Pfizer Library
Chesterfield, MO

CYTOTOXIC CELLS:
Recognition, Effector Function, Generation, and Methods

M. V. Sitkovsky
P. A. Henkart

Editors

Birkhäuser
Boston · Basel · Berlin

Michail V. Sitkovsky
Laboratory of Immunology
Biochemistry and Immunopharmacology Unit
NIH-NIAID
Bethesda, Maryland 20892
USA

Pierre A. Henkart
Experimental Immunology Branch
National Cancer Institute
National Institutes of Health
Bethesda, Maryland 20892
USA

Library of Congress Cataloging-in-Publication Data
Cytotoxic cells: recognition, effector function, generation, and
  methods / M.V. Sitkovsky, P.A. Henkart, editors.
    p.   cm.
   Includes bibliographical references and index.
   ISBN 0-8176-3608-0 (H: alk. paper).
  -- ISBN 3-7643-3608-0 (H: alk. paper)
    1. T cells. 2. Killer cells. 3. Cell-mediated cytotoxicity.
  I. Sitkovsky, M.V. II. Henkart, Pierre.
    [DNLM: 1. T-Lymphocytes, Cytotoxic--physiology. 2. Cytotoxicity,
  Immunologic.   QW 568 C9972 1993]
  QR185.8.T2C99   1993
  616.07'9--dc20
  DNLM/DLC   93-6929
    for Library of Congress   CIP

Printed on acid-free paper.

© 1993 Birkhäuser Boston   *Birkhäuser* ®

Copyright is not claimed for works of U.S. Government employees.
All rights reserved. No part of this publication may be reproduced, stored in a retrieval system or transmitted, in any form or by any means, electronic, mechanical, photocopying, recording or otherwise, without prior permission of the copyright owner.
The use of general descriptive names, trademarks, etc. in this publication even if the former are not especially identified, is not to be taken as a sign that such names, as understood by the Trade Marks and Merchandise Marks Act, may accordingly be used freely by anyone.
While the advice and information in this book are believed to be true and accurate at the date of going to press, neither the authors nor the editors nor the publisher can accept any legal responsibility for any errors or omissions that may be made. The publisher makes no warranty, express or implied, with respect to the material contained herein.
Permission to photocopy for internal or personal use, or the internal or personal use of specific clients, is granted by Birkhäuser Boston for libraries and other users registered with the Copyright Clearance Center (CCC), provided that the base fee of $6.00 per copy, plus $0.20 per page is paid directly to CCC, 21 Congress Street, Salem, MA 01970, U.S.A. Special requests should be addressed directly to Birkhäuser Boston, 675 Massachusetts Avenue, Cambridge, MA 02139, U.S.A.

ISBN 0-8176-3608-0
ISBN 3-7643-3608-0

Typeset by Alden Multimedia, United Kingdom
Printed and bound by Edwards Brothers, Inc., Ann Arbor, MI.
Printed in the United States of America

9 8 7 6 5 4 3 2 1

CONTENTS

Preface ix
Contributors x

SECTION I: INTRODUCTION AND OVERVIEW

1 Introductory Remarks
  Herman N. Eisen 3
2 T-Cell-Mediated Cytotoxicity: A Historical Note
  Jean-Charles Cerottini 5
3 Overview of CTL–Target Adhesion and Other Critical Events in the Cytotoxic Mechanism
  Eric Martz 9

SECTION II: TARGET CELL RECOGNITION

4 $\alpha\beta$ T-Cell Receptor Repertoires Among Cytotoxic and Helper T Lymphocytes
  David M. Kranz and Benjamin Tjoa 49
5 Class I MHC/Peptide/β_2-Microglobulin Interactions: The Basis of Cytotoxic T-Cell Recognition
  David H. Margulies, Lisa F. Boyd, Maripat Corr, Rosemarie D. Hunziker, Sergei Khilko, Steven Kozlowski, Michael Mage, and Randall K. Ribaudo 58
6 The Role of CD8–Class I Interactions in CTL Function
  Anne M. O'Rourke and Matthew F. Mescher 65
7 Interactions Between CD2 and T-Cell Receptor Isoforms in CTL Function
  Shigeo Koyasu and Ellis L. Reinherz 72
8 Triggering Structures on NK Cells
  Lewis L. Lanier and Joseph H. Phillips 84
9 Targeted Cellular Cytotoxicity
  David M. Segal, Carolina R. Jost, and Andrew J. T. George 96

SECTION III: GENERATION OF CYTOTOXIC CELLS

10 Immunobiology and Molecular Characteristics of Peritoneal Exudate Cytotoxic T Lymphocytes (PEL), Their In Vivo IL-2 Dependent Blasts and IL-2 Independent Cytolytic Hybridomas
  Gideon Berke, Dalia Rosen, Denise Ronen, and Barbara Schick 113
11 Regulatory Effects of Cytokines on the Generation of CTL and LAK Cells
  Mark R. Alderson and Michael B. Widmer 128
12 IL-2-Independent Activation of LAK Cells by a Heterodimeric Cytokine, Interleukin-12
  Maurice K. Gately, Aimee G. Wolitzky, Phyllis M. Quinn, and Richard Chizzonite 138
13 Immunobiology of β_2-Microglobulin-Deficient Mice
  J. A. Frelinger, D. G. Quinn, and D. Muller 145

SECTION IV: MOLECULAR MECHANISMS OF CELLULAR CYTOTOXICITY

14 The Granule Exocytosis Model for Lymphocyte Cytotoxicity and Its Relevance to Target Cell DNA Breakdown
  Pierre A. Henkart, Mark P. Hayes, and John W. Shiver 153
15 Subpopulations of Cytotoxic T Lymphocytes with Different Cytotoxic Mechanisms
  David W. Lancki, Maureen McKisic, and Frank W. Fitch 166
16 Multiple Lytic Pathways in Cytototoxic T Lymphocytes
  William R. Clark and Anna Ratner 178

CONTENTS

17 Properties of Cytotoxicity Mediated by CD4$^+$, Perforin-Negative T-Lymphocyte Clones
    Hajime Takayama 190
18 Direct Contact of Cytotoxic T Lymphocyte Receptors with Target Cell Membrane Determinants Induces a Prelytic Rise of [Ca^{2+}]$_i$ in the Target That Triggers Disintegration
    Gideon Berke 196
19 Target Cell Events Initiated by T-Cell Attack
    John H. Russell and Scott I. Abrams 202
20 Apoptosis and Cytotoxic T Lymphocytes
    Richard C. Duke 213
21 Molecular Mechanisms of Lymphocyte Cytotoxicity
    Mark J. Smyth and John R. Ortaldo 223

SECTION V: GRANULE PROTEASES

22 Subtractive and Differential Molecular Biology Approaches to Molecules Preferentially Expressed in Cytotoxic and Other T Cells
    Eric Rouvier and Pierre Golstein 237
23 Structure and Possible Functions of Lymphocyte Granzymes
    Patrick Haddad, Dieter E. Jenne, Olivier Krähenbühl, and Jürg Tschopp 251
24 The Role of Granzyme A in Cytotoxic Lymphocyte-Mediated Lysis
    Mark S. Pasternack 263
25 The Granzyme A Gene: A Marker for Cytolytic Lymphocytes In Vivo
    Gillian M. Griffiths, Susan Alpert, R. Jane Hershberger, Lishan Su, and Irving L. Weissman 273
26 Molecular Analysis and Possible Pleiotropic Function(s) of the T Cell-Specific Serine Proteinase-1 (TSP-1)
    Markus M. Simon, Klaus Ebnet, and Michael D. Kramer 278
27 Serine Protease Control of Lymphocyte-Mediated Cytolysis
    Dorothy Hudig, N. Janine Allison, Gerald R. Ewoldt, Ruther Gault, Dale Netski, Timothy M. Pickett, Doug Redelman, Ming T. Wang, Ulrike Winkler, Susan J. Zunino, Chih-Min Kam, Shinjiro Odake, and James C. Powers 295

SECTION VI: ALTERNATIVE MECHANISMS OF CYTOLYSIS

28 Possible Role of Extracellular ATP in Cell–Cell Interactions Leading to CTL-Mediated Cytotoxicity
    Frank Redegeld, Antonio Filippini, Guido Trenn, and Michail V. Sitkovsky 307
29 Cell-Permeabilizing Properties of Extracellular ATP in Relation to Lymphocyte-Mediated Cytotoxicity
    Francesco Di Virgilio, Paola Zanovello, and Dino Collavo 314
30 The Role of Free Fatty Acids in CTL-Target Cell Interactions
    Alan M. Kleinfeld 321

SECTION VII: BIOCHEMICAL AND IMMUNOPHARMACOLOGICAL MANIPULATIONS OF CYTOTOXIC CELLS

31 Identification of Protein Kinases and Protein Phosphatases Involved in CTL Effector Functions. "ON" and "OFF" Signaling and Immunopharmacological Implications
    Hirotaka Sugiyama, Sergei Apasov, Guido Trenn, Frank Redegeld, and Michail Sitkovsky 331
32 Cytolytic Granules as Targets for Immunosuppressive Therapy: Selective Ablation of CTL by Leucyl-Leucine Methyl Ester
    Dwain L. Thiele and Peter E. Lipsky 340

SECTION VIII: FUNCTIONS OF CYTOTOXIC CELLS IN VIVO

33 Role of CD8$^+$ $\alpha\beta$ T Cells in Respiratory Infections Caused by Sendai Virus and Influenza Virus
    Peter C. Doherty, William Allan, Maryna Eichelberger, Sam Hou, Jacqueline M. Katz, Rudolf Jaenisch, and Maarten Zijlstra 351

CONTENTS

34	CD4+ and CD8+ Cytolytic T Lymphocyte Recognition of Viral Antigens *Vivian Lam Braciale*	358
35	Can CTL Control Virus Infections Without Cytolysis? The Prelytic Halt Hypothesis *Eric Martz*	366
36	Immunologic Control of *Toxoplasma Gondii* Infection by CD8+ Lymphocytes: A Model for Class I MHC- Restricted Recognition of Intracellular Parasites *Ricardo T. Gazzinelli, Eric Denkers, Frances Hakim, and Alan Sher*	370
37	Antigen-Specific Suppression of Antibody Responses by Class II MHC-Restricted CTL *Nobukata Shinohara*	378
38	The Immunosenescence of Cytolytic T Lymphocytes (CTL): Reduction of Pore-Forming Protein and Granzyme Levels *Eda T. Bloom and Judith A. Horvath*	384
39	Bone Marrow Graft Rejection as a Function of T_{NK} Cells *Gunther Dennert*	394
40	Class I MHC Antigens and the Control of Virus Infections by NK Cells *Raymond M. Welsh, Paul R. Rogers, and Randy R. Brutkiewicz*	400
41	Clinical Trials of Immunotherapy of Cancer Utilizing Cytotoxic Cells *Stephen E. Ettinghausen and Steven A. Rosenberg*	407

SECTION IX: MACROPHAGE-MEDIATED CYTOTOXICITY

42	Macrophage-Mediated Cytotoxicity *Penelope J. Duerksen-Hughes and Linda R. Gooding*	439

SECTION X: METHODS

43	The ^{51}Cr-Release Assay for CTL-Mediated Target Cell Lysis *Eric Martz*	457
44	DNA Fragmentation and Cytolysis Assayed by ^{3}H-Thymidine *Eric Martz*	468
45	The JAM Test: An Assay of Cell Death *Polly Matzinger*	472
46	Target Cell Detachment Assay *Scott I. Abrams and John H. Russell*	478
47	Protocol for Assaying CTL Activity Against *Toxoplasma gondii* *Ricardo T. Gazzinelli, Eric Denkers, Frances Hakim, and Alan Sher*	481
48	Granule Exocytosis Assay of CTL Activation *Michail V. Sitkovsky*	482
49	Measurement of Cytolysin Hemolytic Activity *Pierre Henkart*	484
50	SPDP Crosslinking of Antibodies to Form Heteroconjugates Mediating Redirected Cytotoxicity *David M. Segal*	485
51	Derivatization of Cells with Antibody *Anna Ratner and William R. Clark*	487
52	Mixed Lymphocyte Culture for the Generation of Allospecific CTL *Eda T. Bloom*	488
53	Generation of CD4+ and CD8+ Antiinfluenza CTL and Assay of In Vitro Cytotoxicity *Vivian Lam Braciale*	490
54	Generation of Antigen-Specific Murine CTL Under Weakly Immunogenic Conditions *Sergey G. Apasov*	492
55	Commercial Liposomes and Electroporation Can Deliver Soluble Antigen for Class I Presentation in CTL Generation *Weisan Chen and James McCluskey*	494
56	Stimulation of CTLs on Antibody-Coated Plates *Anna Ratner and William R. Clark*	497

57	CTL Recognition of Purified MHC Antigens and Other Cell Surface Ligands *Matthew F. Mescher, Paul Champoux, and Kevin P. Kane*	498
58	Use of Protease Inhibitors as Probes for Biological Functions: Conditions, Controls, and Caveats *Dorothy Hudig and James C. Powers*	502
59	The Murine T Cell-Specific Serine Proteinase-1: Cleavage Activity on Synthetic and Natural Substrates *M. D. Kramer, U. Vettel, K. Ebnet, and M. M. Simon*	516
60	Detection of Specific mRNAs by In Situ Hybridization *Gillian M. Griffiths, Susan Alpert, R. Jane Hershberger, Lishan Su, and Irving L. Weissman*	521
Index		525

PREFACE

Our motivation for putting together this book was the need for a single source reference that could be used as an introduction to cell-mediated cytotoxicity for newcomers to this field, such as students and fellows beginning work in our laboratories. At present no such book is available, and we felt that it would be useful as a teaching tool and as a way of conveying our enthusiasm about recent progress in the cytotoxicity field to our colleagues in allied areas. It was with some hesitation that we approached our colleagues with the proposal for this book, and we were pleased to find them very supportive of the idea and willing to participate. We thought it important to broaden the scope of the book to include historical, molecular, cell biological, and clinical aspects of cell-mediated cytotoxicity. To our knowledge this is the first book on cell-mediated cytotoxicity with such a broad scope.

Historically, studies on cellular cytotoxicity were part of cellular immunology from its origin. One development of tremendous import was the advent of the ^{51}Cr assay, which allowed this arm of the immune response to be measured easily and quantitatively. Thus, a readout of this effector pathway is available within a few hours; other immune effector functions can take days or even longer to assay, and the assays are often less quantitative. For example, using the ^{51}Cr-release assay, studies of cytotoxic T cells played a crucial role in discovering the rules of antigen recognition and in establishing the concept of major histocompatibility complex (MHC) restriction. In turn, the major breakthroughs in the understanding of the structure and functions of target recognition receptors on cytotoxic cells led to the development of such powerful approaches as retargeted cytotoxicity, which is further extending the utility of cytotoxic cells.

Natural killer (NK) cells and cytotoxic T cells are to be considered as two different but related arms of host defense, e.g., against virus infection. Indeed, some interesting interrelations between the class I MHC expression and sensitivity of virus-infection cells to destruction by cytotoxic T lymphocytes (CTL) or NK cells are found. Both positive and negative feedback mechanisms could be at play here, since virus-induced interferon makes cells resistant to NK cells as these infected cells become increasingly sensitive to CTL. The least studied, but most clinically used, cytotoxic cells [lymphokine-activated killer (LAK) cells, tumor-infiltrating lymphocytes (TIL)] may provide the efficient means of antitumor treatments.

The studies of the mechanisms of effector functions are more controversial than other areas (e.g., CTL recognition), since they are linked to broader issues of cell biology and physiology. The diversity of opinion is reflected in the various contributions to this book and provides a flavor of the excitement of this fast-moving area.

We have been immensely pleased with the quality of the chapters we have received for this book. We are confident that it will fulfil the need we have set out to address. We thank the various contributors for their efforts and hope that the readers will get as much out of these chapters as we have.

Michail V. Sitkovsky
Pierre A. Henkart
Editors

CONTRIBUTORS

Scott I. Abrams, Department of Molecular Biology and Pharmacology, Washington University School of Medicine, St. Louis, Missouri 63110, USA. Current Affiliation: Laboratory of Tumor Immunology and Biology, National Cancer Institute, National Institutes of Health, Bethesda, Maryland 20892, USA

Mark R. Alderson, Department of Immunology, Immunex Research and Development Corporation, Seattle, Washington 98101, USA

William Allan, Department of Immunology, St. Jude Children's Research Hosptial, Memphis, Tennessee 38105, USA

N. Janine Allison, Cell and Molecular Biology Program, School of Medicine and College of Agriculture, University of Nevada, Reno, Nevada 89557-0046, USA

Susan Alpert, Department of Pathology and Howard Hughes Medical Institute, Stanford, California 94305, USA

Sergey G. Apasov, Laboratory of Immunology, National Institute of Allergy and Infectious Diseases, National Institutes of Health, Bethesda, Maryland 20892, USA

Gideon Berke, Department of Cell Biology, The Weizmann Institute of Science, 76100 Rehovot, Israel

Eda T. Bloom. Division of Cytokine Biology (HFB-826), Center for Biologics Evaluation and Research, Food and Drug Administration, Bethesda, Maryland 20892, USA

Lisa F. Boyd, Laboratory of Immunology, National Institute of Allergy and Infectious Diseases, National Institutes of Health, Bethesda, Maryland 20892, USA

Vivian Lam Braciale, University of Virginia, Beirne Carter Center for Immunology Research and Department of Microbiology, Health Sciences Center, Charlottesville, Virginia 22908, USA

Randy R. Brutkiewicz, Department of Pathology, University of Massachusetts Medical Center, Worcester, Massachusetts 01655, USA

Jean-Charles Cerottini, Ludwig Institute for Cancer Research, Lausanne Branch, 1066 Epalinges, Switzerland

Paul Champoux, Division of Membrane Biology, Medical Biology Institute, La Jolla, California 92037, USA

Weisan Chen, Department of Clinical Immunology, Flinders Medical Center, Bedford Park, 5042, South Australia, Australia

Richard Chizzonite, Molecular Genetics Department, Hoffmann-La Roche, Inc., Nutley, New Jersey 07110-1199, USA

William R. Clark, Department of Biology, The Molecular Biology Institute, University of California, Los Angeles, California 90024–1606, USA

Dino Collavo, Chair of Immunology, University of Padova, I-35100 Padova, Italy

Maripat Corr, Laboratory of Immunology, National Institute of Allergy and Infectious Diseases, National Institutes of Health, Bethesda, Maryland 20892, USA

Eric Denkers, LPD, NIAID, National Institutes of Health, Bethesda, Maryland 20892, USA

Gunther Dennert, Department of Microbiology, University of Southern California, Los Angeles, California 90033, USA

Penelope J. Duerksen-Hughes, Department of Microbiology and Immunology, Emory University School of Medicine, Atlanta, Georgia 30322, USA

CONTRIBUTORS

Francesco Di Virgilio, Institute of General Pathology, University of Ferrara, I-44100 Ferrara, Italy

Peter C. Doherty, Department of Immunology, St. Jude Children's Research Hospital, Memphis, Tennessee 38105, USA

Richard C. Duke, Immunology Program, University of Colorado Cancer Center, University of Colorado School of Medicine, Denver, Colorado 80262, USA

Klaus Ebnet, Max Planck-Institute für Immunbiologie, D-7800 Freiburg/Br., Germany

Maryna Eichelberger, Department of Immunology, St. Jude Children's Research Hospital, Memphis, Tennessee 38105, USA

Herman N. Eisen, Center for Cancer Research, E17-128, Massachusetts Institute of Technology, Cambridge, Massachusetts 02139, USA

Stephen E. Ettinghausen, Surgery Branch, NCI, National Institutes of Health, Bethesda, Maryland 20892, USA

Gerald R. Ewoldt, Cell and Molecular Biology Program, School of Medicine and College of Agriculture, University of Nevada, Reno, Nevada 89557-0046, USA

Antonio Filippini, Institute of Histology and Embriology, University of Rome "La Sapienza", 00161 Rome, Italy

Frank W. Fitch, Department of Pathology, The University of Chicago, Chicago, Illinois 60637, USA

Jeffrey A. Frelinger, Department of Microbilogy and Immunology, University of North Carolina, Chapel Hill, North Carolina 27599-7290, USA

Maurice K. Gately, Immunopharmacology Department, Hoffmann-La Roche, Inc., Nutley, New Jersey 07110-1199, USA

Ruth A. Gault, Cell and Molecular Biology Program, School of Medicine and College of Agriculture, University of Nevada, Reno, Nevada 89557-0046, USA

Ricardo T. Gazzinelli, LPD, NIAID, National Institutes of Health, Bethesda, Maryland 20892, USA

Andrew J. T. George, Immune Targeting Section, Experimental Immunology Branch, National Cancer Institute, National Institutes of Health, Bethesda, Maryland 20892, USA

Pierre Golstein, Centre d'Immunologie, INSERM-CNRS de Marseille-Luminy, Case 906, 13288 Marseille Cedex 9, France

Linda R. Gooding, Department of Microbiology and Immunology, Emory University School of Medicine, Atlanta, Georgia 30322, USA

Gillian M. Griffiths, Basel Institute of Immunology, Postfach CH-4005 Basel, Switzerland

Patrick Haddad, Institute of Biochemistry, University of Lausanne, 1066 Epalinges, Switzerland

Frances Hakim, EIB, NCI, National Institutes of Health, Bethesda, Maryland 20892, USA

Mark P. Hayes, Division of Cytokine Biology, CBER, FDA, Bethesda, Maryland 20892, USA

Pierre A. Henkart, Experimental Immunology Branch, National Cancer Institute, National Institutes of Health, Bethesda, Maryland 20892, USA

R. Jane Hershberger, Department of Pathology and Howard Hughes Medical Institute, Stanford, California 94305, USA

Judith A. Horvath, Division of Cytokine Biology (HFB–826), Center for Biologics Evaluation and Research, Food and Drug Administration, Bethesda, Maryland 20892, USA

Sam Hou, Department of Immunology, St. Jude Children's Research Hospital, Memphis, Tennessee 38105, USA

CONTRIBUTORS

Dorothy Hudig, Cell and Molecular Biology Program, School of Medicine and College of Agriculture, University of Nevada, Reno, Nevada 89557-0046, USA

Rosemarie D. Hunziker, Laboratory of Immunology, National Institute of Allergy and Infectious Diseases, National Institutes of Health, Bethesda, Maryland 20892, USA

Rudolf Jaenisch, Whitehead Institute, Massachusetts Institute of Technology, Cambridge, Massachusetts 02142, USA

Dieter E. Jenne, Institute of Biochemistry, University of Lausanne, 1066 Epalinges, Switzerland

Carolina R. Jost, Immune Targeting Section, Experimental Immunology Branch, National Cancer Institute, National Institutes of Health, Bethesda, Maryland 20892, USA

Chih-Min Kam, School of Chemistry and Biochemistry, Georgia Institute of Technology, Atlanta, Georgia 30332, USA

Kevin P. Kane, Department of Immunology, University of Alberta, Edmonton, Alberta, Canada T6G 2H7

Jacqueline M. Katz, Department of Immunology, St. Jude Children's Research Hospital, Memphis, Tennessee 38105, USA

Sergei Khilko, Laboratory of Immunology, National Institute of Allergy and Infectious Diseases, National Institutes of Health, Bethesda, Maryland 20892, USA

Alan M. Kleinfeld, Medical Biology Institute, La Jolla, California 92037, USA

Steven Kozlowski, Laboratory of Immunology, National Institute of Allergy and Infectious Diseases, National Institutes of Health, Bethesda, Maryland 20892, USA

Shigeo Koyasu, Laboratory of Immunobiology, Dana-Farber Cancer Institute, Department of Pathology, Harvard Medical School, Boston, Massachusetts 02115, USA

Olivier Krähenbühl, Institute of Biochemistry, University of Lausanne, 1066 Epalinges, Switzerland

Michael D. Kramer, Institut für Immunologie und Serologie der Universität, Immunpatholgie, D-6900 Heidelberg, Germany

David M. Kranz, Department of Biochemistry, University of Illinois, Urbana, Illinois 61801, USA

David W. Lancki, Department of Pathology, The University of Chicago, Chicago, Illinois 60637, USA

Lewis L. Lanier, DNAX Research Institute of Molecular and Cellular Biology, Department of Immunology, Palo Alto, California 94304, USA

Peter E. Lipsky, The University of Texas Southwestern Medical Center at Dallas, Rheumatic Diseases Division, Department of Internal Medicine, Dallas, Texas 75235-8887, USA

Michael Mage, Laboratory of Biochemistry, National Cancer Institute, National Institutes of Health, Bethesda, Maryland 20892, USA

David H. Margulies, Molecular Biology Section, Laboratory of Immunology, National Institute of Allergy and Infectious Diseases, National Institutes of Health, Bethesda, Maryland 20892, USA

Eric Martz, Department of Microbiology and Program in Molecular and Cell Biology, University of Massachusetts, Amherst, Massachusetts 01003, USA

Polly Matzinger, National Institutes of Health, Bethesda, Maryland 20892, USA

James McCluskey, Department of Clinical Immunology, Flinders Medical Center, Bedford Park, 5042, South Australia, Australia

Maureeen McKisic, Department of Pathology, The University of Chicago, Chicago, Illinois 60637, USA

Matthew F. Mescher, Division of Membrane Biology, Medical Biology Institute, La Jolla, California 92037, USA

CONTRIBUTORS

Dan Muller, Division of Rheumatology, Department of Medicine, University of Wisconsin, Madison, Wisconsin 53792, USA

Dale M. Netski, Cell and Molecular Biology Program, School of Medicine and College of Agriculture, University of Nevada, Reno, Nevada 89557-0046, USA

Shinjiro Odake, School of Chemistry and Biochemistry, Georgia Institute of Technology, Atlanta, Georgia 30332, USA

Anne M. O'Rourke, Division of Membrane Biology, Medical Biology Institute, La Jolla, California 92037, USA

John R. Ortaldo, Laboratory of Experimental Immunology, National Cancer Institute, Frederick Cancer Research Facility, Frederick, Maryland 21702–1201, USA

Mark S. Pasternack, Infectious Disease Units (Children's and Medical Services), Massachusetts General Hospital, Boston, Massachusetts 02114, USA

Joseph H. Phillips, DNAX Research Institute of Molecular and Cellular Biology, Department of Immunology, Palo Alto, California 94304, USA

Timothy M. Pickett, Cell and Molecular Biology Program, School of Medicine and College of Agriculture, University of Nevada, Reno, Nevada 89557–0046, USA

James C. Powers, School of Chemistry and Biochemistry, Georgia Institute of Technology, Atlanta, Georgia 30332, USA

Daniel G. Quinn, Department of Microbiology and Immunology, University of North Carolina, Chapel Hill, North Carolina 27599-7290, USA

Phyllis M. Quinn, Immunopharmacology Department, Hoffmann-La Roche, Inc., Nutley, New Jersey 07110-1199, USA

Anna Ratner, Departments of Anatomy and Cell Biology, University of California, Los Angeles, California 90024-1606, USA

Frank Redegeld, Laboratory of Immunology, National Institute of Allergy and Infectious Diseases, National Institutes of Health, Bethesda, Maryland 20892, USA

Doug Redelman, Sierra Cytometry, Reno, Nevada 89509, USA

Ellis L. Reinherz, Laboratory of Immunobiology, Dana-Farber Cancer Institute, Department of Medicine, Harvard Medical School, Boston, Massachusetts 02115, USA

Randall K. Ribaudo, Laboratory of Immunology, National Institute of Allergy and Infectious Diseases, National Institutes of Health, Bethesda, Maryland 20892, USA

Paul R. Rogers, Department of Molecular Genetics and Microbiology, University of Massachusetts Medical Center, Worcester, Massachusetts 01655, USA

Denise Ronen, Department of Cell Biology, The Weizmann Institute of Science, 76100 Rehovot, Israel

Dalia Rosen, Department of Cell Biology, The Weizmann Institute of Science, 76100 Rehovot, Israel

Steven A. Rosenberg, Surgery Branch, NCI, National Institutes of Health, Bethesda, Maryland 20892, USA

Eric Rouvier, Centre d'Immunologie, INSERM-CNRS de Marseille-Luminy, Case 906, 13288 Marseille Cedex 9, France

John H. Russell, Department of Molecular Biology & Pharmacology, Washington University Medical School, St. Louis, Missouri 63110, USA

Barbara Schick, Department of Cell Biology, The Weizmann Institute of Science, 76100 Rehovot, Israel

David M. Segal, Immune Targeting Section, Experimental Immunology Branch, National Cancer Institute, National Institutes of Health, Bethesda, Maryland 20892, USA

CONTRIBUTORS

Alan Sher, LPD, NIAID, National Institutes of Health, Bethesda, Maryland 20892, USA

Nobukata Shinohara, Laboratory of Cellular Immunology, Mitsubishi Kasei Institute of Life Sciences, Tokyo 194, Japan

John W. Shiver, Merck, Sharp and Dohme Research Laboratories, West Point, Pennsylvania 19486, USA

Markus M. Simon, Max Planck-Institut für Immunbiologie, D-7800 Freiburg/Br., Germany

Michail V. Sitkovsky, Chief, Biochemistry and Immunopharmacology Section, Laboratory of Immunology, National Institutes of Health, NIAID, Bethesda, Maryland 20892, USA

Mark J. Smyth, Cellular Cytotoxicity Laboratory, Austin Research Institute, Heidelberg 3084, Victoria, Australia, and Laboratory of Experimental Immunology, National Cancer Institute, Frederick Cancer Research and Development Center, Frederick, Maryland 21702-1201, USA

Lishan Su, Department of Pathology and Howard Hughes Medical Institute, Stanford, California 94305, USA

Hirotaka Sugiyama, Laboratory of Immunology, National Institute of Allergy and Infectious Diseases, National Institutes of Health, Bethesda, Maryland 20892, USA

Hajime Takayama, Laboratory of Cellular Immunology, Mitsubishi Kasei Institute of Life Sciences, Tokyo 194, Japan

Dwain L. Thiele, The University of Texas Southwestern Medical Center at Dallas, Liver Unit, Department of Internal Medicine, Dallas, Texas 75235-8887, USA

Benjamin Tjoa, Department of Biochemistry, University of Illinois, Urbana, Illinois 61801, USA

Guido Trenn, Department of Hematology, University of Essen, D-4300 Essen, Germany

Jürg Tschopp, Institute of Biochemistry, University of Lausanne, 1066 Epalinges, Switzerland

Ulrike Vettel, Institut für Immunologie und Serologie der Universität, Immunpathologie, D-6900 Heidelberg, Germany

Ming T. Wang, Cell and Molecular Biology Program, School of Medicine and College of Agriculture, University of Nevada, Reno, Nevada 89557-0046, USA

Irving L. Weissman, Department of Pathology and Howard Hughes Medical Institute, Stanford, California 94305, USA

Raymond Welsh, Department of Pathology, University of Massachusetts Medical Center, Worcester, Massachusetts 01655, USA

Michael B. Widmer, Department of Immunology, Immunex Research and Development Corporation, Seattle, Washington 98101, USA

Ulrike Winkler, Cell and Molecular Biology Program, School of Medicine and College of Agriculture, University of Nevada, Reno, Nevada 89557-0046, USA

Aimee G. Wolitzky, Immunopharmacology Department, Hoffman-La Roche, Inc., Nutley, New Jersey 07110-1199, USA

Paola Zanovello, Chair of Immunology, University of Padova, I-35100 Padova, Italy

Maarten Zijlstra (Deceased), Department of Molecular Genetics, The Netherlands Cancer Institute, 1066 CX Amsterdam, The Netherlands

Susan J. Zunino, Howard Hughes Medical Institute, University of Texas Southwestern Medical Center, Dallas, Texas 75235-9050, USA

I
Introduction and Overview

1
Introductory Remarks

Herman N. Eisen

In putting together this volume of selected reviews of the cytotoxic activities of T lymphocytes, natural killer (NK) cells, and, to some extent, macrophages, Drs. Michail Sitkovsky and Pierre Henkart deserve the gratitude of the community of concerned research workers. The assembly in one volume of the contemporaneous views of the most active investigators in this field will probably come to be seen in the future as a landmark record of where matters stood in 1992.

Publication of this record at this time is especially propitious, for it seems likely that many of the central issues that have been passionately debated over the past 5–10 years may soon be largely resolved by the powerful techniques that make use of homologous recombination in embryonic stem cells to generate transgenic mice with specifically knocked-out genes. The selective elimination of genes that encode components of the cytotoxic granules of cytotoxic T cells and NK cells is of obvious interest for the present volume. The phenotypes of these "knockout" mice are likely to establish the extent to which products of the knocked-out genes are (or are not) indispensable for cytotoxic or other activities. Indeed, the ability not only to eliminate genes but to replace them with custom-tailored altered versions and to overexpress the wild-type or altered genes now promises to make available for investigators variant mice that might otherwise never have turned up or turned up only by chance operating over millions of years of evolution.

In carrying out the destruction of target cells, cytotoxic cells have to exercise many functions; they have to crawl, to form firm but transient conjugates with target cells, to search with their antigen-specific receptors for antigenic structures on target cells, to trigger lysis of the specifically recognized target cells, to disengage from the damaged target, and to crawl away to hunt for other target cells. Many of these activities are shared with other leukocytes, especially other T cells. Only the lysis of target cells is unique to cytotoxic cells, and so it is understandable that this volume focuses almost entirely on killing mechanisms.

Given the immense amount of effort made in the past decade to understand these mechanisms, it may seem surprising that a consensus about the principal mechanisms is far from having been reached. Is this because key components have yet to be discovered? Or because our understanding of how the known components work is defective? Or is a balanced view of how various mechanisms contribute still to be worked out? The difficulties in reaching a clear concensus can be illustrated by considering the exocytosis model.

According to this model, the cytotoxic activity of T cells and NK cells is attributable to the cytotoxic granules they release into the synapselike space formed by adhesion of these cells to their target cells. As is well known, the granules contain a pore-forming, complement-like protein (perforin, cytolysin) that is capable of forming the destructive channels that appear in membranes of target cells undergoing lysis. That this lytic protein is indeed released by cytotoxic T cells is evident from the extracellular appearance of another granule protein, a serine protease (formerly termed BLT-serine esterase and now called granzyme A) that is easily detected with a sensitive substrate. When isolated from the cytoxic cells, the granules readily lyse all cell types tested except cytotoxic T lymphocytes (especially cultured clones of these cells), which are relatively resistant, accounting in part for the different effects of released granules on the cytotoxic cells and on the target cells to which they adhere. All of this evidence strongly supports the exocytosis model.

CYTOTOXIC CELLS
M. Sitkovsky and P. Henkart, Editors
© 1993 Birkhäuser Boston

There are, nevertheless, several indications that this mechanism may not account for cytotoxic activities in all situations. Thus, lysis of some target cells is not blocked by chelating agents that lower calcium ion concentrations in the extracellular medium to levels too low to support detectable granule exocytosis. There are also some cytotoxic cells that are reported not to produce detectable levels of perforin or granule serine proteases.

Although these negative findings appear to be highly significant, in the absence of further information they are inconclusive, because the lowest levels of perforin that can be detected by current analytical methods and the lowest levels that are active cytolytically have yet to be established. Until they are, the finding of "not detectable" is inconclusive. It is precisely this ambiguity that ought to be dispelled by knockout mice.

And yet, one ought not to expect too much from knockout and other "high-tech" approaches, for it is already evident from work on transgenic mice lacking CD4 or CD8 or other genes that appear vital for immune responses that there is enormous redundancy in the immune system. Thus, compensatory changes sometimes overcome major deficiencies, as is seen in skin allograft rejection by some knockout mice that lack the β^2-microglubin gene (and thereby class IMHC-dependent T cells) and by others that lack class IIMHC genes (and thereby class II-dependent T cells). In interpreting such results it is important to bear in mind that they show what can happen, not what necessarily happens normally.

For many who are interested in cytotoxic cells, $CD8^+$ cytotoxic T lymphocytes (CTL) clones are especially fascinating. These cells not only kill the target cells they recognize, but in their varied responses to antigenic structures and lymphokine signals they exhibit many of the diverse responses exhibited by B lymphocytes and T-helper lymphocytes, i.e., by antigen-recognizing cells in general. Thus, as individual clones (from mice) that can be maintained in culture almost indefinitely, murine $CD8^+$ CTL represent a virtual microcosm of the immune system. It is gratifying that these cells are so extensively represented in this volume.

2
T-Cell-Mediated Cytotoxicity: A Historical Note

Jean-Charles Cerottini

On May 15, 1967, the editor of the British journal *Immunology* found on his desk a manuscript submitted for publication, the Summary section of which read as follows: "The *in vitro* cytotoxic effect of spleen cells of mice immunized by tumor allografts was studied by measuring target cell inactivation as a function of release of radioactive label (^{51}Cr) or loss of cloning efficiency." The first author of this paper was Teddy Brunner, head of the Department of Immunology at the Swiss Institute for Experimental Cancer Research in Lausanne. The manuscript was accepted and was published in February 1968 (Brunner et al., 1968). As I have been fortunate enough to be associated with this work, a book such as this offers me an opportunity to give a brief personal account of the "early days" of T-cell-mediated cytotoxicity.

My introduction to this field began in 1965. I was then working in the Department of Biochemistry, University of Lausanne, under the leadership of Henri Isliker, and my laboratory was located next to (and directly connected to) Teddy Brunner's laboratory. While I was working on the preparation of drug–antibody conjugates ("the magic bullets"), I was introduced to the phenomenon of immunological enhancement by Frank Fitch, on sabbatical leave from the University of Chicago in the Department of Biochemistry. For somebody like me who was mainly familiar with the cytotoxic function of antibodies, it was quite intriguing that antibodies could paradoxically inhibit the immune destruction of allografts. I therefore followed with great interest the attempts made by Teddy Brunner to develop an *in vitro* assay system allowing a quantitative analysis of the effect of major histocompatibility complex (MHC) alloreactive lymphocytes on target cells in the presence or absence of antibodies.

At that time, most of the assays used relied on visual assessment of target cell death and, hence, were subject to severe limitations. In a first attempt to meaure lymphocyte–target cell interactions quantitatively, Teddy Brunner developed an assay based on the loss of the clonogenic potential of tumor cells after incubation with alloimmune lymphocytes. He selected the P815 mastocytoma cell line (derived by Dunn and Potter in 1957) as the source of tumor target cells. This choice, which turned out to be most valuable, was dictated by the results of studies performed by Richard Schindler, a cell biologist working in a laboratory adjacent to that of Teddy Brunner. Richard Schindler, who was primarily interested in the biochemistry of the cell cycle, had developed a cloning assay for P815 cells in fibrin gels that reached 100% efficiency. The availability of a cloning assay with such efficiency allowed for the first time a direct analysis of the dynamics of target cell inactivation by alloimmune lymphocytes. Indeed, by using this assay, Brunner, together with Schindler and Jacques Mauël, a graduate student at that time, was able to demonstrate that the colony-forming potential of P815 cells was significantly reduced as early as 3 hr after incubation with alloimmune lymphocytes (Brunner et al., 1966). These findings were in apparent conflict with the prevailing view based on studies using morphological observation of target cells that the cytotoxic effect of lymphocytes was slow, i.e. that it required 24 to 48 hr. In other experiments, the cloning assay revealed that alloantibodies could inhibit either the generation *in vivo* or the cytotoxic activity *in vitro* of alloimmune lymphocytes (Brunner et al., 1967b), in agreement with previous work performed by Göran and Erna Möller.

As a passive spectator of these studies, I was quite impressed by the sensitivity of the cloning assay, but the idea of using it for further studies did not appeal to me because of the laborious work involved. Therefore, I became quite interested by a report indicating that release of radioactive chromium could be used to measure complement-dependent lysis of target cells by alloantibodies. My first attempts to use P815 cells labeled with ^{51}Cr failed miserably because of a very high spontaneous release of the isotope. It turned out that the procedure published in the literature, which recommended a preincubation of freshly labeled cells in the cold for half an hour before addition of antibodies and complement, was totally inadequate for P815 cells. Fortunately, suitable conditions were worked out relatively rapidly so that a direct comparison between the cloning assay and the ^{51}Cr release assay became feasible.

It was at that time (January 1967) that I left Switzerland to pursue my postdoctoral training with Frank Dixon in the Division of Experimental Pathology at Scripps Clinic. During my stay in La Jolla, I was kept informed about the progress made in Lausanne. Together with Jacques Mauël, Hans Rudolf, and Bernard Chapuis, Teddy Brunner rapidly accumulated quantitative data regarding the kinetics of target cell inactivation *in vitro* by alloimmune spleen cells as well as the time course of the development *in vivo* of cytotoxic effector cells (Brunner et al., 1968). These results, which were first presented at the Vth International Immunopathology Symposium held in Punta Ala in June 1967 (Brunner et al., 1967a), clearly established that the ^{51}Cr release assay was an extremely useful tool for the assessment of cell-mediated cytotoxicity because of its simplicity and sensitivity.

At the time the assay was developed, there was a continuing controversy as to the actual mechanism of cell-mediated cytotoxic reactions *in vitro*. As mentioned before, several studies indicated that alloantibodies inhibited rather than increased the cytotoxic activity of MHC-alloreactive lymphoid cells. It was therefore assumed that antibodies were not involved in cell-mediated cytotoxic reactions *in vitro*. The validity of this assumption, however, was questioned by the observation that the appearance of circulating alloantibodies after immunization temporally coincided with the development of cell-mediated cytotoxic activity in lymphoid cell populations. In addition, studies by the groups of Peter Perlmann and Ian MacLennan showed that target cells coated with appropriate antibodies could be specifically lysed in the presence of lymphoid cells from normal individuals. Based on these findings, it was proposed that the cytotoxic activity of alloimmune lymphoid cell populations reflected an antibody-dependent, cell-mediated phenomenon, whereby alloantibody-producing cells released antibodies which, after binding to target cells, allowed normal lymphocytes (or macrophages) equipped with Fc receptors to bind to and kill target cells.

In an attempt to settle this question, I decided, upon my return to Lausanne, to investigate whether the cytotoxic effector cells in MHC-alloreactive populations were thymus-derived or not. Although such an analysis can be performed easily today by any student in immunology, the experimental tools available in the late sixties were quite limited. In particular, surface markers for T and B cells were not yet identified. My initial approach was based on a series of investigations performed by Jacques Miller and Graham Mitchell which clearly demonstrated in the mouse that thymus-derived cells were not the precursors of antibody-producing cells, although they were needed for the production of antibodies by bone marrow-derived cells. In these experiments, thymus cells had first to be "educated," i.e., they were injected together with the antigen into syngeneic irradiated mice and recovered a week later from the spleen of the irradiated recipients. I reasoned that if alloimmune cytotoxic lymphocytes were thymus-derived, injection of thymus cells into MHC-incompatible mice should lead to the generation of cytotoxic cells of donor origin directed against the recipient MHC antigens.

I had the opportunity to discuss this idea directly with Jacques Miller in September 1969 when he attended a course organized in Lausanne by the World Health Organization for young immunologists from developing countries. Although Jacques was receptive to the idea, he proposed a different experimental protocol based on the use of cells from parental and F1 hybrid mice and H-2 alloantibodies. After a lengthy discussion, it was agreed that Teddy Brunner and I would proceed with the first approach, whereas the second protocol would be developed in Jacques Miller's laboratory. Therefore, we immediately injected lethally irradiated DBA/2 mice with thymus cells from A or C3H mice, and 5 days later recovered the lymphoid cells present in the recipient spleens. To our delight, the vast majority of the recovered cells were of donor origin and were cytotoxic for (DBA/2) P815 cells. In contrast, transfer of bone marrow cells was not accompanied by the appearance of cell-mediated cytotoxicity. We were excited by these findings and it was decided to extend them to another mouse strain combination, namely C57BL/6 cells injected into irradiated DBA/2 mice. The results of these experiments were completely erratic: spleen cells from recipients of thymus cells indeed exhibited cytotoxic activity, but the cytotoxic activity was even greater in spleen cell populations recovered from recipients of bone marrow cells. While we were completely at a loss to explain these results, I fortuitously read about the effect of antilymphocyte serum administered *in vivo* on circulating lymphoid cells. It occurred to me that the unexpected results obtained with C57BL/6 bone marrow cells might be explained by the presence of contaminating thymus-derived cells in the bone marrow preparations. To test this possibility, we rapidly produced a rabbit anti-mouse lymphocyte serum and injected it into C57BL/6 mice one day before the collection of thymus and bone marrow cells. This treatment did not affect the ability of thymus cells to develop into cytotoxic cells upon transfer into irradiated DBA/2 mice, but it abrogated the development of cytotoxic activity in transferred bone marrow cells. (It is noteworthy that subsequent studies using antibodies against T-cell markers have shown that the degree of T-cell contamination is higher in C57BL/6 bone marrow preparations than in other mouse strains.)

Although these studies provided direct evidence that precursors of cytotoxic lymphocytes were present in the thymus, we felt that a critical experiment should include a direct comparison between cytotoxic cells and antibody-forming cells in the same alloimmune populations. At this stage of our work, Al Nordin, on sabbatical leave from the University of Notre Dame, joined the laboratory. He insisted that the purpose of his stay with us was to learn *in vitro* assays for cell-mediated immunity and take a break from several years of testing antibody plaque-forming cells. Fortunately, he quickly realized the potential usefulness of an *in vitro* assay system allowing the detection of alloantibody plaque-forming cells. Within a few weeks, he developed such an assay using P815 cells as the source of indicator cells (Nordin et al., 1971). This remarkable achievement then allowed us to detect independently alloimmune cytotoxic lymphocytes and alloantibody-producing cells with the same target cells. Using the cell transfer system described before, we were able to show that both types of effector cells were generated upon transfer of spleen cells into irradiated allogeneic recipients, whereas thymus cells produced only cytotoxic cells, but no alloantibody plaque-forming cells (Cerottini et al., 1970a). Moreover, by using an anti-Thy-1.2 antiserum prepared in our laboratory, we demonstrated that treatment of alloimmune spleen cells with such an antiserum and complement abrogated cell-mediated cytotoxicity but had no effect on alloantibody plaque formation (Cerottini et al., 1970b, 1971). Conversely, treatment of the same immune spleen cell populations with an anti-plasma cell rabbit serum and complement had no effect on cell-mediated cytotoxicity but completely inhibited the formation of plaque-forming cells.

I presented these results at the VIth International Immunopathology Symposium held in Grindelwald in September 1970 (Cerottini et al., 1970c). Although the evidence looked compelling to me, the lack of participation of antibodies in the cytotoxic activity exhibited by MHC-alloreactive lymphoid cells was still questioned by some participants. Obviously, there was some reluctance to admit that thymus-derived cells could differentiate into helper cells on the one hand and cytotoxic cells on the other. However, the concept of T cell-mediated cytotoxicity became well accepted after a while, so that no objection was raised when Teddy Brunner presented our work at the First International Congress of Immunology held in Washington in 1971.

In these personal recollections, I have concentrated on the work performed in our laboratory. Obviously, many other groups were involved in the area of cell-mediated cytotoxicity in the late sixties (for review, see Cerottini and Brunner, 1974). Most, if not all, of the investigators who participated in the early development of this field are still very active, as shown by their contributions to this book. There is no better evidence that studying cell-mediated cytotoxicity is a healthy addiction!

References

Brunner KT, Mauël J., Cerottini JC, Chapuis B (1968): Quantitative assay of the lytic action of immune lymphoid cells on ^{51}CR-labelled allogeneic target cells *in vitro*; inhibition by isoantibody and by drugs. *Immunology* 14: 181–196

Brunner KT, Mauël J, Cerottini JC, Rudolf H, Chapuis B (1967a): *In vitro* studies of cellular and humoral immunity induced by tumor allografts. In: *Vth International Immunopathology Symposium*, Miescher PA, Grabar P, eds. Basel/Stuttgart: Schwabe & Co., pp. 342–355

Brunner KT, Mauël J, Schindler R (1966): *In vitro* studies of cell-bound immunity; cloning assay of the cytotoxic action of sensitized lymphoid cells on allogeneic target cells. *Immunology* 11: 499–506

Brunner KT, Mauël J, Schindler R (1967b): Inhibitory effect of isoantibody on *in vivo* sensitization and on the *in vitro* cytotoxic action of immune lymphocytes. *Nature* 213: 1246–1247

Cerottini JC, Brunner KT (1974): Cell-mediated cytotoxicity, allograft rejection and tumor immunity. *Adv Immunol* 18: 67–132

Cerottini JC, Nordin AA, Brunner KT (1970a): *In vitro* cytotoxic activity of thymus cells sensitized to alloantigens. *Nature* 227: 72–73

Cerottini JC, Nordin AA, Brunner KT (1970b): Specific *in vitro* cytotoxicity of thymus-derived lymphocytes sensitized to alloantigens. *Nature* 228: 1308–1309

Cerottini JC, Nordin AA, Brunner KT (1970c): Competence of various cell populations for humoral and cell-mediated transplantation immunity. In: *VIth International Immunopathology Symposium*, Miescher PA, ed. Basel/Stuttgart: Schwabe & Co., pp. 97–103

Cerottini JC, Nordin AA, Brunner KT (1971): Cellular and humoral response to transplantation antigens. I. Development of alloantibody-forming cells and cytotoxic lymphocytes in the graft-*becsus*-host reaction. *J Exp Med* 134: 553–564

Nordin AA, Cerottini JC, Brunner KT (1971): The antibody response of mice to allografts as determined by a plaque assay with allogeneic target cells. *Eur J Immunol* 1: 55–56

3
Overview of CTL–Target Adhesion and Other Critical Events in the Cytotoxic Mechanism

Eric Martz

Despite decades of work and much real progress, the mechanisms by which cytotoxic T lymphocytes (CTL) damage target cells remain largely obscure. In addition, the question has been raised whether cytolysis is the primary antiviral effector mechanism employed by CTL *in vivo* (Martz and Howell, 1989). Greater progress has been made in understanding the mechanisms by which CTL adhere to targets. Many of the molecules involved in recognition and adhesion have been identified, although the mechanism of adhesion-strengthening, the recognition-controlled rapid increase in avidity of adhesion molecules for target ligands, remains to be elucidated.

My goal here is to give an overview of the steps in CTL–target interaction. Areas that have been extensively dealt with elsewhere will be summarized briefly, while some areas that have received less attention will be highlighted. I shall also attempt to provide some historical perspective.

CTL Timeline

In this vein, here is a brief timeline for CTL biology. Where references are omitted, details and citations will be found in the text. My attempt to keep this short results in the omission of many important findings, for which I apologize to my colleagues. Also, in general I have tried to list the earliest observation, which in some cases is not the most definitive.

1960
  Cell-mediated killing by lymphocytes immunized *in vivo* (Govaerts, 1960).
  Adhesion, putatively specific, first noted microscopically for sensitized lymphocytes (Rosenau and Moon, 1961).

1960s
  Assays extended over 1 or more days.

1967
  ^{51}Cr begins to be used with cell-mediated killing, leading to shorter assays (a few hours).

1968
  Lymphotoxin discovered (reviewed in Gately and Mayer, 1978; Paul and Ruddle, 1988).

1969
  In vitro generation of specific killer cells (Berke et al., 1969)

1970
  CTL spare antigenically "innocent" bystanders, despite killing adjacent cognate targets (Cerottini et al., 1970).
  Thy-1 alloantisera verify that most allospecific cell-mediated killing activity resides in T cells, hence the term **CTL**.
  Ca^{2+} important for cytolysis (i.e., programming for lysis; Mauel et al., 1970).

1971
  Clonality: CTL specificities separated by specific adhesion to target monolayers (Golstein et al., 1971; Brondz and Snegiröva, 1971).

1972
  Peritoneal exudate lymphocytes (PEL) following intraperitoneal ascites tumor allograft rejection identified as a rich and relatively pure source of highly active CTL (Berke et al., 1972b), often used to this day.

CYTOTOXIC CELLS
M. Sitkovsky and P. Henkart, Editors
© 1993 Birkhäuser Boston

Recycling likely: PEL can kill more target cells than the number of PEL when repeated populations of targets are provided (Berke et al., 1972b).

Assay with ^{51}Cr and PEL at effector-to-target cell ratios near 1 routinely shortened to about 90 min, permitting investigators to get home for dinner! (Berke et al., 1972c).

Apoptosis distinguished from necrosis (Kerr et al., 1972).

1973

Killer cell-independent lysis follows < 1 hr of interaction of CTL with targets (Martz and Benecerraf, 1973).

Mg^{2+} important in CTL adhesion (Stulting and Berke, 1973). Studies of Ca^{2+} and Mg^{2+} throughout the 1970s refined understanding of the roles of divalent cations (Martz, 1980, reviewed in Martz, 1977; Golstein and Smith, 1977). The Mg^{2+} requirement resides in LFA-1 (Shaw et al., 1986; Dustin and Springer, 1989); Ca^{2+} is likely required in several processes that have been more elusive (Martz et al., 1982b; Clark et al., 1988).

1974

MHC restriction of recognition by CTL, reported by Zinkernagel and Doherty (1974), drove home the realization that the physiological function of CTL is to control virus infection; previously, CTL were viewed largely as effectors of allograft rejection.

1974–5

A few minutes of contact suffices for CTL to program target cells to lyse (Wagner and Röllinghoff, 1974; Martz, 1975a).

1975

One minute is sufficient for half-maximal adhesion formation (Martz, 1975a).

Conjugates, discrete and binary, counted microscopically, leading to the ability to enumerate CTL (Martz, 1975; Berke et al., 1975).

CTL unharmed when killing, since each CTL can unequivocally kill multiple target cells sequentially in time (Zagury et al., 1975).

CD8 (Lyt-2) recognized as a CTL marker (Cantor and Boyse, 1975).

Concanavalin A induces CTL to kill nonspecifically (Bevan and Cohn, 1975).

Ca^{++}-independent lysis (Martz, 1975b; Plaut et al., 1976; Martz, 1977; MacLennan et al., 1980).

1976–7

Apoptosis induced in targets by CTL but not complement (Sanderson, 1976; Don et al., 1977).

1977

Golgi/MTOC (microtubule-organizing center) in CTL reorient toward target (Thiernesse et al., 1977).

1978–9

Granule secretion noted from CTL (Bykovskaja et al., 1978a; David et al., 1979).

1979

CD8 (Lyt-2) is the first molecule shown to have a specialized role in CTL function (reviewed in Martz et al., 1982a).

Cloning of CTL (Baker et al., 1979).

1980

DNA fragmentation induced prelytically in targets by CTL but not complement (Russell et al., 1980; Russell and Dobos, 1980).

CD8 (Lyt-2) recognized as first T-cell adhesion molecule (Fan et al., 1980).

1981

Cloned CTL shown capable of controlling virus infection *in vivo* (Lin and Askonas, 1981).

LFA-1 and CD3 shown to have specialized role in CTL function (reviewed in Martz et al., 1982a; Golstein et al., 1982).

LFA-1 recognized as second T-cell adhesion molecule (Davignon et al., 1981a).

CD8 (Lyt-2) matches MHC class I restriction, not CTL function (Swain, 1981).

1982

Secretory granules seen in natural killer (NK)-mediated lysis (Henkart and Henkart, 1982).

$CD4^+$ cloned CTL are MHC class II-restricted (Krensky et al., 1982; Biddison et al., 1982; Meuer et al., 1982).

1983

Rings, reminiscent of complement, seen on CTL-lysed target membranes (Dennert and Podack, 1983).

Antigen receptor protein of CTL identified as a clonotypic protein (Meuer et al., 1983; Haskins et al., 1983).

L3T4 (mouse CD4) expressed on murine MHC class II-restricted CTL (Dialynas et al., 1983).

1984
  Granules from CTL (Podack and Konigsberg, 1984) and NK (Henkart et al., 1984) are cytolytic.
  CD2 and LFA-3 function in adhesion (Krensky et al., 1984).

1985
  Serine esterase in CTL (Pasternak and Eisen, 1985).
  Perforin protein purified from CTL granules (Masson and Tschopp, 1985; Podack et al., 1985).
  Ca^{++}-independence varies with target (Tirosh and Berke, 1985).

1986
  Perforin homologous with 9th component of complement (reviewed in Podack et al., 1991).
  Lymphotoxin induces DNA fragmentation (Schmid et al., 1986).
  Serine esterase-related genes/cDNAs cloned (reviewed in Brunet et al., 1988).
  Granular, esterase-positive CTL induced by LCM virus *in vivo* (Biron et al., 1986; Mcintyre and Welsh, 1986; Simon et al., 1986).
  Talin in CTL but not target concentrates in contact zone Kupfer et al., 1986).
  ICAM-1 identified (Rothlein et al., 1986).
  LFA-1:ICAM-1 avidity regulation. Activation of adhesion does not involve density changes, and hence involves configurational avidity changes (Rothlein and Springer, 1986).

1986-7
  Serine esterase release from CTL triggered by the antigen receptor (Pasternack et al., 1986; Takayama et al., 1987).

1987
  Peptide-binding groove in MHC class I revealed by x-ray crystallography (Bjorkman et al., 1987).
  Ca^{2+} and perforin-independent pathways(s) for cytolysis likely (Berke and Rosen, 1987a; Berke and Rosen, 1987b; Brunet et al., 1987; Trenn et al., 1987; Ostergaard et al., 1987).

1988
  Interleukin-2 (IL-2) *in vitro* induces lytic granules/esterases in PEL lacking them (Berke and Rosen, 1988).
  DNA is not fragmented by CTL granules (Duke et al., 1988; Gromkowski et al., 1988a) except postlytically by high concentrations (Podack, 1986; Munger et al., 1988).
  Immobilized MHC class I binds and activates CTL (Kane et al., 1988).
  Detachment of adherent targets by CTL but not complement (Russell et al., 1988).

1989
  Perforin genes and cDNAs cloned (reviewed in Podack et al., 1991).
  DNA is not fragmented by perforin (Duke et al., 1989).
  Polyoma virus DNA is fragmented during CTL attack (Sellins and Cohen, 1989).
  LFA-1, not ICAM-1, is the molecule that changes during activation to increase mutual avidity (Dustin and Springer, 1989).

1990
  CD8 undergoes avidity regulation for MHC class I (O'Rourke et al., 1990).

1991
  CD2 undergoes avidity regulation for LFA-3 (Hahn et al., 1991).

1992
  Size of target ($\geq 4\,\mu m$) crucial in activating CTL (Mescher, 1992).

What Is a CTL?

"CTL" stands for "cytotoxic T lymphocyte," one of several types of cells that kill other cells. Thus **any lymphocyte which is cytotoxic and thymus-derived** qualifies as a CTL. Most, but not all, killer T cells kill in an antigen-specific, MHC-restricted manner. Culturing of T cells, especially with exogenous cytokines, may encourage the development of nonspecific or NK-like killing (Shortman et al., 1984; Shortman and Wilson, 1986). The role of such nonspecific T killers *in vivo*, if any, is unclear; at the present time, there is no basis for excluding them from the designation "CTL," although the term "CTL" is usually meant to imply antigen-specific, MHC-restricted killing.

Historically, **T cells** were identified by the expression of Thy-1 (earlier called θ) on mouse cells or CD2 (then called the sheep red blood cell receptor, or T11) on human cells. Antigen-specific killer cells were first recognized to be T cells because they could be generated from thymocytes and, more definitively, because the resulting killing activity could be deleted with anti-θ serum (or rabbit antithymocyte serum) and complement (Cerottini et al., 1970; Chapter 2, this volume). Presently, T cells can best be defined by the expression of an $\alpha\beta$ or $\gamma\delta$ antigen-specific receptor resulting from genetic rearrangement of germ-line genes, or the resulting MHC-restricted antigen-specific recognition.

"Cytotoxic" means having a toxic effect on target cells. Formally, "cytotoxicity" includes a wide range of possible effects: mild cytotoxicity, such as a slowing or halting effect on cell growth (cytostasis); impairment of metabolic activities; the permanent loss of the ability of a cell to divide (loss of reproductive potential or clonogenicity); irreversible cessation of respiration and energy metabolism (cytocidal effect); and the most extreme form, physical breaking of the cell membrane and dissolution of cell structure (cytolysis). I and others have at times preferred to consider "CTL" to stand for "cyto**lytic** T lymphocyte" because most work with CTL since the introduction of the ^{51}Cr release assay (Holm and Perlmann, 1967) employed rapid cytolysis as the definition of CTL function. Recently, however, I have questioned whether rapid lysis is the most important function of CTL *in vivo* in controlling virus infections (Martz and Howell, 1989; Martz and Gamble, 1991; Chapter 35, this volume). Hence, the original and broader term "cytotoxic T lymphocyte" remains the most satisfactory general appellation, although "cytolytic T lymphocyte" is perfectly appropriate when cytolysis is the assay employed.

The definition of "CTL" remains largely functional, and hence the term "CTL" embraces **diverse cells** which likely employ several distinct cytolytic mechanisms, and possibly additional antiviral mechanisms or other effector mechanisms as well. Generally, cells that take more than 1 day to lyse target cells would be considered noncytotoxic. If nothing else, times longer than 1 day allow for expression of new genetic programs (differentiation) and proliferation; hence, cells which produce such slow killing may be the *precursors* of CTL. Indeed, it is known that classical allospecific CTL memory cells which lack rapid cytolytic function (assayed within a few hours) can aquire such function in less than a day, and that this requires protein but not DNA synthesis (MacDonald et al., 1975).

In addition to classical CD8$^+$ CTL, some CD4$^+$ CTL have similar effector properties, such as bystander sparing (see below). Other CD4$^+$ T cells with inflammatory helper activity (e.g., IL-2 and interferon-γ) kill targets with different properties (Tite et al., 1985; Jones et al., 1986; Schmid et al., 1986; Nakamura et al., 1986; Ozaki et al., 1987; Gromkowski et al., 1988b; Strack et al., 1990; Tite, 1990a, 1990b; Ju, 1991). These **inflammatory CTL** kill more slowly and are much less sparing of bystanders. They typically use tumor necrosis factor (TNF) as one of their effector molecules, but this is not their only lytic mechanism (Ju et al., 1988, 1990). Inflammatory CTL are beyond the scope of this review.

In view of the heterogeneity of CTL and their similarities to NK cells, it is perhaps not surprising that **no protein markers** have been identified that reliably distinguish CTL, as a group, from noncytotoxic T cells, or from cytotoxic non-T lymphocytes. Under many circumstances, the majority of CTL responding to an infection, alloantigen, or immunization are restricted by class I MHC and hence express CD8. However, class II-restricted CD4$^+$ CTL are typically present as a minority, and can be cloned as readily as can class I-restricted CTL (Wagner et al., 1977; Maimone et al., 1986); also CD8$^+$ T cells lacking cytotoxic activity (often having helper functions) exist (Swain et al., 1980). Cytolytic granules, perforin, and characteristic esterases are present at very low to undetectable levels in some highly efficient CTL (Brunet et al., 1987; Berke and Rosen, 1987c; Dennert et al., 1987) and are present in NK cells.

In view of the **heterogeneity** of CTL, it is important not to think of CTL as a "single" cell type. When analyzing the properties of CTL, it is important to keep clearly separate the different types of CTL. CTL generated *in vivo* have different properties from CTL generated *in vitro*, and time *in vitro* or cloning induces further changes. The amounts of granules (if any), perforin, certain esterases, or cytokines produced also vary and have functional consequences. Even within a single population of CTL multiple mechanisms of lysis are involved. Both calcium-dependent and calcium-independent pathways are typically active. Certain target cells may be insensitive or hypersensitive to certain pathways of damage (Tirosh and Berke, 1985; Ostergaard and Clark, 1989; Sugawara et al., 1990; Landon et al., 1990). Unfortunately, most work to date has tended to lump CTL into a single category. It is now of the utmost importance that attention be paid to discriminating between types of CTL and types of effector mechanisms. Progress will require development and refinement of operational methods to categorize CTL by markers, metabolic pathways, divalent cations, mediators, and target

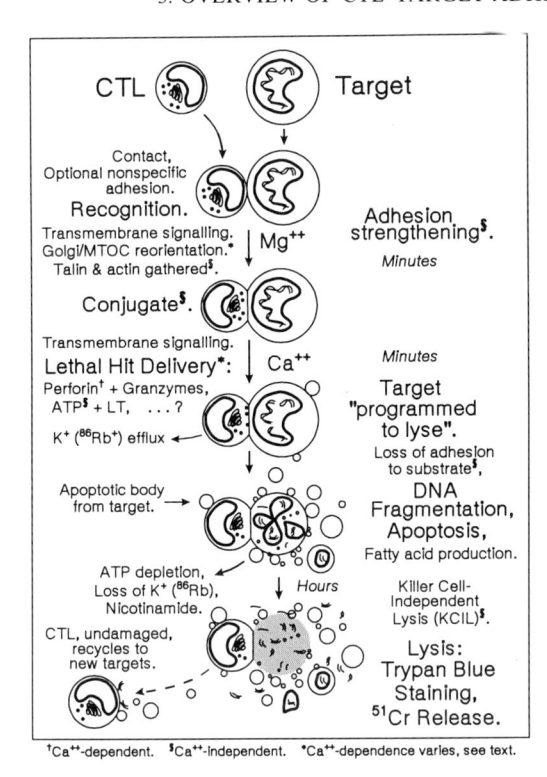

FIGURE 1. Overview of CTL–target interaction.

susceptibility. Ultimately one hopes for the development of clones that express each lytic mechanism, one per clone (Cosgrove et al., 1991), or for drugs that selectively block single lytic pathways.

In view of the limitations of much of the past work on CTL, the following overview will often lapse into presenting generalizations as though they apply to all "CTL." The reader is advised to maintain a healthy skepticism regarding whether any given phenomenon applies to all CTL—*caveat emptor*.

Steps in CTL–Target Interaction

Figure 1 illustrates the steps in CTL–target interaction, each of which is discussed below. The remainder of this chapter will discuss each step, more or less in the order in which they occur during CTL–target interaction.

Requirement for Contact

A premise in the following discussion is that CTL-mediated killing requires intimate contact between the CTL and target cell membranes, at least for efficient killing. It is believed that most, if not all, physiological CTL recognition and lethal hit delivery occurs in rapid succession on one and the same antigen-presenting (e.g., infected) cell, each step requiring intimate contact.[1] *In vitro*, conditions that favor CTL–target contact increase killing. In early experiments using flat dishes, the dishes were rocked to improve killing (Canty and Wunderlich, 1970; Berke et al., 1972b). Conditions that minimize direct contact, such as suspension of the cells in viscous, dextran-containing medium, prevent killing, while centrifugation, which maximizes intercellular contact in a pellet, is routinely used to maximize killing (Martz, 1975a, 1977).

Nonspecific Adhesion

Regardless of whether a target cell bears antigen, a contacting CTL has the opportunity to engage in adhesion with it. Certain types of CTL, especially cloned CTL, form adhesions promiscuously with any cell they contact (reviewed in Martz, 1987; Blakely et al., 1987). This may be an artificial result of prolonged culturing, which may tend to encourage constitutive hyperaction of adhesion molecules such as LFA-1, CD8, and CD2, known to mediate such nonspecific adhesions. On the other hand, nonspecific adhesion may provide the antigen receptors with an optimal chance to engage cognate antigen-bearing MHC molecules on the target cell surface. It is also possible that nonspecific adhesion is activated in response to cell contact by receptor–ligand interaction other than that between TCR and antigen. In any case, mouse allospecific CTL generated *in vivo* show little or no nonspecific adhesion (see below) yet are very efficient specific killers; hence, nonspecific adhesion is not necessary in all cases.

Antigen Recognition and Signaling

If cognate antigen exists on the target in the form of peptide–MHC complexes, these are engaged by the T-cell antigen receptors, and cytosolic metabolic signals are generated within seconds (see Chapter 49, this volume). These signals include tyrosine kinase activation (Rudd, 1990) and polyphosphoinositide hydrolysis (PI pathway) leading to a rise in intracellular Ca^{2+} and production of diacylglycerol, which activate protein kinase C. Release of inositol phosphate intermediates (Treves et al.,

1987; O'Rourke and Mescher, 1988) and a rise in intracellular Ca^{2+} (Gray et al., 1987; Poenie et al., 1987) occur rapidly upon CTL recognition of targets. There is reason to believe that the products of the PI pathway, elevated cytosolic Ca^{2+} (indicated with ionophores) and diacylglycerol (indicated with phorbol esters), are sufficient to trigger both adhesion-strengthening and lethal hit delivery (Munger and Lingquist, 1982; Russell, 1986; Berrebi et al., 1987; Treves et al., 1987; Sitkovsky, 1988; Fortier et al., 1989). Multiple pathways or compartments for Ca_i^{2+} may be required for a full response (Haverstick et al., 1991). However, specific killing of targets can also occur under conditions where neither release of inositol phosphates (O'Rourke and Mescher, 1988) nor rise in Ca_i^{2+} (Ostergaard and Clark, 1987) can be detected, suggesting that alternative signaling events are also sufficient to trigger CTL, perhaps invoking a different lytic mechanism (O'Rourke and Mescher, 1988).

Conjugate Formation

Conjugates form rapidly when CTL contact target cells. A conjugate is a CTL that adheres to a target cell with sufficient strength to resist shear forces that separate control cells. The control depends on the system being tested. In systems where conjugation requires antigen recognition, the controls involve substitution of noncognate target cells (or CTL) for cognate target cells (or CTL). In such cases, a reciprocal protocol is required to determine whether conjugation depends on immunologically specific recognition of antigen (reviewed in Martz, 1987; Hubbard et al., 1990). A reciprocal protocol employs two cognate CTL–target pairs, and test each type of CTL against both types of targets.

Mouse CTL form conjugates **specifically**, that is, adhesion formation depends on antigen recognition (reviewed in Martz, 1987). For both PEL (which contain non-CTL) and cloned CTL, typically the ratio of the number of conjugates with cognate targets to those with noncognate targets is four to eight. This level of specificity is much lower than the specificity of cytolysis, which is about 10-fold greater. Human CTL clones often adhere to potential targets **promiscuously**, in an antigen-nonspecific manner (reviewed in Martz, 1987). Some mouse cloned CTL behave similarly (Blakely et al., 1987). Possibly cloning has rendered adhesion molecules (e.g., LFA-1) constitutively activated. Whether human CTL conjugate specifically *in vivo* is not known.

In the early 1960s, before T-cell markers were available or ^{51}Cr was in use, cytotoxicity was assessed by microscopic observation. Several groups observed apparently specific adhesion to target cells of what were likely CTL (Rosenau and Moon, 1961; Kowproski and Fernandes, 1962; Taylor and Culling, 1963; Wilson, 1963). However, the major interest was in cytotoxicity; the adhesions were not enumerated, studied with reciprocal protocols, nor were the fates of binary conjugates followed. After the first use of ^{51}Cr in 1967 to study cell-mediated cytotoxicity (reviewed in Martz, 1976a), this assay rapidly became popular, resulting in less microscopic observation. Quantitative but indirect evidence for immunologically specific adhesion of mouse CTL to target cell monolayers was provided by several groups, using either microscopy (Brondz, 1968) or ^{51}Cr (Golstein et al., 1971; Berke and Levey, 1972) to estimate lysis.

Not until 14 years after the first report of specific killer–target adhesion were **discrete conjugates enumerated directly by microscopy**. Martz (1975a) and Berke et al. (1975, introducing the term "conjugate") showed that conjugation preceded and could quantitatively account for ^{51}Cr release at the population level. Subsequently, microscopic monitoring of the fates of targets in 1 : 1 CTL : target cell conjugates showed that one CTL can kill a target (indeed, multiple targets sequentially in time, Zagury et al., 1975; see also Martz and Benacerraf, 1976) and permitted uneqivocal **enumeration of CTL** in uncloned lymphocyte populations for the first time (Grimm and Bonavida, 1977; Ryser et al., 1977). Not all CTL form strong conjugates (Ryser et al., 1977), and some conjugating CTL may fail to lyse their targets within a given observation period and conditions. Of course, some non-CTL lymphocytes form conjugates (e.g., NK cells in antibody-independent or Fc-receptor antibody-dependent modes), and helper T cells with cognate antigen-presenting cells (APC) which may be B lymphocytes (see below). Thus, even the most refined methods for enumerating CTL by conjugate fate analysis (Grimm and Bonavida, 1979; Bonavida et al., 1983) are subject to some uncertainty.

Conjugate-based analyses also made possible the demonstration on the single conjugate level that, compared to primary CTL, secondary CTL (two cycles of immunization) are able to adhere to targets faster (Glasebrook, 1978) and also kill targets faster (Grimm and Bonavida, 1977). Ultimately, conjugate-based analysis facilitated the identification of the molecules that mediate T-cell adhesions (reviewed in Martz et al., 1982a). Although initially recognized for CTL, it is now clear that conjugation is a general response of T cells to antigen-presenting cells, e.g., T-helper cells with B cells (Inaba et al., 1984; Ohnishi and Bonavida, 1986; Sanders et al., 1986; Kupfer and Singer, 1989a).

Assays for Conjugation and Its Strength

The most reliable and definitive assay for conjugate formation is by **microscopic counting** of conjugates vs. unconjugated cells following a reproducible level of shearing. This works best with target cells that are low in nonspecific "stickiness," typically transformed leukocyte cell lines. This assay gives accurate results at the expense of being tedious and, since the number of cells sampled is necessarily modest, having some imprecision.

Flow cytofluorometry has also been used to enumerate conjugates (Berke, 1985; Perez et al., 1985) and even to distinguish living from dead targets in conjugates (Lebow et al., 1986). This method has a greater setup cost. It may underestimate weak conjugates, since the instrument subjects the conjugates to shear. It does not lend itself as well as does microscopy to determination of the rate of lysis of conjugated targets (and hence estimation of the percentage of conjugating lymphocytes which are CTL), since CTL may detach after target cell death but before cytometric analysis. However, if conjugation *per se* is the goal, it enables the processing of large numbers of samples with minimal statistical imprecision.

Indirect, population-level assays for conjugate formation have been based on cytolysis (^{51}Cr release) and termed **postdispersion lysis**. The percentage of targets in conjugates at a given point in time is equated to the percentage that lyse following maneuvers (such as dispersion in viscous dextran solutions) to prevent any subsequent CTL–target interactions. This type of assay is capable of excellent quantitation and time resolution, showing half-maximal conjugation of PEL in 1–2 minutes (Martz, 1975sa, 1977; confirmed by flow cytometry, Perez et al., 1985). An adaptation allowed the demonstration of a reversible equilibrium in the adhesions between CTL and targets agitated in suspension (Balk et al., 1981; Balk and Mescher, 1981a). Earlier assays utilized panning of the CTL-containing cell populations on target cell monolayers (reviewed in Martz, 1977; Berke, 1980) but were less satisfactory for most purposes.

Strength of CTL target adhesion can be estimated in relative terms by varying the amount of shear applied before enumeration of conjugates (Bongrand and Goldstein, 1983; Bongrand et al., 1983; cf. Garcia-Peñarrubia and Bankhurst, 1989a). A more elegant method involves the use of laminar shear between parallel plates to remove a population of labeled CTL from a fibroblast target monolayer. Using this method, Hubbard et al. (1990) made the important point that "specificity" of CTL: target adhesion may be overlooked if too low a shear is applied. Both groups noted a wide range (100-fold) in the strengths of conjugates. Absolute strengths have been calculated from the force necessary to separate the CTL from the target using micropipets (Sung et al., 1986; Tozeren et al., 1989); however, the separation may measure tearing of cell membranes rather than a true intermembrane adhesion strength.

Direct quantitation of temperature-dependent CTL **adhesion-strengthening** was achieved with the laminar shear method (Hubbard et al., 1990). LFA-1-dependent adhesion-strengthening has also been quantitated in the interaction between macrophages and tumor cells (Strassman et al., 1986).

Adhesion-Strengthening Controlled by Antigen Recognition: LFA-1, CD8, CD2

Antigen-specific conjugation of CTL to target cells requires that CTL–target adhesion be determined by the T-cell antigen receptor. In principle, the adhesion could be either mediated directly by the **antigen receptor–peptide–MHC bond**, or mediated by other molecules in turn controlled by the antigen receptor. Early observations of CTL–target conjugation showed a large temperature-dependence, energy-dependence, and Mg^{2+}-dependence. These seemed unlikely to represent properties of the antigen receptor–antigen bond. Thus it was proposed that the antigen receptor controlled an adhesion-strengthening process involving other molecules (Martz, 1975a, 1980).

It is now clear that **LFA-1** plays a major role in most CTL–target adhesions. LFA-1 was the second T-cell accessory molecule recognized to have an adhesion function (see timeline at 1981). The binding of CTL LFA-1 to target cell **ICAM-1** contributes a large portion of CTL–target adhesions; CD2–LFA-3 and CD8–MHC class I also contribute. In addition to their mechanical binding function, not only the antigen receptor–CD3 complex, but also LFA-1 (van Seventer et al., 1990;

Moy and Brian, 1992), CD8 (O'Rourke et al., 1990), and CD2 (Bierer and Burakoff, 1989) can each generate costimulatory metabolic signals in the CTL cytoplasm.

LFA-1 is a member of the **integrin** superfamily of cell adhesion molecules (Springer, 1990). All integrins have in common functional dependence on **temperature and divalent cations** (typically Ca^{2+}, but Mg^{2+} in the case of LFA-1). The alpha chain of LFA-1 contains several consensus metal ion-binding sites. It has been shown that the temperature- and Mg^{2+}-dependence of CTL-target adhesion resides in the LFA-1-ICAM-1 component and not in the CD2-LFA-3 component (Shaw et al., 1986). Moreover, it has been shown that the direct binding of purified LFA-1 micelles to immobilized purified ICAM-1 is Mg^{2+}-dependent (Dustin and Springer, 1989). Thus, the Mg^{2+}-dependence and much of the temperature/energy dependence quite likely reside in the LFA-1 component of CTL adhesion.

The detachment and dispersion assay revealed that the formation of specific adhesions to target cells by mouse CTL can be **half maximal within 1 minute** following contact and antigen recognition (Martz, 1975a, 1977). In a system which models this process (Mentzer et al., 1985), stimulation of lymphocytes with phorbol ester results in an increase in LFA-1 dependent adhesion strength of greater than tenfold; this increase begins to occur in less than 80 seconds (Martz, unpublished data). The increase occurs without a change in the densities of either LFA-1 or ICAM-1 (Rothlein and Springer, 1986), and has been shown to be due to a change in LFA-1 (not ICAM-1) which increases the LFA-1:ICAM-1 avidity (Dustin and Springer, 1989).

Elegant studies of T-helper cells, where the density of antigen on the antigen-presenting cell could be controlled, indicate that that activation of LFA-1 adhesion-strengthening requires a less intense signal, or a different kind of signal, than does activation of T-cell effector functions or mitogenesis (Kupfer and Singer, 1989b). Conjugation, with gathering of LFA-1 and talin (see below), was induced at lower densities of antigen than those required for MTOC reorientation (see below) or other T-cell responses.

In summary, recognition of the peptide-MHC complex by the antigen receptor-CD3 complex on the CTL triggers signal pathways which activate LFA-1, increasing its avidity for its ligand ICAM-1. Activation of LFA-1 involves undefined configurational changes which may include changes in the extracellular domains (van Kooyk et al., 1991), increased anchoring to the cytoskeleton (perhaps via talin, see below), or homotypic molecular clustering on the cell surface (known to occur for the closely related leukocyte integrin CR3; Detmers et al., 1987). These changes likely depend on phosphorylation or other events involving the cytoplasmic domains of the LFA-1 chains (Hibbs et al., 1991a, 1991b).

In addition to LFA-1-ICAM-1/2/3 (de Fougerolles and Springer, 1992), **CD8-MHC class I and CD2-LFA-3** contribute to CTL-target adhesion (Bierer and Burakoff, 1991). Rapid antigen-recognition controlled regulation of avidity for ligand occurs not only for LFA-1, but also for CD8 (O'Rourke et al., 1990) and probably for CD2 (Hahn et al., 1991). Thus, rapid avidity regulation is a common pattern of function for CTL accessory adhesion molecules. Recent evidence indicates that further accessory adhesion molecules are likely involved in interactions between T cells and antigen-presenting cells, namely, **CD43-ICAM-1** (Rosenstein et al., 1991). These four pairs of mutual ligands may be involved in most CTL-target interactions, since the target cell ligands ICAM-1, LFA-3 and MHC class I are expressed on most tissues. When CTL interact with certain types of target cells, additional adhesion molecules are likely involved, for example, selectins with endothelial targets (Springer, 1990) and CTL CD28 with BB1/B7 on hematopoietic target cells (Linsley et al., 1990; Koulova et al., 1991).

Adhesion Requirement for Efficient Cytolysis

Much evidence points to the necessity for a strong and extensive adhesion between CTL and target in order for efficient cytolysis to ensue. Although the mediators of cytolysis remain somewhat murky (see below), it seems likely that they are most efficient when delivered into an extensive area (Andre et al., 1990) of close apposition between CTL and target membranes. Without exception, agents that inhibit adhesion also inhibit cytolysis. These agents include drugs and antibodies, which will now be considered in turn.

The best evidence that adhesion is required for efficient cytolysis would be inhibition of cytolysis by a drug that inhibits adhesion but that does not inhibit *programming for lysis* as an isolated step (discussed below). Identification of such drugs has been difficult. If, indeed, efficient programming for lysis requires adhesion, how can the effect of the drug be tested on programming for lysis in isolation (while maintaining adhesion)? One approach was to "glue" the CTL to the targets with the lectin concanavalin A. This produced an adhesion insensitive to cytochalasin B, a drug which inhibits the antigen-specific adhesion. This made possible the demonstration that, at the proper dose, **cytochalasin B** inhibits adhesion but not programming for lysis (Gately et al., 1980). Earlier evidence, based on addition of cytochalasin A at various stages, supported the same conclusion (Golstein et al., 1978; Shortman and Golstein, 1979).

Thus, the fact that cytochalasins inhibit cytolysis is strong evidence that adhesion is required for efficient cytolysis. Independent evidence comes from the use of antibodies to adhesion molecules, notably CD8 and LFA-1.

Antibodies to CD8 were shown to inhibit CTL-mediated killing in 1979 (see timeline). CD8 (then called Lyt-2) was the first T-cell receptor recognized to function in adhesion (shown by inhibition of CTL–target conjugate formation by antibody; Fan et al., 1980). In this original study, the authors showed that although antibody to CD8 significantly reduced the number of conjugates formed, the targets in the conjugates which did form were killed to the same extent as targets in control conjugates. They also showed that the addition of antibody after conjugate formation did not reduce the amount of killing. These findings strongly support the conclusion that antibody to CD8 has no direct inhibitory effect on programming for lysis, but inhibits indirectly by inhibiting adhesion. Thus they support the conclusion that adhesion is necessary for effective killing.

Antibodies to LFA-1 (either the alpha or beta chain) are potent inhibitors of CTL-mediated cytolysis (Sanchez-Madrid et al., 1983; Ware et al., 1983). Indeed, LFA-1 was discovered by screening for monoclonal antibodies to CTL which block cytolysis (Davignon et al., 1981b; Dialynas et al., 1982; Pierres et al., 1982; Sanchez-Madrid et al., 1982a, 1982b; LFA-1 was the first lymphocyte receptor to be discovered in this way). Antibodies to LFA-1 also greatly diminish the strength of adhesion between CTL and targets, reducing conjugate numbers (Davignon et al., 1981a; Krensky et al., 1984).

Antibodies to LFA-1 can generate **costimulatory or inhibitory metabolic signals** (Mishra et al., 1986; Carrera et al., 1988; Geppert and Lipsky, 1988; van Noesel et al., 1988; Pardi et al., 1989), as can binding to ICAM-1 (Braakman et al., 1990; van Seventer et al., 1990; Moy and Brian, 1992). Antibody-induced inhibition of *cytolysis* could result from (1) inhibitory signals, (2) interference with generation of stimulatory signals normally resulting from interaction with ICAM-1, or (3) interference with adhesion *per se*. The signal pathways could diminish the activation of programming for lysis mechanisms and/or the activation of other accessory adhesion molecules such as CD8 and CD2. Diminished activation of CD8 and CD2 would result in less binding of their respective ligands (O'Rourke et al., 1990; Hahn et al., 1991). Since ligand binding results in costimulatory signals from both CD8 (O'Rourke et al., 1990) and CD2 (Bierer and Burakoff, 1991), diminished binding of these receptors could further diminish activation of programming for lysis.

Antibodies to LFA-1 could inhibit *adhesion* by any of several mechanisms. Most simply, they could **sterically** block the ICAM-1 binding site on LFA-1. Alternatively, they could alter the configuration of LFA-1 so as to indirectly reduce its **avidity** for ICAM-1. Such alterations could include both physical alterations of the binding site and alterations of other sites so as to interfere with the ability of the activators (e.g., kinases) to produce or maintain the activated state. In addition to reducing the amount of LFA-1–ICAM-1 binding, the changes in LFA-1-generated **signals** (mentioned above) could reduce the amount of CD8–MHC and/or CD2–LFA-3 binding, thereby further reducing intercellular adhesion.

The possibilities are not greatly simplified in the case of inhibition by **antibodies to ICAM-1**. Anti-ICAM-1 antibodies inhibit both cytolysis (Makgoba et al., 1988) and adhesion (Rothlein et al., 1986). With anti-ICAM-1, the signals known to be generated by the binding of antibody directly to LFA-1, case (1) above, are eliminated. However, effects (2) and (3) remain likely. Additionally, CTL may express ICAM-1, and the binding of antibody to this receptor could generate signals.

In summary, antibodies to LFA-1 and ICAM-1 inhibit cytolysis as a result of inhibition of CTL–target adhesion. However, it cannot presently be distinguished whether the inhibition of cytolysis derives solely from the reduction in physical access (reduced duration and area of adhesion, increased distance between membranes) or also from the resultant alterations in cytoplasmic signals which may diminish the programming-for-lysis activity for the CTL. Perhaps it will be possible through molecular engineering to create LFA-1 capable of being activated but not of generating signals, or vice versa, for testing these hypotheses. For now, adhesion, likely including adhesion-dependent signals, appears to be required for efficient cytolysis.

Is Bystander Lysis Dependent on Active Bystander Adhesion?

One of the striking properties of classical allospecific MHC class I-restricted CTL generated *in vivo* (e.g., the peritoneal exudate lymphocytes or PEL first described by Berke) is their ability to **spare bystander cells**, that is, noncognate targets mixed among cognate targets (Berke et al., 1972a; Cerottini and Brunner, 1974; Martz, 1975b; Wei and Lindquist, 1981). Bystanders are also spared by some *cloned* class I-restricted CTL (Duke, 1989) as well as by some primary (Golding and Singer, 1985) and cloned (Lukacher et al., 1985; Maimone et al., 1986; Ottenhoff and Mutis, 1990) class II-restricted CTL. The cytolytic action of these CTL appears to be narrowly focused/vectorially delivered toward the cognate target cell, sparing noncognate bystander cells that are in close contact with the same CTL during the killing of the cognate targets.

In contrast to CTL generated *in vivo*, CTL **cloned** *in vitro* seem more prone to killing bystanders. Although some clones spare bystanders, bystanders are killed by some class I-restricted human (Fleischer, 1986; Lanzavecchia, 1986) and mouse (Duke, 1989; Skinner et al., 1989a) clones. Fleischer noted the lack of bystander killing with "uncloned multiclonal CTL lines."

Duke made the important observation that in most cases of mouse bystander killing, expression of **self-MHC** on the target is required, evidently reflecting a weak recognition by the T-cell antigen receptor (Duke, 1989). Thus, in future studies, it will be important to distinguish killing of self-MHC bystanders vs. irrelevant third-party MHC or MHC-negative bystanders.

A hypothesis which seems consistent with most of the data is that killing of bystanders requires both activation of CTL lytic machinery (e.g., by recognition of cognate peptide : MHC on a target cell) and simultaneous **activation of adhesion molecules on the CTL** (LFA-1, CD8, CD2, etc.) toward the bystander target. The absence of bystander killing by PEL could then be explained if activation of adhesion molecules by recognition of antigen is localized to the region of the CTL in contact with the cognate target cell; that is, if recognition does not lead to bystander adhesion. The extent to which bystander adhesion occurs has not been tested, so far as I know. Killing of bystanders by cloned CTL could be explained on the basis that cloning tends to change CTL from adhering specifically to adhering promiscuously (reviewed in Martz, 1987), probably by leading to constitutive activation of signals that control functional affinity of LFA-1, CD8, and CD2.

Another factor that may contribute to bystander sparing is the **reorientation of the Golgi/MTOC** toward the cognate target that occurs immediately after recognition (see below). The notion that this may vectorially direct the lethal hit is consistent with the observation that CTL adhering simultaneously to more than one cognate target appear to kill only one target at a time (Zagury et al., 1979). The plausible role of MTOC reorientation in bystander sparing (Kupfer et al., 1986) remains to be tested. Golgi reorientation is obviously directed by signals from the antigen receptor and accessory receptors. A higher degree of activation of accessory adhesion molecules leading to ligand engagement on the bystander cell, even in the absence of antigen receptor engagement on the bystander, could well intensify Golgi-attracting signals. Ligation of LFA-1 and CD2 remote from the site of antigen receptor engagement has been shown capable of generating costimulatory signals (van Seventer et al., 1991).

At least two studies have **directly addressed** the role of bystander adhesion. In one, bystanders were "glued" to CTL with wheat germ agglutinin or antithymocyte serum (which engaged Fc receptor on the targets employed). The bystanders were spared despite being glued to CTL which were simultaneously killing cognate targets (Wei and Lindquist, 1981). In another study, the bystanders were T-helper cells which recognized alloantigens on the CTL, thus producing an adhesion active from the bystander side. The adhering bystander T-helper cells were killed when the CTL engaged cognate targets (Lanzavecchia, 1986). Thus, the two studies used similar strategies but came to opposite conclusions. It can be noted that when the bystanders were not killed, the CTL were mouse PEL generated *in vivo*, whereas when they were killed, the CTL were human clones generated *in vitro*. Human CTL clones are well known for promiscuous adhesion dependent on LFA-1 and CD8 (reviewed in Martz, 1987).

In conclusion, the evidence is consistent with the hypotheses that bystander killing (1) requires engagement of accessory adhesion molecules in an active manner from the CTL side, and (2) is augmented when the CTL antigen receptor recognizes self-MHC even in the absence of cognate peptide (Duke, 1989). Direct tests for the roles of bystander adhesion and of LFA-1, CD8, and CD2 in bystander killing are needed.

The same mechanisms important in control of bystander lysis likely also regulate delivery of help from T cells to bystander B cells (Inaba et al., 1984; Kupfer and Singer, 1989c).

MTOC and Talin Reorientation

CTL, like most cells, have an organelle-rich region of the cytoplasm centered around the Golgi apparatus and the centrosomes or MTOC. In granular CTL, many of the granules are found in this region (Yannelli et al., 1986). When the CTL initially comes into contact with a possible target cell, the MTOC may be in any random position. If, and only if, the target cell is recognized by the CTL as bearing cognate antigen (Kupfer and Dennert, 1984; Yannelli et al., 1986), an active reorientation of the MTOC occurs within a few minutes, resulting in the MTOC facing the region of CTL–target contact (Thiernesse et al., 1977; Bykovskaja et al., 1978b; Geiger et al., 1982; Kupfer and Singer, 1989a). The same occurs when T-helper cells conjugate with antigen-presenting B cells (Kupfer et al., 1987). The reorientation requires microtubule function (Kupfer and Dennert, 1984). The target's MTOC remains randomly orientated (Kupfer et al., 1986, 1987).

The CTL reorientation appears to "aim" the lytic apparatus of the CTL toward the target cell, and may contribute to the **bystander-sparing** characteristics of CTL-mediated killing. It may also be responsible for the fact, mentioned above, that when a CTL adheres to more than one cognate target simultaneously, the targets lyse one at a time, at intervals (Zagury et al., 1979; Kupfer et al., 1985). It is also possible that the CTL has greater **resistance** to its own lytic mediators in this region proximal to the MTOC. When one CTL recognizes a different CTL as a cognate antigen-presenting target, the response is unidirectional; MTOC reorientation occurs only in the recognizing, not the recognized, CTL (Kupfer et al., 1986). In cases where the recognized CTL resists cytolysis, it usually fails to induce reorientation of the MTOC (Ostergaard et al., 1988).

MTOC reorientation was Ca^{2+}-dependent in an allospecific CTL clone conjugated to cognate S194 myeloma cells, a combination in which cytolysis was inhibited >95% by the absence of Ca^{2+} [2.5 mM ethyleneglycol-bis(β-aminoethylether)-N,N,N',N'-tetraacetate (EGTA); Kupfer et al., 1985]. Restoration of Ca^{2+} induced reorientation within minutes, preceding lysis. MTOC reorientation was also found to be Ca^{2+}-dependent in a number of helper T-cell combinations with cognate antigen-presenting cells (Kupfer et al., 1987). These results were confirmed for several additional cloned CTL lines and target cells (Ostergaard and Clark, 1987). However, the cytolysis of certain target cells is Ca^{2+}-independent (Tirosh and Berke, 1985), and in those cases, MTOC reorientation is also Ca^{2+}-independent (Ostergaard and Clark, 1987).

Other cytoplasmic constituents that selectively accumulate in the contact region are **actin and talin** (Ryser et al., 1982; Kupfer et al., 1986; Kupfer and Singer, 1989a); talin is believed to link integrins (such as LFA-1) to the filamentous actin of the cytoskeleton. Other cytoskeletal constituents, such as myosin, alpha-actinin, and vinculin, are not gathered. This gathering of talin is independent of Ca^{2+} (Kupfer et al., 1987; Kupfer and Singer, 1989a), as is the LFA-1-dependent adhesion strengthening. Adhesion and talin-gathering are triggered by lower densities of antigen on the antigen-presenting cell than the densities required to induce MTOC reorientation (Kupfer and Singer, 1989b).

In summary, evidence to date is consistent with the hypothesis that MTOC reorientation is necessary for CTL-mediated cytolysis. It is, however, not sufficient for lysis, since a number of situations or drugs have been identified in which lysis is prevented but MTOC reorientation occurs (Ostergaard et al., 1988; Prochazka et al., 1988; Nowicki et al., 1991). Studies of inhibition by nicotinamide also indicate that phosphatidylinositol bisphosphate hydrolysis and increased Ca_i^{2+} may be unnecesary signals for MTOC reorganization (Nowicki et al., 1991).

Programming for Lysis/Lethal Hit Delivery

Cells (other than erythrocytes) are capable of repairing damage to their membranes mediated, for example, by complement (Burakoff et al., 1975; Ohanian and Schlager, 1981; Ramm et al., 1983; Morgan et al., 1987). Thus, cytolysis mediated by CTL might, in principle, require continued damage by the CTL until the moment of lysis. In fact, however, inactivation of the CTL during its action on the target cell revealed that CTL activity was needed for only a short time, and that lysis would nevertheless follow at a later time. In early experiments, **CTL were inactivated** by antibody plus complement (Ab + C; Martz and Benacerraf, 1973) or heat (Wagner and Röllinghoff, 1974, where the term *lethal hit* was introduced). Early time-lapse studies of CTL had indicated the same conclusion (Ginsburg et al., 1969, where the term *kiss of death* was introduced). The existence of *killer-cell-independent lysis* (*KCIL*) necessitates that the CTL be capable of programming the target cell for lysis at a later time. The phrase *programming for lysis* was introduced (Martz, 1975a) because it makes no assumptions about the nature of the effect of the CTL on the target required for later lysis. In particular, it is not clear whether programming for lysis initially involves patent physical damage (e.g., membrane pores) or a regulatory change (e.g., lymphotoxin or ATP), or both. With many issues about the mechanism of CTL attack remained unresolved, the phrase *programming for lysis* remains appropriate.

Various agents, such as ethylene diaminetetra acetate (EDTA) (to remove the Mg^{2+} needed for LFA-1 function) or cytochalasins, weaken CTL-target adhesion and so allow mild vortex shear to **detach CTL from targets** at will. When such induced detachment is followed by dispersion (e.g., in high-molecular-weight dextran solutions) to prevent new CTL-target interactions, the ensuing KCIL can be quantitated. Using this *detachment and dispersion* assay, the percentage of targets which ultimately lyse was *defined as the percentage of target cells programmed to lyse* (Martz, 1975a, 1977). Using PEL with sensitive target cells (e.g., P815) at effector-to-target-cell ratios > 1, adhesion became strong enough to resist vortex shearing in 1-2 min, but programming for lysis was half-maximal in 3-10 min. Thus, programming for lysis required additional time beyond that required for formation of a strong adhesion, and detachment induced prior to the completion of programming for lysis rescued the targets. These conclusions based on postdispersion ^{51}Cr release were confirmed using the clonogenicity of P815 target cells in soft agar (Martz, 1977).

Some related observations agree well with the original timing of programming for lysis. Fusion of CTL granules with the membrane facing the target cell has been observed within a few minutes after contact (Yannelli et al., 1986). A sharp rise in Ca_i^{2+} in the target cytoplasm occurs within minutes after CTL contact, and follows within about a minute a regulated rise in CTL Ca_i^{2+}.

The detachment and dispersion assay allowed further characterization of programming for lysis (reviewed in Martz, 1977). Programming for lysis is more **temperature**-sensitive than adhesion (which is more sensitive than KCIL). Thus, CTL can form strong, specific adhesions at 15°C, but little or no programming for lysis occurs at this temperature. Programming for lysis is partially Ca^{2+}-dependent (Martz, 1977; Martz et al., 1982b) and may also be Mg^{2+}-dependent (Redegeld et al., 1991). $^{86}Rb^+$, an analog of K^+, is released from target cells at about the time programming for lysis is completed (Martz, 1976b; Sanderson, 1981a). Whether this reflects membrane damage (e.g., perforin pores) or physiological responses to regulatory disturbances (e.g., K^+ release mediated by elevated Ca_i^{2+}; Poenie et al., 1987) is not known. Using ^{51}Cr release as the definition of lysis, some small metabolites (nicotinamide; Martz et al., 1974) are **released prelytically**, while others (nucleotides, amino acids; Sanderson and Thomas, 1977) are not. Possibly small leaks form early and gradually progress to larger leaks (as $^{86}Rb^+$ and nicotinamide suggest); alternatively, the membrane may remain relatively intact until a sudden collapse and lysis (in which case $^{86}Rb^+$ release is a physiologic response and nicotinamide is somehow unique).

Many **drugs** inhibit CTL-mediated killing (e.g., Martz, 1977). The vast majority of these, when tested, inhibit adhesion formation as well (e.g., cAMP; Gray et al., 1988). Exceptions include depletion of calcium with EGTA (Martz et al., 1982b), and certain calcium antagonists, including Mn^{2+} (Gately and Martz, 1981), verapamil (Tirosh and Berke, 1985; Howell and Martz, 1988), and ruthenium red (Howell and Martz, 1988). (Similar results were obtained for NK cells; Solovera et al., 1987.) Thus, programming for lysis is partially (Tirosh and Berke, 1985; Trenn et al., 1987; Clark et al., 1988; Ostergaard and Clark, 1989) calcium-dependent, while adhesion is calcium-independent (in the presence of physiological $[Mg^{2+}]$; Martz, 1980). The degree of Ca^{2+}-dependence can be determined entirely by the target cell employed, regardless of whether the CTL are PEL produced *in vivo* or cloned CTL produced *in vitro* (Tirosh and Berke, 1985; Ostergaard and Clark, 1987, 1989).

A number of agents with no obvious direct effect on Ca^{2+} channels have been shown to inhibit cytolysis *with little or no effect on conjugate formation*, although the mechanisms are unclear. These include antibodies to CD3 (Landegren et al., 1982; Tsoukas et al., 1982; Platsoucas, 1982), certain fatty acids (Richieri and Kleinfeld, 1990, absence of Cl^- or blockade of Cl^- flux with stilbene disulfonates (e.g., "DIDS"; Gray and Russell, 1986; Prochazka et al., 1988), ornithine decarboxylase inhibitors (Chayen et al., 1990), tyrosine ethyl ester (a protease inhibitor; Redelman and Hudig, 1983), and phorbol ester pretreatment (Russell, 1984). It should be clear in view of the discussion above that the absence of an effect on conjugate numbers at a single shear level does not guarantee the absence of an effect on adhesion strength. Ideally, drugs of interest should be tested at a range of shear levels. Nevertheless, the drugs listed in this paragraph are unusual in that the vast majority of CTL-inhibiting drugs have obvious inhibitory effects on conjugate formation.

The following agents may act primarily on programming for lysis, although the lack of effect on conjugate formation was ascertained indirectly: high K^+ (Russell and Dobos, 1981), to syllysine chloromethyl ketone (TLCK) (Chang and Eisen, 1980), inhibitors of the lipoxygenase pathway (Taylor et al., 1985), blockade of protein synthesis or heat shock or chemical stress in the *target* cell (for some target cells but not others, Landon et al., 1990; Sugawara et al., 1990). Nicotinamide may fall into this category—it inhibits cytolysis, serine esterase release, and elevation of cytoplasmic IP_3 and Ca^{2+}, but not reorientation of the Golgi/MTOC (Nowicki et al., 1991; unfortunately, no conjugate quantitation).

Ca^{2+} and Mg^{2+} in CTL Function

Mg^{2+} is required for CTL—target **adhesion** (Martz, 1980), and there is excellent evidence that this requirement resides in the LFA-1–ICAM-1 component (Shaw et al., 1986; Dustin and Springer, 1989), presumably in the consensus metal-binding domains of the LFA-1α chain (Larson et al., 1989). Mg^{2+} is also likely to be required in some components of programming for lysis (Redegeld et al., 1991). As with many Mg^{2+}-dependent processes, Mn^{2+} can substitute for Mg^{2+}, and indeed has higher efficacy (Martz, 1980; Dransfield et al., 1992). Unlike Mg^{2+}, Mn^{2+} inhibits programming for lysis (Gately and Martz, 1981).

Ca^{2+} cannot substitute for Mg^{2+} in supporting CTL–target adhesion. In the presence of several-millimolar Mg^{2+} (physiologic), Ca^{2+} affects neither the rate nor the extent of adhesion (Martz, 1980). When Mg^{2+} is suboptimal, however, Ca^{2+} synergizes dramatically with Mg^{2+} (Martz, 1980; Rothlein and Springer, 1986). Thus, in the absence of Ca^{2+} (presence of EGTA), the concentration of Mg^{2+} required for half-maximal adhesion is about 300 μM, but in the presence of 50 μM Ca^{2+}, this drops more than tenfold (Martz, 1977, 1980). Media made without intentional addition of divalent cations ("calcium- and magnesium-free") typically contain sufficient Mg^{2+} to make Ca^{2+} appear to support adhesion (via synergy with the Mg^{2+}); its inability to do this is revealed when the contaminating Mg^{2+} is removed by pretreating the medium with a chelating ion exchange resin (Martz, 1980).

Ca^{2+} is required for some, but not all, components of **programming for lysis** (Tirosh and Berke, 1985; Trenn et al., 1987; Clark et al., 1988; Ostergaard and Clark, 1989). In the absence of calcium, killing has a longer lag and proceeds at a lower rate. Often, the absence of calcium may slow cytolysis by only a fewfold. In eight experiments using PEL against P815, EGTA increased the time required for half-maximal cytolysis several-fold (Martz, 1977; MacLennan et al., 1980). Some killing of P815 is observed even in the presence of EDTA (Martz, 1975b). The calcium requirement is greatly affected by the type of target cell used; some targets can be killed efficiently in a Ca^{2+}-independent manner (Tirosh and Berke, 1985; Ostergaard and Clark, 1987, 1989).

There are many points at which Ca^{2+} could act during programming for lysis. Ca^{2+} is required in several signaling events (see above) which trigger programming for lysis. In addition, it is strictly required for the exocytosis of perforin/esterase-containing granules (see above), as well as for both the binding and the insertion of perforin into lipid membranes (Podack et al., 1985; Ishiura et al., 1990; Kraut et al., 1990). Finally, calcium leaks rapidly into damaged target cells (Poenie et al., 1987; Berke, 1989a), and there is much evidence to support the idea that excess cytoplasmic calcium may itself be cytolytic (Orrenius et al., 1989; McConkey et al., 1990; Golstein et al., 1991; Berke, 1989a) and contribute to DNA fragmentation (Hameed et al., 1989; McConkey et al., 1989).

Sr^{2+} substitutes for Ca^{2+} in programming for lysis, but not in synergizing with Mg^{2+} in adhesion formation (Gately and Martz, 1979; Martz et al., 1982b). Sr^{2+} alone is not sufficient for perforin-mediated lysis (Kraut et al., 1990).

Killing can be inhibited by **calcium antagonists** and calcium transport blockers. In some cases these can be shown to act selectively on programming for lysis without inhibiting conjugate formation (see Programming for Lysis/Lethal Hit Delivery, above). Secretion by mast cells can similarly be inhibited with calcium transport antagonists, and in some cases this inhibition is relieved by a calcium ionophore. This led to the notion of "bypassing" the transport blocker with the ionophore. On closer scrutiny, however, we were unable to confirm "bypass" in mast cells, and hence it was equally inapplicable to CTL (Parker and Martz, 1982).

Unsuccessful attempts were made to trigger cytolysis of noncognate target cells by admitting Ca^{2+} into the CTL with an **ionophore** (A23187; Parker,

1980; Martz et al., 1982b). To guarantee that the CTL was in intimate adhesive contact with the target at the moment the ionophore was applied, CTL were "glued" to targets with a lectin (wheat germ agglutinin) which did not itself trigger (nor did it inhibit) killing. The CTL employed in these studies were primed *in vivo* and boosted one cycle *in vitro* (1–1) without exogenous IL-2, and hence were likely low in granules and perforin. In more recent experiments using cloned CTL or IL-2-cultured CTL, likely high in granules and perforin and possibly promiscuously adhesive, killing was successfully induced with calcium ionophore (A23187 or ionomycin) plus a phorbol ester (Berrebi et al., 1987; Nishimura et al., 1987) or in some cases with either agent alone (Treves et al., 1987; Russell, 1986). Nongranular PEL were also triggered to lyse noncognate targets with phorbol ester alone (Russell, 1986). Phorbol esters are known to activate strongly LFA-1, and were observed to induce the formation of strong conjugates between the CTL and noncognate targets (Berrebi et al., 1987). In contrast, we have been unable to activate LFA-1 with A23187 (unpublished; also Berrebi et al., 1987). Thus is appears likely that the reason for our early failure to trigger CTL with Ca^{2+} ionophore was the absence of active CTL/LFA-1 adhesion, again underscoring the importance of this in programming for lysis.

Perforin, ATP, and Other Cytolytic Mediators

The discovery of **secretory granules** in NK cells and CTL together with **ring structures** on membranes of lysed target cells (Henkart and Henkart, 1982; Dennert and Podack, 1983) initiated a new era in CTL science. The subsequent discovery and characterization of **perforin/cytolysin** provided the first plausible chemically defined mediator for CTL-mediated cytolysis. This and the accompanying discovery of granule-associated esterases secreted from CTL have been reviewed extensively in this volume and elsewhere (e.g., Henkart and Yue, 1988; Young, 1989; Tschopp and Nabholz, 1990; Podack et al., 1991), so a detailed discussion will not be attempted here. The facts that perforin lyses a wide range of target cell types rapidly, and that its action is Ca^{2+}-dependent, agree well with the properties of CTL-mediated lysis. Despite the appeal of perforin as a CTL lytic mediator, and despite some elegant experimental designs (e.g., Acha-Orbea et al., 1990; Shiver and Henkart, 1991), it has been difficult to obtain compelling direct evidence that perforin plays a major role in CTL-mediating killing (Clark et al., 1988; Berke, 1991a; Hengel et al., 1991; Krähenbühl and Tschopp, 1991). Nevertheless, it seems highly likely that it does so, at least in some situations (Nagler-Anderson et al., 1988a).

Although it has been suggested that perforin binds to phosphorylcholine headgroups on **phospholipids** (Tschopp et al., 1989), any headgroup specificity is outweighed by phospholipid spacing. Perforin binds efficiently to membranes in which the outer leaflet has widely spaced phospholipids; perforin binds poorly to tightly spaced (ordered) phospholipid memebranes (Antia et al., 1992). The lipids in some CTL are more tightly spaced than those in CTL-susceptible target cells such as P815; hence, it has been proposed that resistance of CTL to perforin-induced lysis may be partly explained by lipid spacing (Antia et al., 1992). However, examination of additional cell lines revealed exceptions that cast doubt on this hypothesis (Ojcius et al., 1990).

A variety of **esterases** have been identified as unusually highly expressed in CTL (Pasternak and Eisen, 1985; Brunet et al., 1988; Simon et al., 1989a). Some appear to be colocalized in the same secretory granules with perforin (Ojcius et al., 1991b; Peters et al., 1991) and have been extremely useful as markers of CTL granule exocytosis (Pasternack et al., 1986; Takayama et al., 1987). Recent evidence indicates that the enzymatic activity of some of these esterases is required for the lytic activity of granule contents, including perforin (Simon et al., 1989b; Hudig et al., 1987, 1991; see also Chapter 28, this volume). Remarkably, casein is also expressed in CTL, and may serve to stabilize perforin (Grusby et al., 1990).

It has been hypothesized that CTL **secrete granular contents**, including esterases and perforin, onto target cell membranes, and that perforin is the primary effector of cytolysis. Several problems with this hypothesis became apparent (reviewed in Clark, 1988; Clark et al., 1988; Ferguson et al., 1988; Berke, 1989a, 1989b; Martz and Howell, 1989; Berke, 1991a, 1991b; elsewhere in this volume). In contrast to CTL generated in Il-2-rich cultures, secretory granules are not apparent in classical allospecific PEL generated *in vivo*, and levels of perforin are minuscule in these cells depite their potent lytic activity (see also Dennert et al., 1987; Nagler-Anderson et al., 1989). Moreover, a substantial portion of CTL lytic activity is calcium-independent, while neither the secretion nor the assembly of perforin in target cell membranes occurs appreciably in the absence of calcium.

Killing was blocked in several situations in which esterase release was intact (Clark et al., 1988). Finally, CTL induce **prelytic DNA fragmentation** in many types of target cells, while neither perforin nor complement do so. (Another effect of CTL but not complement is detachment of adherent targets from substrata; Russell et al., 1988.) It is clear that CTL may employ more than one mechanism of lysis, and that at least one mechanism is calcium-independent.

Evidence has been provided to support the idea that granzyme A (an esterase colocalized with perforin in CTL and NK granules) may penetrate into target cells with the help of perforin and induce DNA fragmentation (Munger et al., 1988; Hayes et al., 1989), possibly by cleaving nucleolin (Pasternack et al., 1991). Alternatively, a polyadenylate-binding protein found in CTL granules may induce DNA fragmentation (Tian et al., 1991).

Perhaps the best candidate for a mediator of calcium-independent lysis is **ATP**. It was recently proposed by Sitkovsky and co-workers (Filippini et al., 1990a, 1990b; Redegeld et al., 1991; elsewhere in this volume) and independently by Zanovello, di Virgilio, and co-workers (di Virgilio et al., 1990) that secretion of ATP by CTL may mediate a component of the cytolytic effects on target cells. ATP was well known for many years to be capable of permeabilizing cell membranes. CTL were shown to secrete ATP (ATP$_o$) and to be relatively resistant to its cytotoxic effects. Most interestingly, ATP$_o$ induces DNA fragmentation (Zanovello et al., 1990; Zheng et al., 1991), and both its cytolytic and DNA-fragmenting activities are Ca^{2+}-independent, at least for some target cells (Zanovello et al., 1990). Studies of pH and Mg^{2+}-dependence of ATP- vs. CTL-induced lysis show that if ATP contributes to cytolysis by CTL, it does not utilize the known pathway of pore formation by ATP^{4-} (Redegeld et al., 1991).

A variety of **slow, nonhemolytic cytotoxins** have been described that may play a role in CTL-mediated lysis. These began with lymphotoxin in 1968 (reviewed in Gately and Mayer, 1978; Paul and Ruddle, 1988) and have continued to the present (Liu et al., 1987, 1989a; Young, 1989) with tumor necrosis factor (TNF)-related agents. For the most part, the long time required to induce lysis and the limited target cell range of these mediators argue against their being primary mediators in classical CTL-mediated lysis. TNF does induce calcium-independent (Hasegawa and Bonavida, 1989) DNA fragmentation in some target cells (Schmid et al., 1987; Laster et al., 1988; Gromkowski et al., 1989; Flieger et al., 1989). It seems quite possible that these mediators play an important role by cooperating or synergizing with each other or with other mediators, perhaps with ATP.

Proteins able to neutralize putative lytic mediators have generally failed to inhibit CTL-mediated killing. Agents tested include **antibodies to perforin**/cytolysin (Shiver and Henkart, 1991), **antibodies to lymphotoxin** (Gately et al., 1976; Ju et al., 1988, 1990), and **enzymes that consume ATP** (Redegeld et al., 1991). It seems possible that such agents are unable to penetrate into the space between the closely apposed CTL and target cell membranes, and that this accounts for their failure to inhibit CTL-mediated lysis.

Antiviral and Antimicrobial Mechanisms of CTL

The natural physiologic function of CTL is believed to be primarily to control **virus** infections (Martz and Howell, 1989; Chapter 35 in this volume). Clearly, lysis of an infected cell will stop production of new virions in that cell. But is this the only, or the major, method employed by CTL or immune T lymphocytes to stop virus infections? The prominent DNA fragmentation induced in some target cells by CTL (see below) underscores the possibility that CTL may have prelytic or nonlytic virus control mechanisms. These issues are discussed in more detail Chapter 35 in this volume.

There is little if any evidence that T cells are capable of **directly inactivating** viruses or other microorganisms (Kaufmann, 1988), but this area deserves more exploration. Antimicrobial activity has been sought in extracts of CTL without success (Joag and Young, 1989), unlike neutrophils, which seem replete with antimicrobial factors (Gabay et al., 1989) capable of killing tumor cells as well (Clark and Klebanoff, 1979; Lichtenstein et al., 1988).

The possible involvement of CTL as **killers of cells infected** with, or presenting antigens from, intracellular parasites also needs further exploration. On the one hand, bacteria may benefit from the killing of antigen-presenting cells (Dohlsten et al., 1991); on the other hand, lysis of the host cell may limit infection by exposing bacteria to antibody, complement, and effector cells (Kaufmann, 1988). CTL-mediated lysis of macrophages infected with the protozoan *Leishmania* leaves the protozoa unharmed; however, CTL-produced interferon-γ induces killing of the parasites (Smith et al., 1991). CTL can lyse macrophages infected with yeast-like *Cryptococcus neoformans*

(Huffnagle et al., 1991) or *Toxoplasma* protozoa (Hakim et al., 1991), and rat NK cells contain various factors, including cytolysin/perforin, which inhibit the growth of this organism (Hidore et al., 1990) and have antitrypanosome activity (Albright et al., 1988). Human NK cells do not kill *Candida albicans* (Zunino and Hudig, 1988), but contain extractable factors with potent bactericidal activity (Garcia-Peñarrubia and Bankhurst, 1989a).

It has been reported that T cells can kill *Pseudomonas aeruginosa* bacteria (Markham et al., 1984), but the data seem more likely to represent cytostasis by nutrient depletion than an immunological phenomenon. The lymphocyte-to-bacteria ratio was 2000 (highly unphysiological) and the lymphocyte density was 4×10^6/ml (near the lymphocyte self-starvation level *in vitro*); no controls were provided to rule out nutrient effects; reciprocal specificity controls were missing; and the macrophage contamination level was uncertain.

Do CTL Detach Spontaneously?

It is a popular idea that after a CTL programs a target cell to lyse, it spontaneously detaches to begin looking for additional targets. Despite the popularity of this idea, the evidence presently available provides little direct support for it. Of course, detachment has been *experimentally induced* by agents such as EDTA, antibodies to adhesion molecules, and cytochalasins, and this has allowed the separation of programming for lysis from KCIL. However, experimentally induced detachment must be distinguished from spontaneous detachment.

Early studies indicated that CTL (PEL generated *in vivo*) do not adhere to **dead** cognate target cells (Stulting et al., 1975; Schick and Berke, 1979). Exposure of target cells (EL4) to low concentrations of **formaldehyde** ($<1.2\%$), which leaves them able to exclude dyes (e.g., trypan blue) and take up $^{51}CrO_4^{2-}$, also leaves them able to conjugate specifically with CTL, and indeed to be lysed (^{51}Cr release) by CTL (Bubbers and Henney, 1975; Schick and Berke, 1978; Balk and Mescher, 1981b). This treatment abolishes the ability of the target cells to incorporate thymidine or leucine (probably due to loss of transport) and abolishes capping of H-2. At higher concentrations of formaldehyde, the target cells become dye-stainable, lose about 40% of preloaded ^{51}Cr, are unable to take up ^{51}Cr, and fail to conjugate with CTL (Schick and Berke, 1978).

These results suggest, but do not prove, that a CTL would detach from a target cell which has lysed (Garcia-Peñarrubia and Bankhurst, 1989b). However, we know that cloned CTL can **adhere and respond to purified allo-MHC class I immobilized on plastic** (Kane et al., 1988). Such adhesion involves CD8 (Kane et al., 1989; O'Rourke et al., 1990), but evidently neither LFA-1 nor CD2 could participate. Why would CTL adhere to "dead" plastic but not to dead cells? This distinction may have to do with the immobilization of the antigen on the plastic, which may serve as a super-cytoskeleton. Perhaps the aversion of CTL for dead cells results in part from changes in the cytoskeleton after death. The fact that **membrane vesicles** are not able to compete with living target cells for CTL binding (Balk et al., 1981; Balk and Mescher, 1981a) may result from a combination of changes in the cytoskeleton plus their small size (Mescher et al., 1982; see also Todd et al., 1975). Curiously, **apoptotic vesicles** shed from lysing targets have been observed adhering to CTL in scanning electron micrographs, but these adhesions may be weak or noncompetitive (Grimm et al., 1979; Liepins et al., 1978). Recent experiments with **membrane-coated beads of various sizes** show that cloned CTL respond well to antigen-bearing particles of 4–5 μm diameter, but not to particles $\leqslant 3$ μm in diameter (Mescher, 1992).

The above discussion concerns spontaneous detachment *after* target lysis, but a separate question can be posed: do CTL spontaneously detach as soon as the target is programmed to lyse, and *before* it has lysed? The idea behind **prelytic detachment** is that the CTL would be free to search for another target after programming the first to lyse, thus increasing its efficiency. Detachment might be based on the ability of the CTL to detect whether a target has been programmed to lyse (e.g., has entered apoptosis), or the CTL might simply spend a limited amount of time on each target and then move on, regardless of whether it had succeeded in programming the target.

When detaching, where is the CTL to go? Several **types of detachment** can be envisioned. It could detach (1) in suspension, falling free into the fluid; (2) crawling away attached to a solid substratum or extracellular matrix; (3) crawling away onto alternative but noncognate targets; or (4) crawling away onto alternative cognate targets. In (2–4), CTL adhesions to the test target cell are being exchanged for CTL adhesions to something else.

CTL are **highly motile** cells (Sanderson, 1976; Rothstein et al., 1978; Grimm et al., 1979). If case (4) were observed, it would not be clear whether the CTL had sensed that the first target had been programmed to lyse, or whether to the CTL the various contacting target cells all appeared equal regardless of whether they were programmed. Thus, given motility, detachment of type (4) could be merely accidental exchange, and does not provide evidence for a signal-regulated detachment.

A major portion of the strength of CTL-target adhesion resides in the binding of activated LFA-1 to ICAM-1. Antibody to CD3 was found to **activate LFA-1 transiently**, peaking in about 10 min and waning away by 30 min (Dustin and Springer, 1989). Activation was assayed by adhesion of resting human peripheral blood T cells to purified ICAM-1 immobilized on plastic. While this observation fits beautifully with the notion of spontaneous CTL detachment, caution is necessary since this system is artifical in both the activation and the adhesion processes, and the cells were not CTL. Similarly, it was found that cloned CTL exposed to purified cognate MHC alloantigen immobilized on plastic reached an equilibrium in which only 30% adhered (Mescher et al., 1991). Individual cells **cycled between strong and weak adhesion**, despite being in continuous contact with cognate antigen (which triggers degranulation). This observation also fits the notion of spontaneous detachment beautifully, and employed cloned CTL, but note that the adhesions are unphysiologic in that they involve TCR and CD8 but neither LFA-1 nor CD2.

The above considerations are hypothetical, artificial, or indirect. What about **direct evidence on CTL-target conjugates**? One study widely cited in support of spontaneous prelytic detachment in fact barely touches on the subject and offers no quantitative evidence (Poenie et al., 1987). One of the best ways to look for detachment is by careful observations of time-lapse films. Probably the most quantitative analysis of time-lapse data is that by Sanderson (1976). Using PEL generated *in vivo*, he made complete histories of CTL-induced deaths of 22 P815 cells, 4 macrophages, 1 neutrophil, and 1 lymphocyte. He found that apoptosis (which he called zeiosis) could begin any time from immediately to 3 hr after CTL contact. He found that:

without exception . . . a T cell has remained in contact with the target cell throughout the phase of zeiosis. Previous reports . . . that the T cell detaches from the target cell before target cell death are probably due to confusion between post-mortem changes and cell death. The first analyses in these studies lead to a similar view, but subsequent examination has shown that in each case the phase of zeiosis with the T cell attached had been overlooked.

Using rapid inactivation of CTL by antibody and complement, Sanderson (1981a) concluded that a target cell is programmed to lyse at the beginning of zeiosis. He also concluded that lysis occurs a variable time after the end of zeiosis, from 0 to about 20 min (Sanderson, 1981b). Returning to his original time-lapse study, he found that:

The time of detachment of the T cell from a dead target cell varied considerably. In some cases it occurred soon after zeiosis, but in others the T cell continued to drag the target cell debris around for more than 1 h of observation . . . a second target cell has been observed to be killed while the T cell was in contact with the debris of a previous target.

Sanderson also observed "several cases . . . in which a T cell has detached from a target cell without killing it, and subsequently killed another cell." Judging from the photos in these papers, detachment was usually type (2). In summary, Sanderson found that CTL **did not detach spontaneously prior to lysis**, but sometimes detached without programming the target to lyse.

Another highly quantitative time-lapse study was that of Rothstein et al. (1978). Alloimmune spenocytes produced *in vivo* were observed for about 10 hr on a dense monolayer of EL4 targets which had been attached to plastic with antibodies. In this situation, a CTL could be in contact with several target cells simultaneously, and type (4) exchange occurred frequently (unlike Sanderson's films). In a series of 4 films, 4 CTL contacted 54 targets, 19 of which died. Some CTL remained in contact with targets for an hour or more after they had lysed, and others detached sooner. Rothstein concluded that on the average, CTL moved away from lethally hit targets 14 min before death. However, Sanderson later examined one of these films and raised the possibility that the appearance Rothstein considered to mark "death" was postlytic (Sanderson, 1981b). Thus, Sanderson could not confirm that CTL were detaching before the completion of zeiosis (which was difficult to see in these films which were at lower magnification than Sanderson's). Clearly, it is difficult to be certain about the time-lapse appearance which corresponds to dye uptake or ^{51}Cr release; loss of fura-2 may provide a good marker in future studies (Poenie et al., 1987). What is also clear from Rothstein's films is that in at least half of the cases, **CTL remained in contact with targets until they had lysed**, and often for up to 30 min beyond.

Rothstein observed 35 CTL-target contacts after which the CTL moved away and the target did not lyse. The films generally allowed several hours of postcontact observation, long enough that lysis would have been seen had the target been programmed to lyse (Martz, 1977). It is possible that these nonlethal contacts were qualitatively different than the lethal ones (e.g., less talin accumulation, weaker adhesion, or orientation of the MTOC towards another target). Taken at face value, these cases indicate that a type 4 exchange can occur before a target has been programmed to lyse. Balk's work, mentioned below, supports the same conclusion.

Three other time-lapse studies, although offering no quantitative data, are consistent with Sanderson's conclusions. Grimm et al. (1979) concluded that "CTL usually remained in contact with the zeiosing target for variable lengths of time, but often until the final disintegration." Matter (1979) mentions nothing about detachment, and shows a sequence with the CTL remaining attached throughout the death of the target. Sung et al. (1988) noted that a cloned human CTL remained attached to the target it had killed even after the target had been expelled from a micropipet.

Glasebrook (1978) studied conjugation of primary and secondary CTL generated *in vivo* and *in vitro* using microscopic counts. During the course of 1 hour at 37°C, he noted a decline in the number of conjugates beginning at 15–30 min. This was attributed to postlytic detachment. In contrast, at 23°C, or at 37°C in EGTA, which prevented lysis, no significant decline in conjugate numbers was observed during 1 hour. Type 4 exchange before lysis would have gone undetected with this protocol.

In another study, cloned CTL which displayed nonspecific adhesion to EL4 cells were induced to lyse EL4 by attaching an anti-T-cell-receptor (anti-TCR) antibody to the targets. During the first 30 min of interaction, unconjugated target cells appeared with high Ca_i^{2+}, suggesting postlytic detachment (Allbritton et al., 1988). In both this study and in Glasebrook's study, any detachment occurred in a mixed cell pellet, and thus type 4 exchange cannot be ruled out.

When CTL–tumor cell target conjugates were agitated in suspension (in the absence of Ca^{2+} to minimize lysis), **no spontaneous detachment** of type 1 occurred during several hours of observation. However, if free targets were provided in the suspension, exchange (detachment of type 4) apparently occurred rapidly (Balk and Mescher, 1981b). In contrast to tumor cells, normal splenocytes are killed more slowly, do not bind to CTL in suspension, and function poorly as cold target inhibitors; hence, they appear to bind to CTL more weakly. CTL–splenocyte conjugates agitated in suspension underwent 50% spontaneous type 1 detachment within 2 hr (Balk and Mescher, 1981a). In another study, cloned CTL conjugated to splenocyte targets failed to undergo type 1 detachment within 30 min at 24°C (Perez et al., 1985).

The above-cited direct evidence on CTL detachment from real targets can be summarized as follows:

Type 1 detachment (falling off free into liquid medium) with lysis *blocked*: **against** (Balk, except for poorly adhering targets; Glasebrook, Perez).

Type 1 detachment *after* lysis: **against** (Sung).

Type 2 detachment (crawling away on noncellular substrate) *without* ensuing lysis: **for** (Sanderson).

Type 2 detachment *followed by* lysis: **against** (Sanderson, Grimm?, Matter?).

Type 2 detachment *after* lysis: **for** (Sanderson).

Type 3 detachment (exchange for noncognate target): no data.

Type 4 detachment (exchange for cognate target) *without* ensuing lysis **for** (Rothstein), indirectly for (Balk).

Type 4 detachment *followed by* lysis: **for?** (Rothstein, questioned by Sanderson).

Type 4 detachment *after* lysis: **strongly for** (Rothstein, Glasebrook, Albritton).

In short, the limited evidence available suggests that **CTL do not release targets until after they lyse**, and even then CTL often remain attached for an additional half hour or more. It appears likely that in most cases when a CTL spontaneously detaches from an unlysed target, the target has not been programmed to lyse. The presence of alternative cognate targets probably facilitates the exchange-release of lysed targets, as well as the exchange-release of unlysed targets regardless of whether they have been programmed to lyse.

Apoptosis, DNA Fragmentation, Fatty Acid Production

Programming for lysis is accompanied by several characteristic phenomena which occur in the target cell before lysis, including apoptosis, DNA fragmentation, and fatty acid production.

The term **apoptosis** was introduced to describe a pathway of cell death distinct from so-called accidental death, which had been termed "necrosis" (Kerr et al., 1972; Duvall and Wyllie, 1986; Golstein et al., 1991; Kerr and Harmon, 1991). The distinction was initially largely morphological, apoptosis being defined as "fragmentation into membrane-bound particles which are phagocytosed by other cells". In addition to fragmentation of the cytoplasm, the nucleus often fragments within the cytoplasm, and the chromatin condenses, forming dark masses marginalized against the nuclear envelope.

During embryonic development, masses of tissue die by apoptosis in a regular, programmed manner. So-called **programmed cell death** refers to death induced by a signal which is not directly toxic, usually the increase or decrease of a hormone, which is controlled by a developmental program. This developmental death sculpts out tissue without disrupting the surrounding architecture. In contrast, necrotic death is induced by a toxic condition, disrupts tissue organization, and results in scarring.

Prelytic DNA fragmentation accompanied programmed cell death in mammals wherever it has been examined (Cohen, 1991). Endogenous endonucleases cut exposed internucleosomal DNA links, leaving nucleosomal monomers and oligomers which form a characteristic ladder pattern on agarose gel electrophoresis (multiples of approximately 180 bp). Prelytic DNA fragmentation has thus become accepted as part of the definition of apoptosis. Interestingly, however, developmentally programmed cell death in an invertebrate is not accompanied by DNA fragmentation (Schwartz et al., 1993), although it is otherwise very similar to that in mammals.

The **thymus** is the site of massive apoptotic death associated with the negative selection which deletes T cells capable of producing autoimmunity (Smith et al., 1989; Murphy et al., 1990; Golstein et al., 1991). Thymocyte and lymphocyte apoptosis and DNA fragmentation have been extensively studied. A wide variety of stimuli, including recognition of antigen, corticosteroids, radiation, and withdrawal of various cytokines, induce apoptosis and DNA fragmentation in lymphocytes in the appropriate developmental stages (Martz and Howell, 1989; Shi et al., 1990).

A specific **genetic suicide program** is required to be expressed prior to the onset of DNA fragmentation and apoptosis in developmentally programmed cell death (Schwartz et al., 1990; Golstein et al., 1991; Cohen et al., 1992). Inhibition of transcription or translation prevents or delays these events. TNF also induces DNA fragmentation, often accompanied by apoptosis (Laster et al., 1988). However, TNF-induced death is accelerated by inhibition of transcription or translation (reviewed in Golstein et al., 1991). Thus, the TNF-induced genetic program is protective rather than suicidal. Certain human leukemia cell lines can be induced to undergo apoptotic death by a variety of stimuli. Inhibition of transcription or translation does not delay apoptosis, but rather accelerates it, and indeed is sufficient for the induction of apoptosis (Martin et al., 1990). Thus, expression of a genetic program is evidently not always required for apoptotic suicide.

That **CTL induce apoptosis** in target cells was first reported by the group which introduced the term (Don et al., 1977). Sanderson had also seen this distinction, and appreciated the fact that complement induces necrosis unlike the death induced by CTL, but used the term "zeiosis" to describe what is now called apoptosis (Sanderson, 1976). Indeed, most early ultrastructural studies of CTL-mediated lysis clearly showed what we now recognize as apoptosis before the term and the concept were generally recognized (e.g., Koren et al., 1973; Liepens et al., 1977; Ryser et al., 1977; Liepins et al., 1978).

Russell discovered that **CTL induce the prelytic DNA fragmentation** now recognized as the hallmark of apoptosis (reviewed in Russell, 1983). Russell showed that the fragmentation (solubilization) of DNA occurred prelytically because removing the membranes of unlysed cells with Triton X-100 nonionic detergent released a large amount of soluble DNA (see also example in Martz and Howell, 1989). In contrast, lysis of targets by complement or many other physical stresses or chemical toxins leaves the DNA insoluble in Triton X-100, even long after lysis (reviewed in Martz and Howell, 1989). Generally speaking, CTL induce less extensive DNA fragmentation in human targets than in mouse targets (Christiaansen and Sears, 1985; Gromkowski et al., 1986; cf. Brahmi et al., 1991), although there are exceptions in both directions (Howell and Martz, 1987; Liu et al., 1989c). Hematopoietic cell lines commonly used in CTL research because of their high sensitivity to lysis show, in general, much more dramatic DNA fragmentation than do fibroblasts or epithelial cells, which are less easily lysed (Howell and Martz, 1987; Sellins and Cohen, 1991).

The DNA fragmentation induced by CTL appeared to differ from classical programmed cell death in several respects (reviewed in Martz and Howell, 1989; Golstein et al., 1991). As mentioned above, programmed cell death usually required hours for the expression of a genetic program before DNA fragmentation was initiated; inhibition of gene expression delayed or prevented the DNA fragmentation. However, in some target cells, CTL could initiate DNA fragmentation within minutes, seemingly too rapidly for a genetic program to be expressed. This seemingly unique form of DNA fragmentation suggested that CTL may utilize this mechanism to control virus replication (reviewed in Martz and Howell, 1989). I have discussed this idea further elsewhere in this volume (Chapter 35).

An important question related to the above observations is whether the **target cell needs to play an active role in its own demise** during CTL-induced lysis. Early observations indicated that the target was a passive participant. Poisoning the target cell with brief exposure to low concentrations of aldehyde fixatives sufficient to block nucleotide and amino acid incorporation did not prevent CTL-mediated cytolysis (Bubbers and Henney, 1975). This has been confirmed (Schick and Berke, 1978), although it appears that blockage of precursor *transport* can explain the lack of incorporation, and the degree of internal metabolic blockade is unresolved.

Inhibitors of **transcription and translation** have been reported partially to inhibit CTL-mediated lysis (Brunner et al., 1968; Strom et al., 1975; Tite, 1990a; Zychlinsky et al., 1990) or to produce no inhibition at all[2] (Brondz, 1972; Cerottini et al., 1974; Duke et al., 1983; Landon et al., 1990). In contrast, results with inflammatory CTL (MHC class II-restricted; Tite, 1990a; Ju, 1991; Ju et al., 1990; Strack et al., 1990) show pronounced inhibition of lysis. Actinomycin D appears to have a more pronounced effect on CTL-induced DNA fragmentation than on lysis (Zychlinsky et al., 1990; Zheng et al., 1991), possibly because it interferes directly with the fragmentation mechanism which may involve topoisomerases (Nishioka and Welsh, 1991). An inhibitor of ornithine decarboxylase inhibited lysis (Chayen et al., 1990); the inhibition was attributed to the target, but as presented the data do not rule out a reversible action on the CTL. In all the experiments cited above concerning MHC class I-restricted CTL, the mixture of CTL and target cells was exposed to the drug, so that it is unclear which cell is being affected.

Short pretreatments ($\leqslant 1$ hr) of CTL (MHC class I-restricted) with inhibitors of transcription and translation have produced no inhibition of lysis (Thorn and Henney, 1976; Ostergaard and Clark, 1989; Strack et al., 1990) or of DNA fragmentation (Ju, 1991), while longer ($\geqslant 3$ hr) pretreatments of the CTL have produced a curious cycling of lytic activity (Landon et al., 1990), which may explain some of the discrepant results cited above.

Pretreatment of the **target** cells with inhibitors of transcription and translation, followed by washing so that effects on the CTL can be unequivocally excluded, deserves more study than it has received. One very early study reported no inhibition, but the results must be discounted, since lysis was at plateau levels and hence quite insensitive to inhibition (Clark et al., 1971); this is an important caveat in CTL assays. The only other study I know of showed inhibition on some target cells and less or no inhibition of others (Landon et al., 1990). With lymphokine-activated killer (LAK) cells, clearcut inhibition was observed (Zychlinsky et al., 1990). In contrast, such pretreatment renders targets more sensitive to NK-mediated lysis (Collins et al., 1981; Kunkel and Welsh, 1981). In the aggregate, these experiments, plus those cited above in which the inhibitors had little effect when added to CTL plus target mixtures, suggest that transcription and translation in the target play a minor role, if any, in MHC class I-restricted CTL-mediated killing, and that the importance varies with the type of target.

Heat shock or chemical stress of the target cells before exposure to CTL induces partial resistance to lysis (Sugawara et al., 1990), suggesting that transcription and translation can play a protective role rather than a suicidal one. In contrast, with inflammatory MHC class II-restricted CTL, heat shock induces susceptibility independently of antigen recognition (Ozdemirli et al., 1991).

A crucial issue is whether **CTL-induced DNA fragmentation requires the expression of a suicidal genetic program**, as does developmentally programmed cell death (but unlike TNF-induced DNA fragmentation). When this issue was first examined, a report of two experiments (Duke et al., 1983) was interpreted to mean that neither cycloheximide nor emetine (protein synthesis inhibitors) could prevent CTL-induced DNA fragmentation. However, no target cell pretreatment was employed, so some CTL-induced protein synthesis in the target might have preceded achievement of effective intracellular drug concentrations, and 24% inhibition of DNA fragmentation was in fact observed in one of the two experiments. In a more recent study which pretreated the target for 30 min, inhibition of CTL-induced DNA fragmentation was seen, but action via the effector cell cannot be ruled out (Zychlinsky et al., 1990). Also, the effect of emetine in this study on lysis could not be confimed in another laboratory using the same CTL clone and targets (W.R. Clark, personal communication). In another recent study, the CTL were pretreated and the drugs remained in the assay but had no effect on DNA fragmentation (Ju, 1991). Also LAK-induced DNA fragmentation was apparently independent of protein synthesis (Brahmi et al., 1991). Thus, while the evidence is consistent with the original interpretation, that CTL-induced DNA fragmentation does not require synthesis of suicide proteins in the target, clearly more study is needed to resolve this issue satisfactorily.

Recently, it has been shown that CTL, but not other means of lysis, induce the rapid release of fatty acids in target cells. Fatty acids may contribute to the demise of the target cell, but were also shown capable of inhibiting killing by acting on the CTL. Hence, this may be a self-protective mechanism employed by some target cells, particularly tumor cells. This topic is reviewed in Chapter 30, this volume by Alan Kleinfeld. Very rapid (5 min) labilization of target lysosomes has also been reported in CTL-mediated killing (Pitsillides et al., 1988).

Killer Cell-Independent Lysis (KCIL)

The evidence that the CTL is not required to continue to act on the target until the moment of lysis was introduced above (Programming for Lysis/Lethal Hit Delivery). (Whether CTL detach spontaneously is a separate issue, also discussed above.) KCIL takes a longer time than adhesion formation or programming for lysis, and hence is the rate-limiting event for lysis under ordinary conditions. Since CTL-induced lysis often begins without much of a lag, lysis can occur within minutes, probably for the target cells being acted upon by several CTL simultaneously. After CTL inactivation, however, KCIL can take up to 4 hr to reach completion (Martz and Benacerraf, 1975). KCIL proceeds in the absence of divalent cations, and can be halted at 0°C. It is, however, less temperature-sensitive than adhesion formation, which in turn is much less sensitive than programming for lysis, the latter being nearly halted at 20°C (reviewed in Martz, 1977).

When KCIL is halted at low temperature, no recovery occurs in the cold; rewarming allows lysis to proceed (Martz and Benacerraf, 1975). The mechanism of the temperature sensitivity remains to be elucidated, but when the same types of target cells are damaged by complement, lysis is equally temperature-sensitive (Martz et al., 1974; Burakoff et al., 1975). Since complement is believed to lyse cells solely by permeabilizing the plasma membrane, the temperature effect seems likely to result from a physical gelling of the cytoplasm/cytoskeleton, preventing the exit of isotopic markers (even ^{86}Rb; Burakoff et al., 1975) and perhaps the progression of osmotic water entry and swelling. Erythrocytes are unlike leukocytes in that following Ab + C-induced damage, their lysis cannot be prevented in the cold (Martz et al., 1974; Burakoff et al., 1975).

To separate programming for lysis from KCIL, it is necessary to remove or inactivate the CTL. The method of detachment or inactivation should ideally be one which will not add further damage to the target cell. Otherwise, it could be argued that the target would have been capable of repair, and that the CTL's action would have been sublytic, in the absence of the additional damage. Perhaps the most convincing method was the first, inactivation of the CTL with specific Ab + C (Martz and Benacerraf, 1973). The subsequent methods of inactivation by heat (Wagner and Röllinghoff, 1974) or EDTA to detach the Mg^{2+}-dependent grip of LFA-1 (Martz, 1975a) may be capable of better time resolution,[3] but hasten the lysis of the target. A direct comparison of EDTA with Ab + C showed the acceleration of lysis by EDTA, but lysis plateaued at the same level for both methods (Martz et al., 1974). Less acceleration of lysis might be obtained with detaching agents such as cytochalasins or antibodies to LFA-1.

When adhesion-weakening agents, such as EDTA or cytochalasins, are used to end programming for lysis, it is imperative that mechanical shear be applied physically to separate the CTL from the target cells. In the absence of physical detachment, programming for lysis can continue, albeit at a reduced rate (Martz, 1975b). Following physical detachment, it is imperative to prevent the CTL from forming new contacts with target cells. Dispersion in viscous dextran-containing medium is one of the most effective ways to accomplish this (Martz, 1975a, 1977).

The mechanism of KCIL remains to be elucidated. Early studies on complement-mediated lysis of erythrocytes suggested that colloid osmotic lysis was responsible for lysis following formation of one or more complement pores. Nonerythrocytes, however, cannot be osmotically protected from lysis by complement (Kim et al., 1989) and their lysis is hastened by extracellular Ca^{2+} (Kim et al., 1987), indicating that more complicated processes are involved. In view of the complexity of CTL action, little can presently be concluded about the mechanism of KCIL.

Recycling

The high lytic activity of PEL, harvested from the site of intraperitoneal graft rejection, allowed the demonstration that a multiclonal population rich in CTL could lyse a number of targets exceeding the number of PEL, and therefore necessarily exceeding the number of CTL (Berke et al., 1972b). The rate of lysis did not diminish during 5 hr of incubation. ^{51}Cr release from labeled PEL did not increase during killing. These observations strongly suggested that **CTL were neither inactivated nor lysed during the killing process**, and that each CTL could kill a series of target cells sequentially in time. Confronting PEL populations with successive populations of target cells provided evidence that **each CTL could kill more than 6 target cells**, and could likely do so sequentially in time, even when the PEL were irradiated to block proliferation (Martz and Benacerraf, 1976). There was no diminution in lytic activity during the first three cycles of target populations. However, these population experiments left some uncertainty about the postkilling lytic activity of single CTL, especially in view of the observation that a single CTL could conjugate with more than one target cell simultaneously (Martz, 1975a; Berke et al., 1975).

The first unequivocal demonstration that a single CTL could kill **3 target cells sequentially in time** involved micromanipulation. After the target in a 1:1 conjugate lysed, a second target was pushed into contact with the same CTL. After a second lysed, a third was pushed into contact, and this also lysed (Zagury et al., 1975).

Recently, transient inactivation of human CTL clones during interaction with ordinary target cells was reported (Hommel-Berrey et al., 1991). This appears to conflict with much evidence for mouse CTL generated in vivo. However, I know of no recycling studies on mouse or human cloned CTL. Clearly, more studies will be needed to evaluate this possibility.

All proposed CTL lytic mediators (perforin, ATP, lymphotoxin), once released, seem potentially capable of acting on the CTL as well as the target (but see below for evidence of CTL resistance). A recent alternative proposal is that CTL granules may be released with active recognition molecules on their surfaces, such as the T-cell antigen receptor and LFA-1, so that they interact specifically with the target cell surface (Peters et al., 1990).

Resistance of CTL to CTL

The ability of CTL-containing lymphocyte populations to survive the lytic interaction with undiminished lytic activity raised the likelihood that CTL were **resistant to their own cytolytic mediators**. Experiments to address this issue involve using CTL as both targets and effectors; hence, we shall refer to *target*-CTL and *effector*-CTL. The issue of whether CTL resist other CTL seems a simple one, but in fact is amazingly complicated. In discussing this issue, it is of the utmost importance to distinguish **loss of lytic activity (inactivation)** of target-CTL from **lysis** of target-CTL. Although the early experiments necessarily studied extensively the former, later discussions of these early studies have almost unanimously slipped into referring to the results incorrectly as "lysis," "death," or "killing." More recent work has shown that inactivation does not necessarily mean lysis.

The question of CTL resistance was first addressed by using b-anti-c-lymphocytes as targets for a-anti-b-lymphocytes (where a, b, and c represent H-2 haplotypes). Since the majority of the lymphocytes employed were not CTL, ^{51}Cr release could not be used reliably to assess the fate of the target-CTL subpopulation. Rather, after the CTL:CTL interaction, the residual lytic activity of the target b-anti-c-CTL was quantitated on ^{51}Cr-labeled c-targets. The a-anti-b-lymphocytes substantially **inactivated the lytic activity of the target-CTL** (Golstein, 1974; Martz and Benacerraf, 1976). Similarly, exposing CTL to phytohemagglutinin (PHA), a lectin which causes CTL to bind to and kill targets nonspecifically, inactivated the CTL, presumably due to mutual cytotoxicity (Bradley and Bonavida, 1978). These experiments showed that CTL were not entirely immune to their own cytotoxic mediators. Unanswered was the question of whether the target-CTL were actually lysed or merely inactivated.

Early dye exclusion tests suggested that target-CTL were actually lysed (Fishelson and Berke, 1978), but these tests were not definitive, since the lymphocytes employed were only one-third CTL. Perhaps the first demonstration of CTL-induced lysis of cloned CTL involved the apparent processing of exogenous cognate protein antigen, resulting in antigen-specific ^{51}Cr release from the cloned CTL population by presumed **self-lysis** (Snodgrass et al., 1981). Recently, it has been shown that cloned mouse MHC class I-restricted CTL, when incubated for many hours with their cognate peptide, lose lytic activity and self-lyse (Walden and Eisen, 1990; Pemberton et al., 1990); similar results were obtained for cloned human class I-restricted CTL (Moss et al., 1991) and human class II-restricted CTL (Ottenhoff and Mutis, 1990).

Self-lysis of cloned CTL has **not been universally observed**, however. Some mouse CTL clones incubated with cognate peptide were not lysed (Townsend et al., 1986) but were partially inactivated and released serine esterase (Vitiello et al., 1989). Similarly, anti-fluorescein CTL clones did not self-lyse when conjugated with fluorescein (Skinner and Marbrook, 1987), although they became activated and could kill nonspecific targets (Skinner et al., 1989a). The lectins concanavalin A or PHA induce strong adhesion and nonspecific lysis by CTL (Bevan and Cohn, 1975); however, addition of such lectins to mouse CTL clones has usually not produced self-lysis (Blakely et al., 1987), with a few exceptions (Luciani et al., 1986; including multiclonal: Bradley and Bonavida, 1978), nor did it produce self-inactivation (Luciani et al., 1986).

In cases of self-lysis of clones, it has not been resolved whether a single CTL, in the absence of contact with another CTL, can kill itself (literal **suicide** at the single cell level) or whether one CTL must act on a second (**fractride**). In a study with uncloned CTL in which self-lysis was induced with PHA, incubation of isolated CTL (no cell–cell contact) with PHA produced no lysis, arguing against suicide (Bradley and Bonavida, 1978). In a recent study of peptide-induced self-lysis, the authors favored a suicidal mechanism, although isolated cell studies were not performed (Walden and Eisen, 1990).

A question best treated separately is whether one CTL clone (effector-CTL) can lyse another CTL clone (target-CTL) bearing cognate antigen (**heteroclonal lysis**). In many cases of such heteroclonal confrontations, the target CTL were highly or absolutely resistant, but most studies have found some combinations where lysis occurs (Blakely et al., 1987; Kranz and Eisen, 1987; Skinner and Marbrook, 1987; Luciani et al., 1986; Fink et al., 1984). Skinner and Marbrook noted that target-CTL were susceptible only when they had lower lytic activity than the effector-CTL (Skinner and Marbrook, 1987). Thus, clones could be arranged in a single hierarchy in which effector-CTL could kill all clones lower on the hierarchy, but not themselves nor clones higher in lytic activity. In support of this idea, these authors showed that manipulations which reduced lytic activity (3 days in PMA or 2 hr in colcemid) rendered the clones susceptible to lysis by untreated cells of the same clone or weaker clones (Skinner et al., 1989b). Data from other studies are consistent with the hierarchy idea (Kranz and Eisen, 1987; Luciani et al., 1986). Noncytolytic T-cell lines or MHC class II-restricted clones, even when cytotoxic, seem more susceptible as target-CTL (Kranz and Eisen, 1987; Ottenhoff and Mutis, 1990).

Amazingly, some of the most resistant target-CTL clones are able to be **lysed by multiclonal primary CTL** effectors generated in a 5-day mixed lymphocyte culture (Fink et al., 1984; Luciani et al., 1986; Blakely et al., 1987; Kranz and Eisen, 1987). Despite being lysed, they are still 3-fold to 20-fold less susceptible than highly susceptible standard tumor targets such as EL4 or P815. When they are not lysed, they are inactivated (Gorman et al., 1988; see also Vitiello et al., 1989, for nonlytic self-inactivation with peptide). Interestingly, very few cloned effector-CTL inactivate cloned target-CTL (Kranz and Eisen, 1987; Gorman et al., 1988). Perhaps one lytic mechanism employed by primary CTL is lost during prolonged culture/cloning.

The **mechanism of resistance** of cloned target-CTL to lysis by cloned effector-CTL is unclear. In cases of allorecognition, normal levels of MHC antigen are expressed on the target (Blakely et al., 1987). However, in most cases, the effector-CTL fails to be triggered: there is no elevation of Ca_i^{2+}, no reorientation of the MTOC, and no serine esterase release (Ostergaard et al., 1988). Cloned CTL are **not resistant to complement**-mediated lysis (Blakely et al., 1987; Kranz and Eisen, 1987) nor to mellitin-mediated lysis (Verret et al., 1987).

Cloned CTL are also generally extremely **resistant to lysis by CTL granules** (Blakely et al., 1987; Verret et al., 1987) or purified CTL **perforin** (Liu et al., 1989b; Jiang et al., 1990). Their susceptibility is about 100-fold less than that of tumor target cells (Blakely et al., 1987; Jiang et al., 1990). In one case tested, a clone that resisted lysis was nevertheless inactivated; interestingly, the granule component responsible for inactivation was not itself inactivated by Ca^{2+}, and hence was not the pore-forming activity of perforin (Gorman et al., 1988). This nonperforin inactivator in CTL granules was dubbed **inhibitin**. $CD4^+$ T-helper clones were much less resistant to lysis by granules (Verret et al., 1987). CTL clones are also relatively **resistant to lysis by ATP** (di Virgilio et al., 1989; Filippini et al., 1990a).

The **mechanism of resistance of cloned CTL to granules/perforin** is unclear. The modest resistance of $CD4^+$ T cells is greatly diminished when they are depleted of ATP; however, this has much less effect on the resistance of CTL (Verret et al., 1987). The resistance does not depend on the granule content, since agranular cytoplast fragments retain resistance (Ojcius et al., 1991a). As mentioned above, lipid spacing does not seem to provide the answer (Ojcius et al., 1990). CTL are less able to inactivate granules/perforin that sensitive cells (Verret et al.,

1987; Jiang et al., 1990). This has been interpreted as reduced ability to bind perforin, but direct tests with biosynthetically labeled perforin suggest that resistant human CTL can bind as much as sensitive cells (Jones and Morgan, 1991).

$CD8^+$ lymphocytes, normal or activated during a primary response to alloantigen or lectin, also resist granules much more than $CD4^+$ lymphocytes or activated macrophages (Nagler-Anderson et al., 1988b; Nagler-Anderson and Eisen, 1990). Unseparated target-PEL, however, seemed as susceptible to lysis by cognate effector-CTL as normal lymphocytes (Schick and Berke, 1990). Normal lymphocytes are severalfold more resistant to lysis than tumor targets.

In summary, cloned CTL are often resistant to self-lysis and resist lysis by other cloned CTL, especially if the effector-CTL has weaker lytic activity than the target-CTL. Although often resisting lysis, they may nevertheless lose lytic activity. Curiously, they are more susceptible to lysis and inactivation by primary multiclonal CTL, raising the possibility that cloning produces the qualitative loss of one lytic mechanism. Whether multiclonal primary CTL resist CTL-mediated lysis is very difficult to determine, since many of the cells in the population may not be CTL. They can be inactivated, but this measure of cytotoxicity cannot be compared to non-CTL target cells. Cloned CTL and also multiclonal $CD8^+$ lymphocytes resist granules, perforin, and ATP, but neither complement nor mellitin.

In assessing the resistance of CTL to CTL-mediated lysis, bear in mind that the "standard" target cells (P815, EL4, and other hematopoietic tumor cell lines) are popular because of their exteme sensitivity to CTL-mediated lysis. In fact, unestablished uncloned mouse fibroblast lines are about 20-fold less sensitive to CTL-mediated lysis than P815 (Martz and Gamble, 1991). In comparison, cloned CTL were 3-fold to 20-fold less susceptible to primary CTL than were "standard" targets. Thus, **perhaps cloned CTL have normal susceptibility, while the "standard" targets have abnormally high susceptibility**. Studies of a broader range of target cell types would be useful.

When two CTL attack each other, **perhaps an equilibrium is reached in which each CTL becomes weakened to the point that it cannot lyse the other**. This would require only that (1) lysis be the result of the sum of many sublytic packets of damage delivered over a period of time (even if only 1 min), and (2) the sublytic packets of damage received by a CTL impair its ability to continue to deliver packets of damage in the other direction. This straightforward notion would explain the frequent observation of the absence of self-lysis, accompanied by transient inactivation. It is consistent with the hierarchy observed in heteroclonal confrontations. The observation of self-lysis in some clones raises the possibility of heterogeneity within the clone. Alternatively, CTL may be more resistant near the MTOC than on the opposite side, or only when activated and not when resting (heterogeneity of resistance in space or time). Quantitative studies of the lytic activity of the survivors of self-lysis would be useful to ask whether they are a subpopulation with the highest lytic activity per cell.

The resistance of target-CTL to effector-CTL could result from a "backwards" damage by the target-CTL (cf. Lanzavecchia, 1986) leading to a mutual standoff or equilibrium as described above. The resistance of agranular cytoplasts to perforin (Ojcius et al., 1991a) argues against this, but only if perforin is the sole mechanism of damage, which appears highly unlikely. The often-seen failure of the effector-CTL to be triggered (see above) would seem to argue against this hypothesis when it occurs. Nevertheless, this "CTL–CTL standoff" mechanism of resistance would not be relevant to the interaction of a CTL with a non-CTL target. Thus, open-mindedness is needed about whether the general resistance of cloned CTL to other CTL provides the explanation of the survival of CTL after lysis of non-CTL targets.

References

Acha-Orbea H, Scarpellino L, Hertig S, Dupuis M, Tschopp J (1990): Inhibition of lymphocyte mediated cytotoxicity by perforin antisense oligonucleotides. *EMBO J* 9: 3815–3819

Albright JW, Munger WE, Henkart PA, Albright JF (1988): The toxicity of rate large granular lymphocyte tumor cells and their cytoplasmic granules for rodent and African trypanosomes. *J Immunol* 140: 2774–2778

Allbritton NL, Verret CR, Wolley RC, Eisen HN (1988): Calcium ion concentrations and DNA fragmentation in target cell destruction by murine cloned cytotoxic T lymphocytes. *J Exp Med* 167: 514–527

Andre P, Benoliel AM, Capo C, Foa C, Buferne M, Boyer C, Schmitt-Verhulst AM, Bongrand P (1990): Use of conjugates made between a cytolytic T cell clone and target cells to study the redistribution of membrane molecules in cell contact areas. *J Cell Sci* 97: 335–347

Antia R, Schlegel RA, Williamson P (1992): Binding of perforin to membranes is sensitive to lipid spacing and not headgroup. *Immunol Lett* 32: 153–7

Baker PE, Gillis S, Smith KA (1979): Monoclonal cytolytic T-cell lines. *J Exp Med* 149: 273–278

Balk SP, Mescher MF (1981a): Specific reversal of cytolytic T cell-target cell functional binding is induced by free target cells. *J Immunol* 127: 51–57

Balk SP, Mescher MF (1981b): Specific reversal of cytolytic T lymphocyte-target cell interaction. *J Supramol Struct* 16: 43–52

Balk SP, Walker J, Mescher MF (1981): Kinetics of cytolytic T lymphocyte binding to target cells in suspension. *J Immunol* 126: 2177–2183

Berke G (1980): Interaction of cytotoxic T lymphocytes and target cells. *Prog Allergy* 27: 69–133

Berke G (1985): Enumeration of lymphocyte-target cell conjugates by cytofluorometry. *Eur J Immunol* 15: 337–340

Berke G (1989a): The cytolytic T lymphocyte and its mode of action. *Immunol Lett* 20: 169–178

Berke G (1989b): Functions and mechanisms of lysis induced by cytotoxic T lymphocytes and natural killer cells. In: *Fundamental Immunology*, 2nd ed., Paul WE, ed. New York: Raven Press, pp. 735–764

Berke G (1991a): Lymphocyte-triggered internal target disintegration. *Immunol Today* 12: 396–399, 403

Berke G (1991b): T-cell mediated cytotoxicity. *Curr Opin Immunol* 3: 320–325

Berke G, Levey RH (1972): Cellular immunoabsorbents in transplantation immunity. Specific *in vitro* deletion and recovery of mouse lymphoid cells sensitized against allogeneic tumors. *J Exp Med* 135: 972–984

Berke G, Rosen D (1987a): Are lytic granules and perforin 1 involved in lysis induced by *in vivo*-primed peritoneal exudate cytolytic T lymphocytes? *Transplant Proc* 19: 412–416

Berke G, Rosen D (1987b): Ciruclar lesions detected on membranes of target cells lysed by antibody and complement or natural killer (spleen) cells but not by *in vivo* primed cytolytic T lymphocytes. In: *Membrane Mediated Cytotoxicity*. Bonavida B, Collier RJ, eds. New York: Allan Liss Inc., pp. 367–378

Berke G, Rosen D (1987c): Mechanism and inhibition of target damage. Are lytic granules and perforin 1 involved in lysis induced by *in vivo*-primed peritoneal exudate cytolytic T lymphocytes. *Transplant Proc* 19: 412–416

Berke G, Rosen D (1988): Highly lytic *in vivo* primed cytolytic T lymphocytes devoid of lytic granules and BLT-esterase activity acquire these constituents in the present of T cell growth factors upon blast transformation *in vitro*. *J Immunol* 141: 1429–1436

Berke G, Ax W, Ginsburg H, Feldman M (1969): Graft reduction in tissue culture II. Quantification of the lytic action on mouse fibroblasts by rat lymphocytes sensitized on mouse embryo monolayers. *Immunology* 16: 643–657

Berke G, Gabison D, Feldman M (1975): The frequency of effector cells in populations containing cytotoxic T lymphocytes. *Eur J Immunol* 5: 813–818

Berke G, Sullivan KA, Amos B (1972a): Rejection of ascites tumor allografts I. Isolation, characterization, and *in vitro* reactivity of peritoneal lymphoid effector cells from BALB/c mice immune to EL4 leukosis. *J Exp Med* 135: 1334–1350

Berke G, Sullivan KA, Amos DB (1972b): Tumor immunity *in vitro*: Destruction of a mouse ascites tumor through a cycling pathway. *Science* 177: 433–434

Berke G, Sullivan KA, Amos DB (1972c): Rejection of ascites tumor allografts II. A pathway for cell-mediated tumor destruction *in vivo* by peritoneal exudate lymphoid cells. *J Exp Med* 136: 1594–1604

Berrebi G, Takayama H, Sitkovsky MV (1987): Antigen-receptor requirement for conjugate formation and lethal hit triggering by cytotoxic T lymphocytes can be bypassed by protein kinase C activators and Ca^{++} ionophores. *Proc Natl Acad Sci USA* 84: 1364–1368

Bevan MJ, Cohn M (1975): Cytotoxic effects of antigen- and mitogen-induced T cells on various targets. *J Immunol* 114: 559–565

Biddison WE, Rao PR, Talle MA, Goldstein G, Shaw S (1982): Possible involvement of the OKT4 molecule in T cell recognition of class II HLA antigens: Evidence from studies of cytotoxic T lymphocytes specific for Sb antigens. *J Exp Med* 156: 1065–1076

Bierer BE, Burakoff SJ (1989): T-lymphocyte activation: The biology and function of CD2 and CD4. *Immunol Rev* 111: 267–294

Bierer BE, Burakoff SJ (1991): T cell receptors: Adhesion and signaling. *Adv Cancer Res* 56: 49–76

Biron CA, Natuk RJ, Welsh RM (1986): Generation of large granular T lymphocytes *in vivo* during viral infection. *J Immunol* 136: 2280–2286

Bjorkman PJ, Saper MA, Samraoui B, Bennett WS, Strominger JL, Wiley DC (1987): Structure of the human class I histocompatibility antigen, HLA-A2, *Nature* 329: 506–512

Blakely A, Gorman K, Ostergaard H, Svoboda K, Liu C, Young JD, Clark WR (1987): Resistance of cloned cytotoxic T lymphocytes to cell-mediated cytotoxicity. *J Exp Med* 166: 1070–1083

Bonavida B, Bradley TP, Grimm EA (1983): The single-cell assay in cell-mediated cytotoxicity. *Immunol Today* 4: 196

Bongrand P, Golstein P (1983): Reproducible dissociation of cellular aggregates with a wide range of calibrated shear forces: Application to cytolytic lymphocyte-target cell conjugates. *J Immunol Meth* 58: 209–224

Bongrand P, Pierres M, Golstein P (1983): T cell-mediated cytolysis: On the strength of effector-target cell interaction. *Eur J Immunol* 13: 424–429

Braakman E, Goedegebuure PS, Vreugdenhil RJ, Segal DM, Shaw S, Bolhuis RL (1990): ICAM–melanoma cells are relatively resistant to CD3-mediated T-cell lysis. *Int J Cancer* 46: 475–480

Bradley TP, Bonavida B (1978): Studies on the induction and expression of T cell-mediated immunity. VII. Inactivation of autologous cytotoxic T lymphocytes when used as both effectors and targets in a lectin dependent cellular cytotoxic reaction. *Transplant* 26: 212–217

Brahmi Z, Tomia A, Hommel-Berrey G, Woodard G, Morse P (1991): Mouse targets undergo double-strand DNA fragmentation when exposed to syngeneic or xenogeneic LAK cells whereas human targets undergo single-strand breaks. *Nat Immun Cell Growth Regul* 10: 196–206

Brondz BD (1968): Complex specificity of immune lymphocytes in allogeneic cell cultures. *Folia Biol (Prague)* 14: 115–131

Brondz BD (1972): Lymphocyte receptors and mechanisms of *in vitro* cell-mediated immune reactions. *Transplant Rev* 10: 112–151

Brondz BD, Snegiröva AE (1971): Interaction of immune lymphocytes with the mixtures of target cells possessing selected specificities of the H-2 immunizing allele. *Immunology* 20: 457

Brunet J, Denizot F, Suzan M, Haas W, Mencia-Huerta J, Berke G, Luciani M, Golstein P (1987): CTLA-1 and CTLA-3 serine esterase transcripts are detected mostly in cytotoxic T cells, but not only and not always. *J Immunol* 138: 4102–4105

Brunet JF, Denizot, F, Golstein P (1988): A differential molecular bilology and search for genes preferentially expressed in functional T lymphocytes: The CTLA genes. *Immunol Rev* 103: 21–36

Brunner KT, Mauel J, Cerottini J-C, Chapuis B (1968): Quantitative assay of the lytic action of immune lymphoid cells on ^{51}Cr-labelled allogeneic target cells *in vitro*; inhibition of isoantibody and by drugs. *Immunology* 14: 181–916

Brunner KT, Mauel J, Rudolf H, Chapuis B (1970): Studies of allograft immunity in mice I. Induction, development and *in vitro* assay of cellular immunity. *Immunology* 18: 501–515

Bubbers JE, Henney CS (1975): Studies on the synthetic capacity and antigenic expression of glutaraldehyde-fixed target cells. *J Immunol* 114: 1126–1131

Burakoff SJ, Martz E, Benacerraf B (1975): Is the primary complement lesion insufficient for lysis? Failure of cells damaged under osmotic protection to lyse in EDTA or at low temperature after removal of osmotic protection. *Clin Immunol Immunopathol* 4: 108–126

Bykovskaja SN, Rytenko AN, Rauschenbach MO, Bykovsky AF (1978a): Ultrastructural alteration of cytolytic T lymphocytes following their interaction with target cells. II. Mørphogenesis of secretory granules and intracellular vacuoles. *Cell Immunol* 40: 175–185

Bykovskaja SN, Rytenko AN, Rauschenbach MO, Bykovsky AF (1978b): Ultrastructural alteration of cytolytic T lymphocytes following their interaction with target cells. I Hypertrophy and change of orientation of the golgi apparatus. *Cell Immunol* 40: 164–174

Cantor H, Boyse EA (1975): Functional subclasses of T lymphocytes bearing different Ly antigens I. The generation of functionally distinct T cell subclasses is a differentiative process independent of antigen. *J Exp Med* 141: 1376–1389

Canty TG, Wunderlich JR (1970): Quantitative *in vitro* assay of cytotoxic cellular immunity. *J Natl Cancer Inst* 45: 761–772

Carrera AC, Rincon M, Sanchez-Madrid F, Lopez-Botet M, de Landazuri MO (1988): Triggering of co-mitogenic signals in T cell proliferation by anti-LFA-1 (CD18, CD11a), LFA-3, and CD7 monoclonal antibodies. *J Immunol* 141: 1919–1924

Cerottini J-C, Brunner KT (1974): Cell-mediated cytotoxicity, allograft rejection, and tumor immunity. *Adv Immunol* 18: 67–132

Cerottini J-C, MacDonald HR, Engers HD, Thomas K, Brunner KT (1974): Mechanisms of cell-mediated immunologic injury. *Adv Biosci* 12: 47–56

Cerottini J-C, Nordin AA, Brunner KT (1970): Specific *in vitro* cytotoxicity of thymus-derived lymphocytes sensitized to alloantigens. *Nature* 228: 1308–1309

Chang TW, Eisen HN (1980): Effects of n-tosyl-L-lysl-chloromethylketone on the activity of cytotoxic T lymphocytes. *J Immunol* 124: 1028–1033

Chayen J, Pitsillides AA, Bitensky L, Muir IH, Taylor PM, Askonas BA (1990): T-cell mediated cytolysis: Evidence for target-cell suicide. *J Exp Pathol (Oxford)* 71: 197–208

Christiaansen JE, Sears DW (1985): Lack of lymphocyte-induced DNA fragmentation in human targets during lysis represents a species-specific difference between human and murine cells. *Proc Natl Acad Sci USA* 82: 4482–4485

Clark RA, Klebanoff SJ (1979): Role of the myeloperoxidase-H_2O_2-halide system in concanavalin A-induced tumor cell killing by human neutrophils. *J Immunol* 122: 2605–2610

Clark W, Ostergaard H, Gorman K, Torbett B (1988): Molecular mechanisms of CTL-mediated lysis: A cellular perspective. *Immunol Rev* 103: 37–51

Clark WR (1988): Perforin—a primary or auxiliary lytic mechanism? *Immonol Today* 9: 101–104

Clark WR, Berke G, Feldman M, Sarid S (1971): Macromolecular synthesis during the sensitization of rat lymphocytes on mouse fibroblasts *in vitro*. *Immunochemistry* 8: 487–498

Cohen JJ (1991): Programmed cell death in the immune system. *Adv Immunol* 50: 55–85

Cohen JJ, Duke RC, Fadok VA, Sellins KS (1992): Apoptosis and programmed cell death in immunity. *Annu Rev Immunol* 10: 267–293

Collins, JL, Patek PQ, Cohn M (1981): Tumorigenicity and lysis by natural killers. *J Exp Med* 153: 89–106

Cosgrove JM, Howcroft TK, Tatum SM, Lindquist RR (1991): Membranolytic and nucleolytic activities of cytolytic T lymphocyte clones. *Biochem Biophys Res Commun* 179: 1562–1567

David A, Bernard J, Thiernesse N, Nicolas G, Cerottini J-C, Zagury D (1979): Le processus d'exocytose lysosomale localisée est-il responsable de l'action cytolytique des lymphocytes T tuers? *C R Acad Sci Ser D, Paris* 288: 441

Davignon D, Martz E, Reynolds T, Kürzinger K, Springer TA (1981a): Monoclonal antibody to a novel lymphocyte function-associated antigen (LFA-1). Mechanism of blockade of T lymphocyte-mediated killing and effects on other T and B lymphocyte functions. *J Immunol* 127: 590–595

Davignon D, Martz E, Reynolds T, Kürzinger K, Springer TA (1981b): Lymphocyte function-associated angtigen one (LFA-1): A surface antigen distinct from Lyt-2/3 that participates in T lymphocyte-mediated killing. *Proc Natl Acad Sci USA* 78: 4535–4539

de Fougerolles AR, Springer TA (1992): Intercellular adhesion molecule 3, a third adhesion counter-receptor for lymphocyte function-associated molecule-1 on resting lymphocytes. *J Exp Med* 175: 185–190

Dennert G, Anderson CG, Prochazka G (1987): High activity of N-alpha-benzyloxycarbonyl-L-lysine thiobenzyl ester serine esterase and cytolytic perforin in cloned cell lines is not demonstrable in *in-vivo*-induced cytotoxic effector cells. *Proc Natl Acad Sci USA* 84: 5004–5008

Dennert G, Podack E (1983): Cytolysis by H-2-specific T killer cells. Assembly of tubular complexes on target membranes. *J Exp Med* 157: 1483–1495

Detmers PA, Wright SD, Olsen E, Kimball B, Cohn ZA (1987): Aggregation of complement receptors on human neutrophils in the absence of ligand. *J Cell Biol* 105: 1137–1145

Dialynas D, Loken M, Sarmiento M, Fitch FW (1982): Identification of lysis-relevant molecules on the surface of CTL: Primary screening of monoclonal antibodies for the capacity to block cytolysis by cloned CTL lines. *Adv Exp Med Biol* 146: 547–556

Dialynas DP, Wilde DB, Marrack P, Pierres A, Wall KA, Havran W, Otten G, Loken MR, Pierres M, Kappler J, Fitch FW (1983): Characterization of the murine antigenic determinant designated L3T4a recognized by monoclonal antibody GK1.5: Expression of L3Ta by functional T cell clones appears to correlate primarily with class II MHC antigen-reactivity. *Immunol Rev* 74: 30–56

di Virgilio F, Bronte V, Collavo D, Zanovello P (1989): Responses of mouse lymphocytes to extracellular adenosine 5′-triphosphate (ATP). Lymphocytes with cytotoxic activity are resistant to the permeabilizing effects of ATP. *J Immunol* 143: 1955–1960

di Virgilio F, Pizzo P, Zanovello P, Bronte V, Collavo D (1990): Extracellular ATP as a possible mediator of cell-mediated cytotoxicity. *Immunol Today* 11: 274–277

Dohlsten M, Hedlund G, Kalland T (1991): Staphylococcal-enterotoxin-dependent cell-mediated cytotoxicity. *Immunol Today* 12: 147–150

Don MM, Ablett G, Bishop CJ, Bundesen PG, Donald KJ, Searle J, Kerr JFR (1977): Death of cells by apoptosis following attachment of specifically allergized lymphocytes *in vitro*. *Aust J Exp Biol Med Sci* 55: 407–417

Dransfield I, Cabanas C, Craig A, Hogg N (1992): Divalent cation regulation of the function of the leukocyte integrin LFA-1. *J Cell Biol* 116: 219–26

Duke RC (1989): Self recognition by T cells. I. Bystander killing of target cells bearing syngeneic MHC antigens. *J Exp Med* 170: 59–71

Duke RC, Cohen JJ (1988): The role of nuclear damage in lysis of target cells by cytotoxic T lymphocytes. In: *Cytolytic Lymphocytes and Complement: Effectors of the Immune System*, Podack ER, ed. Boca Raton, FL: CRC Press, 2: 35–37

Duke RC, Chervenak R, Cohen JJ (1983): Endogenous endonuclease-induced DNA fragmentation: An early event in cell-mediated cytolysis. *Proc Natl Acad Sci USA* 80: 6361–6365

Duke RC, Persechini PM, Chang S, Liu CC, Cohen JJ, Young JD (1989): Purified perforin induces target cell lysis but not DNA fragmentation. *J Exp Med* 170: 1451–1456

Duke RC, Sellins KS, Cohen JJ (1988): Cytotoxic lymphocyte-derived lytic granules do not induce DNA fragmentation in target cells. *J Immunol* 141: 2191–2194

Dustin ML, Springer TA (1989): T-cell receptor crosslinking transiently stimulates adhesiveness through LFA-1. *Nature* 341: 619–624

Duvall E, Wyllie AH (1986): Death and the cell. *Immunol Today* 7: 115–119

Fan J, Ahmed A, Bonavida B (1980): Studies on the induction and expression of T cell-mediated immunity. X. Inhibition by Lyt-2,3 antisera of cytotoxic T lymphocyte-mediated antigen-specific and -nonspecific cytotoxicity: Evidence for the blocking of the binding between T lymphocytes and target cells and not the post-binding cytolytic steps. *J Immunol* 125: 2444–2453

Ferguson WS, Verret CR, Reilly EB, Iannini MJ, Eisen HN (1988): Serine esterase and hemolytic activity in human cloned cytotoxic T lymphocytes. *J Exp Med* 167: 528–540

Filippini A, Taffs RE, Agui T, Sitkovsky MV (1990a): Ecto-ATPase activity in cytolytic T-lymphocytes. Protection from the cytolytic effects of extracellular ATP. *J Biol Chem* 265: 334–340

Filippini A, Taffs RE, Sitkovsky MV (1990b): Extracellular ATP in T-lymphocyte activation: Possible role in effector functions. *Proc Natl Acad Sci USA* 87: 8267–8271

Fink PJ, Rammensee H-G, Benedetto JD, Staerz UD, Lefrancois L, Bevan MJ (1984): Studies on the mechanism of suppression of primary cytotoxic responses by cloned cytotoxic T lymphocytes. *J Immunol* 133: 1769–1774

Fishelson Z, Berke G (1978): T lymphocyte-mediated cytolysis: Dissociation of the binding and lytic mechanisms of the effector cell. *J Immunol* 120: 1121–1126

Fleischer J (1986): Lysis of bystander target cells after triggering of human cytotoxic T lymphocytes. *Eur J Immunol* 16: 1021–1024

Fleiger D, Riethmuller G, Ziegler-Heitbrock HW (1989): Zn^{++} inhibits both tumor necrosis factor-mediated DNA fragmentation and cytolysis. *Int J Cancer* 44: 315–319

Fortier AH, Nacy CA, Stikovsky MV (1989): Similar molecular requirements for antigen receptor-triggered secretion of interferon and granule enzymes by cytolytic T lymphocytes. *Cell Immunol* 124: 64–76

Gabay JE, Scott RW, Campanelli D, Griffith J, Wilde C, Marra MN, Seeger M, Nathan CF (1989): Antibiotic proteins of human polymorphonuclear leukocytes [published erratum appears in *Proc Natl Acad Sci USA* 1989 Dec; 86 (24): 10133]. *Proc Natl Acad Sci USA* 86: 5610–5614

Garcia-Peñarrubia P, Bankhurst AD (1989a): Quantitation of effector-target affinity in the human NK cell and K562 tumor cell system. *J Immunol Methods* 122: 177–184

Garcia-Peñarrubia P, Bankhurst AD (1989b): Kinetic analysis of effector cell recycling and effector-target binding capacity in a model of cell-mediated cytotoxicity. *J Immunol* 143: 2101–2111

Gately MK, Martz E (1979): Early steps in specific tumor cell lysis by sensitized mouse T lymphocytes III. Resolution of two distinct roles for calcium in the cytolytic process. *J Immunol* 122: 482–489

Gately MK, Martz E (1981): Early steps in specific tumor cell lysis by sensitized mouse T lymphocytes V. Evidence that manganese inhibits a calcium-dependent step in programming for lysis. *Cell Immunol* 61: 78–89

Gately MK, Mayer MM (1978): Purification and characterization of lymphokines: An approach to the study of molecular mechanisms of cell-mediated immunity. *Prog Allergy* 25: 106–162

Gately MK, Mayer MM, Henney CS (1976): Effect of anti-lymphotoxin cell-mediated cytotoxicity. Evidence for two pathways, one involving lymphotoxin and the other requiring intimate contact between the plasma membranes of killer and target cells. *Cell Immunol* 27: 82–93

Gately MK, Wechter WJ, Martz E (1980): Early steps in specific tumor cell lysis by sensitized mouse T lymphocytes IV. Inhibition of programming for lysis by pharmacologic agents. *J Immunol* 125: 783–792

Geiger B, Rosen D, Berke G (1982): Spatial relationships of microtubule-organizing centers and the contact area of cytoxic T lymphocytes and target cells. *J Cell Biol* 95: 137–143

Geppert TD, Lipsky PE (1988): Activation of T lymphocytes by immobilized monoclonal antibodies to CD3. Regulatory influences of monoclonal antibodies to additional T cell surface determinants. *J Clin Invest* 81: 1497–1505

Ginsburg H, Ax W, Berke G (1969): Graft reaction in tissue culture by normal rat lymphocytes. *Transplant Proc* 1: 551–555

Glasebrook AL (1978): Conjugate formation by primary and secondary populations of murine immune T lymphocytes. *J Immunol* 121: 1870

Golding H, Singer A (1985): Specificity, phenotype, and precursor frequency of primary cytolytic T lymphocytes specific for class II major histocompatibility antigens. *J Immunol* 135: 1610–1615

Golstein P (1974): Sensitivity of cytotoxic T cells to T-cell mediated cytotoxicity. *Nature* 252: 81–83

Golstein P, Foa C, MacLennan ICM (1978): Mechanism of T-cell-mediated cytolysis: The differential impact of cytochalasins at the recognition and lethal hit stages. *Eur J Immunol* 8: 302–309

Golstein P, Goridis C, Schmitt-Verhulst A-M, Hayot B, Pierres A, van Agthoven A, Kaufmann Y, Eshhar Z, Pierres M (1982): Lymphoid cell surface interaction structures detected using cytolysis-inhibiting monoclonal antibodies. *Immunol Rev* 68: 5–42

Golstein P, Ojcius D, Young JD (1991): Cell death mechanisms and the immune system. *Immunol Rev* 121: 29–65

Golstein P, Smith ET (1977): Mechanism of T cell-mediated cytolysis: The lethal hit stage, *Contemp Topics Immunobiol* 7: 273–300

Golstein P, Svedmyr EAJ, Wigzell H (1971): Cells mediating specific *in vitro* cytotoxicity I. Detection of receptor-bearing lymphocytes. *J Exp Med* 134: 1385–1402

Gorman K, Liu CC, Blakely A, Young JD, Torbett BE, Clark WR (1988): Cloned cytotoxic T lymphocytes as target cells. II. Polarity of lysis revisited. *J Immunol* 141: 2211–2215

Govaerts A (1960): Cellular antibodies in kidney homotransplanation. *J Immunol* 85: 516–522

Gray LS, Russel JH (1986): Cytolytic T lymphocyte effector function requires plasma membrane chloride flux. *J Immunol* 136: 3032–3037

Gray LS, Gnarra JR, Englehard VH (1987): Demonstration of a calcium influx in cytolytic T lymphocytes in response to target cell binding. *J Immunol* 138: 63–69

Gray LS, Gnarra J, Hewlett EL, Englehard VH (1988): Increased intracellular cyclic adenosine monophosphate inhibits T lymphocyte-mediated cytolysis by two distinct mechanisms. *J Exp Med* 167: 1963–1968

Grimm EA, Bonavida B (1977): Studies on the induction and expression of T cell mediated immunity VI. Heterogeneity of lytic efficiency exhaused by isolated cytotoxic T lymphocytes prepared from highly enriched populations of effector-target conjugates. *J Immunol* 119: 1041–1047

Grimm E, Bonavida B (1979): Mechanism of cell-mediated cytotoxicity at the single cell level. I. Estimation of cytotoxic T lymphocyte frequency and relative lytic efficiency. *J Immunol* 123: 2861–2869

Grimm E, Price Z, Bonavida B (1979): Studies on the induction and expression of T cell-mediated immunity. VIII. Effector-target junctions and target cell membrane disruption during cytolysis. *Cell Immunol* 46: 77–99

Gromkowski S, Brown TC, Cerutti PA, Cerottini J-C (1986): DNA of human Raji target cells is damaged upon lymphocyte-mediated lysis. *J Immunol* 136: 752–756

Gromkowski SH, Brown TC, Masson D, Tschopp J (1988a): Lack of DNA degraduation in target cells lysed by granules derived from cytolytic T lymphocytes. *J Immunol* 141: 774–778

Gromkowski SH, Hepler KM, Janeway CA, Jr (1988b): Low doses of interleukin 2 induce bystander cell lysis by antigen-specific CD4+ inflammatory T cell clones in short-term assay. *Eur J Immunol* 18: 1385–1389

Gromkowski SH, Yagi J, Janeway CA, Jr (1989): Rapid, prelytic DNA fragmentation can be triggered by TNF. *Eur J Immunol* 19: 1709

Grusby MJ, Mitchell SC, Nabavi N, Glimcher LH (1990): Casein expression in cytotoxic T lymphocytes. *Proc Natl Acad Sci USA* 87: 6897–6901

Hahn WC, Rosenstein Y, Burakoff SJ, Bierer BE (1991):

TCR-CD3 stimulation increases the avidity of CD2 for its ligand LFA-3. *FASEB J* 5: A1377

Hakim FT, Gazzinelli RT, Denkers E, Hieny S, Shearer GM, Sher A (1991): CD8+ T cells from mice vaccinated against *Toxoplasma gondii* are cytotoxic for parasite-infected or antigen-pulsed host cells. *J Immunol* 147: 2310-2316

Hameed A, Olsen KJ, Lee M-K, Lichtenheld MG, Podack ER (1989): Cytolysis by Ca-permeable transmembrane channels. Pore formation causes extensive DNA degradation and cell lysis. *J Exp Med* 169: 765-777

Hasegawa Y, Bonavida B (1989): Calcium-independent pathway of tumour necrosis factor-mediated lysis of target cells. *J Immunol* 142: 2670-2676

Haskins, K, Kubo R, White J, Pigeon M, Kappler, J, Marrack P (1983): The major histocompatibility complex-restricted antigen receptor on T cells. I. Isolation with a monoclonal antibody. *J Exp Med* 157: 1149-1169

Haverstick DM, Engelhard VH, Gray LS (1991): Three intracellular signals for cytotoxic T lymphocyte-mediated killing. Independent roles for protein kinase C, Ca^{2+} influx, and Ca^{2+} release from internal stores. *J Immunol* 146: 3306-3313

Hayes MP, Berrebi GA, Henkart PA (1989): Induction of target cell DNA release by the cytotoxic T lymphocyte granule protease granzyme A. *J Exp Med* 170: 933-946

Hengel H, Wagner H, Heeg K (1991): Triggering of CD8+ cytotoxic T lymphocytes via CD3-epsilon differs from triggering via alpha/beta T cell receptor. CD3-epsilon-induced cytotoxicity occurs in the absence of protein kinase C and does not result in exocytosis of serine esterases. *J Immunol* 147: 1115-1120

Henkart MP, Henkart PA (1982): Lymphocyte mediated cytolysis as a secretory phenomenon. *Adv Exp Med Biol* 146: 227-242

Henkart P, Yue CC (1988): The role of cytoplasmic granules in lymphocyte cytotoxicity. *Prog Allergy* 40: 82-110

Henkart PA, Millard PJ, Reynolds CW, Henkart MP (1984): Cytolytic activity of purified cytoplasmic granules from cytotoxic rat large granular lymphocyte tumors. *J Exp Med* 160: 75-93

Henney CS, Gaffney J, Bloom BR (1974): On the relation of products of activated lymphocytes to cell-mediated cytolysis. *J Exp Med* 140: 837-852

Hibbs ML, Jakes S, Stacker SA, Wallace RW, Springer TA (1991a): The cytoplasmic domain of the integrin lymphocyte function-associated antigen 1 beta subunit: Sites required for binding to intercellular adhesion molecule 1 and the phorbol ester-stimulated phosphorylation site. *J Exp Med* 174: 1227-1238

Hibbs ML, Xu H, Stacker SA, Springer TA (1991b): Regulation of adhesion to ICAM-1 by the cytoplasmic domain of LFA-1 integrin beta subunit. *Science* 251: 1611-1613

Hidore MR, Nabavi N, Reynolds CW, Henkart PA, Murphy JW (1990): Cytoplasmic components of natural killer cells limit the growth of *Cryptococcus neoformans*. *J Leukoc Biol* 48: 15-26

Holm G, Perlmann P (1967): Quantitative studies on phytohaemagglutinin-induced cytotoxicity by human lymphocytes against homologous cells in tissue culture. *Immunology* 12: 525-536

Hommel-Berrey GA, Shenoy AM, Brahmi Z (1991): Receptor modulation and early signal transduction events in cytotoxic T lymphocytes inactivated by sensitive target cells. *J Immunol* 147: 3237-3243

Howell DM, Martz E (1987): The degree of CTL-induced DNA solubilization is not determined by the human vs. mouse origin of the target cell. *J Immunol* 138: 3695-3698

Howell DM, Martz E (1988): Low calcium concentrations support killing by some but not all cytolytic T lymphocytes, and reveal inhibition of a postconjugation step by calcium antagonists. *J Immunol* 140: 1982-1988

Hubbard BB, Glacken MW, Rodgers JR, Rich RR (1990): The role of physical forces on cytotoxic T cell-target cell conjugate stability. *J Immunol* 144: 4129-4138

Hudig D, Allison NJ, Pickett TM, Winkler U, Kam CM, Powers JC (1991): The function of lymphocyte proteases. Inhibition and restoration of granule-mediated lysis with isocoumarin serine protease inhibitors. *J Immunol* 147: 1360-1368

Hudig D, Gregg NJ, Kam CM, Powers JC (1987): Lymphocyte granule-mediated cytolysis requires serine protease activity. *Biochem Biophys Res Commun* 149: 882-888

Huffnagle GB, Yates JL, Lipscomb MF (1991): Immunity to a pulmonary *Cryptococcus neoformans* infection requires both CD4+ and CD8+ T cells. *J Exp Med* 173: 793-800

Inaba K, Wintmer MD, Steinman RM (1984): Clustering of dendritic cells, helper T lymphocytes, and histocompatible B cells during primary antibody responses *in vitro*. *J Exp Med* 160: 858-876

Ishiura S, Matsuda K, Koizumi H, Tsukahara T, Arahata K, Sugita H (1990): Calcium is essential for both the membrane binding and lytic activity of pore-forming protein (perforin) from cytotoxic T-lymphocyte. *Mol Immunol* 27: 803-807

Jiang SB, Ojcius DM, Persechini PM, Young JD (1990): Resistance of cytolytic lymphocytes to perforin-mediated killing. Inhibition of perforin binding activity by surface membrane proteins. *J Immunol* 144: 998-1003

Joag S, Young JD (1989): The absence of direct antimicrobial activity in extracts of cytotoxic lymphocytes. *Immunol Lett* 22: 195-198

Jones B, Tite JP, Janeway CA, Jr (1986): Different phenotypic variants of the mouse B cell tumor A20/2J are selected by antigen- and mitogen-triggered cytotoxicity of L3T4-positive, I-A-restricted T cell clones. *J Immunol* 136: 348-356

Jones J, Morgan BP (1991): Killing of cells by perforin. Resistance to killing is not due to diminished binding of perforin to the cell membrane. *Biochem J* 280: 199-204

Ju S-T (1991): Distinct pathways of CD4 and CD8 cells induce rapid target DNA fragmentation. *J Immunol* 146: 812-818

Ju ST, Ruddle NH, Strack P, Dorf ME, DeKruyff RH (1990): Expression of two distinct cytolytic mechanisms among murine CD4 subsets. *J Immunol* 144: 23–31

Ju ST, Strack P, Stromquist D, Dekruyff RH (1988): Cytolytic activity of Ia-restricted T cell clones and hybridomas: Evidence for a cytolytic mechanism independent of interferon-gamma, lymphotoxin, and tumor necrosis factor-alpha. *Cell Immunol* 117: 399–413

Kane KP, Goldstein SA, Mescher MF (1988): Class I alloantigen is sufficient for cytolytic T lymphocyte binding and transmembrane signaling. *Eur J Immunol* 18: 1925–1929

Kane KP, Sherman L, Mescher MF (1989): Molecular interactions required for triggering alloantigen-specific cytolytic T lymphocytes. *J Immunol* 142: 4153–4160

Kaufmann SHE (1988): $CD8^+$ T lymphocytes in intracellular microbial infections. *Immunol Today* 9: 168–174

Kerr JFR, Harmon BV (1991): Definition and incidence of apoptosis: an historical perspective. In: *Apoptosis: The Molecular Basis of Cell Death*, Tomei D, Coupe F, eds. New York: Cold Spring Harbor Laboratory Press

Kerr JFR, Wyllie AH, Currie AR (1982): Apoptosis: A basis biological phenomenon with wide-ranging implications in tissue kinetics. *B J Cancer* 26: 239–257

Kim SH, Carney DF, Hammer CH, Shin ML (1987): Nucleated cell killing by complement: Effects of C5b-9 channel size and extracellular Ca^{2+} on the lytic process. *J Immunol* 138: 1530–1536

Kim SH, Charney DF, Papadimitriou JC, Shim ML (1989): Effect of osmotic protection on nucleated cell killing by C5b-9: Cell death is not affected by the prevention of cell swelling. *Mol Immunol* 26: 323–331

Koren HS, Ax W, Freund-Moelbert E (1973): Morphological observations on the contact-induced lysis of target cells. *Eur J Immunol* 3: 32–37

Koulova L, Clark EA, Shu G, Dupont B (1991): The CD28 ligand B7/BB1 provides costimulatory signal for alloactivation of $CD4^+$ T cells. *J Exp Med* 173: 759–762

Kowproski H, Fernandes MV (1962): Autosensitization reaction *in vitro*. Contactual agglutination of sensitized lymph node cells in brain tissue culture accompanied by destruction of glial elements. *J Exp Med* 116: 467–476

Krähenbühl O, Tschopp J (1991): Perforin-induced pore formation. *Immunol Today* 12: 399–402

Kranz DM, Eisen HN (1987): Resistance of cytotoxic T lymphocytes to lysis by a clone of cytotoxic T lymphocytes. *Proc Natl Acad Sci USA* 84: 3375–3379

Kraut RP, Bose D, Cragoe EJ, Jr, Greenberg AH (1990): The influence of calcium sodium, and the Na^+/Ca^{2+} antiport on susceptibility to cytolysin/perforin-mediated cytolysis. *J Immunol* 144: 3498–3505

Krensky AM, Clayberger C, Reiss CS, Strominger JL, Buarakoff SJ (1982): Specificity of OKT4 cytotoxic T lymphocyte clones. *J Immunol* 129: 2001–2003

Krensky AM, Robbins E, Springer TA, Burakoff SJ (1984): LFA-1, LFA-2, and LFA-3 antigens are involved in CTL-target conjugation. *J Immunol* 132: 2180–2182

Kunkel LA, Welsh RM (1981): Metabolic inhibitors render "resistant" target cells sensitive to natural killer cell-mediated lysis. *Int J Cancer* 27: 73–79

Kupfer A, Dennert G (1984): Reorientation of the microtubule-organizing center and the golgi apparatus in cloned cytotoxic lymphocytes triggered by binding to lysable target cells. *J Immunol* 133: 2762–2766

Kupfer A, Singer SJ (1989a): Cell biology of cytotoxic and helper T-cell functions. *Ann Rev Immunol* 7: 309–338

Kupfer A, Singer SJ (1989b): The specific interaction of helper T cells and antigen-presenting B cells. IV. Membrane and cytoskeletal reorganizations in the bound T cell as a function of antigen dose. *J Exp Med* 170: 1697–1713

Kupfer A, Singer SJ (1989c): Cell biology of cytotoxic and helper T cell functions: Immunofluorescence microscopic studies of single cells and cell couples. *Annu Rev Immunol* 7: 309–337

Kupfer A, Dennert G, Singer SJ (1985): The reorientation of the golgi apparatus and the microtubule-organizing center in the cytotoxic effector cell is a prerequisite in the lysis of bound target cells. *J Mol Cell Immunol* 2: 37–49

Kupfer A, Singer SJ, Dennert G (1986): On the mechanism of unidirectional killing in mixtures of two cytotoxic T lymphocytes. Unidirectional polarization of cytoplasmic organelles and the membrane-associated cytoskeleton in the effector cell. *J Exp Med* 163: 489–498

Kupfer A, Swain SL, Singer SJ (1987): The specific direct interaction of helper T cells and antigen-presenting B cells. II. Reorientation of the microtubule organizing centre and reorganization of the membrane-associated cytoskeleton inside the bound helper T cells. *J Exp Med* 165: 1565–1580

Landegren U, Ramstedt U, Axberg I, Ullberg M, Jondal M, Wigzell H (1982): Selective inhibition of human T cell cytotoxicity at levels of target recognition or initiation of lysis by monoclonal OKT3 and Leu-2a antibodies. *J Exp Med* 155: 1579–1584

Landon C, Nowicki M, Sugawara S, Dennert G (1990): Differential effects of protein synthesis inhibition on CTL and targets in cell-mediated cytotoxicity. *Cell Immunol* 128: 412–426

Lanzavecchia A (1986): Is the T-cell receptor involved in T-cell killing? *Nature* 319: 778–780

Larson RS, Corbi AL, Berman L, Springer T (1989): Primary structure of the leukocyte function-associated molecule-1 alpha subunit: An integrin with an embedded domain defining a protein superfamily. *J Cell Biol* 108: 703–712

Laster SM, Wood JG, Gooding LR (1988): Tumor necrosis factor can induce both apoptotic and necrotic forms of cell lysis. *J Immunol* 141: 2629–2634

Lebow LT, Stewart CC, Perelson AS, Bonavida B (1986): Analysis of lymphocyte-target conjugates by flow cytometry. I. Discrimination between killer and non-killer lymphocytes bound to targets and sorting of conjugates containing one or multiple lymphocytes. *Nat Immun Cell Growth Regul* 5: 221–237

Lichtenstein AK, Ganz T, Nguyen TM, Selsted ME, Lehrer RI (1988). Mechanism of target cytolysis by peptide defensins. Target cell metabolic activities, possibly involving endocytosis, are crucial for expression of cytotoxicity. *J Immunol* 140: 2686–2694

Liepins A, Faanes RB, Choi YS, de Harven E (1978): T-lymphocyte mediated lysis of tumor cells in the presence of alloantiserum. *Cell Immunol* 36: 331–334

Liepens A, Faanes RB, Lifter J, Choi YS, DeHarven E (1977): Ultrastructural changes during T-lymphocyte-mediated cytolysis. *Cell Immunol* 28: 109–124

Lin Y, Askonas BA (1981): Biological properties of an influenza A virus-specific killer T cell clone inhibition of virus replication *in vivo* and induction of delayed-type hypersensitivity reactions. *J Exp Med* 154: 225–234

Linsley PS, Clark EA, Ledbetter JA (1990): T-cell antigen CD28 mediates adhesion with B cells by interacting with activation antigen B7/BB-1. *Proc Natl Acad Sci USA* 87: 5031–5035

Liu CC, Detmers PA, Jiang SB, Young JD (1989a): Identification and characterization of a membrane-bound cytotoxin of murine cytolytic lymphocytes that is related to tumor necrosis factor/cachectin. *Proc Natl Acad Sci USA* 86: 3286–3290

Liu CC, Jiang S, Persechini PM, Zychlinsky A, Kaufmann Y, Young JD (1989b): Resistance of cytolytic lymphocytes to perforin-mediated killing. Induction of resistance correlates with increase in cytotoxicity. *J Exp Med* 169: 2211–2225

Liu C, Steffen M, King F, Young JD (1987): Identification, isolation, and characterization of a novel cytotoxin in murine cytolytic lymphocytes. *Cell* 51: 393–403

Liu Y, Mullbacher A, Waring P (1989c): Natural killer cells and cytotoxic T cells induce DNA fragmentation in both human and murine target cells *in vitro*. *Scand J Immunol* 30: 31–37

Luciani MF, Brunet JF, Suzan M, Denizot F, Golstein P (1986): Self-sparing of long-term *in vitro*-cloned or uncloned cytotoxic T lymphocytes. *J Exp Med* 164: 962–967

Lukacher A, Morrison L, Braciale V, Malissen B, Braciale T (1985): Expression of specific cytolytic activity by H-2I region-restricted, influenza virus-specific T lymphocyte clones. *J Exp Med* 162: 171–187

MacDonald HR, Sordat B, Cerottini J-C, Brunner KT (1975): Generation of cytotoxic T lymphocytes *in vitro* IV. Functional activation of memory cells in the absence of DNA synthesis. *J Exp Med* 142: 622–636

MacLennan ICM, Gotch FM, Golstein P (1980): Limited specific T-cell mediated cytolysis in the absence of extracellular Ca^{2+}. *Immunol* 39: 109–117

Maimone MM, Morrison LA, Braciale VL, Braciale TJ (1986): Features of target cell lysis by class I and class II MHC-restricted cytolytic T lymphocytes. *J Immunol* 137: 3639–3643

Makgoba MW, Sanders ME, Ginther Luce GE, Gugel EA, Dustin ML, Springer TA, Shaw S (1988): Functional evidence that intracellular adhesion molecule-1 (ICAM-1) is a ligand for LFA-1-dependent adhesion in T cell-mediated cytotoxicity. *Eur J Immunol* 18: 637–640

Markham RB, Goellner J, Pier GB (1984): *In vitro* T cell-mediated killing of *Pseudomonas aeruginosa*. I. Evidence that a lymphocyte mediates killing. *J Immunol* 133: 962–968

Martin SJ, Lennon SV, Bonham AM, Cotter TG (1990): Induction of apoptosis (programmed cell death) in human leukemic HL-60 cells by inhibition of RNA or protein synthesis. *J Immunol* 145: 1859–1867

Martz E (1975a): Early steps in specific tumor cell lysis by sensitized mouse T-lymphocytes I. Resolution and characterization. *J Immunol* 115: 261–267

Martz E (1975b): Inability of EDTA to prevent damage mediated by cytolytic T-lymphocytes. *Cell Immunol* 20: 304–314

Martz E (1976a): Sizes of isotopically-labelled molecules released during lysis of tumor cells labelled with ^{51}Cr and [^{14}C]nicotinamide. *Cell Immunol* 26: 313–321

Martz E (1976b): Early steps in specific tumor cell lysis by sensitized mouse T lymphocytes II. Electrolyte permeability increase in the target cell membrane concomitant with programming for lysis. *J Immunol* 117: 1023–1027

Martz E (1977): Mechanism of specific tumour cell lysis by alloimmune T-lymphocytes: Resolution and characterization of discrete steps in the cellular interaction. *Contemp Top Immunobiol* 7: 301–361

Martz E (1980): Immune lymphocyte to tumor cell adhesion: Magnesium sufficient, calcium insufficient. *J Cell Biol* 84: 584–598

Martz E (1987): LFA-1 and other accessory molecules functioning in adhesions of T and B lymphocytes. *Hum Immunol* 18: 3–37

Martz E, Benacerraf B (1973): An effector-cell independent step in target cell lysis by sensitized mouse lymphocytes. *J Immunol* 111: 1538–1545

Martz E, Benacerraf B (1974): Crucial events in target-cell lysis by sensitized mouse T-lymphocytes: A killer-cell independent, temperature sensitive step. *Adv Biosci* 12: 37–46

Martz E, Benacerraf B (1975): T-lymphocyte mediated cytolysis: Temperature dependence of killer cell dependent and independent phases and lack of recovery from the lethal hit at low temperatures. *Cell Immunol* 20: 81–91

Martz E, Benacerraf B (1976): Multiple target cell killing by the cytolytic T-lymphocyte and the mechanism of cytotoxicity. *Transplant* 21: 5–11

Martz E, Gamble SR (1991): How do CTL control virus infections? Evidence for prelytic halt of *Herpes simplex*. *Virol Immunol* 5: 81–91

Martz E, Howell DM (1989): CTL: Virus control cells first and cytolytic cells second? DNA fragmentation, apoptosis, and the prelytic halt hypothesis. *Immunol Today* 10: 79–86

Marz E, Burakoff SJ, Benacerraf B (1974): Interruption of the sequential release of small and large molecules from tumor cells by low temperature during cytolysis mediated by immune T-cells or complement. *Proc Natl Acad Sci USA* 71: 177–181

Martz E, Davignon D, Kürzinger K, Springer TA (1982a): The molecular basis for cytolytic T lymphocyte function: Analysis with blocking monoclonal antibodies. *Adv Exp Med Biol* 146: 447–465

Martz E, Parker WL, Gately MK, Tsoukas CD (1982b): The role of calcium in the lethal hit of T lymphocyte-mediated cytolysis. *Adv Exp Med Biol* 146: 121–143

Masson D, Tschopp J (1985): Isolation of a lytic, pore-forming protein (perforin) from cytolytic T lymphocytes. *J Biol Chem* 260: 9069–9072

Matter A (1979): Microcinematographic and electron microscopic analysis of target cell lysis induced by cytotoxic T lymphocytes. *Immunology* 36: 179–190

Mauel J, Rudolf H, Chapuis B, Brunner KT (1970): Studies of allograft immunity in mice. II. Mechanism of target cell inactivation *in vitro* by sensitized lymphocytes. *Immunology* 18: 517–535

McConkey DJ, Chow SC, Orrenius S, Jondal M (1990): NK cell-induced cytoxicity is dependent on a Ca^{2+} increase in the target. *FASEB J* 4: 2661–2664

McConkey DJ, Hartzell P, Nicotera P, Orrenius S (1989): Calcium-activated DNA fragmentation kills immature thymocytes. *FASEB J* 3: 1843–1849

Mcintyre KW, Welsh RM (1986): Accumulation of natural killer and cytotoxic T large granular lymphocytes in the liver during virus infection. *J Exp Med* 164: 1667–1681

Mentzer SJ, Gromkowski SH, Krensky AM, Burakoff SJ, Martz E (1985): LFA-1 membrane molecule in the regulation of homotypic adhesions of human B lymphocytes. *J Immunol* 135: 9–11

Mescher MF (1992): Cell-cell interaction and surface contact requirements: Activation of cytotoxic T lymphocytes (submitted for publication)

Mescher MF, Balk SP, Burakoff SJ, Herrmann SH (1982): Cytolytic T lymphocyte recognition of subcellular antigen. In: *Mechanisms of Cell-Mediated Cytotoxicity*, Clark WR, Golstein P, eds. New York: Plenum Press, pp. 41–55

Mescher MF, O'Rourke AM, Champoux P, Kane KP (1991): Equilibrium binding of cytotoxic T lymphocytes to class I antigen. *J Immunol* 147: 36–41

Meuer SC, Fitzgerald KA, Hussey RE, Hodgdon JC, Schlossman SF, Reinherz EL (1983): Clonotypic structures involved in antigen specific human T cell function: Relationship to the T3 molecular complex. *J Exp Med* 157: 705–719

Meuer SC, Schlossman SF, Reinherz EL (1982): Clonal analysis of human cytotoxic T lymphocytes: T4+ and T8+ effector T cells recognize products of different major histocompatibility complex regions. *Proc Natl Acad Sci USA* 79: 4395–4399

Mishra GC, Berton MT, Oliver KG, Krammer PH, Uhr JW, Vitetta ES (1986): A monoclonal anti-mouse LFA-I alpha antibody mimics the biological effects of B cell stimulatory factor-1 (BSF-1). *J Immunol* 137: 1590–1598

Morgan BP, Dankert J, Esser AF (1987): Recovery of human neutrophils from complement attack: Removal of the membrane attack complex by endocytosis and exocytosis. *J Immunol* 138: 246–253

Moss DJ, Burrows SR, Baxter GD, Lavin MF (1991): T cell-T cell killing is induced by specific epitopes: Evidence for an apoptotic mechanism. *J Exp Med* 173: 681–686

Moy VT, Brian AA (1992): Signaling by lymphocyte function-associated antigen 1 (LFA-1) in B cells: Enhanced antigen presentation after simulation through LFA-1. *J Exp Med* 175: 1–7

Munger WE, Lindquist RR (1982): Effect of phorbol esters on alloimmune cytolysis. *Cancer Res* 42: 5023–5029

Munger WE, Berrebi GA, Hankart PA (1988): Possible involvement of CTL granule proteases in target cell DNA breakdown. *Immunol Rev* 103: 99–109

Murphy KM, Heimberger AB, Loh DY (1990): Induction by antigen of intrathymic apoptosis of $CD4^+$ $CD8^+$ TCR^{10} thymocytes *in vivo*. *Science* 250: 1720–1722

Nagler-Anderson C, Eisen HN (1990): Resistance of normal, unstimulated, $CD8^+$ T cells to lysis by cytotoxic granules from cloned T cell lines. *Int Immunol* 2: 99–103

Nagler-Anderson C, Allbritton NL, Verret CR, Eisen HN (1988a): A comparison of the cytolytic properties of murine primary $CD8^+$ cytotoxic T lymphocytes and cloned cytotoxic T cell lines. *Immunol Rev* 103: 111–125

Nagler-Anderson C, Lichtenheld M, Eisen HN, Podack ER (1989): Perforin mRNA in primary peritoneal exudate cytotoxic T lymphocytes. *J Immunol* 143: 3440–3443

Nagler-Anderson C, Verret CR, Firmenich AA, Berne M, Eisen HN (1988b): Resistance of primary $CD8^+$ cytotoxic T lymphocytes to lysis by cytotoxic granules from cloned T cell lines. *J Immunol* 141: 3299–3305

Nakamura M, Ross DT, Briner TJ, Gefter ML (1986): Cytolytic activity of antigen-specific T cells with helper phenotype. *J Immunol* 136: 44–47

Nishimura T, Burakoff SJ, Herrmann SH (1987): Protein kinase C required for cytotoxic T lymphocyte triggering. *J Immunol* 139: 2888–2891

Nishioka WK, Welsh RM (1991): Inhibition of cytotoxic T lymphocyte-induced target cell DNA fragmentation, but not lysis, by inhibitors of DNA topoisomerases I and II. *J Exp Med* 175: 23–27

Nowicki M, Landon C, Sugawara S, Dennert G (1991): Nicotinamide and 3-aminobenzamide interfere with receptor-mediated transmembrane signaling in murine cytotoxic T cells: Independent of Golgi reorientation from calcium mobilization and inositol phosphate generation. *Cell Immunol* 132: 115–126

Ohanian SH, Schlager SI (1981): Humoral immune killing of nucleated cells: Mechanisms of complement-mediated attack and target cell defense. *CRC Crit Rev Immunol* January: 165–209

Ohnishi K, Bonavida B (1986): Specific lymphocyte-target cell conjugate formation between tumor-specific helper T-cell hybridomas and IA-bearing RCS tumors and IE-bearing allogeneic cells. *J Immunol* 137: 3681–3688

Ojcius DM, Jiang SB, Persechini PM, Detmers PA, Young JD (1991a): Cytoplasts from cytotoxic T lymphocytes are resistant to perforin-mediated lysis. *Mol Immunol* 28: 1011–1018

Ojcius DM, Jiang SB, Persechini PM, Storch J, Young JD (1990): Resistance to the pore-forming protein of cytotoxic T cells: Comparison of target cell membrane rigidity. *Mol Immunol* 27: 839–845

Ojcius DM, Zheng LM, Sphicas EC, Zychlinsky A, Young JD (1991b): Subcellular localization of perforin and serine esterase in lymphokine-activated killer cells and cytotoxic T cells by immunogold labeling. *J Immunol* 146: 4427–4432

O'Rourke AM, Mescher MF (1988): T cell receptor-mediated signaling occurs in the absence of inositol phosphate production. *J Biol Chem* 263: 18594–18597

O'Rourke AM, Rogers J, Mescher MF (1990): CD8 binding to class I is activated via the T cell receptor and results in signalling for response. *Nature* 346: 187–189

Orrenius S, McConkey DJ, Bellomo G, Nicotera P (1989): Role of Ca^{2+} in toxic cell killing. *Trends Pharmacol Sci* 10: 281–285

Ostergaard H, Clark WR (1987): the role of Ca^{++} in activation of mature cytotoxic T lymphocytes for lysis. *J Immunol* 139: 3573–3579

Ostergaard HL, Clark WR (1989): Evidence for multiple lytic pathways used by cytotoxic T lymphocytes. *J Immunol* 143: 2120–2126

Ostergaard H, Gorman K, Clark WR (1988): Cloned cytotoxic T lymphocyte target cells fail to induce early activation in effector cytotoxic T lymphocytes. *Cell Immunol* 114: 188–197

Ostergaard HL, Kane KP, Mescher MF, Clark WR (1987): Cytotoxic T lymphocyte mediated lysis can occur in the absence of serine esterase release. *Nature* 330: 71–72

Ottenhoff TH, Mutis T (1990): Specific killing of cytotoxic T cells and antigen-presenting cells by $CD4^+$ cytotoxic T cell clones. A novel potentially immunoregulatory T-T cell interaction in man. *J Exp Med* 171: 2011–2024

Ozaki S, York-Jolley J, Kawamura H, Berzofsky JA (1987): Cloned protein antigen-specific, Ia-restricted T cells with both helper and cytolytic activities: Mechanisms of activation and killing. *Cell Immunol* 105: 301–316

Ozdemirli M, Akdeniz H, El-Khatib M, Ju ST (1991): A novel cytotoxicity of $CD4^+$ Th1 clones on heat-shocked tumor targets. I. Implications for internal disintegration model for target death and hyperthermia treatment of cancers. *J Immunol* 147: 4027–4034

Pardi R, Bender JR, Dettori C, Giannazza E, Engleman EG (1989): Heterogeneous distribution and transmembrane signaling properties of lymphocyte function-associated antigen (LFA-1) in human lymphocyte subsets. *J Immunol* 143: 3157–3166

Parker WL (1980): Studies on the mechanism of T cell-mediated cytolysis. Cambridge MA: Ph.D. Thesis, Harvard University

Parker WL, Martz E (1982): Calcium ionophore A23187 as a secretagogue for rat mast cells: Does it bypass inhibition by calcium flux blockers? *Agents Actions* 12: 276–284

Pasternak MS, Eisen HN (1985): A novel series esterase expressed by cytotoxic T lymphocytes. *Nature* 314: 743–745

Pasternack MS, Bleier KJ, McInerney TN (1991): Granzyme A binding to target cell proteins. Granzyme A binds to an cleaves nucleolin *in vitro*. *J Biol Chem* 266: 14703–14708

Pasternack MS, Verret CR, Liu MA, Eisen HM (1986): Serine esterase in cytolytic T lymphocytes *Nature* 322: 740–743

Paul NL, Ruddle NH (1988): Lymphotoxin. *Ann Rev Immunol* 6: 407–438

Pemberton RM, Wraith DC, Askonas BA (1990): Influenza peptide-induced self-lysis and down-regulation of cloned cytotoxic T cells. *Immunology* 70: 223–229

Perez P, Bluestone JA, Stephany DA, Segal DM (1985): Quantitative measurements of the specificity and kinetics of conjugate formation between cloned cytotoxic T lymphocytes and splenic target cells by dual parameter flow cytometry. *J Immunol* 134: 478–485

Peters PJ, Borst J, Oorschot V, Fukuda M, Krahenbuhl O, Tschopp J, Slot JW, Geuze HJ (1991): Cytotoxic T lymphocyte granules are secretory lysosomes, containing both perforin and granzymes. *J Exp Med* 173: 1099–1109

Peters PJ, Geuze HJ, van der Donk HA, Borst J (1990): A new model for lethal hit delivery by cytotoxic T lymphocytes. *Immunol Today* 11: 28–32

Pierres M, Goridis C, Golstein P (1982): Inhibition of murine T cell-mediated cytolysis and T cell proliferation by a rat monoclonal antibody immunoprecipitating two lymphoid cell surface polypeptides of 94,000 and 180,000 molecular weight. *Eur J Immunol* 12: 60

Pitsillides AA, Tyalor PM, Bitensky L, Chayen J, Muir IH, Askonas BA (1988): Rapid changes in target cell lysosomes induced by cytotoxic T cells: Indication of target suicide? *Eur J Immunol* 18: 1203–1208

Platsoucas CD (1982): Inhibition of specific T-cell-mediated cytotoxicity by the OKT3/anti-Leu 4 and anti-Leu-2a/OKT8 monoclonal antibodies occurs at different stages of the cytolytic process. *Immunobiology* 163: 329

Plaut M, Bubbers JE, Henney CS (1976): Studies on the mechanism of lymphocyte-mediated cytolysis. VII. Two stages in the T cell-mediated lytic cycle with distinct cation requirements. *J Immunol* 116: 150–155

Podack ER (1986): Molecular mechanisms of cytolysis by complement and by cytolytic lymphocytes. *J Cell Biochem* 30: 127–164

Podack ER, Konigsberg PJ (1984): Cytolytic T granules. Isolation, structural, biochemical, and functional characterization. *J Exp Med* 160: 695–710

Podack ER, Hengartner H, Lichtenheld MG (1991): A central role of perforin in cytolysis? *Ann Rev Immunol* 9: 129–157

Podack ER, Young JD, Cohn ZA (1985): Isolation and biochemical and functional characterization of perforin 1 from cytolytic T-cell granules. *Proc Natl Acad Sci USA* 82: 8629–8633

Poenie M, Tsien RY, Schmitt-Verhulst AM (1987): Sequential activation and lethal hit measured by $[Ca^{2+}]_i$ in individual cytolytic T cells and targets. *EMBO J* 6: 2223–2232

Prochazka G, Landon C, Dennert G (1988): Transmembrane chloride flux is required for target cell lysis but not for Golgi reorientation in cloned cytolytic effector cells. Golgi reorientation, N alpha-benzyloxycarbonyl-L-lysine thiobenzyl ester serine esterase release, and delivery of the lethal hit are separable events in target cell lysis. *J Immunol* 141: 1288–1294

Ramm LE, Whitlow MB, Koski CL, Shin ML, Mayer MM (1983): Elimination of complement channels from the plasma membranes of U937, a nucleated mammalian cell line: Temperature dependence of the elimination rate. *J Immunol* 131: 1411–1415

Redegeld F, Filippini A, Sitkovsky M (1991): Comparative studies of the cytotoxic T lymphocyte-mediated cytotoxicity and of extracellular ATP-induced cell lysis. Different requirements in extracellular Mg^{2+} and pH. *J Immunol* 147: 3638–3645

Redelman D, Hudig D (1983): The mechanism of cell-mediated cytotoxicity. III. Protease-specific inhibitors preferentially block later events in cytotoxic T lymphocyte-mediated lysis than do inhibitors of methylation or thiol-reactive agents. *Cell Immunol* 81: 9–21

Richieri GV, Kleinfeld AM (1990): Free fatty acids inhibit cytotoxic T lymphocyte-mediated lysis of allogeneic target cells. *J Immunol* 145: 1074–1077

Rosenau W, Moon HD (1961): Lysis of homologous cells by sensitized lymphocytes in tissue culture. *J Natl Cancer Inst* 27: 471–483

Rosenstein Y, Park JK, Hahn WC, Rosen FS, Bierer BE, Burakoff SJ (1991): CD43, a molecular defective in Wiskott-Aldrich syndrome, binds ICAM-1. *Nature* 354: 233–235

Rothlein R, Springer TA (1986): The requirement for lymphocyte function-associated antigen 1 in homotypic leukocyte adhesion stimulated by phorbol ester. *J Exp Med* 163: 1132–1149

Rothlein R, Dustin ML, Marlin SD, Springer TA (1986): A human intercellular adhesion molecule (ICAM-1) distinct from LFA-1. *J Immunol* 137: 1270–1274

Rothstein TL, Mage M, Jones G, McHugh LL (1978): Cytotoxic T lymphocyte sequential killing of immobilized allogeneic tumor target cells measured by time-lapse microcinematography. *J Immunol* 121: 1652

Rudd CE (1990): CD4, CD8 and the TCR-CD3 complex: A novel class of protein-tyrosine kinase receptor. *Immunol Today* 11: 400–406

Russell JH (1983): Internal disintegration model of cytotoxic lymphocyte-induced target damage. *Immunol Rev* 72: 97–118

Russell JH (1984): Phorbol esters inactivate the lytic apparatus of cytotoxic T lymphocytes. *J Immunol* 133: 907–912

Russell JH (1986): Phorbol-ester simulated lysis of weak and nonspecific target cells by cytotoxic T lymphocytes. *J Immunol* 136: 23–27

Russell JH, Dobos (1980): Mechanisms of immune lysis. II. CTL-induced nuclear disintegration of the target begins within minutes of cell contact. *J Immunol* 125: 1256–1261

Russell JH, Dobos CB (1981): The role of monovalent cations in the interaction between the cytotoxic T lymphocyte and its target. *Eur J Immunol* 11: 840–843

Russell JH, Masakowski VR, Dobos CB (1980): Mechanisms of immune lysis. I. Physiological distinction between target cell death mediated by cytotoxic T-lymphocytes and antibody plus complement. *J Immunol* 124: 1100–1105

Russell JH, Musil L, McCulley DE (1988): Loss of adhesion: A novel and distinct effect of the cytotoxic T lymphocyte-target interaction. *J Immunol* 140: 427–432

Ryser J-E, Rungger-Brändle E, Chaponnier C, Gabbiani G, Vassalli P (1982): The area of attachment of cytotoxic T lymphocytes to their target cells shows high motility and polarization of actin but not myosin. *J Immunol* 128: 1159–1162

Ryser J-E, Sordat B, Cerottini J-C, Brunner KT (1977): Mechanism of target cell lysis by cytolytic T lymphocytes I. Characterization of specific lymphocyte-target cell conjugates separated by velocity sedimentation. *Eur J Immunol* 7: 110–117

Sanchez-Madrid F, Davignon D, Martz E, Springer TA (1982a): Functional screening for antigens associated with mouse T lymphocyte-mediated killing yields antibodies to Lyt-2,3 and LFA-1. *Cell Immunol* 73: 1–11

Sanchez-Madrid F, Krensky AM, Ware CF, Robbins E, Strominger JL, Burakoff SJ, Springer TA (1982b): Three distinct antigens associated with human T lymphocyte-mediated cytolysis: LFA-1, LFA-2, and LFA-3. *Proc Natl Acad Sci USA* 79: 7489–7493

Sanchez-Madrid F, Simon P, Thompson S, Springer TA (1983): Mapping of antigenic and functional epitopes on the alpha- and beta-subunits of two related mouse glycoproteins involved in cell interactions, LFA-1 and Mac-1. *J Exp Med* 158: 586–602

Sanders VM, Snyder JM, Uhr JW, Vitetta ES (1986): Characterization of the physical interaction between antigen-specific B and T cells. *J Immunol* 137: 2395–2404

Sanderson CJ (1976): The mechanism of T cell mediated cytotoxicity II. Morphological studies of cell death by time-lapse microcinematography. *Proc. Roy Soc B* 192: 241–255

Sanderson CJ (1981a): The mechanism of T cell mediated cytotoxicity VIII. Zeiosis corresponds to irreversible phase (programming for lysis) in steps leading to lysis. *Immunology* 42: 201–206

Sanderson CJ (1981b): The mechanism of lymphocyte-mediated cytotoxicity. *Biol Rev* 56: 153–197

Sanderson CJ, Thomas JA (1977): The mechanism of K cell (antibody-dependent) cell mediated cytotoxicity. I. The release of different cell components. *Proc Roy Soc B* 197: 407–415

Schick B, Berke G (1978): Is the presence of serologically-defined target cell antigens sufficient for binding of cytotoxic T lymphocytes? *Transplant* 26: 14–18

Schick B, Berke G (1979): Cellular activities necessary for specific binding of cytotoxic T lymphocytes and target cells. *Transplant Proc* 11: 800–803

Schick B, Berke G (1990): The lysis of cytotoxic T lymphocytes and their blasts by cytotoxic T lymphocytes. *Immunology* 71: 428–433

Schmid DS, Hornung R, McGrath KM, Paul N, Ruddle NH (1987): Target cell DNA fragmentation is mediated by lymphotoxin and tumor necrosis factor. *Lymphokine Res* 6: 195–202

Schmid DS, Tite JP, Ruddle NH (1986): DNA fragmentation: Manifestation of target cell destruction mediated by cytolytic T-cell lines, lymphotoxin-secreting helper T-cell clones, and cell-free lymphotoxin-containing supernatant. *Proc Natl Acad Sci USA* 83: 1881–1885

Schwartz LM, Kosz L, Kay BK (1990): Gene activation is required for developmentally programmed cell death. *Proc Natl Acad Sci USA* 87: 6594–6598

Schwarz LM, Smith SW, Jones MEE, Osborne BA (1993): Do all programmed cell deaths occur by apoptosis? *Proc Natl Acad Sci USA* in press

Sellins KS, Cohen JJ (1989): Polyomavirus DNA is damaged in target cells during cytotoxic T-lymphocyte-mediated killing. *J Virol* 63: 572–578

Sellins KS, Cohen JJ (1991): Cytotoxic T lymphocytes induce different types of DNA damage in target cells of different origins. *J Immunol* 147: 795–803

Shaw S, Luce GE, Quinones R, Gress RE, Springer TA, Sanders ME (1986): Two antigen-independent adhesion pathways used by human cytotoxic T-cell clones. *Nature* 323: 262–264

Shi YF, Szalay MG, Paskar L, Boyer M, Singh B, Green DR (1990): Activation-induced cell death in T cell hybridomas is due to apoptosis. Morphologic aspects and DNA fragmentation. *J Immunol* 144: 3326–3333

Shiver JW, Henkart PA (1991): A noncytotoxic mast cell tumor line exhibits potent IgE-dependent cytotoxicity after transfection with the cytolysin/perforin gene. *Cell* 64: 1175–1181

Shortman K, Golstein P (1979): Target cell recognition by cytolytic T cells: Different requirements for the formation of strong conjugates or for proceeding to lysis. *J Immunol* 123: 833–839

Shortman K, Wilson A (1986): Degradation of specificity in cytolytic T lymphocyte clones. The separate YAC-1-Type (NK-like) and P815 type broad specificity killing patterns are both restricted to the larger cells within a clone but may be expressed independently in clones from different mouse strains. *J Immunol* 137: 798–804

Shortman K, Wilson A, Scollay R (1984): Loss of specificity in cytolytic T lymphocyte clones obtained by limit dilution culture of Ly-c$^+$ T cells. *J Immunol* 132: 584–593

Simon MM, Fruth U, Simon HG, Gay S, Kramer MD (1989a): Evidence for multiple functions of T-lymphocytes associated serine proteinases. *Adv Exp Med Biol* 247A: 609–613

Simon MM, Fruth U, Simon HG, Kramer MD (1986): A specific serine proteinase is inducible in Lyt-2$^+$, L3T4$^-$ and Lyt-2$^-$, L3T4$^+$ T cells *in vitro* but is mainly associated with Lyt-2$^+$, L3T4$^-$ effector cells *in vivo*. *Eur J Immunol* 16: 1559–1568

Simon MM, Prester M, Kramer MD, Fruth D (1989b): An inhibitor specific for the mouse T-cell associated serine proteinase 1 (TSP-1) inhibits the cytolytic potential of cytoplasmic granules but not of intact cytolytic T cells. *J Cell Biochem* 40: 1–13

Sitkovsky MV (1988): Mechanistic, functional and immunopharmacological implications of biochemical studies of antigen receptor-triggered cytolytic T-lymphocyte activation. *Immunol Rev* 103: 127–160

Skinner M, Marbrook J (1987): Communications. The most efficient cytotoxic T lymphocytes are the least susceptible to lysis. *J Immunol* 139: 985–987

Skinner M, Hartley L, Marbrook J (1989a): Hierarchy of cytotoxic T cell clones. II. Intraclonal triggering of lysis by hapten-specific CTL. *Cell Immunol* 122: 471–482

Skinner M, Hartley L, Marbrook J (1989b): Hierarchy of cytotoxic T cell clones. I. Reversal of resistance to lysis and killing between anti-hapten cytotoxic T cells. *Cell Immunol* 122: 461–470

Smith CA, Williams GT, Kingston R, Jenkinson EJ, Owen JJ (1989): Antibodies to CD3/T-cell receptor complex induce death by apoptosis in immature T cells in thymic cultures. *Nature* 337: 181–184

Smith LE, Rodrigues M, Russell DG (1991): The interaction between CD8$^+$ cytotoxic T cells and Leishmania-infected macrophages. *J Exp Med* 174: 499–505

Snodgrass HR, Bosma MJ, Wilson DB (1981): T lymphocytes specific for immunoglobin allotype II. Cloned Igh-1b-specific cytotoxic T cells. *J Exp Med* 154: 491–500

Solovera JJ, Alverex-Mon M, Casas J, Carballido J, Durantez A (1987). Inhibition of human natural killer (NK) activity by calcium channel modulators and a calmodulin antagonist. *J Immunol* 139: 876–880

Springer TA (1990): Adhesion receptors of the immune system. *Nature* 346: 425–434

Strack P, Martin C, Saito S, Dekruyff RH, Ju ST (1990): Metabolic inhibitors distinguish cytolytic activity of CD4 and CD8 clones. *Eur J Immunol* 20: 179–184

Strassman G, Springer TA, Somers SD, Adams DO (1986): Mechanisms of tumor cell capture by activated macrophages: Evidence for involvement of lymphocyte function-associated (LFA)-1 antigen. *J Immunol* 136: 4328–4333

Strom TB, Garovoy MR, Bear RA, Gribik M, Carpenter CB (1975): A comparison of the effects of metabolic inhibitors upon direct and antibody-dependent lymphocyte mediated cytotoxicity. *Cell Immunol* 20: 247–256

Stulting RD, Berke G (1973): Nature of lymphocyte-tumor interaction. A general method for cellular immunoabsorption. *J Exp Med* 137: 932–942

Stulting RD, Todd RF, III, Amos DB (1975): Lymphocyte-mediated cytolysis of allogeneic tumor cells in *vitro* II. Binding of cytotoxic lymphocytes to formaldehyde-fixed target cells. *Cell Immunol* 20: 54–63

Sugawara S, Nowicki M, Xie S, Song H-J, Dennert G (1990): Effects of stress on lysability of tumor targets by cytotoxic T cells and tumor necrosis factor[1]. *J Immunol* 145: 1991–1998

Sung K-LP, Sung LA, Crimmins M, Burakoff SJ, Chein S (1986): Determination of junction avidity of cytolytic T cell and target cell. *Science* 234: 1405–1408

Sung KL, Sung LA, Crimmins M, Burakoff SJ, Chien S (1988): Dynamic changes in viscoelastic properties in cytotoxic T-lymphocyte-mediated killing. *J Cell Sci* 91: 179–189

Swain SL (1981): Significance of Lyt phenotypes: Lyt2 antibodies block activities of T cells that recognize class 1 major histocompatibility complex antigens regardless of their function. *Proc Natl Acad Sci USA* 78: 7101–7105

Swain SL, Dennert G, Dutton RW (1980): Association of Ly phenotypes, T cell function, and MHC recognition. *Fed Proc* 39: 353

Takayama H, Trenn G, Humphrey J, Jr, Bluestone JA, Henkart PA, Sitkovsky MV (1987): Antigen receptor-triggered secretion of a trypsin-type esterase from cytotoxic T lymphocytes. *J Immunol* 138: 566–569

Taylor AS, Howe RC, Morrison AR, Sprecher H, Russell JH (1985): Inhibition of cytotoxic T lymphocyte-mediated lysis by ETYA: Effect independent of arachidonic acid metabolism. *J Immunol* 134: 1130–1135

Taylor HE, Culling CFA (1963): Cytopathic effect *in vitro* of sensitized homologous and heterologous spleen cells on fibroblasts. *Lab Invest* 12: 884–894

Thiernesse N, David A, Bernard J, Jeannesson P, Zagury D (1977): Activité phosphatiasique acide de la cellule T cytolytique au cours du processus de cytolyse. *CR Acad Sci Paris* 285: 713–715

Thorn RM, Henney CS (1976): Studies on the mechanism of lymphocyte-mediated cytolysis. VI. A reappraisal of the requirement for protein synthesis during T cell-mediated lysis. *J Immunol* 116: 146–149

Tian Q, Streuli M, Saito H, Schlossman SF, Anderson P (1991): A polyadenylate binding protein localized to the granules of cytolytic lymphocytes induces DNA fragmentation in target cells. *Cell* 67: 629–639

Tirosh R, Berke G (1985): T-lymphocyte-mediated cytolysis as an excitatory process of the target. I. Evidence that the target cell may be the site of Ca^{++} action. *Cell Immunol* 95: 113–123

Tite JP (1990a): Differential requirement for protein synthesis in cytolysis mediated by class I and class II MHC-restricted cytotoxic T cells. *Immunology* 70: 440–445

Tite JP (1990b): Evidence of a role for TNF-alpha in cytolysis by $CD4^+$, class II MHC-restricted cytotoxic T cells. *Immunology* 71: 208–212

Tite JP, Powell MB, Ruddle NH (1985): Protein-antigen specific Ia-restricted cytolytic T cells: Analysis of frequency, target cell susceptibility, and mechanism of cytolysis. *J Immunol* 135: 25–33

Todd RF, III, Stulting RD, Amos DB (1975): Lymphocyte-mediated cytolysis of allogeneic tumor cells *in vitro* I. Search for target antigens in subcellular fractions. *Cell Immunol* 18: 304–323

Townsend ARM, Rothbard J, Gotch FM, Bahadur G, Wraith D, McMichael AJ (1986): The epitopes of influenza nucleoprotein recognized by cytotoxic T lymphocytes can be defined with short synthetic peptides. *Cell* 44: 959–968

Tozeren A, Sung KL, Chien S (1989): Theoretical and experimental studies on cross-bridge migration during cell disaggregation. *Biophys J* 55: 479–487

Trenn G, Takayama H, Sitkovsky MV (1987): Exocytosis of cytolytic granules may not be required for target cell lysis by cytotoxic T-lymphocytes. *Nature* 330: 72–74

Treves S, Di Virgillio F, Cerundolo V, Zanovello P, Collavo D, Pozzan T (1987): Calcium and inositol-phosphates in the activation of T cell-mediated cytotoxicity. *J Exp Med* 166: 33–42

Tschopp J, Nabholz M (1990): Perforin-mediated target cell lysis by cytolytic T lymphocytes. *Annu Rev Immunol* 8: 279–302

Tschopp J, Schafer S, Masson D, Peitsch MC, Heusser C (1989): Phosphorylcholine acts as a $Ca2^+$-dependent receptor molecule for lymphocyte perforin. *Nature* 337: 272–274

Tsoukas CD, Carson DA, Fong S, Vaughan JH (1982): Molecular interactions in human T cell-mediated cytotoxicity to EBV. II. Monoclonal antibody OKT3 inhibits a post-killer-target recognition/adhesion step. *J Immunol* 129: 1421–1425

van Kooyk Y, Weder P, Hogervorst F, Verhoeven AJ, van Seventer G, Te Velde AA, Borst J, Keizer GD, Figdor CG (1991): Activation of LFA-1 through a $Ca(+)$-dependent epitope stimulates lymphocyte adhesion. *J Cell Biol* 112: 345–354

van Noesel C, Miedema F, Brouwer M, de Rie MA, Aarden LA, van Lier RAW (1988): Regulatory properties of LFA-1 alpha and beta chains in human T-lymphocyte activation. *Nature* 333: 850–852

van Seventer GA, Shimizu Y, Horgan KJ, Luce GE, Webb D, Shaw S (1991): Remote T cell co-stimulation via LFA-1/ICAM-1 and CD2/LFA-3: Demonstration with immobilized ligand/mAb and implication in monocyte-mediated co-simulation. *Eur J Immunol* 21: 1711–1718

van Seventer GA, Schimizu Y, Horgan KJ, Shaw S (1990): The LFA-1 ligand ICAM-1 provides an important costimulatory signal for T cell receptor-mediated activation of resting T cells. *J Immunol* 144: 4579–4586

Verret CR, Firmenich AA, Kranz DM, Eisen HN (1987): Resistance of cytotoxic T lymphocytes to the lytic effects of their toxic granules. *J Exp Med* 166: 1536–1547

Vitiello A, Heath WR, Sherman LA (1989): Consequences of self-presentation of peptide antigen by cytolitic T lymphocytes. *J Immunol* 143: 1512–1517

Wagner H, Röllinghoff M (1974): T cell-mediated cytotoxicity: Discrimination between antigen recognition, lethal hit and cytolysis phase. *Eur J Immunol* 4: 745–750.

Wagner H, Starzinski-Powitz A, Jung H, Rollinghoff M (1977): Induction of I region-restricted hapten-specific cytotoxic T lymphocytes. *J Immunol* 119: 1365

Walden PR, Eisen HN (1990): Cognate peptides induce self-destruction of CD8+ cytolytic T lymphocytes. *Proc Natl Acad Sci USA* 87: 9015–9019

Ware CF, Sanchez-Madrid F, Krensky AM, Burakoff SJ, Strominger JL, Springer TA (1983): Human lymphocyte function associated antigen-1 (LFA-1): Identification of multiple antigenic epitopes and their relationship to CTL-mediated cytotoxicity. *J Immunol* 131: 1182–1188

Wei W-Z, Lindquist RR (1981): Alloimmune cytolytic T lymphocyte activity: Triggering and expression of killing mechanisms in cytolytic T lymphocytes. *J Immunol* 126: 513–516

Wilson DB (1963): The reaction of immunologically activated lymphoid cells against homologous target tissue cells *in vitro*. *J Cell Comp Physiol* 62: 273–286

Yannelli JR, Sullivan JA, Mandell GL, Engelhard VH (1986): Reorientation and fusion of cytotoxic T lymphocyte granules after interaction with target cells as determined by high resolution cinemicrography. *J Immunol* 136: 377–382

Young JD (1989): Killing of target cells by lymphocytes: A mechanistic view. *Physiol Rev* 69: 250–314

Zagury D, Bernard J, Jeannesson P, Thiernesse N, Cerottini J-C (1979): Studies on the mechanism of T cell-mediated lysis at the single effector cell level. I. Kinetic analysis of lethal hits and target cell lysis in multicellular conjugates. *J Immunol* 123: 1604–1609

Zagury D, Bernard J, Thierness N, Feldman M, Berke G (1975): Isolation and characterization of individual functionally relative cytotoxic T lymphocytes: Conjugation, killing and recycling at the single cell level. *Eur J Immunol* 5: 818–822

Zanovello P, Bronte V, Rosato A, Pizzo P, di Virgilio F (1990): Responses of mouse lymphocytes to extracellular adenosine 5'-triphosphate (ATP). II. Extracellular ATP causes cell type-dependent lysis and DNA fragmentation. *J Immunol* 145: 1545–1550

Zheng LM, Zychlinsky A, Liu CC, Ojcius DM, Young JD (1991): Extracellular ATP as a trigger for apoptosis or programmed cell death. *J Cell Biol* 112: 279–288

Zinkernagel RM, Doherty PC (1974): Restriction of *in vitro* T cell-mediated cytotoxicity in lymphocytic choriomeningitis within a syngeneic or semiallogeneic system. *Nature* 248: 701–702

Zunino SJ, Hudig D (1988): Interactions between human natural killer (NK) lymphocytes and yeast cells: Human NK cells do not kill Candida albicans, although C. albicans block NK lysis of K562 cells. *Infect Immun* 56: 564–569

Zychlinsky A, Zheng LM, Liu C-C, Young JD (1990): Cytolytic lymphocytes induce both apoptosis and necrosis in target cells. *J Immunol* 146: 393–400.

Acknowledgments

My heartfelt appreciation for first intriguing me with CTL, and for our early collaborations, goes to Baruj Benacerraf. I am grateful for grant support from the Cellular Biology Program of the National Science Foundation. Thanks are due Matthew Mescher, Anne O'Rourke, and Rick Duke for specific suggestions on portions of this manuscript.

Endnotes

1. Recognition, of course, requires contact because it involves interaction between integral membrane proteins on each cell: the T-cell antigen receptor complex and the antigen-peptide-presenting MHC molecules on the target cell membrane. However, the formal possibility exists that antigen released from a virus-infected cell (perhaps a dying cell) could be presented to CTL by a professional antigen-presenting cell, and that this would activate the CTL to kill cells by action at a distance through soluble mediators. Such mediators would lack antigen specificity, since many investigators have searched for specific T-cell mediators of several kinds without reproducible success. A nonspecific mediator capable of acting at a distance, such as lymphotoxin, would kill indiscriminately, including "innocent" bystander cells lacking cognate antigen. The subject of bystander killing is discussed in more detail below. It appears likely that bystander killing may be an artifactual result of prolonged culture or cloning of CTL. CTL taken directly from mice (PEL) are noted for their extreme degree of bystander sparing.

2. Pactamycin was reported not to inhibit killing at $\leq 2 \times 10^{-6}$ M, which is sufficient to inhibit $\geq 95\%$ of leucine incorporation. At higher doses it inhibits killing which was attributed to nonspecific toxicity (Henney et al., 1974).

3. Years after we first used Ab + C to inactivate CTL, we attempted to refine this method, and to see how quickly we could inactivate CTL in this way. To our surprise, we were unable sufficiently to inactivate CTL using our original alloantisera + C (C.D. Tsoukas and E. Martz, unpublished results). We surmised that our earlier success was possible because in the earlier experiments (Martz and Benacerraf, 1973, 1974; Martz et al., 1974; Martz and Benacerraf, 1975) we employed CTL which had been stored frozen in liquid nitrogen (Brunner et al., 1970). Although a portion of the CTL survived freezing with enough lytic activity to be useful, we later determined that freezing and thawing reduced lytic activity about 90%, whereupon we abandoned this procedure. Despite using a controlled-rate freezing apparatus and DMSO cryoprotectant, recovery of viable CTL is much poorer than for other types of lymphocytes, perhaps because freezing unleashes the lytic apparatus in a suicidal mode. In any case, the freezing damage may have made possible the near-complete inactivation of CTL which we obtained with Ab + C in the early studies. Subsequently, Sanderson succeeded in lysing CTL within 2 minutes with a monoclonal IgM anti-Thy-1 (Sanderson, 1981a; see also Duke and Cohen, 1988). Thy-1 is one of the highest-density antigens on T cells, and IgM is very efficient at fixing complement.

II
Target Cell Recognition

4
$\alpha\beta$ T-Cell Receptor Repertoires Among Cytotoxic and Helper T Lymphocytes

David M. Kranz and Benjamin Tjoa

Introduction

It has become clear, since the discovery of the genes that encode the α chain and the β chain of the T-cell receptor (TCR), that the potential repertoire of receptors that could be generated is quite large ($\sim 10^{18}$ by one estimate). Previous reviews have described the genetic mechanisms that generate such diversity (Kronenberg et al., 1986; Davis and Bjorkman, 1988). The mechanisms include the assorted use of different variable (V) region genes ($\sim 30\,V\beta$ and $\sim 100\,V\alpha$), diversity (D) gene segments ($2\,D\beta$), and joining (J) gene segments ($12\,J\beta$ and $\sim 50\,J\alpha$). Even greater diversity is generated in the CDR3 equivalent of immunoglobulins through the use of three mechanisms that operate at the junctions of the rearranged genes (i.e., $V\beta D\beta$, $D\beta J\beta$, and $V\alpha J\alpha$). These mechanisms involve the template-independent addition of nucleotides by the enzyme terminal deoxynucleotidyl transferase, the variable deletion of bases from the coding ends of genes, and the use of palindromic nucleotides at the ends of full-length gene segments (Lafaille et al., 1989).

The functional repertoire will be shaped by two general types of processes. The first process involves the genetic and structural constraints that lead to a nonrandom expression of receptors. Examples are the bias in rearrangement of $V\alpha$ and $J\alpha$ gene segments (Roth et al., 1991) and the preferential pairing of α chains with β chains (Saito and Germain, 1989). The other process involves the selection of T cells bearing receptors that bind to appropriate major histocompatibility complex (MHC) products and peptide antigens (reviewed by Blackman et al., 1990). During ontogeny thymocytes "learn" to recognize self-MHC products (positive selection) but those that do so with too high an affinity are apparently deleted (negative selection). Peripheral expansion occurs when specific foreign antigens stimulate only those T cells bearing receptors that bind with sufficient affinity. Thus, the functional repertoire against a particular foreign antigen is much smaller than the total repertoire and it could be quite restricted. An apparent exception to this notion are staphylococcal enterotoxins and other "superantigens" which react with a larger fraction of the T-cell repertoire by binding to particular $V\beta$ domains (reviewed by Marrack and Kappler, 1990).

The purpose of this review is to summarize what is known about the $\alpha\beta$ repertoires on different subpopulations of cytotoxic T lymphocytes (CTLs) and T-helper (T_H) cells. Most of the results are from inbred strains of mice, although they probably reflect, to a reasonable approximation, various aspects of the human T-cell repertoire. The current status of the following questions is disussed: Is the $\alpha\beta$ repertoire different between CTLs and T_H? How diverse are the receptors among T cells that react with alloantigens and "foreign" antigens (e.g., haptens, specific peptides, viral antigens, and tumor-specific antigens)? Do T-cell populations undergo a change in repertoires with time, analogous to affinity maturation of antibodies?

CTL and T_H

Most CTL recognize peptides when bound by a class I but not by a class II MHC product. Conversely, T_H cells recognize peptides when presented by a class II but not a class I product. This observation could in part be due to the accessory molecules CD8 and CD4 which are found on CTL and T_H cells and which are known to bind class I and class

II products, respectively. Hence, the TCR could be involved primarily in the binding of peptide, while CD4 or CD8 could bind the MHC product (a version of the so-called two-receptor model). However, several lines of evidence suggest that the $\alpha\beta$ receptor itself contacts the MHC product and so dicates whether class I or class II elements are recognized.

When $\alpha\beta$ TCR transgenes derived from CTL (Kisielow et al., 1988; Sha et al., 1988) and T_H cells (Berg et al., 1989) are expressed in the periphery, the corresponding T cells also express the expected accessory molecule, CD8 or CD4, respectively. This result suggested, indirectly, that the $\alpha\beta$ heterodimer interacts during positive selection with either a class I or a class II product. If this scenario is correct, then it might be expected that TCR on peripheral T cells react with class I or class II product and that the repertoires of CTL and T_H receptors are non-overlapping and distinct.

Attempts to identify the structural features of the TCR that allow it to recognize a class I or class II product have shown that there is not a simple correlation between TCR structure and MHC reactivity. It became clear early on that the constant regions of the TCR could not account for the differences in CTL and T_H cell recognition, since both types of T cells use the same C regions. Thus, many investigators turned to the use of monoclonal antibodies specific for V regions to examine various issues concerning TCR repertoires. A description of antibodies that have been generated to mouse V regions is shown in Table 1. Studies with anti-$V\beta$ antibodies have shown that some $V\beta$ regions are preferentially expressed by CD4$^+$ or CD8$^+$ T cells. In several cases, T cells which express a particular $V\beta$ region together with the appropriate accessory molecule have been shown to have reactivity with a specific MHC product. Furthermore, some $V\beta$ regions are associated with reactivity to the minor lymphocyte-stimulating antigen Mls-1^a. More recently, several anti-$V\alpha$ antibodies have been generated, and one of these, specific for the $V\alpha 3$ region, was used to show that this $V\alpha$ is expressed preferentially by CD8$^+$ cells. In contrast, $V\alpha 11$ appears to be expressed preferentially on CD4$^+$ cells. It is likely that as more $V\alpha$-specific antibodies become available, additional cases of subpopulation specific expression will be identified. From a structural viewpoint, these findings may support the argument that the two helices of the MHC products are contacted by the TCR, perhaps by both the $V\alpha$ and the $V\beta$ regions (Ajitkumar et al., 1988; Davis and Bjorkman, 1988). Although not examined in detail yet, it will also be interesting to see if specific $V\alpha J\alpha$ or $V\beta D\beta J\beta$ combinations or $V\alpha V\beta$ pairs will be selectively expressed on CTL or T_H cells.

Alloreactive T Cells

It has been assumed that the alloreactive repertoire is likely to be larger than an antigen-specific repertoire, since up to 2% of peripheral T cells can react with alloantigen. Initial experiments by Garman et al. (1986) in fact showed that CD8$^+$ populations from various mixed lymphocyte cultures expressed at some level each of the three $V\alpha$ and three $V\beta$ genes examined, although a bias was seen in particular mixed lymphocyte reactions (MLR) (Table 2 summarizes results of various studies that have examined TCR repertoires among different T-cell subpopulations). Other studies have focused on responses to a single class I or class II alloantigen. The alloantigens K^{bm1} (three amino acid differences with K^b) and K^{bm14} (sequence unknown) elicited CTL in H-2^b mice that expressed the $V\beta 8$ chain on approximately 36% and 45% of the cells (Reimann and Bellan, 1986). Similarly, the primary response to I-A^{bm12}, which differs by three amino acids from I-A^b, is characterized by preferential use of some $V\beta$ genes ($V\beta 14$, $V\beta 15$, $V\beta 16$) but diverse use of other TCR gene segments (Bill et al., 1989b). Diverse $V\beta$ rearrangements were observed among five anti-D^b CTL clones (Schilham et al, 1987).

Several examples of preferential $V\beta$ use by human CTL clones have also been reported. These studies examined clones derived from primary anti-DPw2 MLR (Beale et al, 1987), anti-HLA-A2 MLR (Borst et al, 1987), and renal transplant patients (Miceli and Finn, 1989). The expansion of T cells bearing the same $V\beta$ region in transplant patients could be useful in the development of more specific immunosuppression. Among the many issues that remain are the degree of restriction in $V\alpha$ use and the actual size of the alloreactive TCR repertoire *in vivo*. Regardless, the preliminary reports appear to indicate that alloreactive repertoires are indeed more diverse than most of the well-characterized antigen-specific repertoires described below.

Antigen-Specific T Cells

As with antibody responses, it is likely that the degree of TCR diversity that is elicited will vary with different antigens. The size of the specific repertoire will be influenced by various factors: the number of peptides from a particular protein that can be presented, the number of MHC products

TABLE 1. Monoclonal antibodies to mouse TCR V regions

V-region specificity	Designation	Ig type	T-cell[a] preference	Reactivity associated with V region	Reference
Vβ2	B20.6	Rat IgG2a			Grègoire et al., 1991
Vβ3	KJ25	Hamster IgG		Mls2[a]	Pullen et al., 1988
Vβ4	KT4	Rat IgG	CD8[+]	H2[d]	Tomonari et al., 1990
Vβ5.1	MR9-4	Mouse IgG	CD4[+], CD8[+]	I-E	Reich et al., 1989
Vβ6	44-22-1	Rat IgG	CD4[+], CD8[+]		Payne et al., 1988
	46-6B5	Rat IgM	CD4[+], CD8[+]		Payne et al., 1988
	RR4-7	Rat IgG	CD4[+], CD8[+]		Kanagawa et al., 1989
Vβ7	TR310	Rat IgG2b		Mls1[a]	Okada et al., 1990
Vβ8.1 & 8.2	KJ16-133	Rat IgG		Mls1[a]	Haskins et al., 1984
Vβ8	F23.1	Mouse IgG2a		Mls1[a]	Staerz et al., 1985
Vβ8.2	F23.2	Mouse IgG1			Staerz et al., 1985
Vβ9	MR10-2	Mouse IgG		Mls1[a]	Utsunomiya et al. 1991
Vβ10	B21.5	Rat Ig	CD4[+], CD8[+]		Necker et al., 1991
Vβ11	KT11	Rat IgG2b		I-E	Tomonari et al., 1988
	RR3-15	Rat IgG	CD4[+]	I-E	Bill et al., 1989a
Vβ13	MR 12-4	Mouse IgG1			Zaller et al., 1990
Vβ14	14-2	Rat IgM	CD4[+], CD8[+]	K[k]	Liao et al., 1989
Vβ17a	KJ23a	Mouse IgG		I-E[k]	Kappler et al., 1987
Vα2	B20.1	Rat IgG2a			Grègoire et al., 1991
Vα3.2	RR3-16	Rat IgG	CD8[+]		Utsunomiya et al., 1989
Vα8	KT50	Rat IgG2a	CD4[+], CD8[+]		Tomonari et al., 1989
	KT65	Rat IgG2a	CD4[+], CD8[+]		Tomonari et al., 1989
	B21.13	Rat IgG			Necker et al., 1991
	B21.14	Rat IgG			Necker et al., 1991
Vα11.1 & Vα11.2	1.F2	Hamster Ig	CD4[+]	I-E[k]	Jameson et al., 1990, 1991
	RR8-1	Rat Ig	CD4[+]	I-E[k]	Jameson et al., 1991

[a]Where indicated, antibodies were used to examine T-cell preference. CD4[+], CD8[+] indicates that V region was expressed equally by both CD4 and CD8 populations.

TABLE 2. Studies that examined TCR use by specific T cells

	T-cell specificity	V-region preference[a]	Restricted by	Reference
Alloreactive	K^k	$V\beta14$		Liao et al., 1989
	$H-2^d$ anti $H-2^k$	$V\alpha11$		Garman et al., 1986
	$I-A^{bm12}$	$V\beta14, V\beta15, V\beta16$		Bill et al., 1989b
	D^{bm14} and K^{bm1}	$V\beta8$		Reimann and Bellan, 1986
	D^b	Diverse		Schilham et al., 1987
Hapten-specific	p-Azobenzenearsonate	$V\alpha3$	$I-A^k, I-A^d$	Tan et al., 1988
	2, 4, 6-Trinitrophenyl (TNP)	$V\alpha1l2-2, V\beta3$	K^b	Hochgeschwender et al., 1986, 1987
	N-Iodoacetyl-sulfonic-naphthl-ethylene-diamine (AED)	Diverse	K^b & D^b	Iwamoto et al., 1987
Peptide/protein-specific	Cytochrome c	$V\alpha11$	$I-E^k$	Fink et al., 1986, 1990; Winoto et al., 1986
	Sperm whale myoglobin 110–121	$V\beta8.2$	$I-E^d$	Morel et al., 1987
	Beef insulin	Diverse	$I-A^b$	Spinella et al., 1987; Sherman et al., 1987
	Beef insulin	$V\beta6$	$I-A^b$	Falcioni et al., 1990
	Beef insulin	$V\alpha3, V\beta8$	$I-A^d$	Wither et al., 1991
Virus-specific	LCMV (GP2$_{275-289}$)	$V\alpha4, V\beta10$	D^b	Aebischer et al., 1990
	LCMV (278–286)	$V\alpha4$	D^b	Yanagi et al., 1990
	Murine CMV (pp89)	$V\beta8$	L^d	Rodewald et al., 1989
	Influenza virus hemagglutinin	Diverse	$I-E^d$	Taylor et al., 1990
Tumor-specific	FBL-3	$V\alpha1, V\beta10$	D^b	Hirama et al., 1991
	Reticulum cell sarcoma (RCS)	$V\beta17$		Katz et al., 1988

[a] V regions indicated were shown to be used preferentially but, in most cases, not exclusively by the T cell population.

that bind the peptide(s), and the size of the peripheral T-cell repertoire after thymic selection in the particular MHC background of the individual. One should also keep in mind that the observed repertoire might be biased by the methods of immunization and the culture conditions prior to TCR analysis.

Haptens

Based on the structural similarities between TCR and immunoglobulins (Ig) and their similar mechanisms of generating diversity, it might be predicted that the size of the repertoires against the same chemically defined antigen, such as hapten, may also be similar. However, the fundamental differences between recognition by T cells and recognition by antibodies make such a prediction less tenable. For example, a hapten may be recognized by different TCR when coupled to different MHC-bound peptides. This could act to increase the possible TCR repertoire compared to antibodies. On the other hand, the TCR repertoire, unlike antibodies, is first restricted by selection processes that occur in the thymus, and this may reduce the potential repertoire against specific antigens.

Several reports on the diversity of TCR expressed by hapten-specific T cells begin to examine these possibilities. The CTL response to the trinitrophenyl (TNP) group in the context of K^b is remarkably restricted in that almost 50% of the clones expressed the same $V\alpha J\alpha V\beta D\beta J\beta$ combinations (Hochgeschwender et al., 1986, 1987). As some of these clones were derived from different mice, it would be interesting to determine if they exhibit junctional diversity. Four CTL clones to the hapten AED were found to use several different α and β gene segments, although two of the four used the same $J\alpha$ and $J\beta$ (Iwamoto et al., 1987). Because the four clones were restricted by different class I molecules (K^b or D^b), examination of a larger panel will be needed to determine the actual repertoire size. T-helper cells that are reactive with arsonate in the context of either I-A^k or I-A^d frequently use the $V\alpha 3$ region (Tan et al., 1988). The finding that the class I-restricted CTL clone 2C, which also expresses $V\alpha 3$, reacts with arsonate suggested that the $V\alpha 3$ region may in fact contact the arsonate group and not the MHC product.

Peptides

The most thorough analyses of TCR repertoires have been with T cells that are specific for cytochrome c (Fink et al., 1986; Winoto et al., 1986; Hedrick et al., 1988: Fink et al., 1990), insulin (Sherman et al., 1987; Spinella et al., 1987; Falcioni et al., 1990; Wither et al., 1991), and myoglobin (Morel et al., 1987; Danska et al., 1990). T cells specific for pigeon cytochrome c (residues 81 to 104) in the context of $E^{k,b}$ use preferentially the $V\alpha 11$, $J\alpha 84$, $V\beta 3$, and $J\beta 2.1$ gene segments. Cells that recognize the same antigen in the context of $E^{k,s}$ use preferentially the $V\alpha 10$ (or $V\alpha 11$), $V\beta 1$ and $J\beta 1.2$ gene segments. Interestingly, there is a preference for an asparagine residue at the $V\beta D\beta$ junction of the $E^{k,b}$ clones and an aspartic acid residue at the $V\beta D\beta$ junction of the $E^{k,s}$ clones. Such a restricted repertoire could be attributed to selection for binding to either the cytochrome antigen, the MHC product, or both. Results of site-directed mutagenesis suggested that antigen binding is part of the explanation (Engel and Hedrick, 1988), but both positive and negative selection also restrict the final repertoire (Fry and Matis, 1988; Fink et al., 1990).

Initial studies of a limited panel of T cells specific for insulin showed diverse use of $V\alpha$ and $V\beta$ genes (Sherman et al., 1987; Spinella et al., 1987). In contrast, a recent study of insulin-specific, I-A^d-restricted T cells showed a more restricted repertoire, and specific residues at the $V\beta D\beta$ junction appeared to be important in antigen binding (Wither et al., 1991). The insulin-specific repertoire also appears to be influenced by thymic selection processes (Falcioni et al., 1990).

T-cell clones specific for at least two different myoglobin-derived peptides used the $V\beta 8$ region when the antigens were presented by I-E^d but not when presented by IA^d. Differences in the fine specificity of these clones could also be attributed to amino acid residues encoded by bases at $V\alpha J\alpha$ and $V\beta D\beta J\beta$ junctions (Danska et al., 1990). Cumulatively, these data lend support to the model of Davis and Bjorkman (1988), which suggests that the repertoire of peptide-specific T cells may be shaped in large part by interactions of the CDR1 and CDR2 regions ($V\alpha$ and $V\beta$) with the MHC product and the CDR3 regions with the peptide.

Viruses

Restrictions in TCR gene usage have now been reported in the responses to lymphocytic choriomeningitis virus (LCMV) presented by D^b (Aebischer et al., 1990; Yanagi et al., 1990) and pp89 of the murine cytomegalovirus (CMV) presented by L^d (Rodewald et al., 1989). Influenza-specific CTL appeared to express a more diverse array of TCR, but this could be due to the multiple epitopes that are recognized (Taylor et al., 1990). In this regard, Moss et al. (1991) showed that there is conservation in both the α and β chains from human T cells that are specific for influenza matrix peptide in the context of HLA-A2. The restricted use of particular TCR may be useful in the treatment of viral infections which have an accompanying immunopathology by allowing T cells that bear these receptors to be targeted for destruction. However, restricted repertoires might also lead to cases of compromised T-cell activity. The recent report of viral escape by the selection of variants which are no longer recognized by CTL that express a TCR transgene is an elegant example of what could occur in cases of monoclonal T-cell responses to foreign antigens (Pircher et al., 1990).

Tumors

A similar scenario, whereby highly restricted responses might be subject to escape from immune surveillance, might be envisioned for tumor-specific antigens. Restricted TCR usage by tumor-specific CTL and tumor-infiltrating lymphocytes (TIL) has recently been reported. As many as 30% of the CTL that are specific for the FBL-3 tumor antigen in C57BL/6 mice react with a clonotypic monoclonal antibody and express the same α and β gene segments (Hirama et al., 1991). Katz et al. (1988) showed that the response to the reticulum cell sarcoma of SJL/J mice is predominated by T cells that bear the Vβ17 region. The polymerase chain reaction (PCR) was used to show that TIL from patients with uveal melanoma used the Vα7 gene in seven out of eight cases (Nitta et al., 1990). Future studies should be useful in determining whether such restricted responses to tumor-specific antigens may play a role in some cancers.

Primary Versus Secondary T Cells

The question of whether TCR repertoires to the same antigen change over time will impact each of the issues described above. There are now several reports that suggest T cells can undergo a change in avidity (reviewed by MacDonald et al., 1982) and that they may change their specificity during culturing. However, unlike Ig, there is no evidence that TCR genes undergo somatic mutation to produce receptors with higher affinity (reviewed by Vitetta et al., 1991). Recently, we examined the sequences of α chains from hyperimmune alloreactive T cells and showed that these T cells do indeed express the germline genes (Tjoa and Kranz, 1992). Thus the observed absence of mutations does not appear to be due to experimental bias, in that most T cells examined to date have been from primary responses.

Examples of rearrangements at the α and β loci that lead to *in vitro* changes in specificity have been reported (Chen et al., 1986; Epplen et al., 1986). The recent findings that T cells can express more than one α chain (Malissen et al., 1988) and that secondary rearrangements occur at this locus (Marolleau et al., 1988) raise the interesting possibility that TCR with different α chains could be selected for optimal antigen binding over time. We have recently compared long-term and short-term mixed lymphocyte cultures (MLC) derived from spleens of mice that have been hyperimmunized. Using the PCR to examine α chains, we detected the emergence of dominant clones that preferentially express particular VαJα combinations in the long-term cultures (Tjoa and Kranz, 1992). Whether there are similar processes that occur *in vivo* that act to restrict the TCR repertoire remains to be seen. The various effects of interferon, interleukin-2 (IL-2) and interleukin-4 (IL-4) on different populations of T cells have been described (Chen et al., 1986; Gajewski et al., 1989), and it is important to keep in mind that such antigen-independent mechanisms might lead to the expansion of T-cell subsets that express a restricted TCR repertoire. Nevertheless, possible correlations between changes in TCR structure and changes in T-cell "avidity" or specificity need to be examined in more detail.

Concluding Remarks

There are now numerous reports of restrictions in the use of TCR that are specific for various antigens. Cumulatively, the data may suggest more restricted repertoires for T cells than for B cells. This finding is at least in part the result of thymic selection whereby the $\alpha\beta$ repertoire is restricted even prior to exposure to foreign antigens. Through the use of V region-specific antibodies and the PCR, the functional *in vivo* repertoires to a variety of antigens can be examined more thoroughly. The information should be useful in understanding the nature of T-cell responses to viruses, tumor cells, and transplantation antigens. In cases where the repertoire to infectious agents is restricted, it may be desirable to augment the response. In the case of alloreactive T cells, it would be of obvious interest to specifically inhibit the response. Toward this end, the degree of TCR restriction *in vivo* will be of primary importance.

References

Aebischer TS, Oehen S, Hengartner H (1990): Preferential usage of Vα4 and Vβ10 T cell receptor genes by lymphocytic choriomeningitis virus glycoprotein-specific H-2D^b-restricted cytotoxic T cells. Eur J Immunol 20: 523–531

Ajitkumar P, Geier SS, Kesari KV, Borriello F, Nakagawa M, Bluestone JA, Saper MA, Wiley DC, Nathenson SG (1988): Evidence that multiple residues on both the α-helices of the class I MHC molecule are simultaneously recognized by the T cell receptor. Cell 54: 47–56

Beall SS, Lawrence JV, Bradley DA, Mattson DH, Singer DS, Biddison WE (1987): β-chain gene rearrangements and Vβ gene usage in DPw2-specific T cells. J Immunol 139: 1320–1325

Berg LJ, Pullen AM, Fazekas de St Groth B, Mathis D, Benoist C, Davis MM (1989): Antigen/MHC-specific T cells are preferentially exported from the thymus in the presence of their MHC ligand. Cell 58: 1035–1046

Bill J, Kanagawa O, Woodland DL, Palmer E (1989a): The MHC molecule I-E is necessary but not sufficient for the clonal deletion of Vβ-bearing T cells. J Exp Med 169: 1405–1419

Bill J, Yague J, Appel VB, White J, Horn G, Erlich HA, Palmer E (1989b): Molecular genetic analysis of 178 I-A^{bm12} reactive T cells. J Exp Med 169: 115–133

Blackman M, Kappler J, Marrack P (1990): The role of the T cell receptor in positive and negative selection of developing T cells. Science 248: 1335–1341

Borst J, De Vries E, Spits H, De Vries JE, Boylston AW, Matthews EA (1987): Complexity of T cell receptor recognition sites for defined alloantigens. J Immunol 139: 1952–1959

Chen L-K, Mathieu-Mahul D, Bach FH, Dausset J, Bensussan A, Sasportes M (1986): Recombinant interferon α can induce rearrangement of T-cell antigen receptor α-chain genes and maturation to cytotoxicity in T-lymphocyte clones in vitro. Proc Natl Acad Sci USA 83: 4887–4889

Danska JS, Livingstone AM, Paragas V, Ishihara T, Fathman CG (1990): The presumptive CDR3 regions of both T cell receptor α and β chains determine T cell specificity for myoglobin peptides. J Exp Med 172: 27–33

Davis MM, Bjorkman PJ (1988): T-cell antigen receptor genes and T-cell recognition. Nature 334: 395–402

Engel I, Hedrick SM (1988): Site-directed mutations in the VDJ junctional region of a T cell receptor β chain causes changes in antigenic peptide recognition. Cell 54: 473–484

Epplen JT, Bartels F, Becker A, Nerz G, Prester M, Rinaldy A, Simon MM (1986): Change in antigen specificity of cytotoxic T lymphocytes is associated with the rearrangement and expression of a T-cell receptor β-chain gene. Proc Natl Acad Sci USA 83: 4441–4445

Falcioni G, Dembic A, Muller S, Lehmann PV, Nagy ZA (1990): Flexibility of the T cell repertoire: Self tolerance causes a shift of T cell receptor gene usage in response to insulin. J Exp Med 171: 1665–1681

Fink PJ, Blair MJ, Matis LA, Hedrick SM (1990): Molecular analysis of the influences of positive selection, tolerance induction, and antigen presentation on the T cell receptor repertoire. J Exp Med 172: 139–150

Fink PJ, Matis LA, McElligott DL, Bookman MA, Hedrick SM (1986): Correlations between T cell specificity and the structure of the antigen receptor. Nature 321: 219–226

Fry AM, Matis LA (1988): Self-tolerance alters T-cell receptor expression in an antigen-specific MHC restricted immune response. Nature 335: 830–832

Gajewski TF, Schell SR, Nau G, Fitch FW (1989): Regulation of T-cell activation: Differences among T-cell subsets. Immunol Rev 111: 79–110

Garman R, Ko J-L, Vulpe C, Raulet D (1986): T-cell receptor variable gene usage in T-cell populations. Proc Natl Acad Sci USA 83: 3987–3991

Gregoire C, Rebai N, Schweisguth F, Necker A, Mazza G, Auphan N, Millward A, Schmitt-Verhulst A-M, Malissen B (1991): Engineered secreted T-cell receptor αβ heterodimers. Proc Nal Acad Sci USA 88: 8077–8081

Haskins K, Hannum C, White J, Toehm N, Kubo R, Kappler J, Marrack P (1984): The antigen-specific major histocompatibility complex restricted receptor on T cells VI. An antibody to a receptor allotype. J Exp Med 160: 452–471

Hedrick SM, Engel I, McElligott DL, Fink PJ, Hsu M, Hansburg D, Matis LA (1988): Selection based on specificity for amino acid sequences in the putative third hypervariable region of the beta chain of the T cell antigen receptor. Science 239: 1541–1544

Hirama T, Takeshita S, Matsubayashi Y, Iwashiro M, Masuda T, Kuribayashi K, Yoshida Y, Yamagishi H (1991): Conserved V(D)J junctional sequence of cross-reactive cytotoxic T cell receptor idiotype and the effect of a single amino acid substitution. Eur J Immunol 21: 483–488

Hochgeschwender E, Welzien H, Eichmann K, Wallace R, Epplen J (1986): Preferential expression of a defined T-cell receptor β-chain gene in hapten-specific cytotoxic T-cell clones. Nature 322: 376–378

Hochgeschwender U, Simon H-G, Weltzien HU, Bartels F, Becker A, Epplen JT (1987): Dominance of one T-cell receptor in the H-2 K^b/TNP response. Nature 326: 307–309

Iwamoto A, Ohashi PS, Pircher H, Walker CL, Michalopoulos EE, Rupp F, Hengartner H, Mak TW (1987): T cell receptor variable gene usage in a specific cytotoxic T cell response: Primary structure of the antigen-MHC receptor of four hapten-specific cytotoxic T cell clones. J Exp Med 165: 591–600

Jameson SC, Kaye J, Gascoigne NRJ (1990): A T cell receptor Vα region selectively expressed in CD4$^+$ cells. J Immunol 145: 1324–1331

Jameson SC, Nakajima PB, Brooks JL, Heath W, Kanagawa O, Gascoigne NRJ (1991): The T cell receptor Vα11 gene family: Analysis of allelic sequence polymorphism and demonstration of Jα region-dependent recognition by allele-specific antibodies. *J Immunol* 147: 3185-3193

Kanagawa O, Palmer E, Bill J (1989): The T cell receptor Vβ6 domain imparts reactivity to the Mls-1ª antigen. *Cell Immunol* 119: 412-426

Kappler JW, Wade T, White J, Kushnir E, Blackman M, Bill J, Roehm N, Marrack P (1987): A T cell receptor Vβ segment that imparts reactivity to a class II major histocompatibility complex product. *Cell* 49: 263-271

Katz JD, Ohnishi K, Lebow LT, Bonavida B (1989): The SJL/J T cell response to both spontaneous and transplantable syngeneic reticulum cell sarcoma is mediated predominantly by the Vβ17a+ T cell clonotype. *J Exp Med* 168: 1553-1562

Kisielow P, Teh HS, Bluthmann H, von-Boehmer (1988): Positive selection of antigen-specific T cells in thymus by restrictig MHC molecules. *Nature* 335: 730-733

Kronenberg M, Siu G, Hood LE, Shastri N (1986): The molecular genetics of the T-cell antigen receptor and T-cell antigen recognition. *Ann Rev Immunol* 4: 529-591

Lafaille JJ, DeCloux A, Bonneville M, Takagaki Y, Tonegawa S (1989): Junctional sequences of T cell receptor γδ genes: implications for γδ T cell lineages and for a novel intermediate of V-(D)-J joining. *Cell* 59: 859-870

Liao N, Maltzman J, Raulet DH (1989): Positive selection determines T cell receptor Vβ14 gene usage by CD8+ cells. *J Exp Med* 170: 135-143

MacDonald HR, Glasebrook AL, Bron C, Kelso A, Cerottini J-C (1982): Clonal heterogeneity in the functional requirement for Lyt-2/3 molecules on CTL: Possible implications for the affinity of CTL antigen receptors. *Immunol Rev* 68: 69-85

Malissen M, Trucy J, Letourneur F, Rebai N, Dunn DE, Fitch FW, Hood L, Malissen B (1988): A T cell clone expresses two T cell receptor α genes but uses one αβ heterodimer for allorecognition and self MHC-restricted antigen recognition. *Cell* 55: 49-59

Marolleau JP, Fondell JD, Malissen M, Trucy J, Barbier E, Marcu KB, Cazenave DA, Primi D (1988): The joining of germ-line Vα to Jα genes replaces the preexisting VαJα complexes in a T cell receptor αβ positive T cell line. *Cell* 55: 291-300

Marrack P, Kappler J (1990): The staphylococcal enterotoxins and their relatives. *Science* 248: 705-711

Miceli MC, Finn OJ (1989): T cell receptor β-chain selection in human allograft rejection. *J Immunol* 142: 81-86

Morel PA, Livingstone AM, Fathman CG (1987): Correlation of T cell receptor Vβ gene family with MHC restriction. *J Exp Med* 166: 583-588

Moss PAH, Moots RJ, Rosenberg WMC, Rowland-Jones SJ, Bodmer HC, McMichael AJ, Bell JI (1991): Extensive conservation of α and β chains of the human T-cell antigen receptor recognizing HLA-A2 and influenza A matrix peptide. *Proc Natl Acad Sci USA* 88: 8987-8991

Necker AN, Rebai N, Matthes M, Jouvin-Marche E, Cazenave P, Swarnworawong P, Palmer E, MacDonald HR, Malissen B (1991): Monoclonal antibodies raised against engineered soluble mouse T cell receptors and specific for Vα8-, Vβ2- or Vβ10-bearing T cells. *Eur J Immunol* 21: 3035-3040

Nitta T, Oksenberg JR, Rao NA, Steinman L (1990): Predominant expression of T cell receptor Vα7 in tumor-infiltrating lymphocytes of uveal melanoma. *Science* 249: 672-674

Okada CY, Horlzmann B, Guidos C, Palmer E, Weissman IL (1990): Characterization of a rat monoclonal antibody specific for a determinant encoded by the Vβ7 gene segment: Depletion of Vβ7+ T cells in mice with Mls-1ª haplotype. *J Immunol* 144: 3473-3477.

Payne J, Huber BT, Cannon NA, Schneider R, Schilham MW, Acha-Orbea H, MacDonald HR, Hengartner H (1988): Two monoclonal rat antibodies with specificity for the β-chain variable region Vβ6 of the murine T-cell receptor. *Proc Natl Acad Sci USA* 85: 7695-7698

Pircher H, Moskophidis D, Rohrer U, Burki K, Hengartner H, Zinkernagel RM (1990): Viral escape by selection of cytotoxic T cell-resistant virus variants *in vivo*. *Nature* 346: 629-633

Pullen AM, Marrack P, Kappler JW (1988): The T-cell repertoire is heavily influenced by tolerance to polymorphic self-antigens. *Nature* 335: 796-801

Reich EP, Sherwin RS, Kanagawa O, Janeway CA, Jr (1989): An explanation for the protective effect of the MHC class II I-E molecule in murine diabetes. *Nature* 341: 326-328

Reimann J, Bellan A (1986): Use of Vβ8 genes in splenic Lyt-2+ cytotoxic lymphocyte precursors reactive to bm1 or bm14 alloantigen in individual C57BL/6 mice. *Eur J Immunol* 16: 1597-1602

Rodewald HR, Koszinowski UH, Eichmann K, Melchers I (1989): Predominant utilization of Vβ8 T cell receptor genes in the H-2Lᵈ-restricted cytotoxic T cell response against the immediate-early protein pp89 of the murine cytomegalovirus. *J Immunol* 143: 4238-4243

Roth ME, Holman PO, Kranz DM (1991): Non-random use of Jα gene segments: Influence of Vα and Jα gene location. *J Immunol* 147: 1075-1081

Saito T, Germain RN (1989): Marked differences in the efficiency of expression of distinct αβ T cell receptor heterodimers. *J Immunol* 143: 3379-3384

Schilham MW, Lang R, Acha-Orbea H, Benner R, Joho R, Hengartner H (1987): Fine specificity and T cell receptor β-chain rearrangements of five H-2Dᵇ-specific cytotoxic T-cell clones. *Immunogenetics* 25: 171-178

Sha WC, Nelson CA, Newbury RD, Kranz DM, Russell JH, Loh DY (1988): Selective expression of an antigen receptor on CD8-bearing T lymphocytes in transgenic mice. *Nature* 335: 271-274

Sherman DH, Hochman PS, Dick R, Tizard R, Ramachandran KL, Flavell RA, Huber BT (1987): Molecular analysis of antigen recognition by insulin-specific T-cell hybridomas from B6 wild-type and bm12 mutant mice. *Mol Cell Biol* 7: 1865-1872

Spinella DG, Hansen TH, Walsh WD, Behlke MA, Tillinghast JP, Chou HS, Whiteley PJ, Kapp JA, Pierce CW, Shevach EM, Loh DY (1987): Receptor diversity of insulin-specific T cell lines from C57BL (H-2ᵇ) mice. *J Immunol* 138: 3991-3995

Staerz UD, Rammensee HG, Benedetto JD, Bevan MJ (1985): Characterization of a murine monoclonal antibody specific for an allotypic determinant on T cell antigen receptor. *J Immunol* 134: 3994–4000

Tan K-N, Datlof BM, Gilmore JA, Kroman AC, Lee JH, Maxam AM, Rao A (1988): The T cell receptor Vα3 gene segment is associated with reactivity to p-azobenzenearsonate. *Cell* 54: 247–261

Taylor AH, Haberman AM, Gerhard W, Caton AJ (1990): Structure–function relationships among highly diverse T cells that recognize a determinant from influenza virus hemagglutinin. *J Exp Med* 172: 1643–1651

Tjoa B, Kranz DM (1992): Diversity of T cell receptor-α chain transcripts from hyperimmune alloreactive T cells. *J Immunol* 149: 253–259

Tomonari K, Lovering E (1988): T-cell receptor-specific monoclonal antibodies against a Vβ-positive mouse T-cell clone. *Immunogenetics* 28: 445–451

Tomonari K, Lovering E, Fairchild S, Spencer S (1989): Two monoclonal antibodies specific for the T cell receptor Vα8. *Eur J Immunol* 19: 1131–1135

Tomonari K, Lovering E, Spencer S (1990): Correlation between the Vβ4+CD8+ T-cell population and the H-2^d haplotype. *Immunogenetics* 31: 333–339

Utsunomiya TJ, Bill J, Palmer E, Gollob K, Takagaki Y, Kanagawa O (1989): Analysis of a monoclonal rat antibody directed to the α-chain variable region (Vα3) of the mouse T cell antigen receptor. *J Immunol* 143: 2602–2608

Utsonomiya Y, Kosaka H, Kanagawa O (1991): Differential reactivity of Vβ9 T cells to minor lymphocyte stimulating antigen *in vitro* and *in vivo*. *Eur J Immunol* 21: 1007–1011

Vitetta ES, Berton MT, Burger, C, Kepron M, Lee WT, Yin X-M (1991): Memory B and T-cells. *Ann Rev Immunol* 9: 193

Winoto A, Urban JL, Lan NC, Goverman J, Hood L, Hansburg D (1986): Predominant use of a Vα gene segment in mouse T-cell receptors for cytochrome c. *Nature* 324: 679–682

Wither J, Pawling J, Phillips L, Delovitch T, Hozumi N (1991): Amino acid residues in the T cell receptor CDR3 determine the antigenic reactivity patterns of insulin-reactive hybridomas. *J Immunol* 146: 3513–3522

Yanagi Y, Maekawa R, Cook T, Kanagawa O, Oldstone MB (1990): Restricted V-segment usage in T-cell receptors from cytotoxic T lymphocytes specific for a major epitope of lymphocytic choriomeningitis virus. *J Virol* 64: 5919–5926

Zaller DM, Osman G, Kanagawa O, Hood L (1990): Prevention and treatment of murine experimental allergic encephalomyelitis with T cell receptor Vβ-specific antibodies. *J Exp Med* 171: 1943–1955.

5
Class I MHC/Peptide/β_2-Microglobulin Interactions: The Basis of Cytotoxic T-Cell Recognition

David H. Margulies, Lisa F. Boyd, Maripat Corr, Rosemarie D. Hunziker, Sergei Khilko, Steven Kozlowski, Michael Mage, and Randall K. Ribaudo

Abstract

The past decade has witnessed a major revolution in our thinking and understanding of the molecular basis of T-cell recognition. In this brief review we outline the historical development of this knowledge and how it has drawn upon advances in cellular immunology, virology, genetics, molecular biology, and structural biology. The current status of our view of the molecular details is summarized, and an outline of critical molecular and cellular questions for the future is presented.

Introduction

It would be a mark of intellectual arrogance of the highest order to attempt to identify the first suggestion that genetic loci control the immune response to viral or parasitic pathogens, but it is clear that the studies of allograft survival in humans and in animals followed parallel paths that have led to a common understanding. Medieval physicians recognized that autografts of skin had far better survival than allografts (grafts from other human donors), but it was Gorer (1936) who had the insight to recognize the relationship between genetics, transplantation, and serology, and to establish that antisera could define genetic differences that marked the ability of strains of mice to reject or accept skin grafts. Thus, the notion that differences in cell surface antigens between inbred mouse strains defined by antibody reactivity reflected differences in the recognition of nonself by cellular effectors of the immune system became the basis for biochemical studies of the molecules identified by alloantisera. The formal proof that immune response genes, genetic loci that control the ability of particular strains to mount a response (either humoral or cellular), encoded particular structural entities involved in recognition by T cells was initially based on the biochemical description of the class I MHC antigens (Orr et al., 1979; Coligan et al., 1981). The principal facts are that the MHC class I molecules are highly polymorphic, ubiquitously expressed cell surface proteins. These consist of a 45-kDa membrane-anchored heavy chain, known as HLA-A, -B, or -C in the human and H-2K, -D, or -L in the mouse, noncovalently associated with the relatively nonpolymorphic 12-kDa light chain, β_2-microglobulin (β_2-m). The cloning of the genes encoding the class I and class II molecules and their transfection and functional expression were the final proof of the basis of this well-described phenomenon (Germain et al., 1983; Margulies et al., 1983; Ozato et al., 1983; Reiss et al., 1983).

MHC Molecules as Peptide Binders

The classic work of Zinkernagel and Doherty (1974, 1975) and Shearer (1974) outlined the basis of MHC restriction. In the former case, cytolytic T lymphocytes (CTL) primed against the murine viral pathogen, lymphocytic choriomeningitis virus (LCMV), could kill cells infected with the virus only if they shared the MHC type of the effector cells. In the latter, the same phenomenon was observed when nonself was defined chemically, that is when the nonself antigens were trinitrophenylated cell surface molecules. Further insight into the molecular nature of MHC restriction was derived from studies of class II-restricted helper T lymphocytes, cells that proliferate in response to exposure to macromolecular soluble protein antigens such as cytochrome c (Solinger et al., 1979), lysozyme (Maizels et al., 1980), or myoglobin (Berkower et al., 1982). Early observations revealed that such T-cell responses did not discriminate denatured from native protein (Schirrmacher and Wigzell, 1972). Several groups studying the class II-restricted antigens noted (1) the exquisite sensitivity of T-cell responses to structural variants of the antigen encoded by different species; (2) the ability to use proteolytically derived protein fragments or synthetic peptides to stimulate T cells; and (3) the sensitivity of the T-cell response to structural variants of the MHC class II molecules.

It was becoming clear that class II molecules likely formed molecular complexes with peptides derived from the macromolecular protein antigens, but it still remained puzzling how class I molecules could mediate the interaction of T cells with virally infected cells. At this time several different lines of evidence converged, and further experiments outlining part of the answer were performed.

The cloning of MHC class I-encoding genes, the analysis of their function following transfection into appropriate cell types, and the evaluation of the specificity of antigen and T-cell interaction of novel genes generated by *in vitro* manipulation ("exon-shuffling": Margulies and McCluskey, 1985) focused attention on the membrane distal amino terminal aspects of the class I molecule. Both allorecognition and antigen-specific MHC-restricted T-cell recognition relied most heavily on interactions between clonotypic T-cell receptors and the amino-acid terminal $\alpha 1$ and $\alpha 2$ domains of the class I molecule. Studies of class II-restricted antigen presentation revealed functional interactions of the MHC class II molecule with peptide fragments derived from the soluble protein antigens. Therefore, antigen-presenting cells exposed to mild fixation could be used to display synthetic peptides to stimulate specific T cells, but such fixed cells were unable to process whole soluble antigens by a pathway dependent on endocytosis and proteolysis (Shimonkevitz et al., 1983).

The view of MHC molecules as peptide-binding molecules was first convincingly demonstrated in the class II system by Babbit et al. (1985) and Buus et al. (1986), who demonstrated biochemically the interaction of known labeled antigenic peptides with purified MHC class II molecules in detergent solution. Subsequently, Townsend and his colleagues (Bastin et al., 1987; Gould et al., 1989) demonstrated that influenza nucleoprotein-specific MHC-restricted CTL could lyse cells expressing fragments of the nucleoprotein and that synthetic peptides could sensitize target cells for such cytolysis.

Further evidence for the view that MHC class I molecules bind to potentially antigenic peptides derived from the experiments of Townsend and Kärre and colleagues (Townsend et al., 1989) in which the cell surface expression of the MHC class I molecules expressed by defective cells could be augmented or stabilized by exposure to synthetic peptides known to be antigenic for the MHC of the cell. Exposure of the RMA-S cell line (a mutant of this sort) to peptides derived from the influenza nucleoprotein caused a higher steady-state level of H-2D^b on the cell surface. Although this was originally thought to represent the effect of internalized peptide, a number of different lines of investigation demonstrated that it resulted from the binding of MHC molecules at the cell surface to antigenic peptide and the stabilization of the complex there. The behavior of cells like RMA-S, the embryonic cell line transfectant LKD-8 (Otten et al., 1992), and the human mutant cell LBL 721.174 (Cerundolo et al., 1990) is all consistent with the lack of appropriate levels of peptides available for proper folding and assembly of the heavy chain/peptide β_2-m complex in a pre-Golgi compartment. An additional common feature of cells with apparent defects in availability of intracellular peptides is that the cell surface expression of the class I molecules is improved by prolonged incubation of the cells at temperatures lower than those which warm-blooded mammals maintain (Ljunggren et al., 1990). This is consistent with the view that tight interaction with self peptides leads to greater stability of the complex. In the absence of a full repertoire of self peptides, most molecules are thermolabile.

A more precise model for MHC class I/peptide interaction derives from x-ray crystallographic studies. These revealed extra electron density, representing bound self peptides that copurified with the human HLA-A2 MHC class I molecule (Bjorkman et al., 1987). The other striking theme elucidated by the crystallographic studies was that the MHC class I molecules have a peptide-binding cleft consisting of two α helices (one derived from part of each of the two amino terminal $\alpha 1$ and $\alpha 2$ domains), supported by a floor of β-sheet (Bjorkman et al., 1987a, 1987b; Garrett et al., 1989; Saper et al., 1991).

The Role of Putative Peptide Transporters

In addition to the demonstration that peptides control the cell surface expression of class I molecules in mutant cell lines such as those mentioned above, analysis of several different cell lines indicated that the defect was a direct result of the lack of functional peptide transporters or of the macromolecular proteasome complex. Studies by Monaco and his colleagues, in which murine MHC-linked genes under control of the inductive effects of interferon were cloned (Deverson et al., 1990; Monaco et al., 1990), revealed two genes, known as HAM-1 and HAM-2 (now known as TAP-1 and TAP-2), that encoded proteins whose structure was similar to that of members of the multiple drug resistance family. In some cases, a defect in peptide transport correlated with cell surface expression of the class I molecule. Studies of the defective cell surface class I expression of the *cim* locus mutants of the rat (Livingstone et al., 1991: Powis et al, 1991) and of the human LBL 721.174 and its derivatives (Cerundolo et al, 1990) imply that peptides derived from the degradation of cytoplasmic proteins and transported across a pre-Golgi membrane are a normal and often necessary component of the properly folded MHC class I molecule. (An interesting exception to the requirement for transmembrane peptide transport for MHC class I folding is that the human class I molecule HLA-A2 manages to arrive at the surface of a defective cell due to its ability to complex with amino-terminal signal peptides that have entered this compartment via a transporter-independent pathway (Henderson et al, 1992; Wei and Cresswell, 1992).)

Demonstration of Peptide Binding by MHC Molecules

Although the *in vivo* formation of native class I molecules from their constituents, the MHC class I heavy chain, β_2-m, and peptide, is apparently an ordered, vectorial process, dependent on a number of different components of the cellular organizational machinery (proteasome complex, peptide transporters, chaperonins, and prolyl and disulfide isomerases), the individual stages can be reproduced in several *in vitro* models. Thus, systems for analyzing the binding of different MHC class I molecules to cognate peptides in solution (Chen and Parham, 1989; Boyd et al., 1992) or in solid phase (Bouillot et al, 1989; Frelinger et al., 1990) have been developed. The molecules capable of interaction with such peptides can be native molecules from tissue culture cells (Bouillot et al., 1989; Chen and Parham, 1989; Boyd et al., 1992), denatured and refolded molecules synthesized in bacterial expression systems (Parker et al., 1992), or those synthesized in microsome-coupled *in vitro* translation systems (Kvist and Hamann, 1990; Ribaudo and Margulies, 1992).

The Role of β_2-Microglobulin in Peptide Binding

With the original description of β_2-m as a noncovalent subunit of the MHC class I molecule (Smithies and Poulik, 1972; Silver and Hood, 1974) and the subsequent demonstration that most MHC class I molecules could not be expressed at the cell surface in the absence of the coexpression of β_2-m (Hyman and Stallings, 1976), the critical nature of β_2-m became apparent. However, the absolute need for β_2-m remains poorly defined. Allen and colleagues demonstrated that the murine class I molecule H-2D^b could be expressed at the cell surface in the absence of β_2-m, but in a serologically (and thus, structurally) different form (Williams et al., 1989). These H-2D^b molecules had apparently normal expression of the membrane proximal α3 domain, but no detectable epitopes of the peptide-binding domain. Nevertheless, the molecules were the size of normal molecules. Similarly, the β_2-m "knockout" mice (Koller et al., 1990), generated by homologous recombination in embryonic stem cells, though defective in their class I MHC expression, do indeed express the H-2D^b molecule in a similar conformation to that of the β_2-m-defective tissue culture cells.

In evaluating the role of β_2-m in MHC class I/peptide interactions as assessed functionally, several groups demonstrated that the β_2-m component found in the fetal calf serum in which most functional assays are performed played a critical role in the ability of cells or purified class I molecules to bind antigenic peptides such that they could be detected by MHC-restricted, antigen-specific T cells (Rock et al., 1990; Vitiello et al., 1990; Kozlowski et al., 1991). Although the details of the systems employed by these groups differ significantly, the evidence indicates that β_2-m plays a critical role in either preserving MHC class I heavy chains for peptide binding or in stabilizing the MHC/peptide/β_2-m complex.

In evaluating the effect of the addition of excess β_2-m in the solution binding of the MHC class I molecule H-2L^d to an antigenic peptide, our laboratory consistently observed that the addition of β_2-m generated an increasing number of high-affinity peptide-binding sites (Boyd et al., 1992). It is not yet clear whether this is due to the recruitment of otherwise nonfunctional class I molecules so that they could become peptide binders, or a major change in the kinetic dissociation of peptide from the heavy chain/peptide complex as compared to the rate from the free heavy chain.

Role of β_2-Microglobulin and Peptide in Achieving Native Conformation of the MHC I Heavy Chain: Is This a Transition from Molten Globule to Native Structure?

The accumulated data have suggested that the MHC class I heavy chain may be viewed as a molecule that requires both peptide (either self or antigenic) and β_2-m to permit folding into a native structure capable of being transported to the cell surface. One interpretation of this behavior would be that the partially folded, unassembled heavy chain is in fact a "molten globule" (Christensen and Pain, 1991), i.e., a structure that has native secondary structure but has not yet fully achieved its final tertiary structure. Detailed studies, requiring assessment of changes in spectroscopic properties on peptide and MHC binding and development of resistance to proteolysis, have not yet been performed, but it remains intriguing to consider this possibility.

Copurification of Peptides with MHC Class I Molecules

Among the most striking results of the past few years relevant to the interaction of self peptides with MHC class I molecules has been the demonstration by Rammensee and his colleagues (Falk et al., 1991) that immunoadsorbent purified MHC class I molecules can be dissociated in acid, releasing a pool of self peptides that can be fractionated by reverse-phase HPLC and sequenced as the pool. Experiments of this kind have led to the determination that (1) self peptides bound to MHC class I molecules are of about 9 residues in length; (2) "motifs" of the sequences of peptides that copurify with particular MHC molecules can be defined by such sequence analysis of pools of peptides; and (3) some regions of the peptide are conserved and serve as "anchor" points for MHC binding, and other regions lack restraints on the available sequences. This kind of analysis has been extended in several ways: (1) the application of more sensitive mass spectrometry methods to the detection and sequencing of the peptides (Henderson et al., 1992; Hunt et al., 1992); (2) the application of the method to radiolabeled and virally infected cells so that fewer cells might be needed for the analysis (Van Bleek and Nathenson, 1990). Thus, the longer-term hope and goal is that the full panel of self and antigenic peptides restricted to particular MHC molecules will be identified and that the contribution of these peptides to T-cell recognition and, in some cases, to autoimmune disease will be ascertained.

Role of Extracellular Proteolysis in the Processing of Suboptimal Peptides for Presentation by Cell Surface Class I Molecules

The realization that most self peptides eluted from purified MHC class I molecules are of 8 to 10 amino acids in length (Falk et al., 1991), and the demonstration that smaller peptide contaminants in preparations of synthetic peptides can be profoundly more potent than the larger predominant peptide (Rötzschke et al., 1990; Schumacher et al., 1991) led two groups to evaluate the effects of serum proteases in the processing of antigenic peptides for presentation either by purified MHC class I molecules or by cell surface class I molecules (Kozlowski et al., 1992; Sherman et al, 1992). In both cases, although the antigenic peptides and the respective MHC molecules examined were different, a common serum carboxy dipeptidase, angiotensin-converting enzyme (ACE), an enzyme that plays a critical role in the regulation of blood pressure through its control of levels of the potent pressor angiotensin, played a major role in the activation of suboptimal antigenic peptides. It is likely that this particular serum protease is only one of several, and that different antigenic peptides of suboptimal length may have different requirements for proteolytic processing.

Summary

Thus, over the past half-century our understanding of MHC-restricted recognition has taken us from the vague description of a graft and serological phenomenon to a detailed mechanistic molecular view of the interaction of an MHC class I molecule with specific antigenic peptides and β_2-m. Our view of the MHC class I is of a polymorphic membrane-anchored protein that is synthesized on membrane-bound polyribosomes and vectorially discharged into the endoplasmic reticulum, where it folds, forms intrachain disulfide bonds, and assembles with both peptides and the light chain β_2-m prior to its egress from a pre-Golgi compartment and its journey to the cell surface. At the cell surface this ternary complex is available for interaction with, and stimulation of, T cells bearing appropriate $\alpha\beta$ receptors that are capable of binding the complex with sufficient avidity to trigger the T cell.

A Look into The Future

Our understanding of some of the molecular details of classI/peptide/β_2-m interactions is very sophisticated, but there remain a number of critical questions that hopefully will be answered in the next decade. Mechanistically, the most controversial and still unresolved issue is the question of the order of interaction of the three components. Must the heavy chain be free of the light chain β_2-m prior to effective interaction of the heavy chain with peptide, or can heavy chain/β_2-m complexes still interact effectively with peptide? Are these obligate or only favored pathways? What conformational changes accompany the interaction of the MHC class I molecule with peptide? Can specific antigenic peptides be eluted from MHC molecules in inflammatory tissue fluids, permitting specific evaluation of peptide antigens potentiating an ongoing inflammatory reaction and providing a rational basis for development of specific pharmacological intervention? Can peptide vaccines be specifically designed with attention to the role that serum and tissue proteases undoubtedly play in the activation and degradation of antigenic peptide? The answers to these and a host of other important questions await our attention, creativity, and industry.

References

Babbitt BP, Allen PM, Matsueda G, Haber E, Unanue ER (1985): Binding of immunogenic peptides to Ia histocompatibility molecules *Nature* 317: 359–361

Bastin J, Rothbard J, Davey J, Jones I, Townsend A (1987): Use of synthetic peptides of influenza nucleoprotein to define epitopes recognized by class I-restricted cytotoxic T lymphocytes. *J Exp Med* 165: 1508–1523

Berkower I, Buckenmayer GK, Gurd FR, Berzofsky JA (1982): A possible immunodominant epitope recognized by murine T lymphocytes immune to different myoglobins. *Proc Natl Acad Sci USA* 79: 4723–4727

Bjorkman PJ, Saper MA, Samraoui B, Bennett WS, Strominger JL, Wiley DC (1987a): Structure of the human class I histocompatibility antigen, HLA-A2. *Nature* 329: 506–512

Bjorkman PJ, Saper MA, Samraoui B, Bennett WS, Strominger JL, Wiley DC (1987b): The foreign antigen binding site and T cell recognition regions of class I histocompatibility antigens. *Nature* 329: 512–518

Bouillot M, Choppin J, Cornille F, Martinon F, Papo T, Gomard E, Fournie-Zaluski MC, Levy JP (1989): Physical association between MHC class I molecules and immunogenic peptides. *Nature* 339: 472–475

Boyd LF, Kozlowski S, Margulies DH (1992): Solution binding of an antigenic peptide to a major histocompatibility complex class I molecule and the role of β2-microglobulin. *Proc Natl Acad Sci USA* 89: 2242–2246

Buus S, Sette A, Colon SM, Jenis DM, Grey HM (1986): Isolation and characterization of antigen-Ia complexes involved in T cell recognition. *Cell* 47: 1071–1077

Cerundolo V, Alexander J, Anderson K, Lamb C, Cresswell P, McMichael A, Gotch F, Townsend A (1990): Presentation of viral autigen controlled by a gene in the major histocompatibility complex. *Nature* 345: 449–452

Chen BP, Parham P (1989): Direct binding of influenza peptides to class I HLA molecules. *Nature* 337: 743–745

Christensen H, Pain RH (1991): Molten globule intermediates and protein folding. *Eur Biophys J* 19: 221–229

Coligan JE, Kindt TJ, Uehara H, Martinko J, Nathenson SG (1981): Primary structure of a murine transplantation antigen. *Nature* 291: 35–39

Deverson EV, Gow IR, Coadwell WJ, Monaco JJ, Butcher GW, Howard JC (1990): MHC class II region encoding proteins related to the multidrug resistance family of transmembrane transporters. *Nature* 348: 738–741

Falk K, Rötzschke O, Stevanovic S, Jung G, Rammensee HG (1991): Allele-specific motifs revealed by sequencing of self-peptides eluted from MHC molecules. *Nature* 351: 290–296

Frelinger JA, Gotch FM, Zweerink H, Wain E, McMichael AJ (1990): Evidence of widespread binding of HLA class I molecules to peptides. *J Exp Med* 172: 827–834

Garrett TP, Saper MA, Bjorkman PJ, Strominger JL, Wiley DC (1989): Specificity pockets for the side chains of peptide antigens in HLA-Aw68. *Nature* 341: 692–696

Germain RN, Norcross MA, Margulies DH (1983): Functional expression of a transfected murine class II MHC gene. *Nature* 306: 190–194

Gorer P (1936): The detection of antigenic differences in mouse erythrocyte by the employment of immune sera. *Br J Exp Pathol* 17: 42–50

Gould K, Cossins J, Bastin J, Brownlee GG, Townsend A (1989): A 15 amino acid fragment of influenza nucleoprotein synthesized in the cytoplasm is presented to class I-restricted cytotoxic T lymphocytes. *J Exp Med* 170: 1051–1056

Henderson RA, Michel H, Sakaguchi K, Shabanowitz J, Appella E, Hunt DF, Engelhard VH (1992): HLA-A2.1 associated peptides from a mutant cell line: A second pathway of antigen presentation. *Science* 255: 1264–1266

Hunt DF, Henderson RA, Shabanowitz J, Sakaguchi K, Michel H, Sevilir N, Cox AL, Appella E, Engelhard VH (1992): Characterization of peptides bound to the class I MHC molecule HLA-A2.1 by mass spectrometry. *Science* 255: 1261–1263

Hyman R, Stallings V (1976): Characterization of a TL-variant of a homozygous TL⁺ mouse lymphoma. *Immunogenetics* 3: 75–84

Koller BH, Marrack P, Kappler JW, Smithies O (1990): Normal development of mice deficient in beta 2M, MHC class I proteins, and CD8⁺ T cells. *Science* 248: 1227–1230

Kozlowski S, Takeshita T, Boehncke WH, Takahashi H, Boyd LF, Germain RN, Berzofsky JA, Margulies DH (1991): Excess β 2 microglobulin promoting functional peptide association with purified soluble class I MHC molecules. *Nature* 349: 74–77

Kozlowski S, Corr MJ, Takeshita T, Boyd LF, Pendleton CD, Germain RN, Berzofsky JA, Margulies DH (1992): Serum angiotensin-1 converting enzyme activity processes and HIV 1 gp160 peptide for presentation by MHC class I molecules. *J Exp Med* 175: 1417–1422

Kvist S, Hamann U (1990): A nucleoprotein peptide of influenza A virus stimulates assembly of HLA-B27 class I heavy chains and beta 2-microglobulin translated *in vitro*. *Nature* 348: 446–448

Livingstone AM, Powis SJ, Gunther E, Cramer DV, Howard JC, Butcher GW (1991): Cim: An MHC class II-linked allelism affecting the antigenicity of a classical class I molecule for T lymphocytes. *Immunogenetics* 34: 157–163

Ljunggren HG, Stam NJ, Ohlen C, Neefjes JJ, Hoglund P, Heemels MT, Bastin J, Schumacher TN, Townsend A, Karre K, et al. (1990): Empty MHC class I molecules come out in the cold. *Nature* 346: 476–480

Maizels RM, Clarke JA, Harvey MA, Miller A, Sercarz, EE (1980): Epitope specificity of the T cell proliferative response to lysozyme: Proliferative T cells react predominantly to different determinants from those recognized by B cells. *Eur J Immunol* 10: 509–515

Margulies DH, McCluskey J (1985): Exon shuffling: New genes from old. *Surv Immunol Res* 4: 146–159

Margulies DH, Evans GA, Ozato K, Camerini-Otero RD, Tanaka K, Appella E, Seidman JG (1982): Expression of H-2D^d and H-2L^d mouse major histocompatibility antigen genes in L cells after DNA-mediated gene transfer. *J Immunol* 130: 463–470

Monaco JJ, Cho S, Attaya M (1990): Transport protein genes in the murine MHC: Possible implications for antigen processing. *Science* 250: 1723–1726

Orr HT, Lopez de Castro JA, Parham P, Ploegh HL, Strominger JL (1979): Comparison of amino acid sequences of two human histocompatibility antigens, HLA-A2 and HLA-B7: Location of putative alloantigenic sites. *Proc Natl Acad Sci USA* 76: 4395–4399

Otten GR, Bikoff E, Ribaudo RK, Kozlowski S, Margulies DH, Germain RN (1992): Peptide and β_2-microglobulin regulation of cell surface MHC class I conformation and expression. *J Immunol* 148: 80–96

Ozato K, Evans GA, Shykind B, Margulies DH, Seidman JG (1983): Hybrid H-2 histocompatibility gene products assign domains recognized by alloreactive T cells. *Proc Natl Acad Sci USA* 80: 2040–2043

Parker KC, Carreno BM, Sestak L, Utz U, Biddison WE, Coligan JE (1992): Peptide binding to HLA-A2 and HLA-B27 isolated from *Escherichia coli*. Reconstitution of HLA-A2 and HLA-B27 heavy chain/β 2-microglobulin complexes requires specific peptides. *J Biol Chem* 267

Powis SJ, Howard JC, Butcher GW (1991): The major histocompatibility complex class II-linked cim locus controls the kinetics of intracellular transport of a classical class I molecule. *J Exp Med* 173: 913–921

Reiss CS, Evans GA, Margulies DH, Seidman JG, Burakoff SJ (1983): Allospecific and virus-specific cytolytic T lymphocytes are restricted to the N or C1 domain of H-2 antigens expressed on L cells after DNA-mediated gene transfer. *Proc Natl Acad Sci USA* 80: 2709–2712

Ribaudo RK, Margulies, DH (1992): Independent and synergistic effects of disulfide bond formation, β2-microglobulin, and peptides on class I MHC folding and assembly in an in vitro translation system. *J Immunol* 149: 2935–2944

Rock KL, Gamble S, Rothstein L (1990): Presentation of exogenous antigen with class I major histocompatibility complex molecules. *Science* 249: 918–921

Rötzschke O, Falk K, Deves K, Schild H, Norda M, Metzger J, Jung G, Rammensee HG (1990): Isolation and analysis of naturally processed viral peptides as recognized by cytotoxic T cells. *Nature* 348: 195–197

Saper MA, Bjorkman PJ, Wiley DC (1991): Refined structure of the human histocompatibility antigen HLA-A2 at 2.6 Å resolution. *J Mol Biol* 291: 277–319

Schirrmacher V, Wigzell H (1972): Immune responses against native and chemically modified albumins in mice. I. Analysis of non-thymus-processed (B) and thymus-processed (T) cell responses against methylated bovine serum albumin. *J Exp Med* 136: 1616–1630

Schumacher TN, De Bruijn ML, Vernie LN, Kast WM, Melief CJ, Neefjes JJ, Ploegh HL (1991): Peptide selection by MHC class I molecules. *Nature* 350: 703–706

Shearer G (1974): Cell mediated cytotoxicity to trinitrophenyl-modified syngeneic lymphocytes. *Eur J Immunol* 4: 527–533

Sherman LA, Burke TA, Biggs JA (1992): Extracellular processing of peptide antigens that bind class I major histocompatibility molecules. *J Exp Med* 175: 1221–1226

Shimonokevitz R, Kappler J, Marrack P, Grey H (1983): Antigen recognition by H-2-restricted T cells. I. Cell-free antigen processing. *J Exp Med* 158: 303–316

Silver J, Hood L (1974): Detergent-solubilised H-2 alloantigen is associated with a small molecular weight polypeptide. *Nature* 249: 764–765

Smithies O, Poulik MD (1972): Dog homologue of human

2-microglobulin. *Proc Natl Acad Sci USA* 69: 2914–2917

Solinger AM, Ultee ME, Margoliash E, Schwartz RH (1979): T-lymphocyte response to cytochrome c. I. Demonstration of a T-cell heteroclitic proliferative response and identification of a topograhic antigenic determinant on pigeon cytochrome c whose immune recognition requires two complementing major histocompatibility complex-linked immune response genes. *J Exp Med* 150: 830–848

Townsend A, Ohlen C, Bastin J, Ljunggren HG, Foster L, Karre K (1989): Association of class I major histocompatibility heavy and light chains induced by viral peptides. *Nature* 340: 443–448

Van Bleek GM, Nathenson SG (1990): Isolation of an endogenously processed immunodominant viral peptide from the class I H-2K^b molecule. *Nature* 348: 213–216

Vitiello A, Potter TA, Sherman LA (1990): The role of beta 2-microglobulin in peptide binding by class I molecules. *Science* 250: 1423–1426

Wei ML, Cresswell P (1992): HLA-A2 molecules in an antigen-processing mutant cell contain signal sequence-derived peptides. *Nature* 356: 443–446

Williams DB, Barber BH, Flavell RA, Allen H (1989): Role of beta 2-microglobulin in the intracellular transport and surface expression of murine Class I histocompatibility molecules. *J Immunol* 142: 2796–2806

Zinkernagel R, Doherty P (1974): restriction of *in vitro* T cell-mediated cytotoxicity in lymphocytic choriomeningitis within a syngeneic or semiallogeneic system. *Nature* 248: 701–702

Zinkernagel RM, Doherty PC (1975): H-2 compatibility requirement for T-cell mediated lysis of target cells infected with lymphocytic choriomeningitis virus: Different cytotoxic T-cell specificities are associated with structures coded for in H-2K and H-2D. *J Exp Med* 141: 1427–1436

6
The Role of CD8–Class I Interactions in CTL Function

Anne M. O'Rourke and Matthew F. Mescher

With few exceptions, cytotoxic T lymphocytes (CTL) are activated to lyse cells that bear surface antigen displayed in the context of major histocompatibility complex (MHC) class I proteins. Although recognition of antigen *per se* is a function of the polymorphic T-cell receptor complex (TCR/CD3), expression of the CD8 glycoprotein correlates with restriction of the CTL to class I protein, and CD8 has been shown to interact with nonpolymorphic residues in the class I $\alpha 3$ domain (Salter et al., 1990). Considerable evidence has now accumulated to demonstrate the critical role this interaction plays in adhesion and signal generation for CTL effector function (reviewed in Parnes, 1989).

CD8 is composed of α and β chains (termed Lyt-2 and Lyt-3 in the mouse), which are present on the CTL surface as homo- or heterodimers. The intracytoplasmic domain of the α chain is noncovalently associated with the src-related tyrosine kinase p56^{lck} (Veillette et al., 1988), and, as will be discussed later, there is evidence that activation of this enzyme by CD8 engagement contributes to CTL activation.

CD8 is one of several non-antigen-specific, or "accessory," proteins that interact with target cell ligands to allow CTL to form stable conjugates with antigen-bearing target cells (Martz, 1987). The large number of accessory receptor–ligand interactions that can contribute to CTL–target cell (TC) adhesion makes the study of an individual interaction extremely complex. Many laboratoies have therefore taken the approach of immobilizing affinity-purified ligands in artificial membranes, which allows the interaction of a single receptor with its ligand to be studied in detail (see Methods section, this volume). This approach has resulted in evidence that ligand binding by several accessory molecules, including CD8, LFA-1, and the VLA antigens, is acutely regulated by TCR engagement (O'Rourke and Mescher, 1990). Using the purified native ligands in artificial membrane systems, adhesion of T cells was in each case found to be dramatically increased after TCR engagement. Thus, at least two regulatory mechanisms exist by which non-antigen-specific T-cell proteins contribute to function: increased cell surface expression of several molecules, including LFA-1, CD2, and some VLA antigens follows primary T-cell stimulation *in vivo* and expression remains elevated thereafter (Shimizu et al., 1990b); in the short term, antigen binding by the TCR complex leads to a transient increase in the avidities of adhesions mediated by accessory receptor–ligand interactions. The TCR-triggered avidity modulation is termed "activated" binding to distinguish it from ligand binding, which occurs in an antigen-independent fashion, and has so far been demonstrated to occur for interactions between CD8 and class I (O'Rourke et al., 1990), LFA-1 and ICAM-1 (Dustin and Springer, 1989; van Kooyk et al., 1989), VLA-4, -5 and fibronectin, and VLA-6 and laminin (Shimizu et al., 1990a).

A CTL encounter with another cell appears very likely to involve the events outlined in Figure 1. Initial antigen-independent adhesion is necessary to overcome the innate repulsion between cells due to the net surface negative charge and may be mediated by basal (nonactivated) binding by LFA-1 and CD2. In the absence of antigen, this adhesion is probably too weak to sustain conjugate formation, and the cells readily dissociate. In the event that specific antigen/MHC complexes are present on the target cell, TCR engagement occurs and activates ligand binding by CD8, LFA-1, and possibly other molecules, resulting in stable con-

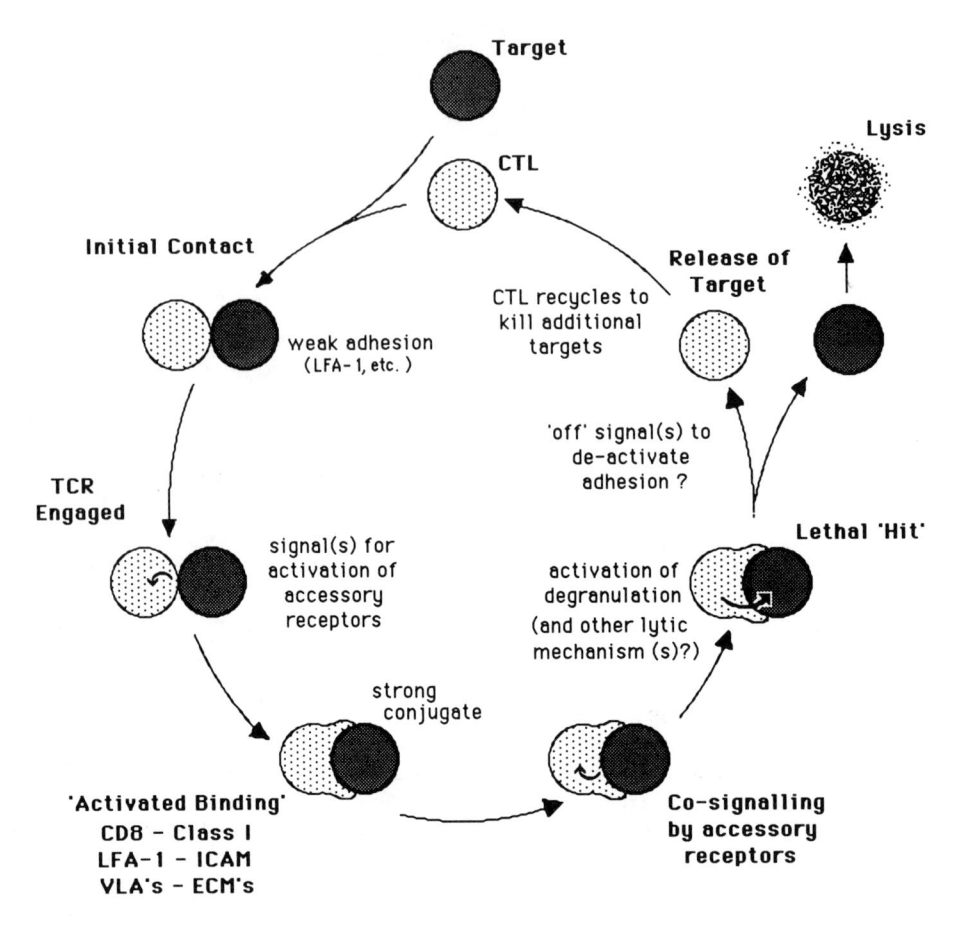

FIGURE 1. Proposed model for antigen-independent and -dependent adhesion systems in CTL–target cell interactions. See text for discussion.

jugate formation. Upon engagement, these coreceptors generate additional signals, possibly involving a feedback mechanism to the TCR/CD3 complex, which combine with direct TCR/CD3-initiated signals, to activate delivery of the lethal hit. Loss of adhesion then occurs and the target cell is released, allowing the CTL to recycle and bind additional target cells. The following discussion will focus on the role of activated CD8–class I interactions in murine CTL function.

Adhesion of CD8 to Class I Protein

Normal CTL–TC interactions involve numerous receptor–ligand binding events, any one of which may be blocked by antireceptor antibodies and lead to a decrease or complete elimination of lysis. However, alloantigen- or peptide-specific CTL clones can also be activated solely by isolated alloantigenic class I protein or peptide-pulsed self class I immobilized in planar membranes (Goldstein and Mescher, 1987; Kane et al., 1988, 1989b). Thus, when class I is the only protein present, engagement of the TCR and CD8 alone can be sufficient for adhesion and CTL activation (as measured by exocytosis of serine esterase enzymes: see Chapter 57). With numerous CTL clones, it was noted that a critical class I density is required, which is comparable to that displayed on normal spleen stimulators (Kane et al., 1989a). It has subsequently been shown that a major component of the density dependence is due to a requirement for effective CD8–class I interactions. Coimmobilization of a suboptimal density of alloantigenic class I protein (which alone is insufficient to activate the CTL) with either third-party or syngeneic class I leads to a very strong degranulation response. Since CTL activation does not occur with any density of irrelevant class I protein alone, the results are consistent with the requirements of a minimal TCR stimulus for CTL activation, providing that sufficient CD8–class I binding also occurs. Interestingly, the total class I density dependence for activation by alloantigen alone is comparable to that by suboptimal alloantigen plus irrelevant class I, suggesting that a major component of the alloantigen stimulus is provided by CD8 interaction with the α3 domain of the class I protein.

Despite the fact that the functional evidence strongly supports the importance of CD8 recognition of conserved class I residues, CTL binding to purified class I protein is only detected when alloantigenic class I protein is present. Thus, binding occurs with optimal densities of alloantigen, or with suboptimal alloantigen coimmobilized with irrelevant class I, but not to irrelevant class I protein alone. These results suggested that the CD8–class I interaction might only occur when a requirement for minimal TCR engagement is satisfied. This hypothesis was directly confirmed by the finding that addition of fluid-phase anti-TCR monoclonal antibodies (mAb) to the CTL in the presence of immobilized nonantigenic class I protein resulted in avid adhesion of the CTL to the class I-bearing surface (O'Rourke et al., 1990). The results are consistent with CD8's normally being in an "inactive" state in which interaction with class I protein is either extremely weak or nonexistent, but upon TCR engagement some event is initiated which now makes CD8 capable of binding avidly to class I.

The specificity of TCR-activated CTL binding to conserved class I residues was demonstrated by the ability of any class I protein (third-party or syngeneic) to mediate adhesion, but not irrelevant proteins such as fetal calf serum (FCS), bovine serum albumin (BSA), or class II protein; binding was blocked by mAb to CD8 but not to other CTL proteins, and class I binding was initiated only by mAb against the TCR/CD3 complex (Vβ or CD3ε). It should be noted that activated adhesion of LFA-1 to ICAM-1 and of VLA-4, -5 and -6 to the extracellular matrix proteins fibronectin and laminin can be elicited by mAb to either the TCR/CD3 complex or to CD2 (Dustin and Springer, 1989; van Kooyk et al., 1989; Shimizu et al., 1990a); we have not yet investigated the ability of anti-CD2 mAb to initiate CD8–class I binding by the murine CTL clones.

Activated binding of CTL to class I is rapid and reaches a plateau which is sustained for several hours at 37°C (Mescher et al., 1991). Interestingly, we have employed CTL with several different specificities and routinely observed that only 30–40% of cells are bound to class I-coated wells at any given time, despite the fact that cloned CTL were used. This did not result from limited availability of class I protein for binding, as 30–40% binding occurs irrespective of the initial cell concentration or class I density employed (Mescher et al., 1991). Moreover, perturbation of the system by addition or removal of cells during incubation results in reestablishment of the level of cells bound to 30–40%. Removal of unbound cells to new wells containing class I protein and measurement of binding has also shown that released or unbound CTL are not refractory to rebinding. These results strongly suggest that binding of CTL to class I protein is governed by an equilibrium between bound and unbound cells. It appears that a TCR stimulus (in the form of either anti-TCR mAb or alloantigen) initiates the binding of CD8 to conserved class I residues, and that once binding has occurred an "off" signal is generated which terminates binding, but does not prevent subsequent rebinding, provided a TCR stimulus is still present.

At a physiological level, the ability of CTL to "recycle" rapidly between antigen-bearing target cells is well characterized (Zagury et al., 1975; Martz, 1976). Death of the target cell is not necessary for the CTL to deadhere once the lethal hit has been delivered, suggesting an active release process. Evaluation of the properties of the CTL–TC interaction has provided strong evidence that conjugate formation is governed by equilibrium binding events (Balk and Mescher, 1981; Balk et al., 1981; Garcia-Penarrubia and Bankhurst, 1989). The demonstration at the molecular level that CTL binding to purified class I protein exhibits such properties is therefore in keeping with a model in which the sum of individual activated adhesion systems determines the net equilibrium level of specific CTL–TC conjugate formation. As the "activation" process involves specific engagement of the TCR by antigen (or anti-TCR antibody in the *in vitro* systems), there exists an inherent fail-safe mechanism to prevent nonproductive tight binding to cells lacking antigen.

The mechanism by which TCR engagement brings about an avidity change in CD8–class I-mediated adhesion is as yet unknown, but several possibilities exist. The cell surface expression of CD8 does not change after addition of soluble anti-TCR antibody to the CTL, as determined by flow cytometry; therefore, rapid movement of internal pools of CD8 to increase surface density does not appear to account for the increased avidity (unpublished observations). A signaled conformational change may occur which increases the affinity of CD8 for class I protein; such a mechanism may be involved in the TCR-activated binding of LFA-1 to ICAM-1. In this case an "activation epitope" is revealed in LFA-1 after cell activation which is not apparent on resting cells (Larson et al., 1990; van Kooyk et al., 1991). Affinity modulation may result from change in the phosphorylation state of CD8, as is thought to occur for several growth-factor receptors (Sibley et al., 1987). Finally, TCR engagement might induce microclustering of CD8, as has been demonstrated to account for the increased capacity of CR3 to bind to particles coated with C3bi (Detmers et al., 1987).

It is very likely that cytoskeletal elements play a role in the events leading to "activated" CD8–class I binding. Cytochalasins and colchicine, which interfere with microfilament and microtubule function, respectively (Plaut et al., 1973; Strom et al., 1973), are able to inhibit CTL binding to alloantigen, or TCR-activated binding to irrelevant class I protein (O'Rourke et al., 1991). Since CTL treated with these compounds do not undergo normal cell deformation and spreading, it may be that they disrupt the initial TCR-derived transmembrane signal, or the CD8 "activation" mechanism itself. However, CTL binding and response to immobilized anti-TCR antibodies are not disrupted by either colchicine or the cytochalasins, suggesting that they do not directly interfere with TCR-mediated signaling events (O'Rourke et al., 1991). It is well known that CTL–TC conjugate formation depends on intact cytoskeleton function (Martz, 1977); it is therefore probable that at least one reason for this is the involvement of cytoskeletal elements in evoking the TCR-activation adhesion systems.

Signaling Events in Activated Adhesion of CTL

The earliest known signaling event following TCR engagement is activation of a tyrosine kinase (TK), possibly $p59^{fyn}$ (Samelson et al., 1990). Activation of phosphatidylinositol-specific phospholipase C (PI-PLC) has been shown to occur subsequent to, and probably depends upon, TK activation (June et al., 1990; Mustelin et al., 1990). More recently, tyrosine phosphorylation has been shown to increase the catalytic activity of (PI-PLC)γ1 in the human T-cell line Jurkat (Secrist et al., 1991; Weiss et al., 1991), as is the case for PI-PLCγ1 activation for several growth factor receptors (Ullrich and Schlessinger, 1990). Although treatment of CTL with soluble anti-TCR antibodies alone does not result in a detectable degranulation response, it seems clear from the results described above that TCR engagement under these conditions does deliver the signal necessary to increase the avidity of CD8 for class I protein. TCR-activated adhesion to class I protein is blocked by pretreatment of the CTL with the broad-range kinase inhibitors K252a and staurosporine (O'Rourke et al., 1990). Although these compounds were initially described as serine/threonine kinase-specific, the TCR-stimulated phosphorylation of tyrosine residues of PI-PLCγ1 in human T cells has recently been shown to be inhibited by staurosporine (Secrist et al., 1991). In recent work, we have found that the TK-specific inhibitors genestein and herbimycin A also block activated CD8 binding, suggesting that TCR-activation of a TK may be the initial signal required for the activation. If so, there are likely to be intermediate steps between TK activation and activation of CD8, since neither the α nor the β chains of murine CD8 contain tyrosine residues in their cytoplasmic domains (Nakauchi et al., 1985; Panaccio et al., 1987).

CD8 binding to class I clearly provides a transmembrane signal, as evidenced by the fact that fluid phase anti-TCR mAb does not trigger degranulation unless the cells also adhere to class I (O'Rourke et al., 1990). We have recently found that PI turnover (by measurement of [^{3}H]inositol phosphate production or changes in intracellular free Ca^{2+} concentrations) does not occur when CTL are treated with soluble anti-TCR antibodies alone. However, PI turnover is readily detectable in the same CTL in response to alloantigen. Initiation of PI turnover when both the TCR and CD8 are engaged (i.e., by alloantigen), but not by a TCR stimulus alone (soluble anti-TCR mAb), therefore raised the possibility that CD8–class I interactions were required to activate this signaling pathway. This was indeed found to be the case: PI turnover was initiated when CTL treated with the same concentrations of soluble anti-TCR mAb were allowed to bind to wells coated with irrelevant class I protein (O'Rourke and Mescher, 1992). The conditions required for initiation of PI turnover in the CTL examined are identical to those required for activated binding to occur. Thus, it appears that PI turnover is not *required* for initiation of CD8–class I binding, but instead *results* from the interaction. Consistent with this, we have found that neither calcium ionophore nor PKC-activating phorbol esters initiate CTL adhesion to class I protein (unpublished results).

How then does CD8 binding to class I protein allow coupling to the PI pathway under conditions

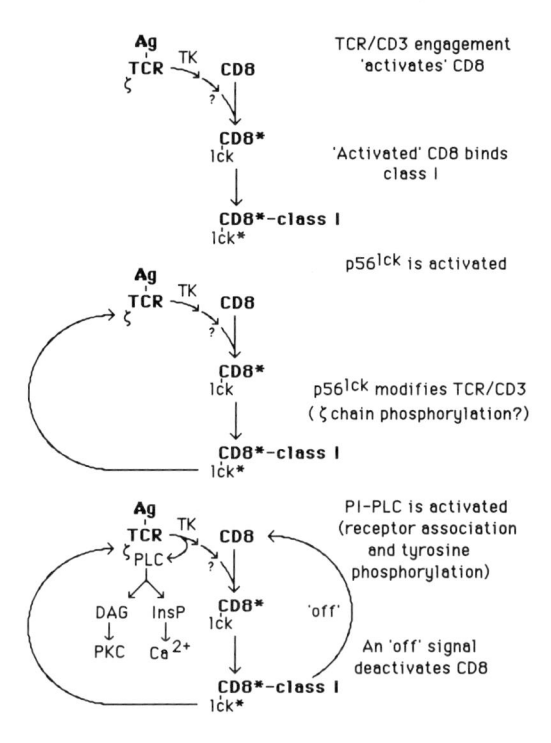

FIGURE 2. Model for TCR- and CD8-mediated signal generation for CTL activation. See text for discussion.

of minimal TCR engagement? As mentioned previously, CD8α is associated intracytoplasmically with the TK p56^{lck} and antibody-mediated crosslinking of CD8 increases the activity of the enzyme. It is therefore reasonable to assume that binding of "activated" CD8 to class I protein also increases the activity of p56^{lck}. The ζ chain of the TCR/CD3 complex has been shown to play a critical role in allowing TCR occupancy to couple to signal generation (Irving and Weiss, 1991); this chain is also a substrate for tyrosine phosphorylation by p56^{lck} (Barber et al., 1989). These observations have led us to suggest a model (Figure 2) by which CD8–class I binding may result in coupling to activation of the PI pathway. TCR engagement would lead to activation of the TCR/CD3-associated TK (p59^{fyn}) which (a) is involved in activation of CD8 so that binding to class I protein may now take place, and (b) phosphorylates PI-PLC on tyrosine residues, which is necessary, but not sufficient, for enzyme activation. Activated CD8 binding to class I protein results in activation of p56^{lck}, which feedback phosphorylates the TCR complex (possibly the ζ chain) such that full PI-PLC enzyme activation may now occur.

An alternative possibility is that CD8 directly couples to activation of PI-PLC, but this mechanism would be difficult to reconcile with the observation that immobilized anti-TCR antibodies alone are very potent initiators of PI turnover in T cells. Under these conditions no CD8 ligand is present, and therefore p56^{lck} activation presumably does not occur. However, immobilized anti-TCR provide a highly multivalent stimulus by which large numbers of cell surface receptors are engaged and may become polarized or clustered. It is possible that, if TCR occupancy is high, basal activity of "trapped" p56^{lck}–CD8 molecules may be sufficient to feedback-activate adjacent TCR coupling to PI-PLC. That this may indeed be the case is supported by the observations that immobilized anti-TCR-stimulated PI turnover in the CTL is still susceptible to at least partial inhibition by anti-CD8 mAb, but not by antibodies specific for other CTL surface proteins (unpublished observations).

Our results thus demonstrate that CD8–class I interactions are required to initiate PI turnover and the degranulation response when TCR occupancy is minimal. Recent estimates indicate that the number of antigen/MHC complexes necessary to activate T cells is very low (Vitiello et al., 1990; Christinck et al., 1991), suggesting that activated adhesion systems may be vital to signal generation for T-cell activation under normal physiological conditions. In keeping with this, there is evidence that other accessory receptors contribute to augmented functional responses (Shimizu et al., 1990c; van Seventer et al., 1990) and may couple to second messenger generation (Pardi et al., 1989; Hahn et al., 1991). If activated adhesion systems generate intracellular second messengers which contribute to cell activation, it seems reasonable that the final stage of CTL–TC interactions, i.e., the deadhesion step, may also result from a signaled termination of binding. This may be due either to decay of an existing signal necessary to sustain binding, or, alternatively, to generation of an active "off" signal. Distinguishing between these possibilities must await a more detailed understanding of the mechanism by which CD8, LFA-1, and other receptors are activated to bind their ligands.

Conclusion

The "adhesion-strengthening hypothesis," first expounded by Martz (1987 and references therein) suggested that the properties of murine CTL–TC interactions were such that antigen–TCR bonds were unlikely to contribute to adhesion, but that "adhesion strengthening" was mediated by accessory receptors, possibly triggered by antigen recognition. Such a mechanism would also explain the specificity of the interaction, in that sustained conjugates are not formed if specific antigen is not present on the TC. Evidence from the last few years suggests that this, indeed, is the likely mechanism of CTL-TC conjugate formation. TCR-activated adhesion has now been shown at the molecular level for CD8, LFA-1, and the VLA antigens; it is probable that additional systems will be identified as the purified ligands of other T-cell surface receptors become available. The extent to which TCR-activated binding systems are involved in CTL function will presumably depend upon the spectrum of ligands displayed by the particular TC involved. However, all nucleated cells bear class I protein, and it therefore seems reasonable to assume that activated CD8-class I interactions make a contribution to CTL activation in most, if not all, cases.

References

Balk SP, Mescher MF (1981): Specific reversal of cytolytic T lymphocyte-target cell binding is induced by free target cells. *J Immunol* 127: 51-57

Balk SP, Walker J, Mescher MF (1981): Kinetics of cytolytic T lymphocyte binding to target cells. *J Immunol* 126: 2177-2183

Barber EK, Dev Dasgupta J, Schlossman SF, Trevllyan JM, Rudd CE (1989): The CD4 and CD8 antigens are coupled to a protein-tyrosine kinase (p56[lck]) that phosphorylates the CD3 complex. *Proc Natl Acad Sci USA* 86: 3277-3281

Christinck ER, Luscher MA, Barber BH, Williams DB (1991): Peptide binding to class I MHC on living cells and quantitation of complexes required for CTL lysis. *Nature* 352: 67-70

Detmers PA, Wright SD, Olsen E, Kimball B, Cohn ZA (1987): Aggregation of complement receptors on human neutrophils in the absence of ligand. *J Cell Biol* 105: 1137-1145

Dustin ML, Springer TA (1989): T-cell receptor crosslinking transiently stimulates adhesiveness through LFA-1. *Nature* 341: 619-624

Garcia-Penarrubia P, Bankhurst AD (1989): Kinetic analysis of effector cell recylcing and effector-target binding capacity in a model of cell-mediated cytotoxicity. *J Immunol* 143: 2101-2111

Goldstein SAN, Mescher MF (1987): Cytotoxic T cell activation by class I protein on cell-size artificial membranes: Antigen density and Lyt-2/3 function. *J Immunol* 138: 2034-2043

Hahn WC, Rosenstein Y, Burakoff SJ, Bierer BE (1991): Interaction of CD2 with its ligand lymphocyte function-associated antigen-3 induces adenosine 3′,5′-cyclic monophosphate production in T lymphocytes. *J Immunol* 147: 14-21

Irving BA, Weiss A (1991): The cytoplasmic domain of the T cell receptor ζ chain is sufficient to couple to receptor-associated signal transduction pathways. *Cell* 64: 891-901

June CH, Fletcher MC, Ledbetter JA, Samelson LE (1990): Increases in tyrosine phosphorylation are detectable before phospholipase C activation after T cell receptor stimulation. *J Immunol* 144: 1591-1599

Kane KP, Champoux P, Mescher MF (1989a): Solid-phase binding of class I and class II MHC proteins: Immunoassay and T cell recognition. *Mol Immunol* 26: 759-768

Kane KP, Goldstein SAN, Mescher MF (1988): Class I alloantigen is sufficient for cytolytic T lymphocyte binding and transmembrane signalling. *Eur J Immunol* 18: 1925-1929

Kane KP, Vitiello A, Sherman LA, Mescher MF (1989b): Cytolytic T lymphocyte response to isolated class I H-2 proteins and influenza peptides. *Nature* 340: 157-159

Larson RS, Hibbs ML, Springer TA (1990): The leukocyte integrin LFA-1 reconstituted by cDNA transfection in a nonhematopoietic cell line is functionally active and not transiently regulated. *Cell Reg* 1: 359-367

Martz E (1976): Multiple target cell killing by cytolytic T lymphocytes and the mechanism of cytotoxicity. *Transplantation* 21: 5-11

Martz E (1977): Mechanism of specific tumor cell lysis by alloimmune T-lymphocytes: Resolution and characterization of discrete steps in the cellular interaction. *Contemp Top Immunobiol* 7: 301-361

Martz E (1987): LFA-1 and other accessory molecules functioning in adhesions of T and B lymphocytes. *Human Immunol* 18: 3-37

Mescher MF, O'Rourke AM, Champoux P, Kane KP (1991): Equilibrium binding of cytotoxic T lymphocytes to class I antigen. *J Immunol* 147: 36-41

Mustelin T, Coggeshall KM, Isakov N, Altman A (1990): T cell antigen receptor-mediated activation of phospholipase C requires tyrosine phosphorylation. *Science* 247: 1584-1587

Nakauchi H, Nolan GP, Hsu C, Huang HS, Kavathas P, Herzenberg LA (1985): Molecular cloning of Lyt2, a membrane glycoprotein marking a subset of mouse T lymphocytes: Molecular homology to its human counterpart, Leu-2/T8, and to immunoglobulin variable regions. *Proc Natl Acad Sci USA* 82: 5126-5130

O'Rourke AM, Mescher MF (1990): T-cell receptor-activated adhesion systems. *Curr Opinion Cell Biol* 2: 888-893

O'Rourke AM, Mescher MF (1992): Cytotoxic T lymphocyte activation involves a cascade of signalling and adhesion events. *Nature* 358: 253-255

O'Rourke AM, Apgar JR, Kane KP, Martz E, Mescher MF (1991): Cytoskeletal function in CD8- and T cell receptor-mediated interaction of cytotoxic T lymphocytes with class I protein. *J Exp Med* 173: 241-249

O'Rourke AM, Rogers J, Mescher MF (1990): Activated

CD8 binding to class I protein mediated by the T-cell receptor results in signalling. *Nature* 346: 187–189

Panaccio M, Gillespie, MT, Walker ID, Kirszbaum L, Sharpe JA, Tobias GH, McKenzie IFC, Deacon NJ (1987): Molecular characterization of the murine cytotoxic T-cell membrane glycoprotein Ly-3 (CD8). *Proc Natl Acad Sci USA* 84: 6874–6878

Pardi R, Bender JR, Dettori C, Giannazza E, Engleman EG (1989): Heterogeneous distribution and transmembrane signalling properties of lymphocyte function-associated antigen (LFA-1) in human lymphocyte subsets. *J Immunol* 143: 3157–3166

Parnes JR (1989): Molecular biology and function of CD4 and CD8. *Adv Immunol* 44: 265–311

Plaut M, Lichtenstein LM, Henney CS (1973): Studies on the mechanisms of lymphocyte mediated cytolysis III. The role of microfilaments and microtubules. *J Immunol* 110: 771–780

Salter RD, Benjamin RJ, Wesley PK, Buxton SE, Garrett TP, Clayberger C, Krensky AM, Norment AM, Littmann DR, Parham P (1990): A binding site for the T-cell co-receptor CD8 on the α3 domain of HLA-A2. *Nature* 345: 41–46

Samelson LE, Phillips AF, Luong ET, Klausner RD (1990): Association of the fyn protein-tyrosine kinase with the T-cell antigen receptor. *Proc Natl Acad Sci USA* 87: 4358–4362

Secrist JP, Karnitz L, Abraham RT (1991): T-cell antigen receptor ligation induced tyrosine phosphorylation of phospholipase C-γ1. *J Biol Chem* 266: 12135–12139

Shimizu Y, van Seventer GA, Horgan KJ, Shaw S (1990a): Regulated expression and binding of three VLA (β1) integrin receptors on T cells. *Nature* 345: 250–253

Shimizu Y, van Seventer GA, Horgan KJ, Shaw S (1990b): Roles of adhesion molecules in T-cell recognition: Fundamental similarities between four integrins on resting human T cells (LFA-1, VLA-4, VLA-5, VLA-6) in expression, binding, and constimulation. *Immunol Rev* 114: 109–143

Shimizu Y, van Seventer GA, Horgan KJ, Shaw S (1990c): Costimulation of proliferative responses of resting CD4+ T cells by the interaction of VLA-4 and VLA-5 with fibronectin or VLA-6 with laminin. *J Immunol* 145: 59–67

Sibley DR, Benovic JL, Caron MG, Lefkowitz RJ (1987): Regulation of transmembrane signalling by receptor phosphorylation. *Cell* 48: 913–922

Strom TB, Garovoy MR, Carpenter CB, Merrill JP (1973): Microtubule function in immune and non-immune lymphocyte mediated cytotoxicity. *Science* 181: 171–173

Ullrich A, Schlessinger J (1990): Signal transduction by receptors with tyrosine kinase activity. *Cell* 61: 203–212

van Kooyk Y, van de Wiel-van Kemenade P, Weder P, Kuijpers TW, Figdor CG (1989): Enhancement of LFA-1-mediated cell adhesion by triggering through CD2 or CD3 on T lymphocytes. *Nature* 342: 811–813

van Kooyk Y, Weder P, Hogervorst F, Verhoeven AJ, van Seventer G, te Velde AA, Borst J, Keizer GD, Figdor CG (1991): Activation of LFA-1 through a Ca^{2+}-dependent epitope stimulates lymphocyte adhesion. *J Cell Biol.* 112: 345–354

van Seventer GA, Shimizu Y, Horgan KJ, Shaw S (1990): The LFA-1 ligand ICAM-1 provides an important costimulatory signal for T cell receptor-mediated activation of resting T cells. *J Immunol* 144: 4579–4586

Veillette A, Bookman MA, Horak EM, Bolen JB (1988): The CD4 and CD8 T cell surface antigens are associated with the internal membrane tyrosine-protein kinase p56[lck]. *Cell* 55: 301–308

Vitiello A, Potter TA, Sherman LA (1990): The role of β2-microglobulin in peptide binding by class I molecules. *Science* 250: 1423–1426

Weiss A, Koretzky G, Schatzman RC, Kadlecek T (1991): Functional activation of the T-cell antigen receptor induces tyrosine phosphorylation of phospholipase C-γ1. *Proc Natl Acad Sci USA* 88: 5484–5488

Zagury D, Bernard J, Thierness N, Feldman M, Berke G (1975): Isolation and characterization of individual functionally reactive cytotoxic T lymphocytes: Conjugation, killing and recycling at the single cell level. *Eur J Immunol* 5: 818–822

7
Interactions Between CD2 and T-Cell Receptor Isoforms in CTL Function

Shigeo Koyasu and Ellis L. Reinherz

Introduction

The recognition of specific target cells by cytotoxic T lymphocytes (CTL) involves a set of cell surface proteins. The physical interaction of a T-cell receptor (TCR) with a nominal peptide antigen bound to a specific major histocompatibility complex (MHC) molecule confers specificity on the CTL. The clonally unqiue antigen-specific binding component has been termed Ti and exists as a disulfide-linked Tiα-β heterodimer on the majority of peripheral T cells, although approximately 5% of T cells express Tiγ-δ heterodimers (Meuer et al., 1984a; Marrack and Kappler, 1987; Brenner et al., 1988; Davis and Bjorkman, 1988; Raulet, 1989). The immunoglobulin-like Ti structure is noncovalently associated in a molecular complex with the CD3 subunits γ, δ, ε, ζ, and η, the latter molecules being involved in signal transduction (Clevers et al., 1988; Ashwell and Klausner, 1990; Koyasu et al., 1991a). The TCR has therefore been referred to as the CD3-Ti complex.

The antigen recognition process is not singularly dictated by the CD3-Ti complex itself, but rather is dependent on other accessory molecules, including CD4 and CD8. CD4 (L3T4 in mouse) and CD8 (Lyt-2/3 in mouse) molecules are expressed on mature T lymphocytes in a mutually exclusive manner (Reinherz and Schlossman, 1980; Swain, 1983; Parnes, 1989). CD4$^+$8$^-$ T lymphocytes were found to be restricted by class II MHC molecules, whereas CD4$^-$8$^+$ T lymphocytes demonstrated a class I MHC restriction (Meuer et al., 1982; Greenstein et al., 1984; Gabert et al., 1987; Gay et al., 1987; Ratnofsky et al., 1987). Subsequently, it was discovered that CD4 and CD8 structures themselves bind to monomorphic regions of class II and class I MHC structures, respectively, thereby facilitating the interaction of the TCR with the MHC-restricting element (Doyle and Strominger, 1987; Norment et al., 1988; Clayton et al., 1989). In addition to the MHC-binding accessory structures, a set of adhesion structures facilitates antigen-dependent and -independent interactions between T lymphocytes and their cognate partners. These include CD2/LFA-3 and LFA-1/ICAM-1 receptor–ligand pairs (Shaw et al., 1986; Spits et al., 1986). The purpose of this article is to summarize the current information about the structure and function of the CD2 molecule and the TCR complex on CTL.

Adhesion Function of CD2

The CD2 molecule was originally identified on the basis of its ability to participate in spontaneous aggregates between T cells and sheep erythrocytes (SRBC) (Howard et al., 1981; Kamoun et al., 1981; Bernard et al., 1982). CD2 on CTL interacts with lymphocyte function-associated antigen 3 (LFA-3; CD58) on target cells (Hunig et al., 1986; Selvaraj et al., 1987; Takai et al., 1987). Unlike the LFA-1/ICAM-1 pair, the interaction between CD2 and LFA-3 is temperature- and divalent cation-independent. The adhesion domain of CD2 is encoded by the second exon of the CD2 gene and consists of the N-terminal 102 amino acids (Sewell et al., 1986; Sayre et al., 1987; Seed and Aruffo, 1987; Diamond et al., 1988; Lang et al., 1988; Richardson et al., 1988). Mutagenesis and biochemical analysis of a recombinant CD2 protein localized the T11$_1$ epitope associated with LFA-3 binding and a distinct set of epitopes including T11$_2$ to this protein segment (Peterson and Seed, 1987; Richardson et al., 1988; Sayre et al., 1989; Recny et al., 1990). By utilizing a soluble monomeric form of CD2, the dissociation constant (K_d) between CD2 and LFA-3 was determined to be 0.4 μM (Sayre et al., 1989; Recny et al., 1990). The affinity between CD2 and LFA-3 is independent of the cytoplasmic domain of CD2 (Moingeon et al., 1989, 1991), which contrasts to results obtained with different adhesion structures such as E-cadherin and LFA-1 (Nagafuchi and Takeichi, 1988; Hibbs et al., 1991).

CD2/LFA-3 Interaction Augments CTL Recognition of Target Cells

Certain monoclonal antibodies (mAb) directed against CD2 or LFA-3 inhibit proliferation of T cells as well as cytotoxic effector function and antigen-independent cell adhesion between CTL and target cells (Palacios and Martinez-Maza, 1982; Krensky et al., 1984; Tadmori et al., 1985; Bolhuis et al., 1986; Shaw et al., 1986; Spits et al., 1986; Denning et al., 1987), indicating the importance of CD2/LFA-3 interaction in cell adhesion. CD2 also plays an important role in antigen-dependent T-cell recognition. Several studies in addition to antibody-blocking experiments have suggested that CD2/LFA-3 interaction is important for the T-cell recognition process (Bierer et al., 1988a, 1988b; Moingeon et al., 1989). For example, transfection of human CD2 into alloreactive murine T-cell hybridomas that recognize human class II MHC enhances interleukin 2 (IL2) production on exposure to human LFA-3-expressing stimulator cells (Bierer et al., 1988b). Likewise, transfection of human CD2 molecules into murine T-cell hybridoma's and human LFA-3 into murine antigen presenting cells (APC) enhances IL-2 production induced by such antigen-pulsed APC (Moingeon et al., 1989). Further evidence of the role of CD2/LFA-3 interaction in recognition of nominal peptide antigen has been shown in a defined model system utilizing a human T-cell clone whose precise antigen specificity and MHC restriction element are known, and murine fibroblast cells transfected with human molecules as APC (Koyasu et al., 1990; Koyasu and Reinherz, 1991). The latter findings are particularly relevant given that the model utilizes CTL.

A human $CD4^+$ CTL clone specific for HIV-gp120 and restricted to DR4, termed Een 217, kills a murine L-cell target transfected with human DR4 molecules in the presence of a specific peptide antigen corresponding to a fragment of HIV H3DCG gp120 (Siliciano et al., 1988). When the L cell is overtransfected with human LFA-3, the CTL clone kills the target in the presence of a lower concentration of peptide antigen (Koyasu and Reinherz, 1991). Comparable proliferation of Een 217 is induced by a 50-fold lower concentration of peptide antigen with L cells expressing both DR4 and LFA-3 than with L cells expressing only DR4 molecules (Koyasu et al., 1990). It is thus indicated that the CD2/LFA-3 interaction augments the recognition process by lowering the concentration of nominal antigen required for effective CTL recognition of the target cells.

Mechanism of the Augmenting Effect of CD2/LFA-3 Interaction

The CD2 molecule is known to transmit activation signals as described later. One possible mechanism for augmenting the effect of the CD2/LFA-3 interaction is binding of LFA-3 to the extracellular segment of CD2, thus triggering the signal transduction function of CD2. However, transmembrane CD2 molecules that have been truncated so that they lack a cytoplasmic tail still enhance antigen-induced production of IL2 when transfected into murine T-cell hybridomas stimulated by human LFA-3-expressing murine APC (Moingeon et al., 1989). Thus, the adhesion function *per se* contributes, at least in part, to the augmenting effect as CD2/LFA-3 interaction facilitates physical interaction between CTL and target cells.

Antigen-dependent and -independent conjugate formation was directly examined in the system described above, and the following results were obtained (Koyasu et al., 1990): (1) Conjugate formation between CTL and a $DR4^+ LFA-3^-$ L-cell target was marginal in the absence of peptide antigen but was enhanced with increasing concentration of the antigen. (2) Stable binding of CTL to a $DR4^+ LFA-3^+$ L-cell target was observed, but the binding was further enhanced with increasing concentration of the peptide antigen. (3) Treatment of the $DR4^+ LFA-3^+$ L-cell target cells with mAb against LFA-3 resulted in a binding pattern of CTL indistinguishable from that with $DR4^+ LFA-3^-$ target cells. These findings indicated that coexpression of LFA-3 with the MHC-restricting element stabilizes the conjugate formation between CTL and target cells, even if the expression of CD2 alone or MHC alone is not sufficient for the stable conjugate formation. Furthermore, they show that CD2/LFA-3 interaction contributes to conjugate formation independently of the interaction between TCR and Ag/MHC. The fact that only the MHC-restricting element, and not the irrelevant MHC molecule, contributes to stable conjugate formation in this system suggests that the TCR has a measurable affinity for the correct MHC molecule (i.e., restricting element), even in the absence of peptide antigen.

The K_d between CD2 and LFA-3 has been shown to be only 0.4 μM (Sayre et al., 1989; Recny et al., 1990). This might appear to be lower than that required for stable conjugate formation, particularly since no significant changes in the affinity between CD2 and LFA-3 have been observed after TCR triggering (Moingeon et al., 1991). However, immunofluorescence microscopic analysis revealed that CD2, but not TCR, rapidly reorganized to the site of conjugate formation between CTL and target cells (Koyasu et al., 1990; Moingeon et al., 1991). This reorganization was observed even in the absence of peptide antigen, indicating that the reorganization does not require an activation process through TCR. The reorganization of CD2 into the site of conjugate formation which occurs as a consequence of CD2/LFA-3 binding likely permits multimeric interaction at the cell–cell junction that enhances the avidity of the CTL for their target cells, despite the low-affinity interaction between monomeric CD2 and LFA-3.

The other major T-cell adhesion system, LFA-1/ICAM-1, utilizes a different mechanism to stablize the conjugate formation. The interaction between LFA-1 and ICAM-1 is of low affinity before T-cell activation, but the affinity increases dramatically after TCR triggering (Dustin and Springer, 1989; van Kooyk et al., 1989). This affinity conversion involves the phosphorylation of the cytoplasmic domain of the β subunit of LFA-1 (Hibbs et al., 1991). In contrast to the CD2/LFA-3 system, no significant reorganization of LFA-1 molecules was observed after TCR triggering (Moingeon et al., 1991). Although such reorganization of LFA-1 has been reported (Kupfer and Singer, 1989), the fraction of conjugates with redistribution was very low, and LFA-1 molecules were localized in pseudopodia-like structure in such cases (Moingeon et al., 1991). Collectively, two adhesion pathways play complementary roles in the interaction between CTL and target cells. It is likely that CD2/LFA-3 interaction in conjunction with TCR/MHC interaction initiates the conjugate formation between CTL and target cells, and that subsequent TCR stimulation activates the LFA-1-mediated adhesion system, thereby optimizing both T-cell activation and the effector function of CTL.

Signal Transduction Function of CD2

Perturbation of the adhesion domain of CD2 by LFA-3 or certain mAb reactive with the adhesion segment, including anti-T11$_2$ in conjunction with an antibody to a more membrane-proximal epitope such as anti-T11$_3$, results in T-cell stimulation (Meuer et al., 1984b; Brottier et al., 1985; Huet et al., 1986; Yang et al., 1986; Alcover et al., 1987; Hunig et al., 1987: Bierer et al., 1988b). A combination of anti-T11$_2$ and anti-T11$_3$ mAb induces a series of activation events indistinguishable from that induced by TCR stimulation. These events include phosphatidylinositol (PI) turnover, calcium mobilization, activation of protein tyrosine kinase(s) (PTK), and IL-2 gene activation (Pantaleo et al., 1987; Alcover et al., 1988a; Samelson et al., 1990a). Calcium mobilization involves both the release of intracellular calcium and the opening of voltage-insensitive plasma membrane calcium channels (Alcover et al., 1986; Gardner et al., 1989). Importantly, such activation also induces cytotoxic activity of CD2$^+$ CD3$^+$ CTL and CD2$^+$ CD3$^-$ natural killer (NK) cells (Siliciano et al., 1985; Bolhuis et al., 1986; Scott et al., 1989).

The signal transduction function of CD2 is dependent on its cytoplasmic tail (Bierer et al., 1988b; He et al., 1988; Chang et al., 1989; 1990). Truncation analysis as well as site-directed mutagenesis has pinpointed a region between amino acid residues 253 and 287 which is necessary for calcium mobilization and IL-2 gene induction (Chang et al., 1989). Furthermore, disruption of a single histidine and arginine at positions 264 and 265 within the first of two tandem PPPGHR motifs can abrogate or dramatically reduce the T-cell activation process (Chang et al., 1990). Activation of T cells through CD2 is also dependent on other cellular structures, namely TCR (Alcover et al., 1987). T-cell variants lacking TCR because of a mutation in Tiβ subunit cannot be stimulated through CD2, but the signal transduction function can be restored by transfection of a cDNA encoding the Tiβ subunit, leading to the surface expression of TCR (Alcover et al., 1988b). Two other indirect pieces of evidence also support this notion in T cells. First, prior crosslinking of TCR inhibits subsequent CD2-mediated activation (Fox et al., 1985; Breitmeyer et al., 1987; Pantaleo et al., 1987). Second, submitogenic concentrations of anti-TCR mAb and anti-CD2 mAb, which independently fail to activate T cells, in combustion induce T-cell activation (Yang et al., 1986). The importance of TCR expression for CD2 signal transduction function has been further emphasized by the failure of CD2 to trigger activation events when transfected and expressed in *Spodoptera frugiperda* worm gut epithelial cells or mouse fibroblasts (Alcover et al., 1988b; Clipstone and Crumpton, 1988).

Despite the dependence of CD2 function on TCR surface expression in T cells, both thymocytes and NK cells lacking TCR can be triggered upon CD2 stimulation (Siliciano et al., 1985; Fox et al., 1985; Bolhuis et al., 1986; Scott et al., 1989). A likely candidate that functionally interacts with the CD2 pathway in these $CD2^+CD3^-$ NK cells is CD16 (FcγRIII). CD16 is known to be physically associated with CD3ζ (Anderson et al., 1989; Lanier et al., 1989) and is also associated with the γ subunit of FcεRI (FcεRIγ) (Hibbs et al., 1989). The latter is a component of the high-affinity IgE receptor expressed on mast cells and basophils (Blank et al., 1989) and is now known to be present in various tissues (Ra et al., 1989). FcεRIγ has significant structural homology to CD3ζ and CD3η and seems to belong the same gene family (Huppi et al., 1988; Weissman et al., 1988b; Blank et al., 1989; Ra et al., 1989; Jin et al., 1990a; Clayton et al., 1991).

The facts that TCR contain CDζ/η subunits, and that CD16 and FcεRI contain FcεRIγ, raised the possibility that CD2 signal transduction is linked to either the CD3ζ/η or FcεRIγ structure. Two lines of evidence support this hypothesis. First, CD2 can transmit signals to induce IL6 production as well as histamine release in mast cells transfected with human CD2 (Arulanandam et al., 1991). Mast cells express FcεRI and FcγRIII but do not express TCR or CD3ζ/η, indicating that expression of TCR or CD3ζ/η is not essential for CD2 signal transduction function in mast cells. Second, transfection of CD16 into the Tiβ^-TCR$^-$ T-cell line described above restored CD2 signaling function in the absence of FcεRIγ (Moingeon et al., 1992). It is thus likely that CD2 transmits signals through Fc receptors in the absence of TCR, and that this mechanism is operative in $CD2^+CD3^-$ NK cells and thymocytes.

Structure and Function of the TCR Complex on CTL

The TCR is composed of a clonotypic Ti heterodimer and CD3 subunits. The CD3γ, CD3δ, and CD3ε subunits are structurally related polypeptides encoded by closely linked genes located on human chromosome 11 and mouse chromosome 9 and exist primarily as monomeric components in the TCR (van den Elsen et al., 1985; Krissansen et al., 1986; Gold et al., 1987). In contrast, the structurally distinct CD3ζ and CD3η molecules are encoded on chromosome 1 in both human and mouse and are derived from the same gene locus by alternative splicing events (Weissman et al., 1988a, 1988b; Baniyash et al., 1989; Jin et al., 1990a; Ohno and Saito, 1990; Clayton et al., 1991). The CD3ζ and CD3η subunits form disulfide-linked homo- or heterodimers within the CD3-Ti complex, thereby adding complexity to the TCR structure (Baniyash et al., 1988a; Clayton et al., 1990; Bauer et al., 1991; Koyasu et al., 1991a). Furthermore, recent studies have shown that a fraction of CD3ε subunits exists as disulfide-linked dimers possibly within a single TCR complex, or linking two TCR complexes (Jin et al., 1990b). In addition, other studies suggest that individual TCR complexes likely contain two CD3ε monomers (Blumberg et al., 1990; Koning et al., 1990). Several studies raise the possibility that CD3γ and CD3δ may exist in separate TCR complexes (Alarcon et al., 1991; Kappes and Tonegawa, 1991). Immunoprecipitation of surface TCR with anti-CD3γ, co-precipitated Tiα-β, CD3ε, and CD3ζ/η, but not CD3δ and reciprocally anti-CD3δ failed to co-precipitate CD3γ but identified Tiα-β, CD3ε, and CD3ζ/η. Transfection analysis also demonstrated that TCR can be expressed on HeLa cells by transfection of Tiα, Tiβ, CD3ε, CD3ζ/η, and either CD3γ or CD3δ. Thus the precise stoichiometry of TCR is yet to be defined, but clearly several possibilities have to be considered, including Tiα-βCD3$\gamma\delta\varepsilon_2\zeta/\eta$, Ti$\alpha$-$\beta$CD3$\gamma_2\varepsilon_2\zeta/\eta$, Ti$\alpha$-$\beta$CD3$\delta_2\varepsilon_2\zeta/\eta$, and (Ti$\alpha$-$\beta$)$_2$CD3$\gamma\delta\varepsilon_2\zeta/\eta$ etc. To avoid confusion, we refer to only one of the possible models, namely Tiα-βCD3$\gamma\delta\varepsilon_2\zeta/\eta$, in this chapter.

Because the CD3 subunits possess extensive cytoplasmic tails relative to the Tiα and Tiβ subunits, they are almost certainly important components of the signal transduction machinery. In particular, a critical role of CD3ζ in signal transduction as well as regulation of surface TCR expression has been documented (Weissman et al., 1989; Frank et al., 1990). As described above, FcεRIγ is also encoded on the same chromosome (mouse chromosome 1), suggesting that CD3ζ/η and FcεRIγ are derived from a common ancestral gene (Baniyash et al., 1989; Huppi et al., 1989; Kuster et al., 1990). Here we focus on the structure and signal transduction function of TCR isoforms generated by the differential usages of members of the CD3ζ/η-FcεRIγ family.

Regulation of CD3ζ and CD3η Expression

Most of the TCR contain CD3ζ/η components and thus are composed of either Tiα-βCD3$\gamma\delta\varepsilon_2\zeta$-$\zeta$ or Tiα-βCD3$\gamma\delta\varepsilon_2\zeta$-$\eta$ complexes (Baniyash et al., 1988a; Clayton et al., 1990, 1991). It is not known whether helper T cells and CTL utilize different isoforms. Transfection studies show that a high level of CD3η expression results in the expression of Tiα-βCD3$\gamma\delta\varepsilon\eta$-$\eta$, but this isoform is rarely observed in normal tissues, implying that it is present in low amounts and/or expressed in rare cell types (Bauer et al., 1991). The expression of CD3ζ and CD3η is apparently coordinate during T-cell development (Clayton et al., 1991). Both CD3ζ and CD3η mRNA expression appears after day 14 and before or on day 16 of fetal murine gestation. The steady-state level of CD3ζ mRNA is always $\geq$ 40 to 60-fold that of CD3η mRNA in any cell population tested. Where in immature CD4$^+$CD8$^+$CD3^{low} double-positive thymocytes and CD4$^+$CD8$^-$CD3^{high} or CD4$^-$CD8$^+$CD3^{high} single-positive thymocytes, the respective steady-state CD3ζ and CD3η mRNA levels are equivalent, the CD3ζ and CD3η mRNA levels in splenic T cells are at least one order of magnitude lower than those in thymocytes. The same discordance in mRNA levels in thymocytes and splenic T cells is found in other TCR subunits. The mRNA levels of Tiα, Tiβ, CD3γ, CD3δ, and CD3ε are also higher in thymocytes than in splenic T cells (Maguire et al., 1990). Despite the distribution of steady-state CD3ζ and CD3η mRNA noted above, the levels of both CD3ζ and CD3η proteins increase during T-cell development (Clayton et al., 1991). Thus, the amount of CD3ζ and CD3η proteins is lowest in CD4$^+$CD8$^+$CD3^{low} double-positive thymocytes and highest in CD3^{high} single-positive thymocytes and splenic T cells. The amount of receptor-associated CD3ζ and CD3η proteins in double-positive thymocytes is $\sim$ 10-fold less than in single-positive thymocytes and splenic T cells. Furthermore, discordance between mRNA and protein levels for CD3ζ and CD3η is also observed in splenic T cells which express amounts of CD3ζ and CD3η proteins equivalent to those of single-positive thymocytes, whereas the steady-state level of CD3ζ and CD3η mRNAs is one order of magnitude less in the same splenic T cells than those of single-positive thymocytes. Thus, posttranscriptional as well as transcriptional regulatory mechanisms control CD3ζ and CD3η expression during T-cell development.

Transfectants expressing only CD3η-η always express lower surface TCR than those expressing CD3ζ-ζ (Clayton et al., 1990; Bauer et al., 1991). CD3ζ and CD3η double transfectants which express lower amounts of CD3η mRNA express CD3ζ-ζ and CD3ζ-η, but not CD3η-η, while those expressing high levels of CD3η mRNA express CD3ζ-ζ, CD3ζ-η, and CD3η-η dimers in association with the other TCR experiments. Based on these observations, we have suggested that the preference for subunit assembly into stable TCR complexes is likely to be CD3ζ-ζ > CD3ζ-η > CD3η-η. Perhaps this occurs as a consequence of differences in affinities of the different domains for the incomplete Tiα-βCD3$\gamma\delta\varepsilon_2$ complex or their ability to salvage partially assembled TCR complexes from degradation. In addition, it should be noted that the intrinsic stability of these protein dimers likely affects the steady-state level of the TCR isoforms. In this regard, the stabilities of the different dimers are dissimilar when examined by pulse-chase experiments in a transfectant expressing all three TCR isoforms. The half-lives of CD3ζ-ζ, CD3ζ-η, and CD3η-η were observed to be > 4, 3, and 1.5 hr, respectively (Koyasu et al., 1991a). The relatively low abundance of CD3η protein in most T cells, as well as its relatively short half-life as a homodimer, may account for the general lack of detectable CD3η-η homodimers in bulk populations of thymocytes, splenic T cells, and T cell lines where the level of CD3ζ mRNA clearly dominates over that of CD3η mRNA.

Signal Transduction Through TCR Isoforms

Transfection studies with CD3ζ and/or CD3η cDNAs into a variant of a T-cell hybridoma lacking CD3ζ and CD3η revealed that all the resulting TCR isoforms transmitted signals indistinguishable with regard to Ca^{2+} mobilization, PI turnover, and IL-2 production resulting from TCR stimulation by anti-Ti, anti-CD3 mAbs, or Ag/MHC (Bauer et al., 1991). Although direct mobilization of intracellular free Ca^{2+} and activation of PKC can function in lieu of TCR triggering to activate IL-2 production (Kaibuchi et al., 1985; Truneh et al., 1985; Koyasu et al., 1987), recent studies suggest that a PTK pathway seems critical for IL-2 production. The time course of the PTK activation is more rapid than either PI turnover or Ca^{2+} mobilization, suggesting an important role for the PTK activation pathway at an early stage of T cell activation (Hsi et al., 1989; June et al., 1990a). Consistent with this notion are reports showing that PTK inhibitors affected IL-2 production (June et al., 1990b; Mustelin et al., 1990; Stanley et al., 1990). Recent studies have shown that TCR stimulation induces the tyrosine phosphorylation of phospholipase C (PLC)-γ1 (Park et al., 1991; Secrist et al., 1991; Weiss et al., 1991). Since tyrosine phosphorylation of PLC-γ activates its enzymatic activity (Nishibe et al., 1990; Kim et al., 1991), the PTK pathway is likely the primary step to induce PI turnover. It is known that the TCR is physically associated with p59^{fyn} PTK (Samelson et al., 1990b) as CD4 and CD8 are associated with p56^{lck} (Rudd et al., 1988; Viellette et al., 1988, 1989). It is thus likely that crosslinking of TCR as well as CD4/CD8 by Ag/MHC results in the activation of these PTK, which triggers a series of signal transduction events. The importance of tyrosine phosphorylation was also shown in the activation of NK cell function (Einspahr et al., 1991).

Whereas TCR stimulation results in tyrosine phosphorylation of a common set of proteins in all the above transfectants, thus indicating linkage of each TCR isoform to PTK pathways, tyrosine phosphorylation of the CD3ζ subunit but not the CD3η subunit was observed (Baniyash et al., 1988b; Bauer et al., 1991). The lack of tyrosine phosphorylation of the CD3η subunit is not due to the absence of PTK in CD3η-containing TCR isoforms, since association of p56^{fyn} is observed in CD3η-containing TCR isoforms as well as CD3ζ-containing TCR isoforms (Koyasu et al., 1992b). The clear differential in tyrosine phosphorylation between CD3ζ and CD3η subunits stands in marked contrast to the ability of all CD3ζ/η isoforms (CD3ζ-ζ, CD3ζ-η, and CD3η-η) to transmit other early signals, such as Ca^{2+} mobilization and PI turnover as well as the more complex responses required for the subsequent IL-2 gene activation upon antigen/MHC stimulation. Thus, it is unlikely that CD3ζ phosphorylation directs the primary signaling of these responses. The observed slowness of the CD3ζ phosphorylation kinetics relative to other substrates (June et al., 1990a) also supports this notion. Rather, given the regulatory role of tyrosine phosphorylation in many receptor systems (Hunter and Cooper, 1985; Sibley et al., 1987; Ullrich and Schlessinger, 1990), phosphorylation of CD3ζ could subsequently affect stimulation of these same pathways or modify other transduction signals. One possible function of CD3ζ phosphorylation is to desensitize TCR triggering by antigen/MHC. If so, TCR expressing the CD3η subunit will not be subjected to the same effects as TCR expressing CD3ζ. Differential expression of TCR isoforms during T lineage development or in T-cell subsets could thereby qualitatively modify the cellular response. In this context, it is worth noting again that in contrast to the situation in mature T cells, a fraction of CD3ζ subunits are constitutively phosphorylated on at least one tyrosine residue in thymocytes (Nakayama et al., 1989). Such a difference could account for functional distinctions in the responsiveness of thymocytes as opposed to mature T cells in TCR crosslinking, as shown previously (Ramarli et al., 1987).

Further Complexities in TCR Isoforms

Recent molecular cloning of the FcεRIγ subunit identified this receptor subunit as another member of the CD3ζ/η gene family (Kuster et al., 1990). Structural and functional similarity between FcεRIγ and CD3ζ was further shown by the ability of CD3ζ to complement the formation of a high-affinity IgE receptor (FcεRI) on *Xenopus* oocytes by injecting mRNAs of FcεRIα, FcεRIβ, and CD3ζ in the absence of FcεRIγ (Howard et al., 1990). FcεRIγ is distributed in cells of a number of different lineages, including mast cells, basophils, macrophages, and NK cells. Importantly, recent studies have shown that FcεRIγ can coassociate to form homo- or heterodimeric structures with CD3ζ and CD3η in certain TCR complexes.

The mouse CTL line, CTLL, expresses TCR of at least four different isoforms (Orloff et al., 1990), including Tiα-βCD3γδε$_2$ζ-ζ, Tiα-βCD3γδε$_2$ζ-η, Tiα-βCD3γδε$_2$ζ-FcεRIγ, and Tiα-βCD3γδε$_2$η-FcεRIγ. In addition, TCRs containing FcεRIγ homodimers in lieu of CD3ζ/η have also been observed in certain mouse T-cell populations (Koyasu et al., 1992a). Large granular lymphocytes (LGL) can be induced by cultivation of total splenocytes with IL-2 (Koyasu et al., 1991b). Such cells show a CD2$^+$, CD3/Tiα-β$^+$, CD4$^-$, CD8$^-$, CD16$^+$, NK1.1$^+$ surface phenotype and represent approximately 0.5% of the splenocyte population (Levitsky et al., 1991). These LGL display NK-like cytotoxic activity and are cytotoxic for B-cell hybridomas producing anti-CD3ε and anti-CD16 mAb, demonstrating the signaling capacity of both TCR and CD16 (Koyasu et al., 1992a). They express CD3ζ and CD3η mRNAs of normal size but do not express CD3ζ/η proteins. The TCR of this subset contain disulfide-linked homodimers of FcεRIγ in lieu of CD3ζ/η proteins, which is similar to the discordance between mRNAs and proteins during T-cell development (Clayton et al., 1991). Thus, both TCR and CD16 complexes, which are mutually exclusive, with rare exceptions, in the human system (Perussia et al., 1984; Lanier et al., 1985), share the same FcεRIγ subunit as a component likely involved in signal transduction. Presumably, such signaling is important for the cytotoxic function of the LGL population.

Although it is not known whether FcεRIγ-containing TCR possesses different signal transduction properties than "conventional" TCR, various TCR isoforms consisting of CD3ζ, CD3η, and FcεRIγ may play an important role in T-cell function. Because transfection of the FcεRIγ cDNA into CD3ζ$^-$ η$^-$ MA5.8 cells leads to surface expression of a Tiα-βCD3γδε$_2$FcεRIγ-γ TCR which transmits signals resulting in IL-2 production (Rodewald et al., 1991) without imparting cytotoxic activity to the transfectant (A.R.N. Arulanandam, personal communication), it is clear that FcεRIγ itself is not uniquely linked to the cytotoxic program of cells. As described above, differential usage of CD3γ and CD3δ subunits in TCR complexes creates another level of TCR isoforms. It should also be examined whether such different TCR isoforms have distinct signal transduction mechanisms in T-cell function.

It is noteworthy that members of the CD3ζ/η-FcεRIγ family can dimerize differentially in other receptor complexes. For example, human NK cells express CD3ζ as well as FcεRIγ in association with CD16 in the absence of other TCR components (Tiα, Tiβ, CD3γ, CD3δ, CD3ε) (Anderson et al., 1989, 1990; Lanier et al., 1989). Although it is not known whether CD3η is expressed in these cells or forms a complex with CD16, it seems likely that members of this gene family may play an important role in various types of cells. By analogy with the TCR, there may exist different isoforms of CD16. Again, it will be important to determine if different signal transduction pathways occur through these putative CD16 isoforms. Furthermore, there is a possibility that there are additional members of the CD3ζ/η-FcεRIγ gene family and that these structures participate in such related and other receptor systems. The functional attributes of the various CD3ζ/η-FcεRIγ dimers are now critical to ascertain.

Perspectives

Here we have discussed the role of the CD2 molecule in both the process of mitogenic recognition and the triggering function of CTL. In addition, the structure and signaling activation of TCR isoforms generated by alternative usage of CD3ζ/η-FcεRIγ dimers and their functional linkage to CD2 are reviewed. The roles of the CD2 adhesion function and the signal transduction function in T-cell development are yet to be defined. However, since CD2 is expressed on virtually all thymocytes, including CD4$^+$CD8$^+$ double-positive cortical thymocytes (Reinherz and Scholssman, 1980), which are subject to the thymic selection process, it is likely that CD2/LFA-3 interaction will modulate the process of selection by affecting the overall avidity of the thymocyte for the thymic epithelial cells or APC. Analysis of the role of CD2/LFA-3 interaction in modulating T-cell repertoire selection should provide a fertile area for further research.

Differential expression of the members of the CD3ζ/η-FcεRIγ gene family might quantitatively modify signal transduction events not only in the effector phase but also during the course of development. Various developmental programs of the thymic selection process might be dependent on such differential utilization of receptor signaling units. Thus, overexpression of the CD3ζ/η-FcεRIγ family in transgenic mice or production of CD3ζ/η-FcεRIγ family-deficient mice by gene targeting with homologous recombination in the embryonic stem cell system (Frohman and Martin, 1989) should result in an alteration in T-cell development and/or thymic selection processes if the CD3ζ/η-FcεRIγ family plays an important role in these events. These experimental approaches will provide a definitive answer regarding the function of these proteins *in vivo*.

References

Alarcon B, Ley SC, Sanchez-Madrid F, Blumberg RS, Ju ST, Fresno M, Terhorst C (1991): The CD3-γ and CD3-δ subunits of the T cell antigen receptor can be expressed within distinct functional TCR/CD3 complexes. *EMBO J* 10: 903–912

Alcover A, Alberini C, Acuto O, Clayton LK, Transy C, Spagnoli G, Moingeon P, Lopez P, Reinherz EL (1988a): Interdependence of T3-Ti and T11 activation pathways in human T lymphocytes. *EMBO J* 7: 1973–1977

Alcover A, Chang H-C, Sayre PH, Hussey RE, Reinherz EL (1988b): The T11 (CD2) cDNA encodes a transmembrane protein which expresses T11$_1$, T11$_2$ and T11$_3$ epitopes but which does not independently mediate calcium influx: Analysis by gene transfer in a baculovirus system. *Eur J Immunol* 18: 363–367

Alcover A, Ramarli D, Richardson NE, Chang H-C,

Reinherz EL (1987): Functional and molecular aspects of human T lymphocyte activation via T3-Ti and T11 pathways. *Immunol Rev* 95: 5–36

Anderson P, Caligiuri M, O'Brien C, Manley T, Ritz J, Schlossman SF (1990): Fcγ receptor type III (CD16) is included in the ζ NK receptor complex expressed by human natural killer cells. *Proc Natl Acad Sci USA* 87: 2274–2278

Anderson P, Caligiuri M, Ritz J, Schlossman SF (1989): CD3-negative natural killer cells express ζ TCR as part of a novel molecular complex. *Nature* 341: 159–162

Arulanandam ARN, Koyasu S, Reinherz EL (1991): T cell receptor-independent CD2 signal transduction in FcR+ cells. *J Exp Med* 173: 859–868

Ashwell JD, Klausner RD (1990): Genetic and mutational analysis of the T cell antigen receptor. *Ann Rev Immunol* 8: 139–167

Baniyash M, Garcia-Morales P, Bonifacino JS, Samelson LE, Klausner RD (1988a): Disulfide linkage of the ζ and η chains of the T cell receptor: Possible identification of two structural classes of receptors. *J Biol Chem* 263: 9874–9878

Baniyash M, Garcia-Morales P, Luong E, Samelson LE, Klausner RD (1988b): The T cell antigen receptor ζ chain is tyrosine phosphorylated upon activation. *J Biol Chem* 263: 18225–18230

Baniyash M, Hsu VW, Seldin MF, Klausner RD (1989): The isolation and characterization of the murine T cell antigen receptor ζ chain gene. *J Biol Chem* 264: 13252–13257

Bauer A, McConkey DJ, Howard FD, Clayton LK, Novick D, Koyasu S, Reinherz EL (1991): Differential signal transduction via T cell receptor $CD3\zeta_2$, $CD3\zeta\text{-}\eta$ and $CD3\eta_2$ isoforms. *Proc Natl Acad Sci USA* 88: 3842–3846

Bernard A, Gelin C, Raynal B, Pham D, Gosse C, Boumsell L (1982): Phenomenon of human T cells rosetting with sheep erythrocytes analyzed with monoclonal antibodies: "Modulation" of partially hidden epitope determining the conditions of interaction between T cells and erythrocyes. *J Exp Med* 155: 1317–1333

Bierer BE, Peterson A, Barbosa J, Seed B, Burakoff SJ (1988a): Expression of the T-cell surface molecule CD2 and an epitope-loss CD2 mutant to define the role of lymphocyte function-associated antigen 3 (LFA-3) in T-cell activation. *Proc Natl Acad Sci USA* 85: 1194–1198

Bierer BE, Peterson A, Gorga JC, Herrmann SH, Burakoff SJ (1988b): Synergistic T cell activation via the physiological ligands for CD2 and the T cell receptor. *J Exp Med* 168: 1145–1156

Blank U, Ra C, Miller L, White K, Metzger H, Kinet J-P (1989): Complete structure and expression in transfected cells of high affinity IgE receptor. *Nature* 337: 187–189

Blumberg RS, Ley S, Sancho J, Lonberg N, Lacy E, McDermott F, Schad V, Greenstein JL, Terhorst C (1990): Structure of the T-cell antigen receptor: Evidence for two CD3 ε subunits in the T-cell receptor-CD3 complex. *Proc Natl Acad Sci USA* 87: 7220–7224

Bolhuis RLH, Roozemond RC, van de Griend RJ (1986): Induction and blocking of cytolysis in $CD2^+$, $CD3^-$ NK and $CD2^+$, $CD3^+$ cytotoxic T lymphocytes via CD2 50KD sheep erythrocyte receptor. *J Immunol* 136: 3939–3944

Breitmeyer JB, Daley JF, Levine HB, Schlossman SF (1987): The T11 (CD2) molecule is functionally linked to the T3/Ti T cell receptor in the majority of T cells. *J Immunol* 139: 2899–2905

Brenner MB, Strominger JL, Krangel MS (1988): The γδ T cell receptor. *Adv Immunol* 43: 133–192

Brottier P, Boumsell L, Gelin C, Bernard A (1985): T cell activation via CD2 (T, gp50) molecules: Accessory cells are required to trigger T cell activation via CD2-D66 plus CD2-9.6/T11₁ epitopes. *J Immunol* 135: 1624–1631

Chang H-C, Moingeon P, Lopez P, Krasnow H, Stebbins C, Reinherz EL (1989): Dissection of the human CD2 intracellular domain: Identification of a segment required for signal transduction and interleukin 2 production. *J Exp Med* 169: 2073–2083

Chang H-C, Moingeon P, Pederson R, Lucich J, Stebbins C, Reinherz EL (1990): Involvement of the PPPGHR motif in T cell activation via CD2. *J Exp Med* 172: 351–355

Clayton LK, Bauer A, Jin Y-J, D'Adamio L, Koyasu S, Reinherz EL (1990): Characterization of thymus-derived lymphocytes expressing Tiα-βCD3γδεζ-ζ, Tiα-βCD3γδεη-η or Tiα-βCD3γδεζ-ζ/ζ-η antigen receptor isoforms: Analysis by gene transfection. *J Exp Med* 172: 1243–1253

Clayton LK, D'Adamio L, Howard FD, Sieh M, Hussey RE, Koyasu S, Reinherz EL (1991): CD3η and CD3ζ are alternatively spliced products of a common genetic locus and are transcriptionally and post-transcriptionally regulated during T-cell development. *Proc Natl Acad Sci USA* 88: 5202–5206

Clayton LK, Sieh M, Pious DA, Reinherz EL (1989) Identification of human CD4 residues affecting class II MHC versus HIV-1 gp120 binding. *Nature* 339: 548–551

Clevers H, Alarcon B, Wileman T, Terhorst C (1988): The T cell receptor/CD3 complex: A dynamic protein ensemble. *Ann Rev Immunol* 6: 629–662

Clipstone NA, Crumpton MJ (1988): Stable expression of the cDNA encoding the human T lymphocyte-specific CD2 antigen in murine L cells. *Eur J Immunol* 18: 1541–1545

Davis MM, Bjorkman PJ (1988): T cell antigen receptor genes and T cell recognition. *Nature* 334: 395–402

Denning SM, Tuck DT, Vollger LW, Springer TA, Singer KH, Haynes BF (1987): Monoclonal antibodies to CD2 and lymphocyte function-associated antigen 3 inhibit human thymic epithelial cell-dependent mature thymocyte activation. *J Immunol* 139: 2573–2578

Diamond DJ, Clayton LK, Sayre PH, Reinherz EL (1988): Exon-intron organization and sequence comparison of human and murine T11 (CD2) genes. *Proc Natl Acad Sci USA* 85: 1615–1619

Doyle C, Strominger JL (1987): Interaction between CD4 and class II MHC molecules mediates cell adhesion. *Nature* 330: 256–259

Dustin ML, Springer TA (1989): T-cell receptor cross-linking transiently stimulates adhesiveness through LFA-1. *Nature* 341: 619–624

Einspahr KJ, Abraham RT, Binstadt BA, Uehara Y, Leibson PJ (1991): Tyrosine phosphorylation provides an early and requisite signal for the activation of natural killer cell cytotoxic function. *Proc Natl Acad Sci USA* 88: 6279–6283

Fox DA, Hussey RE, Fitzgerald KA, Bensussan A, Daley JF, Schlossman SF, Reinherz EL (1985): Activation of human thymocytes via the 50KD T11 sheep erythrocyte binding protein induces the expression of interleukin 2 receptors on both T3$^+$ and T3$^-$ populations. *J Immunol* 134: 330–335

Frank SJ, Niklinska BB, Orloff DG, Mercep M, Ashwell JD, Klausner RD (1990): Structural mutations of the T cell receptor ζ chain and its role in T cell activation. *Science* 249: 174–177

Frohman MA, Martin GR (1989): Cut, paste, and save: New approaches to altering specific genes in mice. *Cell* 56: 145–147

Gabert J, Langlet C, Zamoyska R, Parnes JR, Schmitt-Verhulst A-M, Malissen B (1987): Reconstitution of MHC class I specificity by transfer of the T cell receptor and Lyt-2 genes. *Cell* 50: 545–554

Gardner P, Alcover A, Kuno M, Moingeon P, Weyland CM, Goronzy J, Reinherz EL (1989): Triggering of T-lymphocytes via either T3-Ti or T11 surface structures opens a voltage-insensitive plasma membrane calcium-permeable channel: Requirement for interleukin-2 gene function. *J Biol Chem* 264: 1068–1076

Gay D, Maddon P, Sekaly R, Talle MA, Godfrey M, Long E, Goldstein G, Ches L, Axel R, Kappler J, Marrack P (1987): Functional interaction between human T-cell protein CD4 and the major histocompatibility complex HLA-DR antigen. *Nature* 328: 626–629

Gold DP, van Dongen JJM, Morton CC, Bruns GAP, van den Elsen P, van Kessel AHMG, Terhorst C (1987): The gene encoding the epsilon subunit of the T3/T cell receptor complex maps to chromosome 11 in humans and to chromosome 9 in mice. *Proc Natl Acad Sci USA* 84: 1664–1668

Greenstein JL, Kappler J, Marrack P, Burakoff SJ (1984): The role of L3T4 in recognition of Ia by a cytotoxic, H-2D^d-specific T cell hybridoma. *J Exp Med* 159: 1213–1224

He Q, Beyers AD, Barclay AN, Williams AF (1988): A role in transmembrane signaling for the cytoplasmic domain of the CD2 T lymphocyte surface antigen. *Cell* 54: 979–984

Hibbs ML, Selvaraj P, Carpen O, Springer TA, Kuster H, Jouvin M-HE, Kinet J-P (1989): Mechanisms for regulating expression of membrane isoforms of FcγRIII (CD16). *Science* 246: 1608–1611

Hibbs ML, Xu H, Stacker SA, Springer TA (1991): Regulation of adhesion of ICAM-1 by the cytoplasmic domain of LFA-1 integrin beta subunit. *Science* 251: 1611–1613

Howard FD, Ledbetter JA, Wong J, Bieber CP, Stinson EB, Herzenberg LA (1981): A human T lymphocyte differentiation marker defined by monoclonal antibodies that block E-rosette formation. *J Immunol* 126: 2117–2122

Howard FD, Rodewald H-R, Kinet J-P, Reinherz EL (1990): CD3ζ subunit can substitute for the γ subunit of Fcε receptor type I in assembly and functional expression of the high affinity IgE receptor: Evidence for interreceptor complementation. *Proc Natl Acad Sci USA* 87: 7015–7019

Hsi ED, Siegel JN, Minami Y, Luong ET, Klausner RD, Samelson LE (1989): T cell activation induces rapid tyrosine phosphorylation of a limited number of cellular substrates. *J Biol Chem* 264: 10836–10842

Huet S, Wakasugi H, Sterkers G, Gilmour J, Tursz T, Boumsell L, Bernard A (1986): T cell activation via CD2 (T, gp50): The role of accessory cells in activating resting T cells vai CD2. *J Immunol* 137: 1420–1428

Hunig T, Mitnacht R, Tiefenthaler G, Kohler C, Miyasaka M (1986): T11TS, The cell surface molecule binding to the "erythrocyte receptor" of T lymphocytes: Cellular distribution, purification to homogeneity and biochemical properties. *Eur J Immunol* 16: 1615–1621

Hunig T, Tiefenthaler G, zum Buschenfelde K-HM, Meuer SC (1987): Alternative pathway activation of T cells by binding of CD2 to its cell-surface ligand. *Nature* 326: 298–301

Hunter T, Cooper JA (1985): Protein-tyrosine kinases. *Ann Rev Biochem* 54: 897–930

Huppi K, Mock BA, Hilgers J, Kochan J, Kinet J-P (1988): Receptors for Fcε and Fcγ are linked on mouse chromosome 1. *J Immunol* 141: 2807–2810

Huppi K, Siwarski D, Mock BA, Kinet J-P (1989): Gene mapping of the three subunits of the high affinity FcR for IgE to mouse chromosomes 1 and 19. *J Immunol* 143: 3787–3791

Jin Y-J, Clayton LK, Howard FD, Koyasu S, Sieh M, Steinbrich R, Tarr GE, Reinherz EL (1990a): Molecular cloning of the CD3η subunit identifies a CD3ζ-related product in thymus-derived cells. *Proc Natl Acad USA* 87: 3319–3323

Jin Y-J, Koyasu S, Moingeon P, Steinbrich R, Tarr GE, Reinherz EL (1990b): A fraction of CD3ε subunits exists as disulfide-linked dimers in both human and murine T lymphocytes. *J Biol Chem* 265: 15850–15853

June CH, Fletcher MC, Ledbetter JA, Samelson LE (1990a): Increases in tyrosine phosphorylation are detectable before phospholipase C activation after T cell receptor stimulation. *J Immunol* 144: 1591–1599

June CH, Fletcher MC, Ledbetter JA, Schieven GL, Siegel JN, Phillips AF, Samelson LE (1990b): Inhibition of tyrosine phosphorylation prevents T cell receptor-mediated signal transduction. *Proc Natl Acad Sci USA* 87: 7722–7726

Kaibuchi K, Takai Y, Nishizuka Y (1985): Protein kinase

C and calcium ion in mitogenic response of macrophage-depleted human peripheral lymphocytes. *J Biol Chem* 260: 1366–1369

Kamoun M, Martin PJ, Hansen JA, Brown MA, Siadak AW, Nowinski RC (1981): Identification of a human T lymphocyte surface protein associated with the E-rosette receptor. *J Exp Med* 153: 207–212

Kappes DJ, Tonegawa S (1991): Surface expression of alternative forms of the TCR/CD3 complex. *Proc Natl Acad Sci USA* 88: 10169–10173

Kim HK, Kim JW, Zilberstein A, Margoilis B, Kim JG, Schlessinger J, Rhee SG (1991): PDGF stimulation of inositol phospholipid hydrolysis requires PCL-γl phosphorylation on tyrosine residues 783 and 1254. *Cell* 65: 435–441

Koning F, Maloy WL, Coligan JE (1990): The implications of subunit interactions for the structure of the T cell receptor-CD3 complex. *Eur J Immunol* 20: 299–305

Koyasu S, Reinherz EL (1991): Contribution of accessory molecules to MHC/antigen presentation: Role of CD2/LFA-3 interaction in T-cell recognition of nominal antigen: In: *Antigen Processing and Recognition*, McClusky J, ed. Boca Raton, FL: CRC Press, pp. 229–247

Koyasu S, D'Adamio L, Arulanandam ARN, Abraham S, Clayton LK, Reinherz EL (1992a): T cell receptor complexes containing FcεRIγ homodimers in lieu of CD3ζ and CD3η components: A novel isoform expressed on large granular lymphocytes. *J Exp Med* 175: 203–209

Koyasu S, D'Adamio L, Clayton LK, Reinherz EL (1991a): T cell receptor isoforms and signal transduction. *Curr Opinion Immunol* 3: 32–39

Koyasu S, Lawton T, Novick D, Recny MA, Siliciano RF, Wallner BP, Reinherz EL (1990): Role of interaction of CD2 molecule with lymphocyte function-associated antigen 3 in T-cell recognition of nominal antigen. *Proc Natl Acad Sci USA* 87: 2603–2607

Koyasu S, McConkey DJ, Clayton LK, Abrahams S, Yandava B, Katagiri T, Moingeon P, Yamamoto T, Reinherz EL (1992b): Phosphorylation of multiple CD3ζ tyrosine residues leads to formation of pp21 *in vitro* and *in vivo*: Structural changes upon T cell receptor stimulation. *J Biol Chem* 267: 3375–3381

Koyasu S, Suzuki G, Asano Y, Osawa H, Diamantstein T, Yahara I (1987): Signals for activation and proliferation of murine T lymphocyte clones. *J Biol Chem* 262: 4689–4695

Koyasu S, Tagaya Y, Sugie K, Yonehara S, Yodoi J, Yahara I (1991b): The expression of IL-2R α-chain is enhanced by activation of adenylate cyclase in large granular lymphocytes and natural killer cells. *J Immunol* 146: 233–238

Krensky AM, Robbins E, Springer TA, Burakoff SJ (1984): LFA-1, LFA-2 and LFA-3 antigens are involved in CTL-target conjugation. *J Immunol* 132: 2180–2182

Krissansen GW, Owen MJ, Verbi W, Crumpton MJ (1986): Primary structure of the T3 γ subunit of the T3/T cell antigen receptor complex deduced from cDNA sequences: Evolution of the T3 γ and δ subunits. *EMBO J* 5: 1799–1808

Kupfer A, Singer SJ (1989): The specific interaction of helper T cells and antigen presenting B cells. IV. Membrane and cytoskeletal reorganization in the bound T cell as a function of antigen dose. *J Exp Med* 170: 1697–1713

Kuster H, Thompson H, Kinet J-P (1990): Characterization and expression of the gene for the human Fc receptor γ subunit. *J Biol Chem* 265: 6448–6452

Lang G, Wotton D, Owen MJ, Sewell WA, Brown MH, Mason DY, Crumpton MJ, Kioussis D (1988): The structure of the human CD2 gene and its expression in transgenic mice. *EMBO J* 7: 1675–1682

Lanier LL, Kipps TJ, Phillips JH (1985): Functional properties of a unique subset of cytotoxic CD3+ T lymphocytes that express Fc receptors for IgG (CD16/Leu-11 antigen). *J Exp Med* 162: 2089–2106

Lanier LL, Yu G, Phillips JH (1989): Co-association of CD3ζ with a receptor (CD16) for IgG Fc on human natural killer cells. *Nature* 342: 803–805

Levitsky HI, Golumbek PT, Pardoll DM (1991): The fate of CD4$^-$8$^-$ T cell receptor-$\alpha\beta$ + thymocytes. *J Immunol* 146: 1113–1117

Maguire JE, McCarthy S, Singer A, Singer DS (1990): Inverse correlation between steady-state RNA and cell surface T cell receptor levels. *FASEB J* 4: 3131–3134

Marrack P, Kappler J (1987): The T cell receptor. *Science* 238: 1073–1079

Meuer SC, Acuto O, Hercend T, Schlossman SF, Reinherz EL (1984a): The human T cell receptor. *Ann Rev Immunol* 2: 23–50

Meuer SC, Hussey RE, Fabbi M, Fox D, Acuto O, Fitzgerald KA, Hodgdon JC, Protentis JP, Schlossman SF, Reinherz EL (1984b): An alternative pathway of T-cell activation: A functional role for the 50KD T11 sheep erythrocyte receptor protein. *Cell* 36: 897–906

Meuer SC, Schlossman SF, Reinherz EL (1982): Clonal analysis of human cytotoxic T lymphocytes: T4$^+$ and T8$^+$ effector T cells recognize products of different major histocompatibility complex regions. *Proc Natl Acad Sci USA* 79: 4395–4399

Moingeon P, Chang H-C, Wallner BP, Stebbins C, Frey AZ, Reinherz EL (1989): CD2-mediated adhesion facilitates T lymphocyte antigen recognition function. *Nature* 339: 312–314

Moingeon P, Lucich JL, McConkey DJ, Letourneur F, Malissen B, Kochan J, Chang H-C, Rodewald H-R, Reinherz EL (1992): CD3ζ dependence of the CD2 pathway of activation in T Lymphocytes and NK cells. *Proc Natl Acad Sci USA* 89: 1492–1496

Moingeon PE, Lucich JL, Stebbins CC, Recny MA, Wallner BP, Koyasu S, Reinherz EL (1991): Complementary roles for CD2 and LFA-1 adhesion pathways during T cell activation. *Eur J Immunol* 21: 605–610

Mustelin T, Coggeshall KM, Isakov N, Altman A (1990): T cell antigen receptor-mediated activation of phospholipase C requires tyrosine phosphorylation. *Science* 247: 1584–1587

Nagafuchi A, Takeichi M (1988): Cell binding function of E cadherin is regulated by the cytoplasmic domain. *EMBO J* 7: 3679–3684

Nakayama T, Singer A, Hsi ED, Samelson LE (1989): Intrathymic signaling in immature CD4+ CD8+ thymocytes results in tyrosine phosphorylation of the T cell receptor zeta chain. *Nature* 341: 651–653

Nishibe S, Wahl MI, Hernandez-Sotomayor SMT, Tonks NK, Rhee SG, Carpenter G (1990): Increase of the catalytic activity of phospholipase C-γ1 by tyrosine phosphorylation. *Science* 250: 1253–1256

Norment AM, Salter RD, Parham P, Engelhard VH, Littman DR (1988): Cell-cell adhesion mediated by CD8 and MHC class I molecules. *Nature* 336: 79–81

Ohno H, Saito T (1990): CD3ζ and η chains are produced by alternative splicing from a common gene. *Int Immunol* 2: 1117–1119

Orloff DG, Ra C, Frank SJ, Klausner RD, Kinet J-P (1990): Family of disulfide-linked dimers containing the ζ and η chains of the T cell receptor and the γ chain of Fc receptors. *Nature* 347: 189–191

Palacios R, Martinez-Maza O (1982): Is the E receptor on human T lymphocytes a "negative signal receptor"? *J Immunol* 129: 2479–2485

Pantaleo G, Olive D, Poggi A, Kozumbo WJ, Moretta L, Moretta A (1987): Transmembrane signaling via the T11-dependent pathway of human T cell activation. Evidence for the involvement of 1,2-diacylglycerol and inositol phosphates. *Eur J Immunol* 17: 55–62

Park DJ, Rho HW, Rhee SG (1991): CD3 stimulation causes phosphorylation of phospholipase C-γ1 on serine and tyrosine residues in a human T-cell line. *Proc Natl Acad Sci USA* 88: 5453–5456

Parnes JR (1989): Molecular biology and function of CD4 and CD8. *Adv Immunol* 44: 265–311

Perussia B, Trinchieri G, Jackson A, Warne NL, Faust J, Rumpold H, Kraft D, Lanier LL (1984): The Fc receptor for IgG on human natural killer cells: Phenotypic, functional and comparative studies with monoclonal antibodies. *J Immunol* 133: 180–189

Peterson A, Seed B (1987): Monoclonal antibody and ligand binding sites of the T cell erythrocyte receptor (CD2). *Nature* 329: 842–846

Ra C, Jouvin M-HE, Kinet J-P (1989): Complete structure of the mouse mast cell receptor for IgE (FcεRI) and surface expression of chimeric receptors (rat-mouse-human) on transfected cells. *J Biol Chem* 264: 15323–15327

Ramarli D, Fox DA, Reinherz EL (1987): Selective inhibition of interleukin 2 gene function following thymocyte antigen/major histocompatibility complex receptor crosslinking: Possible thymic selection mechanism. *Proc Natl Acad Sci USA* 84: 8598–8602

Ratnofsky SE, Peterson A, Greenstein JL, Burakoff SJ (1987): Expression and function of CD8 in a murine T cell hybridoma. *J Exp Med* 166: 1747–1757

Raulet DH (1989): The structure, function, and molecular genetics of the γ/δ T cell receptor. *Ann Rev Immunol* 7: 175–207

Recny MA, Neidhardt EA, Sayre PH, Ciardelli TL, Reinherz EL (1990): Structural and functional characterization of the CD2 immunoadhesion domain. Evidence for inclusion of CD2 in an α-β protein folding class. *J Biol Chem* 265: 8542–8549

Reinherz EL, Schlossman SF (1980): The differentiation and function of human T lymphocytes. *Cell* 19: 821–827

Richardson NE, Chang H-C, Brown NR, Hussey RE, Sayre PH, Reinherz EL (1988): Adhesion domain of human T11 (CD2) is encoded by a single exon. *Proc Natl Acad Sci USA* 85: 5176–5180

Rodewald H-R, Arulanandam ARN, Koyasu S, Reinherz EL (1991): The high affinity Fcε receptor γ subunit (FcεRIγ) facilitates T cell receptor expresion and antigen/major histocompatibility complex-driven signaling in the absence of CD3ζ and CD3η. *J Biol Chem* 266: 15974–15978

Rudd CE, Trevillyan JM, Dasgupta JD, Wong LL, Schlossman SF (1988): The CD4 receptor is complexed in detergent lysates to a proetin-tyrosine kinase (pp58) from human T lymphocytes. *Proc Natl Acad Sci USA* 85: 5190–5194

Samelson LE, Fletcher MC, Ledbetter JA, June CH (1990a): Activation of tyrosine phosphorylation in human T cells via the CD2 pathway. Regulation by the CD45 tyrosine phosphatase. *J Immunol* 145: 2448–2454

Samelson LE, Phillips AF, Luong ET, Klausner RD (1990b): Association of the fyn protein-tyrosine kinase with the T cell antigen receptor. *Proc Natl Acad Sci USA* 87: 4358–4362

Sayre PH, Chang H-C, Hussey RE, Brown NR, Richardson NE, Spagnoli G, Clayton LK, Reinherz EL (1987): Molecular cloning and expression of T11 cDNAs reveal a receptor-like structure on human T lymphocytes. *Proc Natl Acad Sci USA* 84: 2941–2945

Sayre PH, Hussey RE, Chang H-C, Ciardelli TL, Reinherz EL (1989): Structural and binding analysis of a two domain extracellular CD2 molecule. *J Exp Med* 169: 995–1009

Scott CF, Jr, Bolender S, McIntyre GD, Holldack J, Lambert JM, Venkatesh YP, Morimoto C, Ritz J, Schlossman SF (1989): Activation of human cytolytic cells through CD2/T11: Comparison of the requirements for the induction and direction of lysis of tumor targets by T cells and NK cells. *J Immunol* 142: 4105–4112

Secrist JP, Karnitz L, Abraham RT (1991): T-cell antigen receptor ligation induces tyrosine phosphorylation of phospholipase C-γ1. *J Biol Chem:* 266: 12135–12139

Seed B, Aruffo A (1987): Molecular cloning of the CD2 antigen, the T cell erythrocyte receptor, by a rapid immunoselection procedure. *Proc Natl Acad Sci USA* 84: 3365–3369

Selvaraj P, Plunkett ML, Dustin M, Sanders ME, Shaw S, Springer TA (1987): The T lymphocyte glycoprotein CD2 binds the cell surface ligand LFA-3. *Nature* 326: 400–403

Sewell WA, Brown MH, Dunne J, Owen MJ, Crumpton MJ (1986): Molecular cloning of the T lymphocyte surface CD2 (T11) antigen. *Proc Natl Acad Sci USA* 83: 8718-8722

Shaw S, Luce GEG, Quinones R, Gress RE, Springer TA, Sanders ME (1986): Two antigen-independent adhesion pathways used by human cytotoxic T cell clones. *Nature* 323: 262-264

Sibley DR, Benovic JL, Caron MG, Lefkowitz R (1987): Regulation of transmembrane signaling by receptor phosphorylation. *Cell* 48: 913-922

Siliciano RF, Lawton T, Knall C, Karr RW, Berman P, Gregory T, Reinherz EL (1988): Analysis of host-virus interactions in AIDS with anti-gp120 T cell clones: Effect of HIV sequence variation and a mechanism for CD4+ cell depletion. *Cell* 54: 561-575

Siliciano RF, Pratt JC, Schmidt RE, Ritz J, Reinherz EL (1985): Activation of cytolytic T lymphocyte and natural killer cell function through the T11 sheep erythrocyte binding protein. *Nature* 317: 428-430

Spits H, van Schooten W, Keizer H, van Seventer G, van de Rijn M, Terhorst C, de Vries JE (1986): Alloantigen recognition is preceded by nonspecific adhesion of cytotoxic T cells and target cells. *Science* 232: 403-405

Stanley JB, Gorczynski R, Huang C-K, Love J, Mills GB (1990): Tyrosine phosphorylation is an obligatory event in IL-2 secretion. *J Immunol* 145: 2189-2198

Swain SL (1983): T cell subsets and the recognition of the MHC class. *Immunol Rev* 74: 129-142

Tadmori W, Reed JC, Nowell PC, Kamoun M (1985): Functional properties of the 50KD protein associated with the E-receptor on human T lymphocytes: Suppression of IL2 production by anti-p50 monoclonal antibodies. *J Immunol* 134: 1709-1716

Takai Y, Reed ML, Burakoff SJ, Herrmann SH (1987): Direct evidence for a receptor-ligand interaction between the T cell surface antigen CD2 and lymphocyte-function-associated antigen 3. *Proc Natl Acad Sci USA* 84: 6864-6868

Truneh A, Albert F, Goldstein P, Schmitt-Verhulst A-M (1985): Early steps of lymphocyte activation bypassed by synergy between calcium ionophores and phorbol ester. *Nature* 313: 318-320

Ullrich A, Schlessinger J (1990): Signal transduction by receptors with tyrosine kinase activity. *Cell* 61: 203-212

van den Elsen P, Bruns G, Gerhard DS, Pravtcheva D, Jones C, Housman D, Ruddle FH, Orkin S, Terhorst C (1985): Assignment of the gene coding for the T3 delta subunit of the T3/T cell receptor complex to the long arm of the human chromosome 11 and to mouse chromosome 9. *Proc Natl Acad Sci USA* 82: 2920-2924

van Kooyk Y, van de Wiel-van Kemenade P, Weder P, Kujipers TW, Figdor CG (1989): Enhancement of LFA-1-mediated cell adhesion by triggering through CD2 or CD3 on T lymphocytes. *Nature* 342: 811-813

Veillette A, Bookman MA, Horak EM, Bolen JB (1988) The CD4 and CD8 T cell surface antigens are associated with the internal membrane tyrosine-protein kinase p56^{lck}. *Cell* 55: 301-308

Veillette A, Bookman MA, Horak EM, Samelson LE, Bolen JB (1989): Signal transduction through the CD4 receptor involves the activation of the internal membrane tyrosine-protein kinase p56^{lck}. *Nature* 338: 257-259

Weiss A, Koretzky G, Schazman RC, Kadlecek T (1991): Functional activation of the T-cell antigen receptor induces tyrosine phosphorylation of phospholipase C-γ1. *Proc Natl Acad Sci USA* 88: 5484-5488

Weissman AM, Baniyash M, Hou D, Samelson LE, Burgess WH, Klausner RD (1988a): Molecular cloning of the zeta chain of the T cell antigen receptor. *Science* 239: 1018-1021

Weissman AM, Frank SJ, Orloff DG, Mercep M, Ashwell JD, Klausner RD (1989): Role of the zeta chain in the expression of the T cell antigen receptor. Genetic reconstitution studies. *EMBO J* 8: 3651-3656

Weissman AM, Hou D, Orloff DG, Modi WS, Seuznez H, O'Brien SJ, Klausner RD (1988b): Molecular cloning and chromosomal localization of the human T cell receptor ζ chain: Distinction from the molecular CD3 complex. *Proc Natl Acad Sci USA* 85: 9709-9713

Yang SY, Chouaib S, Dupont B (1986): A common pathway for T lymphocyte activation involving both the CD3-Ti complex and CD2 sheep erythrocyte receptor determinants. *J Immunol* 137: 1097-1100.

8
Triggering Structures on NK Cells

Lewis L. Lanier and Joseph H. Phillips

Introduction

Natural killer (NK) cells are a subpopulation of lymphocytes distinct from both T and B cells (Lanier et al., 1986d). In man, NK cells are identified as lymphocytes with the antigenic phenotype membrane CD3ε^-, CD16$^+$ and/or CD56$^+$ (Lanier et al., 1986d). In the mouse, NK cells are membrane CD3ε^-, and in some strains NK1.1$^+$. Unlike T lymphocytes, NK cells do not rearrange T-cell receptor (TCR) α, β, γ, or δ genes (Lanier et al., 1986a, 1986b; Tutt et al., 1986, 1987; Loh et al., 1988). NK cell function and maturation, moreover, is normal in *scid* mice (Hackett et al., 1986; Tutt et al., 1987), in which the development of T and B lymphocytes is arrested due to a defect in the process necessary for rearrangement of immunoglobulin and T-cell antigen receptor (Bosma et al., 1983; Schuler et al., 1986; Bosma et al., 1991). These observations suggest that NK cells constitute a distinct lineage of lymphocytes, and they indicate that the recombinase mechanisms present in B and T lymphocytes are not required for NK development or function. A relationship between T and NK progenitor cells prior to TCR rearrangement is nonetheless quite possible, given the remarkable similarities in the functional and antigenic phenotypes of these lymphocytes.

In the quest for tumor-specific cytolytic T lymphocytes (CTL), NK cells were discovered independently by several laboratories as an annoying "background" activity. Unlike antigen-specific CTL, NK cells lysed both autologous and allogeneic tumor cell targets, were present in normal, unimmunized hosts, and did not demonstrate a "memory" response. Although originally identified as "natural killers" of tumors, it is now clear that NK cells also produce cytokines in response to stimulation. These NK cell-derived soluble factors may regulate the immune and hematopoietic systems (reviewed in Trinchieri, 1989). Moreover, NK cells may play an important role in early defense against certain viral infections (reviewed in Welsh, 1986).

The membrane receptors that are responsible for NK-cell recognition of "foreign" cells are presently unknown. The most obvious and direct approach to identify these receptors is to produce monoclonal antibodies (mAb) that react with these structures and to use these reagents to characterize and clone the molecules. Based on prior studies of the B- and T-cell antigen receptors, there are several criteria that putative NK-cell receptors should satisfy: (1) anti-receptor mAb against relevant epitopes should have agonist or antagonist activity (i.e., trigger or inhibit NK cytotoxicity or cytokine production), (2) the receptors should be present on cells mediating NK-cell recognition, but generally absent on cells without these functions; (3) cells that have lost the receptor should not mediate the function; and (4) transfection of the receptor genes or cDNA into cytotoxic cells without the receptor should enable these cells to recognize new target antigens.

It is now generally appreciated that interaction of the TCR with specific antigen may be insufficient to initiate effector functions. For example, alloantigen-specific helper T cells fail to produce cytokines in response to HLA-DR7 antigens expressed in murine L cells. However, cotransfection of HLA-DR7 and CD54 (ICAM-1), a cellular ligand for CD11a/18 (LFA-1), restores T-cell responsiveness (Altmann et al., 1989). Similar requirements for accessory molecules are necessary for alloantigen-specific CTL to kill murine L cells bearing HLA-A2 antigens (H. Spits, personal communication). In these experimental systems, mAb against either TCR or CD11a/18 are able to block T-cell effector function, and there is evidence that both TCR and CD11a/18 may transmit signals necessary for cellular activation. With T-cell effectors, it is clear that TCR is the antigen receptor and that CD11a/18 is mediating an accessory function, since antigen specificity and MHC restriction can be conferred only by TCR gene transfection (Dembic et al., 1986). However, this issue is more problematic with NK cells. Since many of the cytotoxic activities of NK cells are not influenced by MHC expression on the target and as yet no "fine antigen specificity" has been defined, it becomes very difficult to distinguish a primary receptor necessary for NK recognition from an accessory or costimulatory molecule. Nonetheless, several candidate molecules that may participate in NK cell recognition have been identified based on the ability of mAb against these structures to either trigger or inhibit a cytolytic response. Properties of these molecules are described below.

The CD16 (FcγRIII)/ζ Complex

NK cells can efficiently lyse otherwise resistant tumors provided that the cells are coated with an IgG antibody of the appropriate isotype (e.g., human IgG1 or IgG3, mouse IgG3 or IgG2a), a process referred to as antibody-dependent cellular cytotoxicity. Although in man three distinct IgG Fc receptors exist (FcγRI-CD64, RII-CD32, and RIII-CD16), NK cells express only FcγRIII and do not express Fc receptors for IgM, IgA, or IgE. FcγRIII is a low-affinity IgG receptor that was initially identified by several independently generated mAb, e.g., 3G8, B73.1, VEP-13, and Leu 11 (Perussia et al., 1984). These mAb reacted with granulocytes, macrophages, and a small subset of T cells in addition to NK cells (Fleit et al., 1982; Lanier et al., 1983; Perussia et al., 1983; Lanier et al., 1985). Depending on experimental conditions, mAb against CD16 either inhibit ADCC (Lanier et al., 1983; Perussia et al., 1983) or trigger cytolytic function (van de Griend et al., 1987; Lanier et al., 1988; Werfel et al., 1989). Perussia and colleagues (Anegon et al., 1988; Cassatella et al., 1989) demonstrated that engagement of CD16 on NK cells stimulates the inositol phosphate activation pathway, mobilizes intracellular stores of Ca^{2+}, and initiates transcription of γ-interferon and GM-CSF. Leibson and colleagues (Ting et al., 1991) have further demonstrated that CD16 signal transduction involves a tyrosine kinase activation, independent of G-protein coupling.

A cDNA encoding a glycoprotein reactive with anti-CD16 mAb was initially cloned from a placenta library (Simmons and Seed, 1988). COS-7 cells transfected with this cDNA (now referred to as CD16-I or FcγRIII-1) expressed a membrane glycoprotein that was susceptible to cleavage with phosphatidylinositol-specific phospholipase C (PI-PLC). Similarly, the CD16 glycoprotein expressed on human neutrophils was cleaved by PI-PLC treatment (Huizinga et al., 1988; Selvaraj et al., 1988). By contrast, CD16 expressed on NK cells was resistant to PI-PLC cleavage, and the polypeptide was substantially larger than the granulocyte isoform (Lanier et al., 1988, 1989b). Subsequent isolation of cDNA from NK cells (designated FcγRIII-2 or CD16-II) revealed the existence of a distinct structure containing a longer cytoplasmic tail and several amino acid differences in the extracellular domain (Ravetch and Perussia, 1989; Scallon et al., 1989). CD16-I and CD16-II are encoded by two distinct but closely linked genes on chromosome 1 that arose by gene duplication (Pletz et al., 1989; Ravetch et al., 1989; Qiu et al., 1990). Unlike CD16-I, CD16-II is anchored to the plasma membrane by a membrane-spanning domain. The mode of membrane attachment (PI-glycan versus transmembrane) is determined by a single amino acid at codon 203 in CD16-I (Ser) and CD16-II (Phe) (Hibbs et al., 1989; Kurosaki and Ravetch, 1989; Lanier et al., 1989a). The core polypeptide of CD16-I is ~29 kDa, whereas CD16-II is ~32–34 kDa and both contain N-linked carbohydrates (Lanier et al., 1988). CD16-I is exclusively expressed by granulocytes, whereas CD16-II is expressed by NK cells, macrophages, and a small subset of T lymphocytes (Ravetch and Perussia, 1989; Perussia and Ravetch, 1991; Phillips et al., 1991).

A biallelic polymorphism of the CD16-I gene exists that was first identified as the NA1/NA2 antigen using serum from multiparous women and transfusion recipients (Werner et al., 1986; Tetteroo et al., 1987; Huizinga et al., 1990). The NA1 and NA2 gene products differ at codons in a manner that provides additional sites for N-linked glycosylation in NA2, resulting in a molecular weight difference in the two glycoproteins (Ory et al., 1989a, 1989b; Ravetch and Perussia, 1989; Trounstine et al., 1990; Ory et al., 1991).

Initial attempts to express CD16-II efficiently on the cell surface of COS-7 were unsuccessful, suggesting that additional subunits might be required. Both the IgE-binding FcεRI-α and murine IgG-binding FcγRIIα glycoproteins require coassociation with the FcεRI-γ subunit for membrane expression (Blank et al., 1989; Ra et al., 1989). The transmembrane domain of CD16-II demonstrates significant homology to the transmembrane segments of FcεRI-α and murine FcγRIIα. Subsequent studies revealed that cotransfection with either FcεRI-γ or a related molecule, the human ζ chain of the TCR complex (Weissman et al., 1988), permitted membrane expression of CD16-II (Hibbs et al., 1989; Kurosaki and Ravetch, 1989; Lanier et al., 1989d). ζ and FcεRI-γ form disulfide-linked homodimers and ζγ heterodimers. Human NK cells can express CD16-II-ζζ, CD16-II-γγ, and CD16-II-γζ complexes (Anderson et al., 1989; Hibbs et al., 1989; Lanier et al., 1989d; Anderson et al., 1990b; Letourneur et al., 1991; Vivier et al., 1991c). Homologs of CD16-II have been identified in rat and mouse NK cells (Perussia et al., 1989; Zegler et al., 1990). In mice, the homolog is designated FcγRIIα. Murine ζ cannot associate with FcγRIIα, suggesting that only FcεRI-γ is used in murine NK cells (Kurosaki anD Ravetch, 1989). Mice do not express a gene equivalent to human CD16-I that encodes a PI-glycan-anchored molecule.

In T lymphocytes, ζ is required for efficient membrane expression and function of the CD3/TCR complex (Sussman et al., 1988; Frank et al., 1990). Most T lymphocytes also express the γ subunit of FcεRI in association with CD3/TCR. FcεRI-γ can functionally replace ζ in CD3/TCR-mediated functions (Orloff et al., 1990; Rodewald et al., 1991). Engagement of TCR results in tyrosine phosphorylation of ζ (Samelson et al., 1987, 1989), a signal presumably necessary for the subsequent cascade of activation events that result in cytokine production and other effector functions. Similarly, it has been demonstrated that engagement of the CD16-II complex on NK cells results in rapid tyrosine phosphorylation of ζ (O'Shea et al., 1991; Vivier et al., 1991a). Although FcεRI-γ subunits are also coassociated with CD16-II, there has been no indication of phosphorylation of FcεRI-γ; however, tyrosine phosphorylation of this subunit has recently been reported to occur when IgE binds to the high-affinity IgE receptor on basophils (Paolini et al., 1991). CD16-II can be induced on monocytes by TGF-β (Welch et al., 1990). The CD16-II complex on macrophages is competent to mediate ADCC, even though macrophages express FcεRI-γ, but not ζ (Perussia and Ravetch, 1991; Phillips et al., 1991).

While engagement of the CD16-II complex clearly activates the tyrosine kinase and inositol phosphate-associated pathways in NK cells and triggers cytotoxicity and cytokine secretion, the role of CD16-I in granulocytes is more controversial. Engagement of CD16-I does not initiate a cytolytic response (Graziano and Fanger, 1987; Lanier et al., 1988); however, it has been suggested that the PI-glycan CD16-I glycoprotein can initiate phagocytosis and release of granule-associated proteins from granulocytes (Huizinga et al., 1988; Anderson et al., 1990a; Salmon et al., 1991). Analysis of Fc receptor function on granulocytes is complicated by the presence of FcγRII (CD32), since this receptor may cooperate with CD16-I in mediating Fc-dependent functions.

The existence of CD16-II as a ζ-associated receptor on NK cells that behaves functionally in a manner similar to the TCR initially suggested CD16-II as a potential candidate for "the NK receptor." Is is conceivable that certain targets susceptible to NK lysis may possess membrane antigens that are structurally similar to the Fc region of immunoglobulins and which thereby permit NK cell activation. Although this is an attractive model, we have been unable to obtain any supporting evidence for this hypothesis. Whilst most NK cells in human peripheral blood express the CD16-ζ complex, we have identified a subset of NK cells (comprising ~10% of total peripheral blood NK cells) that do not express CD16 on the cell surface or transcribe CD16 mRNA (Lanier et al., 1986c; Nagler et al., 1989). These CD16$^-$ NK cells lyse an NK-sensitive target, K562, and can kill a broad range of tumors following IL-2 activation (Lanier et al., 1986c; Nagler et al., 1989). We have prepared a panel of CD16$^-$ and CD16$^+$ NK clones from the same individual and tested them on a large panel of target cells, but we have been unable to distinguish any obvious differences in "target repertoire" (unpublished observations). As expected, however, the CD16$^+$ NK clones mediated ADCC, whereas the CD16$^-$ clones did not. Thus, the major role of the CD16-ζ complex in NK function is apparently dependent on interaction with complexed IgG, and CD16 does not mediate the specific recognition of most (or all) NK targets.

CD2/CD58 (LFA-3)

CD2 is a 50-kDa glycoprotein expressed on T cells and NK cells. It is responsible for sheep erythrocyte rosette formation (Kamoun et al., 1981). CD2 specifically binds to another membrane glycoprotein, CD58 (LFA-3), that is broadly distributed in the hematopoietic system (Shaw et al., 1986; Dustin et al., 1987). Monoclonal antibodies against appropriate epitopes of CD2 can induce a rise in intracellular free Ca^{2+}, initiate cytokine production, and trigger cell-mediated cytotoxicity (Meuer et al., 1984; Weiss et al., 1984; Siliciano et al., 1985).

Several lines of evidence indicate that CD2 can participate in NK-cell-mediated cytotoxicity. Anti-CD2 mAb against certain epitopes induce human NK cells to lyse tumor cell targets (Schmidt et al., 1985; Siliciano et al., 1985; Bolhuis et al., 1986; Schmidt et al., 1987; van de Griend et al., 1987; Schmidt et al., 1988; Scott et al., 1989), and under certain circumstances to undergo autolysis (Schmidt et al., 1988). Additionally, anti-CD2 mAb can inhibit, albeit weakly, murine and human NK-cell-mediated cytotoxicity against some, but not all, tumor cell targets (Bolhuis et al., 1986; Nakamura et al., 1990; Robertson et al., 1990). Seaman and colleagues have shown that lysis of the NK-sensitive murine cell line YAC-1 by the rat NK leukemia cell line RNK-16 is efficiently inhibited by soluble anti-rat CD2 mAb OX-34 (Seaman et al., 1987). In addition to inhibiting killing, crosslinking CD2 on RNK-16 cells induced the generation of inositol trisphosphate (IP3) and an increase in intracellular free Ca^{2+} (Seaman et al., 1987). However, anti-CD2 fails to generate IP3 or to affect cytotoxicity mediated by NK cells from normal rats (Seaman et al., 1991). Interestingly, Imboden and colleagues have demonstrated that CD2 coassociates with gp35 (OX-44), a membrane glycoprotein on all mature rat lymphocytes that is homologous to human CD37 (Imboden et al., 1991). Anti-gp35 mAb also induces inositol phosphate generation in rat RNK-16 and blocks lysis of YAC-1 targets, suggesting cooperation with CD2 in this process. The possibility that CD2 is involved in NK cell-mediated cytotoxicity is further supported by the observation that transfection of CD2 cDNA into a $CD2^-$ murine NK clone enabled these cells to lyse certain otherwise resistant targets (Nakamura et al., 1991). By contrast, transfection with a CD2 cDNA lacking the cytoplasmic domain did not confer cytolytic function, implicating CD2 in possible signal transduction. There is a report indicating that transfection of murine L cells with CD58 (LFA-3) renders these cells susceptible to lysis by human NK cells (Anasetti et al., 1989). We have been unable to verify this in our laboratory (unpublished observation).

It is impossible to discriminate whether CD2 serves as a "target receptor" or an "accessory molecule." However, several lines of evidence argue against a role for CD2 as a primary receptor. First, extensive studies of CD2 in antigen-specific T cells suggest that CD2 works primarily in conjunction with, and is regulated by, the TCR. Modulation of CD3 inhibits anti-CD2 mAb-induced T-cell activation (Meuer et al., 1984), and stimulation of Jurkat T leukemia cells via the CD2/CD58 interaction is dependent upon expression of the CD3/TCR complex (Bockenstedt et al., 1988). The cascade of events in T cells accompanying CD2 activation appears to be similar, if not identical, to TCR stimulation (Ley et al., 1991). The role of CD2 as an "accessory" molecule in TCR-dependent activation is also supported by the finding that expression of the CD2 ligand, CD58, on class II MHC-bearing antigen-presenting cells decreases the concentration of peptide necessary for T-cell cytokine production, but CD58 is not absolutely required (Koyasu et al., 1990). Similarly, expression of CD58 on target cells does not correlate well with their susceptibility to natural killing, and many $CD58^+$ targets are resistant to lysis by NK cells. Thus, if CD2 is a "primary" receptor on NK cells, it appears to play only a minor role in conferring target specificity.

There are suggestions that the CD16-ζ complex may regulate the function of CD2 in NK cells. Anderson and colleagues have recently demonstrated that stimulation with anti-CD2 mAb induces tyrosine phosphorylation of ζ in NK cells (Vivier et al., 1991b). Additionally, it has been observed that signal transduction via CD2 in NK cells may require CD16 (Anasetti et al., 1987; Spruyt et al., 1991). Collectively, these findings suggest that CD2 may participate in initiation of NK cell signal transduction and cytotoxicity in certain limited circumstances, but the primary signal may be delivered by other receptor molecules.

Disulfide-Linked Homodimers of the Ly-49 Family

A family of disulfide-linked homodimers that are preferentially expressed on NK cells that have been identified in mouse, rat, and man. In mouse, the prototypic marker for the identification of NK cells is the NK1.1 antigen (Glimcher et al., 1977). This antigen was originally identified by allo-antiserum and later by mAb (Koo and Peppard, 1984). The NK1.1 allele is expressed in B6 and related mouse strains (Glimcher et al., 1977); however, a second allele of the gene has not as yet been identified by serology. NK1.1 is expressed on all NK cells in appropriate stains, but it is not NK-specific, since a small subset of T lymphocytes also expresses this antigen (Ballas and Rasmussen, 1990; Levitsky et al., 1991). Yokoyama and colleagues have demonstrated that anti-NK1.1 mAb can trigger a cytolytic response in NK cells (Karlhofer and Yokoyama, 1991), suggesting that the molecule may participate in NK cytolytic function. Although NK1.1 has not been cloned, the gene maps to murine chromosome 6, near the Ly-49 gene family (Yokoyama et al., 1990).

The Ly-49 gene (also known as A1 or YE1/48) (Chan and Takei, 1989; Yokoyama et al., 1989) encodes a disulphide-linked homodimer composed of 43-kDa subunits. Ly-49 is expressed on ~20% of NK cells, as well as some T lymphocytes, but does not trigger NK cytotoxicity (Yokoyama et al., 1990). Southern blot analysis using a Ly-49 probe has revealed remarkable complexity, suggesting the existence of numerous related genes and polymorphisms (Yokoyama et al., 1990). Additional family members have been recently cloned (Wong et al., 1991). The sequence of Ly-49 indicates that it encodes a type II membrane protein containing animal lectin domains that may bind specific carbohydrate determinants (Yokoyama et al., 1989). Related genes are CD23, a low-affinity receptor for IgE, and Lyb2 (CD72), a B-cell-associated antigen.

Chambers and colleagues (1989) have described a mAb, 3.2.3, that reacts with a disulfide-linked homodimer (designated NKR-P1) that is expressed on rat NK cells, granulocytes, and a small subset of T lymphocytes. Like NK1.1, 3.2.3 triggers NK cell-mediated cytotoxicity. Seaman and colleagues (Ryan et al., 1991) have demonstrated that 3.2.3 mAb stimulates NK cells via the inositol phosphate activation pathway. As yet, ligands have not been identified for NKR-P1; however, Seaman (Niemi et al., 1991) has reported that a NKR-P1 loss variant of the RNK-16 NK leukemia cell line has diminished cytotoxic activity against the YAC-1 target. Although these results suggest that NKR-P1 may be involved in RNK-16-mediated cytotoxicity, genetic reconstitution of NKR-P1 expression is necessary to verify this.

Molecular cloning of the cDNA for NKR-P1 revealed that it shares structural features with Ly-49, including an extracellular domain that has features of known C-type lectins (Giorda et al., 1990). A relationship between NKR-P1 and the Ly-49 gene family was further supported by genetic mapping of a mouse homolog for rat NKR-P1; the gene lies only ~0.4 cM telomeric to the Ly-49 gene family (Yokoyama et al., 1991). The mouse NKR-P1 locus may contain more than one gene, as there are at least two cross-hybridizing and highly homologous cDNAs (Giorda and Trucco, 1991). Despite the genetic linkage and structural similarities between NKR-P1 and Ly-49, they share only weak amino acid homology and are quite different in their cytoplasmic domains, suggesting that they may be functionally distinct molecules.

By differential and subtractive hybridization, Houchins has cloned a series of closely related cDNAs from NK cells that are structurally similar to the murine Ly-49 and NKR-P1 genes (Houchins et al., 1990, 1991). Like the rodent genes, the human NKG2 gene family contains at least three members that encode type II membrane proteins containing lectin-binding domains (Houchins et al., 1991). While the NKG2-encoded proteins have not yet been characterized, it seems likely that a mAb generated by López-Botet and colleagues (Aramburu et al., 1990, 1991) against the human Kp43 antigen may be related to this family. The Kp43 antigen is a disulfide-linked homodimer composed of glycosylated 43-kDa subunits (Aramburu et al., 1990). This antigen is expressed on NK cells, as well as a subset of T lymphocytes (Aramburu et al., 1990). Anti-Kp43 mAb inhibits the IL-2-dependent proliferation of NK cells and $\gamma\delta$-TcR$^+$ T lymphocytes (Aramburu et al., 1990). Interestingly, F(ab')$_2$ fragments of anti-Kp43 induce IL-2-activated NK cells to lyse autologous and allogeneic T-cell blasts, but do not affect cytotoxicity against several tumor cell targets. Fresh, resting NK cells do not mediate this function. The mechanism of this phenomenon is presently unknown. We also have produced a mAb, designated DX-1, that is similar to anti-Kp43. DX1 mAb preferentially reacts with NK cells and a subset of T cells and immunoprecipitates a disulfide-linked homodimer with 43-kDa subunits. Removal of N-linked carbohydrates reveals the presence of a 24-kDa core polypeptide. Unlike anti-Kp43, anti-DX1 does not affect IL-2-induced NK cell proliferation (unpublished observation). Although it seems likely that DX1 and Kp43 may be members of a human Ly49-like family based on similarities in distribution and biochemical properties, this will need to be confirmed by cloning the cDNA encoding these antigens.

The presence of lectin-binding domains in the Ly49-related molecules suggests that these structures may serve as membrane receptors for either soluble factors or other cell surface antigens. The possibility that cell surface antigens may be ligands for Ly-49 proteins is supported by the recent observation that CD72/Lyb2, which also contains lectin-binding domains, specifically interacts with CD5 (van de Velde et al., 1991).

Other NK Signal-Transducing Molecules

The molecules or gene families described above have been extensively studied by numerous laboratories in several species. Several additional molecules that can initiate signal transduction in NK cells have been recently identified and are currently under investigation. In the rat, Imboden and Seaman have generated mAb against a PI-glycan anchored membrane glycoprotein on the RNK-16 leukemia cell line that stimulates the inositol phosphate activation pathway (Imboden et al., 1989). gp42 is absent on resting NK cells but is present on all NK cells following activation by culture in IL-2 (Imboden et al., 1989; Seaman et al., 1991). Unlike all other IL-2 induced activation antigens (e.g., transferrin receptors, CD25, CD69, and MHC class II), expression of gp42 is restricted to NK cells (Imboden et al., 1989; Seaman et al., 1991). Although anti-gp42 mAb induces an increase in intracellular $[Ca^{2+}]$ and inositol trisphosphate in the RNK-16 cell line, there is no effect on IL-2-activated NK cells from normal rats, similar to the situation with anti-CD2 stimulation of rat NK cells (Seaman et al., 1991). The role of gp42 in NK cell function is presently unknown.

Ortaldo and colleagues have used an interesting strategy to identify NK-cell-associated membrane antigens involved in the recognition of K562 cells. First, they produced a mAb against K562, designated mAb 36, that partially blocks NK binding and cytotoxicity. They then used this mAb to generate rabbit anti-idiotype serum to identify NK cell surface antigens reactive with the K562 antigen (Frey et al., 1991). One such antiserum has been shown to preferentially bind human NK cells, to induce NK cell-mediated cytokine production and cytotoxicity against Fc receptor-bearing targets, and to partially block conjugate formation and killing of K562 (Frey et al., 1991). Glycoproteins of 80, 110, and 150 kDa were immunoprecipitated from the surface of ^{125}I-labeled NK cells by using this antiserum, and a partial cDNA has been cloned from an NK-cell library by bacterial expression of the protein. The sequence of this cDNA is unusual for a membrane protein, containing cyclophilin- and histone-like domains. Transcripts of this mRNA are quite rare in NK cells, requiring the use of poly A-selected RNA for detection. These investigators have not yet been able to transfect this cDNA in appropriate host cells to demonstrate that it confers the ability to recognize and induce lysis of K562. The biochemical properties of the K562 antigen recognized by mAb 36 are also unknown, since this mAb is unable to immunoprecipitate any radiolabeled structures from these cells (Frey et al., 1991). Therefore, further studies will be necessary to confirm the role of this molecule in K562 recognition and to determine the structural nature of its ligand.

In addition to mediating lysis of tumor or virus-infected cells, NK cells also possess the ability to recognize alloantigens present on normal hematopoietic cells. Bennett and colleagues have extensively described the ability of NK cells to reject allogeneic bone marrow grafts based upon incompatibility at *Hh*, a locus tightly linked to the murine major histocompatibility complex *H-2D* gene (Bennett, 1987). Susceptibility to *Hh*-mediated rejection is conferred as an autosomal recessive trait. Sentman et al., 1989, 1991) have recently reported that mAb 5E6 reacts with a subset of murine NK cells that can adoptively transfer *Hh* specificity to appropriate strains of mice that lack this recognition.

Moretta and co-workers have described human NK cell lines and clones that mediate alloantigen-specific lysis of normal PHA lymphoblasts using *in vitro* cytotoxicity assays (Ciccone et al., 1988). Like murine *Hh*, target cell susceptibility is transmitted in an autosomal recessive fashion by a locus linked to human leukocyte antigens (HLA) (Ciccone et al., 1990). Although these target antigens have not been identified by serology, Moretta (Moretta et al., 1990a) has described two mAb, GL183 and EB6, generated against NK clones that identify subsets of NK cells and expression of these antigens correlates with allo-antigen specificity. Although GL183 and EB6 do not affect allo-antigen recognition, these mAb can trigger cell-mediated cytotoxicity against several NK resistant human tumor cell lines and induce TNF-α secretion (Moretta et al., 1990a, 1990b). Paradoxically, GL183 mAb inhibits NK clone lysis of murine P815 cells and prevents NK cell activation via CD2 and CD16 (Moretta et al., 1990b). Anti-GL183 induces an increase in intracellular $[Ca^{2+}]$, suggesting possible involvement of the inositol phosphate activation pathway (Moretta et al., 1990b). Further insight into the role of the murine 5E6 and human GL183/EB6 molecules awaits cloning of these genes and transfection studies to determine whether they are necessary and sufficient to confer allo-antigen recognition.

Harris and Evans (Harris et al., 1991) have reported that a murine mAb 5C6, generated against catfish NK-like cells, specifically reacts with mouse, rat, and human NK cells, blocks NK cell-mediated cytotoxicity against a variety of targets, and triggers cytolytic function. At the Seventh International Workshop on Natural Killer Cells (June 4–7, 1991, Stockholm, Sweden), these investigators reported a partial protein sequence of the antigen reactive with mAb 5C6. They find that 5C6 reacts specifically with vimentin on the surface of NK cells and that commercially available anti-vimentin mAb demonstrate the same binding and functional properties as 5C6. We have obtained mAb reactive with human vimentin, and these mAb efficiently bind to intracellular vimentin proteins. However, in our hands these reagents did not bind to the surface of viable NK cells and did not affect NK cytotoxicity.

Griffin and colleagues (1983) generated one mAb, NKH1, initially raised against a myeloid leukemia, that preferentially reacts with human NK cells. The NKH1 antigen (CD56) is expressed on essentially all NK cells and on a small subset of peripheral blood T lymphocytes that can mediate MHC-unrestricted cytotoxicity (Lanier et al., 1986c; Schmidt et al., 1986). Subsequent studies have demonstrated that NKH1 and similar mAb (e.g., Leu 19) react with the neural cell adhesion molecule (N-CAM) (Lanier et al., 1989c). N-CAM was originally identified as a glycoprotein expressed on neural tissues. It can be expressed as several isoforms which are generated by alternative mRNA splicing. Transmembrane-anchored polypeptides of 140 and 180 kDa and a 120-kDa phosphatidylinositol-glycan anchored polypeptide have been reported (reviewed in Cunningham et al., 1987). NK cells predominantly express the 140-kDa isoform, and the NK cell-derived cDNA sequence is essentially identical to that reported for the neural 140-kDa species (Lanier et al., 1991). While N-CAM has been implicated in homophilic binding in neural tissues (reviewed in Edelman, 1986; Cunningham et al., 1987; Edelman, 1988), we have recently reported that transfection of tumor cell targets with N-CAM cDNA does not affect the interaction with NK cell effectors (Lanier et al., 1991). Moreover, anti-N-CAM mAB NKH1 and Leu 19 do not affect NK cell-mediated cytotoxicity against N-CAM$^+$ targets (Lanier et al., 1991). Recently, Suzuki et al. (1991) have reported that two mAb against N-CAM (2-13 and 5-38) block alloantigen-specific NK cell-mediated lysis of B lymphoblastoid cell lines. Moreover, they find that these mAb signal the induction of NK cell-mediated cytotoxicity (Suzuki et al., 1991). These investigators have generously provided us with these reagents; however, we have been unable to reproduce their results. Thus, these findings await independent confirmation.

Conclusions

The most extensively characterized receptor expressed on NK cells is FcγRIII (CD16-II), an Fc receptor for IgG. In this case, the ligand is known. The subunits of the receptor have been characterized biochemically, cloned, and confirmed by transfection. Signal transduction via this receptor involves both tyrosine kinase-dependent and inositol phosphate-associated activation pathways. FcγRIII is exclusively and specifically responsible for the antibody-dependent cellular cytotoxicity (ADCC) function mediated by NK cells. However, it is clear that NK cells must possess other receptors for recognition of target cells that do not involve Ig.

The role of the other NK cell membrane structures that can influence NK function is less clear. As mentioned previously, it is difficult to distinguish whether these molecules are recognition receptors or accessory molecules. It is conceivable that NK cells will not possess a single, dominant receptor for target antigens. A "slot-machine hypothesis" of NK recognition would predict that initiation of the cytolytic response may result from engagement of an appropriate array of "adhesion" molecules with their ligands. Interference with individual receptor/ligand pairs could partially or totally inhibit the cytolytic response; however, no one of the receptors alone would be sufficient for the initial "recognition" event. Since it is now appreciated that many of the "cellular adhesion molecules" not only can mediate physical binding but also can transmit intracellular signals, there is no *a priori* reason to assume that "NK recognition" could not be a consequence of cooperative interactions between these structures. This model is experimentally testable by genetic transfection.

A great deal of controversy, speculation, and false leads has accompanied the quest to identify NK recognition receptors. Undoubtedly such molecules exist, but identification, confirmation, and an understanding of their function have thus far remained elusive and will certainly provide a challenge for the near future. Careful, detailed studies of signal-transducing molecules expressed on NK cells should provide valuable insight into this problem.

Acknowledgments

We thank Dr. Bill Seaman for helpful discussion,

critical review of the manuscript, and contribution to the discussion of Ly-49 molecules. DNAX Research Institute is supported by Schering Plough Corporation.

References

Altmann DM, Hogg N, Trowsdale J, Wilkinson D (1989): Cotransfection of ICAM-1 and HLA-DR reconstitutes human antigen-presenting cell function in mouse L cells. *Nature* 338: 512–514

Anasetti C, Hansen JA, Martin PJ, Guralski D, Barbosa JA (1989): Activation of natural killer cells by LFA-3 binding to CD2. *Tissue Antigens* 33: 73

Anasetti C, Martin PJ, June CH, Hellstrom KE, Ledbetter JA, Rabinovitch PS, Morishita Y, Hellstrom I, Hansen JA (1987): Induction of calcium flux and enhancement of cytolytic activity in natural killer cells by cross-linking of the sheep erythrocyte binding protein (CD2) and the Fc-receptor (CD16). *J Immunol* 139: 1772–1779

Anderson CL, Shen L, Eicher DM, Wewers MD, Gill JK (1990a): Phagocytosis mediated by three distinct Fcg receptor classes on human leukocytes. *J Exp Med* 171: 1333–1345

Anderson P, Caligiuri M, O'Brien C, Manley T, Ritz J, Schlossman SF (1990b): FcγRIII (CD16) is included in the ζ NK receptor complex expressed by human natural killer cells. *Proc Natl Acad Sci USA* 87: 2274–2278

Anderson P, Caligiuri M, Ritz J, Schlossman SF (1989): CD3-negative natural killer cells express ζ TCR as part of a novel molecular complex. *Nature* 341: 159–162

Anegon I, Cuturi MC, Trinchieri G, Perussia B (1988): Interaction of Fc receptor (CD16) ligands induces transcription of interleukin 2 receptors (CD25) and lymphokine genes and expression of their products in human natural killer cells. *J Exp Med* 167: 452–472

Aramburu J, Balboa MA, Izquierdo M, Lopez-Botet M (1991): A novel functional cell surface dimer (Kp43) expressed by natural killer cells and γ/δ TCR+ T lymphocytes II. Modulation of natural killer cytotoxicity by anti-Kp43 monoclonal antibody. *J Immunol* 147: 714–721

Aramburu J, Balboa MA, Ramirez A, Silva A, Acevedo A, Sanchez-Medrid F, DeLandazuri MO, Lopez-Botet M (1990): A novel functional cell surface dimer (Kp43) expressed by natural killer cells and T cell receptor-γ/δ+ T lymphocytes. I. Inhibition of the IL-2 dependent proliferation by anti-Kp43 monoclonal antibody. *J Immunol* 144: 3238–3247

Ballas ZK, Rasmussen W (1990): NK1.1+ thymocytes: Adult murine CD4–, CD8– thymocytes contain an NK1.1+, CD3+, CD5^{hi}, CD44^{hi}, TCR-Vβ8+ subset. *J Immunol* 145: 1039–1045

Bennett M (1987): Biology and genetics of hybrid resistance. *Adv Immunol* 41: 333–445

Blank U, Ra C, White K, Metzger H, Kinet J-P (1989): Complete structure and expression in transfected cells of high affinity IgE receptor. *Nature* 337: 187–189

Bockenstedt LK, Goldsmith MA, Dustin M, Olive D, Springer TA, Weiss A (1988): The CD2 ligand LFA-3 activates T cells but depends on the expression and function of the antigen receptor. *J Immunol* 141: 1904–1911

Bolhuis RLH, Roozemond RC, van de Griend RJ (1986): Induction and blocking of cytolysis in CD2+, CD3– and CD2+, CD3+ cytotoxic T lymphocytes via CD2 50 kd sheep erythrocyte receptor. *J Immunol* 136: 3939

Bosma GC, Cluster RP, Bosma MJ (1983): A severe combined immunodeficiency mutation in the mouse. *Nature* 301: 527–530

Bosma MJ, Carroll AM (1991): The SCID mouse mutant: Definition, characterization, and potential use. *Annu Rev Immunol* 9: 323–350

Cassatella MA, Angeon I, Cuturi MC, Griskey P, Trinchieri G, Perussia B (1989): FcγR (CD16) interaction with ligand induces Ca^{2+} mobilization and phosphoinositide turnover in human natural killer cells. Role of Ca^{2+} in FcγR (CD16)-induced transcription and expression of lymphokine genes. *J Exp Med* 169: 549–567

Chambers WH, Vujanovic NL, DeLeo AB, Olszowy MW, Herberman RB, Hiserodt JC (1989): Monoclonal antibody to a triggering structure expressed on rat natural killer cells and adherent lymphokine-activated killer cells. *J Exp Med* 169: 1373–1389

Chan P-Y, Takei F (1989): Molecular cloning and characterization of a novel murine T cell surface antigen, YE1/48. *J Immunol* 142: 1727–1736

Ciccone E, Colonna M, Viale O, Pende D, Di Donato C, Reinharz D, Amoroso A, Jeannet M, Guardiola J, Moretta A, Spies T, Strominger J, Moretta L (1990): Susceptibility or resistance to lysis by alloreactive natural killer cells is governed by a gene in the human major histocompatibility complex between BF and HLA-B. *Proc Natl Acad Sci USA* 87: 9794–9797

Ciccone E, Viale O, Pende D, Malnati M, Biassoni R, Melioli G, Moretta A, Long EO, Moretta L (1988): Specific lysis of allogeneic cells after activation of CD3– lymphocytes in mixed lymphocyte culture. *J Exp Med* 168: 2403–2408

Cunningham BA, Hemperly JJ, Murray BA, Prediger EA, Brackenbury R, Edelman GM (1987): Neural cell adhesion molecule: Structure, immunoglobulin-like domains, cell surface modulation, and alternative RNA splicing. *Science* 236: 799–806

Dembic Z, Haas W, Weiss S, McCubrey J, Kiefer H, von Boehmer H, Steinmetz M (1986): Transfer of specificity by murine α and β T-cell receptor genes. *Nature* 320: 232–238

Dustin ML, Sanders ME, Shaw S, Springer TA (1987): Purified lymphocyte function-associated antigen 3 binds to CD2 and mediates T lymphocyte adhesion. *J Exp Med* 165: 677–692

Edelman GM (1986): Cell adhesion molecules in the regulation of animal form and tissue pattern. *Ann Rev Cell Biol* 2: 81–116

Edelman GM (1988): Morphoregulatory molecules. *Biochemistry* 27: 3533–3543

Fleit HB, Wright SD, Unkeless JC (1982): Human neutrophil Fcγ receptor distribution and structure. *Proc Natl Acad Sci USA* 79: 3275–3279

Frank SJ, Niklinska BB, Orloff DG, Mercep M, Ashwell JD, Klausner RD (1990): Structural mutations of the T cell receptor ζ chain and its role in T cell activation. *Science* 249: 174–177

Frey JL, Bino T, Kantor RRS, Segal DM, Giardina SL, Roder J, Anderson S, Ortaldo JR (1991): Mechanism of target cell recognition by natural killer cells: Characterization of a novel triggering molecule restricted to CD3− large granular lymphocytes. *J Exp Med* 174: 1527–1536

Giorda R, Trucco M (1991): Mouse NKR-P1: A family of genes selectively coexpressed in adherent lymphokine-activated killer cells. *J Immunol* 147: 1701–1708

Giorda R, Rudert WA, Vavassori C, Chambers WH, Hiserodt JC, Trucco M (1990): NKR-P1, a signal transduction molecule on natural killer cells. *Science* 249: 1298–1300

Glimcher L, Shen FW, Cantor H (1977): Identification of a cell-surface antigen selectively expressed on the natural killer cell. *J Exp Med* 145: 1–9

Graziano RF, Franger MW (1987): FcγRI and FcγRII on monocytes and granulocytes are cytotoxic trigger molecules for tumor cells. *J Immunol* 139: 3536–3541

Griffin JD, Hercend T, Beveridge R, Schlossman SF (1983): Characterization of an antigen expressed by human natural killer cells. *J Immunol* 130: 2947–2951

Hackett J, Jr, Bosma GC, Bosma MJ, Bennett M, Kumar V (1986): Transplantable progenitors of natural killer cells are distinct from those of T and B lymphocytes. *Proc Natl Acad Sci USA* 83: 3427–3431

Harris DT, Jaso-Friedmann L, Devlin RB, Koren HS, Evans DL (1991): Identification of an evolutionary conserved, function-associated molecule on human natural killer cells. *Proc Natl Acad Sci USA* 88: 3009–3013

Hibbs ML, Selvaraj P, Carpen O, Springer TA, Kuster H, Jouvin M-HE, Kinet J-P (1989): Mechanisms for regulating expression of membrane isoforms of FcγRIII (CD16). *Science* 246: 1608–1611

Houchins JP, Yabe T, McSherry C, Bach FH (1991): DNA sequence analysis of NKG2, a family of related cDNA clones encoding type II integral membrane proteins on human natural killer cells. *J Exp Med* 173: 1017–1020

Houchins JP, Yabe T, McSherry C, Miyokawa N, Bach FH (1990): Isolation and characterization of NK cell or NK/T cell-specific cDNA clones. *J Mol Cell Immunol* 4: 295–306

Huizinga TWJ, Kleijer M, Tetteroo PAT, Roos K, Kr. von dem Borne AEG (1990): Biallelic neutrophil Na-antigen system is associated with a polymorphism on the phospho-inositol-linked Fcγ receptor III (CD16). *Blood* 75: 213–217

Huizinga TWJ, van der Schott CE, Jost C, Klaassen R, Kleijer M, Kr. von dem Borne AEG, Roos D, Tetteroo PAT (1988): The PI-linked receptor FcRIII is released on stimulation of neutrophils. *Nature* 333: 667–669

Imboden JB, Bell G, Seaman WE (1991): Characterization of signal-transducing molecules on natural killer cells. *Biochem Soc Trans* 19: 265–268

Imboden JB, Eriksson EC, McCutcheon M, Reynolds CW, Seaman WE (1989): Identification and characterization of a cell-surface molecule that is selectivity induced on rat lymphokine-activated killer cells. *J Immunol* 143: 3100–3103

Kamoun M, Martin PJ, Hansen JA, Brown MA, Siakak AW, Nowinski RC (1981): Identification of a human T lymphocyte surface protein associated with the E-rosette receptor. *J Exp Med* 153: 207–212

Karlhofer FM, Yokoyama WM (1991): Stimulation of murine natural killer (NK) cells by a monoclonal antibody specific for the NK1.1 antigen: IL-2-activated NK cells possess additional specific stimulation pathways. *J Immunol* 146: 3662–3673

Koo GC, Peppard JR (1984): Establishment of monoclonal anti-NK1.1 antibody. *Hybridoma* 3: 301

Koyasu S, Lawton T, Novick D, Recny MA, Siliciano RF, Wallner BP, Reinherz EL (1990): Role of interaction of CD2 molecules with lymphocyte function-associated antigen 3 in T-cell recognition of nominal antigen. *Proc Natl Acad Sci USA* 87: 2603–2607

Kurosaki T, Ravetch JV (1989): A single amino acid in the glycosyl phosphatidylinositol attachment domain determines the membrane topology of FcγRIII. *Nature* 342: 805–807

Lanier LL, Chang C, Azuma M, Ruitenberg JJ, Hemperly JJ, Phillips JH (1991): Molecular and functional analysis of human NK cell-associated neural cell adhesion molecule (N-CAM/CD56). *J Immunol* 146: 4421–4426

Lanier LL, Cwirla S, Federspiel N, Phillips JH (1986a): Human natural killer cells isolated from peripheral blood do not rearrange T cell antigen receptor β chain genes. *J Exp Med* 163: 209–214

Lanier LL, Cwirla S, Phillips JH (1986b): Genomic organization of T cell γ genes in human peripheral blood natural killer cells. *J Immunol* 137: 3375–3377

Lanier LL, Cwirla S, Yu G, Testi R, Phillips JH (1989a): A single amino acid determines phosphatidylinositol-glycan membrane anchoring of a human Fc receptor for IgG (CD16). *Science* 246: 1611–1613

Lanier LL, Kipps TJ, Phillips JH (1985): Functional properties of a unique subset of cytotoxic CD3+ T lymphocytes that express Fc receptors for IgG (CD16/Leu-11 antigen). *J Exp Med* 162: 2089–2106

Lanier LL, Le AM, Civin CI, Loken MR, Phillips JH (1986c): The relationship of CD16 (Leu-11) and Leu-19 (NKH-1) antigen expression on human peripheral blood NK cells and cytotoxic T lymphocytes. *J Immunol* 136(12): 4480–4486

Lanier LL, Le AM, Phillips JH, Warner NL, Babcock GF (1983): Subpopulations of human natural killer cells defined by expression of the Leu-7 (HNK-1) and Leu-11 (NK-15) antigens. *J Immunol* 131: 1789–1796

Lanier LL, Phillips JH, Hackett J, Jr, Tutt M, Kumar V (1986d): Natural killer cells: Definition of a cell type rather than a function. *J Immunol* 137: 2735–2739

Lanier LL, Phillips JH, Testi R (1989b): Membrane-anchoring and spontaneous release of CD16 (FcR III) by natural killer cells and granulocytes. *Eur J Immunol* 19: 775–778

Lanier LL, Ruitenberg JJ, Phillips JH (1988): Functional and biochemical analysis of CD16 antigen on natural killer cells and granulocytes. *J Immunol* 141: 3478–3485

Lanier LL, Testi R, Bindl J, Phillips JH (1989c): Identity of Leu 19 (CD56) leukocyte differentiation antigen and neural cell adhesion molecule (N-CAM). *J Exp Med* 169: 2233–2238

Lanier LL, Yu G, Phillips JH (1989d): Co-association of CD3ζ with a receptor (CD16) for IgG Fc on human natural killer cells. *Nature* 342: 803–805

Letourneur O, Kennedy ICS, Brini AT, Ortaldo JR, O'Shea JJ, Kinet J-P (1991): Characterization of the family of dimers associated with Fc receptors (FcεRI and FcγRIII). *J Immunol* 147: 2652–2656

Levisky HI, Golumbek PT, Pardoll DM (1991): The fate of CD4− 8− T cell receptor-αβ+ thymocytes. *J Immunol* 146: 1113–1117

Ley SC, Davies AA, Druker B, Crumptom MJ (1991): The T cell receptor/CD3 complex and CD2 stimulate the tyrosine phosphorylation of indistinguishable patterns of polypeptides in the human T leukemic cell line Jurkat. *Eur J Immunol* 21: 2203–2209

Loh EY, Cwirla S, Serafini AT, Phillips JH, Lanier LL (1988): Human T-cell receptor δ chain: Genomic organization, diversity, and expression in populations of cells. *Proc Natl Acad Sci USA* 85: 9714–9718

Meuer SC, Hussey RE, Fabbi M, Fox D, Acuto O, Fitzgerald KA, Hodgdon JC, Protentis JP, Schlossman SF, Reinherz EL (1984): An alternative pathway of T-cell activation: A functional role for the 50 kD T11 sheep erythrocyte receptor protein. *Cell* 36: 897–906

Moretta A, Bottino C, Pende D, Tripodi G, Tambussi G, Viale O, Orengo A, Barbaresi M, Merli A, Ciccone E, Moretta L (1990a): Identification of four subsets of human CD3− CD16+ natural killer (NK) cells by the expression of clonally distributed functional surface molecules: Correlation between subset assignment of NK clones and ability to mediate specific alloantigen recognition. *J Exp Med* 172: 1589–1998

Moretta A, Tambussi G, Bottino C, Tripodi G, Merli A, Ciccone E, Pantaleo G, Moretta L (1990b): A novel surface antigen expressed by a subset of human CD3− CD16+ natural killer cells. Role in cell activation and regulation of cytolytic function. *J Exp Med* 171: 714

Nagler A, Lanier LL, Cwirla S, Phillips JH (1989): Comparative studies of human FcRIII-positive and negative NK cells. *J Immunol* 143: 3183–3191

Nakamura T, Takahashi K, Fukazawa T, Koyanagi M, Yokoyama A, Kato H, Yagita H, Okumura K (1990): Relative contribution of CD2 and LFA-1 to murine T and natural killer cell functions. *J Immunol* 145: 3628–3634

Nakamura T, Takahashi K, Koyanagi M, Yagita H, Okumura K (1991): Activation of a natural killer clone upon target cell binding via CD2. *Eur J Immunol* 21: 831–834

Niemi EC, Ryan JC, Seaman WE (1991): Mutational loss of NKR-P1 from RNK-16 cells is accompanied by loss of cytotoxicity against YAC-1 targets. *N Immun Cell Growth Regul* 10: 146–147

Orloff DG, Ra C, Frank SJ, Klausner RD, Kinet J-P (1990): Family of disulphide-linked dimers containing the ζ and η chains of the T-cell receptor and the γ chain of Fc receptors. *Nature* 347: 189–191

Ory PA, Clark MR, Kwoh EE, Clarkson SB, Goldstein IM (1989a): Sequences of complementary DNAs that encode the NA1 and NA2 forms of Fc receptor III on human neutrophils. *J Clin Invest* 84: 1688–1691

Ory PA, Clark MR, Talhouk AS, Goldstein IM (1991): Transfected NA1 and NA2 forms of human neutrophil Fc receptor III exhibit antigenic and structural heterogeneity. *Blood* 77: 2682–2687

Ory PA, Goldstein IM, Kwoh EE, Clarkson SB (1989b): Characterization of polymorphic forms of Fc receptor III on human neutrophils. *J Clin Invest* 83: 1676–1681

O'Shea J, Weissman AM, Kennedy ICS, Ortaldo JR (1991): Engagement of the natural killer cell IgG Fc receptor results in tyrosine phosphorylation of the ζ chain. *Proc Natl Acad Sci USA* 88: 350–354

Paolini R, Jouvin M-H, Kinet J-P (1991): Phosphorylation and dephosphorylation of the high-affinity receptor for immunoglobulin E immediately after receptor engagement and disengagement. *Nature* 353: 855–858

Perussia B, Ravetch JV (1991): FcγRIII (CD16) on human macrophages is a functional product of FcγRIII-2 gene. *Eur J Immunol* 21: 425–429

Perussia B, Starr S, Abraham S, Fanning V, Trinchieri G (1983): Human natural killer cells analyzed by B73.1, a monoclonal antibody blocking Fc receptor functions. I. Characterization of the lymphocyte subset reactive with B73.1. *J Immunol* 130: 2133–2141

Perussia B, Trinchieri G, Jackson A, Warner NL, Faust J, Rumpold H, Kraft D, Lanier LL (1984). The Fc receptor for IgG on human natural killer cells: Phenotypic, functional, and comparative studies with monoclonal antibodies. *J Immunol* 133: 180–189

Perussia B, Tutt MM, Qui WQ, Kuziel WA, Tucker PW, Trinchieri G, Bennett M, Ravetch JV, Kumar V (1989): Murine natural killer cells express functional Fcγ receptor II encoded by the FcγRα gene. *J Exp Med* 170: 73–86

Phillips JH, Chang C, Lanier LL (1991): Platelet-induced expression of FcγRIII (CD16) on human monocyte. *Eur J Immunol* 21: 895–899

Pletz GA, Grundy HO, Lebro RV, Yssel H, Barsh GS, Moore KW (1989): Human FcγRIII: Cloning, expression, and identification of the chromosomal locus of two Fc receptors for IgG. *Proc Natl Acad Sci USA* 86: 1013–1017

Qiu WQ, de Bruin D, Brownstein BH, Pearse R, Ravetch JV (1990): Organization of the human and mouse low-affinity FcgR genes: Duplication and recombination. *Science* 248: 732–735

Ra C, Jouvin M-HE, Blank U, Kinet J-P (1989): A macrophage Fcγ receptor and the mast cell receptor for immunoglobulin E share an identical subunit. *Nature* 341: 752–754

Ravetch JV, Perussia B (1989): Alternative membrane forms of FcγRIII (CD16) on human natural killer cells and neutrophils: Cell type-specific expression of two genes that differ in single nucleotide substitutions. *J Exp Med* 170: 481–497

Robertson MJ, Caligiuri MA, Manley TJ, Levine H, Ritz J (1990): Human natural killer cell adhesion molecules: Differential expression after activation and participation in cytolysis. *J Immunol* 145: 3194–3201

Rodewald H-R, Arulanandam ARN, Koyasu S, Reinherz EL (1991): The high affinity Fcε receptor γ subunit (FcεRIγ) facilitates T cell receptor expression and antigen/major histocompatibility complex-driven signaling in the absence of CD3ζ or CD3η. *J Biol Chem* 266: 15974–15978

Ryan RC, Niemi EC, Goldfien RD, Hiserodt JC, Seaman WE (1991): NKR-P1, an activating molecule on rat natural killer cells, stimulates phosphoinositide turnover and a rise in intracellular calcium. *J Immunol* 147: 3244–3250

Salmon JE, Brogle NL, Edberg JC, Kimberly RP (1991): Fcγ receptor III induces actin polymerization in human neutrophils and primes phagocytosis mediated by Fcγ receptor II. *J Immunol* 146: 997–1004

Samelson LE, O'Shea JJ, Luong H, Ross P, Urdahl KB, Klausner RD, Bluestone J (1987): T cell antigen receptor phosphorylation induced by an anti-receptor antibody. *J Immunol* 139: 2708–2714

Samelson LE, Patel MD, Weissman AM, Harford JB, Klausner RD (1989): Antigen activation of murine T cells induces tyrosine phosphorylation of a polypeptide associated with the T cell antigen receptor. *Cell* 46: 1083–1090

Scallon BJ, Scigliano E, Freedman VH, Miedel MC, Pan Y-CE, Unkeless JC, Kochan JP (1989): A human immunoglobulin G receptor exists in both polypeptide-anchored and phosphatidylinositol-glycan-anchored forms. *Proc Natl Acad Sci USA* 86: 5079–5083

Schmidt RE, Caulfield JP, Michon J, Hein A, Kamada MM, MacDermott RP, Stevens RL, Ritz J (1988): T11/CD2 activation of cloned human natural killer cells results in increased conjugate formation and exocytosis of cytolytic granules. *J Immunol* 140: 991–1002

Schmidt RE, Hercend T, Fox DA, Bensussan A, Bartley G, Daley JF, Schlossman SF, Reinherz EL, Ritz J (1985): The role of interleukin 2 and T11 E rosette antigen in activation and proliferation of human NK clones. *J Immunol* 135: 672–678

Schmidt RE, Michon JM, Woronicz J, Schlossman SF, Reinherz EL, Ritz J (1987): Enhancement of natural killer function through activation of the T11 E rosette receptor. *J Clin Invest* 79: 305–308

Schmidt RE, Murray C, Daley JF, Schlossman SF, Ritz J (1986): A subset of natural killer cells in pheripheral blood displays a mature T cell phenotype. *J Exp Med* 164: 351–356

Schuler W, Weiler IJ, Schuler A, Phillips RA, Rosenberg N, Mak TW, Kearney JF, Perry RP, Bosma MJ (1986): Rearrangement of antigen receptor genes is defective in mice with severe combined immune deficiency. *Cell* 46: 963–972

Scott CF, Boloender S, McIntyre GD, Holldack J, Lambert JM, Venkatesh YP, Morimoto C, Ritz J, Schlossman SF (1989): Activation of human cytolytic cells through CD2/T11: Comparison of the requirements for the induction and direction of lysis of tumor targets by T cells and NK cells. *J Immunol* 142: 4105–4112

Seaman WE, Eriksson E, Dobrow R, Imboden JB (1987): Inositol trisphosphate is generated by a rat natural killer cell tumor in response to target cells or to cross-linked monoclonal antibody OX-34: Possible signaling role for the OX-34 determinant during activation by target cells. *Proc Natl Acad Sci USA* 84: 4239–4243

Seaman WE, Niemi EC, Stark MR, Goldfien RD, Pollock AS, Imboden JB (1991): Molecular cloning of gp42: A cell-surface molecule that is selectively induced on rat natural killer cells by interleukin 2: Glycolipid membrane anchoring and capacity for transmembrane signaling. *J Exp Med* 173: 251–260

Selvaraj P, Rosse WF, Silber R, Springer TA (1988): The major Fc receptor in blood has a phosphatidylinositol anchor and is deficient in paroxysmal nocturnal haemoglobinuria. *Nature* 333: 565–567

Sentman CL, Hackett J, Kumar V, Bennett M (1989): Identification of a subset of murine natural killer cells that mediates rejection of $Hh\text{-}1^d$ but not $Hh\text{-}1^b$ bone marrow grafts. *J Exp Med* 170: 191–202

Sentman CL, Kumar V, Bennett M (1991): Rejection of bone marrow cell allografts by natural killer cell subsets: 5E6+ cell specificity for Hh-1 determinant 2 shared by $H\text{-}2^d$ and $H\text{-}2^f$. *Eur J Immunol* 21: 2821–2828

Shaw S, Luce GEG, Quinones R, Gress RE, Springer TA, Sanders ME (1986): Two antigen-independent adhesion pathways used by human cytotoxic T-cell clones. *Nature* 323: 262–264

Siliciano RF, Pratt JC, Schmidt RE, Ritz J, Reinherz EL (1985): Activation of cytolytic T lymphocyte and natural killer cell function through the T11 sheep erythrocyte binding protein. *Nature* 317: 428–430

Simmons D, Seed B (1988): The Fcγ receptor of natural killer cells is a phospholipid-linked membrane protein. *Nature* 333: 568–570

Spruyt LL, Glennie MJ, Beyers AD, Williams AF (1991): Signal transduction by the CD2 antigen in T cells and natural killer cells: Requirement for expression of a functional T cell receptor or binding of antibody Fc to the Fc receptor, FcγRIIIA (CD16). *J Exp Med* 174: 1407–1415

Sussman JJ, Bonifacino JS, Lippincott-Schwartz J, Weissman AM, Saito T, Klausner RD, Ashwell JD (1988): Failure to synthesize the T cell CD3-ζ chain: Structure and function of a partial T cell receptor complex. *Cell* 52: 85–95

Suzuki N, Suzuki T, Engleman EG (1991): Evidence for the involvement of CD56 molecules in alloantigen-specific recognition by human natural killer cells. *J Exp Med* 173: 1451–1461

Tetteroo PAT, van der Schoot CE, Visser FJ, Bos MJE,

von dem Borne AEGK (1987): Three different types of Fcγ receptors on human leucocytes defined by Workshop antibodies; FcγR$_{low}$ of neutrophils, FcγR$_{low}$ of K/NK lymphocytes, and FcγRII. In: *Leucocyte Typing III. White Cell Differentiation Antigens*, McMichael AJ, ed. Oxford: Oxford University Press, pp. 702–706

Ting AT, Einspahr KJ, Abraham RT, Leibson PJ (1991): Fcγ receptor signal transduction in natural killer cells: Coupling to phospholipase C via a G protein-independent, but tyrosine kinase-dependent pathway. *J Immunol* 147: 3122–3127

Trinchieri G (1989): Biology of natural killer cells. *Adv Immunol* 47: 187–376

Trounstine ML, Peltz GA, Yssel H, Huizinga TWJ, Kr. von dem Borne AEG, Spits H, Moore KW (1990): Reactivity of cloned, expressed human FcγRIII isoforms with monoclonal antibodies which distinguish cell-type-specific and allelic forms of FcγRIII. *Int Immunol* 2: 303–310

Tutt MM, Kuziel WA, Hackett J, Jr, Bennett M, Tucker PW, Kumar V (1986): Murine natural killer cells do not express functional transcripts of the α-, β-, or γ-chain genes of the T cell receptor. *J Immunol* 137: 2998–3001

Tutt MM, Schuler W, Kuziel WA, Tucker PW, Bennett M, Bosma MJ, Kumar V (1987): T cell receptor genes do not rearrange or express functional transcripts in natural killer cells of *scid* mice. *J Immunol* 138: 2338–2344

van de Griend RJ, Bolhuis RLH, Stoter G, Roozemond RC (1987): Regulation of cytolytic activity in CD3− and CD3+ killer cell clones by monoclonal antibodies (anti-CD16, anti-CD2, anti-CD3) depends on subclass specificity of target cell IgG-FcR. *J Immunol* 138: 3137–3144

van de Velde H, von Hoegen I, Luo W, Parnes JR, Thielemans K (1991): The B-cell surface protein CD72/Lyb-2 is the ligand for CD5. *Nature* 351: 662–665

Vivier E, Morin P, O'Brien C, Druker B, Schlossman SF, Anderson PA (1991a): Tyrosine phosphorylation on the FcγRIII (CD16): ζ complex in human natural killer cells. Induction by antibody dependent cytotoxicity but not by natural killing. *J Immunol* 146: 206–210

Vivier E, Morin PM, O'Brien C, Schlossman SF, Anderson P (1991b): CD2 is functionally linked to the ζ-natural killer receptor complex. *Eur J Immunol* 21: 1077–1080

Vivier E, Rochet N, Kochan JP, Presky DH, Schlossman SF, Anderson P (1991c): Structural similarity between Fc receptors and T cell receptors: Expression of the γ-subset of FcεRI in human T cells, natural killer cells, and thymocytes. *J Immunol* 147: 4263–4270

Weiss MJ, Daley JF, Hodgdon JC, Reinherz EL (1984): Calcium dependency of antigen-specific (T3-Ti) and alternative (T11) pathways of human T-cell activation. *Proc Natl Acad Sci USA* 81: 6836–6840

Weissman AM, Hou D, Orloff DG, Modi WS, Seuanez H, O'Brien SJ, Klausner RD (1988): Molecular cloing and chromosomal localization of the human T-cell receptor ζ chain: Distinction from the molecular CD3 complex. *Proc Natl Acad Sci USA* 85: 9709–9713

Welch GR, Wong HL, Wahl SM (1990): Selective induction of FcγRIII on human monocytes by transforming growth factor-β. *J Immunol* 144: 3444–3448

Welsh RM (1986): Regulation of virus infections by natural killer cells: A review. *Nat Immun Cell Growth Regul* 5: 169–199

Werfel T, Uchiechowski P, Tetteroo PAT, Kurrie R, Deicher H, Schmidt RE (1989): Activation of cloned human natural killer cells via FcγRIII. *J Immunol* 142: 1102–1106

Werner G, von dem Borne AEGK, Bos MJE, Tromp JF, van der Plas-van Dalen CM, Visser FJ, Engelfriet CP, Tetteroo PAT (1986): Localization of the human NA1 alloantigen on neutrophil Fcγ receptors. In: *Leucocyte Typing II*, Reinherz EL, Haynes BF, Nadler LM, Bernstein ID, eds. New York: Springer-Verlag, p. 109

Wong S, Freeman JD, Kelleher C, Mager D, Takei F (1991): Ly-49 multigene family: New members of a superfamily of type II membrane proteins with lectin-like domains. *J Immunol* 147: 1417–1423

Yokoyama WM, Jacobs LB, Kanagawa O, Shevach EM and Cohen DI (1989): A murine T lymphocyte antigen belongs to a supergene family of type II integral membrane proteins. *J Immunol* 143: 1379–1386

Yokoyama WM, Kehn PJ, Cohen DI, Shevach EM (1990): Chromosomal location of the Ly-49 (A1. YE1/48) multigene family: Genetic association with the NK1.1 antigen. *J Immunol* 145: 2353–2358

Yokoyama WM, Ryan JC, Hunter JJ, Smith HRC, Stark M, Seaman WE (1991): cDNA cloning of mouse NKR-P1 and genetic linkage with Ly-49: Identification of a natural killer cell gene complex on mouse chromosome 6. *J Immunol* 147: 3229–3236

Zegler DL, Hogarth PM, Sears DW (1990): Characterization and expression of an Fcγ receptor cDNA cloned from rat natural killer cells. *Proc Natl Acad Sci USA* 87: 3425–3429

9
Targeted Cellular Cytotoxicity

David M. Segal, Carolina R. Jost, and Andrew J.T. George

Introduction

Cell-mediated cytolysis occurs when a cytotoxic cell binds and delivers a "lethal hit" to a target cell. Several types of cells commonly found in blood are capable of performing cytolysis, including monocytes, neutrophils, eosinophils, natural killer (NK) cells, platelets, and T lymphocytes. The T-cell receptors (TCR) on T cells and the Fc$_\gamma$ receptors (Fc$_\gamma$R) on myeloid cells, NK cells, and platelets are two well-characterized families of cell surface glycoproteins that are involved in binding target cells and triggering lysis.

The TCR on cytotoxic T cells recognizes a peptide bound to a major histocompatibility complex (MHC) molecule on the surface of the target cell. Subsequently several other cell surface molecules such as CD4, CD8, and a number of adhesion molecules participate in the cell–cell interaction (Springer, 1990), resulting in a strong and specific binding between the cells. TCR crosslinking at the cell–cell interface most likely provides the cytotoxic cell with a signal that triggers it to deliver a lethal hit to the target cell. Similarly, cells mediating antibody-dependent cellular cytotoxicity (ADCC) recognize antibody-coated target cells via their Fc$_\gamma$ receptors (Lovchick and Hong, 1977; Segal, 1990). Again, crosslinking of Fc$_\gamma$R at the cell–cell interface is thought to trigger the effector cells to deliver a lethal hit. The TCR and the various Fc$_\gamma$R are therefore termed "triggering molecules". (See other chapters in Section II, this volume).

Cytolytic cells can be redirected against cells they would not normally lyse using antibodies against triggering molecules on their surfaces. For example, hybridoma cells producing antibodies against either TCR or Fc$_\gamma$R are lysed by appropriate cytotoxic cells (Ertl et al., 1982; Lancki et al., 1984; Graziano and Fanger, 1987a, 1987b; Graziano et al., 1989). This lysis is induced by binding of cell surface antibody (which is present on most hybridoma cells) to the triggering molecule on the cytotoxic cell, thus forming an effector:target cell conjugate, and triggering a lytic response (Figure 1A). Hybridoma cell killing has provided a convenient means of identifying triggering molecules on the surfaces of various types of cytolytic cells. Another type of antibody-mediated retargeted lysis occurs when anti-TCR IgG antibodies are present in mixtures containing cytotoxic T lymphocytes and target cells that express Fc$_\gamma$R, (Leeuwenberg et al., 1985): (Figure 1B). By a process known as "reverse ADCC," the anti-TCR antibodies link the target cells (via their Fc$_\gamma$R) to TCR on the effector cells and trigger lysis of the target cells.

Finally, cytotoxicity can be induced by using bispecific antibodies that incorporate the binding specificities of two different antibodies into one single molecule, one specificity being against a triggering molecule on a cytotoxic cell, the other being against a cell surface determinant on the target cell (Figure 1C). Addition of bispecific antibodies to effector cell–target cell mixtures allows the killing of a broad spectrum of target cells by numerous types of cytolytic cells (Karpovsky et al., 1984; Perez et al., 1985; Staerz et al., 1985; Titus et al., 1987b). The ability of bispecific antibodies to thus redirect cytolysis to a variety of cells, including tumor cells and virally infected cells, has resulted in numerous studies that aim ultimately at using bispecific antibodies in clinical protocols.

The general features of redirected cytolysis, the various aspects of producing bispecific antibodies, and the clinical application of redirected cytolysis will be discussed in this chapter.

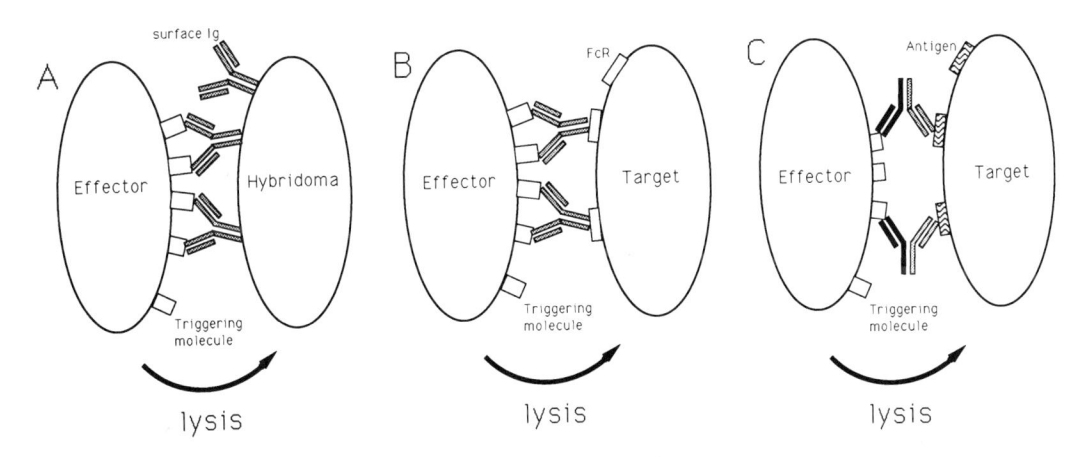

FIGURE 1. Types of targeted cytotoxicity. Three ways in which antibodies can redirect cell-mediated lysis are illustrated in this figure. (A) A hybridoma cell expressing on its surface an antibody against a triggering molecule binds an effector cell and induces its own lysis. (B) An IgG antibody against a triggering molecule mediates lysis of cells expressing Fc_γ receptors by bridging the triggering molecule on the effector cells with the Fc_γ receptors on the target cells. (C) A bispecific antibody mediates lysis of any desired target cell by binding antigens on target cells to triggering molecules on effector cells.

Effector Cells

$Fc_\gamma R$-Bearing Effector Cells

Three classes of human $Fc_\gamma R$ have been identified, first by using monoclonal antibodies, and later by gene cloning. These receptors, termed $Fc_\gamma RI$ (CD64), $Fc_\gamma RII$ (CD32), and $Fc_\gamma RIII$ (CD16), are present on different cell types, and by using bispecific antibodies or hybridoma cells as targets, the abilities of the various $Fc_\gamma R$ to serve as triggering molecules have been determined. Most, if not all, $Fc_\gamma R$ are able to trigger lysis of erythrocytes, regardless of the leukocyte type expressing them. However, the ability of $Fc_\gamma R^+$ cells to lyse tumor cells is much more restricted. Using hybridoma cell killing as a model, Fanger and colleagues showed that $Fc_\gamma RI$ on monocytes, macrophages, or interferon-γ (IFN-γ)-treated neutrophils, but not on U937 or HL-60 cell lines, is capable of triggering lysis of tumor cells. $Fc_\gamma RII$ is also capable of mediating hybridoma cell lysis when present on monocytes (Shen et al., 1989), but not on freshly prepared neutrophils or eosinophils. These latter cell types require activation with granulocyte colony-stimulating factor (G-CSF) to generate lytic activity through $Fc_\gamma RII$. Such treatment does not alter the number of $Fc_\gamma RII$ on the cell surface but does enable these cells to lyse the hybridoma cells (Graziano et al., 1989; Erbe et al., 1990). These results suggest that cytolytic ability depends not only on the presence of a triggering molecule, but also on the activation and/or differentiation state of the effector cell and the intrinsic lysability of the target cell.

Human $Fc_\gamma RIII$ exists in two forms produced by separate but closely related genes. NK cells bear a transmembrane form of $Fc_\gamma RIII$ ($Fc_\gamma RIII$-2) and are capable of mediating cytolysis (Titus et al., 1987b), whereas activated neutrophils possess a phosphatidylinositol glycan (PI)-linked form ($Fc_\gamma RIII$-1) and are incapable of killing hybridoma cells. This is not due to the fact that activated neutrophils are not cytolytic, as the same cells do trigger lysis through $Fc_\gamma RII$ (Graziano et al., 1989). Apparently, the PI-linked form of $Fc_\gamma RIII$ is incapable of triggering lysis of tumor cells.

Cytotoxic T Cells

TCR on cytotoxic T lymphocytes are also capable of triggering redirected cytolysis. In freshly isolated human blood T lymphocytes, a small subset of cells [approximately 2% of peripheral blood lymphocytes (PBL)], which is exclusively $CD8^+$ and $CD56^+$, mediates all redirected cytolysis, and these cells require interleukin-2 (IL-2) to maintain their optimal activity (Perez et al., 1986a). The great majority of PBL, however, are incapable of mediating lysis, and therefore require activation in order to generate targetable lytic activity. This has been achieved by several approaches, for example by crosslinking the TCR with immobilized anti-CD3 antibody and adding IL-2 (Garrido et al., 1990a) or by stimulating with lectins plus IL-2 (Staerz et al., 1985). In addition, T cells have been activated by using two different bispecific antibodies, anti-target × anti-CD3 plus anti-target × anti-CD28 (Jung et al., 1987; Jung and Muller-Eberhard, 1988). Presumably these two bispecific antibodies cause both CD3 and CD28 to coaggregate at the cell–cell interface when the bispecific antibodies bind to the target cells, a process that results in T-cell activation. Recently, T cells have also been activated by using trispecific antibodies that crosslink CD3 to CD2 on the surface of T cells while simultaneously binding the target cell through the third arm of the trispecific antibody (Jung et al., 1991; Tutt et al., 1991b).

Mechanisms of Targeted Lysis

Retargeting studies are usually done to develop strategies to target cytotoxic cells against tumor cells or virally infected cells. When using cytotoxic T cells as effector cells, bispecific antibodies provide the most convenient agents for (1) binding cytotoxic cells to target cells and (2) triggering lysis. In order to mediate these functions, both specificities of a bispecific antibody must be physically linked, as expected; administration of the two parental antibodies separately or together in an unlinked mixture does not result in either conjugate formation or target cell death. In the case of T cells, a bispecific antibody can alter the specificity of a particular clonotype, preventing it from recognizing its natural target, and causing it to lyse a surrogate target cell (Figure 2).

In $Fc_\gamma R$ mediated-cytolysis, convention IgG antibodies with specificity for the target cell can mediate ADCC. However, *in vivo*, circulating IgG (which is present at concentrations greater than the K_d of monomeric IgG for most $Fc_\gamma R$) will compete with antibody-coated target cells for $Fc_\gamma R$, and will therefore down-modulate ADCC (Segal et al., 1983). When the anti-$Fc_\gamma R$ portion of a bispecific antibody binds outside the ligand binding site or has a very high affinity for the $Fc_\gamma R$, then circulating IgG will not be able to block ADCC. Therefore, bispecific antibodies can potentially enhance ADCC in a physiological setting.

Not all cell surface molecules are capable of transducing lytic signals, and therefore not all bispecific antibodies are capable of mediating redirected cytolysis. For example, bispecific antibodies with anti-class I and anti-hapten specificities (anti-class I × anti-hapten) will link a hapten-modified target cell to a cytotoxic T lymphocyte expressing class I molecules, but the target cell will not be lysed. By contrast, an anti-TCR × anti-hapten bispecific antibody will both form effector–target conjugates and trigger targeted cell death (Perez et al., 1986a). In fact, most structures on effector cell surfaces are not targeting molecules. Molecules known to be capable of triggering redirected cytolysis by cytotoxic cells are indicated in Table 1. Myeloid cells are triggered exclusively by $Fc_\gamma R$. NK cells can be triggered both by $Fc_\gamma RIII$-2 and by CD2. T-cell triggering normally occurs through the TCR–CD3 complex consisting of an α-β (or γ-δ) heterodimer of the TCR, responsible for antigen binding, and several invariant chains of CD3 responsible for signal transduction. The CD3 complex is thought to be composed of at least two parallel signal transducing units: the $\gamma,\delta,\varepsilon$ subunits and the ζ-ζ or ζ-η dimers (Wegener et al., 1992). The ζ-ζ and ζ-η dimers are also associated with $Fc_\gamma RIII$-2 on NK cells (Lanier et al., 1989) and thus can act as transducing elements in those cells as well. Other cell surface structures can serve as cytotoxic triggers (Table 1), but their role in the lytic process remains unclear. Some of these molecules, such as CD2 and CD44, are adhesion molecules, normally involved in strengthening the binding of effector cells to target cells. In addition, the crosslinking of these molecules at the cell–cell interface may enhance the triggering signal.

Transduction of a lytic signal causes the cytotoxic cell to deliver a lethal hit to the target cell. Lethal hit delivery is known to involve the exocytosis of cytolytic granules at the cell–cell interface. This, in turn, exposes cells to several proteins, including a number of proteases and perforin (also known as cytolysin), a homolog of the C9 pore-forming protein of the complement system. Perforin is thought to be responsible for the formation of multiple pores in the plasma membranes of target cells (Henkart, 1985; Tschopp and Nabholz, 1990; Podack and Kupfer, 1991). NK cells are also known to use perforin in killing (Henkart et al., 1984; Henkart, 1985), but the exact mechanism by which $Fc_\gamma R$-mediated killing by other cells occurs is less well documented.

An important characteristic of cell-mediated lysis (either targeted or natural) is that only the bound target cells are lysed. Bystander cells in close proximity to both the effector and target cells are not lysed in the standard 4-hr ^{51}Cr-release assay (Perez et al., 1986a). This suggests that degranulation results in a localized release of cytolytic substances,

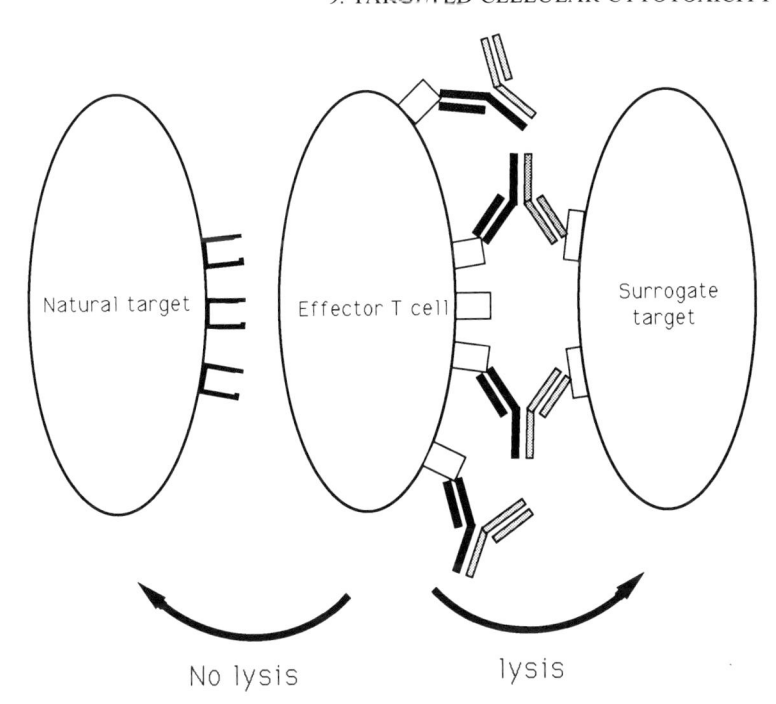

FIGURE 2. Bispecific antibodies induce cytotoxic T cells to lyse surrogate target cells by linking their TCR with target cell antigens. In addition, these same bispecific antibodies can inhibit lysis of the natural target cells, thus, in effect, changing the specificity of the cytotoxic T cell.

Tumor Growth Inhibition

When T cells, tumor, and bispecific antibody are mixed and injected subcutaneously into murine hosts, tumor growth is strongly blocked in a manner that is dependent upon the bispecific antibody (Titus et al., 1987a). However, in contrast to the situation *in vitro*, the growth of tumor cells that are not recognized by the bispecific antibody (i.e. bystander tumor cells) is inhibited in the subcutaneous environment, provided that bona fide target cells (i.e. cells that do bind the bispecific antibody) are also present. This suggests that targeted T cells may affect tumor growth *in vivo* by a mechanism(s) different from direct lysis. The most likely mechanism for inhibiting the growth of bystander cells would involve a local release of tumoristatic factors by T cells as a result of TCR crosslinking. Such factor release was in fact observed in an *in vitro* tumor growth inhibition assay in which tumor and targeted T cells were cultured together for 7–9 days. In this assay, bystander tumor growth was blocked and factors released into the medium were shown to contain tumor growth inhibitory activity. Antibodies against TNF-α and IFN-γ reversed the inhibition of tumor growth, suggesting that these cytokines were necessary for mediating this antitumor activity (Qian et al., 1991).

Figure 3 illustrates the mechanism by which the growth of bystander tumor cells is blocked by targeted T cells. Thus, targeted T lymphocytes are able to eradicate tumor in at least two ways: firstly by direct cytolysis, and secondly by growth inhibition via factors secreted into the milieu. The latter mechanism has the potential of being able to block, *in vivo*, the growth of tumor cells that are inaccessible to the targeted T cells, or that have lost their tumor antigen and so would escape detection by the targeted cytotoxic T lymphocytes.

TABLE 1. Effector cells and their triggering molecules

Effector cells	Triggering molecules
T cells	TcR/CD3, CD44, CD69, Ly-6.2C (mouse)
NK cells	CD2, CD16 (FcRIII-2)
Monocytes	CD64 (FcRI), CD32 (FcRII), CD16 (FcRIII-2)
Neutrophils	CD64 (FcRI), CD32 (FcRII)
Eosinophils	CD64 (FcRI), CD32 (FcRII)

Data taken from the following references: TcR/CD3 (Leeuwenberg et al., 1985; Staerz et al., 1985; Perez et al., 1985), CD2 (Siliciano et al., 1985), CD44 (Seth et al., 1991), CD69 (Moretta et al., 1991), Ly-6.2C (Leo et al., 1987), CD64 (Graziano and Fanger, 1987b; Shen et al., 1989; Erbe et al., 1990), CD32 (Graziano et al., 1989; Shen et al., 1989; Erbe et al., 1990), CD16 (Titus et al., 1987b; Shen et al., 1989; Erbe et al., 1990).

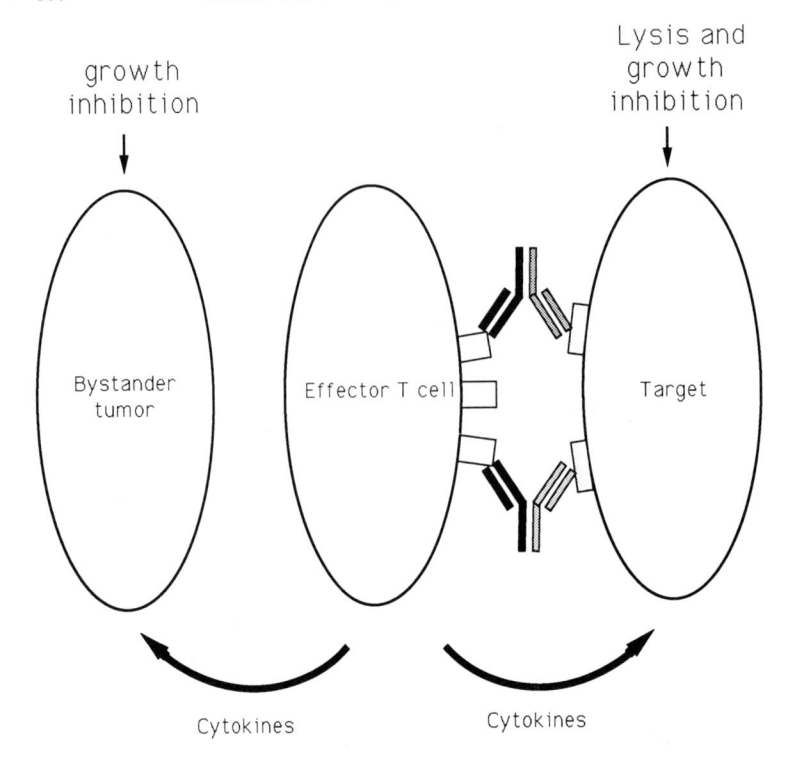

FIGURE 3. Targeted cytokine release. Bispecific antibodies, in the presence of target cells, crosslink TCR on T cells, inducing them to release cytokines into the medium that can block the growth of both target and bystander cells.

Types of Bispecific Antibodies

Several methods exist for the manufacture of bispecific antibodies. However, none of them is optimal, and it has proved difficult to produce the large amounts of bispecific antibody needed for clinical trials. At present there are two major ways to produce bispecific antibodies: by fusion of two hybridomas to produce a hybrid hybridoma (or "quadroma") capable of secreting antibodies with dual specificity, and by chemical conjugation of antibodies or their fragments.

Hybrid Hybridomas

Fusion of two hybridomas to produce a hybrid hybridoma is often used to generate bispecific antibodies (Milstein and Cuello, 1983). A major advantage of this approach is that the hybrid hybridoma can be a stable source, producing the bispecific antibody on a continual basis. However, the production of a stable hybrid hybridoma is not an insignificant task. In addition, the combination of the two heavy and two light chains produced by the hybrid hybridomas can lead to assembly of up to 10 different antibody species, only one of which has bispecific activity, both reducing the yield of bispecific antibody, and necessitating complicated purification protocols.

Fusion and Selection of Hybrid Hybridomas

Fusion of two hybridoma cells to produce a hybrid hybridoma can be carried out by conventional methods, such as polyethylene glycol fusion. However, it is necessary to select the fused hybrid hybridoma cells from the unfused parental lines, which is most often done by the introduction of selectable drug markers. The most common such marker is the loss of the enzyme hypoxanthine phosphoribosyl transferase (HPRT), which makes the cell line unable to grow in hypoxanthine, aminopterin, and thymidine (HAT) medium. Selection of HPRT-negative cells can be accomplished by growing the hybridomas in increasing concentrations of 8-azaguanine or 6-thioguanine.

In order to select a hybrid hybridoma from each parental line, two selectable markers must be used. For example, HAT sensitivity can also be produced by selecting thymidine kinase (TK)-negative varients using bromodeoxyuridine. Fusion of two hybridomas, each with a complementary enzyme deficiency [(HPRT$^-$, TK$^+$) and (HPRT$^+$, TK$^-$)] will allow selection on HAT medium of the (HPRT$^+$, TK$^+$) hybrid hybridoma cells (Wong and Colvin, 1987; Urnovitz et al., 1988).

If one hybridoma cell line is to be used as a fusion partner with a number of hybridomas, then it may be advantageous to carry all the selection markers in that line, obviating the need to introduce markers into every hybridoma to be fused. The introduction of a positive selection marker, resistance to a particular drug, together with a second negative selection marker, such as sensitivity to HAT medium, will allow, following fusion with a wild-type hybridoma (HAT-resistant, drug-sensitive), the selection of hybrid hybridomas in medium containing both HAT and the drug. Such drug resistance can be obtained either by selecting variants that grow in the drug (e.g., ouabain; Staerz and Bevan, 1986a) or by introduction of a plasmid containing a gene that confers resistance to a particular drug (e.g., geneticin, a neomycin analog) (Lanzavecchia and Scheidegger, 1987; De Lau et al., 1989).

Rather than using a dual drug selection system, one can metabolically poison one of the hybridoma cell lines immediately prior to fusion by treatment with iodoacetamide (Clark et al., 1987). The killed cell line is unable to grow but will confer HAT resistance when fused to a HAT-sensitive hybridoma, allowing selection of the hybrid hybridoma in HAT medium. An extension of this approach treats both hybridoma cell lines with metabolic inhibitors that have distinct actions, so that both types of cell die unless fused to the other. Such an approach has been used with actinomycin D to inhibit RNA synthesis and with emetine to block protein synthesis (Suresh et al., 1986).

An alternative strategy is to use just one established hybridoma (HAT-sensitive) and fuse it directly with spleen cells from an animal immunized with an appropriate antigen (Milstein and Cuello, 1983, 1984). The splenic B cells will be unable to grow *in vitro*, and the resulting "triomas" can be screened directly for the production of bispecific antibodies of the desired specificities. The obvious disadvantage of this approach is that the specificity of one arm of the bispecific has to be characterized from scratch.

Rather than using drug markers to select the hybrid hybridomas, one can directly pick the fused cells using fluorescent cell sorting techniques. Thus the two parental hybridoma cells can be labeled with different fluorescent dyes and the fused, doubly labeled cells selected from singly labeled parents by fluorescence-activated cell sorting (Karawajew et al., 1987). A number of dye pairs have been used, for example, hydroethidine and rhodamine 123, two intracellular stains which produce, respectively, red and green fluorescent signals (Shi et al., 1991), and octadecylamine fluorescein isothiocyanate (FITC) and octadecylamine tetramethyl rhodamine isothiocyanate (TRITC), whose long hydrophobic tails anchor the molecules in the cell membrane (Koolwijk et al., 1988). Alternatively, one can directly label the cell surface using FITC or TRITC (Karawajew et al., 1987). It is essential that all the dyes used are stable and will not be passed from one cell to the other under the extreme conditions of polyethylene glycol (PEG) fusion. In addition, the fluorescent signals of the two dyes must be well separated to allow unequivocal determination of the cell populations. Isolating the small percentage of fused cells while preventing contamination with the parental cells can be difficult, but the efficiency of the process may be improved by Percoll gradient centrifugation (Koolwijk et al., 1988), or by sorting a second time, using a vital DNA stain to identify the octaploid (i.e., fused) cells (Karawajew et al., 1990).

Fusion of two hybridomas that make either pentameric IgM or dimeric IgA molecules can lead to the production of bispecific polymeric immunoglobulins that incorporate the 8S subunits of both antibody molecules (Urnovitz et al., 1988). While the large size of these molecules may be disadvantageous in many settings, their multivalency may be useful if the antibodies are of low affinity.

A final approach to the production of heterohybridomas is to transfect into a cell the relevant genes encoding the heavy and light chains of the two antibodies (Songsivilai et al., 1989). This method allows the use of humanized or chimerized immunoglobulin molecules, and so may have advantages in therapeutic situations.

Purification of Bispecific Antibodies

A total of up to 10 different immunoglobulin species can be produced by a hybrid hybridoma, only one of which is the desired bispecific antibody (Figure 4) (Milstein and Cuello, 1983). While three of the species produced are "nonsense" molecules, whose heavy and light chains are totally mispaired and so have none of the specificities of the parental antibodies, six of the molecules share the same reactivity as one of the parental antibodies. These antibodies may be deleterious in therapy by blocking the binding of the bispecific antibody and by modulating the target antigen. In addition, antibodies directed against effector cells may target them for lysis via ADCC mechanisms (Clark et al., 1987), and, in the case of anti-CD3, be directly immunosuppressive (Ortho Multicenter Transplant Study Group, 1985).

The proportion of the bispecific species secreted by the hybrid hybridoma cells depends on a number

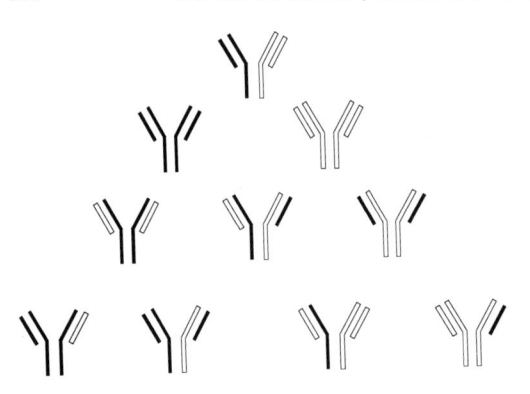

FIGURE 4. Production of immunoglobulin molecules by hybrid hybridoma cells. Up to 10 different immunoglobulin species can be produced by a hybrid hybridoma. Only one (top row) is the desired bispecific antibody. Two of the molecules (second row) represent the parental antibodies, three (third row) are "nonsense" molecules, while four (fourth row) are various univalent forms of the parental antibodies.

of factors, including the relative rates of synthesis of the various heavy (H) and light (L) chains, and any preferential associations that might occur between chains. In the situation where the synthesis of the chains is equal and their combination totally random, 12.5% of the total immunoglobulin produced would be bispecific (Staerz and Bevan, 1986a). However, the actual yield can vary between 0 and 50% (Corvalan and Smith, 1987; Songsivilai and Lachmann, 1990). The ability of the H and L chains to correctly pair greatly influences the yield of bispecific antibody, and it has been suggested that the homologous H and L chains will show preferential pairing in a quadroma (Milstein and Cuello, 1984). If this were the case, then the yield of bispecific could approach a maximum theoretical level of 50% (assuming random recombination of H chains). However, recent studies have indicated that while, in some cases, there can be a propensity for homologous H-L chain pairing, preferential heterologous, or random, pairing often occurs (De Lau et al., 1991). Thus, although in some quadromas preferential H-L chain pairing will boost the yield of bispecific antibody, this cannot be relied on, and in some cases the yield will be lower than that theoretically predicted, due to heterologous H-L chain preferences.

The association of the two H chains should be random if both antibodies are of the same immunoglobulin subclass. However, if two different subclasses are used, then the yield of mixed H-H chain combinations may drop (Milstein and Cuello, 1984), and, if two different classes of H chain are used, association between the heterologous H chains may not be possible (Urnovitz et al., 1988). While the formation of mixed subclass bispecific antibodies may seem disadvantageous, the lower yield may be offset by an increased ability to purify the mixed H chain bispecific molecule (Milstein and Cuello, 1984). A further advantage of mixed isotype bispecific antibodies may be that they are often incapable of mediating ADCC of effector cells bound by the bispecific antibody (Clark and Waldmann, 1987; Clark et al., 1988).

The best method for isolation of a bispecific antibody is dependent on the nature of the antibodies used. In most cases a combination of different methods is necessary to achieve full purification. In the rare cases when the antigen recognized by one arm of the bispecific antibody is available in sufficient quantities, one can use it to affinity purify the antibody (Corvalan and Smith, 1987; Brissinck et al., 1991; Demanet et al., 1991). Otherwise one has to rely on differences between the physical properties of the two parental antibodies. Thus, differential adsorption to protein A (Demanet et al., 1991), various types of ion exchange chromatography (Milstein and Cuello, 1983; Staerz and Bevan, 1986a), preparative isoelectric focusing (Wong and Colvin, 1987) and hydrophobic interaction chromatography (Weiner and Hillstrom, 1991) have all been used to purify hybrid hybridoma antibodies.

Chemical Conjugation

A number of methods exist for chemical conjugation of antibodies or their fragments. Some produce a crude mix of heteroaggregates which, while suitable for experimental *in vitro* use, cannot be used in patients, while others produce suitable, well-defined molecular species suitable for *in vivo* applications.

Simple conjugation of two antibodies can be achieved using a linker such as succinimidyl-3-(2-pyridyldithiol)-propionate (SPDP) to crosslink ε-amino groups on lysine residues, forming disulfide bonds between the molecules (Figure 5) (Karpovsky et al., 1984). Such linkages form randomly, and the resulting bispecific molecules can consist of more than two antibodies linked together at random sites in the molecules. The formation of monospecific homoaggregates is prevented by labeling both antibodies with SPDP, reducing one, and mixing it with the non-reduced molecule (Karpovsky et al., 1984) (Figure 5). Such

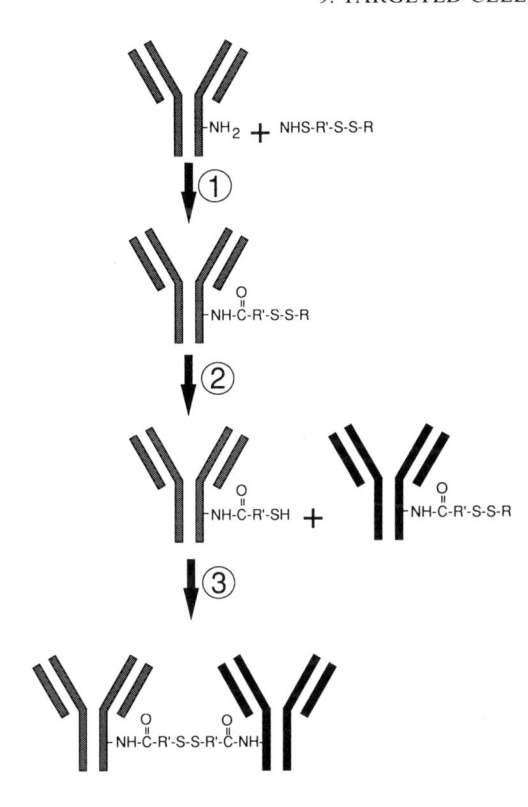

FIGURE 5. Conjugation of two antibodies via lysine residues. N-Hydroxysuccinimide (NHS) containing crosslinkers, such as N-succinimidyl 3-(2-pyridyldithio)propionate (SPDP), introduce dithiol groups onto ε-amino groups of lysine residues (1). Subsequent reduction of the disulfide bonds forms free SH groups (2) that will exchange with a dithiol derivatized molecule to form disulfide bonds between the two antibodies (3). In the case of SPDP R' = CH2CH2, R = pyridine.

directed coupling can also be achieved by labeling one antibody with iminothiolane (Traut's reagent), which introduces sulfhydryl groups directly onto the lysine residues (King et al., 1978), followed by reaction with a SPDP-coupled antibody (Liu et al., 1985). Bispecific antibodies formed in this manner have reducible disulfide bonds, which may be unstable in certain situations. An alternative is to link maleimide-containing compounds to ε-amino residues of lysine residues of one antibody, using compounds such as m-maleimidobenzoyl-N-hydroxysuccinimide ester (MBS); reaction with either reduced SPDP-linked or iminothiolane-linked antibody leads to the formation of nonreducible thioether bonds between the two molecules (Segal and Snider, 1989).

For many applications, bispecific antibodies made from Fab or Fab' fragments of antibodies may be preferable to those made from whole IgG. The lack of an Fc region will prevent bispecific antibodies from binding to Fc receptors, thus removing their ability to target effector cells for lysis by ADCC mechanisms and decreasing their chances of being cleared by the reticuloendothelial system. In addition, removal of Fc fragments from the antibodies might decrease their immunogenicity and improve their pharmacokinetic behavior. It is possible to randomly link antibody Fab fragments at their lysine residues, as described for whole antibody molecules (Karpovsky et al., 1984; Perez et al., 1986a; Shen et al., 1986).

As an alternative to crosslinking at lysine residues, one can utilize the hinge region cysteines made available by the reduction of F(ab')$_2$ fragments (normally produced by pepsin digestion of whole IgG) into Fab' fragments (Figure 6). A mixture of Fab' fragments can be reoxidized via the hinge region cysteines, as was first shown for polyclonal rabbit antisera (Nisonoff and Mandy, 1962). However, such a procedure leads to the formation of both homodimers and heterodimers. For a more defined product, the two Fab' can be crosslinked by disulfide exchange. For this, it is necessary to form disulfide groups on one of the Fab' using reagents such as 5,5'-dithiobis(2-nitrobenzoic acid) (Ellman's reagent). Mixing reduced Fab' with the dithiol derivative produces only the bispecific molecule, uncontaminated by parental type F(ab')$_2$ (Brennan et al, 1985).

Stable thioether bonds between the cysteine residues can be produced by use of a dimaleimide reagent such as o-phenylenedimaleimide (o-PDM) to crosslink two Fab' fragments (Glennie et al., 1987). This reaction can be highly specific, as the hinge region SH groups on one of the Fab' molecules can by alkylated with o-PDM to provide free maleimide groups that can react with the hinge region SH groups on the other Fab' species (Glennie et al., 1987) (Figure 7). In addition, as most Fab' have more than a single SH in their hinge region, it is possible, by altering the reaction conditions, to generate bispecific F(ab')$_3$ constructs in which two of the arms have the same specificity, while the third is derived from the other antibody (Tutt et al., 1991a). Bivalent recognition of one of the antigens confers a higher avidity on the F(ab')$_3$ reagent to that antigen, which can lead to more efficient targeting of cytotoxicity (Tutt et al., 1991a). By using a modification of the same procedure, trispecific F(ab')$_3$ molecules can be produced, in which each of the Fab' arms has a different specificity (Tutt et al., 1991b).

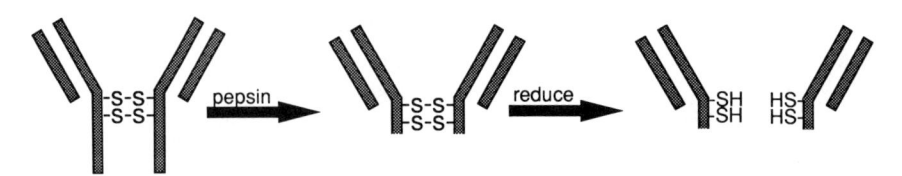

FIGURE 6. Formation of antibody Fab' fragments. Cleavage of the Fc regions of antibodies with pepsin forms F(ab')$_2$ fragments which, upon reduction, yield the Fab' species with free hinge region SH groups available for crosslinking.

Methods such as the directed disulfide exchange or the use of dimaleimide crosslinkers are capable of producing well-defined bispecific antibodies, suitable for *in vivo* applications. However, they rely on the production of F(ab')$_2$ fragments from the parental antibodies. This can lead to a substantial loss of material and a decrease in yield. Such problems may be overcome by producing recombinant F(ab')$_2$ molecules in bacterial expression systems and chemically crosslinking them as for conventional F(ab')$_2$ fragments (Shalaby et al., 1992). The recombinant F(ab')$_2$ molecules can be produced in large amounts and do not need digestion prior to crosslinking.

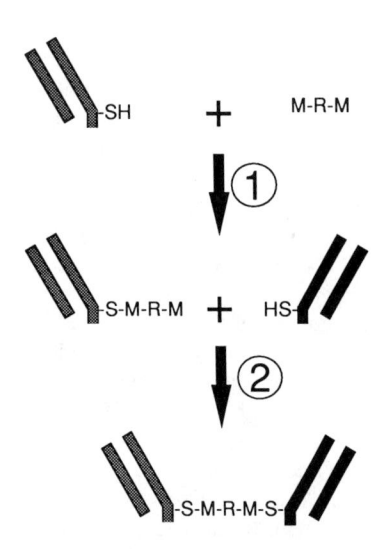

FIGURE 7. Reaction of hinge region sulfhydryl groups with a dimaleimide compound (M-R-M) leads to the formation of a maleimidated derivative (1), which can react with a second Fab'S-H to form the bispecific F(ab')$_2$ molecule, where the two Fab' fragments are linked by tandem thioether bonds.

Genetic Engineering

In recent years genetic engineering techniques have allowed the development of a number of novel recombinant antibody molecules (Winter and Milstein, 1991). These antibodies can be designed to have improved properties for therapeutic or diagnostic applications. The ability to produce bispecific antibodies by such techniques is an attractive prospect, as it should be possible to produce large amounts of highly pure material relatively cheaply, and in addition, one could modify the molecule to make it more suitable for clinical use, for example, by humanization (Winter and Milstein, 1991).

One advance that may be of considerable benefit to the development of recombinant bispecific antibodies has been the development of single-chain Fv (sFv) constructs, which consist of just the two variable domains of the antibody molecule linked by a peptide spacer (Bird et al., 1988; Huston et al., 1988). The sFv molecules contain the antigen-binding sites of antibodies, and so show similar specificities and affinities as the Fab fragments derived from the parental antibodies. The small size (25–30 kDa) of sFv as compared to the intact IgG (about 150 kDa) can lead to increased tissue penetrance and more rapid clearance of unbound antibody. In addition, sFvs are likely to prove less immunogenic than whole antibodies. One enticing prospect would be to link two such sFv together to form a single-chain bispecific antibody (Andrew et al., 1991). The fusion of sFv with other proteins, such as toxins (Chaudhary et al., 1989), protein A (Tai et al., 1990), and CD4 (Traunecker et al., 1991) suggests that such an approach is feasible. Though it has yet to be determined what approach will be best for the production of recombinant bispecific molecules, it is to be hoped that future technological advances will make possible the production of large quantities of bispecific antibody suitable for clinical use.

Clinical Uses of Targeted Lymphocytes

Background

It is now well established that bispecific antibodies can target T cells (Segal and Snider, 1989), NK cells (Titus et al., 1987b), and myeloid cells (Graziano and Fanger, 1987b; Fanger et al., 1989; Fanger and Guyre, 1991) against tumor and virally infected cells *in vitro*. In cases where the effector cells exhibit no natural cytotoxicity against the target cells, bispecific antibodies induce a totally new lytic activity, while in cases where the effector cells are naturally cytotoxic, bispecific antibodies can enhance lysis.

Virally infected cells often express viral proteins on their surfaces. Therefore, antibodies against viral coat proteins when linked to anti-CD3 or anti-CD16 cause T or NK cells to specifically lyse infected cells. Examples include cells infected with human immunodeficiency virus (Zarling et al., 1988), herpes simplex virus (Paya et al., 1989), and influenza virus (Moran et al., 1991). Blockage of viral replication by targeted T cells has also been reported (Paya et al., 1989; Moran et al., 1991). When mAb that bind preferentially to tumor cells are coupled to anti-CD3 or anti-Fc$_\gamma$R mAb, the resulting bispecific antibodies cause T (Perez et al., 1986b; Segal and Snider, 1989), NK (Titus et al., 1987b), and myeloid cells (Graziano and Fanger., 1987b; Fanger et al., 1989; Fanger and Guyre, 1991) to lyse tumor cells *in vitro*. Antitumor antibodies recognize tumor-associated antigens (Jung et al., 1986; Perez et al., 1986b; Segal and Wunderlich, 1988; Segal and Snider, 1989), many of which are also present to small extents on normal cells, while others, such as idiotypic determinants of surface Ig on B-cell neoplasms (Brissinck et al., 1991; Demanet et al., 1991; Weiner and Hillstrom, 1991) are expressed exclusively by the tumor cells. Cross reactions of antitumor bispecific antibodies on normal cells are likely to be small (depending upon the specificity of the antitumor component) and could be compensated for in at least two ways when targeted lymphocytes are directed against tumor cells. First, a threshold for triggering T cells may exist such that normal cells expressing low antigen levels may not trigger a T-cell response. Second, targeted cytotoxic lymphocytes may act by releasing cytokines such as tumor necrosis factor-α (TNF-α) that are much more toxic to tumor than to normal cells (Old, 1985; Larrick and Wright, 1990). Thus targeted T or NK cells might respond to both tumor and normal cells, but kill only the tumor cells.

Mouse Models

A major problem in studying targeted lymphocytes in tumor-bearing mice has been the relative lack of availability of mAb suitable for targeting murine T cells against murine tumors. Accordingly, models have been developed in which human tumors are transplanted into immunodeficient mice, followed by treatment with targeted human lymphocytes. Early studies showed that targeted T (Titus et al., 1987a) and NK (Titus et al., 1987b) cells could mount a potent antitumor response in a Winn-type assay in which effector cells, tumor cells, and bispecific antibody were mixed and injected subcutaneously into nude mice. Experiments showing that the growth of bystander tumor cells is blocked in such assays strongly suggested that the release of cytokines by lymphocytes played an important role in blocking tumor growth in the subcutaneous environment (Qian et al., 1991).

In subsequent investigations (Garrido et al., 1990b; Mezzanzanica et al., 1991), human ovarian carcinoma cells were injected intraperitoneally into nude mice, and tumors were allowed to become established for 4 days, by which time the number of tumor cells in ascites had more than doubled, and solid tumor had begun to metastasize to the pancreas and mesenteric lymph nodes (Mezzanzanica et al., 1991). On day 4, mice were given intraperitoneal injections of human T cells targeted with bispecific antibodies containing anti-CD3 linked to an antitumor antibody. When mice were examined for tumor growth by peritoneal lavage on day 15, a large majority of mice had no detectable tumor. Moreover, mean survival times of treated mice were dramatically increased when compared with those of untreated controls, although most of the treated mice eventually died of tumor. Thus, treatment of tumor-bearing mice with targeted human T cells on day 4 presumably destroyed most, but not all, tumor cells, and the residual cells eventually grew out and killed the animals.

A more realistic mouse model for studying the effects of targeted cytotoxic cells on tumors would involve murine effector cells targeted against murine tumors. Toward this end, Staerz and Bevan (1986b) treated Thy-1.2$^+$ mice bearing a Thy-1.1$^+$ lymphoma with activated host-type T cells and a bispecific antibody directed against Vβ8$^+$ T-cell receptors and Thy-1.1. Preliminary results showed that treated mice were significantly protected against the tumor when compared with untreated controls. More recently, two laboratories have reported that targeted T cells have a remarkably powerful ability to block the growth of BCL_1 (Brissinck et al., 1991) and 38C13 (Demanet et al., 1991; Weiner and Hillstrom, 1991) syngeneic lymphoma transplants in mice. Both tumors were treated with hybrid hybridoma antibodies having specificity for CD3 and an idiotypic determinant on the surface IgM of the lymphoma. Preliminary *in vitro* results showed that such bispecific antibodies had the ability to specifically redirect T cell lysis against the lymphoma cells. BCL_1 is a relatively slow-growing lymphoma that kills essentially all mice receiving a 500-tumor-cell inoculum by day 80. Brissinck et al. (1991) treated mice that had received 500 tumor cells on day 0 with a single 5-μg intraperitoneal injection of bispecific antibody on day 9. By day 150, all mice that had been given the bispecific antibody remained tumor-free, whereas all of the untreated mice had died of lymphoma. Effective therapy was also observed when mice were given higher tumor doses on day 0, and in all of these experiments, the bispecific antibody was found to be much more effective in blocking tumor growth than either parental antibody alone.

In contrast to BCL_1, 38C13 is an extremely aggressive tumor; mice receiving 100 tumor cells intraperitoneally have a mean survival time of about 20 days. When mice were given 5 or 10 μg of bispecific antibody intravenously (Demanet et al., 1991) or intraperitoneally (Weiner and Hillstrom, 1991) 2 days after tumor, highly significant cure rates were observed (Demanet et al., 1991; Weiner and Hillstrom, 1991) and *in vivo* activation of T cells by inoculation with allogeinic cells enhanced the efficacy of the treatment (Demanet et al., 1991). However, in some of the animals treated with bispecific antibodies, tumors arose that had lost the idiotypic determinant to which the bispecific antibody bound (Weiner and Hillstrom, 1991), suggesting that the treatment had selected for antigen-loss variants.

Lymphocyte Targeting in Cancer Patients

A major obstacle to using targeted cells in patients has been the lack of availability of sufficient amounts of clinical-grade bispecific antibodies. Nevertheless, suitable bispecific antibodies are beginning to become available, and as a result, two clinical trials are currently underway and others are in the planning stage. In one study (Nitta et al., 1990), targeted T cells are being used to treat patients suffering from malignant glioma. Tumors are first removed surgically, leaving a passage with access to the tumor site through an Ommaya reservoir. Patients' peripheral blood mononuclear cells are activated *ex vivo* with IL-2, incubated with 100 μg of a hetero-F(ab')$_2$ anti-CD3 × antitumor bispecific antibody, and infused six times over a 3-week period into the tumor site. In an initial publication (Nitta et al., 1990), 10 patients were treated with bispecific antibody together with activated lymphocytes; four patients showed tumor regression 10–18 months following therapy, and another 4 showed eradication of tumor remnants left after surgery. By contrast, in the historical control group, 9 of the 10 patients had recurrences within 1 year of therapy and 8 patients had died within 4 years. In the latest report (Nitta et al., 1991), 77% of patients treated with targeted cells were alive and free of tumor 2 years after treatment, whereas only 33% of patients treated with activated lymphocytes alone were alive. The authors emphasize that their results are preliminary, but promising.

A phase I–II clinical trial to treat ovarian carcinoma with activated, targeted T cells is currently underway in the Netherlands and in Italy (R. Bolhuis and G. Stoter, Daniel den Hoed Cancer Center, Rotterdam, personal communication). In this study, ovarian cancer patients are given multiple intraperitoneal injections of autologous activated T cells in the presence of recombinant IL-2 and a murine hybrid hybridoma antibody with anti-CD3 and anti-ovarian cancer specificities (Lanzavecchia and Scheidegger, 1987; Miotti et al., 1987). Adverse side reactions in patients have been minimal, although some human anti-mouse antibody (HAMA) response has been observed and may prove to be a limiting factor in the future. Early results concerning the efficacy of therapy are encouraging but must be considered anecdotal due to the low number of patients treated. Other clinical trials are currently being developed, and we should know within the next few years whether targeted lymphocytes can be used beneficially to treat cancers.

Conclusions

Bispecific antibodies change the specificities of cytotoxic cells by linking triggering molecules on cytotoxic cells to cell surface components on target cells. As a result, effector cell–target cell conjugates are formed, and triggering molecules are cross-linked at the cell–cell interface, thus inducing a cytotoxic response from the effector cell. Two major classes of triggering molecules, T-cell receptors on all populations of T cells, and Fc$_\gamma$ receptors on NK cells and myeloid cells, can be used to target effector cells.

Lymphocytes can mediate at least two targetable responses: (1) direct lysis of bound target cells and (2) cytokine release which can block the growth of both bystander and bound tumor cells. Since bispecific antibodies can focus all T cells, NK cells, and myeloid cells against a particular antigen, they have great potential for therapeutic use, especially in treating tumors and virally infected cells. Indeed, a number of studies have demonstrated that targeted T cells can eradicate established tumors in experimental animals. Bispecific antibodies can be produced by chemical methods, by fusing hybridoma lines, and by genetic engineering. However, the difficulty in producing large amounts of clinical-grade bispecific antibody has hampered efforts to establish clinical trials. In spite of this problem, preliminary clinical trials using targeted lymphocytes against glioblastomas and ovarian carcinoma have been initiated and already show some encouraging results. Thus, targeting experiments have provided valuable insights into the mechanisms by which effector cells mediate their immune functions, and the clinical use of targeted lymphocytes remains an exciting and active field of investigation.

References

Andrew SM, Perez P, Nicholls PJ, George AJT, Huston JS, Oppermann H, Seal DM (1991): Production of single chain bispecific antibody by recombinant DNA technology. In: *Bispecific Antibodies and Targeted Cellular Cytotoxicity; Second International Conference, Seillac, France. October 9–13, 1990*, Romet-Lemonne JL, Fanger MW, Segal DM, eds. Fondation Nationale de Transfusion Sanguine, Les Ulis, France. 197–199

Bird RE, Hardman KD, Jacobson JW, Johnson S, Kaufman BM, Lee SM, Lee T, Pope SH, Riordan GS, Whitlow M (1988): Single-chain antigen-binding proteins. *Science* 242: 423–426

Brennan M, Davison PF, Paulus H (1985): Preparation of bispecific antibodies by chemical recombination of monoclonal immunoglobulin G1 fragments. *Science* 229: 81–83

Brissinck J, Demanet C, Moser M, Leo L, Thielemans K (1991): Treatment of mice bearing BCL$_1$ lymphoma with bispecific antibodies. *J Immunol* 147: 4019–4026

Chaudhary VK, Queen C, Junghans RP, Waldmann TA, FitzGerald DJ, Pastan I (1989): A recombinant immunotoxin consisting of two antibody variable domains fused to pseudomonas exotoxin. *Nature* 339: 394–397

Clark MR, Waldmann H (1987): T-cell killing of target cells induced by hybrid antibodies: Comparison of two bispecific monoclonal antibodies. *J Natl Cancer Inst* 79: 1393–1401

Clark M, Gilliland L, Waldmann H (1988): Hybrid antibodies for therapy. *Prog Allergy* 45: 31–49

Corvalan JR, Smith W (1987): Construction and characterisation of a hybrid-hybrid monoclonal antibody recognising both carcinoembryonic antigen (CEA) and vinca alkaloids. *Cancer Immunol Immunother* 24: 127–132

De Lau WBM, Heije K, Neefjes JJ, Oosterwegel M, Rozemuller E, Bast BJEG (1991): Absence of preferential homologous H/L chain association in hybrid hybridomas. *J Immunol* 146: 906–914

De Lau WB, Van Loon AE, Heije K, Valerio D, Bast BJ (1989): Production of hybrid hybridomas based on HAT(s)-neomycin(r) double mutants. *J Immunol Methods* 117: 1–8

Demanet C, Brissinck J, Van Mechelen M, Leo O, Thielemans K (1991): Treatment of murine B cell lymphoma with bispecific monoclonal antibodies (anti—idiotype × anti-CD3). *J Immunol* 147: 1091–1097

Erbe DV, Collins JE, Shen L, Graziano RF, Fanger MW (1990): The effect of cytokines on the expression and function of Fc receptors for IgG on human myeloid cells. *Mol Immunol* 27: 57–67

Ertl HC, Greene MI, Noseworthy JH, Fields BN, Nepom JT, Spriggs DR, Finberg RW (1982): Identification of idiotypic receptors on retrovirus-specific cytotoxic T cells. *Proc Natl Acad Sci USA* 79: 7479–7483

Fanger MW, Guyre PM (1991): Bispecific antibodies for targeted cellular cytotoxicity. *Trends Biotechnol* 9: 375–380

Fanger MW, Graziano RF, Shen L, Guyre PM (1989): Fc$_\gamma$R cytotoxicity exerted by mononuclear cells. *Chem Immunol* 47: 214–253

Garrido MA, Perez P, Titus JA, Valdayo MJ, Winkler DA, Barbieri SA, Wunderlich JR, Segal DM (1990a): Targeted cytotoxic cells in human peripheral blood lymphocytes. *J Immunol* 144: 2891–2898

Garrido MA, Valdayo MJ, Winkler DF, Titus JA, Hecht TT, Perez P, Segal DM, Wunderlich JR (1990b): Targeting human T lymphocytes with bispecific antibodies to react against human ovarian carcinoma cells in nu/nu mice. *Cancer Res* 50: 4227–4232

Glennie MJ, McBride HM, Worth AT, Stevenson GT (1987): Preparation and performance of bispecific F(ab)'$_2$ antibody containing thioether-linked Fab'$_\gamma$ fragments. *J Immunol* 139: 2367–2375

Graziano RF, Fanger MW (1987a): Human monocyte-mediated cytotoxicity: The use of Ig-bearing hybridomas as target cells to detect trigger molecules on the monocyte cell surface. *J Immunol* 138: 945–950

Graziano RF, Fanger MW (1987b): Fc$_\gamma$RI and Fc$_\gamma$RII on monocytes and granulocytes are cytotoxic trigger molecules for tumor cells. *J Immunol* 139: 3536–3541

Graziano RF, Looney RJ, Shen L, Fanger MW (1989): Fc$_\gamma$R-mediated killing by eosinophils. *J Immunol* 142: 230–235

Henkart PA (1985): Mechanism of lymphocyte-mediated cytotoxicity. *Annu Rev Immunol* 3: 31–58

Henkart PA, Millard PJ, Reynolds CW, Henkart MP (1984): Cytolytic activity of purified cytoplasmic granules from cytotoxic rat large granular lymphocyte tumors. *J Exp Med* 160: 75–93

Huston JS, Levinson D, Mudgett-Hunter M, Tai MS, Novotny J, Margolies SN, Ridge RJ, Bruccoleri RE, Haber E, Crea R (1988): Protein engineering of antibody binding sites: Recovery of specific activity in an anti-digoxin single-chain Fv analogue produced in Escherichia colie. *Proc Natl Acad Sci USA* 85: 5879–5883

Jung G, Muller-Eberhard HJ (1988): An *in vitro* model for tumor immunotherapy with antibody heteroconjugates. *Immunol Today* 9: 257–260

Jung G, Freimann U, Von Marschall Z, Reisfeld RA, Wilmanns W (1991): Target cell-induced T cell activation with bi- and trispecific antibody fragments. *Eur J Immunol* 21: 2431–2435

Jung G, Honsik CJ, Reisfeld RA, Muller-Eberhard HJ (1986): Activation of human peripheral blood mononuclear cells by anti-T3: Killing of tumor target cells coated with anti-target-anti-T3 conjugates. *Proc Natl Acad Sci USA* 83: 4479–4483

Jung G, Ledbetter JA, Muller-Eberhard HJ (1987): Induction of cytotoxicity in resting human T lymphocytes bound to tumor cells by antibody heterconjugates. *Proc Natl Acad Sci USA* 84: 4611–4615

Karawajew L, Micheel B, Behrsing O, Gaestel M (1987): Bispecific antibody-producing hybrid hybridomas selected by a fluorescence activated cell sorter. *J Immunol Methods* 96: 265–270

Karawajew L, Rudchenko S, Wlasik T, Trakht I, Rakitskaya V (1990): Flow sorting of hybrid hybridomas using the DNA stain Hoechst 33342. *J Immunol Methods* 129: 277–282

Karpovsky B, Titus JA, Stephany DA, Segal DM (1984): Production of target-specific effector cells using heterocross-linked aggregates containing anti-target cell and anti-Fc$_\gamma$ receptor antibodies. *J Exp Med* 160: 1686–1701

King TP, Li Y, Kochoumian L (1978): Preparation of protein conjugates via intermolecular disulfide bond formation. *Biochemistry* 17: 1499–1506

Koolwijk P, Rozemuller E, Stad RK, De Lau WB, Bast BJ (1988): Enrichment and selection of hybrid hybridomas by Percoll density gradiant centrifugation and fluorescent-activated cell sorting. *Hybridoma* 7: 217–225

Lancki DW, Ma DI, Havran WL, Fitch FW (1984): Cell surface structures involved in T cell receptor complex but involved in T cell activation. *Immunol Rev* 81: 65–94

Lanier LL, Yu G, Phillips JH (1989): Co-association of CD3 zeta with a receptor (CD16) for IgG Fc on human natural killer cells. *Nature* 342: 803–805

Lanzavecchia A, Scheidegger D (1987): The use of hybrid hybridomas to target human cytotoxic T lymphocytes. *Eur J Immunol* 17: 105–111

Larrick JW, Wright SC (1990): Cytotoxic mechanism of tumor necrosis factor-alpha. *FASEB J* 4: 3215–3223

Leeuwenberg JTM, Spits H, Tax WJM, Capel PJA (1985): Induction of nonspecific cytotoxicity by monoclonal anti-T3 antibodies. *J Immunol* 134: 3770–3775

Leo O, Foo M, Segal DM, Shevach E, Bluestone JA (1987): Activation of murine T lymphocytes with monoclonal antibodies: Detection on Lyt2+ cells of an antigen not associated with the T cell activation. *J Immunol* 139: 1214–1222

Liu MA, Kranz DM, Kurnick JT, Boyle LA, Levy R, Eisen HN (1985): Heteroantibody duplexes target cells for lysis by cytotoxic T lymphocytes. *Proc Natl Acad Sci USA* 82: 8648–8652

Lovchik JC, Hong R (1977): Antibody-dependent cell-mediated cytolysis (ADCC): analyses and projections. *Prog Allergy* 22: 1–44

Mezzanzanica D, Garrido MA, Noblock DS, Daddona PE, Andrew SM, Zurawski VR, Segal DM, Wunderlich JR (1991): Human T-lymphocytes targeted against an established ovarian carcinoma with bispecific F(ab')$_2$ antibody prolong host survival in a murine xenograft model. *Cancer Res* 51: 5716–5721

Milstein C, Cuello AC (1983): Hybrid hybridomas and their use in immunohistochemistry. *Nature* 305: 537–540

Milstein C, Cuello AC (1984): Hybrid hybridomas and production of bispecific monoclonal antibodies. *Immunol Today* 5: 299–304

Miotti S, Canevari S, Menard S, Mezzanzanica D, Porro G, Pupa SM, Regazzoni M, Tagliabue E, Colnaghi MI (1987): Characterization of human ovarian carcinoma-associated antigens defined by novel monoclonal antibodies with tumor-restricted specificity. *Int J Cancer* 39: 297–303

Moran TM, Usuba O, Kuzu Y, Schulman J, Bona CA (1991): Inhibition of multicycle influenza virus replication by hybrid antibody-directed cytotoxic T lymphocyte lysis. *J Immunol* 146: 321–326

Moretta A, Poggi A, Pende D, Tripodi G, Orengo AM, Pella N, Augugliaro R, Bottino C, Ciccone E, Moretta L (1991): CD69-mediated pathway of lymphocyte activation: Anti-CD69 monoclonal antibodies trigger the cytolytic activity of different lymphoid effector cells with the exception of cytolytic T lymphocytes expressing T cell receptor α/β. *J Exp Med* 174: 1393–1398

Nisonoff A, Mandy WJ (1962): Quantitative estimation of the hybridization of rabbit antibodies. *Nature* 194: 355–359

Nitta T, Ikeda M, Azuma A, Yagita H, Sato K, Okumura K, Steinman L (1991): Clinical results of specific targeting therapy against human malignant glioma and prospects for future reagents based on the restricted T cell receptor repertoire in tumor infiltrating lymphocytes. In: *Bispecific Antibodies and Targeted Cellular Cytoxicity. Second International Conference, Seillac, France October 9–13, 1990*, Romet-Lemonne JL, Fanger MW, Segal DM, eds. Les Ulis, France: Foundation Nationale de Transfusion Sanguine 233–235

Nitta T, Sato K, Yagita H, Okumura K, Ishii S (1990): Preliminary trial of specific targeting therapy against malignant glioma. *Lancet* 335: 368–371

Old LJ (1985): Tumor necrosis factor (TNF). *Science* 230: 630–632

Ortho Multicenter Transplant Study Group (1985): A randomized clinical trial of OKT3 monoclonal antibody for acute rejection of cadaveric renal transplants. *N Engl J Med* 313: 337–342

Paya CV, McKean DJ, Segal DM, Schoon RA, Schowalter SD, Leibson PJ (1989): Heteroconjugate antibodies enhance cell-mediated anti-herpes simplex virus immunity. *J Immunol* 142: 666–671

Perez P, Hoffman RW, Shaw S, Bluestone JA, Segal DM (1985): Specific targeting of cytotoxic T cells by anti-T3 linked to anti-target cell antibody. *Nature* 316: 354–356

Perez P, Hoffman RW, Titus JA, Segal DM (1986a): Specific targeting of human peripheral blood T cells by heteroaggregates containing anti-T3 crosslinked to anti-target cell antibodies. *J Exp Med* 163: 166–178

Perez P, Titus JA, Lotze MT, Cuttitta F, Longo DL, Groves ES, Rabin H, Durda PJ, Segal DM (1986b): Specific lysis of human tumor cells by T cells coated with anti-T3 crosslinked to anti-tumor antibody. *J Immunol* 137: 2069–2072

Podack ER, Kupfer A (1991): T-cell effector functions: Mechanisms for delivery of cytotoxicity and help. *Annu Rev Cell Biol* 7: 479–504

Qian JH, Titus JA, Andrew SM, Mezzanzanica D, Garrido MA, Wunderlich JR, Segal DM (1991): Human PBL targeted with bispecific antibodies release cytokines that are essential for inhibiting tumor growth. *J Immunol* 146: 3250–3256

Segal DM (1990): Antibody-mediated killing by leukocytes. In: *Fc Receptors and the Action of Antibodies* Metzger H, ed. Washington DC, American Society for Microbiology: 291–301

Segal DM, Snider DP (1989): Targeting and activation of cytotoxic lymphocytes. *Chem Immunol* 47: 179–213

Segal DM, Wunderlich JR (1988): Targeting of cytotoxic cells with heterocrosslinked antibodies. *Cancer Invest* 6: 83–92

Segal DM, Dower SK, Titus JA (1983): The role of nonimmune IgG in controlling IgG-mediated effector functions. *Mol Immunol* 20: 1177–1189

Seth A, Gote L, Nagarkatti M, Nagarkatti PS (1991): T-cell-receptor-independent activation of cytolytic activity of cytotoxic T lymphocytes mediated through CD44 and gp90^{MEL-14}. *Proc nat Acad Sci USA* 88: 7877–7881

Shalaby MR, Shepard HM, Presta L, Rodrigues ML, Beverley PCL, Feldmann M, Carter P (1992): Development of humanized bispecific antibodies reactive with cytotoxic lymphocytes and tumor cells overexpressing the HER2 protooncogene. *J Exp Med* 175: 217–225

Shen LR, Graziano RF, Fanger MW (1989): The functional properties of Fc gamma RI, II and III on myeloid cells: A comparative study of killing of erythrocytes and tumor cells mediated through the different Fc receptors. *Mol Immunol* 26: 959–969

Shen L, Guyre PM, Anderson CL, Fanger MW (1986): Heteroantibody-mediated cytotoxicity: Antibody to the high affinity Fc receptor for IgG mediates cytotoxicity by human monocytes that is enhanced by interferon-γ and is not blocked by human IgG. *J Immunol* 137: 3378–3382

Shi T, Eaton AM, Ring DB (1991): Selection of hybrid hybridomas by flowing cytometry using a new combination of fluorescent vital stains. *J Immunol Methods* 141: 165–175

Siliciano RF, Pratt JC, Schmidt RE, Ritz J, Reinherz EL (1985): Activation of cytotoxic T lymphocyte and natural killer cell function through the T11 sheep erthrocyte binding protein. *Nature* 317: 428–429

Songsivilai S, Lachmann PJ (1990): Bispecific antibody: A tool for diagnosis and treatment of disease. *Clin Exp Immunol* 79: 315–321

Songsivilai S, Clissold PM, Lachmann PJ (1989): A novel strategy for producing chimeric bispecific antibodies by gene transfection. *Biochem Biophys Res Commun* 164: 271–276

Springer TA (1990): Adhesion receptors of the immune system. *Nature* 346: 425–434

Staerz UD, Bevan MJ (1986a): Hybrid hybridoma producing a bispecific monoclonal antibody that can focus effector T-cell activity. *Proc Natl Acad Sci USA* 83: 1453–1457

Staerz UD, Bevan MJ (1986b): Use of anti-receptor antibodies to focus T cell activity. *Immunol Today* 7: 241–245

Staerz UD, Kanagawa O, Bevan MJ (1985): Hybrid antibodies can target sites for attack by T cells. *Nature* 314: 628–631

Suresh MR, Cuello AC, Milstein C (1986): Advantages of bispecific hybridomas in one-step immunocytochemistry and immunoassays. *Proc Natl Acad Sci USA* 83: 7989–7993

Tai MS, Mudgett-Hunter M, Levinson D, Wu GM, Haber E, Oppermann H, Huston JS (1990): A bifunctional fusion protein containing Fc-binding fragment B of staphylococcal protein A amino terminal to antidigoxin single-chain Fv. *Biochemistry* 29: 8024–8030

Titus JA, Garrido MA, Hecht TT, Winkler DF, Wunderlich JR, Segal DM (1987a): Human T cells targeted with anti-T3 crosslinked to anti-tumor antibody prevent tumor growth in nude mice. *J Immunol* 138: 4018–4022

Titus JA, Perez P, Kaubisch A, Garrido MA, Segal DM (1987b): Human K/NK cells targeted with heterocrosslinked antibodies specifically lyse tumor cells *in vitro* and prevent tumor growth *in vivo*. *J Immunol* 139: 3153–3158

Traunecker A, Lanzavecchia A, Karjalainen K (1991): Bispecific single chain molecules (Janusins) target cytotoxic lymphocytes on HIV infected cells. *EMBO J* 10: 3655–3659

Tschopp J, Nabholz M (1990): Perforin-mediated target cell lysis by cytolytic T lymphocytes. *Annu Rev Immunol* 8: 279–302

Tutt A, Greenman J, Stevenson GT, Glennie MJ (1991a): Bispecific F(ab'γ)$_3$ antibody derivatives for redirecting unprimed cytotoxic T cells. *Eur J Immunol* 21: 1351–1358

Tutt A, Stevenson GT, Glennie MJ (1991b): Trispecific F(ab')$_3$ derivatives that use cooperative signaling via the TCR/CD3 complex and CD2 to activate and redirect resting cytotoxic T cells. *J Immunol* 147: 60–69

Urnovitz HB, Chang Y, Scott M, Fleischmann J, Lynch RG (1988): IgA:IgM and IgA:IgA hybrid hybridomas secrete heteropolymeric immunoglobulins that are polyvalent and bispecific. *J Immunol* 140: 558–563

Wegener A-MK, Letourneur F, Hoeveler A, Brocker T, Luton F, Malissen B (1992): The T cell receptor/CD3 complex is composed of at least two autonomous transduction molecules. *Cell* 68: 83–95

Weiner GJ, Hillstrom JR (1991): Bispecific anti-idiotype/anti-CD3 antibody therapy of murine B cell lymphoma. *J Immunol* 147: 4035–4044

Winter G, Milstein C (1991): Man-made antibodies. *Nature* 349: 293–299

Wong JT, Colvin RB (1987): Bispecific monoclonal antibodies: Selective binding and complement fixation to cells that express two different surface antigens. *J Immunol* 139: 1369–1374

Zarling JM, Moran PA, Grosmarie LS, McClure J, Shriver K, Ledbetter JA (1988): Lysis of cells infected with HIV-1 by human lymphocytes targeted with monoclonal antibody heteroconjugates. *J Immunol* 140: 2609–2613

III
Generation of Cytotoxic Cells

10
Immunobiology and Molecular Characteristics of Peritoneal Exudate Cytotoxic T Lymphocytes (PEL), Their In Vivo IL-2-Dependent Blasts and IL-2-Independent Cytolytic Hybridomas

Gideon Berke, Dalia Rosen, Denise Ronen, and Barbara Schick

Major histocompatibility complex (MHC)-restricted cytolytic T lymphocytes (CTL) are generated in response to allogeneic tissues (normal and malignant), tumors (autologous and syngeneic), viruses, and certain bacteria and self-antigens. CTL can be derived directly from spleen or lymph nodes after immunization, but these cells often require a secondary *in vitro* stimulation (Cerottini and Brunner, 1974). Peritoneal exudate CTL (PEL), collected during or shortly after a primary intraperitoneal (i.p.) immunization of rats and mice with allogeneic or irradiated syngeneic tumors, usually yield a highly potent, specific population of CTL (Berke et al., 1972a, 1972b, 1972c; Fishelson and Berke, 1978; Schick and Berke, 1977; Berke and Schick, 1980). Generation of PEL can be augmented by, but does not require, a secondary stimulation *in vivo* or *in vitro*.

PEL effectively bind to and lyse target cells *in vitro* and can retard syngeneic tumor growth *in vivo* (reviewed in Berke and Amos, 1973; Berke and Schick, 1980; Berke, 1980, 1989). The small to medium-sized, nondividing PEL can be somatically hybridized to yield immortalized CTL-hybridomas (Kaufmann et al., 1981a; Kaufmann and Berke, 1983) or can, upon incubation in interleukin-2 (IL-2), be transformed into large dividing granular CTL blasts [PEL blasts (PEB); Berke and Rosen, 1988]. This chapter describes our studies on the generation, composition, mode of action, and function of PEL, PEB, and PEL hybridomas. The mechanism by which CTL lyse target cells will be discussed separately (Berke, Chapter 18, this volume).

Initial Studies on Cell-Mediated Immunity to Allogeneic and Syngeneic Tumors in the Peritoneal Cavity

The abundance of lymphatic tissues and connections within the peritoneal cavity, such as lymphoid patches, mesenteric nodes, retroperitoneal nodes, and the spleen, probably facilitates the migration of immunocompetent host cells to and from the peritoneal cavity (Straube et al., 1955). This migration may promote the infiltration of intraperitoneal tumors by immunocompetent cells, homing of activated cells to a milieu conducive to effector cell differentiation, and migration of mature effector cells to the tumor site. The variety of lymphocytes present in the peritoneal cavity, in conjunction with the lymphatic plexi, probably creates an *in vivo* environment promoting CTL differentiation in the peritoneal cavity. An *in vitro* correlate of such an environment may be mixed lymphocyte cultures.

Tumors capable of multiplying as single cells within the peritoneal or pleural cavities are referred to as ascites tumors. Since most established ascites tumors grow as single cells, not subject to the intense vascularization and necrosis that occurs with solid tumors (Gorer, 1956), their use to examine antitumor and allograft responses has been widespread. Pioneering studies of immune responses within the peritoneal cavity consisted mainly of visual observations. Ascites tumors, upon i.p. injection into syngeneic hosts, multiply

CYTOTOXIC CELLS
M. Sitkovsky and P. Henkart, Editors
© 1993 Birkhäuser Boston

rapidly (Patt and Straube, 1956). The growth of i.p. injected ascites tumors in allogeneic hosts is initially similar to that observed in syngeneic ones, but after several days a dramatic decrease in the number of tumor cells occurs (Amos and Wakefield, 1959; Baker et al., 1962; Berke and Amos, 1973). No change or only small changes in host cell populations are visually observed in syngeneic ascites tumor systems (Baker et al., 1962). In contrast, in most allogeneic ascites tumor systems, variations in the number of peritoneal macrophages occur, corresponding to the number of tumor cells present (Amos, 1960, 1962; Baker et al., 1962). This correlation suggested that macrophages are responsible for the tumor rejection.

Other findings also indicate that macrophages are involved in the i.p. rejection of allogeneic ascites tumors. At the peak of allogeneic ascites tumor growth, peritoneal macrophage populations consist primarily of small cells with compact cytoplasm and rounded nuclei, and the macrophages exhibit only slight phagocytic activity for the tumor cells (Amos, 1960). As the number of tumor cells decreases, larger macrophages, which contain irregular lipid granules and tumor cell fragments (Amos, 1960) and which adhere to the tumor cells *in vivo* (Baker et al., 1962), are observed. In the allogeneic sarcoma I system studied by Baker and co-workers, no other host cell type exhibiting affinity for the tumor cells was detected. Furthermore, the i.p. injection of sarcoma I cells caused tumor cell–macrophage clusters within 30 min, and the generation of cellular debris and large macrophage–tumor cell clumps within several hours *in vivo*. Immunity to the sarcoma I could be adoptively transferred by spleen cells or macrophage-rich peritoneal exudates from alloimmunized mice, but not by alloimmune sera, cell-free ascites, or sonic extracts of alloimmune cells (Baker et al., 1962). Also, alloimmune peritoneal cells, capable of phagocytosing iron particles, obtained after the i.p. injection of the DBA/2 lymphoma L1210, prevented L1210 growth in allogeneic C3H mice (Amos, 1962).

The Peritoneal Exudate CTL (PEL)

Elucidation of the actual host cells involved in the response to ascites tumors required the fractionation and *in vitro* testing of peritoneal exudate cells (PEC). Peritoneal exudates of Donryu rats, primed subcutaneously (s.c.) with nitrogen-mustard-treated Yoshida sarcoma and challenged i.p. with untreated Yoshida cells, exhibited antitumor activity, as determined by adoptive transfer, tumor neutralization, and diffusion chamber experiments (Hashimoto et al., 1965). The antitumor effect was due to intact host effector cells, since lysed exudate preparations exhibited no antitumor activity and sarcoma cells inside diffusion chambers in the peritoneal cavities of alloimmune mice were not lysed unless alloimmune PEC were present within the chambers. Adherent cells were removed from the alloimmune exudates to obtain lymphocyte-enriched nonadherent fractions (Hashimoto and Sudo, 1968), now referred to as peritoneal exudate lymphocytes (PEL). When these alloimmune PEL were incubated with tumor cells *in vitro* at low effector-to-target ratios (30 : 1), more than 90% of the tumor cells were destroyed within 7 hr. These alloimmune PEL adhered to Yoshida cells *in vitro* within 1 hr and the number of bound PEL correlated with the antitumor activity. Yoshida cells to which two or more lymphocytes bound were lysed within several hours, and even one tumor-bound PEL could cause tumor cell destruction after 24 to 48 hr of incubation.

Several groups have demonstrated that upon i.p. injection of allogeneic tumors, the resulting PEL exhibit greater cytocidal activity than the corresponding splenic, lymph node, or peripheral blood lymphocytes (Berke et al., 1972a) (Figure 1). The

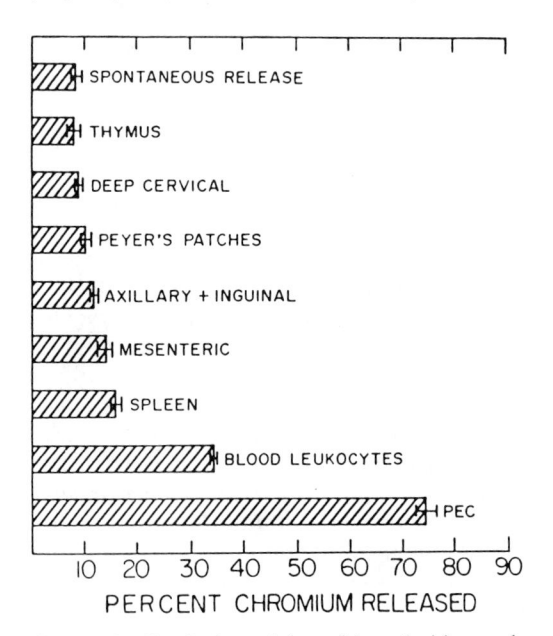

FIGURE 1. Cytolytic activity of lymphoid populations derived from alloimmunized BALB/c mice. After the i.p. rejection of EL4, lymphocytes from the indicated sources were obtained and incubated with ^{51}Cr-labeled EL4 cells for 2.5 hr at 37°C (Berke et al., 1972a).

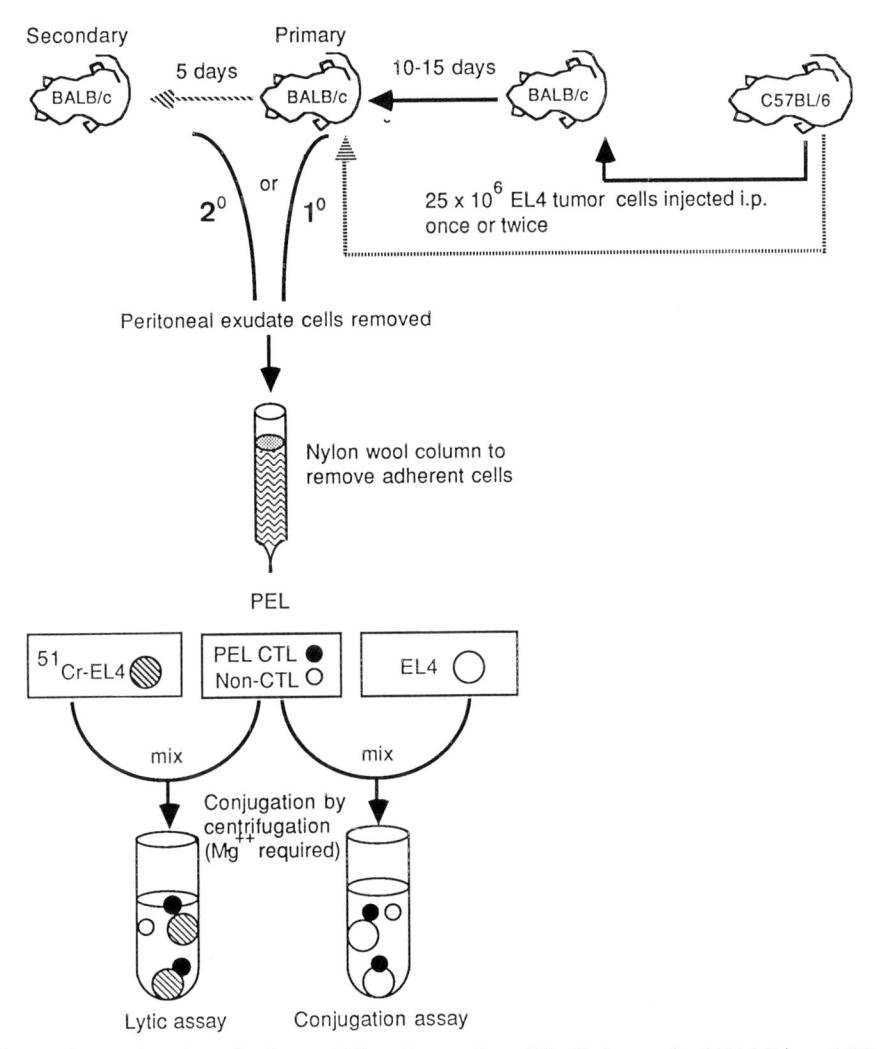

FIGURE 2. Generation and testing of primary (1°) and secondary (2°) alloimmunized BALB/c anti-EL4 PEL (from Berke, 1989).

standard procedures for inducing and testing the binding (conjugation) and lysis of target tumor cells by PEL are depicted for the allogeneic BALB/c (H-2^d) anti-leukemia EL4 (H-2^b) system (Berke et al., 1972 a,c,b) (Figure 2). By injecting γ-irradiated tumor cells, the protocol can be modified to generate and examine allogeneic PEL at earlier times after immunization and to induce PEL in syngeneic systems (Schick and Berke, 1977; Hurt et al., 1979; Denizot et al., 1989). Upon removal of macrophages and other adherent cells, by incubating unfractionated allogeneic or syngeneic PEC on nylon wool columns, nonadherent PEL are obtained, which exhibit potent cytotoxicity (^{51}Cr release) (Table 1) and conjugation (target binding) (Figure 3). Target cell conjugation and lysis by PEL

TABLE 1. Lytic activity of BALB/c anti-EL4 nonadherent PEC (PEL), unfractionated crude PEC, and spleen cells

Effector/target cell ratio	Immune effector cells [specific ^{51}Cr release (%/hr)]		
	Nonadherent PEC (PEL)	Crude PEC	Spleen
0:1	2.7	3.3	3.0
1:1	23.1	10.0	3.2
5:1	48.4	22.5	5.1
10:1	60.0	30.0	7.8
25:1	NT	42.7	9.5
50:1	NT	44.7	17.5
100:1	NT	NT	30.2

NT, Not tested.

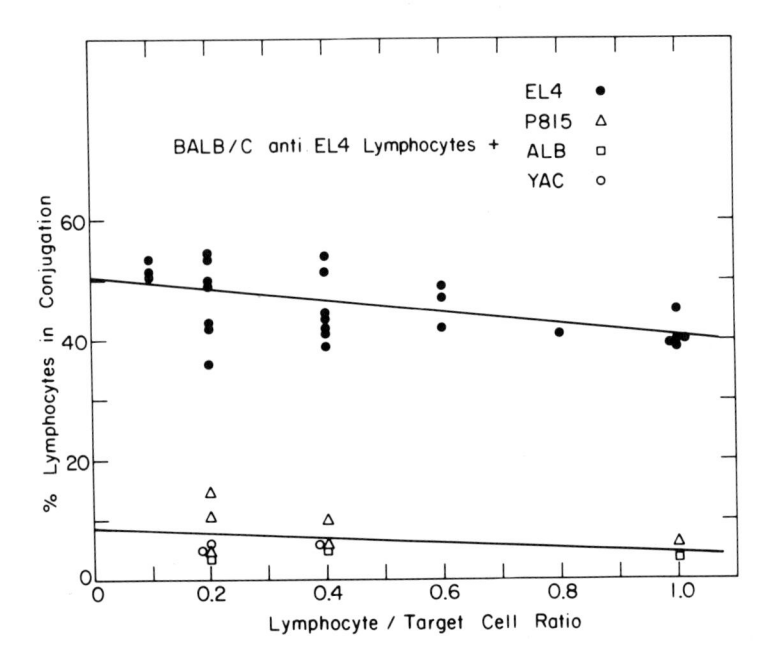

FIGURE 3. Specificity and quantification of conjugation. EL4, P815, ALB, and YAC are transplantable ascites tumors of C57BL/6 (H-2^b), DBA/2 (H-2^d), BALB/c (H-2^d), and A (H-2^a) mouse strains, respectively. Tumor cells (10^6) were conjugated with various quantities of BALB/c anti-EL4 PEL. Extrapolation to zero on the x-axis indicates the percent of effector cells in the PEL population ($\sim 50\%$), a fifth of which may not be specific (from Berke, 1980).

are specific and exhibit a correlation of 85% at both the single-cell and the population levels (Berke et al., 1972a, 1972b, 1975; Zagury et al., 1975). However, slight, but consistent, nonspecific cross-reactivity of PEL-mediated conjugation and lysis has been observed (see Berke, 1980; Martz, 1987). The cellular composition and surface membrane markers of PEL, obtained 11 days after primary alloimmunization, of which 35–55% can enter into specific conjugation with target cells (Figure 3), are shown in Table 2. Interestingly, although PEL are small to medium-sized and nondividing, they incorporate significant amounts of ^{3}H-thymidine (Table 2), probably due to extensive DNA repair. PEL consist primarily of Thy-$1,2^+$ and Lyt-2^+ cells. The morphology of target cell binding (conjugate formation) and lysis by PEL is shown in Figures 4 and 5, respectively.

A major drawback of ascites tumor systems is the tremendous tumor overgrowth, which hampers the study of host cell populations associated with the tumor at early and intermediate phases of the antitumor response. Tumor cells can be specifically lysed with antibodies plus complement. However, large amounts of specific antisera are required, complete removal of tumor cells is not guaranteed,

TABLE 2. Cellular composition and thymidine incorporation of PEL, PEB, and PEL hybridomas

Cell type	Morphology (% of total)					Surface markers [positive cells (%)]			Thymidine incorporation (cpm/no. of cells)
	Small to medium	Large	Histiocytes	PMN	Other	Thy-1.2	Lyt-2	L3T4	
PEC crude	38	<1	47	5	9	NT	NT	NT	NT
PEL nonadherent	91	<2	2	2	3	95	81.1	19.5	11,170
PEL naive	60	3	31	3	4	NT	NT	NT	2,040
PEL blasts (PEB)	25	75	NT	NT	NT	99.5	94.1	0	High
PEL hybridoma	5	95	NT	NT	NT	99	0	NT	High

NT, Not tested; PMN, Polymorphonuclear leucocytes

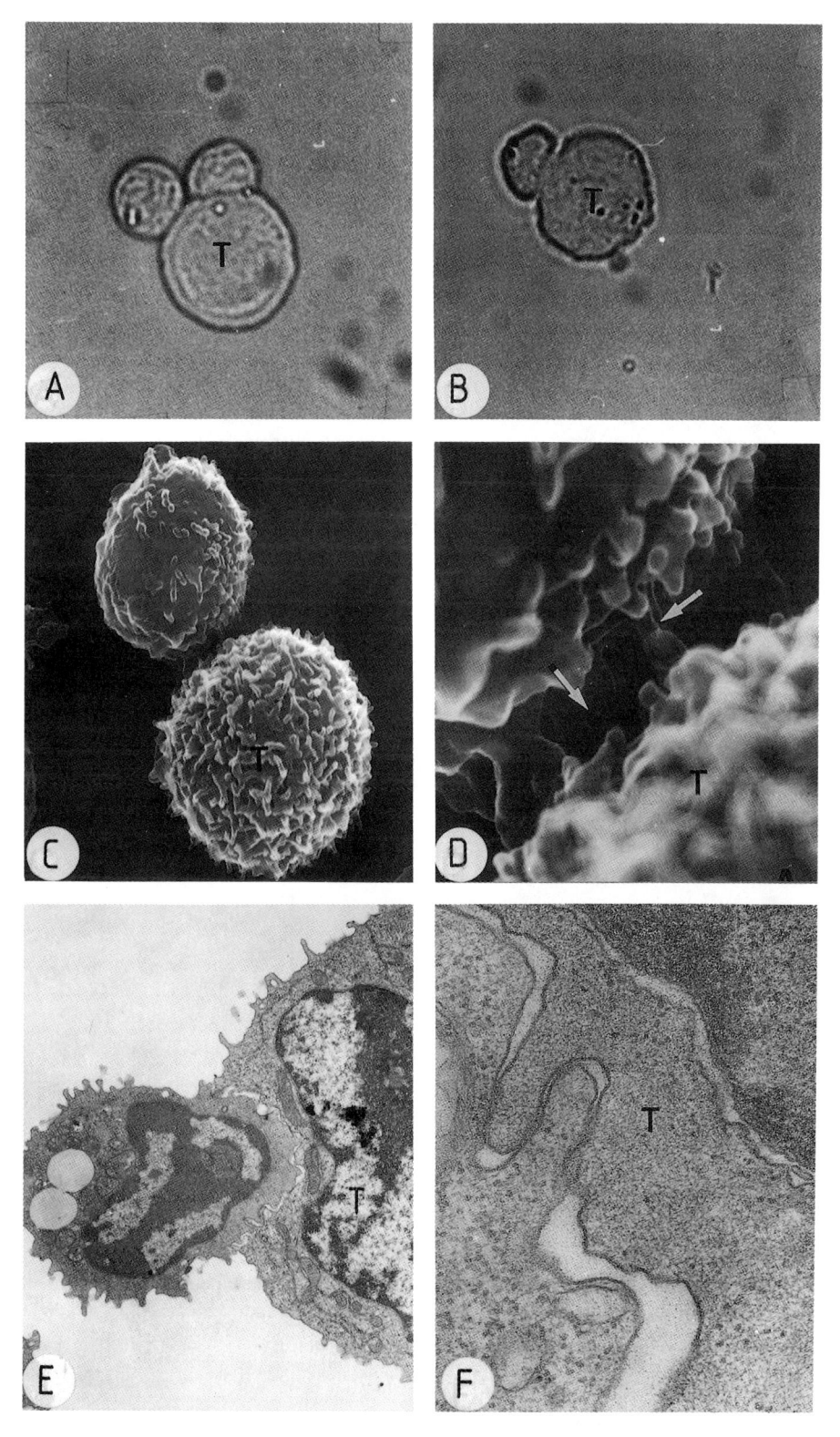

FIGURE 4. BALB/c anti-EL4 PEL conjugated with EL4 target cells. Light microscope (A and B), scanning (C and D), and transmission (E and F) electromicrographs. T, EL4 target cell. Arrows in panel D point at adhesion sites (from Berke, 1989).

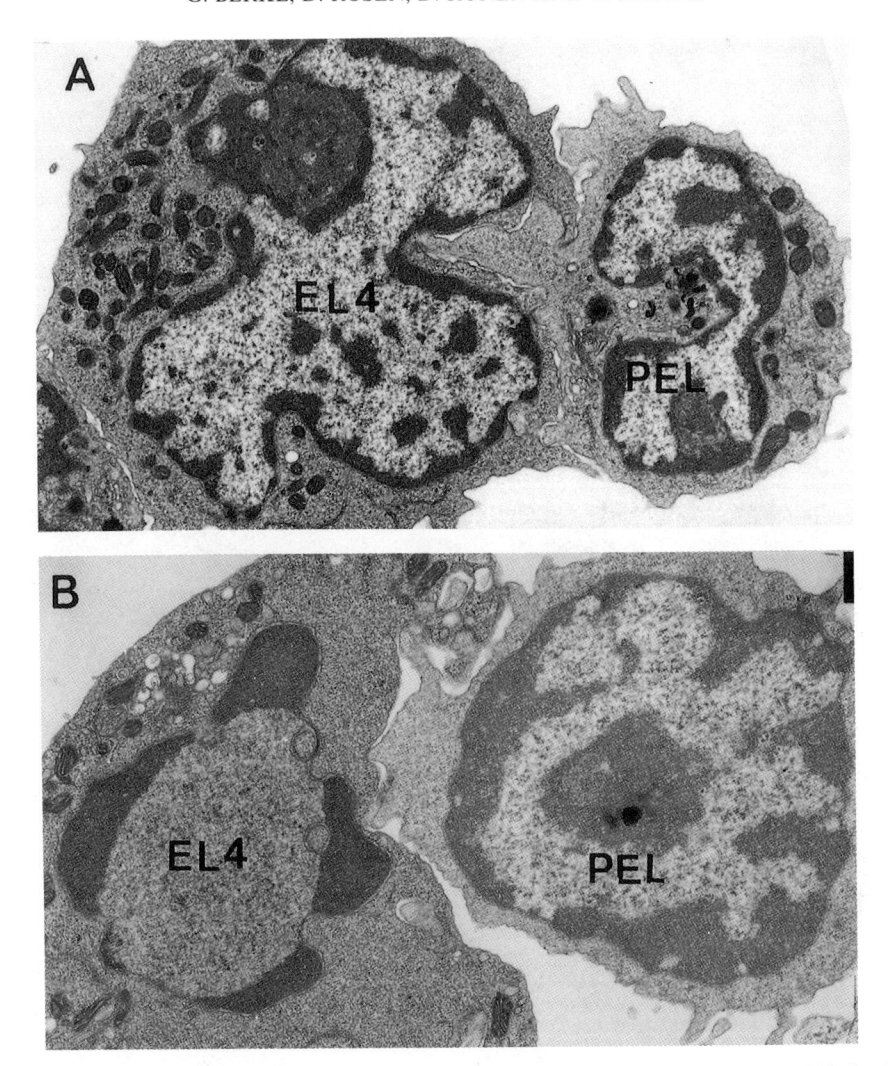

FIGURE 5. Conjugation (binding) and lysis of EL4 target cells by BALB/c anti-EL4 PEL. (A) Conjugation; (B) prelytic chromatin condensation in EL4; (see next page) (C) "zeiosis" of EL4; (D) lysis of the conjugated EL4.

and host cells may be affected by the complement. Host cells can also be separated from tumor cells on the basis of size with gradients, but complete separation by this method of different subpopulations is difficult, if not impossible, due to the heterogeneity of the cell populations involved. This drawback can be circumvented by using specific antibodies to tumor or lymphocyte cell surface markers linked to magnetic particles. Successful separation of tumor-infiltrating lymphocytes (TIL) by such a method has been demonstrated (Lotze and Finn, 1990).

Alternatively, tumors whose capacity to replicate has been inhibited can be injected for immunization. A single i.p. injection of leukemia EL4 cells, exposed to ^{60}Co irradiation (irr-EL4), induced immunity to non-irr-EL4 in syngeneic C57BL/6 and allogeneic BALB/c mice (Schick and Berke, 1977). EL4 were lysed in *in vitro* cytolytic assays by tumor-associated syngeneic and allogeneic PEC as early as 4 days after irr-EL4 injection, with peak activity occurring on days 5–6 (Figure 6). Adoptive immunity was exhibited by syngeneic PEC obtained 5 days after the i.p. administration of irr-EL4. Therefore, a single i.p. injection of irr-tumor cells can induce a cell-mediated immune response at the tumor inoculation site, which can be measured *in vivo* and *in vitro*, in addition to conferring resistance *in vivo* against the nonirradiated tumor.

The cytolytic capacity of effector PEL induced by irr-EL4 is lower than that induced by non-irr-EL4 (Schick and Berke, 1977). Since irradiation

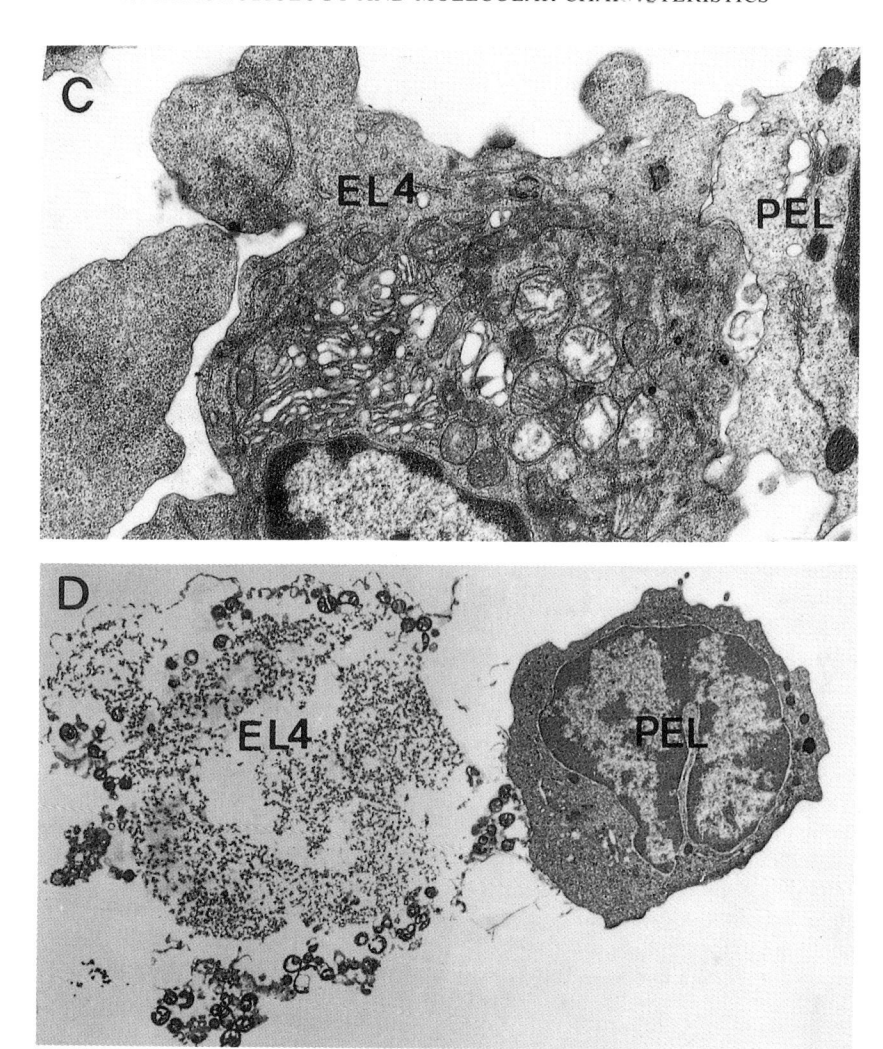

FIGURE 5. Continued.

(5,000 R) did not affect the cellular antigenicity of EL4, as determined by *in vitro* cell-mediated lytic assays, the weaker PEL response elicited by irr-EL4 cannot be attributed to gross antigenic changes caused by irradiation (see also Herberman, 1974). The differences in the PEL responses may be related to the amount of cellular antigen available for sensitization in the peritoneal cavity, since non-irr-EL4 proliferate profusely in the peritoneal cavity (Berke and Amos, 1973), while irr-EL4 do not multiply. That repeated injections of irr-tumor cells increase the potency of PEL (Hurt et al., 1979) supports this assumption. The immune response induced by irr-EL4 is probably identical to the early response induced by nonirradiated EL4, and is useful for studying early stages of the response. The time course of appearance of syngeneic and allogeneic effector activity generated against irr-EL4 *in vivo* and in mixed lymphocyte reactions (MLR) *in vitro* was similar (Wagner et al., 1973; Cerottini and Brunner, 1974; Andersson and Häyry, 1975). Thus, in this system, as in several others, it appears that cytolytic effectors differentiate at the tumor site in a process involving replication of their precursors (Denizot et al., 1986). The decline in PEL activity, which occurs earlier after irr-EL4 than non-irr-EL4 injection, may be due to migration of immune effectors from the peritoneal cavity once the sensitizing cellular antigen is no longer present. No differences in the characteristics of PEL induced by irr-tumor and by non-irr-tumor cells have been observed.

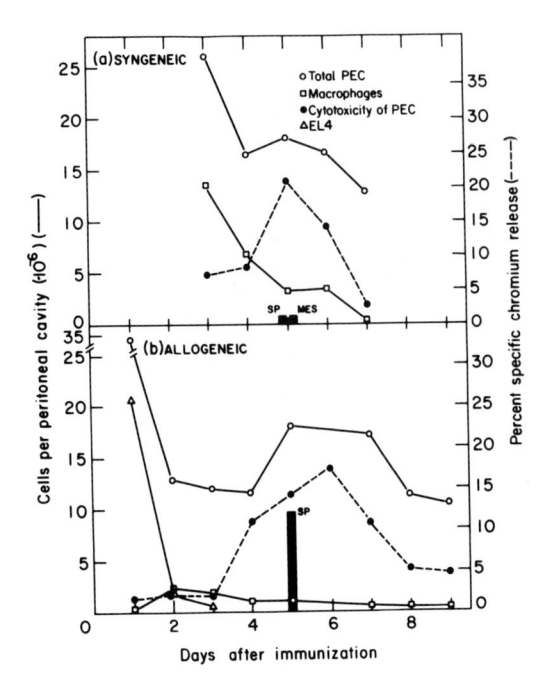

FIGURE 6. Time course of the cell populations in and cytolytic activity of PEC from C57BL/6 (a) and BALB/c (b) mice inoculated i.p. with 150 and 25 × 10⁶ irr-EL4, respectively. Effectors: ^{51}Cr-EL4 ratios of 20:1 with syngeneic PEC, 4:1 with allogeneic PEC, 100:1 with spleen (SP), and 50:1 with mesenteric node lymphocytes (MES) were employed in a 2-hr ^{51}Cr-release cytotoxic assay. Heavily granulated cells were considered to be macrophages. (From Schick and Berke, 1977.)

PEL-Blasts (PEB) and PEL Hybridomas

Upon stimulation *in vitro* in the presence of T-cell growth factors (TCGF) or recombinant interleukin-2 (rIL-2), the small, nondividing PEL transform into large, IL-2-dependent, dividing PEB that in short assays express specific lytic activity similar to that of the original *in vivo*-primed PEL (Table 3; Berke and Rosen, 1988). PEB, in contrast to PEL, possess massive quantities of typical intracytoplasmic granules (Figure 7; Berke and Rosen, 1988). IL-2-independent CTL hybridomas have been generated from PEL and other CTL by polyethylene glycol-induced fusion with BW5147 lymphoma cells (Kaufmann et al., 1981a, 1981b; Kaufmann and Berke, 1983). The CTL hybridomas exhibit MHC-specific lytic activity (Kaufmann and Berke, 1983). The cellular surface markers and lytic activity of PEL, PEB, and PEL hybridomas are depicted in Tables 2 and 3. They consist almost entirely of Thy-1,2⁺ and Lyt-2⁺ cells. Some of the CTL hybridomas represent "memory" CTL (Moscovitch et al., 1984), since their lytic capacity is stimulated upon exposure to antigen or mitogen [concanavalin A (Con A)] in a dose- and time-dependent manner (Figure 8). The PEL hybridomas, unlike PEL or PEB, did not respond when stimulated by rIL-2 (Berke, Rosen, and Ronen, submitted for publication). Also, unlike PEL or PEB, which exhibit a 5 to 10 min lag before the onset of lysis (Berke, 1989), lysis mediated by CTL

TABLE 3. Lytic activity and specificity of *in vivo*-primed BALB/c anti-EL4 PEL, rIL-2-dependent PEL blasts (PEB), and PEL hybridoma[a]

Effector cells	Effector/target ratio	Assay duration (min)	EL4 (K^bD^b)	L1210 (K^dD^d)	YAC (K^kD^d)	BW (K^kD^k)
PEL	5:1	75	40.5	0.6	7.0	4.2
	5:1	240	95.6	3	14.8	9.8
PEB[b]	1:1	90	54	8.4	8.8	NT
	1:1	180	84.6	16.3	19	NT
PEL hybridoma	10:1	75	6	NT	0	NT
	10:1	240	46	NT	0	0

[a] The various H-2^d anti-H-2^b effectors were mixed with ^{51}Cr-labeled target cells in the indicated effector/target ratios and cocentrifuged at room temperature to promote conjugate formation. Specific ^{51}Cr release was determined after incubation at 37°C for the indicated times. Spontaneous ^{51}Cr release was less than 10% at 75 and 90 min and less than 25% after 240 min.
[b] PEB were cultured in IL-2 prior to assay.
NT, Not tested.

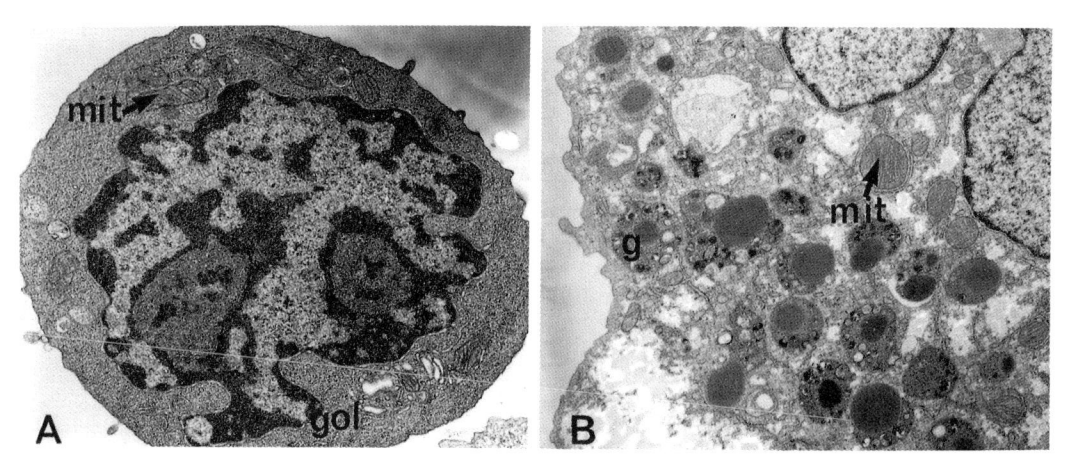

FIGURE 7. Electronmicrographs of BALB/c anti-EL4 PEL (A) and PEB (B). Gol, Golgi; g, intracytoplasmic granules; mit, mitochondria.

hybridomas is characterized by a 2 hr lag period, which is shortened by stimulation with antigen or mitogen (Figure 9).

Lytic Granules, Perforin, and Granzymes in PEL, PEB, and PEL Hybridomas

A unified mechanism for the lethal phase of both lymphocyte-mediated and antibody-plus-complement-induced immune cytolysis has been suggested (Henkart, 1985; Krähenbühl and Tschopp, 1991; Podack et al., 1991). According to this mechanism, lymphocyte-mediated lysis is the result of target cell membrane perforation by perforin-lined transmembrane pores (10–20 nm in diameter) that traverse the membrane. CTL secretory lytic granules, which contain the Ca^{2+}-dependent lytic protein perforin/cytolysin (ca. 65 kDa) and several serine proteases (granzymes), have been implicated in this mechanism. However, extensive electron microscopy studies employing negative staining techniques have not confirmed the presence of complement membrane attack complex (MAC)-like structures, proposed for perforin-induced lysis, on membranes of cells lysed by the *in vivo*-primed PEL (Berke and Rosen, 1987b) (Figure 10).

In early studies on PEL–target interaction (Zagury et al., 1975; Kalina and Berke, 1976), a few single-membrane-bound lysosomal granules and a well-developed Golgi were observed in PEL, but no lytic granule, perforin, or granzyme activity has been detected (Berke and Rosen, 1987a, 1988). Massive granule induction occurs in PEL as they

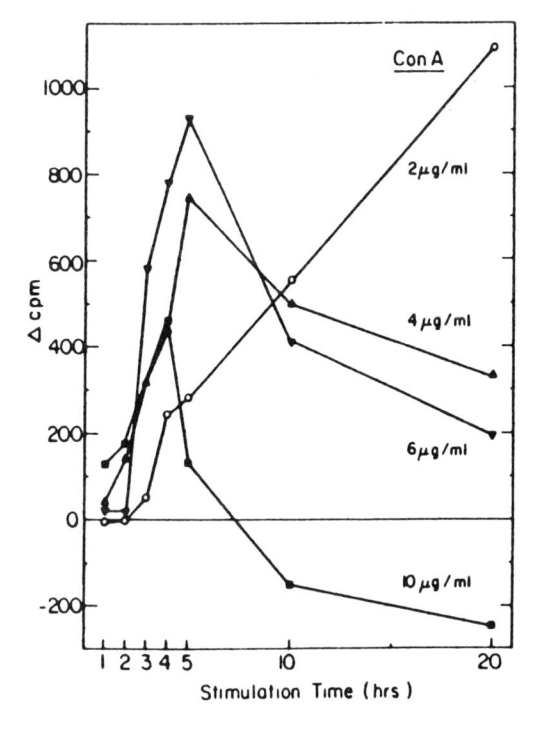

FIGURE 8. Time course and dose dependence of the enhancement of the lytic capacity of CTL hybridoma stimulated by Con A. BALB/c anti-EL4 hybridomas were preincubated with Con A (0–10 μg/ml for 1–20 hr). After incubating with α-methylmannoside to neutralize unreacted Con A, the hybridoma cells were incubated with ^{51}Cr-EL4 at a 1:1 ratio for 3.5 hr at 37°C. Enhancement due to Con A stimulation of hybridoma-mediated lysis is expressed as the Δcpm of ^{51}Cr release from EL4 by stimulated and nonstimulated cells.

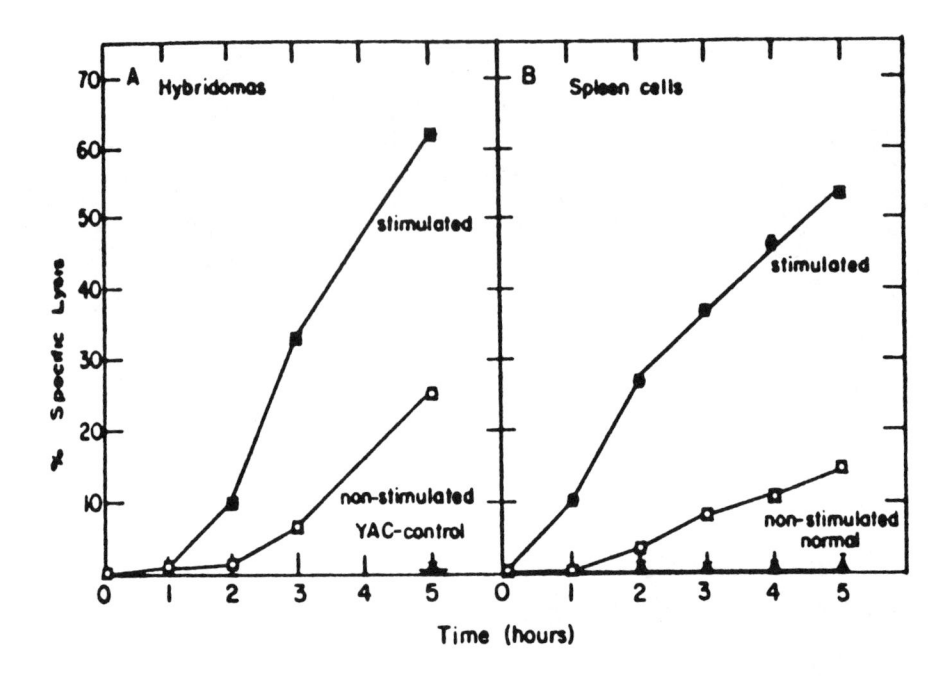

FIGURE 9. Kinetics of the lytic activity of BALB/c anti-EL4 CTL-hybridomas and memory spleen CTL stimulated by Con A. Hybridoma cells (A) or normal and immunized (60 days before assay with EL4) spleen cells (B) of BALB/c mice were cultured with (□) or without (■) Con A (2 μg/ml) for 20 hr. Excess unreacted Con A was then neutralized by addition of α-methylmannoside. Lysis of ^{51}Cr-labeled EL4 and YAC was examined at hybridoma/target cells ratios of 3:1 (A) and 10:1 (B), respectively, in triplicate.

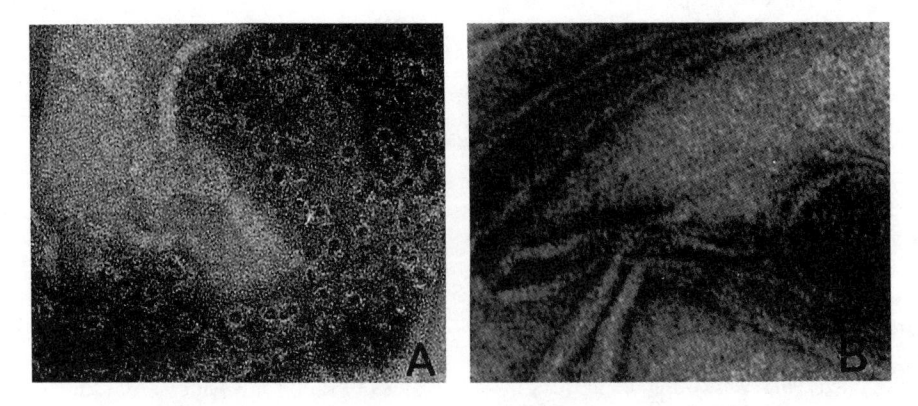

FIGURE 10. Negatively stained electronmicrographs of plasma membranes from EL4 lysed by antibody plus complement (A) or by BALB/c anti-EL4 PEL (B).

PEL

PEB

HYBRIDOMA

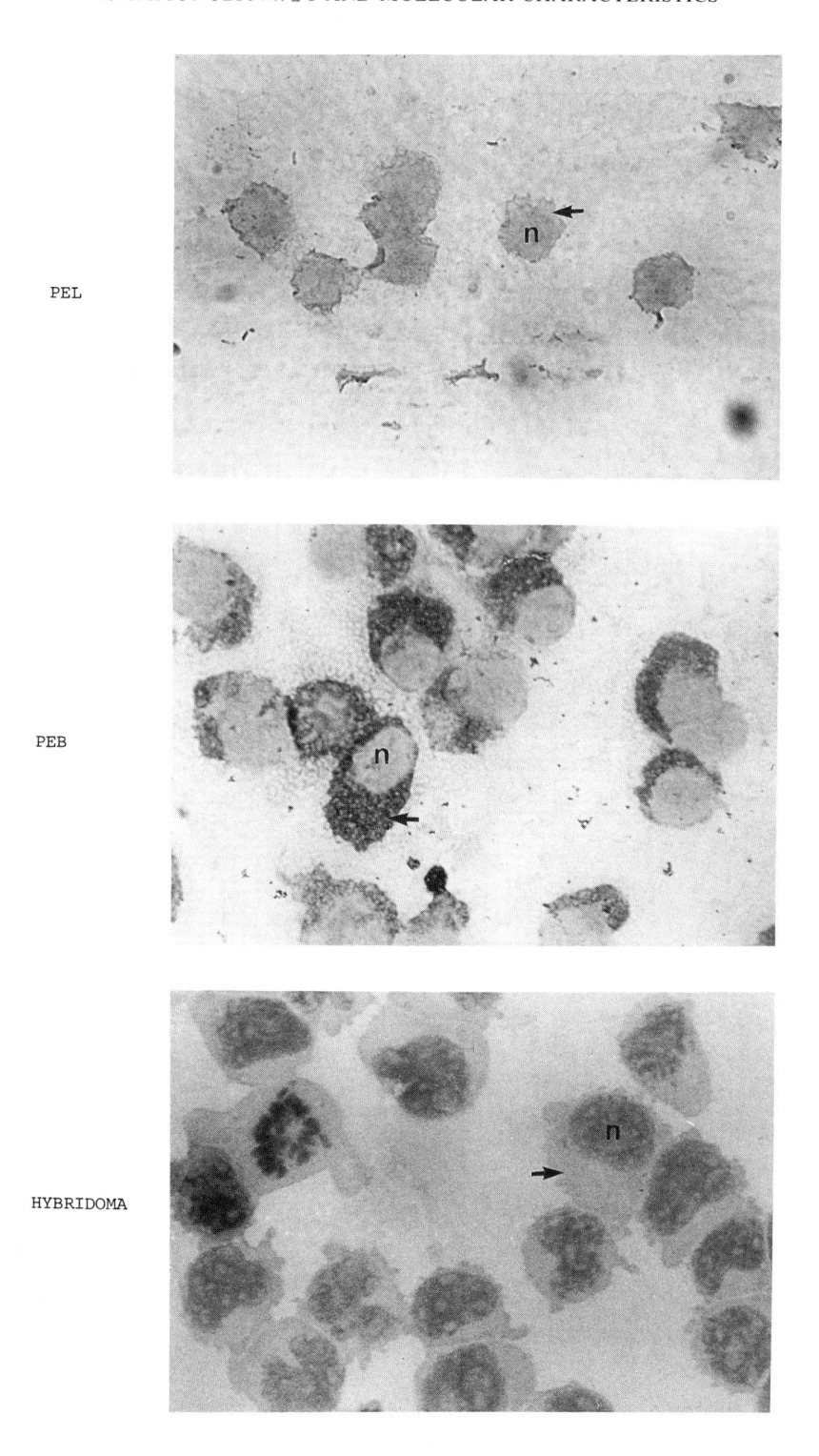

FIGURE 11. Immunocytostaining of PEL, PEB, and CTL hybridoma with monoclonal antibody for perforin. n, nucleus, counterstained blue with hematoxylin on the original slide. The arrow indicates the cytoplasm which was red on the positively stained original slide of PEB and pale pink on the negatively stained PEL and hybridoma (from Berke, Rosen, and Ronen, 1993).

TABLE 4. Lytic (granule and cellular) and BLT-esterase activities of BALB/c anti-EL4 CTL

Cell	Stimulant	Lytic activity (% specific ^{51}Cr release) Cell-mediated against ^{51}Cr-EL4	of Lytic granules against ^{51}Cr-SRBC	BLT-esterase activity (ΔOD 412 nm/10^6 cells/15 min at RT)
PEL	EL4 in vivo	84	−2.6	0.32
PEB	rIL-2	82.6	100	9.06
Hybridoma[a]	None	4.3	−0.7	0.30
	Con A	22.4	−0.7	0.30
	irr-EL4	47.5	−0.9	0.32

Lytic granule activity (perforin/cytolysin) was monitored against ^{51}Cr-labeled sheep red blood cells, in the presence and absence of Ca^{2+} (Henkart, 1985). BLT-esterase (granzyme) activity was determined by the cleavage of a synthetic substrate spectrophotometrically. SRBC, Sheep red blood cells.
[a] PMMI.

transform to PEB upon IL-2 stimulation *in vitro* (Berke and Rosen, 1988) (see Figure 7). The lytic granule and granzyme [*N*-α-bezyloxycarbonyl-L-lysine thiobenzyl (BLT)-esterase] activities in cellular extracts of PEL, PEB, and PEL hybridomas and the cell-mediated lytic activity of the intact cells from which the extracts were derived are depicted in Table 4. Whereas PEL exhibited potent

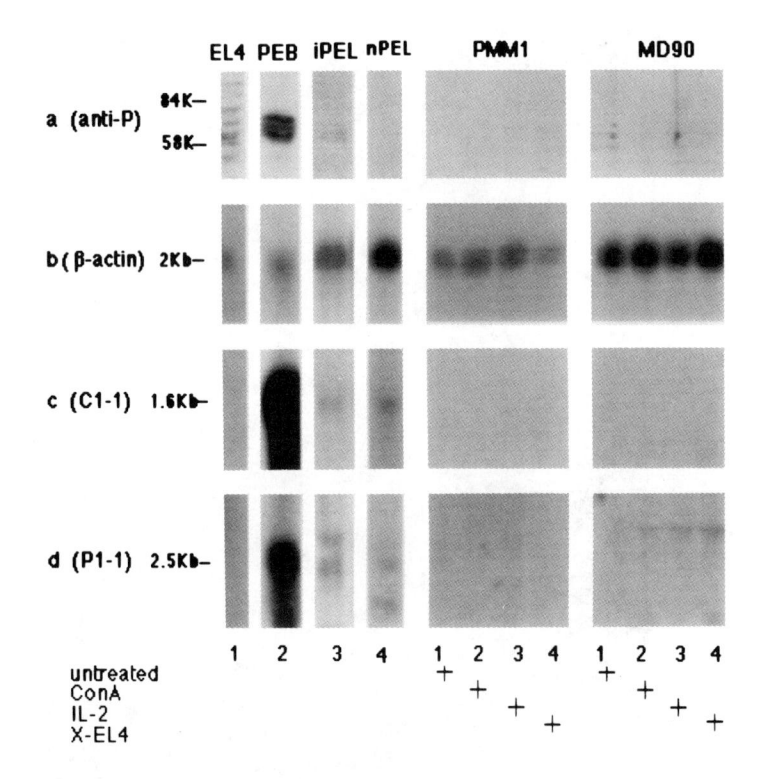

FIGURE 12. Western blot screening for perforin (a) and Northern blot screening for β-actin (b), granzyme B (c), and perforin (d) mRNAs of EL4 (negative control), PEB (positive control), PEL [normal (n) and immune (i)], and CTL-hybridomas (PMM1 and MD90, derived from PEL and MLC CTL, respectively). Cl-1 (CTLA-1) and P1-1 are probes for granzyme B and perforin mRNA, respectively. Hybridomas were stimulated with Con A (2 μg/ml), rIL-2 (300–500 U/ml), or equivalent numbers of irradiated (4000 R) (x) EL4, as indicated (From Berke, Rosen, and Ronen, 1993.)

and specific cell-mediated cytocidal activity, only background levels of lytic granules (i.e., perforin) and BLT-esterase were detected in them (Dennert et al., 1987; Berke and Rosen, 1988; Nagler-Anderson et al., 1988). In contrast, PEB derived from PEL exhibited massive quantities of BLT-esterase, perforin, and lytic granules, although usually not exhibiting more cell-mediated activity than the parental PEL (Table 4). Like PEL, but in contrast to PEB, PEL hybridomas showed no trace of lytic granules or BLT-esterase activity prior to or after stimulation with Con A or irr-EL4, treatments which would activate memory CTL (Table 4).

The presence or absence of perforin and serine esterases in PEL, PEB, and PEL hybridomas as determined by functional assays (see above) has been confirmed using monoclonal antibody against perforin (kindly provided by K. Okumura, Tokyo) and cDNA probes for perforin and granzymes (kindly provided by K. Okumura and by P. Golstein of Marseilles, respectively). Immunocytostaining for perforin of fixed preparations of PEL and PEL hybridomas revealed no perforin, while large granular PEB stained profusely (Figure 11), corresponding to the detection of cytoplasmic lytic granules by lysis of sheep red blood cells (Table 4). Western blotting (Figure 12) confirmed a strong perforin band (stained by anti-perforin plus peroxidase) in PEB and only faintly stained bands, equivalent to those often seen in nonimmunized lymphoid cells, in PEL. Cytolytic PEL hybridomas were as negative for perforin as noncytocidal cells, such as EL4 (Figure 12).

Northern blotting and hybridization with a ^{32}P-labeled cDNA probe of perforin (P1-1) demonstrated a strong perforin mRNA signal in PEB, faint signals in PEL and nonimmunized lymphoid cells, and no signal in CTL hybridomas (Figure 12). Expression of granzyme B mRNA, monitored with the C1-1 (CTLA-1) probe, paralleled perforin mRNA expression (Figure 12). No perforin or granzyme mRNA expression in the cytolytic hybridomas was detected by polymerase chain reaction (PCR) analysis (C. Bleackley, personal communication), corroborating our Northern blot, immunostaining, and Western blot analyses (Berke and Rosen, 1988, Berke, Rosen, and Ronen, 1993).

Summary

This chapter describes the generation, isolation, and cellular and molecular characterization of *in vivo*-primed peritoneal exudate CTL (PEL), involved in intraperitoneal allograft rejection and syngeneic tumor immunity; the PEL-blasts (PEB), obtained from PEL by culturing in IL-2; and cytocidal CTL hybridomas, derived by somatic hybridization of CTL with a T-cell lymphoma. Typically, the CTL used in most studies are large lymphoblastoid cells or cell lines, cultured in the presence of IL-2 and requiring repeated restimulation with antigen. In contrast, the *in vivo*-derived PEL-CTL are small to medium-sized, nondividing cells exhibiting potent and specific MHC-restricted binding and lytic activities. PEB are probably characteristic of CTL/NK cells cultured *in vitro*, whereas the immortalized PEL hybridomas are more characteristic of *in vivo*-derived CTL, like PEL, with regard to expression of lytic granules, perforin, and granzymes. The presence of lytic granules, perforin, and granzymes in PEL, PEB, and PEL hybridomas was examined because of their proposed roles in lymphocyte-mediated cytotoxicity. Their roles, if any, in the lytic mechanism of CTL, such as PEL, is discussed in Chapter 18 of this volume.

Acknowledgments

We would like to thank Ms. Malvine Baer for typing the manuscript.

Supported in part by funds from the U.S.-Israel Binational Science Foundation, Jerusalem, Israel; the German-Israeli Fund; the Deutsche Krebsforschungzentrum, Heidelberg, Germany; the Israel Academy of Sciences and Humanities; and the Israel Cancer Association.

References

Amos DB (1960): Possible relations between the cytotoxic effects of isoantibody and host cell function. *Ann NY Acad Sci* 87: 273

Amos DB (1962): The use of simplified systems as an aid to the interpretation of mechanisms of graft rejection. *Prog Allergy* 6: 648

Amos DB, Wakefield JD (1959): Growth of ascites tumor cells in diffusion chambers. II. Lysis and growth inhibition by diffusible isoantibody. *J Natl Cancer Inst* 22: 1077

Andersson LC, Häyry P (1975): Clonal isolation of alloantigen-reactive T cells and characterization of their memory. *Transplant Rev* 25: 121

Baker P, Weiser RS, Jutila J, Evans CA, Blandau RJ (1962): Mechanisms of tumor homograft rejection: The behavior of Sarcoma I ascites tumor in the A/Jax and C57BL/6K mouse. *Ann NY Acad Sci* 101: 46

Berke G (1980): Interaction of cytotoxic T lymphocytes and target cells *Prog Allergy* 27: 69–133

Berke G (1989): Functions and mechanisms of lysis induced by cytotoxic T-lymphocytes and natural killer cells. In: *Fundamental Immunol.* Paul WE ed. Raven Press. pp 735–764

Berke G (1993): Direct contact of cytotoxic T-lymphocyte receptors with target cell membrane determinants induces a prulytic rise of [CA^{2+}]i in the target that triggers disintegration. In this volume p. 194

Berke G, Amos DB (1973): Mechanism of lymphocyte-mediated cytolysis: The LMC cycle and its role in transplantation immunity. *Transplant Rev* 17: 71–107

Berke G, Rosen D (1987a): Are lytic granules, and perforin 1 thereof, involved in lysis induced by *in vivo* primed, peritoneal exudate CTL? *Transplant Proc* 19: 412–416

Berke G, Rosen D (1987b): Circular lesions detected on membranes of target cells lysed by antibody and complement or natural killer (spleen) cells but not by *in vivo* primed cytolytic T lymphocytes. In: *Membrane Mediated Cytotoxicity*, Bonavida B, Collier RJ, eds. (UCLA Symposia, Park City), 1986, New York: Alan Liss, pp. 367–378

Berke G, Rosen D (1988): Highly lytic *in vivo* primed CTL devoid of lytic granules and BLT-esterase activity acquire these constituents in the presence of T cell growth factors upon blast transformation *in vivo*. *J Immunol* 141: 1429–1436

Berke G, Schick B (1980): Tumor immunity in the peritoneal cavity. *Cont Top. Immunobiol.* 10: 297–315

Berke G, Gabison D, Feldman M (1975): The frequency of effector cells in populations containing cytotoxic T lymphocytes. *Eur J Immunol* 5: 813–818

Berke G, Rosen D, Ronen D (1993): Mechanism of lymphocyte-medialed cytolysis: Functional cytolylic T cells lacking perforin and granzymes. *Immunology* (in press)

Berke G, Sullivan KA, Amos DB (1972a): Rejection of ascites tumor allografts. I. Isolation, characterization and *in vitro* reactivity of peritoneal lymphoid effector cells from BALB/c mice immune to EL4 leukosis. *J Exp Med* 135: 1334–1350

Berke G, Sullivan KA, Amos DB (1972b): Rejection of ascites tumor allograft. II. A pathway for cell-mediated tumor destruction *in vitro* by peritoneal exudate lymphoid cells. *J Exp Med* 136: 1594–1604

Berke G, Sullivan KA, Amos DB (1972c): Tumor immunity *in vitro*: Destruction of a mouse ascites tumor through a cycling pathway *Science* 177: 433–434

Cerottini J-C, Brunner KT (1974): Cell-mediated cytotoxicity, allograft rejection and tumor immunity. *Adv Immunol* 18: 67

Denizot F, Brunet J-F, Roustan P, Harper K, Suzan M, Luciani M-F, Mattéi M-G, Golstein P (1989): Novel structures CTLA-2 alpha and CTLA-2 beta expressed in mouse activated T cells and mast cells and homologous to cysteine proteinase proregions. *Eur J Immunol* 19: 631–635

Denizot F, Wilson A, Battye F, Berke G, Shortman K (1986): Clonal expansion of T cells: A cytotoxic T cell response *in vivo* that involves precursor cell proliferation. *Proc Natl Acad Sci USA* 83: 6089–6092

Dennert G, Anderson C, Prochazka G (1987): High activity of Nα-benzyloxycarbonyl-L-lysine thiobenzyl ester serine esterase and cytolytic perforin in cloned cell lines is not demonstrable in *in-vivo*-induced cytotoxic effector cells. *Proc Natl Acad Sci USA* 184: 5004–5008

Fishelson Z, Berke G (1978): T lymphocyte-mediated cytolysis: Dissociation of the binding from the lytic mechanism of the effector cells. *J Immunol* 120: 1121–1126

Gorer D (1956): Some recent work on tumor immunity. *Adv Cancer Res* 4: 149

Hashimoto Y, Sudo H (1968): Studies on acquired transplantation resistance. III. Cytocidal effect of sensitized peritoneal lymphocytic cells of Donryu rats against the target Yoshida sarcoma cells in vitro. *Gann* 59: 7

Hashimoto Y, Ishidate M, Takaku M (1965): Studies on acquired transplantation resistance. II. Action of peritoneal exudate cells of Donryu rats immune to the tumor against Yoshida sarcoma. *Gann* 56: 23

Henkart PA (1985): Mechanisms of lymphocyte-mediated cytotoxicity. *Ann Rev Immunol* 3: 31

Herberman RB (1974): Cell-mediated immunity to tumor cells. *Adv Cancer Res* 19: 207

Hurt SN, Berke G, Clark WR (1979): A rapid method for generating cytotoxic effector cells *in vivo*. *J Immunol Methods* 28: 321–329

Kalina M, Berke G (1976): Contact regions of cytotoxic T lymphocyte-target cell conjugates. *Cell Immunol* 25: 41–51

Kaufmann Y, Berke G (1980): Enucleated cytotoxic T lymphocytes bind specifically to target cells *in vitro*. *Transplantation* 29: 374–378

Kaufmann Y, Berke G (1981): Cell surface glycoproteins of cytotoxic T lymphocytes induced *in vivo* and *in vitro*. *J Immunol* 126: 1443–1446

Kaufmann Y, Berke G (1983): Monoclonal cytotoxic T lymphocyte hybridoma capable of specific killing activity, antigenic responsiveness and inducible interleukin(s) secretion. *J Immunol* 131: 50–56

Kaufmann Y, Berke G, Eshhar Z (1981a): Cytotoxic T lymphocyte hybridomas that mediate specific tumor cell lysis *in vitro*. *Proc Natl Acad Sci USA* 78: 2502–2506

Kaufmann Y, Berke G, Eshhar Z (1981b): Functional cytotoxic T lymphocyte hybridomas. *Transplant Proc* 13: 1170–1174

Krähenbühl O, Tschopp J (1991): Perforin-induced pore formation. *Immunol Today* 12: 399

Lotze MT, Finn OJ (1990): Recent advances in cellular immunology: Implications for immunity to cancer. *Imm Today* 11: 190

Martz E (1987): LFA-1 and other accessory molecules functioning in adhesions of T and B lymphocytes. *Hum Immunol* 18: 3

Moscovitch M, Kaufmann Y, Berke G (1984): Memory CTL-hybridoma: A model system to analyze the anamnestic response of CTL. *J Immunol* 133: 2369–2374

Nagler-Anderson C, Allbritton NL, Verret CR, Eisen HN (1988): A comparison of the cytolytic properties of murine primary CD8$^+$ cytotoxic T lymphocytes and cloned cytotoxic T cell lines. *Immunol Rev* 103: 111–125

Patt HM, Straube RL (1956): Measurement and nature of ascites tumor growth. *Ann NY Acad Sci* 63: 728

Podack ER, Hengartner H, Lichtenheld M (1991): A central role of perforin in cytolysis? *Ann Rev Immunol* 9: 129

Schick B, Berke G (1977): Activity of tumor-associated lymphoid cells at short intervals after administration of irradiated syngeneic and allogeneic tumor cells. *J Immunol* 118: 986

Straube RL, Hill MS, Patt HM (1955): Vascular permeability and ascites tumor growth. *Proc Am Assoc Cancer Res* 2: 49

Wagner H, Röllinghoff M, Nossal GJV (1973): T cell-mediated immune responses induced *in vitro*: A probe of allograft and tumor immunity. *Transplant Rev* 17: 3

Zagury D, Bernard J, Thierness N, Feldman M, Berke G (1975): Isolation and characterization of individual functionally reactive cytotoxic T lymphocytes: Conjugation, killing and recycling at the single cell level. *Eur J Immunol* 5: 818–822

11
Regulatory Effects of Cytokines on the Generation of CTL and LAK Cells

Mark R. Alderson and Michael B. Widmer

Introduction

It has long been recognized that the growth and differentiation of resting mature T lymphocytes into functional helper or cytolytic cells is dependent upon not only an antigenic stimulus but also soluble factors (Plate, 1976; Ryser et al., 1978). Consequently, recent attention has been focused upon the identification and further characterization of these soluble regulators. Following the description of interleukin-2 (IL-2) as a T-cell growth factor that can influence the generation of cytolytic T lymphocytes (CTL) (Gillis et al., 1978), it became apparent that several other cytokines exist which have similar activity. However, many of the early studies on T cell-active factors were clouded by the use of impure cytokine preparations and the possibility of indirect effects occurring in bulk lymphocyte cultures due to cytokine cascades. Thus, it was possible that a reputed T-cell growth factor or CTL induction factor was in fact affecting the secondary production of IL-2 or other T-cell growth factors, rather than itself directly stimulating T cells.

Our understanding of the role of cytokines in T-cell growth and differentiation has been aided over the past several years by the application of several important technologies. First, biochemical purification and molecular cloning have been of paramount importance in ascribing the biological activity present in a crude culture supernatant to a purified, defined cytokine of known amino acid sequence. Once the cytokine is available in purified form, it can be tested for its effects on T-cell growth and differentiation in the absence of contaminating proteins. In some cases, cytokines originally described based upon their effects on cells outside the T lineage have subsequently been found to influence the behavior of T lymphocytes. For example, the cDNAs for both IL-4 and IL-7 were molecularly cloned based upon the activity of the corresponding native protein on B-lineage cells (Noma et al., 1986: Namen et al., 1988), and it was not until these cytokines were available in recombinant form that their ability to cause the growth and differentiation of T cells was realized. Second, limiting dilution analysis has allowed the determination of the effect of cytokines at the precursor level and has permitted the quantification of cytokine-responsive cells in bulk populations (MacDonald et al., 1980). Finally, the increasing availability of cytokine antagonists, both neutralising antibodies and soluble cytokine receptors, has helped to distinguish direct from indirect effects of cytokines and has facilitated descriptions of the role played by endogenously produced cytokines in immune responses.

In this chapter, we will summarize recent literature, including studies performed in our laboratory, in which it has been reported that IL-2, IL-4, and IL-7 are all major T-cell growth and differentiation factors, particularly with respect to the ability of these and other cytokines to influence the generation of cytolytic cell populations. We will focus our discussion on human cytolytic cells, though important similarities and differences between human and murine studies will also be addressed. Two cell types will be discussed, CTL and lymphokine-activated killer (LAK) cells, which can be distinguished based upon their lytic specificity, their requirement for antigen, and, in part, their cell surface phenotype. For the purposes of this review, CTL are operationally defined as cytolytic T lymphocytes that require the presence of antigen during their generation from inactive precursors and that exhibit antigen-specific lytic activity. We have used mixed leukocyte cultures (MLC) as an *in vitro* model for defining the role of cytokines in the generation of CTL. Alloreactive CTL generated in such cultures are detected based upon their lytic activity for phytohemagglutinin (PHA)-stimulated blasts derived from the stimulating cell donor, but not the responding cell donor. On the other hand, LAK cells are defined as killer cells with broad specificity which are generated in the absence of an antigenic stimulus. LAK cells kill a variety of tumor cell targets, including natural killer (NK) cell-insensitive tumor lines and fresh tumor cells. We have used cultures of peripheral blood mononuclear cells (PBMC) supplemented with exogenous cytokines to stimulate the induction of LAK cells, and the NK-insensitive tumor cell line, Daudi, as a target cell to measure their lytic activity. Our laboratory has approached the analysis of the effects of cytokines on the regulation of cytolytic cells in two ways. First, purified, recombinant cytokines are added to culture systems in which a cytolytic response occurring in the absence of added cytokines is relatively minimal. Cytolytic activity present at the end of the culture period would be influenced presumably by any cytokine which affects CTL cell growth and/or differentiation. Second, to determine whether endogenously produced cytokines are involved in a particular response, specific cytokine antagonists are added to cultures in which a relatively strong cytolytic response occurs in the absence of exogenous cytokine. These two complementary approaches allow a determination of whether a cytokine (a) has the capability to stimulate a cytolytic response and (b) is normally involved in the response.

Interleukin-2

IL-2 is the prototype T-cell growth factor, and its ability to induce T-cell growth and enhance the generation of CTL has been extensively documented. Early studies suggested that IL-2 is one of the factors that is essential for the induction of CTL, as neutralizing antibodies to IL-2 virtually eliminate CTL generation (Gillis et al., 1981; Kern et al., 1981). LAK activity was first described based upon the ability of IL-2 to induce nonspecific killer cells from human peripheral blood (Lotze et al., 1981; Grimm et al., 1982). Thus, IL-2 had become the cytokine of choice for the induction and numerical expansion of cytolytic cell populations, and it is currently being used to generate cytolytic cells for adoptive transfer into patients with certain malignancies (reviewed by Rosenberg and Lotze, 1986). However, during the early to mid-1980s, numerous reports suggested the existence of at least two other cytokines that could play a role in the generation of CTL (Raulet and Bevan, 1982; Kaieda et al., 1982; Wagner et al., 1982; Mannel et al., 1983; Takai et al., 1986).

Interleukin-4

Prior to its molecular cloning, IL-4 was characterized by its ability to affect mature B cell function, including B-cell growth and immunoglobulin isotype switching (reviewed by Paul and Ohara, 1987). With the availability of highly pure natural and recombinant IL-4, several laboratories reported on the ability of IL-4 to induce T-cell growth (Fernandez-Botran et al., 1986; Mosmann et al., 1986; Grabstein et al., 1987; Hu-Li et al., 1987). Simultaneously, our laboratory demonstrated that IL-4 enhanced the generation of alloreactive CTL in both murine and human primary MLC (Widmer and Grabstein, 1987; Widmer et al., 1987), and Pfeifer and co-workers (1987) reported that IL-4 induced the proliferation and differentiation of lectin-costimulated murine CTL precursors.

The activity of IL-4 on human cytolytic cells could be differentiated from that of IL-2 in a number of ways (Widmer et al., 1987). First, IL-4 was most potent at enhancing CTL generation if its addition was delayed relative to the initiation of culture. Second, in spite of its ability to enhance antigen-specific CTL development, IL-4 had no detectable LAK cell-inducing activity. On the contrary, IL-4 inhibited the development of CTL and LAK effector cells that occurred in the presence of exogenous IL-2. The inhibitory effect of IL-4 on the generation of human LAK cells has been confirmed by several other investigators (Nagler et al., 1988; Spits et al., 1988; Kawakami et al., 1989). These findings are in stark contrast to the results of studies on murine lymphoid cells in which IL-4 has been demonstrated to enhance LAK activity in the presence of IL-2 (Mulé et al., 1987). The mechanism by which IL-4 precludes the ability of IL-2 to induce human LAK activity is unclear, though it has been suggested that it may be related to the ability of IL-4 to inhibit cytokine production by both monocytes and T cells. In this regard, it has been demonstrated that two of the cytokines whose secretion is inhibited by IL-4, namely interferon-γ (IFN-γ) and tumor necrosis factor-α (TNF-α), are themselves positive regulators of LAK induction and are able to partially reverse the inhibitory effect of IL-4 (Swisher et al., 1990). Alternatively, the inhibitory effect of IL-4 may be related to its ability to down-regulate the β chain of the IL-2 receptor on human mononuclear cells (Lindqvist et al., 1991).

Interleukin-7

IL-7 is a stromal cell-derived cytokine that was originally purified and molecularly cloned based upon its ability to stimulate the growth of murine pre-B cells (Namen et al., 1988). Subsequent studies by our laboratory and others revealed that IL-7 is a potent growth factor for immature (Chantry et al., 1989; Conlon et al., 1989; Okazaki et al., 1989; Watson et al., 1989) as well as mature (Chazen et al., 1989; Morrissey et al., 1989; Alderson et al., 1990; Armitage et al., 1990; Grabstein et al., 1990; Londei et al., 1990) murine and human T cells, the effect on mature cells being particularly evident when IL-7 was used in conjunction with antigen or with suboptimal doses of T-cell mitogens. IL-7 was also found to enhance expression of the α chain of the IL-2 receptor, suggesting a possible involvement of IL-2 in the IL-7 response (Armitage et al., 1990; Grabstein et al., 1990). However, the addition of anti-IL-2 serum to cultures of T cells stimulated with IL-7 plus mitogen only partially inhibited their proliferative response, suggesting the existence of both IL-2-dependent and IL-2-independent components of IL-7 activity (Armitage et al., 1990).

Simultaneously, we analyzed the effect of IL-7 on the generation of CTL and LAK cells from human peripheral blood (Alderson et al., 1990). The addition of either IL-2 or IL-7 to MLC induced a 10 to 30-fold increase in alloantigen-specific CTL activity. As observed for T-cell proliferation, the generation of CTL in MLC supplemented with IL-7 contained both IL-2-dependent and IL-2-independent components. Subsequently, IL-7 was shown to enhance the generation of antiviral CTL from human T cells (Hickman et al., 1990). IL-7 was also found to induce LAK activity in human peripheral blood leukocyte cultures, though the maximum lytic activity induced by IL-7 was approximately 10-fold less than that induced by IL-2 (Alderson et al., 1990). However, in contrast to the partial dependence of IL-7 upon endogenously produced IL-2 in the generation of CTL, the effect of IL-7 on LAK cell induction was found to be virtually entirely IL-2-independent. The induction of LAK cells by IL-7, like IL-2, was inhibited by the addition of IL-4. Stotter et al. (1991) have confirmed that IL-7 induces human LAK cells in an IL-2-independent manner, and that IL-4 inhibits the induction of LAK cells by IL-7. IL-7 also induces LAK activity among murine lymphoid cells, though the majority of cytolytic cells induced in this fashion are of T-cell phenotype, rather than a heterogeneous mixture of T and non-T LAK cells which are induced by IL-2 (Lynch and Miller, 1990).

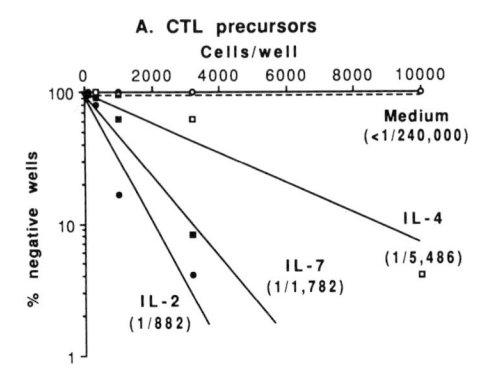

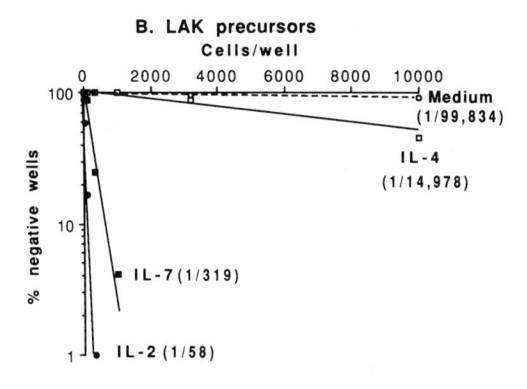

FIGURE 1. Limiting dilution analysis of the effect of exogenous IL-2, IL-4, and IL-7 on the generation of (A) CTL and (B) LAK cells. Numbers in parentheses represent precursor frequencies.

To define more precisely the relationship among CTL and LAK cell precursors responsive to IL-2, IL-4, and IL-7, limiting dilution assays were performed under conditions in which these cytokines were used to promote cellular expansion and/or differentiation. Microcultures were established with limiting numbers of responding cells, 10^5 irradiated stimulating cells, and excess cytokine (30 ng/ml). After 14 days, cultures were assayed for lytic capacity against PHA blasts derived from the stimulating cell donor. A high frequency of CTL precursors (CTL-p) was detected in cultures supplemented with IL-2, IL-4, or IL-7 (Figure 1A).

% Lysis of Allogeneic Stimulator Targets

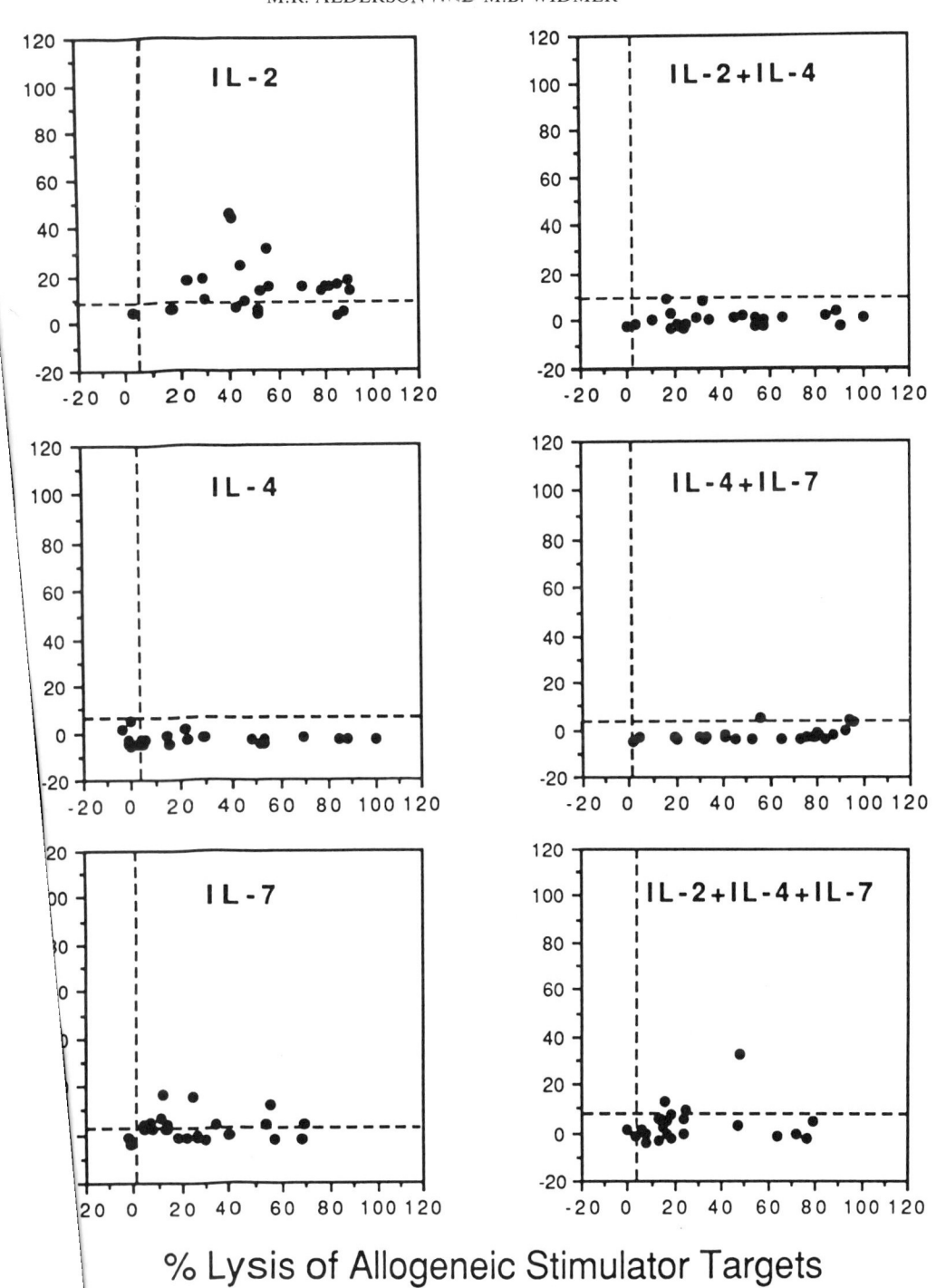

it-well limiting dilution analysis of the effect of exogenous IL-2, IL-4, and IL-7 on the loreactive versus autologous cytolytic cells from PBMC. Data were obtained from cultures PBMC were seeded per well with 24 replicate wells.

Moreover, clone size estimates have suggested that CTL clones generated in IL-2, IL-4, or IL-7 have similar lytic activity (Alderson et al., 1990). We also analyzed the frequency of LAK cell precursors (LAK-p) responsive to these cytokines by culturing limiting numbers of PBMC with irradiated autologous feeder cells and assaying lytic activity against ^{51}Cr-labeled Daudi targets. In contrast to CTL-p, the frequency of LAK-p detectable among PBMC differed markedly depending upon the cytokine added to the culture (Figure 1B). A very high frequency of LAK-p was responsive to IL-2, whereas IL-4 stimulated no significant LAK outgrowth compared to medium alone. A five-fold lower frequency of LAK-p was responsive to IL-7 than to IL-2. The differences in potency of IL-2, IL-4, and IL-7 for inducing LAK cells in bulk cultures are thus reflected at the level of the frequency of LAK-p responsive to these cytokines.

More recently we have used split-clone limit-dilution analysis to demonstrate that the majority of alloreactive CTL clones generated in MLC supplemented with IL-2 exhibit lytic activity against autologous targets (Figure 2). A lower, though significant, proportion of clones generated in IL-7 also exhibit such autologous lytic activity. In contrast, CTL clones generated in IL-4 are acutely specific for target cells from the stimulating donor and exhibit very little, if any, self-reactivity. Interestingly, the inclusion of IL-4 in culture together with IL-2 and/or IL-7 inhibits the development of autologous lytic activity that would otherwise occur, yet the generation of antigen-specific CTL is virtually unaffected. The occurrence of autologous lytic activity in limiting dilution microcultures may be partially due to an inherent degradation of lytic specificity of some CTL clones exposed to IL-2. Indeed, transfer of an alloantigen-specific CTL clone generated in the presence of IL-4 into IL-2 resulted in a loss of antigen specificity, which could be reversed by the simultaneous addition of IL-4 (Paliard et al., 1989). Alternatively, the concomitant presence of LAK cells in cultures generated in the presence of IL-2 may mask the antigen specificity of CTL clones.

Other Cytokines in the Generation of CTL and LAK Cells

Several other cytokines have been reported to act on mature T cells and LAK cell precursors, though not with the same potency as IL-2, IL-4, and IL-7. For example, both IL-5 and IL-6 to enhance the generation of hur have been reported to require tl to exert their effect (Okada et al 1988). A neutralizing monoclo IL-6 had little or no effect o alloreactive CTL in non-cy MLC or in MLC supplement IL-7 (Alderson et al, unpubli that endogenous IL-6 plays such responses. Addition o IFN-α has been reported to tion (Chen et al., 1986), against IFN-γ, Simon showed a role for endogen IL-2-dependent CTL requirement for both IFN reported for the inductic cell-free limiting dilution sky et al., 1989).

IFN-α and IFN-γ h human LAK cells, but this respect (Ellis et al and IL-6 have been human natural killer duction of IL-2 (A 1989). As previously can partially reverse on LAK cell induc cribed and cloned affect the generatic full spectrum of a other cells involve further analysis. factor (IL-12) sti (Wolf et al., 199 recently reporte generation of r tion with exog

Role of Produce Genera

To examir cytokines agonists v MLC fo selected absence bodies IL-7, ar were

FIGURE 2. Sp induction of a in which 3200

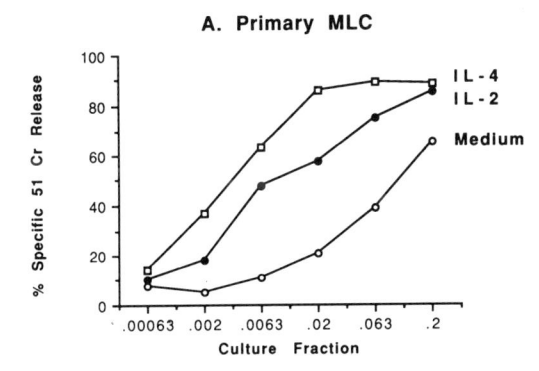

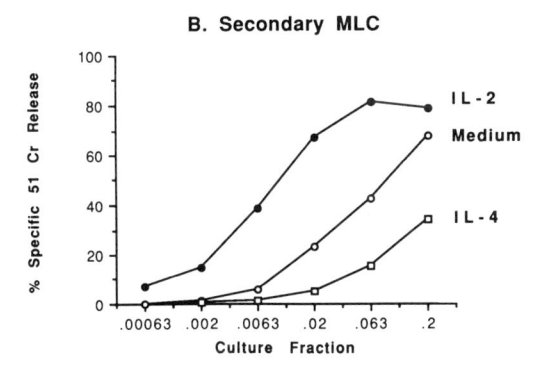

FIGURE 3. Effect of exogenous IL-2 and IL-4 on CTL generation in (A) primary versus (B) secondary MLC.

ongoing CTL response. Of these antagonists, only an IL-2-neutralizing antiserum consistently and significantly inhibited the generation of CTL in primary human MLC, demonstrating once again that IL-2 occupies a central role in CTL generation. IL-1 antagonists (both anti-IL-1 antibodies and soluble type I IL-1 receptor) had variable effects on the generation of CTL. In secondary MLC, cytokine antagonists revealed a role for endogenously produced IL-2 and IL-4, although the role of IL-4 differed markedly depending on whether human or murine cultures were employed. When used in conjunction with an antibody against IL-2, soluble IL-4 receptor completely blocked CTL induction in murine secondary MLC (Widmer et al., unpublished data). In contrast, exogenously added IL-4 was found to inhibit CTL generation in human secondary MLC (Figure 3); furthermore, addition of an anti-IL-4 serum to human secondary MLC resulted in enhanced CTL generation. These two results suggest an inhibitory role for endogenously produced IL-4 in such cultures (Alderson et al., manuscript in preparation).

Interestingly, systemic administration of soluble IL-4 receptor to mice receiving heterotopic heart allografts significantly prolonged graft survival, demonstrating a role for this cytokine in the allogeneic response *in vivo* (Fanslow et al., 1991).

Discussion

The increasing availability of purified recombinant cytokines and cytokine antagonists has paved the way for detailed analysis of the roles of individual cytokines in the generation of CTL and LAK cells. For the most part, studies to define the role of cytokines in these responses have been conducted by employing *in vitro* model systems, as described herein. Studies in which exogenous cytokines are added to cultures of lymphoid cells have demonstrated that IL-2, IL-4, and IL-7 are major growth factors for T cells and can exert significant effects on the generation of CTL and LAK cells. Other cytokines may also influence the generation of cytolytic cells, but their effects in various model systems have been found in many cases to depend upon the concomitant presence of a second cytokine, such as IL-2. Furthermore, studies involving the use of cytokine antagonists have identified IL-2 and IL-4 as endogenous regulators of CTL. In the case of IL-4, both stimulatory and inhibitory effects on cytolytic cell generation have been observed, depending on (1) its time of addition to culture, (2) the presence of other cytokines, (3) prior exposure to antigen (primary or secondary cultures), and (4) the species of origin of the lymphoid cells.

Despite the relatively large amount of data generated by employing *in vitro* systems, the identification of cytokines that regulate cytolytic responses *in vivo* and the mechanism by which they may act is a relatively unexplored area of cytokine biology. Studies in our laboratory have demonstrated that systemic administration of soluble receptors for either IL-1 or IL-4 prolongs the survival of allografts in mice and thus point to a role for the endogenous production of these two cytokines in allograft rejection (Fanslow et al., 1990, 1991), although their involvement in the generation of CTL *in vivo* has yet to be established. A related area of intense interest concerns the manner by which the production of cytokines may be regulated *in vivo*. Studies by Mosmann et al. (1986) have demonstrated the existence of two major subsets of murine CD4+ long-term T-helper cell lines (termed T_H1 and T_H2) which are distinguished by their profiles of cytokine secretion. Although cells with the characteristics of T_H1 and

T_H2 clones have been difficult to demonstrate in naive animals and in humans, in vivo studies have suggested that different modes of challenge to the immune system may result in the preferential production of certain cytokines, and that these cytokines subsequently determine the nature of the immune response. For example, the immune response to parasitic infections is typically characterized by IgE production and eosinophilia, responses normally associated with the presence of IL-4 and IL-5, which are both T_H2 products. In contrast, viral infections are characteristically associated with IgG2a production and delayed hypersensitivity, two responses directly influenced by T_H1 products. It is thus of interest that two of the cytokines that play a major role in CTL generation, namely IL-2 and IL-4, are made by T_H1 and T_H2 cells, respectively (Mosmann et al., 1986). Notably, these two cytokines often modulate the effect of one another on both LAK cell and CTL precursors.

In contrast to IL-2 and IL-4, which are usually produced by T cells, IL-7 appears to be primarily a stromal cell product, and thus factors involved in the regulation of IL-7 production remain to be determined. The presence of IL-7 mRNA in spleen and thymus (Namen et al., 1988) suggests that CTL precursors may be exposed to IL-7 in vivo. Studies of murine fetal thymus cultures have demonstrated that exogenous IL-7 can stimulate proliferation of developing cells in the cytolytic lineage, but expression of lytic activity requires subsequent exposure to IL-2 (Widmer et al., 1990). Interestingly, the development of such lytic activity is inhibited by exogenous IL-4.

Another approach to assessing the role of cytokines in vivo involves the generation of transgenic mice in order to determine the effect of constitutive expression of a particular cytokine or cytokine receptor gene. Alternatively, a more recent and elegant approach involves the elimination of a particular cytokine or cytokine receptor in genetically deficient animals generated by targeted recombination. Using this second approach, Schorle et al. (1991) recently demonstrated that elimination of a functional IL-2 gene in mice has virtually no effect on T-cell development, though the ability of these mice to fight infection remains to be determined.

Another expanding area of CTL biology concerns the regulation of generation and function of T cells expressing the $\gamma\delta$ T-cell receptor (Borst et al., 1987). These T cells have been demonstrated to mediate non-MHC-restricted lytic activity after culture in IL-2, especially $\gamma\delta$ T cells that fail to express either CD4 or CD8 molecules (Moingeon et al., 1987; Koide et al., 1989). Recently it was demonstrated that $\gamma\delta$ T-cell clones could be generated in the presence of IL-4, though unlike $\gamma\delta$ clones generated in IL-2, the majority of clones generated in IL-4 were antigen-specific and failed to lyse NK-sensitive tumor cells (Paliard et al., 1989). The effect of IL-7 and other cytokines on $\gamma\delta$ T-cells remains to be examined, though our initial studies indicate that IL-7 induces $\gamma\delta$ cells with LAK activity from human PBMC (Alderson et al., 1990). The effects of cytokines on $\gamma\delta$ T cells coexpressing either CD4 or CD8 also await analysis.

The ability of cytokines to influence the generation of CTL and LAK cells in vitro has prompted attempts to use cells grown in cytokine-supplemented cultures as immunotherapeutic agents for the treatment of certain cancers. Initial studies focused upon the effect of IL-2 on the induction of LAK cells and expansion of tumor-infiltrating lymphocytes (TIL) obtained from cancer patients and early results have been promising, though only in a minority of patients (Rosenberg et al., 1986, 1989). Thus, current treatments for enhancing cytolytic responses in vivo are generally nonspecific; similarly, treatments for the prevention of graft rejection are generally, not selectively, immunosuppressive. If our aim is to take advantage of the specificity of the immune system by treating patients with tumor-specific killer cells that do not simultaneously lyse normal cells, then IL-4 or IL-7 or combinations of these cytokines may prove more effective for the in vitro expansion of cytolytic cells than IL-2 alone. Our data from split-clone limiting dilution analysis (Figure 2) suggest that the combination of IL-2 plus IL-4 or IL-7 plus IL-4 may prove to be the best stimulus for generating antigen-specific CTL in the absence of self-reactive cytolytic cells. However, the fact that IL-4 has the ability to both induce and suppress the generation of CTL, depending upon its time of addition to culture or its costimulus, suggests some caution should be exercised in its use as an immunotherapeutic. In addition, the relationship between lytic activity of cells generated by cytokine treatment in vitro and their ability to eliminate tumors in vivo remains to be determined. A considerable amount of work remains to be done before it is possible to manipulate the generation of specific cytolytic populations at will in both a positive and a negative manner. However, we have seen considerable progress in this area in the last decade; the next decade promises to be just as fruitful.

Acknowledgments

We gratefully acknowledge Drs. Ken Grabstein,

Richard Armitage, William Fanslow, David Lynch, and Phil Morrissey for sharing their data and ideas, and Rob Voice and Teresa Tough for expert technical assistance.

References

Alderson MR, Sassenfeld HM, Widmer MB (1990): Interleukin 7 enhances cytolytic T lymphocyte generation and induces lymphokine-activated killer cells from human peripheral blood. *J Exp Med* 172: 577–587

Aoki T, Kikuchi H, Miyatake S, Oda Y, Iwasaki K, Yamasaki T, Kinashi T, Honjo T (1989): Interleukin 5 enhances interleukin 2-mediated lymphokine-activated killer activity. *J Exp Med* 170: 583–588

Armitage RJ, Namen AE, Sassenfeld HM, Grabstein KH (1990): Regulation of human T-cell proliferation by interleukin 7. *J Immunol* 144: 938–941

Borst J, van de Griend RJ, van Oostveen JW, Ang S, Melief CJ, Seidman JG, Bolhuis RLH (1987): A T-cell receptor γ/CD3 complex found on cloned functional lymphocytes. *Nature* 325: 683–688

Chantry D, Turner M, Feldmann M (1989): Interleukin 7 (murine pre-B cell growth factor/lymphopoietin 1) stimulates thymocyte growth: Regulation by transforming growth factor beta. *Eur J Immunol* 19: 783–786

Chazen GD, Pereira GMB, LeGros G, Gillis S, Shevach EM (1989): Interleukin 7 is a T-cell growth factor. *Proc Natl Acad Sci USA* 86: 5923–5927

Chen L-K, Tourvieille B, Burns GF, Bach FH, Mathieu-Mahul D, Sasportes M, Bensussan A (1986): Interferon: A cytotoxic T lymphocyte differentiation signal. *Eur J Immunol* 16: 767–770

Chen W-F, Zlotnick A (1991): IL-10: A novel cytotoxic T cell differentiation factor. *J Immunol* 147: 528–534

Conlon PJ, Morrissey PJ, Nordan RP, Grabstein KH, Prickett KS, Reed SG, Goodwin R, Cosman D, Namen AE (1989): Murine thymocytes proliferate in direct response to interleukin-7. *Blood* 74: 1368–1373

Ellis TM, McKenzie RS, Simms PE, Helfrich BA, Fisher RI (1989): Induction of human lymphokine-activated killer cells by IFN-α and IFN-γ. *J Immunol* 143: 4282–4286

Fanslow WC, Clifford KN, Park LS, Rubin AS, Voice RF, Beckmann MP, Widmer MB (1991): Regulation of alloreactivity in vivo by IL-4 and the soluble IL-4 receptor. *J Immunol* 147: 535–540

Fanslow WC, Sims JE, Sassenfeld H, Morrissey PJ, Gillis S, Dower SK, Widmer MB (1990): Regulation of alloreactivity in vivo by a soluble form of the interleukin-1 receptor. *Science* 248: 739–742

Fernandez-Botran R, Krammer PH, Diamantstein T, Uhr JW, Vitetta ES (1986): B cell-stimulatory factor 1 (BSF-1) promotes growth of helper T cell lines. *J Exp Med* 164: 580–593

Gillis S, Ferm MM, Ou W, Smith KA (1978): T cell growth factor: Parameters of production and a quantitative microassay for activity. *J Immunol* 120: 2027–2032

Gillis S, Gillis AE, Henney CS (1981): Monoclonal antibody directed against interleukin-2. I. Inhibition of T lymphocyte mitogenesis and the in vitro differentiation of alloreactive cytolytic T cells. *J Exp Med* 154: 983–988

Grabstein KH, Park LS, Morrissey PJ, Sassenfeld H, Price V, Urdal DL, Widmer MB (1987): Regulation of murine T cell proliferation by B cell stimulatory factor-1. *J Immunol* 139: 1148–1153

Grabstein KH, Namen AE, Shanebeck K, Voice RF, Reed SG, Widmer MB (1990): Regulation of T cell proliferation by IL-7. *J Immunol* 144: 3015–3020

Grimm EA, Mazumder A, Zhang HZ, Rosenberg SA (1982): Lymphokine-activated killer cell phenomenon. Lysis of natural killer-resistant fresh solid tumor cells by interleukin-2-activated autologous human peripheral blood lymphocytes. *J Exp Med* 155: 1823–1841

Hickman CJ, Crim JA, Mostowski HS, Siegal JP (1990): Regulation of human cytotoxic T lymphocyte development by IL-7. *J Immunol* 145: 2415–2420

Hu-Li J, Shevach EM, Mizuguchi J, Ohara J, Mosmann T, Paul WE (1987): B cell stimulatory factor 1 (interleukin 4) is a potent costimulant for normal resting T lymphocytes. *J Exp Med* 165: 157–172

Kaieda T, Okada M, Yoshimura N, Kishimoto S, Yamamura Y, Kishimoto T (1982): A human helper T cell clone secreting both killer helper factor(s) and T cell-replacing factor(s). *J Immunol* 129: 46–51

Kawakami Y, Custer MC, Rosenberg SA, Lotze MT (1989): IL-4 regulates IL-2 induction of lymphokine-activated killer activity from human lymphocytes. *J Immunol* 142: 3452–3461

Kern DE, Gillis S, Okada M, Henney CS (1981): The role of interleukin-2 (IL-2) in the differentiation of cytotoxic T cells: The effect of monoclonal anti-IL-2 antibody and absorption with IL-2 dependent T cell lines. *J Immunol* 127: 1323–1328

Koide J, Rivas A, Engleman EG (1989): Natural killer (NK)-like cytotoxic activity of allospecific T cell receptor-γ,δ^+ T cell clones. Distinct receptor-ligand interactions mediate NK-like and allospecific cytotoxicity. *J Immunol* 142: 4161–4168

Lindqvist C, Nihlmark EL, Nordstrom T, Andersson LC (1991): Interleukin-4 downregulates the p70 chain of the IL-2 receptor on peripheral blood mononuclear cells. *Cell Immunol* 136: 62–68

Londei M, Verhoef A, Hawrylowicz C, Groves J, De Berardinis P, Feldmann M (1990): Interleukin 7 is a growth factor for mature human T cells. *Eur J Immunol* 20: 425–428

Lotze MT, Grimm EA, Mazumder A, Strausser JL, Rosenberg SA (1981): Lysis of fresh and cultured autologous tumor by human lymphocytes cultured in T cell growth factor. *Cancer Res* 41: 4420–4425

Luger TA, Krutmann J, Kirnbauer R, Urbanski A, Schwarz T, Klappacher G, Kock A, Micksche M, Malejczyk J, Schauer E, May LT, Sehgal PB (1989): IFN-β2/IL-6 augments the activity of human natural killer cells. *J Immunol* 143: 1206–1209

Lynch DH, Miller RE (1990): Induction of murine lymphokine-activated killer cells by recombinant IL-7. *J Immunol* 145: 1983–1990

MacDonald HR, Cerottini J-C, Ryser J-E, Maryanski JL, Taswell C, Widmer MB, Brunner KT (1980): Quantitation and cloning of cytolytic T lymphocytes and their precursors. *Immunol Rev* 51: 93–123

Mannel DN, Falk W, Droge W (1983): Induction of cytotoxic T cell function requires sequential action of three different lymphokines. *J Immunol* 130: 2508–2510

Maraskovsky E, Chen W-F, Shortman K (1989): IL-2 and IFN-γ are two necessary lymphokines in the development of cytolytic T cells. *J Immunol* 143: 1210–1214

Moingeon P, Jitsukawa S, Faure F, Troalen F, Triebel F, Graziani M, Forestier F, Bellet D, Bohuon C, Hercend T (1987): A γ-chain complex forms a functional receptor on cloned human lymphocytes with natural killer-like activity. *Nature* 325: 723–726

Morrissey PJ, Goodwin RG, Nordan RP, Anderson D, Grabstein KH, Cosman D, Sims J, Lupton S, Acres B, Reed SG, Mochizuki D, Eisenman J, Conlon PJ, Namen AE (1989): Recombinant interleukin 7, pre-B cell growth factor, has costimulatory activity on purified mature T cells. *J Exp Med* 169: 707–716

Mosmann TR, Cherwinski H, Bond MW, Giedlin MA, Coffman RL (1986): Two types of murine helper T cell clone. I. Definition according to profiles of lymphokine activities and secreted proteins. *J Immunol* 136: 2348–2357

Mulé JJ, Smith CA, Rosenberg SA (1987): Interleukin 4 (B cell stimulatory factor 1) can mediate the induction of lymphokine-activated killer cell activity directed against fresh tumor cells. *J Exp Med* 166: 792–797

Nagler A, Lanier LL, Phillips JH (1988): The effects of IL-4 on human natural killer cells. A potent regulator of IL-2 activation and proliferation. *J Immunol* 141: 2349–2351

Namen AE, Lupton S, Hjerrild K, Wignall J, Mochizuki DY, Schmierer A, Mosley B, March CJ, Urdal D, Gillis S, Cosman D, Goodwin RG (1988): Stimulation of B-cell progenitors by cloned murine interleukin-7. *Nature* 6173: 571–573

Noma Y, Sideras P, Naito T, Bergstedt-Lindquist S, Azuma C, Severinson E, Tanabe T, Kinashi T, Matsuda F, Yaoita Y, Honjo T (1986): Cloning of cDNA encoding the murine IgG1 induction factor by a novel strategy using SP6 promoter. *Nature* 319: 640–646

Okada M, Kitahara M, Kishimoto S, Matsuda T, Hirano T, Kishimoto T (1988): IL-6/BSF-2 functions as a killer helper factor in the in vitro induction of cytotoxic T cells. *J Immunol* 141: 1543–1549

Okazaki H, Ito M, Sudo T, Hattori M, Kano S, Katsura Y, Minato N (1989): IL-7 promotes thymocyte proliferation and maintains immunocompetent thymocytes bearing $\alpha\beta$ or $\gamma\delta$ T-cell receptors in vitro: Synergism with IL-2 *J Immunol* 143: 2917–2922

Paliard X, Yssel H, Blanchard D, Waitz JA, de Vries JE, Spits H (1989): Antigen–specific and MHC nonrestricted cytotoxicity of T cell receptor $\alpha\beta^+$ and $\gamma\delta^+$ human T cell clones isolated in IL-4. *J Immunol* 143: 452–457

Paul WE, Ohara J (1987): B-cell stimulatory factor-1/interleukin 4. *Annu Rev Immunol* 5: 429–459

Pfeifer JD, McKenzie DT, Swain SL, Dutton RW (1987): B cell stimulatory factor 1 (interleukin-4) is sufficient for the proliferation and differentiation of lectin-stimulated cytolytic T lymphocyte precursors. *J Exp Med* 166: 1464–1470

Plate JMD (1976): Soluble factors substitute for T-T-cell collaboration in generation of T-killer lymphocytes. *Nature* 260: 329–331

Raulet DH, Bevan MJ (1982): A differentiation factor required for the expression of cytotoxic T-cell function. *Nature* 296: 754–757

Rosenberg SA, Lotze MT (1986): Cancer immunotherapy using interleukin-2 and interleukin-2-activated lymphocytes. *Annu Rev Immunol* 4: 681–709

Rosenberg SA, Lotze MT, Yang JC, Aebersold PM, Linehan WM, Seipp CA, White DE (1989): Experience with the use of high-dose interleukin-2 in the treatment of 652 cancer patients. *Ann Surg* 210: 474–484

Rosenberg SA, Speiss P, Lafreniere R (1986): A new approach to the adoptive immunotherapy of cancer with tumor-infiltrating lymphocytes. *Science* 233: 1318–1321

Ryser J-E, Cerottini JC, Brunner KT (1978): Generation of cytolytic T lymphocytes in vitro. IX. Induction of secondary CTL responses in primary long-term MLC by supernatants from secondary MLC. *J Immunol* 120: 370–377

Schorle H, Holtschke T, Hunig T, Schimpl A, Horak I (1991): Development and function of T cells in mice rendered interleukin-2 deficient by gene targeting. *Nature* 352: 621–624

Simon MM, Hochgeschwender U, Brugger U, Landolfo S (1986): Monoclonal antibodies to interferon-γ inhibit interleukin 2-dependent induction of growth and maturation in lectin/antigen-reactive cytolytic T lymphocyte precursors. *J Immunol* 136: 2755–2762

Spits H, Yssel H, Paliard X, Kastelein R, Figdor C, de Vries JE (1988): IL-4 inhibits IL-2-mediated induction of human lymphokine-activated killer cells, but not the generation of antigen-specific cytotoxic T lymphocytes in mixed leukocyte cultures. *J Immunol* 141: 29–36

Stotter H, Custer MC, Bolton, ES, Guedez L, Lotze MT (1991): IL-7 induces human lymphokine-activated killer cell activity and is regulated by IL-4. *J Immunol* 146: 150–155

Swisher SG, Economou JS, Carmack Holmes E, Golub SH (1990): TNF-α and IFN-γ reverse IL-4 inhibition of lymphokine-activated killer cell function. *Cell Immunol* 128: 450–461

Takai Y, Herrmann SH, Greenstein JL, Spitainy GL, Burakoff SJ (1986): Requirement for three distinct lymphokines for the induction of cytotoxic T lymphocytes from thymocytes. *J Immunol* 137: 3494–3500

Takai Y, Wong GG, Clark SC, Burakoff SJ, Herrmann SH (1988): B cell stimulatory factor-2 is involved in the differentiation of cytotoxic T lymphocytes. *J Immunol* 140: 508–512

Wagner H, Hardt C, Rouse BT, Rollinghoff M, Scheurich

P, Pfizenmaier K (1982): Dissection of the proliferative and differentiative signals controlling murine cytotoxic T lymphocyte responses. *J Exp Med* 155: 1876–1881

Watson JD, Morrissey PJ, Namen AE, Conlon PJ, Widmer MB (1989): Effect of IL-7 on the growth of fetal thymocytes in culture. *J Immunol* 143: 1215–1222

Widmer MB, Grabstein KH (1987): Regulation of cytolytic T-lymphocyte generation by B-cell stimulatory factor. *Nature* 326: 795–798

Widmer MB, Acres RB, Sassenfeld HM, Grabstein KH (1987): Regulation of cytolytic cell populations from human peripheral blood by B cell stimulatory factor 1 (interleukin 4). *J Exp Med* 166: 1447–1455

Widmer MB, Morrissey PJ, Namen AE, Voice RF, Watson JD (1990): Interleukin 7 stimulates growth of fetal thymic precursors of cytolytic cells: Induction of effector function by interleukin 2 and inhibition by interleukin 4. *Int Immunol* 2: 1055–1061

Wolf SF, Temple PA, Kobayashi M, Young D, Dicig M, Lowe L, Dzialo R, Fitz L, Ferenz C, Hewick RM, Kelleher K, Herrmann SH, Clark SC, Azzoni L, Chan SH, Trinchieri G, Perussia B (1991): Cloning of cDNA for natural killer cell stimulatory factor, a heterodimeric cytokine with multiple biologic effects on T and natural killer cells. *J Immunol* 146: 3074–3081

12
IL-2-Independent Activation of LAK Cells by a Heterodimeric Cytokine, Interleukin-12

Maurice K. Gately, Aimee G. Wolitzky, Phyllis M. Quinn and Richard Chizzonite

Introduction

Interleukin-12 (IL-12) is a heterodimeric cytokine which was originally isolated from cultures of activated human B lymphoblastoid cells and called natural killer cell stimulatory factor (Kobayashi et al., 1989) or cytotoxic lymphocyte maturation factor (Stern et al., 1990). IL-12 has been shown to stimulate the proliferation of activated T cells and natural killer (NK) cells (Gately et al., 1991) and to cause interferon-γ (IFN-γ) production (Kobayashi et al., 1989; Chan et al., 1991) and enhanced NK lytic activity (Kobayashi et al., 1989; Wolf et al., 1991) by resting peripheral blood mononuclear cells (PBMC). IL-12, unlike IL-2, does not cause resting PBMC to proliferate, but it can stimulate enhanced proliferation of PBMC cultured in suboptimal concentrations of IL-2 (Gately et al., 1991). IL-12 is composed of two disulfide-bonded subunits with molecular masses of 35 and 40 kDa (Kobayashi et al., 1989; Stern et al., 1990). The cDNA encoding each of these two subunits has recently been cloned and bioactive recombinant IL-12 (rIL-12) expressed (Gubler et al., 1991; Wolf et al., 1991). Coexpression of the two subunits is required for biologically active IL-12 to be produced (Gubler et al., 1991: Wolf et al., 1991).

We previously reported that IL-12 synergized with suboptimal concentrations of IL-2 to induce lymphokine-activated killer (LAK) cells in the presence of hydrocortisone, but that IL-12 by itself was ineffective in causing LAK cell generation (Stern et al., 1990; Gubler et al., 1991). Hydrocortisone was included in those assays to inhibit endogenous cytokine production and the triggering of cytokine cascades so that the effects of individual added cytokines could be clearly discerned. However, these results appeared discordant with observations that in the absence of hydrocortisone, IL-12 could enhance NK lytic activity in 24-hr cultures and that combining IL-2 with IL-12 resulted, at best, in additive, not synergistic, effects (Kobayashi et al., 1989; M.K. Gately, unpublished results). Therefore, we have examined the ability of IL-12 to induce LAK cell activity in the absence of hydrocortisone. As shown below, we found that in cultures lacking hydrocortisone, IL-12 could cause the generation of human LAK cells independently of IL-2. Furthermore, IL-12 may play a role in the induction of LAK cell responses by suboptimal amounts of IL-2.

Results

IL-12 Causes LAK Cell Activation Independently of IL-2

When the ability of IL-12 to activate LAK cells was compared in the presence and absence of hydrocortisone, dramatically different results were observed (Figure 1). As previously reported, in the presence of 0.1 mM hydrocortisone, IL-12 alone caused little LAK cell generation but synergized with 5 U/ml IL-2 to cause a substantial LAK cell response. However, in cultures to which hydrocortisone had not been added, IL-12 by itself caused a strong LAK response, and combining IL-12 and IL-2 at the concentrations indicated in Figure 1 did not result in enhancement of the response above the maximum response caused by either cytokine alone. Nevertheless, synergy between IL-2 and IL-12 in causing LAK cell activation could be observed in the absence of hydrocortisone when lower concentrations of cytokines and/or longer

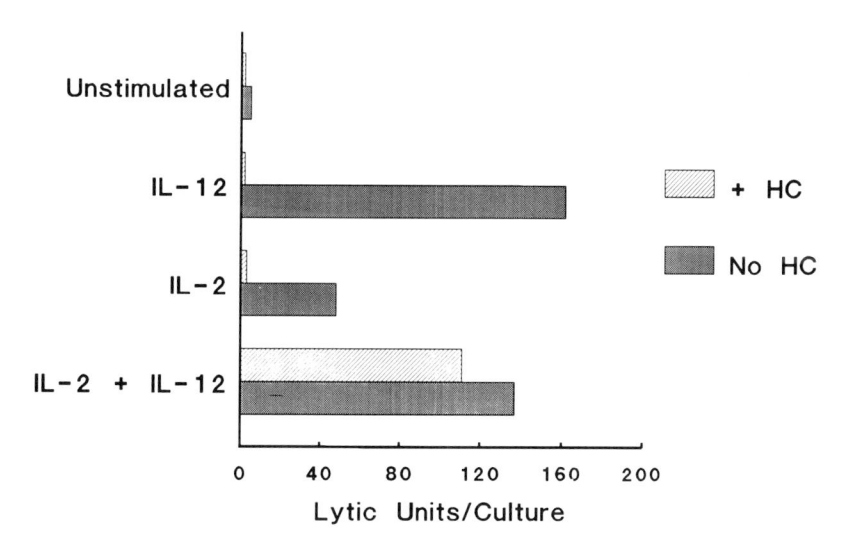

FIGURE 1. Effects of IL-12 ± IL-2 on LAK cell induction in the presence or absence of hydrocortisone (HC). The induction of LAK cells from low-density human peripheral blood lymphocytes (PBL) was performed as previously described (Stern et al., 1990) in the presence (+ HC) or absence (no HC) of 0.1 mM hydrocortisone sodium succinate. IL-12 and IL-2 were added to the cultures at final concentrations of 10 and 5 U/ml, respectively. Cells were harvested from the cultures on day 4 and tested for their ability to lyse NK-resistant Daudi target cells in a 5-hr ^{51}Cr-release assay in which each effector population was tested at several different effector/target ratios. One lytic unit was defined as the number of effectors required to cause 30% specific ^{51}Cr release from the targets and was calculated as described (Gately et al., 1982). The spontaneous ^{51}Cr release in each of the experiments shown in this manuscript was ≤ 22%. IL-12 was produced by activated NC-37 B lymphoblastoid cells and purified as previously described (Stern et al., 1990).

culture times (≥ 6 days) were examined (data not shown).

The ability of IL-12 to induce LAK cell activation was dose-dependent (Figure 2). In four separate experiments, the concentration of IL-12 which induced half-maximum activation of LAK cells in cultures of low-density peripheral blood lymphocytes (PBL) varied from 1 to 20 pM, similar to the concentration of IL-2 which was required for half-maximum LAK cell activation under these culture conditions. However, the maximum LAK cell response induced by IL-12 was generally only 20–30% of the maximum LAK cell activity which could be stimulated by IL-2 (data not shown). In kinetic studies, both IL-12- and IL-2-induced LAK cell activity peaked at days 4 to 6 of culture and declined thereafter. Phenotyping studies indicated that LAK cell activity induced by each of these two cytokines was predominantly mediated by CD56$^+$ NK cells. CD3$^+$ T cells isolated from cultures of IL-12-activated low-density lymphocytes displayed little or no LAK activity (data not shown).

One possible explanation for the difference in the ability of IL-12 to activate LAK cells in the

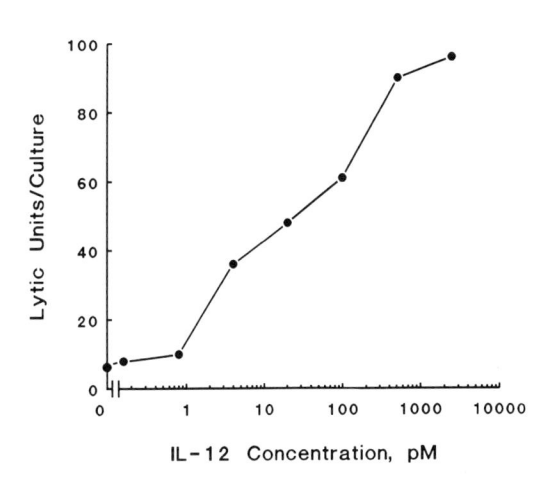

FIGURE 2. Dose-response for the activation of LAK cells by purified IL-12 in the absence of hydrocortisone. LAK cells were induced from low-density human PBL by culture for 4 days in the presence of IL-12 at the indicated concentrations. One picomolar purified NC-37-derived IL-12 is equal to 6 U/ml.

TABLE 1. Activation of LAK Cells by IL-12 is inhibited by anti-TNF-α but not anti-IL-2[a]

Culture contents			% Specific ^{51}Cr release from Daudi at E/T:	
Cytokine[b]	Anti-IL-2 or NGS[c]	Anti-TNF-α or NRbS[c]	10/1	2.5/1
None	NGS	NRbS	11 ± 1	2 ± 1
IL-2	NGS	NRbS	64 ± 4	30 ± 1
IL-2	Anti-IL-2	NRbS	12 ± 1	2 ± 1
IL-2	NGS	Anti-TNF-α	40 ± 1	12 ± 1
IL-12	NGS	NRbS	50 ± 5	22 ± 2
IL-12	Anti-IL-2	NRbS	53 ± 2	18 ± 1
IL-12	NGS	Anti-TNF-α	27 ± 1	5 ± 2
IL-12	Anti-IL-2	Anti-TNF-α	28 ± 2	5 ± 1

[a] Low-density PBL were cultured for 4 days with the indicated cytokines and sera. The cultures were then harvested and the cells tested for their ability to lyse ^{51}Cr-labeled Daudi cells at the indicated E/T ratios. All values are means ± 1 SEM of quadruplicates.
[b] The concentrations of IL-2 and IL-12 were 4 and 100 U/ml, respectively.
[c] Goat anti-human rIL-2, normal goat serum (NGS), rabbit anti-human rTNF-α, and normal rabbit serum (NRbS) were all used at a final dilution of 1/1000.

presence or absence of added hydrocortisone is that IL-12-induced LAK cell activation is dependent on *in situ* production of other cytokines in the cultures. To examine this possibility, we have tested the ability of neutralizing antisera to various cytokines to inhibit IL-12-induced LAK cell generation. In three separate experiments, a potent neutralizing goat anti-human rIL-2 antiserum, which neutralizes natural as well as recombinant IL-2, had no effect on IL-12-induced LAK cell generation (for example, see Table 1). Hence, IL-12 appeared to activate LAK cells independently of IL-2. However, in five experiments, such as the one shown in Table 1, a neutralizing rabbit antiserum to human recombinant tumor necrosis factor-α (rTNF-α) consistently inhibited (50–100%) IL-12-induced LAK cell activation. Likewise, it caused substantial inhibition of IL-2-induced LAK cell activation (Table 1). Studies examining whether a neutralizing mouse monoclonal antibody to human IFN-γ can inhibit LAK cell generation induced by IL-12 are in progress.

Anti-IL-12 Partially Inhibits LAK Cell Responses and PBMC Proliferation Induced by Suboptimal Concentrations of IL-2

An antiserum against human IL-12 was produced by repetitive immunization of a rat with highly purified IL-12 using an immunization protocol similar to that previously described (Chizzonite et al., 1991). This antiserum was very potent in its ability to neutralize human IL-12, and its specificity was indicated by its lack of activity against human IL-2, IL-4, or IL-7 (Table 2). We examined the ability of this antiserum, alone or in combination with rabbit anti-human rTNF-α, to cause inhibition of human LAK cell activation and PBMC proliferation caused by suboptimal (≤ 25 U/ml) concentrations of human rIL-2. Two experiments were performed with similar results, and one of these is illustrated in Figure 3. On day 3 of culture, anti-IL-12, unlike anti-TNF-α, had no effect on the LAK cell response or the proliferative response induced by IL-2. Hence IL-12 does not appear to play an essential role in the initial steps in IL-2-induced LAK cell activation. However, when LAK cell activity and PBMC proliferation were examined on or after day 6 of culture, anti-IL-12, like anti-TNF-α, caused substantial inhibition of both LAK cell activity and PBMC proliferation induced by IL-2. Moreover, in cultures which received both anti-IL-12 and anti-TNF-α, the levels of LAK cell activity and PBMC proliferation were similar to those seen in control cultures without added cytokines (Figure 3). Hence, in these experiments IL-12 produced *in situ* appeared to be required for the maintenance of the IL-2-induced LAK response beyond day 3 and for the generation of an optimal proliferative response to IL-2.

Discussion

A number of cytokines in addition to IL-2 have been reported to activate LAK cells, either alone or in combination with IL-2. These include IFN-γ (Ellis et al., 1989), TNF-α (Espevik et al., 1988;

TABLE 2. Specificity of rat anti-human IL-12 antiserum[a]

Cytokine[b]	Culture contents Anti-IL-12 or NRtS[c]	[3]Thymidine incorporation by PHA-activated human lymphoblasts
None	NRtS	14,253 ± 406
IL-12	NRtS	71,557 ± 866
IL-12	Anti-IL-12	18,844 ± 212
IL-2	NRtS	74,552 ± 4,086
IL-2	Anti-IL-12	95,471 ± 2,896
IL-4	NRtS	37,040 ± 765
IL-4	Anti-IL-12	39,933 ± 1,568
IL-7	NRtS	57,255 ± 4,558
IL-7	Anti-IL-12	64,929 ± 1,486

[a] The ability of anti-IL-12 to inhibit cytokine-induced proliferation of human PHA-activated lymphoblasts in a 48-hr assay was assessed as previously described (Gubler et al., 1991). All values are mean cpm ± 1 SEM of triplicates.
[b] The concentrations of IL-12, rIL-2, rIL-4, and rIL-7 were 100 U/ml, 8 U/ml, 100 U/ml, and 3 ng/ml, respectively.
[c] Rat anti-human IL-12 and normal rat serum (NRtS) were used at a final dilution of 1/2000.

Owen-Schaub et al., 1988), IL-1 (Crump et al., 1989), IL-7 (Alderson et al., 1990: Stotter et al., 1991), and, under some circumstances, IL-4 (Kawakami et al., 1989). See also chapter C2. Clearly, IL-12 may be added to this list. In the presence of 0.1 mM hydrocortisone, IL-12 synergized with IL-2 in the induction of LAK cells but did not by itself activate LAK cells. However, in cultures to which hydrocortisone was not added, IL-12 by itself was active in eliciting LAK cell responses. The concentrations of IL-2 and IL-12 required to elicit LAK cell responses were similar. However, the maximum response elicited by IL-12 was generally only 20–30% of the LAK cell activity that could be induced by optimal concentrations of IL-2. In part, this could reflect the fact that IL-2 can induce proliferation of CD56[+] NK/LAK cells (Gately et al., 1991; Naume et al., 1991), whereas IL-12 does not stimulate proliferation of resting PBMC in the absence of an additional mitogenic stimulus (Gately et al., 1991). Activation of LAK cells by IL-12 appeared to be independent of IL-2, since it was not affected by the addition of a potent neutralizing anti-IL-2 antiserum. However, IL-12-induced LAK generation was inhibited to a variable extent (50–100%) by addition of an antiserum to human TNF-α. Hence, activation of LAK cells by IL-12 appeared to be at least partially dependent upon *in situ* production of TNF-α in the cultures. Variability in the degree of inhibition seen from experiment to experiment using cells from different donors likely reflects heterogeneity in the spectrum of cytokines released in culture by PBL from different individuals. In addition to inhibiting IL-12-induced LAK cell generation, the anti-TNF-α also inhibited IL-2-induced LAK cell generation, consistent with the reports of others (Owen-Schaub et al., 1988; Naume et al., 1991). Hence, it appears that TNF-α may play a central role in the process of LAK cell activation.

Glucocorticoids have been reported to inhibit the production of a variety of cytokines both *in vitro* (Gillis et al., 1979; Daynes and Araneo, 1989) and *in vivo* (Mier et al., 1990; Alegre et al., 1991). Dexamethasone was reported to inhibit IL-2-induced TNF-α release in human cancer patients receiving IL-2 therapy (Mier et al., 1990). It seems possible that the failure of IL-12 to induce LAK activity in the presence of hydrocortisone may have been due in part to inhibition of endogenous TNF-α production. Conceivably, cytokines other than TNF-α may also play a role in IL-12-induced LAK cell activation. IL-12 is known to induce the production of substantial amounts of interferon-γ (Kobayashi et al., 1989; Chan et al., 1991; M.K. Gately, unpublished results), and the possible role of interferon-γ in IL-12-induced LAK cell responses is under study.

The results of studies using a neutralizing rat antiserum prepared against purified human IL-12 suggested that IL-12 produced *in situ* may play a role in the LAK cell response and proliferation induced when human PBMC are cultured with low concentrations (≤ 25 U/ml) of IL-2. Although anti-IL-12 had no effect on the LAK cell activity or proliferation observed on day 3 of culture, anti-IL-12 partially inhibited both IL-2-induced proliferation and LAK cell activity on days 6 and 9. In contrast, anti-TNF-α caused partial inhibition of both IL-2-induced proliferation and LAK cell

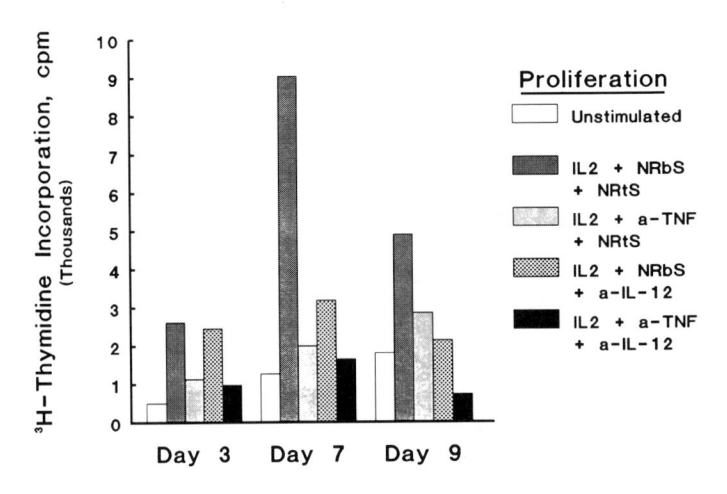

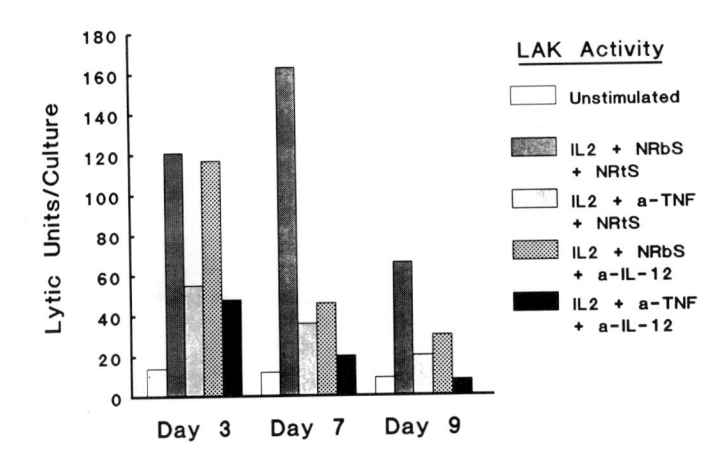

FIGURE 3. Effects of anti-TNF-α and anti-IL-12 on proliferation and LAK cell activity induced by culture of unfractionated human PBMC with IL-2 (5 U/ml). Proliferation of PBMC was measured as previously described (Gately et al., 1991). LAK cell activity was assayed on Daudi targets as described above. Rabbit anti-human rTNF-α (a-TNF), control normal rabbit serum (NRbS), rat anti-human IL-12 (a-IL-12), and control normal rat serum (NRtS) were all added to cultures at a final dilution of 1/2000.

activity at all observation times. The effects of anti-IL-12 and anti-TNF-α were additive on days 6 and 9. We have previously shown that the cells which proliferate in response to IL-2 under these culture conditions are predominantly CD56$^+$ NK/LAK cells (Gately et al., 1991). The results of the present studies suggest that IL-12 is not required for the initial activation of LAK cells but is required at later times (at least when activation is induced by low concentrations of IL-2) for the maintenance of an IL-2-induced LAK response and for the generation of an optimal proliferative response. These results are consistent with our previous observations that added IL-12 could enhance the prolifera-

tive response of PBMC to low concentrations of IL-2 and that the IL-12-mediated enhancement of IL-2-induced proliferation was only observed on or after day 6 of culture (Gately et al., 1991). Likewise, the observation that anti-IL-12 suppressed IL-2-induced LAK cell activity and PBMC proliferation only after day 3 of culture is consistent with the results of recent studies on the kinetics of IL-2-induced up-regulation of IL-12 receptors on resting PBMC (B.B. Desai et al., submitted for publication). Recent preliminary results with neutralizing monoclonal antibodies to IL-12 are similar to the results reported herein (M.K. Gately, unpublished results). In one experiment in which LAK cells were activated by ≥ 100 U/ml IL-2, anti-IL-12 appeared to have no effect (data not shown).

Hence, these studies suggest not only that IL-12 can induce LAK cell activation, but also that IL-12 produced by normal PBMC may, at least in some circumstances, contribute to the LAK cell activity and proliferation induced by IL-2. Further studies to delineate the role(s) of IL-12 in cell-mediated immune responses, including cytotoxic T-cell responses, are in progress.

Summary

IL-12 is a heterodimeric cytokine which has been shown to stimulate the proliferation of activated T and NK cells, to induce IFN-γ production by resting T and NK cells, and to enhance NK cell lytic activity. We previously reported that in the presence of hydrocortisone, IL-12 synergized with suboptimal concentrations of IL-2 to induce LAK cell activity; however, IL-12 by itself, was inactive. We now demonstrate that IL-12 by itself, in the absence of hydrocortisone, can cause dose-dependent generation of human LAK cells from populations of low-density PBL. The maximum lytic activity generated in the presence of optimal concentrations of IL-12 was generally 20–30% of the lytic activity which could be induced by culture of the same lymphocyte populations with optimal amounts of IL-2. IL-12 appeared to activate LAK cells independently of IL-2 inasmuch as IL-12-mediated LAK cell induction was not inhibited by a potent neutralizing goat anti-human IL-2 antiserum. However, IL-12- as well as IL-2-induced LAK cell generation could be partially inhibited by a rabbit antiserum against human TNF-α. Anti-IL-12, unlike anti-TNF-α, did not inhibit LAK cell activation induced by suboptimal (≤ 25 U/ml) IL-2 when lytic activity was measured after day 3 of culture. Nevertheless, when LAK cell activity was measured at later times, anti-IL-12 by itself was partially inhibitory, and the combination of anti-IL-12 plus anti-TNF-α resulted in additive inhibition. Hence IL-12 can activate LAK cells independently of IL-2 and may play a role in LAK cell activation caused by suboptimal concentrations of IL-2.

Acknowledgments

We wish to thank Mr. Frank Podlaski, Department of Protein Biochemistry, Hoffmann-La Roche Inc., for supplying the purified NC-37-derived IL-12 used in these studies. We are also grateful to Dr. Manfred Brockhaus, F. Hoffmann-La Roche and Co. Ltd., Basel, Switzerland, for providing the rat anti-human rTNF-α used in these studies.

References

Alderson MR, Sassenfeld HM, Widmer MB (1990): Interleukin 7 enhances cytolytic T lymphocyte generation and induces lymphokine-activated killer cells from human peripheral blood. *J Exp Med* 172: 577–587

Alegre M-L, Vandenabeele P, Depierreux M, Florquin S, Deschodt-Lanckman M, Flamand V, Moser M, Leo O, Urbain J, Fiers W, Goldman M (1991): Cytokine release syndrome induced by the 145-2C11 anti-CD3 monoclonal antibody in mice: Prevention by high doses of methylprednisolone. *J Immunol* 146: 1184–1191

Chan SH, Perussia B, Gupta JW, Kobayashi M, Pospisil M, Young HA, Wolf SF, Young D, Clark SC, Trinchieri G (1991): Induction of interferon γ production by natural killer cell stimulatory factor: Characterization of the responder cells and synergy with other inducers. *J Exp Med* 173: 869–879

Chizzonite R, Truitt T, Podlaski FJ, Wolitzky AG, Quinn PM, Nunes P, Stern AS, Gately MK (1991): IL-12: Monoclonal antibodies specific for the 40-kDa subunit block receptor binding and biologic activity on activated human lymphoblasts. *J Immunol* 147: 1548–1556

Crump WL III, Owen-Schaub LB, Grimm EA (1989): Synergy of human recombinant interleukin 1 with interleukin 2 in the generation of lymphokine-activated killer cells. *Cancer Res* 49: 149–153

Daynes RA, Araneo BA (1989): Contrasting effects of glucocorticoids on the capacity of T cells to produce the growth factors interleukin 2 and interleukin 4. *Eur J Immunol* 19: 2319–2325

Ellis TM, McKenzie RS, Simms PE, Helfrich BA, Fisher RI (1989): Induction of human lymphokine-activated killer cells by IFN-α and IFN-γ. *J Immunol* 143: 4282–4286

Espevik T, Figari IS, Ranges GE, Palladino MA, Jr (1988): Transforming growth factor-β_1 (TGF-β_1) and recombinant human tumor necrosis factor-α reciprocally regulate the generation of lymphokine-activated killer cell activity. Comparison between natural porcine platelet-derived TGF-β_1 and TGF-β_2, and recombinant human TGF-β_1. *J Immunol* 140: 2312–2316

Gately MK, Desai BB, Wolitzky AG, Quinn PM, Dwyer CM, Podlaski FJ, Familletti PC, Sinigaglia F, Chizzonite R, Gubler U, Stern AS (1991): Regulation of human lymphocyte proliferation by a heterodimeric cytokine, IL-12 (cytotoxic lymphocyte maturation factor). *J Immunol* 147: 874–882

Gately MK, Glaser M, Dick SJ, Mettetal RW, Jr, Kornblith PL (1982): In vitro studies on the cell-mediated immune response to human brain tumors. I. Requirement for third-party stimulator lymphocytes in the induction of cell-mediated cytotoxic responses to allogeneic cultured gliomas. *JNCI* 69: 1245–1254

Gillis S, Crabtree GR, Smith KA (1979): Glucocorticoid-induced inhibition of T cell growth factor production. I. The effect on mitogen-induced lymphocyte proliferation. *J Immunol* 123: 1624–1631

Gubler U, Chua AO, Schoenhaut DS, Dwyer CM, McComas W, Motyka R, Nabavi N, Wolitzky AG, Quinn PM, Familletti PC, Gately MK (1991): Coexpression of two distinct genes is required to generate secreted, bioactive cytotoxic lymphocyte maturation factor. *Proc Natl Acad Sci USA* 88: 4143–4147

Kawakami Y, Custer MC, Rosenberg SA, Lotze MT (1989): IL-4 regulates IL-2 induction of lymphokine-activated killer activity from human lymphocytes. *J Immunol* 142: 3452–3461

Kobayashi M, Fitz L, Ryan M, Hewick RM, Clark SC, Chan S, Loudon R, Sherman F, Perussia B, Trinchieri G (1989): Identification and purification of natural killer cell stimulatory factor (NKSF), a cytokine with multiple biological effects on human lymphocytes. *J Exp Med* 170: 827–845

Mier JW, Vachino G, Klempner MS, Aronson FR, Noring R, Smith S, Brandon EP, Laird W, Atkins MB (1990): Inhibition of interleukin-2-induced tumor necrosis factor release by dexamethasone: Prevention of an acquired neutrophil chemotaxis defect and differential suppression of interleukin-2-associated side effects. *Blood* 10: 1933–1940

Naume B, Shalaby R, Lesslauer W, Espevik T (1991): Involvement of the 55- and 75-kDa tumor necrosis factor receptors in the generation of lymphokine-activated killer cell activity and proliferation of natural killer cells. *J Immunol* 146: 3045–3048

Owen-Schaub LB, Gutterman JU, Grimm EA (1988): Synergy of tumor necrosis factor and interleukin 2 in the activation of human cytotoxic lymphocytes: Effect of tumor necrosis factor α and interleukin 2 in the generation of human lymphokine-activated killer cell cytotoxicity. *Cancer Res* 48: 788–792

Stern AS, Podlaski FJ, Hulmes JD, Pan YE, Quinn PM, Wolitzky AG, Familletti PC, Stremlo DL, Truitt T, Chizzonite R, Gately MK (1990): Purification to homogeneity and partial characterization of cytotoxic lymphocyte maturation factor from human B-lymphoblastoid cells. *Proc Natl Acad Sci USA* 87: 6808–6812

Stotter H, Custer MC, Bolton ES, Guedez L, Lotze MT (1991): IL-7 induces human lymphokine-activated killer cell activity and is regulated by IL-4. *J Immunol* 146: 150–155

Wolf SF, Temple PA, Kobayashi M, Young D, Dicig M, Lowe L, Dzialo R, Fitz L, Ferenz C, Hewick RM, Kelleher K, Herrmann SH, Clark SC, Azzoni L, Chan SH, Trinchieri G, Perussia B (1991): Cloning of cDNA for natural killer cell stimulatory factor, a heterodimeric cytokine with multiple biologic effects on T and natural killer cells. *J Immunol* 146: 3074–3081

13
Immunobiology of β_2-Microglobulin-Deficient Mice

J.A. Frelinger, D.G. Quinn, and D. Muller

Introduction

The ability to disrupt the function of a given protein molecule is a powerful technique. This approach has been extremely useful in modern biology. Monoclonal antibodies have been used to disrupt molecular function *in vitro* and *in vivo*. In addition, synthetic antisense oligonucleotides have been useful in disrupting the function of genes at the pretranslational level. *In vitro* depletion of particular cell types by treatment with monoclonal antibodies and complement has played a major role in elucidating the activities of cells expressing the appropriate target molecules. Combinations of these approaches led to the definition of lymphocyte subsets and their unequivocal characterization *in vivo* and *in vitro*. However, definition of the precise functions of these molecules has been much more difficult to attain. In prokaryotes and lower eukaryotes, the ability to produce null mutations has been important in defining the function of numerous gene products, ultimately allowing the determination of the function of each gene in the intact organism. Until recently, it has proved extremely difficult to obtain null mutants in mammals. Although there are examples of spontaneous null mutants *in vivo* and, in principle, such mutants can be readily selected *in vitro*, relatively little information has been derived from these.

In the case of β_2-microglobulin (β_2-m), a null mutant was initially noted in the human B-cell lymphoma, Daudi. This led to the realization that β_2-m is required for the cell surface expression of MHC class I molecules (Arce-Gomez et al., 1978). Subsequently, a number of immunoselected class I-deficient mutations have turned out to be β_2-m-negative. Several *in vitro* experiments have suggested that β_2-m plays an important role in the binding of exogenous peptides to class I molecules (Vitiello and Sherman, 1990). These studies demonstrated decreased binding of exogenously added peptide antigen, as assessed by cytotoxic T lymphocyte (CTL) killing, to target cells cultured in β_2-m-free medium compared to target cells cultured in the presence of this molecule. This led to the concept that exchange of β_2-m might be important in binding of exogenous peptide. Apart from the implications for class I expression, these experiments provide no information on the importance of β_2-m *in vivo*, for example in the thymic selection of CD8[+] T lymphocytes.

Several investigators have postulated that, in addition to the known immunologic function of K and D class I molecules, they might also serve as recognition structures in cell-cell interactions during fetal development (Artzt et al., 1974). Indeed, early experiments had even suggested that the F9 antigen might be the product of the wild-type allele of one of the developmentally lethal *t*-mutants in the mouse (Artzt et al., 1973). Thus it was of considerable interest to determine whether expression of any class I molecules was required for normal murine development. In addition, in studies using immature rat lymphoid cells, β_2-m was shown to be identical to the prothymocyte chemotactic factor, thymotaxin (Dargemont et al., 1989). This suggested that this molecule might play a role in thymic development quite distinct from any subsequent role in T-cell selection by inducing thymic colonization by T-cell precursors. These experiments, however, provide no information on the role of β_2-m *in vivo*, for example, in the thymic selection of CD8[+] T lymphocytes.

Biologic Effects of Interruption of the β_2-m Gene

Recently the technology has been developed to create null mutants in any known sequence by homolous recombination *in vitro*. This technique relies on the ability to select embryonic stem (ES) cells which have incorporated antibiotic resistance genes interrupting the coding sequence of the target gene. Following identification of the mutant cells, the stem cells bearing this mutation can be reincorporated into the germ line following injection into the Blastocoel cavity of a morula. Chimeric animals can be recovered and a mouse which has incorporated the mutant stem cell into the germ line may be detected by progeny testing. These animals are then intercrossed to produce a homozygous dilution line in the manner of classical Mendelian genetics. Two groups have utilized this gene targeting approach to prodice mice lacking expression of β_2-m (Koller et al., 1990; Zijlstra et al., 1990) using ES cells derived from mouse strain 129 which were selected for interruption of the β_2-m gene. The 129 chimeras were crossed to C57BL/6 (B6) mice to obtain progeny from the ES-derived germ cells. The fact that homozygous β_2-m-deficient (β_2-m$^-$) mice were viable demonstrated unequivocally that the expression of class I proteins on the cell surface is not required for normal murine development. Thus, while it does not preclude a role for class I in development, it strongly argues that such a role is not necessary.

Based on the studies of β_2-m$^-$ cell lines, it was expected that class I molecules would not be expressed on the surface of cells from β_2-m$^-$ mice. Strain 129 (H-2^b) mice, the ES cell donor, and strain B6, the other parental strain in the cross, both express K^b and D^b cell surface molecules. As expected, β_2-m$^-$ mice are deficient in cell surface expression of these molecules throughout all their tissues, including the thymus. This led to a direct test of the proposition that class I was involved in both positive and negative selection of CD8$^+$ T cells. If expression of class I was not required for thymic selection, then β_2-m$^-$ mice ought to produce numerous CD8$^+$ T cells which were not negatively selected for class I. In contrast, if positive selection was required, then β_2-m$^-$ mice ought to be deficient in CD8$^+$ T cells. That β_2-m$^-$ mice express very few CD8$^+$ T cells (Koller et al., 1990; Zijlstra et al., 1990) established unequivocally the role of class I expression in the positive selectionof CD8$^+$ T cells.

The second surprising outcome was that β_2-m$^-$ mice are healthy. Kept in a clean but not germ-free facility, they show reasonable lifespan (>6 months) and no obvious sensitivity to bacterial or viral diseases. The mice breed well and produce large, healthy litters. These mice, therefore, are not grossly immunocompromised as a result of this lesion. This has also been more recently observed with mice lacking class II and CD4 (Cosgrove et al., 1991; Grusby et al., 1991; Rahemtulla et al., 1991).

Allograft Response in β_2-m$^-$ Mice

As expected from the lack of CD8$^+$ T cells, lymphocytes from β_2-m$^-$ mice which were stimulated *in vitro* by allogeneic spleen cells produced no strong allospecific CTL (Zijlstra et al., 1990). In contrast, a strong proliferative response mediated by CD4$^+$ T cells was elicited. Further, as expected, concanavalin A-induced T cell blasts from β_2-m$^-$ mice are poor targets for normal alloreactive CTL, although they are efficiently lysed by NK cells (see later).

CTL Activity in β_2-m$^-$ Mice

As it has been proposed that CD8$^+$ and CTL are crucial for recovery from and clearance of virus infection, it was of interest to test the prediction that β_2-m$^-$ mice would be highly susceptible to viral infections due to the absence of CD8$^+$ T lymphocytes. We have begun a study of the immune response to lymphocyte choriomeningitis virus (LCMV) in β_2-m$^-$ mice. In normal mice the pathobiology of the disease is mediated directly by CD8$^+$ CTL attack on virus-infected cells in the brain (reviewed in Doherty et al., 1990). If LCMV infection occurs outside the central nervous system (CNS), there is little CNS pathology and the infection is readily cleared. This infection produces large numbers of CD8$^+$, class I-restricted, LCMV-specific CTL. These CTL, upon intracranial injection, are capable of transferring disease to mice with a persistent CNS infection. Thus the disease is immunopathological in its basis, although the immune response observed is normal.

LCMV is also unusual in that, unlike most viruses, it induces CTL which are readily detectable in the spleens of infected animals without the requirement for a secondary stimulation *in vitro*. We reasoned that LCMV infection would be an interesting model to test the supposition that β_2-m$^-$ mice are indeed deficient in CTL. When normal

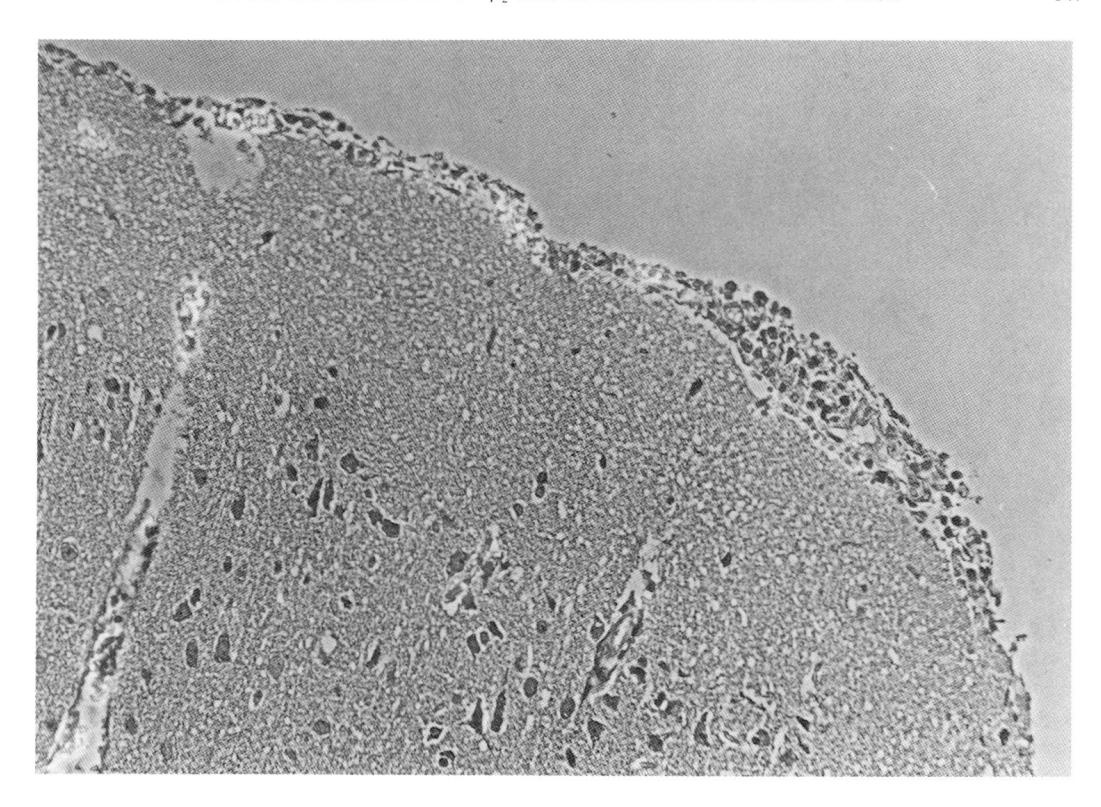

FIGURE 1. Section through the cortex of a β_2 m-deficient mouse infected intracerebrally 6 days previously with 10^3 plaque-forming units of the Armstrong-3 strain of lymphocytic choriomeningitis virus. A mononuclear cell leptomeningeal infiltrate is seen. H & E, phase contrast, original magnification 200 ×.

B6 or (B6 × 129) F_1 mice are infected intracranially by LCMV, the animals die between days 7 and 9 postinfection. Death is associated with a massive lymphocytic infiltrate into the CNS and obvious leptomeningitis. This disease can be blocked by administration of anti-CD8 monoclonal antibody, cyclophosphamide, or other immunosuppressive drugs. Mice which are immunosuppressed do not succumb to lethal meningitis, but instead become persistently infected with LCMV. We predicted that β_2-m$^-$ mice, when infected with LCMV, should respond like immunosuppressed mice, i.e., they would not succumb to acute disease and would become persistently infected. Contrary to our expectations, β_2-m$^-$ mice that were injected intracranially with LCMV began to die 12 days after infection. Indeed, mice at this time have an active leptomeningitis that is histologically very similar to that seen in normal mice (Figure 1). Thus β_2-m$^-$ mice do show an altered kinetics of response to LCMV in that they succumb more slowly than wild-type mice; however, the outcome is the same.

It is also interesting to note that virus loads in the CNS and the periphery do not differ between β_2-m$^-$ and control mice during the course of the disease.

To investigate further the identity of the cells mediating LCM-induced immunopathology in these β_2-m$^-$ mice, we reexamined splenocytes from mice innoculated i.p. 7 days previously with LCMV. Splenocytes from these animals were unable to lyse LCMV-infected L929 cells transfected with the gene for D^b, or LCMV-infected MC-57 cells, an $H-2^b$ target. These target cells were efficiently lysed by effector cells obtained from virally infected B6 spleens. In contrast, when LCMV-infected CH.B2 cells were used as targets, there was significant killing by β_2-m$^-$ as well as by B6 effector cells (Figure 2). CH.B2 is an $H-2^b$ B-cell lymphoma expressing both class I and class II antigens. Lysis of CH.B2 by β_2-m$^-$ effectors could be blocked by anti-class II but not by anti-class I monoclonal antibodies. This killing is H-2-restricted, since an $H-2^a$ B-cell lymphoma, CH12, is not susceptible to

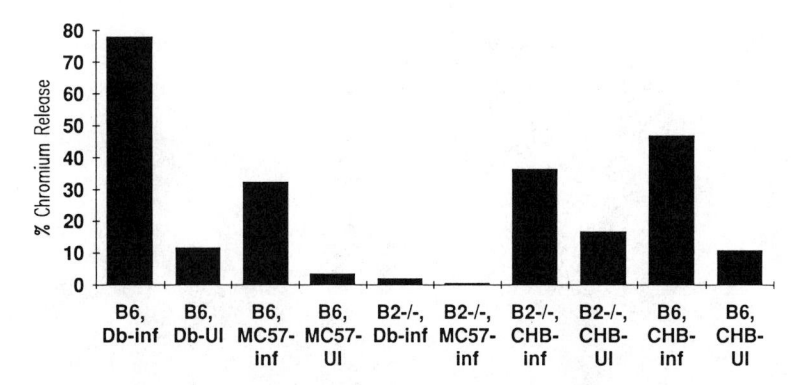

FIGURE 2. Cytotoxic activity of spleen cells from LCMV-infected B6 and β_2-m$^-$ mice against a range of LCMV-infected (inf) or uninfected (UI) target cells at an effector to target ratio of 50:1. For CTL assays, target cells which had been infected with LCMV for 24–48 hr were washed and resuspended in serum-free RPMI 1640. Target cells were incubated for 60–90 min with ^{51}Cr and then washed 3 times in RPMI 1640 containing 5% (v/v) fetal calf serum (FCS) and supplements. Each individual well contained 2×10^4 target cells. After 5 hr incubation at 37°C, supernatants were harvested and counted. D^b : L929 fibroblast cell line (H-2^k) transfected with H-D^b.

lysis by β_2-m$^-$ CTL. Complement depletion experiments (Figure 3) demonstrated that the LCMV-specific effector population in these mice was indeed CD4$^+$.

Based on these results, we were curious to determine if there was a significant increase in the number of class II$^+$ cells in the brains of these mice, thus providing potential targets for the class II-restricted CTL. We infected mice with LCMV intracranially and examined the expression of I-A$_\beta$ on Western blots of brain homogenates. This demonstrated that class II is significantly induced in LCMV-infected brains of both normal B6 mice and β_2-m$^-$ mice (Muller et al., submitted for publication). It is therefore possible that LCMV-specific, class II-restricted CTL may be important in the immunopathology of this disease in β_2-m$^-$ mice.

Effect of Class I Deficiency on NK Activity

It has recently been reported that bone marrow transplants from β_2-m$^-$ mice fail to survive in irradiated major histocompatibility complex

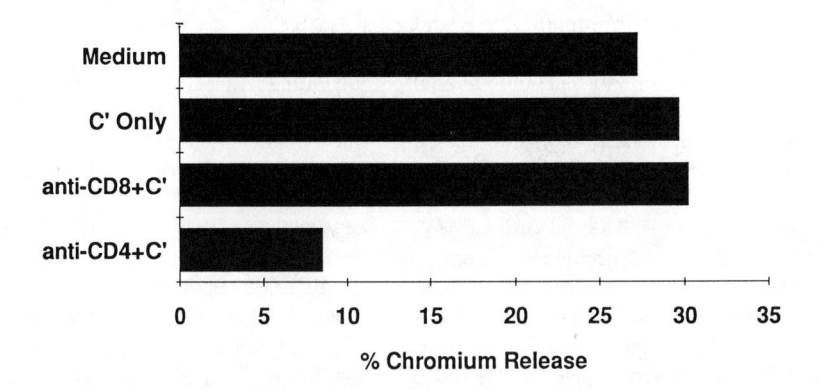

FIGURE 3. Cytotoxic activity of spleen cells from LCMV-infected β_2-m$^-$ mice against infected CH.B2 target cells after depletion of effector population with anti-CD4 + complement, anti-CD8 + complement, or complement alone, compared with medium control. Effector to target ratio was 100:1. Splenocytes (10^7/ml) were incubated with antibody for 15 min at 37°C and washed. Baby rabbit complement was added (1/15 dilution) and the incubation was continued for a further 45 min. After washing, the cells were resuspended to the original volume and assayed as described in Figure 2.

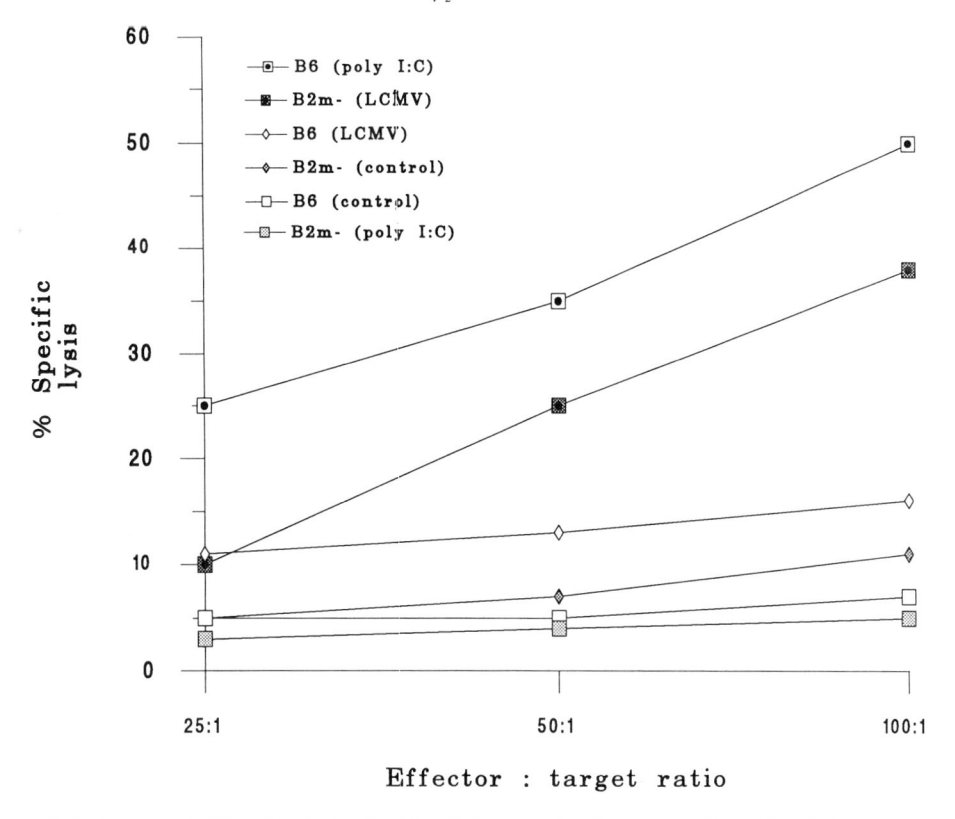

13. IMMUNOBIOLOGY OF β_2-MICROGLOBULIN-DEFICIENT MICE

FIGURE 4. Induction of NK cell activity in B6 and β_2-m⁻ mice that were injected i.p. 7 days previously with LCMV or 2 days previously with poly I:C (100 μg), compared with untreated mice (control). Cytotoxic activity was assayed against YAC-1 tumor cells as described under Figure 2 except that the assay was incubated for 4 hr at 37°C.

(MHC)-matched normal mice (Bix et al., 1991). This is most likely because β_2-m⁻ cells, lacking class I expression, are susceptible to lysis by NK cells. In response to poly I:C induction, however, β_2-m⁻ mice themselves express low levels of NK activity compared to normal mice as assessed by lysis of YAC-1 target cells (Liao et al., 1991). Unlike NK cells from normal mice, β_2-m⁻ NK cells were unable to kill concanavalin A-induced T-cell blasts from β_2-m⁻ mice. There was no discernible difference in the number of NK1.1⁺ cells between β_2-m⁻ and normal mice (2–3% of unfractionated splenocytes). Results in our laboratory suggest, however, that NK activity is efficiently induced in β_2-m⁻ mice following viral infection. Lysis of YAC-1 cells by splenocytes from LCMV-infected β_2-m⁻ mice is comparable to that observed using spleen cells from LCMV or poly I:C-injected B6 mice (Figure 4). There would therefore appear to be some difference in NK induction between β_2-m⁻ and normal mice. Poly I:C, functioning by interferon-α and -β induction, seems to be unable to generate a good NK response in β_2-m⁻ mice, whereas direct viral infection appears to be much more effective. This may indicate that LCMV infection is a more potent interferon-αβ inducer than poly I:C, or alternatively that virally induced interferon-γ is more effective than interferon-α and -β in inducing NK activity. We have, however, been unable to detect NK activity in β_2-m⁻ mice following i.p. injection of interferon-γ, suggesting that NK induction by LCMV is mediated by a different mechanism.

Conclusions

The study of viral infection of β_2-m⁻ mice has revealed significant immunological responses, demonstrating the profound plasticity of the immune system. In mice which are devoid of conventional CD8⁺, class I-restricted CTL, we can readily detect virus-specific, CD4⁺, class II-restricted CTL *in vitro*. The immunopathology of LCMV disease in these mice suggests that these class II-restricted CTL function *in vivo* similarly to the virus-specific, class I-restricted CTL in normal mice. β_2-m$^-$ mice are, therefore, not obviously immunodeficient due to a lack of CD8$^+$ CTL; rather there is a compensatory increase in the level of CD4$^+$ activity.

References

Arce-Gomez B, Jones EA, Barnstable CJ, Solomon E, Bodmer WF (1978): The genetic control of HLA-A and B antigens in somatic cell hybrids: requirement for β_2 microglobulin. *Tissue Antigens* 11: 96–112

Artzt K, Bennet D, Jacob F (1974): Primitive teratocarcinoma cell express a differentiation antigen specified by a gene at the *T*-locus in the mouse. *Proc Natl Acad Sci USA* 71: 811–814

Artzt K, Dubois P, Bennet D, Condamine H, Babinet C, Jacob F (1973): Surface antigens common to mouse cleavage embryos and primitive teratocarcinoma cells in culture. *Proc Natl Acad Sci USA* 70: 2988–2992

Bix M, Liao N-S, Zijlstra M, Loring J, Jaenisch R, Raulet D (1991): Rejection of class I MHC-deficient haemopoietic cells by irradiated MHC-matched mice. *Nature* 349: 329–331

Cosgrove D, Gray D, Dierich A, Kaufman J, Lemeur M, Benoist C, Mathis C (1991): Mice lacking MHC class II molecules. *Cell* 66: 1051–1066

Dargemont C, Dunon D, Deugnier M-A, Denoyelle M, Girault J-M, Lederer F, Le K, Godeau F, Thierry JP, Imhof BA (1989): Thymotaxin, a chemotactic protein, is identical to β_2 microglobulin. *Science* 246: 803–806

Doherty PC, Allen JE, Lynch F, Ceredig R (1990): Dissection of an inflammatory process induced by CD8$^+$ T cells. *Immunol Today* 11: 55–59

Grusby MJ, Johnson RS, Papaioannou VE, Glimcher LH, (1991): Depletion of CD4$^+$ T cells in major histocompatibility complex class II-deficient mice. *Science* 253: 1417–1420

Koller BH, Marrack P, Kappler JW, Smithies O (1990): Normal development of mice deficient in β_2M, MHC class I proteins, and CD8$^+$ T cells. *Science* 248: 1227–1230

Liao N-S, Bix M, Zijlstra M, Jaenisch R, Raulet D (1991): MHC class I deficiency: Susceptibility to natural killer (NK) cells and impaired NK activity. *Science* 253: 199–202

Rahemtulla A, Fung-Leung WP, Schilham MW, Kundig TM, Sambhara SR, Narendran A, Arabian A, Wakeham A, Paige CJ, Zinkernagel RM, Miller RG, Mak TW (1991): Normal development and function of CD8$^+$ cells but markedly decreased helper cell activity in mice lacking CD4. *Nature* 353: 180–184

Vitiello A, Potter TA, Sherman LA (1990): The role of β_2 microglobulin in peptide binding by class I molecules. *Science* 250: 1423–1426

Zijlstra M, Bix M, Simister NE, Loring J, Raulet DH, Jaenisch R (1990): β_2-Microglobulin deficient mice lack CD4$^-$8$^+$ cytolytic T cells. *Nature* 344: 742–746

IV
Molecular Mechanisms of Cellular Cytotoxicity

14
The Granule Exocytosis Model for Lymphocyte Cytotoxicity and Its Relevance to Target Cell DNA Breakdown

Pierre A. Henkart, Mark P. Hayes[1], and John W. Shiver[2]

The granule exocytosis model for lymphocyte cytotoxicity was proposed some years ago (Henkart and Henkart, 1982, Henkart, 1985) and has become generally accepted as one pathway which cytotoxic lymphocytes use to kill target cells (Krahenbuhl and Tschapp, 1991). The model is appealing because it can account for most of the classically known properties of lymphocyte cytotoxicity, and it utilizes a basic process well known to cell biology: the regulated pathway of protein secretion (Henkart et al., 1987b). Indeed, it remains the only well-defined model which accounts for the rapid, lymphocyte-mediated death of target cells *in vitro*. However, there remain questions as to whether it is the major pathway of cytotoxicity, as it seems likely that cytotoxic T lymphocytes (CTL) can sometimes kill when this pathway is inoperative (Ostergaard et al., 1987; Trenn et al., 1987).

Some of these questions arise from the phenomenon of target cell "internal disintegration" accompanying death mediated by cytotoxic lymphocytes, as discussed in other chapters in this book. The hallmark property of this mode of cell death is DNA fragmentation, which is well correlated with an "apoptotic" morphology of cells undergoing programmed cell death. It is our contention that the granule exocytosis model accounts for the delivery of signals to the target cell which result in both membrane damage and internal disintegration. We will summarize in this chapter the current status of the granule exocytosis model and describe experiments which show that the granule exocytosis model accounts for the triggering of a target cell disintegration pathway which includes target DNA breakdown.

The Granule Exocytosis Model

Development of the Model

A number of groups carried out comprehensive studies on the mechanism of CTL cytotoxicity in the 1970s. They developed the general concept of an initial killer–target adhesion step, followed by a calcium-dependent lethal hit step in which the target was irreversibly injured, and finally a more prolonged lysis step in which the killer cell played no role. The nature of the lethal damage inflicted on the target cell remained elusive. Although some workers supported the hypothesis that lymphotoxin or related cytotoxic lymphokines damaged the target cells (Ruddle and Schmid, 1987), this mediator did not have the required rapid kinetics, nor could it account for the lack of damage to bystander cells near the antigenic target cells. We became interested in the idea that cytotoxic lymphocytes inflicted membrane damage on target cells, and utilized an antigenic lipid bilayer membrane to demonstrate a striking increase in ionic permeability inflicted by large granular lymphocytes (LGL) in a model for antibody-dependent cytotoxicity (Henkart and Blumenthal, 1975). We then carried out a series of experiments with red cell ghost targets, again with LGL effectors (Simone and Henkart, 1980). These experiments showed a clear sieving behaviour of released marker proteins from these ghosts, leading to an inference of pore formation as the mechanism of membrane damage. The subsequent electron micrographs showing these ghost membranes with porelike structures strikingly similar to those known with complement

[1] Current address: Division of Cytokine Biology, CBER, FDA, Bethesda, MD 20892.
[2] Current address: Merck, Sharp and Dohme Research Laboratories, WP16-306, West Point, PA 19486.

gave strong support to the membrane-damage hypothesis (Dourmashkin et al., 1980).

If membrane damage was indeed the lethal injury to the target cell, this left open the question of molecular mediator and how the killer cell delivered it to the target cell. For some time, a secretory process for delivery of the lethal hit had seemed unlikely, because it was difficult to account for the sparing of bystanders if soluble toxic molecules were secreted (Henney, 1977). However, electron micrographic studies of natural killer (NK) cell killing in our laboratory (Henkart and Henkart, 1982) and of CTL-target interactions in Zagury's laboratory (Zagury, 1982) suggested that the effector cell became polarized and secreted the contents of preexisting granules into the narrow gap between the killer cell and its bound target. These studies suggested that isolation of granules from cytotoxic lymphocytes could reveal the molecular mediators of target cell damage, and led to the description of cytolysin/perforin in NK tumors by our laboratory (Henkart et al., 1984) and from CTL clones by Podack's laboratory (Dennert and Podack, 1983).

Target Cell Binding and Effector Polarization

As described above, the model has proved to be applicable to LGL-mediated NK cell killing, antibody-dependent killing, cytotoxicity by CTL, and, to the extent tested, cytotoxicity by LAK cells.

As diagrammed in Figure 1, the initial steps in this model are a binding of the target cell by the killer cell and the polarization of the killer cell cytoplasm as a result of this interaction Although the diagram shows the binding to be mediated by a single type of receptor-ligand interaction, this is an oversimplification for the sake of the illustration. A substantial number of adhesion molecules have now been described which often play a critical role in both adhesion and triggering events, which can greatly influence subsequent processes. Some of these, e.g., the LFA-1-ICAM interaction and the CD2-LFA-3 interaction, were originally defined on the basis of their role in CTL killing. These adhesion molecules often mask the specificity of killer-target binding, so that cells that are not lysed can be seen to bind as readily as target cells that are lysed.

The polarization of the effector cell was first described by Zagury and colleagues in EM studies of CTL killing in which they noted a preponderance of granules on the side of the CTL where the target cell was bound (David et al., 1979). This asymmetry in cytoplasmic organelles was subsequently confirmed for both NK cells and CTL bound to their targets, by fluorescence microscopy using both Golgi and microtubule organizing center markers (Kupfer and Singer, 1989). Because detecting and quantitating this polarization is tedious, relatively few studies of this phenomenon hae been reported. This rearrangement of lymphocyte cytoplasm in response to the surface-bound antigenic stimulus is both rapid and calcium-requiring (Kupfer et al., 1985), and is a remarkable phenomenon whose mechanism has not been defined. However, it is in many ways parallel to phagocytosis, where the phagocyte membrane completely surrounds a particle before excytosing its lysosomal granules into the phagosome. It should also be noted that a similar polarization has been observed for cloned helper T lmphocytes whose newly synthesized lymphokines have been localized in the cytoplasm en route to being secreted by the cell, apparently by the constitutive secretory pathway (Kupfer and Singer, 1989).

Secretion

The polarized secretion of preformed and stored granule components is the critical step in the delivery of the lethal hit by the effector cell in this model. The secretion of various granule components cytolysin, lysosomal enzymes, granzyme A, and proteoglycans) in response to receptors recognizing target cells has been documented for both CTL and NK cells for a number of granule components. Most studies of this secretion have measured the release of soluble granule components into the medium. The stimuli shown to trigger cytotoxic lymphocyte degranulation have included target cells (Schmidt et al., 1985; Pasternack et al., 1986), purified membrane antigens (Kane et al., 1988), and antibodies against the T-cell receptor (TcR) complex, CD16, and other LGL surface antigens (Williams et al., 1985; Chambers et al., 1989). In general, the secretion requires a polyvalent stimulus such as immobilized or membrane-bound ligands; it is quite rapid, occurring within several hours of triggering, and it requires calcium in the medium. It is difficult experimentally to measure the degree of polarity of the secretion, but high-resolution phase-contrast cinematography images have been interpreted as compatible with a highly polarized secretion (Yannelli et al., 1986).

Properties of Cytolysin

The functional and biochemical properties of cyto-

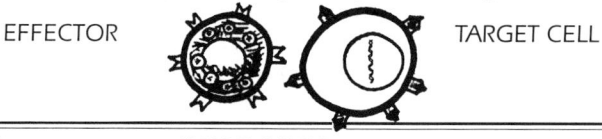

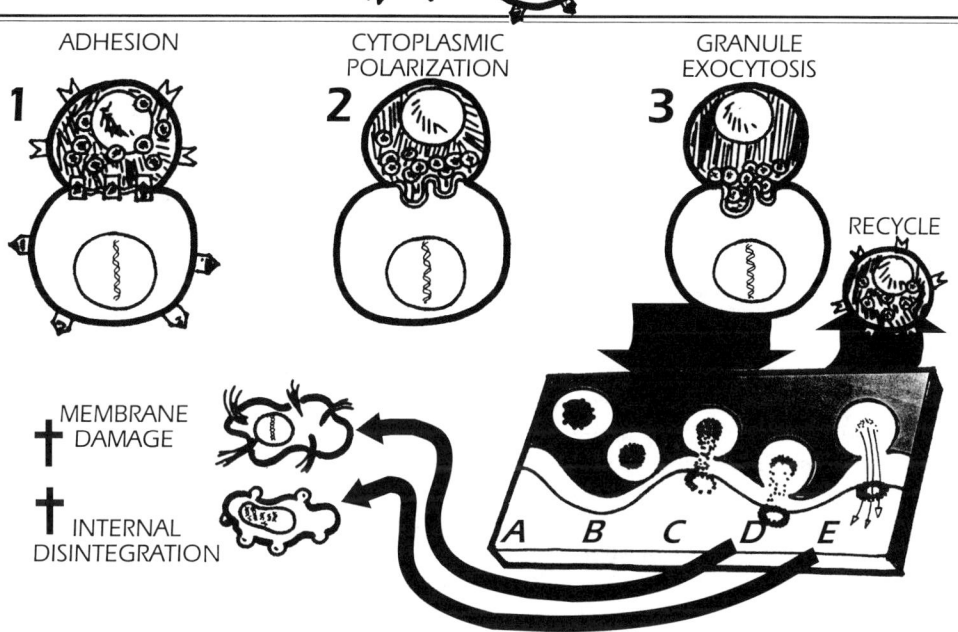

FIGURE 1. The granule exocytosis model of lymphocyte cytotoxicity. After binding the target cell, the effector cell receives activating signals from triggering receptors on its surface, and the next step of cytoplasmic polarization occurs. This orients the secretory apparatus so that the critical secretion step occurs selectively into the synapse-like junctional region between the target and effector. Exocytosis results in the release of granule components into this region while the granule membrane becomes part of the plasma membrane. The initial pathway of target cell damage is the insertion of cytolysin into the target membrane, aggregation, and pore formation. Although this can result in target death, it is also repairable by endocytosis. However, the cytolysin allows permeabilization of the target membrane to granzymes and other granule mediators which inflict a second, "internal disintegration," pathway of target cell death involving DNA degradation. The effector cell has not exocytosed all its granules and can immediately recycle to begin another round of killing.

lysin have been reviewed previously (Young et al., 1986a; Henkart and Yue, 1988: Podack et al., 1991), and only a brief summary will be given here. The most striking features of cytolysin's lytic activity are the following: (1) This agent permeabilized lipid bilayers (Blumenthal et al., 1984; Young et al., 1986b) and is relatively nonselective with respect to target cells, although red blood cells are about 50 times more sensitive than most tumor cells (Henkart et al., 1984). (2) Target lysis is rapid, occurring within seconds or minutes of exposure. (3) The lytic activity requires Ca^{2+} in the 10^{-4}–10^{-3} M range (Henkart et al., 1984). (4) The lytic activity is inhibited by lipids and lipoproteins (Yue et al., 1987; Tschopp et al., 1989).

The cytolysin molecule has been purified as a soluble protein from granules of rat LGL tumors (Millard et al., 1987) and cloned murine CTL (Dennert and Podack, 1983; Podack et al., 1985). It is a single protein on sodium dodecyl sulfate (SDS) gels with a MW_{app} of 65–75 kDA. The protein aggregates in the presence of calcium and lipid to form large cylindrical structures inserted into the lipid bilayer (Dennert and Podack, 1983; Blumenthal et al., 1987). These have the appearance of structures observed on target membranes after cytotoxic lymphocyte attack (Dourmashkin et al.,

1980). These structures have a definite pore-like appearance, and the internal diameter correlates well with the maximal size of released markets, from resealed red cell ghosts attacked by large granular lymphocytes (Simone and Henkart, 1980). It should be emphasized, however, that smaller cytolysin aggregates undoubtedly occur and are probably capable of causing smaller functional pores, which may predominate in physiological killer cell–target interactions.

Cytolysin cDNA sequences have been reported from mouse, human, and rat sources (Lichtenheld et al., 1988; Shinkai et al., 1988b: Ishikawa et al., 1989). The deduced protein sequences are quite homologous to each other, and correspond to mature proteins of 534 amino acids, or about 60 kDa. Protein homology searches with the databases show that the protein is related to the complement proteins C6, C8α, C7, C9, and C8β, in that order. The homology to these complement proteins occurs over a central region of roughly 300 amino acids and is in the range of 20–25% identity, indicating evolutionary relationships among these proteins. It is interesting that these proteins comprise the complement "membrane attack" complex along with C5b, but the functional implications for the cytolysin protein are unclear. From protein functional studies, the C9 protein has the most analogous behavior, including aggregation, metal binding, membrane insertion, and pore formation. Unfortunately, the basic protein structure–function relationship in C9 is still obscure. Cytolysin contains no hydrophobic domain typical of integral membrane proteins, and there is no calmodulin-type calcium-binding motif. When mouse, human, and trout C9 sequences were examined for periodic hydrophobic "moments," one short conserved region of possible amphipathic helix was revealed (Stanley and Herz, 1987); the amphipathic character of this stretch is conserved in all three cytolysin sequences. Such an amphipathic helix could be a region of membrane interaction, perhaps as the side of a transmembrane channel.

The actual mechanism by which pore-forming molecules kill nucleated cells is not clear. It has been shown that complement-attacked tumor cells can repair membrane damage, apparently by endocytosis of the inserted pore complexes (Carney et al., 1985). A similar repair process was demonstrated with cytolysin-attacked tumor cells (Bashford et al., 1988). In many cases, degranulation of a CTL may release a high local concentration of cytolysin, creating massive pore formation which would be difficult to repair, as suggested by the EM images (Dourmashkin et al., 1980). In other cases, where less cytolysin is delivered, the net damage to the target membrane is the difference between a pore-formation process and a repair process. Based on older literature about complement, there has been a tendency to ascribe the cytolysin-mediated death of nucleated target cells to a colloid osmotic mechanism. This type of lysis results from the equilibration of ions across the membrane due to their enhanced permeability, followed by an osmotic swelling of the cell due to the unbalanced internal osmotic pressure of the "colloids," i.e., macromolecules. The unchecked swelling then leads to membrane rupture. Experimental support for this model has come from the blocking of complement-mediated lysis by high concentrations of macromolecules in the medium (Green et al., 1959). However, recent studies have shown that, unlike complement-mediated hemolysis, complement-mediated cytotoxicity of tumor cells is not blocked when the colloid osmotic pressure is equalized on the outside of the cell (Kim et al., 1989). Attempts to apply this approach to lymphocyte-mediated cytotoxicity have been complicated by the viscosity of the blocking solutions. Thus, at present the detailed mechanism leading to the death of nucleated cells attacked by pore-forming agents like complement and cytolysin is still unknown. As discussed below, this pathway of lysis does not normally result in DNA breakdown.

Why Are the Killer Cells Not Killed by Degranulation?

The issue of whether CTL are immune to their own cytotoxic mechanism has been addressed by a number of groups over many years. Early studies using polyclonal CTL populations showed that allospecific CTL could functionally inactivate CTL bearing the appropriate major histocompatibility complex (MHC) antigens on their surface (Kuppers and Henney, 1977). These results were interpreted as indicating that CTL could kill other CTL, and that the lytic mechanism needed to be highly polarized, since the effector cell had no "armor plating" preventing its own destruction. When cloned CTL became available, labeling them with ^{51}Cr allowed their lysis to be measured directly. Using this approach, CTL were found to be resistant to lysis by other cloned allospecific or hapten-specific CTL with appropriate recognition (Blakely et al., 1987; Skinner and Marbrook, 1987). When lectin was added to trigger the effector CTL or the specificity was redirected by anti-TCR, a similar resistance was found (Blakely et al, 1987; Kranz and Eisen,

1987:, but exceptions were noted (Blakely et al., 1987). Thus the general view has emerged that CTL are resistant to being lysed by their own cytotoxic mechanism, and that the earlier results reflected a functional inactivation rather than lysis. In accord of this, the cloned CTL have been shown to be resistant to lysis by the granule cytolysin (Nagler-Anderson et al., 1988; Shinkai et al., 1988a; Jiang et al., 1990).

The mechanism for this has been suggested to be due to a resistant property of the CTL plasma membrane (Ojcius et al., 1991) which has not yet been defined.

Controversies Involving the Granule Exocytosis Model

Since its development in the early 1980s, the granule exocytosis model has been the only reasonably well-defined model for lymphocyte cytotoxicity to gain acceptance. As indicated above, it has been well supported experimentally in a number of important respects, but there is no clear evidence that it represents a dominant pathway of target cell damage during lymphocyte cytotoxicity. There are three lines of evidence that can be considered as challenges to this model:

1. Cytolysin may not be well expressed in some CTL. There are two situations where this has been found:

   (a) Granule components generally are expressed at lower levels in *in vivo* CTL than in the cloned *in vitro*-grown CTL generally used for biochemical studies. This point has been prominently made by Berke and colleagues and is covered in Chapter 18 of this volume. Although their studies have failed to detect various granule components in *in vivo*-grown CTL, other laboratories using more sensitive techniques have routinely done so (Podack et al., 1991). Thus the controversy revolves on the issue of how much cytolysin is necessary to inflict lethal damage on target cells. One can only point out that the cytolysin is extremely potent when added to the medium and that the polarized delivery postulated by the model would result in a high local concentration of the secreted material. Thus, while a definitive assessment of this issue is not possible, it cannot be a strong argument against the model.

   (b) In some cytotoxic CD4$^+$ T-cell clones, cytolysin does not seem to be expressed when sensitive detection methods are used. The status of the granule exocytosis pathway in these cells has not been reported, and it remains possible that the cells kill by a granule exocytosis model in which some other mediator serves to permeabilize the target membrane to secreted granzymes, which cause an internal disintegration of the target cell. More mechanistic studies need to be done on this type of cytotoxicity.

2. Granule exocytosis and cytolysin activity require calcium, but a few target cells can by lysed by CTL in its absence. Early studies of lymphocyte cytotoxicity recognized that the process was calcium-dependent, and in the 1970s several laboratories defined a calcium-dependent lethal hit stage of the CTL mechanism when the effector irreversibly injured the target cell. Later it was found that some target cells had a calcium-independent component of the CTL lethal effect, and with a few targets this accounts for the majority of the damage (MacLennan et al., 1980; Tirosh and Berke, 1985). When these targets were used with cloned CTL, it was found that they were killed in the absence of calcium even though degranulation was completely blocked (Ostergaard et al., 1987; Trenn et al., 1987). These results make it likely that another cytotoxic pathway exists in CTL in addition to granule exocytosis; unfortunately, such a pathway has not been molecularly defined (see Section VI, this volume).

3. A third question regarding the granule exocytosis model involves the target cell DNA breakdown phenomenon, which is covered in detail below. In our view there is no real controversy in this regard.

In summary, the granule exocytosis model clearly has much evidence in its favor and explains most of the phenomena of lymphocyte cytotoxicity. However, since other undefined killing pathways probably exist, it is hard to conclusively assess their relative importance.

Target Cell DNA Breakdown and the Granule Exocytosis Model

The original description of DNA breakdown accompanying CTL-mediated killing of tumors was in a series of elegant studies by Russell and his collaborators (Russell, 1983). They pointed out that cytolysis by complement or by osmotic lysis was not accompanied by DNA breakdown in the same target cells where it was dramatically observed during CTL-induced lysis. The former damage corresponded to the morphological syndrome of "necrosis," describing cell death by agents causing membrane injury, while the latter fell under the description of "apoptosis," characterized by striking condensation of nuclear chromatin and cytoplasmic shrinking. Apoptosis has been associated with developmentally programmed cell death in a number of systems (Wyllie et al., 1980). The challenge to the granule exocytosis model came from the fact that the original evidence for the model emphasized the lytic properties of cytolysin. Since this molecule has a number of similarities with the pore-forming complement protein C9, it appeared unlikely that it could account for the DNA breakdown accompanying cell death mediated by lymphocytes. We have considered for some time that other granule components in addition to cytolysin might play a role in target cell damage, and have approached this possibility using several different experimental approaches. We will review here three different lines of evidence which argue for a role for the effector cell granule protease granzyme A in target DNA breakdown during cytolysis mediated by CTL.

In considering these studies, it must be appreciated that there is no defined biochemical pathway for programmed cell death in any system. Although it has been proposed that the DNA breakdown itself is critical to cell death, the evidence for this is weak. The biochemistry of the DNA breakdown process is also not worked out. In the best-studied system, the steroid-induced death of immature thymocytes, it has been difficult to characterize the enzyme responsible for the dramatic DNA breakdown (Gaido and Cidlowski, 1991). We have been hampered by this lack of biochemical knowledge in dealing with target DNA breakdown induced by cytotoxic lymphocytes.

Cytolysin Alone Induces Minimal DNA Breakdown During Lysis

Although complement and osmotic lysis were among the original samples of necrotic cell death in which DNA breakdown did not occur, it remained possible that the cytolysin itself was capable of causing target DNA breakdown in some cells. The literature contains examples of many agents inducing DNA breakdown in certain cell types, e.g., thymocytes, which seem primed to undergo the apoptotic type of death, but most of these agents do not have this effect on other cells such as the typical CTL tumor targets. In the case of cytolysin, both we (Munger et al., 1988; Hayes et al., 1989) and Young's laboratory (Duke et al, 1989) have reported that purified cytolysin did not cause appreciable DNA fragmentation during cytolysis, while another laboratory reported the opposite result (Hameed et al., 1989). However, the report that staphylococcal alpha toxin, a bacterial protein causing small membrane pores (Fusshe et al., 1981), can induce DNA fragmentation reinforces the possibility that some conditions of pure membrane injury can trigger DNA breakdown (Hameed et al., 1989).

Given the possibility that some pore-formers can trigger DNA breakdown, it was important to test the effect of cytolysin when delivered by degranulation of a bound effector cell, where the target membrane receives a high local concentration of cytolysin over one patch of membrane rather than the global attack resulting from cytolysin in solution. We have recently been able to do this using a model system we devised to test the granule exocytosis model for lymphocyte cytotoxicity. We have employed the mast cell tumor line, rat basophilic leukemia (RBL), which has a well-defined degranulation pathway triggered by the IgE Fc receptor (Fewtrell and Metzger, 1981). While RBL has no cytolysin and is not cytotoxic itself, our strategy has been to transfect this line with CTL granule components and thus attempt to create a model killer cell. We can hapten-modify the plasma membrane of our target cell and use an IgE anti-haptem antibody to trigger a kind of antibody-dependent cellular cytotoxicity (ADCC). We have shown (Shiver and Henkart, 1991a) that red blood cells were readily killed by RBL cells transfected with cytolysin, giving a potent cytotoxicity comparable to that of cloned CTL. The properties of this cytotoxicity were essentially similar to that mediated by CTL, and the general predictions of the granule exocytosis model were confirmed. However, in our original series of transfectants we were unable to obtain good killing of nucleated target cells (Shiver and Henkart, 1991b). This result was rationalized by the fact that a larger dose of cytolysin is required to kill such cells compared to red cells. Subsequently, we have modified our transfection protocol so that by using electroporation we obtain much better cytolysin expression, and we chose parental RBL lines with much better overall secretion efficiency. The RBL-cy transfectants obtained in this way are able to give reasonable killing of tumor target cells, although not as potent as cloned CTL. The ability to kill tumor cells with these transfectants has allowed us to ask if target DNA breakdown accompanies death in this cell-delivered lethal hit. The answer is clearly negative, as shown in the experiment in Figure 2. These

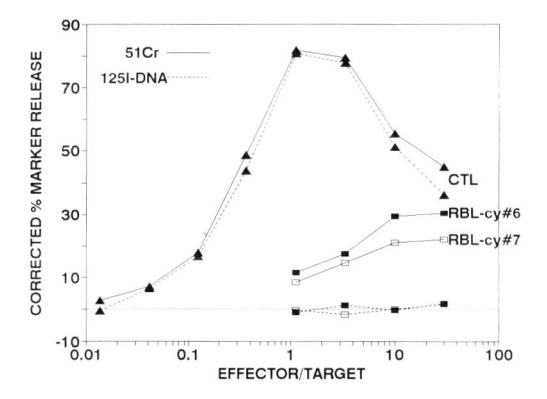

FIGURE 2. Lack of target DNA breakdown in target cells killed by cytolysin delivered by degranulating RBL cells. TNP-EL4 mouse lymphoma cells were labeled with ^{51}Cr and iododeoxyuridine-125 (^{125}I-UR) and incubated for 4 hr with either murine cloned CTL in the presence of α-CD3xα-TNP heteroconjugate (triangles), or two different clones of cytolysin-transfected RBL cells in the presence of IgE α-TNP (squares). The parental untransfected RBL cells are not cytolytic (Figure 4).

results confirm our previous findings with purified cytolysin in the medium, but are more significant because the cell-delivered cytolysin replicates the physiological case. We next turned to the question of whether or not other granule components may play a role in target DNA breakdown.

Pretreatment of CTL to Inactivate Granule Proteases Compromises Their Ability to Mediate Target DNA Breakdown but Not Cytolysis

Ever since the recognition that granzymes are major components of cytotoxic lymphocyte granules, there has been a debate about their physiological role. One idea which was considered was that they processed the cytolysin protein so that it became lytically active. This hypothesis predicted that higher-molecular-weight inactive cytolysin precursors would exist and that inactivation of granule proteases would inactivate cytolytic function. Neither of these predictions has been borne out; although some workers have interpreted a loss of CTL activity after treatment with protease inhibitors as evidence in favor of this, it has never been established that the granule proteases are a target for these inhibitors. Indeed, degranulation itself in mast cells is blocked by protease inhibitors (Ishizaka and Ishizaka, 1984). The cytolysin pro-

cessing hypothesis is also not easily compatible with the results described above, in which lytic activity was conferred to RBL cells which do not contain lymphocyte granzymes by transfection with the gene for intact cytolysin. Although one could argue that other proteases in RBL granules carry out this function, it seems hard to accept this explanation, since the known mast cell granzymes appear to have different specificities than those in lymphocyte granules (Serafin and Reynolds, 1990).

A large number of inhibitors of various serine proteases have been developed; many of these are reagents that mimic substrates and form a covalent intermediate with the active site serine. Instead of forming a transient intermediate, the inhibitor-serine bond is stable, thus clogging the active site. The reaction of this type of inhibitor with the active site is pH-dependent, with a marked falloff at lower pH, similar to the formation of the catalytic intermediate (Gold and Fahrney, 1964). This means that previous attempts to inactivate CTL granule proteases *in situ* with these reagents would have been ineffective if the granules had a pH of around 5, as is typical of many secretory granules (Orci et al., 1987). In confirmation of this, we showed that the reaction of the classical serine protease inhibitor diisopropyl fluorphosphate (DFP) with granzyme A in intact CTL was greatly enhanced by coincubation of DFP with agents that raise the pH of acidic intracellular compartments (Henkart et al., 1987a). This approach showed that it was possible to distinguish the granule proteases that responded to lysosomal agents from others whose DFP reaction was unaffected (i.e., those in neutral compartments such as the cytoplasm). We also found that inactivation of CTL granzyme A *in situ* by PMSF was greatly enhanced by agents raising the pH of acidic intracellular compartments; this treatment protocol did not kill the cells. When the lytic activity of these CTL was analyzed (Figure 3), phenylmethylsulfonyl fluoride (PMSF) treatment caused a minimal inactivation, and this was not enhanced by the lysosotropic agents (Henkart et al., 1987a). These results suggested a nongranule protease was required for optimal lytic activity, but failed to implicate granule enzymes; obviously such results do not rule out the processing hypothesis for granzyme function.

An alternative hypothesis for granzyme function is that these enzymes play a role in secondary target cell damage after gaining access to the target cytoplasm via pores made by the cytolysin. We considered DNA breakdown as one candidate for such damage, even though it was not clear how a protease would be involved in DNA cleavage.

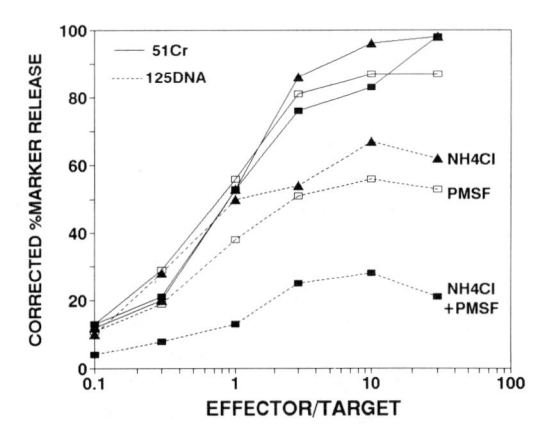

FIGURE 3. Pretreatment of CTL with PMSF and a lysosomotropic agent does not alter lytic activity but inactivates their ability to mediate target DNA release. Cloned murine anti-H2K^b specific CTL were pretreated with 0.01 M NH$_4$Cl (triangles), a lysosomotropic agent which reversibly raises the pH of acidic intracellular compartments; 0.5 mM PMSF (open squares), a reagent which irreversibly reacts with active site serines of serine proteases in a pH-dependent reaction; or a combination of the above NH$_4$Cl and PMSF (closed squares). Control CTL pretreated with buffer alone were identical to the NH$_4$Cl-treated CTL. After the pretreatment, the CTL were washed and tested for their ability to cause lysis (^{51}Cr release, solid lines) and DNA release (^{125}-DNA) of the murine H2K^b-bearing EL4 lymphoma cells in a standard 4-hr assay. In these experiments a maximum of 60% DNA release was achieved even when CTL caused complete lysis.

When measuring target DNA breakdown after a granzyme inactivation, we obtained a rather different result from ^{51}Cr release (Munger et al., 1988; Hayes et al., 1989). There was a clear loss of target DNA breakdown when the CTL were pretreated with both PMSF and agents raising the intragranular pH, as shown in Figure 3. In this case, both agents were required in order to see a significant effect. These results showed that a PMSF-sensitive enzyme located in a low-pH intracellular compartment of the CTL was critically required for target DNA breakdown. The granzymes were obvious candidates, and granzyme A in particular was suggested since its N-α-benzyloxycarbonyl-L-lysine thiobenzyl (BLT)-esterase activity was diminished in parallel with the ability of the CTL to mediate DNA breakdown.

Granzyme A Accounts for the Ability of CTL Granules to Break Down DNA in Permeabilized Cells

Pursuing the hypothesis that some granule component was capable of triggering target DNA breakdown after gaining access to the target cell cytoplasm, we looked for DNA breakdown when cells were lysed by granule extracts. We found that this was indeed observed with several different target cells, but required a much larger granule dose than was required for ^{51}Cr release. Indeed, we could find a dose of granule extract which would cause ^{51}Cr release but little DNA release (Munger et al., 1988). However, when higher doses of granule extract were used, both markers were released. This dose–response raised the question of whether DNA release by superlytic doses of granule extract was relevant to the cell-delivered lethal hit, expecially considering that two groups had reported that granule extracts did not cause target DNA breakdown accompanying lysis (Gromkowski et al., 1988; Selmaj et al., 1991). Our rationale for thinking that this was relevant was that addition of granule extracts to cell suspensions is a poor imitation of secretion by a killer cell in two respects: (1) The cell secretion probably causes many cytolysin pores to form in the target membrane where the local granzyme A concentration is high, thus allowing efficient diffusion of the enzyme to intracellular sites; in order to achieve this degree of permeabilization, a superlytic dose of cytolysin is probably required. (2) When added to a bulk solution containing calcium, the cytolysin probably acts like an affinity reagent for membranes because of its amphipathic nature, allowing it to act on the cell suspension at a very low concentration. An enzyme, on the other hand, would be unlikely to have this benefit, and might require a higher relative concentration when added to a cell suspension. However, when delivered by a degranulating lymphocyte adherent to the target membrane, the affinity effect of the cytolysin becomes negligible.

We devised an experimental system to characterize the granule component causing DNA breakdown in permeabilized cells, routinely using detergent rather than cytolysin to provide access to the nucleus. With this assay, CTL and LGL tumor granule extracts showed good DNA degradation activity, which could be blocked by protease inhibitors (Munger et al., 1988; Hayes et al., 1989). When the CTL granule activity was purified to homogeneity, it proved to be identical to granzyme A.

E:T   0.1   1   10   NG/ML   50   150   500

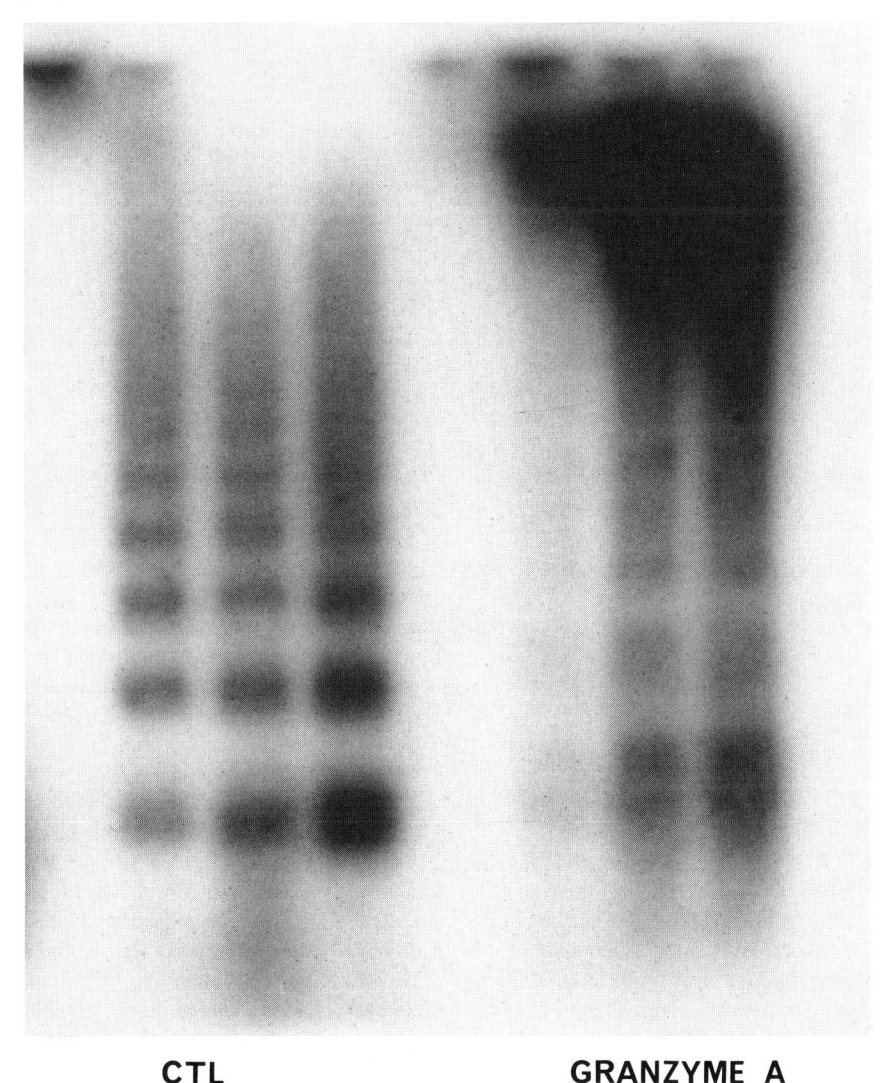

FIGURE 4. Granzyme A induces DNA breakdown in permeabilized target cells. Autoradiograms of 1% agarose gels on supernatants of ^{125}I-DNA-labeled EL4 cells treated 6 hr with granzyme A at the concentrations shown (right) or CTL at the E/T ratio shown (left). The lowest band corresponds to approximately 200 bp.

Purified granzyme A was subsequently shown to induce DNA breakdown in combination with purified cytolysin (Hayes et al., 1989). Thus, preparations of cytolysin which are contaminated by granzyme A could induce DNA breakdown in permeabilized cells. In causing DNA release from nuclei, granzyme A induced the characteristic nucleosomal ladder pattern of DNA fragments, as shown in Figure 4. Thus, by a completely independent approach, we found further evidence for a role for granzyme A in target DNA fragmentation.

A Combination of Cytolysin and Granzyme A Delivered by Transfected RBL Cells Causes Tumor Target Lysis Accompanied by DNA Breakdown

We have recently completed a study of RBL cells transfected with combinations of cytolysin and granzyme A as an additional means of testing the latter's physiological role. After transfection with

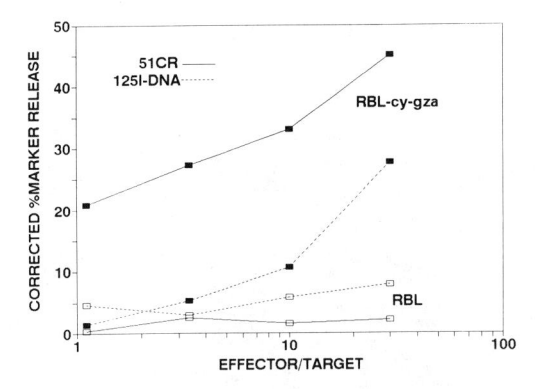

FIGURE 5. Degranulating RBL cells delivering both granzyme A and cytolysin mediate target lysis with accompanying DNA breakdown. RBL double transfectants expressing both granzyme A and cytolysin were used in this experiment (solid squares), which is otherwise similar to that shown in Figure 1. Lack of lytic or DNA release activity by control untransfected parental RBL cells is also shown (open squares).

the granzyme A gene, RBL cells produce granzyme A mRNA and protein at levels comparable to cloned CTL. The protein is properly targeted to the secretory granules, and is secreted in response to crosslinking the IgE receptor. As expected, RBL transfected with granzyme A alone are not cytolytic and do not cause target DNA breakdown. The double transfectants are not strikingly more cytolytic than the cytolysin single transfectants, arguing against a role for granzyme A in the lytic process or the cyolysin processing hypothesis. Most striking, however, is the result in Figure 5, clearly showing that the double transfectants do cause target DNA breakdown accompanying lysis of their targets. While this does not occur as strongly as we see it with CTL (where the levels of DNA and ^{51}Cr release are typically equal), it is clearly positive and thus in striking contrast to the single transfectants, as described in Figure 2. We would interpret the less complete target DNA breakdown by the RBL double transfectants compared to CTL as due to a less efficient delivery system and lack of appropriate adhesion molecules to give a tight compartment to concentrate the polarized secretion products.

Granule Exocytosis and the Target Cell "Internal Disintegration" Pathway

We have described three separate lines of investigation which point to a role for effector cell granzyme A in target DNA breakdown during lymphocyte-mediated cytotoxicity. We believe that all three approaches together argue convincingly that the granule exocytosis mechanism can account for target death accompanied by DNA breakdown. While the RBL transfection experiments are in many ways the most convincing demonstration of the plausibility of this model, they do not address the issue of whether the DNA breakdown which normally accompanies CTL killing is attributable to this pathway. However, the results from PMSF pretreatment of CTL strongly suggest that this is the predominant means of inducing DNA breakdown, at least for some target cells. The most significant conclusion we draw from the transfection experiments is that the granule exocytosis model can account for target DNA breakdown. While we do not rule out the possibility of other pathways playing a role in target injury and death, there is no need to hypothesize another delivery pathway or molecular mediators beyond those already described in order to achieve target DNA breakdown.

Thus, granule exocytosis of cytolysin and granzyme A (and probably other granule components) triggers two damage pathways in target cells. This is schematically shown in Figure 1. One pathway is that of cytolysin-mediated membrane damage, which in many cases may be sufficient to kill the target cell. This pathway dominates with RBL-cy, PMSF-treated CTL (Henkart et al., 1987a), in the presence of topoisomerase inhibitors (Coffman et al., 1989), or with certain targets such as fibroblasts (Sellins and Cohen, 1991). However, since this pathway is vulnerable to defeat by target repair via endocytosis, it is plausible that another damage pathway has evolved involving other granule components; this second pathway leads to DNA breakdown and possibly alternative membrane damage by other agents. This could be the dominant pathway in target cells capable of repairing the initial cytolysin damage. The second pathway depends on the initial phase of the first pathway to permeabilize the target membrane and allow access of other granule components to the target cell cytoplasm. We clearly need to learn more about the biochemical nature of this second, "internal disintegration" pathway. Most obviously, we would like to know the physiologically relevant substrate for granzyme A and how its cleavage leads to DNA breakdown. This substrate appears to be inside the cell, as treatment of cells with granzyme A in solution or degranulating granzyme A onto target cells does not trigger DNA breakdown. Certainly the cytolysin has the ability to induce pores of sufficient diameter to allow granzyme A penetration into the target cell (Simone and Henkart, 1980). The granzyme A substrate inside the cell could be cytoplasmic, or quite possibly nuclear. Since access to the nucleus for proteins > 30 kDa is controlled by nuclear pores, granzyme A may be too large to diffuse in freely unless the disulfide dimer is reduced. It is worth pointing out, however, that the short basic "nuclear localization sequence" motifs required for translocation of proteins through the nuclear pore are present in murine granzyme A, and this may assure its access to the nucleus. Recent studies by Pasternack et al. (Chapter 24, this volume; Pasternack et al., 1991) have shown that granzyme A has a number of nuclear protein substrates, including nucleolin, which shuttles between the nucleus and cytoplasm.

The above discussion has emphasized our findings implicating granzyme A as being involved in target DNA breakdown. Recently two other groups have produced evidence for a role for other granule components in this process. Shi et al. (1992) have found an NK-cell granule serine protease they have termed fragmentin which acts to cause DNA breakdown in the presence of a sublytic concentration of cytolysin. Fragmentin is similar to the NK-cell protease RNKP-1 (Zunino et al., 1990) and granzyme B, and thus probably has a very different specificity than granzyme A. Furthermore, this group showed that another granule protease enhanced the DNA breakdown by cytolysin and fragmentin, suggesting a cooperative digestion of target proteins leading to DNA breakdown. Tian et al. (1991) have shown that a nonprotease polyadenate-binding protein from CTL granules can induce DNA fragmentation in permeabilized target cells. It seems quite plausible that such proteins collaborate with proteases to trigger the internal damage pathway leading to DNA breakdown.

The major conclusion from these studies is that the granule exocytosis model should not be viewed as being limited to the secretion of cytolysin. Other granule components can cause internal damage to the target cell after its membrane has been permeabilized by the cytolysin, and this damage includes DNA breakdown. Future studies will seek to define this internal damage pathway on a molecular basis.

References

Bashford CL, Menestrina G, Henkart PA, Pasternak CA (1988): Cell damage by cytolysin: Spontaneous recovery and reversible inhibition by divalent cations. *J Immunol* 141: 3965-3974

Blakely A, Gorman K, Ostergaard H, Svoboda K, Liu CC, Young JD, Clark WR (1987): Resistance of cloned cytotoxic T lymphocytes to cell-mediated cytotoxicity. *J Exp Med* 166: 1070-1083

Blumenthal R, Millard PJ, Henkart MP, Reynolds CW, Henkart PA (1984): Liposomes as targets for granule cytolysin from cytotoxic LGL tumors. *Proc Natl Acad Sci USA* 81: 5551-5555

Carney DF, Koski CL, Shin ML (1985): Elimination of terminal complement intermediates from the plasma membrane of nucleated cells: The rate of disappearance differs for cells carrying C5b-7 or C5b-8 or a mixture of C5b-8 with a limited number of C5b-9. *J Immunol* 134: 1804-1809

Chambers WH, Vujanovic NL, Deleo AB, Olszowy MW, Herberman RB, Hiserodt JC (1989): Monoclonal antibody to a triggering structure expressed on rat natural killer cells and adherent lymphokine-activated killer cells. *J Exp Med* 169: 1374-1389

Coffman FD, Green LM, Godwin A, Ware CF (1989): Cytotoxicity mediated by tumor necrosis factor in variant subclones of the ME-180 cervical carcinoma line: Modulation by specific inhibitors of DNA topoisomerase II. *J Cell Biochem* 39: 95-105

David A, Bernard J, Thiernesse N, Nicolas G, Cerottini JC, Zagury D (1979): Le processus d'exocytose lysomale localisee: Est-il responsable de l'action cytolytique des lymphocytes T tuers? *CR Acad Sci Paris Ser D* 288: 441-444

Dennert G, Podack ER (1983): Cytolysis by H-2 specific T killer cells. Assembly of tubular complexes on target membranes. *J Exp Med* 157: 1483-1495

Dourmashkin RR, Deteix P, Simone CB, Henkart PA (1980): Electron microscopic demonstration of lesions on target cell membranes associated with antibody-dependent cytotoxicity. *Clin Exp Immunol* 43: 554-560

Duke RC, Persechini PM, Chang S, Liu CC, Cohen JJ, Young JD (1989): Purified perforin induces target cell lysis but not DNA fragmentation. *J Exp Med* 170: 1451-1456

Fewtrell C, Metzger H (1981): Stimulus-secretion coupling in rat basophilic leukemia cells. *KROC Found Ser* 14: 295-314

Fussle R, Bhakdi S, Sziegoleit A, Tranum-Jensen J, Kranz T, Wellensiek H-J, (1981): On the mechanism of membrane damage by S. aureus alpha-toxin. *J Cell Biol* 91: 83-90

Gaido ML, Cidlowski JA (1991): Identification, purification and characterization of a calcium-dependent endonuclease (NUC18) from apoptotic rat thymocytes. NUC18 is not histone H_2B. *J Biol Chem* 266: 18580-18585

Gold AM, Fahrney D (1964): Sulfonyl fluorides as inhibitors of enzymes. II. Formation and reactions of phenylmethanesulfonyl alpha-chymotrypsin. *Biochemistry* 3: 783-791

Green H, Fleischer RA, Barrow P, Goldberg B (1959): The cytotoxic action of immune gamma globulin and complement on Krebs ascites tumor cells. II. Chemical studies. *J Exp Med* 109: 510-511

Gromkowski SH, Brown TC, Masson D, Tschopp J (1988): Lack of DNA degradation in target cells lysed by granules derived from cytolytic T lymphocytes. *J Immunol* 141: 77

Hameed A, Olsen KJ, Lee MK, Lichtenheld MG, Podack ER (1989): Cytolysis by Ca-permeable transmembrane channels. *J Exp Med* 169: 765-777

Hayes MP, Berrebi GA, Henkart PA (1989): Induction of target cell DNA release by the cytotoxic T lymphocyte granule protease granzyme A. *J Exp Med* 170: 933-946

Henkart MP, Henkart PA (1982): Lymphocyte mediated cytolysis as a secretory phenomenon. *Adv Exp Med Biol* 146: 227-242

Henkart PA (1985): Mechanism of lymphocyte-mediated cytotoxicity. *Annu Rev Immunol* 3: 31-58

Henkart P, Blumenthal R (1975): Interaction of lymphocytes with lipid bilayer membranes: A model for the lymphocyte-mediated lysis of target cells. *Proc Natl Acad Sci USA* 72: 2789-2793

Henkart P, Yue CC (1988): The role of cytoplasmic granules in lymphocyte cytotoxicity. *Prog Allergy* 40: 44-81

Henkart PA, Berrebi GA, Takayama H, Munger WE, Sitkovsky MV (1987a): Biochemical and functional properties of serine esterases in acidic cytoplasmic granules of cytotoxic T lymphocytes. *J Immunol* 139: 2398-2405

Henkart P, Henkart M, Hodes R, Taplits M (1987b): Secretory processes in lymphocyte function. *Biosci Rep* 7: 345-353

Henkart PA, Millard PJ, Reynolds CW, Henkart MP (1984): Cytolytic activity of purified cytoplasmic granules from cytotoxic rat LGL tumors. *J Exp Med* 160: 75-93

Henney CS (1977): T cell-mediated cytolysis: An overview of some current issues. *Contemp Top Immunobiol* 7: 245-272

Ishikawa H, Shinkai YI, Yagita H, Yue CC, Henkart PA, Sawada S, Young HA, Reynolds CW, Okumura K (1989: Molecular cloning of rat cytolysin. *J Immunol* 143: 3069-3073

Ishizaka T, Ishizaka K (1984): Activation of mast cells for mediator release through IgE receptors. *Prog Allergy* 34: 188-236

Jiang S, Ojcius DM, Persechini PM, Young, JD (1990): Resistance of cytolytic lymphocytes to perforin-mediated killing. *J Immunol* 144: 998-1003

Kane KP, Goldstein SAN, Mescher MF (1988): Class I alloantigen is sufficient for cytolytic T lymphocyte binding and transmembrane signaling. *Eur J Immunol* 18: 1925-1929

Kim SH, Carney DF, Papadimitriou JC, Shin ML (1989): Effect of osmotic protection on nucleated cell killing by C5b-9. Cell death is not affected by the prevention of cell swelling *Mol Immunol* 26: 323-331

Krahenbuhl O, Tschopp J (1991): Debate: The mechanism of lymphocyte-mediated killing. Perforin-induced pore formation. *Immunol Today* 12: 399-401

Kranz DM, Eisen HN (1987): Resistance of cytotoxic T lymphocytes to lysis by a clone of cytotoxic T lymphocytes. *Proc Natl Acad Sci USA* 84: 3375-3379

Kupfer A, Singer SJ (1989): Cell biology of cytotoxic and helper T cell functions: Immunofluorescence microscopic studies of single cells and cell couples. *Ann Rev Immunol* 7: 309-337

Kupfer A, Dennert G, Singer SJ (1985): The reorientation of the Golgi apparatus and the microtubule-organizing center in the cytotoxic effector cell is a prerequisite in the lysis of bound target cells. *J Mol Cell Immunol* 2: 37-49

Kuppers RC, Henney CS (1977): Studies on the mechanism of lymphocyte-mediated cytolysis. IX. Relationships between antigen recognition and lytic expression in killer T cells. *J Immunol* 118: 71-76

Lichtenheld MG, Olsen KJ, Lu P, Lowrey DM, Hameed A, Hengartner, Podack ER (1988): Structure and function of human perforin. *Nature* 335: 448-451

MacLennan ICM, Gotch FM, Golstein P (1980): Limited specific T-cell mediated cytolysis in the absence of extracellular Ca^{+2}. *Immunology* 39: 109-117

Millard PJ, Henkart MP, Reynolds CW, Henkart PA (1984): Purification and properties of cytoplasmic granules from cytotoxic rat LGL tumors. *J Immunol* 132: 3197-3204

Munger WE, Berrebi GA, Henkart PA (1988): Possible involvement of CTL granule proteases in target cell DNA breakdown. *Immunol Rev* 103: 99-109

Nagler-Anderson C, Verret CR, Firmenich AA, Berne M, Eisen HJ (1988): Resistance of primary $CD8^+$ cytotoxic T lymphocytes to lysis by cytotoxic granules from cloned T cell lines. *J Immunol* 141: 3299-3305

Ojcius DM, Jiang S, Persechini PM, Detmers PA, Young JD-E (1991): Cytoplasts from cytotoxic T lymphocytes are resistant to perforin-mediated lysis. *Mol Immunol* 28: 1011-1018

Orci L, Ravazzola M, Anderson RG (1987): The condensing vacuole of exocrine cells is more acidic than the mature secretory vesicle. *Nature* 326: 77-79

Ostergaard HL, Kane KP, Mescher MF, Clark WR (1987): Cytotoxic T lymphocyte mediated lysis without release of serine esterase. *Nature* 330: 71-72

Pasternack MS, Blier KJ, McInerney TN (1991): Granzyme A binding to target cell proteins. Granzyme A binds to and cleaves nucleolin in vitro./ *J Biol Chem* 266: 14703-14708

Pasternack MS, Verret CR, Liu MA, Eisen HN (1986): Serine esterase in cytolytic T lymphocytes. *Nature* 322: 740-743

Podack ER, Hengartner H, Lichtenheld MG (1991): A central role of perforin in cytolysis? *Ann Rev Immunol* 9: 129-157

Podack ER, Young JD, Cohn ZA (1985): Isolation and biochemical and functional characterization of perforin

1 from cytolytic T-cell granules. *Proc Natl Acad Sci USA* 82: 8629–8633

Ruddle NH, Schmid DS (198): The role of lymphotoxin in T-cell-mediated cytotoxicity. *Ann Inst Pasteur Immunol* 138: 314–320

Russell JH (1983): Internal disintegration model of cytotoxic lymphocyte-induced target damage. *Immunol Rev* 72: 97–118

Schmidt RE, MacDermott RP, Bartley G, Bertovich M Amato DA, Austen KF, Schlossman SF, Stevens RL, Ritz J (1985): Specific release of proteoglycans from human natural killer cells during target lysis. *Nature* 318: 289–291

Sellins KS, Cohen JJ (1991): Cytotoxic T lymphocytes induce different types of DNA damage in target cells of different origins *J Immunol* 147: 795–803

Selmaj K, Raine CS, Farooq M, Norton WT, Brosnan CF (1991): Cytokine cytotoxicity against oligodendrocytes: Apoptosis induced by lymphotoxin. *J Immunol* 147: 1522–1529

Serafin WE, Reynolds DS (1990): Neutral proteases of mouse mast cells. *Monogr Allergy* 27: 31–50

Shi L, Kraut RP, Aebersold R, Greenberg AH (1992): A natural killer cell granule protein that induces DNA fragmentation and apoptosis. *J Exp Med* 175: 553–566

Shinkai Y, Ishikawa H, Hattori M, Okumura K (1988): Resistance of mouse cytolytic cells to pore-forming protein-mediated cytolysis. *Eur J Immunol* 18: 29–33

Shinkai Y, Takio K, Okumura K (1988b): Homology of perforin to the ninth component of complement (C9). *Nature* 334: 525–527

Shiver JW, Henkart PA (1991a): A noncytotoxic mast cell tumor line exhibits potent IgE-dependent cytotoxicity after transfection with the cytolysin/perforin gene. *Cell* 62: 1174–1181

Shiver JW, Henkart PA (1991b): Cytolytic activity of RBL cells transfected with the cytolysin/perforin gene. In: *NK Cell-mediated Cytotoxicity: Receptors, Signalling and Mechanisms*, Herberman RB, Lotzova E, eds. Miami: CRC Press. pp 341–348

Simone CB, Henkart P (1980): Permeability changes induced in erythrocyte ghost targets by antibody-dependent cytotoxic effector cells: Evidence for membrane pores. *J Immunol* 124: 954–963

Skinner M, Marbrook J (1987): The most efficient cytotoxic T lymphocytes are the least susceptible to lysis. *J Immunol* 139: 985–987

Stanley KK, Herz J (1987): Topological mapping of complement component C9 by recombinant DNA techniques suggests a novel mechanism for its insertion into target membranes. *EMBO J* 6: 1951–1957

Tian Q, Streuli M, Saito H, Schlossman SF, Anderson P (1991): A polyadenylate binding protein localized to the granules of cytolytic lymphocytes induces DNA fragmentation in target cells. *Cell* 67: 629–639

Tirosh R, Berke G (1985): T-lymphocyte-mediated cytolysis as an excitatory process of the target. I. Evidence that the target cell may be the site of CA^{2+} action. *Cell Immunol* 95: 113–123

Trenn G, Takayama H, Sitkovsky MV (1987): Exocytosis of cytolytic granules may not be required for target cell lysis by cytotoxic T lymphocytes. *Nature* 330: 72–74

Tschopp JP, Schfer S, Masson D, Peitsch MC, Heusser C (1989): Phosphorylcholine acts as a Ca^{2+}-dependent receptor molecule for lymphocyte perforin. *Nature* 337: 272–274

Williams JM, Deloria D, Hansen JA, Dinarello CA, Loertscher R, Shapiro HM, Strom TB (1985): The events of primary T cell activation can be staged by use of Sepharose-bound anti-T3 (64:1) monoclonal antibody and purified interleukin 1. *J Immunol* 135: 2249–2255

Wyllie AH, Kerr JFR, Currie AR (1980): Cell death: The significance of apoptosis. *Int Rev Cytol* 68: 251–306

Yannelli JR, Sullivan JA, Mandell GL, Engelhard VH (1986): Reorientation and fusion of cytotoxic T lymphocyte granules after interaction with target cells as determined by high resolution cinemicrography. *J Immunol* 136: 377–382

Young JD, Cohn ZA, Podack ER (1986a): The ninth component of complement and the pore-forming protein (perforin 1) from cytotoxic T cells: Structural immunological, and functional similarities. *Science* 233: 184–190

Young JD, Hengartner H, Podack ER, Cohn ZA (1986b): Purification and characterization of a cytolytic pore-forming protein from granules of cloned lymphocytes with natural killer activity. *Cell* 44: 849–859

Yue C-C, Reynolds CW, Henkart PA (1987): Inhibition of cytolysin activity in granules for a membrane insertion mechanism of lysis. *Mol Immunol* 24: 647–653

Zagury D (1982): Direct analysis of individual killer T cells: Susceptibility of target cells to lysis and secretion of hydrolytic enzymes by CTL. *Adv Exp Med Biol* 146: 149–163

Zunino SJ, Bleackley RC, Martinez J, Hudig D (1990): RNKP-1, a novel natural killer associated serine protease gene cloned from RNK-16 cytotoxic lymphocytes. *J Immunol* 144: 2001–2009

15
Subpopulations of Cytotoxic T Lymphocytes with Different Cytotoxic Mechanisms

David W. Lancki, Maureen McKisic, and Frank W. Fitch

Introduction

Subpopulations of T lymphocytes have been identified based on (1) expression of CD4 or CD8 accessory molecules on the cell surface [CD4$^+$ cells generally recognize antigen (Ag) presented in the context of class II major histocompatibility complex (MHC) molecules, while CD8$^+$ cells generally recognize Ag presented in the context of class I MHC molecules], (2) the structure of their T-cell receptor (TCR) (α/β versus γ/δ), (3) the array of lymphokines that the cells produce, and (4) functional characteristics. T-cell functions in heterogeneous populations of T cells appeared to be correlated with the expression of CD4 or CD8 molecules on the T-cell surface. Thus, T-cell-mediated cytotoxicity was associated with CD8$^+$ T cells, while T-cell-mediated helper functions appear to reside primarily in the CD4$^+$ T cells. However, it is now generally accepted that cytotoxic activity is not restricted to CD8$^+$ T cells. The lytic activity of cells bearing the γ/δ TCR is well established. The extent of expression of cytotoxicity among the members of CD4$^+$ subsets is still controversial. In this review, we summarize recent observations regarding the expression of cytotoxic activity among T-cell subsets bearing CD4 molecules. In addition, we explore the growing body of evidence indicating that cytotoxic activity expressed by lymphocytes within different T-cell subpopulations may be accomplished through distinct and independent mechanisms.

Subpopulations of CD4$^+$ T lymphocytes have been defined based on the array of lymphokines produced. Thus, T-helper lymphocytes (T$_H$) cells type 1 (T$_H$1) produce interferon-γ (IFN-γ) and interleukin-2 (IL-2) but not IL-4, while T$_H$ type 2 (T$_H$2) produce IL-4 but not IFN-γ or IL-2 (Mosmann et al., 1986). Still other subsets of T cells exist based on other patterns of lymphokine production (Glasebrook and Fitch, 1980; Gajewski et al., 1989). Other differences appear to exist between the T$_H$ subsets, including distinct signaling pathways and costimulatory requirements (Gajewski et al., 1990, 1991). CD8$^+$ T cells also have distinct subsets, since they too can differ in the array of lymphokines produced. Some CD8$^+$ cells produce an array of lymphokines similar to those produced by T$_H$1 (Swain and Panfili, 1979; Fong and Mosmann, 1990); these appear to represent the helper-independent CD8$^+$ T-cell populations previously described (Klarnet et al., 1989). Collectively, these observations indicate that a complex array of functionally distinct subpopulations of T cells exists.

This chapter reviews evidence indicating that cytotoxic activity is expressed in representatives of each of the subpopulations of T cells described above. Differences in the lytic properties of these effector cells suggest that distinct and dissociable lytic mechanisms are involved in the destruction of target cells.

Caveats for the Measurement of T-Cell-Mediated Injury to Target Cells

Cytotoxicity has frequently been assessed by measuring the release of radiolabeled molecules from target cells. These techniques provide accurate and sensitive means of measuring the extent of at least one parameter of target cell injury, but they provide little information about how this injury occurred.

For example, the cytotoxicity mediated by T cells or by complement leads to the rapid release of labeled ^{51}Cr from the target cells, but conventional T-cell-mediated cytotoxicity also can involve rapid degradation of target cell DNA (presumably due to activation of DNA endonucleases in the target cell) (Duke et al., 1983; Russell, 1983). Nuclear damage has also been observed in very different cytotoxic systems, such as withdrawal from growth factors (Duke and Cohen, 1986) or activation of T-cell hybridomas (Mercep et al., 1989). In addition, it is not clear how the release of ^{51}Cr-labeled macromolecules from the target cell or nuclear degradation, as measures of target cell injury, relate to cytotoxic function in vivo.

Clonal populations of T cells have been useful in assessing the cytotoxic potential of cells within distinct T-cell subpopulations and for determining the expression and relative contribution of molecules thought to be involved in the lytic process. The expression of these characteristics in well-defined clonal populations of antigen-reactive T cells can then be correlated with the lytic activity against particular target cells. The extent to which cloned T cells are representative of functionally active cells in vivo is not known, but they provide a useful system for correlating the different functional properties of these cells.

Many tumor cell targets used in early studies of cytotoxicity express only class I molecules. Such target cells are suitable for assaying cytotoxic activity of CD8$^+$ T cells, which are generally restricted by class I MHC molecules. However, they are not suitable for assaying direct cytotoxicity of CD4$^+$ T cells, which are generally restricted by class II MHC antigens. These limitations may in part account for the impression that CD4$^+$ cells are not cytotoxic. Two significant developments have facilitated the study of cytotoxic activity among CD4$^+$ subsets of T cells. One of these was the derivation of target cell lines that express class II MHC molecules and are also capable of processing and presenting Ag efficiently to the appropriate effector cells (Kim et al., 1979; Glimcher et al., 1983).

Another important advance came from the observation that cytotoxicity can be activated by anti-TCR/CD3 monoclonal antibody (mAb) in a "redirected lytic assay" (Chapter 9, this volume; Lancki et al., 1984; Spits et al., 1985; Staerz et al., 1985; Leo et al., 1986). The rationale for using the target cells coated with anti-CD3 mAb to redirect lysis has been described previously (Lancki et al., 1991a). Briefly, it is based on the observation that the CD3-ϵ component of the TCR appears to be nonpolymorphic. The effects of possible differences in affinity of the TCR may be avoided, since anti-CD3-ϵ mAb should bind all T cells with the same affinity. In addition, anti-CD3-ϵ mAb provides an efficient stimulus for activation of lymphokine production and mRNA expression, therefore cytotoxicity can be readily compared with the other functional responses. This system provides a sensitive means of assessing cytotoxic function of T cells against a range of target cells that do not bear the relevant antigen or MHC molecules on their surface. Although redirected lysis of nucleated target cells has properties similar to those of Ag-specific lysis (Leo et al., 1986; Lancki et al., 1989), stimulation with anti-CD3-ϵ mAb may not be identical to stimulation via the TCR by Ag-presenting cells (APC) (Hengel et al., 1991). In addition, some target cells may not bear the ligands for T-cell surface molecules such as CD2, CD4, CD8, or CD28 that may normally contribute to activation.

Cytotoxic Activity and Lytic Mechanisms in T-Cell Subpopulations

Cytotoxic Activity Among Members of the CD8$^+$ T-Cell Populations

Antigen-specific T cell-mediated cytotoxicity has been investigated using primarily CD8$^+$ populations of T cells as effector cells. The characteristics of T cell-mediated lysis by these "conventional" cytotoxic T cells have been reviewed extensively (Möller, 1988; Young, 1989). Briefly, this process requires direct cell contact between the effector cell and the target cell. Surface molecules, including CD8 and LFA-1, contribute to the Mg^{2+} cation-dependent formation of conjugates between the effector and target cells that leads to the functional activation of the T cells (Spits et al., 1986). Antigen specificity is provided by the TCR, while intracellular signaling leading to the triggering of the "lethal hit" is mediated via the TCR-associated CD3 complex. The delivery of the lethal hit is a rapid event, occurring within minutes. These characteristics distinguish conventional T cell-mediated lysis from other forms of lysis, such as that mediated by soluble lymphokines.

Two predominant models have emerged to explain the lytic activity observed with CD8$^+$ T cells against their targets: (1) a granule release or exocytosis-dependent model (Chapter 14, this volume) and (2) a granule/exocytosis-independent model.

In the present review, we focus on the evidence suggesting that one or more of these proposed mechanisms may be involved in the cytotoxic activity present in certain T-cell populations. In the following section we examine the characteristics of the cytotoxic activity of conventional CD8+ cells.

The Perforin/Exocytosis-Dependent Mechanism of Lysis in CD8+ T Cells

Many CD8+ CTL contain large cytoplasmic granules. Purified granules prepared from CD8+ T cells and NK cells have potent lytic activity against certain target cells (Millard et al., 1984; Podack and Konigsberg, 1984; Leo et al., 1987). The characteristics of these granules are discussed in greater detail in Chapter 14 and Section V. Pore-forming proteins (also referred to as perforin or cytolysin) (Masson and Tschopp, 1985; Podack et al., 1985; Young et al., 1986), serine esterases (also referred to as granzymes A through H) (Pasternack and Eisen, 1985; Bleackley et al., 1988; Jenne and Tschopp, 1988), and tumor necrosis factor (TNF)-like material (Liu et al., 1987) are among the molecules that have been identified in these granules. Perforin and serine esterase appear to be co-localized within the granules of CTL and other cytotoxic cells (Peters et al., 1991). Furthermore, CTL can be stimulated via the TCR to degranulate and exocytose the contents of these granules, usually measured as the release of N-α-benzyloxycarbonyl-L-lysine thiobenzyl (BLT)-esterase activity (Takayama et al., 1987). The BLT-esterase activity in T-cell clones appears to be predominantly due to the expression of granzyme A in these cells (Jenne and Tschopp, 1988). BLT-esterase activity is expressed at high levels in most, but not all, cytotoxic CD8+ T-cell populations (Brunet et al., 1987; Berke and Rosen, 1988). Lysis and exocytosis are triggered by target cells bearing antigen, and this activation can be mimicked by anti-TCR/CD3-ϵ mAb, or by stimulation with the combination of phorbol esters (such as PMA) plus a calcium ionophore, pharmacologic reagents that activate T cells but bypass TCR stimulation (Lancki et al., 1989). Neither CD8+ T cell-mediated lysis nor exocytosis of BLT-esterase activity is inhibited by cycloheximide (Takayama and Sitkovsky, 1987; Lancki et al., 1989). These and other observations are the basis for the perforin/exocytosis model for lysis by T cells in which release of perforin and other molecules that are "prepackaged" in cytoplasmic granules is accomplished by the TCR-mediated triggering of exocytosis (Henkart and Henkart, 1982; Takayama and Sitkovsky, 1987). Upon exocytosis, the granule contents somehow cause target cell injury. This model for the induction of membrane damage by T cells is discussed in greater detail in Chapter 14.

Because of the difficulty in measuring perforin or granzyme B proteins directly, several investigators have examined T cells for their expression of perforin and granzyme B mRNA. Although perforin or other granule-associated molecules are thought to be synthesized and stored for use, the cells that utilize these molecules must have a mechanism of replacing them upon depletion. A likely time for the expression of mRNA for these molecules would be after the cells are activated through the TCR. Northern analysis of perforin mRNA expression indicates that CD8+ T cells do indeed express readily detectable levels of perforin mRNA under these conditions (Mueller et al., 1988; Liu et al., 1990; Lancki et al., 1991a). Unstimulated cells express low or undetectable perforin mRNA levels. CTLA-1 (granzyme B) mRNA expression is also selectively induced by stimulation of T cells via the TCR (Liu et al., 1990). These findings suggest that perforin and CTLA-1 mRNA expression are linked in many T-cell clones. However, under certain conditions the expression of CTLA-1 mRNA can be dissociated from the expression of perforin mRNA. Thus, cloned T cells stimulated with high concentrations of rIL-2 (1000 U/ml for 18 hrs) expressed levels of CTLA-1 (granzyme B) mRNA comparable to those induced by anti-CD3-ϵ mAb, but perforin mRNA levels remained at background levels (Lancki et al., 1991a).

Target cells differ widely in their susceptibility to cell-mediated cytotoxicity, but the basis for differences is generally not well understood. However, certain target cells have defined properties that can be useful in distinguishing effector mechanisms. The role of the perforin/exocytosis process in T cell-mediated lysis has been investigated using conventional nucleated target cells and anucleated erythrocyte target cells. The erythrocyte targets provide a sensitive system for measuring the destructive activity of effector molecules, such as complement or perforin, only on cell membranes. Mammalian erythrocytes do not have nuclei, and therefore the potential contribution of nuclear degradation to cytotoxicity is avoided (Millard et al., 1984; Podack and Konigsberg, 1984). T cell-mediated lysis of erythrocyte targets occurs by a process that presumably involves membrane injury. There appears to be a good correlation between exocytosis of BLT-esterase activity and the capacity of T-cell clones to lyse erythrocyte target cells.

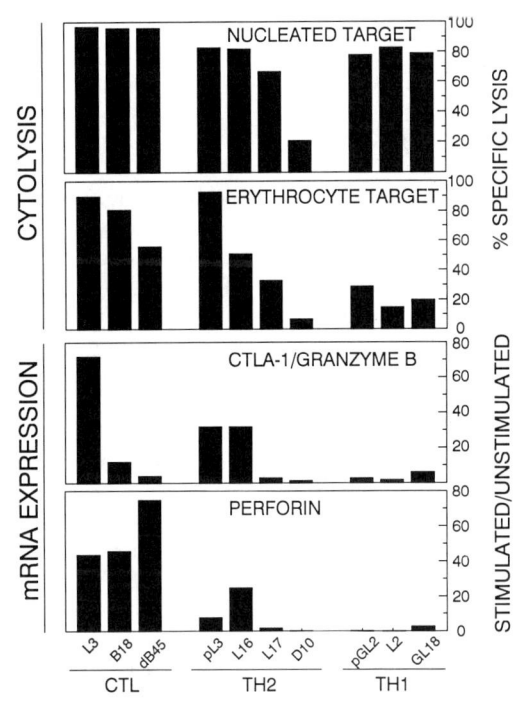

FIGURE 1. T-cell clones appear to mediate lysis of target cells coated with anti-CD3 mAb by perforin-dependent and perforin-independent mechanisms. CD8$^+$ T-cell clones having properties of CTL (L3, B18, and dB45), or CD4$^+$ clones having properties of either T$_H$2 (L16, PL3, and D10), or T$_H$1 (L2, pGL2, and GL18) were tested for their capacity to lyse TNP-derivatized, ^{51}Cr-labeled erythrocytes coated with anti-CD3/anti-TNP heteroconjugated antibody at a saturating concentration as previously described (Lancki et al., 1991a). The ^{51}Cr-labeled FcR-bearing P-815 mastocytoma target cells were incubated with effector cells either as a control in culture medium alone (not shown) or in the presence of hybridoma supernatants containing anti-CD3 mAb. The effector and target cells were combined at an effector-to-target cell ratio of 25:1 and incubated for 6 hr. The percent specific lysis was determined as described above. Cloned T cells (1 × 10^7) having properties of CD8$^+$ and CD4$^+$ T$_H$1 and T$_H$2 subsets were examined for their capacity to express perforin or granzyme B mRNA as described previously (Lancki et al., 1991a). Unstimulated cloned cells expressed low or undetectable levels of mRNA; cells were stimulated with anti-CD3 mAb immobilized on microspheres (at a ratio of 5 beads per responder cell) for 18 hr. Total cell mRNA was harvested and subjected to Northern analysis using specific probes for perforin as described previously (Lancki et al., 1991a). Band peak areas were assessed using an LKB UltroScan XL densitomiter.

Thus, cloned CD8$^+$ T cells that exocytose BLT-esterase activity are capable of lysing efficiently the sheep erythrocyte targets coated with anti-CD3-ϵ mAb (Lanzavecchia and Staerz, 1987; Lancki et al., 1989). Inhibitors of exocytosis [ethyleneglycolbis(β-aminoethylether)-N, N, N', N'-tetraacetate (EGTA) or cyclosporin A (CsA)] also inhibit significantly the capacity of the T-cell clones to lyse erythrocyte targets (Lancki et al., 1989; Lancki et al., 1991b; Ratner and Clark, 1991). All of the CD8$^+$ T-cell clones that efficiently lysed erythrocytes coated with anti-CD3-ϵ mAb also expressed readily detectable levels of BLT esterase, perforin, and granzyme B mRNA (see Figure 1 and Lancki et al., 1991a, 1991b). Although the perforin exocytosis model appears to account for membrane damage, the relative contribution of the several molecules implicated in this process cannot be readily discerned in typical CD8$^+$ cells. Further evidence for the involvement of cytolysin/perforin in at least some forms of lysis has now been provided by studies in which the cytolysin/perforin gene has been transfected into cells previously lacking expression. Thus, a noncytotoxic mast cell tumor line, transfected with the cytolysin/perforin gene, acquired the capacity to lyse erythrocytes coated with IgE (Shiver and Henkart, 1991). Also, cytotoxicity has been inhibited in cells treated with perforin antisense oligonucleotides (Acha-Orbea et al., 1990).

Perforin/Exocytosis-Independent Mechanism(s) of Lysis in CD8+ T Cells

Several observations indicate that the perforin exocytosis model alone does not account for all of the observed characteristics of cytotoxicity. CD8+ T-cell-mediated cytotoxicity is characterized by release of labeled macromolecules from the target cell, presumably resulting from loss of membrane integrity, but there is also degradation of target cell DNA (Russell, 1983). In contrast, complement-mediated cytotoxicity causes membrane damage, but DNA degradation is not observed (Russell, 1983). Purified perforin does not appear to be capable of inducing efficient DNA degradation (Gromkowski et al., 1988; Duke et al., 1989), though conflicting results have been reported (Hameed et al., 1989). Another granule-associated molecule, a novel cytotoxin, has been reported to cause DNA degradation (Liu et al., 1987). Other findings suggest the existence of perforin/exocytosis-independent mechanisms of lysis.

Certain highly lytic populations of T cells appear to lack appreciable expression of perforin or cytotoxic granules (Dennert et al., 1987; Berke and Rosen, 1988). In addition, the exocytosis process can be inhibited significantly using EGTA (Ostergaard et al., 1987; Trenn et al., 1987) or CsA (Lancki et al., 1989) without appreciably affecting the lytic activity of some T cells against certain target cells. These results suggest strongly that a mechanism of lysis independent of perforin/exocytosis is likely to be involved in the lysis of some target cells. The role of soluble or membrane-bound mediators of perforin/exocytosis-independent mechanisms of cytotoxicity is discussed in greater detail in the sections dealing with CD4+ T-cell mechanisms of lysis below, and in Chapters 16 and 17.

Cytotoxic Activity Among Members of the CD4+ T-cell Subpopulations

The cytotoxic potential of some CD4+ cells has been demonstrated with clonal populations of T cells that specifically lyse Ag-bearing APC (Tite and Janeway, 1984; Spits et al., 1985; Tite et al., 1985; Lancki et al., 1991a). Cytotoxicity among CD4+ cells was first shown with T-cell clones reactive with nominal antigen, but lysis is not uniquely a feature of such cells, since several alloreactive CD4+ clones have also been reported (Lancki et al., 1991a; McKisic et al., 1991). Concern about the effects of long-term culture on the expression of cytotoxic activity in T cell clones has stimulated efforts to assess cytotoxic function of freshly derived populations of CD4+ T cells. These studies have indicated that both nominal and alloreactive cytotoxic CD4+ T cells are detectable early in culture (Lancki et al., 1991a, 1991b).

Limiting dilution analysis allows T cells to be assessed early in culture for (1) a specific pattern of lymphokine production and (2) the expression of cytotoxicity. Using this approach, a high proportion of ovalbumin (OVA)-reactive murine T-lymphocyte clones derived from different strains of mice were found to have the ability to lyse target cells bearing antigen as well as target cells coated with anti-CD3 mAb (Lancki et al., 1991a). The array of lymphokines produced by freshly derived clones was compared with their capacity to lyse target cells coated with anti-CD3 mAb in a short-term (5-h) ^{51}Cr-release cytotoxicity assay. The results, summarized in Figure 2, show that the majority of OVA-reactive clones from five different strains of mice lysed A20 B lymphoma hybridoma tumor target cells coated with anti-CD3 mAb with varying degrees of efficiency (ranging from 10 to 100% lysis). Approximately the same proportion of T_H1 cells from each of the five strains were lytic. In most strains, a high proportion of the T_H2 cells also lysed the A20 B lymphoma target cells in redirected lytic assays.

Surprisingly, none of the T_H2 clones derived from BALB/c mice lysed the A20 target cells. This characteristic does not appear to be determined by genes within the MHC, since the ability of T_H2 clones derived from other H-2^d haplotype strains, including DBA/2 and B10.D2 mice, were able to lyse target cells. Most of the BALB/c OVA-specific clones that produce IFN-γ *were* able to lyse A20 target cells coated with anti-CD3 mAb. The results, summarized in Figure 3 and Table 1 indicate that BALB/c OVA-specific clones that failed to produce IFN-γ, regardless of the other lymphokines secreted, were not lytic (see also Lancki et al., 1991b). Thus, the cytotoxic activity of OVA-reactive clones derived from BALB/c mice did *not* correlate with the capacity of those clones to produce IL-2 or IL-4. This notion is supported by earlier studies using T_H1 and T_H2 clones derived from BALB/c mice (Erb et al., 1990). The correlation between IFN-γ production and the expression of cytotoxic activity is most evident in the BALB/c strain, but there may be a similar tendency among several of the other strains examined (Figure 3). These findings are consistent with an earlier observation that nonlytic clones derived from AKR mice failed to produce lymphotoxin (LT) or IFN-γ, while lytic clones derived from BALB/c or B10.S

PROPORTION OF OVA-REACTIVE CLONES LYTIC IN AN ANTI-CD3 MAb REDIRECTED LYSIS ASSAY

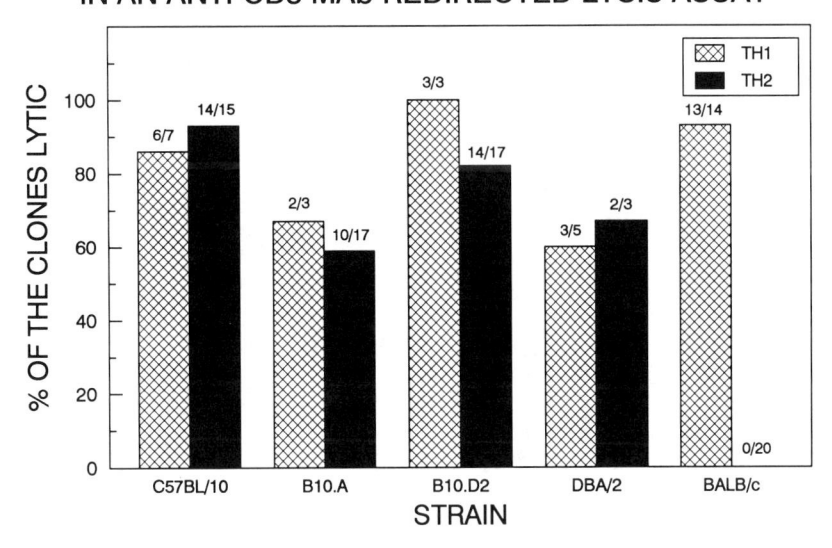

FIGURE 2. Murine T_H1 and T_H2 clones derived from different strains are cytotoxic but strain differences exist. Murine T-cell clones derived by limiting dilution and having properties of either T_H1 cells (producing IL-2 and IFN-γ) or T_H2 cells (producing IL-4 and not IL-2 or IFN-γ) were combined in V-bottom microtiter plates (Linbro) with the ^{51}Cr-labeled FcR-bearing A20.2J tumor target cells (2×10^3) in the presence of hybridoma supernatants containing anti-CD3 mAb (1/8 dilution) in a total volume of 200 μl. Cultures were centrifuged at 600 $\times$ g for 5 min and allowed to incubate for 5 hr. The percent specific lysis was calculated as:

$$\% \text{ Specific } ^{51}\text{Cr release} = \frac{\text{cpm (exp)} - \text{cpm (spont)}}{\text{cpm (maximum)} - \text{cpm (spont)}} \times 100$$

mice produced LT and IFN-γ (Tite et al., 1985). The factors controlling lymphokine gene expression and cytotoxic activity are not known, but these results suggest these factors may be associated in some strains.

If multiple mechanisms of cytotoxicity exist, it is possible that not all of these mechanisms are expressed in each of the cytotoxic cells. Thus cytotoxic CD4$^+$ clones representative of distinct T-cell subpopulations may share the perforin/exocytosis-dependent and/or -independent cytotoxic mechanisms that appear to be present in CD8$^+$ cytotoxic T cells. In the following section we examine evidence for differences in the effector mechanisms used by CD8$^+$ and CD4$^+$ T-cell subsets indicated by the differential capacity of effector cells to lyse certain target cells.

Evidence for Perforin/Exocytosis-Dependent Lysis in CD4$^+$ Cytotoxic T Cells

We have previously used anucleated erythrocytes to demonstrate mechanism(s) of lysis that involve membrane, but not nuclear, damage. T-cell clones from different subsets differ in their capacity to lyse anucleated erythrocytes, even though they have similar capacities to lyse nucleated target cells. While CD8$^+$ T-cell clones indeed were able to lyse erythrocytes coated with anti-CD3-ϵ mAb, a number of CD4$^+$ T$_H$1 clones were unable to lyse the erythrocyte targets (Lanzavecchia and Staerz, 1987; Lancki et al., 1989). Some of these CD4$^+$ clones *could* lyse nucleated target cells coated with anti-CD3-ϵ mAb or bearing Ag. Erythrocytes coated

THE RELATIONSHIP BETWEEN THE EXPRESSION OF LYTIC ACTIVITY AND IFN-g PRODUCTION IN OVA-REACTIVE CLONES FROM DIFFERENT STRAINS

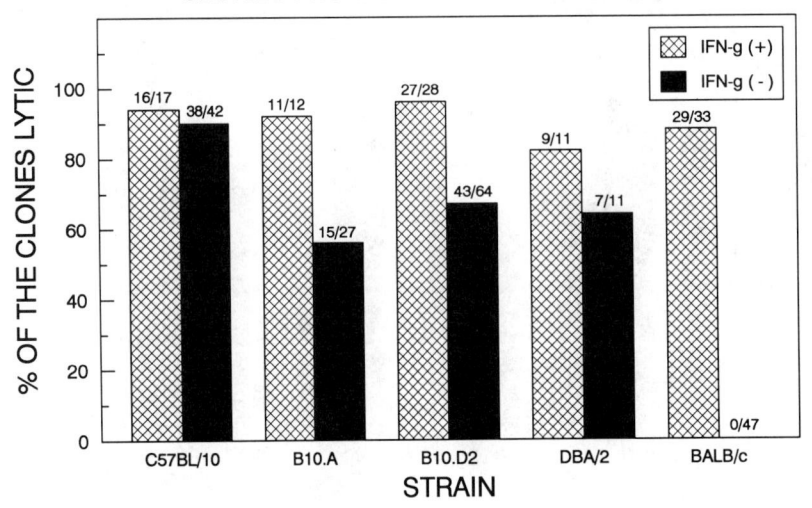

FIGURE 3. The strain differences in the expression of cytotoxic activity by murine OVA-reactive clones are associated with the production of IFN-γ. Murine T-cell clones derived by limiting dilution were screened for the array of lymphokines produced, including IL-2, IL-4, and IFN-γ, and were assessed for cytotoxic potential as described above. The percent specific lysis was calculated as described above.

with heteroconjugated anti-TNP/anti-CD3-ϵ antibody (Lancki et al., 1991a) were lysed efficiently by CD8$^+$ CTL clones (Figure 1) but not by CD4$^+$ T$_H$1 cells (also shown in Figure 1). Interestingly, some T$_H$2 cells lysed the erythrocyte targets as efficiently as the CD8$^+$ clones, while other T$_H$2 cells appeared to have no lytic activity. Several of the T-cell clones that did *not* lyse efficiently erythrocytes *could* lyse nucleated cells that were coated with anti-CD3 mAb. A few T$_H$2 clones lysed neither the erythrocytes nor the nucleated target cells coated with anti-CD3 mAb. Differences in the efficiency of lysis of nucleated target cells did not correlate with the capacity of these clones to lyse erythrocytes (Lancki et al., 1991a, 1991b). Collectively, the results indicate that CD8$^+$ and some T$_H$2 clones utilize a mechanism of lysis involving membrane damage that does not appear to be utilized by the T$_H$1 clones.

The expression of cytotoxic granules and/or granule-associated molecules in populations of T cells other than CD8$^+$ cells appears to depend on the CD4$^+$ subset. Granzyme A (BLT-esterase activity) or CTLA-3 transcripts have been observed in CD4$^+$ T cells (Brunet et al., 1987; Garcia-Sanz et al., 1987; Lancki et al., 1991a). BLT-esterase activity was found predominantly, though not exclusively, in the T$_H$2 subset of CD4$^+$ cells (Lancki et al., 1991a), but levels of expression are variable. BLT-esterase expression could not be detected in some T$_H$2 clones, while expression in others occurred at levels comparable to the highest levels

TABLE 1. Most OVA-specific clones derived from BALB/c mice that produce IFN-γ lyse target cells coated with anti-CD3 mAb

Number of lytic clones/total number of clones[a] producing the indicated lymphokines

IL-4	IL-2	IL-2 + IL-4	IL-4 + IFN-γ	IL-2 + IFN-γ	IL-2 + IL-4 + IFN-γ
0/22	0/3	0/22	2/3	13/14	14/16

[a] Murine T-cell clones producing the indicated lymphokines were combined in V-bottom microtiter plates (Linbro) with the ^{51}Cr-labeled FcR-bearing A20.2J tumor target cells (2 × 10^3) in the presence of hybridoma supernatants containing anti-CD3 mAb (1/8 dilution) in a total volume of 200 μl. Cultures were centrifuged at 600 × g for 5 min and allowed to incubate for 5 hr. The percent specific lysis was calculated as described in Figure 1.

observed among the CD8+ cells. BLT-esterase activity expressed by members of the T_H1 subset tended to be much lower; modest levels of activity were observed in some of these clones. Other T_H1 clones did not express detectable levels of BLT-esterase activity, although they were able to lyse nucleated target cells (Lancki et al., 1991a). Thus, the expression of granzyme A does not correlate with the capacity of cloned T cells to lyse conventional target cells.

Although perforin and granzyme B gene products are expressed by CD8+ T cells (Brunet et al., 1987; Bleackley et al., 1988; Podack et al., 1988), it appears that perforin and granzyme B mRNA expression is also inducible by stimulation via the TCR in some, but not all, T_H2 cells. The T_H2 clones that were induced to express perforin mRNA did so at somewhat lower levels than that observed with the CTL clones. Expression of perforin mRNA among the T_H2 clones correlated with the capacity of those clones to lyse efficiently erythrocytes coated with anti-CD3 mAb. T cells of the T_H1 subset did not produce appreciable perforin or granzyme B mRNA, and none of these clones efficiently lysed erythrocytes coated with anti-CD3 mAb. The expression of granzyme B generally correlates with perforin mRNA expression among the clones tested from the different CD4+ T-cell subpopulations (Figure 2) (Lancki et al., 1991a).

The role of exocytosis has also been investigated in the cytotoxic function of CD4+ cells. Cytotoxic granules have not been generally observed in CD4+ cells, but it appears that exocytosis may occur in at least some CD4+ cells. At least some CD4+ cells can express BLT-esterase activity (Garcia-Sanz et al., 1987; Ju et al., 1990; Lancki et al., 1991b). Helper T cells that express appreciable levels of serine esterase activity also release this activity when stimulated via the TCR (Lancki et al., 1991b). Thus, CD4+ clones of the T_H2 subset exocytose BLT-esterase activity in response to stimulation through the TCR. As with CD8+ clones, the release of this activity by CD4+ clones is inhibited by CsA (Schumacher et al., 1990; Lancki et al., 1991b). Furthermore, lysis of erythrocyte targets by these clones is also blocked by CsA (Lancki et al., 1989, 1991b). Collectively, the results suggest that exocytosis is not an exclusive property of CD8+ CTL but also occurs in T cells of the CD4+ subset. Interestingly, T_H1 clones expressing BLT-esterase activity, but not perforin or granzyme B mRNA, did not lyse erythrocyte target cells coated with anti-CD3 mAb efficiently, even though they also were able to exocytose BLT-esterase activity. Thus exocytosis, as measured by the release of BLT-esterase activity, may be required but is not sufficient for the lysis of erythrocyte target cells, which occurs presumably through membrane damage. The significance *in vivo* of the capacity to lyse erythrocyte target cells by a process that involves exocytosis is not clear, but it may be an important means of facilitating alternative mechanisms of lysis.

Evidence for Perforin/Exocytosis-Independent Lysis in CD4+ Cytotoxic T Cells

Several CD4+ T_H1 clones could lyse efficiently nucleated target cells bearing Ag or coated with anti-CD3-ε mAb, even though they did not express appreciable levels of BLT-esterase activity, or perforin and granzyme B mRNA. Furthermore, these cells did not lyse erythrocytes coated with anti-CD3-ε mAb. This suggests that the lysis of nucleated target cells by the T_H1 effector cells occurs by a mechanism other than the perforin/exocytosis-dependent mechanism. Lysis of the nucleated target cells by CD4+ cells, including T_H1 cells, also involved the rapid induction of target cell nuclear damage that is characteristic of CD8+ T-cell-mediated cytotoxicity (Strack et al., 1990; Abrams and Russell, 1991; Ju, 1991; Lancki et al., 1991b). Cytolytic T_H2 cells also appear to share this capacity with CTL and T_H1 cells (J. Russell, personal communication). While the mechanism by which T cells induce target cell nuclear damage is not known, these findings indicate that the capacity of T cells to induce nuclear damage can be independent of the perforin/exocytosis mechanism of lysis. Several molecules, including soluble TNF/LT, have been thought to be involved (Tite et al., 1985; Liu et al., 1987). As described previously, the role of these soluble mediators in the rapid lysis by effector cells from the T-cell subpopulations is controversial. These molecules appear to induce a pattern of nuclear damage similar to that induced by CTL and T_H1 clones. Most studies suggest that synthesis *de novo* of molecules with this activity is not required for lysis by CD8+ clones. This also appears to be true for at least some CD4+ cells, since inhibitors of protein (cycloheximide) or RNA (actinomycin D) synthesis, or of lymphokine production (CsA) did not inhibit lysis by some CD4+ T-cell clones (Russell et al., 1988; Abrams and Russell, 1991; Lancki et al., 1991b). The involvement of soluble TNF and perhaps other proteins in some forms of lysis has been suggested by studies indicating that lysis by some CD4+ cells requires synthesis of proteins *de novo* (Ju et al., 1990). In that system, TNF-independent as well as TNF-dependent forms of lysis were blocked by inhibitors of protein synthesis (Ju et al., 1990). The properties of soluble cytotoxic molecules produced by T cells will only be discussed briefly here.

Many cytotoxic T cells have the capacity to produce TNF/LT activity when stimulated and can lyse target cells that are susceptible to these molecules (Schmid et al., 1986). At least some lytic activity involving CD4$^+$ effector cells can be blocked by anti-TNF antibodies (Ju et al., 1990). The kinetics of soluble TNF/LT-mediated lysis are relatively slow, usually requiring in excess of 12 hr to produce these effects. For this reason, some investigators consider these molecules unlikely candidates for rapid T cell-mediated lysis. Moreover, target cells that are not sensitive to soluble TNF/LT can be readily lysed by cytotoxic cells from each of the T-cell subpopulations. These observations are consistent with the findings showing that *de novo* synthesis of TNF/LT activity alone does not appear to account for the cytotoxicity of CD8$^+$ cells and some CD4$^+$ cells.

Membrane-bound molecules having properties of cell surface TNF, including a novel cytotoxin, have been reported to be present in the granules of CD8$^+$ cytotoxic T cells or on the surface of cells (Liu et al., 1987; Bakouche et al., 1988). It is possible that these or similar molecules may also play a role in nuclear damage observed in CD4$^+$ T-cell cytolysis of target cells (Liu et al., 1987), but the expression of such molecules among the subsets of CD4$^+$ cytotoxic cells has not been studied.

Other Effector Molecules or Functions Associated with T Cell-Mediated Cytotoxicity

Extracellular ATP has been proposed as an effector molecule which is secreted by cytotoxic T cells (Chapters 28 and 29; Filippini et al., 1990). However, CD8$^+$ or CD4$^+$ subsets have not been examined with respect to this lytic mechanism. The relationship of this activity to the expression of distinct cytotoxic patterns described in this review needs to be determined.

T cell-induced loss of adhesion in target cells has been described as another distinct effector function of CD4$^+$ as well as CD8$^+$ T cells (Russell et al., 1988; Abrams and Russell, 1991). This interesting effector function appears to be a prerequisite for the efficient lysis of some adherent cells by CD4$^+$ cells (Abrams and Russell, 1991). It is not clear if this requirement exists for all CD4$^+$ subsets or for CD4$^+$ cytotoxic cells that express the perforin/exocytosis-dependent mechanism of lysis.

Directions for Future Studies

The functional significance of the different patterns of lysis observed *in vitro* among distinct subpopulations of T cells may only be determined once we have a more complete understanding of effector mechanisms. Apparent differences in lytic potential may be due to alternative signaling pathways activated through the TCR or to selective expression of particular effector molecules; ultimately both will determine the functional responses of T cells. Signaling differences among the T-cell subsets, including CD8$^+$ cells and CD4$^+$ cells of the T_H1 and T_H2 subset, may contribute to differences in the activation of certain functional responses.

Although nuclear damage appears to be induced in the target cell as an event associated with lysis by distinct T-cell subsets, it is not clear if this is an obligate part of perforin/exocytosis-independent cytotoxic mechanisms. The capacity to efficiently inflict membrane lesions in target cells may enhance the capacity of other molecules released by the effector cells that express lytic function to lyse certain target cells. Certain effector cell and target cell combinations appear to be predisposed by the expression of class I versus class II MHC molecules by distinct APC/target cells. Distinct accessory molecules and their ligands appear to be important in the activation of the different T-cell subsets. These too may play a role in determining the efficiency of functional interactions between a particular effector cell and the antigen-bearing target cell.

Summary

Cytotoxic activity *in vitro* and probably *in vivo* is not absolutely restricted to a particular subpopulation of T cells. This is true regardless of whether subsets are defined on the basis of TCR configuration, expression of CD4 or CD8 surface molecules, the pattern of lymphokines produced, or other functional responses. Strain differences do exist in the expression of cytotoxic activity (as measured by redirected lysis) in cells lacking the capacity to produce IFN-γ regardless of whether they produce IL-2 or IL-4. However, the molecular basis for these differences has not been determined. Cytotoxicity can be mediated by at least two distinct mechanisms. One of these mechanisms appears to depend on expression of perforin and exocytosis of granule contents. This mechanism involves lysis of target cells through direct cell membrane damage

TABLE 2. Cell-mediated cytotoxic mechanisms currently identified among α/β T-cell subsets

Phenotype		Cytotoxic activity target cells		Proposed mechanism: perforin/exocytosis	
		Nucleated	Erythrocytes	Dependent PFP/CTLA-1	Independent
I.	CD8$^+$	+	+	+	+
II.	CD4$^+$ T$_H$2	+	+	+	+
III.	CD4$^+$ T$_H$2	+	−	−	+
IV.	CD4$^+$ T$_H$1	+	−	−	+

but does not involve nuclear degradation. Other mechanisms involve DNA degradation and appear to be independent of perforin or exocytosis of granule contents. The expression of these mechanisms in various T-cell subsets is summarized in Table 2. CD8$^+$ cytotoxic cells express both of these mechanisms. Some CD4$^+$ T$_H$2 cytotoxic effector cells express both mechanisms, while other T$_H$2 cells and T$_H$1 cells have a perforin-independent mechanism of lysis. In the latter case, release of radiolabeled macromolecules and target cell nuclear damage can occur in the absence of effector cell release of perforin through exocytosis. Some CD4$^+$ cells, including T$_H$1, T$_H$2, and intermediate phenotypes (based on the pattern of lymphokine production), appear to lack both mechanisms.

References

Abrams SI, Russel JH (1991): CD4$^+$ T lymphocyte-induced target cell detachment. A model for T cell-mediated lytic and nonlytic inflammatory processes. *J Immunol* 146: 405–413

Acha-Orbea H, Scarpellino L, Hertig S, Dupuis M, Tschopp J (1990): Inhibition of lymphocyte mediated cytotoxicity by perforin antisense oligonucleotides. *EMBO J* 9: 3815–3819

Bakouche O, Ichinose Y, Hiecappel R, Fidler IS, Lachman LB (1988): Plasma membrane-associated tumor necrosis factor. A non-integral membrane protein possibly bound to its own receptor. *J Immunol* 140: 1142–1147

Berke G, Rosen D (1988): Highly lytic *in vivo* primed cytolytic T lymphocytes devoid of lytic granules and BLT-esterase activity acquire these constituents in the presence of T cell growth factors upon blast transformation in vitro. *J Immunol* 141: 1429–1436

Bleackley RC, Lobe CG, Duggan B, Ehrman N, Fregeau C, Meier M, Letellier M, Havele C, Shaw J, Paetkau V (1988): The isolation and characterization of a family of serine protease genes expressed in activated cytotoxic T lymphocytes. *Immunol Rev* 103: 5–19

Brunet JF, Denizot F, Suzan M, Haas W, Mencia-Huerta JM, Berke G, Luciani MF, Golstein P (1987): CTLA-1 and CTLA-3 serine esterase transcripts are detected mostly in cytotoxic T cells, but not only and not always. *J Immunol* 138: 4102–4105

Dennert G, Anderson CG, Prochazka G (1987): High activity of Nα-benzylocarbonyl-L-lysine thiobenzyl ester serine esterase and cytolytic perforin in cloned cell lines is not demonstrable in *in vivo*-induced cytotoxic effector cells. *Proc Natl Acad Sci USA* 84: 5004–5008

Duke RC, Cohen JJ (1986): IL-2 addiction: Withdrawal of growth factor activates a suicide program in dependent T cells. *Lymphokine Res* 5: 289–293

Duke RC, Chervenak R, Cohen JJ (1983): Endogenous endonuclease-induced DNA fragmentation: An early event in cell-mediated cytolysis. *Proc Natl Acad Sci USA* 80: 6361–6365

Duke RC, Persechini PM, Chang S, Liu C-C, Cohen JJ, Young JD-E (1989): Purified perforin induces target cell lysis but not DNA fragmentation. *J Exp Med* 170: 1451–1456

Erb P, Grogg D, Troxler M, Kennedy M, Fluri M (1990): CD4$^+$ T cell-mediated killing of MHC class II-positive antigen-presenting cells. I. Characterization of target cell recognition by *in vivo* or *in vitro* activated CD4$^+$ killer T cells. *J Immunol* 144: 790–795

Filippini A, Taffs RE, Sitkovsky MV (1990): Extracellular ATP in T-lymphocyte activation: Possible role in effector functions. *Proc Natl Acad Sci USA* 87: 8267–8271

Fong TA, Mosmann TR (1990): Alloreactive murine CD8$^+$ cell clones secrete the Th1 pattern of cytokines. *J Immunol* 144: 1744–1752

Gajewski TF, Joyce J, Fitch FW (1989): Anti-proliferative effect of IFN-γ in immune regulation. III. Differential selection of T$_H$1 and T$_H$2 murine helper T lymphocyte clones using recombinant IL-2 and recombinant IFN-γ. *J Immunol* 143: 15–22

Gajewski TF, Pinnas M, Wong T, Fitch FW (1991): Murine T$_H$1 and T$_H$2 clones proliferate optimally in response to distinct antigen presenting cell populations. *J Immunol* 146: 1750–1758

Gajewski TF, Schell SR, Fitch FW (1990): Evidence implicating utilization of different T cell receptor-associated signalling pathways by T_H1 and T_H2 clones. *J Immunol* 144: 4110–4120

Garcia-Sanz JA, Plaetinck G, Velotti F, Masson D, Tschopp J, MacDonald HR, Nabholtz M (1987): Perforin is present only in normal activated Lyt2 + T lymphocytes and not in L3T4 + cells, but serine protease granzyme A is made by both subsets. *EMBO J* 6: 933–936

Glasebrook AL, Fitch FW (1980): Alloreactive cloned T cell lines. I. Interactions between cloned amplified and cytolytic T cell lines. *J Exp Med* 151: 876–895

Glimcher LH, Sharrow SO, Paul WE (1983): Serologic and functional characterization of a panel of antigen-presenting cell lines expressing mutant I-A class II molecules. *J Exp Med* 158: 1573–1588

Gromkowski SH, Brown TC, Masson D, Tschopp J (1988): Lack of DNA degradation in target cells lysed by granules derived from cytolytic T lymphocytes. *J Immunol* 141: 774–778

Hameed A, Olsen KJ, Lee M-K, Lichtenheld MG, Podack ER (1989): Cytolysis by Ca-permeable transmembrane channels: Pore formation causes extensive DNA degradation and cell lysis. *J Exp Med* 767: 765–767

Hengel H, Wagner H, Heeg K (1991): Triggering of CD8+ cytotoxic T lymphocytes via CD3-ϵ differs from triggering via α/β T cell receptor. CD3-ϵ-induced cytotoxicity occurs in the absence of protein kinase C and does not result in exocytosis of serine esterases. *J Immunol* 147: 1115–1120

Henkart MP, Henkart PA (1982): Lymphocyte mediated cytolysis as a secretory phenomenon. *Adv Exp Med Biol* 146: 227–247

Jenne DE, Tschopp J (1988): Granzymes, a family of serine proteases released from granules of cytolytic T lymphocytes upon T cell receptor stimulation. *Immunol Rev* 103: 53

Ju S-T (1991): Distinct pathways of CD4 and CD8 cells induce rapid target DNA fragmentation. *J Immunol* 146: 812–818

Ju S-T, Ruddle NH, Strack P, Dorf ME, DeKruyff RH (1990): Expression of two distinct cytolytic mechanisms among murine CD4 subsets. *J Immunol* 144: 23–31

Kim KJ, Kanellopoulos-Langevin C, Merwin RM, Sachs DH, Asofsky R (1979): Establishment and characterization of BALB/c lymphona lines with B cell properties. *J Immunol* 122: 549–554

Klarnet JP, Kern DE, Dower SK, Matis LA, Cheever MA, Greenberg PD (1989): Helper-independent CD8+ cytotoxic T lymphocytes express IL-1 receptors and require IL-1 for secretion of IL-2. *J Immunol* 142: 2187–2191

Lancki DW, Hsieh C-S, Fitch FW (1991a): Mechanisms of lysis by cytotoxic T lymphocyte clones: Lytic activity and gene expression in cloned antigen-specific CD4+ and CD8+ T lymphocytes. *J Immunol* 146: 3242–3249

Lancki DW, Kaper BP, Fitch FW (1989): The requirements for triggering of lysis by cytolytic T lymphocyte clones. II. Cyclosporin A inhibits TCR-mediated exocytosis but only selectively inhibits TCR-mediated lytic activity by cloned CTL. *J Immunol* 142: 416–424

Lancki DW, Ma DI, Havran WL, Fitch FW (1984): Cell surface structures involved in T cell activation. *Immunol Rev* 81: 65–94

Lancki DW, McKisic M, Hsieh C-S, Fitch FW (1991b): Mechanism of lysis by murine cytolytic and helper T cell clones. In: *NK Cell Mediated Cytotoxicity: Receptors, Signalling and Mechanisms*, Lotzová E, Herberman RB, eds. Boca Raton, FL: CRC Press

Lanzavecchia A, Staerz UD (1987): Lysis of nonnucleated red blood cells by cytotoxic T lymphocytes. *Eur J Immunol* 17: 1073–1074

Leo O, Foo M, Henkart P, Perez P, Shinohara N, Segal D, Bluestone JA (1987): Activation of murine T lymphocytes with monoclonal antibody: Detection of Lyt-2 + cells of an antigen not associated with the T cell receptor but involved in T cell activation. *J Immunol* 139: 1214–1222

Leo O, Sachs DH, Samelson LE, Foo M, Quinones R, Gress R, Bluestone JA (1986): Identification of monoclonal antibodies specific for the T cell receptor complex by Fc receptor-mediated CTL lysis. *J Immunol* 137: 3874–3880

Liu C-C, Joag SV, Kwon B, Young JD (1990): Induction of perforin and serine esterases in a murine cytotoxic T lymphocyte clone. *J Immunol* 144; 1196–1201

Liu C-C, Steffan M, King F, Young JD-E (1987): Identification, isolation, and characterization of a novel cytotoxin in murine cytolytic lymphocytes. *Cell* 51: 393–403

Masson D, Tschopp J (1985): Isolation of a lytic, pore-forming protein (perforin) from cytolytic T-lymphocytes. *J Biol Chem* 260: 9069–9072

McKisic MD, Sant AJ, Fitch FW (1991): Some cloned murine CD4+ T cells recognize H-2L^d Class I MHC determinants directly; other cloned CD4+ T cells recognize H-2L^d Class I MHC determinants in the context of Class II MHC molecules. *J Immunol* 147: 2868–2874

Mercep M, Weissman AM, Frank SJ, Klausner RD, Ashwell JD (1989): Activation-driven programmed cell death and T cell receptor $\zeta\eta$ expression. *Science* 246: 1162–1165

Millard PJ, Henkart MP, Reynolds CW, Henkart PA (1984): Purification and properties of cytoplasmic granules from cytotoxic granules from cytotoxic rat LGL tumors. *J Immunol* 132: 3197–3204

Möller G, ed. (1988): Molecular mechanisms of T cell-mediated lysis. *Immunol Rev* 103: 1–211

Mosmann TR, Cherwinski H, Bond MW, Giedlin MA, Coffman RL (1986): Two types of murine helper T cell clone. I. Definition according to profiles of lymphokine activities and secreted proteins. *J Immunol* 136: 2348–2357

Mueller C, Gershenfeld HK, Weissman IL (1988): Activation of CTL-specific genes during cell-mediated cytolysis *in vivo*: Expression of the HF gene analyzed by *in situ* hybridization. *Immunol Rev* 103; 73–85

Ostergaard HL, Kane KP, Mescher MF, Clark WR

(1987): Cytotoxic T lymphocytes mediated lysis without release of serine esterase. *Nature* 330: 71–72

Pasternack MS, Eisen HN (1985): A novel serine esterase espressed by cytotoxic T lymphocytes. *Nature* 314: 743–745

Peters PJ, Borst J, Oorschot V, Fukuda M, Krähenbühl O, Tschopp, J, Slot JW, Geuze HJ (1991): Cytotoxic T lymphocyte granules are secretory lyosomes, containing both perforin and granzymes. *J Exp Med* 173: 1099–1109.

Podack ER, Konigsberg PJ (1984): Cytolytic T cell granules. Isolation, structural, biochemical, and functional characterization. *J Exp Med* 160: 695–710

Podack ER, Lowrey DM, Lichtenheld M, Olsen K, Aebischer T, Binder D, Rupp F, Hengartner H (1988): Structure, function, and expression of murine and human perforin 1 (P1). *Immunol Rev* 103: 203–211

Podack ER, Young JD-E, Cohn ZA (1985): Isolation and biochemical and functional characterization of perforin 1 from cytolytic T-cell granules. *Proc Natl Acad Sci USA* 82: 8629–8633

Ratner A, Clark WR (1991): Lack of target cell participation in cytotoxic T lymphocyte-mediated lysis. *J Immunol* 147: 55–59

Russell JH (1983): Internal disintegration model of cytolytic lymphocyte-induced target damage. *Immunol Rev* 72: 97–118

Russell JH, Musil L, McCulley DE (1988): Loss of adhesion. A novel and distinct effect of the cytotoxic T lymphocyte-target interaction. *J Immunol* 140: 427–432

Schmid DS, Tite JP, Ruddle NH (1986): DNA fragmentation: Manifestation of target cell destruction mediated by cytotoxic T-cell lines, lymphotoxin-secreting helper T-cell clones, and cell-free lymphotoxin-containing supernatant. *Proc Natl Acad Sci USA* 83: 1881–1885

Schumacher TNM, Heemels M-T, Neefjes JJ, Kast WM, Melief CJM, Ploegh HL (1990): Direct binding of peptide to empty MHC class I molecules on intact cells and *in vitro*. *Cell* 62: 563–567

Shiver JW, Henkart PA (1991): A noncytotoxic mast cell tumor line exhibits potent IgE-dependent cytotoxicity after transfection with the cytolysin/perforin gene. *Cell* 64: 1175–1181

Spits H, van Schooten W, Keizer H, van Seventer G, van de Rijn M, Terhorst C, De Vries JE (1986): Alloantigen recognition is preceded by nonspecific adhesion of cytotoxic T cells and target cells. *Science* 232: 403–405

Spits H, Yssel H, Leeuwenberg J, De Vries JE (1985): Antigen-specific cytotoxic T cell and antigen-specific proliferating T cell clones can be induced to cytolytic activity by monoclonal antibodies against T3. *Eur J Immunol* 15: 88–91

Staerz UD, Kanagawa O, Bevan MJ (1985): Hybrid antibodies can target sites for attack by T cells. *Nature* 314: 628–631

Strack P, Martin C, Saito S, DeKruyff RH, Ju ST (1990): Metabolic inhibitors distinguish cytolytic activity of CD4 and CD8 clones. *Eur J Immunol* 20: 179–184

Swain SL, Panfili PR (1979): Helper cells activated by allogeneic H-2K or H-2D differences have a Ly phenotype distinct from those responsive to I differences. *J Immunol* 122: 383–391

Takayama H, Sitkovsky MV (1987): Antigen receptor-regulated exocytosis in cytoxic T lymphocytes. *J Exp Med* 166: 725–743

Takayama H, Trenn G, Sitkovsky MV (1987): A novel cytotoxic T lymphocyte activation assay: Optimized conditions for antigen receptor triggered granule enzyme secretion. *J Immunol Methods* 104: 183–185

Tite JP, Janeway CA, Jr (1984): Cloned helper T cells can kill B lymphoma cells in the presence of specific antigen: Ia restriction and cognate vs. non cognate interactions in cytolysis. *Eur J Immunol* 14: 878–886

Tite JP, Powell MB, Ruddle NH (1985): Protein-antigen specific Ia-restricted cytolytic T cells: Analysis of frequency, target cell susceptibility, and mechanism of cytolysis. *J Immunol* 135: 25–33

Trenn G, Takayama H, Sitkovsky MV (1987): Exocytosis of cytolytic granules may not be required for target cell lysis by cytotoxic T-lymphocytes. *Nature* 330: 72–74

Young JD (1989): Killing of target cells by lymphocytes: A mechanistic view. *Physiol Rev* 69: 250–314

Young JD-E, Podack ER, Cohn ZA (1986): Properties of a purified pore-forming protein (perforin 1) isolated from H-2-restricted cytotoxic T cell granules. *J Exp Med* 164: 144–155.

16
Multiple Lytic Pathways in Cytotoxic T Lymphocytes

William R. Clark and Anna Ratner

Introduction

The study of the mechanism of cell-mediated cytotoxicity received renewed interest with the introduction of a rather elegant and simple model: exocytosis of toxin-containing granules from cytotoxic effector cells upon encounter with and activation by their specific target cells (Dennert and Podack, 1983; Henkart et al., 1984; Podack and Konigsberg, 1984; Young and Cohn, 1986). A pore-forming protein, perforin (also called cytolysin), was identified first in the cytoplasmic granules of natural killer (NK) cells, and subsequently in cytotoxic T lymphocytes (CTL). Binding of CTL and NK cells to their target cells is proposed to trigger a degranulation event, releasing perforin monomers which assemble in the presence of Ca^{2+} into complement-like pore structures that insert into the target membrane, thereby causing damage to the target cell. Initially it seemed that this model could adequately account for the known features of antigen-specific, CTL-mediated cytotoxicity. That perforin exists is beyond question, the gene having been cloned and sequenced by several laboratories. A role for perforin in cell-mediated cytolysis has been greatly strengthened by recent antisense (Acha-Orbea et al., 1990) and gene transfection (Shiver and Henkart, 1991) experiments.

However, a number of observations have also been reported, since this model was first proposed, that are inconsistent with perforin's being the exclusive lytic mechanism for CTL. In this review we summarize this evidence, examine recent data suggesting multiple lytic pathways in CTL, and speculate on what these pathways may be. A distinction will be made throughout this article between "classical" or "acute" lysis, and what we will refer to as "slow" lysis. Acute lysis is normally activated in CTL upon interaction of the T-cell receptor (TCR) complex with either antigen–major histocompatibility complex (MHC) complexes, or TCR complex antibodies, on the target cell surface. In acute CTL lysis, target cell DNA fragmentation and release of target cell-entrapped ^{51}Cr begins in minutes and is complete in 1–2 hr. "Slow lysis" is typical of lysis of target cells not recognized via interaction of the CTL–TCR complex with the target cell surface, and may take on the order of 18–36 hr to completion. We do not know how cytolysis is activated in such cases; it is clearly not via the TCR. The possible involvement of other T-cell activation pathways is largely unexplored, although recent experiments with mouse CD-2 antibodies suggest that that particular pathway may not be involved (Nakamma et al., 1990). Acute and slow pathways are found in both $CD4^+$ and $CD8^+$ CTL.

Experimental Findings Inconsistent with the Perforin Model

Dissociation of Exocytosis and Killing

Several groups have observed that granule exocytosis and acute target cell lysis are dissociable events. For example, acute CTL-mediated cytolysis has been demonstrated in the absence of Ca^{2+} in the assay medium (MacLennan et al., 1980; Tirosh and Berke, 1985; Ostergaard et al., 1987; Trenn et al., 1987; Ostergaard and Clark, 1989). Under this condition, no exocytosis of secretory granules should occur, and perforin, unable to assemble (Henkart et al., 1984), should not be lytic. In most of these experiments, serine esterase was used as a marker for degranulation. Recent experiments have shown that serine esterase and perforin are indeed colocalized in individual granules, both in T cells (Peters et al., 1991) and in lymphokine-activated killer (LAK) cells (Ojcius et al., 1991). Release of serine esterase should therefore be a valid marker for release of perforin. On the other hand, it has been found repeatedly that there is little or no quantitative correlation between serine esterase release and extent of cytotoxicity (Dennert et al., 1987; Berke, 1988; Shiver and Henkart, 1991).

As expected, there is no measurable release of serine esterase from CTL in the absence of extracellular Ca^{2+}, whether they are stimulated directly by the relevant antigen-bearing target cell, in redirected lysis via lectin or TCR antibody, or by the synergistic action of protein kinase C (PKC) activators and Ca^{2+} inophores (Ostergaard and Clark, 1989). We do not know at what level in the overall lytic sequence Ca^{2+} is acting. Although omission of Ca^{2+} would certainly block degranulation, it is not obvious that this is the sole, or even primary, locus of the Ca^{2+} effect. Ca^{2+} could very well be involved in any of a range of primary or auxiliary receptor–ligand interactions, or in some aspect of signal transduction. Since both the CTL and the target cell are exposed to ethylene-bis (β-aminoethylether)-N, N, N', N-tetraacetate (EGTA), the locus could as well be the target cell (Tirosh and Berke, 1985). Of course we cannot say on the basis of undetectable serine esterase release that no perforin molecules are released, but the absence of Ca^{2+} would also prevent assembly of perforin subunits into pore structures. It therefore seems extremely unlikely that perforin is mediating acute lysis in the absence of Ca^{2+}.

Similar results can be obtained using the drug cyclosporin A (CsA). Lancki et al. (1989) found that CsA is a potent inhibitor of degranulation at concentrations that have no effect on acute CTL-mediated lysis of nucleated target cells. (These same CsA concentrations, however, significantly decrease lysis of red blood cells. We will return to this point later.)

It could be argued that cytolysis seen in the absence of extracellular Ca^{2+} is due to leakage of intracellular Ca^{2+} from either the effector cell or the target cell. The existence of closed pockets formed by interdigitations between target cell and effector cell membranes has been documented (Kalina and Berke, 1976), and it has been suggested that the concentration of Ca^{2+} in these pockets from intracellular sources could rise to levels in excess of the exogenous EGTA. It seems unlikely (but would be difficult to prove) that intracellular Ca^{2+} could be released in sufficient amounts to saturate extracellular EGTA of up to 11mM (MacLennan et al., 1980). It is also possible that perforin molecules are still released in the presence of EGTA by a Ca^{2+}-independent constitutive secretion pathway. Such a pathway should be highly sensitive to inhibitors of protein synthesis (Kelly, 1985), but in fact there is no inhibition of lysis during Ca^{2+}-independent killing in the presence of sufficient emetine to completely shut down constitutive secretion (Ostergaard and Clark, 1989). And again, perforin released under such conditions should still be unable to assemble into lytic pore structures in the absence of external Ca^{2+}.

Interestingly, the requirement for Ca^{2+} in acute lysis was found to be dictated by the target cell used: the same antigen-specific CTL can kill one target cell in the absence of Ca^{2+}, yet absolutely require Ca^{2+} for lysis of another target (Tirosh and Berke, 1985; Ostergaard and Clark, 1989; see also Lancki et al., 1989). The implications of this for putative cytolytic mechanisms are unclear. The major difference seen between lysis in the presence and absence of Ca^{2+} is an approximately 45-min delay in both DNA fragmentation and ^{51}Cr release. However, once initiated, both of these processes proceed with kinetics that are indistinguishable in the two cases (Ostergaard and Clark, 1989). We cannot say from these data alone whether this represents a delay, in the absence of Ca^{2+}, in initiating the same lytic program that operates in the presence of Ca^{2+}, or whether lysis in the absence of Ca^{2+} is due to initiation of a completely different program.

The fact that substantial, even apparently normal, levels of target cell killing can be seen in the absence of degranulation in no way invalidates perforin as a cytolytic mechanism in acute lysis. But it does suggest that mechanisms other than the release of granule-associated perforin may exist.

Paucity of Perforin Expression in Primary CTL

One concern about perforin from the beginning is that is has been described largely in transformed cells or in cloned cells grown in the presence of exogenously supplied growth factors such as interleukin-2 (IL-2). The presence of perforin in primary CTL has been much more difficult to demonstrate. Dennert et al. found that in primary CTL generated in mixed lymphocyte culture (MLC), variable amounts of serine esterase activity could be found in the resultant CTL, but no detectable hemolytic activity (Dennert et al., 1987). Nagler-Anderson et al. (1988) reported low but detectable levels of hemolytic activity in MLC-generated CTL. Since it is possible that variable amounts of IL-2 and other growth/differentiation factors may be produced in MLC, depending on strain combination, cell concentration, and other factors, it is possible that the presence, absence, or level of such factors may influence perforin expression in a variable way.

Much of the confusion about the role of perforin in primary CTL has come from work with peritoneal exudate lymphocytes (PEL) produced by alloimmunization *in vivo*. These cells are an extremely potent source of antigen-specific CD8+ CTL, but they have no detectable cytoplasmic granules or hemolytic activity (Dennert et al., 1987; Berke, 1988; Nagler-Anderson et al., 1988) and only modest amounts of perforin mRNA (Nagler-Anderson et al., 1989). Yet after a few days of culture in the presence of IL-2, PELs accumulate cytoplasmic granules and readily detectable levels of perforin (Berke and Rosen, 1988). Such IL-2 induction of granular activity has also been seen in cloned CTL lines (Zanovello et al., 1989).

Results in our laboratory generally agree with the published results for PEL. We have never been able to detect serine esterase activity in PEL, or its release from PEL during target cell lysis, even when the colorimetric reaction for serine esterase detection is allowed to develop for up to 4 hr. (Serine esterase released by cloned CTL can usually be detected within 5 min under the same assay conditions). Yet PEL are as lytically potent as, and in some cases more lytically potent than, our cloned CTL lines. PEL thus represent an extreme example of the generally observed phenomenon that CTL lytic efficiency does not correlate well with serine esterase release. Because serine esterases and perforin have been shown to localize the same individual granules (Peters et al., 1991) these results strongly suggest that perforin is not a major lytic mechanism in PEL.

It has been suggested that the acute lysis exerted by PEL, which is as strong as that exhibited by cloned CTL, could be due to extremely small amounts of endogenous perforin, undetectable even in the sensitive red blood cell (RBC) hemolytic assay. This is certainly a formal possibility. However, in view of the fact that PEL are also able to lyse target cells in the absence of Ca^{2+} (Tirosh and Berke, 1985; see also Figure 1), where degranulation would be blocked and perforin could not assemble, it would seem more reasonable to

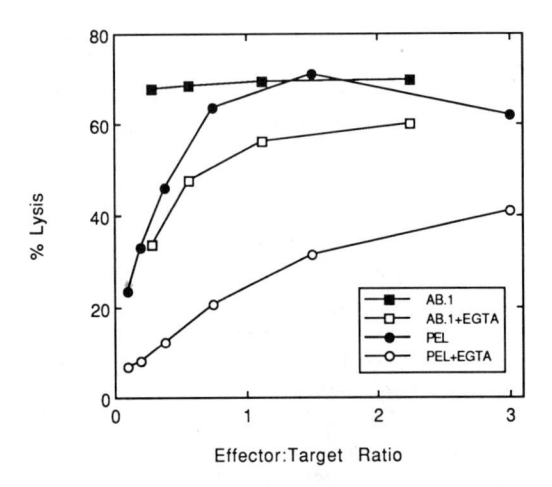

FIGURE 1. PEL can lyse target cells in the absence of Ca^{2+}. Lysis of EL-4 by AB.1 and by primary H-2^b-specific PEL is compared in the presence or absence of EGTA in the external medium.

conclude that PEL may be expressing an alternate lytic pathway.

Failure of Perforin to Cause DNA Fragmentation

One of the earliest events in target cells damaged by CTL is DNA fragmentation, accompanied by permeabilization of the nuclear membrane (Russell and Dobos, 1980). These events are also seen in systems involving apoptotic cell death, with the difference that after target cell damage by CTL they occur in minutes, rather than hours, as is typical in most apoptotic systems. As just noted, DNA fragmentation also occurs during lysis mediated by CTL in the absence of Ca^{2+}. It has been argued that if perforin mediates acute lysis of target cells, then it should also cause target cell DNA fragmentation.

The role of perforin in mediating DNA fragmentation has been investigated by several laboratories, with mixed results. Several investigations have failed to find DNA fragmentation in target cells treated with either crude granule preparations or purified perforin (Duke et al., 1988; Gromkowski et al., 1988; Duke et al., 1989). Other researchers did see fragmentation (Albritton et al., 1988; Hameed et al., 1989). The most definitive study (Duke et al., 1989) suggests that highly purified perforin, while able to lyse target cells, does not cause DNA fragmentation. These authors suggest that previous results showing DNA fragmentation with purified perforin could be explained by residual contamination of the perforin with other granule components which actually cause

DNA breakdown. This introduces the possibility that perforin itself may not induce DNA fragmentation, but rather facilitates entry into the target cell of other granule components that do. At any rate, we do not consider any of the results so far as a *a priori* inconsistent with a role for perforin in cytolysis. Target cell lysis and DNA fragmentation may well be dissociable events.

Evidence in Favour of Multiple Pathways

One of the major problems plaguing the analysis of lytic pathways in CTL has been the fact that different laboratories study different effector cells, different target cells, and different effector–target cell combinations. As a result, different laboratories often come up with seemingly contradictory results when studying what appears to be, at least superficially, the same kind of lysis. Nevertheless, a common thread that seems increasingly to run through many of these studies is that cytotoxic effector cells probably use more than one lytic pathway to kill target cells. We review here some of the recent evidence for multiple lytic pathways. It should be borne in mind when comparing these systems that cytolysis consists of at least three phases: (1) recognition/binding, (2) delivery of the lethal hit, and (3) dissolution of the target cell. When we say "lytic pathway" in the following sections, we refer to all of these components of lysis, not just lytic mechanisms *per se* used by effector cells.

Our own evidence for multiple pathways in antigen-specific CTL can be briefly summarized as follows. We identified a Ca^{2+}-dependent acute lytic pathway and a Ca^{2+}-independent acute lytic pathway in CTL. As noted above, the same CTL can kill by either pathway, depending on the target cell. It is extremely unlikely that perforin is the lytic mechanism in the absence of Ca^{2+}, and therefore to the extent that perforin *is* a lytic mechanism, the Ca^{2+} data are strong evidence of more than one acute lytic pathway.

Evidence for multiple pathways in acute CTL killing can also be obtained with other drugs. We found that Ca^{2+}-dependent lysis is inhibited by prolonged exposure to PMA, a treatment which has been proposed to deplete (PKC) activity (Russell and Coggeshall, 1987). Ca^{2+}-independent lysis, on the other hand, is completely refractory to PMA treatment. As noted earlier, Lancki et al., (1989) showed that exocytosis could be inhibited in $CD8^+$ CTL by the drug (CsA), with little effect on cytolysis. Trenn et al. (1989) dissected this effect further. They found that in systems where an antigen-specific CTL appears to be killing a target cell using both Ca^{2+}-dependent and a Ca^{2+}-independent pathway, the Ca^{2+}-dependent killing is in fact sensitive to CsA, whereas Ca^{2+}-independent killing is completely resistant to CsA. Thus two quite different reagents, PMA and CsA, can be used to distinguish Ca^{2+}-dependent and Ca^{2+}-independent killing. In a very recent paper, Hangel et al. (1991) related the requirement for PKC activation (and hence PMA sensitivity) to the mode of CTL activation. In the CTL–target cell systems they studied, CTL activation via the α/β portion of the TCR complex resulted in serine esterase release, and cytolysis was blocked by prolonged exposure to PMA. When CTL were activated by CD3 antibodies attached to the target cell surface, PKC quenching with PMA had no effect on cytolysis; moreover, cytolysis occurred in the absence of serine esterase release. It is not completely clear how these results relate to those just described (Lancki et al., 1989; Ostergaard and Clark, 1989; Trenn et al., 1989). The slight differences may well be due to the use of different effector and target cells, but overall the data from all of these studies are most consistent with distinctive lytic pathways within the same cell. Whether the end lytic *mechanism* is different, we cannot say, nor can we conclude that perforin is or is not involved.

Zanovello et al. (1989) reported evidence for multiple pathways in antigen-specific CTL clones driven to a promiscuous, LAK-like state by high concentrations of IL-2. Lysis of antigen-specific target cells by these CTL is accompanied by serine esterase release, whereas lysis of nonspecific targets by the same CTL proceeds in the absence of serine esterase release. Similarly, killing by these promiscuous CTL of specific target cells is accompanied by Ca^{2+} influx into the CTL, whereas no such influx is observed when these same CTL kill nonspecific target cells. We had previously reported a similar observation with cloned CTL made promiscuous by short-term activation with PMA (Ostergaard and Clark, 1989). These results strongly suggest the simultaneous expression of distinct lytic pathways in at least IL-2-driven, LAK-like CTL.

A quite different study suggesting multiple lytic pathways is that of Zychlinsky et al. (1991). These investigators carried out a careful histological analysis of target cells killed by CTL or LAK cells, and concluded that the target cells as a population showed ultrastructural changes associated with both apoptosis and necrosis. Previous investigators have provided evidence that apoptosis takes place in target populations acted on by CTL and LAK cells, but these studies rarely examined necrotic cell death, and certainly not comparatively in the same populations. Inasmuch as apoptosis and necrosis are assumed to arise by distinct mechanisms, the results of Zychlinsky et al. are most compatible with multiple lytic pathways in the effector cells.

Returning to the studies of Lancki et al. (1989) cited earlier, these authors found that levels of CsA that completely inhibited degranulation (serine esterase release) had little or no effect on direct lysis of nucleated target cells in a 5-hr assay. However, these same concentrations completely blocked CTL lysis of RBC derivatized with CD3 antibody in a 5-hr assay. This strongly suggests that distinct lytic pathways are involved in the acute lysis of nucleated vs. RBC target cells. In a follow-up study (Lancki et al., 1991), these authors showed that the ability of both $CD8^+$ and $CD4^+$ CTL to lyse sheep red blood cells (SRBC) target cells correlates strongly with expression of perforin mRNA. These results collectively raise the possibility that perforin may well be involved as a lytic pathway when RBC are the target cells. Our own experiments using RBC ghosts are consistent with this conclusion (Ratner and Clark, 1991). We found that any treatment that blocks degranulation of CTL always blocks lysis of anti-CD3-derivated RBC ghosts. On the other hand, the physiological relevance of a lytic mechanism selectively directed against RBC is unclear.

The possibility that lymphokines such as tumor necrosis factor-α (TNF-α), TNF-β, or interferon-γ (IFN-γ) might play a role in CTL-mediated cytolysis has been advocated for a number of years, but the relation of such a mechanism to acute CTL lysis (complete lysis of target cells recognized via the TCR in 2–4 hr) has been difficult to establish. Cytotoxic lymphokines clearly represent a potential pathway for slow lysis, although some have proposed them as the primary mediator of acute lysis as well. We will discuss this particular point further in the next section.

The study of TNF-α and TNF-β as lytic mediators received renewed interest with the serious study of $CD4^+$ CTL. In recent years we have come to appreciate that cell-mediated cytotoxicity by $CD4^+$ T cells is not an *in vitro* artifact, but an important physiological immune defense mechanism against viruses and perhaps other pathogens as well. Like $CD8^+$ CTL, $CD4^+$ CTL express both acute and slow lysis, depending on whether or not the TCR is engaged. Tite et al. (1985) presented strong (albeit indirect) evidence that at least some CD4 CTL kill in acute, antigen-specific assays by soluble mediators such as TNF-β and IFN-γ. They found that antigen-nonspecific bystander target cells were killed during lysis of antigen-specific targets; separate experiments showed that their $CD4^+$ CTL were capable of releasing TNF-β, although not necessarily within the time frame of their cytolytic assays. However, it was not shown that the bystander cells were in fact TNF-sensitive, nor was the ability of TNF antibodies to block the killing of innocent bystander cells tested.

In a follow-up study, Tite (1990) showed that TNF-α-resistant forms of normally TNF-α-sensitive targets were also resistant to acute killing by antigen-specific, class II-restricted, $CD4^+$ CTL. The TNF-resistant target cells retained their full sensitivity to acute lysis by class I-restricted $CD8^+$ CTL.

Ju et al. (1990) studied killing by both T-helper type 1 (T_H1) and T-helper type 2 (T_H2) CTL clones. They described what they believe to be TNF-α/β-dependent and -independent pathways in these cells. In short-term, antigen-specific or concanavalin A (Con A)-mediated cytolytic assays, killing was unaffected when TNF (and other lymphokine) production was blocked by CsA or cholera toxin (ChT), and was insensitive to TNF-α antibody. Yet killing was sensitive to blockers of RNA and protein synthesis, suggesting that the synthesis of some sort of protein mediator is required for lysis. (This contrasts with killing mediated by $CD8^+$ CTL in short-term assays, where it has been shown repeatedly that protein synthesis is not required for lysis; (see, e.g., Lancki et al., 1989; Ostergaard and Clark, 1989; Strack et al., 1990). Slow killing by $CD4^+$ CTL against non-specific targets was also blocked by actinomycin D and cycloheximide, and could be blocked by CsA and TNF antibody as well. Thus it seems likely that in $CD4^+$ CTL, acute and slow target cell killing are mediated by different pathways.

Blockage of killing by inhibition of RNA and protein synthesis in $CD4^+$ CTL in acute lytic assays would argue against release of a preformed mediator such as perforin. Consistent with this, Ju et al. (1990) found no correlation of cytolytic activity with serine esterase release, and blockers of serine esterase release (CsA and ChT) had no effect on cytolysis. Thus it seems unlikely that the acute, TNF-independent pathway in $CD4^+$ CTL involves perforin.

Similar results were recently reported by Abrams and Russell (1991). They found that antibodies to TNF-α and IFN-γ had no effect on direct antigen-specific lysis by primary $CD4^+$ CTL, but could at least partially inhibit the lysis of TNF-sensitive bystander cells in the same assay. The simultaneous expression of TNF-dependent and -independent pathways by the same effector cells further strengthens the notion of multiple lytic pathways in the same CTL.

The results of Ju et al. (1990) and Abrams and Russell (1991) are quite at odds with those of Tite and colleagues regarding acute lysis by CD4$^+$ CTL (Tite et al., 1985; Tite, 1990), and at present it is difficult to reconcile their respective findings. This may be a classic case of the use of different target cells confounding what otherwise seem like identical experiments.

By comparing the effect of metabolic inhibitors on acute cytolytic activity in CD4$^+$ and CD8$^+$ clones, Strack et al. (1990) concluded that distinct lytic pathways must be used by these cells. The CD4$^+$ T$_H$1 clones used by these authors appear not to use a perforin-dependent mechanism of lysis; no perforin mRNA was detected in these clones (see also Lancki et al., 1991). Acute lysis by CD4$^+$ but not CD8$^+$ CTL clones was sensitive to RNA or protein synthesis inhibition, while lysis mediated by CD8$^+$ but not CD4$^+$ clones was sensitive to ChT. Importantly, these authors found that whereas pretreatment of CD8$^+$ CTL with actinomycin D and cycloheximide had no effect on cytolysis, this pretreatment did halt production of lymphokines. This argues against lymphokines secreted via the constitutive pathway as mediators of rapid lysis by CD8$^+$ CTL.

The possibility that CTL are not themselves lytic agents, but rather induce programmed cell death in target cells, has been raised by numerous authors. If true, this would represent yet another cytolytic pathway. This topic is reviewed elsewhere in this volume and will not be discussed here. However, it is worth noting that our studies with CTL-mediated lysis of RBC ghosts (Ratner and Clark, 1991), which could not possibly execute an endogenous cell death program, make it clear that CTL do possess one or more entirely self-contained lytic pathways.

Candidate Mechanisms for Alternate Pathways in Acute Lysis

Regardless of whether or not perforin turns out to be a major player in the cytotoxicity arena, it seems unlikely that it will be the only player. In a sense, we are back where we were before the perforin model was put forward. Who are the other players?

Before setting off on a quest to identify unknown molecules and pathways mediating acute CTL lysis, it would seem appropriate to obtain definitive evidence for the involvement or lack thereof of known effector cell products that have already been implicated in cytotoxicity. For CTL, these products would include TNF-α, TNF-β, and, to a lesser extent, IFN-γ. Detailed reviews have been written on each of these lymphokines, and we will not describe them extensively here. However, a few brief comments about their possible role in acute CTL-mediated lysis will be useful for focussing our discussion.

IFN-γ. The evidence that IFN-γ may be directly cytotoxic is not impressive, and hence it has not generally been considered a candidate molecule for CTL-mediated target cell lysis. Among the well-recognized roles of IFN-γ are modulation of the levels of surface MHC products, facilitation of inflammation, and inhibition of viral replication (for review see De Maeyer-Guignard and De Maeyer, 1985). Virtually all of the studies carried out with IFN-γ are based on systemically administered IFN-γ. A major difference between IFN-γ administered systemically and IFNγ delivered by a CTL bound to a specifically recognized target cell would again be effective local concentration. When IFN-γ secretion is activated by contact of CTL with target cells, the concentration of cytokine accruing in the intracellular pockets formed by membrane interdigitation is likely to be much higher than that which can act on the same cell when exogenous (even recombinant) IFN-γ is supplied *in vitro*. When a cloned line of human tumor-specific CTL was augmented with retrovirally introduced IFN-γ genes, the *in vitro* cytotoxicity expressed by this line was greatly enhanced (Miyatake et al., 1990). It is unclear whether IFN-γ was acting as a lytic agent or modifying target cell susceptibility to other lytic mechanisms. Abrams and Russell (1991) found a slight reduction in bystander (slow) cytolysis in the presence of IFN-γ antibodies. It is entirely possible that an important role for IFN-γ in CTL-mediated cytotoxicity may have been overlooked in previous studies.

TNF-β, or lymphotoxin (LT), was one of the earliest proposed mechanisms for CTL-mediated cytolysis. Soluble, purified TNF-β has a variety of effects on susceptible target cells, being only cytostatic for some, but strongly cytolytic for others (Paul and Ruddle, 1988). In those cases where TNF-β is cytolytic, target cell death involves apoptosis, at least as indicated by DNA fragmentation (Schmid et al., 1987). It is produced by both CD8$^+$ and CD4$^+$ T cells (usually the T$_H$1 subset), either as

TABLE 1. Presence of the specific target cell EL-4 causes enhanced lysis of WEHI-164 by AB.1 cells

CTL	^{51}Cr-labeled target cell	Third-party cell	Assay length (hr)	%^{51}Cr release at E/T: 1	0.5	0.25	0.12
AB.1	EL4	–	4		48.9	37.8	22.2
	WEHI-164	–	18	14.8	11.2	6.7	3.6
	EL4	WEHI-164	4		48.3	38.5	19.9
	WEHI-164	EL4	18	60.9	51.3	44.8	36.5

AB.1 cells were incubated with target cells at the indicated E/T ratios. Third-party cells, when present, were added in equal numbers to the target cells.

a result of specific antigen presentation, crosslinking of the TCR/CD3 complex, or mitogen stimulation. It is also produced by both CTL types upon coincubation with certain TNF-sensitive target cells. *TNF-α* (identical with cachectin) was long thought to be a macrophage cytokine. However, there is now clear evidence that it is a T-cell lymphokine as well (Sung et al., 1988; Rawle et al., 1989; Kinkabhwala et al., 1990). In particular, TNF-α has been directly implicated in cytolysis mediated by both CD4$^+$ and CD8$^+$ CTL (Bakouche et al., 1988; Liu et al., 1989; Penezetal, 1990; Strack et al., 1990) and by LAK cells (Chong et al., 1989).

One major reservation about both TNF-α and TNF-β as lytic agents in acute CTL killing has been the kinetics of lysis. When TNF-sensitive, antigen-nonspecific target cells are exposed to either form of TNF, or to effector cells secreting TNF, maximal lytic effects are observed only at times on the order of 24 (even up to 72) hr. Killing in these slow lytic reactions can almost always be completely blocked with TNF antibodies, and few would argue that lymphokines are the exclusive lytic agent in these situations. Acute CTL-mediated killing, on the other hand, can be completed in 2–4 hr, and is never blocked by TNF antibodies. Inasmuch as the same individual CTL can often mediate both acute and slow killing reactions (Table 1), it is clear that CTL do possess multiple lytic pathways. But the problem currently occupying many cellular immunologists is whether there is more than one lytic pathway mediating acute CTL lysis. For example, does acute killing seen in the presence of inhibitors of degranulation (e.g., EGTA, CsA) involve the same pathway or different pathways as lysis seen in the absence of these inhibitors? The real question is therefore whether lymphokines such as TNF-α or -β, despite apparent differences in kinetics of lysis, could mediate acute killing as well, *either* as the primary pathway *or* as an alternate acute pathway. Could triggering of CTL via the TCR complex, perhaps aided by accessory molecule interactions, cause TNF-α/β to induce a more rapid form of lysis than is seen when these lymphokines are released via other pathways, and picked up only by target cells displaying TNF-α/β receptors? In antigen-specific (acute) lysis, CTL and target cells form numerous, tightly bounded intercellular "pockets" as a result of extensive membrane interdigitations (Kalina and Berke, 1976). Lymphokines released by CTL into such pockets will be accompanied by a range of other CTL products, such as perforin or serine esterases, that may well modulate TNF-α or -β activity, either directly or by affecting the target cell surface. These molecules could, for example, enhance target cell uptake of coreleased cytotoxic lymphokines. Ruddle's laboratory showed that artificially facilitated entry of TNF-β into target cells changes the kinetics of target cell lysis from the typically slow lymphokine-mediated lysis to a rate of killing very similar to acute CTL killing (Schmid et al., 1985).

The fact that many target cells lysed readily by CTL in acute CTL assays cannot be lysed (or can be lysed only poorly) by soluble lymphokines would seem to argue against the involvement of soluble lymphokines as a lytic mechanism in such assays. But it may not be appropriate to compare the action of purified (even recombinant) TNF-α or -β, introduced into assay medium, with the same lymphokines released by CTL during target cell lysis. If lymphokines are released vectorially into the intercellular "pockets" formed by CTL–target cell membrane interdigitations, then the concentration of TNF-α or -β within these pockets would almost certainly far exceed (financially) practical levels of purified lymphokines that could be added to an assay mixture. The work by Strack et al. (1990) cited earlier (see also Ju et al., 1986) using biosynthetic inhibitors makes it unlikely that lymphokines released by the constitutive secretion pathway are involved in classical CTL killing. However, the use of already synthesized, stored forms of TNF, or of membrane-bound forms of TNF, has not been convincingly eliminated. Liu et al. (1987) reported the presence of a TNF-like molecule stored in CTL granules. Release of such material would not be blocked by inhibitors of macromolecular synthesis. Membrane-bound forms of TNF-α, originally found in macrophages (Bakouche et al., 1988) and of TNF-β (Browning et al., 1991) have now been described in T cells as well. These would also be relatively unaffected by biosynthetic inhibitors, especially in short-term assays. The membrane form of CTL-associated TNF-α may be the 26-kDa precursor of secreted (17-kDa) TNF-α (Bakouche et al., 1988; Penez et al., 1990) or a somewhat larger (polymeric?) form (Liu et al., 1989). The parameters regulating the involvement of membrane-associated TNF-α or -β in CTL-mediated cytolysis have not yet been fully explored, but it is entirely possible that the kinetics of killing via membrane-associated forms of these lymphokines may be different from that of the soluble forms. Whether all antibodies prepared against secreted TNF-α or -β cross-react with the membrane forms also remains to be clarified.

Another objection to the involvement of TNF-α or -β in acute CTL lysis is that antibodies to these molecules do not block acute CTL killing (but see Tite, 1990). Again, if these lymphokines are released vectorically into membrane-enclosed spaces, or if membrane-associated forms are used, exogenously added antibodies (other than those enclosed within intercellular spaces during CTL–target cell conjugation) may not have access to the entrapped lymphokines. This argument is often used to explain the failure of perforin antibodies to block CTL killing (Podack et al., 1988; Shiver and Henkart, 1991). Whether such pockets form when TNF-producing effector cells, including CTL, interact with antigen-nonspecific target cells has never been established.

We recently began a series of experiments to examine the role of TNF-α in acute and slow lysis mediated by the CD8+ CTL clone AB.1. As shown in Table 1, AB.1 can kill its specific target, EL-4, in an acute reaction at the same time it is engaged in lysing the TNF-α-sensitive target, WEHI-164, in a slow lytic reaction. When unlabeled EL-4 cells are added to a mixture of AB.1 and ^{51}Cr-labeled WEHI-164, lysis of the WEHI cells is increased severalfold. As shown in Figure 2, AB.1 cells produce large amounts of TNF-α activity when stimulated by EL-4 cells. However, as shown in Table 2, TNF-α antibodies have very little effect on lysis of WEHI-164 cells by AB.1, either in the presence or absence of EL-4 cells. These same concentrations of TNF-α antibodies strongly block the lysis of WEHI-164 by L10A2.J cells, a "natural cytotoxicity" effector cell purported to lyse TNF-sensitive target cells via a membrane-bound form of TNF-α (Vanderslice and Collins, 1991); they also block the soluble TNF-α activity released by AB.1 in response to EL-4 (Table 3). We tentatively conclude that radially released, soluble TNF-α plays at best a minor role in lysis of WEHI-164 by AB.1. We interpret the enhanced lysis of WEHI-164 by EL-4-stimulated AB.1 as most likely due to enhanced cell (rather than soluble lymphokine-) mediated cytotoxicity. [This could also explain the results of Tite et al. (1985), who also observed "bystander lysis" of target cells during acute lysis of antigen-specific targets.] Whether this form of cell-mediated cytotoxicity is mediated by vectorially released, or membrane-associated, forms of TNF-α or other lymphokines is the subject of current investigations. It is interesting that the killing of WEHI-164 by L1OA2.J is readily blocked by TNF-α antibodies (Vanderslice and Collins, 1991; Table 2), while killing of WEHI-164 by AB.1 cells is not blocked. If L10A2.J do indeed kill targets using a membrane-associated form of TNF-α, then presumably some sort of conjugate must form between effector and target cells. Clearly the intercellular plane of this conjugate is accessible to TNF antibody.

At this stage, the bulk of published evidence does not favor the involvement of lymphokines such as TNF-α or -β in acute lysis by either CD4+ or CD8+ CTL. However much of this evidence is directed to *de novo* synthesized, soluble forms of TNF. In our opinion, the involvement of stored forms of TNF, membrane-associated TNF, and perhaps isoforms (oligomers?) of TNF with altered immunological properties has not been vigorously ruled out. Again, before beginning a search for completely new mechanisms to explain degranulation-independent acute lysis, or acute lysis by cells with no apparent granule/perforin capacity, we believe a definitive statement about all possible modes for involvement of known cytotoxic CTL lymphokines should be obtained. We are pursuing this line of investigation at the cellular level with the system just described.

Future Directions

A great deal of evidence now exists suggesting that perforin is unlikely to be the exclusive pathway used by CTL in mediating acute ("classical") cytolysis. However, the powerful attractiveness of the perforin model has dampened enthusiasm for

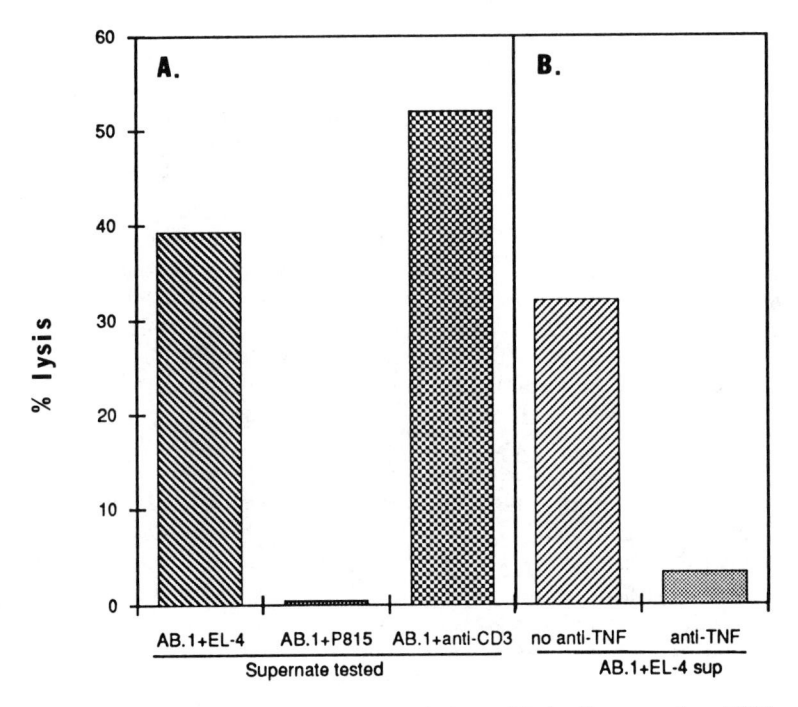

FIGURE 2. Stimulation via the T-cell receptor complex induces AB.1 cells to produce TNF-α. (A) AB.1 cells were stimulated for 5 hr either with an equal number of EL-4 cells, an equal number of P815 (H-2^d) cells, or in wells containing immobilized CD3 antibodies (2C11). Undiluted supernates from these reactions were then incubated with ^{51}Cr-labeled WEHI-164 cells for 18 hr. (B) AB.1 cells were incubated for 18 hr with an equal number of EL-4 cells. The supernate was harvested, diluted 1:12, and incubated for 18 hr with ^{51}Cr-labeled WEHI-164 in the presence or absence of a polyclonal TNF-α antiserum.

exploring alternative lytic pathways and mechanisms. Several laboratories are now engaged in generating cloned CTL lines, and inbred mouse strains, completely lacking the perforin gene, using the techniques of gene disruption by homologous recombination (Johnson et al., 1989). Very likely, by the time this volume appears these materials will be ready for analysis. They will allow immediate clarification of which lytic pathways do or do not require perforin. Similarly, cloned CTL lines lacking each of the lymphokine genes (TNF-α or -β; IFN-γ) will further define the role of these gene

TABLE 2. Enhanced killing of WEHI-164 is not due to soluble TNF-α

Effector cell	Third-party cell	Target cell	Anti-TNF-α	%^{51}Cr release at E/T: 20	10	5
L10A2.J	–	WEHI-164	–	15.4	13.9	17.3
L10A2.J	–	WEHI-164	+	1.3	4.4	3.8
AB.1	–	WEHI-164	–	35.8	23.7	15.7
AB.1	–	WEHI-164	+	28.9	18.8	9.5
AB.1	EL4	WEHI-164	–	74.5	80.1	79.6
AB.1	EL4	WEHI-164	+	76.5	70.7	56.6

The indicated effector cells were incubated with labeled WEHI-164 target cells for 18 hr at the indicated E/T ratios, in the presence or absence of a polyclonal TNF-α antibody. Unlabeled EL-4 cells, where present, were added in equal numbers to the target cells. Assays carried out in the absence of TNF-α antibody contained an equivalent concentration of control (normal rabbit) serum.

TABLE 3. Cytotoxicity present in the 18-hr supernates of AB.1 + EL-4 + WEHI-164 assays is fully inhibitable by TNF-α antibody

Supernate from culture of:	Target cell	% ^{51}Cr release at dilution of supernate: Undil.	1:2	1:4
AB.1 + EL-4 + WEHI + control serum	WEHI-164	79.5	63.9	40.2
AB.1 + EL-4 + WEHI + anti-TNF-α	WEHI-164	3.7	3.5	4.4

Supernates from 18-hr assays of AB.1 + EL-4 + WEHI-164, in the presence or absence of TNF-α antibody, exactly parallel to those in Table 2 except that WEHI-164 was not ^{51}Cr-labeled, were tested against labeled WEHI-164 in an 18-hr assay.

products in acute CTL killing, either alone or with perforin as a possible "facilitator" molecule. For example, if perforin in combination with one or more of the lymphokine gene products plays a direct lytic role in CTL-mediated cytolysis, then elimination of each such gene should lead to a partial, but not complete, reduction in lysis. Only elimination of both would reduce cytolysis to zero. But if perforin plays a nonlytic, facilitating role for entry of one of the other molecules, which is responsible for lysis *per se*, then elimination of *either* perforin *or* the lytic lymphokine should lead to complete loss of cytolysis, except when these targets display specific receptors for the lytic lymphokine. Studies of this type, combined with previous and ongoing studies at the cellular level, should greatly clarify our understanding of the process of cell-mediated cytotoxicity.

References

Abrams SL, Russell JH (1991): CD4+ T lymphocyte-induced target cell detachment: A model for T cell-mediated lytic and nonlytic inflammatory processes. *J. Immunol.* 146: 405

Acha-Orbea H, Scarpellino L, Hertzig S, Dupris S, Tschopp J (1990): Inhibition of lymphocyte-mediated cytotoxicity by perforin anti-sense oligonucleotides. *EMBO J* 9: 3815

Allbritton NL, Verret CR, Wolley RC, Eisen HN (1988): Calcium ion concentrations and DNA fragmentation in target cell destruction by murine cloned cytotoxic T lymphocytes. *J Exp Med* 167: 514

Bakouche O, Ichinose Y, Heicappell R, Fidler IJ, Lachman LB (1988): Plasma membrane-associated tumor necrosis factor: A non-integral membrane protein possibly bound to its own receptor. *J Immunol* 140: 1142

Berke G (1988): Lymphocyte-mediated cytolysis. *Ann NY Acad Sci* 532: 314

Berke G, Rosen D (1988): Highly lytic *in vivo* primed cytolytic T lymphocytes devoid of lytic granules and BLT-esterase activity acquire these constituents in the presence of T cell growth factors upon blast transformation *in vitro*. *J Immunol* 141: 1429

Browning J, Androlewicz M, Ware C (1991): Lymphotoxin and an associated 33 κDa glycoprotein are expressed on the surface of an activated human T cell hybridoma. *J Immunol* 147: 1230

Chong A, Scuderi P, Grimes W, Hersch E (1989): Tumor targets stimulated IL-2-activated killer cells to produce IFN-α and TNF. *J Immunol* 142: 2133

De Maeyer-Guignard J, De Maeyer E, Gresser I, eds. (1985): Immunomodulation by interferons: Recent developments. *Interferon* 6: 69

Dennert G, Podack ER (1983): Cytolysis by H-2-specific T killer cells: Assembly of tubular complexes on target membranes. *J Exp Med* 157: 1483

Dennert G, Anderson C, Prochazka G (1987): High activity of serine esterase and cytolytic perforin in cloned cell lines is not found in *in vivo* induced cytolytic effector cells. *Proc Natl Acad Sci USA* 84: 5004

Duke RC, Persechini PM, Chang S, Liu C-C, Cohen JJ, Young JD-E (1989): Purified perforin induces target cell lysis but not DNA fragmentation. *J Exp Med* 170: 1451

Duke RC, Sellins KS, Cohen JJ (1988): Cytotoxic lymphocyte-derived lytic granules do not induce DNA fragmentation in target cells. *J Immunol* 141: 2191

Gromkowski SH, Brown TC, Masson D, Tschopp J (1988): Lack of DNA degradation in target cells lysed by granules derived from cytolytic T lymphocytes. *J Immunol* 141: 774

Hameed A, Olsen KJ, Lee M-K, Lichtenheld MG, Podack ER (1989): Cytolysis by Ca-permeable transmembrane channels: Pore formation causes extensive DNA degradation and cell lysis. *J Exp Med* 169: 765

Hangel H, Wagner H, Heeg K (1991): Triggering of CD8+ cytotoxic T lymphocytes via CD3-ε fidders from triggering via α/β T cell receptor: CD3-ε-induced cytotoxicity occurs in the absence of protein kinase C and does not result in exocytosis of serine esterases. *J Immunol* 147: 1115

Henkart P, Millard PJ, Reynolds CW, Henkart MP (1984): Cytolytic activity of purified cytoplasmic granules from cytotoxic rat large granular lymphocyte tumors. *J Exp Med* 160: 75

Johnson RS, Sheng M, Greenberg ME, Kolonder RD, Papaioannou VE, Spiegelman BM (1989): Targeting of nonexpressed genes in embryonic stem cells via homologous recombination. *Science* 245: 1234

Ju S., DeKruyff R, Dort M (1986): Inducer T cell-mediated killing of antigen presenting cells. *Cell Immunol* 101: 613

Ju S-T, Ruddle NH, Strack P, Dorf ME, and DeKruyff RH (1990): Expression of two distinct cytolytic mechanisms among murine CD4 subsets. *J Immunol* 144: 23

Ju S-T, Strack P, Stromquist D, DeKruyff RH (1988): Cytolytic activity of a Ia-restricted T cell clones and hybridomas: Evidence for a cytolytic mechanism independent of interferon-γ, lymphotoxin, and tumor necrosis factor-α. *Cell Immunol* 117: 399

Kalina M, Berke G (1976): Contact regions of cytotoxic T lymphocyte target cell conjugation. *Cell Immunol* 25: 41

Kelly R (1985): Pathways of protein secretion in eukaryotes. *Science* 230: 25

Kinkabhwala M, Sehajpal P, Skolnik E, Smith D, Sharma VK, Vlassara H, Cerami A, Suthanthiran M (1990): A novel addition to the T cell repertoire: Cell surface expression of tumor necrosis factor/cachectin by activated normal human T cells. *J Exp Med* 171: 941

Lancki DW, Hsieh CS, Fitch FW (1991): Mechanisms of lysis by cytotoxic T lymphocyte clones: Lytic activity and gene expression in cloned antigen-specific CD4[+] and CD8[+] T lymphocytes. *J Immunol* 146: 3242

Lancki DW, Kaper BP, Fitch FW (1989): The requirements for triggering of lysis by cytolytic T lymphocyte clones. *J Immunol* 142: 416

Liu C, Detmers P, Jiang S, Young J (1989): Identification and characterization of a membrane-bound cytotoxin of murine CTLs that is related to TNF/cachectin. *Proc Natl Acad Sci USA* 86: 3286

Liu C, Steffen M, King F, Young J (1987): Identification, isolation and characterization of a novel cytotoxin in murine cytotoxic lymphocytes. *Cell* 51: 393

MacLennan ICM, Gotch FM, Golstein P (1980): Limited specific T-cell mediated cytolysis in the absence of extracellular Ca^{2+}. *Immunology* 39: 109

Miyatake SI, Nishihara K, Kikucki H, Yamashita J, Namba Y, Hanaoka M, Watanabe Y (1990): Efficient tumor suppression by glioma-specific murine cytotoxic T lymphocytes transfected with interferon-γ gene. *J Nat'l Cancer Inst* 82: 217

Nagler-Anderson C, Allbriton N, Verrett C, Eisen H (1988): A comparison of the cytolytic properties of murine primary CD8[+] CTLs and cloned CTL lines. *Immunol Rev* 103: 111

Nagler-Anderson C, Lichtenheld M, Eisen H, Podack E (1989): Perforin mRNA in primary peritoneal exudate cytotoxic T lymphocytes. *J Immunol* 143: 33440

Nakamura T, Takahashi K, Fukazawa T, Koyanagi M, Yokoyama A, Kato H, Yagita H, Okumura K (1990): Relative contribution of CD2 and LFA-1 to murine T and natural killer cell functions. *J Immunol* 145: 3628

Ojcius D, Zheng L, Sphicas E, Zychlinsky A, Young J (1991): Subcellular localization of perforin and serine esterase in LAK cells and CTLs by immunogold labeling. *J Immunol* 146: 4427

Ostergaard HL, Clark WR (1989): Evidence for multiple lytic pathways used by cytotoxic T lymphocytes. *J Immunol* 143: 2120

Ostergaard HL, Kane KP, Mescher MF, Clark WR (1987): Cytotoxic T lymphocyte mediated lysis without release of serine esterase. *Nature* 330: 71

Paul N, Ruddle N (1988): Lymphotoxin. *Annu Rev Immunol* 6: 407

Perez C., Albert I, DeFay K, Zachariades N, Gooding L, Kriegler M (1990): A nonsecretable cell surface mutant or tumor factor (TNF) kills by cell-to-cell contact. *Cell* 63: 251

Peters PJ, Brost J, Oorschot V, Fukuda M, Krahenbuhl O, Tschopp J, Slot JW, Geuze H (1991): Cytotoxic T lymphocyte granules are secretory lysosomes, containing both perforin and granzymes. *J Exp Med* 173: 1099

Podack ER, Konigsberg PJ (1984): Cytolytic T cell granules. Isolation, structural, biochemical, and functional characterization. *J Exp Med* 160: 695

Podack E, Lowrey D, Lichtenfeld M, Harmeed A (1988): Function of granule perforin and esterases in T cell mediated reactions. *Ann NY Acad Sci* 532: 292

Ratner A, Clark W (1991): Lack of target cell participation in CTL-mediated lysis. *J Immunol* 147: 55

Rawle FC, Tollefson AE, Wold WSM, Gooding LR (1989): Mouse anti-advenovirus cytotoxic T lymphocytes: Inhibition of lysis by E3 gp19K but not E3 14.7K. *J Immunol* 143: 2031

Russell J, Coggeshall K (1987): The role of protein kinase C in the cytotoxic T cell response. *Ann Inst Pasteur Immunol* 138: 320

Russell JR, Dobos CB (1980): Mechanisms of immune lysis. II. CTL-induced nuclear disintegration of the target begins within minutes of cell contact. *J Immunol* 125: 1256

Schmid DS, Hornung R, McGrath KM, Paul N, Ruddle N (1987): Target cell DNA fragmentation is mediated by lymphotoxin and tumor necrosis factor. *Lymphokine Res* 6: 195

Schmid DS, Powell MB, Mahoney KA, Ruddle NH (1985): A comparison of lysis mediated by Lyt 2[+] TNP-specific cytotoxic-T-Lymphocyte (CTL) lines with that mediated by rapidly internalized lymphotoxin-containing supernatant fluids: Evidence for a role of soluble mediators in CTL-mediated killing. *Cell Immunol* 93: 68

Shiver J, Henkart P (1991): A noncytotoxic mast cell tumor line exhibits potent IgE-dependent cytotoxicity after transfection with the cytolysin/perforin gene. *Cell* 64: 1175

Strack P, Martin C, Saito S, DeKruyff RH, Ju S-T (1990): Metabolic inhibitors distinguish cytolytic activity of CD4 and CD8 clones. *Eur J Immunol* 20: 179

Sung S-SJ, Bjorndahl JM, Wang CY, Kao HT, Fu SM (1988): Production of tumor necrosis factor/cachectin by human T cell lines and peripheral blood T lymphocytes stimulated by phorbol myrisate acetate and anti-CD3 antibody. *J Exp Med* 167: 937

Tirosh R, Berke G (1985): T lymphocyte mediated cytolysis as an excitatory process of the target. I. Evidence that the target cell may be the site of Ca^{2+} action. *Cell Immunol* 95: 113

Tite J (1990): Evidence for a role for TNF-α in cytolysis by $CD4^+$, class II MHC-restricted cytotoxic T cells. *Immunol* 71: 208

Tite JP, Powell MB, Ruddle NH (1985): Protein-antigen specific Ia-restricted cytolytic T cells: Analysis of frequency, target cell susceptibility, and mechanism of cytolysis. *J Immunol* 135: 25

Trenn GR, Taffs R, Hohman R, Kincaid R, Shevach EM, Sitkovsky M (1989): Biochemical characterization of the inhibitory effect of CsA on cytolytic T lymphocyte effector functions. *J Immunol* 142: 3796

Trenn G, Takayama H, Sitkovsky M (1987): Exocytosis of cytolytic granules may not be required for target cell lysis by cytotoxic T-lymphocytes. *Nature* 330: 72

Vanderslice W, Collins J (1991): Differences in TNF-α-mediated lysis by fixed NC cells and fixed cytotoxic macrophages. *J Immunol* 146: 156

Young JD-E, Cohn ZA (1986): Cell-mediated killing: A common mechanism? *Cell* 46: 641

Zanovello P, Rosato A, Bronte V, Cerundolo V, Treves S, Di Virgilio F, Pozzan T, Biasi G, Collavo D (1989): Interaction of lymphokine-activated killer cells with susceptible targets does not induce second messenger generation and cytolytic granule excoytosis. *J Exp Med* 170: 655

Zychlinsky A, Zheng LM, Liu CC, Young JD-E (1991): Cytolytic lymphocytes induce both apoptosis and necrosis in target cells. *J Immunol* 146: 393

17
Properties of Cytotoxicity Mediated by CD4$^+$, Perforin-Negative T-Lymphocyte Clones

Hajime Takayama

Introduction

There has been a long-lasting debate about the involvement of perforin in antigen-specific, directional killing of target cells mediated by cytotoxic T lymphocytes (CTL). Perforin, a pore-forming protein found in cytolytic granules of natural killer (NK) cells and CTL lines, must be secreted from cytosolic granules by exocytosis, according to the "exocytosis model" for lymphocyte-mediated cytotoxicity. Indeed, it is reported that noncytotoxic rat basophilic leukemia cell line (RBL) became cytolytic against sheep erythrocyte target (SRBC) after transfection with perforin gene (Shiver and Henkart, 1991).

However, there are several reports suggesting the existence of perforin-independent mechanisms in CTL-mediated cytotoxicity (Table 1). If such mechanisms are indeed involved in CTL-mediated cytotoxicity, one would expect cytotoxic activity in perforin-deficient CTL. Conversely, if perforin is the sole major cytolytic mechanism, no cytotoxicity would be expected in perforin-deficient T-cell populations.

Expecting the latter case, Acha-Orbea et al. (1990) have tried to abrogate the expression of perforin by means of antisense oligonucleotide treatment. In their report, however, the suppression of both perforin expression and cytolytic activity was partial. Thus, again, it is not clear whether the remaining cytotoxic activity was due to remaining perforin activity or to unidentified cytolytic mechanisms other than perforin. With this line of approach, therefore, the perforin gene itself must be disrupted. Recently developed gene targeting technology (Zijlstra et al., 1990) has made it feasible to construct perforin-deficient mice and perforin-deficient CTL clones.

On the other hand, it would also be reasonable to search the T-cell population of normal mice for the perforin-negative CTL. Recently, we and others have reported that subsets of the CD4$^+$ T-cell population were found to be cytolytic against antigen-presenting cells without expressing perforin (Strack et al., 1990; Lancki et al., 1991; Takayama et al., 1991). These cells are conventionally classified as helper T cells, and the cytotoxicity by these CD4$^+$ T cells is thought to be mediated by soluble factor(s), since it is usually associated with bystander cytolysis; thus it is different from that of CD8$^+$ CTL. Nevertheless, we found that cytolysis and DNA fragmentation of target cells (antigen-presenting cells) induced by these CD4$^+$ T-cell clones were highly antigen-specific and directional, and that bystander cytolysis was mediated by totally different mechanisms from directional cytotoxicity. Thus, the cytotoxic mechanism used by these T-cell clones was not different from that of CD8$^+$ CTL-mediated cytotoxicity. These observations not only indicate the existence of cytotoxic T-cell subsets that do not utilize perforin, but also suggest that perforin is not mandatory for specific target lysis by T cells.

Methods

Cells

CD4$^+$ T-cell clones specific for soluble protein, such as keyhole limpet hemocyanin (KLH) or ovalbumin (OVA), were prepared as described elsewhere (Shinohara et al., 1991). The CD8$^+$, class I MHC-specific CTL clone, OE4, was kindly provided by Dr. Osami Kanagawa; OE4 and the tumor cell lines, P815, EL4, BW5147, were maintained in culture as previously described

TABLE 1. Reports suggesting the existence of perforin-independent mechanisms in CTL-mediated cytotoxicity

Reported observation	Reference
Extracellular Ca^{2+}-independent, exocytosis-independent CTL-mediated cytotoxicity was detected.	Ostergaard et al., 1987 Trenn et al., 1987
CTL hybridoma which did not express detectable perforin activity was isolated.	Berke & Rosen, 1988
Nucleated target cells were less sensitive to perforin-mediated cytolysis than erythrocyte target.	Henkart et al., 1984 Shiver & Henkart, 1991
Antiperforin antibody failed to inhibit CTL-mediated cytolysis.	Reynolds et al., 1987 Shiver & Henkart, 1991
Several cytolytic $CD4^+$ T-cell clones were found to be perforin-negative.	Strack et al., 1990 Lancki et al., 1991 Takayama et al., 1991
Purified perforin failed to induce DNA fragmentation in target cell.	Duke et al., 1989

(Takayama et al., 1991). The Ia^+ B lymphoma A20HL, which is transfected with the structural gene of the IgM anti-trinitrophenyl (TNP) antibody and thus capable of taking up and presenting TNP–protein conjugates, were kindly provided by Dr. Novumichi Hozumi (Watanabe et al., 1987).

Assay of Cytotoxic Activity of T-Cell Clones

Cytolysis and DNA fragmentation were measured by ^{51}Cr release and detergent-soluble ^{125}I release from ^{51}Cr-labeled or [^{125}I] iododeoxyuridine ($^{125}IUdR$)-labeled target cells, respectively, as described in the previous report (Takayama et al., 1991).

Detection of Perforin Expression in T-Cell Clones

Perforin mRNA and protein were detected as described elsewhere (Takayama et al., 1991). Briefly, perforin mRNA was detected by Northern hybridization with a probe which contains a 731-bp fragment of mouse perforin gene (position −72 to 659). The perforin protein content of T-cell clones was quantitatively measured by sandwich enzyme-linked immunosorbent assay (ELISA) as described previously (Kawasaki et al., 1990).

Results

Antigen-Specific Cytotoxicity Mediated by Soluble Protein-Specific, Class II MHC-Restricted $CD4^+$ T-Cell Clones

Two CTL clones named BK1 and BK2 were isolated from a bulk cell line established from lymph node cells of KLH-immunized BALB/c mice. These KLH-specific clones were $CD4^+$, $V\beta 8^+$, and antigen recognition was restricted by $I-E^d$. The OVA-specific $CD4^+$ clone, BO1, which is $I-A^d$-restricted, was also isolated from OVA-immunized BALB/c mice by a similar procedure.

The cytolysis mediated by these $CD4^+$ T-cell clones was antigen-specific, since the antigen-presenting target cell, A20HL, presensitized with TNP–KLH, was lysed by BK1 and BK2, but not by BO1, whereas A20HL presensitized with TNP–OVA was lysed only by BO1 but not by BK1 or BK2 (Shinohara et al., 1991). These $CD4^+$ T-cell clones caused not only target cell lysis but also extensive and rapid DNA fragmentation (Figure 1). Thus, the cytotoxic effect by these $CD4^+$ clones against target cells was quite similar to that of conventional $CD8^+$ CTL.

Mechanisms of Bystander Cytolysis Mediated by $CD4^+$ T-Cell Clones

Since the cytolysis of antigen-presenting cells induced by $CD4^+$ T cells has been claimed to be due to nondirectional secretion of cytotoxic substances such as tumor necrosis factor (TNF or TNF-α) or lymphotoxin (LT or TNF-β), whereas cytotoxicity induced by $CD8^+$ CTL is highly directional, the cytotoxic mechanisms of these two types of T cells are though to be totally different (Tite and Janeway, 1984; Tite et al., 1985; Tite, 1990). However, a previous report showed data suggesting that both class I MHC-restricted CTL and class II MHC-restricted CTL cause cytolysis by an essentially similar direct mechanism (Maimone et al., 1986). In addition, several reports showed that involvement of TNF in the cytolysis of target cells was dependent on the susceptibility of target cells to the factor (Chang and Moorhead, 1986; Ju et al., 1988, 1990).

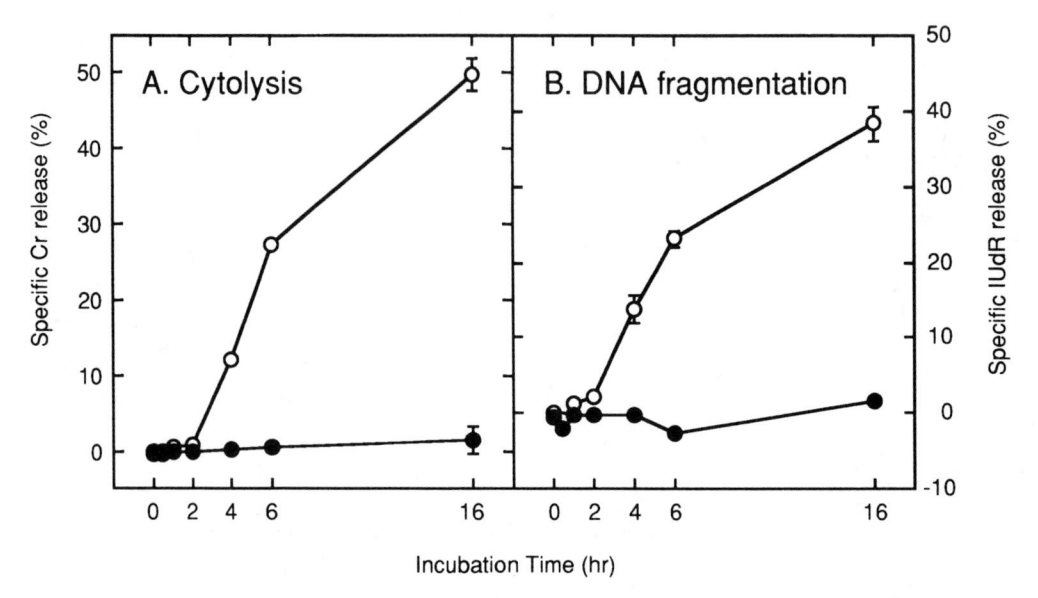

FIGURE 1. Cytotoxicity mediated by CD4[+] T-cell clone BK1. (A) Cytolysis measured by ^{51}Cr release. (B) DNA fragmentation measured by detergent soluble [^{125}I]-iododeoxyuridine (^{125}IUdR)-labeled DNA fragment release. Target cells were A20HL pretreated with TNP-KLH (○) and A20HL pretreated with TNP-OVA (●). (Modified with permission of Oxford University Press from Takayama H, et al. (1991): Antigen-specific directional target cell lysis by perforin-negative T lymphocyte clones. *Int Immunol* 3(11): 1149–1156.)

Our CD4[+] T-cell clones did not show bystander lysis in short-term incubations (6 hr), whereas antigen-specific directional target lysis was readily detectable. Bystander lysis was detectable in longer incubations (16 hr). In the experiment, the CD4[+] T-cell clone BK1 was incubated with ^{51}Cr-labeled A20HL pretreated with TNP–OVA (irrelevant antigen) in the presence of A20HL pretreated with TNP–KLH (specific antigen). Since both the antigen-specific target and the bystander cells were the same A20HL, susceptibility to any cytotoxic effector should be the same in antigen-specific target and bystander cells. Nevertheless, bystander lysis required a longer incubation period than the lysis of antigen-specific target. Moreover, bystander lysis was inhibitable with certain serum preparations, including anti-TNF, while antigen-specific target lysis was not inhibited by the antibody. Nevertheless, the specificity of the inhibition was not clear, since control rabbit immunoglobulin G (IgG) was also found to be inhibitory in the same condition. Even though TNF/LT-like activity was found in culture supernatants of BK1 by a standard fibroblast cytotoxicity assay, the evidence does not show that bystander lysis caused by BK1 is mediated by such a factor, since A20HL was resistant to the TNF/LT-like activity in BK1 culture supernatant. It is clear, however, that antigen-specific target lysis and bystander lysis caused by BK1 are mediated by totally different mechanisms, since the kinetics and effect of anti-TNF antibody were completely different.

Expression of Perforin in Soluble Protein-Specific, Class II MHC-Restricted CD4[+] T-Cell Clones

Our CD4[+] T-cell clone, BK1, was capable of inducing antigen-specific directional cytolysis and DNA fragmentation of target cells as described above. Since such a feature of cytotoxicity is similar to that of conventional CD8[+] class I MHC-restricted CTL, and since perforin has been a major candidate as an effector molecule responsible for CTL-mediated cytotoxicity, the expression of perforin in these T-cell clones was examined.

Perforin mRNA expression was examined by Northern blot hybridization (Figure 2). Whereas the CD8[+] CTL clone OE4 and the NK cell line SPB2.4 expressed the perforin message, the CD4[+] T-cell clones BK1, BK2, and BO1 did not express a detectable amount of perforin message. Since

8
Direct Contact of Cytotoxic T-Lymphocyte Receptors with Target Cell Membrane Determinants Induces a Prelytic Rise of $[Ca^{2+}]_i$ in the Target That Triggers Disintegration

Gideon Berke

Background

Because of the central role of cytotoxic T lymphocytes (CTL) (both CD4$^+$ and CD8$^+$ cells) in immune responses against viruses, tumors, and transplants, in AIDS and in autoimmunity, it is important to understand the precise molecular mechanism(s) whereby CTL destroy target cells (TC). No single mechanism proposed till now provides a satisfactory explanation of the entire process (Clark et al., 1988; Berke, 1991; Krähenbühl and Tschopp, 1991; Podack et al., 1991). CTL–TC interactions are initiated by specific binding of the TC by CTL (Brondz, 1968; Golstein et al., 1971; Berke and Levey, 1972), which results in conjugate formation (Berke et al., 1975; Martz, 1975, 1977; Berke, 1985; Dustin and Springer, 1991). The Mg^{2+}-dependent binding step (Stulting and Berke, 1973) is frequently, but not always (Tirosh and Berke, 1985b) followed by a Ca^{2+}- and temperature-dependent delivery of the effector's lethal hit. The lethal hit [also referred to as "kiss of death" or "programming for lysis" (Ginsburg et al., 1969; Martz, 1977; Golstein and Smith, 1977)], induces the prelytic fragmentation of TC DNA into [fragments] consisting of multiples of 180 bp (Russell, 1983; Duke and Cohen, 1988). The subsequent temperature-dependent but killer-independent TC [dissol]ution (Berke et al., 1972b; Martz, 1977) completes one round of the lytic process. The effector [can] then recycle to start a new lytic interaction (Berke et al., 1969, 1972a; Zagury et al., 1975). For [extensi]ve reviews of the process of TC lysis by CTL, [see Wil]son and Billingham (1967), Berke and Amos (1973), Cerottini and Brunner (1974), Martz (1977), Golstein and Smith (1977), Henney (1977), Sanderson (1981), Henkart (1985), Berke (1980, 1989, 1991), and Podack et al. (1991).

Lytic Mechanisms

Although CTL–TC recognition and binding (conjugate formation) have been defined at the cellular and molecular levels (Berke et al., 1975; Dustin and Springer, 1991; Martz, 1977), the precise molecular signal(s) by which TC are irreversibly damaged is still controversial and multiple lytic pathways may exist (Ostergaard and Clark, 1989; Berke, 1991; Krähenbühl and Tschopp, 1991). Any lymphocytotoxicity theory must account for the following facts: (1) CTL-mediated TC lysis is an apoptotic, not a necrotic, process (Wyllie et al., 1984; Duke and Cohen, 1988; Golstein et al., 1991); (2) fragmentation of TC DNA, a feature characteristic of CTL-mediated lysis, occurs prior to ^{51}Cr release from the affected TC (Russell, 1983); (3) effective CTL-induced lysis of certain TC can occur when there is no secretion (either regulated or constitutive) of CTL granule constituents (Tirosh and Berke, 1985b; Ostergaard et al., 1987; Trenn et al., 1987); and (4) effector CTL are not self-annihilated and can recycle repeatedly to lyse additional TC (Berke et al., 1969, 1972a; Zagury et al., 1975), even though CTL themselves are susceptible to their own lytic mechanism. In fact, lysis of CTL by other CTL (e.g., lysis of B anti-C by A anti-B CTL) has been demonstrated in several laboratories (Golstein,

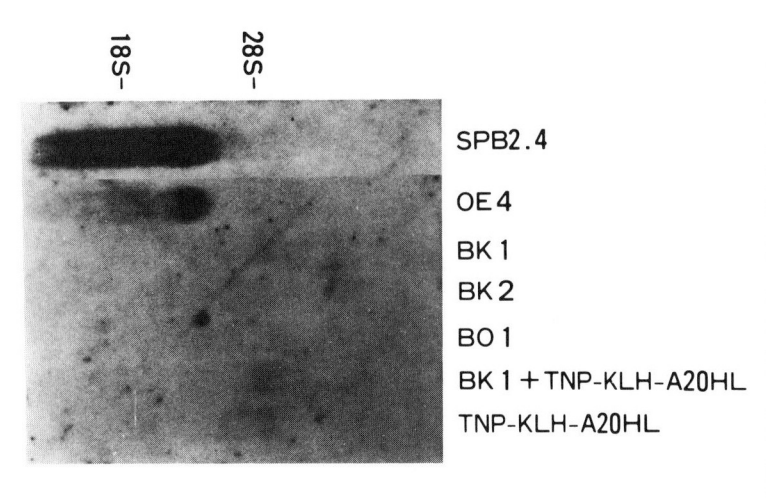

FIGURE 2. Northern blot analysis of expression of perforin mRNA in CTL clones OE4, BK1, BK2, BO1.BK1 stimulated with TNP-KLH in the presence of A20HL was also analyzed in comparison with A20HL alone in the presence of TNP-KLH. Cytoplasmic RNA from 1×10^6 cells was analyzed. (Modified with permission of Oxford University Press from Takayama H, et al. (1991): Antigen-specific directional target cell lysis by perforin-negative T lymphocyte clones. *Int Immunol* 3(11): 1149–1156.)

perforin is known to be inducible by appropriate stimulation (Liu et al., 1989; Smyth et al., 1990; Lancki et al., 1991), BK1 was stimulated with the antigen TNP–KLH in the presence of the antigen-presenting cell A20HL. Even after such stimulation, perforin mRNA was not incduced in BK1 (Figure 2).

Perforin protein levels in these cells were also examined by recently developed quantitative immunoassay (Kawasaki et al., 1990). Perforin in SPB2.4 was abundant (3.2–7.5×10^6 molecules/cell), and OE4 contained a detectable amount of perforin (2.5×10^5 molecules/cell), whereas BK1 as well as mastocytoma P815 did not express any detectable amount of perforin (less than 4×10^4 molecules/cell). Thus, the content of perforin protein was well correlated with mRNA expression, and our CD4+ T-cell clones did not express detectable amounts of perforin mRNA or of protein, suggesting that our CD4+ T-cell clones were capable of killing the antigen-presenting target cells by perforin-independent mechanisms.

Discussion

The involvement of perforin in CTL-mediated cytotoxicity has been controversial, mainly because there has been no reliable report showing the secretion of perforin from CTL corresponding with target cytolysis. Perforin gene transfection experiments provided direct evidence showing that perforin indeed mediated cytolysis upon stimuli-regulated granule secretion (Shiver and Henkart, 1991). However, CTL-mediated cytotoxicity has been found not to be inhibited by anti-perforin antibody, even though the same antibody preparation did inhibit perforin-RBL-mediated hemolytic activity (Reynolds et al., 1987; Shiver and Henkart, 1991). In addition, nucleated tumor cell lines were much less susceptible to perforin-RBL-mediated cytotoxicity than SRBC. Thus, even though these observations support the idea that perforin indeed works as an effector molecule upon secretion, they do not rule out the possibility that effector mechanisms other than perforin are involved in CTL-mediated cytotoxicity.

It has been reported that a subset of CD4+ T cells are cytotoxic against antigen-presenting cells (Nakamura et al., 1986; Ozaki et al., 1987; Watanabe et al., 1987; Hancock et al., 1989; Erb et al., 1990) and that the mechanism of lysis mediated by these CD4+ "helper"-type T cells involves soluble cytotoxic factor(s) such as TNF/LT (Tite and Janeway, 1984; Tite et al., 1985; Tite, 1990). However, several reports suggested that the involvement of TNF/LT in cytotoxicity depended on the susceptibility of target cells to the factor (Chang and Moorhead, 1986; Ju et al., 1988, 1990). We have also found that even though BK1 is able to produce TNF/LT-like activity in the culture supernatant, this factor may not be involved in antigen specific target cell lysis or in bystander lysis, since A20HL used in the assay was found to be highly resistant to the factor and the antibody against murine TNF/LT failed to inhibit BK1-mediated cytotoxicity (Takayama et al., 1991). Thus, such a factor seems not to be the sole major factor in directional target cell lysis induced by these T-cell clones.

The involvement of perforin in CD4+ T-cell-mediated cytotoxicity has been examined with several independent clones (Strack et al., 1990; Lancki et al., 1991; Takayama et al., 1991). Strack et al. have found cytotoxic CD4$^+$ T cells in the T helper type 1 (T_H1) subset but not in the T helper type 2 (T_H2) subset, and they did not detect perforin mRNA expression in T_H1 clones. BK1, the CD4$^+$ clone described here, is also classified as T_H1 because of its ability to produce IL-2 upon activation. Lancki et al. have found cytotoxic cells in both the T_H1 and the T_H2 subset, and perforin-positive cells were found only in T_H2 subsets. Interestingly, perforin expression in T_H2 was correlated with hemolytic activity against SRBC (Lancki et al., 1991). This observation may correspond with the hemolytic activity of RBL acquired after transfection with the perforin gene (Shiver and Henkart, 1991).

To date, however, there is no report that directly describes a mechanism other than perforin for CTL-mediated cytotoxicity. Pharmacological studies with metabolic inhibitors showed that cytolysis mediated by CD4$^+$ T cells was RNA- and protein synthesis-dependent, whereas cytolysis mediated by CD8$^+$ CTL was not (Strack et al., 1990). Preliminary experiments showed, however, that BK1-mediated cytolysis was not RNA- or protein synthesis-dependent (data not shown), suggesting that the mechanism of CD4$^+$ T-cell-mediated cytotoxicity is also heterogeneous. In addition, even though it has been speculated that perforin-independent cytotoxicity must be mediated by extracellular Ca^{2+}-independent mechanisms (Ostergaard et al., 1987; Trenn et al., 1987), BK1-mediated cytotoxicity was found to be extracellular Ca^{2+}-dependent (data not shown).

Possible involvement of N-α-benzyloxycarbonyl-L-lysine thiobenzyl (BLT) esterase in CTL-induced DNA fragmentation in target cells was recently reported (Hayes et al., 1989). This mechanism was shown to be dependent upon permeabilization of target cell membrane, and thus perforin might be actively involved in the process. BK1, however, was found to be able to induce target cell DNA fragmentation in the absence of perforin, as perforin-positive CD8$^+$ CTL do. In addition, even though BLT esterase activity was detectable in BK1 lysate, it was far less than that of the CD8$^+$ CTL clone OE4, and 2C (enzyme activity as absorbence at 412 nm for 20 min was 0.02 for BK1, 0.79 for OE4, and 0.46 for 2C, respectively). Thus, DNA fragmentation induced by these T-cell clones seems to be independent of perforin and BLT-esterase. Alternatively, it is also possible that DNA fragmentation induced by CD4$^+$ T-cell clones and CD8$^+$ CTL clones were mediated by different mechanisms, as suggested in recent report (Ju, 1991). The involvement of other potential effector molecules, such as CTL-specific lipase (Grusby et al., 1990) or ATP (Filippini et al., 1991), in the cytotoxicity mediated by these perforin-negative CD4$^+$ T-cell clones has not been investigated yet.

As we described here and as reports have shown (Maimone et al., 1986; Takayama et al., 1991), the mode of cytotoxicity of some CD4$^+$ T-cell clones was not different from that of CD8$^+$ CTL. Hence, we believe that these cytolytic CD4$^+$ T-cell clones can be classified as CTL irrespective of their CD4 phenotype and MHC restriction, and that these clones provide a good model to analyze the perforin-independent cytolytic mechanism involved in CTL-mediated cytotoxicity.

Acknowledgments

This work was carried out in collaboration with Drs. Yo-ichi Shinkai, Akemi Kawasaki, Hideo Yagita, and Ko Okumura (Department of Immunology, Juntendo University School of Medicine). The author thanks Yoshiko Someya and Satoko Hanaoka (Laboratory of Cellular Immunology, Mitsubishi Kasei Institute of Life Sciences) for excellent technical assistance, and Dr. Nobukata Shinohara for his helpful discussion, collaboration, and continuous support of the study.

References

Acha-Orbea H, Scarpellino L, Hertig S, Dupuis M, Tschopp J (1990): Inhibition of lymphocyte mediated cytotoxicity by perforin antisense oligonucleotides. *EMBO J* 9: 3815–3819

Berke G, Rosen D (1988): Highly lytic *in vivo* primed cytolytic T lymphocytes devoid of lytic granules and BLT-esterase activity acquire these constituents in the presence of T cell growth factors upon blast transformation *in vitro*. *J Immunol* 141: 3440–3445

Chang JCC, Moorhead JW (1986): Hapten-specific, class II-restricted killing by cloned T cells: Direct lysis and production of a cytotoxic factor. *J Immunol* 136: 2826–2831

Duke RC, Persechini PM, Chang S, Liu C-C, Cohen JJ, Young JD-E (1989): Purified perforin induces target cell lysis but not DNA fragmentation. *J Exp Med* 170: 1451–1456

Erb P, Grogg D, Troxler M, Kennedy M, Fluri M (1990): CD4$^+$ T cell-mediated killing of MHC class II-positive antigen-presenting cells. I. Characterization of target cell recognition by *in vivo* or *in vitro* activated CD4$^+$ killer cells. *J Immunol* 144: 790–795

Filippini A, Taffs RE, Sitkovsky MV (1990): Extracellular ATP in T-lymphocyte activation: Possible role in effector functions. *Proc Natl Acad Sci USA* 87: 8267–

8271, with clarification (1991): *Proc Natl Acad Sci USA* 88: 6899

Grusby MJ, Nabavi N, Wong H, Dick RF, Bluestone JA, Schotz MC, Glimcher LH (1990): Cloning of an interleukin-4 inducible gene from cytotoxic T lymphocytes and its identification as a lipase. *Cell* 60: 451–459

Hancock GE, Cohn ZA, Kaplan G (1989): The generation of antigen-specific, major histocompatibility complex-restricted cytotoxic T lymphocytes of the CD4+ phenotype. Enhancement by the cutaneous administration of interleukin 2. *J Exp Med* 169: 909–919

Hayes PM, Berrebi GA, Henkart PA (1989): Induction of target cell DNA release by the cytotoxic T lymphocyte granule protease granzyme A. *J Exp Med* 170: 933–946

Henkart PA, Millard PJ, Reynolds CW, Henkart M (1984): Cytolytic activity of purified cytoplasmic granules from cytotoxic rat large granular lymphocyte tumors. *J Exp Med* 160: 75–93

Ju S-T (1991): Distinct pathways of CD4 and CD8 cells induce rapid target DNA fragmentation. *J Immunol* 146: 812–818

Ju S-T, Ruddle NH, Strack P, Dorf ME, DeKruyff RH (1990): Expression of two distinct cytolytic mechanisms among murine CD4 subsets. *J Immunol* 144: 23–31

Ju S-T, Strack P, Stromquist D, DeKruyff RH (1988): Cytolytic activity of Ia-restricted T cell clones and hybridomas: Evidence for a cytolytic mechanism indepdent of interferon-γ, lymphotoxin, and tumor necrosis factor-α. *Cell Immunol* 117: 399–413

Kawasaki A, Shinkai Y, Kuwana Y, Furuya A, Iigo Y, Hanai N, Itoh S, Yagita H, Okumura K (1990): Perforin, a pore-forming protein detectable by monoclonal antibodies, is a functional marker for killer cells. *Int Immunol* 2: 677–684

Lancki DW, Hsieh C-S, Fitch FW (1991): Mechanisms of lysis by cytotoxic T lymphocyte clones. Lytic activity and gene expression in cloned antigen-specific CD4+ and CD8+ T lymphocytes. *J Immunol* 146: 3242–3249

Liu C-C, Rafii S, Granelli-Piperno A, Trapani JA, Young JD-E (1989): Perforin and serine esterase gene expression in stimulated human T cells. *J Exp Med* 170: 2105–2118

Maimone MM, Morrison LA, Braciale VL, Braciale TJ (1986): Features of target cell lysis by class I and class II MHC-restricted cytotoxic T lymphocytes. *J Immunol* 137: 3639–3643

Nakamura M, Ross DT, Briner TJ, Gefter ML (1986): Cytolytic activity of antigen-specific T cells with helper phenotype. *J Immunol* 136: 44–47

Ostergaard HL, Kane KP, Mescher MF, Clark WR (1987): Cytotoxic T lymphocyte-mediated lysis without release of serine esterase: *Nature* 330: 71–72

Ozaki S, York-Jolley J, Kawamura H, (1987): Cloned protein antigen-specific, cells with both helper and cytolytic activi isms of activation and killing. *Cell Immu* 316

Reynolds CW, Reichardt D, Henkart M, Henkart PA (1987): Inhibition of NK ar activity by antibodies against purified cy granules from rat LGL tumors. *J Leukocyte* 642–652

Shinohara N, Huang Y-Y, Muroyama A (1991): suppression of antibody response by soluble pr specific, class II-restricted cytolytic T lymph clones. *Eur J Immunol* 21: 23–27

Shiver JW, Henkart PA (1991): A noncytotoxic mast tumor line exhibits potent IgE-dependent cytotoxic after transfection with the cytolysin/perforin gene. C 64: 1175–1181

Smyth MJ, Ortaldo JR, Shinkai Y-I, Yagita H, Nakata M, Okumura K, Young HA (1990): Interleukin-2 induction of pore-forming protein gene expression in human peripheral blood CD8+ T cells. *J Exp Med* 171: 1269–1281

Strack P, Martin C, Saito S, DeKruyff RH, Ju S-T (1990): Metabolic inhibitors distinguish cytolytic activity of CD4 and CD8 clones. *Eur J Immunol* 20: 179–184

Takayama H, Shinohara N, Kawasaki A, Someya Y, Hanaoka S, Kojima H, Yagita H, Okumura K, Shinkai Y-I (1991): Antigen-specific directional target cell lysis by perforin-negative T lymphocyte clones. *Int Immunol* 3: 1149–1156

Tite JP (1990): Evidence of a role for TNF-α in cytolysis by CD4+, class II MHC-restricted cytotoxic T cells. *Immunology* 71: 208–212

Tite JP, Janeway CA, Jr (1984): Cloned helper T cells can kill B lymphoma cells in the presence of specific antigen: Ia restriction and cognate vs. noncognate interaction in cytolysis. *Eur J Immunol* 14: 878–886

Tite JP, Powell MB, Ruddle NH (1985): Protein-anti specific Ia-restricted cytolytic T cells: Analysis of quency, target cell susceptibility, and mechani cytolysis. *J Immunol* 135: 25–33

Trenn G, Takayama H, Sitkovsky MV (1987): E of cytolytic granules may not be required for lysis by cytotoxic T-lymphocytes. *Nature* 3

Watanabe M, Yoshikawa M, Hozumi N (toxic function of a cloned helper T cell *Lett* 15: 133–138

Zijlstra M, Bix M, Simister NE, Loring Jaenisch R (1990): β2-Microglobulin CD4-8+ cytolytic T cells. *Nature*

1974; Kuppers and Henney, 1977; Fishelson and Berke, 1978; Schick and Berke, 1990).

Currently, there are three proposed mechanisms for lymphocyte-mediated lysis: (1) CTL-triggered internal disintegration (apoptosis) of the target upon contact with its membrane (Berke and Clark, 1982; Russell, 1983; Berke, 1991); (2) formation of pores (10–20 nm in diameter) in the TC membrane by the lytic protein perforin, secreted from CTL (Henkart, 1985; Podack et al., 1991); and (3) lysis induced by ATP secreted from CTL (Filippini et al., 1990; Zanovello et al., 1989). The first proposed mechanism accounts best for prelytic TC DNA disintegration and subsequent TC lysis under conditions in which granule exocytosis does not occur and perforin is not lytic (Tirosh and Berke, 1985a, 1985b; Ostergaard et al., 1987; Trenn et al., 1987). Contact-induced apoptosis is also consistent with self-sparing and recycling of effector CTL (Berke et al., 1969, 1972c; Berke, 1991) and lysis mediated by CD8+ or CD4+ CTL that do not express perforin (Allbritton et al., 1988a; Berke and Rosen, 1988; Takayama et al., 1991; Berke et al., 1993).

Contact-Induced Elevation of Cytosolic Ca^{2+} in the Target Triggers Its Disintegration (Apoptosis): A Theory and Supporting Evidence

Lymphocyte-triggered disintegration of TC by an autolytic cascade initiated by prelytic nuclear fragmentation was first suggested by Russell (1983). However, the triggering mechanism(s) and mediator(s), for example, contact with the killer alone or a secreted lytic agent(s), respectively, and the "second messenger" in the TC responsible for inducing such a lytic cascade, have not been delineated. At the First International Workshop on Mechanisms in Cell-Mediated Cytotoxicity, held in Marseilles in 1981, we proposed that the effector's lethal signal may be delivered directly by cell surface receptors, later shown to be the T-cell receptor (TCR) complex and associated components, including adhesion molecules, upon contact with the TC cell surface determinants, including major histocompatibility complex (MHC) and other TC molecules involved in the interaction (Berke and Clark, 1982; Clark and Berke, 1982). That contact alone of lymphocytes and targets is sufficient to induce target membrane depolarization, as measured by ^{86}Rb efflux, was demonstrated (Sanderson, 1981). Redistribution, aggregation, and cross-linking of TC MHC and non-MHC surface proteins, upon interaction with mobile effector TCR, associated surface receptors, and adhesion molecules, could in itself alter the TC membrane permeability to critical ions. Non-MHC TC surface proteins involved could include adhesion proteins, such as ICAM-1 or -2 and LFA-1 or -3 (Martz, 1987).

CTL-induced prelytic elevation of free cytosolic Ca^{2+} ($[Ca^{2+}]_i$) in TC, preceding TC membrane damage (^{51}Cr release), was initially observed by us (Tirosh and Berke, 1985a, 1985b) in the peritoneal exudate CTL (PEL) EL4 system and then confirmed by others (Hassin et al., 1987; Poenie et al., 1987; Allbritton et al., 1988b; McConkey et al., 1990a) in various in vitro CTL and natural killer cell (NK) systems. The CTL-induced, sustained elevation of intracellular Ca^{2+} levels observed in conjugated TC could trigger, perhaps in conjunction with other factors, a multitude of internal degradation processes in the TC, ultimately leading to its dissolution (Tirosh and Berke, 1985a, 1985b; Berke, 1989, 1991). These degradative processes could include Ca^{2+}-mediated activation of endonucleases, proteases, phospholipases, and ATPases, which cause prelytic DNA fragmentation; cytoskeleton disruption resulting in bleb formation; cytoplasmic streaming; metabolic exhaustion; and finally, cytolysis. That nuclear disintegration, followed by cytolysis, is actually related to elevated $[Ca^{2+}]_i$ in the target has recently been demonstrated (McConkey et al., 1990a, 1990b).

Whereas CTL binding to TC surface moieties can trigger processes resulting in TC dissolution, the effector CTL involved is only stimulated. This unidirectional lysis (survival rather than self-annihilation of the effector in the course of killing the TC) may be due to specialization and orientation of CTL surface receptors (see Geiger et al., 1982) or to the effector's ability to regulate stimulatory membrane signals which cause a significant but transient rise in $[Ca^{2+}]_i$ in the CTL (Poenie et al., 1987). The relative refractoriness of CTL, when used as TC (Kranz and Eisen, 1987; Shinkai et al., 1988; Zanovello et al., 1989), supports this explanation.

The induction of cell death through CTL-induced crosslinking or redistribution of TC surface receptors, not involving the formation of 10- to 20-nm pores by a CTL lytic protein, such as perforin, is not intuitively obvious. However, programmed cell death induced by crosslinking of cell surface receptors has recently been demonstrated in the following systems: (1) apoptosis of immature thymocytes or leukemia cells induced by CD3 antibodies in the absence of complement, or by antigen or "superantigen" (Jenkinson et al., 1989; Takahashi et al., 1989; Nieto et al., 1990; Tadakuma et al., 1990); (2) apoptosis of tumor cells mediated by high-affinity antibody, but not by F(ab)$_2$ fragments, against the surface membrane determinant Apo-1 (Trauth et al., 1989); (3) apoptosis of T-cell hybridomas and mature T cells induced by antigen (Mercep et al., 1989; Ucker et al., 1989; Odaka et al., 1990; Shi et al., 1990); (4) glutamate-induced neurotoxicity mediated by N-methyl-D-aspartate (NMDA)-glutamate receptors (McCaslin and Smith, 1990); (5) tumor necrosis factor (TNF)-like cytotoxicity induced by antibodies against membrane-bound TNF receptors (Engelmann et al., 1990); (6) mature T-cell lysis mediated by crosslinking of CD4 and TCR (Newell et al., 1990); and (7) apoptosis of T cells induced by the interaction of TCR and the α_3 domain of class I MHC. Interestingly, a sustained, prelytic elevation of $[Ca^{2+}]_i$, which may induce lysis, has also been demonstrated in these systems.

Although fulfilling a pivotal role as a second messenger, an uncontrolled rise of $[Ca^{2+}]_i$ has long been implicated in the induction of cell lysis by various drugs, toxins, and viruses (Campbell, 1987; Orrenius et al., 1989). Ca^{2+} transport across intact biological membranes is regulated by: (1) voltage-dependent Ca^{2+} channels, which open upon membrane depolarization; (2) Na^+-Ca^{2+} exchangers; and (3) ATP-fueled Ca^{2+} pumps, which exchange intracellular Ca^{2+} ions and extracellular protons. Direct or, more likely, indirect, via membrane transducers such as G proteins, interaction of CTL receptors with any of the constituents involved in these three mechanisms could influence Ca^{2+} flux across the TC membrane, although other mechanisms are possible. Fluxes of ions, including Ca^{2+}, across the TC membrane may result from the mere crosslinking of TC membrane constituents, including MHC, induced by binding of mobile CTL receptors. That voltage-regulated Ca^{2+} transport may be involved in excitable myocardial cells under CTL attack has been demonstrated (Hassin et al., 1987; see also Binah, 1992, 1993). Interestingly, even in the absence of medium Ca^{2+}, $[Ca^{2+}]_i$ elevation occurs in some TC upon CTL attack (Tirosh and Berke, 1985b), possibly due to Ca^{2+} release from internal stores (mitochondria, endoplasmic reticulum) upon CTL-induced cytoplasmic acidification (Russell, 1983). Hence, elevation of $[Ca^{2+}]_i$ upon CTL attack may be a secondary, but crucial, event in the course of the lytic process. As previously mentioned, a sustained increase in $[Ca^{2+}]_i$ can potentially induce DNA fragmentation due to Ca^{2+}-dependent activation of endonucleases and/or bleb formation due to activation of proteases, which cleave the cytoskeleton. Irreversible failure of selective membrane permeability, culminating in lysis, probably results from the inability of ATP-fueled ionic pumps to maintain ionic gradients due to ATP wastage caused by Ca^{2+} activation of actomyosin ATPases. These, however, are late-stage effects preceded by the prelytic rise in $[Ca^{2+}]_i$ and nuclear fragmentation.

Summary

Currently there are three theories to explain the mechanism of lymphocytotoxicity: (1) formation of pores in the target cell membrane by perforin, believed to be a lytic protein secreted by some CTL and NK cells, (2) target cell lysis induced by ATP secreted by CTL, and (3) direct contact of CTL receptors with target cell membrane determinants triggering internal disintegration, mediated by a sustained increase of cytosolic Ca^{2+} in the target. This chapter deals with the third theory. Prelytic elevation of target intracellular Ca^{2+} ($[Ca^{2+}]_i$) and its relation to the onset of TC lysis, first described with PEL (Tirosh and Berke, 1985a, 1985b), has now been confirmed and extended to other CTL/NK systems. Receptor-mediated triggering of TC disintegration mediated by prelytic elevated $[Ca^{2+}]_i$ has recently been implicated in apoptosis triggered by antibodies in the absence of complement or by ligands interacting with cell surface components such as CD3, TNF receptors, superantigen receptors, APO-1 antigen, CD4–TCR, and the α_3 domain of MHC class I, supporting the feasibility of the proposed mechanism. No pore formation has been proposed for any of these processes. Furthermore, the most potent CTL, in vivo primed PEL, only contain amounts of perforin and granzymes comparable to those in nonlytic lymphocytes, and these PEL can lyse TC under conditions where perforin is neither secreted nor lytic. Recently, we also generated (Chapter 10, this volume) several functional and homogeneous CTL hybridomas, which are completely devoid of lytic granules, perforin, and granzymes, and of mRNA for perforin and granzymes. Possible pathways whereby a sustained increase in $[Ca^{2+}]_i$ can bring about target cell apoptosis are discussed. Although a lytic mechanism involving receptor-triggered internal disintegration of the TC is in agreement with the characteristics of CTL/NK-mediated TC lysis, it is still feasible that under certain circumstances lysis effected by certain CTL and NK cells can be mediated by pore formation or secreted ATP.

Acknowledgments

I would like to thank Mrs. Malvine Baer for typing the manuscript, Dr. Barbara Schick for reviewing it, Dr. Dalia Rosen and Denise Ronen for technical assistance, and Moshe Kushnir for some critical comments.

Supported in part by funds from the U.S.-Israel Binational Science Foundation, Jerusalem, Israel; the German-Israeli Fund; the Deutsche Krebsforschungzentrum, Heidelberg, Germany; the Israel Academy of Sciences and Humanities; and the Israel Cancer Association.

References

Albritton NL, Nagler-Anderson C, Elliot TJ, Verret CR, Eisen HN (1988a): Target cell lysis by cytolytic lymphocytes that lack detectable hemolytic perforin activity. *J Immunol* 141: 3243

Albritton NL, Verret CR, Wolley RC, Eisen HN (1988b): Calcium ion concentrations and DNA fragmentation in target cell destruction by murine cloned cytotoxic T lymphocytes. *J Exp Med* 167: 514-527

Berke G (1980): Interaction of cytotoxic T lymphocytes and target cells. *Prog Allergy* 27: 69-133

Berke G (1985): How T lymphocytes kill infected cells. *Microbiol Sci* 2: 44-48

Berke G (1989) Functions and mechanisms of lysis induced by cytotoxic T lymphocytes and natural killer cells. *Fundamental Immunology*, 2nd ed, Paul W, ed. New York: Raven Press, pp. 735-764

Berke G (1991): Lymphocyte-triggered Internal Target Disintegrations. *Immunol Today* 12: 396-399

Berke G, Amos DB (1973): Mechanism of lymphocyte-mediated cytolysis. The LMC cycle and its role in transplantation immunity. *Transplant Rev* 17: 71-107

Berke G, Clark WR (1982): T lymphocyte-mediated cytolysis: A comprehensive theory. I. The mechanism of CTL-mediated cytolysis. In: *Mechanism of Cell Mediated Cytotoxicity*, Clark WR, Golstein P, eds, New York: Plenum Publ. Corp., pp. 57-69

Berke G, Levey R (1972): Cellular immunoadsorbents in transplantation immunity. Specific in vitro deletion and recovery of mouse lymphoid cells sensitized against allogeneic tumors. *J Exp Med* 135: 972-984

Berke G, Ax W, Ginsburg H, Feldman M (1969): Graft reaction in tissue culture. II. Quantification of the lytic action on mouse fibroblasts by rat lymphocytes sensitized in mouse embryo monolayers. *Immunology* 16: 643-657

Berke G, Gabison D, Feldman M (1975): The frequency of effector cells in populations containing cytotoxic T lymphocytes. *Eur J Immunol* 5; 813-818

Berke G, Rosen D, Ronen D (1993): Mechanism of lymphocyte-mediated cytolysis: Functional cytolytic T cells lacking perforin and granzymes. *Immunol* (in press)

Berke G, Sullivan KA, Amos DB (1972a): Tumor immunity *in vitro*: Destruction of a mouse ascites tumor through a cycling pathway. *Science* 177: 433-434

Berke G, Sullivan KA, Amos DB (1972b) Rejection of ascites tumor allografts. II. A pathway for cell-mediated tumor destruction *in vitro* by peritoneal exudate lymphoid cells. *J Exp Med* 136: 1594-1604

Binah O, Kline R, Berke G, Hoffman B (1993): Lytic granules from cytotoxic T lymphocytes damage guinea pig ventricular myocytes by opening channels with large conductance. *Pflugers Arch Scand J Immunol* (in press)

Binah O, Marom S, Rubinstein I, Robinson RB, Berke G, Hoffman BF (1992): Immunological rejection of heart transplant: How lytic granules from cytotoxic T lymphocytes damage guinea pig ventricular myocytes. *Pflugers Arch* 420: 172-179

Brondz BD (1968): Complex specificity of immune lymphocytes in allogeneic cell cultures. *Folia Biol Praha* 14: 115

Campbell AK (1987): Intracellular calcium: Friend or foe? *Clin Sci* 72: 1-10

Clark WR, Berke G (1982): T lymphocyte-mediated cytolysis. II. Lytic vs. nonlytic interactions of T lymphocytes. *Immunol Rev* 103: 69-79

Clark WR, Ostergaard HL, Gorman K, Torbett BE (1988): Molecular mechanisms of CTL-mediated lysis: A cellular perspective. *Immunol Rev* 103: 37-51

Duke RC, Cohen JJ (1988): The role of nuclear damage in lysis of target cells by cytotoxic T lymphocytes. In: *Cytolytic Lymphocytes and Complement as Effectors of the Immune System*, Podack ER, ed., Boca Raton, FL: CRC Press, p. 235

Dustin ML, Springer TA (1991): Role of lymphocyte adhesion receptors in transient interactions and cell locomotion. *Annu Rev Immunol* 9: 27-66

Engelmann H, Holtmann H, Brakebusch C, Avni YS, Sarov I, Nophar Y, Hadas E, Leitner O, Wallach D (1990): Antibodies to a soluble form of a tumor necrosis factor (TNF) receptor have TNF-like activity. *J Biol Chem* 265: 14497-14504

Filippini A, Taffs RE, Agui T, Sitkovsky MV (1990): Ecto-ATPase activity in cytolytic T-lymphocytes. *J Biol Chem* 265: 334-340

Fishelson Z, Berke G (1978): T lymphocyte-mediated cytolysis: Dissociation of the binding from the lytic mechanism of the effector cells. *J Immunol* 120: 1121-1126

Geiger B, Rosen D, Berke G (1982): Spatial relationships of microtuble-organizing centers and the contact area of cytotoxic T lymphocytes and target cells. *J Cell Biol* 95: 137-143

Ginsburg H, Ax W, Berke G (1969): Graft reaction in tissue culture by normal rat lymphocytes. *Transplant Proc* 1: 551-555

Golstein P (1974): Sensitivity of cytotoxic T cells to T-cell mediated cytotoxicity. *Nature* 252: 81

Golstein P, Smith ET (1977): Mechanism of T cell-mediated cytolysis: The lethal hit stage. *Contemp Top Immunobiol* 7: 273–300

Golstein P, Ojcius DM, Young JD-E (1991): Cell death mechanisms and the immune system. *Immunol Rev* 121: 29–65

Golstein P, Svedmyr EAJ, Blomgren H (1971): Cells mediating specific *in vivo* cytotoxicity. I. Detection of receptor-bearing lymphocytes. *J Exp Med* 134: 1385–1402

Hassin D, Fixler R, Shimoni Y, Rubinstein E, Ras S, Gotsman MS, Hasin Y (1987): Physiological changes induced in cardiac myocytes by cytotoxic T lymphocytes. *Am J Physiol* 252: C10–C16

Henkart PA (1985): Mechanism of lymphocyte-mediated cytotoxicity. *Annu Rev Immunol* 3: 31–58

Henney CS (1977): T-cell-mediated cytolysis: An overview of some current issues. *Contemp Top Immunobiol* 7: 245–272

Jenkinson EJ, Kingston R, Smith CA, Williams GT, Owen JJT (1989): Antigen-induced apoptosis in developing T cells: A mechanism for negative selection of the T cell receptor repertoire. *Eur J Immunol* 19: 2175–2177

Krähenbühl O, Tschopp J (1991): Perforin-induced pore formation. *Immunol Today* 12: 399

Kranz DM, Eisen HN (1987): Resistance of cytotoxic T lymphocytes to lysis by a clone of cytotoxic T lymphocytes. *Proc Natl Acad Sci USA* 84: 3375–3379

Kuppers RC, Henney CS (1977): Studies on the mechanism of lymphocyte-mediated cytolysis. IX. Relationships between antigen recognition and lytic expression in killer T cells. *J Immunol* 118: 71

Martz E (1975): Early steps in specific tumor cell lysis by sensitized mouse T-lymphocytes. I. Resolution and characterization. *J Immunol* 115: 261

Martz E (1977): Mechanism of specific tumor-cell lysis by alloimmune T lymphocytes: Resolution and characterization of discrete steps in the cellular interaction. *Contemp Top Immunobiol* 7: 301–361

Martz E (1987): LFA-1 and other accessory molecules functioning in adhesions of T and B lymphocytes. *Semin T-Cell Immunobiol* 18: 3–37

McCaslin PP, Smith TG (1990): Low calcium-induced release of glutamate results in autotoxicity of cerebellar granule cells. *Brain Res* 513: 280–285

McConkey DJ, Chow SC, Orrenius S, Jonda IM (1990a): NK cell-induced cytotoxicity is dependent on a Ca^{2+} increase in the target. *FASEB J* 4: 2661–2664

McConkey DJ, Orrenius S, Jondal M (1990b): Cellular signalling in programmed cell death (apoptosis). *Immunol Today* 11: 120–121

Mercep M, Weissman AM, Frank SJ, Klausner RD, Ashwell JD (1989): Activation-driven programmed cell death and T cell receptor zeta eta expression. *Science* 246: 1162–1165

Newell MK, Haughn LJ, Maroun CR, Julius MH (1990): Death of mature T cells by separate ligation of CD4 and T-cell receptor for antigen. *Nature* 347: 286–289

Nieto MA, Gonzalez A, Lopez-Rivas A, Diaz-Espada F, Gambon F (1990): IL-2 protects against anti-CD3-induced cell death in human medullary thymocytes. *J Immunol* 145: 1364–1368

Odaka C, Kizaki H, Tadakuma T (1990): T cell receptor-mediated DNA fragmentation and cell death in T cell hybridomas. *J Immunol* 144: 2096–2101

Orrenius S, McConkey DJ, Bellomo G, Nicotera P (1989): Role of Ca^{2+} in toxic cell killing. *TIPS* 10: 281–285

Ostergaard H, Clark WR (1989): Evidence for multiple lytic pathways used by cytotoxic T lymphocytes. *J Immunol* 143: 2120–2126

Ostergaard HL, Kane KP, Mescher MF, Clark WR (1987): Cytotoxic T lymphocyte-mediated lysis without release of serine esterase. *Nature* 330: 71

Podack ER, Hengartner H, Lichtenheld M (1991): A central role of perforin in cytolysis? *Annu Rev Immunol* 9: 129

Poenie M, Tsien RY, Schmitt-Verhulst AM (1987): Sequential activation and lethal hit measured by $[Ca^{2+}]_i$ individual cytolytic T cells and targets. *EMBO J* 6: 2223–2232

Russell J (1983): Internal disintegration model of cytotoxic lymphocyte-induced target damage. *Immunol Rev* 72: 97–118

Sanderson CJ (1981): The mechanism of lymphocyte-mediated cytotoxicity. *Biol Rev* 56: 153–197

Schick B, Berke G (1990): The lysis of cytotoxic T lymphocytes and their blasts by cytotoxic T lymphocytes. *Immunol* 71: 428–433

Shi Y, Szalay MG, Paskar L, Boyer M, Singh B, Green DR (1990): Activation-induced cell death in T cell hybridomas is due to apoptosis: Morphologic aspects and DNA fragmentation. *J Immunol* 144: 3326–3333

Shinkai Y, Takio K, Okumura K (1988): Homology of perforin to the ninth component of complement. *Nature* 334: 525

Stulting RD, Berke G (1973): The use of ^{51}Cr release as a measure of lymphocyte-mediated cytolysis in vitro. *Cell Immunol* 9: 474–476

Tadakuma T, Kizaki H, Odaka C, Kubota R, Ishimura Y, Yagita H, Okumura K (1990): $CD4^+CD8^+$ thymocytes are susceptible to DNA fragmentation induced by phorbol ester, calcium ionophore and anti-CD2 antibody. *Eur J Immunol* 20: 779–784

Takahashi S, Maecker HT, Levy R (1989): DNA fragmentation and cell death mediated by T cell antigen receptor/CD3 complex on a leukemia T cell line. *Eur J Immunol* 19: 1911–1919

Takayama H, Shinohara N, Kawasaki A et al. (1991): Antigen-specific directional target cell lysis by perforin-negative T lymphocyte clones. *Int Immunol* 3: 1149–1156

Tirosh R, Berke G (1985a): Immune cytolysis viewed as a stimulatory process of the target. In: *Proc. 2nd CMC Conference*, Henkart P, Martz E, eds. New York: Plenum, pp. 473–492

Tirosh R, Berke G (1985b): T lymphocyte mediated cytolysis as an excitatory process of the target: I. Evidence that the target may be the site of Ca^{2+} action. *Cell Immunol* 95: 113–123

Trauth BC, Klas C, Peters AMJ, Matzku S, Moller P,

Falk W, Debatin K-M, Krammer PH (1989): Monoclonal antibody-mediated tumor regression by induction of apoptosis. *Science* 245: 301–305

Trenn G, Takayama H, Sitkovsky MV (1987): Antigen-receptor regulated exocytosis of cytolytic granules may not be required for target cell lysis by cytotoxic T lymphocytes. *Nature* 330: 72

Ucker DS, Ashwell JD, Nickas G (1989): Activation-driven T cell death: I. Requirements for de novo transcription and translation and association with genome fragmentation. *J Immunol* 143: 3461–3469

Wilson DB, Billingham RE (1967): Lymphocytes and transplantation immunity. *Adv Immunol* 7: 189

Wyllie AH, Morris RG, Smith AL, Dunlop D (1984): Chromatin cleavage in apoptosis: Association with condensed chromatin morphology and dependence on macromolecular synthesis. *J Pathol* 142: 67–77

Zagury D, Bernard J, Thierness N, Feldman M, Berke G (1975): Isolation and characterization of individual functionally reactive cytotoxic T lymphocytes. Conjugation, killing and recycling at the single cell level. *Eur J Immunol* 5: 818–822

Zanovello P, Cerundolo V, Bronte V, Giunta M, Panozzo M, Biasi G, Collavo D (1989): Resistance of lymphokine-activated T lymphocytes to cell-mediated cytotoxicity. *Cell Immunol* 122: 450–460

19
Target Cell Events Initiated by T-Cell Attack

John H. Russell and Scott I. Abrams

The interaction between the thymus-derived lymphocyte (T cell) and its antigen-bearing target cell represents a model of contact-initiated intercellular communication. The unique feature of T cells is that they have evolved to distinguish self from nonself at the cellular level. That is, they do not recognize foreign antigens such as viruses as an entity alone, but rather they recognize aberrant neighboring cells as a perturbation of what they have "learned" as self. The basis of this self-perception is the conformation of complexes of peptides bound to a set of cell surface MHC (major histocompatibility complex) proteins unique to each individual. Activated T cells constantly sample neighboring cells in their environment for alterations in the conformation of these MHC complexes (Chang et al., 1979). Detection of "foreign" peptide-altered MHC complexes (e.g., on a virally infected cell) triggers the activation of a program of differentiated function within the T cell. In this chapter we will outline how the choice of antigen-bearing target cells and methods of assay have been used to explore the spectrum of functional T-cell activities resulting from the T cell–target interaction.

The earliest notion of an antibody-independent, functional activity by lymphocytes arose from the classical delayed-type hypersensitivity experiments of Landsteiner and Chase (Landsteiner and Chase, 1942; Chase, 1945). The possibility of a direct cytotoxic function for lymphocytes is usually credited to the transplantation experiments of Billingham and Medawar (Billingham et al., 1954). The first *ex vivo* assay for cytotoxicity was developed by Winn (1959) who demonstrated that lymphocytes from mice undergoing a tumor allograft response could suppress tumor formation if mixed with tumor cells before reimplantation. Subsequently, using cell culture assays, it was shown that the failure of tumor growth was due to a cytocidal rather than a cytostatic mechanism (Wilson, 1965; Brunner et al., 1968). These experiments constituted the first direct demonstration of contact-induced functional activity in T cells. The efficacy of the lymphocyte response to challenges with large tumor cell numbers stimulated the earliest enthusiasm for immunotherapy of neoplastic disease (Klein et al., 1960).

Morphological Descriptions of Cell Death

The requirement for cell–cell contact between the T cell and the target, and the low frequency of specific effectors in mixed cell populations, have greatly complicated study of functional mechanisms in this two-cell system. Thus, the earliest studies suggesting that T-cell-mediated lysis was distinct from other forms of death utilized morphological observations. In a remarkable series of early observations on tumor rejection, Kidd suggested that lymphocytes could be directly cytotoxic. He also concluded that direct killing by lymphocytes and death by antibody-fixed complement were distinct mechanisms based on a comparison of the morphology of tumor cells dying *in situ* in close apposition to lymphocytes with similar tumor cells dying in culture after treatment with hyperimmune serum (Kidd, 1950).

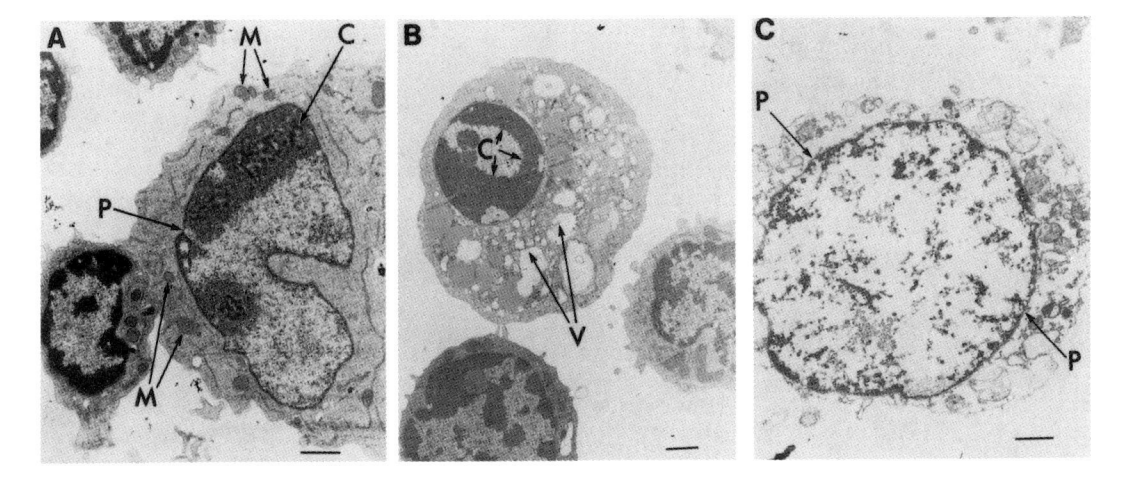

FIGURE 1. Ultrastructure of target cells being attacked by CTL and complement. (A) CTL/target conjugate within 10 min of interaction. The small attacking lymphocyte (bottom left) has already induced chromatin (C) condensation without mitochondrial (M) swelling. Also at this point the nuclear envelope appears unharmed, as indicated by the presence of pore (P) structures. (B) A CTL target (apoptotic body, Don et al., 1977) 20 min after interaction with CTL. Note condensed chromatin (C) and vacuoles (V). Also at this time the integrity of the nuclear membrane is lost. (C) A similar tumor target 20 min after attack with complement. Note the lack of chromatin condensation and integrity of the nuclear membrane, indicated by complete pore (P) structures. Bars indicate 1.0 μm. (Adapted from Russell et al., 1982.)

Sanderson described the "boiling" of the plasma membrane and the blebbing and shedding of cytoplasmic particles of targets dying after T-cell attack. He termed the phenomenon "zeosis" and drew analogies to "natural" death occurring in tumor populations (Sanderson, 1976). Independently, Kerr and colleagues concluded that the morphology of cells dying after T-cell attack was similar to examples of "natural" cell death in populations which they had previously described as "apoptosis" (Don et al., 1977). Apoptosis was characterized by cytoplasmic blebbing, but more importantly, by the condensation of chromatin material within the nucleus of a cell which otherwise appeared morphologically normal. Especially important was that mitochondria, which are exquisitely sensitive to osmotic changes, should remain normal (see Figure 1A). Wyllie extended these studies and proposed a unified "apoptotic" mechanism for a "natural" or "programmed" death from a variety of causes (Wyllie et al., 1980).

Nuclear Lesion

Henney was one of the first to take advantage of cloned tumor cells as uniform targets, "asking" them about the nature of the T-cell-induced lesion by differential labeling of cellular compartments. His results indicated that with time, progressively larger particles, from ions to DNA, were released from the cell. This was analogous to similar experiments using antibody and complement and measuring the release of different red cell constituent molecules. Thus, he concluded that the primary lesion was an osmotic lesion in the plasma membrane similar to that of antibody and complement (Henney, 1973).

We later confirmed Henney's results, but found that the release of nuclear material occurred with T-cell effectors but not with antibody plus complement (Russell et al., 1980). Upon closer examination (Russell and Dobos, 1980), we observed that the nuclear lesion began at the same time as damage to the cytoplasmic membrane (within 10 min), but actually went to completion within the dying cell faster (15 min) than cytoplasmic components leaked from the plasma membrane (90–120 min). In contrast, DNA required treatment with proteases or nucleases to be released from cells killed by complement and a variety of other agents, including osmotic lysis induced by hypotonicity. Analysis of the DNA at early times (10 min) after T-cell attack revealed that initially high-molecular-weight forms can be extracted by nonionic detergent but the DNA is cleaved over the next 30–60 min into lower-molecular-weight species (Russell et al., 1982). Eventually, repeating units of 180–200 bp are observed (see Figure 2; Russell, 1983; Duke et al., 1983), which suggests endonuclease attack at exposed sites between nucleosomal particles (Keene and Elgin, 1981).

The cytotoxic T lymphocyte (CTL)-initiated nuclear lesion is caused not only by the cleavage of the DNA by endonucleases, but also by the breakdown of the inner nuclear membrane (see Figure 1B; Russell et al., 1982). The lamin structure of the inner nuclear membrane resists DNA extraction by nonionic detergents in undamaged interphase cells (Ottaviano and Gerace, 1985). However, the loss of integrity of the inner nuclear membrane is an early event during CTL-initiated death, but not during complement-mediated death (Figure 1C).

Significance of the Nuclear Lesion

There is no question that a cell whose DNA has been reduced to 200-bp fragments is no longer viable. However, there have been two schools of thought as to the significance of the phenomenon for the target cell being attacked. The first way of thinking is that the nuclear lesion is a marker for events which occur in the cytoplasm of the cell during CTL lysis, but not during lysis initiated by other agents such as complement (Russell and Dobos, 1980). Before the discovery of the CTL-induced nuclear lesion, it had already been shown that CTL could "kill" enucleated cells (Berke and Fishelson, 1977; Siliciano and Henney, 1978). Thus damage to the plasma membrane does not require a nucleus. This has recently been confirmed in experiments with redirected lysis which demonstrate that mammalian red blood cells can also be lysed by CTL triggered by antibodies to the T-cell antigen–receptor complex (Lanzavecchia and Staerz, 1987). Ratner and Clark (1991) have recently demonstrated that this form of lysis requires no active participation of the red cell target.

Other experiments also indicate that the nucleus need not be damaged to produce plasma membrane damage within the same cell. Topoisomerase inhibitors (Nishioka and Welsh, 1991), intracellular Ca^{2+} chelators within the target (Hameed et al., 1989), and appropriate pretreatment of effectors with alkylating agents (Hayes et al., 1989) have been reported to prevent or reduce the nuclear lesion with no effect on the plasma membrane lesion. Using neurons as targets, we have found that physically moving the point of attack away from the nucleus allows rupture of distal membrane

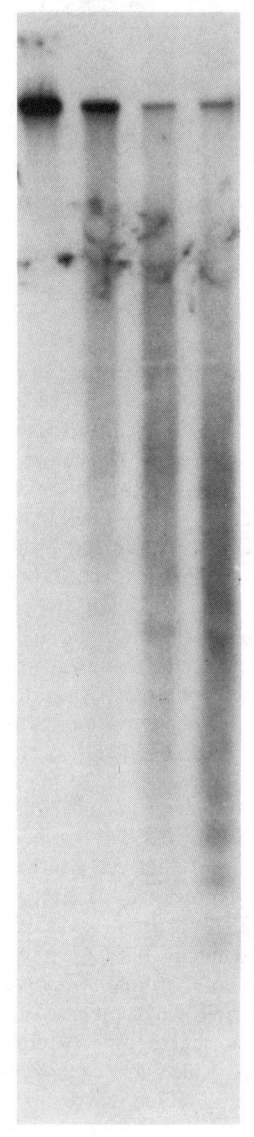

FIGURE 2. Time-dependent DNA fragmentation in CTL targets. ^{125}IUdR released from targets by nonionic detergent at various times after interaction with CTL. Shown is an autoradiograph of an agarose gel. Standards indicate the bands at lower end of the gel are at 180- to 200-bp intervals. Total DNA extracted from control cells or complement-lysed cells remains at the top of the gel indistinguishable from the band at the 10-min time point. (Adapted from Russell, 1983.)

elements without affecting the nucleus (Manning et al., 1987).

Alternatively, Duke and Cohen have promoted the centrality of the nuclear lesion within the CTL target (Duke et al., 1983 and Chapter 20, this volume). In large part this hypothesis was based on experiments using Zn^{2+}, which is a potent inhibitor of endonucleases. Their experiments have found that Zn^{2+} blocks both nuclear and plasma membrane lesions. While a major known effect of Zn^{2+} is endonuclease inhibition, Zn^{2+} also inhibits perforin-mediated lytic activity of granules isolated from cytolytic cells (Podack et al., 1985; Bashford et al., 1988). Ju (1991) has reported that Zn^{2+} has a general effect on T-cell activation using a variety of assays.

The nuclear lesion does accompany all CTL-target interactions to one degree or another (see below). Therefore, regardless of its role in the lytic process, the nuclear lesion may reflect a biological response which has significance beyond the individual CTL-target interaction. The primary role of classical $CD8^+$ CTL is probably protection from intracellular parasites such as viruses and the nuclear lesion may reflect a response that would eliminate potentially infectious material from within the target which would otherwise be released into the surrounding tissue by other forms of lysis. Martz and Howell (1989) have enunciated the antiviral "prelytic halt" hypothesis as the major function of $CD8^+$ cells. Sellins and Cohen (1989) have found that the DNA of some nuclear viruses is in fact destroyed during CTL attack. In contrast, Howell and Martz (1987a) found no effect of CTL on an RNA virus. Thus it is unclear whether this antiviral function can be generalized to all viruses and be demonstrated to actually reduce the release of infectious material through a mechanism other than through interferon-γ (IFN-γ) production.

Initiation of the Nuclear Lesion

The evidence is very strong that the primary plasma membrane lesion in $CD8^+$ cell-mediated lysis is the result of the insertion into the target membrane of a molecule termed perforin (Podack et al., 1985; Acha-Orbea et al., 1990; Podack et al., 1991) or cytolysin (Henkart et al., 1984). This molecule has homology to the terminal complement components (Shinkai et al., 1988; Lowrey et al., 1989) and is vectorially secreted from the CTL in a Ca^{2+}-dependent (Plaut et al., 1976; Golstein and Smith, 1977; Martz, 1977) and protein kinase C-dependent (Russell et al., 1986; Nishimura et al., 1987) fashion by exocytosis of intracellular granules (Henkart, 1985; Haverstick et al., 1991).

The initiation of the nuclear lesion is much less well understood. Gromkowski and colleagues found that in some target cells the unique, CTL-mediated nuclear lesion was present, but less dramatic than in other cells (Gromkowski et al., 1986). This work has been extended by others using the same cloned effector cells with various targets. These experiments have demonstrated that the amount of DNA cleavage is determined by the target cell lineage (Sellins and Cohen, 1991) and not by the species of origin (Howell and Martz, 1987b), as had been suggested by others (Christiaansen and Sears, 1985). While not conclusive, the target cell dependence of the severity of the nuclear lesion is evidence that it is a response of the target to CTL attack rather than a passive function of an activity injected into the target by the CTL.

There is some biochemical support in the literature for the role of granules or perforin in initiating the nuclear lesion (Albritton et al., 1988; Hameed et al., 1989). However, the balance of the evidence would indicate that neither crude granule preparations (Duke et al., 1988; Gromkowski et al., 1988) nor purified perforin/cytolysin (Duke et al., 1989; Hayes et al., 1989) is capable of initiating the nuclear lesion. Morphological analysis of cells lysed by perforin/cytolysin indicates that these cells exhibit similar morphological characteristics to those seen in Figure 1C with complement (Golstein et al., 1991). Contaminating enzymes in various preparations of pore-forming molecules may be the source of the biochemical discrepancies.

Munger and colleagues first suggested that proteases or nucleases within the cytolytic granule might be involved in mediating the nuclear lesion (Munger et al., 1988 and Chapter 14, this volume). High concentrations of granules or purified enzymes in the presence of perforin/cytolysin can release some DNA from damaged cells (Hayes et al., 1989), although the kinetics of ^{51}Cr and iododeoxyuridine ($^{125}IUdR$) release are very different from the lesion initiated by intact CTL. The protease hypothesis was attractive, as broad-specificity proteases were known to release the DNA from cells damaged by other agents (Klein and Perlmann, 1963). In fact, given the ease with which proteases like trypsin release DNA from nuclei of cells lysed by complement, it is quite striking that nuclei lysed by granules are relatively resistant to the plethora of proteases that have been associated with the granules (Jenne and Tschopp, 1988). It remains possible that these granule proteases may in fact contribute to the nuclear lesion in cells with a poor endogenous response. However, it seems unlikely that sufficient granule contents can be transferred in the time required to account for the rapid nuclear damage observed in many target cells.

Tirosh and Berke (1985) suggested that increased intracellular Ca^{2+} in the target was responsible for activating an endogenous pathway resulting in the nuclear lesion. Certainly Ca^{2+} has been implicated in the suicide pathways of a variety of systems (Jewell et al., 1982; McConkey et al., 1989), and Poenie and colleagues (1987) have measured such increases in targets being attacked by CTL. The suggestion that intracellular Ca^{2+} chelators can inhibit DNA damage (Hameed et al., 1989) also supports the Ca^{2+} hypothesis. The recent suggestion that Ca^{2+}-requiring topoisomerases are effectors of the DNA cleavage (Nishioka and Welsh, 1991) is also consistent with such a hypothesis. While it is likely that the effector of the nuclear lesion does in fact require Ca^{2+}, it is not possible to correlate absolute increases in intracellular Ca^{2+} with a nuclear lesion or even lysis (Mercep et al., 1989; Jones et al., 1990; Duke et al., 1991).

A variety of other agents, such as TNF (tumor necrosis factor), lymphotoxin (Schmid et al., 1986), and ATP (Chapters 28 and 29, this volume; Filippini et al., 1990; Zanovello et al., 1989; Zheng et al., 1991), have also been proposed over the last several years to account for the nuclear lesion. TNF and the TNF-like molecule lymphotoxin have been especially extensively studied and clearly can initiate a similar lesion in some cells (Laster et al., 1988; Schmid et al., 1986). However, both TNF and lymphotoxin fail to satisfy the kinetic and target cell spectrum of sensitivity requirements to convincingly demonstrate that they are the agents *generally* responsible for initiating the nuclear lesion induced by CTL.

Based on available evidence, the nuclear lesion either is initiated by a labile granule component which has resisted detection, or is the result of an activity that is independent of granule secretion (Trenn et al., 1987; Ostergaard et al., 1987; Young et al., 1987). Related to the granule-independent mechanism, Young and colleagues have described, on some cytolytic lines, a membrane-bound peptide that is structurally related to TNF (Liu et al., 1989). Further experiments are required to determine the role of this peptide as a general cytotoxic element distinct from activities already ascribed to secreted TNF.

Another working hypothesis has been that the nuclear lesion is the result of a CTL-induced triggering of a molecule at the target cell surface which transmits the suicide signal from the CTL into the target (Russell, 1983). There is increasing evidence that such suicide "triggers" exist on the surface of a variety of cell types (Mercep et al., 1988; Englemann et al., 1990; Trauth et al., 1989; Yonehara et al., 1989). It is possible that CTL activate such endogenous "triggers" on all cell types. A difference between most of these cell-type-specific triggers and the CTL-induced nuclear lesion is the absence of a requirement for macromolecular synthesis in the targets of CTL (Golstein et al., 1991). Therefore, either a cell surface trigger is not the mechanism, or there are differences in the "arming" of the suicide mechanism in the CTL-induced pathway. The recent experiments of Strack et al. (1990) and Lancki et al. (1991) indicating that some $CD4^+$ clones are lytic, but appear unable to activate the perforin gene, may provide the key to understanding the molecular basis of perforin-independent lytic mechanisms (see also Chapters 17 and 18, this volume).

Other Target Cell Events Associated with CTL Attack

Protein Synthesis

The two-cell nature of the CTL–target interaction has limited the capacity to determine contributions of the individual participants to those processes that can be examined morphologically or through pharmacological or radiolabeled studies that can be performed on pretreatment of one cell or the other. There continue to be occasional reports of a requirement for target cell protein synthesis for lysis to occur with $CD8^+$ effectors (Zychlinsky et al., 1991). However, the consensus remains that $CD8^+$ CTL lyse targets in the absence of protein synthesis by either the effector or the target (Thorn and Henney, 1976; Golstein et al., 1991). This does not exclude the necessity of target cells to retain sensitivity to CTL through protein synthesis (Landon et al., 1990). It is difficult to know at this point whether such proteins are required in an active target cell suicide response or at some earlier step in the CTL–target interaction.

Some $CD4^+$ cytolytic effectors can be inhibited by protein synthesis blockade (Ju et al., 1990; Abrams and Russell, 1991). This argues against a preformed granule mechanism in these cells, but does not necessarily eliminate a possible perforin participation. In addition, a subset of $CD4^+$ effectors require target cell protein synthesis when the targets are in an adherent state, but not when

Immune Effector Mechanisms

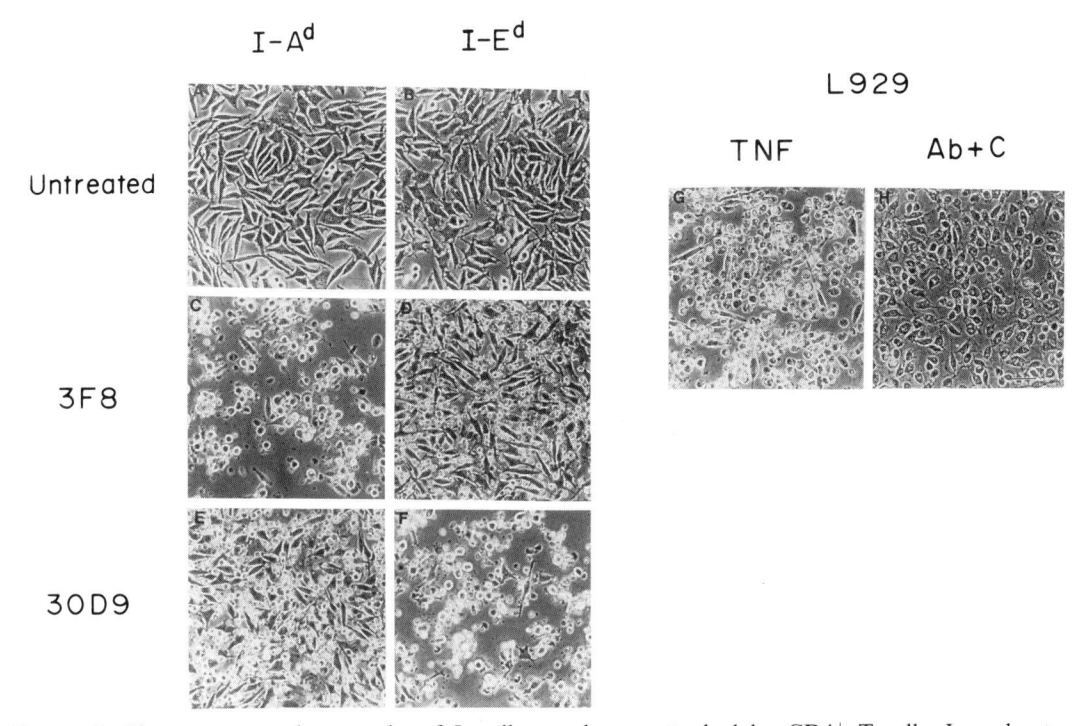

FIGURE 3. Phase-contrast photographs of L cell monolayers attacked by CD4+ T cells. In order to demonstrate antigen specificity and separate the effect from TNF/lymphotoxin activity, we have used MHC class II-specific CD4+ effector cells with TNF-resistant, transfected L cell targets, but a similar phenomenon occurs with MHC class I-specific CD8+ effectors. The target monolayer is disrupted by detaching large, refractile L cells analogous to those mitotic cells in the control monolayer. In the absence of extracellular Ca^{2+}, these detached cells are viable as assessed by cloning assays (Russell et al., 1988). TNF-sensitive targets are detached as dead cells, indicated by the dark appearance of the cells under phase contrast. Complement produces cell ghosts with a retention of the basic form of the attached fibroblast. Bar = 50 μm.

the same targets are attacked as cells in suspension (Abrams and Russell, 1991). The requirement for target protein synthesis in this instance appears more related to a required detachment of the targets from the substrate rather than to the lytic process itself.

Lysosomes

One potential target for an autolytic cascade is the intracellular disruption of lysosomes (DeDuve, 1969). Using microcytochemical techniques, Pitsillides and colleagues found evidence for the intracellular release of a lysosomal enzyme in targets following CTL–target interaction (Pitsillides et al., 1988). The postulated mechanism of this release of lysosomal enzyme is the labilization of lysosomal membranes by activation of polyamine synthesis (Chayen et al., 1990). The mechanistic significance of these findings remains to be established. A lysosomal model would predict a rather nonselective catabolism of intracellular proteins. Other than chromatin and DNA, there is no evidence for general intracellular proteolysis. Rather, the evidence suggests that the intracellular events are a disorganization of structure, with little or no proteolysis. Thus the alterations suggested for changes in the compartmentalization of lysosomal enzymes as a result of interaction with CTL may reflect alterations in the target cytoskeleton responsible for the disorganization of other membrane structures. Attempts to detect a reduction of intracellular pH necessary for optimal catabolic activity by lysosomal components have been unsuccessful (Russell, 1983).

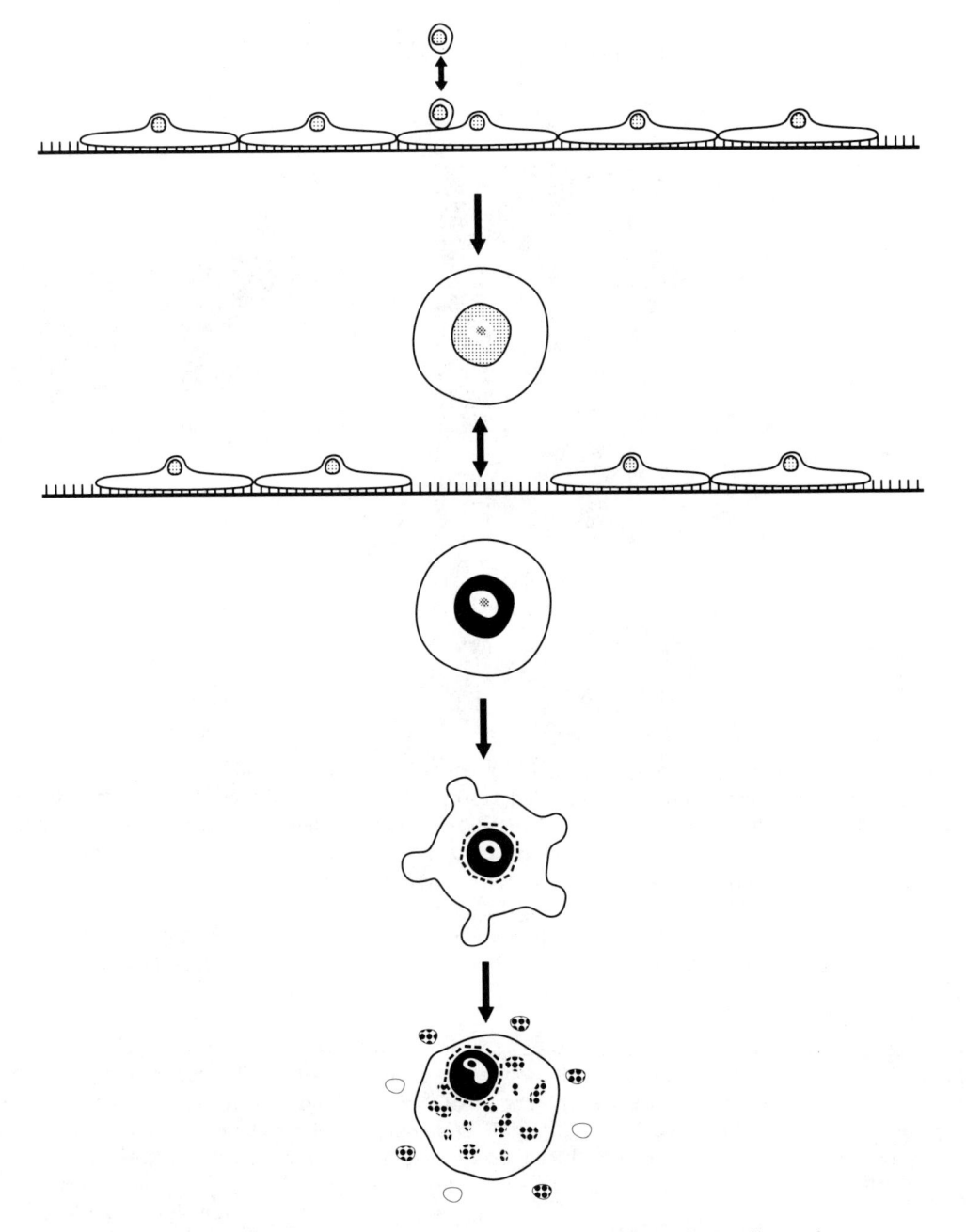

FIGURE 4. Schematic representation of target cell events unique to cytolytic T-cell attack.

Smith and colleages (1991) recently examined the fate of the intralysosomal parasite *Leishmania* in macrophage targets attacked by a CD8⁺ CTL clone. They found that macrophages were rapidly lysed, but the parasites not only remained viable, but could reinfect bystander macrophage. The reduction of viable parasite numbers required a subsequent macrophage activation reaction to IFN-γ produced by the CTL in response to the initial interaction with the infected macrophage. These experiments also argue against a rapid, general lysosomal activation response in targets under CTL attack.

Detachment from Substrate

Most assays for the CTL–target interaction involve

target cells from suspension culture. However, *in vivo* the interaction between T cells and antigen-bearing cells in organized tissue is biologically very important. Therefore we have begun to analyze that interaction using fibroblast and macrophage targets. Figure 3 demonstrates the morphological differences in fibroblast monolayers attacked by different immunological effectors. An early event in T-cell attack is the disruption of the monolayer by detachment of live cells from their substrate. This is an antigen-specific event, as evidenced in Figure 3 by the MHC class II specificity of the activity of CD4$^+$ effectors on transfected targets.

Detachment appears to be a distinct event in the sequence of T cell–target interaction as it can occur in the absence of lysis and degranulation using either Ca^{2+}-free conditions (Russell et al., 1988) or PMA-pretreated effectors (Abrams et al., 1989). Even under physiological conditions, some T cells have the capacity to induce detachment with little target cell lysis (Wang et al., 1993). Furthermore, the event is independent of soluble mediators such as TNF and granule proteases, based on target cell sensitivity as well as bystander, supernatant transfer, and antibody inhibition assays (Abrams and Russell, 1991).

Death induced by TNF or lymphotoxin causes simultaneous target detachment and death (Abrams and Russell, 1991) in the absence of a nuclear lesion in many cell lines (Laster et al., 1988; Abrams and Russell, 1991). In contrast, pore-forming elements such as complement (Figure 3) produce lysis *in situ* in which the remaining ghost retains the general morphology of the adherent cell.

The capacity of T cells, particularly CD4$^+$ cells, to initiate detachment of adherent targets from substrate may play an important role in the inflammatory response and the regulation of immune responses. Attack and denuding of vascular endothelium by CD4$^+$ cells with or without lysis would be equally effective at initiating an inflammatory response. Similarly CD4$^+$ cell-induced activation and migration of antigen-bearing macrophages could be important in recruiting additional T-cell clones and mobilization of activated macrophages to inflammatory sites.

Summary

Figure 4 is a schematic representation of the features of the CTL–target interaction that distinguish it from other forms of death. Initially there is a reversible binding of the T cell to an individual cell in the tissue. This is followed by a signal for detachment of the target cell from the surrounding tissue. Detachment appears to reflect early changes in the target cytoskeleton but can also be a reversible phenomenon. Through additional signals or increased intensity of the same signals, the target becomes irreversibly committed ("programmed") to the lytic pathway. Accompanying the "programming" or "lethal hit", morphological alterations appear in the target cell chromatin and the membrane becomes unable to partition ion gradients. Additional alterations appear in the target cytoskeleton reflected by membrane blebbing, intracellular vacuolation ("apoptotic" bodies), and the loss of integrity of the inner nuclear membrane. Nucleases either from within the nucleus (topoisomerases?) or from the cytoplasm complete the degradation of the cellular genome to ≈ 200-bp fragments.

Acknowledgments

This work has been supported by a Public Health Service Grant CA28533 from the National Cancer Institute. We would also like to express our appreciation to Dr. David Ucker for comments on the manuscript.

References

Abrams SI, McCulley DE, Melcedy-Rey P, Russell JH (1989): Cytotoxic T lymphocyte-induced loss of target cell adhesion and lysis involve common and separate signalling pathways. *J Immunol* 142: 1789–1796

Abrams SI, Russell JH (1991): CD4$^+$ T lymphocyte-induced target cell detachment. A model for T cell-mediated lytic and nonlytic inflammatory processes. *J Immunol* 146: 405–413

Acha-Orbea H, Scarpellino L, Hertig S, Dupuis M, Tschopp J (1990): Inhibition of lymphocyte-mediated cytotoxicity by perforin antisense oligonucleotides. *EMBO J* 9: 3815–3819

Albritton NL, Verret CR, Wolley RC, Eisen HN (1988): Calcium ion concentrations and DNA fragmentation in target cell destruction by murine cloned cytotoxic T lymphocytes. *J Exp Med* 167: 514–527

Bashford CL, Memestrina G, Henkart PA, Pasternak CA (1988): Cell damage by cytolysin: Spontaneous recovery and reversible inhibition by divalent cations. *J Immunol* 141: 3965–3974

Berke G, Fishelson Z (1977): T-lymphocyte mediated cytolysis: Contribution of intracellular components to target cell susceptibility. *Transplant Proc* 9: 671–674

Billingham RE, Brent L, Medawar PB (1954): Quantitative studies on tissue transplantation immunity. II. The origin, strength and duration of actively and adoptively acquired immunity. *Proc Roy Soc London B* 143: 58–80

Brunner KT, Mauel J, Cerottini J-C, Chapus B (1968):

Quantitative assay of the lytic action of immune lymphoid cells on ^{51}Cr-labelled allogeneic target cells *in vitro*. *Immunology* 14: 181–196

Chang TW, Celis E, Eisen HN, Solomon F (1979): Crawling movements of lymphocytes on and beneath fibroblasts in culture. *Proc Natl Acad Sci USA* 76: 2917–2921

Chase MW (1945): The cellular transfer of cutaneous hypersensitivity to tuberculin. *Proc Soc Exp Biol Med* 59: 134–135

Chayen J, Pitsillides AA, Bitensky L, Muir IH, Taylor PM, Askonas BA (1990): T-cell mediated cytolysis: Evidence for target-cell suicide. *J Exp Pathol* 71: 197–208

Christiaansen JE, Sears DW (1985): Lack of lymphocyte-induced DNA fragmentation in human targets during lysis represents a species-specific difference between human and murine cells. *Proc Natl Acad Sci USA* 82: 4482–4485

DeDuve C (1969): The lysosome in retrospect. In: *Lysosomes in Biology and Pathology, Vol. 1*, Dingle JT, Fell HB, eds. Amsterdam: North Holland, p. 3

DiVirgilio F, Pizzo P, Zanovello P, Bronte V, Collavo D (1990): Extracellular ATP as a possible mediator of cell-mediated cytotoxicity. *Immunol Today* 11: 274–277

Don MM, Ablett G, Bishop CJ, Bundesen PG, Konald KJ, Searle J, Kerr JF (1977): Death of cells by apoptosis following attachment of specifically allergized lymphocytes *in vitro*. *Aust J Exp Biol Med Sci* 55: 407–417

Duke RC, Chervenak R, Cohen JJ (1983): Endogenous endonuclease-induced DNA fragmentation: An early even in cell-mediated cytolysis. *Proc Natl Acad Sci USA* 80: 6361–6365

Duke RC, Persechini PM, Chang S, Liu C-C, Cohen JJ, Young JD-E (1989): Purified perforin induces target cell lysis but not DNA fragmentation. *J Exp Med* 170: 1451–1456

Duke RC, Sellins KS, Cohen JJ (1988): Cytotoxic lymphocyte-derived lytic granules do not induce DNA fragmentation in target cells. *J Immunol* 141: 2191–2194

Duke RC, Zulauf R, Cohen JJ, Zheng LM, Young JD-E, Ojcius D. (1991): A rise in intracellular calcium concentrations is not sufficient for the induction of apoptosis. *J Cell Biol* (submitted for publication)

Englemann H, Holtmann H, Brakebusch C, Avni YS, Sarov I, Nophar Y, Hadas E, Leitner O, Wallach D (1990): Antibodies to a soluble form of a tumor necrosis factor (TNF) receptor have TNF-like activity. *J Biol Chem* 265: 14497–14504

Filippini A, Taffs RE, Sitkovsky MV (1990): Extracellular ATP in T-lymphocyte activation: Possible role in effector functions. *Proc Natl Acad Sci USA* 87: 8267–8271

Golstein P, Smith ET (1977): Mechanisms of T-cell mediated cytolysis: The lethal hit stage. *Contemp Top Immunol* 4: 273–300

Golstein P, Ojcius DM, Young JD-E (1991): Cell death mechanisms and the immune system. *Immunol Rev* 121: 29–65

Gromkowski SH, Brown TC, Cerutti PA, Cerottini J-C (1986): DNA of human raji target cells is damaged upon lymphocyte-mediated lysis. *J Immunol* 136: 752–756

Gromkowski SH, Brown TC, Masson D, Tschopp J (1988): Lack of DNA degradation in target cells lysed by granules derived from cytolytic T lymphocytes. *J Immunol* 141: 774–778

Hameed A, Olsen KJ, Lee M-K, Lichtenheld MG, Podack ER (1989): Cytolysis by Ca-permeable transmembrane channels. Pore formation causes extensive DNA degradation and cell lysis. *J Exp Med* 169: 765–777

Haverstick DM, Englehard VH, Gray LS (1991): Three intracellular signals for cytotoxic T lymphocyte-mediated killing. Independent roles for protein kinase C, Ca^{++} influx, and Ca^{++} release from intracellular stores. *J Immunol* 146: 3306–3313

Hayes MP, Berrebi GA, Henkart PA (1989): Induction of target cell DNA release by the cytotoxic T lymphocyte granule protease granzyme A. *J Exp Med* 170: 933–946

Henkart PA (1985): Mechanism of lymphocyte-mediated cytotoxicity. *Annu Rev Immunol* 3: 31–58

Henkart PA, Millard PJ, Reynolds CW, Henkart MP (1984): Cytolytic activity of purified cytoplasmic granules from cytotoxic rat LGL tumors. *J Exp Med* 160: 75–93

Henney CS (1973): Studies on the mechanism of lymphocyte-mediated cytolysis. II. The use of various target cell markers to study cytolytic events. *J Immunol* 110: 73–84

Howell DM, Martz E (1987a): Intracellular reovirus survives CTL-mediated lysis of its host cell. *J Gen Virol* 68: 2899–2907

Howell DM, Martz E (1987b): The degree of CTL-induced DNA solubilization is not determined by the human vs mouse origin of the target cell. *J Immunol*. 138: 3695–3698

Jenne DE, Tschopp J (1988): Granzymes: A family of serine proteases in granules of cytolytic T lymphocytes. *Curr Top Microbiol Immunol* 140: 33–47

Jewell SA, Bellomo G, Thor H, Orrenius S, Smith M (1982): Bleb formation in hepatocytes during drug metabolism is caused by disturbances in thiol and calcium ion homeostasis. *Science* 217: 1257–1259

Jones J, Hallett MB, Morgan BP (1990): Reversible cell damage by T-cell perforins. Calcium influx and propidium iodide uptake into K562 cells in the absence of lysis. *Biochem J* 267: 303–307

Ju S-T (1991): Distinct pathways of CD4 and CD8 cells induce rapid target DNA fragmentation. *J Immunol* 146: 812–818

Ju S-T, Ruddle NH, Strack P, Dorf ME, DeKruyff RH (1990): Expression of two distinct cytolytic mechanisms among murine CD4 subsets. *J Immunol* 144: 23–31

Keene MA, Elgin SCR (1981): Micrococcal nuclease as a probe of DNA sequence organization and chromatin structure. *Cell* 27: 57–64

Kidd JG (1950): Experimental necrobiosis—a venture in cytobiology. *Proc Inst Med Chicago* 18: 50–60

Klein G, Perlmann P (1963): *In vitro* cytotoxic effect of

isoantibody measured as isotope release from labelled target cell DNA. *Nature* 199: 451–453

Klein G, Sjögren HO, Klein E, Hellström KE (1960): Demonstration of resistance against methylcholanthrene-induced sarcomas in the primary autochthonous host. *Cancer Res* 20: 1561–1572

Lancki DW, Hsieh C-S, Fitch FW (1991): Mechanisms of lysis by cytotoxic T lymphocyte clones. Lytic activity and gene expression in cloned antigen-specific CD4+ and CD8+ T lymphocytes. *J Immunol* 146: 3242–3249

Landon C, Nowicki M, Sugawara S, Dennert G (1990): Differential effects of protein synthesis inhibition on CTL and targets in cell-mediated cytotoxicity. *Cell Immunol* 128: 412–426

Landsteiner K, Chase MW (1942): Experiments on transfer of cutaneous sensitivity to simple compounds. *Proc Soc Exp Biol Med* 49: 688–690

Lanzavecchia A, Staerz U (1987): Lysis of non-nucleated red blood cells by cytotoxic T lymphocytes. *Eur J Immunol* 17: 1073–1074

Laster SM, Wood JG, Gooding LR (1988): Tumor necrosis factor can induce both apoptotic and necrotic forms of cell lysis. *J Immunol* 141: 2629–2634

Liu C-C, Detmers PA, Jiang S, Young JD-E (1989): Identification and characterization of a membrane-bound cytotoxin of murine cytolytic lymphocytes that is related to tumor necrosis factor/cachectin. *Proc Natl Acad Sci USA* 86: 3286–3290

Lowrey DM, Aebischer T, Olsen K, Lichtenheld M, Rupp F, Hengartner H, Podack ER (1989): Cloning analysis and expression of murine perforin 1 cDNA, a component of cytolytic T-cell granules with homology to complement component C9. *Proc Natl Acad Sci USA* 86: 247–251

Manning PT, Johnson EM, Jr, Wilcox CL, Palmatier MA, Russell JH (1987): MHC-specific cytotoxic T lymphocyte killing of dissociated sympathetic neuronal cultures. *Am J Pathol* 128: 395–409

Martz E (1977): Mechanisms of specific tumor-cell lysis by alloimmune T lymphocytes: Resolution and characterization of discrete steps in the cellular interaction. *Contemp Top Immunobiol* 7: 301–361

Martz E, Howell DM (1989): CTL: Virus control cells first and cytolytic cells second? DNA fragmentation, apoptosis and the prelytic halt hypothesis. *Immunol Today* 10: 79–86

McConkey DJ, Hartzell P, Nicotera P, Orrenius S (1989): Calcium-activated DNA fragmentation kills immature thymocytes. *FASEB J* 3: 1843–1849

Mercep M, Bluestone JA, Noguchi PD, Ashwell JD (1988): Inhibition of transformed T cell growth *in vitro* by monoclonal antibodies directed against distinct activating molecules. *J Immunol* 140: 324–335

Mercep M, Noguchi PD, Ashwell JD (1989): The cell cycle block and lysis of an activated T cell hybridoma are distinct processes with different Ca^{++} requirements and sensitivity to cyclosporin A. *J Immunol* 142: 4085–4092

Munger WE, Berribi GA, Henkart PA (1988): Possible involvement of CTL granule proteases in target cell DNA breakdown. *Immunol Rev* 103: 99–109

Nishimura T, Burakoff SJ, Herrmann SA (1987): Protein kinase C required for cytotoxic T lymphocyte triggering. *J Immunol* 139: 2888–2891

Nishioka WK, Welsh RM (1991): Inhibition of cytotoxic T lymphocyte-induced target cell DNA fragmentation, but not lysis, by inhibitors of DNA topoisomerases I and II. *J Exp Med* (in press)

Ostergaard HL, Kane KP, Mescher MF, Clark WR (1987): Cytotoxic T lymphocyte mediated lysis without release of serine esterase. *Nature* 330: 71–72

Ottaviano Y, Gerace L (1985): Phosphorylation of the nuclear lamins during interphase and mitosis. *J Biol Chem* 260: 624–632

Pitsillides AA, Taylor PM, Bitensky L, Chayen J, Muir IH, Askonas BA (1988): Rapid changes in target cell lysosomes induced by cytotoxic T-cells: Indication of target suicide. *Eur J Immunol* 18: 1203–1208

Plaut M, Bubbers JE, Henney CS (1976): Studies on the mechanisms of lymphocyte-mediated cytolysis. VII. Two stages in the T cell-mediated lytic cycle with distinct cation requirements. *J Immunol* 116: 150–155

Podack ER, Hengartner H, Lichtenheld MG (1991): A central role of perforin in cytolysis. *Annu Rev Immunol* 9: 129–157

Podack ER, Young JD-E, Cohn ZA (1985): Isolation and biochemical and functional characterization of perforin 1 from cytolytic T-cell granules. *Proc Natl Acad Sci USA* 82: 8629–8633

Poenie M, Tsien RY, Schmitt-Verhulst A-M (1987): Sequential activation and lethal hit measured by $(Ca^{++})_i$ in individual cytolytic T cells and targets. *EMBO J* 6: 2223–2232

Ratner A, Clark WR (1991): Lack of target cell participation in CTL-mediated lysis. *J Immunol* 147: 55–59

Russell JH (1983): Internal disintegration model of cytotoxic lymphocyte-mediated target damage. *Immunol Rev* 72: 97–118

Russell JH, Dobos CB (1980): CTL-induced nuclear disintegration of the target begins within minutes of cell contact. *J Immunol* 125: 1256–1261

Russell JH, Masakowski VR, Dobos CB (1980): Physiological distinction between target cell death mediated by cytotoxic T lymphocytes and antibody plus complement. *J Immunol* 124: 1100–1105

Russell JH, Masakowski VR, Rucinsky T, Phillips G (1982): Characterization of the nature and kinetics of the cytotoxic T lymphocyte-induced nuclear lesion in the target. *J Immunol* 128: 2087–2094

Russell JH, McCulley DE, Taylor AS (1986): Antagonistic effects of phorbol esters on lymphocyte activation. Evidence that protein kinase C provides an early signal associated with lytic function. *J Biol Chem* 261: 12643–12648

Russell JH, Musil L, McCulley DE (1988): Loss of adhesion. A novel and distinct effect of the cytotoxic T lymphocyte-target interaction. *J Immunol* 140: 427–432

Sanderson CJ (1976): The mechanism of T cell mediated cytotoxicity. II. Morphological studies of cell death by time-lapse microcinematography. *Proc Roy Soc London B* 192: 241–255

Schmid DS, Tite JP, Ruddle NH (1986): DNA fragmentation: Manifestation of target cell destruction mediated by cytotoxic T-cell lines, lymphotoxin-secreting helper T-cell clones, and cell-free lymphotoxin-containing supernatant. *Proc Natl Acad Sci* 83: 1881–1885

Sellins KS, Cohen JJ (1989): Polyomavirus DNA is damaged in target cells during cytotoxic T-lymphocyte-mediated killing. *J Virol* 63: 572–578

Sellins KS, Cohen JJ (1991): Cytotoxic T lymphocytes induce different types of DNA damage in target cells of different origins. *J Immunol* 147: 795–803

Shinkai Y, Takio K, Okumura K (1988): Homology of perforin to the ninth component of complement (C9). *Nature* 334: 525–527

Siliciano RF, Henney CS (1978): Enucleated cells as targets for cytotoxic attack. *J Immunol* 121: 186–191

Smith LE, Rodrigues M, Russell DG (1991): The interaction between CE8$^+$ cytotoxic T cells and *Leishmania*-infected macrophages. *J Exp Med* 174: 499–505

Strack P, Martin C, Saito S, DeKruyff RH, Ju S-T (1990): Metabolic inhibitors distinguish cytolytic activity of CD4 and CD8 clones. *Eur J Immunol* 20: 179–184

Thorn RM, Henney CS (1976): Studies on the mechanism of lymphocyte-mediated cytolysis. VI. A reappraisal of the requirements for protein synthesis during cell-mediated lysis. *J Immunol* 116: 146–149

Tirosh R, Berke G (1985): T lymphocyte-mediated cytolysis as an excitatory process of the target. I. Evidence that the target may be the site of Ca^{++} action. *Cell Immunol* 95: 113–123

Trauth BC, Klas C, Peters AMJ, Matzku S, Moller P, Falk W, Debatin K-M, Krammer PH (1989): Monoclonal antibody-mediated tumor regression by induction of apoptosis. *Science* 245: 301–305

Trenn G, Takayama H, Sitkovsky MV (1987): Exocytosis of cytolytic granules may not be required for target cell lysis by cytotoxic T-lymphocytes. *Nature* 330: 72–74

Wang R, Abrams SI, Loh DY, Hsieh C-S, Murphy KM, Russell JH (1993): Separation of CD4$^+$ functional responses by peptide dose in Th1 and Th2 subsets expressing the same transgenic antigen receptor. *Cell Immunol* (in press)

Wilson DB (1965): Quantitative studies on the behavior of sensitized lymphocytes *in vitro*. *J Exp Med* 122: 143–166

Winn H (1959): The immune response and the homograft reaction. *Natl Cancer Inst Monogr* 2: 113–138

Wyllie AH, Kerr JFR, Currie AR (1980): Cell death. The significance of apoptosis. *Int Rev Cytol* 68: 251–306

Yonehara S, Ishii A, Yonehara M (1989): A cell-killing monoclonal antibody (anti-fas) to a cell surface antigen co-downregulated with the receptor of tumor necrosis factor. *J Exp Med* 169: 1747–1756

Young JD-E, Clark WR, Liu C-C, Cohn ZA (1987): A calcium- and perforin-independent pathway of killing mediated by murine cytolytic lymphocytes. *J Exp Med* 166: 1894–1899

Zanovello P, Rosato A, Bronte V, Cerundolo V, Treves S, DiVirgilio F, Pozzan T, Biasi G, Collavo D (1989): Interaction of lymphokine-activated killer cells with susceptible targets does not induce second messenger generation and cytolytic granule exocytosis *J Exp Med* 170: 665–677

Zheng LM, Zychlinsky A, Liu C-C, Ojcius DM, Young JD-E (1991): Extracellular ATP as a trigger for apoptosis of programmed cell death. *J Cell Biol* 112: 279–288

Zychlinsky A, Zheng LM, Liu C-C, Young JD-E (1991): Cytolytic lymphocytes induce both apoptosis and necrosis in target cells. *J Immunol* 146: 393–400

20
Apoptosis and Cytotoxic T Lymphocytes

Richard C. Duke

Introduction: Apoptosis in the Immune System

Apoptosis is a mode of cell death that occurs under normal physiological conditions. *In vivo*, it is most often found during normal tissue turnover, embryogenesis, metamorphosis, and endocrine-dependent tissue atrophy (Kerr et al., 1972; Wyllie et al., 1980). The inference has been made that cell death in these instances is "programmed," meaning that it is a normal function of the cell, tissue, or organ. This has led some investigators to use the terms *apoptosis* and *programmed cell death* (*PCD*) interchangeably, although it is becoming apparent that there are examples of PCD which are not apoptotic and examples of apoptosis which are not truly programmed (Cohen, 1991; Cohen et al., 1992). Nevertheless, the fact that the morphological and biochemical changes associated with apoptosis are highly conserved in various cell types and systems suggests that an apoptotic mechanism (cell death program) exists in all cells that can be activated under the appropriate physiological conditions.

Some of the morphological and biochemical criteria that have been employed to assess apoptosis are described in Wyllie et al., 1980. In brief, the morphological appearance of apoptosis is characterized by progressive loss of cell volume and widespread condensation and fragmentation of nuclear chromatin (Wyllie, 1981). Cytoplasmic organelles, including mitochondria, appear normal. Biochemically, loss of cell volume translates as an increase in cell density, allowing viable apoptotic cells to be separated from nonapoptotic cells by density gradient centrifugation (Wyllie and Morris, 1982). Disruption of chromatin structure is often correlated with internucleosomal DNA cleavage in apoptotic cells (Skalka et al., 1976; Wyllie, 1980; Cohen et al., 1985; Duvall and Wyllie, 1986, Cohen et al., 1992). It should be stressed that apoptosis is only poorly understood and must be approached empirically at the present time, since it is unclear whether all cells undergoing apoptosis will undergo the morphological and biochemical changes described above (reviewed in Cohen et al., 1992).

The cells of the immune system provide numerous examples of apoptotic cell death (Cohen et al., 1985, 1992). In fact, it appears that the majority of lymphocytes, macrophages, and granulocytes are extremely short-lived. The circumstances in which apoptosis may be induced in these cells are as diverse as the cells themselves. For example, immature T lymphocytes can be induced to undergo apoptosis upon exposure to physiological concentrations of glucocorticoids (Wyllie, 1980; Vanderbilt et al., 1982; Cohen and Duke, 1984), low-dose gamma irradiation (Shalka et al., 1976; Zvonareva et al., 1983; Sellins and Cohen, 1987), certain toxins including dioxin (Bell and Jones, 1982; McConkey et al., 1988), heatshock (Sellins and Cohen, 1991), and antibodies that crosslink their antigen receptors (McConkey et al., 1989a; Shi et al., 1989; Smith et al., 1989). The mechanisms controlling programmed cell death in leukocytes are equally complex; apoptosis in cells of the immune system can be either dependent on, independent of, or induced by cessation of protein synthesis (reviewed in Cohen, 1991; Cohen et al., 1992).

While apoptosis can occur at any time during the lifetime of a cytotoxic T lymphocyte (CTL), it is beyond the scope of this discussion to describe apoptosis during T-cell development, and so we will concentrate on two examples that occur during a typical CTL-mediated immune response. These examples are (1) death of dependent T cells upon removal of growth factor (Bishop et al., 1985; Duke and Cohen, 1986) and (2) death of target cells induced by CTL (Battersby et al., 1974; Don et al., 1977; Duke et al., 1983; Russell, 1983; Stacey et al., 1985; Duke et al., 1986).

Apoptosis in CTL following growth factor deprivation

As a result of the initial binding interaction with an antigen-presenting cell, CTL become "activated" and express high-affinity receptors for interleukin-2 (IL-2) (Sharon et al., 1986; Weissman et al., 1986; Robb et al., 1987; Teshigawara et al., 1987). IL-2 secreted by activated T helper (T_H) cells, or by the CTL itself, binds to the high-affinity receptor on the CTL, which eventually proliferates. Gillis and Smith were among the first investigators to describe how antigen-specific CTL could be maintained *in vitro* with IL-2 derived from T_H cells. In their initial report they indicated that these CTL were not merely responsive to the growth factor, but were totally addicted, in that they died if IL-2 was withheld (Gillis et al., 1978).

Growth factors induce proliferation following activation of certain genes and the subsequent macromolecular synthesis necessary for cell growth and division (Cantrell and Smith, 1984). Removal of growth factors would be expected to inhibit proliferation; however, the majority of antigen-specific cells actually die rather than become quiescent (Gillis et al., 1978). Cells deprived of growth factors should die eventually from a lack of appropriate stimulation. However, the biochemical characteristics of growth factor withdrawal-induced death indicate that apoptosis, rather than acquiescent expiration, is involved.

The first cells tested were the CTLL-2 and HT-2 cell lines, which were derived from antigen-specific T cell clones that had lost the ability to respond to antigen but grew continuously in the presence of exogenously added growth factor. This growth pattern is unlike that of typical cytotoxic and helper T cells, which require continued antigenic stimulation in order to maintain the necessary level of expression of high-affinity IL-2 receptors. CTLL-2 and HT-2 cells are exquisitely sensitive to growth factor removal and begin to undergo apoptosis as early as 6 hr after withdrawal (Duke and Cohen, 1986). Antigen-specific T cells, in contrast, are slightly more recalcitrant to IL-2 deprivation and do not begin to fragment their DNA, and die, until 24 hr or longer.

IL-2 seems both to promote proliferation and to prevent induction of apoptosis. This is best illustrated with the antigen-specific CTL. As antigen concentration declines, less IL-2 is produced by T_H cells, and high-affinity IL-2 receptor levels decrease on antigen-specific CTL and other effector T cells. However, if growth factor, but not antigen, is provided to antigen-specific CTL, they neither proliferate nor die (Duke and Nash, in preparation). It seems that IL-2 binding to the low-affinity receptor that is constitutively expressed on T cells may be sufficient to allow the cells to survive in the absence of antigenic stimulation. Under these *in vitro* conditions, IL-2 acts to inhibit apoptosis. Further investigation is required to understand the molecular basis for this observation.

Physiologically, apoptosis of antigen-responsive CTL upon antigen clearance is advantageous for at least two reasons. First, induction of programmed cell death assures rapid termination of an immune response without the metabolic costs involved in sustaining a large number of effector cells. Second, death of the expanded clones of lymphokine-dependent effector cells prevents interference with subsequent immune responses to unrelated antigens by avoiding competition for limiting concentrations of growth factors.

Target Cell Apoptosis Induced by CTL

At present, the most popular models of lymphocyte-mediated cytotoxicity suggest that killer lymphocytes kill in a manner similar to that of complement. In these models, the killer cell becomes triggered following binding to the target cell such that it secretes the contents of "lytic granules" into the intercellular space between itself and its target (Young and Cohn, 1986). Lytic granules isolated from CD8-bearing T-cell clones (Podack et al., 1985a, 1985b) or from rat, natural killer (NK)-like, large granular lymphocyte (LGL) tumors (Henkart et al., 1984) contain a pore-forming protein termed perforin or cytolysin and various enzymatic activities, including serine esterases (Masson et al., 1985; Pasternak and Eisen, 1985). It has been shown that perforin molecules insert into cell membranes and, in the presence of calcium, assemble into porelike structures (Podack, 1985).

The pore-forming models postulate that once a threshold amount of membrane damage is attained, the target cell dies by colloid osmotic lysis, just as is observed during complement-mediated killing. This type of cell death is classifed as "necrotic" (Kerr et al., 1972; Wyllie et al., 1980). Indeed, when cells are treated with lytic granules or purified perforin in the absence of extracellular calcium, they do undergo colloid osmotic lysis if calcium is added subsequently (Duke et al., 1988, 1989). However, the same cells killed by CTL do not appear to be passive victims of a pore-forming process; rather, the morphology of their death suggests they are dying by apoptosis.

The idea that target cells are actively participating in their own death is based on observations of cell-mediated killing both *in vitro* and *in vivo* during viral infections, autoimmune reactions, and graft rejection (Battersby et al., 1974; Slavin and Woodruff, 1974; Don et al., 1977; Sanderson and Glauert, 1977; Searle et al., 1977; Matter, 1979; Russel et al., 1980; Wyllie et al., 1980; Russell et al., 1982; Duke et al., 1983; Duke and Cohen, 1988). By morphological and biochemical criteria, lymphocyte-mediated killing is an example of apoptosis.

There are several prominent features of lymphocyte-mediated killing that distinguish it from colloid osmotic lysis. First, unlike cells undergoing colloid osmotic lysis, targets of CTL, like apoptotic cells, do not swell initially but actually lose volume. The reason why target and apoptotic cells shrink is unknown. Second, dramatic changes in chromatin organization and structure can be detected in cells undergoing apoptosis prior to their lysis.

Normal cell nuclei, or nuclei of complement-lysed cells, have "structure"; there are variations in intensity reflecting the distribution of euchromatin and heterochromatin. The nuclei of apoptotic cells, including target cells which have interacted with CTL, contain highly condensed chromatin. Condensation first takes the form of crescents around the periphery of the nucleus, and eventually the entire nucleus can present as one or a group of featureless, bright spherical beads. Changes in nuclear structure temporally precede cell lysis, as detected by release of vital dyes.

Biochemically, killer lymphocyte-induced nuclear damage is manifested as extensive internucleosomal DNA cleavage, which can be readily detected in centrifugal sedimentation assays or by gell electrophoresis (Russell et al., 1980, 1982; Duke et al., 1983). The degree of DNA cleavage correlates very closely with the number of cells with apoptotic nuclei, such that a measured value of 50% fragmented DNA corresponds with 50% of the cells having apoptotic nuclei and vice-versa (Duke and Cohen, 1988, and unpublished observations). In agreement with morphological data, target cell DNA fragmentation precedes loss of membrane integrity (lysis), as measured by chromium release, by 30–120 min (Duke et al., 1986). It is important to stress that none of the morphological or biochemical changes in chromatin structure associated with lymphocyte-mediated killing are observed during killing mediated by complement or purified perforin (Duke et al., 1983; Russell, 1983; Duvall and Wyllie, 1986).

Apoptotic-like morphological and biochemical changes are also observed when susceptible tumor cells are killed by NK and lymphokine-activated killer (LAK) cells (Stacey et al., 1985; Duke et al., 1986; Zychlinski et al., 1990, 1991), tumor necrosis factor (TNF-α) (Laster et al., 1988), lymphotoxin (TNF-β) (Schmid et al., 1986), natural killer cell cytotoxic factor (NKCF) (Duke et al., 1986), or "leukalexin" (Liu et al., 1987). In each of these instances, the mode of tumor cell death is the same and is distinct from that observed when the same cells are killed with complement, purified perforin, or other pore-forming molecules (Duke et al., in preparation). These distinct changes provide the most compelling evidence that the "pore-forming" models of lymphocyte-mediated cytotoxicity need to be modified and that new molecules and mechanisms to explain how killer cells function need to be addressed.

How Might the Killer Lymphocyte Induce Target Cell Apoptosis?

It is an intriguing idea that killer lymphocytes hold the secret to activating an intrinsic cell death program in any cell. Despite much research, there is no evidence to support any particular mechanism; however, two possibilities seem likely: (1) the killer cell transfers the relevant enzymes or enzyme activators in the target cell, or (2) the killer cell transfers an activating factor or signals the target cell in such a way that an intrinsic cell death program involving nuclear damage is initiated.

Although the killer lymphocyte could directly transfer the nuclear-modifying enzymes to the target, there is little evidence to support such an idea. There have been several reports that certain preparations of lytic granules contain an endonuclease or a factor that can induce endonuclease activity in target cells (Munger et al., 1985; Konigsberg and Podack, 1986; Munger et al., 1988; Hameed et al., 1989; Hayes et al., 1989); however, these reports have not been verified (Duke et al., 1988; Gromkowski et al., 1988; Duke et al., 1989, and unpublished observations). Purified perforin is unable to induce the type of DNA fragmentation and apoptotic morphology observed during CTL-mediated killing (Duke et al., 1989). In addition, several killer lymphocytes that have potent cytolytic activity associated with DNA fragmentation do not contain detectable perforin at the protein or mRNA levels (Dennert et al., 1987; Berke and Rosen, 1988; Nagler-Anderson et al., 1989; Strack et al., 1990; Welsh et al., 1990; Berke, 1991). These observations suggest that mediators other than perforin are likely to be involved in the induction of nuclear damage.

TNF-α (Laster et al., 1988), TNF-β (Schmid et al., 1986), and a TNF-like molecule termed leukalexin (Liu et al., 1987) can induce apoptosis in tumor cells but are poor candidates for the mechanism of lymphocyte-mediated DNA fragmentation. First, these molecules are cytotoxic for only a limited number of tumor cells, whereas CTL can kill any normal (e.g., allogeneic), virally infected, or tumor cell to which they can bind. Second, TNF-related factors are slow-acting, inducing apoptosis in susceptible cells only after 24 hr or more of exposure. CTL can induce DNA fragmentation in minutes. Finally, while some CTL express TNF-β or leukalexin, some cells with potent cytolytic activity do not (Ju et al., 1990; Ju, 1991).

A soluble, rapidly acting, apoptosis-inducing factor cannot be ruled out, although the experiments describing such a molecule(s) (Munger et al., 1988; Duke et al., 1989) have been difficult to reproduce. ATP, secreted by the CTL, has been implicated as a possible mediator of CTL-induced apoptosis (Filippini et al., 1990; Zanovello et al., 1990); however, whereas some tumor cells can be induced to undergo apoptosis when exposed to ATP, others are quite resistant (Zanovello et al., 1990; Zheng et al., 1991). The data obtained thus far suggest that secreted ATP does not represent a general mechanism to account for CTL-induced apoptosis.

A very recent report by Shi and colleagues (1992) describing a highly labile soluble factor isolated from rat, NK-like, LGL tumor cell granules provides the best evidence to date in support of a soluble apoptosis-inducing molecule. This factor, termed "fragmentin," shares a high degree of homology with granzyme A (see chapters in Section V, this volume) and, in the presence of purified perforin, induces rapid DNA fragmentation in cells. Curiously, fragmentin-induced DNA cleavage seems to follow, rather than precede, cytolysis; however, this result could be due to an underestimate of DNA damage inherent in the DNA solubilization assay used by Shi et al. (Duke and Cohen, 1988; Martz and Howell, 1989).

How the CTL would deliver an apoptosis-inducing factor is unclear, but it probably would not involve granule exocytosis. It has been shown that CTL-mediated killing can occur under conditions that preclude exocytosis (Ostergaard et al., 1987; Trenn et al., 1987; Takayama et al., 1988; Trenn et al., 1989; Nash and Duke, 1991). Under these conditions, granule exocytosis is not detected, yet the CTL induce lysis *and* DNA fragmentation (Ostergaard and Clark, 1989). Thus, it seems most likely that the CTL might directly signal activation of a suicide program.

A Working Model of CTL-Mediated Cytolysis

Given the available data, the following model can be constructed to explain how CTL kill (Figure 1). Engagement of the CTL antigen receptor triggers the CTL to deliver the so-called lethal hit (Wagner and Rollinghoff, 1974; Martz, 1975). The lethal hit is operationally defined as the point from which the target cell is irreversibly "programmed for lysis" (Golstein and Smith, 1977; Martz, 1977) and is associated with secretion of lytic granule contents from the CTL (Henkart, 1985; Pasternak and Eisen, 1985; Podack, 1985), a rise in intracellular calcium concentration in the target cell (Martz et al., 1983; Tsien and Poenie, 1986), and induction of DNA fragmentation (Russell and Dobos, 1980; Duke et al., 1983; Duke and Cohen, 1988). Thus, triggering of the CTL results in at least two events: (1) direct damage to the target cell membrane and (2) initiation of the process(es) that leads to induction of DNA cleavage and breakdown of overall nuclear structure.

Induction of Target Cell Membrane Damage

The prevailing data suggest that damage to the target cell membrane occurs as a result of granule exocytosis with subsequent insertion of perforin monomers in the target cell membrane (Podack, 1985). Granule exocytosis can be shown to be polarized towards the tight junction formed between the killer lymphocyte and its target (Yannelli et al., 1986; Gray et al., 1988). After degranulation and perforin insertion, calcium causes the perforin to aggregate into porelike structures that cause the cell to become leaky. This form of membrane damage, leakiness, has been detected at lytic granule or

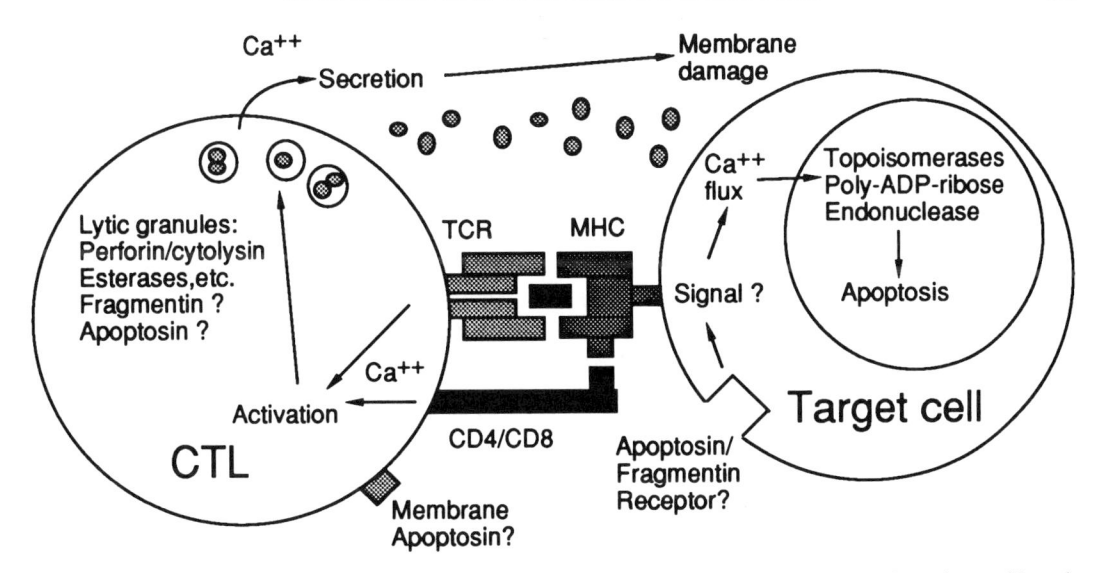

FIGURE 1. Model of CTL-mediated cytolysis. After binding to the target cell, the CTL is "triggered" such that (1) the CTL causes direct damage to the target cell's plasma membrane, and (2) the CTL activates an intrinsic pathway in the target cell which leads to apoptotic-like nuclear damage.

perforin concentrations which are sublethal (Young et al., 1986).

Induction of Apoptotic-Like Nuclear Damage in the Target Cell

It seems likely that the enzymes which cause DNA fragmentation and disrupt chromatin structure are already present in all cells as proenzymes. The proenzymes, when activated, encompass the final common pathway of apoptotic cell death. The CTL interacts with the target via specific target-cell membrane proteins, including, but not limited to, major histocompatibility complex (MHC) molecules. Recently, a membrane-associated factor(s) has been obtained from non-specific LAK cells (Felgar and Hiserodt, 1990). This factor induces apoptosis in any cell to which it is added and would certainly make an attractive candidate for an apoptosis-inducing molecule. Perhaps the binding of this molecule to, or the proteolytic action of fragmentin (Shi et al., 1992) on, some ubiquitously expressed receptor results in the transduction of a death signal to the target cell.

The evidence for such a signal transduction mechanism may be provided by the rise in target cell intracellular calcium concentration $[Ca^{2+}]_i$ which temporally coincides with the target cell's being irreversibly programmed to die. Using Quin-2 or BAPTA to chelate target cell intracellular free calcium prevents DNA fragmentation but does not always block killing (Hameed et al., 1989;

McConkey et al., 1991). This observation is a reminder that the cytotoxic cell employs multiple cytolytic mechanisms to assure killing (Young et al., 1987; Ostergaard and Clark, 1989), but more importantly, it suggests that a rise in target cell intracellular calcium is required for the CTL to induce apoptosis. It should be noted that a rise in $[Ca^{2+}]_i$ is not sufficient to induce apoptosis; a number of agents, including the calcium ionophores A23187 and ionomycin, as well as sublethal and lethal concentrations of the pore-forming proteins mellitin, staphylococcal α-toxin, complement, and perforin, do not mediate DNA fragmentation, even though they all cause a detectable rise in intracellular calcium (Duke et al., in preparation). How the CTL induces apoptosis remains to be elucidated.

Roles of Membrane Damage and Apoptotic-Like Nuclear Damage in Target Cell Lysis

The relative contributions of membrane and apoptotic-like internal damage to target cell lysis are unknown at the present time. Nevertheless, two questions can be addressed:

In the Absence of Induction of Nuclear Damage, Can Lysis Occur?

At very high effector-to-target ratios (greater than 200:1 with highly enriched effector cells), NK cells and CTL are able to kill some of the targets without inducing DNA fragmentation (Duke et al., 1986; Gromkowski et al., 1988). It has also been reported that lymphocyte-mediated killing can sometimes present a mixed morphology of both necrotic and apoptotic cells (Zychlinsky et al., 1991). Therefore, if several killer lymphocytes are bound to a single target, or if these killer lymphocytes are able to rapidly deliver a lethal amount of perforin, the target cell may die without the induction of nuclear damage.

In contrast, it has been reported that target cell mutants that apparently lack the suicide program cannot be killed by CTL (Ucker, 1987). Zinc and aurintricarboxylic acid, potent inhibitors of the endogenous endonuclease thought to be responsible for DNA fragmentation (Cohen and Duke, 1984), block CTL-mediated DNA fragmentation as well as lysis (Duke et al., 1983; Cohen et al., 1985; McConkey et al., 1989b). The most plausible conclusion that can be drawn from these observations is that the amount of membrane damage induced by killer lymphocytes in most cases will be sublethal for cells that have not been induced to fragment their DNA.

In the Absence of Induction of Membrane Damage, Will Lysis Occur?

Inhibition of granule exocytosis does not block CTL-mediated DNA fragmentaiton or lysis (Ostergaard et al., 1987; Trenn et al., 1987; Takayama et al., 1988; Trenn et al., 1989; Nash and Duke, 1991). Lack of granule exocytosis is defined as inhibition of esterase release into the culture medium. It remains entirely possible that enough perforin is being delivered to the target cell under these conditions to result in lysis, and so the question cannot be answered with this observation. Antigranule antibodies block NK cell but not CTL-mediated cytolysis (Henkart et al., 1985; Berke, 1991). This also does not answer the question, and in order for this to be truly resolved, killer cells which either cannot produce or secrete perforin need to be developed (Krahenbuhl and Tschopp, 1991).

If it is possible for a CTL to induce nuclear damage without inducing membrane damage, we would expect the target cell to die with the kind of kinetics associated with truly programmed cell death; however, the experimental evidence does not support this idea, as CTL-mediated killing occurs with the same kinetics under conditions where granule exocytosis is blocked. In addition, even though the similarities between programmed cell death and CTL-mediated cytolysis suggest that killer cells activate an intrinsic apoptotic process in their targets, it is also clear that CTL-mediated killing is more complex than typical apoptosis. In support of this, cell death (lysis) occurs with faster kinetics following induction of DNA fragmentation in CTL-mediated killing than in truly programmed cell death (e.g., IL-2 withdrawal from dependent T cells; Duke and Cohen, 1986). This is a real difference in mechanism, and not just due to differences in target cell phenotype, as is readily apparent when the kinetics of lysis of the same cells (CTLL-2) are compared during CTL-mediated cytolysis and following growth factor removal. CTLL-2 die approximately 60 min after induction of CTL-mediated DNA fragmentation versus 4–6 hr after growth factor withdrawal-induced nuclear damage (Duke and Cohen, 1988).

Perhaps Apoptotic-Like Nuclear Damage Synergizes with Membrane Damage Leading to Rapid Lysis?

It is the ultimate goal of the killer lymphocyte to kill its target. It seems likely, therefore, that killer lymphocytes use multiple cytolytic mechanisms to assure rapid killing (Young et al., 1987; Ostergaard and Clark, 1989; Ju et al., 1990). In a very simplistic model, we propose that the induction of DNA fragmentation renders the target cell susceptible to what would otherwise be a sublethal amount of membrane damage caused by perforin. If the combination of nuclear damage and perforin-induced membrane damage is insufficient to rapidly kill the target cell, then perhaps TNF-α, TNF-β, NKCF, leukalexin, or some other slow-acting cytotoxic factor provides the necessary effector function.

What is the Role of Perforin in Lymphocyte-Mediated Cytolysis?

In an attempt to address this question, Henkart's group transfected the perforin gene into rat basophilic leukemia cells and showed that the transfectants could lyse red cells but not nucleated cells (Shiver and Henkart, 1991). They suggested that the failure of these cells to lyse nucleated cells was due to their inability to deliver a lethal amount of perforin. An alternative explanation is that nucleated cells are only sensitive to low amounts of perforin if they have been induced to undergo apoptosis. Erythrocytes are cells that have undergone apoptosis as part of their differentiation process and are therefore sensitive to perforin secreted by the transfectants. A similar conclusion can be drawn from the recent results of Ratner and Clark (1991).

Tschopp and colleagues tried a different approach to determine whether perforin was required for killing. This approach involved the use of perforin antisense oligonucleotides (Acha-Orbea et al., 1990). These oligos were unable to prevent killing mediated by activated CTL clones but were able to partially inhibit the development of cytotoxic effector function from CTL precursors. Tschopp's group is currently attempting to create perforin-ablated mice, which may yield definitive proof concerning the role of perforin in cytolysis (Krahenbuhl and Tschopp, 1991).

Even though the exact role of perforin in lymphocyte-mediated killing cannot be determined at the present time, it should be noted that it is expressed by the majority of lymphocytes with cytolytic capability (Liu et al., 1989; Joag et al., 1990). Thus, it seems likely that perforin is involved in the cytolytic process at some level. It could be involved in the events leading to induction of apoptosis or, as proposed above, could accelerate the kinetics of lysis in target cells that had been induced to undergo apoptosis.

Acknowledgments

The author is grateful to Drs. J. John Cohen and Karen Sellins and to Paul Nash for their innumerable contributions to the ideas presented in this text. This work was supported by grants from the USPHS-NIH (AI-11661 and AI-29953), the American Cancer Society, and the Pauline A. Morrison Charitable Trust.

References

Acha-Orbea H, Scarpellino L, Hertig S, Dupuis M, Tschopp J (1990): Inhibition of lymphocyte mediated cytotoxicity by perforin antisense oligonucleotides. *EMBO J* 9: 3815

Battersby C, Egerton WS, Balderson G, Kerr JF, Burnett W (1974): Another look at rejection of pig liver homografts. *Surgery* 76: 617

Bell PA, Jones CN (1982): Cytotoxic effects of butyrate and other "differentiation inducers" on immature lymphoid cells. *Biochem Biophys Res Commun* 104: 1202

Berke G (1991): T-cell-mediated cytotoxicity. *Curr Opinion Immunol* 3: 320

Berke G, Rosen D (1988): Highly lytic *in vivo* primed CTL devoid of lytic granules and BLT-esterase activity acquire these constituents in the presence of T cell growth factors upon blast transformation *in vitro*. *J Immunol* 141: 1429

Bishop CJ, Moss DJ, Ryan JM, Burrows SR (1985): T lymphocytes in infectious mononucleosis. II. Response *in vitro* to interleukin-2 and establishment of T cell lines. *Clin Exp Immunol* 60: 70

Cantrell, DA, Smith KA (1984): The interleukin-2 T-cell system: A new cell growth model. *Science* 24: 1312

Cohen JJ (1991): Programmed cell death in the immune system. *Adv Immunol* 50: 55

Cohen JJ, Duke RC (1984): Glucocorticoid activation of a calcium-dependent endonuclease in thymocyte nuclei leads to cell death. *J Immunol* 132: 38

Cohen JJ, Duke RC, Chervenak R, Sellins KS, Olson LK (1985): DNA fragmentation in targets of CTL: An example of programmed cell death in the immune system. *Adv Exp Med Biol* 184: 439

Cohen JJ, Duke RC, Fadok V, Sellins KS (1992): Apoptosis and programmed cell death in immunity. *Annu Rev Immunol* 10: 267

Dennert G, Anderson C, Prochazka G (1987): High activity of Nα-benzyloxycarbonyl-L-lysine thiobenzyl ester serine esterase and cytolytic perforin in cloned cell lines is not demonstable in *in vivo*-induced cytotoxic effector cells. *Proc Natl Acad Sci USA* 84: 5004

Don MM, Ablett G, Bishop CJ, Bundesen PG, Searle KJ, Kerr JF (1977): Death of cells by apoptosis following attachment of specifically allergized lymphocytes *in vitro*. *Aust Exp Biol Med Sci* 55: 407

Duke RC, Cohen JJ (1986): IL-2 addiction: Withdrawal of growth factor activates a suicide program in dependent T cells. *Lymphokine Res* 5: 289

Duke RC, Cohen JJ (1988): The role of nuclear damage in lysis of target cells by cytotoxic T lymphocytes. In: *Cytolytic Lymphocytes and Complement: Effectors of the Immune System*, Podack ER, ed. Boca Raton, FL: CRC Press, p. 35

Duke RC, Sellins KS (1989): Target cell nuclear damage in addition to DNA fragmentation during cytotoxic T lymphocyte-mediated killing. *Prog Cell Biol* 9: 311

Duke RC, Chervenak R, Cohen JJ (1983): Endogenous endonuclease-induced DNA fragmentation: An early event in cell-mediated cytolysis. *Proc Natl Acad Sci USA* 80: 6361

Duke RC, Cohen JJ, Chervenak R (1986): Differences in target cell DNA fragmentation induced by mouse cytotoxic T lymphocytes and natural killer cells. *J Immunol* 137: 1442

Duke RC, Persechini PM, Chang S, Liu C-C, Cohen JJ, Young JD-E (1989): Purified perforin induces target cell lysis but not DNA fragmentation. *J Exp Med* 170: 1451

Duke RC, Sellins KS, Cohen JJ (1988): Cytotoxic lymphocyte-derived lytic granules do not induce DNA fragmentation in target cells. *J Immunol* 141: 2191

Duvall E, Wyllie AH (1986): Death and the cell. *Immunol Today* 7; 115

Felgar R, Hiserodt J (1990): A novel cytotoxic activity isolated from the plasma membrane of highly purified lymphokine activated killer cells and the LGL tumor CRNK-16. *FASEB J* 4: A1900

Filippini A, Taffs RE, Agui T, Sitkowski MV (1990): Ecto-ATPase activity in cytolytic T-lymphocytes. *J Biol Chem* 265: 334

Gillis S, Ferm MM, Ou W, Smith KA (1978): T cell growth factor: Parameters of production and a quantitative microassay for activity. *J Immunol* 120: 2027

Golstein P, Smith ET (1977): Mechanism of T-cell-mediated cytolysis: The lethal hit stage. *Contemp Top Immunobiol* 7: 273

Gray LS, Gnarra JR, Sullivan JA, Mandell GL, Engelhard VH (1988): Spatial and temporal characteristics of the increase in intracellular Ca^{2+} induced in cytotoxic T lymphocytes by cellular antigen. *J Immunol* 141: 2424

Gromkowski SH, Brown TC, Masson D, Tschopp J (1988): Lack of DNA degradation in target cells lysed by granules derived from cytolytic T lymphocytes. *J Immunol* 141: 774

Hameed A, Olsen KJ, Lee M-K, Lichtenheld MG, Podack ER (1989): Cytolysis by Ca-permeable transmembrane channels. Pore formation causes extensive DNA degradation and cell lysis. *J Exp Med* 169: 765

Hayes MP, Berrebi GA, Henkart PA (1989): Induction of target cell DNA release by the cytotoxic T lymphocyte granule protease Granzyme A. *J Exp Med* 170: 933

Henkart PA (1985): Mechanism of lymphocyte-mediated cytotoxicity. *Annu Rev Immunol* 3: 31

Henkart P, Henkart M, Millard P, Frederikse P, Bluestone J, Blumenthal R, Yue C, Reynolds C (1985): The role of cytoplasmic granules in cytotoxicity by large granular lymphocytes and cytotoxic T lymphocytes. *Adv Exp Med Biol* 184: 121

Henkart PA, Millard PJ, Reynolds CW, Henkart MP (1984): Cytolytic activity of purified cytoplasmic granules from cytotoxic rat LGL tumors. *J Exp Med* 160: 75

Joag SV, Liu C-C, Kwon BS, Clark WR, Young JD-E (1990): Expression of mRNAs for pore-forming protein and two serine esterases in murine primary and cloned effector lymphocytes. *J Cell Biochem* 31: 81

Ju S-T (1991): District pathways of CD4 and CD8 cells induce rapid target DNA fragmentation. *J Immunol* 146: 812

Ju S-T, Ruddle NH, Struck P, Dorf ME, DeKruyff RH (1990): Expression of two distinct cytolytic mechanisms among murine CD4 subsets. *J Immunol* 144: 23

Kerr JFR, Wyllie AH, Currie AR (1972): Apoptosis: A basic biological phenomenon with wide-ranging implications in tissue kinetics. *Br J Cancer* 26: 239–257

Konigsberg PJ, Podack ER (1985): Target cell DNA fragmentation induced by cytolytic T-cell granules. *J Leukocyte Biol* 38: 109

Krahenbuhl O, Tschopp J (1991): Perforin-induced pore formation. *Immunol Today* 12: 399

Laster SM, Wood JG, Gooding LR (1988): Tumor necrosis factor can induce both apoptotic and necrotic forms of cell lysis. *J Immunol* 141: 2629

Liu C-C, Rafii S, Granelli-Piperano A, Trapani JA, Young JD-E (1989): Perforin and serine esterase gene expression in stimulated human T cells. *J Exp Med* 170: 2105

Liu C-C, Steffen M, King F, Young JD-E (1987): Identification, isolation, and characterization of a novel cytotoxin in murine cytolytic lymphocytes. *Cell* 51: 393

Martz E (1975): Early steps in specific tumor cell lysis by sensitized mouse T-lymphocytes. I. Resolution and characterization. *J Immunol* 115: 261

Martz E (1977): Mechanism of specific tumor cell lysis by alloimmune T lymphocytes: Resolution and characterization of discrete steps is in the cellular interaction. *Contemp Top Immunobiol* 7: 301

Martz E, Howell DM (1989): CTL: Virus control cells first and cytolytic cells second. DNA fragmentation, apoptosis and the prelytic halt hypothesis. *Immunol Today* 10: 79–86

Martz E, Heagy W, Gromkowski H (1983): The mechanism of CTL-mediated killing: Monoclonal antibody analysis of the roles of killer and target cell membrane proteins. *Immunol Rev* 72: 73

Masson D, Corthesy P, Nabholz M, Tschopp J (1985): Appearance of cytolytic granules upon induction of cytolytic activity in CTL-hybrids. *EMBO J* 4: 2533

Matter A (1979): Microcinematographic and electron microscopic analysis of target cell lysis induced by cytotoxic T lymphocytes. *Immunology* 36: 179

McConkey DJ, Chow SC, Orrenius S, Jondal M (1991): NK cell-induced cytotoxicity is dependent on a Ca^{2+} increase in the target. *FASEB J* (in press)

McConkey DJ, Hartzell P, Duddy SK, Hakansson H, Orrenius S (1988): 2,3,7,8-Tetrachlorodibenzo-p-dioxin kills immature thymocytes by Ca^{2+} mediated endonuclease activation. *Science* 242: 256

McConkey, DJ, Hartzell P, Jondal M, Orrenius S (1989a): Calcium-dependent killing of immature thymocytes by stimulation via the CD3/T cell receptor complex. *J Immunol* 143: 1801

McConkey DJ, Hartzell P, Nicotera P, Orrenius S (1989b): Calcium-activated DNA fragmentation kills immature thymocytes. *FASEB J* 3: 1843

Munger WE, Berrebi GA, Henkart PA (1988): Possible involvement of CTL granule proteases in target cell DNA breakdown. *Immunol Rev* 103: 99

Munger WE, Reynolds CW, Henkart, PA (1985): DNase activity in cytoplasmic granules of cytotoxic lymphocytes. *Fed Proc* 44: 1284

Nagler-Anderson C, Allbritton NL, Verret CR, Eisen HN (1989): A comparison of the cytolytic properties of murine $CD8^+$ cytolytic T lymphocytes and cloned cytotoxic T cell lines. *Immunol Rev* 103: 111

Nash PB, Duke RC (1991): Certain requirements for delivery of the lethal hit can be overcome by "triggering" the effector cells prior to interaction with their targets. *FASEB J* 5: A600

Ostergaard HL, Clark WR (1989): Evidence for multiple lytic pathways used by cytotoxic T lymphocytes. *J Immunol* 143: 2120

Ostergaard HL, Kane KP, Mescher MF, Clark WR (1987): Cytotoxic T lymphocyte mediated lysis without release of serine esterase. *Nature* 330: 71

Pasternak MS, Eisen HN (1985): A novel serine esterase expressed by cytotoxic T lymphocytes. *Nature* 314: 743

Podack ER (1985): The molecular mechanism of lymphocyte-mediated tumor cell lysis. *Immunol Today* 6: 21

Podack ER, Konigsberg PJ, Acha-Orbea H, Pircher H, Hengartner H (1985a): Cytolytic T-cell granules: Biochemical properties and functional specificity. *Adv Exp Med Biol* 184: 99

Podack ER, Young JD-E, Cohn ZA (1985b) Isolation and biochemical and functional characterization of perforin 1 from cytolytic T cell granules. *Proc Natl Acad Sci USA* 82: 8629

Ratner A, Clark WR (1991): Lack of target cell participation in cytotoxic T lymphocyte-mediated lysis. *J Immunol* 147: 55

Robb RJ, Rusk CM, Yodoi J, Greene WC (1987): Interleukin 2 binding molecule distinct from the Tac protein: Analysis of its role in formation of high affinity receptors. *Proc Natl Acad Sci USA* 84: 2002

Russell JH (1983): Internal disintegration model of cytotoxic lymphocyte-induced target damage. *Immunol Rev* 72: 97

Russell JH, Dobos CB (1980): Mechanisms of immune lysis. II. CTL-induced nuclear disintegration of the target begins within minutes of cell contact. *J Immunol* 125: 1256

Russell JH, Masakowski VR, Dobos CB (1980): Mechanisms of immune lysis. I. Physiological distinction between target cell death mediated by cytotoxic T lymphocytes and antibody plus complement. *J Immunol* 124: 1100

Russell JH, Masakowski V, Rucinsky T, Phillips G (1982): Mechanisms of immune lysis. III. Characterization of the nature and kinetics of the cytotoxic T lymphocyte-induced nuclear lesion in the target. *J Immunol* 128: 2087

Sanderson CJ, Glauert AM (1977): The mechanism of T-cell mediated cytolysis. V. Morphological studies by electron microscopy. *Proc R Soc London Ser B* 198: 315

Schmid DS, Tite JP, Ruddle NH (1986): DNA fragmentation: A manifestation of target cell destruction mediated by cytotoxic T cell lines, lymphotoxin-secreting helper T cell clones, and cell-free lymphotoxin containing supernatant fluids. *Proc Natl Acad Sci USA* 83: 1881

Searle J, Kerr JFR, Battersby C, Egerton WS, Balderson G, Burnett W (1977): An electron microscopic study of the mode of donor cell death in unmodified rejection of pig liver allografts. *Aust J Exp Biol Med Sci* 55: 401

Sellins, KS, Cohen JJ (1987): Gene induction by gamma-irradiation leads to DNA fragmentation in lymphocytes. *J Immunol* 139: 3199

Sellins KS, Cohen JJ (1991): Hyperthermia induces apoptosis in thymocytes. *Radiat Res* 126: 88–95

Sharon M, Klausner RD, Cullen BR, Chizzonite R, Leonard WJ (1986): Novel interleukin 2 receptor subunit detected by cross-linking under high-affinity conditions. *Science* 234: 859

Shi L, Kraut RP, Aebersold R, Greenberg AH (1992): A natural killer (NK) cell granule protein that induces DNA fragmentation and apoptosis. *J Exp Med* 175: 553

Shi YF, Sahai BM, Green DR (1989): Cyclosporin A inhibits activation-induced cell death in T-cell hybridomas and thymocytes. *Nature* 339: 625

Shiver JW, Henkart PA (1991): A noncytotoxic mast cell tumor line exhibits potent IgE-dependent cytotoxicity after transfection with the cytolysin/perforin gene. *Cell* 64: 1175

Skalka RE, Mattyasova J, Cejkova M (1976): DNA in chromatin of irradiated lymphoid tissues degrades *in vivo* into regular fragments. *FEBS Lett* 72: 271

Slavin RE, Woodruff JM (1974): The pathology of bone marrow transplantation. *Pathol Ann* 9: 291

Smith CA, Williams GT, Kingston R, Jenkinson EJ, Owen JJT (1989): Antibodies to CD3/T-cell receptor complex induce death by apoptosis in immature T cells in thymic cultures. *Nature* 337: 181

Stacey NH, Bishop CJ, Halliday JW, Halliday WJ, Cooksley WGE, Powell LW, Kerr JFR (1985): Apoptosis as the mode of cell death in antibody-dependent lymphocytotoxicity. *J Cell Sci* 74: 169

Strack P, Martin C, Saito S, DeKruff RH, Ju S-T (1990): Metabolic inhibitors distinguish cytolytic activity of CD4 and CD8 clones. *Eur J Immunol* 20: 179

Takayama H, Trenn G, Sitkovsky MV (1988): Locus of inhibitory action of cAMP-dependent protein kinase in the antigen receptor-triggered cytotoxic T lymphocyte activation pathway. *J Biol Chem* 263: 2330

Teshigawara K, Wang HM, Kato K, Smith KA (1987): Interleukin 2 high-affinity receptor expression requires two distinct binding proteins. *J Exp Med* 165: 223

Trenn G, Takayama H, Sitkowski MV (1987): Exocytosis of cytolytic granules may not be required for target cell lysis by cytotoxic T-lymphocytes. *Nature* 330: 72

Trenn G, Taffs R, Hohman R, Kincaid R, Shevach EM, Sitkovsky MV (1989): Biochemical characterization of the inhibitory effect of CsA on cytolytic T lymphocyte effector functions. *J Immunol* 142: 3796

Tsien RY, Poenie M (1986): Fluorescence ratio imaging: A new window into intracellular ionic signaling. *Trends Biochem Sci* 11: 450

Ucker DS (1987): Cytotoxic T lymphocytes and glucocorticoids activate an endogenous suicide process in target cells. *Nature* 327: 62

Vanderbilt JN, Bloom KS, Anderson JN (1982): Endogenous nuclease. Properties and effects on transcribed genes in chromatin. *J Biol Chem* 257: 13009

Wagner H, Rollinghoff M (1974): T cell mediated cytotoxicity: Discrimination between antigen recognition, lethal hit and cytolysis phase. *Eur J Immunol* 4: 745

Weissman AM, Harford JB, Svetlik PB, Leonard WJ, Depper JM, Waldmann, TA, Greene WC, Klausner RD (1986): Only high affinity receptors for interleukin-2 mediate internalization of ligand. *Procl Natl Acad Sci USA* 83: 146

Welsh RM, Nishioka WK, Antia R, Dundon PL (1990): Mechanism of killing of virus-induced cytotoxic T lymphocytes induced *in vitro*. *J Virol* 64: 3726

Wyllie AH (1980): Glucocorticoid-induced thymocyte-apoptosis is associated with endogenous endonuclease activation. *Nature* 284: 555

Wyllie AH (1981): Cell death: A new classification separating apoptosis from necrosis. In: *Cell Death in Biology and Pathology*, Bowen ID, Lockshin RA eds.) London: Chapman and Hall, pp. 9–34

Wyllie AH, Morris RG (1982): Hormone-induced cell death. Purification and properties of thymocytes undergoing apoptosis after glucocorticoid treatment. *Am J Pathol* 109: 78–87

Wyllie AH, Kerr JFR, Currie AR (1980): Cell death: The significance of apoptosis. *Int Rev Cytol* 68: 251

Yannelli JR, Sullivan JA, Mandell GL, Engelhard LH (1986): Reorientation and fusion of cytotoxic T lymphocyte granules after interaction with target cells as determined by high resolution cinemicrography. *J Immunol* 136: 377

Young JD-E, Cohn ZA (1986): Cell-mediated killing: A common mechanism? *Cell* 46: 641

Young JD-E, Clark WR, Liu CC, Cohn ZA (1987): A calcium- and perforin-independent pathway of killing mediated by murine cytolytic lymphocytes. *J Exp Med* 160: 1894

Young, JD-E, Hengartner H, Podack ER, Cohn ZA (1986): Purification and characterization of a cytolytic pore-forming protein from granules of cloned lymphocytes with natural killer activity. *Cell* 44: 849

Zanovello P, Bronte V, Rosato A, Pizzo P, DiVigilio F (1990): Responses of mouse lymphocytes to extracellular ATP. II. Extracellular ATP causes cell type-dependent lysis and DNA fragmentation. *J Immunol* 145: 1545

Zheng LM, Zychlinsky A, Liu C-C, Ojcius DM, Young JD-E (1991): Extracellular ATP as a trigger for apoptosis or programmed cell death. *J Cell Biol* 112: 279

Zvonareva NB, Zhitovsky BD, Hanson KP (1983): Distribution of nuclease attack sites and complexity of DNA in the products of post-irradiation degradation of rat thymus chromatin. *Int J Radiat Biol* 44: 261

Zychlinsky A, Joag SV, Liu C-C, Young JD-E (1990): Cytotoxic mechanisms of murine lymphokine-activated killer cells: Functional and biochemical characterization of homogeneous populations of spleen LAK cells. *Cell Immunol* 126: 377

Zychlinsky A, Zheng LM, Liu C-C, Young JD-E (1991): Cytolytic lymphocytes induce both apoptosis and necrosis in target cells. *J Immunol* 146: 393

21
Molecular Mechanisms of Lymphocyte Cytotoxicity

Mark J. Smyth and John R. Ortaldo

Introduction

Natural killer (NK) cells are $CD3^-$, T-cell receptor $(TCR)^-$, large granular lymphocytes (LGL) that express spontaneous lytic activity against tumor cells, virally infected cells, and perhaps certain hematopoietic progenitor cells (Trinchieri, 1989). Understanding the cellular and molecular mechanisms by which NK cells recognize and destroy target cells has become an area of considerable interest. Previous studies (Henney, 1973; Herberman et al., 1986) with NK cells have proposed to divide the process of NK killing into four identifiable stages, consisting of: (1) target cell binding (adhesion), (2) effector cell activation (recognition/signal transduction), (3) delivery of the lethal signal to the target (lethal hit), and (4) effector cell detachment and recycling. Ca^{2+} plays a central role in the killing process, yet only recently has it been possible to delineate more clearly the site(s) of Ca^{2+} requirements in the lytic mechanism. Early studies involving cytotoxic T lymphocyte (CTL) models and more recent studies with NK cells demonstrated that Ca^{2+} was required at a point in the lytic process distal to Mg^{2+}-dependent target cell adhesion but proximal to target cell disintegration (Roder and Haliotis, 1980; Quan et al., 1982; Martz et al., 1983; Berke, 1989). Recent evidence supported a role for Ca^{2+} in the activation of a stimulus–secretion response by killer cells. It also suggested that Ca^{2+} can be a potent toxic agent if allowed to accumulate at high concentrations in target cells. Therefore, Ca^{2+} appears to be required not only to activate killer cell function but must also enter the killer cell to activate additional processes related to stimulus–secretion coupling. Analysis of the lethal hit by which killer lymphocytes mediate target cell damage has focused on CTL and their interactions with specific target cells (Roder and Haliotis, 1980; Quan et al., 1982). It suggested, but did not prove, that during "programming for lysis" the killer cells may deposit (or secrete) materials onto the target cell that mediate the lytic signals. The hypothesis that material(s) were transferred from a killer cell to a target cell was a requisite for this target cell lysis model (Henkart and Henkart, 1982). Collectively, numerous data support this hypothesis as a general phenomenon common to numerous types of killer cells. The observation that NK cells and in vitro-activated CTL contain intracytoplasmic azurophilic granules suggested that these granules may be released during the Ca^{2+}-dependent stage of killing and may contain materials capable of mediating target cell lysis.

Based on the above, Henkart et al. (1984) proposed a unified hypothesis of cell-mediated killing that included granule exocytosis as the fundamental mechanism of the lytic process. This model was supported by the isolation of potent membranolytic molecules from purified cytoplasmic granules of NK cells and in vitro interleukin-2 (IL-2)-propagated CTL clones, but was inconsistent with the studies of Berke and Rosen (1987). These membranolytic molecules were termed cytolysin and perforin, respectively, and were later shown to be identical proteins [herein termed pore-forming protein (PFP)]. PFP is a 68-kDa protein stored in the matrix of the granule as a nonlytic protomer. It is postulated that during granule exocytosis these protomers are released and rapidly polymerize in the membrane of the target cell to form transmembrane pores. These "pores" can be visualized by negative staining electron microscopy and have uniform diameters of 10 nm I.D. and

CYTOTOXIC CELLS
M. Sitkovsky and P. Henkart, Editors
© 1993 Birkhäuser Boston

20 nm O.D. The process of polymerization to form transmembrane pores is specifically dependent upon Ca^{2+} ions (Henkart et al., 1984; Podack and Konigsberg, 1984). Polymerization is also temperature- and pH-dependent. Although Ca^{2+} is required for the polymerization of membranolytic PFP protomers, in the absence of an acceptor membrane, Ca^{2+} will also cause a rapid inactivation of PFP, presumably by inducing irreversible polymerization of the PFP. This control mechanism is an integral part of the granule exocytosis/polymerization hypothesis, since it suggests that if PFP monomers do not immediately react with the target cell membrane, they will rapidly be inactivated and be unable to diffuse as an active lytic moiety. This phenomenon restricts the lytic activity to the vicinity of effector–target contact where granule exocytosis occurs and thus prevents innocent bystander lysis.

Pore-Forming Protein

PFP is an excellent candidate for a role in lymphocyte-mediated cytolysis; however, the expression, rather than the function, of PFP has been most comprehensively studied. A number of publications highlight a correlation between PFP expression and cytolytic activity in lymphocyte populations (Liu et al., 1989; Smyth et al., 1990b). PFP transcription is tightly regulated and specific for CTL and NK cells. The mature polyadenylated PFP transcript is 2.6–2.9 kbp, as determined by Northern analysis. The cDNA sequences for human, murine, and rat PFP show considerable homology (Lichtenheld et al., 1988; Shinkai et al., 1988; Ishikawa et al., 1989; Lichtenheld and Podack, 1989). Studies of PFP expression have been restricted to PFP mRNA (Northerns) and protein (hemolysis, immunohistochemistry). Detection of PFP mRNA has correlated with PFP protein measurements where tested (Muller et al., 1989; Young et al., 1989).

PFP expression in unstimulated peripheral blood lymphocytes (PBL) is restricted to $CD3^-$ LGL, $\gamma\delta^+$ T cells, and $CD11b^+$ $CD8^+$ T cells (Nakata et al., 1990; Smyth et al., 1990b; Table 1). In $CD3^-$ LGL and $\gamma\delta^+$ T cells, constitutive PFP expression does not appear to be regulated by cytokines. All other resting PBL subpopulations, B cells, and monocytes are PFP-negative. T-cell PFP mRNA induction occurs within 6–12 hr, followed by PFP protein expression and functional cytotoxicity 18–30 hr later (Smyth et al., 1990b; Lichtenheld et al., 1988). PFP expression can be induced in $CD8^+$ $CD11b^-$ T cells independently of TCR occupation by stimulation with IL-2 or IL-7 alone (Smyth et al., 1990b, 1991a; Table 1). It is curious that a single stimulus (e.g., IL-2 via the p75 IL-2 receptor) can induce PFP and cytolytic potential in $CD8^+$ T lymphocytes, particularly when IL-6, which alone is ineffective, augments the effects of low levels of IL-2 (Smyth et al., 1990a). $CD4^+$ T cells do not express PFP in response to these stimuli; however, both $CD4^+$ and $CD8^+$ T cells produce PFP upon culture with solid phase OKT3 monoclonal antibody (mAb) ± IL-2 (Table 1). Interestingly, TGF-β_1 reversibly inhibits IL-2 induction of $CD8^+$ T-cell PFP and cytotoxicity, suggesting that transforming growth factor-β_1 (TGF-β_1) may be an important regulator of CTL induction (Smyth et al., 1991a). In addition, nonphysiological stimulation of T cells with PMA/phytohemagglutinin (PHA) or PBL with calcium-ionophore/accessory cells induces high levels of PFP in an IL-2-independent manner (Lu et al., 1989b; Lu et al., 1990). In PBL, PFP mRNA induction is independent of *de novo* protein synthesis, and its half-life is 2–3 hr (Lu et al., 1990). However, upon effectory–target cell interaction, there is a rapid abrogation of PFP mRNA, followed by a recovery (Hommel-Berrey et al., 1990). The 3′ untranslated A-T-rich sequence that is responsible for the short half-life of many lymphokine messages is not present in PFP. In addition, protein synthesis inhibitors and anti-CD28 mAb do not increase PFP mRNA levels (Muller et al., 1989; Thompson et al., 1989). Acha-Orbea et al. (1990) demonstrated that PFP antisense oligonucleotide partially inhibited the cytolytic activity of anti-CD3 mAb-activated PBL. These experiments did not take into account the contribution of contaminating LGL that constitutively express PFP, nor did they rule out merely a role for PFP in the CD3-mediated differentiation of CTL.

All NK and $CD8^+$ CTL express PFP (Lichtenheld et al., 1988; Podack et al., 1988; Shinkai et al., 1988; Ishikawa et al., 1989; Lowrey et al., 1989). Initial reports claiming the absence of PFP in CTL and peritoneal exudate lymphocytes (PEL) (Berke and Rosen, 1987; Allbritton et al., 1988a; Nagler-Anderson et al., 1988) were based on both insensitive and unreliable assays. Recently the presence of PFP mRNA has been confirmed in these highly lytic cells (Nagler-Anderson et al., 1989). The majority of $CD4^+$ T-helper type 2 (T_H2) and some T-helper type 1 (T_H1) effectors express PFP and lyse both nucleated targets and antibody-coated red blood cells (RBC). Clearly, several lytic $CD4^+$ cell lines do not express PFP or lyse RBC, yet can lyse

TABLE 1. PFP expression in human PBL subsets

Stimulus	CD3⁻ LGL	γδ⁺ T	αβ⁺ T	CD8⁺ T	CD4⁺ T
—	++++[a]	++	—	−/+	—
IL-2[b]	++++	++	++	+++	—
IL-6	++++	N.D.	—	−/+	—
IL-2[c]/IL-6	++++	N.D.	++	+++	N.D.
IL-4	N.D.	N.D.	—	−/+	—
IL-7	++++	++	+	++	—
IFN-α/γ	++++	N.D.	—	−/+	—
OKT-3mAb[d]	N.D.	N.D.	++	+++	++
OKT-3mAb/IL-2	N.D.	N.D.	+++	+++	+++
OKT-3mAb[e]/PMA	N.D.	N.D.	++	N.D.	N.D.
PMA	++++	N.D.	—	−/+	—
Ionomycin	N.D.	N.D.	—	−/+	—
PMA/ionomycin	++++	N.D.	+	+	—
PHA	++++	N.D.	—	−/+	—
TGF-β₁	++++	N.D.	—	−/+	—
TGF-β₁[f]/IL-2	N.D.	N.D.	—	−/+	—
TGF-β1/IL-2/IL-6	N.D.	N.D.	—	−/+	—
TGF-β1/IL-2/OKT-3 mAb	N.D.	N.D.	—	−/+	—

[a] (++++; high expression) → (−; no expression); N.D. = not determined.
[b] High dose IL-2 (1000 U/ml).
[c] Low dose IL-2 (10 U/ml).
[d] Immobilized OKT-3 mAb.
[e] Fluid phase OKT-3 mAb.
[f] TGF-β₁ pretreatment for ≥ 3 hr.

nucleated targets (Strack et al., 1990). Transfection of PFP cDNA into a rat basophilic leukemia and triggering via the IgE receptor enabled the transfected cells to lyse RBC (Shiver and Henkart, 1991). However, PFP involvement in lymphocytotoxicity against nucleated target cells was not supported by these experiments.

The demonstrated ability of purified PFP to mediate a potent lytic attack on membranes of a wide range of targets by forming a tubular lesion is the basis for its putative role in lysis. Recently, we have examined the exocytosis model using mAb reactive with human PFP (Table 2). Although some staining of PFP was seen over the nucleus and in the Golgi region, the vast majority of PFP was concentrated in the cytoplasmic granules. These data emphasize the localization of PFP at the interaction site of NK cells and the NK tumor target, K562. We also demonstrated that the highly concentrated PFP molecule present in the granules of NK cells is released to the target in a time-dependent fashion. The initial interactions involve the rearrangement of cytoskeletal elements to align the PFP-rich granules to the site of target interaction (Table 2). Upon effector–target interaction, PFP molecules are seen with increased frequency on the surface of the effector cell, in the microenvironment, and subsequently on the surface of the target. The localization of PFP is

TABLE 2. Summary of immunogold microscopy with anti-PFP antibody

Characteristic location of PFP staining	Comment
In resting LGL	Staining over nucleus, Golgi, with concentration in cytoplasmic granules
After target interaction	Concentrated staining at effector/target junction localized: 1. On effector membrane pseudopods 2. In intracellular space 3. On target membranes
After target interaction without Ca²⁺	Same as resting LGL

selective, but not exclusive to the effector–target interaction site. The increase in PFP localization at other target sites could reflect: (1) PFP release that was not totally limited to the effector binding site, (2) numerous effector contacts with the target cell, and (3) PFP that was deposited and integrated into the target membrane lipid bilayer but diffused to the other target areas due to high membrane fluidity at 37°C. It should be noted that no intact granules were seen in the microenvironment or on the target surface. Overall, these data provide direct visual evidence for support of PFP involvement in the "granule exocytosis model" of lysis by NK cells. The temporal deposition of PFP was paralleled by loss of target cell integrity and release of chromium. These studies, although consistent with a role of PFP in cellular lysis, do not exclude the possibility that other molecules are released and involved in lysis by NK cells.

Expression of PFP is expected in conditions where specific CTL-mediated cytotoxicity is observed. Indeed, *in vivo*, PFP is detected in the immune response to viral infection, in antitumor immunity, in autoimmune disorders, and in transplant rejection (Muller et al., 1989; Nagler-Anderson et al., 1989; Young et al., 1989). Taken together, these studies convincingly indicate that PFP can be produced by CTL and NK cells primed *in vivo*.

Efforts are currently being focused on analyzing how various stimuli or inhibitors regulate PFP gene transcription. Several putatively imporant 5' noncoding sequences are not conserved between human and mouse; however, the immediate 5' flanking sequences of the human and murine PFP genes are highly homologous and novel (Lichtenheld and Podack, 1989; Youn et al., 1991). Therefore, it appears likely that the PFP gene transcription is regulated, at least in part, by novel transcription factors. Specific *cis*-acting elements are suspected because of the highly restricted PFP expression in killer lymphocytes. By contrast, the conservation of a response element TRE tempts speculation that a complex interaction of protooncogene products regulates PFP expression. It will be important to establish the role of the putative enhancer/silencer elements for PFP expression. In particular, the PFP gene represents an excellent model in which to compare transcriptional control in T cells and LGL and to study IL-2 regulation of genes.

Granzymes/Proteoglycans

Data indicate that, with few exceptions, expression of granzyme genes is restricted to T cells and their thymic precursors (Garcia-Sanz et al., 1990). The granules of CTL contain a number of granzymes (human = 3, murine = 7) that represent about 90% of the total granule proteins (Bleackley et al., 1988). All the above granzymes are serine proteases, and their cDNAs have been cloned (see review in Tschopp, 1990). The members of the granzyme family can be divided into two groups: granzymes B, C, D, E, F, and G are highly homologous (~ 60–94%), whereas granzyme A is less related (~ 54–62%). The biological functions of granzymes are unknown. Granzyme A (Pasternack et al., 1986) exhibits a trypsin-like activity and in particular cleaves fibrin and other extracellular proteins (Young et al., 1986; Simon et al., 1988). Granzyme B preferentially hydrolyzes substrates after serine and aspartic acid (Poe et al., 1991), and granzyme D displays a weak trypsin-like activity (Jenne and Tschopp, 1988). The substrates for the remaining granzymes are unknown. Of the granzymes, granzyme B is most tightly associated with CTL; however, the observation of all the granzymes in noncytotoxic activated $CD4^+$ T cells argues against their direct involvement in the lethal hit (Garcia-Sanz et al., 1987, 1990). Alternative functions have been proposed for granzyme A, which can stimulate the growth of B cells and thymic lymphoma cells (Simon et al., 1986), including the modulation of PFP activity (Munger, 1988) or DNA fragmentation of target cells (Munger et al., 1988). We suggest that families of granzymes with different substrate specificities are involved in the migration of activated lymphocytes into tissues by interacting with different endothelial-derived matrix proteins. Similarly, granzymes may be required for the detachment of effector lymphocytes from target cells following granule exocytosis (Tschopp and Nabholz, 1987). Future studies should clarify the function of granzymes. PFP and granzyme mRNAs do not appear to be coordinately regulated in activated T cells or CTL. While IL-2 induces the expression of PFP and several granzymes (Liu et al., 1989b, 1990), the kinetics of granzyme induction are somewhat slower and the putative promoter elements in PFP and granzyme genes are quite distinct (Lichtenheld and Podack, 1989).

LGL and CTL synthesize chondroitin sulfate-A proteoglycans, and their exocytosis from the granules is stimulated by target cells (Stevens et al., 1988). Granzymes all possess a basic isoelectric point and are ionically bound to proteoglycans at the acid pH of the granules. Thus, packaging in this manner may prevent degradation or inactivation of PFP/granzymes prior to exocytosis. Alternatively, interaction of proteoglycans with extracellular endothelial matrix proteins may release granzymes and thereby activate granzyme function.

Other Lytic Molecules

Several lines of evidence indicate that cell lysis can occur independently of granule exocytosis and involve molecules other than those described above. First, lysis sometimes proceeds in the absence of Ca^{2+} (Ostergaard et al., 1987; Trenn et al., 1987). Second, the DNA of CTL targets is degraded into ~200-bp fragments, suggesting the activation of an internal T-cell endonuclease (Russell, 1983). Third, the nuclear blebbing in the T cell that precedes effector exocytosis is not consistent with the pore-formation model. These observations have prompted alternative models of lymphocyte-mediated cytolysis (Ojcius and Young, 1990). Lymphotoxin (LT) is a prototype calcium-independent mediator of lysis secreted by antigen- and mitogen-activated CTL (Gray et al., 1984; Aggarwal et al., 1985b). LT is closely related to tumor necrosis factor-α (TNF-α) (~35% homology), which was discovered because of its ability to cause hemorrhagic necrosis of tumors (Old, 1988). The biological properties of LT and TNF-α are similar, and both molecules share the same receptor (Aggarwal et al., 1985a). LT fragments DNA in some tumor cells (Ruddle and Schmid, 1987). Additionally, a number of calcium-independent, DNA-fragmenting cytotoxins have been described in the supernatants of CTL and activated T cells (Green et al., 1986; Yamamoto et al., 1986; Liu et al., 1987). Several of these are TNF-related, although only one, leukalexin, has been well characterized (Liu et al., 1987). Leukalexin is a 50-kDa protein that has been detected in the granules, in the cytosol, and on the surface of CTL (Liu et al., 1987, 1989a). DNA fragmentation by purified leukalexin is quite inefficient, therefore suggesting that it may only act in the context of other factors. The cytolytic properties of another molecule, natural killer cytotoxic factor (NKCF), have been described; however, this factor has not been biochemically purified or characterized as a distinct entity (Wright and Bonavida, 1987).

Thus, the search for other potential lytic factors that might be involved in lymphocyte-mediated lysis remains a high priority. In light of this fact, we examined the subcellular fractions of the rat natural killer (RNK) cell leukemia percoll gradients used to isolate PFP. These fractions demonstrated a potent lytic activity that was associated with the membrane. The characteristics of this RNK-derived lytic factor (RNK-LF) are summarized in Table 3. In preliminary experiments "RNK-LF" activity was also isolated from spontaneously cytotoxic LGL but was absent in resting noncytotoxic T lymphocytes. Therefore the RNK-LF activity appears restricted to LGL that exhibited cytotoxic activity. The kinetics of RNK-LF lysis were similar to those of LGL; significant lysis occurred after several hours and maximal lysis after 4–6 hr. The specificity of RNK-LF activity appears similar to that of PFP in that a broad range of target cells were susceptible. Interestingly, although sheep red blood cells (SRBC) were lysed by RNK-LF, they were less susceptible than nucleated targets, such as K562 and YAC. In contrast, SRBC are several logs more susceptible to PFP than are nucleated target cells. Studies also were performed to determine the Ca^{2+} requirement of this RNK-LF. In the presence of MgEGTA or after pretreatment with Ca^{2+}, PFP was inactivated, whereas the RNK-LF lysed targets efficiently and independently of Ca^{2+} concentration prior to or during cytoassay. In addition, a rabbit anti-PFP antiserum was unable to abrogate the lytic activity of the RNK-LF. Presently we are continuing to purify and characterize the RNK-LF present in RNK cells and in fresh human LGL to eliminate identity with PFP. The presence of RNK-LF in cytolytic cells of $CD8^+$ and $CD4^+$ phenotype should determine whether RNK-LF is present in CTL that do or do not contain PFP. The presence of a potential cytolytic molecule that is associated with the membrane of the cytolytic effector cell may offer an alternative mechanism for cell-mediated lysis.

Lymphocyte Subsets: Different Mechanisms of Cytolysis?

CTL, like other T lymphocytes, respond to antigens via cell surface TCR. This response is highly specific, sometimes with single amino acids conferring recognition. The TCR is associated with the invariant CD3 complex, which functions in transmembrane signaling. The antigen must be proteolytically processed by the target cell and displayed as peptides with major histocompatibility complex (MHC) proteins. The ability to recognize antigen in the context of class I and class II MHC correlates with the expression of CD8 and CD4, respectively. LGL, unlike CTL, are not MHC-restricted and do not have functional TCR, despite expressing the ζ chain of the CD3 complex (Anderson et al., 1989). Furthermore, LGL are spontaneously cytotoxic to

TABLE 3. Comparison of PFP and RNK membrane-associated lytic factor (RNK-LF)

Characteristic	PFP	RNK-LF
Cellular location	Cytoplasmic granules	Membrane-rich
Cell source	NK cells and cell lines in vitro activated CTL	NK cells and cell lines
Kinetics of lysis	Rapid (minutes)	Rapid (hours)
Specificity of lysis	None	None
	SRBC > YAC = K562 > leukocytes	K562 = YAC > SRBC ≃ leukocytes
Ca^{2+} requirement	Absolute	None
Inhibited by anti-PFP	Yes	No
Causes nuclear degradation	No	?

a wide variety of target cells and have Fc receptors that allow them to lyse antibody-coated target cells (Ortaldo and Herberman, 1984; Trinchieri, 1989). LGL thus provide an early, memory-independent immunity at a stage when the antigen-specific CTL have not yet expanded clonally. The majority of studies examining the lytic mechanism of CTL have utilized T-cell clones. In particular, a panel of metabolic inhibitors have demonstrated some differences in the cytotoxicity of CD4[+] and CD8[+] T-cell clones (Strack et al., 1990; Ju, 1991). Furthermore, we have recently used these metabolic inhibitors to compare the antibody-redirected cytolytic potential and the direct cytotoxicity of IL-2/anti-CD3 mAb-activated human peripheral blood CD4[+] and CD8[+] T cells (Table 4). In both T-cell clones and in our study with activated peripheral blood T lymphocytes, it was found that CD4[+] cells, but not CD8[+] cells, require de novo synthesis of RNA and proteins to express cytotoxicity. Conversely, cholera toxin, which ADP-ribosylates G proteins and increases intracellular cAMP, inhibited the cytotoxicity of CD8[+] T cells, but not CD4[+] T cells. These observations may suggest that in CD8[+] T cells the cytolytic mediator is stored and secreted upon interaction with the target cell. In addition, the apparently different cytolytic mechanisms of CD4[+] and CD8[+] T cells used to induce target ^{51}Cr release also lead to target DNA fragmentation, as judged by their distinct sensitivity to specific inhibitors (Ju, 1991). Therefore, an emerged common mechanism for target cell lysis may be employed by CD4[+] and CD8[+] T cells. Both CD4[+] and CD8[+] T cells do not require the synthesis of lymphokines [TNF-α, LT, interferon-γ (IFN-γ), etc.] to lyse TNF-resistant targets; however both subsets are directly cytotoxic to TNF-sensitive targets in an 18-h assay. Interestingly, CD4[+] or CD8[+] T cells may lyse the same TNF-sensitive target by TNF-dependent and TNF-independent

TABLE 4. Effect of metabolic inhibitors on human PBL subset cytotoxicity[a]

Inhibitor	CD4[+] T activated[b]	CD8[+] T activated[c]	LGL Spontaneous[d]	LGL Activated[e]
Emetine	↓	–	↓	↓
Cycloheximide	↓	–	–	↓
Actinomycin D	↓	–	–	↓
Cyclosporin A	–	–	?	?
Mitomycin C	–	–	–	–
Dibutyryl cAMP	–	↓	↓	↓
Cholera toxin	–	↓	↓	↓
EGTA	↓	↓	↓	↓

[a] 4-hr ^{51}Cr-release assay (T-redirected cytotoxic potential using an anti-CD3 × anti-NP heteroconjugate and NP-labeled targets; LGL-direct vs K562).
[b] IL-2/anti-CD3 mAb, 72 hr.
[c] IL-2/anti-CD3 mAb, 36 hr.
[d] Untreated.
[e] IFN-α (1000 U/ml).
[f] (–) No effect.

mechanisms (Smyth et al., 1992). Despite their use of metabolically distinct pathways of lysis, both subsets of activated peripheral blood T cells expressed PFP and several granzymes (Garcia-Sanz et al., 1990; Smyth et al., 1992). Therefore it would appear that the potentially distinct molecules involved in CD4$^+$ and CD8$^+$ T-cell cytolysis remain to be isolated.

The sensitivity of the NK activity of CD3$^-$ LGL to metabolic inhibitors distinguishes it from the cytotoxicity of T lymphocytes. Inhibition of RNA synthesis had no effect on the spontaneous NK activity, but did abrogate IFN stimulation of NK activity. The effect of protein synthesis inhibitors was dependent on the agent used: some inhibited basal NK activity (emetine/puromycin), others (cycloheximide) only blocked enhancement of NK activity (Ortaldo et al., 1980). In other studies with LGL, cAMP and cholera toxin have been demonstrated to inhibit NK activity (Katz et al., 1982), and therefore CD8$^+$ T cells and CD3$^-$ LGL may share some common lytic pathway distinct from that in CD4$^+$ T cells. With the current expansion of our knowledge of intracellular signaling following ligation of the TCR or IL-2 stimulation, the specific pathways/effector mechanisms utilized by different lymphocyte subsets should become apparent.

Models of Lymphocyte-Mediated Cytotoxicity

There are two major theories to explain the mechanism(s) by which CTL and NK cells lyse target cells: granule exocytosis of PFP and contact-induced internal disintegration. However, distinct effector mechanisms are expressed by different killer cells toward targets with various susceptibilities.

In the presence of Ca^{2+}, isolated PFP reversibly interacts with the phospholipid headgroups of the membrane (Tschopp et al., 1989). Only at 37°C, PFP inserts and polymerizes in the membrane irreversibly; this mechanism is reminiscent of the interaction of C9 with C5b-8 (Podack, 1984). Upon complete polymerization of approximately 20 PFP molecules in the membrane, poly-PFP forms a tubular complex that creates a large functional transmembrane channel. Such pores enable transfer of even large cellular proteins and may lead to eventual target cell lysis by osmosis, Ca^{2+} influx, and energy depletion (Figure 1A). It also is reasonable to assume that PFP pores initiate an endocytic repair process in the target cell. Pinocytosis of extracellular fluid, potentially containing concentrated LT and TNF-α, by the target cell may contribute to the observed DNA degradation in the target cell (Figure 1B). In support of this hypothesis, the addition of TNF to PFP greatly increases the subsequent DNA degradation in the target cell (Hameed et al., 1989). Experiments examining isolated granule degradation of target cell DNA have produced conflicting data (Allbritton et al., 1988b; Gromkowski et al., 1988). Similarly, no consensus has been reached regarding the ability of PFP to cause target cell DNA degradation (Duke et al., 1989; Hameed et al., 1989). A sustained Ca^{2+} rise and acidic lysosomes (Hameed et al., 1989) are required in the target cell for DNA fragmentation to occur. Lymphokine-activated killer (LAK) cells mediating DNA degradation and ^{51}Cr release from target cells may lyse targets without DNA fragmentation (Podack and Lee, 1988). Therefore, Ca^{2+} influx through PFP lesions may be required for target DNA fragmentation without being the sole mechanism of lysis.

Early findings led to the suggestion that PFP-containing vesicles fused with the target cell membrane resulting, in PFP polymerization and subsequent cell death (Dennert and Podack, 1983; Podack and Dennert, 1983). More recently a refined model was proposed based on ultrastructural studies indicating that granules have the properties of secretory granules and lysosomes (Burkhardt et al., 1989; Peters et al., 1989, 1990). In designating this organelle a granulosome, it has been postulated that communication with the plasma membrane via an endosomal pathway enables the membrane proteins CD3, CD8, LFA-1, and MHC to localize to the outer surface of the vesicles that contain PFP and granzymes in their interior (Peters et al., 1989, 1990) (Figure 1C). Secretion of granulosomes is thought to allow self-targeting of PFP and granzymes to the target cell via the vesicle-associated CD3/TCR complex and other membrane proteins. Because granule cores also are enclosed by a membrane containing these proteins, they also are targeted to the target cell (Figure 1D). This mechanism may explain how the effector is spared during lytic interactions. This model predicts MHC-restricted lysis of target cells by isolated granules; however, to date isolated granules appear relatively unspecific (Henkart et al., 1984).

Although the granule exocytosis/PFP model of cytotoxicity is conceptually very appealing (particularly for granule-rich NK cells), several key observations are inconsistent with this model. First, target cells killed by purified granules or purified PFP fail to demonstrate nuclear DNA fragmenta-

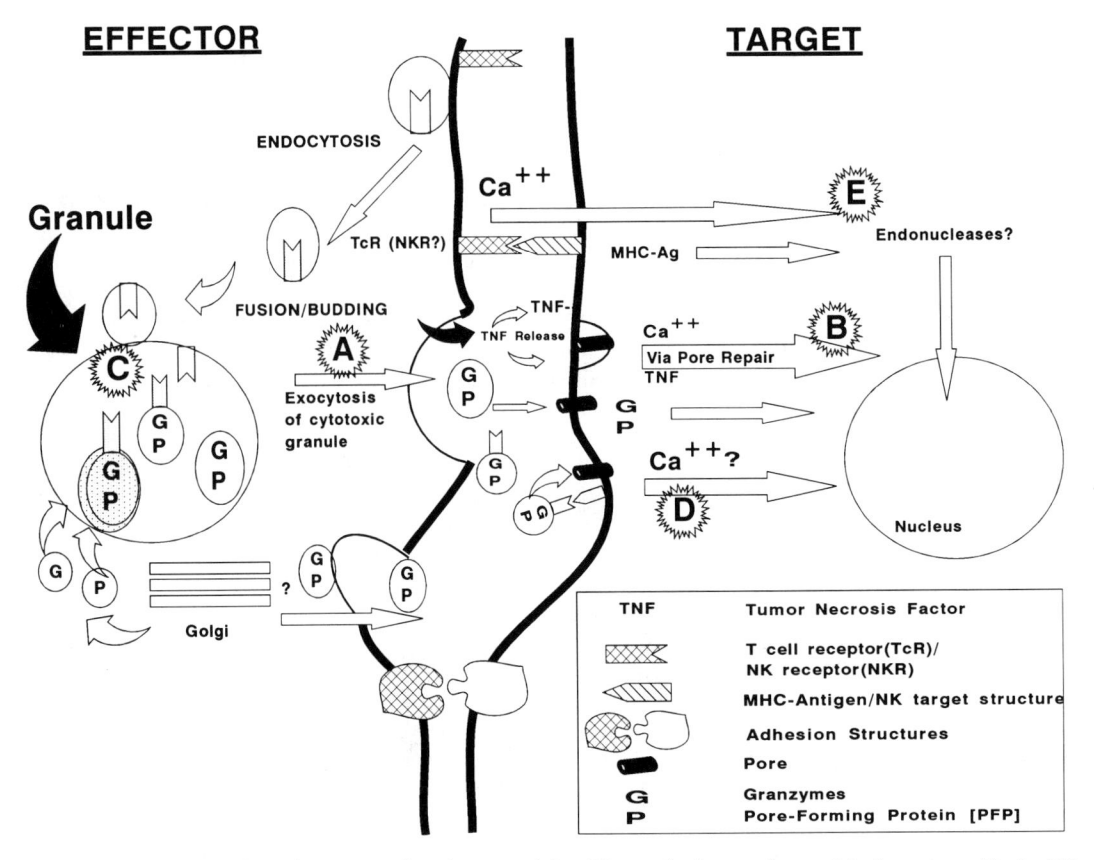

FIGURE 1. Models for lymphocyte-mediated cytotoxicity. Figure depicts various critical aspects of lysis. (A) Exocytosis of granule contents such as PFP/granzymes. (B) TNF uptake either via pore (pinocytic) or by direct binding to TNF receptor. Receptor-target granulosomes are shown for effector (C) and for target (D). (E) Receptor-mediated target cell disintegration.

tion, a characteristic feature of the lytic process induced by both NK cells and CTL. Second, Berke and Rosen (1987) have demonstrated a lack of detectable intracytoplasmic granules in *in vivo*-elicited CTL [peritoneal exudate T lymphocytes (PEL)]. Small agranular CD3−, CD56+ lymphocytes also have been shown to mediate non-MHC-restricted lysis (Inverardi et al., 1991). Third, cytolysis sometimes proceeds in the absence of Ca^{2+}. Most investigators studying NK–target interactions have utilized a cytoplasmic marker (^{51}Cr) to measure cell disintegration. An important CTL-mediated observation made initially by Russell (1983) was that nuclear DNA appeared to fragment in targets treated with either NK cells or CTL prior to the observable release of ^{51}Cr. Therefore, study of killer lymphocytes must consider the possible mechanisms that induce DNA fragmentation.

Unless the effector CTL is resistant to PFP, as has been suggested (Kranz and Eisen, 1987; Jiang et al., 1990), another difficulty with the PFP exocytosis model stems from the finding that CTL escape self-lysis and recycle after killing their targets (Zagury et al., 1975). The recent demonstration of specific killing of CTL by CTL (Ottenhoff and Mutis, 1990) and of antigen-induced self-lysis of CTL (Tadakuma et al., 1990) confirms that CTL are not refractory to their own lytic mechanisms, thereby questioning the involvement of PFP in CTL-induced lysis.

Lymphocyte-triggered internal disintegration of the target via an autolytic cascade was originally proposed by Russell (1983). An alternative pathway of receptor-mediated triggering of target cell internal disintegration may not require pore formation, although these two mechanisms are not necessarily mutually exclusive and may be used by the same effector cell at different stages of activation. The nature of the trigger and what mediates the lytic effect in the target cell have not been elucidated. It is clear, however, that receptor-mediated contact of the effector and its target is necessary. The lethal hit delivered by the effector may stem from the interaction of its TCR complex (or the NK cell equivalent) and associated molecules with target cell molecules (possibly MHC determinants) that transduce the signal (Figure 1E). Redistribution/crosslinking of the target cell surface may alter membrane permeability to Ca^{2+}, enabling the signaling of internal activation processes such as stimulation of endonuclease and protease activity. There are numerous examples of cell death signaled through ligand-mediated crosslinking that do not appear to involve pore formation. This apoptotic death can be induced by antibodies to cell surface receptors (e.g., CD3, TNF receptor, Apo-1) (Trauth et al., 1989; Engelman et al., 1990; Tadakuma et al., 1990) antigen (Jenkison et al., 1989; Mercep et al., 1989) or ligation of receptors (Newell et al., 1990), and is characterized by DNA fragmentation. Additional lytic mechanisms involving membrane-bound cytotoxin (Liu et al., 1987) and CTL-secreted ATP (Filippini et al., 1990; Zanovello et al., 1990) also have been proposed; however, these mechanisms require further investigation.

Conclusion

Despite efforts to equate PFP with lymphocytotoxicity, more recent studies have indicated that lymphocytes can utilize several different mechanisms and/or cytotoxins to lyse target cells. None of these need be mutually exclusive, and indeed together they may all contribute to damage target cells. Evidence against PFP participation in lymphocyte-mediated lysis is currently difficult to verify, and the molecular details of alternative mechanisms remain to be illustrated. With the determination of the structure and *in vitro* function of various components of the lytic mechanism and elucidation of their relationship to one another, it will be possible to distinguish different cytotoxic effector–target cell interactions.

References

Acha-Orbea H, Scarpellino L, Hertig S, Dupuis M, Tschopp J (1990): Inhibition of lymphocyte mediated cytotoxicity by perforin antisense oligonucleotides. *EMBO J* 9: 3815–3819

Aggarwal BB, Eessalu TE, Hass PE (1985a): Characterization of receptors for human tumor necrosis factor and regulation by γ-interferon. *Nature* 318: 665–667.

Aggarwal BB, Henzel WJ, Moffat B, Kohr WJ, Harkins RN (1985b): Primary structure of human lymphotoxin derived from 1788 lymphoblastoid cell line. *J Biol Chem* 260: 2334–2344

Allbritton NL, Nagler-Anderson C, Elliott TJ, Verret CR, Eisen HN (1988a): Target cell lysis by cytotoxic T lymphocytes that lack detectable hemolytic perforin activity. *J Immunol* 141: 3243–3248

Allbritton NL, Verret CR, Wolley RC, Eisen HN (1988b): Calcium ion concentrations and DNA fragmentation in target cell destruction by murine cloned cytotoxic T lymphocytes. *J Exp Med* 167: 514–527

Anderson P, Caligiuri M, Ritz J, Schlossman SF (1989): CD3-negative natural killer cells express ζ TCR as part of a novel molecular complex. *Nature* 341: 159–162

Berke G (1989): Functions and mechanisms of lysis induced by cytotoxic T lymphocytes and natural killer cells. In: *Fundamental Immunology*, Paul WE, ed. New York: Raven Press

Berke C, Rosen D (1987): Are lytic granules and perforin 1 involved in lysis induced by *in vivo* primed peritoneal exudate cytolytic T lymphocytes? *Transplant Proc* 19: 412–416

Bleackley RC, Lobe CG, Duggan B, Ehrman N, Fregeau C, Meier M, Letellier M, Havele C, Shaw J, Paetkau V (1988): The isolation and characterization of a family of serine protease genes expressed in activated cytotoxic T lymphocytes. *Immunol Rev* 103: 5–19

Burkhardt JK, Hester S, Argon Y (1989): Two proteins targeted to the same lytic granule compartment undergo very different posttranslational processing. *Proc Natl Acad Sci USA* 86: 7128–7132

Dennert G, Podack ER (1983): Cytolysis by H-2 specific killer cells: Assembly of tubular complexes on target membranes. *J Exp Med* 157: 1483–1495

Duke RC, Persechini PM, Chang S, Liu C-C, Cohen JJ, Young JD-E (1989): Purified perforin induces target cell lysis but not DNA fragmentation. *J Exp Med* 170: 1451–1456

Engelmann H, Holtmann H, Brakebusch C, Shemer-Avni Y, Sarov I, Nophar Y, Hadas E, Leitner O, Wallach D (1990): Antibodies to a soluble form of a tumor necrosis factor (TNF) receptor have TNF-like activity. *J Biol Chem* 265: 14497–14504

Filippini A, Taffs RE, Agui T, Sitkovsky M (1990): Ecto-ATPase activity in cytolytic T-lymphocytes. *J Biol Chem* 265: 334–340

Garcia-Sanz JA, MacDonald HR, Jenne DE, Tschopp J, Nabholz M (1990): Cell specificity of granzyme gene expression. *J Immunol* 145: 3111–3118

Garcia-Sanz JA, Plaetinck, G, Velotti F, Masson D, Tschopp J, MacDonald HR, Nabholz M (1987): Perforin is present only in normal activated Lyt 2$^+$ T lymphocytes and not in L3T4$^+$ cells, but the serine protease granzyme A is made by both subsets. *EMBO J* 6: 933–938

Gray PW, Aggarwal BB, Benton CV, Bringman TS, Henzel WS, Jarrett JA, Leung DW, Mofat B, Ng P, Sverdsky LP, Palladino MA, Nedwin GA (1984): Cloning and expression of cDNA for human lymphotoxin: A lymphokine with tumor necrosis activity. *Nature* 312: 721–724

Green LM, Reade JL, Ware CF, Devlin PE, Liang C-M, Devlin JJ (1986): Cytotoxic lymphokines produced by cloned human cytotoxic T lymphocytes. II. A novel CTL-produced cytotoxin that is antigenically distinct from tumor necrosis factor and α-lymphotoxin. *J Immunol* 137: 3488–3493

Gromkowski SH, Brown TC, Masson D, Tschopp J (1988): Lack of DNA degradation in target cells lysed by granules derived from cytolytic T lymphocytes. *J Immunol* 141: 774–778

Hameed A, Olsen KJ, Lee M-K, Lichtenheld MG, Podack ER (1989): Cytolysis by Ca-permeable transmembrane channels: Pore formation causes extensive DNA degradation and cell lysis. *J Exp Med* 169: 765–777

Henkart MP, Henkart PA (1982): Lymphocyte mediated cytolysis as a secretory process. In: *Mechanisms in Cell Mediated Cytotoxicity*, Clark and Goldstein, eds. New York: Plenum

Henkart PA, Millard PJ, Reynolds CW, Henkart MP (1984): Cytolytic activity of purified cytoplasmic granules from cytotoxic rat large granular lymphocyte tumors. *J Exp Med* 160: 75–93

Henney CS (1973): On the mechanism of T-cell mediated cytolysis. *Transplant Rev* 17: 37–41

Herberman RB, Reynolds CW, Ortaldo JR (1986): Mechanism of cytotoxicity by natural killer (NK) cells. *Annu Rev Immunol* 4: 651–680

Hommel-Berrey G, Goebel WS, Bajpai A, Shenoy AM, Brahmi Z (1990): Loss of serine esterase activity and granule protein messages from CTL inactivated by a sensitive target. 4th International Cell-Mediated Cytotoxicity Workshop, Ogelbay Park, WV (Abstract)

Inverardi L, Witson JC, Fuad SA, Winkler-Pickett RT, Ortaldo JR, Bach FH (1991): CD3 negative "small granular lymphocytes" are natural killer cells. *J Immunol* 146: 4048–4052

Ishikawa H, Shinkai Y-I, Yagita H, Yue CC, Henkart PA, Sawada S, Young HA, Reynolds CW, Okumura K (1989): Molecular cloning of rat cytolysin. *J Immunol* 143: 3069–3073

Jenkison EJ, Kingston R, Smith CA, Williams GT, Owen JJT (1989): Antigen-induced apoptosis in developing T cells. A mechanism for negative selection of the T cell receptor repertoire. *Eur J Immunol* 19: 2175–2177

Jenne DE, Tschopp J (1988): Granzymes, a family of serine proteases released from granules of cytolytic T lymphocytes upon T cell receptor stimulation. *Immunol Rev* 103: 53–71

Jiang S, Ojcius DM, Persechini PM, Young JD-E (1990): Inhibition of perforin binding activity by surface membrane proteins. *J Immunol* 144: 998–1003

Ju S-T (1991): Distinct pathways of CD4 and CD8 cells induce rapid target DNA fragmentation. *J Immunol* 146: 812–818

Katz P, Zaytoun AM, Fauci AS (1982): Mechanisms of human cell-mediated cytotoxicity. I. Modulation of natural killer cell activity by cyclic nucleotides. *J Immunol* 129: 287–296

Kranz DM, Eisen HN (1987): Resistance of cytotoxic T lymphocytes to lysin by a clone of cytotoxic T lymphocytes. *Proc Natl Acad Sci USA* 84: 3375–3379

Lichtenheld MG, Podack ER (1989): Structure of the human perforin gene: A simple organization with interesting potential regulatory sequences. *J Immunol* 143: 4267–4274

Lichtenheld MG, Olsen KJ, Lu P, Lowrey DM, Hameed A, Hengartner H, Podack ER (1988): Structure and function of human perforin. *Nature* 335: 448–451

Liu C-C, Detmers PA, Kiang S, Young JD-E (1989a): Identification and characterization of a membrane-bound cytotoxin of murine cytolytic lymphocytes that is related to tumor necrosis factor/cachectin. *Proc Natl Acad Sci USA* 86: 3286–3290

Liu C-C, Joag SV, Kwon BS, Young JD-E (1990): Induction of perforin and serine esterase in a murine cytotoxic T lymphocyte clone. *J Immunol* 144: 1196–1201

Liu C-C, Rafii S, Granelli-Piperno A, Trapani JA, Young JDE (1989b): Perforin and serine esterase gene expression in stimulated human T cells. *J Exp Med* 170: 2105–2188

Liu C-C, Steffen M, King F, Young JD-E (1987): Identification, isolation and characterization of a novel cytotoxin in murine cytolytic lymphocytes. *Cell* 51: 393–403

Lowrey DM, Aebischer T, Olsen K, Lichtenheld M, Rupp F, Hengartner H, Podack ER (1989): Cloning analysis and expression of murine perforin 1 cDNA, a component of cytolytic T-cell granules with homology to complement component C9. *Proc Natl Acad Sci USA* 86: 247–251

Lu P, Garcia-Sanz JA, Lichtenheld MG, Podack ER (1990): Calcium ionophore can induce perforin gene expression in human peripheral blood lymphocytes. *FASEB J* 4: 1898 (Abstract)

Martz E, Heagy W, Gromkowski SH (1983): The mechanism of CTL-mediated killing: Monoclonal antibody analysis of the roles of killer and target cell membrane proteins. *Immunol Rev* 72: 73–94

Mercep M, Weissman AM, Frank SJ, Klausner RD, Ashwell JD (1989): Activation-driven programmed cell death and T cell receptor ζη expression. *Science* 246: 1162–1165

Muller C, Kagi D, Aebischer T, Odermatt B, Held W, Podack ER, Zinkernagel RM, Hengartner H (1989): Detection of perforin and granzyme A mRNA in infiltrating cells during infection of mice with lymphocytic choriomeningitis virus. *Eur J Immunol* 19: 1253–1259

Munger WE (1988): LGL secretory granule-associated 60-kd BLT esterase augments the lytic activity of cytolysin against nucleated target cells. *Nat Immun Cell Growth Regul* 7: 61–62

Munger WE, Berrebi GA, Henkart PA (1988): Possible involvement of CTL granule proteases in target cell DNA breakdown. *Immunol Rev* 103: 99–109

Nagler-Anderson C, Allbritton NL, Verret CR, Eisen HN (1988): A comparison of the cytolytic properties of murine primary CD8+ cytotoxic T lymphocytes and cloned cytotoxic T cell lines. *Immunol Rev* 103: 111–124

Nagler-Anderson C, Lichtenheld M, Eisen HN, Podack ER (1989): Perforin mRNA in primary peritoneal exudate cytotoxic T lymphocytes. *J Immunol* 143: 3440–3443

Nakata M, Smyth MJ, Norihisa Y, Kawasaki A, Shinkai Y, Okumura K, Yagita H (1990): Constitutive expression of pore-forming protein in peripheral blood γ/δ T cells: Implication for their cytotoxic role in vivo. *J Exp Med* 172: 1877–1880

Newell MK, Haughn LJ, Maroun CR, Julius MH (1990): Death of mature cells by separate ligation of CD4 and T-cell receptor for antigen. *Nature* 347: 286–289

Ojcius DM, Young JD-E (1990): Cell-mediated killing: Effector mechanisms and mediators. *Cancer Cells* 2: 138–145

Old LJ (1988): Tumor necrosis factor. *Sci Am* 258: 59–75

Ortaldo JR, Herberman RB (1984): Heterogeneity of natural killer cells. *Annu Rev Immunol* 2: 359–394

Ortaldo JR, Phillips W, Wasserman K, Herberman RB (1980): Effects of metabolic inhibitors on spontaneous and interferon-boosted human natural killer cell activity. *J Immunol* 125: 1839–1844

Ostergaard HL, Kane KP, Mescher MF, Clark WR (1987): Cytotoxic T lymphocyte mediated lysis without release of serine esterase. *Nature* 330: 71–72

Ottenhoff THM, Mutis T (1990): Specific killing of cytotoxic T cells and antigen-presenting cells by CD4$^+$ cytotoxic T-cell clones. A novel potentially immunoregulatory T-T cell interaction in man. *J Exp Med* 171: 2011–2024

Pasternack MS, Verret CR, Liu MA, Eisen HN (1986): Serine esterase in cytolytic T lymphocytes. *Nature* 322: 740–743

Peters PJ, Geuze HJ, Van der Donk HA, Borst J (1990): A new model for lethal hit delivery by cytotoxic T lymphocytes. *Immunol Today* 11: 28–32

Peters PJ, Geuze HJ, Van der Donk HA, Slot JW, Griffith JM, Stam NJ, Clevers HC, Borst J (1989): Molecules relevant for T cell-target cell interaction are present in cytolytic granules of human T lymphocytes. *Eur J Immunol* 19: 1469–1475

Podack ER (1984): Molecular composition of the tubular structure of the membrane attack complex complement. *J Biol Chem* 259: 8641–8647

Podack ER, Dennert G (1983): Assembly of two types of tubules with putative cytolytic function by cloned natural killer cells. *Nature* 302: 442–445

Podack ER, Konigsberg PJ (1984): Cytolytic T cell granules: Isolation, structural, biochemical and functional characterization. *J Exp Med* 160: 695–710

Podack ER, Lee MK (1988): Mechanism of lymphocyte-mediated tumor cell lysis: Selective inhibition of DNA breakdown and release from target cells does not interfere with chromium release and cell death. In: *Tumor Necrosis Factor/Cachectin and Related Cytokines*, Bonavida B, Gifford GE, Kirchner H, Old LJ, eds. Basel: Karger

Podack ER, Lowrey DM, Lichtenheld MG, Olsen KJ, Aebischer T, Binder D, Rupp F, Hengartner H (1988): Structure, function, and expression of murine and human perforin 1 (P1). *Immunol Rev* 103: 203–211

Poe M, Blake JT, Boulton DA, Gammon M, Sigal NH, Wu JK, Zweerink HJ (1991): Human cytotoxic lymphocyte granzyme B. *J Biol Chem* 266: 98–103

Quan PC, Ishizaka T, Bloom BR (1982): Studies on the mechanism of NK cell lysis. *J Immunol* 128: 1786–1791

Roder JC, Haliotis TA (1980): Comparative analysis of the NK cytolytic mechanism and regulatory genes. In: *Natural Cell-Mediated Immunity Against Tumors*, Herberman RB, ed. New York: Academic Press

Ruddle NH, Schmid DS (1987): The role of lymphotoxin in T-cell mediated cytotoxicity. *Ann Inst Pasteur Immunol* 138: 314–320

Russell JH (1983): Internal disintegration model of cytotoxic lymphocyte-induced target damage. *Immunol Rev* 72: 97–118

Shinkai Y, Takio K, Okumura K (1988): Homology of perforin to the ninth component of complement (C9). *Nature* 33: 525–527

Shiver JW, Henkart PA (1991): A noncytotoxic mast cell tumor line exhibits potent IgE-dependent cytotoxicity after transfection with the cytolysin/perforin gene. *Cell* 64: 1175–1181

Simon MM, Hoschutzky, H, Fruth U, Simon HG, Kramer MD (1986): Purification and characterization of a T cell specific serine proteinase (TSP-1) from cloned cytolytic T lymphocytes. *EMBO J* 5: 3267–3274

Simon MM, Prester, M, Nerz G, Kramer MD, Fruth U (1988): Release of biologically active fragments from human plasma-fibronectin by murine T cell-specific proteinase 1 (TSP-1). *Biol Chem Hoppe Seiler* 369: 107–112

Smyth MJ, Norihisa Y, Gerard JR, Young HA, Ortaldo JR (1991a): IL-7 regulation of cytotoxic lymphocytes. Pore-forming protein gene expression, interferon-γ production and cytotoxicity of human peripheral blood lymphocyte subsets. *Cell Immunol* 138: 390–403

Smyth MJ, Norihisa Y, Ortaldo JR (1992): Multiple cytolytic mechanisms displayed by activated human peripheral blood T cell subsets. *J Immunol* 148: 55–62

Smyth MJ, Ortaldo JR, Bere W, Yagita H, Okumura K, Young HA (1990a): IL-2 and IL-6 synergize to augment the pore-forming protein gene expression and cytotoxic potential of human peripheral blood T cells. *J Immunol* 145: 1159–1166

Smyth MJ, Ortaldo JR, Shinkai Y-I, Yagita H, Nakata M, Okumura K, Young HA (1990b): Interleukin 2 induction of pore-forming protein gene expression in human peripheral blood CD8$^+$ T cells. *J Exp Med* 171: 1269–1281

Smyth MJ, Strobl SL, Young HA, Ortaldo JR, Ochoa AC (1991b): Regulation of lymphokine-activated killer activity and pore-forming protein gene expression in human peripheral blood CD8$^+$ T lymphocytes. Inhibition by transforming growth factor-β. *J Immunol* 146: 3290–3297

Stevens RL, Kamada MM, Serafin WE (1988): Structure and function of the family of proteoglycans that reside in the secretory granules of natural killer cells and other effector cells of the immune response. *Curr Top Microbiol Immunol* 140: 93–108

Strack P, Martin C, Saito S, DeKruyff RH, Ju S-T (1990): Metabolic inhibitors distinguish cytolytic activity of CD4 and CD8 clones. *Eur J Immunol* 20: 179–184

Tadakuma T, Harutoshi K, Odaka, C, Kubota R, Ishimura Y, Tagita H, Okumura K (1990): CD4+, CD8+ thymocytes are susceptible to DNA fragmentation induced by phorbol ester, calcium ionophore and anti-CD3 antibody. *Eur J Immunol* 20: 779–784

Thompson CB, Lindsten T, Ledbetter JA, Kunkel SL, Young HA, Emerson SG, Leiden JM, June CH (1989): CD28 activation pathway regulated the production of multiple T-cell-derived lymphokines/cytokines. *Proc Natl Acad Sci USA* 86: 1333–1337

Trauth BC, Klas C, Peters AMJ, Matzku S, Moller P, Falk W, Debatin K-M, Krammer PH (1989): Monoclonal antibody-mediated tumor regression by induction of apoptosis. *Science* 245: 301–305

Trenn G, Takayama H, Sitkovsky M (1987): Exocytosis of cytolytic granules may not be required for target cell lysis by cytotoxic T-lymphocytes. *Nature* 330: 72–74.

Trinchieri G (1989): The biology of natural killer cells. *Adv Immunol* 47: 187–376

Tschopp J (1990): Perforin-mediated target cell lysis by cytolytic T lymphocytes. *Annu Rev Immunol* 8: 279–302

Tschopp J, Nabholz M (1987): The role of cytoplasmic granule components in cytolytic lymphocyte-mediated cytolysis. *Ann Inst Pasteur Immunol* 138: 290–295

Tschopp J, Schafer S, Masson D, Peitsch MC, Heusser C (1989): Phosphorylcholine acts as a Ca^{2+}-dependent receptor molecule for lymphocyte perforin. *Nature* 337: 272–274

Wright SC, Bonavida B (1987): Studies on the mechanism of natural killer cell-mediated cytotoxicity. VIII. Functional comparison of human natural killer cytotoxic factors with recombinant lymphotoxin and tumor necrosis factor. *J Immunol* 138: 1791–1798

Yamamoto RS, Ware CF, Granger GA (1986): The human LT system. XI. Identification of LT and "TNF-like" forms from stimulated natural killers, specific and non-specific cytotoxic human T cells in vitro. *J Immunol* 137: 1878–1884

Youn BS, Liu C-C, Kim K-K, Young JD-E, Kwon MH, Kwon BS (1991): Structure of the mouse pore-forming protein (perforin) gene: Analysis of transcription initiation site, 5′-flanking sequence, and alternative splicing of 5′ untranslated regions. *J Exp Med* 173: 813–822

Young JD-E, Leong LG, Liu C-C, Damiano A, Wall DA, Cohn ZA (1986): Isolation and characterization of a serine esterase from cytolytic T cell granules. *Cell* 47: 183–194

Young LH, Klavinskis LS, Oldstone MBA, Young JD-E (1989): In vivo expression of perforin by CD8+ lymphocytes during an acute viral infection. *J Exp Med* 169: 2159–2171

Zagury D, Bernard J, Thierness N, Feldman M, Berke G (1975): Isolation and characterization of individual functionally reactive cytotoxic T lymphocytes. Conjugation, killing and recycling at the single cell level. *Eur J Immunol* 5: 818–822

Zanovello P, Bronte V, Rosato A, Pizzo P, DiVirgilio F (1990): Responses of mouse lymphocytes to extracellular ATP. II. Extracellular ATP causes cell type-dependent lysis and DNA fragmentation. *J Immunol* 145: 1545–1550

V
Granule Proteases

22
Subtractive and Differential Molecular Biology Approaches to Molecules Preferentially Expressed in Cytotoxic and Other T Cells

Eric Rouvier and Pierre Golstein

Introduction

After a phase of mostly phenomenological characterization (reviewed in Golstein and Smith, 1977; Henney, 1977; Martz, 1977; Berke, 1989), research on T-cell-mediated cytotoxicity is focusing on its molecular mechanism(s). Some of the already isolated candidate molecules, their functional significance, and the likely conclusion that they may not fully account for the observed mechanisms (encouraging one to continue the search) are also discussed in other chapters of this volume.

Among the few experimental methodologies that may give access to functionally relevant molecules, the preparation and selection of cytotoxicity-inhibiting monoclonal antibodies have led to the isolation of molecules (reviewed in Golstein et al., 1982) involved in recognition, but probably not at the lethal hit stage of T-cell-mediated cytotoxicity. The lethal hit stage is precisely the one for which function-related molecules are now sought. We and others therefore turned to other approaches, based in particular on molecular biology techniques. Among these, functional complementation would seem most appropriate: one may try to complement a cytotoxicity-negative variant of a cytotoxic cell by shotgun transfection of a population of cDNAs, to isolate a cytotoxicity-reestablishing one. A major difficulty in this case is that no good system for the selection, after transfection, of a few cytotoxic cells from a large population of noncytotoxic cells seems currently available.

Another approach makes use of the postulate that some of the molecules involved in a given function may be expressed mostly, if not only, in cells exerting this function. A first step, then, is to try to isolate molecules expressed preferentially in cytotoxic cells; this has usually entailed subtractive or differential molecular biology methodologies. A necessary second step is to demonstrate the functional role of the molecules thus isolated.

We shall here review published work aiming at the isolation of molecules preferentially expressed in cytotoxic T or natural killer (NK) cells, or other T cells, using subtractive or differential methodologies; comment on the CTLA molecules thus isolated and studied in this laboratory; and, finally, consider what may have been learned using these approaches and which ways might be used to test functional relevance.

Subtractive or Differential Approaches Applied to T Cells, and the Molecules Thus Obtained

Identifying preferentially or specifically expressed molecules may be achieved through the construction and differential screening of cDNA libraries. Both these libraries and the probes used for screening them can be enriched as to their specificity and sensitivity by subtractive hybridization.

Until recently, the methodology of subtraction was based on the hybridization of cDNA of one cell type with an excess of mRNA of another cell type; the unhybridized cDNA (considered enriched in cDNA corresponding to mRNA specifically expressed in the first cell type) was subsequently isolated using hydroxyapatite columns, and cloned. In these protocols, subtraction came before cloning. New technologies are now available (Duguid et al., 1988; Sive and St John, 1988; Rubenstein et al., 1990), which take advantage of phagemid and avidin-biotin systems: the cDNA to subtract is first cloned in a phagemid, which is used to produce single-stranded DNA. This DNA is then hybridized to an excess of biotinylated mRNA. Avidin is used to remove biotinylated sequences. The remaining unhybridized single-stranded phagemids can be double-stranded and used to transform bacteria. With these technologies, cloning precedes subtraction, circumventing the problem of cloning very small amounts of material encountered in the previous approaches. This technological improvement should allow easier access to differences between closely related cells, entailing very narrow subtraction windows (as for example between variants of the same original cell, provided these differences are transcriptional).

Subtractive hybridization has been used in immunology mainly in two circumstances; first, to try to clone cDNAs corresponding to already known molecules, by enriching either the library or the probe for these molecules, using cells expressing or not expressing the corresponding mRNAs (see, e.g., Alt et al., 1979; Hedrick et al., 1984a; Kavathas et al., 1984; Sims et al., 1984; Littman et al., 1985; Maddon et al., 1985; Sugimura et al., 1985; Raschke, 1987); second, and now perhaps more frequently, to try to clone cDNAs corresponding to unknown molecules by difference between cells exerting or not exerting a specific function, with the hope that differentially expressed molecules may be involved in the function.

Differential screening (plus/minus) has been carried out in many different ways: screening of a library with total cDNA probes from two types of cells (known/expected to express and not to express, respectively, the desired molecules); subtracted probes; autosubtracted probes (Jongstra et al., 1987); "negative selection" of clones (based on their absence of signal with a "minus" probe) or a combination of these screenings; or screening of a library by competitive hybridization of a "plus" radioactive cDNA probe with an excess "minus" mRNA (Sugimura et al., 1985; Gershenfeld and Weissman, 1986).

The use of a subtracted cDNA library, rather than a total library subjected to differential screening, has one main advantage. The clones of interest are in principle enriched by the subtraction process, which enables one to then work with a more limited number of still-relevant clones. This smaller number allows one to prepare an ordered set of these clones, easily amenable to repeated screenings, ultimately allowing the detailed characterization of all of the clones of the ordered set. The drawbacks are that subtraction is technically demanding, and that it tends to enrich in specifically expressed structures only for medium-abundance transcripts: it may incompletely remove cDNA corresponding to high-abundance transcripts, and tends to ignore cDNAs corresponding to very low-abundance transcripts. Some of these aspects have been discussed before (Brunet et al., 1988), and will also be considered below.

Tables 1 and 2 review some of the molecules that have been isolated, using subtractive and/or differential approaches, from cytotoxic cells (either CTL or NK clones) and from other T-cell types. From cytotoxic cells, the construction and screening of at least four subtracted and six total cDNA libraries have led to the isolation (either as a discovery or as a rediscovery) of molecules of the immunoglobulin (Ig) superfamily (Thy-1, various chains of the T-cell receptor/CD3 complex, CTLA-4, LAG3), IL-2 receptor, serine esterases, a molecule homologous to proregions of cysteine proteinase (CTLA-2), α-casein, pancreatic lipase, metallothionein, several molecules from a family of small induced secreted (SIS) proteins, and a certain number of differentially expressed molecules with no homologies to any known molecule in the databases at the time of study. The function of many of these molecules is still unknown. In addition, CD4, CD8, T200, MAL, and a number of cytokines/interleukins and DNA-binding proteins were isolated from noncytotoxic T cells.

Table 2 also features a rapidly growing group of experiments aiming at isolating not only molecules represented in some (T) cells and not in others, but also molecules appearing at particular stages of cell life. In the examples shown, the subtraction partners were either activated vs. nonactivated T cells, or dying (here corticoid-treated) vs. healthy thymocytes (Harrigan et al., 1989; Owens et al., 1991). The aim in these experiments is not to find tissue-specific molecules, but to find differentiation stage-specific ones. Some of the molecules appearing upon activation (reviewed in Crabtree, 1989; Ullman et al., 1990) or upon induction of cell death (see for recent reviews Arends and Wyllie, 1991; Cohen, 1991; Golstein et al., 1991; Kerr and Harmon, 1991) are expected not to be tissue-specific. They would be found under similar circumstances in different cell types, and T cells are used here mainly because of the quality of the available experimental systems.

The tables do not comment on tissue distribution, which has been studied thoroughly (making it both more difficult and more legitimate to affirm tissue specificity) for only some of the

TABLE 1. Some of the molecules isolated by subtractive or differential methods from cytotoxic T or NK cells

Authors[a]	Cell combination[b]	Library/screening	Molecules
Saito et al. (1984a) Saito et al. (1984b) Koyama et al. (1987) Haser et al. (1987)	**T_c-B** (2C)-(A20.2.J) (2C)-(CH1)	Subtracted/subtracted	TCR-α, -β, and -γ; CD3γ; Thy-1; pT49 (36% homology with fibrinogen β and γ in the COOH terminal part); 10 other molecules[e]
Brunet et al. (1986) Brunet et al. (1987) Brunet et al. (1988) Denizot et al. (1989)	**T_c-B** (KB5C20)-(M12.4.1)	Subtracted/differential Subtracted/subtracted	CTLA-1[c]; CTLA-2; CTLA-3[c] CTLA-4
Gershenfeld and Weissmann (1986)	**T_c-T_h** (1E4)-(VL3)	Total/competitive	Hanukkah factor[c]
Lobe et al. (1986a) Lobe et al. (1986b)	**T_c-T_h** (MTL2.8.2)-(CH1)	Total/differential	B10[c]; C11[c]
Kwon et al. (1987) Kwon et al. (1989) Kwon et al. (1988)	**T_c-B** **T_c-T_h** (ConA-L3)-(A20.2.J) (ConA-L3)-(ConA-L2)	Total/differential	TCR-β; L3G10[c]; L3.1[c]; 2 other molecules[e]
Jongstra et al. (1987) Schall et al. (1988)	***T_c-B*** (AH2)-(LB)	Subtracted/subtracted	519; RANTES[d]; 2 other molecules[e]
Grusby et al. (1990a) Grusby et al. (1990b)	*IL-4-induced T_c -* *IL-2-induced T* [CT4.R(IL4)]-[CTEV(IL2)]	Total/differential	Pancreatic lipase; α casein
Houchins et al. (1990)	***NK-B***	Subtracted/differential	TCR-α and -β; metallothionein; 519-like; granzymes A[c] and B[c], Act-2[d]; 25H8; a member of supergene family of type II integral membrane proteins; 3 other molecules[e]
Yabe et al. (1990) Dahl et al. (1990) Houchins et al. (1991)	(B22)-(FJO)	Total/differential	Granzyme B[c]; rIL-2α Act 2[d]; 25H8.
Baixeras et al. (1990) Triebel et al. (1990)	NK-(***T + B + erythromyeloid + histiocytic***)(F5)- (Jurkat + Laz388 + K562 + U937)	Total/subtracted	LAG1[d]; RANTES[d]; LAG3 (related to CD4)

[a] Data regrouped by laboratory.
[b] The indicated cells (from mouse in **bold** letters, from man in ***bold italic*** letters) are either subtraction partners for a subtracted library or cell combinations used for the screening of a total library with total or subtracted probes. This subtraction window is reflected in general in the tissue distribution of the isolated molecules.
[c] Molecules belonging to the serine esterases superfamily; within this family, the following are equivalent: CTLA-1 = CCP1 = granzyme B = 1-3E; CTLA-3 = Hanukkah factor = granzyme A = L3G10.
[d] Molecules belonging to a family of small secreted proteins; within this family, the following are presumably identical in mouse or man: TCA3.0 = 1309 = SIS-ε; G26 = Act2 = LAG-1 = HC21 = SIS-γ; SIS-α = pLD78.
[e] Refers to sequences, corresponding to the desired differential pattern of expression, and showing no homology to any known sequence in the databases at the time of study. For some of these, the tissue distribution has been studied in detail. Some others have not been further investigated.

TABLE 2. Some of the molecules isolated by subtractive or differential methods from noncytotoxic T cells

Authors[a]	Cell combination[b]	Library/screening	Molecules
Yanagi et al. (1984)	**T-B**	Total/differential	TCR-α and -β; one molecule with 92% homology with p56lck; 18 other molecules[e]
Yanagi et al. (1985)	(MOLT3)-(HSC58)	Subtracted/differential	
Koga et al. (1986)			
Takihara et al. (1989)			
Hedrick et al. (1984a)	T$_h$-B; T$_h$-T$_h$ (3.3T)-(Bal 17)	Subtracted/subtracted	TCR-β; Thy-1; 8 other molecules[e]
Hedrick et al. (1984b)	(M12)-(L10A) (2B4)-(L10A)	Total/subtracted	
Chien et al. (1984)	(2B4)-(C10) (2B4)-(EL4)		
Kavathas et al. (1984)	T8 L transfectant-L	Total/subtracted	CD8
Littman et al. (1985)	T8(T4) L transfectant-L	Total PBL/subtracted	CD8; CD4
Maddon et al. (1985)			
Sugimura et al. (1985)	T$_s$-T	Total/competitive	PC-T cell suppressing factor
Raschke (1987)	T200 L transfectant-L	Total thymocyte/subtracted	T200
Obaru et al. (1986)			pLD78[d]
Schmid and Weissmann (1987)	*Activated PBL-PBL*	Total/differential	1-3E[c]; 3-10C[d]: 42% homology with β-thromboglobulin; 2 other molecules[e]
Hata et al. (1987)	**T-B**	Total/subtracted	TCR-α, -β, and -δ; CD3 δ/ε; I309[d]; G26[d]; 10 other molecules[e]
Miller et al. (1989)	(IDP2)-(JY)		
Alonso et al. (1987)	*T-T?* (MOLT4)-(CCRF HSB.2) (HPB ALL)-(CRF HSB.2)	Subtracted/subtracted	MAL
Kwon et al. (1987)	Th-B; Th-Tc (ConA-L2)-(A20.2.J)	Total/differential	GM-CSF; IL-3; IL-2; TCR-α and -β; c-myc; preproenkephalin; T-cell replacing factor; L2G25B[d]
Kwon and Weissman (1989)	(ConA-L2)-(ConA-L3)		
Burd et al. (1987)	Th-B; Th-L (Con A-C1.Lyl.T1)- (L, 2PK3, MOPC315, C1.Lyl.T1)	Total/subtracted	TCA3.0[d]
Lipes et al. (1988)	*Activated PBL-PBL*	Total/differential	Act-2[d]
Chang and Reinherz (1989)	*Activated PBL-PBL*	Subtracted/subtracted	HC21[d]

Zipfel et al. (1989) Bours et al. (1990)	*Activated PBL-PBL*	Subtracted/differential	c-myc; IL-3; IL-4; rIL-2; preproenkephalin; early growth response gene-2; a member of the steroid receptor family; a DNA binding factor interacting with NF-κb sites; a zinc-finger-containing DNA-binding protein; Act-2[d]; 56 other molecules[e]
Brown et al. (1989)	**Activated T-T**	Total/differential	SIS-α[d]; SIS-ε[d]; SIS-γ[d]; 1 other molecule[e]
Harrigan et al. (1989)	**Glucocorticoid T-T (WEHI-7TG)-(CXG56D3)**	Total/subtracted	One molecule 85% homologous to the chondroitin sulfate proteoglycan core protein; VL30 element; 9 other molecules[e]
MacLeod et al. (1990a) MacLeod et al. (1990b)	**T-T (SL12.4)-(SL12.3)**	Total/subtracted	One molecule homologous to the murine ecotropic virus receptor; 5 other molecules[e]
Owens et al. (1991)	**Glucocorticoid T-T**	Subtracted/subtracted	A zinc-finger-containing protein; 1 other molecule[e]

Footnotes as for Table 1.

ged described molecules. Also, since it has emerged that several distinct mechanisms of cytotoxicity may well exist in the same or in different cells (MacLennan et al., 1980; Tirosh and Berke, 1985; Ostergaard et al., 1987; Trenn et al., 1987; Berke, 1989; Lancki et al., 1989; Young et al., 1989; Ju et al., 1990; Tite, 1990; Ju, 1991; Lancki et al., 1991), some of the deviations from a strict CTL- or NK-specific tissue distribution may, rather than exclude a candidate molecule from playing a functional role, reflect the limits of a given mechanism of cytotoxicity. From another point of view, for instance, perforin, purified from the cytoplasmic granules of CTL and NK cell lines (Henkart, 1985; Podack, 1985; Tschopp and Jongeneel, 1988; Young et al., 1989), has not been cloned by subtraction technology, in spite of its apparently appropriate differential expression in most of the cellular combinations used to construct the libraries cited in Table 1. This may reflect a bias in the subtraction technique, or more likely a particular behavior of this sequence in subtractive hybridization and cloning. This example underlines the possibility that functionally important and differentially expressed molecules may go undetected using this approach.

The CTLA Molecules

The following molecules were isolated in this laboratory using the early subtractive cDNA cloning methodology mentioned above:

Two serine esterases, CTLA-1 and CTLA-3 (Brunet et al., 1986). CTLA-1 has also been described under other names, notably CCP1 (Lobe et al., 1986a) and granzyme B (Jenne and Tschopp, 1988). CTLA-3 has also been described as H factor (Gershenfeld and Weissman, 1986) and granzyme A (Jenne and Tschopp, 1988).

CTLA-2, in fact two very homologous cDNAs called CTLA-2α and CTLA-2β, coding for putative proteins homologous to proregions of cysteine proteinases, and expressed upon T-lymphocyte activation (Denizot et al., 1989).

CTLA-4, a member of the Ig superfamily (Brunet et al., 1987a).

In retrospect, it is perhaps no wonder that the subtraction procedure led to the isolation of three proteases or protease-related molecules and a member of the Ig superfamily; subtraction just provides a slice within the world of the molecules of the immune system, which seems rich in precisely these types of molecules. Rather than reviewing all of the known various properties of the CTLA molecules, we shall concentrate for CTLA-1, -2, and -4 on only a few of their characteristics, which perhaps are of general significance or applicability.

CTLA-1 and the Charms of Molecular Neighborhood

To our initial surprise, the *Ctla-1* gene in mouse turned out to map, by *in situ* hybridization, to the D region of chromosome 14 (Brunet et al., 1986), where the gene for the alpha chain of the T-cell receptor had previously been mapped (Kranz et al., 1985). Using restriction fragment length polymorphisms and a backcross progeny, the number of recombination events detected between the *Tcrα* and the *Ctla-1* genes was 0 in 100 meioses (Harper et al., 1988), 1 in 198 meioses (Ceci et al., 1990), or 1 in 75 meioses (Crosby et al., 1990); the locus order thus determined was *NP-1-Tcrα/Ctla-1-Gdh-X* (Harper et al., 1988), centromere–[*Psp-2,Rib-1*]–*Tcrα–Ctla-1* (Ceci et al., 1990), and *Np-1–Tcrα–Ctla-1–Rb-1* (Crosby et al., 1990). Most interestingly, not only *Ctla-1*, but also three other serine–esterase genes (called *Ctla-5, -6, and -7*, corresponding most probably to granzymes D, E, and F, respectively) were detected in the same area, with no recombination detected between them and *Ctla-1* (Crosby et al., 1990).

Similarly, in man the CTLA-1 gene was localized by *in situ* hybridization to the q11-q12 region of chromosome 14 (Harper et al., 1988; Klein et al., 1989; Lin et al., 1990), where the *TCRα* gene had been mapped (Caccia et al., 1985; Croce et al., 1985). The use of a cell line bearing a convenient inversion enabled to establish the order centromere–NP-1–TCRα–CTLA-1 (Harper et al., 1988). Also, another serine esterase gene (Meier et al., 1990; Hanson et al., 1990), most probably identical to granzyme H (Haddad et al., 1991), and the cathepsin G gene (Hanson et al., 1990) were detected in the same chromosomal area. More precisely, as determined by cosmid cloning, the CTLA-1, granzyme H, and cathepsin G genes reside within the same 50-kbp chromosomal segment (Hanson et al., 1990).

A main message from these findings is that a serine esterase gene cluster, expressed mostly in activated T cells, is located near the TCRα locus in both mouse and man. The differences between both species are most probably related to the finding that there are some homologous, but also some nonhomologous, serine esterase genes in mouse and man (Haddad et al., 1991). How far exactly is the serine esterase cluster from the TCRα locus? Co-mapping by *in situ* hybridization provides relatively poor discrimination, indicating that the genes may not be farther apart than, say, one or two megabases. These values would be consistent with the genetic distance (cumulated result from the data mentioned above: 2 recombinations/373 meioses). Attempts to find by pulse-field gel electrophoresis bands hybridizing with both probes were unsuccessful, indicating that there is at least one CpG-rich island between the serine esterase and the TCRα genes (Harper et al., 1988).

Since most of these clustered serine esterases, and TCRα, show preferential expression in T cells, these results may point at a possible coregulation of gene expression on a sizable stretch of chromosome. Also, these results provide an example of proximity of members of the serine esterase and of the immunoglobulin superfamilies (of which there are other examples, see Harper et al., 1988). As discussed before (Harper et al., 1988), this proximity could occur by chance if the family members are in high numbers, or if there had been accidental proximity of the two ancestor genes before amplification/divergence. Or the proximity could be functionally significant, for instance, in the case of coregulation of expression. From a practical point of view, these results, together with the proximity of genes coding for several members of the Ig superfamily (see references in Harper et al., 1991), for instance, in regions such as the 11q23 segment (see references in McConville et al., 1990), suggest approaches based on "proximity cloning," i.e., the cloning of genes residing in the immediate chromosomal neighborhood of a gene of interest.

CTLA-2 Molecules as Pieces of Unsolved Puzzles

There are at least two such puzzles. The first one stems from considerations on the possible role of CTLA-2. We like the idea that the CTLA-2 proteins, which are homologous to cysteine protease proregions and are induced upon lymphocyte activation, might then combine with a cysteine protease, thus functionally inactivating it and perhaps taking part in the down-regulation of lymphocyte activation (there are several examples of proregion peptides inhibiting the corresponding mature enzyme, see Denizot et al., 1989). CTLA-2 molecules would thus be inhibitors of cysteine proteinases. We feel that the possible role within the immune system of cysteine proteases and of their inhibitors has perhaps not been considered as much as that of, say, serine esterases. There are serious hints that cysteine proteinases and/or their natural inhibitors might be involved in processes as diverse as not only cell-mediated cytotoxicity (Redelman and Hudig, 1980) (interestingly, the amounts of the cysteine proteinase dipeptidyl peptidase I are increased in cytotoxic lymphocytes, see Chapter 32, Thiele and Lipsky, 1990a, 1990b), but also antigen processing and presentation (Buus and Werdelin, 1986; Puri and Factorovitch, 1988), some aspects of the abnormal phenotype of the beige mouse (Ito et al., 1989; Sato et al., 1990), stroke occurrence in patients with hereditary amyloid angiopathy (Levy et al., 1989), and transforming activity of the p21 *ras* oncogene products (Hiwasa et al., 1990).

The second puzzle is that of the relationship between T cells and mast cells. This longstanding question is now being tackled at the molecular level. In recent years, a growing number of molecules known to be produced by activated T cells have been shown to be also produced by mast cells. These include interleukin-4 (IL-4) (Brown et al., 1987), proteoglycans (Razin et al., 1982; MacDermott et al., 1985), tumor necrosis factor-α (TNF-α) (Young et al., 1987; Richards et al., 1988; Steffen et al., 1989; Gordon and Galli, 1990; Ohno et al., 1990; Walsh et al., 1991), CTLA-1 (Brunet et al., 1987b) and other serine esterases (Reynolds et al., 1990; Vanderslice et al., 1990), and indeed the cysteine proteinase proregion homolog CTLA-2 (Denizot et al., 1989). As noted before (Denizot et al., 1989), the sharing of these molecules between activated T cells and mast cells may reflect common cellular derivation and/or activation mechanisms.

CTLA-4 and Its Strikingly Close Relationship to CD28

It was progressively realized that CTLA-4 bore striking similarities to CD28, also a single-V-domain member of the Ig superfamily. This is of special interest considering what is known about the functional role of CD28. In the presence of co-activators, antibodies directed against CD28 induce T-cell activation (Hara et al., 1985; Moretta et al., 1985; Lesslauer et al., 1986; June et al., 1987; Pierres et al., 1988; Turka et al., 1990). CD28 is therefore able to transduce an activation signal. However, the activation induced by CD28 has peculiar features (reviewed in June et al., 1990). CD28 is expressed mostly on T cells. The demonstration (Linsley et al., 1990, 1991) that

B7/BB1, yet another member of the same superfamily, which is, however, mostly expressed on B cells, is a receptor for CD28 and for CTLA-4 suggests an essential role of the CD28-B7/BB1 interaction in T-cell help to B cells, and very significantly adds to the similarities between CD28 and CTLA-4.

The organization map of the human CTLA-4 gene (Dariavach et al., 1988; Harper et al., 1991), the human CD28 gene (Lee et al., 1990), and the mouse CTLA-4 gene (Harper et al., 1991) showed that both CTLA-4 genes and the human CD28 gene share the same overall intron/exon pattern. The corresponding primary structures of CD28 became known through the cloning and analysis of the human cDNA (Aruffo and Seed, 1987) and gene (Lee et al., 1990) and of the mouse cDNA (Gross et al., 1990), while for CTLA-4 the mouse cDNA (Brunet et al., 1987a) and the human genomic DNA (Dariavach et al., 1988) sequences were determined. The nucleic acid sequence data, recently completed (Harper et al., 1991), allowed a direct comparison of the four putative complete protein sequences of CD28 and CTLA-4 in mouse and man, showing striking homologies especially in some stretches (such as a MYPPPY hexamer in the hinge region) conserved across molecules and across species (Harper et al., 1991).

Not only are CTLA-4 and CD28 structurally homologous, but also the CD28 and CTLA-4 genes map to the same chromosomal region in both mouse and man: in situ hybridization studies on metaphase chromosomes showed that the CTLA-4 and CD28 genes mapped at the same site, in man in the 2q33 region (Dariavach et al., 1988; Lafage-Pochitaloff et al., 1990) and in mouse on band C of chromosome 1 (Brunet et al., 1987a; Harper et al., 1991). Recently, restriction fragment length polymorphism (RFLP) segregation studies showed no recombination event between CTLA-4 and CD28 among 114 mouse meioses (Howard et al., 1991). Experiments using yeast artificial chromosomes indicate a hecto–kilobase range for the distance between the human CTLA-4 and CD28 genes (N. Buonavista et al., in preparation).

Thus, CD28 and CTLA-4 were found very similar in two species in terms of structure, sequence, expression (not discussed here), receptor, and gene location, strongly suggesting that their genes are the direct products of a duplication event, and raising the possibility of functional homologies between the corresponding proteins.

What Have We Learned, and What Have We Not Learned, in the Process of Isolating and Studying the CTLA Molecules?

We did gather a number of pieces of methodological information. For instance, although there is no question that the subtractive approaches are powerful ones, in the sense that they have helped to isolate a number of interesting molecules, their degree of relevance to the initial question asked is more problematic. It could be argued that functional questions cannot be answered readily with a tissue-distribution-axed methodology (more about this below); however, even the tissue distribution of the isolated molecules does not correspond *exactly* to the desired pattern. One way to improve this situation would be to narrow the subtraction window, i.e., to use for subtraction a pair of cell types ideally differing only as to the desired character (for instance, cytotoxic cells minus a cytotoxicity-negative variant of the same cells, provided that this negativity results from transcript loss). As discussed before, narrowing the subtraction window requires an improved methodology, for instance, of the cloning-before-subtraction type.

Another consideration is the sensitivity threshold. When screening a subtracted library with a radiolabeled cDNA probe, even if this probe had been itself subtracted following the same combination as the subtracted library, only about one out of 10 clones would give a detectable signal (Brunet et al., 1988). This is most probably due to the fact that most of the cDNA clones correspond to low-abundance transcripts (and especially so after subtraction); these transcripts are insufficiently represented in the radiolabeled probe to give a detectable signal on the corresponding cDNA clones. These silent clones can still be investigated using more sensitive techniques, for instance, by using them as probes on Northern blots (Milner and Sutcliffe, 1983; Palazzolo et al., 1989), or they can be directly sequenced (Adams et al., 1991). However, these clones, if they correspond to low-abundance transcripts, would be precisely those in a cDNA library that would be the least efficiently subtracted away, and thus the ones with the highest chance of not fulfilling the specificity pattern corresponding to the subtraction combination. One is therefore led to the conclusion that in a subtracted cDNA library only about one clone in 10, which can be labeled with a radiolabeled cDNA probe, is of practical use; the silent ones are both more cumbersome to work with and less specific. Therefore, unless new techniques are developed to analyze them, they may be, and perhaps should be, disposed of as soon as possible during the screening process. Some of these observations now serve as a basis for new rounds of subtractive cloning in this laboratory.

Another note of caution stems from the very fact that at least some of the CTLA molecules, and many of the other molecules isolated through the approaches considered here, seem to be interesting in their own right. Although they were isolated as possible answers to a given question, they may provide fascinating answers to other questions; once molecules have been isolated, it is only too easy to follow *them* rather than the initial question.

Indeed, what we have not learned so far with the CTLA and other subtraction-derived molecules is new information about the mechanism(s) of cytotoxicity. We do not know, as yet, whether these molecules play a role in at least some types of T-cell-mediated cytotoxicity. More generally, it turns out to be rather difficult to establish the functional role of a given molecule isolated through reverse genetics. What we know from a functional point of view, for instance, for the CTLAs but also for other molecules such as perforin, was deduced from the putative protein sequences, and usually not through tests deliberately aiming at elucidating function (although the first results of such tests are now forthcoming; see below).

To demonstrate the functional role of a candidate molecule, one may try to confer cytotoxicity to a given cell by transfection with this molecule. For example, after transfection of perforin cDNA into a mast cell was performed, given the appropriate stimulus, the resulting transfected cells were found to lyse red blood cells, but to lyse nucleated cells much less (Shiver and Henkart, 1991).

Practically all other currently available techniques rely on inhibition of a given function, here cytotoxicity, using probes specific for a given candidate molecule. These probes can interfere with this molecule at several distinct levels. At the protein level, specific antibodies could be used to try to inhibit the function, obviously only if the candidate molecule is expressed at the membrane or secreted; anti-perforin antibodies were generally found not to block cytolysis, while, for instance, anti-TNF-α antibodies could block natural cytotoxic cell-mediated cytotoxicity.

At the messenger RNA level, blocking could be realized with antisense RNA. In the cytotoxicity field, the only published experiment of this sort that we are aware of deals again with perforin; antisense perforin oligonucleotides were able to inhibit the induction of T-cell-mediated cytotoxicity in populations of mouse spleen cells or human peripheral blood lymphocytes (Acha-Orbea et al., 1990).

At the gene level, one would try to eliminate or to inactivate the corresponding gene, thus preventing the synthesis of the product, with subsequent testing of the impact on cytotoxicity. *In vitro*, using as starting material a cytotoxic T-cell hybridoma (Conzelmann et al., 1982), we have derived a number of clones that are still cytotoxic, although they have lost a sizable amount of genetic material. In this "gene pruning" type of experiment, candidate molecules not expressed would be demonstrably not required for this type of cytotoxicity (manuscript in preparation). *In vivo* gene targeting experiments ultimately leading to mice homozygous for a given inactivated gene, which have been very conclusive for some immunologically relevant molecules (see for instance Koller et al., 1990; Zijlstra et al., 1990), are now under way in several laboratories for some of the candidate molecules mentioned above.

Acknowledgments

We thank INSERM, CNRS, and ARC for support.

References

Acha-Orbea H, Scarpellino L, Hertig S, Dupuis M, Tschopp J (1990): Inhibition of lymphocyte mediated cytotoxicity by perforin antisense oligonucleotides. *EMBO J* 9: 3815–3819

Adams MD, Kelley JM, Gocayne JD, Dubnick M, Polymeropoulos MH, Xiao H, Merril CR, Wu A, Olde B, Moreno RF, Kerlavage AR, McCombie WR, Venter JC (1991): Complementary DNA sequencing: Expressed sequence tags and human genome project. *Science* 252: 1651–1656

Alonso MA, Weissman SM (1987): cDNA cloning and sequence of MAL, a hydrophobic protein associated with human T-cell differentiation. *Proc Natl Acad Sci USA* 84: 1997–2001

Alt FW, Enea V, Bothwell ALM, Baltimore D (1979): Probes for specific mRNAs by subtractive hybridization: Anomalous expression of immunoglobulin genes. In: *Eucaryotic Gene Regulation, UCLA Symposia on Molecular and Cellular Biology.* New York: Academic Press, pp. 407–419

Arends MJ, Wyllie AH (1991): Apoptosis: Mechanisms and roles in pathology. *Int Rev Exp Pathol* 32: 223–254

Aruffo A, Seed B (1987): Molecular cloning of a CD28 cDNA by a high-efficiency COS cell expression system. *Proc Natl Acad Sci USA* 84: 8573–8577

Baixeras E, Roman-Roman S, Jitsukawa S, Genevee C, Mechiche S, Viegas-Pequignot E, Hercend T, Triebel F

(1990): Cloning and expression of a lymphocyte activation gene (LAG-1). *Mol Immunol* 27: 1091–1102

Berke G (1989): Functions and mechanisms of lysis induced by cytotoxic T lymphocytes and natural killer cells. In: *Fundamental Immunology*, Paul WE, ed. New York: Raven Press, pp. 735–764

Bours V, Villalobos J, Burd PR, Kelly K, Siebenlist U (1990): Cloning of a mitogen-inducible gene encoding a κB DNA-binding protein with homology to the *rel* oncogene and to cell-cycle motifs. *Nature* 348: 76–80

Brown KD, Zurawski SM, Mosmann TR, Zurawski G (1989): A family of small inducible proteins secreted by leukocytes are members of a new superfamily that includes leukocyte and fibroblast-derived inflammatory agents, growth factors, and indicators of various activation processes. *J Immunol* 142: 679–687

Brown MA, Pierce JH, Watson CJ, Falco J, Ihle JN, Paul WE (1987): B cell stimulatory factor-1/interleukin-4 mRNA is expressed by normal and transformed mast cells. *Cell* 50: 809–818

Brunet J-F, Denizot F, Golstein P (1988): A differential molecular biology search for genes preferentially expressed in functional T lymphocytes: The CTLA genes. *Immunological Rev* 103: 21–36

Brunet J-F, Denizot F, Luciani M-F, Roux-Dosseto M, Suzan M, Mattéi M-G, Golstein P (1987a): A new member of the immunoglobin superfamily—CTLA-4. *Nature* 328: 267–270

Brunet J-F, Denizot F, Suzan M, Haas W, Mencia-Huerta JM, Berke G, Luciani M-F, Golstein P (1987b): CTLA-1 and CTLA-3 serine-esterase transcripts are detected mostly in cytotoxic cells, but not only and not always. *J Immunol* 138: 4102–4105

Brunet J-F, Dosseto M, Denizot F, Mattéi M-G, Clark WR, Haqqi TM, Ferrier P, Nabholz M, Schmitt-Verhulst A-M, Luciani M-F, Golstein P (1986): The inducible cytotoxic-T-lymphocyte-associated gene transcript CTLA-1 sequence and gene localization to mouse chromosome 14. *Nature* 322: 268–271

Burd PR, Freeman GJ, Wilson SD, Berman M, Dekruyff R, Billings PR, Dorf ME (1987): Cloning and characterization of a novel T cell activation gene. *J Immunol* 139: 3126–3131

Buus S, Werdelin O (1986): A group-specific inhibitor of lysosomal cysteine proteinases selectively inhibits both proteolytic degradation and presentation of the antigen dinitrophenyl-poly-L-lysine by guinea pig accessory cells to T cells. *J Immunol* 136: 452–458

Caccia N, Bruns GAP, Kirsch IR, Hollis GF, Bertness V, Mak TW (1985): T cell receptor alpha chain genes are located on chromosome 14 at 14q11-14q12 in humans. *J Exp Med* 161: 1255–1260

Ceci JD, Kingsley, DM, Silan CM, Copeland NG, Jenkins NA (1990): An interspecific backcross linkage map of the proximal half of mouse chromosome 14. *Genomics* 6: 673–678

Chang H-C, Reinherz EL (1989): Isolation and characterization of a cDNA encoding a putative cytokine which is induced by stimulation via the CD2 structure on human T lymphocytes. *Eur J Immunol* 19: 1045–1051

Chien Y-H, Becker DM, Lindsten T, Okamura M, Cohen DI, Davis MM (1984): A third type of murine T-cell receptor gene. *Nature* 312: 31–35

Cohen JJ (1991): Programmed cell death in the immune system. *Adv Immunol* 50: 55–83

Conzelmann A, Corthézy P, Cianfriglia M, Silva A, Nabholz M (1982): Hybrids between rat lymphoma and mouse T cells with inducible cytolytic activity. *Nature* 298: 170–172

Crabtree GR (1989): Contingent genetic regulatory events in T lymphocyte activation. *Science* 243: 355–361

Croce CM, Isobe M, Palumbo A, Puck J, Ming J, Tweardy D, Erikson J, Davis M, Rovera G (1985): Gene for alpha-chain of human T-cell receptor: Location on chromosome 14 region involved in T-cell neoplasms. *Science* 227: 1044–1047

Crosby JL, Bleackley RC, Nadeau JH (1990): A complex of serine protease genes expressed preferentially in cytotoxic T-lymphocytes is closely linked to the T-cell receptor alpha- and delta-chain genes on mouse chromosome 14. *Genomics* 6: 252–259

Dahl CA, Bach FH, Chan W, Huebner K, Russo G, Croce CM, Herfurth T, Cairns JS (1990): Isolation of a cDNA clone encoding a novel form of granzyme B from human NK cells and mapping to chromosome 14. *Hum Genet* 84: 465–470

Dariavach P, Mattéi M-G, Golstein P, Lefranc M-P (1988): Human Ig superfamily CTLA-4 gene: Chromosomal localization and identity of protein sequence between murine and human CTLA-4 cytoplasmic domains. *Eur J Immunol* 18: 1901–1905

Denizot F, Brunet J-F, Roustan P, Harper K, Suzan M, Luciani M-F, Mattéi M-G, Golstein P (1989): Novel structures CTLA-2 alpha and CTLA-2 beta expressed in mouse activated T cells and mast cells and homologous to cysteine proteinase proregions. *Eur J Immunol* 19: 631–635

Duguid JR, Rohwer RG, Seed B (1988): Isolation of cDNAs of scrapie-modulated RNAs by subtractive hybridization of a cDNA library. *Proc Natl Acad Sci USA* 85: 5738–5742

Gershenfeld HK, Weissman IL (1986): Cloning of a cDNA for a T cell-specific serine protease from a cytotoxic T lymphocyte. *Science* 232: 854–858

Golstein P, Smith ET (1977): Mechanism of T cell-mediated cytolysis: The lethal hit stage. *Contemp Top Immunobiol* 7: 273–300

Golstein P, Goridis C, Schmitt-Verhulst A-M, Hayot B, Pierres A, Van Agthoven A, Kaufmann Y, Eshhar Z, Pierres M (1982): Lymphoid cell surface interaction structures detected using cytolysis-inhibiting monoclonal antibodies. *Immunol Rev* 68: 5–42

Golstein P, Ojcius DM, Young JD-E (1991): Cell death mechanisms and the immune system. *Immunol Rev* 121: 29–65

Gordon JR, Galli SJ (1990): Mast cells as a source of both preformed and immunologically inducible TNF-alpha/cachectin. *Nature* 346: 274–276

Gross JA, St. John T, Allison JP (1990): The murine homologue of the T lymphocyte antigen CD28: Molecular cloning and cell surface expression. *J Immunol* 144: 3201–3210

Grusby MJ, Mitchell SC, Nabavi N, Glimcher LH (1990a): Casein expression in cytotoxic T lymphocytes. *Proc Natl Acad Sci USA* 87: 6897–6901

Grusby MJ, Nabavi N, Wong H, Dick RF, Bluestone JA, Schotz MC, Glimcher LH (1990b): Cloning of an interleukin-4 inducible gene from cytotoxic T lymphocytes and its identification as a lipase. *Cell* 60: 451–459

Haddad P, Jenne DE, Tschopp J, Clément M-V, Mathieu-Mahul D, Sasportes M (1991): Structure and evolutionary origin of the human granzyme H gene. *Int Immunol* 3: 57–66

Hanson RD, Hohn PA, Popescu NC, Ley TJ (1990): A cluster of hematopoietic serine protease genes is found on the same chromosomal band as the human α/δ T-cell receptor locus. *Proc Natl Acad Sci USA* 87: 960–963

Hara T, Fu SM, Hansen JA (1985): Human T cell activation II. A new activation pathway used by a major T cell population via a disulfide-bond dimer of a 44 kilodalton polypeptide (9.3 antigen). *J Exp Med* 161: 1513–1524

Harper K, Balzano C, Rouvier E, Mattéi M-G, Luciani M-F, Golstein P (1991): CTLA-4 and CD28 activated lymphocyte molecules are closely related in both mouse and human as to sequence, message expression, gene structure and chromosomal location. *J Immunol* 147: 1037–1044

Harper K, Mattéi M-G, Simon D, Suzan M, Guénet J-L, Haddad P, Sasportes M, Golstein P (1988): Proximity of the CTLA-1 serine-esterase and Tcr alpha loci in mouse and man. *Immunogenetics* 28: 439–444

Harrigan MT, Baughman G, Campbell NF, Bourgeois S (1989): Isolation and characterization of glucocorticoid- and cyclic AMP-induced genes in T lymphocytes. *Mol Cell Biol* 9: 3438–3446

Haser WG, Saito H, Koyama T, Tonegawa S (1987): Cloning and sequencing of murine T3 gamma cDNA from a subtractive cDNA library. *J Exp Med* 166: 1186–1191

Hata S, Brenner MB, Krangel MS (1987): Identification of putative human T cell receptor delta complementary DNA clones. *Science* 238: 678–682

Hedrick SM, Cohen DI, Nielsen EA, Davis MM (1984a): Isolation of cDNA clones encoding T cell-specific membrane-associated proteins. *Nature* 308: 149–153

Hedrick SM, Nielsen EA, Kavaler J, Cohen DI, Davis MM (1984b): Sequence relationships between putative T-cell receptor polypeptides and immunoglobulins. *Nature* 308: 153–158

Henkart PA (1985): Mechanism of lymphocyte-mediated cytotoxicity. *Annu Rev Immunol* 3: 31–58

Henney CS (1977): T-cell-mediated cytolysis: An overview of some current issues. *Contemp Top Immunobiol* 7: 245–272

Hiwasa T, Sawada T, Sakiyama S (1990): Cysteine proteinase inhibitors and *ras* gene products share the same biological activities including transforming activity toward NIH3T3 mouse fibroblasts and the differentiation-inducing activity toward PC12 rat pheochromocytoma cells. *Carcinogenesis* 11: 75–80

Houchins JP, Yabe T, McSherry C, Bach FH (1991): DNA sequence analysis of NKG2, a family of related cDNA clones encoding type II integral membrane proteins on human natural killer cells. *J Exp Med* 173: 1017–1020

Houchins JP, Yabe T, McSherry C, Miyokawa N, Bach FH (1990): Isolation and characterization of NK cell or NK/T cell-specific cDNA clones. *J Mol Cell Immunol* 4: 295–306

Howard TA, Rochelle JM, Seldin MF (1991): *Cd28* and *Ctla-4*, two related members of the *Ig* supergene family, are tightly linked on proximal mouse chromosome 1. *Immunogenetics* 33: 74–76

Ito M, Sato A, Tanabe F, Ishida E, Takami Y, Shigeta S (1989): The thiol proteinase inhibitors improve the abnormal rapid down-regulation of protein kinase C and the impaired natural killer cell activity in (Chediak-Higashi syndrome) beige mouse. *Biochem Biophys Res Commun* 160: 433–440

Jenne DE, Tschopp J (1988): Granzymes, a family of serine proteases released from granules of cytolytic T lymphocytes upon T cell receptor stimulation. *Immunol Rev* 103: 53–71

Jongstra J, Schall TJ, Dyer BJ, Clayberger C, Jorgensen J, Davis MM, Krensky AM (1987): The isolation and sequence of a novel gene from a human functional T cell line. *J Exp Med* 165: 601–614

Ju S-T (1991): Distinct pathways of CD4 and CD8 cells induce rapid target DNA fragmentation. *J Immunol* 146: 812–818

Ju S-T, Ruddle NH, Strack P, Dorf ME, Dekruyff RH (1990): Expression of two distinct cytolytic mechanisms among murine CD4 subsets *J Immunol* 144: 23–31

June CH, Ledbetter JA, Gillespie MM, Lindsten T, Thompson CB (1987): T-cell proliferation involving the CD28 pathway is associated with cyclosporin-resistant interleukin 2 gene expression. *Mol Cell Biol* 7: 4472–4481

June CH, Ledbetter JA, Linsley PS, Thompson CB (1990): Role of the CD28 receptor in T-cell activation. *Immunol Today* 11: 211–216

Kavathas P, Sukhatme VP, Herzenberg LA, Parnes JR (1984): Isolation of the gene encoding the human T-lymphocyte differentiation antigen Leu-2 (T8) by gene transfer and cDNA subtraction. *Proc Natl Acad Sci USA* 81: 7688–7692

Kerr JFR, Harmon BV (1991): Definition and incidence of apoptosis: An historical perspective. In: *Apoptosis: The Molecular Basis of Cell Death*, Tomei LD, Cope FO, eds. New York: Cold Spring Harbor Laboratory Press 11: 5–29

Klein JL, Shows TB, Dupont B, Trapani JA (1989): Genomic organization and chromosomal assignment for a serine protease gene (CSPB) expressed by human cytotoxic lymphocytes. *Genomics* 5: 110–117

Koga Y, Caccia N, Toyonaga B, Spolski R, Yanagi Y, Yoshikai Y, Mak TW (1986): A human T cell-specific cDNA clone (YT16) encodes a protein with extensive homology to a family of protein-tyrosine kinases. *Eur J Immunol* 16: 1643–1646

Koller BH, Marrack P, Kappler JW, Smithies O (1990): Normal development of mice deficient in β_2M, MHC Class I proteins, and CD8+ T cells. *Science* 248: 1227–1230

Koyama T, Hall LR, Haser WG, Tonegawa S, Saito H (1987): Structure of a cytotoxic T-lymphocyte-specific gene shows a strong homology to fibrinogen β and γ chains. *Proc Natl Acad Sci USA* 84: 1609–1613

Kranz DM, Saito H, Disteche CM, Swisshelm K, Pravtcheva D, Ruddle FH, Eisen HN, Tonegawa S (1985): Chromosomal locations of the murine T-cell receptor alpha-chain gene and the T-cell gamma gene. *Science* 229: 941–945

Kwon BS, Kestler D, Lee E, Wakulchik M, Young JD-E (1988): Isolation and sequence analysis of serine protease cDNAs from mouse cytolytic T lymphocytes. *J Exp Med* 168: 1839–1854

Kwon BS, Kim GS, Prystowsky MB, Lancki DW, Sabath DE, Pan J, Weissman SM (1987): Isolation and initial characterization of multiple species of T-lymphocyte subset cDNA clones. *Proc Natl Acad Sci USA* 84: 2896–2900

Kwon BS, Weissman SM (1989): cDNA sequences of two inducible T-cell genes. *Proc Natl Acad Sci USA* 86: 1963–1967

Lafage-Pochitaloff M, Costello R, Couez D, Simonetti J, Mannoni P, Mawas C, Olive D (1990): Human *CD28* and *CTLA-4* Ig superfamily genes are located on chromosome 2 at bands q33-q34. *Immunogenetics* 31: 198–201

Lancki DW, Hsieh C-S, Fitch FW (1991): Mechanisms of lysis by cytotoxic T lymphocyte clones: Lytic activity and gene expression in cloned antigen-specific CD4+ and CD8+ T lymphocytes. *J Immunol* 146: 3242–3249

Lancki DW, Kaper BP, Fitch FW (1989): The requirements for triggering of lysis by cytolytic T lymphocyte clones. II. Cyclosporin A inhibits TCR-mediated exocytosis but only selectively inhibits TCR-mediated lytic activity by cloned CTL. *J Immunol* 142: 416–424

Lee KP, Taylor C, Petryniak B, Turka LA, June CH, Thompson CB (1990): The genomic organization of the CD28 gene: Implications for the regulation of CD28 mRNA expression and heterogeneity. *J Immunol* 145: 344–352

Lesslauer W, Koning F, Ottenhoff T, Giphart M, Goulmy E, van Rood JJ (1986): T90/44 (9.3 antigen). A cell surface molecule with a function in human T cell activation. *Eur J Immunol* 16: 1289–1296

Levy E, Lopez-Otin C, Ghiso J, Geltner D, Frangione B (1989): Stroke in Icelandic patients with hereditary amyloid angiopathy is related to a mutation in the cystatin C gene, an inhibitor of cysteine proteases. *J Exp Med* 169: 1771–1778

Lin CC, Meier M, Sorensen O, Sasi R, Tainaka T, Bleackley RC (1990): Chromosome localization of two human serine protease genes to region 14q11.2→q12 by in situ hybridization. *Cytogenet Cell Genet* 53: 169–171

Linsley PS, Brady W, Urnes M, Grosmaire LS, Damle NK, Ledbetter JA (1991): CTLA-4 is a second receptor for the B cell activation antigen B7. *J Exp Med* 174: 561–569

Linsley PS, Clark EA, Ledbetter JA (1990): T-cell antigen CD28 mediates adhesion with B cells by interacting with activation antigen B7/BB-1. *Proc Natl Acad Sci USA* 87: 5031–5035

Lipes MA, Napolitano M, Jeang K-T, Chang NT, Leonard WJ (1988): Identification, cloning, and characterization of an immune activation gene. *Proc Natl Acad Sci USA* 85: 9704–9708

Littman DR, Thomas Y, Maddon PJ, Chess L, Axel R (1985): The isolation and sequence of the gene encoding T8: A molecule defining functional classes of T lymphocytes. *Cell* 40: 237–246

Lobe CG, Finlay BB, Paranchych W, Paetkau VH, Bleackley RC (1986a): Novel serine proteases encoded by two cytotoxic T lymphocyte-specific genes. *Science* 232: 858–861

Lobe CG, Havele C, Bleackley RC (1986b): Cloning of two genes that are specifically expressed in activated cytotoxic T lymphocytes. *Proc Natl Acad Sci USA* 83: 1448–1452

MacDermott RP, Schmidt RE, Caulfield JP, Hein A, Bartley GT, Ritz J, Schlossman SF, Austen KF, Stevens RL (1985): Proteoglycans in cell-mediated cytotoxicity. Identification, localization, and exocytosis of a chondroitin sulfate proteoglycan from human cloned natural killer cells during target cell lysis. *J Exp Med* 162: 1771–1787

MacLennan ICM, Gotch FM, Golstein P (1980): Limited specific T-cell mediated cytolysis in the absence of extracellular Ca^{++}. *Immunology* 39: 109–117

MacLeod CL, Finley K, Kakuda D, Kozak CA, Wilkinson MF (1990a): Activated T cells express a novel gene on chromosome 8 that is closely related to the murine ecotropic retroviral receptor. *Mol Cell Biol* 10: 3663–3674

MacLeod CL, Fong AM, Seal BS, Walls L, Wilkinson MF (1990b): Isolation of novel complementary DNA clones from T lymphoma cells: One encodes a putative multiple membrane-spanning protein. *Cell Growth Differ* 1: 271–279

Maddon PJ, Littman DR, Godfrey M, Maddon DE, Chess L, Axel R (1985): The isolation and nucleotide sequence of a cDNA encoding the T cell surface protein T4: A new member of the immunoglobulin gene family. *Cell* 42: 93–104

Martz E (1977): Mechanism of specific tumor-cell lysis by alloimmune T lymphocytes: Resolution and characterization of discrete steps in the cellular interaction. *Contemp Top Immunobiol* 7: 301–361

McConville CM, Formstone CJ, Hernandez D, Thick J, Taylor AMR (1990): Fine mapping of the chromosome 11q22-23 region using PFGE, linkage and haplotype analysis; localization of the gene for ataxia telangectasia to a 5cM region flanked by NCAM/DRD2 and STMY/CJ52.75,q2.22. *Nucl Acids Res* 18: 4335–4343

Meier M, Kwong PC, Frégeau CJ, Atkinson EA, Burrington M, Ehrman N, Sorensen O, Lin CC, Wilkins J, Bleackley RC (1990): Cloning of a gene that encodes a new member of the human cytotoxic cell protease family. *Biochemistry* 29: 4042–4049

Miller MD, Hata S, De Waal Malefyt R, Krangel MS (1989): A novel polypeptide secreted by activated human T lymphocytes. *J Immunol* 143: 2907–2916

Milner R, Sutcliffe JG (1983): Gene expression in rat brain. *Nucl Acids Res* 11: 5497–5520

Moretta A, Pantaleo G, Lopez-Botet M, Moretta L (1985): Involvement of T44 molecules in an antigen-independent pathway of T cell activation. Analysis of the correlations to the T cell antigen-receptor complex. *J Exp Med* 162: 823–838

Obaru K, Fukuda M, Maeda S, Shimada K (1986): A cDNA clone used to study mRNA inducible in human tonsillar lymphocytes by a tumor promoter. *J Biochem* 99: 885–894

Ohno I, Tanno Y, Yamauchi K, Takishima T (1990): Gene expression and production of tumour necrosis factor by a rat basophilic leukaemia cell line (RBL-2H3) with IgE receptor triggering. *Immunology* 70: 88–93

Ostergaard HL, Kane KP, Mescher MF, Clark WR (1987): Cytotoxic T lymphocyte mediated lysis without release of serine esterase. *Nature* 330: 71–72

Owens GP, Hahn WE, Cohen JJ (1991): Identification of mRNAs associated with programmed cell death in immature thymocytes. *Mol Cell Biol* 11: 4177–4188

Palazzolo MJ, Hyde DR, VijayRaghavan K, Mccklenburg K, Benzer S, Meyerowitz E (1989): Use of a new strategy to isolate and characterize 436 Drosophila cDNA clones corresponding to RNAs detected in adult heads but not in early embryos. *Neuron* 3: 527–539

Pierres A, Lopez M, Cerdan C, Nunes J, Olive D, Mawas C (1988): Triggering CD28 molecules synergize with CD2 (T11.1 and T11.2)-mediated T cell activation. *Eur J Immunol* 18: 685–690

Podack ER (1985): Molecular mechanism of lymphocyte-mediated tumor cell lysis. *Immunol Today* 6: 21–27

Puri J, Factorovitch Y (1988): Selective inhibition of antigen presentation to cloned T cells by protease inhibitors. *J Immunol* 141: 3313–3317

Raschke WC (1987): Cloned murine T200 (Ly-5) cDNA reveals multiple transcripts within B- and T-lymphocyte lineages. *Proc Natl Acad Sci USA* 84: 161–165

Razin E, Stevens RL, Akiyama F, Schmid K, Austen KF (1982): Culture from mouse bone marrow of a subclass of mast cells possessing a distinct chondroitin sulfate proteoglycan with glycosaminoglycans rich in N-acetylgalactosamine-4,6-disulfate. *J Biol Chem* 257: 7229

Redelman D, Hudig D (1980): The mechanism of cell-mediated cytotoxicity. I. Killing by murine cytotoxic T lymphocytes requires cell surface thiols and activated proteases. *J Immunol* 124: 870–878

Reynolds DS, Stevens RL, Lane WS, Carr MH, Austen KF, Serafin WE (1990): Different mouse mast cell populations express various combinations of at least six distinct mast cell serine proteases. *Proc Natl Acad Sci USA* 87: 3230–3234

Richards AL, Okuno T, Takagaki Y, Djeu JY (1988): Natural cytotoxic cell-specific cytotoxic factor produced by IL-3-dependent basophilic/mast cells: Relationship to TNF. *J Immunol* 141: 3061–3066

Rubenstein JLR, Brice AEJ, Ciaranello RD, Denney D, Porteus MH, Usdin TB (1990): Subtractive hybridization system using single-stranded phagemids with directional inserts. *Nucl Acids Res* 18: 4833–4842

Saito H, Kranz DM, Takagaki Y, Hayday AC, Eisen HN, Tonegawa S (1984a): A third rearranged and expressed gene in a clone of cytotoxic T lymphocytes. *Nature* 312: 36–40

Saito H, Kranz DM, Takagaki Y, Hayday AC, Eisen HN, Tonegawa S (1984b): Complete primary structure of a heterodimeric T-cell receptor deduced from cDNA sequences. *Nature* 309: 757–762

Sato A, Tanabe F, Ito M, Ishida E, Shigeta S (1990): Thiol proteinase inhibitors reverse the increased protein kinase C down-regulation and concanavalin A cap formation in polymorphonuclear leukocytes from Chediak-Higashi syndrome (beige) mouse. *J Leukocyte Biol* 48: 377–381

Schall TJ, Jongstra J, Dyer BJ, Jorgensen J, Clayberger C, Davis MM, Krensky AM (1988): A human T cell-specific molecule is a member of a new gene family. *J Immunol* 141: 1018–1025

Schmid J, Weissmann C (1987): Induction of mRNA for a serine protease and a beta-thromboglobulin-like protein in mitogen-stimulated human leukocytes. *J Immunol* 139: 250–256

Shiver JW, Henkart PA (1991): A noncytotoxic mast cell tumor line exhibits potent IgE-dependent cytotoxicity after transfection with the cytolysin/perforin gene. *Cell* 64: 1175–1181

Sims JE, Tunnacliffe A, Smith WJ, Rabbitts TH (1984): Complexity of human T-cell antigen receptor beta-chain constant- and variable-region genes. *Nature* 312: 541–545

Sive HL, St John T (1988): A simple subtractive hybridization technique employing photoactivatable biotin and phenol extraction. *Nucl Acids Res* 16: 10937

Steffen M, Abboud M, Potter GK, Yung YP, Moore MAS (1989): Presence of tumor necrosis factor or a related factor in human basophil/mast cells. *Immunology* 66: 445–450

Sugimura K, Yamasaki N, Matsuura M, Watanabe T (1985): Antigen-specific T cell suppressor factor (TsF): Isolation of a cDNA clone encoding for a functional polypeptide chain of a phosphorylcholine-specific TsF. *Eur J Immunol* 15: 873–880

Takihara Y, Caccia N, Yanagi Y, Mak TW (1989): Isolation of a T cell specific cDNA clone possibly involved in the T cell activation pathway. *Int Immunol* 1: 59–65

Thiele DL, Lipsky PE (1990a): Mechanism of L-leucyl-L-leucine methyl ester-mediated killing of cytotoxic lymphocytes: Dependence on a lysosomal thiol protease, dipeptidyl peptidase I, that is enriched in these cells. *Proc Natl Acad Sci USA* 87: 83–87

Thiele DL, Lipsky PE (1990b): The action of leucyl-leucine methyl ester on cytotoxic lymphocytes requires uptake by a novel dipeptide-specific facilitated transport system and dipeptidyl peptidase I-mediated conversion to membranolytic products. *J Exp Med* 172: 183-194

Tirosh R, Berke G (1985): T lymphocyte-mediated cytolysis as an excitatory process of the target. I. Evidence that the target may be the site of Ca^{++} action. *Cellular Immunol* 95: 113-123

Tite JP (1990): Evidence of a role for TNF alpha in cytolysis by $CD4^+$, class II MHC-restricted cytotoxic T cells. *Immunology* 71: 208-212

Trenn G, Takayama H, Sitkovsky MV (1987): Exocytosis of cytolytic granules may not be required for target cell lysis by cytotoxic T-lymphocytes. *Nature* 330: 72

Triebel F, Jitsukawa S, Baixeras E, Roman-Roman S, Genevée C, Viegas-Péquignot E, Hercend T (1990): LAG-3, a novel lymphocyte activation gene closely related to CD4. *J Exp Med* 171: 1393-1405

Tschopp J, Jongeneel CV (1988): Cytotoxic T lymphocyte mediated cytolysis. *Biochemistry* 27: 2641-2646

Turka LA, Ledbetter JA, Lee K, June CH, Thompson CB (1990): CD28 is an inducible T cell surface antigen that transduces a proliferative signal in CD3+ mature thymocytes. *J Immunol* 144: 1646-1653

Ullman KS, Northrop JP, Verweij CL, Crabtree GR (1990): Transmission of signals from the T lymphocyte antigen receptor to the genes responsible for the cell proliferation and immune function: The missing link. *Annu Rev Immunol* 8: 421-452

Vanderslice P, Ballinger SM, Tam EK, Goldstein SM, Craik CS, Caughey GH (1990): Human mast cell tryptase: Multiple cDNAs and genes reveal a multigene serine protease family. *Proc Natl Acad Sci USA* 87: 3811-3815

Walsh LJ, Trinchieri G, Waldorf HA, Whitaker D, Murphy GF (1991): Human dermal mast cells contain and release tumor necrosis factor alpha, which induces endothelial leukocyte adhesion molecule 1. *Proc Natl Acad Sci USA* 88: 4220-4224

Yabe T, McSherry C, Bach FH, Houchins JP (1990): A cDNA clone expressed in natural killer and T cells that likely encodes a secreted protein. *J Exp Med* 172: 1159-1163

Yanagi Y, Chan A, Chin B, Minden M, Mak TW (1985): Analysis of cDNA clones specific for human T cells and the α and β chains of the T-cell receptor heterodimer from a human T-cell line. *Proc Natl Acad Sci USA* 82: 3430-3434

Yanagi Y, Yoshikai Y, Leggett K, Clark SP, Aleksander I, Mak TW (1984): A human T cell-specific cDNA clone encodes a protein having extensive homology to immunoglobulin chains. *Nature* 308: 145-149

Young JD-E, Liu C-C, Butler G, Cohn ZA, Galli SJ (1987): Identification, purification, and characterization of a mast cell-associated cytolytic factor related to tumor necrosis factor. *Proc Natl Acad Sci USA* 84: 9175-9179

Young LHY, Klavinskis LS, Oldstone MBA, Young JD-E (1989): In vivo expression of perforin by $CD8^+$ lymphocytes during an acute viral infection. *J Exp Med* 169: 2159-2171

Zijlstra M, Bix M, Simister NE, Loring JM, Raulet DH, Jaenisch R (1990): β2-microglobulin deficient mice lack $CD4^- 8^+$ cytolytic T cells. *Nature* 344: 742-746

Zipfel PF, Irving SG, Kelly K, Siebenlist U (1989): Complexity of the primary genetic response to mitogenic activation of human T cells. *Mol Cell Biol* 9: 1041-1048

23
Structure and Possible Functions of Lymphocyte Granzymes

Patrick Haddad, Dieter E. Jenne, Olivier Krähenbühl, and Jürg Tschopp

Introduction

The participation of cellular serine proteinases in many physiological processes has been widely documented. These enzymes are most active at neutral, i.e., physiological pH. A number of studies have suggested that they may be involved in cellular chemotaxis, protein turnover in tissues, endocytosis and exocytosis, peptide hormone processing, cellular invasiveness or tumorigenesis, and cell proliferation (for reviews, see Mullins and Rohrlich, 1983; Neurath, 1984). These suggestions were based on the interference of proteinase inhibitors with cellular function or proliferation and the detection of cell-associated esterolytic, amidolytic, or proteolytic activities, notably in hemopoietic cells of the immune defense system, including lymphocytes, natural killer (NK) cells, mast cells, and phagocytes. This review will focus on a subfamily of serine proteases expressed in granules of lymphocytes and NK cells.

Lymphocytes fulfill their immune defense task by penetrating through subendothelial basement membranes of blood vessel walls, interstitial matrices, in particular after infection with pathogens. The ability of a serine esterase expressed and secreted by effector T cells and their tumorigenic variants to degrade high-molecularweight proteins and components of the extracellular matrix was taken as an indication of participation of the enzyme extravasation (Kramer et al., 1985). There is also considerable indirect evidence that lymphocyte-associated proteases play an important role in cell mediated cytotoxicity: cytolytic activity elicited by intact cytotoxic lymphocytes (CTl) or NK cells can be strongly inhibited by proteinase inhibitors such as diisopropylfluorophosphate (DFP), phenylmethylsulfonyl fluoride (PMSF), tosyllysine chloromethyl ketone (TLCK), and isocoumarins (Ferluga and Allison, 1974; Chang and Eisen, 1980; Redelman and Hudig, 1980; Quan et al., 1982; Redelman and Hudig, 1983; Hudig et al., 1984; Lavie et al., 1985; Hudig et al., 1991). Antibody-dependent cellular cytotoxicity (ADCC) is abolished by protease substrates such as acetyl tyrosine ester in a competitive manner, and by chloromethyl ketone derivatives of amino acids. Furthermore, plasma proteinase inhibitors such as α_1-antitrypsin and α_1-antichymotrypsin also suppress lysis when present in NK cell assays (Hudig et al., 1981, 1984), suggesting inactivation of functionally important serine proteinases. Protease inhibitors have no effect on the lytic activity of NK cells when the effector cells are pretreated, but cytotoxicity is highly sensitive to the presence of such inhibitors for a short period after the addition of the target cell (Lavie et al., 1985). This, it appears that target cell binding triggers the exposure of proteolytic enzymes to the external environment.

These observations provided fuel for the stimulus–secretion hypothesis of CTL- and NK-cell-mediated target cell lysis (Henkart, 1985; Podack, 1986; Tschopp and Jongeneel, 1988). According to this model, initially cryptic proteases stored in dense-core cytoplasmic granules are released upon effector–target cell contact and could contribute to the lytic mechanism by an as yet undefined pathway. In support of this notion, it was shown that granule-mediated catalysis can be completely inhibited by mechanism-based isocoumarin serine protease inhibitors (Hudig et al., 1987, 1991).

A large number of highly homologous serine esterases, termed granzymes (for *gran*ule-associated en*zymes*), have been shown to be expressed by

CTL and NK cells of murine, rat, and human origin. Although these enzymes are localized in secretory cytoplasmic granules, their relevance to cytotoxicity is still not known. The isolation, structure, genomic organization, expression, and putative functions of human and murine granzymes are discussed in the following sections.

Isolation of Murine and Human Granzymes

Pasternack and Eisen (1985) first described a serine esterase activity in cloned murine CTL cell lines by using the compound N-α-benzyloxcarbonyl-L-lysine thiobenzyl ester (BLT) as a chromogenic substrate. This trypsin-like protease could be labeled with the affinity reagent for serine esterases, DFP, and was initially reported to be CTL-specific. Following this report, several serine esterases, designated granzymes A and B (Masson et al., 1986a), T-cell-specific serine proteinase 1 (TSP-1) (Fruth et al., 1987), serine esterases 1 and 2 (SE1 and SE2) (Young et al., 1986), cytotoxic cell protease I (CCPI) (Redmond et al., 1987), BLT esterase, or simply serine esterase (Pasternack et al., 1986), were identified in cytolytic granules of CTL and NK cell lines (see Table 1 for nomenclature).

Subsequently, Masson and Tschopp (1987) purified to homogeneity the major proteins present in cytolytic granules of mouse CTL. Besides perforin and proteoglycans (see Chapter 14, this volume), a family of at least six serine esterases, which they termed granzymes A to F, was isolated from granules of a murine CTL cell line.

The identification and structural analysis of the granzymes was greatly facilitated by the application of cDNA subtraction library techniques. A pool of mRNA transcripts preferentially expressed in cytolytically active CTL was identified by subtracting from a CTL-derived total cDNA library those transcripts which hybridized with the total cDNA from noncytotoxic T cells. Three murine serine proteases expressed in T lymphocytes have thus been cloned and sequenced by various investigators: H factor (Gershenfeld and Weismann, 1986) or CTLA-3 (Brunet et al., 1987) which is identical to granzyme A (Masson et al., 1986b) CTLA-1 (Brunet et al., 1986) or CCPI (Lobe et al., 1986), which is identical to granzyme B (Masson and Tschopp, 1987); and CCPII (Lobe et al., 1986), which corresponds to granzyme C (Jenne et al., 1988b) (see Table 1 for the nomenclature of granzymes and their corresponding cDNA clones).

The subtractive screening approach described above is limited by the fact that rare mRNA transcripts are difficult to detect and that an unknown number of expressed genes are related to the activation rather than to the effector functions of CTL. Another approach using antibody screening of cDNA expression libraries resulted in the isolation of cDNA clones for further mouse granzymes. In addition to granzymes A and B, the murine granzymes C, D, E, and F have been sequenced and their complete covalent structure described (Jenne et al., 1988a, 1988b). Furthermore, a seventh murine serine esterase cDNA clone, granzyme G, has been isolated, while the encoded protein was shown by N-terminal sequence analysis to be correctly processed and sorted into CTL granules (Kwon et al., 1988; Jenne et al., 1989; Prendergast et al., 1991).

In human CTL and LAK cell granules, there are at least three, possible four, serine proteinases besides perforin and other undefined proteins. Two of these have been shown to be the human homologs of the murine granzymes A and B (Table 1). Like murine granzyme A, which elutes from the strong cation exchanger mono S at NaCl concentrations above 650 mM (Masson and Tschopp, 1987; Simon et al., 1986), human granzyme A is a covalently linked homodimer of 50 kDa. Human granzyme A is identical to the T cell-associated serine proteinase (HuTSP) isolated from a human CD8$^+$ CTL line (Fruth et al., 1987), and to the Q31 tryptase isolated from an essentially (90%) CD4$^+$ population of cells generated by repeated allogeneic stimulation of human peripheral blood lymphocytes (Poe et al., 1988). Human granzyme A is encoded by the HuHF cDNA clone isolated by Gershenfeld et al. (1988).

Human granzyme B was identified on the basis of its M_r and its crossreactivity with an anti-murine granzyme A/B antibody, as well as by its N-terminal sequence. The first 12 N-terminal amino acids of human granzyme B are identical to those deduced for three independently described cDNA clones encoding the HLP (Schmid and Weissmann, 1987). SECT (Caputo et al., 1988), or CSP-B (Trapani et al., 1988) serine esterase. The HLP clone was isolated from a cDNA library constructed from human peripheral blood leukocytes stimulated with the mitogen staphylococcal enterotoxin A. This mitogen induces a 10-fold increase in the levels of the mRNA corresponding to the HLP cDNA clone. The SECT cDNA clone, which was isolated from phytohenagglutinin (PHA)-stimulated human peripheral blood lymphocytes (Caputo et al., 1988), is identical to HLP

TABLE 1. Nomenclature of lymphocyte granzymes

Human lymphocyte granzymes	Synonyms	Murine lymphocyte granzymes	Alternative names	Rat granzymes
Granzyme A (Krahenbuhl et al., 1988)	HuTSP (Fruth et al., 1987) H factor (Gershenfeld et al., 1988 CTL Q31 tryptase (Poe et al., 1988) Granzyme 1 (Hameed et al., 1988)	Granzyme A (Masson and Tschopp, 1987)	H factor (Gershenfeld and Weissman, 1986) TSP-1 (Simon et al., 1986) BLT-esterase (Pasternack and Eisen, 1985) CTLA-3 (Brunet et al., 1987) SE1 (Young et al., 1986)	Tryptase (Hudig et al., 1991)
Granzyme B (Krahenbuhl et al., 1988)	HLP (Schmid and Weissmann, 1987) CSP-B (Trapani et al., 1988) SECT (Caputo et al., 1988) Granzyme 2 (Hamed et al., 1988) CGL-1 (Hanson et al., 1990)	Granzyme B (Masson and Tschopp, 1987)	CCP-I (Lobe et al., 1986) CTLA-1 (Brunet et al., 1986)	Aspase (Hudig et al., 1991) RNKP-1 (Zunino et al., 1990)
		Granzyme C (Jenne et al., 1988b)	CCP-II (Lobe et al., 1986)	
		Granzyme D (Jenne et al., 1988a)		
		Granzyme E (Jenne et al., 1988a)	CCP-3 (Bleackley et al., 1988)	
		Granzyme F (Jenne et al., 1988a)	CCP-4 (Bleackley et al., 1988)	
		Granzyme G (Jenne et al., 1989)	MSCP- (Kwon et al., 1988)	
Granzyme H (Haddad et al., 1991)	CGL-2 (Hanson et al., 1990)			
Granzyme 3 (Hameed et al., 1988)	CSP-C (Klein et al., 1990) CCP-X (Meier et al., 1990)			

[a] Proper assignment of synonyms to granzymes B to G is based on cDNA sequence or partial protein sequence data. Other names for human and murine granzyme A are also listed when the identity could be ascertained from the unique physical, biochemical, and functional properties.

except for five nucleotide substitutions in the coding sequence, three of which are silent.

In addition to granzymes A and B, a third member of this family has been cloned in man and designated granzyme H, CSP-C, or CCP-X (Klein et al., 1990; Meiler et al., 1990; Haddad et al., 1991). The complete gene and cDNA sequence expressed in activated T cells have been described; however, the predicted 246-amino-acid protein in T cells has not yet been isolated. Transcripts were detected only in cytotoxic cells. Granzyme H shows the highest degree (greater than 54%) of amino acid sequence homology with granzyme B and cathepsin G, and does not appear to represent a human counterpart of the known murine granzymes A to G. In contrast to murine CTL granules, which contain seven granzymes in approximately equal concentrations (Masson and Tschopp, 1987; Jenne et al., 1989), Hameed et al. (1988) detected only three serine esterases (termed granzymes 1–3) in human lymphokine-activated killer (LAK) cells, which can be labeled with DFP. On the other hand, Ferguson et al. (1988) have only found two proteins in cell extracts of human CTL clones that are labeled with DFP. These two DFP-labeled proteins had the same M_r as granzymes A and B, determined by sodium dodecylsulfate–polyacrylamide gel electrophoresis (SDS–PAGE) under reducing and nonreducing conditions.

Granzymes A and B are likely to be identical to the granzymes 1 and 2, respectively, described by Hameed et al. (1988). A potential third granzyme, with BLT esterase activity, has been isolated from CD3-peripheral blood mononuclear cells (PBMC) and LAK cells (O. Krähenbühl et al., unpublished results). This activity probably corresponds to the LAK cell-derived granzyme 3 described by Hameed et al. (1988). It is not clear, however, whether granzyme 3 is only expressed in LAK cells and anti-CD3 monoclonal antibody (mAb)-activated PBMC, which, in fact, contain a large number of LAK cells. Indeed, anti-CD3 mAb stimulation has been shown to induce a massive production of interleukin-2 (IL-2) by peripheral blood lymphocytes (Stankova et al., 1989; van Lier et al., 1989), which in turn generates LAK cells. It cannot be excluded, however, that granzyme 3 is not found in T cells, but rather in another cell type present in the PBMC/LAK cell population.

There is no evidence that any of the other granule proteins correspond to murine granzymes C, D, E, F, or G. However, the anti-murine granzyme C–G antibody may not crossreact with the corresponding human proteins. The murine granzyme probes used in Northern blot analyses did not crosshybridize to other transcripts, even under reduced stringency conditions, but this may be due to species variation, as hinted by the very weak signal observed with the murine granzyme A probe.

Structural Features of Granzymes

Granzyme A has a structure which is unique among serine proteases in that it forms a disulfide-linked homodimer of 60 kDa, while granzymes B–G all consist of a single polypeptide chain with M_r ranging between 27 and 55 kDa. The granzymes are antigenically related and, except for granzyme C, highly glycosylated, although the carbohydrate content of each granzyme is heterogeneous, ranging from 10 to 50% to the total molecular mass.

Granzymes are highly homologous with each other (at least 39% amino acid identities in the most unfavorable case, between granzymes A and D). Alignment of the amino acid sequences of granzymes to those of well-characterized members of the serine protease family, such as rat mast cell proteases (RMCP I and II), elastase, or trypsin, reveals the typical features of serine proteases (Figures 1 and 2). All granzymes contain, at equivalent positions, the three residues His, Asp, and Ser, which form the catalytic center of serine proteases. Adjacent residues are also highly conserved, as well as six cysteine residues which are involved in disulfide bond formation in a 1-2, 3-6, and 4-5 pattern by analogy to RMCP II. The structure of granzyme A, however, shows two peculiarities in that it contains a fourth cysteine bond at the carboxy terminus analogous to that of chymotrypsin, trypsin, and elastase (Woodbury and Neurath, 1980), and that it occurs as a disulfide-linked homodimer via a free cysteine residue at position 76. Since both catalytic centers of the granzyme A dimer appear to be accessible and actually cleave substrates (Masson and Tschopp, 1988), it appears that granzyme A resembles factor XI of the coagulation system.

All predicted sequences, including granzyme A, whose initially reported 5′ terminal cDNA sequence was incomplete (Gershenfeld and Weissman, 1986; Brunet et al., 1987), but which has since been extended (Jenne and Tschopp, 1988), start with a typical hydrophobic signal peptide indicating that granzymes are translocated across the lipid membrane into the rough endoplasmatic reticulum (Figure 2). Two sequence features appear to characterize the granzyme subfamily of serine proteases. First, the mature enzymes found in granules

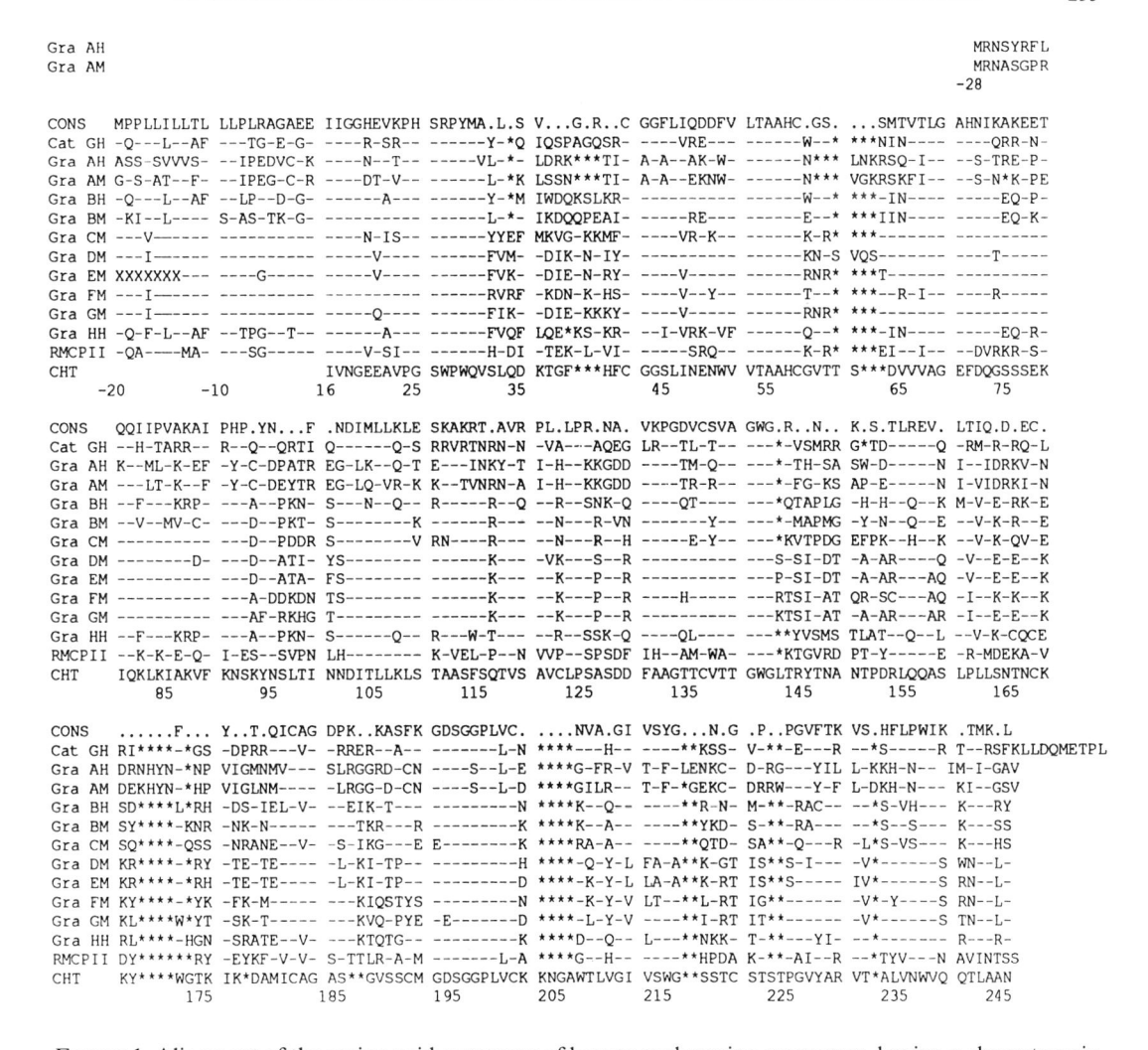

23. STRUCTURE AND POSSIBLE FUNCTIONS OF LYMPHOCYTE GRANZYMES

```
Gra AH                                                                                                                    MRNSYRFL
Gra AM                                                                                                                    MRNASGPR
                                                                                                                          -28

CONS      MPPLLILLTL LLPLRAGAEE IIGGHEVKPH SRPYMA.L.S V...G.R..C GGFLIQDDFV LTAAHC.GS. ...SMTVTLG AHNIKAKEET
Cat GH    -Q---L--AF ---TG-E-G- ----R-SR-- ------Y-*Q IQSPAGQSR- ----VRE--- ------W--* ***NIN---- ----QRR-N-
Gra AH    ASS-SVVVS- --IPEDVC-K ----N--T-- ------VL-*- LDRK***TI- A-A--AK-W- ------N*** LNKRSQ-I-- --S-TRE-P-
Gra AM    G-S-AT--F- --IPEG-C-R ----DT-V-- ------L-*K LSSN***TI- A-A--EKNW- ------N*** VGKRSKFI-- --S-N*K-PE
Gra BH    -Q---L--AF --LP--D-G- ------A--- ------Y-*M IWDQKSLKR- ---------- ------W--* ***-IN---- -----EQ-P-
Gra BM    -KI--L---- S-AS-TK-G- ---------- ------L-*- IKDQQPEAI- ----RE---- ------E--* ***IIN---- -----EQ-K-
Gra CM    ---V------ ---------- ----N-IS-- ------YYEF MKVG-KKMF- ----VR-K-- ------K-R* ***------- ----------
Gra DM    ---I------ ---------- ---------- ------V--- ------FVM- -DIK-N-IY- ---------- ------KN-S VQS------- ----T-----
Gra EM    XXXXXXX--- ----G----- ---------- ------V--- ------FVK- -DIE-N-RY- ----V----- ------RNR* ***T------ ----------
Gra FM    ---I------ ---------- ---------- ---------- ------RVRF -KDN-K-HS- ----V--Y-- ------T--* ***--R-I-- ----R-----
Gra GM    ---I------ ---------- ---------- ------Q--- ------FIK- -DIE-KKKY- ----V----- ------RNR* ***------- ----------
Gra HH    -Q-F-L--AF --TPG--T-- ------A--- ------FVQF LQE*KS-KR- ---I-VRK-VF ------Q--* ***-IN---- -----EQ-R-
RMCPII    -QA------MA- ---SG----- ------V-SI-- ------H-DI -TEK-L-VI- ----SRQ-- ------K-R* ***EI--I-- --DVRKR-S-
CHT                             IVNGEEAVPG SWPWQVSLQD KTGF***HFC GGSLINENWV VTAAHCGVTT S***DVVVAG EFDQGSSSEK
          -20        -10        16         25         35                   45         55         65         75

CONS      QQIIPVAKAI PHP.YN...F .NDIMLLKLE SKAKRT.AVR PL.LPR.NA. VKPGDVCSVA GWG.R..N.. K.S.TLREV. LTIQ.D.EC.
Cat GH    --H-TARR-- R-Q--QRTI  Q------Q-S RRVRTNRN-N -VA----AQEG LR--TL-T-- ---*-VSMRR G*TD-----Q -RM-R-RQ-L
Gra AH    K---ML-K-EF -Y-C-DPATR EG-LK--Q-T E---INKY-T I-H--KKGDD ----TM-Q-- ---*-TH-SA SW-D-----N I--IDRKV-N
Gra AM    ---LT-K--F -Y-C-DEYTR EG-LQ-VR-K K--TVNRN--A I-H--KKGDD ----TR-R-- ---*-FG-KS AP-E-----N I-VIDRKI-N
Gra BH    --F---KRP- ---A--PKN- S---N--Q-- R-----R--Q --R--SNK-Q ----QT---- ---*QTAPLG -H-H--Q--K M-V-E-RK-E
Gra BM    --V--MV-C- ---D--PKT- S-------A- ------R--- --N---R-VN ------Y--- ---*-MAPMG -Y-N--Q--K --V-K-R--E
Gra CM    ---------- ---D--PDDR S--------V RN----R--- --N---R--H -----E-Y-- ---*KVTPDG EFPK--H--K --V-K-QV-E
Gra DM    --------D- ---D--ATI- YS-------- ------K--- -VK---S--R ---------- ---S-SI-DT -A-AR----Q -V--E-E--K
Gra EM    --------D- ---D--ATA- FS-------- ------K--- --K---P--R ---------- ---P-SI-DT -A-AR----AQ -V--E-E--K
Gra FM    --------A- ---A-DDKDN TS-------- ------K--- --K---P--R ----H----- ---RTSI-AT QR-SC---AQ -I--K-K--K
Gra GM    ---------- ---AF-RKHG T--------- ------K--- --K---P--R ---------- ---KTSI-AT -A-AR---AR -I--E-E--K
Gra HH    --F---KRP- ---A---PKN- S-------Q- R--W-T--- --R--SSK-Q -----QL---- ---*YVSMS  TLAT--Q--L --V-K-CQCE
RMCPII    --K-K-E-Q- I-ES--SVPN LH-------- K-VEL--P--N VVP--SPSDF IH--AM-WA-- ---*KTGVRD PT-Y-----E -R-MDEKA-V
CHT       IQKLKIAKVF KNSKYNSLTI NNDITLLKLS TAASFSQTVS AVCLPSASDD FAAGTTCVTT GWGLTRYTNA NTPDRLQOAS LPLLSNTNCK
          85         95         105        115        125        135        145        155        165

CONS      ......F... Y..T.QICAG DPK..KASFK GDSGGPLVC. ....NVA.GI VSYG...N.G .P..PGVFTK VS.HFLPWIK .TMK.L
Cat GH    RI****-*GS -DPRR---V- -RRER--A-- -------L-N ****---H-- ----**KSS- V-**-E---R --*S-----R T--RSFKLLDQMETPL
Gra AH    DRNHYN-*NP VIGMNMV--- SLRGGRD-CN ----S--L-E ****G-FR-V T-F-LENKC- D-RG---YIL L-KHN-N-- IM-I-GAV
Gra AM    DEKHYN-*HP VIGLNM---- -LRGG-D-CN ----S--L-D ****GILR-- T-F-*GEKC- DRRW---Y-F L-DKH-N--- KI--GSV
Gra BH    SD****L*RH -DS-IEL-V- --EIK-T--- ---------N ****K--Q-- ----**R-N- M-**-RAC-- --*S-VH--- K---RY
Gra BM    SY****-KNR -NK-N----- ---TKR---R --------K  ****K--A-- ----**YKD- S-**-RA--- --*S--S--- K---SS
Gra CM    SQ****-QSS -NRANE--V- -S-IKG---E E--------K ****RA-A-- ----**QTD- SA**-Q---R -L*S-VS--- K---HS
Gra DM    KR****-*RY -TE-TE---- -L-KI-TP-- ---------H ****-Q-Y-L FA-A**K-GT IS**S-I--- -V*------S WN--L-
Gra EM    KR****-*RH -TE-TE---- -L-KI-TP-- ---------D ****K-Y-L LA-A**K-RT IS**S----- IV*------S RN--L-
Gra FM    KY****-*YK -FK-M----- ---KIQSTYS ---------N ****K-Y-V  LT--**L-RT IG**----- -V*-Y----S RN--L-
Gra GM    KL****W*YT -SK-T----- --KVQ-PYE  -E-------D ****L-Y-V  ----**I-RT IT**----- -V*------S TN--L-
Gra HH    RL****-HGN -SRATE--V- --KTQTG--- ---------K ****D--Q-- L---*NKK- T-*----YI- --*------N K--V-CQCE
RMCPII    DY******RY -EYKF-V-V- S-TTLR-A-M -------L-A ****G--H-- ----**HPDA K-**-AI--R --*TYV---N AVINTSS
CHT       KY****WGTK IK*DAMICAG AS**GVSSCM GDSGGPLVCK KNGAWTLVGI VSWG**SSTC STSTPGVYAR VT*ALVNWVQ QTLAAN
          175        185        195        205        215        225        235        245
```

FIGURE 1. Alignment of the amino acid sequences of human and murine granzymes, bovine α-chymotrypsin (CHT) and rat mast cell protease II. The asterisk (*) represents a gap that has been inserted to maximum sequence similarity between the aligned 12 serine protease sequences. The top line, named CONS, shows a consensus amino acid when at least five residues in the aligned sequences are identical. The residues are printed at those positions that differ from the consensus sequence; the residues that agree with the consensus sequence are indicated by a dash. The numbering of aa residues shown in the bottom line is that of CHT. Sequence data sources are: human cathepsin G (Cat GH) (Salvesen et al., 1987); human granzyme A (Gra AH) (Gershenfeld et al., 1988); murine granzyme A (Gra AM) (Gershenfeld and Weissmann, 1986), human granzyme B (Gra BH) (Schmid and Weissmann, 1987); murine gransyme B (Gra BM) (Brunet et al., 1987; Lobe et al., 1986); murine granzyme C (Gra CM) (Jenne et al., 1988b); murine granzyme D (Gra DM), E (Gra EM), and F (Gra FM) (Jenne et al., 1988a); murine granzyme G (Gra GM) (Jenne et al., 1989); human granzyme H (Gra HH) (Meier et al., 1990; Klein et al., 1990; Haddad et al., 1991); RMCP II (Benfey et al., 1987); CHT, NBRF identification code KYBOA.

share a strictly conserved N-terminal sequence from positions +1 to +4 (I I G G) and positions +9 to +16 (P H S R P Y M A). Second, they are synthesized as inactive precursor molecules with a very short acidic propeptide at the N-terminus consisting of either Gly-Glu or Glu-Glu (Figure 1).

It is not known in which cellular compartment the propeptide is cleaved, but the endoplasmic reticulum and the Golgi apparatus are likely candidates. The propeptide Glu-Arg and Glu-Lys of human and mouse granzyme A, respectively, appears to be an exception, since it ends with a

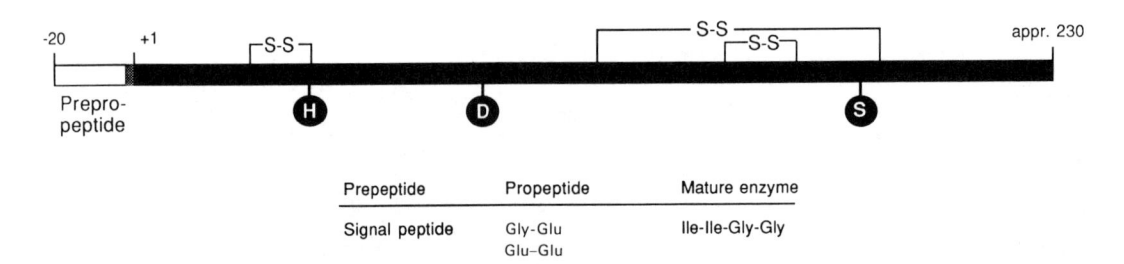

FIGURE 2. Structural features and biosynthetic processing of mouse granzymes B–G and human granzymes B and H. The thick black bar represents the polypeptide chain of the mature, fully processed granzymes, the open bar the hydrophobic signal peptide followed by the short dotted bar standing for the propeptide. The approximate location of the three residues forming the catalytic triad (H, D, S) is shown as well as the pattern of the disulfide bridges. The sequences of the prepeptides is highly conserved among granzymes; a glutamic acid residue is always found at the second position. Granzyme A is not included in this figure, since the number of disulfide bridges and the sequences of the propeptide are different.

positively charged residue. According to these criteria, RMCP I and II (Beney et al., 1987; Le Trong et al., 1987; Woodbury et al., 1978), cathepsin G (Salvesen et al., 1987), and elastase (Takahashi et al., 1988), enzymes which are present in granules of differentiated mast cells or neutrophilic granulocytes, have also been included in this family. In contrast to serine proteases of the coagulation and complement system, granzymes lack additional regulatory domains.

Genomic Organization and Chromosomal Localization of Granzymes

The genomic organization, sequence, and chromosomal localization of murine granzymes B, C (Lobe et al., 1988), and F (Jenne et al., 1991) and of human granzyme B and H genes have been reported (Klein et al., 1989; Caputo et al., 1990; Haddad et al., 1990; Meier et al., 1990; Haddad et al., 1991).

Mouse granzyme B has been mapped to band D of chromosome 14 (Brunet et al., 1986), while close physical kinkage of the granzymes B, C, E, and F to the T-cell receptor-α (TCR-α)-chain locus was demonstrated by the very low recombination frequencies between polymorphic markers of the TCR-α-chain locus and the murine granzyme B/C/E/F locus in interspecific mouse backcross progeny (Harper et al., 1988; Crosby et al., 1990). By using high-resolution fluorescence *in situ* hybridization, granzyme F was found in close proximity distally to the TCR-α-chain locus on mouse chromosome 14 (Jenne et al., 1991). In contrast, the granzyme A gene has been localized to chromosome 13 in mice (Mattei et al., 1987).

Similarly, human granzyme B is located close to the TCR-α/δ-chain locus on chromosome 14 (Harper et al., 1988; Lin et al., 1990) and displays the exon-intron structure typical of the granzyme family. The gene for human granzyme A has been localized to chromosome 5 (Gershenfeld et al., 1988).

Furthermore, Hanson et al., have established a continuous genomic restriction map for cathepsin G and two other cathepsin G-like serine esterase genes (CGL-1 and 2), showing that all three genes are tightly clustered in a 50-kbp region close to the TCR-α/δ locus (Hanson et al., 1990). CGL-1 is identical to granzyme B, while CGL-2 corresponds to granzyme H (Klein et al., 1990; Haddad et al., 1991) and hCCP-X (Meier et al., 1990). The order of the entire locus on the long arm of human chromosome 14 has been determined as: centromere–TCR-α/δ–granzyme B–granzyme H–cathepsin G. Granzyme H (CSP-C/hCCP-X) was shown to be a functional gene specifically expressed in activated T cells (Klein et al., 1990; Haddad et al., 1991; Heusel et al., 1991). Interestingly, the three proximal genes, i.e., the TCR-α/δ locus, and the granzyme B and H genes, are specifically expressed in T cells, while the distal gene, cathepsin G, is only transcribed in the myelo-monocyte lineage (Hohn et al., 1989). Granzyme H shows the highest degree of amino acid sequence homology (> 54%) with granzyme B and cathepsin G, and like these genes, consists of five exons separated by introns at equivalent positions. Nevertheless, evolutionary analyses suggest that granzyme H is unlikely to be the human counterpart of one of the known murine granzymes A to G, and that it should exist in other mammalian lineages as well. Likewise, a human counterpart for the murine granzymes C to G has yet to be identified.

All granzyme genes studied to date, including RMCP I and II (Benfey et al., 1987), neutrophil cathepsin G (Hohn et al., 1989), and elastase (Takahashi et al., 1988), share a common exon-intron organization, i.e., five exons are separated by introns situated at homologous positions. Exon 1 codes for the signal peptide and the first nucleotide of the pro-dipeptide, while exons 2, 3, and 5 contain the codons for the three catalytic site residues His^{57}, Asp^{102}, and Ser^{195} (chymotrypsin numbering), respectively. This genomic organization is coherent with the hypothesis of a common ancestral granzyme gene which has evolved via duplication and reshuffling of the different exons, resulting in the formation of different active sites (for more details concerning the evolution of granzymes, see Jenne et al., 1989; Haddad et al., 1991; Prendergast et al., 1991).

Studies aimed at understanding the mechanisms underlying the differences in tissue-specific and developmental regulation of the three linked human serine protease genes, cathepsin G, granzyme B, and granzyme H, have identified two regulatory 5′ and cis-acting sequences in granzyme B which act as upstream promoter elements (Hanson and Ley, 1990; Heusel et al., 1991). The reorganization of chromatin structure also appears to be involved in granzyme B trancriptional activation.

Substrates and Inhibitors of Granzymes

Granzymes A and D are strongly labeled by the serine esterase affinity label DFP, whereas the other granzymes are either weakly labeled (granzymes B and E) or completely fail to bind with DFP (granzymes C, F and G) (Masson and Tschopp, 1987; Jenne et al., 1989). A panel of synthetic peptide substrates has been tested, but substrates have only been found for granzymes A, B, and D. Both granzymes A and D show trypsin-like activity in that they cleave best after Arg or Lys residues, although the activity of granzyme D is very low compared to that of granzyme A (Simon et al., 1986a; Masson and Tschopp, 1987; Odake et al., 1991). Granzyme A cleaves Pro-Phe-Arg-7-amino-4-methyl-coumarin, Pro-Phe-Arg-nitroanilide, and BLT most efficiently (Masson et al., 1986b; MM Simon et al., 1987b).

More recently, Odake et al. (1991) have shown that granzyme B has Asp-ase activity and efficiently hydrolyzes Boc-Ala-Ala-Asp-SBzl. Granzyme B also has significant activity towards Boc-Ala-Ala-X-Sbzl substrates where X is Asn or Ser. The pH optimum of granzyme A and B for the cleavage of the various substrates is around 8.

Inhibitors of serine proteases like PMSF, aprotinin, DFP, and benzamidine are all potent inhibitors for the esterolytic activity of granzyme A (Masson et al., 1986b; Simon et al., 1986; Young et al., 1986). TLCK and TPCK, excellent inhibitors for trypsin and chymotrypsin, respectively, have no or little effect on granzyme A, whereas they inhibit the activity of granzyme D efficiently (Jenne and Tschopp, 1988). The plasma inhibitor antithrombin III (Masson and Tschopp, 1988) and the tissue inhibitor protease nexin-1 (Gurwitz et al., 1989) inhibit granzyme A by forming a heparin-dependent, covalently linked 2:1 complex with granzyme A, each subunit interacting with one molecule of antithrombin III or nexin-1. A specfic inhibitor of granzyme A, H-D-Pro-Phe-Arg-chloromethyl-ketone (PFR-CK), designed according to the preferred amino acid sequence cleaved by this protease, has been shown to inhibit granzyme A (Simon et al., 1989).

The esterolytic activity of granzyme B is efficiently blocked by isocoumarins (Odake et al., 1991).

Expression of Granzymes

Since the mature mRNA transcripts for granzymes C, D, E, F, and G have about the same size of 1.1 kb and show high nucleotide sequence homologies, unambiguous identification of the different granzyme mRNAs (except for granzymes A and B) is difficult to achieve. Garcia-Sanz et al. (1990) performed RNAse protection assays, which, in contrast to the Northern blot technique, allow discrimination between the transcripts of the different members of the granzyme family. The analysis of a wide range of lymphoid and nonlymphoid cell populations and tissues shows that, with few exceptions, expression of granzyme genes is restricted to T cells and their thymic precursors. In mature T cells, granzymes are expressed only upon activation, although their expression is not strictly correlated with cytolytic activity. Worthy of note is the observation that peritoneal exudate lymphocytes (PEL) express granzyme A and B genes but none of the other granzymes, and that only the granzyme F gene appears to be exclusively expressed in the CD4$^-$CD8$^+$ subset of peripheral T cells (Garcia Sanz et al., 1990).

In the mouse thymus, expression of the granzyme A and B genes is almost exclusively associated with CD4⁻ CD8⁻ (double negative) immature T cells, as clearly demonstrated by *in situ* hybridization (Held et al., 1990a).

Granzymes A, B, and C have been detected in various CTL lines, in nude mouse spleen (which contains many NK cells), in some helper T cell lines (CD4⁺ CD8⁻), and in NL-cell-derived tumor cell lines (Gershenfeld and Weissman, 1986; Lobe et al., 1986; Brunet et al., 1987). Granzyme A was also identified in approximately equal amounts in CD4⁺ CD8⁻ and CD4⁻ CD8⁺ subsets of mixed leukocyte cultures that had been maintained in IL-2 for 3–5 days (Garcia Sanz et al., 1987). However, when the frequency of resting precursor CD4⁺ and CD8⁺ lymphocytes which can be induced *in vitro* by antigen or lectins to express granzyme A was determined, practically all of the growing lectin/antigen-reactive CD8⁺ lymphocytes, but only 12–27% of the CD4⁺ lymphocytes, were positive for the enzyme (Fruth et al., 1988).

As in mouse, the granzyme A and B genes are expressed in human lymphocytes, including normal peripheral blood lymphocytes (PBL) activated by (IL-2) or PHA, alloreactive CTL clones, noncytolytic T-cell lines, NK cells, and large granular lymphocytes (LGL) (Schmid and Weissmann, 1987; Caputo et al., 1988; Gershenfeld et al., 1988; Trapani et al., 1988; Liu et al., 1989; Clément et al., 1990a). Moreover, granzyme A mRNA is present in the human T-cell leukemia line Jurkat (Gershenfeld et al., 1988), and low levels of granzyme B mRNA have been detected in Jurkat cells and in an Epstein–Barr virus-transformed B-cell line (Caputo et al., 1988). By using the *in situ* hybridization technique, Clément et al. (1990a, 1990b) showed that, in addition to inducing LAK activity, IL-2 stimulation increased the amount of perforin and granzyme B mRNA at the single-cell level in 40–100% of the total CD3⁻ human LGL population. Granzyme B mRNA appeared within 30 min after addition of IL-2. The stimulatory effect of IL-2 on both LAK activity and granzyme B/perforin gene expression was down-regulated by IL-4. In a noncytotoxic T-cell clone (BJ15) which becomes cytotoxic upon treatment with recombinant α or γ interferon (rIFN), the expression of granzyme B gene was increased 10- to 15-fold during the acquisition of the cytotoxicity (Haddad et al., 1990). Interestingly, Wagner et al. (1989) demonstrated an increased number of BLT esterase-positive T cells in hairy-cell leukemia patients after rIFN-α treatment. Granzyme B mRNA has also been detected in peripheral blood mononuclear cells (PBMC) of a patient having a marked proliferation of LGL (80% of PBMC are NK cells); in the same manner, variable levels of perforin mRNA have been found in PBMC of patients having LGL-proliferative disorders (Oshimi et al., 1990).

To overcome the problem of a likely cross-hybridization with the granzyme B gene, granzyme H gene expression was analyzed by S1 nuclease protection assays (Haddad et al., 1991; Heusel et al., 1991). The results of this approach have shown granzyme H expression in activated PBL, CTL clones, LAK cells, and NK cells, but at much lower levels relative to the granzyme B transcript.

More recently, Ebnet et al. (1991) have analyzed mouse granzyme A–G gene expression using polymerase chain reaction (PCR) techniques. A panel of four CD8⁺ and six CD4⁺ long-term-cultured T-lymphocyte clones were shown to express all seven granzymes, although at greatly differing concentrations. mRNA species specific for granzymes A and B, and to a lesser extent for granzyme C, were detected in both CD8⁺ and CD4⁺ T cells primed *in vivo* and in T cells activated and cultured at short-term *in vitro*. In contrast, no transcripts for granzymes D–G were detected by PCR in either of these effector cell populations, arguing against their participation in T-cell-mediated functions *in vivo*.

The biological relevance of murine granzymes A and B *in vivo* has been highlighted by *in situ* hybridization and immunohistological investigations using specific riboprobes and polyclonal antibodies (Griffiths and Mueller, 1991). Granzyme A transcripts have been detected in normal tissues such as intestine, lung, spleen (Garcia Sanz et al., 1990), and the metrial glands of the uterus during pregnancy and the stages of the menstrual cycle (Zheng et al., 1991), and in a small population of CD4⁻,CD8⁻ (double negative) thymocytes (Garcia Sanz et al., 1990).

Other analyses of *in vivo*-activated T cells reveal that granzyme A is expressed preferentially by virus-specific CD8⁺CD4⁻ CTL of mice infected with lymphocytic choriomeningitis virus (LCMV), but not by CD8⁻CD4⁺ T cells of mice sensitized with either *Listeria monocytogenes* or I-A alloantigens (HG Simon *et al.*, 1987; Kramer et al., 1989; Müller et al., 1989; Welsh et al., 1990), suggesting a possible involvement of CTL granules in the control of the viral infection or in the development of the immunopathological damage of the meninges and choroid plexus. In heart allograft rejections, the granzyme A and B genes have been shown to be preferentially expressed by infiltrating CD8⁺CD4⁻ T cells (Müller et al., 1988a, 1988b), while the granzyme A gene has been shown to be expressed by infiltrating lymphocytes during the spontaneous development of autoimmune diabetes mellitus in nonobese diabetic mice (Held et al., 1990b).

Granzyme B mRNA-expressing lymphocytes were also detected in renal biopsies of allografted patients irrespective of whether the patients were undergoing acute rejection or not (Clément et al., 1990a). This observation points out the presence of activated infiltrating T cells in allografts even in the absence of clinical rejection, thus suggesting that granzyme B expression could be used as an early diagnostic marker of T-cell activation for monitoring the outcome of an allograft in humans. Confirming the possible usefulness of perforin, granzyme A, or granzyme B as early markers of allograft rejection, the perforin and granzyme A (Griffiths et al., 1991) or granzyme B (Clément et al., 1991) genes were found to be mainly expressed in T cells found in endomyocardial cellular infiltrates of patients who had undergone heart transplantation. More recently, the presence of granzyme B has been detected in cytotoxic cells infiltrating human cardiac allografts using purified polyclonal anti-granzyme B antibody (Hameed et al., 1991). In bone marrow transplantation, donor T cells interact with host antigens, generally resulting in a graft versus host disease (GVHD). In acute GVHD, skin involvement is early and frequent and presents lymphoid infiltration of the dermis and epidermis. Analysis of skin biopsies from patients with GVHD revealed that granzyme B and perforin gene expression correlates with the presence of a large number of CD8$^+$ T cells (CTL) in the biopsy infiltrates (Clément et al., submitted for publication).

These results show that granzyme (and perforin) genes constitute good markers of inflammatory processes and suggest a possible involvement of granzymes in cytotoxic function (and/or its activation) *in vivo*.

Functions of Granzymes

Several physiological functions of granzymes have been proposed (Kramer and Simon, 1987); however, the evidence for none of them is without any flaws. Although one or several of them may well be the proteases that are inactivated by the panel of protease inhibitors that abrogate CTL activity, it is not clear whether granzymes are the targets of these inhibitors. PMSF blocks granzyme A activity but does not reduce the lytic activity of isolated granules (Masson et al., 1986a; Henkart et al., 1987). A direct cytolytic effect of granzyme A on target cells was excluded when the purified enzyme was tested on a panel of target cells (Simon et al., 1986). Granzymes are also present in granules of noncytotoxic CD4$^+$ T cells, further arguing against their direct involvement in the lethal hit. There is no evidence that granzymes act in a proteolytic cascade leading to the activation of perforin (or another molecule), as is the case in the complement cascade system, with the possible exception of inhibition of granzyme A by H-D-Pro-Phe-Arg-chloromethyl-ketone, as reported by Simon et al. (1989), which may be interpreted to mean that CTL-mediated lysis of target cells involves activation of granzyme A upon exocytosis.

Several alternative roles have been proposed for granzymes. One observation reported by Henkart's laboratory (Munger et al., 1988; Hayes et al., 1989) suggests that granzyme A is a potent mediator of target cell DNA degradation if the cells are concomitantly permeabilized with perforin. This result has, however, not been confirmed, and in fact, several reports have described the lack of DNA degradation in cells lysed with granules (Duke et al., 1988; Gromkowski et al., 1988).

A role in DNA degradation was also proposed by Pasternack et al. (1991) who demonstrated the selective binding of granzyme A to several proteins present in the target cell P815. Subcellular fractionation of target cells showed that the nuclear fraction contained proteins that interact with granzyme A even when its proteolytic activity is blocked by inhibitors. A protein with $M_r = 100,000$ and too closely migrating proteins with $M_r = 35,000$ and $38,000$ were the predominant reactive moieties, and the N-terminal sequence of the 100-kDa protein showed that this protein was murine nucleolin. Incubation of granzyme A with nucleolin generated a discrete proteolytic cleavage product of $M_4 = 88,000$. Since nucleolin is known to shuttle between nucleus and cytoplasm, it was suggested that the interaction of granzyme A and nucleolin may be important in the process of apoptosis which accompanies CTL-mediated lysis of target cells.

Secretion of granule contents, including granzymes, is triggered by incubating normal CTL or CTL lines with antigen-bearing cells or immobilized anti-TCR mAb which mimics TCR occupation (Pasternack et al., 1986; Takayama et al., 1987; Velotti et al., 1987). Granzymes may act after the actual delivery of the lethal protein, for example, in facilitating the recycling event by mediating the dissociation of effector and target cells. There is, however, no evidence at present to support this hypothesis.

Granzyme A has been shown to cleave endothelial cell-derived extracellular matrix, suggesting that granzymes may allow activated T cells to invade and penetrate blood vessel walls, thereby facilitating their migration (Jenne and Tschopp, 1988; MM Simon et al., 1987b).

The involvement of granzyme A in B-cell growth has been postulated on the grounds that thrombin and trypsin had previously been shown to drive antigen-independent B-cell proliferation. Indeed, granzyme A can stimulate the growth of B cells and thymic lymphoma cells (HG Simon et al., 1986; MM Simon et al., 1987a).

Furthermore, time-dependent activation of recombinant human pro-urokinase, as well as natural pro-urokinase derived from human melanoma cells, accompanied by the conversion of single-chain into active two-chain urokinase, mediated by granzyme A, has been demonstrated (Brunner et al., 1990). These data suggest a novel pathway for plasmin generation during T-cell-mediated immune responses and extravasation of immune cells. Yet another function suggested for granzyme A is the inactivation of retroviral replication through limited proteolysis (Simon et al., 1987).

Human neutrophil lysosomal cathepsin G exerts broad-spectrum antibacterial action *in vitro* against Gram-negative and Gram-positive bacteria independent of its serine protease activity. An internal peptide of cat G (HPQYNQR), obtained after digestion of cathepsin C with clostripain, possesses broad-spectrum antibacterial action *in vitro*. The cathepsin G bactericidal peptide displays similarity to granzyme sequences. An internal peptide of one human granzyme (granzyme B) with the sequence of HPAYNPK also displays bactericidal action *in vitro*. Shafer et al. (1991) suggested that an internal antibacterial domain in the human serine proteases cathepsin G and granzyme B has been functionally conserved through evolution, perhaps for the purpose of host defense against microbial pathogens and targets of CTL killing.

Since the initial discovery of granzymes, this new class of serine proteases has drawn much attention. Yet their true biological functions still need to be defined more precisely and remain speculative. There is some circumstantial evidence for the hypothesis that one or several granzymes are involved in cell killing. Several important issues remain to be solved. Which of the many proteases are really required for the lytic response? Are some of the proteases alone capable of triggering apoptotic cell death of certain cells? Is the membrane-inserted lytic form of perforin a preolytically activated form of the released perforin molecule? Can certain mast cell proteases substitute for lymphocyte proteases, which may explain the lytic capability of a rat basophilic leukemia cell line transfected with the killer cell-specific perforin gene (Shiver and Henkart, 1991)? The actions of granzymes apparently are not restricted to cytolytic cells, and some of them may exert a role in the induction of proliferative responses. As the breadth of our knowledge of granzymes continuously increases in the future, we are confident that the biological niches that granzymes occupy will be discovered soon.

Acknowledgments

This work was supported by grants of the Swiss National Science Foundation and the French Association de Recherche sur le Cancer.

References

Benfey PN., Yin FH, Leder P (1987): *J Biol Chem* 262: 5377–5384

Bleackley RC, Duggan B, Ehrman N, Lobe CG (1988): *FEBS Lett* 234: 153–159

Brunet JF, Denizot F, Suzan M, Haas W, Mencia Huerta JM, Berke G, Luciani MF, Goldstein P (1987): *J Immunol* 138: 4102–4105

Brunet JF, Dosseto M, Denizot F, Mattei MG, Clark WR, Haqqu TM, Ferrier P, Nabholz M, Schmitt Verhulst AM, Luciani MF, Golstein P (1986): *Nature* 322: 268–271

Brunner G, Simon MM, Kramer MD (1990): *FEBS Lett* 260: 141–144

Caputo A, Fahey D, Lloyd C, Vozab R, McCairns E, Rowe PB (1988): *J Biol Chem* 263: 6363–6369

Caputo A, Sauer DE, Rowe PB (1990): *J Immunol* 145: 737–744

Chang TW, Eisen HM (1980): *J Immunol* 124: 1028–1033

Clément M-V, Haddad P, Ring GH, Pruna A, Sasportes M (1990a) *Hum Immunol* 28: 159–166

Clément M-V, Haddad P, Soulié A, Benevenuti C, Lichtenheld MG, Podack ER, Sigaux N, Sasportes M (1991): *Int Immunol* (in press)

Clément M-V, Haddad P, Soulié A, Legros-Maida S, Guillet J, Cesar E, Sasportes M (1990b): *Res Immunol* 141: 477–489

Crosby JL, Bleackley RC, Nadeau JH (1990): *Genomics* 6: 252–259

Duke RC, Sellins KS, Cohen JJ (1988): *J Immunol* 141: 2191–2194

Ebnet K, Chluba de Tapia J, Hurtenbach U, Kramer MD, Simon MM (1991): *Int Immunol* 3: 9–19

Ferguson WS, Verret CR, Reilly EB, Iannini MJ, Eisen HN (1988): *J Exp Med* 167: 528–540

Ferluga J, Allison AC (1974): *Nature* 250: 673–675

Fruth U, Nerz G, Prester M, Simon HG, Kramer MD, Simon MM (1988): *Eur J Immunol* 18: 773–781

Fruth U, Sinigaglia F, Schlesier M, Kilgus J, Kramer MD, Simon MM (1987): *Eur J Immunol* 17: 1625–1633

Garcia Sanz JA, MacDonald, HR, Jenne DE, Tschopp J, Nabholz M (1990): *J Immunol* 145: 3111–3118

Garcia Sanz, JA Plaetinck G, Velotti F, Masson D, Tschopp J, MacDonald HR, Nabholz M (1987): *EMBO J* 6: 933–938

Gershenfeld HK, Weissman IL (1986): *Science* 232: 854–858

Gershenfeld HK, Hershberger RJ, Shows TB, Weissman IL (1988): *Proc Natl Acad Sci USA* 85: 1184–1188

Griffiths G, Mueller C (1991): *Immunol Today* 12: 415–419

Griffiths GM, Namikawa R, Müller C, Liu CC, Young JD, Billingham M, Weissman I (1991): *Eur J Immunol* 21: 687–693

Gromkowski SH, Brown TC, Masson D, Tschopp J (1988): *J Immunol* 141: 774–778

Gurwitz D, Simon MM, Fruth U, Cunningham DD (1989): *Biochem Biophys Res Commun* 161: 300–304

Haddad P, Clement, MV, Bernard O, Larsen CJ, Degos L, Sasportes M, Mathieu-Mahul D (1990): *Gene* 87: 265–271

Haddad P, Jenne D, Tschopp J, Clement MV, Mathieu Mahul D, Sasportes M (1991): *Int Immunol* 3: 57–66

Hameed A, Lowrey DM, Lichtenheld M, Podack ER (1988): *J Immunol* 141: 3142–3147

Hameed A, Truong LD, Price V, Kruhenbul, O, Tschopp J (1991): *Am J Pathol* 138: 1069–1075

Hanson RD, Ley TJ (1990): *Molec Cell Biol* 10: 5655–5662

Hanson RD, Hohn PA, Popescu NC, Ley TJ (1990): *Proc Natl Acad Sci USA* 87: 960–963

Harper K, Mattei MG, Simon D, Suzan M, Guenet JL, Haddad P, Sasportes M, Golstein P (1988): *Immunogenetics* 28: 439–444

Hayes MP, Berrebi GA, Henkart PA (1989): *J Exp Med* 170: 933–946

Held W, MacDonald, HR Müller C (1990a): *Int Immunol* 2: 57–62

Held W., MacDonald HR, Weissman IL, Hess MW, Müller C (1990b): *Proc Natl Acad Sci USA* 87: 2239–2243

Henkart PA (1985): *Annu Rev Immunol* 3: 31–58

Henkart PA, Berrebi GA, Takayama H, Munger WE, Sitkovsky MV (1987): *J Immunol* 139: 2398–2405

Heusel JW, Hanson RD, SIlverman GA, Ley TJ (1991): *J Biol Chem* 266: 6152–6158

Hohn PA, Popescu NC, Hanson RD, Salvesen G, Ley TJ (1989): *J Biol Chem* 264: 13412–13419

Hudig D, Allison NJ, Pickett TM, Winkler U, Kam CM, Powers JC (1991): *J Immunol* 147: 1360–1368

Hudig D, Gregg NJ, Kam CM, Powers JC (1987): *Biochem Biophys Res Commun* 149: 882–888

Hudig D, Haverty T, Fulcher C, Redelman D, Mendelsohn J (1981): *J Immunol* 126: 1569–1574

Hudig D, Redelman D, Minning LL (1984): *J Immunol* 133: 2647–2654

Jenne DE, Tschopp J (1988): *Immunol Rev* 103: 53–71

Jenne DE, Masson D, Zimmer M, Haefliger JA, Li W-H, Tschopp J (1989): *Biochemistry* 28: 7953–7961

Jenne D, Rey C, Haefliger JA, Qiao BY, Groscurth P, Tschopp J (1988a): *Proc Natl Acad Sci USA* 85: 4814–4818

Jenne D, Rey C, Masson D, Stanley KK, Herz J, Plaetinck G (1988b): *J Immunol* 140: 318–323

Jenne D, Zimmer M, Garcia-Sanz JA, Tschopp J, Lichter P (1991): *J Immunol* 147: 1045–1052

Klein JL, Selvakumar A, Trapani JA, Dupont B (1990): *Tissue Antigens* 35: 220–228

Klein JL, Shows, TB, Dupont B, Trapani JA (1989): *Genomics* 5: 110–117

Krahenbuhl O, Rey C, Jenne D, Lanzavecchia A, Groscurth P, Carrel S, Tschopp J (1988): *J Immunol* 141: 3471–3477

Kramer, MD, Simon MM (1987): *Immunol Today* 8: 140–142

Kramer MD, Fruth U, Simon HG, Simon MM (1989): *Eur J Immunol* 19: 151–156

Kramer MD, Robinson P, Vlodavsky I, Barz D, Friberger P, Fuks Z, Schirrmacher V (1985): *Eur J Cancer Clin Oncol* 21: 307–316

Kwon BS, Kestler D, Lee E, Wakulchik M, Young JDE (1988): *J Exp Med* 168: 1839–1854

Lavie G, Leib Z, Servadio C (1985): *J Immunol* 135: 1470–1476

Le Trong H, Parmelee DC, Walsh KA, Neurath H, Woodbury RG (1987): *Biochemistry* 26: 6988–6994

Lin CC, Meier M, Sorensen O, Sasi R, Tainaka T, Bleackley RC (1990): *Cytogenet Cell Genet* 53: 169–171

Liu CC, Rafii S, Granelli Piperno A, Trapani JA, Young JD (1989): *J Exp Med* 170: 2105–2118

Lobe CG, Havele C, Bleackley RC (1987): *Proc Natl Acad Sci USA* 83: 1448–1452

Lobe CG, Upton, C, Duggan B, Ehrman N, Letellier M, Bell J, Bleackley RC (1988): *Biochemistry* 27: 6941–6946

Masson D, Tschopp J (1987): *Cell* 49: 679–685

Masson D, Tschopp J (1988): *Mol Immunol* 25: 1283–1289

Masson D, Nabholz M, Estrade C, Tschopp J (1986a): *EMBO J* 5: 1595–1600

Masson D, Zamai M, Tschopp J (1986b): *FEBS Lett* 208: 84–88

Mattei MG, Harper K, Brunet J-F, Denizot F, Mattei JF, Golstein P, Giraud F (1987): *Cytogenet Cell Genet* 46: 657–658

Meier M, Kwong PC, Frégeau CJ, Atkinson EA, Burrington M, Ehrman N, Sorenson O, Lin CC, Wilkins J, Bleackley RC (1990): *Biochemistry* 29: 4042–4049

Müller C, Gershenfeld HR, Lobe CG, Okada CY, Bleackley RC, Weissman IL (1988a): *J Exp Med* 167: 1124–1136

Müller C, Gershenfeld HK, Lobe CG, Okada CY, Bleackley RC, Weissman IL (1988b): *Transplant Proc* 20: 251–253

Müller C, Kägi D, Aebischer T, Odermatt B, Held W, Podack, ER, Zinkernagel RM, Hengartner H (1989): *Eur J Immunol* 19: 1253–1259

Mullins DE, Rohrlich ST (1983): *Biochim Biophys Acta* 695: 177–214

Munger WE, Berrebi GA, Henkart PA (1988): *Immunol Rev* 103: 99–109

Neurath H (1984): *Science* 224: 350–357

Odake S, Kam CM, Narasimhan L, Poe M, Blake JT, Krahenbuhl O, Tschopp J, Powers JC (1991): *Biochemistry* 30: 2217–2227

Oshimi K, Shinkai Y, Okumura K, Oshimi Y, Mizoguchi H (1990): *Blood* 75: 704–708

Pasternack MS, Eisen HN (1985): *Nature* 314: 743–745

Pasternack MS, Bleier KJ, McInerney TN (1991): *J Biol Chem* 266: 14703–14708

Pasternack MS, Verret CR, Liu MA, Eisen HN (1986): *Nature* 322: 740–743

Podack ER (1986): *J Cell Biochem* 30: 133–170

Poe M, Bennet CD, Biddison WE, Blake JT, Norton GP, Rodkey JA, Sigal NH, Turnover RV, Wu JK, Zweerink HJ (1988): *J Biol Chem* 263: 13215–13222

Prendergast JA, Pinkoski M, Wolfenden A, Bleackley RC (1991): *J Mol Biol* 220: 867–875

Qan P-C, Ishizaka T, Bloom BR (1982): *J Immunol* 128: 1786–1791

Redelman D, Hudig D (1980): *J Immunol* 124: 870–878

Redelman D, Hudig D (1983): *Cell Immunol* 81: 9–21

Redmond MJ, Letellier M, Parker JM, Lobe C, Havele C, Paetkau V, Bleackley RC (1987): *J Immunol* 139: 3184–3188

Salvesen G, Farley D, Shuman J, Przbyla A, Reilly C (1987): *Biochemistry* 26: 2289–2293

Schmid, J, Weissmann C (1987): *J Immunol* 139: 250–256

Shafer WM, Pohl J, Onuka VC, Bangalore N, Travis J (1991): *J Biol Chem* 266: 112–116

Shiver JW, Henkart PA (1991): *Cell* 1175–1181

Simon HG, Fruth U, Kramer MD, Simon MM (1987): *FEBS Lett* 223: 352–360

Simon MM, Fruth U, Simon HG, Kramer MD (1987a): *Ann Inst Pasteur Immunol* 138: 309–314

Simon MM, Hoschutzky H, Fruth U, Simon HG, Kramer MD (1986): *EMBO J* 5: 3267–3274

Simon MM, Prester M, Kramer MD, Fruth U (1989): *J Cell Biochem* 40: 1–13

Simon MM, Simon HG, Fruth U, Epplen J, Muller Hermelink HK, Kramer MD (1987b): *Immunology* 60: 219–230

Stankova J, Hoskin DW, Roder JC (1989): *Cell Immunol* 121: 13–29

Takahashi H, Nukiwa T, Yoshimura K, Quick CD, States DJ, Whang-Peng J, Knutsen T, Crystal RG (1988); *J Biol Chem* 263: 14739–14747

Takayama H, Trenn G, Humphrey JW, Bluestone JA, Henkart PA, Sitkovsky MV (1987): *J Immunol* 138: 566–569

Trapani JA, Klein JL, White PC, Dupont B (1988): *Proc Natl Acad Sci USA* 85: 6924–6928

Tschopp J & Jongeneel CV (1988): *Biochemistry* 27: 2641–2646

van Lier RAW, Brouwer M, Rebel VI, van Noesel CJM, Aarden LA (1989): *Immunology* 68: 45–50

Velotti F, MacDonald HR, Nabholz M (1987): *Eur J Immunol* 17: 1095–1099

Wagner L, Goldstone AH, Worman CP (1989): *Leukemia* 3: 373–379

Welsh RM, Nishioka WK, Antia R, Dundon P (1990): *J Virol* 64: 3726–3733

Woodbury RG, Neurath H (1980): *FEBS Lett* 114: 189–196

Woodbury RG, Katunuma N, Kobayashi K, Titani K, Neurath H (1978): *Biochemistry* 17: 811–819

Young JDE, Leong LG, Liu CC, Damiano A, Wall DA, Cohn ZA (1986): *Cell* 47: 183–194

Zheng, LM, Ojcius DM, Liu CC, Kramer MD, Simon MM, Parr EL, Young JD (1991): *FASEB J* 5: 79–85

Zunino SJ, Bleackley RC, Martinez J, Hudig D (1990): *J Immunol* 144: 2001–2009

24
The Role of Granzyme A in Cytotoxic Lymphocyte-Mediated Lysis

Mark S. Pasternack

Introduction

Despite considerable information regarding the structure and expression of granzymes, the serine proteases that reside in the cytotoxic granules of cytotoxic T lymphocytes (CTL), little is known about their physiologic substrates or their role in the economy of these effector cells. From a conceptual framework, such proteases may act on the effector cells themselves ("*cis*" action), on the target cells ("*trans*"), or in a more nonspecific fashion; current information remains sketchy and controversial. In this chapter, the functional properties and possible physiologic roles of CTL granzymes will be reviewed, with particular reference to granzyme A.

This enzyme, the first identified granule serine esterase and quantitatively the most abundant of the cytotoxic granule serine esterases, is a basic, homodimeric protein with esterolytic activity against benzyloxycarbonyl lysyl thioester (BLT), a synthetic trypsin substrate (Pasternack and Eisen, 1985), and was defined as a serine esterase because of its inhibition by phenylmethyl sulfonyl fluoride (PMSF) and diisopropylfluorophosphate (DFP). This enzyme was abundantly expressed in cloned CTL lines and could be induced following activation of thymocytes *in vitro* with concanavalin A and interleukin-2 (IL-2), conditions which lead to the differentiation of CTL from thymocytes (Pasternack and Eisen, 1985). Screening studies of several CD4[+] cell lines showed that serine esterase activity was selectively expressed only by cytolytic cells. This serine esterase was characterized by several additional laboratories and given a variety of names, including granzyme A (Masson et al., 1986), Hanukkah factor (Gershenfeld and Weissman, 1986), BLT esterase (Henkart et al., 1987), serine esterase 1 (Young et al., 1986b), and T-cell-specific proteinase (TSP-1) (Simon et al., 1986). Granzyme A was localized to a dense granular fraction within CTL extracts by fractionation studies (Pasternack et al., 1986); (Henkart et al., 1987), (Young et al., 1986b) as well as by histochemical analysis (Peters et al., 1991); Ojcius et al., 1991). It had an optimum of esterolytic activity at pH 8 (Simon et al., 1986; Young et al. 1986b) and was inhibitable by a variety of small molecule and protein inhibitors (benzamidine, aprotinin, soybean trypsin inhibitor) but not by the conventional irreversible trypsin inhibitor tosyllysine chloromethyl ketone (TLCK) (Pasternack and Eisen, 1985). It cleaved a number of model peptide substrates with basic amino acid residues at the carboxyl terminus (Simon et al., 1986). Nonspecific protease activity against casein and fibrin was reported (Masson et al., 1986; Simon et al., 1986; Young et al., 1986b), but in other laboratories granzyme A showed little nonspecific proteolytic activity (Pasternack, Verret, and Eisen, unpublished observations) or only slight activity (Simon et al., 1987b). Further studies by Tschopp (as discussed in Chapter 23, this volume) demonstrated the existence of a family of seven highly related granule-associated serine esterases, or "granzymes" A–G, in murine CTL (Masson and Tschopp, 1987). The remaining granzymes are monomeric proteins with limited if any reactivity against a conventional panel of protease substrates and inhibitors: only granzymes A and D react significantly with DFP.

Granzyme A and Granule Exocytosis

As one of the predominant components of cytotoxic granules, the role of granzyme A in CTL-mediated lysis should be considered within the framework of granule-mediated killing, a topic reviewed at greater length in Chapter 14, this volume. The hypothesis that CTL lyse target cells via a secretory mechanism was originally proposed on the basis of light and electron microscopic analyses of conventional heterogeneous lymphocyte populations (Henkart and Henkart, 1982). Over the past several years, this hypothesis has been largely substantiated at the biochemical level: discrete complement-like transmembrane pores were observed on target cell membranes following CTL-mediated lysis (Dennert and Podack, 1983), and dense cytotoxic granules isolated from CTL and natural killer cells were shown to mediate lysis of nucleated targets and even erythrocytes in a Ca^{2+}-dependent manner (Henkart et al., 1984; Podack and Konigsberg 1984). The apparently restricted presence of these granules in CTL and the similarities between cell-mediated and granule-mediated lysis supported their role in CTL-mediated lysis.

Cytotoxic granules contain three distinct substances: perforin (cytolysin) (Henkart et al., 1984; Masson and Tschopp, 1985; Podack et al., 1985), the cytolytically active moiety that polymerizes to form transmembrane pores; sulfated proteoglycans (MacDermott et al., 1985); and serine esterases (reviewed in Jenne and Tschopp, 1988). It is likely that these basic serine esterases are bound to the highly acidic proteoglycans within cytotoxic granules. As specialized lysosomal organelles (Peters et al., 1991), these granules probably contain conventional acid hydrolytic activity as well (Henkart et al., 1987). All three major granule substituents were shown to be secreted into the extracellular milieu with rapid kinetics after antigenic stimulation (Schmidt et al., 1985; Pasternack et al., 1986; Takayama et al., 1987) or ionophore treatment of CTL (Young et al., 1986a), suggesting that granule exocytosis is the central physiologic event during CTL-mediated lysis. The detection of secreted granzyme A into the medium has been used as a rapid and sensitive assay of CTL activation (Pasternack et al., 1986; Takayama et al., 1987) (see Chapter 48, this volume).

Evidence Supporting a Role for Granzyme A in CTL-Mediated Lysis

The apparently limited expression of granzyme A correlated with cell-mediated cytotoxicity in initial screens of granzyme A expression among different T-cell populations. Moreover, the inhibition of cytotoxicity in a cloned CTL line by the granzyme A inhibitor PMSF implicated granzyme A in the cytolytic process (Pasternack and Eisen, 1985). In addition, when isolated cytotoxic granules were pretreated with an irreversible (chloromethyl ketone) inhibitor of granzyme A, cytotoxicity against erythrocytes and nucleated target cells was abrogated (Simon et al., 1987a). This inhibitor did not affect esterase activity or cytotoxicity of intact cells, suggesting that it did not penetrate (or inactivate granzyme A within) the acid environment of the cytotoxic granules. The role of granzyme proteases in cytotoxic granule-mediated lysis of erythrocytes has been studied in great detail by Hudig and co-workers (Hudig et al., 1987, 1988, 1989, 1991), and is reviewed in Chapter 27, this volume. These analyses, utilizing novel isocoumarin serine protease inhibitors and peptidyl thioester substrates (rather than simpler amino acid-based thiester substrates such as BLT), demonstrate that granule protease activity is indeed necessary for hemolysis. Interestingly, it appears that chymotrypsin-like activity plays a more important role in such lysis than the trypsin-like activity of granzyme A. Since these studies monitor hemolysis rather than the lysis of nucleated target cells, they suggest that granule proteases play a role in the *membrane-associated events* accompanying target cell lysis which cannot be ascribed to the action of perforin alone. As will be discussed below, the ability of isolated cytotoxic granules (but not purified perforin) to induce nucleosomal degradation of target cell chromatin suggests that additional granule moieties, possibly granzymes, are necessary for nuclear injury as well.

Although some cytolytic populations appear to lack granzyme A (Dennert et al., 1987; Berke and Rosen, 1988), hybridization studies using specific cDNA probes have demonstrated that granzyme A and perforin mRNAs are present in CTL even when levels of the expressed proteins are too low to detect by enzymatic or functional assays (Garcia-Sanz et al., 1990) (see below). Furthermore, *in situ* analyses of infiltrating lymphocytes in experimental models of viral infection (Muller et al., 1989) and autoimmune disease (Young et al., 1989; Held et al., 1990), as well as in clinical specimens such as myocardial biopsies obtained from patients with acute myocarditis (Young et al., 1990) and synovial fluid obtained from rheumatoid arthritis patients (Griffiths et al., 1992) have demonstrated that granzyme A and/or perforin are expressed in $CD8^+$ cells *in vivo*. Thus, cytotoxic granules should not simply be dismissed as an artifact of cloned CTL lines but may be physiologically important in the intact host.

Evidence Challenging a Role for Granzyme A in CTL-Mediated Lysis

Several analyses of a variety of lymphoid subpopulations have demonstrated that in fact there is very poor correlation between granzyme A expression and cytotoxicity. For example, murine peritoneal exudate CTL (PEL), generated in vivo by in situ allogeneic sensitization, are highly lytic and yet possess no more detectable BLT-esterase activity than nonlytic target cells (38,39); yet, when cultured in vitro with exogenous IL-2, PEL had a >25-fold increase in granzyme A activity (Berke and Rosen, 1988). When short-term alloreactive mixed lymphocyte cultures were fractionated into minimally cytotoxic CD4$^+$ and highly cytotoxic CD8$^+$ subpopulations, perforin was expressed only in the CD8$^+$ subset (Garcia-Sanz et al., 1987), but granzyme A was present in both subsets and was secreted into the medium upon antigenic stimulation in both subsets (Garcia-Sanz et al., 1987; Velotti et al., 1987). Granzyme A expression in L3T4$^+$ cells was further studied by limiting dilution analysis, and approximately 20% of such clones expressed granzyme A (Fruth et al., 1988). In an analysis of the R8i cell line, an inducibly cytolytic T-cell clone, granzyme A content differed only threefold between the lytic (IL-2 induced) and nonlytic (uninduced) populations (Liu et al., 1989). The virtual absence of granzyme A in highly lytic cytotoxic effectors such as PEL, and its apparently nonselective presence in both CTL and T helper (T$_H$) cells, and more generally in lytic and nonlytic lymphocyte populations, argues against a specific role for granzyme A in CTL-mediated lysis.

Additional inhibition experiments also possibly challenge the role of granzyme A in lysis. Inhibition studies with serine protease inhibitors are limited by the neutral pH optimum of granzyme A and the localization of granzyme A within the acidic compartment of cytotoxic granules, as well as by the possibly restricted locus of action of the enzyme to the intercellular cleft during CTL–target cell interaction. However, when Henkart and colleagues (Henkart et al., 1987) pretreated CTL with PMSF and/or lysosomotropic agents to facilitate granzyme A inactivation, PMSF pretreatment alone inactivated 89% of granzyme A activity and reduced cytotoxicity to 55% of control levels. When CTL were dually pretreated with PMSF and ammonium chloride, cytotoxicity was reduced to 36% of control, yet 98.5% of granzyme A activity was inactivated. Under these conditions, granzyme A activity is far more susceptible to PMSF inhibition than is cytolysis, suggesting that granzyme A activity may not be necessary for cytolysis. However, such experiments do not resolve the threshhold quantity of granzyme A needed for cytolysis, especially since PEL contain only trace quantities of granzyme A and yet have excellent cytotoxic activity.

A more significant obstacle for a cytolytic role for granzyme A (and for the granule exocytosis model more generally) is the demonstration of calcium-independent lysis of target cells by cloned CTL lines (Tren et al., 1987; Ostergaard et al., 1987; Young et al., 1987). In the presence of Mg$_2$EGTA (which chelates free Ca^{2+}), detectable granule exocytosis is abolished, as measured by the absence of granzyme A release into the medium following CTL activation, yet CTL-mediated lysis of several different target cells is variably inhibited, if at all, depending on the effector cell studied (Ostergaard et al., 1987; Trenn et al., 1987; Young et al., 1987). Such observations are most compatible with the concept that CTL possess a multiplicity of cytolytic mechanisms, and that inhibition of granule exocytosis does not preclude target cell killing by an independent mechanism or mechanisms (Klein et al., 1990).

Cytotoxic Lymphocyte (*cis*) Effects of Granules

The sequence of murine perforin cDNA encodes a leader peptide sequence, typical of secreted proteins (Lowrey et al., 1989), which is cleaved during granule biogenesis, but the mature protein does not appear to require further proteolytic cleavage for activation. Thus, granzymes do not appear to be necessary for perforin action. It has been suggested that granzymes may act on the CTL to facilitate disengagement from target cells following delivery of the lethal hit. No data are available to support or refute this hypothesis. In addition to such a theoretical beneficial interaction with cytotoxic effector cells, the potentially lethal effects of cytotoxic granules on the effector cells themselves must be considered. CTL have consistently been found to be resistant to the lytic effects of intact CTL, cytotoxic granules, as well as purified perforin; and for a given CTL population, cytotoxic potential generally parallels resistance to cell- or granule-mediated lysis.

Target Cell (*trans*) Effects of Granules: Granules and Target Cell Nuclear Injury

One of the distinctive features of CTL-mediated lysis of target cells is the rapid fragmentation of target cell chromatin into a series of oligonucleosomes resulting in a characteristic ladder pattern of oligonucleotides differing by approximately 180 bp when analyzed by agarose gel electrophoresis (see Chapter 20, this volume). This phenomenon, part of the programmed cell death mechanism labeled apoptosis (Wyllie et al., 1984; Arends et al., 1990), is observed in a variety of physiologic conditions and is not specific for CTL-mediated killing, but is characteristic of such lysis. Does granule-mediated lysis inflict nuclear injury as does CTL-mediated lysis? Experimental observations to date are somewhat conflicting. Cytotoxic granules isolated from murine CTL have been reported to release ^{51}Cr and ^{3}H-thymidine from suitably labeled EL-4 target cells after a 20-min incubation, associated with typical oligonucleosomal fragmentation (Podack, 1986), and similarly YAC-1 cells underwent oligonucleosomal fragmentation after 60 min (Allbritton et al., 1988). In support of these findings with CTL-derived cytotoxic granules, granules isolated from a large granular lymphocyte (LGL) natural killer (NK) tumor cell line also released ^{51}Cr and ^{125}iododeoxyuridine (^{125}IUdR) from YAC-1 targets (Munger et al., 1988). In one report, purified perforin induced DNA degradation in several target cells (Hameed et al., 1989). In contrast to these observations, there is one report that fails to confirm the nuclear effects of isolated CTL granules (Gromkowski et al., 1988), and in two additional reports there was no evidence that purified perforin induced DNA fragmentation (Gromkowski et al., 1988; Munger et al., 1988; Duke et al., 1989). Thus, although there is no uniform consensus, there is some support that cytotoxic granules possess an effector molecule, or molecules, other than perforin, which may be responsible for target cell DNA degradation. The evidence implicating CTL-derived lymphotoxin (tumor necrosis factor-β) or related molecules in the process of target cell lysis generally, and target cell nuclear injury in particular, is reviewed in Chapter 16, this volume.

Henkart and his colleagues first presented evidence supporting a role for granzyme A in the nuclear events accompanying CTL-mediated lysis. As noted above, LGL cytotoxic granules triggered membrane and nuclear injury in YAC-1 targets (Munger et al., 1988). When granule perforin was inactivated by prior incubation in Ca^{2+}, treated granules lost both activities against intact target cells, but such treated granules could still trigger ^{125}IUdR release from target cells permeabilized by 0.2% Triton X-100. Such detergent treatment of target cells provides an alternative means for granule effector molecules to gain access to the target cell nucleus and participate in the DNA release process. The nuclear DNA release (NDR) activity was inhibitable by the serine protease inhibitors PMSF and DFP, suggesting a role for granzymes in this assay (Munger et al., 1988). More recent studies (Hayes et al., 1989) have shown that pretreatment of CTL by PMSF together with lysosomotropic NH_4Cl disproportionally inhibited ^{125}IUdR release when compared to the modest inhibition of ^{51}Cr release, suggesting that a PMSF-sensitive moiety localized within an acidic intracellular compartment was participating in chromosomal fragmentation. NDR activity copurified with granzyme A, and this activity was inhibited by the serine esterase inhibitors aprotinin and benzamidine. In addition, a combination of perforin and granzyme A augmented ^{125}IUdR release from labeled EL-4 target cells when compared to the ^{125}IUdR release observed using purified perforin alone. These findings support the hypothesis that granzyme A gains access to the internal milieu of the target cell via the membrane pores created by perforin, enabling its participation in the process of nuclear damage in an as yet unknown fashion.

Two additional studies demonstrate that other cytotoxic granule moieties, distinct from granzyme A and perforin, may be implicated in target cell nuclear chromatin degradation. An assay for target cell DNA fragmentation ("fragmentin" activity) based on the release of ^{125}IUdR from YAC target cells treated with rat NK cell granule proteins together with sublytic quantities of cytolysin demonstrated the presence of a discrete DNA-fragmenting activity (Shi et al., 1992). ^{51}Cr release assays demonstrated that fragmentin acts on the target cell membrane in addition to its intranuclear role. Purification and tryptic peptide sequence analysis of this 32-kDa species has shown that fragment is highly homologous to the NK protease RNK-1 as well as murine granzyme B, and that its activity is inhibited by the serine protease inhibitor 3,4-dichloroisocoumarin (Shi et al., 1992). It is of interest that a copurifying N-α-benzyloxycarbonyl-

L-lysine (BLT)-esterase activity acted synergistically with fragmentin when assayed in the ^{125}IUdR release assay (Shi et al., 1992). An independent series of experiments has demonstrated an independent, cytostatic (rather than cytocidal) activity for the rat NK granule protease RNK-1 (Sayers et al., 1992). Since in this report identification of the active moiety was based on N-terminal sequencing, while in the preceding report identification was based on sequence data derived from several internal peptides (Shi et al., 1992), it is possible that both activities may be ascribed to a single granzyme moiety highly homologous to, but distinct from, RNK-1.

Anderson and co-workers have recently shown that a nongranzyme protein present in the cytotoxic granules of human CTL, TIA-1, can also induce DNA fragmentation in target cells (Tian et al., 1991). This protein, which exists in 15- and 40-kDa isoforms, directly induces DNA fragmentation in digitonin-permeabilized thymocytes. The nucleolytic activity requires permeabilization of the target cell consistent with the concept that nucleolytic factors require an independent means to gain access to the intracellular milieu (e.g., via perforin). The 40-kDa isoform of TIA-1 has three tandem polyadenylate-binding domains, which may target this factor to bind not only to polyadenylated mRNA but also to poly-ADP-ribosylated nuclear proteins. The high-molecular-weight isoform can undergo proteolytic processing to generate a 15-kDa carboxy-terminal fragment which retains nucleolytic activity. It is possible that granule proteases may participate in this proteolysis.

Analysis of Granzyme A-Binding Proteins Suggests Granzyme A Acts Within the Nucleus

Nitrocellulose filter-binding experiments, akin to immunoblotting procedures but using soluble murine granzyme A as primary ligand, have been performed in an attempt to identify which target cell proteins may serve as possible substrates for granzyme A during CTL-mediated killing (Pasternack et al., 1991). Filters were prepared from target cell extracts separated by sodium dodecyl sulfate–polyacrylamide gel electrophoresis (SDS–PAGE), then incubated with microgram quantities of granzymes and developed by incubation with a monospecific rabbit anti-granzyme A antiserum and a biotin-avidin-peroxidase system. A restricted

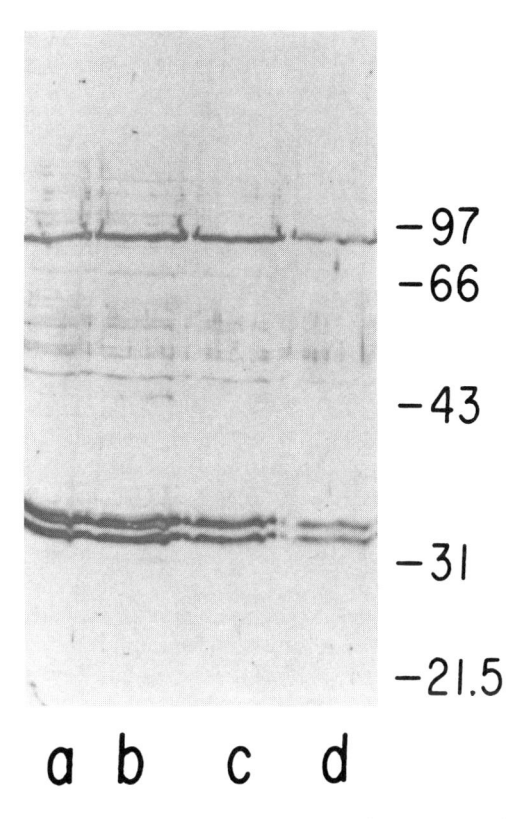

FIGURE 1. Quantitative aspects of granzyme A binding. Lanes a–d contain 20, 15, 10, and 5 µg, respectively, of a P815 nuclear salt wash preparation transferred to nitrocellulose after separation by SDS–PAGE. The filter was then incubated sequentially with granzyme A and monospecific anti-granzyme A antiserum, and developed with biotinylated anti-rabbit immunoglobulin and avidin-biotinylated horseradish peroxidase complex. (Reprinted with permission of The American Society for Biochemistry and Molecular Biology from Pasternack MS, et al. (1991): Granzyme A binding to target cell proteins: Granzyme A binds to and cleaves nucleolin in vitro. J Biol Chem 266(22): 14703–14708.)

number of binding proteins were visualized (Figure 1), and cell fractionation experiments demonstrated that most of the binding activity present in such detergent extracts derived from the nuclear fraction. The specificity of binding was studied by competitive binding experiments: although soluble target cell extracts reduced the binding of granzyme A to the filters, approximately 1000-fold molar excess of bovine albumin (pI 4.9), lysozyme (pI 11.0), and RNase (pI 9.6) failed to inhibit granzyme A binding. The failure of the extremely basic lysozyme and RNase to inhibit binding was particularly noteworthy, suggesting that granzyme A binding was not simply due to trivial electrostatic considerations. Granzyme A binding was not inhibited by pretreatment with aprotinin and was only variably inhibited by pretreatment with PMSF, which are covalently bound to the catalytically active site of the enzyme, but granzyme A binding was abrogated under conditions which denatured the enzyme, such as heating or treatment with SDS. These findings suggest granzyme A has discrete sites for substrate binding distinct from the proteolytic active site.

The most consistently observed binding proteins had M_r = 100, 38, and 35 kDa, respectively. The M_r = 100 kDa moiety was purified, and N-terminal sequence analysis confirmed it to be nucleolin. The latter two proteins are presumed to be B23 on the basis of their size, charge, and nuclear localization. The susceptibility of these proteins to proteolysis by granzyme A and trypsin was assessed by incubating an aliquot of nuclear salt wash with equivalent activities of granzyme A and trypsin (based on their esterolytic activity towards BLT) and monitoring proteolysis by silver staining and granzyme A binding. In contrast to trypsin, which rapidly and universally degraded the starting material to small peptide fragments, granzyme A demonstrated highly restricted proteolytic activity (compare Figures 2a and 2b). Granzyme A binding analysis (compare Figures 2c and 2d) revealed that nucleolin undergoes rapid, restricted cleavage to an 88-kDa fragment which retains granzyme A-binding activity and is largely refractory to further proteolysis. In contrast, the 38- and 35-kDa moieties no longer bind granzyme A after cleavage by this enzyme.

The identification of nuclear proteins as major granzyme-A-binding proteins is certainly surprising, since a nuclear locus of action requires that the enzyme traverse two membrane compartments. During "conventional" cell-mediated cytotoxicity, where granule exocytosis is the dominant cytolytic mechanism, perforin-induced membrane lesions might facilitate the entry of granzymes into the target cell cytoplasm. Nucleolin and B23 have been shown to shuttle between the nucleus and cytoplasm (Borer et al., 1989) and may function in the intranuclear transport of ribosomal proteins; in the present instance, they may serve to carry granzyme A into the nuclear compartment, where it may act directly or via proteolytically modified nucleolin or other intranuclear moieties. In addition, recent studies have shown that a small quantity of nucleolin is present at the cell surface and can serve as a specific receptor for low-density lipoproteins (Semekovich et al., 1990). If nucleolin is generally present on the cell surface, it may directly bind granzyme A for intracellular transport. Additional observations suggest that nucleolin and B23 may serve more generally as transporters for a variety of proteins. For example, in cells infected with human immunodeficiency virus (HIV), B23 forms a highly stable complex with the HIV Rev protein and is believed to shuttle cytoplasmic Rev into the nucleus of infected cells, which in turn is necessary for the export of viral mRNAs from the nucleus to the cytoplasm (Fankhauser et al., 1991).

Additional Effects of Granzyme A

The release of granzyme A into the extracellular environment may have biological effects that transcend the specific CTL–target cell interaction. For example, like trypsin, granzyme A has been shown to act as a polyclonal B-cell activator (inducing the incorporation of ^{3}H-thymidine into C3H nu/nu spleen cells) at concentrations ranging from approximately 200 ng/ml to 1.3 μg/ml *in vitro* (Simon et al., 1986). This activity was abrogated by PMSF pretreatment of granzyme A, confirming that the proteolytic activity of the enzyme was responsible for the proliferative signal. The significance of this observation is unclear, however, since the concentration of granzyme A required for B-cell activation *in vitro* is high compared to the quantity of enzyme secreted by CTL under physiologic conditions, particularly by CTL *in vivo*, where observed granzyme A levels have been quite modest (typically < 10% of cloned CTL) or even virtually undetectable. Furthermore, the half-life of enzymatically active granzyme A in extracellular fluid may be quite brief, since antiprotease macroglobulins presumably bind irreversibly any free serine proteases. Granzyme A has also been shown to degrade the extracellular sulfated proteoglycan matrix deposited by endothelial cells, and this activity is also inhibitable by the serine esterase inhibitor aprotinin (Simon et al., 1987b). Such activity *in vitro* suggests that immune attack by CTL may lead to degradation of the extracellular matrix, perhaps promoting ingress of additional inflammatory cells, etc.

An extracellular role for granzyme A may be analogous to the extracellular functions of tryptase. In possible support of such an hypothesis, tryptase, a granule serine protease structurally related to granzyme A but derived from mast cells rather than

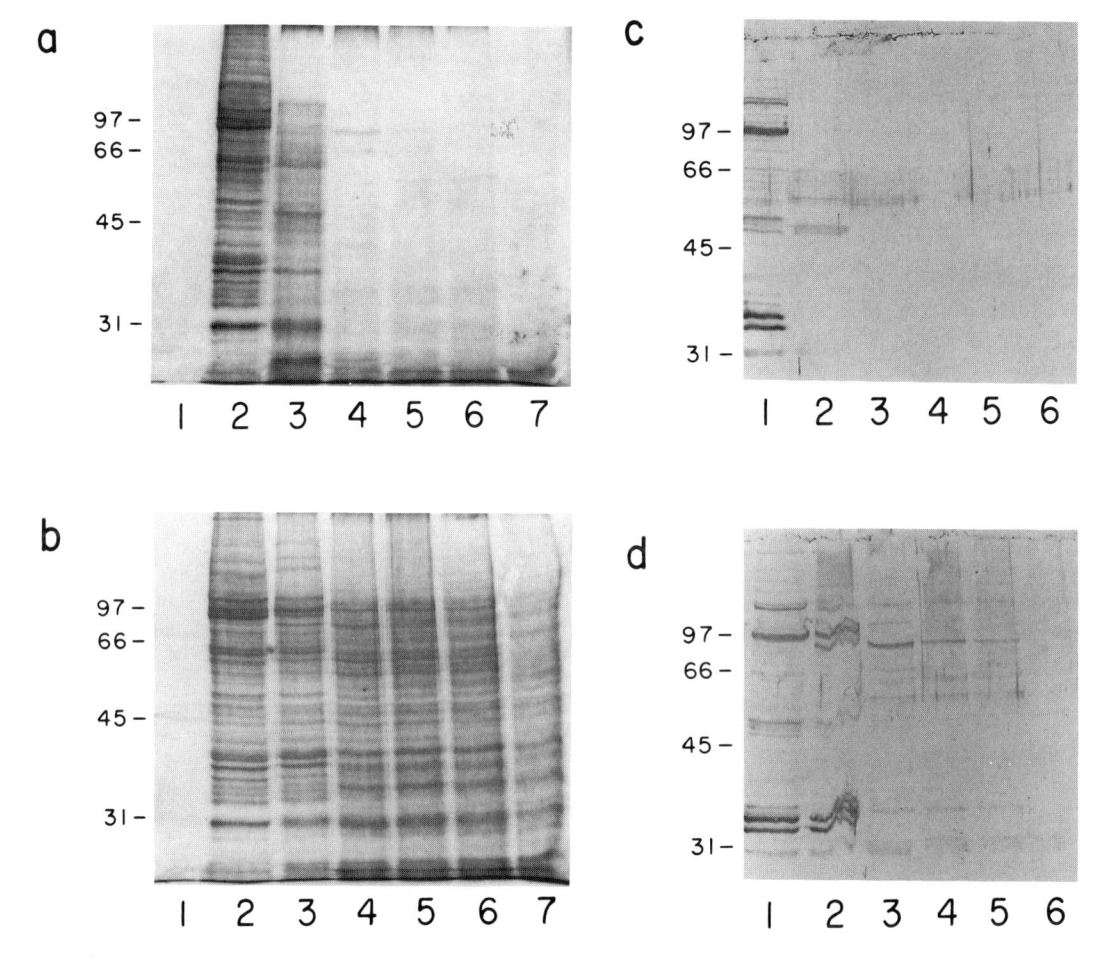

FIGURE 2. Proteolysis of nuclear salt wash proteins by trypsin (a, c) and granzyme A (b, d). a and b: Coomassie Blue staining of proteolytic digest after treatment with trypsin or granzyme A for 1 min (lane 3), 1 hr (lane 4), 2 hr (lane 5), 4 hr (lane 6), and 25 h (lane 7). The added protease is shown in lane 1 and the undigested sample in lane 2. c and d: Aliquots of the digests were separated by SDS–PAGE and processed for granzyme A binding as described in Figure 1. Undigested material, 1-min, 1-hr, 2-hr, 4-hr, and 25-hr samples are shown in lanes 1–6, respectively. (Reprinted with permission of The American Society for Biochemistry and Molecular Biology from Pasternack MS, et al. (1991): Granzyme A binding to target cell proteins: Granzyme A binds to and cleaves nucleolin in vitro. *J Biol Chem* 266(22): 14703–14708.)

from CTL, is secreted upon mast cell activation. *In vitro* studies show that tryptase can activate pro-collagenase to collagenase, which is believed to contribute to cartilage erosion and collagen degradation in rheumatoid arthritis (Gruber et al., 1988). Similar studies of mast cell chymase, a chymotrypsin-like granule enzyme related to granzyme B, show that chymase can cleave the functionally inactive precursor IL-1β (M_r = 31 kDa) to an active form with M_r = 18 kDa (Mizutani et al., 1991), which can amplify a local inflammatory process. Although these studies do not shed light regarding events *in vivo*, it is difficult to exclude these nonspecific amplifying effects of cytotoxic granule serine proteases. In addition, secreted mast cell chymase itself potentiates mast cell granule release, providing a positive feedback or amplifying signal to mast cells (Schick et al., 1984).

Are Proteases Generally Involved in Apoptosis?

Studies of target cell lysis triggered by tumor necrosis factor (TNF) may shed some light on the role of granzymes in cytolysis, since TNF-mediated lysis produces apoptosis (Schmid et al., 1987), and since lymphotoxin (TNF-β) has been implicated as a lytic mediator in CTL-mediated lysis (Schmid et al., 1986; Paul and Ruddle, 1988). In assay systems where soluble TNF is added to susceptible target cells either as culture supernatant or as recombinant TNF, several protease inhibitors (both competitive and noncompetitive) were able to markedly inhibit TNF-mediated lysis (Kirstein and Baglioni, 1986; Ruggiero et al., 1987; Suffys et al., 1988; Kumar and Baglioni, 1991). Furthermore, one of the target cell responses following TNF exposure is the expression of plasminogen-activator inhibitor (PAI), a serine protease inhibitor (Kumar and Baglioni, 1991). When this protein was transfected into TNF-susceptible target cells, they acquired relative resistance to TNF, and among different transfected clones, TNF resistance was proportional to the expression of PAI (Kumar and Baglioni, 1991). These studies suggest that a protease activity or activities within target cells is necessary for TNF-mediated lysis and activation of the apoptotic program.

Concluding Remarks

Tremendous strides have been made toward understanding the mechanisms of target cell lysis by CTL over the past two decades. The granule exocytosis model of target cell lysis continues to provide the fundamental theoretical framework for studies in this area, but newer data suggest that there are a multiplicity of effector mechanisms, presumably utilizing discrete effector molecules to lyse target cells. The analytical problems are heightened by the ability of a particular cytolytic population to utilize multiple lytic pathways, as well as by the surprisingly broad distribution of possible lytic effector molecules among different populations (perforin, TNF-β, granzymes, etc.). The widely disparate quantities of these effector moieties in different cytolytic populations (in vitro vs in vivo, mixed populations vs. IL-2-maintained clones, etc.) further complicate the emerging analysis. Despite these obstacles, the importance of CTL in host defense, immune surveillance, and a variety of pathophysiologic states insures that these mechanisms will be better understood with the goals of developing therapeutic modulation of cytolytic activity and elucidating this instructive area of cellular biology.

References

Allbritton NK. Verret CR, Wolley RC, Eisen HN (1988): Calcium ion concentrations and DNA fragmentation in target cell destruction by murine cloned cytotoxic T lymphocytes. J Exp Med 167: 514–527

Arends MJ, Morris RG, Wyllie AH (1990): Apoptosis: The role of the endonuclease. Am J Pathol 136: 593–608

Berke G, Rosen D (1988): Highly lytic in vivo primed cytolytic T lymphocytes devoid of lytic granules and BLT-esterase activity acquire these constituents in the presence of T cell growth factors upon blast transformation in vitro. J Immunol 141: 1429–1436

Borer RA, Lehner CF, Eppenberger HM, Nigg EA (1989): Major nucleolar proteins shuttle between nucleus and cytoplasm. Cell 56: 379–390

Dennert G, Podack ER (1983): Cytolysis by H-2-specific T killer cells: Assembly of tubular complexes on target membranes. J Exp Med 157: 1483–1495

Dennert G, Anderson CG, Prochazka G (1987): High activity of N-alpha-benzyloxycarbonyl-L-lysine thiobenzyl ester serine esterase and cytolytic perforin in cloned cell lines is not demonstrable in in vivo-induced cytotoxic effector cells. Proc Natl Acad Sci USA 84: 5004–5008

Duke RC, Persechini PM, Chang S, Liu C, Cohen JJ, Young JD (1989): Purified perforin induces target cell lysis but not DNA fragmentation. J Exp Med 170: 1451–1456

Fankhauser C, Izaurralde E, Adachi Y, Wingfield P, Laemmli UK (1991): Specific complex of human immunodeficiency virus type 1 Rev and nucleolar B23 proteins: Dissociation by the Rev response element. Mol Cell Biol 11: 2567–2575

Fruth U. Nerz G, Prester M, Simon HG, Kramer MD, Simon MM (1988): Determination of T cells expressing the T-cell specific serine proteinase 1 (TSP-1) reveals two types of L3T4+ T lymphocytes. Eur J Immunol 18: 773–781

Garcia-Sanz JA, MacDonald HR, Jenne DE, Tschopp J, Nabholz M (1990): Cell specificity of granzyme gene expression. J Immunol 145: 3111–3118

Garcia-Sanz JA, Plaetinck G, Velotti F, Masson D, Tschopp J, MacDonald HR, Nabholz M (1987): Perforin is present only in normal activated Lyt2+ T lymphocytes and not in L3T4+ cells, but the serine protease granzyme A is made by both subsets. EMBO J 6: 933–938

Gershenfeld HK, Weissman IL (1986): Cloning of a cDNA for a T cell-specific serine protease from a cytotoxic T lymphocyte. Science 232: 854–858

Griffiths GM, Alpert S, Lambert E, McGuire J, Weissman IL (1992): Perforin and granzyme A expression identifying cytolytic lymphocytes in rheumatoid arthritis. Proc Natl Acad Sci USA 89: 549–553

Gromkowski SH, Brown TC, Masson D, Tschopp J (1988): Lack of DNA degradation in target cells lysed by granules derived from cytolytic T lymphocytes. J Immunol 141: 774–778

Gruber BL, Schwartz LB, Ramamurthy NS, Irani A, Marchese MJ (1988): Activation of latent rheumatoid synovial collagenase by human mast cell tryptase. J Immunol 140: 3936–3942

Hameed A, Olsen KJ, Lee M, Lichtenheld MG, Podack ER (1989): Cytolysis by Ca-permeable transmembrane channels: Pore formation causes extensive DNA degradation and cell lysis. *J Exp Med* 169: 765–777

Hayes MP, Berrebi GA, Henkart PA (1989): Induction of target cell DNA release by the cytotoxic T lymphocyte granule protease granzyme A. *J Exp Med* 170: 933–946

Held W, MacDonald HR, Weissman IL, Hess MW, Mueller C (1990): Genes encoding tumour necrosis factor alpha and granzyme A are expressed during development of autoimmune diabetes. *Proc Natl Acad Sci USA* 87: 2239–2243

Henkart MP, Henkart PA (1982): Lymphocyte mediated cytolysis as a secretory phenomenon. *Adv Exp Med Biol* 146: 227–242

Henkart PA, Berrebi GA, Takayama H, Munger WE, Sitkovsky MV (1987): Biochemical and functional properties of serine esterases in acidic cytoplasmic granules of cytotoxic T lymphocytes. *J Immunol* 139: 2398–2405

Henkart PA, Millard PJ, Reynolds CW, Henkart MP (1984): Cytolytic activity of purified cytoplasmic granules from cytotoxic rat large granular lymphocyte tumors. *J Exp Med* 160: 75–93

Hudig D, Allison NJ, Kam C, Powers JC (1989): Selective isocoumarin serine protease inhibitors block RNK-16 lymphocyte granule-mediated cytolysis. *Mol Immunol* 26: 793–798

Hudig D, Allison NJ, Pickett TM, Winkler U, Kam C, Powers JC (1991): The function of lymphocyte proteases: Inhibition and restoration of granule-mediated lysis with isocoumarin serine protease inhibitors. *J Immunol* 147: 1360–1368

Hudig D, Callewaert DM, Redelman D, Allison NJ, Krump M, Tardieu B (1988): Lysis by RNK-16 cytotoxic lymphocyte granules: Rate assays and conditions to study control of cytolysis. *J Immunol Methods* 115: 169–177

Hudig D, Gregg, NJ, Kam C, Powers JC (1987): Lymphocyte granule-mediated cytolysis requires serine protease activity. *Biochem Biophys Res Commun* 149: 882–888

Jenne DE, Tschopp J (1988): Granzymes, a family of serine proteases released from granules of cytolytic T lymphocytes upon T cell receptor stimulation. *Immunol Rev* 103: 53–71

Kirstein M, Baglioni C (1986): Tumor necrosis factor induces synthesis of two proteins in human fibroblasts. *J Biol Chem* 261: 9565–9567

Klein JL, Selvakumar A, Trapani JA, Dupont B (1990): Characterization of a novel, human cytotoxic lymphocyte-specific serine protease cDNA clone (CSP-C). *Tissue Antigens* 35: 220–228

Kumar S, Baglioni C (1991): Protection from tumor necrosis factor-mediated cytolysis by overexpression of plasminogen activator inhibitor type-2. *J Biol Chem* 266: 20960–20964

Liu C, Jiang S, Persechini PM, Zychlinsky A, Kaufmann Y, Young, JD (1989): Resistance of cytolytic lymphocytes to perforin-mediated killing. Induction of resistance correlates with increase in cytotoxicity. *J Exp Med* 169: 2211–2225

Lowrey DM, Aebischer T, Olsen K, Lichtenheld M, Rupp F, Hengartner H, Podack ER (1989): Cloning, analysis, and expression of murine perforin 1 cDNA, a component of cytolytic T-cell granules with homology to complement component C9. *Proc Natl Acad Sci USA* 86: 247–251

MacDermott RP, Schmidt RE, Caulfield JP, Hein A, Bartley GT, Ritz J, Schlossman SF, Austen KF, Stevens RL (1985): Proteoglycans in cell-mediated cytotoxicity: Identification, localization, and exocytosis of a chondroitin sulfate proteoglycan from human cloned natural killer cells during target cell lysis. *J Exp Med* 162: 1771–1787

Masson D, Tschopp J (1985): Isolation of a lytic, pore-forming protein (perforin) from cytolytic T-lymphocytes. *J Biol Chem* 260: 9069–9072

Masson D, Tschopp J (1987): A family of serine esterases in lytic granules of cytolytic T lymphocytes. *Cell* 49: 679–685

Masson D, Nabholz M, Estrade C, Tschopp J (1986): Granules of cytolytic T-lymphocytes contain two serine esterases. *EMBO J* 5: 1595–1600

Mizutani H, Schechter N, Lazarus G, Black RA, Kupper TS (1991): Rapid and specific conversion of precursor interleukin 1-beta (IL-1 beta) to an active IL-1 species by human mast cell chymase. *J Exp Med* 174: 821–825

Muller C, Kagi D, Aebischer T, Odermatt B, Held W, Podack ER, Zinkernagel RM, Hengartner H (1989): Detection of perforin and granzyme A mRNA in infiltrating cells during infection of mice with lymphocytic choriomeningitis virus. *Eur J Immunol* 19: 1253–1259

Munger WE, Berrebi GA, Henkart PA (1988): Possible involvement of CTL granule proteases in target cell DNA breakdown. *Immunol Rev* 103: 99–109

Ojcius DM, Zheng LM, Sphicas EC, Zychlinsky A, Young JD (1991): Subcellular localization of perforin and serine esterase in lymphokine-activated killer cells and cytotoxic T cells by immunogold labeling. *J Immunol* 146: 4427–4432

Ostergaard HL, Kane KP, Mescher MF, Clark WR (1987): Cytotoxic T lymphocyte mediated lysis without release of serine esterase. *Nature* 330: 71–72

Pasternack MS, Eisen HN (1985): A novel serine esterase expressed by cytotoxic T lymphocytes. *Nature* 314: 743–745

Pasternack MS, Bleier KJ, McInerney TN (1991): Granzyme binding to target cell proteins: Granzyme A binds to and cleaves nucleolin in vitro. *J Biol Chem* 266: 14703–14708

Pasternack MS, Verret CR, Liu MA, Eisen HN (1986): Serine esterase in cytolytic T lymphocytes. *Nature* 322: 740–743

Paul NL, Ruddle NH (1988): Lymphotoxin. *Annu Rev Immunol* 6: 407–438

Peters PJ, Borst J, Oorschot V, Fukada M, Krahenbuhl O, Tschopp J, Slot JW, Geuze HJ (1991): Cytotoxic T lymphocyte granules are secretory lysosomes, containing both perforin and granzymes. *J Exp Med* 173: 1099–1109

Podack ER (1986): Molecular mechanisms of cytolysis by complement and by cytolytic lymphocytes. *J Cell Biochem* 30: 133–170

Podack ER, Konigsberg PJ (1984): Cytolytic T cell granules: Isolation, structural, biochemical, and functional characterization. *J Exp Med* 160: 695–710.

Podack ER, Young JD, Cohn ZA (1985): Isolation and biochemical and functional characterization of perforin 1 from cytolytic T-cell granules. *Proc Natl Acad Sci USA* 82: 8629–8633

Ruggiero V, Johnson SE, Baglioni C (1987): Protection from tumor necrosis factor cytotoxicity by protease inhibitors. *Cell Immunol* 107: 317–325

Sayers TJ, Wiltrout TA, Sowder R, Munger WL, Smyth MJ, Henderson LE (1992): Purification of a factor from the granules of a rat natural killer cell line (RNK) that reduces tumor cell growth and changes tumor morphology: Molecular identity with a granule serine protease (RNKP-1). *J Immunol* 148: 292–300

Schick B, Austen KF, Schwartz LB (1984): Activation of rat serosal mast cells by chymase, an endogenous secretory granule protease. *J Immunol* 132: 2571–2577

Schmid DS, McGrath KM, Hornung RL, Paul N, Ruddle NH (1987): Target cell DNA fragmentation is mediated by lymphotoxin and tumor necrosis factor. *Lymphokine Res* 6: 195–200

Schmid DS, Tite JP, Ruddle NH (1986): DNA fragmentation: Manifestation of target cell destruction mediated by cytotoxic T-cell lines, lymphotoxin-secreting helper T-cell clones, and cell-free lymphotoxin-containing supernatant. *Proc Natl Acad Sci USA* 83: 1881–1885

Schmidt RE, MacDermott RP, Bartley G, Bertovich M, Amato DA, Austen KF, Schlossman SF, Stevens RL, Ritz J (1985): Specific release of proteoglycans from human natural killer cells during target lysis. *Nature* 318: 289–291

Semenkovich CF, Ostlund RE, Olson MOJ, Yang JW (1990): A protein partially expressed on the surface of HepG2 cells that binds lipoproteins specifically is nucleolin. *Biochemistry* 29: 9708–9713

Shi L, Kraut RP, Aebersold R, Greenberg AH (1992): A natural killer cell granule protein that induces DNA fragmentation and apoptosis. *J Exp Med* 175: 553–566

Simon MM, Fruth U, Simon HG, Kramer MD (1987a): Evidence for the involvement of a T cell associated serine proteinase (TSP-1) in cell killing. *Ann Inst Pasteur* 138: 285–289

Simon MM, Hoschutzky H, Fruth U, Simon H, Kramer MD (1986): Purification and characterization of a T cell specific serine proteinase (TSP-1) from cloned cytolytic T lymphocytes. *EMBO J* 5: 3267–3274

Simon MM, Simon HG, Fruth, U, Epplen J, Muller-Hermelink HK, Kramer MD (1987b): Cloned cytolytic T-effector cells and their malignant variants produce an extracellular matrix degrading trypsin-like serine proteinase. *Immunology* 60: 219–230

Suffys P, Beyaert R, van Roy F, Fiers W (1988): Involvement of a serine protease in tumour-necrosis-factor-mediated cytotoxicity. *Eur J Biochem* 178: 257–265

Takayama H, Trenn G, Humphrey W, Bluestone JA, Henkart PA, Sitkovsky MV (1987): Antigen receptor-triggered secretion of a trypsin-like esterase from cytotoxic T lymphocytes. *J Immunol* 138: 566–569

Tian Q, Streuli M, Saito H, Schlossman SF, Anderson P (1991): A polyadenylate binding protein localized to the granules of cytotoxic lymphocytes induces DNA fragmentation in target cells. *Cell* 67: 629–639

Trenn G, Takayama H, Sitkovsky MV (1987): Exocytosis of cytolytic granules may not be required for target cell lysis by cytotoxic T-lymphocytes. *Nature* 330: 72–74

Velotti F, MacDonald HR, Nabholz M (1987): Granzyme A secretion by normal activated Lyt 2+ and L3T4+ T cells in response to antigenic stimulation. *Eur J Immunol* 17: 1095–1099

Wyllie AH, Morris RG, Smith AL, Dunlop D (1984): Chromatin cleavage in apoptosis: Association with condensed chromatin morphology and dependence on macromolecular synthesis. *J Pathol* 142: 67–77

Young, JD, Clark WR, Liu C, Cohn ZA (1987): A calcium- and perforin-independent pathway of killing mediated by murine cytolytic lymphocytes. *J Exp Med* 166: 1894–1899

Young JD, Leong LG, Liu C, Damiano A, Cohn, ZA (1986a). Extracellular release of lymphocyte cytolytic pore-forming protein (perforin) after ionophore stimulation. *Proc Natl Acad Sci USA* 83: 5668–5772

Young JD, Leong LG, Liu C, Damiano A, Wall DA, Cohn ZA (1986b): Isolation and characterization of a serine esterase from cytolytic T cell granules. *Cell* 47: 183–194

Young LHY, Joag SV, Zheng L, Lee C, Lee Y, Young JD (1990): Perforin-mediated myocardial damage in acute myocarditis. *Lancet* 2: 1019–1021

Young LHY, Peterson LB, Wicker LS, Persechini PM, Young JD (1989): In vivo expression of perforin by CD8+ lymphocytes in autoimmune disease: Studies on spontaneous and adoptively transferred diabetes in nonobese diabetic mice. *J Immunol* 143: 3994–3999

25
The Granzyme A Gene: A Marker for Cytolytic Lymphocytes In Vivo

Gillian M. Griffiths, Susan Alpert, R. Jane Hershberger, Lishan Su, and Irving L. Weissman

Introduction

Granzyme A, initially termed Hanukkah factor (HF) in this laboratory, is a serine protease contained within specialized cytoplasmic granules of activated cytolytic lymphocytes (CTL). These granules contain other granzymes, proteoglycans, and a pore-forming protein termed perforin (Podack, 1989). The granules and the proteins contained within them appear only after T-cell activation occurs, and their exocytosis seems to account for at least one mechanism of lymphocyte cytotoxicity. Whether this is the only mechanism remains controversial (Chapters 16, 18, 28, this volume; Berke, 1991; Krahenbuhl and Tschopp, 1991).

The exact role of the granzymes in killing is still unknown. Even so, there is a strong correlation between the expression of mRNA encoding both granzyme A and perforin, and lymphocyte cytotoxicity. This suggested that granzyme A gene expression might provide a marker for activated CTL *in vivo* (Griffiths and Mueller, 1991). In this chapter we will summarize work from this laboratory which has focused on the gene structure and the use of these genes as markers of CTL *in vivo*.

Granzyme A Gene Structure

The genes encoding both the mouse and the human granzyme A genes have now been cloned by a number of laboratories. Initially these genes were cloned as CTL-specific DNAs by subtractive hybridization techniques (Gershenfeld and Weissman, 1986; Gershenfeld et al., 1988). Two independent cDNAs encoding granzyme A have been isolated from a mouse CTL library (Hershberger et al., 1991). These cDNAs, termed GraA1 and GraA2, differ by a 126-bp insertion in the GraA2 (Figure 1a). This insertion creates a termination codon when transcription is initiated from the GraA1 initiation methionine. A second methionine within the GraA2 sequence would result in the production of the same active protein with an entirely different leader sequence. The leader sequence encoded by GraA1 is highly homologous to that of other murine granzymes and the human granzyme A, while the GraA2 leader sequence is very different. Both mRNAs appear to encode functional proteins; using *in vitro* transcription and translation systems, products of the predicted sizes can be generated from cDNA clones encoding GraA1 and GraA2. When transfected into COS-7 cells the GraA1 clone produces detectable granzyme A enzyme activity.

A polymerase chain reaction (PCR) analysis was used to confirm the expression of the two cDNAs isolated. In these experiments, mRNA was isolated from cytolytic cell lines and animals primed *in vivo* to generate CTL. Primers specific for GraA1 and GraA2 were used to amplify the isolated mRNA, and the specificity was confirmed in each case by DNA hybridization using leader sequence-specific probes. Both GraA1 and GraA2 could be detected in mRNA isolated from cytolytic cell lines. In the hybridoma PC-60, in which granzyme A activity can be induced with interleukin-2 (IL-2), both GraA1 and GraA2 mRNA species were induced upon IL-2 treatment. As in the cell lines, GraA1 and GraA2 could be detected *in vivo*. For these studies skin patches from donor mice were grafted onto either allogeneic or syngeneic recipient mice. Both GraA1 and GraA2 mRNAs were detected in lymphocytes isolated from the rejecting allogeneic grafts. In contrast, neither GraA1 nor GraA2

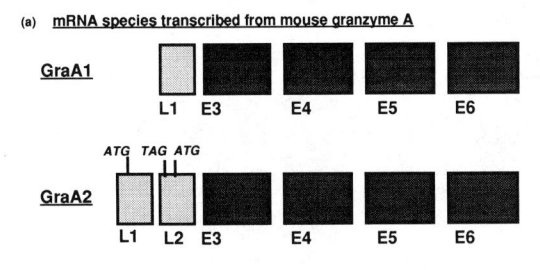

(a) mRNA species transcribed from mouse granzyme A

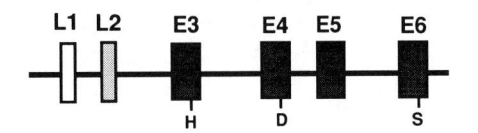

FIGURE 1. Structure and transcription of the mouse granzyme A gene. (A) mRNA species transcribed from mouse granzyme A. (B) Genomic organization of mouse granzyme A.

mRNAs were detectable by PCR amplification in syngeneic grafts.

The genomic organization of the mouse granzyme A gene confirms the existence of two alternative leader sequences. In general, the structure of the mouse granzyme A gene is typical of a serine protease gene, with three active site residues encoded in three separate exons, and the leader sequence in another (Figure 1b). In fact, in the case of the mouse granzyme A gene, there are six exons, and the extra exon encodes the second leader sequence (L2). Analyses of the sequences at the putative splice junctions indicated that the second leader sequence encoded by GraA2 possesses the consensus splice acceptor site at its 5' end and a splice donor sequence at its 3' end, suggesting that the GraA2 mRNA is derived by an alternative splicing mechanism.

Granzyme A Expression As a Marker of *in vivo* CTL

Granzyme A, like perforin, is expressed after T cells have been stimulated by target cell recognition, and the appearance of these markers correlates with the cytolytic ability of T cells in short-term assays (Masson and Tschopp 1985; Garcia et al., 1988; Hayes et al., 1989; Velotti et al., 1989). In addition, granzyme A mRNA has been detected only in normal tissues that might contain activated lymphoid cells, such as spleen, intestine, and lung, but not in heart, kidney, and brain (Garcia et al., 1990). This relationship suggests that the expression of granzyme A, or perforin, might be used to identify activated cells *in vivo*. In order to detect the cells that express granzyme A *in vivo*, *in situ* hybridization can be performed (see Methods). This method allows the detection of single cells expressing a given gene in a tissue section. Consequently it is possible to correlate the presence of cells expressing granzyme A and perforin with histological damage.

Animal Experiments

Transplantation

In an animal model of allograft rejection, where myocardial tissue from newborn BALB/c mice was transplanted under the kidney capsule of allogeneic C57BL/Ka mice, a massive mononuclear infiltration occurred (Mueller et al., 1988). These grafts were completely rejected within 8 to 12 days. Cells with detectable granzyme A transcripts were present at only 2 days posttransplantation, and at 4 days posttransplantation the frequency of cells expressing granzyme A increased dramatically, as compared to syngeneic control grafts, which did not reject. Throughout the course of the experiments, no more than 7% of the inflammatory cells expressed granzyme A, which is consistent with the theory that the majority of infiltrating inflammatory cells in the allograft are not alloantigen-specific. When the infiltrating lymphocytes were characterized for their CD4/CD8 phenotype, it was apparent that although the majority of the infiltrate was $CD8^+$, only 10% of the $CD8^+$ cells expressed granzyme A. This result indicates that the majority of $CD8^+$ cells are not activated killers, and may indicate why CD4/CD8 ratios do not correlate with rejection (McWhinnie et al., 1985).

A recent study of a rat heterotropic heart transplant model showed that the number of cells expressing granzyme A correlated with graft function, as measured by palpatation, survival, and histological analysis. Granzyme A expression was detected 2 days before rejection monitored by the other criteria, indicating that granzyme A may be used as an early marker for rejection in this animal model. In this transplantation system, cyclosporin A and Ox-38, and anti-CD4 monoclonal antibody, prevented both the accumulation of granzyme A^+ cells and graft rejection (Chen et al. 1993. By using granzyme A as a rejection marker, it was possible to see that allografts could be "rescued" with cyclosporin A infusion when treatment was initiated prior to the onset of tissue damage (Chen et al., unpublished data). The results show that immunosuppressive regimens decrease granzyme A expression *in vivo*. In light of these results, it has been suggested that it may be possible not only to use the expression of granzyme A as an early marker for transplantation rejection, but to predict the efficiency of immunosuppressive treatments in animal models monitored by granzyme A expression.

Infectious Disease: LCMV

One system for studying the cytolytic response to virally infected cells is to infect mice with lymphotropic choriomeningitis virus (LCMV). Six days after injection of hepatotropic LCMV, granzyme A-expressing lymphocytes are found in the liver infiltrate. The timing of appearance of these cells in the hepatic infiltrate correlates with the onset of a detectable anti-LCMV response. Granzyme A-expressing cells appear in the central nervous system 6 to 7 days after intracerebral inoculation of LCMV-Armstrong (Muller et al., 1989). Immunohistochemical staining of serial sections indicates that these granzyme A cells are probably the CD8$^+$ cells that are found in the LCMV—infected meninges and choroid plexus. Similar results have been found in immunohistochemical experiments with monoclonal antibodies to granzyme A (Kramer et al., 1989). Thus granzyme A is involved in the cytolytic response to LCMV, and is likely to be involved in the response to other viruses.

Autoimmune Disease: The NOD Mouse

The NOD-Obese Diabetic Mouse (NOD) is one animal model for studying insulin-dependent diabetes mellitus. Mice of 4 to 6 weeks of age exhibit infiltration of the pancreas islets of Langerhans, and by 12 to 16 weeks of age the islets are destroyed. *In situ* hybridization showed that activated T cells were present in the infiltrate, however, the cells that express granzyme A were not near the insulin-producing cells (Held et al., 1990). A significant fraction of the cells in contact with the insulin-producing cell expressed tumor necrosis factor alpha (TNE-α) transcripts. Most of these cells were CD4$^+$. As both CD8$^+$ and CD4$^+$ cells are required to produce the disease in adoptive transfer experiments, it is possible that both granzyme A and TNF-α-producing cells are necessary to induce the disease state. It should be noted that cells expressing perforin are found in the islet infiltrate. These perforin-expressing cells are also found in the islets of recipient animals upon adoptive transfer. It is possible that the production of all three molecules —granzyme A, perforin, and TNF-α—is required to produce the diabetic state in this animal model of autoimmunity.

Human Studies

It has been possible to use the method of *in situ* hybridization to detect CTL that seem to be involved in transplantation rejection and autoimmunity in human patients. These observations suggest an important predictive role for granzyme A expression as a diagnostic marker.

Transplantation

Cardiac transplant patients are constantly monitored postoperatively for signs of imminent rejection. This is usually carried out by examining cardiac biopsies for signs of myocyte necrosis and lymphocytic infiltration. Since CTL are implicated in the rejection process, a series of cardiac biopsies were examined for the presence of granzyme A- and perforin-expressing lymphocytes. There was a very clear correlation between biopsies showing histological damage indicative of rejection and the presence of granzyme- and perforin-expressing lymphocytes: rejecting biopsies always contained granzyme- and perforin-expressing lymphocytes.

In several patients who underwent rejection during monitoring by biopsy, the presence of granzyme- and perforin-expressing lymphocytes could be detected prior to the rejection episode, when standard histological examination could only demonstrate the presence of an infiltrate too sparse to necessarily predict a rejection episode (Griffiths et al., 1991). By identifying a protein that is likely to have a functional role in the cytolytic process, granzyme A expression may provide an accurate predictive marker of rejection. Since the aim of this diagnosis would be to have a better idea of when to increase immunosuppression, it was clear that it was important to understand the effects of immunosuppressive drugs on granzyme expression in transplantation.

Autoimmunity

The effects of immunosuppressive drugs also seemed to be apparent when granzyme A and perforin expression was examined in the synovial fluid lymphocytes from rheumatoid arthritis patients. *In situ* hybridization was used to detect granzyme and perforin expression in lymphocytes isolated directly from the joints of patients with active rheumatoid arthritis. Patients receiving no immunosuppression or low doses of steroidal immunosuppressives were found to contain granzyme- and perforin-expressing lymphocytes. For three patients, samples were obtained both before and after immunosuppressive therapy. In the untreated samples, granzyme A- and perforin-expressing cells were present. However, these granzyme A- and perforin-expressing cells were completely absent following immunosuppressive treatment (Griffiths et al., 1992). These results suggest not only that cytolytic lymphocytes may play a role in the pathogenesis of this disease, but also that granzyme A and perforin may be regulated by immunosuppressive drugs *in vivo*.

Perspectives

There is a good correlation between the expression of granzyme A and the ability of cells to function cytolytically. Thus, *in vivo* it is likely that granzyme A has a role in cell-mediated cytotoxicity. Although the mechanism of cytotoxicity is still unclear, it is very likely that granzyme A could be used as a marker for graft rejection and for certain autoimmune phenomena.

Clearly, more disease states need to be investigated. There are several autoimmune phenomena, such as Sjogren's syndrome, Crohn's disease, and primary biliary cirrhosis, where cytotoxic T cells could potentially play a role in the pathogenesis of the disease. If granzyme A-expressing cells are found in the autoimmune "lesions," it is likely that CTL are directly involved in the disease state.

In bacterial and viral infections, cytolytic cells respond by recognizing and destroying the infected cells. To determine if granzyme A is involved in the response to agents other than LCMV, both animal and human samples need to be studied. Experiments are currently underway to determine if granzyme A-expressing cells are present in Lyme disease and in reactive arthritis.

As granzyme A can be used as a marker of activated CTL *in vivo*, it may be a potential diagnositic tool. Granzyme A expression could be used as a marker for transplantation rejection, an acute phase of infection, or an autoimmune disease.

The effect of immunosuppressive drugs on granzyme A expression should also be studied further. Cylosporin A is known to decrease expression of granzyme A *in vitro*. Murine studies have now shown that cyclosporin A also affects granzyme A expression, and the limited human studies indicate that the number of granzyme A-expressing cells declines after immunosuppressive treatment. Because animal studies allow the effect of a particular treatment on granzyme A expression to be monitored in a defined system, these studies will probably be conducted first. Greater numbers of patients must be studied, in a variety of disease states, for the role of immunosuppressive agents in granzyme A expression to be firmly established.

References

Berke G (1991): Lymphocyte-triggered internal target disintegration. *Immunology Today* 12: 396–399

Chen RH, Ivens KW, et al. (1993): The use of granzyme A as a marker of heart transplant rejection in cyclosporine or anti-CD4 monoclonal antibody-treated rats transplantation (in press)

Garcia SJ, Macdonald HR, et al. (1990): Cell specificity of granzyme gene expression. *J Immunol* 145: 3111–3118

Garcia SJ, Velotti F, et al. (1988): Appearance of granule-associated molecules during activation of cytolytic T-lymphocyte percursors by defined stimuli. *Immunology* 64: 129–134

Gershenfeld HK, Hershberger KJ, et al. (1988): Cloning and chromosomal assignment of a human cDNA encoding a T cell- and natural killer cell-specific trypsin-like protease. *Proc Natl Acad Sci USA* 85: 1184–1188

Gershenfeld NK, Weissman IL (1986): Cloning of a cDNA for a T cell-specific serine protease from a cytotoxic T lymphocyte. *Science* 232: 854–858

Griffiths GM, Alpert S, et al. (1992): Perforin and granzyme A expression identifying cytolytic lymphocytes in rheumatoid arthritis. *Proc Natl Acad Sci USA* 89: 549–553

Griffiths GM, Mueller C (1991): Expression of perforin and granzymes in vivo: Potential diagnostic markers for activated cytotoxic cells. *Immunol Today* 12: 415–419

Griffiths GM, Namikawa R, et al. (1991): Granzyme A and perforin as markers for rejection in cardiac transplantation. *Eur J Immunol* 21: 687–692

Hayes, MP, Berrebi GA, et al. (1989): Induction of target cell DNA release by the cytotoxic T lymphocyte granule protease granzyme A. *J Exp Med* 170: 933–946

Held W, MacDonald HR, et al. (1990): Genes encoding tumor necrosis factor alpha and granzyme A are expressed during development of autoimmune diabetes. *Proc Natl Acad Sci USA* 87: 2239–2243

Hershberger RJ, Gershenfeld HG, et al. (1992): Genomic organization of the mouse granzyme A gene. *J Biol Chem* (in press)

Krahenbuhl O, Tschopp J (1991): *Immunol Today* 12: 399–402

Kramer MD, Fruth U, et al. (1989): Expression of cytoplasmic granules with T cell associated serine protease 1 activity in Ly2 (CD8+) T lymphocytes responding to lymphocytic choriomeningitis virus in vivo. *Eur J Immunol* 19: 151–156

Masson D, Tschopp J (1985): Isolation of a lytic, porforming protein (perforin) from cytolytic T-lymphocytes. *J Biol Chem* 260: 9069–9072

McWhinnie DL, Carter NP, et al. (1985): Is the T4/T8 ratio an irrelevance in renal transplantation monitoring? *Transplant Proc* 17: 2548–2549

Mueller C, Gershenfeld HG, et al. (1988): A high proportion of T lymphocytes that infiltrate H-2 incompatible heart allografts in vivo express genes encoding cytotoxic cell specific serine proteases, but do not express the MEL-14 defined homing receptor. *J Exp Med* 167: 1124–1136

Muller C, Kagi D, et al. (1989): Detection of perforin and granzyme A mRNA in infiltrating cells during infection of mice with lymphocytic choriomeningitis virus. *Eur J Immunol* 19: 1253–1259

Podack ER (1989): Killer lymphocytes and how they kill. *Curr Opinions Cell Biol* 1: 929–933

Velotti F, Palmieri G, et al. (1989): Granzyme A expression by normal rat natural killer (NK) cells in vivo and by interleukin 2-activated NK cells in vitro. *Eur J Immunol* 19: 575–578

26
Molecular Analysis and Possible Pleiotropic Function(s) of the T Cell-Specific Serine Proteinase-1 (TSP-1)

Markus M. Simon, Klaus Ebnet and Michael D. Kramer

Introduction

Recovery from infection requires the induction of relevant effector functions of lymphocytes and their focus to appropriate sites. T lymphocytes play a pivotal role in these complex biological processes. They act not only as regulator and effector cells during the development of both humoral and cellular immune responses, but are also able to interact with and control the physiology of nonimmune cells such as endothelial cells and fibroblasts (Cavender, 1989, 1990; Springer, 1990; Shimizu et al., 1989, 1990, 1991) during extravasation and their performance within infectious foci (Abraham et al., 1989).

T lymphocytes recirculate between blood and lymph, and in response to antigenic stimulation they invade tissues to exert their effector function *in situ*. Recirculation which depends on specific and regulated interactions between receptors expressed on lymphocyte cell surfaces and ligands on the surface of endothelial cells (Woodruff et al., 1987; Shimizu et al., 1991), is thought to increase the likelihood of interaction between lymphocytes and antigen-presenting cells in secondary lymphoid organs leading to the induction of immune responses. On the other hand, recruitment of activated T lymphocytes into inflammatory foci is accomplished by their binding, via a set of antigenically induced cell surface receptors, to appropriate ligands on microvascular endothelial cells. Expression of the latter ligands is induced by nonspecific inflammatory stimuli (Cavender, 1989, 1990; Kupper, 1990; Barker et al., 1991). Extravasation of activated T lymphocytes also seems to involve adhesion to and focal degradation of vascular basement membrane, as shown for polymorphonuclear leukocytes that traverse the vascular basement membrane by virtue of proteolytic enzymes that locally digest the basement membrane (Huber and Weiss, 1989). By analogy, it is assumed that migration of T lymphocytes into inflammatory foci via microvascular capillaries also involves as yet undefined proteolytic processes.

Lymphocyte-associated proteases have long been implicated in basic lymphocyte effector functions such as proliferation and cytolysis. It has been shown that proteolytic enzymes like trypsin and thrombin induce B-cell proliferation (Vischer, 1974; Kaplan and Bona, 1974; Glenn and Cunningham, 1979) and that this process may be suppressed by protease inhibitors (Hart and Streilein, 1976). Moreover, it has been demonstrated that protease inhibitors interfere with lymphocyte-mediated killing (Chang and Eisen, 1980; Redelman and Hudig, 1980). However, in spite of much indirect evidence for the presence of T-lymphocyte-associated proteolytic enzymes, candidate proteases responsible for the various biological activities were not characterized in detail until the mid-1980s. With the advent of T-lymphocyte growth factors (Smith, 1984), it became possible to expand monoclonal populations of T-effector lymphocytes with defined function *in vitro* on a large scale. The availability of cloned $CD4^+$ helper (T_H) and cloned $CD8^+$ cytolytic T lymphocytes (T_c) has allowed the investigation of the molecular mechanisms underlying T-lymphocyte functions, such as B-cell activation, cytolysis, control of virus replication, and migration.

In recent years, we and others (Pasternack and Eisen, 1985; MM Simon et al., 1986; Kramer et al., 1986; Masson et al., 1986a, 1986b; Young et al., 1986b; Pasternack et al., 1986; Fruth et al., 1987b; Kramer and Simon, 1987; Krähenbühl et al., 1988;

TABLE 1. Murine T-cell-associated proteases

Protein	cDNA	M_r red/nred	Substrate	pH optimum
MTSP-1 BLT-esterase Granzyme A SE-1	MTSP-1 H factor CTLA-3	34/67	HD-Pro-Phe-Arg-pNA BLT Pro-Phe-Arg-AMC	7.0–8.0
MTSP-2 Granzyme B SE-2	CCP-I CTLA-1	29/27	Boc-Ala-Ala-Asp-SBzl	7.0–8.0
MTSP-3 Granzyme C	Granzyme C CCP-II	27/24	N.I.	
MTSP-4 Granzyme D	Granzyme D	33–50	N.I.	
MTSP-5 Granzyme E	Granzyme E CCP 3	35–45	N.I.	
MTSP-6 Granzyme F	Granzyme F CCP 4	35–40	Suc-Phe-Leu-Phe-SBzl	N.I.
MTSP-7 Granzyme G	Granzyme G MCSP-1	30	N.I.	

Data are summarized from the following references: Pasternack and Eisen, 1985; Pasternack et al., 1986; Gershenfeld and Weissman, 1986; Kramer et al., 1986; MM Simon et al., 1986; HG Simon 1988; Brunet et al., 1986; Young et al., 1986b; Masson et al., 1986a, 1986b; Masson and Tschopp, 1987; Lobe et al., 1986a, 1986b; Jenne et al., 1988a, 1988b, 1989; Bleackley et al., 1988; Kwon et al., 1988; Jenne et al., 1989; Odake et al., 1991.
BLT = N^α-benzyloxycarbonyl-L-lysine thiobenzyl ester; N.I., Not identified.

Poe et al., 1988) have isolated from cloned mouse and human T_C two analogous trypsin-like serine proteinases termed "murine T-lymphocyte-associated serine proteinase-1 (MTSP-1) and "human T lymphocyte-associated serine proteinase-1 (HuTSP-1)". To date, two families of highly homologous serine proteinases including MTSP-1 and HuTSP-1 have been isolated from cloned mouse and human T_C, named independently BLT esterases (Pasternack and Eisen, 1985; Pasternack et al., 1986), TSPs (Kramer et al., 1986; MM Simon et al., 1986; Fruth et al., 1987b), granzymes A–G (Masson et al., 1986a, 1986b; Masson and Tschopp, 1987), and SE 1 and 2 (Young et al., 1986b; See also Chapters 23, 24, 25, this volume). For the sake of clarity in this review we will be using the following abbreviations for the above synonyms: for serine proteases derived from mouse T_C "MTSP-1–MTSP-7" (Table 1) and for those derived from human T_C, "HuTSP-1, -2, -8, and -X" (Table 2).

It is the purpose of this article to review our recent studies on molecular and functional aspects of MTSP-1 and HuTSP-1 and to put forward our concepts of their possible pleiotropic activities in T-lymphocyte physiology.

MTSP-1 Is a Member of a Multigene Family Encoding Highly Homologous T Lymphocyte-Associated Serine Proteinases

To understand the molecular mechanism of T_C function, one approach was to identify mRNAs specifically expressed in mouse T_C (Gershenfeld and Weissman, 1986b; Lobe et al., 1986b; Brunet et al., 1986). Sequence analysis of different cDNA clones, including that specific for MTSP-1, derived from

TABLE 2. Human T-cell-associated proteases

Designated			Activity	
Protein	cDNA	M_r red/nred	Substrate	pH optimum
HuTSP-1 Granzyme 1 CTL-Tryptase Granzyme A	H factor	30/56	Tos-Gly-Pro-Arg-NA BLT HD-(ε-Cbz)-Lys-Pro-Arg-pNA Gly-Pro-Arg-AMC	7.0–9.0
HuTSP-2 Granzyme 2 Granzyme B	HLP CSP-B SECT CGL-1 CTLA-1	35/30	Boc-Ala-Ala-Asp-SBzl	7.0–8.0
HuTSP-X Granzyme 3	N.I.	25/28	BLT	
HuTSP-8 Granzyme H	N.I. CSP-C CGL-2			

Data have been taken from the following references: Fruth et al., 1987b; Schmid and Weissmann, 1987; Gershenfeld et al., 1988; Krähenbühl et al., 1988; Hameed et al., 1988; Poe et al., 1988, 1991; Trapani et al., 1988; Caputo et al., 1988; Klein et al., 1989, 1990; Hanson et al., 1990; Haddad et al., 1990, 1991; Hanson and Ley, 1990; Meier et al., 1990. N.I., Not identified.

mRNAs of mouse T_C, revealed a set of molecules with extensive homology (Brunet et al., 1986; Lobe et al., 1986a; Gershenfeld and Weissman, 1986; HG Simon, 1988; Jenne and Tschopp, 1988; Jenne et al., 1988a,b, 1989). The predicted proteins resemble serine proteinases in that they express the amino acid residues His, Asp, and Ser, which form the active site of such enzymes (Kraut, 1977). In the mouse, this family now includes seven members, referred to as MTSP-1 to MTSP-7 (see Table 1). In humans, four homologous structures, HuTSP-1 (Fruth et al., 1987b; Krähenbühl et al., 1988; Poe et al., 1988; Hameed et al., 1988; Gershenfeld et al., 1988), HuTSP-2 (Schmid and Weissmann, 1987; Krähenbühl et al., 1988; Hameed et al., 1988; Trapani et al., 1988; Caputo et al., 1988; Klein et al., 1989; Hanson and Ley, 1990; Haddad et al., 1990; Poe et al., 1991), HuTSP-8 (Meier et al., 1990; Klein et al., 1990; Haddad et al., 1991), and HuTSP-X (Hameed et al., 1988), have been described. That their corresponding genes are closely related is strongly supported by recent studies from other laboratories and our own.

The genomic organization of MTSP-1 (Figure 1; Ebnet et al., 1992) revealed a striking similarity with that of five other TSP, i.e. MTSP-2, MTSP-3 (Lobe et al., 1988), MTSP-6 (Jenne et al., 1991), HuTSP-2 (Klein et al., 1989), and HuTSP-8 (Meier et al., 1990), studied so far. First, all enzymes are encoded by four exons. Second, the sizes of the corresponding exons are very similar, and the introns, although greatly differing in size, are located at very similar positions. Third, all six TSP genes contain the nucleotide triplets encoding the catalytic triad (His, Asp, Ser) of serine proteinases, at corresponding positions on exons 2, 3, and 5, respectively. Fourth, nucleotide residues adjacent to the three amino acid residues His, Asp, and Ser are also conserved, as well as six cysteine residues which are involved in intramolecular disulfide bonding. Finally, the splice phase, i.e., the position of a nucleotide within a codon after which an intron is inserted, is identical for the respective exons in all six genes. These features are also characteristic for other serine proteinase genes such as rat mast cell protease II (RMCP II; Benfey et al., 1987) and mouse mast cell protease-1 (MMCP-1; Huang et al., 1991) and suggest that these genes may have evolved from a common ancestral gene via duplication (Rogers, 1985).

As with the other MTSP, the transcribed MTSP-1-specific mRNA encodes a typical signal peptide (Jenne and Tschopp, 1988; HG Simon, 1988) required for the translocation of proteins into the endoplasmic reticulum. Moveover, it encodes the propeptide "Glu-Arg," indicating that MTSP-1 is

Genomic Organisation of Serine Protease Genes

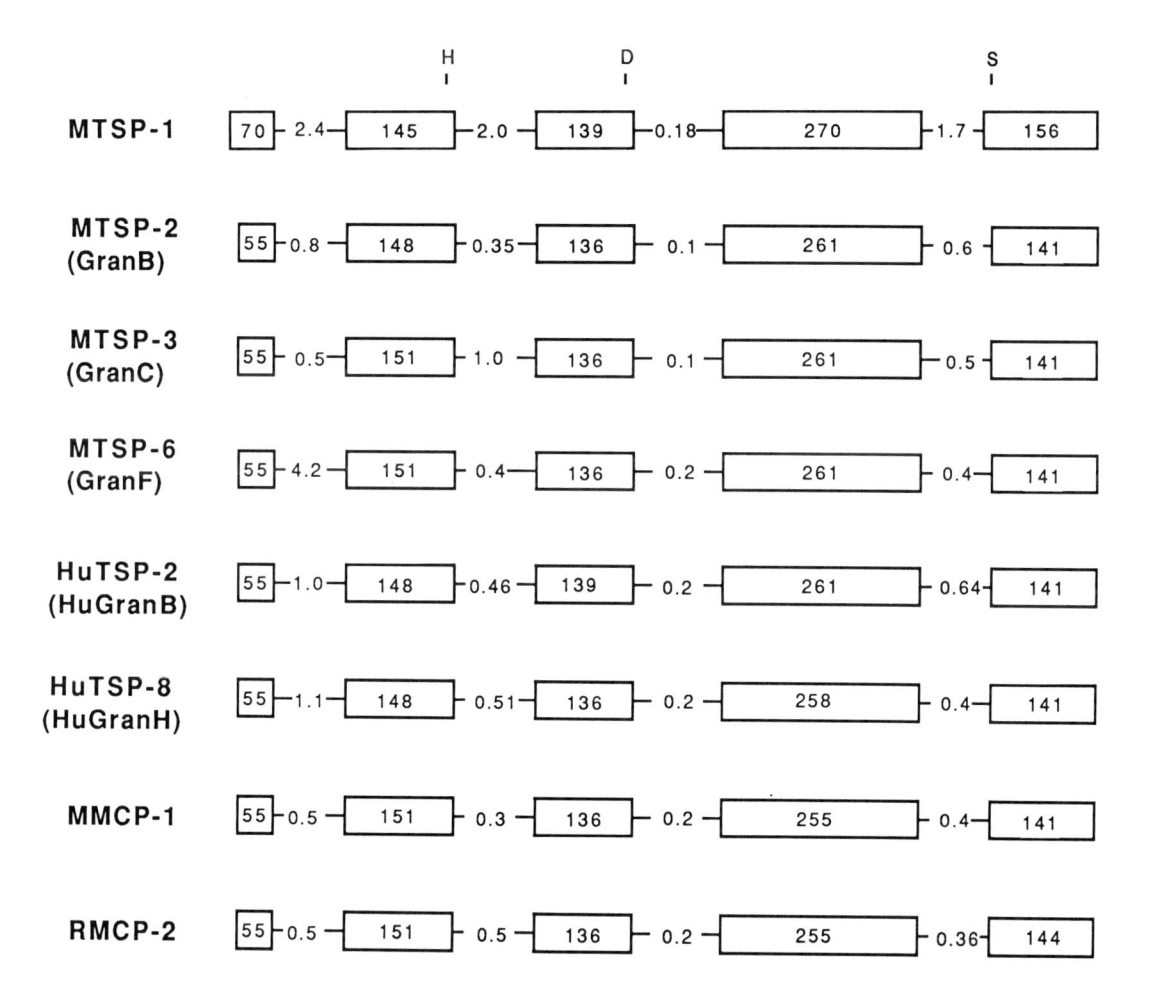

	H	D	S

MTSP-1 [70]─2.4─[145]─2.0─[139]─0.18─[270]─1.7─[156]

MTSP-2 (GranB) [55]─0.8─[148]─0.35─[136]─0.1─[261]─0.6─[141]

MTSP-3 (GranC) [55]─0.5─[151]─1.0─[136]─0.1─[261]─0.5─[141]

MTSP-6 (GranF) [55]─4.2─[151]─0.4─[136]─0.2─[261]─0.4─[141]

HuTSP-2 (HuGranB) [55]─1.0─[148]─0.46─[139]─0.2─[261]─0.64─[141]

HuTSP-8 (HuGranH) [55]─1.1─[148]─0.51─[136]─0.2─[258]─0.4─[141]

MMCP-1 [55]─0.5─[151]─0.3─[136]─0.2─[255]─0.4─[141]

RMCP-2 [55]─0.5─[151]─0.5─[136]─0.2─[255]─0.36─[144]

FIGURE 1. Comparison of the MTSP-1 gene with other lymphoid (MTSP-2, -3, -6; HuTSP-2, -8; see also Tables 1 and 2) and nonlymphoid (MMCP-1; RMCP-2) protease genes. The numbers within (exon) and between (intron) the boxes indicate the sizes of exons (base pairs) and introns (kilobases), respectively. The positions of the amino acids forming the active sites of the predicted enzymes are indicated by H (histidine), D (aspartic acid), and S (serine). MMCP-1, mouse mast cell protease-1; RMCP-2, rat mast cell protease-2. Data taken from: Benfey et al., 1987; Lobe et al., 1988; Klein et al., 1989; Meier et al., 1990; Huang et al., 1991; Jenne et al., 1991).

synthesized as an inactive precursor, which is stored, however, in its active form within cytoplasmic granules (see also Figure 6C Inset: cytochemical detection of MTSP-1 activity in cytoplasmic granules of *in vivo*-sensitized virus-specific T$_c$). Most probably, the propeptide, which is also common to other granule-associated serine proteinases of T lymphocytes and other cells, such as HuTSP-1, -2, and -8 (Gershenfeld et al., 1988; Schmid and Weissman 1987; Meier et al., 1990),

RMCP I and II (Benfey et al., 1987), cathepsin G (Hohn et al., 1989), and polymorphonuclear neutrophil elastase (Takahashi et al., 1988), is removed during transport of the enzymes into the cytoplasmic granules.

MTSP-1 and HuTSP-1, in contrast to all other monomeric T-lymphocyte-associated serine proteinases, occur as disulfide-linked homodimers (Simon et al., 1986; Masson et al., 1986b; Fruth et al., 1988a) with both catalytic sites being active

(Masson and Tschopp, 1988; Gurwitz et al., 1989; MM Simon et al., 1990). Whether these features are of particular relevance for T-lymphocyte-mediated functions is not known at present.

From the Seven MTSP Known So Far, only MTSP-1 and MTSP-2 Are Expressed to a Significant Level in Immune T Lymphocytes *in vivo*

Originally, MTSP-1 to MTSP-7 were isolated as proteins and cDNAs from long-term-cultured mouse T_C. However, their *in vivo* relevance remained unclear. Further analyses of Northern blots using total cytoplasmic RNA from lymphoid and nonlymphoid tissues and cDNA probes specific for MTSP-1 and MTSP-2 revealed that corresponding transcripts were expressed by *in vitro*- and *in vivo*-activated T lymphocytes but not in naive T lymphocytes and most other tissues, suggesting that both enzymes are specific markers for activated T lymphocytes (Gershenfeld and Weissman, 1986; Brunet et al., 1986; Lobe et al., 1986a; Garcia-Sanz et al., 1987, 1988; HG Simon et al., 1988). Due to the high degree of sequence homology between MTSP-3 to MTSP-7, the similar size of their transcripts, as well as the relative insensitivity of Northern blot analysis, the expression pattern of the latter four serine proteases could not unambiguously be determined by this method.

Using the sensitive polymerase chain reaction (PCR), we were able to detect mRNA species from as little as 2 pg of total cRNA (Ebnet et al., 1991). We have found that acute lymphocytic choriomeningitis virus (LCMV)-immune T_C and acute anti-major histocampatibility complex (MHC) class II immune T_H taken *ex vivo*, as well as short-term *in vitro*-activated T cells, express mRNA species specific for MTSP-1 and MTSP-2 and weakly, if at all, also that for MTSP-3 (Table 3). Naive T lymphocytes express only marginal amounts of the respective mRNA species. In all cases, T_H populations expressed transcripts for MTSP-1, MTSP-2, and MTSP-3 much less than T_C populations. Whether this is due to a lower frequency of MTSP-1-producing T lymphocytes in the T_H population generated *in vivo*, as suggested by *in vitro* studies (Fruth et al., 1988b), or to an overall reduced synthesis of specific mRNA in all T_H is unclear. In none of the above T-lymphocyte populations could transcripts for MTSP-4 to MTSP-7 be detected (Table 3). However, similar analyses of a panel of *in vitro*-propagated T_C and T_H clones revealed the expression of all seven MTSP-specific mRNAs, though at greatly differing concentrations (Ebnet et al., 1991). These findings, which in essence corroborate a similar study applying the RNase protection assay (Garcia-Sanz et al., 1990), argue against participation of MTSP-3 to MTSP-7 in T-lymphocyte-mediated functions *in vivo*.

In situ hybridization could demonstrate that MTSP-1- and/or MTSP-2-specific transcripts are expressed in LCMV-specific CD8$^+$ T-lymphocytes (Müller et al., 1989), in T lymphocytes infiltrating allografts (Müller et al., 1988a, 1988b), and in pancreatic islets of Nonobese Diabetic (NOD) mice in the course of development of autoimmune diabetes (Held et al., 1990b). Using the same technique, MTSP-1 and MTSP-2 transcripts were also found to be expressed in CD4$^-$ CD8$^-$ thymocytes (Held et al., 1990a).

The availability of monoclonal antibodies with specificity for MTSP-1 allowed the identification of the enzyme in cytoplasmic granules of T_C during LCMV infection (Kramer et al., 1989; see Figure 6), most probably stored together with perforin (Peters et al., 1991), as well as in T_H infiltrating skin lesions of *Leishmania major*-infected BALB/c mice (Moll et al., 1991). It is interesting to note that MTSP-1 was also found in intracytoplasmic granules of natural killer (NK) cell-like "granulated metrial gland cells" of the murine placenta (Zheng et al., 1991), which are thought to be involved in immune surveillance at the interphase of materno-fetal tissue (Stewart and Mukhtar, 1988). However, a more detailed study is required to reveal whether MTSP-1 may also be expressed by non-T lymphocytes, including NK cells, under certain physiological or pathological conditions.

Similar studies in humans described the presence of HuTSP-1- and HuTSP-2-specific transcripts in infiltrating T lymphocytes of various dermatoses (Wood et al., 1988), and rejected renal (Clement et al., 1990) and cardiac (Hameed et al., 1991; Griffiths et al., 1991) allografts.

The presence of both MTSP-1 and MTSP-2 in mouse CD4$^-$ CD8$^-$ thymocytes as well as in antigen-sensitized T_C and T_H recruited under different physiological and pathological conditions in both man and mouse emphasizes their putative involvement in various immune reactions *in vivo*.

TABLE 3. Expression of T-cell-associated serine proteinases MTSP-1–MTSP-7 in T lymphocytes *in vivo* and *in vitro*

T-cell population	MTSP						
	1	2	3	4	5	6	7
T lymphocytes *in vivo*							
Naive splenic T lymphocytes	(+)	(+)	(+)	−	−	−	−
Naive lymph node T lymphocytes	(+)	(+)	(+)	−	−	−	−
Acute LCMV immune T_C	+++	+++	+	−	−	−	−
Acute anti-MHC class II immune T_H	++	+	+	−	−	−	−
T lymphocytes *in vitro*							
Con A-activated T lymphocytes	+++	+++	−	−	−	−	−
$CD4^+$ T_H clones	+++	+++	++/+	++	++	++/+	++/+
$CD8^+$ T_C clones	+++	+++	++/+	++/+	++/+	++	++/+
El4-F15 thymoma cells	−	(+)	−	−	−	−	−

Expression of MTSP-1–MTSP-7 was studied by using the polymerase chain reaction. Data have been taken from Ebnet et al., 1991.
Strength of expression: +++ very strong, ++ strong, + weak, (+) marginal, − nondetectable.

MTSP-1 and HuTSP-1 Express Specific and Limited Proteolytic Activities Which Are Only Operative in Extracellular Environments

MTSP-1 and HuTSP-1 display detailed enzymatic activities on a panel of model peptide substrates (Table 1 and 2; MM Simon et al., 1986; Masson et al., 1986b; Fruth et al., 1987b; Poe et al., 1988; Krähenbühl et al., 1988), indicating that both enzymes do not exert merely degradative but rather protein-processing functions (Neurath and Walsh, 1976; Neurath, 1989). The finding that MTSP-1 shows optimal activity on the chromogenic substrate HD-Pro-Phe-Arg-pNA at neutral to alkaline pH but is inactive at pH 5–6 (MM Simon et al., 1986), the pH range of lysosomal and endosomal compartments (Young et al., 1986b), suggests that the enzyme, which is stored in cytoplasmic granules, is operative only after exocytosis into the extracellular space. Several groups have shown that enzymatically active MTSP-1 is released from cloned T_C and *in vivo*-activated T_C populations as a result of stimulation by the appropriate antigen, by lectins, or by crosslinking of the T-cell receptor (TCR)/CD3 complex with anti-CD3 monoclonal antibodies (Takayama and Sitkovsky, 1987; Fruth et al., 1987a; Garcia-Sanz et al., 1987; MM Simon et al., 1987a, 1989; Garcia-Sanz et al., 1987). More recently it has been found that MTSP-2 has a pH optimum similar to MTSP-1, but a different substrate specificity, i.e., for Asp residues (Odake et al., 1991). MTSP-6 appears to cleave the substrate Suc-Phe-Leu-Phe-SBzl. Substrates for MTSP-3, -4, -5, and -7 remain to be identified (Odake et al., 1991).

In analogy to MTSP-1 and MTSP-2, HuTSP-1 and HuTSP-2 cleave model peptides with high efficiency also at neutral pH, but with different substrate specificities (Table 2; Fruth et al., 1987b; Krähenbühl et al., 1988; Poe et al., 1988, 1991; Odake et al., 1991). The differential fine specificities of MTSP-1/HuTSP-1 and MTSP-2/HuTSP-2, which are concomitantly expressed in activated T lymphocytes *in vivo*, should be helpful in the search for relevant target structures and suggest an independent contribution of the two enzymes at distinct sites in T-lymphocyte-mediated processes.

MTSP-1 May Enable Activated T Lymphocytes to Penetrate Vascular Basement Membranes and to Reach Appropriate Targets in Tissues

In order to migrate to certain targets and to exert their immune function(s) *in situ*, activated T lymphocytes have initially to bind to and interact with the vascular endothelium (Woodruff et al., 1987; Stoolman, 1989; Cavender, 1989, 1990). It has been shown that this process is guided by a variety of specific receptors and ligands expressed on both activated T lymphocytes (LFA-1, VLA-4) and activated endothelial cells (ICAM-1, VCAM, ELAM-1) (Hynes, 1987; Woodruff et al., 1987; Stoolman, 1989; Shimizu et al., 1990, 1991). After adhesion and egress from the lumen of the vessel, activated T lymphocytes have to traverse vascular basement

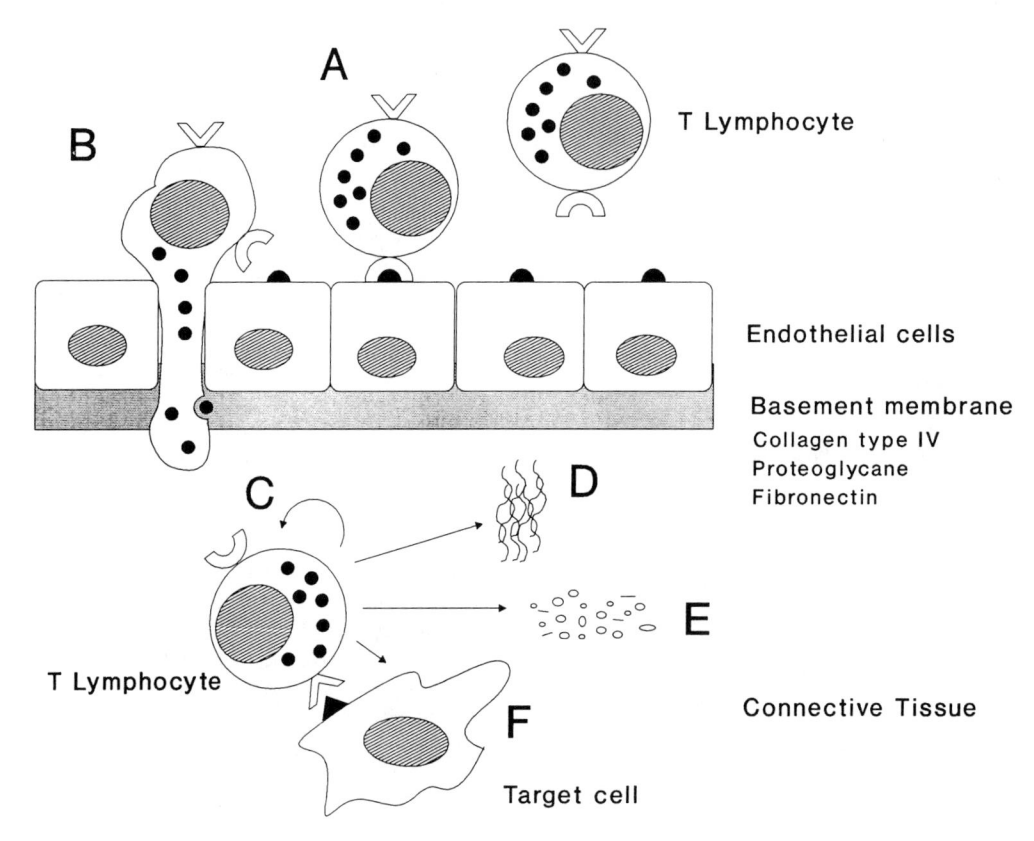

FIGURE 2. Role of MTSP-1 in extravasation and interaction with target cells. (A) Extravasation of T lymphocytes on their way to inflammatory foci requires adhesion to "activated" capillary endothelial cells. (B) Traversing of subendothelial basement membranes is thought to require focal degradation by proteolytic processes, probably including TSP. (C) Interaction of T lymphycytes with appropriate antigen-presenting target cells in inflamed tissues results in degranulation of cytoplasmic granules, release of TSP, and their interaction with various putative extracellular target structures: (C) cell surface structures of T_H and T_C, (D) structures of the ECM, (E) solute proteins of the interstitial fluid, and (F) structures expressed on the target cell. For details see text.

membranes. Although little is known about the actual pathways involved in the latter process, it is assumed that it involves specific cell adhesion to basal lamina structures and subsequent degradation of its constituents, most probably by proteolytic enzymes (Figure 2).

In fact, it has previously been shown that activated T lymphocytes are able to adhere to cell-binding regions of fibronectin, laminin, and vitronectin (Hynes, 1987; Hemler, 1988; Abraham et al., 1989) via specific receptors known to be members of the VLA/integrin family (Hynes, 1987). We have now found that cloned T_C, but not normal T lymphocytes, preferentially attach to the basement membrane collagen type IV, but not, or only marginally, to other collagen types (MM Simon et al., 1991). At present, the functional role of the corresponding fibronectin, laminin, and collagen receptors on T lymphocytes in the immune response is not clear. However, it has recently been shown that under appropriate culture conditions a mixture of anti-CD3 antibody and fibronectin can induce the activation of $CD4^+$ T lymphocytes, whereas neither alone could do it (Matsuyama et al., 1989). These data suggest that fibronectin and probably also other integrin receptor ligands may contribute to activation of T lymphocytes and possibly also to the release of their effector molecules, including granule-associated compounds such as MTSP-1 and MTSP-2 under physiological conditions.

In order to elucidate the direct effect of MTSP-1 on major structural constituents of connective tissue, we have analyzed its capacity to cleave struc-

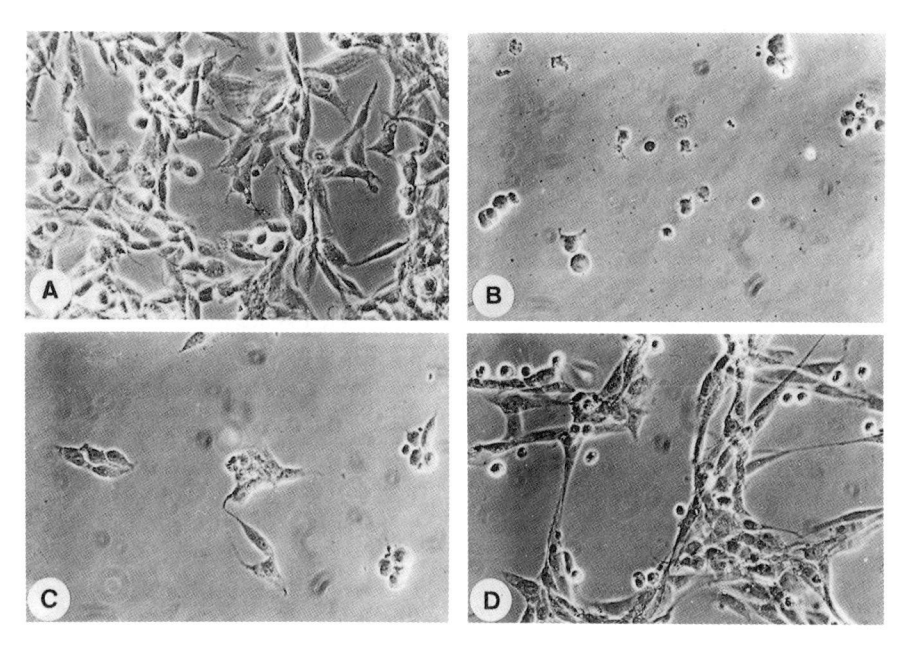

FIGURE 3. Destruction of cell-binding function of fibronectin by MTSP-1. Human plasma fibronectin was coated to microtiter plates (5 μg/ml). Afterwards the plates were treated with (A) buffer only, (B) MTSP-1 (24 μg/ml), (C) MTSP-1 (6 μg/ml), (D) MTSP-1 (24 μg/ml) previously incubated with the synthetic MTSP-1 inhibitor PFR-CK (10 mM). Afer incubation of 3T3 fibroblasts for 12 hr at 37°C, the plates were washed and cell adhesion was monitored. Adhesion of 3T3 fibroblasts was prevented by pretreatment of coated fibronectin with active MTSP-1.

tures such as fibronectin, the distinct types of collagen (types I–VI), and proteoglycans. We could show that human plasma-derived fibronectin, a large glycoprotein made up of various domains with distinct biological functions (Proctor, 1987), is a suitable substrate for MTSP-1 (MM Simon et al., 1988). Its proteolysis leads to various stable peptides with the preservation of two of the major functional properties of the molecule, i.e., its ability to bind to collagen and heparin (Proctor, 1987), located on distinct fragments. However, preincubation of human plasma fibronectin with MTSP-1 resulted in the loss of a site(s) relevant for fibroblasts–fibronectin interaction (see Figure 3; MM Simon et al., unpublished observations). These findings indicate that activated T lymphocytes may have profound effects on the overall architecture of basement membranes by regulating, via MTSP-1 and possibly also other hydrolytic enzymes, cell–matrix interactions. Moreover, the peptide fragments released from fibronectin may compete with intact fibronectin for binding sites on cells or other components of the extracellular matrix (ECM).

MTSP-1 was also shown to directly degrade one of the three bands of collagen type IV, i.e., the a_2 (IV) M_r 80,000 component, into small peptides (MM Simon et al., 1991). This result established that collagen type IV is sensitive not only to typical metalloproteinases (Garbisa et al., 1986), i.e., the classical collagenases, but also to serine proteinases (Mainardi et al., 1980) including MTSP-1.

Preliminary experiments from our laboratory indicated that MTSP-1 might be involved in the degration of ECM-associated proteoglycans (MM Simon et al., 1987b). Moreover, it was shown by others that activated, but not naive, T lymphocytes express an enzyme inhibitable by heparin and able to degrade heparan sulfate side chains of the proteoglycan scaffold of subendothelial ECM (Naparstek et al., 1984; Fridman et al., 1987). Subsequently, we have demonstrated that purified MTSP-1 is able to release only high-molecular-weight split products from sulfated proteoglycans of *in vitro*-derived ECM, but that molecules secreted by cloned T_C upon interaction with target cells were capable of releasing low-molecular-weight products from the same ECM (Vettel et al., 1991). The additional finding that conversion of the MTSP-1-derived high-molecular-weight fragments by supernatants of previously restimulated cloned T_C was inhibited by heparin, an inhibitor for heparanase-like enzymes, but not by H-D-Pro-Phe-Arg-chloromethyl ketone (PFR-CK), a specific inhibitor for MTSP-1 (MM Simon et al., 1989), suggested that activated T lymphocytes contain at least one endoglycosidase (heparanase), which is secreted together with MTSP-1 after restimulation of activated T lymphocytes. It is therefore most probable that the degradation of ECM-associated heparan sulfate by the two T-lymphocyte-associated enzymatic activities proceeds in a sequential manner, such that the product of one enzyme, MTSP-1, provides a more accessible substrate for the following (heparanase). Most interesting in this context is our finding that interaction of MTSP-1 with sulfated glycosaminoglycans, i.e., heparin and chondroitin sulfate, led to increased enzymatic activity as well as an altered fine specificity for peptide substrates (MM Simon et al., 1990). The specific structure of sulfated proteoglycans in individual tissues may therefore differentially regulate MTSP-1 activity, resulting in distinct proteolytic processes confined to distinct organ sites.

It has been shown before that removal of glycosaminoglycans from ECM results in the exposure of the collagen network and its higher susceptibility to collagenolytic enzymes (David and Bernfield, 1979). The fact that T lymphocytes are able to bind collagen type IV and are able to degrade it by virtue of MTSP-1 (MM Simon et al., 1991) supports the hypothesis that this enzyme is used by T_C and at least some T_H to traverse vascular basement membranes.

MTSP-1 may not only cleave structures of the ECM directly but may also accelerate its degradation indirectly via recruitment of additional proteolytic activities. This hypothesis is supported by our recent finding that MTSP-1 is able to activate pro-urokinase-type plasminogen activator (pro-uPA; Brunner et al., 1990), which can then activate plasminogen to plasmin. Plasmin in turn is known to activate matrix-associated metalloproteinases (MMPs) (Okada et al., 1988), including collagenases which can act on various collagens, including collagen type IV (Eeckhout and Vaes, 1977).

It thus seems that activated T lymphocytes express and secrete a variety of potent hydrolytic enzymes, including MTSP-1, which allow them to reach inflammatory foci by directly or indirectly degrading constituents of vascular basement membranes (see Figure 4).

MTSP-1 May Play a Crucial Role in the Control of Viral Infection by Cleaving Proteins Relevant for Virus Replication and Infectivity

One of the most important functions of $CD8^+$ T_C is to control viral infection (Blanden, 1974). It has been widely assumed that T lymphocytes control virus load by lysing infected target cells and thereby inhibiting viral replication (Zinkernagel and Doherty, 1979). Accordingly, it was shown that T_C secrete a variety of effector molecules contained in their cytoplasmic granules, such as perforin (Henkart, 1985; Podack, 1985; Young et al., 1986a) and TSP onto target cells. It has been clearly demonstrated that perforin on its own is hemolytic and cytolytic for various tumour cell lines *in vitro* at low concentrations (Young et al., 1986a; Lichtenheld et al., 1988).

Indirect evidence for the involvement of cellular proteases, including MTSP, in cell-mediated cytotoxicity derived from previous studies demonstrating that proteinase inhibitors can inhibit cytolytic activity of intact cytotoxic T-lymphocytes (Chang and Eisen, 1980; Redelman and Hudig, 1980). We have recently shown that the MTSP-1 inhibitor PFR-CK was able to inhibit, in a dose-dependent manner, the cytolytic potential of isolated cytoplasmic granules, but not of intact cloned T_C (Figure 5; MM Simon et al., 1987a, 1989). Together with the fact that MTSP-1 is not cytolytic on its own (MM Simon et al., 1986), these results are in line with the assumption that MTSP-1 participates in target cell lysis by acting on relevant structures of either the T_C themselves or the target cells. This is also supported by the finding that incubation of target cells with both perforin and MTSP-1, but not with either alone, leads to degradation of their DNA (Hayes et al., 1989).

It was therefore relevant that genes encoding MTSP-1 and perforin were expressed *in vivo* in LCMV-specific T-lymphocyte infiltrates of liver and brain tissues (Müller et al., 1989), and that granule-associated MTSP-1 was associated with the majority of LCMV-specific T_C present in splenic tissue at the peak of virus infection (Figure 6; Kramer et al., 1989). These findings suggest the participation of both MTSP-1 and perforin in T_C-mediated control of virus replication *in vivo* by either contributing to lysis of infected target cells or by inactivating intracellular viruses via DNA and/

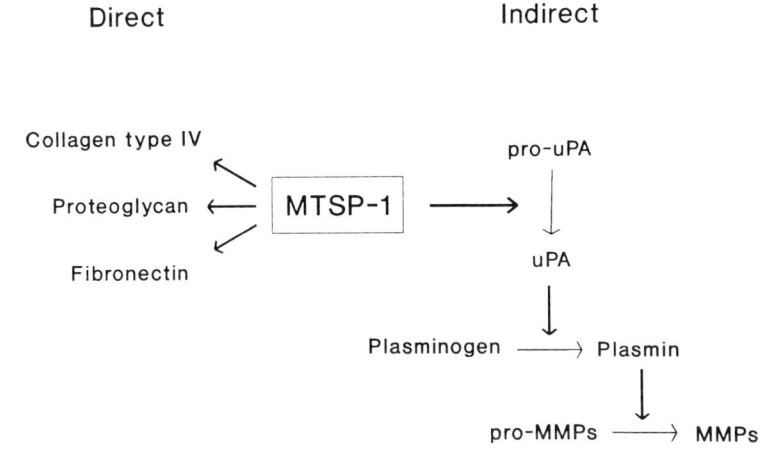

FIGURE 4. Direct and indirect cleavage of basement membrane constituents by MTSP-1. MTSP-1 directly cleaves various structural proteins of basement membranes: collagen type IV (MM Simon et al., 1991), sulfated proteoglycans (MM Simon et al., 1987b; Vettel et al., 1991) and fibronectin (MM Simon et al., 1988). MTSP-1 is able to recruit the broad-specific proteolytic potential of plasmin via activation of pro-uPA (Brunner et al., 1990). Plasmin can activate a variety of so-called matrix-associated metalloproteinases (MMPs) (Okada et al., 1988).

or RNA fragmentation, as proposed recently (Müllbacher and Ada, 1987; Martz and Howell, 1989).

Previous findings that T_C can inhibit virus replication by soluble mediators such as interferon-γ (IFN-γ) (Stewart, 1981) and other less well-defined T-lymphocyte-derived molecules (Walker et al., 1986) have indicated that T_C can limit viral replication by still other noncytolytic mechanism(s). Support for this assumption comes from our finding that viral proteins are suitable substrates for MTSP-1. We have found that the enzyme reverse transcriptase (RT) from the retrovirus Moloney murine leukemia virus (MoMuLV) is inactivated by MTSP-1 (HG Simon et al., 1987), and that pretreatment of intact MoMuLV with the enzyme leads to cleavage of the viral envelope protein gp 70 and concomitant loss of viral infectivity on 3T3 mouse fibroblast cells *in vitro* (B. Maier et al., unpublished). Although the physiological significance of these findings is not known at present, one could speculate that MTSP-1 has a direct impact on the infectivity of mature virus particles by cleaving relevant envelope structures (see Figure 7). It is also possible that MTSP-1 may be bound to and endocytosed by relevant target cells either directly or complexed with intact viruses, thereby interfering with virus replication. The latter hypothesis is at least in part supported by the finding that an exogenous serine proteinase, trypsin, is able to induce cleavage and normal intracellular maturation of viral precursor proteins in a temperature-sensitive mutant of MoMuLV incapable of replicating on its own in 3T3 cells (Traktman and Baltimore, 1982).

T lymphocytes may also combat viral infections by inducing, via MTSP-1, immune functions in bystander lymphocytes. We have recently shown that MTSP-1, like other serine proteinases such as

FIGURE 5. Effect of the inhibitor H-D-Pro-Phe-Arg-chloromethyl ketone (PFR-CK) on the cytolytic and proteolytic activity associated with either intact T_C or with isolated cytoplasmic granules thereof. PFR-CK was synthesized according to the substrate specificity of MTSP-1. PFR-CK inhibited proteolytic and cytolytic activity of cytoplasmic granules but not of intact CTL. SRBC, Sheep red blood cells. (From MM Simon et al., 1989.)

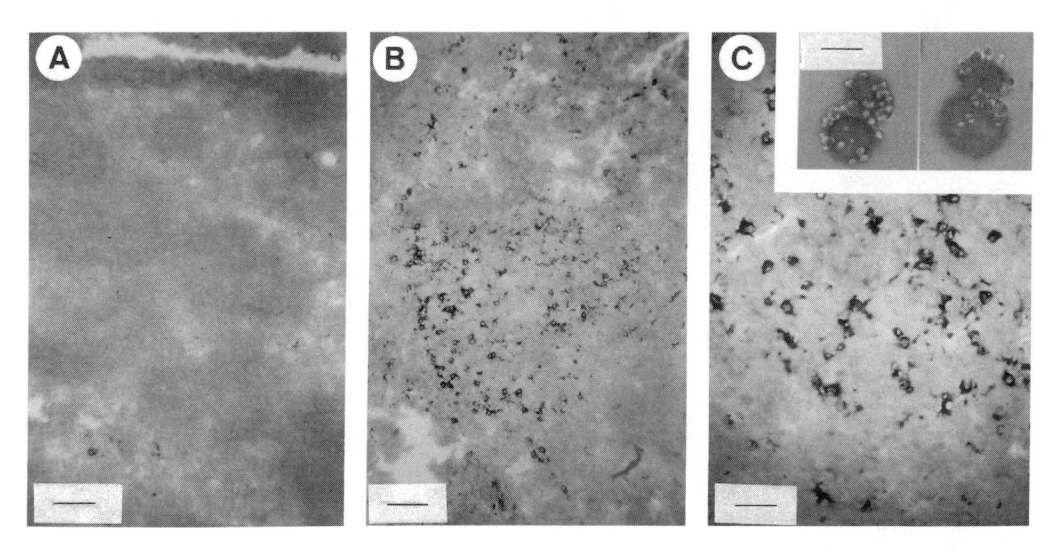

FIGURE 6. Detection of MTSP-1 expressing T lymphocytes *in situ* during an acute viral infection. Frozen sections of spleens of (A) uninfected and (B, C) LCMV-infected mice (7 days postinfection at the peak of antiviral T lymphocyte response) were stained using the MTSP-1-specific monoclonal antibody HD-7.1 (Kramer et al., 1989). Bound antibody was visualized using the Biotin-Streptavidin-peroxidase technique. Note the staining of numerous splenic T lymphocytes in spleens of virus-infected mice, (C) Higher-power micrograph of (B): staining is apparently intracellular. Inset: Staining of enzyme activity in intracellular granules of CD8+ lymphocytes isolated from spleens of LCMV-infected mice by fluorescence-activated cell sorting. Enzyme cytochemistry was performed using the esterase substrate N-α-benzyloxycarbonyl-L-lysine thiobenzyl ester (BLT). Similar granules could be stained using the anti-MTSP-1 antibody HD-6.2 (data not shown; see Kramer et al., 1989). Scale bars: (A,B) 65 μm, (C) 20 μm, (C, inset) 5 μm.

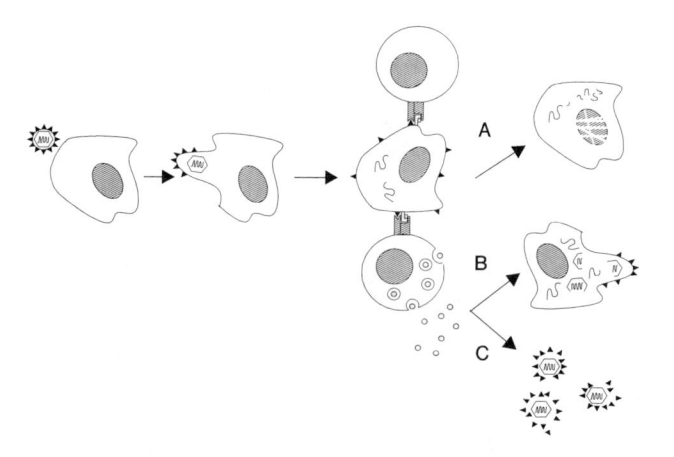

FIGURE 7. Possible T_C mediated control mechanisms of virus infections. (A) T_C may interfere with viral infections by inducing intracellular processes, including DNA fragmentation (Müllbacher and Ada, 1987; Martz and Howell, 1989), or by lysing infected target cells (Zinkernagel and Doherty, 1979). (B) T_C may induce—by virtue of TSP—cleavage of viral proteins relevant for viral replication (HG Simon, 1987). (C) T_C-derived TSP may interfere with viral infectivity by cleaving relevant virus envelope proteins (B. Maier et al., unpublished).

TABLE 4. Physiological inhibitors of MTSP-1

Inhibitor	Mode of action	Residual enzyme activity
Aprotinin	Reversible	No
Antithrombin III	Irreversible	No
Protease nexin-1	Irreversible	No
α_2-Macroglobulin	Irrefersible	Yes, on proteins < 20 kDa

Summarized from Gurwitz et al., 1989; MM Simon et al., 1990.

trypsin and thrombin (Vischer, 1974; Kaplan and Bona, 1974), can induce proliferation of B lymphocytes in the absence of antigen (MM Simon et al., 1986). The findings that various serine proteinases can stimulate B lymphocytes *in vitro* have led to the proposition that induction of immune effector cells by proteolytic enzymes, including MTSP-1, in inflammatory foci may further aid in defending the host against infection (Scher, 1987).

MTSP-1 Activity Is Controlled by Vascular and Extravascular Proteinase Inhibitors

Extracellular proteases play a key role in the fine regulation of inflammatory processes. However, when released into the microenvironment in an uncontrolled fashion, their enzymatic activity may have devastating effects on the integrity of extracellular protein constituents (Neurath and Walsh, 1976; Neurath, 1989). It is for this reason that extracellular proteinases have to be tightly regulated *in vivo* by several means (Neurath and Walsh, 1976; Neurath, 1989). First, proteinases may be synthesized as inactive precursors which require limited proteolysis by other proteases. Second, mature proteinases may be stored in cytoplasmic granules under conditions (e.g., acidic pH) that do not allow the expression of their enzymatic activity *in situ*. However, they become active after appropriate stimulation of effector cells and their secretion into the extracellular milieu. This type of control has been demonstrated for serine proteinases of polymorphonuclear leukocytes or mast cells (Serafin et al., 1990) and now also for proteinases of activated T lymphocytes (Takayama and Sitkovsky, 1987; Fruth et al., 1987a; MM Simon et al., 1989).

The third way to control serine proteinases is provided by proteinase inhibitors. These are abundant in vascular and extravascular fluids (Laskowski and Kato, 1980; Travis and Salvesen, 1983), and some of them have now been shown to inactivate MTSP-1 *in vitro* (Table 4). We demonstrated that interaction of MTSP-1 with the broad-specific protease inhibitor α_2-macroglubulin leads to formation of a complex between the two components, resulting in the inhibition of MTSP-1 activity on large proteins such as fibronectin, but not on small peptides (MM Simon et al., 1990). This finding is notable in light of a study demonstrating the inactivation of IL-2 by α_2-macroglobulin–protease complexes (Borth and Teodorescu, 1986), and it indicates that MTSP-1 could still play a role in T-lymphocyte-induced processes *in situ*, even after its complexation with α_2-macroglobulin.

It has also been shown that the serine proteinase inhibitors antithrombin III and C1-esterase inhibitor, which are present in the vascular system (Travis and Salvesen, 1983), as well as protease nexin-1 (PN-1), an inhibitor associated with extracellular matrices (Baker et al., 1980), irreversibly inhibit MTSP-1 by forming covalent complexes with both catalytic site residues of the homodimeric enzyme (Masson and Tschopp, 1988; Gurwitz et al., 1989; MM Simon et al., 1990). The fact that the respective complexes between MTSP-1 and the individual serpins generated *in vitro* were formed at physiologically relevant association rates (Masson and Tschopp, 1988; Gurwitz et al., 1989; MM Simon et al., 1990) argues for a tight control of MTSP-1 activity in the extracellular milieu, thus limiting the induction of unrelated protease-driven regulatory pathways and damage to surrounding tissue.

Conclusion

The most intriguing finding derived from these and similar studies on proteolytic activities in immune cells of mouse and man is the fact that activated but not naive T lymphocytes express genes encoding a family of highly homologous and tissue-specific serine proteinases (TSP) with significant differences in their substrate specificities. The fact that to date gene expression and enzymatic activity *in vivo* have been demonstrated only for some of the TSP, mainly for MTSP-1/HuTSP-1 and MTSP-2/HuTSP-2, but not for others, does not necessarily mean that the remaining TSP are not induced under different physiological and pathophysiological conditions. Furthermore, the data do not exclude the possibility that at least some of the TSP may also be expressed by cells other than T lymphocytes.

The biochemical properties of TSP, as well as their storage in an enzymatically inert form within secretory cytoplasmic granules, suggest that the serine proteinases are only operative after their release into the extracellular space. In fact, we have shown that MTSP-1 is able to cleave various components of the ECM as well as viral proteins responsible for virus replication and infectivity. These data lead us to propose that antigen-sensitized T lymphocytes use TSP-1, and possibly also TSP-2, probably together with additional T-lymphocyte-derived molecules to traverse vascular endothelial basement membranes, to migrate to target cells, and to exert their immune function(s).

At present it is not known why the murine and human genomes contain so many highly related genes encoding TSP. One point that comes to mind is the variety of polymorphic structures used by the host to overcome strategies developed by infectious agents to evade immune elimination. Accordingly, evolutionary pressure from the pathogen would enforce the generation of T-cell-associated enzymes with different specificities for optimal protection. The fact that MTSP-1/HuTSP-1 and MTSP-2/HuTSP-2 indeed have quite different substrate specificities gives this hypothesis credence.

Acknowledgments

The authors thank all their colleagues who contributed to the development of research on TSP-1. We are especially grateful to numerous individuals in our laboratories for their valuable contributions. These people include Gabi Nerz, Marlot Prester, Uta Schirmer, Thomas Stehle, and Thao Tran. We especially thank Arno Müllbacher for many discussions and the exchange of ideas that color our views and Gordon Ada, Robert Blanden, and Dietmar Vestweber for their critical comments on the manuscript. The authors are especially indebted to J. Reinartz for preparing the line drawings. Our work is supported by the Deutsche Forschungsgemeinschaft (Si 214/5-7; Kr 931/2-2).

Abbreviations

ECM	extracellular matrix
HF	Hanukah factor
LCMV	lymphocytic choriomeningitis virus
MMCP	mouse mast cell protease
MMPS	matrix-associated metalloproteinases
MoMuLV	moloney murine leukemia virus
NK	natural killer cells(s)
PCR	polymerase chain reaction
RMCP	rat mast cell protease
TSP	T lymphocyte-associated serine proteinase
RT	reverse transcriptase
SE	serine esterase
T_C	$CD8^+$ cytolytic T lymphocyte(s)
T_H	$CD4^+$ helper T lymphocyte(s)

Note

Since this article was written, the organizations of the genes coding for MTSP-4 and MTSP-5 have been published (Prendergast et al., 1991). The principal organization of these genes corresponds to that described for the other TSP (see Figure 1).

References

Abraham D, Ince T, Muir H, Olsen I (1989): Fibroblast matrix and surface components that mediate cell-to-cell interaction with lymphocytes. *J Invest Dermatol* 93: 335–340

Baker JB, Low DA, Simmer RL, Cunningham DD (1980): Protease-nexin: A cellular component that links thrombin and plasminogen activator and mediates their binding to cells. *Cell* 21: 37–45

Barker JNWN, Mitra RS, Griffiths CEM, Dixit VM, Nickoloff BU (1991): Keratinocytes as initiators of inflammation. *Lancet* 337: 211–214

Benfey PN, Yin FH, Leder P (1987): Cloning of the mast cell protease, RMCP II. Evidence for cell-specific expression and a multi-gene family. *J Biol Chem* 262: 5377–5384

Blanden RV (1974): T cell response to viral and bacterial infection. *Transplant Rev* 19: 56–88

Bleackley C, Duggan B, Ehrmann N, Lobe CG (1988): Isolation of two cDNA sequences which encode cytotoxic cell proteases. *FEBS Lett* 234: 153–159

Borth W, Teodorescu M (1986): Inactivation of human interleukin-2 (IL-2) by α2-macroglubulin-trypsin complexes. *Immunology* 57: 367–371

Brunet JF, Dosseto M, Denizot F, Mattei MG, Clark WR, Haqqi TM, Ferrier P, Nabholz M, Schmitt-Verhulst A, Luciano MF Golstein F (1986): The inducible cytotoxic T-lymphocyte-associated gene transcript CTLA-1 sequence and gene localization to mouse chromosome 14. *Nature* 322: 268–271

Brunner G, Simon MM, Kramer MD (1990): Activation of pro-urokinase by the human T cell-associated serine proteinase HuTSP-1. *FEBS Lett* 260: 141–144

Caputo A, Fahey D, Lloyd C, Vozab R, McCairns E, Rowe PB (1988): Structure and differential mechanisms of regulation of expression of a serine esterase gene in activated human T lymphocytes. *J Biol Chem* 263: 6363–6369

ulation of hamster lymphoid cells. *Exp* 53-263

bi GA, Henkart P (1989): Induction of A release by the cytotoxic T lymphocyte se granzyme A. *J Exp Med* 170: 933-946

nald HR, Müller C (1990a): Expression ling cytotoxic cell-associated serine pro-ocytes. *Int Immunol* 2: 57-62

onald HR, Weissman IL, Hess MW, 90b): Genes encoding tumor necrosis ranzyme A are expressed during develop-mmune diabetes. *Proc Natl Acad Sci USA*

88): Adhesive protein receptors on hema-. *Immunol Today* 9: 109-113

985): Mechanism of lymphocyte-mediated *Annu Rev Immunol* 3: 31-58

escu NC, Hanson RD, Salvesen G, Ley TJ mic organization and chromosomal locali-human cathepsin G gene. *J Biol Chem* 264:

n T, Hellman L (1991): Cloning and struc-is of MMCP-1, MMCP-4 and MMCP-5, mast cell-specific serine proteases. *Eur J* : 1611-1621

Veiss SJ (1989): Disruption of the suben-basement membrane during neutrophil n an in vitro construct of a vessel wall. *J Clin* 122-1136

1987): Integrins: A family of cell surface *Cell* 48: 549-554

schopp J (1988): Granzymes, a family of eases released from granules of cytolytic T es upon T cell receptor stimulation. *Immunol* 3-71

hter P, Zimmer M, Garcia-Sanz JA, Tschopp Genomic organization and subchromosomal atization of the murine granzyme F, a serine expressed in CD8+ T cells. *J Immunol* 147:

asson D, Zimmer M, Haefliger JA, Li WH, J (1989): Isolation and complete structure of hocyte serine protease granzyme G, a novel of the granzyme multigene family in mrine T lymphocytes. Evolutionary origin of the yte proteases. *Biochemistry* 28: 7953-7961

Rey C, Haefliger JA, Qiao BY, Groscurth P, J (1988a): Identification and sequencing of lones encoding the granule-associated serine s granzymes D, E, and F of cytolytic T lym-s. *Proc Natl Acad Sci USA* 85: 4841-4848

Rey C, Masson D, Stanley KK, Herz J, Plae-5, Tschopp J (1988b): cDNA cloning of ne C, a granule-associated serine protease of c T lymphocytes. *J Immunol* 140: 318-323

G, Bona C (1974): Proteases as mitogens. The of trypsin and pronase on mouse and human cytes. *Exp Cell Res* 88: 389-394

, Selvakumar A, Trapan JA, Dupont B (1990): terization of a novel, human cytotoxic lym-phocyte-specific serine protease cDNA clone (CSP-C). *Tiss Antigens* 35: 220-228

Klein JL, Shows TB, Dupont B, Trapani JA (1989): Genomic organization and chromosomal assignment for a serine protease gene expressed by human cytotoxic lymphocytes. *Genomics* 5: 110-117

Krähenbühl O, Rey, Jenne D, Lanzavecchia A, Groscurth P, Carrel C, Tschopp J (1988): Characterization of granzymes A and B isolated from granules of cloned human cytotoxic T lymphocytes. *J Immunol* 141: 3471-3477

Kramer MD, Simon MM (1987): Are proteinases functional molecules of T lymphocytes? *Immunol Today* 8: 140-142

Kramer M, Binninger L, Schirrmacher V, Moll H, Prester M, Nerz G, Simon MM (1986): Characterization and isolation of a trypsin-like serine protease from a long-term culture cytolytic T cell line and its expression by functionally distinct T cells. *J Immunol* 136: 4644-4651

Kramer MD, Fruth U, Simon H-G, Simon MM (1989): Expression of cytoplasmic granules with T cell-associated serine proteinase-1 activity in Ly-2+(CD8+) T lymphocytes responding to lymphocytic choriomeningitis virus in vivo. *Eur J Immunol* 19: 151-156

Kraut J (1977): Serine proteases: Structure and mechanism of catalysis. *Annu Rev Biochem* 46: 331-358

Kupper TS (1990): Immune and inflammatory processes in cutaneous tissues. Mechanisms and speculations. *J Clin Invest* 86: 1783-1789

Kwon BS, Kestler D, Lee E, Wakulchik M, Young JDE (1988): Isolation and sequence analysis of serine protease cDNAs from mouse cytolytic T lymphocytes. *J Exp Med* 168: 1839-1854

Laskowski MJ, Kato I (1980): Protein inhibitors of proteinases. *Annu Rev Biochem* 49: 593-626

Lichtenheld MG, Olsen KJ, Lu P, Lowrey DM, Hameed A, Hengartner H, Podack ER (1988): Structure and function of human perforin. *Nature* 335: 448-451

Lobe CG, Finaly BB, Paranchych W, Paetkau VH, Bleackley RC (1986a): Novel serine proteases encoded by two cytotoxic T lymphocyte-specific genes. *Science* 232: 858-861

Lobe CG, Havele C, Bleackley RC (1986b): Cloning of two genes that are specifically expressed in activated cytotoxic T lymphocytes. *Proc Natl Acad Sci USA* 83: 1448-1452

Lobe CG, Upton C, Duggan B, Ehrman N, Letellier M, Bell J, Bleackley RC (1988): Organization of two genes encoding cytotoxic T lymphocyte-specific serine proteases CCPI and CCPII. *Biochemistry* 27: 6941-6946

Mainardi CL, Dixit SN, Kang AH (1980): Degradation of type IV (basement membrane) collagen by proteinase isolated from human polymorphonuclear leukocyte granules. *J Biol Chem* 255: 5433-5440

Martz E, Howell DM (1989): CTL: Virus control cell first and cytolytic cells second? *Immunol Today* 10: 79-86

Masson D, Tschopp J (1987): A family of serine esterases in lytic granules of cytolytic T lymphocytes. *Cell* 49: 679-685

Masson D, Tschopp J (1988): Inhibition of lymphocyte

Cavender DE (1989): Lymphocyte adhesion to endothelial cells in vitro: Models for the study of normal lymphocyte recirculation and lymphocyte emigration into chronic inflammatory lesions. *J Invest Dermatol* 93 (Suppl): 88–95

Cavender DE (1990): Organ-specific and non-organ-specific lymphocyte receptors for vascular endothelium. *J Invest Dermatol* 94: 41S–48S

Chang TW, Eisen HN (1980): Effects of N-tosyl-L-chloromethylketone on the activity of cytotoxic T lymphocytes. *J Immunol* 124: 1028–1033

Clement MV, Haddad P, Ring GH, Pruna A, Sasportes M (1990): Granzyme B-gene expression: A marker of human lymphocytes "activated" in vitro or in renal allografts. *Hum Immunol* 28: 159–166

David G, Bernfield MR (1979): Collagen reduced glycosaminoglycan degradation by cultured mammary epithelial cells: Possible mechanism for basal lamina formation. *Proc Natl Acad Sci USA* 76: 786–790

Ebnet K, Chluba-deTapia J, Hurtenbach U, Kramer MD, Simon MM (1991): In vivo primed mouse T cells selectively express T cell-specific serine proteinase-1 and the proteinase-like molecules granzyme B and C. *Int Immunol* 3: 9–19

Ebnet K, Kramer MD, Simon MM (1992): Organization of the gene encoding the mouse T-cell-specific serine proteinase "granzyme A". *Genomics* 13: 502–508

Eeckhout Y, Vaes G (1977): Further studies on the activation of procollagenase, the latent precursor of bone collagenase. Effects of lysosomal cathepsin B, plasmin and kallikrein, and spontaneous activation. *Biochem J* 166: 21–31

Fridman R, Lider O, Naparstek Y, Fuks Z, Vlodavsky I, Cohen IR (1987): Soluble antigen induces T lymphocytes to secreta an endoglycosidase that degrades the heparan sulfate moiety of subendothelial extracellular matrix. *J Cell Physiol* 130: 85–92

Fruth U, Eckerskorn C, Loftspeich F, Kramer MD, Prester M, Simon MM (1988a): Human T cell specific proteinase (HuTSP) is encoded by the T cell and natural killer cell specific human hanukah factor (HuHF) gene. *FEBS Lett* 237: 45–48

Fruth U, Nerz G, Prester M, Simon HG, Kramer MD, Simon MM (1988b): Determination of fequency of T cells expressing the T cell-specific serine proteinase-1 (TSP-1) reveals two types of L3T4+ T lymphocytes. *Eur J Immunol* 18: 773–781

Fruth U, Prester M, Golecki J, Hengartner H, Simon HG, Kramer MD, Simon MM (1987a): The T cell-specific serine proteinase TSP-1 is associated with cytoplasmic granules of cytolytic T lymphocytes. *Eur J Immunol* 17: 613–622

Fruth U, Sinigaglia F, Schlesier M, Kilgus J, Kramer MD, Simon MM (1987b): A novel serine proteinase (HuTSP) isolated from a cloned human CD8+ cytolytic T cell line is expressed and secreted by activated CD4+ and CD8+ lymphocytes. *Eur J Immunol* 17: 1625–1633

Garbisa S, Ballin M, Daga Gordini D, Fastelli G, Naturale M, Negro A, Semenzato G, Liotta LA (1986):

protease granzyme A by antithrombin III. *Mol Immunol* 25: 1283–1289

Masson D, Nabholz M, Estrade C, Tschopp J (1986a): Granules of cytolytic T-lymphocytes contain two serine esterases. *EMBO J* 5: 1595–1600

Masson D, Zamai M, Tschopp J (1986b): Identification of granzyme A isolated from cytotoxic T lymphocyte granules as one of the proteases encoded by CTL-specific genes. *FEBS Lett* 208: 84–88

Matsuyama T, Yamada A, Kay J, Yamada KM, Akiyama SK, Schlossman SF, Morimoto C (1989): Activation of CD4 cells by fibronectin and anti-CD3 antibody. A synergistic effect mediated by the VLA-5 fibronectin receptor complex. *J Exp Med* 170: 1133–1148

Meier M, Kwong PC, Frégeau CJ, Atkinson EA, Burrington M, Ehrman N, Sorenson O, Lin CC, Wilkins J, Bleackley RC (1990): Cloning of a gene that encodes a new member of the human cytotoxic T cell protease family. *Biochemistry* 29: 4042–4049

Moll H, Müller C, Gillitzer R, Fuchs H, Röllinghoff M, Simon MM, Kramer MD (1991): Expression of T cell-associated serine proteinase-1 during murine Leishmania major infection correlates with susceptibility to disease. *Infect Immun* 59: 4701–4705

Müllbacher A, Ada GL (1987): How do cytotoxic T lymphocytes work in vivo? *Microb Pathog* 3: 315–318

Müller C, Gershenfeld HK, Lobe CG, Okada CY, Bleackley RC, Weissman IL (1988a): A high proportion of T lymphocytes that infiltrate H-2-incompatible heart allografts in vivo express genes encoding cytotoxic cell-specific serine proteases, but do not express the MEL-14-defined lymph node homing receptor. *J Exp Med* 167: 1124–1136

Müller C, Gershenfeld HK, Lobe CG, Okada CY, Bleackley RC, Weissman IL (1988b): Expression of two serine esterase genes during an allograft rejection in the mouse. *Transplant Proc* 20: 251–253

Müller C, Kägi D, Aebischer T, Odermatt B, Held W, Podack ER, Zinkernagel RM, Hengartner H (1989): Detection of perforin and granzyme A mRNA in infiltrating cells during infection of mice with lymphocytic choriomeningitis virus. *Eur J Immunol* 19: 1253–1259

Naparstek Y, Cohen IR, Fuks Z, Vlodavsky I (1984): Activated T lymphocytes produce a matrix-degrading heparan sulfate endoglycosidase. *Nature* 310: 241–244

Neurath H (1989): Proteolytic processing and physiological regulation. *Trends Biochem Sci* 14: 268–271

Neurath H, Walsh KA (1976): Role of proteolytic enzymes in biological regulation (A review). *Proc Natl Acad Sci USA* 73: 3825–3832

Odake S, Kam CM, Narasimhan L, Poe M, Blake JT, Kraehenbuehl O, Tschopp J, Powers JC (1991): Human and murine cytotoxic T lymphocyte serine proteases: Subsite mapping with peptide thioester substrates and inhibition of enzyme activity and cytolysis by inhibition of enzyme activity and cytolysis by isocoumarins. *Biochemistry* 30: 2217–2227

Okada Y, Harris ED, Nagase H (1988): The precursor of metallendopeptidase from human rheumatoid synovial fibroblasts. Purification and mechanism of activation by endopeptidases and 4-aminophenylmercuric acetate. *Biochem J* 254: 731–741

Pasternack MS, Eisen HN (1985): A novel serine esterase expressed by cytotoxic T lymphocytes. *Nature* 314: 743–745

Pasternack MS, Verret CR, Liu MA, Eisen HN (1986): Serine esterase in cytolytic T lymphocytes. *Nature* 322: 740–743

Peters PJ, Borst J, Oorschot V, Fukuda M, Krähenbühl O, Tschopp J, Geuze HJ (1991): Cytotoxic T lymphocyte granules are secretory lysosomes, containing both perforin and granzymes. *J Exp Med* 173: 1099–1109

Podack ER (1985): The molecular mechanism of lymphocyte-mediated tumor cell lysis. *Immunol Today* 6: 21–27

Poe M, Bennett CD, Biddison WE, Blake JT, Norton GP, Rodkey JA, Sigal NH, Turner RV, Wu JK, Zweerink HJ (1988): Human cytotoxic lymphocyte tryptase. Its purification from granules and characterization of inhibitor and substrate specificity. *J Biol Chem* 263: 13215–13222

Poe M, Blake JT, Boulton DA, Gammon M, Sigal, NH, Wu JK, Zweerink HJ (1991): Human cytotoxic lymphocyte granzyme B. Its purification from granules and the characterization of substrate and inhibitor specificity. *J Biol Chem* 266: 98–103

Prendergast JA, Pinkoski M, Wolfenden A, Bleackley RC (1991): Structure and evolution of the cytotoxic cell proteinase genes CCP3, CCP4, and CCP5. *J Mol Biol* 220: 867–875

Proctor RA (1987): Fibronectin: A brief overview of its structure, function, and physiology. *Rev Infect Dis.* 9(Suppl 4): S317–S321

Redelman D, Hudig D (1980): The mechanism of cell-mediated cytotoxicity. I. Killing by murine cytotoxic T lymphocytes requires cell surface thiols and activated proteases. *J Immunol* 124: 870–878

Rogers J (1985): Exon shuffling and intron insertion in serine protease genes. *Nature* 315: 458–459

Scher W (1987): Biology of disease. The role of extracellular proteases in cell proliferation and differentiation. *Lab Invest* 57: 607–633

Schmid J, Weissman C (1987): Induction of mRNA for a serine protease and a β-thromboglobulin-like protein in mitogen-stimulated human leukocytes. *J Immunol* 139: 250–256

Serafin WE, Reynolds DS, Rogelj S, Lane WS, Conder GA, Johnson SS, Austen KF, Stevens RL (1990): Identification and molecular cloning of a novel mouse mucosal mast cell serine protease. *J Biol Chem* 265: 423–429

Shimizu Y, Newman W, Gopal TV, Horgan KJ, Graber N, Beall LD, Van Seventer GA, Shaw S (1991): Four molecular pathways of T cell adhesion to endothelial cells: Roles of LFA-1, VCAM-1 and ELAM-1 and changes in pathway hierarchy under different activation conditions. *J Cell Biol* 113: 1203–1212

Shimizu Y, Van Seventer GA, Horgan KJ, Shaw S (1990):

Roles of adhesion molecules in T cell recognition: Fundamental similarities between four integrins on resting human T cells (LFA-1, VLA-4, VLA-5, VLA-6) in expression, binding and costimulation. *Immunol Rev* 114: 109–143

Shimizu Y, Van Seventer GA, Siraganian R, Wahl L, Shaw S (1989): Dual role of the CD44 molecule in T cell adhesion and activation. *J Immunol* 143: 2457–2463

Simon HG (1988): Molekulargenetische Untersuchung zur T-Zell assoziierten Serinprotease 1 (TSP-1) aus der Maus. Doctoral thesis, Freiburg, Germany

Simon HG, Fruth U, Eckerskorn C, Lottspeich F, Kramer MD, Nerz G, Simon MM (1988): Induction of T cell serine proteinase 1 (TSP-1)-specific mRNA in mouse T lymphocytes. *Eur J Immunol* 18: 855–861

Simon HG, Fruth U, Kramer MD, Simon MM (1987): A secretable serine proteinase with highly restricted specificity from cytolytic T lymphocytes inactivates retrovirus-associated reverse transcriptase. *FEBS Lett* 223: 352–358

Simon MM, Fruth U, Simon HG, Kramer MD (1987a): Evidence for the involvement of a T cell-associated serine proteinase-1 (TSP-1) in cell killing. *Ann Pasteur Inst* 17: 309–314

Simon MM, Hoschützky H, Fruth U, Simon HG, Kramer MD (1986): Purification and characterization of a T cell specific serine proteinase (TSP-1) from cloned cytolytic T lymphocytes. *EMBO J* 5: 3267–3274

Simon MM, Kramer MD, Gay S (1991): CD8+ activated T cells secrete MTSP-1, a collagen-degrading enzyme: Potential pathway for the migration of lymphocytes through vascular basement membranes. *Immunology* 73: 117–119

Simon MM, Prester M, Kramer MD, Fruth U (1989): An inhibitor specific for the mouse T-cell associated serine proteinase 1 (TSP-1) inhibits the cytolytic potential of cytoplasmic granules but not of intact cytolytic T cells. *J Cell Biochem* 40: 1–13

Simon MM, Prester M, Nerz G, Kramer MD, Fruth U (1988): Release of biologically active fragments from human plasma-fibronectin by murine T cell-specific proteinase-1 (TSP-1). *Biol Chem Hoppe-Seyler* 369(Suppl): 107–112

Simon MM, Simon H-G, Fruth U, Epplen J, Müller-Hermelink HK, Kramer MD (1987b): Cloned cytolytic T-effector cells and their malignant variants produce an extracellular matrix degrading trypsin-like serine proteinase. *Immunology* 60: 219–230

Simon MM, Tran T, Fruth U, Gurwitz D, Kramer MD (1990): Regulation of mouse T cell-associated serine proteinase-1 (MTSP-1) by proteinase inhibitors and sulfated proteoglycans. *Biol Chem Hoppe-Seyler* 371(Suppl): 81–87

Smith KA (1984): Interleukin 2. *Annu Rev Immunol* 2: 319–333

Springer TA (1990): Adhesion receptors of the immune system. *Nature* 346: 425–434

Stewart I, Mukhtar DDY (1988): The killing of mouse trophoblast cells by granulated metrial glands cells in vitro. *Placenta* 9: 417–425

Stewart WE (1981): *The Interferon System*. Heidelberg: Springer-Verlag

Stoolman LM (1989): Adhesive molecules controlling lymphocyte migration. *Cell* 56: 907–910

Takahashi H, Nukiwa T, Yoshimura K, Quick CD, States DJ, Holmes MD, Whang Peng J, Knutsen T, Crystal RG (1988): Structure of the human neutrophil elastase gene. *J Biol Chem* 263: 14739–14747

Takayama H, Sitkovsky MV (1987): Antigen receptor-regulated exocytosis in cytotoxic T lymphocytes. *J Exp Med* 166: 725–743

Traktman P, Baltimore D (1982): Protease bypass of temperature-sensitive murine leukemia virus maturation mutants. *J Virol* 44: 1039–1046

Trapani JA, Klein JL, White PC, Dupont B (1988): Molecular cloning of an inducible serine esterase gene from human cytotoxic lymphocytes. *Proc Natl Acad Sci USA* 85: 6924–6928

Travis J, Salvesen GS (1983): Human plasma proteinase inhibitors. *Annu Rev Biochem* 52: 655–709

Vettel U, Bar-Shavit R, Simon MM, Brunner G, Vlodavsky I, Kramer MD (1991): Coordinate secretion and functional synergism of T cell-associated serine proteinase-1 (MTSP-1) and endoglycosidase(s) of activated T cells. *Eur J Immunol* 21: 2247–2251

Vischer TL (1974): Stimulation of mouse B lymphocytes by trypsin. *J Immunol* 113: 58–62

Walker CM, Moody DJ, Stites DP, Levy JA (1986): CD8+ lymphocytes can control HIV infection in vitro by suppressing virus replication. *Science* 234: 1563–1566

Wood GS, Müller C, Warnke RA, Weissman I (1988): In situ localization of HUHF serine protease mRNA and cytotoxic cell-associated antigens in human dermatoses. A novel method for the detection of cytotoxic cells in human tissues. *Am J Pathol* 133: 218–225

Woodruff JJ, Clarke LM, Chin YH (1987): Specific cell-adhesion mechanisms determining migration pathways of recirculating lymphocytes. *Annu Rev Immunol* 5: 201–222

Young JDE, Hengartner H, Podack ER, Cohn ZA (1986a): Purification and characterization of a cytolytic pore-forming protein from granules of cloned lymphocytes with natural killer activity. *Cell* 44: 849–859

Young JDE, Leong LG, Liu CC, Damiano A, Wall DA, Cohn ZA (1986b): Isolation and characterization of a serine esterase from cytolytic T cell granules. *Cell* 47: 183–194

Zheng LM, Ojcius DM, Liu CC, Kramer MD, Simon MM, Parr EL, Young JDE (1991): Immunogold labeling of perforin and seine esterases in granulated metrial gland cells. *FASEB J* 5: 79–85

Zinkernagel RM, Doherty PC (1979): MHC-restricted cytotoxic T cells: Studies on the biological role of polymorphic major transplantation antigens determining T-cell restriction-specificity, function, and responsiveness. *Adv Immunol* 27: 51–177

27
Serine Protease Control of Lymphocyte-Mediated Cytolysis

Dorothy Hudig, N. Janine Allison, Gerald R. Ewoldt, Ruth Gault, Dale Netski, Timothy M. Pickett, Doug Redelman, Ming T. Wang, Ulrike Winkler, Susan J. Zunino, Chih-Min Kam, Shinjiro Odake, and James C. Powers

Introduction

Serine proteases initiate several complex biochemical responses to sudden, life-threatening situations. The immediate and localized activation of coagulant serine proteases prevents mammals from bleeding to death. The immediate activation of complement serine proteases kills common bacteria and prevents bacterial sepsis. The granules of cytotoxic lymphocytes (CTL) are packages which contain a unique subfamily of serine proteases and the pore-forming protein, perforin. These sequestered serine proteases are induced during the differentiation of lymphocytes to become competent killer cells (Lobe et al., 1986; Gershenfeld and Weisman, 1986). *In vivo*, lymphocytes laden with toxic granules accumulate at sites of graft rejection (Mueller et al., 1988). Exocytosis exposes the granule contents to high extracellular concentrations of calcium, the cation required for cell lysis. *In vitro*, freshly isolated natural killer (NK) lymphocytes and lymphokine-activated T lymphocytes release these protease-packed granules during killing. Proteases alone do not mediate lysis. Perforin is necessary for cell lysis. The proteases undoubtably perform significant functions immediately associated with cytolysis.

We hypothesize that several serine proteases are essential for cytolysis and that they cleave perforin to activate lysis. This review summarizes data supporting this hypothesis. It evaluates seemingly contradictory conclusions from several other laboratories. Where possible, we correlate results from three different systems: living cells (cell-mediated lysis), granule extracts (granule-mediated lysis), and purified perforin. A separate section of this book (Chapter 58) offers guidelines for the use of protease inhibitors as biological probes. For consideration of other physiological roles of lymphocyte serine proteases, refer to Kramer and Simon (1987) and Simon et al. (1989a).

The lymphocyte serine proteases, or "granzymes" (Tschopp and Jongeneel, 1988), have an unglycosylated molecular weight of approximately 25 kDa. They have a trypsin-like catalytic site. A uniquely reactive Ser residue, together with strategically situated His and Asp residues, forms a characteristic catalytic triad. The active site Ser is a target for inhibition by acylation, phosphonylation, and formation of complexes with transition state analogs, all chemical reactions that can be reversed for subsequent recovery of enzyme activity. The active site His is a target for enzyme inhibition by alkylation, an essentially irreversible chemical reaction. General serine protease inhibitors inactivate almost all members of this family, whereas "specific" inhibitors inactivate only some, depending on the substrate affinity of the protease.

The biochemical characteristics of many of the lymphocyte serine proteases are still unknown. Even the total number of the proteases is uncertain. Seven genes have been sequenced for the mouse (Jenne et al., 1989) and three for the human. No murine gene homologous to the most recently discovered human gene (Meier et al., 1990, Haddad et al., 1991) has yet been sequenced. This observation suggests that even the rodent genetic characterization may be unfinished. Only two gene products have been matched with active, specific proteases. Hanukah Factor or granzyme A is a tryptase, and CCPI or granzyme B is an Asp-ase. Five genetically defined rodent serine proteases need to be assigned substrate specificities. All of these proteases are found in granules where multiple protease activities are discernible. We have recently detected that the

CYTOTOXIC CELLS
M. Sitkovsky and P. Henkart, Editors
© 1993 Birkhäuser Boston

following five peptide thiobenzyl (SBzl) esters, Z-Gly-Arg-SBzl, Suc-Phe-Leu-Phe-SBzl, Boc-Ala-Ala-Met-SBzl, Boc-Ala-Ala-Asp-SBzl, and Boc-Ala-Ala-Ser-SBzl (Hudig et al., 1991; Odake et al., 1991), are good substrates for the following five different lymphocyte protease activities: tryptase (trypsin-line protease, cleaving at Lys or Arg residues), chymase (chymotrypsin-like, cleaving at aromatic amino acid residues), Met-ase, Asp-ase, and Ser-ase actitivies, respectively. Z-Lys-SBzl, the first substrate identified for lymphocyte serine proteases ("BLT," Pasternack et al., 1986), and Z-Gly-Arg-SBzl are both hydrolyzed only by tryptases. Thus, additional protease activities can now be evaluated for their role in cell-mediated killing.

Before the 1986 discovery of the serine protease genes expressed only in lymphocytes (Gershenfeld and Weissman, 1986; Lobe et al, 1986), we and many others found that serine protease inhibitors block NK and T-cell lysis (Matter, 1975; Trinchieri and de Marchi, 1976; Hatcher et al., 1987; Chang and Eisen, 1980; Redelman and Hudig, 1980; Hudig et al., 1981; Goldfarb et al., 1982; Gravagna et al., 1982; Lavie, 1982; Quan et al., 1982; Ristow et al., 1983; Hudig et al., 1984; Petty et al., 1984; Okumura et al., 1985; Poe et al., 1991). To extend the information acquired from cellular experiments, we and other investigators turned to granule-mediated lysis, which is also blocked by protease inhibitors and can be restored when the inhibitors are uncoupled from the proteases. Now two issues are immediate: which particular proteases are required for lysis, and whether perforin must undergo cleavage in order to mediate lysis.

Materials and Methods

Methods for cell-mediated lysis are standard and can be found in the specific citations and in Section X of this book. General methods to measure the granule proteases and lysis by the granule extracts can be found in papers from the Hudig laboratory (Hudig et al., 1987, 1988, 1989, 1991; Zunino et al., 1988). Use of granules bypasses the requirements for certain controls for live cell-mediated lysis, such as effector and target cell viability and lymphocyte adhesion to targets. In addition, proteases inside the granules are often inaccessible to reagents. We have recently modified two methods. Hemoglobin release from red cells (0.5% v/v) works as well as ^{51}Cr release for monitoring granule-mediated lysis. A Molecular Devices (Redwood City, CA) ThermoMax kinetic microplate reader assays proteases faster, more accurately, and less wastefully than standard spectrophotometric assays. To accommodate the detection range of the reader, Ellman's reagent is used to measure the thiobenzyl leaving group at 412 nm instead of the more sensitive dithiodipyridine, which absorbs in the UV range.

Serine Proteases Activate Lysis by Living Lymphocytes, by Their Granules, and by Enriched Perforin

The first evidence that proteases activate killing came from experiments with living killer cells. The killer cells included human and mouse antibody-dependent cellular cytotoxic (ADCC), NK, and T effectors (see citations above). Different laboratories came to contradictory conclusions, despite obtaining similar results with the same reagents. Several investigators even concluded that the proteases are not important for lysis. However, many of the protease inhibitors are problematic reagents which at the time of the experiments had not received the additional characterization that would help interpret the experimental results. It now appears that only highly reactive serine protease inhibitors will block lysis and that only nontoxic reagents can generate interpretable results.

Macromolecular serine protease inhibitors block lymphocyte-mediated lysis. Among the most reliable cellular reagents are the protein inhibitors, such as α_1-antichymotrypsin ($\alpha_1 X$, a blood plasma protein) and soybean trypsin inhibitor (SBTI, a plant protein). These proteins are nontoxic to cells in culture. $\alpha_1 X$ was important because it blocked human NK (Figure 1) and rodent T-cell-mediated lysis and prompted a search for lymphocyte chymases. This 52-kDa protease inhibitor, like other plasma "serpin" (*ser*ine *p*rotease *in*hibitor) inhibitors, couples tightly to serine proteases. $\alpha_1 X$ is selective for chymases and does not react with tryptases. Chymases have now been identified in cytotoxic lymphocytes of rats (Hudig et al., 1987), mice (J Tschopp and JC Powers, unpublished results), and humans (M Poe and JC Powers, unpublished results). α_1-Protease inhibitor (α_1PI, formerly termed α_1-antitrypsin), like α_1-antichymotrypsin, is also a plasma serpin. Even though α_1PI inactivates a wider spectrum of serine proteases than $\alpha_1 X$, we observed that purified α_1-PI does not inhibit lymphocyte-mediated lysis (Hudig et al., 1981).

Controversy exists as to whether $\alpha_1 X$ (Redelman

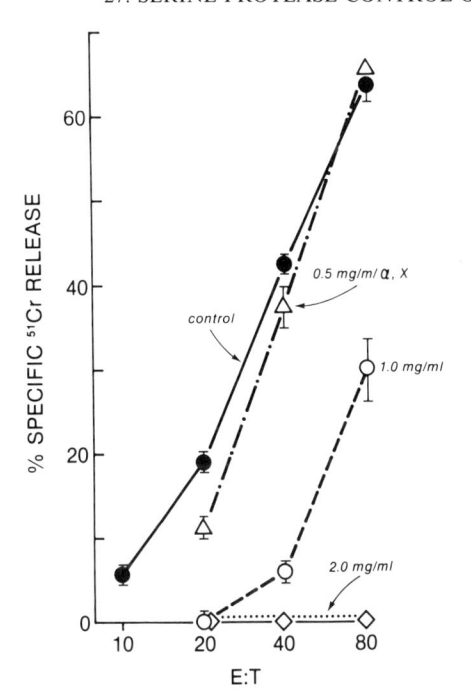

FIGURE 1. α_1-Antichymotrypsin (α_1X) inhibition of human NK to T24 bladder carcinoma cells. A preparation of α_1X with no detectable antitrypsin activity was included in the assay medium at the concentrations indicated. The percentage of specific ^{51}Cr counts released over background release indicates the percentage of dead cells. The effector lymphocyte to tumor target (E:T) ratios are indicated on the abscissa. There were 2×10^4 tumor cells per well in these 12-hr assays. (Reprinted with permission from Hudig et al., 1981.)

and Hudig, 1980; Hudig et al., 1981; Gravagna et al., 1982) or α_1PI (Okumura et al., 1985; Laine et al., 1990) blocks cell-mediated lysis. We submit that it may be possible for both serpins to block cell-mediated lysis if the serpin preparations have fully retained their antiprotease activity. Considerable serpin activity can be lost during purification. The lower molar antiprotease reactivities of some serpin preparations can explain their failure to block lysis. Alternatively, trace amounts of other, highly reactive serpins within the α_1X and α_1PI preparations could augment the inhibition of lysis. Ultimately, recombinant bacterial sources of each serpin will incontrovertibly identify those serpin(s) which specifically inactivate lymphocyte-mediated lysis.

Another nontoxic macromolecular serine protease inhibitor, soybean trypsin inhibitor (Kunitz) (SBTI, M_r 22,000), slows the *rate* of NK cell-mediated lysis but does not prevent final lysis (Hudig et al., 1981). With time, inhibited cells are still capable of attaining lysis equal to that of unhibited control cells. Thus, SBTI inhibition can be obscured in end-point cellular assays. Granule-mediated lysis is also slowed, though not prevented, by SBTI (Hudig et al., 1987).

Low-molecular-weight serine protease inhibitors also block lymphocyte-mediated lysys. It was orginally hoped that low-M_r inhibitors might gain access to "sequestered sites" which macromolecular inhibitors could not enter. In general, the view was that reversible inhibitors would cause fewer serious side effects because they would react noncovalently. However, in practice both reversible and irreversible low-M_r inhibitors are difficult reagents to use with cells. Their potential to cross the lymphocyte plasma membrane requires careful controls to differentiate effects due to toxicity. Low-M_r reagents with a common reactive group but differing in protease specificity were compared. The rationale was that toxicity should be nonspecific and equivalent among compounds with a common reactive group. Selective effects should be associated with the protease specificity of the inhibitor and should be consistently associated with particular protease specificities.

For example, the fungal peptide aldehyde inhibitors of serine proteases share similar reactive aldehyde groups but differ in their peptide sequences. Chymostatin (A,N-[[(S)-1-carboxy-2-phenylethyl]carbamoyl)-α-[2-iminohexahydro-4(S)-pyrimidyl]-L-glycyl-L-leucyl-phenylalaninal) reacts with chymases, and leupeptin reacts with tryptases. We suspected that the aldehyde groups might be toxic to lymphocytes. However, such toxicity does not occur. Chymostatin blocks NK lysis, whereas leupeptin does not (Hudig et al., 1984). We attribute the loss of lysis to the same chymase that is also inactivated when α_1X blocks lysis. This "pair" of aldehyde compounds reacts similarly with granule extracts: granule-mediated lysis is slowed by chymostatin and is not affected by leupeptin (Hudig et al., 1987).

Conjoint comparison of chemically similar compounds, as with the preceding fungal aldehydes, is a reasonable but not infallible approach. Recent findings dictate reexamination of our use of amino acid esters as inhibitors in lymphocyte-mediated cytotoxic assays (Hudig et al., 1984). At the time we considered these reagents to be alternative substrates that would compete with, and thereby reduce proteolysis of, natural substrates. Aromatic (Phe, Tyr) esters, such as AcTyrOEt, block lysis, whereas basic (Lys, Arg) esters do not. This chymase specificity resembles the $\alpha_1 X$ and chymostatin inactivation of lysis. However, lymphocytes have the ability to form oligopeptides from peptide esters. Both mono- and dipeptide amino acid esters serve as substrates. These oligopeptides are formed in lymphocyte lysosomes by dipeptidase I, an enzyme that catalyzes peptide synthesis as well as peptide hydrolysis. The oligopeptides are toxic only when the contain hydrophobic amino acids (Thiele and Lipsky, 1990). Thus the aromatic amino acid esters can ultimately kill lymphocytes whereas the charged amino acids are nontoxic by this mechanism (Thiele and Lipsky, 1990). Consequently, we do not know if the Phe and Tyr ethyl esters that block cell-mediated killing are effective alternative substrates for serine proteases or if they are just toxic.

New peptide ketoester inhibitors such as Z-Phe-Nvl-COOEt (Powers, unpublished), are less likely than peptide ester inhibitors to form oligopeptides in lysosomes, as they are transition-state analogs and will form ketoacids upon cleavage. These reversible inhibitors offer an opportunity to reexamine the earlier cellular experiments. Ketoesters with peptide sequencs similar to the synthetic peptide substrates hydrolyzed by the lymphocyte proteases can be compared with ketoesters with sequences that are not substrates for these enzymes.

Irreversible low-M_r inhibitors present other experimental problems. With the irreversible chloromethyl ketones, problems were recognized with the initial lymphocyte experiments. In 1975, Alex Matter concluded that the toxicity of tosyllysine chloromethyl ketone (TLCK) was too great to support any conclusions as to the role of serine proteases in killing. TLCK inhibition was interpreted by other investigators to indicate a role for proteases (Chang and Eisen, 1980; Lavie, 1982). However, additional data support the initial reservations. Redelman and Hudig (1980) and Ristow et al. (1983) found that the *single* amino acid chloromethyl ketones block cell-mediated lysis without specificity and are needed in unusually high (mM) concentrations. These high concentrations of chloromethyl ketones probably act nonspecifically by binding to thiol groups (Redelman and Hudig, 1980). This reactivity reduces the adhesion of killer lymphocytes to their targets (Ristow et al., 1983; Hudig et al., unpublished results), which is needed for lysis to proceed. TLCK does not inactivate rat lymphocyte serine proteases or block granule-mediated lysis (Hudig et al., 1987).

Recently, more reactive and selective *tripeptide* chloromethyl ketone inhibitors have been identified which block lymphocyte proteases and granule-mediated lysis (Simon et al., 1987; Zunino et al., 1988). Since the *peptide* chloromethyl ketone inhibitors are effective at much lower concentrations than single amino acid chloromethyl ketone inhibitors, it is important to test peptide chloromethyl ketone reagents with lymphocytes. The first peptide chloromethyl ketone to be tested with cells, D-Pro-Phe-Arg-CH_2CH_2Cl, failed to block cell-mediated lysis, whereas it did block granule extract-mediated lysis (Simon et al., 1989b). Numerous factors may have contributed to this apparent inconsistency. The natural substrates in the granules and in the granule extracts compete with inhibitors for access to the proteases. It is possible to pretreat granule extracts with inhibitors, whereas with live cells the granules are exposed to the inhibitor only after exocytosis. Higher concentrations of the natural substrates are present in granules right after exocytosis than in dilute granule extracts. Exocytosis favors natural substrate competition with inhibitors, slower inactivation of the proteases, and natural substrate hydrolysis to permit lysis. Alternatively, chemical decomposition of the Arg-chloromethyl ketones can significantly reduce the effective concentrations during the longer times required for the cellular assays (4 hr vs 20 min for granule-mediated lysis).

Contradictory conclusions have been drawn from highly reproducible experiments from different laboratories using low M_r, irreversible general protease inhibitors. *General* inhibitors should, in theory, inactivate all serine protease-dependent functions. But the general serine protease inhibitors phenylmethylsulfonyl fluoride (PMSF) and diisopropylfluorophosphate (DFP) fail to block lysis. For example, PMSF fails to block T-cell lytic function, even though manipulation to raise lysosomal pH results in good protease inhibition after PMSF pretreatment of the lymphocytes (Henkart et al., 1987). Although these results suggest that proteases may not be important for lysis, further examination indicates that PMSF is not as "general" an inhibitor as might have been supposed. The tryptase measured by Henkart et al. is the most susceptible granule protease to PMSF. PMSF does not react with all lymphocyte proteases and reacts poorly with most of those that are susceptible (Ewoldt et al., 1992).

The PMSF paradox will undoubtedly reappear, even with stronger serine protease inhibitors. Although the tryptase may not be important for lysis (Henkart et al., 1987), the quandry may reappear when lysis is unaffected and other lymphocyte proteases are inhibited. How can lytic activity be protease-dependent and still remain when most protease activity is inhibited? This paradox is a consequence of biology and methodology. Biological functions may be catalyzed by protease activities too low to be detected by standard protease assays. For protease concentrations so low as to escape direct detection, the fraction of surviving protease can be calculated from the second-order rate constants of the inhibitor with the individual proteases, the molar concentration of the inhibitor, and the time of treatment. For example, after treatment with 1 mM PMSF for 20 min, with a second-order inhibition rate constant of $7.7 \, M^{-1} sec^{-1}$ for rat lymphocyte tryptase, 0.01% of the tryptase would remain.

Although two general serine protease inhibitors, DFP and PMSF, do not block cytoxicity, a new, low-M_r, irreversible general protease inhibitor, 3,4-dichloroisocoumarin (DCI), does block lymphocyte-mediated cytotoxicity. The low DCI concentration (5 μM) that inhibits lysis by a human T-cell clone, Q31 (Odake et al., 1991), is remarkable, particularly since this inhibitor has an approximately 20-min aqueous half-life. Comparison of second-order inhibition rate constants indicates that DCI is > 100-fold more reactive than PMSF with all currently detectable lymphocyte proteases (Ewoldt et al., 1992). More reactive isocoumarin serine protease inhibitors that are selective for specific proteases block cell-mediated lysis even more effectively than DCI (Odake et al., 1991).

The first experiments with live cells can be viewed as reconnaissance. They indicated that serine proteases control lymphocytic destruction of cells. Inhibition of lysis by macromolecular, nonpenetrating protein inhibitors is consistent with exocytosis of proteases to activate lysis. The low-M_r inhibitors led to recognition that extremely potent inhibitors are needed to block lysis. Side effects, such as the generation of toxic oligopeptides and interference with lymphocyte adhesion to targets, were also noted. Careful monitoring of potential toxic effects and measurement of the rate of reactivity with lymphocyte proteases should promote development of better inhibitors to block lytic function.

Serine Proteases Also Activate Granule-Mediated Lysis

Serine protease inhibitors block granule-mediated lysis as well as cell-mediated lysis. We began with unfractionated granule extracts, and now we continue further, to determine whether perforin is lytic when completely separated from the proteases. Our rationale was that if perforin was directly activated by protease cleavage (positive regulation), all active perforin preparations would be sensitive to protease inhibitors. On the other hand, if perforin pore formation was controlled indirectly by proteases, lysis could be retarded by additional restrictive granule proteins (negative regulation) and accelerated when these proteins were cleaved. Pure perforin might bypass such indirect control to remain lytic and an indirect control mechanism would not be detected. By measuring lytic rates (Hudig et al., 1988), we found that *reversible* protease inhibitors will slow granule-mediated lysis, probably because they dissociate from the proteases. In contrast, *highly reactive* protease inhibitors that are considered "irreversible" because they form covalent bonds will completely block granule-mediated lysis (Hudig et al., 1987), probably because of their low dissociation from the proteases. The new general, irreversible, serine protease inhibitor, DCI, which blocks cell-mediated killing (Odake et al., 1991), blocks lysis by granule extracts (Hudig et al., 1987, 1991). More reactive and more specific irreversible isocoumarin inhibitors that inactivate chymases, Met-ase and/or Ser-ase and not tryptase or Asp-ase block lysis at lower concentrations than DCI.

Representative isocoumarin and other types of inhibitors that block granule-mediated lysis are listed in Table 1, with their abilities to inactivate chymase, Met-ase, Ser-ase, and/or tryptase activities. Asp-ase inhibition is not listed, because of all inhibitors that we have found to block lysis, only DCI and SBTI significantly inactivate Asp-ase. Fractional inhibition of protease activity listed in Table 1 reflects complete inhibition of one protease, with little inhibition of a second protease that cleaves the same substrate. Biphasic second-order inhibition rate constants [with Z-Gly-Leu-Phe-CH$_2$Cl (Zunino et al., 1988), with 4-chloro-7-(n-phenylcarbamoylamino)-3-(2-phenylethoxy)isocoumarin (Hudig et al., 1989) and the substituted benzenesulfonyl fluoride (Ewoldt et al., 1992)] and biochemical separations, indicate that there are multiple chymases that cleave the Suc-Phe-Leu-Phe-SBzl substrate, some refractory to a particular inhibitor. We conclude from the specific reactivity of those reagents which inactivate lysis that one or more of the chymase, Met-ase, or Ser-ase proteases activates lysis. Purification of perforin should differentiate between direct and indirect activation of lysis.

Highly purified preparations [greater than 98% by sodium dodecyl sulfate–polyacrylamide gel elec-

TABLE 1. Inactivation of granule-mediated lysis and lymphocyte proteases by seven types of serine protease inhibitors[a]

| | | | % inhibition of: | | | |
| | | | | Protease[c] | | |
Inhibitor type (Reference)	Compound (concentration inhibiting lysis)[a]	Lysis[b]	Chym	Met	Ser	Trypt
Irreversible inhibitors that arrest lysis						
Isocoumarin (IC) (Hudig et al., 1989)	4-Cl-7-(N-phenylcarbamoylamino)-3-(2-phenyl ethoxy)-IC (400 µM)	100%	25	24	22	9
Peptide phosphonate (Hudig et al., unpub)	Z-Phe-Pro-Phe-(OPh)$_2$ (1 mM)	97%	89	65	24	4
Peptide CMK (Zunino et al., 1988)	Z-Gly-Leu-Phe-CH$_2$Cl (0.4 mM)	100%	72	56	31	41
Sulfonylfluoride (Ewoldt et al., 1992)	2-Z-(H(CH$_2$)$_2$CONH)C$_6$H$_4$SO$_2$F (0.8 mM)	50%	84	61	30	1
Reversible inhibitors that slow lysis						
Peptide ketoester (Hudig et al., unpub)	Z-Leu-Nvl-COOEt (0.1 mM)	20%	27	22	32	ND
Fungal aldehyde (Zunino et al., 1988)	Chymostatin	50%	97	63	29	8
Plant inhibitor (Hudig et al., 1987)	Soybean trypsin inhibitor (0.022 mM)	71%	79	76	33	0

[a] Granule extracts were pretreated with the irreversible inhibitors for 20 min and then assayed for lysis. The reversible inhibitors were included in the lytic assays at the concentrations indicated.
[b] Percent inhibition of control lysis at 20 min was calculated as [1-(exp lysis/control lysis)] × 100.
[c] Percent remaining protease activity after pretreatment with 100 µM irreversible compounds for 20 min at room temperature or with 100 µM reversible compounds directly in the protease assay.
The sources of the reagents are: Z-Gly-Leu-Phe-CH$_2$Cl, Enzyme Systems Products, Dublin, CA; SBTI, Sigma Chemical Co; chymostatin, Peninsula Labs, Belmont, CA; isocoumarin, peptide phosphonate, sulfonyl fluoride, and ketoester, the Powers laboratory.

trophoresis (SDS–PAGE)] of α_1X and α_1PI inactive granule chymases, Met-ase, and Ser-ase. Significantly, highly purified preparations of both serpins alone do not inactivate lysis even at 2×10^{-5} M, though less pure preparations (with lower concentrations of the serpins) inactivate both proteases and lysis. Second-order inhibition rate constants and characterization of the impurities should clarify these unexpected results.

Many potent inhibitors of the lymphocyte tryptases do not affect granule-mediated lysis. For example, 4-chloro-7-guanidino-3-methoxyisocoumarin (Hudig et al., 1989), aprotinin, Z-Arg-CH$_2$Cl, and the serpin antithrombin III with heparin selectively inactivate tryptases but do not affect lysis. Nonetheless, there *are* tryptase-directed inhibitors that inactivate granule-mediated lysis. D-Pro-Phe-Arg-CH$_2$Cl inhibits both mouse T-cell granule-mediated lysis (Simon et al., 1987) and rat NK-granule mediated lysis. D-Pre-Phe-Arg-CH$_2$Cl and some other peptide chloromethyl ketone inhibitors containing P1*Arg uncharacteristically also inactivate a chymase fraction (Wang and Netski, in preparation). (*Nomenclature is illustrated in our methods chapter, 58, in this volume.) At this time, we cannot distinguish whether (1) extremely high reactivity with a key tryptase component or (2) cross-reactivity with another protease accounts for the effects on lysis. *Reactivation of inhibited granule-mediated lysis.* Lysis is restored when the proteases are chemically reactivated. When deacylation (with hydroxylamine) removes the inhibitor DCI from the serine residues of lymphocyte proteases, substantial activities of several proteases are regenerated, most notably 100% of the inhibited Met-ase. Simultaneously, substantial, though not all, lytic activity is also restored (Figure 2) (Hudig et al., 1991). Inhibitors that acylate the active Ser residue can be uncoupled to reactivate a protease, unless the protease is also alkylated at its active-site His. Alkylation cannot be uncoupled. For example, DCI, the general serine protease inhibitor that blocks lysis, acylates active-site Ser residues and may or may not react with the active-site His residues, depending on the individual protease (Harper et al., 1985). Partial restoration of protease activity is consistent with the mechanism of DCI inhibition of proteases. Partial restoration

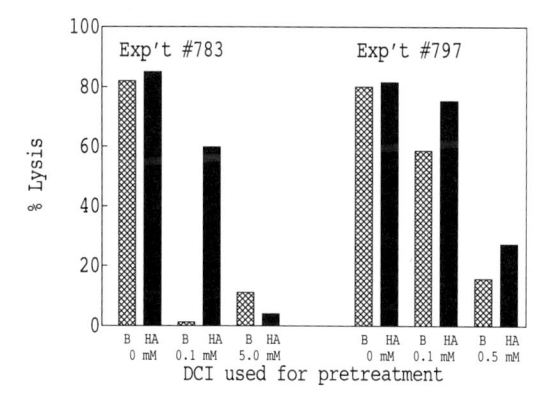

FIGURE 2. Protease-dependent reactivation of granule-mediated lysis. After pretreatment with the general serine protease inhibitor DCI, inactivated proteases and lysis can be concomitantly restored. Two different granule extracts were pretreated with 0.1, 0.5, or 5 mM DCI, dialyzed, then deacylated with 0.4 M hydroxylamine (HA) or treated with buffer (B), and dialyzed again prior to lytic assay. Controls were treated with the diluant, dimethylsulfoxide (DMSO). In these experiments, 100% of inhibited Met-ase was restored by hydroxylamine treatment. Lesser percentages of other granule protease activities were restored. (Reprinted with permission from Hudig et al., 1991.)

Perforin Lysis Depends Upon Serine Proteases

Active perforin preparations contain serine proteases. Proteases are detectable within concentrated perforin preparations, even though purification separates most protease activity from perforin. Protease activities are too low to detect at the perforin concentrations that support lytic activity. However, if one assays preparations containing 50- to 100-fold greater amounts of perforin, it is possible to detect three protease activities, chymase, Ser-ase, and Met-ase at $\sim 10\,\text{milliOD}_{412}/\text{min/ml}$ of perforin. The perforin used was prepared by anion exchange (Podack et al., 1985) followed by size exclusion in 0.5 M NaCl. Tryptases and Asp-ase are not detectable in these perforin preparations and are resolved with little loss of activity in peaks widely separated from perforin.

Perforin lytic activity is inactivated by serine protease inhibitors. Every time we inactivate the three trace proteases within our perforin preparations, lysis is lost. Lysis is blocked when preforin preparations are pretreated with DCI, chloromethylketone, and sulfonyl fluoride inhibitors (Table 2; Winkler and Pickett, in preparation).

of lytic activity is thus fully consistent with protease activation of lysis.

Since lysis is perforin-dependent, one has to take into account potential effects of DCI on perforin. DCI is a mechanism-based or "suicide" serine protease inhibitor. Without activation by the uniquely reactive catalytic serine of proteases, DCI is a very poor acylating reagent (Rusbridge and Beynon, 1990) and is most likely to react with lysine residues. Thus serines of perforin are extremely unlikely to be acylated by DCI, though some lysine residues may be nonspecifically n-acylated. Acylated lysine residues might functionally impair perforin, but since they will remain acylated after hydroxylamine treatment, lost function would not be restored by hydroxylamine treatment. Since hydroxylamine restores lysis to granule extracts inactivated by DCI (Figure 2), it appears that proteases are affected and perforin is unaffected by DCI. Since more reactive and specific (Met-ase and chymase-directed) isocoumarin inhibitors block lysis at lower concentrations than DCI, and this lysis can also be regenerated with hydroxylamine, the effects on lysis appear to be consistently protease-specific.

Discussion and Hypotheses

Must perforin be cleaved for lysis to occur? If so, homogeneous preparations of perforin should be inactive until appropriate proteases are added to them. Experimental data suggest this direct activation. Munger (1988) found that addition of 60-kDa proteases accelerated the rate of lysis by rat perforin. The 60-kDa proteases were identified as "BLT" tryptase but could well have contained other serine proteases, including chymases of identical M_r, since no other proteases were assayed. The gene for perforin has been transfected into rat basophilic leukemia cells to allow these cells to express perforin, which is found in granules. Perforin also made the basophils able to lyse haptenated red cells by anti-hapten IgE-directed granule exocytosis (Shiver and Henkart, 1991). No conclusions as to the requirements for proteases in lysis can be drawn from these experiments, because chymases and tryptases (which are resident in basophil granules) may have activated perforin.

Three hypotheses for protease activation of lysis conform with the available data. Immunologists frequently submit a choice of hypotheses instead of data. We sincerely hope that additional data will

TABLE 2. Inactivation of perforin-mediated lysis and perforin-associated proteases by three types of irreversible serine protease inhibitors[a]

| | | | % inhibition of: | | |
| | | | | Perforin-associated proteases: | |
Compound	Concentration	Lysis	Chymase	Met-ase	Ser-ase
Dichloroisocoumarin (DCI)[b]	1.000 mM	88%	92%	84%	ND
Z-Gly-Leu-Phe-CH$_2$Cl	0.125 mM	96%	100%	100%	100%
2-Z-(NH(CH$_2$)$_2$CONH)C$_6$H$_4$SO$_2$F	0.800 mM	50%	93%	38%	ND

[a] Perforin was purified from RNK-16 granules by MonoQ anion exchange (Podack et al., 1985), followed by Sephacryl S200 gel exclusion chromatography.
[b] DCI was obained from Boehringer Mannheim; see Table 1 for the other sources.
The concentrations listed above were used to inactivate lysis. The concentrations used to inactivate the proteases were 1 mM DCI, 0.5 mM Z-Gly-Leu-Phe-CH$_2$Cl, and 0.8 mM 2-Z-(NH(CH$_2$)$_2$CONH)C$_6$H$_4$SO$_2$F.

help us eliminate only two of three models that hypothesize natural substrates for protease activation of lysis. *Model one* hypothesizes that the natural protease substrate is a "co-factor" for perforin, serving a function analogous to complement C5. Functional and structural comparisons between perforin and C9 have been drawn many times (Tschopp et al., 1986; Young et al., 1986; Shinkai et al., 1988; Peitsch et al., 1990). In this model, proteolysis generates a C5b-like granule protein which facilitates entry of the pore-forming (C9-like) protein into lipid membranes. Faster entry into the target cell membrane (which is calcium-dependent) would reduce the fraction of perforin that is inactivated by calcium in a simultaneous process. *Model two* hypothesizes that perforin is inactive in the presence of a "controlling" protein (i.e. subject to negative regulation). Addition of calcium induces proteolytic destruction of the controlling protein, dissociating perforin and facilitating its access to membranes. For either of these models to be correct, both proteases and auxiliary molecules would have eluded detection on SDS–PAGE gels of perforin. *Model three*, which we favor, hypothesizes that perforin is activated by cleavage (positive regulation). Once cleaved, perforin might serve as a nucleus for a polymeric pore (Podack and Hengartner, 1990). The pore itself could even be largely formed of uncleaved perforin molecules. For this third model to be correct, only catalytic proteases would also have to be undetected after SDS–PAGE electrophoresis, whereas controlling proteins would also have to be undetected in models one and two.

Evaluation of these hypotheses will demand purification of active proteases and preparation of perforin in sufficient quantities to detect minor contaminants. Quantitative assays for inactive as well as active perforin protein are needed. Ultimately, it will be necessary to determine whether full-length recombinant perforin and exhaustively purified native perforin are equally lytic, mole for mole, and at identical rates, with and without lymphocyte proteases. Classical biochemistry is difficult with perforin because of its low abundance in lymphocyte granules and its lability. Molecular biology techniques alone will not identify the natural substrates of the proteases. Combination of the two approaches will promote understanding of the complex mechanism by which lymphocytes kill only selected cells within an organism.

Acknowledgments

We thank Drs. John Adlish, William H. Welch, Jr., and Terry Woodin of the University of Nevada, Reno, and R. Christopher Bleackley of the University of Alberta for their helpful suggestions and critique of the manuscript. The research was made possible by NIH grants T32 CA09563 (D.H.) which supports the predoctoral fellow G.R.E., and R01 awards CA38942 (D.H.), GM42212 (D.H. and J.C.P.) and HL29307 (J.C.P.).

Abbreviations

α_1X, alpha-1 antichymotrypsin; α_1PI, alpha-1-protease inhibitor; ADCC, antibody-dependent cell-mediated cytotoxicity; Asp-ase, protease activity cleaving Boc-Ala-Ala-Asp-SBzl; BLT, Benzyloxycarbonylysine thiobenzyl ester; Boc, butyloxylcarbonyl; DCI, 3,4-dichloroisocoumarin; DMSO, dimethylsulfoxide; IC, isocoumarin; Met-ase, protease activity cleaving Boc-Ala-Ala-Met-SBzl; NK, natural killer lymphocyte; OEt, ethyl ester; SBzl, thiobenzyl ester; SBTI, soybean trypsin inhibitor; Ser-ase, protease activity cleaving Boc-

Ala-Ala-Ser-SBzl; T cell, thymus-dependent lymphocyte; Z, benzyloxycarbonyl.

References

Chang TW, Eisen HN (1980): Effects of N^a-tosyl-L-lysyl-chloromethylketones on the activity of cytotoxic T lymphocytes. *J Immunol* 124: 1028–1033

Gershenfeld HK, Weissman IL (1986): Cloning of a cDNA for a T cell-specific serine protease from a cytotoxic T lymphocyte. *Science* 232: 854–857

Goldfarb R, Timonen TT, Herberman RB (1982): The role of neutral serine proteases in the mechanism of tumour cell lysis by human natural killer cells. In: *NK Cells and Other Natural Effector Cells*, Herberman RB, ed. New York: Academic Press, pp. 931–938

Gravagna P, Gianazza E, Arnaud P, Neels M, Ades EW (1982): Modulation of the immune response by plasma inhibitors. III. Alpha-1-antichymotrypsin inhibits human natural killing and antibody-dependent cell-mediated cytotoxicity. *J Reticuloendothel Soc.* 32: 125–130

Haddad P, Jenne D, Tschopp J, Clement M-V, Mathieu-Mahul D, Sasportes M (1991): Structure and evolutionary origin of the human granzyme H gene. *Int Immunol.* 3: 57–66

Harper JW, Hemmi K, Powers JC (1985): Reaction of serine proteases with substituted isocoumarins: Discovery of 3,4-dichloroisocoumarin, a new general mechanism-based serine protease inhibitor. *Biochemistry* 24: 1831–1841

Hatcher VB, Oberman MS, Lazarus GS, Grayzel AI (1978): A cytotoxic proteinase isolated from human lymphocytes. *J Immunol* 120: 665–670

Henkart PA, Berrebi GA, Takahama H, Munger WE, Sitkovsky M (1987): Biochemical and functional properties of serine esterases in acidic cytoplasmic granules of cytotoxic T lymphocytes. *J Immunol* 139: 2398–2405

Hudig D, Allison NJ, Pickett TM, Kam C-M, Powers JC (1991): The function of lymphocyte proteases: Inhibition and restoration of granule-mediated lysis with isocoumarin serine protease inhibitors. *J Immunol* 147: 1360–1368

Hudig D, Callewaert DM, Redelman D, Gregg NJ, Krump M, Tardieu B (1988): Lysis by RNK-16 cytotoxic–lymphocyte granules: Rate assays and conditions to study control of cytolysis. *J Immunol Methods* 115: 169–177

Hudig D, Gregg NJ, Kam C-M, Powers JC (1987): Lymphocyte granule-mediated cytolysis requires serine protease activity. *Biochem Biophys Res Commun* 149: 882–888

Hudig D, Haverty T, Fulcher C, Redelman D, Mendelsohn J (1981): Inhibition of human natural cytoxicity by macromolecular antiproteinases. *J Immunol* 126: 1569–1574

Hudig D, Powers JC, Allison NJ, Kam C-M (1989): Selective isocoumarin protease inhibitors block RNK-16 lymphocyte granule-mediated cytolysis. *Mol Immunol* 26: 793–798

Hudig D, Redelman D, Minning L (1984): The requirement for proteinase activity for human lymphocyte-mediated natural cytotoxicity (NK): Evidence that the proteinase is serine dependent and has aromatic amino acid specificity of cleavage. *J Immunol* 133: 2647–2654

Jenne DE, Masson D, Zimmer M, Haefliger JA, Li WH, Tschopp J (1989): Isolation and complete structure of the lymphocyte serine protease granzyme G, a novel member of the granzyme multigene family in murine cytolytic T lymphocytes. Evolutionary origin of lymphocyte proteases. *Biochemistry* 28: 7953–7961

Kramer MD, Simon MM (1987): Are proteinases functional molecules of T lymphocytes? *Immunol Today* 8: 140–142

Laine A, Leroy A, Hachulla E, Davril M, Dessaint J-P (1990): Comparison of the effects of purified human α_1-antichymotrypsin and α_1-proteinase inhibitor on NK cytotoxicity: only α_1-proteinase inhibitor inhibits natural killing. *Clin Chim Acta* 1900: 163–174

Lavie G, (1982): The role of surface-associated proteases in human NK cell-mediated cytotoxicity. Evidence suggesting a mechanism by which concealed surface enzymes become exposed during cytolysis. In: *NK Cells and Other Natural Effector Effector Cells*, ed. Herberman RB, ed. New York: Academic Press, pp. 939–947

Lobe CG, Finlay BB, Paranchych W, Paetkau VH, Bleackley RC (1986): Novel serine proteases encoded by two cytotoxic T lymphocyte-specific genes. *Science* 232: 858–861

Matter A (1975): A study of proteolysis as a possible mechanism for T-cell-mediated target cell lysis. *Scand J Immunol* 4: 349–356

Meier M, Kwong PC, Fregeau CJ, Atkinson EA, Burrington M, Ehrman N, Sorenson O, Lin CC, Wilkins J, Bleackley RC (1990): Cloning of a gene that encodes a new member of the human cytotoxic cell protease family. *Biochemistry* 29: 40042–40049

Mueller C, Gershenfeld HK, Lobe CG, Okada CY, Bleackley RC, Weissman IL (1988): A high proportion of T lymphocytes that infiltrate H-2-incompatible heart allografts in vivo express genes encoding cytotoxic cell-specific serine proteases, but do not express the MEL-14-defined lymph node homing receptor. *J Exp Med* 167: 1124–1136

Munger WE (1988): LGL-secretory granule-associated 60-kd BLT esterase augments the lytic activity of cytolysin against nucleated target cells. *Nat Immun Cell Growth Regul* 7: 61–70

Odake S, Kam C-M, Narasimhan L, Poe M, Blake JT, Krahenbuhl O, Tschopp J, Powers JC (1991): Human and murine cytotoxic T lymphocyte serine proteases: Subsite mapping with peptide thioester substrates and inhibition of enzyme activity and cytolysis by isocoumarins. *Biochemistry* 30: 2217–2227

Okumura Y, Kuro J, Ikata T, Kurokawa S, Ishibashi H, Okubo H (1985): Influence of acute phase proteins on the activity of natural killer cells. *Inflammation* 9: 211–219

Pasternack M, Verret CR, Liu MA, Eisen HN (1986):

Serine esterase in cytolytic T lymphocytes. *Nature* 322: 740–743

Peitsch MC, Amiguet P, Guy R, Brunner J, Maizel JV, Jr, Tschopp J (1990): Localization and molecular modelling of the membrane-inserted domain of the ninth component of human complement and perforin. *Mol Immunol* 27: 589–602

Petty HR, Hermann W, Dereski W, Frey T, McConnell HM (1984): Activatable esterase activity of murine natural killer cell–YAC tumour cell conjugates. *J Cell Sci* 72: 1–13

Podack ER, Hengartner H (1990): Structure of perforin and its role in cytolysis. *Year Immunol* 6: 245–261

Podack ER, Young JD-E, Cohn ZA (1985): Isolation and biochemical and functional characterization of perforin 1 from cytolytic T-cell granules. *Proc Natl Acad Sci USA* 82: 8629–8633

Poe M, Wu JK, Blake JT, Zweerink HJ, Sigal NH (1991): The enzymatic activity of human cytotoxic T-lymphocyte granzyme A and cytolysis mediated by cytotoxic T-lymphocytes are potently inhibited by a synthetic antiprotease, FUT-175. *Arch Biochem Biophys* 284: 215–218

Quan P-C, Ishizaka T, Bloom BR (1982): Studies on the mechanism of NK cell lysis. *J Immunol* 128: 1786–1791

Redelman D, Hudig D (1980): The mechanisms of cell-mediated cytoxicity I. Killing by murine cytotoxic T lymphocytes requires cell surface thiols and activiated protease. *J Immunol* 124: 870–878

Ristow SS, Starkey JR, Hass GM (1983): In vitro effects of protease inhibitors on murine natural killer cell activity. *Immunology* 48: 1–8

Rusbridge NM, Beynon RJ (1990): 3,4-Dichloroisocoumarin, a serine protease inhibitor, inactivates glycogen phosphorylase b. *FEBS Lett* 268: 133–136

Shinkai Y, Takio K, Okumura K (1988): Homology of perforin to the ninth component of complement. *Nature* 334: 525–527

Shiver JW, Henkart PA (1991): A noncytotoxic mast cell tumor line exhibits potent IgE-dependent cytotoxicity after transfection with the cytolysin/perforin gene. *Cell* 64: 1175–1181

Simon MM, Fruth U, Simon HG, Gay S, Kramer MD (1989a): Evidence for multiple functions of T-lymphocyte associated serine proteinases. *Adv Exp Med Biol* 247A: 609–613

Simon MM, Fruth U, Simon HG, Kramer MD (1987): Evidence for the involvement of a T-cell-associated serine protease (TSP-1) in cell killing. *Ann Inst Pasteur* 138: 309–314

Simon MM, Prester M, Kramer MD, Fruth U (1989b): An inhibitor specific for the mouse T-cell associated serine proteinase 1 (TSP-1) inhibits the cytolytic potential of cytoplasmic granules but not of intact cytolytic T cells. *J Cell Biochem* 40: 1–13

Thiele DL, Lipsky PE (1990): The action of leucyl-leucine methyl ester on cytotoxic lymphocytes requires uptake by a novel dipeptide-specific facilitated transport system and dipeptidyl peptidase I-mediated conversion to membranolytic products. *J Exp Med* 172: 183–194

Tschopp J, Jongeneel CV (1988): Cytotoxic T lymphocyte mediated cytolysis *Biochemistry* 27: 2641–2646

Tschopp J, Masson D, Stanley KK (1986): Structural/functional similarity between proteins involved in complement- and cytotoxic T lymphocyte-mediated cytolysis. *Nature* 322: 831–833

Trinchieri G, de Marchi M (1976): Antibody-dependent cell-mediated cytotoxicity in humans III. Effect of protease inhibitors and substrates. *J Immunol* 116: 885–891

Young JD, Cohn ZA, Podack ER (1986): The ninth component of complement and the pore-forming protein (perforin 1) from cytotoxic T cells: Structural, immunological, and functional similarities. *Science* 233: 184–190

Zunino S, Allison NJ, Kam C-M, Powers JC, Hudig D (1988): Localization, implications for function, and gene expression of chymotrypsin-like proteinases of cytotoxic RNK-16 lymphocytes. *Biochim Biophys Acta* 967: 331–340

VI
Alternative Mechanisms of Cytolysis

28
Possible Role of Extracellular ATP in Cell–Cell Interactions Leading to CTL-Mediated Cytotoxicity

Frank Redegeld, Antonio Filippini, Guido Trenn, and Michail V. Sitkovsky

The nature of the so-called lethal hit that causes target cells (TC) to die after a relatively short encounter with cytotoxic T lymphocytes (CTL) is of great interest not only to immunologists. This problem is also very challenging to those who realize that CTL–TC interactions present an interesting model system in which contacts between two different cells–(CTL) and antigen (Ag)-bearing (TC)–result in dramatic consequences for both participants: the CTL is activated due to the crosslinking of the T-cell receptor (TCR) with the Ag, whereas the TC eventually dies (Goldstein, 1987).

Crosslinking of the TCR by the Ag on the surface of the TC leads to multiple changes in biochemical processes and functions of the CTL, including the triggering of the exocytosis of cytolytic granules (Chapter 14, this volume; Sitkovsky, 1988). The release of the content of these cytotoxic granules from the CTL toward the TC has been considered to be an important part of the "lethal hit" delivery, since cytolytic, pore-forming polypeptides (cytolysin/perforin) were found to be localized in granules (Bykovskaya et al., 1978; Podack et al., 1984; Podack et al., 1991; Tschopp and Nabholz, 1990; Krahenbuhl and Tschopp, 1991; Chapter 14, this volume). Although exocytosis of cytolytic granules has been widely accepted as a mechanism of cell-mediated cytotoxicity (challenges from G. Berke not withstanding: see Chapter 10, this volume; Berke and Rosen, 1987; Berke, 1991), no information was available concerning the molecular requirements of this process in the CTL, mostly due to the lack of the convenient assay. The description of the trypsin-like protease, N-α-benzyloxycarbonyl-L-lysine thiobenzyl (BLTesterase, in the CTL (Chapter 25, this volume; Pasternack and Eisen, 1985; Masson and Tschopp, 1987), which was shown to be located in the CTL granules (Takayama and Sitkovsky, 1987; Henkart et al., 1987; Peters et al., 1991), helped to establish such an assay (Pasternack et al., 1986; Takayama et al., 1987b; Taffs and Sitkovsky, 1991). The role of BLTesterase in CTL functions is still not clear, but studies of BLT esterase served to accelerate studies of CTL activation and exocytosis. Especially useful were findings that BLT esterase was located in granules and its secretion could be triggered by incubating CTL with Ag-bearing TC or with immobilized anti-TCR mAb (Takayama and Sitkovsky, 1987a; Takayama et al., 1987a). Biochemical and morphological data obtained in experiments using other markers of granules supported the validity of this enzyme for studies of exocytosis in CTL (Takayama and Sitkovsky, 1987; Peters et al., 1991).

The intensity of secretion of BLT esterase depends on the surface density of activating anti-TCR mAb, which made it possible to study different inhibiting or activating agents in conditions most favorable for the detection of their modulating properties. During biochemical studies of exocytosis of cytolytic granules in CTL (see Chapter 31, this volume), we encountered several situations that questioned the validity of the granule exocytosis model of CTL-mediated cytotoxicity (Takayama and Sitkovsky, 1987; Trenn et al., 1987). For example, exocytosis of cytolytic granules from CTL was found to be absolutely dependent on extracellular Ca^{2+}, while TC lysis by CTL could proceed even without the Ca^{2+}, although lower levels of cytotoxicity were observed. When CTL were incubated with Ag-bearing TC or with anti-TCR mAb-modified TC in the presence of 2 mM Mg_2-[ethylenebis(oxyethylenenitrilo)]tetraacetic acid (Mg_2-EGTA), we found that removal of extracellular Ca^{2+} abolished the TCR-triggered exocytosis of granules. Under the same conditions, in parallel assays significent lysis of target cells (30.6%) was still observed (Trenn et al., 1987). A similar conclusion was reached by Ostergaard et al. (1987) using essentially the same approach.

Lytic activity of CTL could be dissociated from the exocytosis of granules in other experimental conditions, including CTL pretreatment with cholera toxin, which induced an increase of intracellular cAMP levels and a strong, almost complete inhibition of exocytosis, which cytotoxicity was only moderately affected (Takayama et al., 1988).

The possibility of dissociating CTL-mediated lysis of cells from cytolytic granules exocytosis led to the conclusion that CTL may not employ the exocytosis of cytolytic granules for this effector function, but may have a different mechanism for the lethal hit delivery.

The Search for an Alternative Model of Cytotoxicity

The following main criteria are to be satisfied for a correct model of cell-mediated cytotoxicity: (1) it should explain the Ca^{2+}-independent TC lysis; (2) it should explain the survival of the CTL during a lethal hit; (3) removal of a suspected "lethal hit" intermediate from the assay should result in the inhibition of CTL-mediated cytotoxicity; and (4) a lethal hit intermediate should cause DNA fragmentation in TC.

In the search for an alternative (to perforin-mediated) mechanism of TC lysis, we first considered the role of resident, cell surface membrane-bound molecules (Sitkovsky, 1988). Consideration of extracellular ATP as the lytic intermediate was prompted by a control experiment that we designed in our parallel studies of the role of calmodulin-dependent phosphatase, calcineurin, in CTL activation. Since we found calcineurin to be the major Ca^{2+}/CaM-binding protein in lymphocytes (Kincaid et al., 1987), it immediately suggested that protein dephosphorylation by calcineurin may be the mechanism of propagation and the actual result of Ca^{2+} increase. Accordingly, to provide the experimental evidence for that, we tested phosphorylation of proteins by protein kinase using γ-S-ATP. The advantage of using γ-S-ATP was that thiophosphorylated proteins were not expected to be efficiently dephosphorylated by protein phosphatases, and thus would be "locked in" the phosphorylated state. Experiments with γ-S-ATP required the use of permeabilized lymphocytes, and control experiments with unpermeabilized cells and ATP were needed. During these control experiments we had to deal with the cell-permeabilizing properties of extracellular ATP (Rosengurt and Heppel, 1979; Steinberg et al., 1987; Dubyak and Fedan, 1990), and these properties of ATP_o prompted us to consider extracellular ATP (ATP_o) as a potential lytic intermediate in CTL-mediated cytotoxicity (Filippini et al., 1989; 1990; Chapter 29, this volume). The role of ATP_o in cytotoxicity is also being investigated by P. Zanovello, F Di Virgilio, and colleagues Di Virgilio et al., Zanovello et al., 1989; 1990; Chapter 29, this volume).

Below we describe experiments that support the model based on the assumption that extracellular ATP (which is released by CTL due to TCR cross-linking by Ag on the surface of TC in CTL–TC conjugates) is possibly involved in the lethal hit delivery.

Cytolytic Properties of Extracellular ATP

Incubation of cultured cell lines with ATP or ATP analogs (Filippini et al., 1990) resulted in the destruction of EL4 thymoma cells and P815 mastocytoma cells. Significant cytoxicity was observed even at 0.5 mM ATP with EL4 cells. CTL were resistant, even when the concentration of ATP was raised to 8 mM. It is remarkable that CTL clones were not lysed by ATP at concentrations that caused a high degree of EL4 or P815 cell death. The same results were obtained using the trypan blue exclusion test and the lactate dehydrogenase (LDH) release assay, which were used independently to confirm the cytotoxic effects of extracellular ATP. The time course of ATP-induced cell lysis is similar to that described for CTL-induced ^{51}Cr release, since significant EL4 and P815 lysis was already observed after 1 hr., and practically all cells were dead after 7 hr. Among other adenine nucleotides tested, only ATP analogs and ADP had any measurable lytic effect toward EL4, although it was always less than that of ATP itself. ADP had no effect on P815 cells, and AMP had no effect in any tested cellular system.

The Effect of ATP$_o$ Is Extracellular Ca^{2+}-Independent

The lytic effects of ATP were not affected by the addition of EGTA to the incubation medium to chelate the extracellular Ca^{2+}. Both sublines of P815 (P815 and P815.2) as well as EL4 cells released approximately the same proportion of intracellular radiolabeled material with or without EGTA in the incubation medium.

High Level of Ecto-ATPase Activity on the CTL surface as a Possible Explanation of Their Survival During the Lethal Hit Delivery to TC

It is a long-standing mystery that CTL is protected from lysis during their lethal hit delivery to TC. The ATP$_o$ model may help to explain the mystery by the fact that CTL have an unusually high level of ecto-ATPase actitity. The resistance of CTL clones to high concentrations of extracellular ATP suggested that it could be due partly to their increased cell surface-associated ATP-degrading activity, possible mediated by the expression of high amounts of ecto-ATPase. Experiments were therefore designed to test is CTL do indeed have the capacity to lower the extracellular concentration of this lytic agent.

1. The basis for the first, indirect approach was the expectation that the cytotoxicity of extracellular ATP for ATP-sensitive cells would be greatly diminished in the presence of CTL (as a source of ATP degradation). Such as experiment could be performed using a CTL clone and ^{51}Cr-labeled cells that are *not* recognized by that clone. Thus, CTL were present in the assay only as a source of the ATP-degrading activity, but not in their capacity to kill target cells, causing ^{51}Cr release. Using such an assay, we demonstrated that CTL can protect (rescue) EL4 cells from lysis by ATP, since much less ^{51}Cr release was detected when CTL OE4 were present during incubation of ^{51}Cr-labeled EL4 with ATP.

2. Measurement and biochemical characterization of ecto-ATPase activity in intact CTL. Since ecto-ATPase activity is measured by incubation of cells with ATP, and since we demonstrated that extracellular ATP can kill cells, causing release of intracellular enzymes and possibly intracellular ATPase, considerable effort was made to localize the measured activity to a plasma membrane ectoenzyme with its nucleotide-hydrolyzing site facing into the extracellular space (Filippini et al., 1990). Several independent assays for ecto-ATPase activity were used. Luminometry provided a sensitive method to measure ATP concentration using a luciferin–luciferase assay system. This method was applied to determine ecto-ATPase activity on CTL by addition of these cells to a reaction mixture containing ATP. Degradation of ATP was evident in the first seconds after addition of CTL to the luminometer chamber. In contract to CTL OE4, very little change in luminescence intensity (ATP degradation) could be detected when tumor cells EL4 and P815 were added to the luminometer.

3. Effect of inhibitors on ecto-ATPase activity. Fluorescein 5′-isothiocyanate (FITC) and 5′-p-fluorosulfonylbenzoyladenosine (FSBA) have been used before for the affinity labeling of the ATP-binding sites of various enzymes. FSBA is a synthetic adenosine analog, which is capable of covalently reacting (alkylation) with proteins. It was shown that FSBA labels two different sites on the subunit of the dog kidney Na^{2+}/K^{+}-transporting ATPase, and the amino acid sequences of the labeled tryptic peptides were described. We made an assumption that ecto-ATPases may contain one or more ATP-binding sites reactive with these affinity labels.

We found that FSBA strongly inhibited ecto-ATPase activity of cloned CTL, whereas FITC and vanadate pretreatments had no effect. Addition of ATP together with FSBA during CTL pretreatment completely protected ecto-ATPase activity from FSBA inactivation, indicating that the molecular target of FSBA is an extracellular ATP-binding site on CTL. We found that even 10 mM ATP was not able to lyse nontreated CTL. However, when the CTL ecto-ATPase was inactivated by pretreatment with FSBA, they became susceptible to the lytic effects of ATP. In contrast, when 3 mM ATP was present during pretreatment of CTL with FSBA (thus protecting ATP-binding sites on CTL), no ecto-ATPase inactivation was observed and CTL were still resistant to the lytic effects of ATP.

These results suggested a functional role of ecto-ATPase: protection of effector cells of the immune system in areas where local concentrations of extracellular ATP could be quite high due to specific release by effector cells or to cell death. When cells are damaged by physical injury or by other means that are not accompanied by extensive degradation,the neighbors of damaged cells may be transiently exposed to concentrations of ATP in the millimolar range. It is also an attractive hypothesis that ecto-ATPase of CTL protects them from the lytic intermediate (ATP) that is involved in the delivery of the lethal hit to the target cell.

TCR-triggered CTL as the source of ATP$_o$

Our hypothesis was based on the assumption that extracellular ATP (ATP$_o$) is available and present during the interactions between CTL and TC. ATP$_o$ has been detected previously in the blood and lymph *in vivo*, and a physiological role for ATP$_o$ has been proposed (reviewed in G. Dubyak and J. Fedan, 1990). We tested if ATP$_o$ is released by the CTL in response to the crosslinking of the TCR by the antigen-bearing TC (A. Filippini et al., 1991). To avoid the presence of the large ATP-rich tumor TC in the assay, we used a model system for CTL–TC interactions in which cells were activated by immobolized anti-TCR mAb. We confirmed in parallel experiments that the crosslinking of the CTL TCR by either anti-CD3 or anti-TCR mAb is sufficient to trigger delivery of the lethal hit by the CTL. No significant extracellular ATP accumulation was detected when peritoneal exudate lymphocytes (PEL) were incubated alone, whereas up to 15% of total cellular ATP content (in some experiments) was available for the luciferin-luciferase reaction after incubation of CTL with anti-CD3 mAb. Similar results were obtained with anti-TCR mAb and concanavalin A (con A). Cloned CTL also accumulated ATP$_o$ after incubation with con A or anti-TCR mAb. Immobolized anti-LFA-1 mAb was much less efficient in triggering ATP$_o$ accumulation, as was shown with both the CTL clone and PET CTL. The accumulation of cell surface-associated ATP$_o$ in these experiments is most likely underestimated because of the presence of ATP-depending enzymes on the CTL surface. Incubation of helper T-cell clones with con A or with immobolized anti-TCR mAb did result in the accumulation of ATP$_o$, suggesting that ATP$_o$ accumulation could be the general property of T lymphocytes in response to the TCR crosslinking.

Since an alternative mechanism of CTL-mediated cytotoxicity should also explain the ability of CTL to kill TC in the absence of extracellular Ca^{2+}, we tested the effect of Mg$_2$EGTA in order to reduce the extracellular Ca^{2+} concentration. We found that the addition of Mg$_2$EGTA to the incubation medium did not prevent the accumulation of ATP$_o$ by anti-CD3 mAb-activated CTL.

One of the criteria for validation of any cytotoxicity model is the need to demonstrate the inhibiting effect of the removal of a lytic intermediate in CTL-mediated cytotoxicity. We attempted to do this by using ATP-degrading enzymes. We failed to provide conclusive data using this approach.

Caveats in the Use of ATP-Degrading Enzymes in the Studies of Cytotoxicity

We originally attempted to add different ATP-degrading enzymes (apyrase, hexokinase, glycerokinase, Na$^+$/K$^+$-ATPase) and found that these enzymes do block cytotoxicity (Filippini et al., 1991). To rule out the possibility that such inhibition could be due to contamination in the enzyme preparations we tested the effects of ATP-degrading enzymes in parallel in ^{51}Cr-release cytotoxicity and BLT esterase release assays. It was found that cytotoxicity was inhibited but granule exocytosis was not. it was technically difficult to perform the most straightforward control experiment by testing the effect of heat-inactivated enzymes in a cytotoxicity assay, since they became gel-like and the observed inhibition could be due to the increased viscosity of the incubation medium. However, in several experiments we did observe that vanadate-inactivated Na$^+$/K$^+$-ATPase was less inhibitory to CTL in the ^{51}Cr-release assay. This combination of suggestive data led us to the mistaken conclusion that an ATP-degrading activity was the major reason for the observed inhibition of CTL-mediated cytotoxicity by these enzyme preparations.

Subsequently, our attention was attracted (S. Gromkovski, personal communication) to the possibility that ammonium sulfate (which was present in the hexokinase preparations) might be the inhibitory factor. Indeed, this was the case for at least one preparation of hexokinase. Subsequently, we reevaluated the effects of all other ATP-degrading enzymes after extensively dialyzing them before used to remove any traces of sales that could be inhibitory to the CTL lethal hit. It was also found that citrate salts can account for a large part of the inhibition of cytotoxicity by hexokinase, which was obtained as a lyophilized preparation.

Clarification of the effect of salts in the experiments with hexokinase is now published (Filippini et al., 1991; Redegeld et al., 1991), but two interesting questions have been raised by these results. (1) Why is ammonium sulfate strongly inhibitory for cytotoxicity, but has no effect on granule exocytosis? (2) Is the need for relatively high concentrations of ATP-degrading enzymes due to the highly localized, closed areas of contact between CTL and TC? Further experiments will be needed to shed light on these questions.

The data described above also attract attention to another paradoxical observation: both extracellular ATP and highly active ecto-ATPase coexist on the surface of the CTL (Filippini et al., 1990, 1991). The obvious explanation is to assume that ATP$_o$-enriched domain(s) on the surface of the CTL are spatially separated from ecto-ATPase containing domain(s) of the CTL plasma membrane. This awaits further experimental confirmation, but it does point to the possible exquisite steric regulation of ATP$_o$ accumulation and of ATP$_o$ degradation processes. Accordingly, we explore the possibility that ATP$_o$ may be accumulated in specialized, limited areas of contact between STL and TC, where highly active protein kinases utilize ATP$_o$ to phosphorylate surface proteins. This model is in agreement with such observations as the coexistence of ATP$_o$ and ecto-ATPase, and the presence of highly active cellular protein kinases (and phosphatases) on CTL (F.R. and M.V.S., work in progress). The important and yet unanswered question is which surface proteins on TC (or on CTL?) are targeted for phosphorylation by extracellular protein kinases and how such modification of the extracellular portions of the surface proteins may lead to gross changes in target cell physiology.

Is ATP$_o$ a "Hit" Molecule or a "Messenger"?

As we described above, the emergence of the extracellular ATP cytotoxicity model was prompted by the known abilities of ATP (as ATP^{4-} species) to permeabilize cells (and to function as a "hit" molecule) and by the important role of MgATP^{2-} in the functioning of protein kinases (ATP$_o$ functions as a "messenger").

The two opposite "hit" and "messenger" models of the participation of ATP in CTL effector functions differ in their requirements for additional molecular intermediate(s): (1) according to a "hit" model, ATP$_o$ can act alone due to its cell-permeabilizing properties; (2) according to a "messenger" model, ATP$_o$ acts in concert with other molecules, e.g., ectokinases, ecto-ATP-receptors, ecto-ATPases, and/or yet other unidentified participants.

The requirements for the increase of the fraction of ATP^{4-} include alkaline pH, but it was out experience that CTL can destroy TC even at pH 6.5. Therefore, we felt that it was unlikely that the cell-permeabilizing properties of ATP$_o$ are fully responsible for its possible role in CTL-mediated cytotoxicity. Direct experiments were designed to test the possibility of dissociating conditions for ATP$_o$-induced and CTL-induced TC lysis, by testing the pH dependence and Mg^{2+} dependence of the lysis of different TC by CTL and ATP$_o$.

It was shown (Redegeld et al., 1991) that ATP$_o$ did not efficiently lyse P815 cells at pH 6.0, 6.5, 7.0, and 7.5. A significant ATP-induced lysis was observed only at pH 8.0. The same TC which are routinely used as Ag-bearing TC for CTL clone OE4 were efficiently killed by these CTL at pH 7.0, 7.5, and 8.0. At pH 6.0 no lysis was observed; at ph 6.5 strong lysis was observed, though less than at pH 7.0. Thus, clear dissociation of ATP$_o$-mediated and CTL-mediated lysis could be demonstrated at pH 6.5, 7.0, and 7.5 when P815 cells were used as TC. Similar results were obtained in the pH-dependence studies with YAC-1 and EL4 cells. Thus, it was possible to demonstrate that the pH requirements of the assay media were different for CTL—mediated and ATP$_o$-induced TC lysis. These results point to the possibility that if ATP$_o$ is indeed involved in extracellular, Ca^{2+}-dependent, CTL-mediated cytotoxicity, it may act not through its ATP^{4-} molecular species.

The differences in requirements for ATP$_o$-mediated and CTL-mediated lysis were confirmed in experiments using different concentration of Mg^{2+}. These experiments were designed to take advantage of the fact that the relative concentrations of ATP^{4-} and MgATP^{2-} are dependent on the Mg^{2+} concentration.

Thus, as the MgATP^{2-} concentration was decreased in the media with low Mg^{2+} concentration, we observed a marked decrease in the lysis of Ag-bearing TC (P815). To bypass the requirements for the accessory molecule-dependent conjugate formation, we used a CTL-activating hybrid antibody to bridge CTL and syngeneic TC. It was found that the removal of Mg^{2+} blocked the CTL-mediated cytotoxicity in a retargeting assay, also raising the interesting possibility that Mg^{2+} is required for a step other than accessory molecule-dependent conjugate formation between CTl and TC. Thus, there is a strong inhibition of CTL-mediated cytotoxicity with the sharpest drop in cytotoxicity between 50 and 25 μM Mg^{2+} in the retargeting assay (Redegeld et al., 1991). When the same experiments were performed in parallel Mg^{2+}-dependence studies of ATP-dependent cytotoxicity, the opposite dependence on Mg^{2+} was observed. Lowering of Mg^{2+} concentration resulted in an actual increase of ATP-mediated lysis, while the presence of 1 mM Mg^{2+} was inhibitory, suggesting that cytotoxic effects of ATP are most likely due to the ATP^{4-} form. Additional evidence that ATP does not act solely as a "hit" molecule comes from the observation that tumor cells grown as ascites are susceptible to lysis by ATP, while after long propagation *in vitro* they lose their susceptibility to ATP. On the other hand, both *in vitro*-grown and *in vivo*-maintained tumor cells are susceptible to lysis by CTL.

The latter experiments provided evidence that it is possible to dissociate ATP_o-induced lysis from Ca^{2+}-dependent, CTL-mediated lysis. Thus, it seems that if ATP_o participates in Ca^{2+}-dependent CTL-mediated cytoxicity, it would not be solely (or maybe not at all) as a "hit" molecule, but also as a "messenger," which requires the involvement of yet another molecular participant. It is also an important observation that Mg^{2+} is needed for TC lysis beyond the step of conjugate formation as demonstrated in the retargeting assay. The most straightforward explanation involved the assumption that Mg^{2+} is required for the functioning of $MgATP^{2-}$-dependent extracellular enzymes, which include ectoprotein kinases, and possibly some yet unidentified $MgATP^{2-}$ receptors.

The inability of ATP_o alone to lyse erythrocytes that are lysed by CTL, which was found during the same studies, could be explained, for example, by the hypothesis that erythrocytes do not express the gap junction protein connexin-43, which was recently suggested to be involved in ATP-induced pore formation (Beyer and Steinberg, 1991). Similarly, differential susceptibility of EL4, P815, and YAC-1 cells to lysis by ATP_o could also be due to differential expression of connexin. It is of interest to test if the loss of susceptibility of P815 cells to ATP during *in vivo* cultur is correlated with the loss of connexin. it could be more that coincidence that *in vivo*-maintained TC are more susceptible both to ATP^{4-}-induced lysis and to extracellular Ca^{2+}-independent, CTL-mediated lysis (unpublished observations). it will be of great interest to correlate the expression of connexin and the susceptibility of TC to lysis by CTL or ATP in the absence of Ca^{2+}.

We are currently exploring the possibility that ATP_o requires ectoprotein kinases and/or other polypeptides which may be shed (released, secreted) from the CTL during interaction with the TC. It is our view that independently of its validity as a mechanism of CTL-mediated cytotoxicity, the ATP_o model is valuable to consider, since it attracted our attention to previously unappreciated proteins and low-molecular-weight intermediates that are localized on the surface of lymphocytes and whose functions are not yet clear.

References

Berke G (1991): Lymphocyte-triggered internal target disintegration. *Immunol Today* 12: 396–399

Berke G, Rosen D (1987): Are lytic granules, and perforin 1 thereof involved in lysis induced by in vivo primed peritoneal exudate CTL? *Transplant Proc* 19: 412–416

Beyer EC, Steinberg TH (1991): Evidence that the gap junction protein connexin-43 is the ATP-induced pore of mouse macrophages *J Biol Chem* 266: 7971–7974

Bykovskaja SN, Rytenko AN, Rauschenbach MO, Bykovsky AF (1978): Ultrastructural alteration of cytolytic lymphocytes following their interaction with target cells. II. Morphogenesis of secretory granules and intracellular vacuoles. *Cell Immunol* 40: 175

Di Virgilio F, Bronte V, Callavo D, Zanovello P (1989): Responses of mouse lymphocytes to extracellular adenosine 5-triphosphate (ATP). Lymphocytes with cytotoxic activity are resistant to the permeabilizing effects of ATP. *J Immunol* 143: 1995–1960

Dubyak GR, Fedan JS, eds. (1990): The biological actions of extracellular ATP. Proceedings of a New York Academy of Sciences Conference, November 27–29, 1989

Filippini A, Taffs RE, Agui T, Sitkovsky M (1990): Ecto-ATPase activity in cytolytic T-lymphocytes protection from the cytolytic effects of extracellular ATP. *J Biol Chem* 265: 334–340

Filippini A, Taffs R, Sitkovsky MV (1989): Alternative molecular mechanisms of T-cell receptor regulated effector functions of cytolytic T lymphocytes. *J Immunol Immunopharmacol* 9: 118–119

Filippini A, Taffs R, Sitkovsky M (1991): Extracellular ATP in T-lymphocytes activation: Possible role in effector functions. *Proc Natl Acad Sci USA* 87: 8267–8271

Golstein P (1987): Cytolytic T-cell melodrama. *Nature* 327: 12

Henkart PA, Berrebi GA, Takayama H, Munger WE, Sitkovsky MV (1987): Biochemical and functional properties of serine esterases in acid cytoplasmic granules of cytotoxic T lymphocytes. *J Immunol* 139: 2398–2405

Kincaid R, Takayama H, Billingsley ML, Sitkovsky MV (1987): Differential expression of calmodulin-binding proteins in B, T lymphocytes and thymocytes. *Nature* 330: 176–178

Krahenbuhl O, Tschopp J (1991): Perforin-induced pore formation. *Immunol Today* 12: 399–402

Masson D, Tschopp J (1987): A family of serine esterases in lytic granules of cytolytic T lymphocytes. *Cell* 49: 679–685

Ostergaard HL, Kane KP, Mesher MF, Clark WE (1987): Cytotoxic T-lymphocyte mediated lysis without release of serine esterase. *Nature* 330: 71–73

Pasternack MS, Eisen HN (1985): A novel serine esterase expressed by cytotoxic T lymphocytes. *Nature* 314: 743–745

Pasternack M, Verret CR, Liu MA, Eisen HN (1986): Serine esterase in cytolytic T lymphocytes. *Nature* 322: 740–743

Peters PJ, Borst J, Oorschot V, Fukuda M, Krahenbuhl O, Tschopp J, Slot JW, Geuze HJ (1991): Cytotoxic T lymphocyte granules are secretory lysosomes, containing both perforin and granzymes. *J Exp Med* 173: 1099–1109

Podack ER, Konigsberg PJ (1984): Cytolytic T cell granules. Isolation, structural, biochemical and functional characterization. *J Exp Med* 160: 695–710

Podack ER, Henhgartner H, Lichtenheld MG (1991): A central role of perforin in cytolysis? *Ann Rev Immunol* 9: 129–157

Redegeld F, Filippini A, Sitkovsky M (1991): Comparative studies of the cytotoxic T-lymphocyte-mediated cytotoxicity and of extracellular ATP-induced cell lysis. Different requirements in extracellular Mg^{2+} and pH. *J Immunol* 147: 3638–3645

Rozengurt E, Heppel LA (1979): Reciprocal control of membrane permeability of transformed cultures of mouse cell lines by external and internal ATP. *J Biol Chem* 254: 708

Sitkovsky MV (1988): Mechanistic, functional and immunopharmacological implications of biochemical studies of antigen receptor-triggered cytolytic T-lymphocyte activation. *Immunol Rev* 103: 127–160

Steinberg TH, Newman AS, Swanson JA, Silverstein SC (1987): ATP^{4-}-permeabilizes the plasm membrane of mouse macrophages to fluorescent dyes. *J Biol Chem* 262: 8884–8888

Taffs RE, Sitkovsky MV (1991): Granule enzyme exocytosis assay for cytotoxic T lymphocyte activation. In: *Current Protocols in Immunology* Unit 3.16. Coligan J, Kruisbeck AM, Margulies DH, Shevach EM, Storber W, eds. New York: John Wiley & Sons

Takayama H, Sitkovsky MV (1987): Antigen receptor-regulated exocytosis in cytotoxic T lymphocytes. *J Exp Med* 166: 725–743

Takayama H, Trenn G, Humphrey W, Bluestone JA, Henkart PA, Sitkovsky MV (1987a): Antigen receptor-triggered secretion of a trypsin-type esterase from cytotoxic T-lymphocytes. *J Immunol* 138: 566–569

Takayama H, Trenn G, Sitkovsky MV (1987b): A novel cytotoxic T lymphocyte activation assay. Optimized conditions for antigen receptor triggered granule enzyme secretion. *J Immunol Methods* 104: 183–190

Takayama H, Trenn G, Sitkovsky MV (1988): Locus of inhibitory action of cAMP-dependent protein kinase in the antigen-receptor triggered cytotoxic T-lymphocyte activation pathway. *J Biol Chem* 263: 2330–2336

Trenn G, Takayama H, Sitkovsky MV (1987): Exocytosis of cytolytic granules may not be required for target cell lysis by cytotoxic T-lymphocytes. *Nature* 330: 72–74

Tschopp J, Nabholz M (1990): Perforin-mediated target cell lysis by cytolytic T lymphocytes. *Annu Rev Immunol* 8: 279–302

Zanovello P, Bronte V, Rosato A, Pizzo P, Di Virgilio F (1990): Responses of mouse lymphocytes to extracellular ATP. II. Extracellular ATP causes cell type-dependent lysis and DNA fragmentation. *J Immunol* 145: 1545–1550

29
Cell-Permeabilizing Properties of Extracellular ATP in Relation to Lymphocyte-Mediated Cytotoxicity

Francesco Di Virgilio, Paola Zanovello, and Dino Collavo

Introduction

Although the phenomenon of lymphocyte-mediated cytotoxicity is well known, the underlying mechanisms are at present a hotly debated issue because the molecular participants involved still elude definition (Berke, 1991; Krahenbuhl and Tschopp, 1991). Although perforin is a well-characterized cytocidal mediator, much evidence indicates that other molecules are implicated (Trenn et al., 1987). In the last three years, we have explored the possibility that extracellular ATP (ATP$_o$) might constitute a mediator of lymphocyte-mediated cytotoxicity (Di Virgilio et al., 1989; Zanovello et al., 1990; Di Virgilio et al., 1990).

That ATP is released by several cell types and displays lytic activity on the outer cell membrane has been long known. Furthermore, we and others observed that ATP$_o$ can cause death by apoptosis in a number of targets (Zanovello et al., 1990; Zheng et al., 1991), while lymphocytes with cytotoxic activity are resistant to its lytic effect (Di Virgilio et al., 1989; Filippini et al., 1990a; Chapter 28, this volume). These latter findings fulfill two of the most stringent requirements for a soluble mediator of cell-mediated killing. Here we will briefly review the sources, extracellular concentrations, and surface receptors for ATP$_o$, as well as the mechanisms by which this molecule causes changes in plasma membrane permeability, and its involvement in lymphocyte-mediated cytotoxicity.

Multifunctional ATP$_o$ Activity and Cellular Receptors

Functional responses to ATP$_o$ span from cell proliferation to cell death, smooth muscle contraction to neurosecretion, and inflammatory cell priming to phagocytosis inhibition (Dubyak and Fedan, 1990). ATP is known to be released into the extracellular milieu from several sources, including platelets, endothelial and chromaffin cells, and adrenergic, cholinergic, and purinergic nerve terminals. Furthermore, ATP can also leak out as a consequence of plasma membrane breakage, for example, during tissue damage. While basal plasma ATP levels are believed to be low (1–5 μM), peaks of up to 20 μM were measured in the effluent of vessels in which extensive platelet aggregation had taken place (Born and Kratzer, 1984). These determinations, however, help little in estimating the local ATP$_o$ level that is reached during thrombus formation, or in protected environments, such as those formed at sites of platelet–endothelial cell, or cytotoxic–target cell interaction. Direct measurement of ATP$_o$ within these secluded pouches is not yet possible, but its levels in a zone of close cell-to-cell contact can be roughly estimated from a calculation of its concentration in the synaptic cleft during nerve terminal stimulation. ATP in fact is co-stored with acetylcholine within the synaptic vesicles at a ratio of 1 to 5; thus, given that acetylcholine reaches a concentration of 0.5–1 mM after an action potential, it is not unlikely that ATP levels as high as 100–200 μM could build up in the synaptic space (Fu and Poo, 1991).

ATP$_o$ effects are mediated by at least three different subtypes of specific cell surface receptors (purinergic P2): P$_{2y}$; P$_{2x}$, and P$_{2z}$ (Gordon, 1986). P$_{2z}$ is responsible for ATP$_o$-dependent changes in plasma membrane permeability to low-molecular-weight aqueous solutes (see below), and therefore we will also refer to it as the "ATP-gated pore" or "pore-receptor." An additional P$_2$ receptor,

TABLE 1. Classification of ATP receptors

Subtype	Signal transduction mechanism
P2y	Receptor coupled to Ins1,4,5P$_3$ formation and Ca^{2+} mobilization, often via a G-protein
P2x	Ligand-gated cation channel (permeable to Na$^+$, Ca^{2+}, and solutes up to 200 daltons)
P2z	Nonselective membrane pore permeable to ions and metabolites up to 900 daltons

ATP-Dependent Plasma Membrane Permeability Changes

Activation of a Nonspecific Aqueous Pore

Rozengurt et al. (1977) and Cockcroft and Gomperts (1979) were the first to show that ATP$_o$ induces the permeabilization of the plasma membrane to normally impermeant low-molecular-weight hydrophilic solutes. It is known that several molecules are freely permeant through the ATP-gated pore, including ethylenediaminetetraacetate (EDTA), ethyleneglycol-bis (β-aminoethylether)-N,N,N',N-tetraacetate (EGTA), ethidium bromide, GTPγS, lucifer yellow, eosin yellowish, brilliant blue, fura-2 free acid, 1-(4-trimethylammonium-phenyl)-6-phenyl 1-1,3,5–hexatriene (TMA-DPH), and ATP itself. Thus, intracellular nucleotides will also leak out through the pore. Although common to several cell types, the molecular mechanism by which ATP$_o$ causes such pores in the plasma membrane is poorly understood (Figure 1). Studies with the patch-clamp technique showed that changes in the plasma membrane potential (the earliest measurable alteration in plasma membrane permeability) occur within a few tenths of a millisecond following ATP$_o$ addition, with no lag (Buisman et al., 1988). The absence of lag was also confirmed by measurement of uptake of the fluorescent tracers ethidium bromide or TMA-DPH (Tatham and Lindau, 1990). The general consensus is that the active ATP species is ATP^{4-}, i.e., the fully dissociated tetraanionic form. ATP can be chelated by Ca^{2+}, Mg^{2+}, and H$^+$; therefore, since biological fluids contain millimolar amounts of Ca^{2+} and Mg^{2+} at a pH close to neutrality, ATP^{4-} will obviously be in equilibrium with the chelated forms, ATPCa^{2-}, ATPMg^{2-}, and HATP^{3-} (Di Virgilio and Steinberg, 1992). Hence, the "active" ATP$_o$ concentration in most experiments usually refers to "total" added ATP.

The picture is further confused by the various experimental conditions used by different workers: presence or absence of divalent cations, pH ranging from 7.4 to 8, and different cell concentrations. Moreover, it is clear that different cell populations have a dissimilar sensitivity to the pore-forming activity of ATP$_o$. Despite these uncertainties, it seems reasonable to say that pore activation requires *total* ATP$_o$ concentrations that are at least one order of magnitude higher than those sufficient

activated by ATP$_o$ as well as other extracellular nucleotides (UTP, GTP, CTP, ITP) has been postulated (Dubyak, 1991), but whether this molelcule has a separate identity or instead coincides with the P$_{2y}$ receptor is controversial. The functional, biochemical, and molecular characterization of the P$_2$ receptors has been hindered by a lack of specific synthetic agonists or antagonists, and none has thus far been isolated and cloned.

ATP$_o$ surface receptors are associated with a variety of different signal transduction mechanisms (see Table 1): Ca^{2+} and Na$^+$ fluxes; plasma membrane depolarization; phospholipase C and D stimulation; Ca^{2+} release from intracellular stores; and activation of protein kinase C. However, ATP$_o$ also triggers other unkown transduction pathways, as has been clearly demonstrated by specific cellular responses, e.g., insulin secretion in RINm5 cells (Li et al., 1991), in the absence of known intracellular second messenger generation.

Once in the extracellular environment, ATP$_o$ is rapidly hydrolyzed by powerful ecto-ATPases found on the plasma membrane of all cell types. Interestingly enough, these enzymes have a K_m for ATP of about 200–300 μM. i.e., at least one to two orders of magnitude higher than the ATP$_o$ concentration in blood. Ecto-ATPase activity in the various cell populations differs, and this characteristic has constituted an explanation for dissimilar susceptibilities to the cytotoxic effects of ATP$_o$ (Filippini et al., 1990a); however, Steinberg and Silverstein (1987) clearly dissociated ecto-ATPase activity from membrane permeabilization in macrophages. The function of plasma membrane ecto-ATPases is unknown, but based on the recognized harmful effect of high ATP$_o$ concentrations, these workers advanced the hypothesis that their role was the rapid hydrolysis of ATP$_o$, much like acetylcholinesterase in the cholinergic synapse.

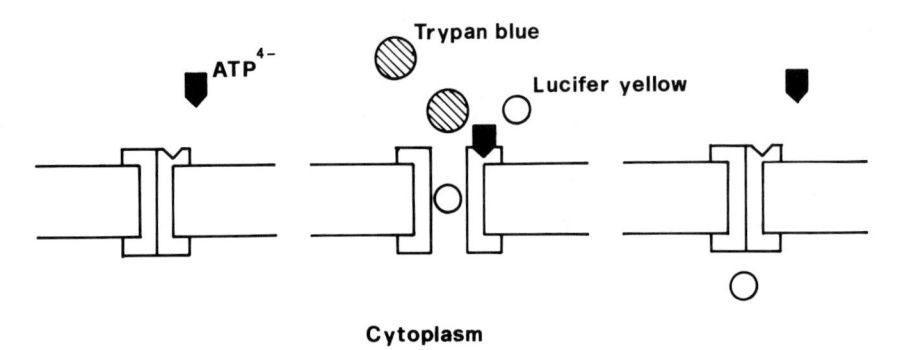

FIGURE 1. Hypothetical model of reversible permeabilization of the plasma membrane by ATP_o. ATP_o, in its fully dissociated form, interacts with a receptor on the outer plasma membrane. Receptor activation causes the opening of an aqueous pore (probably coinciding with the receptor) that allows permeation of water-soluble molecules of M_r up to 900. Thus, the extracellular marker lucifer yellow (M_r 463) can freely permeate through the pore, while trypan blue (M_r 960) is excluded. Removal of ATP_o within a few minutes of its addition causes the pore to close, thus trapping any molecule that permeated through the plasma membrane during the incubation with ATP_o within the cytoplasm. A prolonged incubation in the presence of ATP_o eventually causes an irreversible permeabilization of the plasma membrane, followed by efflux of large soluble cytoplasmic molecules and cell death.

to maximally activate additional cellular responses dependent on the activation of other P_2 purinergic receptors, such as P_{2y} and P_{2x}. All other tri- or diphosphorylated nucleotides thus far tested are inactive. ADP and the nonhydrolyzable ATP analogs are moderately active in some cell types and completely inactive in others. A very interesting aspect of the ATP_o-gated pore is its reversibility. In fact, provided ATP_o is removed within 15–20 min of its addition, cells will reseal and reestablish a normal plasma membrane potential. However, despite the reversibility of the plasma membrane pore, ATP_o-pulsed cells are doomed, even if they are washed free of the nucleotide after 10–15 min (Di Virgilio et al., 1989; Hogquist et al., 1991). The ATP_o-gated pore shares some similarities with gap junctions, including molecular cutoff, temperature dependence, and regulation by ATP (studied in the gap junctions by varying the nucleotide concentration on the intracellular side of the membrane) (Sugiura et al., 1990). These common features prompted studies by Beyer and Steinberg (1991) that showed that the ATP_o-gated pore of mouse macrophages coincides with the gap junctional protein connexin-43. As it is quite probable that significant ATP_o concentrations build up in the extracellular milieu, it is not clear why cells during their evolution would have conserved such a dangerous way to activate gap junctions from the outside, unless there exists a specific role for this mechanism in cytotoxicity.

Activation of a Ligand-Gated Ion Channel

Krishtal et al. (1983) first observed an ATP_o-activated conductance in mammalian sensory neurons. Although electrophysiological properties seem to vary, several other cell types have been shown to possess ATP_o-activatable cation channels. Thus, the ATP_o-gated channel of sensory neurons, for example, shows very little inactivation, while that of atrial or vas deferens cells inactivates within 2 sec. ATP_o EC_{50} varies from 0.3–10 μM in sensory neurons, to 50–60 μM in atrial muscle cells (Bean and Friel, 1990). We very recently characterized a similar ATP_o-gated channel in mouse lymphocytes and determined an EC_{50} of about 200 μM (Pizzo et al., 1991). Like other congeners in other cells, this channel is permeable to both Na^+ and Ca^{2+}, as well as to other cations up to 200 daltons. ATP is the only active ligand, very probably in its fully dissociated form (ATP^{4-}), and opening of the channel does not seem to require generation of known intracellular second messengers. We recently linked this ligand-gated ion channel to cytotoxicity to explain the mechanism of ATP_o-triggered cell death in cell types lacking the ATP_o-

gated pore (mouse lymphocytes, P-815 mastocytoma cells, YAC-1 lymphoma cells) (Di Virgilio et al., 1989; Murgia et al., 1992). We advanced the hypothesis that a prolonged collapse of the ionic gradient across the plasma membrane could eventually cause cell death even in the absence of short-term permeabilization to higher-molecular-weight solutes. A known example of a ligand-gated ion channel involved in cytotoxicity is the glutamate/*N*-methyl-D-aspartate (NMDA) receptor, which is responsible for neuronal cell death triggered by excitatory amino acids (McDermott and Dale, 1987).

ATP$_o$-Mediated Target Cell Killing

In vitro exposure of several cell lines or primary cultures to ATP$_o$ causes death (see Steinberg and Di Virgilio, 1991, for a recent review). The incubation time and the ATP$_o$ dose needed to achieve the cytotoxic effect are cell-type-dependent features, as rat mast cells or YAC-1 mouse lymphocytes are killed by 10–20 min exposure to a few tenths micromolar ATP$_o$, while L929 mouse fibroblasts or P-815 mastocytoma cells require several hours at higher concentrations (500 μM to 1–2 mM) (Di Virgilio et al., 1989; Zanovello et al., 1990; Pizzo et al., submitted). The presence of divalent cations in the culture medium also influences the effective toxic ATP$_o$ concentration, confirming that cell death is specifically triggered by ATP$_o$ in its tetraanionic form. Furthermore, pulse-chase experiments showed that the continuous presence of ATP$_o$ is not required for lysis; in mouse thymocytes, a 15 min exposure followed by washing and further incubation in ATP$_o$-free medium is sufficient to induce irreversible damage. This observation was also confirmed by Hogquist and colleagues (1991) in lipopolysaccharide (LPS)-stimulated murine peritoneal macrophages, in which 30 min incubation with ATP$_o$ sufficed to cause cell death 20 hr later.

We demonstrated that the pathways leading to ATP$_o$-dependent cell death involve not only a process of cell membrane damage (revealed by ^{51}Cr-release assay), but also nuclear DNA degradation into 200-bp fragments (detectable by agarose gel electrophoresis), the kinetics and pattern of which recall the "programmed target cell death" process occurring during lymphocyte-mediated cytotoxicity (Zanovello et al., 1990). Zheng et al. (1991) also confirmed that the most characteristic traits of apoptosis (plasma membrane blebbing, chromatin condensation, and cytoplasmic boiling) are observed in thymocytes and EL-4 cells during ATP$_o$-induced killing. ATP$_o$-induced apoptosis requires that cells be capable of *de novo* protein and RNA synthesis, as inhibitors of macromolecular synthesis blocked morphological changes typical of apoptosis and DNA degradation (Zheng et al. 1991, and our unpublished results). Interestingly, Zychlinsky et al. (1991) recently reported that cytotoxic lymphocytes mediated apoptosis only of metabolically active cells.

An inverse correlation between lysis and DNA fragmentation was observed in several cell lines; in fact, cells that are rapidly killed by ATP$_o$ exhibit minimal DNA degradation, while cells that require longer incubation times also undergo extensive DNA fragmentation (Zanovello et al., 1990). These two patterns are best documented in the YAC-1 lymphoma and P-815 mastocytoma cell lines. In YAC-1 cells, massive ^{51}Cr-release accompanied by very little DNA fragmentation is promoted by low ATP$_o$ concentrations (less than 50 μM). On the contrary, P-815 mastocytoma cells, which are sensitive to rather high ATP$_o$ concentrations (above 500 μM), exhibit massive DNA degradation. Time-course experiments also showed that DNA fragmentation was evident in P-815 cells as early as 2 hr after ATP$_o$ challenge and preceded lysis, while ^{51}Cr-release by YAC-1 cells started without lag after ATP$_o$ addition and reached near plateau levels within 2 hr (Zanovello et al., 1990). Since rapid damage to the target would prevent initiation of the DNA degradation program (Hameed et al., 1989), it is likely that in YAC-1 cells the early lysis occurring at low ATP$_o$ doses prevents the initiation of a DNA fragmentation program.

The Ca^{2+} requirement for ATP$_o$-triggered lysis and DNA fragmentation is not absolute, but is rather related to the cell type investigated, thus suggesting that the target cell contributes to its own killing even when a soluble lytic mediator is used. In fact, ^{51}Cr-release and DNA fragmentation in YAC-1 cells are inhibited in the absence of extracellular Ca^{2+} ([Ca^{2+}]$_o$), whereas these phenomena are [Ca^{2+}]$_o$-independent in several cell lines (e.g., P-815 mastocytoma cells and L929 fibroblasts) (Zanovello et al., 1990; Pizzo et al., submitted). The possibility that [Ca^{2+}]$_o$ independence was due to ATP$_o$ ability to release Ca^{2+} from intracellular stores was excluded in P-815 cells, since the ATP-dependent rise in [Ca^{2+}]$_i$ was suppressed by (Ca^{2+})$_o$ chelation with excess EGTA, thus indicating that it was entirely due to Ca influx from the extracellular medium (Zanovello et al., 1990). These observations support the hypothesis that ATP$_o$ might represent one of the mediators of Ca^{2+}-independent

TABLE 2. Lysis and DNA fragmentation in different cell lines after incubation in the presence of ATP_o, TNF, or a combination of ATP_o and TNF

	Lysis			DNA fragmentation		
	ATP_o	TNF	ATP + TNF	ATP_o	TNF	ATP_o + TNF
P-815	28	<1	26	25	<1	39
L929[a]	33	<1	30	8	<1	24
WEHI 164	5	8	40	2	15	22
L1210	47	<1	45	49	<1	98
BW5147	15	37	49	10	49	60
YAC-1	80	<1	75	2	<1	3

Target cells, labeled in their cytoplasm with ^{51}Cr and in their DNA with ^3H-TdR, were incubated for 4 hr in culture medium containing ATP_o (4 mM), rTNF (500 U/ml), or ATP_o plus rTNF. Data are expressed as percent of specific label release.
[a] L929 cells were labeled with iododeoxyuridine-125 ($^{125}IUdR$) as DNA release marker.

cell lysis in addition to other soluble factors, such as tumor necrosis factor (TNF) and lymphotoxin (LT) (Hasegawa and Bonavida, 1989).

ATP_o cytotoxic activity can be synergistically potentiated by TNF and LT in a Ca^{2+}-independent fashion (Bronte et al., submitted). It is well known that TNF or LT can cause target cell death (by apoptosis, necrosis, or an intermediate form of death) only after several hours of incubation (Schmid et al., 1986; Laster et al., 1988); however, when ATP_o was present, we observed a dramatic increase in TNF/LT-dependent cytotoxicity in a number of cell lines (Table 2). The effects of the combined application of ATP_o plus TNF/LT were clearly target cell-dependent, as some targets (e.g., P-815, L929 and L1210 cells) exhibited a synergistic increase only in DNA fragmentation, while others showed a synergistic (WEHI 164) or an additive (BW 5147) rise in both DNA degradation and ^{51}Cr-release. No synergism was observed in YAC-1 lymphocytes.

Resistance of CTL to ATP_o-Mediated Killing

Cytotoxic lymphocytes (CTL) are known to be highly resistant to cell-mediated lysis; these cytotoxic effectors are also resistant to the membrane-permeabilizing effects of ATP_o as well as to its lytic effects (Di Virgilio et al., 1989; Zanovello et al., 1990; Filippini et al., 1990a; Zheng et al., 1991). The "ATP-resistant" phenotype was observed both in CTL derived from established CTL clones and in those derived from *in vivo*-primed peritoneal exudate lymphocytes (PEL). Moreover, lymphokine-activated killer (LAK) cells generated by *in vitro* culture for 5 days in the presence of high interleukin-2 (IL-2) doses were also resistant to ATP_o. In particular, during the generation of LAK cells we observed a close temporal correlation between the acquisition of the lytic activity and the development of resistance to ATP_o and CTL-dependent lysis in the emerging cytotoxic cell population (Di Virgilio et al., 1989). Moreover, ATP_o was unable to induce apoptosis of CTL clone and LAK cells even when it was associated with TNF/LT (unpublished results).

The mechanism involved in cytotoxic T-cell resistance to ATP_o-mediated lysis is unknown. CTL and LAK cells have a much higher ecto-ATPase activity than non-cytotoxic lymphocytes (Filippini et al., 1990a; Zambon and Zanovello, unpublished observations). It is thus conceivable that CTL are protected against the cytotoxic effects of ATP_o by the ability to hydrolyze ATP_o rapidly. However, high levels of ecto-ATPase activity did not correlate with ATP_o resistance in some tumor cell lines (Zambon and Zanovello, unpublished observations); moreover, ecto-ATPases are strongly Ca^{2+}/Mg^{2+}-dependent enzymes, so resistance to ATP_o should be abrogated in the presence of EDTA. However, CTL remain fully refractory to ATP_o when incubated in several-millimolar EDTA concentrations, suggesting that other mechanisms must be operating. We suggest that a possible alternative explanation might be the lack of functional ATP_o surface receptors responsible for activating the lytic process.

ATP_o as a Possible Cytotoxic Mediator

Target cell killing by cytotoxic lymphocytes is due to two distinct mechanisms: alteration of plasma

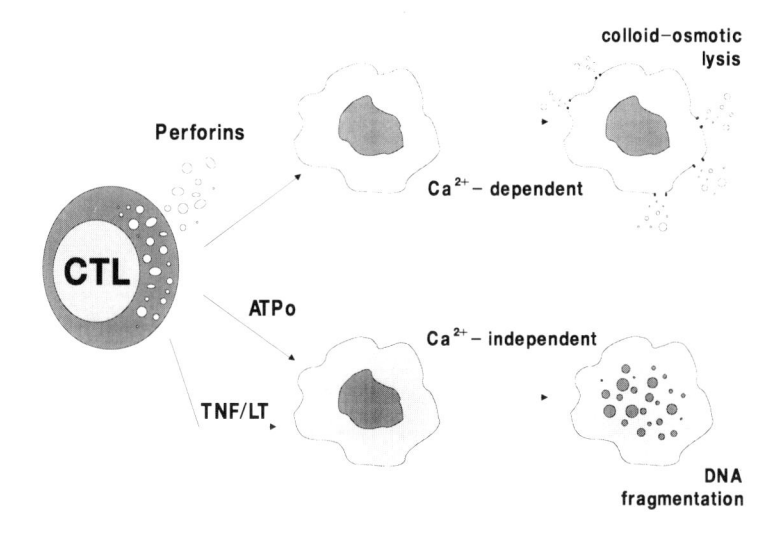

FIGURE 2. Mechanisms of lysis mediated by cytotoxic lymphocytes. Two different phenomena are responsible for target cell killing: osmotic cell lysis and DNA degradation. Osmotic cell lysis is mediated by perforin release and is strictly Ca^{2+}-dependent, as Ca^{2+} is necessary for inducing both lytic granule exocytosis and perforin polymerization. DNA degradation, which may proceed in the absence of extracellular Ca^{2+}, is most likely mediated by the synergistic effect of TNF, LT (or other secretory or membrane-associated proteins closely related to these molecules), and ATP_o.

membrane permeability leading to osmotic lysis, and DNA fragmentation leading to apoptotic death. These mechanisms are not mutually exclusive but, in most cases, complementary. Moreover, while target cell lysis is strictly Ca^{2+}-dependent and principally mediated by secretion of perforins by the effector cells, apoptosis is mainly due to an active participation of the target cell and can occur even in the virtual absence of Ca^{2+} (Figure 2). It is noteworthy that ATP_o is endowed with both properties: the capacity to alter cell membrane permeability and to induce DNA fragmentation. The latter phenomenon is independent of membrane alteration, as, in several targets, it can precede the lytic event. TNF and LT were considered as candidates for the mediators responsible for apoptosis, although the slow kinetics of TNF-induced DNA degradation (12–18 hr) are in contrast with the rapid (< 1 hr) effect mediated by intact effector cells (Figure 2). Interestingly, ATP_o and TNF/LT synergic activity results in a faster (< 2–4 hr) DNA breakdown (Bronte et al. submitted).

However, the observation that effector–target cell interaction leads to ATP_o secretion (Filippini et al. 1990b) requires further investigation; in any case, the ATP_o level reached in the bulk phase seems too low compared to the concentration required to induce *in vitro* cell lysis. It is possible that ATP_o accumulates (Filippini et al., 1990b) in the restricted area of tight contact between effector and target cells, and other mediators, such as TNF/LT and interferon-γ (IFN-γ) (Blanchard et al., (1991), could enhance its activity.

One last unresolved question is how the CTL induce the target cell disintegration program. Berke (1991) recently suggested that effector-target cell interaction activates receptors that transduce a signal across the target cell membrane leading to a persistent Ca^{2+} rise that ultimately triggers the disintegration process. This model is appealing, since purinergic receptors induce an internal Ca^{2+} rise, and ATP_o at low concentration is endowed with cell activation properties. However, the Ca^{2+} requirement for apoptosis is controversial (Cohen, 1991), and both ATP_o (Zanovello et al., 1990) and TNF (Hasegawa and Bonavida, 1989) can cause cell death in the absence of a Ca^{2+} increase.

In conclusion, ATP_o fulfills most of the requirements to be considered a mediator of perforin-independent target cell killing.

References

Bean BP, Friel DD (1990): ATP-activated channels in excitable cells. In: *Ion Channels*, Narahashi T, ed. New York: Plenum Press, pp. 169–203

Berke G (1991): Lymphocyte-triggered internal disintegration. *Immunol Today* 12: 396–399

Beyer EC, Steinberg TH (1991): Evidence that the gap junction protein connexin-43 is the ATP-induced pore of mouse macrophages. *J Biol Chem* 266: 7971–7974

Blanchard DK, McMillen S, Djeu JY (1991): IFN-gamma enhances sensitivity of human macrophages to extracellular ATP-mediated lysis. *J Immunol* 147: 2579–2585

Born GVR, Kratzer MAA (1984): Source and concentration of extracellular adenosine triphosphate during haemostasis in rats, rabbits and man. *J Physiol* 354: 419–429

Buisman HP, Steinberg TH, Fischbarg J, Silverstein SC, Vogelzag SA, Ince C, Vpey DL, Leijh PCJ (1988): Extracellular ATP induces a large, non selective conductance in macrophage plasma membrane. *Proc Natl Acad Sci USA* 85: 7988–7992

Cockcroft S, Gomperts BD (1979): ATP induced nucleotide permeability in rat mast cells. *Nature* 279: 541–542

Cohen JJ (1991): Programmed cell death in the immune system. *Adv Immunol* 50: 55–85

Di Virgilio F, Bronte V, Collavo D, Zanovello P (1989): Responses of mouse lymphocytes to extracellular adenosine 5'-triphospate (ATP). Lymphocytes with cytotoxic activity are resistant to the permeabilizing effects of ATP. *J Immunol* 143: 1955–1960

Di Virgilio F, Pizzo P, Zanovello P, Bronte V, Collavo D (1990): Extracellular ATP as a possible mediator of cell-mediated cytotoxicity. *Immunol Today* 11: 274–277

Di Virgilio F, Steinberg TH (1992): ATP permeabilization. *Protocols in Cell and Tissue Culture* (in press)

Dubyak GR (1991): Signal transduction by P_2-purinergic receptors for extracellular ATP. *Am J Respir Cell Mol Biol* 4: 295–300

Dubyak GC, Fedan JS (1990): Biological actions of extracellular ATP. *Ann NY Acad Sci* 603

Filippini A, Taffs RE, Agui T, Sitkovsky MV (1990a): Ecto-ATPase activity in cytolytic T-lymphocytes. Protection from the cytolytic effects of extracellular ATP. *J Biol Chem* 265: 334–340

Filippini A, Taffs RE, Stikovsky MV (1990b): Extracellular ATP in T-lymphocyte activation: Possible role in effector function. *Proc Natl Acad Sci USA* 87: 8267–8271

Fu W-M, Poo M-M (1991): ATP potentiates spontaneous transmitter release at developing neuromuscular synapses. *Neuron* 6: 837–843

Gordon JL (1986): Extracellular ATP: Effects, sources and fate. *Biochem* 233: 309–319

Hameed A, Olsen KJ, Lee M-K, Lichtenheld MG, Podack ER (1989): Cytolysis by Ca^{2+}-permeable transmembrane channels. Pore formation causes extensive DNA degradation and cell lysis. *J Exp Med* 169: 765–769

Hasegawa H, Bonavida B (1989): Calcium-independent pathway of tumor necrosis factor-mediated lysis of target cells. *J Immunol* 142: 2670–2676

Hogquist KA, Nett MA, Unanue ER, Chaplin DD (1991): Interleukin 1 is processed and released during apoptosis. *Proc Natl Acad Sci USA* 88: 8485–8489

Krahenbuhl O, Tschopp J (1991): Perforin-induced pore formation. *Immunol Today* 12: 399–402

Krishtal O, Marchenko SM, Pidoplichko VI (1983): Receptor for ATP in the membrane of mammalian sensory neurons. *Neurosci Lett* 35: 41–45

Laster SM, Wood JG, Gooding LR (1988): Tumor necrosis factor can induce both apoptotic and necrotic forms of cell lysis. *J Immunol* 141: 2629–2634

Li GD, Milani D, Dunne MJ, Pralong WF, Theler JM, Peterson OH, Wollheim CB (1991): Extracellular ATP causes Ca^{2+}-dependent and independent insulin secretion in RINm5F cells. Phospholipase C mediates Ca^{2+} mobilization but not Ca^{2+} influx and membrane depolarixation. *J Biol Chem* 266: 3449–3457

McDermott AB, Dale N (1987): Receptors, ion channels and synaptic potential underlying the integrative actions of excitatory aminoacids. *Trends Neurosci* 10: 280–284

Murgia M, Pizzo, Zanovello P, Zambon A, Di Virgilio F (1992): In vitro cytotoxic effects of extracellular ATP. *ATLA (Frame)* 20: 66–70

Pizzo P, Zanovello P, Bronte V, Di Virgilio F (1991): Extracellular ATP causes lysis of mouse thymocytes and activates a plasma membrane ion channel. *Biochem J* 274: 139–144

Rozengurt E, Heppel LA, Friedberg I (1977): Effect of exogenous ATP on the permeability properties of transformed cultures of mouse cell lines. *J Biol Chem* 252: 4584–4590

Schmid DS, Tite JP, Ruddle NH (1986): DNA fragmentation: Manifestation of target cell destruction mediated by cytotoxic T cell lines, lymphotoxin secreting helper T cell clones and cell free lymphotoxin containing supernatant. *Proc Natl Acad Sci USA* 83: 1881–1885

Steinberg TH, Di Virgilio F (1991): Cell-mediated cytotoxicity: ATP as an effector and the role of the target cells. *Curr Op Immunol* 3: 71–75

Steinberg TH, Silverstein SC (1987): Extracellular ATP^{4-} promotes cation fluxes in the J774 mouse macrophage cell line. *J Biol Chem* 262: 3118–3122

Sugiura H, Toyama J, Tsuboi N, Kamyia K, Kodama I (1990): ATP directly affects junctional conductance between paired ventricular mycocytes isolated from guinea pig heart. *Circ Res* 66: 1095—1102

Tatham PER, Lindau M (1990): ATP-induced pore formation in the plasma membrane of rat peritoneal mast cells. *J Gen Physiol* 95: 459–476

Trenn GH, Takayama H, Sitkovsky MV (1987): Exocytosis of cytolytic granules may not be required for target cell lysis by cytotoxic T lymphocytes. *Nature* 330: 72–74

Zanovello P, Bronte V, Rosato A, Pizzo P, Di Virgilio F (1990): Responses of mouse lymphocytes to extracellular ATP. II. Extracellular ATP causes cell type-dependent lysis and DNA fragmentation. *J Immunol* 145: 1545–1550

Zheng LM, Zychlinsky A, Liu C-C, Ojcius DM, Young JD-E (1991): Extracellular ATP as a trigger for apoptosis or programmed cell death. *J Cell Biol* 112: 279–288

Zychlinsky A, Zheng LM, Liu C-C, Young JD-E (1991): Cytolytic lymphocytes induce both apoptosis and necrosis in target cells. *J Immunol* 146: 393–400

30
The Role of Free Fatty Acids in CTL–Target Cell Interactions

Alan M. Kleinfeld

Introduction

Fatty acids (FA) are critical to the maintenance of physiological homeostasis. They are the major source of energy, provide essential building blocks of membrane structure, and play a variety of modulatory functions either as FA themselves or as precursors of hormone-like molecules. An immunomodulatory role for dietary FA has long been the subject of investigation (see, e.g., Meade and Mertin, 1978; Cave, 1991). A variety of *in vivo* and *in vitro* studies involving long-term exposure to FA have obtained evidence suggesting both inhibition and stimulation of cytolytic T lymphocyte (CTL) activity by *cis*-unsaturated FA treatment (Meade and Mertin, 1976; Bialick et al., 1984).

Modification of the immune response in these studies, when cells are exposed to altered FA environments for long periods (hours to days), may be due to a complex combination of chemical (including FA oxidation, esterification, and eicosanoid conversion) and physical alteration in cellular components. The resulting cytolytic activity will be difficult to predict. The situation is quite different under conditions of *acute* exposure (seconds to minutes) to free fatty acids (FFA). Under these conditions FA are not chemically modified, and their effects can be shown to result from purely physical perturbation of cellular structural elements.

To assess the potential physiological significance of this type of FA modulation it is necessary to determine the level of FFA presented to the cells of the immune system. While we discuss this in some detail below, it is important to recognize that the chemical species mediating this type of perturbation is the FFA, the FA monomer in aqueous phase. FFA levels in plasma are regulated by albumin and, as a consequence, FFA concentrations are generally quite low (Spector and Fletcher, 1978) compared to the perturbative levels of *in vitro* studies (see, e.g., Karnovsky et al., 1982). In comparison to these other systems, evidence will be discussed in this chapter that CTL may be particularly sensitive to FA perturbation. Moreover, upon conjugation, CTL initiate a process that stimulates FA production within and the secretion of FFA out of cognate target cells. Target cell release of FA appears to be a general phenomenon of the CTL–target cell interaction that, on the one hand, may play a critical role in the killing mechanism and, on the other hand, may be inhibitory for subsequent rounds of CTL-mediated killing and could therefore have importance in tumor suppression of the immune response.

FA Definitions and Nomenclature

The most important FA, as regards the modulatory effects to be discussed, are the long-chain *cis*-unsaturated FA, and these are also most abundant physiologically. The molecular species of the FA are generally given common names and are distinguished by specifying chain length, number of double (*cis*) bonds, and, for polyunsaturated FA, the position of the double bond closest to the terminal (ω) methyl group (Table 1). For example, palmitic acid (16:0) is a 16-carbon-long saturated FA, oleic acid (18:1), $\Delta 9$ is 18 carbons long with a single *cis* bond at the 9–10 position, and α-linoleic acid (18:2)(ω-6) is 18 carbons long with 2 *cis* bonds, the last one at the 11–12 position or 6 carbons from the ω methyl group.

CYTOTOXIC CELLS
M. Sitkovsky and P. Henkart, Editors
© 1993 Birkhäuser Boston

TABLE 1. Common long-chain fatty acids

Common name	Number of carbons and double bonds (n:m)	Position of last double bound (x) (ω–x)	Positions of double bonds Δ
Myristic	14:0	–	–
Palmitic	16:0	–	–
Palmitoleic	16:1	–	9
Stearic	18:0	–	–
Oleic	18:1	–	9
α-Linoleic*	18:2	6	9,12
γ-Linoleic*	18:2	9	6,9
α-Linolenic*	18:3	3	9,12,15
γ-Linolenic*	18:3	6	6,9,12
Dihomo-γ-linolenic*	20:3	6	8,11,14
Arachidonic*	20:4	6	5,8,11,14
Docosahexaenoic*	22:6	3	4,7,10,13,16,19

*Essential fatty acids.

Acute FFA Perturbation of CTL Function

The effects to be described involve perturbations of CTL structure that occur within seconds of FA addition. During this short time, FA metabolism (oxidation or esterification) or indeed any chemical modification is negligible. The added FA can be completely and immediately extracted simply by adding FA-free bovine serum albumin (BSA) (FAFBSA) to the cells, thereby reversing the perturbing effects of the FA.

Cis-unsaturated FA Inhibit the Ag-Stimulated Rise in Intracellular Ca^{2+} but Have No Effect on Phosphatidylinositol Turnover

One of the earliest events to be observed in the interaction of CTL and cognate target cell is a rise in intracellular calcium ([Ca^{2+}]$_i$) (Ostergaard and Clark, 1987; Gray et al., 1987; Poenie et al., 1987). The initial [Ca^{2+}]$_i$ increase (~100–500 nM) occurs within a few seconds and remains elevated above basal levels for more than 20 min. The total response results from a combination of release from intracellular stores as well as flux through the plasma membrane. In murine CTL, at least, most of the rise (~90%) is due to the plasma membrane flux (Richieri and Kleinfeld, 1989; Haverstick et al., 1991).

The rise in [Ca^{2+}]$_i$ is completely inhibited if cis-unsaturated, but not saturated, long-chain FA are added to either the CTL or target cell shortly before conjugation (Richieri and Kleinfeld, 1989). Although the levels of FA necessary for inhibition are discussed in more detail below, in general, less than 5 μM of FA such as oleic, linoleic, or arachidonic are sufficient for complete inhibition. In addition to preventing the rise in [Ca^{2+}]$_i$ by pretreatment, addition of cis-FA to CTL–target cell pairs after conjugation, immediately (within seconds) causes the elevated [Ca^{2+}]$_i$ level to return to the prestimulated value. Moreover, and this is critical to understanding the mechanism of perturbation, if FAFBSA is added either to the CTL pretreated with FA or to the CTL–target cell pair treated with FA after conjugation, [Ca^{2+}]$_i$ levels are restored to control values. Thus the FA inhibition of the calcium response is reversible. FA are inhibitory to the increase in CTL [Ca^{2+}]$_i$ generated by lectin treatment and crosslinking of antibodies to surface proteins, as well as by antigen recognition (Richieri and Kleinfeld, 1989; Richieri et al., 1990).

Inhibition of the calcium response does not represent some general toxic effect of FA. Although more than a single CTL component may be affected by FA perturbations, not all functions are altered. Specifically, the turnover of phosphatidylinositol, which releases inositol phosphates upon CTL activation, is not affected by FA treatment (Richieri and Kleinfeld 1989). In addition, FA treatment does not alter the number of CTL–target cell conjugate pairs, providing further evidence that T-cell receptors (TCR)–antigen recognition is not affected (Richieri et al., 1990). In addition to effects on [Ca^{2+}]$_i$, FA also induce a rapid and substantial acidification of the CTL cytosol (Richieri and Kleinfeld, 1989). Although its relationship to CTL function is less clear than for the [Ca^{2+}]$_i$ response, this acidification may play an important role in modulating CTL function, since it is probably not a simple consequence of the effect on [Ca^{2+}]$_i$.

Cis-Unsaturated FA Inhibit CTL Degranulation

Release of lytic granules from CTL, following target cell recognition, is calcium-dependent (Ostergaard et al., 1987, Trenn et al., 1987). It would not be surprising, therefore, if cis-FA treatment under conditions that inhibit the rise in $[Ca^{2+}]_i$ also inhibited degranulation. Indeed, FA treatment does inhibit alloantigen-mediated degranulation from CTL at cis-FA concentrations similar to those that are inhibitory for the $[Ca^{2+}]_i$ response (Richieri et al., 1990). Moreover, as discussed above, if cis-FA are added after stimulation, $[Ca^{2+}]_i$ falls to basal levels within seconds and the same behavior is exhibited in degranulation. Thus, adding cis-FA at any time during the continuous phase of granule secretion causes its immediate abrogation. Degranulation can be restored if FAFBSA is added to cis-treated cells. Although these results seem to suggest that inhibition of degranulation results directly from the FA inhibition of the $[Ca^{2+}]_i$ response, cis-FA also effectively inhibit degranulation when CTL are activated by phorbol-12-myristate-13-acetate diester (PMA) plus ionomycin (G.V. Richieri and A.M. Kleinfeld, unpublished results). Since $[Ca^{2+}]_i$ levels are elevated under these conditions, it is likely that an element(s) in addition to the $[Ca^{2+}]_i$ response is disrupted by cis-FA, and this factor is critical for degranulation.

Cis-Unsaturated FA Inhibit CTL-Mediated Lysis of Target Cells

Inhibition of degranulation demonstrates that the inhibitory effects have direct functional significance, and it may appear that as a direct consequence, CTL-mediated killing of target cells would also be inhibited. Short-term treatment with cis-unsaturated, but not saturated, FA does, in fact, inhibit killing by alloreactive CTL clones as well as by in vivo-generated effector cells (Samlaska, 1979; Taylor et al., 1985; Richieri and Kleinfeld, 1990). Equivalent inhibition was obtained whether FA were added to effector or to target cells, and inhibition could be prevented by adding FAFBSA to FA-treated CTL before conjugation (Richieri and Kleinfeld, 1990). These results are consistent with a physical perturbation mechanism and with the rapid exchange of FA between membrane and aqueous medium (Richieri and Kleinfeld, 1990; Kleinfeld, 1990). In contrast to the continuous nature of the inhibitory effect of FA on the $[Ca^{2+}]_i$ response and degranulation, inhibition of killing is only effective in a specific temporal window (Richieri and Kleinfeld 1990). Thus, addition of cis-FA before, 2–5 min after, and 15 min after conjugation inhibits lysis completely, 50%, and not at all, respectively. These results indicate that FA inhibit an event specifically related to lethal hit delivery (Martz, 1977).

Although inhibition of killing may be related to inhibition of the rise in intracellular calcium and/or inhibition of degranulation, a number of factors suggest that FA affect additional elements in the lethal hit delivery mechanism. First, although addition of FA within 5 min of conjugation only partially inhibits killing, degranulation is completely abolished. Second, cis-FA inhibit calcium-independent killing. Third, killing is not restored by treating with ionomycin sufficient to raise intracellular calcium. Finally, FA inhibit degranulation stimulated by PMA plus ionomycin, indicating that even when $[Ca^{2+}]_i$ levels are unaffected, another element of the signaling pathway must be impaired. Thus it seems likely that inhibition of killing occurs by inhibiting both granule-dependent and -independent pathways.

The Mechanism(s) of FA Inhibition of the CTL Response

The results outlined in the preceding discussion indicate that cis-FA alter one or more elements of the CTL signaling pathways. It is likely that the site of FA perturbative action is the plasma membrane and, more specifically, the lipid phase of the outer bilayer leaflet. That the action of the FA is confined to the extracellular face of the plasma membrane is suggested by the facts that inhibition occurs within seconds of FA addition, that the rate of FA esterification is quite slow (Richieri and Kleinfeld, 1989), and that FA flip-flop across the plasma membrane is probably also slow on this time scale (Kleinfeld, 1990). Inhibition at the site of the plasma membrane might result from direct (allosteric) perturbation of specific proteins by cis-FA, or the effect might be mediated through the lipid phase. One of the hallmarks of acute FA perturbation, exhibited not only by CTL but by many cells and functions, is that cis-unsaturated long-chain, but not saturated, FA are perturbants. It was previously suggested that this dichotomy is a consequence of the ability of the cis-FA kink to alter lipid acyl chain order, in contrast to the straight-chain saturated FFA, which has no effect (Karnovsky et al., 1982; Richieri and Kleinfeld, 1989).

A more rigorous determination of whether functional modification proceeds through perturbation of the lipid phase of the membrane is now possible with a new technology that allows direct determination of [FFA] (Kleinfeld et al., 1991; Richieri, Ogata and Kleinfeld, 1992). The method uses a fluorescent probe consisting of the fatty acid-binding protein from rat intestine (I-FABP) covalently modified with the fluorescent compound Acrylodan, and is designated ADIFAB. Using this technology it has been possible to show that: (1) the degree of FA partition from the aqueous phase into lipid bilayers or CTL plasma membrane is a sensitive function of the molecular species of FA (the partition of palmitic acid, for example, is much greater than that of linoleic or arachidonic acids); (2) the ability of cis-FA to reduce lipid acyl chain order is proportional to the partition coefficient and related to the FA molecular species; and (3) inhibition of degranulation is a direct function of the decrease in phospholipid acyl chain order (A. Anel, G. Richieri, and A. M. Kleinfeld, 1993). These results strongly support the hypothesis that FA act through alteration of the lipid phase of the membrane.

How does perturbation of the lipid phase of the plasma membrane's outer leaflet inhibit CTL activity? Presumably, the actions of one or more proteins are sensitive to even small changes in global lipid acyl chain order or cis-FA concentrate in particular lipid domains (Karnovsky et al., 1982), and these lipid changes are translated into alterations of one or more elements in the signaling pathways. Although FA perturbation of plasma membrane lipid might affect a wide variety of cellular properties, the CTL FA effects are quite selective. Thus, the lack of effect on phosphatidylinositol turnover and CTL-target cell conjugate number shows that antigen recognition and at least some of the signaling pathways are not affected by FA perturbation. Which specific pathways are affected is much less clear, and at least part of the complication arises because, as discussed above, each of the affected CTL responses involved the perturbation of more than one event. Thus, inhibition of killing appears to involve elements other than degranulation, and inhibition of degranulation involves elements other than those to do with the calcium response.

Inhibitory Concentrations of FFA

To assess the physiological relevance of FA perturbation, it is essential, and surprisingly difficult, to determine the levels of FFA necessary to inhibit CTL function. Inhibition of CTL responses in albumin-free medium occurs at *added* concentrations of $2-10\ \mu M$. Long-chain FA are highly insoluble in the aqueous phase, and therefore the FFA level is not equal to the total amount of FA added to a cell suspension, since much of the FA will bind to the cells and the container walls. This contrasts with the physiological situation, in which a constant FFA level is maintained by the FA-albumin complexes, which serve as a buffer of the aqueous phase monomeric concentration of FA.

Typical normal plasma levels of [total FA] range between 200 and $1000\ \mu M$. Most plasma FA is bound to albumin, but a small fraction of the total dissociates from albumin and becomes hydrated in the aqueous phase ([FFA]) (Spector and Fletcher, 1978). Plasma [FFA] levels are determined by the ratio of total FA to total albumin. To assess whether *in vitro* levels are physiological relevant, it is necessary to determine the *in vitro* and plasma levels of [FFA]. Although estimates of [FFA] have been made in the FFA/albumin system, because of their low solubility direct measurements of [FFA] have not been possible previously. Using the new FA probe ADIFAB, however, these values can now be measured. Initial studies indicate that FFA levels below $1\ \mu M$ inhibit degranulation (A. Anel, G. Richieri, and A. Kleinfeld, 1993. Preliminary results using ADIFAB suggest, in agreement with the earlier estimates of the FA/albumin buffering capacity (Spector and Fletcher, 1978), that [FFA] levels increase exponentially above ratios of 3:1, where oleic acid levels may exceed $1\ \mu M$. In several disease states, including a variety of cancers, ratios greater than 3 have been reported (Brown et al., 1983; Legaspi et al., 1987), raising the possibility that elevated plasma FA levels may suppress cytolytic activity directed against tumor cells.

Target Cell Release of FA

Perhaps the central finding of the results described above is that cellular behavior can be seriously disrupted at concentrations of FFA that are low enough to be physiologically relevant. Two interesting possibilities are raised by these findings. First, since FA are disruptive, and in fact can be lytic (Richieri et al., 1990) when added extracellularly, FA generation *within* the target cell may play an important role in lethal hit delivery. In this event CTL would be expected to activate lipolytic enzymes within the target cell very early after conjugation. Second, because of their high level of metabolic activity, tumor cells might elevate FFA levels locally, and/or upon CTL attack they might

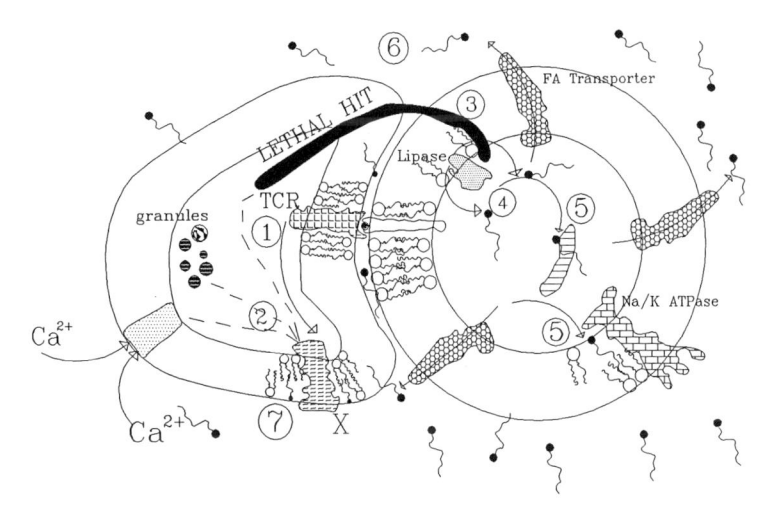

FIGURE 1. The fatty acid view of the CTL–target cell interaction. This cartoon summarizes many of the results described in the text. In this sequence of events (specifically designated with circled numbers), engagement of the TCR and major histocompatibility complex (MHC) class I (1) initiates a series of signaling events (2), such as phosphatidylinositol turnover (not shown), an increase in $[Ca^{2+}]_i$, degranulation, and lethal hit delivery (3). In order to simplify the figure, signal transmission from the TCR engagement and/or other accessory molecules is shown to occur through a membrane protein (X). If extracellular *cis*-FA is present, then this intermediate (X) in the activation pathway is perturbed, and as the dashed lines emanating from X are supposed to indicate, the various CTL responses are inhibited. Recent evidence suggests that X is a tyrosine phosphatase and that tyrosine phosphorylation of a 100 kDa protein, regulated by this phosphate, is a critical intermediate for the CTL responses (1), (2), and (3) (A. Anel and A.M. Kleinfeld, submitted for publication). Very early after its generation, the lethal hit (3) activates target cell lipases that result in the intracellular liberation of large quantities of FFA (4). These intracellular FA will eventually be toxic to the target cell by mechanisms that are as yet unknown, but that may include decoupling of oxidative phosphorylation in the mitochondria and inhibition of the plasma membrane Na/K ATPase (5). Once produced, these intracellular FA will be secreted from the target cell through plasma membrane FA transporters, resulting in substantial extracellular concentrations of FA, especially in regions closest to the target cell (6). Finally the elevation of the *cis*-FA concentration in the vicinity of the CTL will result in the inhibition of further signaling events, including possibly the "off signal" necessary for CTL–target cell disengagement (7).

increase these levels to aid in tumor escape from CTL surveillance. This effect might be especially important when plasma FFA levels are already elevated by the disease.

Stimulation of Target Cell FA Release Is a General Property of the CTL Interaction

FA release has now been measured from a variety of target cells bearing various class I alloantigens (Richieri and Kleinfeld, 1991). Target cells were incubated, long term, with radiolabeled FA so that virtually all the radiolabel was incorporated into esterified FA products, predominantly phospholipids. Conjugation of these targets with cognate effector cells resulted in an increase in supernatant radiolabel within minutes after conjugation. Lipid analysis of the supernatant and pellet revealed that while the FA content of cells is generally less than 0.5% before conjugation, it is greater than 1.5% within 15 min after conjugation, and the FA content of the supernatant is eventually enriched nearly 100-fold. Both radiolabel and gas chromatography studies indicate that although FA release occurs for a wide variety of saturated and *cis*-unsaturated FA, the greatest increase occurs for the *cis*-unsaturated FA, predominantly oleic and linoleic acids. Stimulation of FA release was observed with 7 different murine CTL clones having alloreactivity directed against a variety of target cells (Richieri and Kleinfeld, 1991). Release in these studies was also observed with CTL directed against concanavalin A (Con A) blasts and with mixed lymphocyte culture (MLC) effectors, consistent with an early study of Koren et al. (1971).

CTL-Stimulated Target Cell Release of FA Occurs Early and Is Highly Correlated with Target Cell Lysis

FA release from target cells is highly correlated with target cell lysis, measured by ^{51}Cr release, suggesting a causal relationship between the two events (Richieri and Kleinfeld, 1991). Thus, as the degree of target cell lysis increases, either as a result of CTL lytic potential, effector to target ratio, or time of CTL-target cell incubation, so also does the degree of FA release. Also consistent with a causal relationship, FA are produced in the target cell earlier than 5 min after conjugation. Both production and secretion of FA from the cell increase monotonically with time until about 1 to 2 hr after conjugation when, presumably, cell lysis has proceeded so far that the lipolytic enzymes can no longer function and FA production ceases.

FA release does not occur simply because the cell membranes fragment as the cells are dying. As discussed above, cells themselves contain very little FA, so that production of FA requires activation of specific lipolytic enzymes. These enzymes are specifically activated by CTL; other means of cell disruption, including cycles of freeze/thaw, detergent solubilization, or massive ionophore treatment, do not result in FA production. Moreover, treatment of target cells with lytic granules, sufficient to completely lyse cells, also does not generate FA (Richieri and Kleinfeld, 1991).

Target Cell Release of FA Is Stimulated by Wide-Spectrum Lipases

Although it is clear that activation is CTL-specific and requires antigen recognition, it is not known how CTL activate FA release, which enzymes are involved, or what substrates are hydrolyzed. Examination of the FA distribution and the lipolytic products produced after CTL-target cell conjugation raises the possibility that at least some FA release may be due to phospholipase C and/or phospholipase D, followed by di- and monoacylglycerol lipase hydrolysis. These results, together with the fact that a substantial fraction of the cells' total lipid is converted to FA over the 1-2 hr that FA production is active, suggests that either a wide spectrum of lipases is involved, or those that are activated have little substrate specificity (Richieri and Kleinfeld, 1991).

Is FA Release Crucial for Lethal Hit Delivery?

As the articles in this volume reveal, the mechanism of CTL-mediated killing remains a subject of continuing investigation. Thus, while granule-mediated killing is clearly indicated in a variety of systems (Kolber et al., 1989), considerable evidence suggest that other pathways of CTL-mediated lysis exist (Russell et al., 1982; Ostergaard and Clark, 1989; Filippini et al., 1990). Our own studies of FA inhibition of CTL responses, in fact, argue for additional pathways. Thus, as we described, addition of FA shortly after conjugation leads to only partial inhibition of killing (20-50%), yet addition of FA at these early times (2-5 min) almost completely inhibits degranulation. FA treatment also inhibits killing of calcium-independent targets under conditions in which degranulation does not occur, indicating that an additional factor must be critical to killing.

Several proposed alternative mechanisms of killing involve suicide-like events within the target cell (Russell et al., 1982; Ostergaard and Clark, 1989). Consistent with such a mechanism, early biochemical changes in the target cell, including DNA fragmentation and a rise in intracellular free calcium ($[Ca^{2+}]_i$), have been reported (Russell et al., 1982; Howell and Martz, 1987; Poenie et al., 1987). It seems unlikely, however, that these events are the direct mediators of death. In contrast, the CTL-mediated induction of target cell production of FA provides an attractive and perhaps more direct link between lethal hit delivery and target cell demise. FA are themselves known to be potent modulators of cellular behavior (inhibition of the CTL response is an example), and they have been directly implicated in several different types of cell death (Kelly et al., 1986; Chow et al., 1989; Jones et al., 1989). Moreover, copious amounts of FA are produced upon CTL induction, and while it is not known what levels are obtained intracellularly, the total amounts secreted suggest that FA levels will easily exceed those necessary to be deleterious for the targets.

Could Target Cell Secretion of FA Suppress CTL Killing *In Vivo*?

FA release of the kind and level observed is unusual, and for the reasons enumerated above is very likely to be a specifically activated process having some important functional role(s). While FA production within the target cell may mediate killing, FA secreted from the target cell may inhibit CTL activity. The quantities of FA secreted could, depending upon local concentrations, plasma FA levels, and tissue type, easily exceed those found to inhibit CTL–mediated killing *in vitro*. The target cell that secretes FA will not be spared; the correlation between secretion and death suggests that a secreting cell is committed to death. However, the CTL that has stimulated this release and therefore becomes exposed to high FFA levels may be inhibited from further rounds of lytic activity, thereby allowing neighboring tumor cells to escape CTL attack. Whether this occurs *in vivo* would depend upon the rate at which FA are removed from the CTL–target cell conjugate environment, and this in turn will depend upon the FA-buffering capacity of the surrounding medium. Thus, if total plasma levels of FA are elevated or albumin levels are depressed, then removal of FA will be less effective. This suggests that FA release might be particularly important in neuronal tissue, since cerebrospinal fluid (CSF) has very little albumin (< 1% of plasma levels), and in most body tissue where the surrounding interstitial fluid may also have reduced albumin levels as compared to plasma (Jain, 1987; Krystal et al., 1989).

Acknowledgments

The author thanks Drs. G.V. Richieri and A. Anel for sharing unpublished results and for helpful discussions. This work was supported in part by grants from the American Cancer Society (BE-11) and the NIH (GM44171 and HL44818).

References

Anel A, Richieri GV, Kleinfeld AM (1993): Membrane partition of fatty acids and T cell function. *Biochemistry*, in press.

Bialick R, Gill R, Berke G, Clark WR (1984): Modulation of cell mediated cytoxicity function after alteration of fatty acid composition *in vitro*. *J Immunol* 132: 81

Brown RE, Steele RW, Marmer DJ, Hudson JL, Brewster MA (1983): Fatty acids and the inhibition of mitogen-induced lymphocyte transformation by leukemic serum. *J Immunol* 131: 1011

Cave WT, Jr (1991): Dietary n-3 (ω-3) polyunsaturated fatty acid effects on animal tumorigenesis. *FASEB J* 5: 2160

Chow SC, Sisfontes L, Bjorkhem I, Jondal M (1989): Suppression of growth in a leukemic T cell line by n-3 and n-6 polyunsaturated fatty acids. *Lipids* 24: 700

Filippini A, Taffs RE, Sitkovsky MV (1990): Extracellular ATP in T-lymphocyte activation: Possible role in effector functions. *Proc Natl Acad Sci USA* 87: 8267

Gray LS, Gnarra JR, Engelhard VH (1987): Demonstration of a calcium influx in cycolytic T lymphocytes in response to target cell binding. *J Immunol* 138: 63

Haverstick DM, Engelhard VH, Gray LS (1991): Three intracellular signals for cytotoxic T lymphocyte-mediated killing: Independent roles for protein kinase C, Ca^{2+} influx, and Ca^{2+} release from internal stores. *J Immunol* 146: 3306

Howell DM, Martz E (1987): The degree of CTL induced DNA solubilization is not determined by the human vs mouse origin of the target cell. *J Immunol* 138: 3695

Jain RK (1987): Transport of molecules in the tumor interstitium: A review. *Cancer Res* 47: 3039

Jones RL, Miller JC, Hagler HK, Chien KR, Willerson JT, Buja LM, Bellotto D, Buja D, Williams PK, Yang E (1989): Association between inhibition of arachidonic acid release and prevention of loading during ATP depletion in cultured rat cardiac myocytes. *Am J Pathol* 135: 541

Karnovsky MJ, Kleinfeld AM, Hoover RL, Klausner RD (1982): The concept of lipid domains in membranes. *J Cell Biol* 49: 1

Kelly RA, O'Hara DS, Mitch WE, Smith TW (1986): Identification of NaK-ATPase inhibitors in human plasma as non-esterified fatty acids and lysophospholipids. *J Biol Chem* 261: 11704

Kleinfeld AM (1990): Transport of free fatty acids across membranes. *Comm Molec Cell Biophys* 6: 361

Kleinfeld AM, Ogata RT, Richieri GV (1991): Assay for long chain unbound free fatty acids with fluorescently labeled intestinal FABP. *Biophys J* 59: 637a

Kolber MA, Quinones RR, Henkart PA (1989): Target cell lysis by cytotoxic T lymphocytes redirected by antibody-polystyrene beads. *J Immunol* 143: 1461

Koren HS, Ferber E, Fischer H (1971): Changes in phospholipid metabolism of a tumour target cell during a cell-mediated cytotoxic reaction. *Biochim Biophys Acta* 231: 520

Krystal G, Lam V, Schreiber WE (1989): Application of a silver-binding assay to the determination of protein in cerebrospinal fluid. *Clin Chem* 35: 860

Legaspi A, Jeevanandam M, Starnes HF, Jr, Brennan MF (1987): Whole body lipid and energy metabolism in the cancer patient. *Metabolism* 36: 958

Martz E (1977): Mechanism of specific tumor-cell lysis by alloimmune T lymphocytes: Resolution and characterization of discrete steps in the cellular interaction. In: *Contemporary Topics in Immunobiology*. Vol 7, Stutman, O, ed. New York: Plenum Press, p. 301

Meade C, Mertin J (1978): Fatty acids and immunity. *Adv Lipid Res* 16: 127

Mead CJ, Mertin J (1976): The mechanism of immunoinhibition by arachidonic and linoleic acid: Effects on the lymphoid and reticulo-endothelial systems. *Int Arch Allergy Appl Immunol* 51: 2

Ostergaard H, Clark WR (1987): The role of Ca^{2+} in activation of mature cytotoxic T lymphocytes for lysis. *J Immunol* 139: 3573

Ostergaard HL, Clark WR (1989): Evidence for multiple lytic pathways used by cytotoxic T lymphocytes. *J Immunol* 143: 2120

Ostergaard HL, Kane KP, Mescher MF, Clark WR (1987): Cytotoxic T lymphocyte mediated lysis without release of serine esterase. *Nature* 330: 71

Poenie M, Tsien YR, Schmitt-Verhulst AM (1987): Sequential activation and lethal hit measured by [Ca++]i in individual cytolytic T cells and targets. *EMBO J* 6: 2223

Richieri GV, Kleinfeld AM (1989): Free fatty acid perturbation of transmembrane signaling in cytotoxic T lymphocytes. *J Immunol* 143: 2302

Richieri GV, Kleinfeld AM (1990): Free fatty acids inhibit cytotoxic-T-lymphocyte mediated lysis of allogeneic target cells. *J Immunol* 145: 1074

Richieri GV, Kleinfeld AM (1991): Free fatty acids are produced in secreted from target cells very early in cytotoxic T lymphocyte-mediated killing. *J Immunol* 147: 2809

Richieri GV, Mescher MF, Kleinfeld AM (1990): Short term exposure of cytotoxic-T-lymphocytes to *cis* unsaturated free fatty acids inhibits antigen stimulation of cytosolic calcium increases and degranulation. *J Immunol* 144: 671

Richieri GV, Ogata RT, Kleinfeld AM (1992): A fluorescently labeled intestinal fatty acid binding protein: Interactions with fatty acids and its use in monitoring free fatty acids. *J Biol Chem* 267: 23495

Russel JH, Masakowski V, Rucinsky T, Phillips G (1982): Mechanisms of immune lysis: III. Characterization of the nature and kinetics of the cytotoxic T lymphocyte-induced nuclear lesion in the target. *J Immunol* 128: 2087

Samlaska CP (1979): Linoleic acid inhibition of naturally occurring lymphocytotoxicity to breast cancer-derived cells measured by a chromium-51 release assay. *JNCL* 62: 1427

Spector AA, Fletcher JE (1978): Transport of fatty acid in the circulation. In: *Disturbances in Lipid and Lipoprotein Metabolism*, Dietschy JM, Gotto AM, Ontoko A, eds. pp. 229–249

Taylor AS, Howe RC, Morrison AR, Sprecher H, Russell JH (1985): Inhibition of cytotoxic T lymphocyte-mediated lysis by ETYA: Effect independent of arachidonic acid metabolism. *J Immunol* 134: 1130

Trenn G, Takayama H, Sitkovsky MV (1987): Exocytosis of cytolytic granules may not be required for target cell lysis by cytotoxic T-lymphocytes. *Nature* 330: 72

VII
Biochemical and Immunopharmacological Manipulations of Cytotoxic Cells

31
Identification of Protein Kinases and Protein Phosphatases Involved in CTL Effector Functions. "ON" and "OFF" Signaling and Immunopharmacological Implications

Hirotaka Sugiyama, Sergey Apasov, Guido Trenn, Frank Redegeld and Michail Sitkovsky

Biochemical studies of cytotoxic T-lymphocyte (CTL) effector functions, the description of molecular requirements, and the enzymes needed for CTL-induced target cell (TC) death are likely to shed light on the mechanisms and individual participants of practically all steps of CTL–TC interactions (Lancki et al., 1986). These mechanisms include conjugate formation, lethal hit delivery, and events leading to TC death. Identification of important regulatory molecules will also provide key targets for immunomodulation and allow for design of specific pharmacological agents.

In our studies, we have tried to take advantage of the ability to test several different functional responses of cloned CTL to the same stimuli. These responses, which can be triggered by the interaction of the T-cell receptor (TCR) with the antigen (Ag) on the TC (or with the anti-TCR monoclonal antibody, mAb), or by pharmacological agents, include cytotoxicity (lethal hit delivery), TCR-regulated granule exocytosis, and synthesis and secretion of interferon-γ (IFN-γ) (among the few most easily tested). During these studies, we implicated protein kinase C(PK-C), Ca^{2+}/CaM-dependent phosphatase (calcineurin), and PPIA protein phosphatase as part of the "on" signaling, while PK-A and yet another, still unidentified Ser/Thr phosphatase are linked to an "off", down-regulating pathway. The important role of phosphorylation/dephosphorylation reactions in CTL responses is confirmed by both biochemical and molecular biological experiments. This confirmation forms the basis for the development of a new approach in immunomodulation by using synthetic peptide substrates, pseudosubstrates, and inhibitors of protein kinases. In addition, during experiments on the role of PK-A and of extracellular Ca^{2+} in CTL-mediated cytotoxicity and exocytosis, we have a situation where one can block exocytosis and still observe cytotoxicity. Those studies provided crucial evidence against the granule exycytosis model and resulted in the formulation of an alternative model. Thus, identification of key enzymes involved in lethal hit delivery by CTL and in TC death promises to significantly improve our understanding of cell-mediated cytotoxicity, and provide potentially useful molecular targets for immunomodulation.

In this review we concentrate on studies of proteins implicated in the effector phase of the CTL response, which follows early events in T-cell activation after TCR crosslinking by Ag on the Ag-bearing cells. The studies of early events [tyrosine phosphorylation by tyrosine protein kinases, polyphosphoinositide hydrolysis (PI pathway)] have been extensively reviewed (e.g. June, 1991; Rudd, 1990) as applied to T-cell activation in general. It was confirmed that the PI pathway is triggered in CTL soon after engagement of CTL with target cells (Treves et al., 1987). It was also shown that TCR—mediated signaling may occur even in the absence of inositol phosphate production (O'Rourke and Mesher, 1988). Calcium influx into CTL was demonstrated in response to target cell conjugate formation (Gray et al., 1987). It is, however, quite possible that these early events are nor necessary or sufficient for the execution of CTL effector functions (Sitkovsky, 1988). Thus, we feel that studies of the molecular participants in the later stages of the CTL response may bring more relevant information.

PK-C Is Required for Conjugate Formation, Lethal Hit Delivery, and TCR-Triggered Exocytosis

PK-C and Ca^{2+} increases in lymphocytes have been shown (Trunch et al., 1985; Albert et al., 1985; Utsunomiya et al., 1986) to be involved in triggering of CTL development. To evaluate if PK-C is involved in CTL effector functions, the ability of PK-C-activating phorbol esters to trigger CTL-mediated lysis of non-Ag-bearing target cells was tested. It was demonstrated earlier that phorbol esters alone are able to trigger nonspecific lysis (Russell, 1986). However, when different CTL clones were mixed with non-Ag bearing ^{51}Cr-labeled target cells in the presence of various phorbol esters and Ca^{2+}-ionophore concentrations in a standard ^{51}Cr-release cytotoxicity assay, we found that phorbal miristate acetate (PMA) and Ca^{2+} ionophores synergistically induced a dose-dependent activation of CTL clones, resulting in the lysis of the nonspecific target cells (Berrebi et al., 1987). The Ca^{2+} ionophores A23187 and ionomycin were similar in their effect on CTL-mediated lysis. In most experiments, PMA alone induced a modest increase in nonspecific cytotoxicity, which agrees with the report of Russell (1986). Occasionally PMA-induced, CTL-mediated ^{51}Cr-release from the target cell was high even in the absence of Ca^{2+} ionophores. In such experiments the combined effects of PMA and A23187 could not be considered synergistic.

It is most likely that the effect of PMA is mediated by PK-C, since PMA and 4β-phorbol 12,13-didecanoate, which activate PK-C in vitro (100% and 80%, respectively), also activate CTL either alone or in synergy with Ca^{2+} ionophore. In contrast, 4α-phorbol 12,13-didecanoate, which does not bind or activate this enzyme, has only a small effect on CTL-mediated nonspecific lysis (Berrebi et al., 1987).

These results had interesting implications for understanding how CTL execute their screening of surrounding cells, and supported the view that during the nonspecific adhesion of CTL to target cells all of the requirements (except one) for triggering of lysis are met. The principal role of Ag recognition is simply to trigger a biochemical cascade, which includes PK-C and Ca^{2+}-dependent reactions. Thus, the TCR is not needed for "bridging" of CTL and TC, but only for CTL triggering. This conclusion is in agreement with the results of Lanzaveccia (1986) and of Spits et al. (1986), who used an entirely different approach. The explanation of nonspecific target cells cytolysis by CTL in the presence of PMA and Ca^{2+} ionophore assumes that the immunological function of CTL requires them to engage in contact (nonspecific engagement) with every cell to distinguish Ag-bearing, specific target cells from the surrounding syngeneic cells.

The involvement of PK-C in conjugate formation between CTL and TC was also demonstrated in a separate experiments using fluorescence-activated cell sorter (FACS) analysis of different fluorescent-labeled CTL and TC. It was found that PMA has the ability to promote conjugate formation with non-Ag-bearing target cells, excluding the necessity for Ag receptor–ligand links in the formation and stregthening of the CTL–target cell conjugate. The results described above point to the importance of the accessory cell surface molecules and their phosphorylation/dephosphorylation in CTL–target cell contact interactions (Berrebi et al., 1987). These observations gained support by later detailed studies of the role of LFA-1 phosphorylation in adhesion (Dustin and Springer, 1989). Future studies may reveal a correlation between the effects of Ag recognition and different combinations of PK-C activators and Ca^{2+} ionophores on the organization of the cytoskeleton and contractile system of CTL. Synergistic effects of PMA and Ca^{2+} ionophores in the induction of nonspecific cytotoxicity were also independently found by Lancki and coauthors (1986).

It is still possible that besides providing a triggering signal to CTL, ligand binding to the Ag receptor on the CTL surface could participate in the formation and strengthening of CTL–target cell conjugates. Although the latter remains to be demonstrated, the data obtained so far suggest that Ag receptor occupancy is required only for triggering the CTL lytic function of the CTL and is not a mandatory requirement for conjugate formation.

Role of PK-C in CTL Granule Exocytosis

Granule exocytosis in CTL was studies using an assay (Takayama et al., 1987; Taffs and Sitkovsky, 1991), that takes advantage of the presence of easily detectable enzyme in CTL granules (Pasternack et al., 1986; Masson and Tschopp, 1987; Takayama et al., 1987a). It was found that incubation of CTL clones with soluble anti-TCR mAb does not stimulate the secretion of granule-associated enzyme, whereas solid-phase immobilized anti-TCR mAb do trigger secretion (Takayama et al., 1987b). Neither the tested PK-C activators (phorbol esters or bryostatin I and bryostatin II) nor Ca^{2+} ionophore had a significant effect on N-α-benzyloxycarbonyl-L-lysine (BLT) esterase release from CTL when added alone. However, when added together, Ca^{2+} ionophore and PK-C activators exerted a strong synergistic effect in triggering secretion (Takayama and Sitkovsky, 1987). A maximal response, similar in its magnitude to that elicited by Ag-bearing target cells or by immobilized anti-TCR mAb, was produced by 10 ng/ml PMA in the presence of 0.5 μl/ml A23187. A similar effect of PMA and A23187 was observed when the release of granule-located, affinity-labeled proteins was analyzed.

These data suggest that PK-C activation is involved in the triggering of CTL exocytosis. When different phorbol esters that have different potencies of Pk-C activation *in vitro* (Takayama and Sitkovsky, 1989) were tested with the CTL activation and BLT esterase secretion assays, the same hierarchy was found in their ability to stimulate exocytosis: PMA was the most potent, while the less potent 4-β-phorbol-12,13-didecanoate, (4-β-PDD) still induced strong exocytosis. 4-α-PDD, which does not activate PK-C *in vitro*, also does not exocytosis in CTL.

Role of Extracellular Ca^{2+} in TCR-Triggered Exocytosus

Addition of 2.5 mM [ethylenebis(oxyethylenenitrilo)]tetraacetic acid (EGTA) or ethylenediaminetetraacetic acid (EDTA) alone or in different combinations with 3.0 mM Ca^{2+} or 3.0 mM Mg^{2+} did not affect the basal level of secretion. However, secretion of BLT esterase triggered by immobilized anti-TCR mAb was completely inhibited by 2.5 mM EGTA or EDTA. Addition of excess Ca^{2+} (3.0 mM), but not of Mg^{2+} (3.0 mM), reversed inhibition by EGTA and EDTA almost completely. These results (Takayama and Sitkovsky, 1987) reflect the necessity of extracellular Ca^{2+} in exocytosis, which is triggered by the interactions of solid-phase anti-TCR mAb with TCR on the CTL surface.

EGTA and EDTA also completely blocked BLT esterase secretion, which was triggered by the synergistic action of PMA and A23187. The addition of excess Ca^{2+}, but not Mg^{2+}, reversed the inhibitory effect of EDTA to a large degree, and the inhibitory effect of EGTA was partially reversed by Ca^{2+}. Thus, the presence of Ca^{2+} in incubation buffers is obligatory for the triggering of exocytosis in CTL. This conclusion is supported by experimental results in which Mn^{2+} inhibited (PMA/A23187)-induced exocytosis, providing an important cation specificity control.

Effect of Ca^{2+} Channel Blockers on TCR-Triggered Exocytosis in CTL

The studies described above implicate extracellular Ca^{2+} in TCR-triggered exocytosis in the CTL, and suggest that extracellular Ca^{2+} may be transported through the plasma membrane. To investigate the possibility that Ca^{2+} channels are involved in TCR-regulated exocytosis, we used two known Ca^{2+} channel blockers, nifedipine and verapamil (Takayama and Sitkovsky, 1987). These agents exerted no effect on the basal level of secretion in the CTL, but both inhibited TCR-triggered secretion. A higher level of cxocytosis was less inhibited by both Ca^{2+} antagonists, so that practically complete inhibition was observed with 50 μM of nifedipine when anti-TCR mAb triggered 26% BLT esterase secretion, and 64% inhibition was observed when anti-TCR mAb triggered 36% BLT esterase secretion. The results of these experiments suggest that Ca^{2+} channels are involved in TCR-triggered exocytosis in the CTL.

Role of Calmodulin (CaM) and of CaM-Binding Proteins in TCR-Triggered Exocytosis in the CTL

The obligatory role of the extracellular Ca^{2+} in TCR-triggered exocytosis and reported increases in intracellular Ca^{2+} concentration in CTL (Utsunomiya et al., 1986) suggest that Ca^{2+} and Ca^{2+}-dependent proteins are directly involved in CTL activation. Many Ca^{2+}-dependent reactions in eukaryotic cells are mediated by CaM, which is considered the main intracellular acceptor of Ca^{2+}. To investigate the possibility that CaM is involved in the exocytosis of granules from CTL, we tested the effect of CaM antagonists on TCR-triggered secretion. To exclude side effects of the reagents on TCR-triggered exocytosis, two different CaM antagonists were used in studying CaM involvement in CTL exocytosis. Trifluoperazine (TFP), which prevents Ca^{2+} binding to CaM, and an unrelated CaM antagonist, N-(6-aminohexyl)-5-chloro-1-naphthalene sulfonamide (W-7), were used. We found that both TFP and W-7 were efficient in blocking anti-TCR mAb-triggered exocytosis. The data implicated CaM and CaM-binding proteins in CTL exocytosis granules, as was demonstrated earlier using CaM antagonists in the studies of exocytosis in other cells (Takayama and Stikovsky, 1987).

The strict dependence of both TCR-triggered and PMA/A23187-triggered exocytosis in CTL on extracellular Ca^{2+} suggests that a sustained increase of $[Ca^{2+}]_i$ must be obligatory for maintenance of exocytosis. Such a sustained $[Ca^{2+}]_i$ increase achieved by transmembrane signaling through TCR may be maintained through plasma membrane Ca^{2+} channels, since the Ca^{2+} channel inhibitors nifedipine and verapamil block TCR-mediated exocytosis. Ca^{2+} channels were implicated in stimulation mechanisms of human T lymphoma cells by mAb to the CD3/TCR complex (Oettgen et al., 1985). The obligatory requirements of external Ca^{2+} for both TCR-triggered and PMA/A23187-triggered exocytosis suggest that inositol triphosphate-induced release of calcium from intracellular stores may not be sufficient to support activation of the effector functions of CTL.

The experiments reviewed here established the main features of TCR-regulated exocytosis in CTL. They form a basis for a model of activation of CTL exocytosis through TCR crosslinking and triggering of the phosphoinositide/Ca^{2+} pathway, where a sustained increase of intracellular Ca^{2+} level is maintained by the translocation of extracellular Ca^{2+} through Ca^{2+} channels in the plasma membrane. However, additional detailed studies are needed to understand the role of CaM and CaM-binding proteins in activation and exocytosis in CTL.

CaM-Dependent Protein Phosphatase Calcineurin

This enzyme was recently identified as a molecular target for cyclosporin A (CsA), which also is known to be a major CaM-binding protein in T lymphocytes (Kincaid et al., 1987), suggesting that dephosphorylation may be important in the propagation of the Ca^{2+} "on" signal in CTL. Since Ca^{2+} was important for exocytosis of granules and contributed to the cytotoxic potential of CTL, it was important to determine the Ca^{2+}-dependent processes in CTL. It is believed that Ca^{2+}/CaM-dependent processes are involved in the propagation of the Ca^{2+} signal. Therefore, the identification of Ca/CaM-binding proteins was expected to give a clue to the understanding of the Ca^{2+}-dependent pathways. This was attempted by the analysis of CaM-binding proteins which were purified from lymphocyte extracts using CaM-Separose affinity chromatography (Kincaid et al., 1987). it was found that the major CaM-binding protein in lymphocytes is a polypeptide with M_r 59 kDa, which we identified (using immunochemical and biochemical methods) as the α-chain of a well-known CaM-dependent phosphatase, calcineurin. This finding led us to conclude that calcineurin and the protein dephosphorylation it mediates play an important role in the lymphocyte activation events which follow Ca^{2+} increase. Recently this conclusion was further confirmed by the discovery that the inhibitory CsA mechanism is mediated by calcineurin (Lu et al., 1991). This is in agreement with our earlier results (Trenn et al., 1989) in which CsA was shown to inhibit Ca^{2+}-independent CTL-mediated cytotoxicity and exocytosis of granules in the CTL. In these studies, we concluded that peptydil-prolyl isomerase (a prevailing target for CsA binding) and its assumed mechanism of action at the time (reviewed by Freedman, 1989) were not likely to be part of the lethal hit delivery. Thus, the available data implicate calcineurin, a CaM-dependent phosphatase, in such protein synthesis-independent events as cytotoxicity and granule exocytosis in the CTL. It is of great interest to further explore the precise role of this serine/threonine phosphatase in CTL function.

Serine/Threonine Protein Phosphatases Other than Calcineurin

Other protein phosphatases are also involved in both "on" and "off" signaling. The reversible processes of phosphorylation (by protein kinases) and dephosphorylation (by protein phosphatases) of intracellular proteins are considered to be of major importance in the regulation of the intracellular biochemical pathways in mammalian cells, including T lymphocytes (reviewed in Alexander and Cantrell, 1989). It was suggested that almost all known activities of Ser/Thr phosphatases are accounted for by four major catalytic subunits with overlapping substrate specificities. The Ser/Thr phosphatases can be classified by their ability to specifically dephosphorylate the β subunit of the phosphorylase kinase (Type 1 and type 2) and by their susceptibility to inhibition by two different inhibitory proteins (inhibitor 1 and inhibitor 2).

Further subclassification of protein phosphatases is based on their Ca^{2+} and Mg^{2+} dependence (Cohen et al., 1990). PP1, PP2A, and PP2C protein phosphatases can dephosphorylate a wide variety of important substrates *in vitro*, however, not much is known about their specific substrates *in vivo*.

More direct evidence for the dephosphorylation role in T-cell activation required specific inhibitors of Ser/Thr phosphatases, but no such reagents were available until recently.

The recent discovery of an okadaic acid (OA) (reviewed in Cohen et al., 1990), a specific PP1 and PP2A phosphatase inhibitor, allowed us to address this issue. The effect of OA on CTL functions was examined in cytotoxicity and granule exocytosis assays by using OA-pretreated or OA-TME-pretreated CTL (Taffs et al., 1991) (OA-TME is an inactive analog of OA), or by the addition of OA directly to the assay. It was found that the effects of OA were stronger and more reproducible if evaluated after pretreatment of CTL. Several concentrations of OA were tested in all experiments, because we noticed that the dose dependence varied slightly with different preparations of CTL clones. It was shown that pretreatment of CTL with different concentrations of OA for 4 hr, followed by washing and testing in a 4-hr cytotoxicity assay, profoundly changed the ability of CTL to lyse target cells. Pretreatment of CTL with 50, 100, or 150 nM OA enhanced lysis of TC, whereas higher concentrations of OA were inhibitory in this experiment.

The enhancement of cytotoxicity by OA was an unexpected finding because it is, to our knowledge, the first example of the ability of a reagent to increase Ag-specific cytotoxicity observed at the plateau of high E/T ratios. Pretreatment at concentrations of 75 and 150 nM OA dramatically enhanced plateau-level cytotoxicity from about 45% of specific ^{51}Cr release in the control assay to 65% and 75%, respectively, in the phosphatase-inhibitor treatment group.

Inhibition by OA could be demonstrated with all tested cells and assays of cytotoxicity. Another hydrophobic compound, the inactive analog of PMA, 4-α-phorbol 12,13-didceanoate, was not able to inhibit functional responses in cytotoxicity and granule exocytosis assays even at concentrations up to 1 μM/

Thus, we found that low concentrations of OA enhanced both Ag-specific cytotoxicity and TC-induced exocytosis of CTL clones. At higher concentrations, OA inhibited all tested CTL responses. We interpret the inhibition by OA as most likely due to the interference with intracellular processes of TCR-mediated transmembrane signaling in the CTL.

It seems that OA enhances cytotoxicity by facilitating early stages of CTL–TC interaction, because TC-triggered granule exocytosis, another cellular response of CTL that is triggered by TCR-crosslinking during CTL–TC confugate formation, was also enhanced by OA. The enhancement of TC-induced exocytosis by OA-pretreated CTL eliminates the concern that the enhancement of cytotoxicity by OA is an artifact caused by OA's affecting the permeability of TC membranes for ^{51}Cr during the assay. The enhancing effect of OA may include the facilitation of CTL motility or the alignment of complementary adhesion structures such as LFA-1 and ICAM-1. Enhancement may be mediated at the level of an accessory molecule-dependent step of CTL–TC interactions. Indeed, OA enhanced the Ag-specific CTL–TC interactions that involved accessory molecules CD8 and CD11a, but was not able to enhance cytotoxicity against anti-TCR mAb-modified TC, which does not require CD8 and CD11a. Facilitation by protein phosphatase of CTL–TC accessory molecule interactions may involve their increased expression or changes in their affinity to complementary ligands on the adjacent cell surface, since it is possible that conformational changes in accessory molecules may be affected by OA. It was discussed earlier that such changes could be mediated by phosphorylation processes (Dustin and Springer, 1989). Further studies with OA may point to yet another functionally important protein whose phosphorylation controls a regulatory activity.

These results are consistent with the model that functional responses of CTL are regulated by at lease two Ser/Thr protein phosphatases. The phosphatase that was inhibited when CTL were incubated with lower concentrations of OA could be PP2A protein phosphatase, whereas the phosphatase that was inhibited at high concentrations of OA may belong to the class of PP1 phospatases.

cAMP-Dependent Protein Kinase as an "off," Down-Regulating Enzyme in CTL Effector Functions

If one makes the assumption that the TCR triggers interactions between two cells (CTL and TC), then it follows that there should be biochemical second messengers for "on" signals from the TCR.

Moreover, there should be biochemical second messengers for the "off" disengagement singal. Indeed, it is hard to imagine spontaneous dissociation of CTL adhesion proteins from their corresponding ligands on the TC surface. It is not clear what triggers the unlocking of multiple receptor–ligand complexes between the CTL and the TC surfaces in the CTL–TC conjugates, thereby allowing the eventual disengagement of CTL from TC and their cyclic activities.

We considered two possible disengagement mechanisms. The first is based on the assumption that, if an "on" signal results in TCR crosslinking by the Ag, then the termination of TCR crosslinking would have the effect of a "stop" signal. An attractive candidate for the role of a "stop" signal would be an enzyme (e.g., granule-located protease) that would be released and active in the space between the CTL and the TC in response to TCR crosslinking. This enzyme would possibly cleave off the TCR, thereby terminating the crosslinking signal.

A second possibility is that there is a biochemical "off" signal that by itself or in combination with the "stop" signal will counteract the TCR-triggered "on" signal, resulting in the termination of the "lethal hit" delivery and allowing the disengagement of the CTL from the TC when the intensity of the "on" signal is decreased. The existence of an "off" signal would make biological sense, since it increases the immunologic efficiency of CTL by preventing their irreversible engagement with the first encountered TC. Previously (Sitkovsky et al., 1988; Takayama et al., 1988) we suggested a role for the cAMP-dependent protein kinase (PK-A) in the down-regulating biochemical pathway that is involved in the inhibition of TCR-triggered T-cell activation. This conclusion was based on detailed studies involving the locus of inhibitory PK-A action in the Ca^{2+}/PK-C-mediated, TCR-triggered activating pathway ("on" signal).

Locus of Inhibitory Action of cAMP-Dependent Protein Kinase in TCR-triggered CTL Activation

Using several independent biochemical criteria, we demonstrated that PK-A inhibits both early and late stages of the TCR-triggered activation pathway in CTL, which makes this enzyme a part of a high-fidelity inhibitory, regulatory pathway.

Pretreatment of the CTL with cholera toxin induced cAMP accumulation in the CTL, partially inhibited TCR-triggered "lethal hit" delivery to the target cell, and almost completely blocked TCR-triggered exocytosis of granules from the CTL. Other agents that raise the intracellular level of cAMP, including forskolin and isobutylmethylxanthine (IBMX) also inhibited TCR-triggered CTL activation. The potential role of cyclic nucleotides in the regulation of lymphocyte function was proposed and studied earlier (reviewed in Takayama et al., 1988).

Failure to inhibit the CTL by pretreatment with forskolin or IBMX followed by washing of the CTL suggested that sustained increases in cAMP levels are required to efficiently counteract the TCR-triggered activation pathway.

The ability of CTL clones to respond by granule exocytosis to two such different stimuli as anti-TCR mAb or PMA and A23187 provided an opportunity to investigate the locus of inhibitory PK-A action in the TCR-triggered pathway of CTL activation. Since PMA- and A23187-triggered secretion is inhibited by the cAMP level-enhancing drug IBMX, it suggested that PK-A inhibits CTL activation at stages distant from the TCR-crosslinking and early transmembrane signaling pathway. These data did not, however, rule out the ability of PK-A to inhibit both early and late stages of TCR-triggered CTL activation. The ability of PK-A to inhibit early stages of TCR-triggered CTl activation was directly tested in experiments that studied the effects of cAMP increases on TCR-mediated $P1$-P_2 turnover (Takayama et al., 1988). Strong inhibition of inositol phosphate accumulation by forskolin and IMBX suggested that PK-A can also affect the early stages of the TCR-transmembrane signaling and block diacyglycerol and inositol phosphate formation, thereby inhibiting PK-C- and Ca^{2+}-mediated pathways.

The existence of more than one locus for the inhibitory action of cAMP-dependent biochemical reactions suggests an important regulatory role of cAMP in lymphocyte regulation. It is still not known which particular events of TCR-triggered exocytosis are affected by PK-A. It is possible that different cAMP levels are required to inhibit different steps in exocytosis, thereby providing a mechanism of quantitative regulation by cAMP. The drawback of these studies was the need to induce cAMP in the CTL to be able to demonstrate their inhibition. It was also possible that the effect of cAMP-raising agents was not entirely specific, and this made the interpretation and the results not conclusive.

Antisense RNA Studies of the Role of PK-A in CTL Activation

To study the role of PK-A in CTL function using more specific reagents, we tested the effect of sense, antisense, and nonsense RNA oligonucleotides on the expression of the PK-A catalytic subunit (C) in CTL and on their effector functions. Such pretreatment of CTL resulted in significant inhibition of cAMP-dependent PK-A activity and in enhancement of TCR-triggered granule exocytosis and cytotoxicity. Antisense RNA treatment is only the second example of CTL treatment (except low concentrations of the phosphatase inhibitor OA) that causes the enhancement of the plateau level of cytotoxicity, and these data support the view that PK-A is a down-regulating, inhibitory enzymen. Unexpectedly, antisense C-subunit PK-A pretreatment strongly inhibited the basal level of PK-A activity (cAMP-independent) and did not block any cAMP-inducible PK-A activity. These data suggest that the enhancement of CTL functions that we observed could be due to the inhibition of the free cytoplasmic C subunit. This has interesting implications not only for the CTL biochemical studies but also for the still unclear relationship between the free C subunit and the PK-A holoenzyme in the CTL. It also points to the likeliness that no cAMP increases are required for PK-A-mediated down-regulation.

Extracellular Ca^{2+}-Independent Cell Lysis by CTL

As we stated, cAMP-raising agents allowed the dissociation of granule exocytosis from lethal hit delivery. To obtain a clearer dissociation of cytolytic granule exocytosis from lethal hit delivery, conditions had to be chosen that allowed the TC to be killed, but allowed no possibility of the release and action of granule-associated cytolytic molecules.

Thus, we designed experiments to unambiguously dissociate exocytosis of lytic granules from TC cytolysis by comparing TCR-triggered lysis by CTL and TC-triggered exocytosis in the same experiment without Ca^{2+} (Trenn et al., 1987). We took advantage of the earlier observations by E. Martz's and P. Goldstein's groups (reviewed in Treen et al., 1987) of extracellular Ca^{2+}-independent killing, which was interpreted as an indication that the target cell is the site of Ca^{2+} action (Chapter 18, this volume) in T-lymphocyte-mediated cytolysis. We demonstrated that the addition of 2 mM Mg_2EGTA nearly abolished granule exocytosis, as shown by 0.3 ± 0.7% specific BLT esterase release at 1:1 E/T ratio. In a parallel experiment, the removal of extracellular Ca^{2+} also completely blocked TCR-triggered secretion of another CTL granule enzyme, β-glucuronidase.

The same results were obtained in studies of Con A-induced cytotoxicity and with another CTL clone specific for $H-2^d$-bearing target cells. Similar results were obtained by W. Clark and his colleagues (Ostergaard et al., 1987) who also could not detect release of serine esterases during CTl-mediated lysis in the absence of Ca^{2+}.

Thus, in three different assays of CTL-mediated cytotoxicity using "calcium-free' medium or Mg_2EGTA to decrease the Ca^{2+} concentration, it was possible to achieve TC lysis without detectable exocytosis of lytic granules from the CTL. TC lysis in the absence of Ca^{2+} required occupation and crosslinking of the Ag receptor, as shown by the retargeting assay and the inhibition of cytotoxicity by soluble mAb to TCR on the CTL. We have also found that Ca^2-independent cytotoxicity requires cell-to-cell contact, and soluble mAb to lymphocyte function-associated antigen (LFA-1) completely blocks CTL-induced TC lysis in the presence of Mg_2EGTA.

The following alternative explanations for TCR-regulated lethal hit delivery in Ca^{2+}-free media were considered (Sitkovsky, 1988) and deemed to be unlikely: (1) CTL could be in a preactivated state after stimulation with ag-bearing cells during *in vitro* culture, such that "preactivated" cells could be triggered to kill TC even in the absence of Ca^{2+}. (2) There could be a small subpopulation of very aggressive CTl, able to recycle efficiently between TC, and in the absence of Ca^{2+} only these effector cells would kill TC. Since they would comprise only a small proportion of all the CTL, no exocytosis would be detected even though they still would be able to release granules and employ these as a mechanism for lethal hit delivery.

These data indicate that exocytosis of cytolytic granules from the CTL cannot account for TC lysis in a Ca^{2+}-free medium, not only because exocytosis does not occur in the absence of extracellular Ca^{2+}, but also because Ca^{2+} is required for secretion and pore formation by perforin (Young et al., 1986; Podack and Konigsberg, 1984) (also see Chapter 14).

Extracellular Ca^{2+}-Dependent Secretion of Newly Synthesized Polypeptides by CTL

It is still, however, possible that the secretory event underlies the lytic mechanism, since some as yet unidentified, functional protein(s) could be secreted by CTL in a TCR-regulated Ca^{2+}-independent manner. Description of such a TCR-regulated biochemical pathway of secretion would be an important advantage in understanding T-lymphocyte activation. This question was addressed in experiments where secreted proteins from ^{35}S-methionine-labeled CTL were compared after incubation of cloned CTL with solid-phase anti-TCR mAb in the absence or presence of extracellular Ca^{2+} (G. Trenn, manuscript in preparation).

It was shown that such newly synthesized secreted proteins could be observed only if TCR was crosslinked in the presence of Ca^{2+}, thereby excluding the exocytic mechanism of cytoxicity. This is, however, an interesting methodological approach, since it may allow for detection of novel intermediates of lymphocyte functions.

Taken together, the data described above allow for the following model of CTL activation. According to this model: (1) CTL engage in "nonspecific" conjugates with all surrounding cells as part of their immune surveillance functions (Sitkovsky, 1988). (2) TCR crosslinking results in transmembrane signaling with involvement of GTP-binding proteins (Schrezenmeier et al., 1988) and P1-P$_2$ and Ca^{2+} increases and causes the strengthening of the conjugate due to involvement of PK-C (Berrebi et al., 1987) and another yet unidentified enzyme (such as Ser/Thr protein phosphatase) (Taffs et al., 1987). (3) Extracellular Ca^{2+} translocation through Ca^{2+} channels and Ca^{2+}/CaM-dependent proteins is implicated by the effects of nifedipine and CaM antagonists (Takayama and Sitkovsky, 1987). (4) Propagation of Ca^{2+}/CaM-mediated signal is most likely mediated by protein dephosphorylation processes, as shown by our finding that the major Ca^{2+}/CaM-binding protein in T lymphocytes is the CaM-dependent protein phosphatase, calcineurin (Kincaid et al., 1987), which is a target for CsA and FK506 (Friedman and Weissman, 1991; Liu et al., 1991), and by our finding that CsA blocks CTL-mediated cytotoxicity and granule exocytosis (Trenn et al., 1989). "Off" signaling and/or down-regulation of CTL activities is mediated by the combined action of PK-A and Ser/Thr protein phosphatase. This is evident from observations of (1) inhibition of both early and late stages of TCR transmembrane signaling in CTL by cAMP-raising agents and by cAMP analogs (Takayama et al., 1988), (2) enhancement of CTL response after blocking the expression of the PK-A C subunit by antisense RNA oligonucleotides (T. Sugiyama, 1992), and (3) enhancement of CTL response by low concentrations of protein phosphatase inhibitors (Taffs et al., 1991).

Thus, biochemical studies of CTL have proven the advantages of this system. The system provides a homogeneous source of cloned cells with well-controlled activating stimuli and multiple responses, and the ability to transfect cloned CTL to target genes of interest. These studies are expected to provide new insights not only into CTL, but also into the mechanism of cell activation in general.

References

Albert F, Hua C, Truneh A, Pierre M, Schmitt-Verhulst A-M (1985): Distinction between antigen receptor and IL-2 receptor triggering events in the activation of alloreactive T cell clones with calcium ionophores and phorbol esters. *J Immonol* 134: 3649-3655

Alexander DR, Cantrell DA (1989): Kinases and phosphatases in T-cell activation. *Immunol Today* 10: 200-205

Berrebi G, Takayama H, Sitkovsky MV (1987): Antigen-receptor interaction requirement for conjugate formation and lethal-hit triggering by cytotoxic T lymphocytes can be bypassed by protein kinase C activators and Ca^{2+} ionophores. *Proc Natl Acad Sci USA* 84: 1364-1368

Cohen P, Holmes CFB, Tsukitan Y (1990): Okadaic acid: A new probe for the study of cellular regulation. *Trends Bionchem Sci* 15: 98-102

Dustin ML, Springer TA (1989): T-cell receptor crosslinking transiently stimulates adhesiveness through LFA-1. *Nature* 341: 619-624

Freedman R (1989): A protein with many functions? *Nature* 337: 407

Friedman J, Weissman I (1991): Two cytoplasmic candidates for immunophilin action are revealed by affinity for a new cyclophilin: One in the presence and one in the absence of CsA. *Cell* 66: 799-806

Gray LS, Gnarra JR, Engelhard VM (1987): Demonstration of a calcium influx in cytolytic T lymphocytes in response to target cell binding. *J Immonol* 138: 63-69

June CH (1991): Signal transduction in T cells. *Curr Opinion Immunol* 3: 287-293

Kincaid R, Takayama H, Billingsley ML, Sitkovsky MV (1987): Differential expression of calmodulin-binding proteins in B, T lymphocytes and thymocytes. *Nature* 330: 176-178

Lancki DW, Weiss A, Fitch FW (1986): Requirements for triggering of lysis by cytolytic T lymphocyte clones. *J Immunol* 138: 3646-3653

Lanzavecchia A (1986): Is the T-cell receptor involved in T-cell killing? *Nature* 319: 778–780

Liu J, Farmer JD, Lane WS, Friedman J, Weissman I, Schreiber SL (1991): Calcineurin is a common target of cyclophilin-cyclosporin A and FKBP-FK506 complexes. *Cell* 66: 807–815

Masson D, Tschopp J (1987): A family of serine esterases in lytic granules of cytolytic T lymphocytes. *Cell* 49: 679–685

Oettgen HC, Terhost C, Cantley LC, Rosoff PM (1985): Stimulation of the T3-T cell receptor complex induces a membrane-potential sensitive calcium influx. *Cell* 40: 583–590

O'Rourke AM, Mesher MF (1988): T-cell receptor-mediated signaling occurs in the absence of inositol phosphate production. *J Biol Chem* 263: 18594–18597

Ostergaard HL, Kane KP, Mesher MF, Clark WE (1987): Cytotoxic T-lymphocyte mediated lysis without release of serine esterase. *Nature* 330: 71–73

Pasternack M, Verret CR, Liu MA, Eisen HN (1986): Serine esterase in cytolytic T lymphocytes. *Nature* 322: 740–743

Podack ER, Konigsberg PJ (1984): Cytolytic T cell granules. Isolation, structural, biochemical, and functional characterization. *J Exp Med* 160: 695–710

Rudd CE (1990): CD4, CD8 and the TCR-CD3 complex: A novel class of protein-tyrosine kinase receptor. *Immunl Today* 11: 400–406

Russell JH (1986): Phorbol-ester stimulated lysis of weak and nonspecific target cells by cytotoxic T lymphocytes. *J Immunol* 136: 23–27

Schrezenmeier H, Ahnert-Hilger G, Fleischer B (1988): A T-cell receptor associated GTP-binding protein triggers TCR-mediated granule exocytosis in cytotoxic T lymphocytes. *J Immunol* 141: 3785–3790

Sitkovsky MV (1988): Mechanistic, functional and immunopharmacological implications of biochemical studies of antigen receptor-triggered cytolytic T-lymphocyte activation. *Immunol Rev* 103: 127–160

Sitkovsky MV, Trenn G, Takayama H (1988): Cyclic AMP-dependent protein kinase as part of the possible down-regulating pathway in the antigen receptor-regulated cytotoxic T lymphocyte conjugate formation and granule exocytosis. *Ann NY Acad Sci* 532: 350–358

Spits H, Schooten W, Keizer H, Seventer G, deRijn M, Terhorst C, deVries JE (1986): Alloantigen recognition is preceded by non-specific adhesion of cytotoxic T-cells and target cells. *Science* 232: 403–405

Taffs RE, Redegeld F, Sitkovsky MV (1991): Modulation of the effector functions of cytolytic T-lymphocytes by an inhibitor of serine/threonine phosphatase, okadaic acid. *J Immunol* 147: 722–729

Taffs RE, Sitkovsky MV (1991): Granule enzyme exocytosis assay for cytotoxic T lymphocyte activation. In: *Current Protocols in Immunology*, Coligan J, Kruisbeek AM, Margulies DH, Shevach EM, Strober W, eds. New York: Wiley Interscience

Takayama H, Sitkovsky MV (1987): Antigen receptor-regulated exocytosis in cytotoxic T lymphocytes. *J Exp Med* 166: 725–743

Takayama H, Trenn G, Humphrey W, Bluestone JA, Henkart PA, Sitkovsky MV (1987a): Antigen receptor —triggered secretion of a trypsin-type esterase from cytotoxic T-lymphocytes. *J Immunol* 138: 566–569

Takayama H, Trenn G, Sitkovsky MV (1987b): A novel cytotoxic T lymphocyte activation assay. Optimized conditions for the antigen receptor triggered granule enzyme secretion. *J Immunol Methods* 104: 183–190

Takayama H, Trenn G, Sitkovsky MV (1988): locus of inhibitory action of cAMP-dependent protein kinase in the antigen receptor triggered cytotoxic T-lymphocyte activation pathway. *J Biol Chem* 263: 2330–2336

Trenn G, Taffs R, Hohmann R, Kincaid R, Shevach EM, Sitkovsky M (1989): Biochemical characterization of the inhibitory effect of CsA on cytolytic T-lymphocyte effector functions. *J Immunol* 142: 3796–3802

Trenn G, Takayama H, Sitkovsky MV (1987): Exocytosis of cytolytic granules may not be required for target cell lysis by cytotoxic T-lymphocytes. *Nature* 330: 72–74

Treves S, Di Virgilio F, Cerundolo V, Zanovello P, Collavo D, Pozzan T (1987): Calcium and inositol phosphates in the activation of T-cell mediated cytotoxicity. *J Exp Med* 166: 33–42

Trunch A, Albert F, Goldstein P, Schmidt-Verhulst AM (1985): Early steps of lymphocyte activation bypassed by synergy between calcium ionophores and phorbol ester. *Nature* 313: 318–320

Utsunomiya N, Tsuboi M, Nakanishi M (1986): Early transmembrane events in alloimmune cytotoxic T lymphocyte activation as revealed by stopped-flow fluorometry. *Proc Natl Acad Sci USA* 83: 1877–1880

Young JD-E, Leong LG, Liu CC, Damiano A, Cohn ZA (1986): Extracellular release of lymphocyte cytolytic pore—forming protein (perforin) after ionophore stimulation. *Proc Natl Acad Sci USA* 83: 5668–5672

32
Cytolytic Granules as Targets for Immunosuppressive Therapy: Selective Ablation of CTL by Leucyl-Leucine Methyl Ester

Dwain L. Thiele and Peter E. Lipsky

Introduction

Cytotoxic lymphocytes constitute a subset of immune effector cells defined by their unique capacity for cell-mediated cytotoxicity. A number of observations have suggested that these cells play a crucial role in the mediation of allograft rejection, elimination of virally infected cells, and control of tumor cell growth. Pharamacologic agents capable of selectively modulating cytotoxic lymphocyte function are therefore likely to be of value not only in delineating the role of these effector cells in such immune responses but also in ameliorating a variety of disease states. Whereas a variety of currently available immunosuppressive agents are capable of impairing generation of activated cytotoxic T lymphocytes (CTL), such effects appear to relate largely to nonspecific impairment of cell proliferation or T-helper cell function necessary for CTL activation, and not to selective impairment of this lymphocyte effector function.

Development of immunosuppressive agents specific for cytotoxic lymphocytes is likely to be dependent upon improved understanding of the functional and phenotypic characteristics of this unique population of immune effector cells. Although sensitive assays for the functional characterization of these cells have been available for many years, specific cell surface phenotypic markers of this population of lymphocytes have not been clearly delineated. It has long been recognized that the majority of CTL are $CD8^+$, but $CD4^+$ CTL capable of lysing allogeneic or virally infected targets have been identified as well (Jacobson et al., 1984; Golding and Singer, 1985). It is now apparent that CD4/CD8 expression is more closely related to the role of class II or class I major histocompatibility complex (MHC) molecules, respectively, in antigen presentation than to the effector function of individual T lymphocytes. Similarly, phenotypic markers uniquely expressed by natural killer (NK) cells have been difficult to discern. The overall pattern of cell surface receptor proteins exhibited by NK cells is distinct from that of other leukocyte lineages, but virtually all individual cell surface markers present on NK cells have also been noted to be displayed on other cells of either T lymphocyte or myeloid lineage (Chapter 8, this volume; Lanier et al., 1986). Alternative approaches to identification of CTL and effector mechanisms employed by these cells have utilized differential screening of cDNA libraries or more direct techniques for isolation and characterization of putative effector molecules stored within the specialized granules of these cells. A family of serine proteases termed granzymes, the pore-forming protein, perforin, and a variety of other granule proteins selectively expressed by CTL and NK cells have been identified (Chapters 23–27, this volume; Bleackley et al., 1988; Tschopp and Nabholz, 1990). The granules of CTL and the putative CTL-specific effector molecules contained within these granules, therefore, serve to identify lymphocytes with cytolytic potential and present potential targets for pharamacologic intervention designed to modulate this aspect of immune effector function selectively.

A variety of serine protease inhibitors have been found to impair CTL-mediated cytotoxicity (Chapters 27 and 58, this volume; Redelman and Hudig, 1980; Poe et al., 1991; Odake et al., 1991). The enzyme inhibitors used in most such studies are active against broad classes of serine esterases and, thus, may not necessarily impair cytolytic function by specific inhibition of granule serine proteases.

CYTOTOXIC CELLS
M. Sitkovsky and P. Henkart, Editors
© 1993 Birkhäuser Boston

However, recent analysis of mechanism-based inhibitors of granzyme activity has suggested a correlation between inhibition of granzyme activity and impairment of cell-mediated cytotoxicity (Odake et al., 1991; Poe et al., 1991). The specialized granules of cytotoxic lymphocytes are secretory lysosomes (Peters et al., 1991). As a result, a variety of lysosomotropic agents capable of neutralizing lysosomal pH also have been noted to inhibit the cytolytic effector function of these cells (Thiele and Lipsky, 1985a). Among the lysosomotropic agents noted to impair cytotoxic lymphocyte effector function, L-leucine methyl ester (Leu-OMe), L-leucyl-L-leucine methyl ester (Leu-Leu-OMe), and related dipeptide esters have been noted to exhibit unique spectra and mechanisms of immunosuppressive activities (Thiele and Lipsky, 1985b; Thiele and Lipsky, 1986).

Effects of Leu-OMe and Leu-Leu-OMe on Leukocyte Function

Amino acid and dipeptide methyl esters were initially proposed as nontoxic substituted amines that inhibited lysosomal enzyme function within mammalian cells (Goldman and Kaplan, 1973; Reeves et al., 1981). Like other substituted amines, these agents become protonated upon entry into the acidic environment of lysosomes. In addition, these agents are degraded by various esterases and proteases present within these organelles, thus enhancing their osmotic effects. These attributes appear to be responsible for loss of lysosomal integrity when isolated lysosomes or intact cells are incubated with 10^{-3}–10^{-2} M concentrations of these agents (Goldman and Kaplan, 1973; Reeves et al., 1981). However, when human peripheral monocytes are exposed to Leu-OMe, there is rapid killing of these cells (Thiele et al., 1983), an effect quite different from the reversible effects seen with exposure of cells of nonmyeloid origin to Leu-OMe. Moreover, following exposure of mixed populations of human peripheral blood leukocytes to this agent, neutrophils, NK cells, CTL, and pre-CTL are killed, while eosinophils, B cells, and T-helper cells remain functionally intact (Thiele et al., 1983; Thiele and Lipsky, 1985a, 1985b, 1986; Thiele et al., 1987a; Thiele and Lipsky, 1991). However, NK and CTL are killed only when incubated with Leu-OMe in the presence of monocytes or neutrophils (Thiele and Lipsky, 1985a, 1985b). In the absence of myeloid cells, exposure of NK or CTL to Leu-OMe or related amino acid methyl esters results in only transient loss of cytolytic activity, presumably secondary to reversible disruption of cytolytic granules (Thiele and Lipsky, 1985a). In addition, no adverse effects on monocyte survival were noted after incubation with D-leucine methyl ester or the methyl ester derivatives of other structurally similar L-amino acids, such as isoleucine and valine (Thiele and Lipsky, 1985a). These observations suggested that the monocyte toxicity of Leu-OMe was unlikely to relate only to the lysosomotropic properties of this compound.

Mechanisms of Leu-OMe- and Leu-Leu-OMe-Mediated Killing of Cytotoxic Lymphocytes and Myeloid Cells

As summarised in Figure 1, a series of enzymatically mediated events has been demonstrated to be responsible for the unique spectrum of toxicity mediated by this agent. The toxicity of Leu-OMe for myeloid cells and bystander cytotoxic lymphocytes is dependent upon conversion to the dipeptide ester condensation product Leu-Leu-OMe by the action of a lysosomal enzyme present in neutrophils and freshly isolated peripheral blood monocytes, but not in more differentiated macrophages or cells of nonmyeloid origin (Thiele and Lipsky, 1985b). Leu-Leu-OMe is in turn taken up by NK cells and CTL via a dipeptide-specific facilitated transport process that is severalfold more active in these cells than in Leu-Leu-OMe-resistant lymphocytes (Thiele and Lipsky, 1990a, 1990b). Within the granules of cytotoxic lymphocytes and myeloid cells, Leu-Leu-OMe is converted to higher-molecular-weight polymerization products of general structure (Leu-Leu)$_n$-OMe by the action of the lysosomal thiol protease, dipeptidyl peptidase I (DPPI). The essential role of such (Leu-Leu)$_n$-OMe metabolites in toxicity mediated by Leu-Leu-OMe has been suggested by a number of observations.

As detailed in Figure 2, when ^{51}Cr-labeled red blood cells (RBC) are incubated with Leu$_2$-OMe, Leu$_4$-OMe, or exogenous DPPI alone, no RBC lysis is observed. However, Leu$_6$-OMe or exogenous DPPI and its substrate, Leu-Leu-OMe, cause significant lysis of RBC. In addition, as summarized in Table 1, screening of a diverse panel of dipeptide ester and amide analogs of Leu-Leu-OMe has demonstrated that all NK-toxic dipeptide derivatives act as competitive inhibitors of [^{3}H]Leu-Leu-OMe uptake by human lymphocytes and serve as

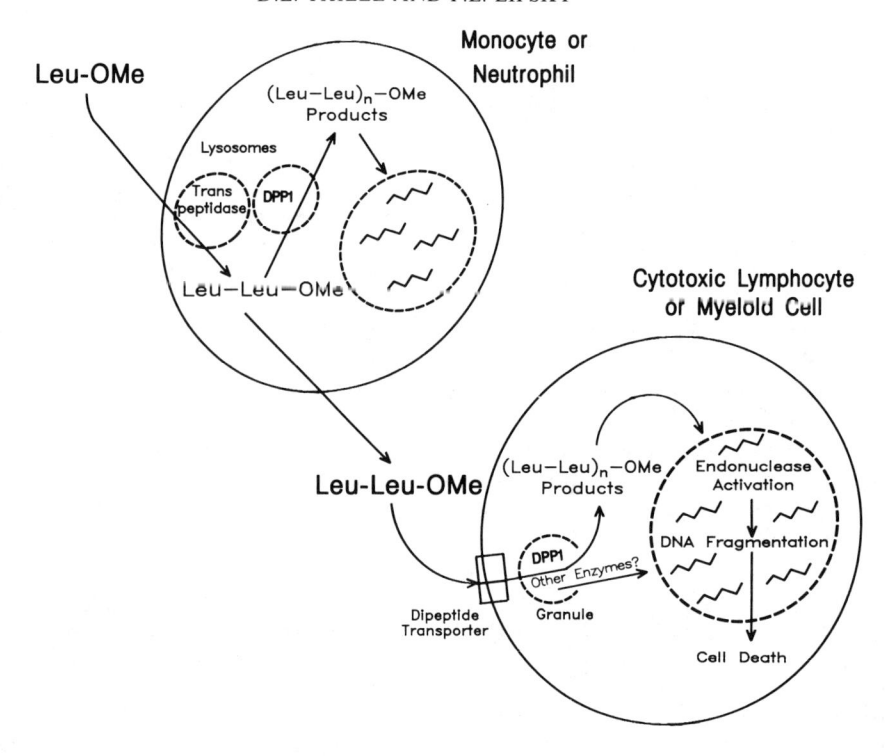

FIGURE 1. Mechanisms of Leu-OMe- and/or Leu-Leu-OMe-mediated killing of cytotoxic lymphocytes and myeloid cells.

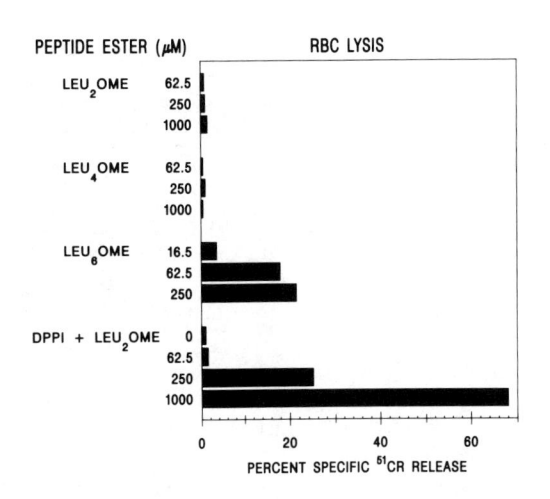

FIGURE 2. RBC lysis is mediated by Leu$_6$-OMe or by DPPI-generated metabolites of Leu-Leu-OMe. ^{51}Cr-labeled human RBC were incubated in the presence or absence of 8×10^{-4} U/ml of purified bovine DPPI and the indicated dipeptide esters for 4 hr at 37°C and were then assessed for percent specific lysis. (From Thiele and Lipsky, 1990a).

substrates for DPPI-catalyzed lysis of RBC. Moreover, treatment of human NK cells (Table 2), CTL precursors (Thiele and Lipsky, 1990a), or myeloid cells (Thiele and Lipsky, 1991) with the specific DPPI inhibitor, Gly-Phe-CHN$_2$, prevents toxicity mediated by any of these agents. Of note, some well-characterized dipeptide ester substrates of DPPI, such as Leu-Tyr-OMe and Ser-Leu-OMe, that contain amino acids with polar R groups are converted to $(R_1-R_2)_n$-OMe polymerization products without discernible membranolytic or NK toxic properties (Thiele and Lipsky, 1990b). In addition, while Leu-Leu-NH$_2$ has been noted to be converted to membranolytic metabolites by the acyl transferase activity of DPPI, this compound does not compete with Leu-Leu-OMe for uptake by lymphocytes and has no adverse effect on NK function (Thiele and Lipsky, 1990b). These findings, therefore, indicate that at least three properties are required for mediation of NK toxicity by Leu-Leu-OMe and related dipeptide esters or amides. Such agents must (1) be suitable ligands for uptake by a unique dipeptide-specific facilitated transporter present in leukocytes, (2) serve as substrates for the acyl transferase activity of DPPI,

TABLE 1. Comparison of the avidity for dipeptide transport, conversion by acyl transferase activity of DDPI to membranolytic products, and NK toxicity of various dipeptide derivatives[a]

Compound	Competitive Inhibition of [³H]Leu-Leu-OMe Uptake[b]	DPPI catalyzed lysis of human RCB[c]	Toxicity for human NK cells[d]	Toxicity for Gly-Phe-CHN₂⁻ treated NK cells[e]
Leu-Leu-OMe	+ +	+ +	+ +	−
Leu-Leu	+	−	−	−
Leu-Leu-NH₂	−	+ +	−	−
Leu-Leu-NHCH₃	−	−	−	−
Leu-Leu-OBenzyl	+ +	+ +	+ +	−
D-Leu-L-Leu-OMe	−	−	−	−
L-Leu-D-Leu-OMe	−	−	−	−
Leu-Phe-OMe	+ +	+ +	+ +	−
Leu-Tyr-OMe	+ +	−	−	−
Val-Phe-OMe	+ +	+ +	+ +	−
Pro-Leu-OMe	−	−	−	−
Asp-Phe-OMe-	−	−	−	−
Ser-Leu-OMe	+ +	−	−	−
Gly-Phe-OMe	+	+	+	−
Gly-Phe-β-naphthylamide	+ +	+ +	+ +	−
Gly-Arg-β-naphthylamide	−	−	−	−

[a] From Thiele and Lipsky, 1990b.
[b] Uptake of 10 μM [³H] Leu-Leu-OMe was assessed in the presence of 250 μM concentratins of unlabeled compounds. >66% inhibition, + +; >33% inhibition, +; <33% inhibition, −.
[c] Assayed as detailed in Figure 2. >33% RBC lysis at 250 μM, + +; <33% RBC lysis at 250, >33% lysis at 1 mM, +; <33% lysis at 1 mM, −.
[d] Human PBL were cultured for 15 min at 22°C with varying concentrations of dipeptide derivatives and then assayed for lysis of K562 targets. >50% loss of NK function, E/T 10:1, dipeptide concentration 50 μM, + +; <50% loss of NK function at 50 μM but >50% at 250 μM, +; <50% loss of NK function at 500 μM, −.
[e] Toxicity assessed as above for cells preincubated for 30 min at 37°C with 10⁻⁵ M Gly-Phe-CHN₂ before exposure to 50–500 μM concentrations of the indicated compound.

and (3) form DPPI-generated hydrophobic polymerization products that exhibit membranolytic properties.

Delineation of the role of a dipeptide transporter and the granule enzyme DPPI in the generation of Leu-Leu-OMe-mediated killing of cytotoxic lymphocytes and myeloid cells has led to a better understanding of the selective nature of Leu-Leu-OMe toxicity for these cells. Dipeptide-specific membrane transporters have been described in mammalian intestine and kidney. However, these previously described transport pathways differ in energy dependence, pH optima, effects of extracellular Na⁺ concentration, and transport kinetics from the mechanism whereby human leuokcytes internalize Leu-Leu-OMe (Thiele and Lipsky, 1990b). The K_m of Leu-Leu-OMe uptake by various subsets of human peripheral blood leukocytes has been found in all cases to be approximately 150 μM (Thiele and Lipsky, 1990b). However, as shown in Figure 3, the V_{max} for Leu-Leu-OMe uptake by various leukocyte subsets varies widely. Thus, monocytes or NK and CD8⁺ T-cell-enriched lymphocyte populations exhibit a V_{max} severalfold higher than that observed in predominantly Leu-Leu-OMe-resistant CD4⁺ T-cell populations. Moreover, the V_{max} for Leu-Leu-OMe uptake by lymphocytes previously depleted of Leu-Leu-OMe-sensitive cells is only half of that observed for control peripheral blood leukocytes (PBL). These finding suggest that rates of Leu-Leu-OMe uptake could in part account for the differential sensitivity of these populations of PBL to Leu-Leu-OMe-mediated toxicity.

The DPPI content of various leukocyte populations varies even more widely than does rate of

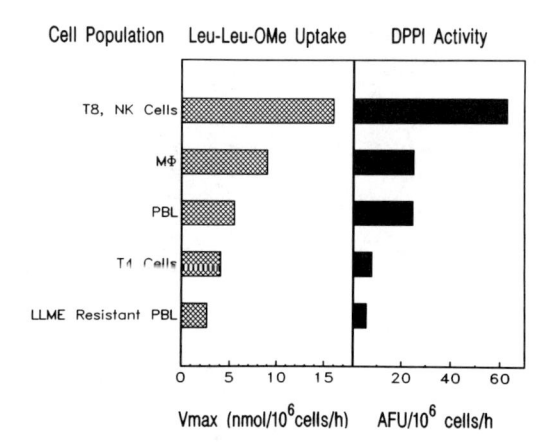

FIGURE 3. Rates of Leu-Leu-OMe uptake and DPPI content of leukocyte subpopulations correlate with sensitivity to Leu-Leu-OMe toxicity. (Data from Table 3 of Thiele and Lipsky, 1990b).

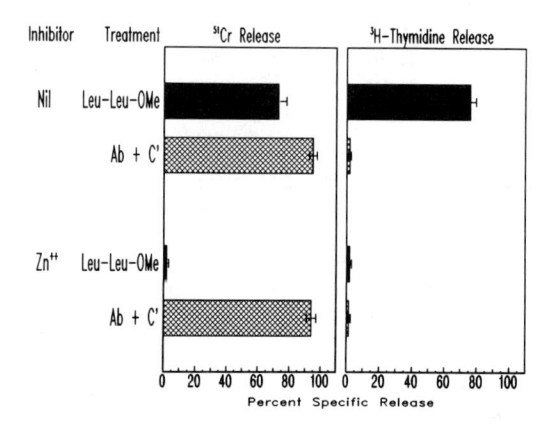

FIGURE 4. The endonuclease inhibitor, Zn^{2+}, prevents Leu-Leu-OMe but not antibody + C mediated lysis of U937 cells. U937 cells were labeled with ^{51}Cr or ^{3}H-thymidine and then treated with 1 mM Leu-Leu-OMe or anti-class I MHC (W6/32) + C in the presence or absence of 5 mM Zn^{2+}.

dipeptide ester uptake (right panel, Figure 3). As illustrated by the data detailed in this figure, T8- and NK-enriched lymphocyte populations contain > 10-fold higher levels of this lysosomal enzyme than do Leu-Leu-OMe-resistant lymphocytes. In other studies, the DPPI content of purified CD16+ human NK cells has been shown to be 20-fold higher than that of Leu-Leu-OMe-resistant cell populations such as purified CD19+ B cells, fibroblasts, or endothelial cells. Thus, while differential rates of Leu-Leu-OMe uptake are likely to account for differences in levels of sensitivity to the toxic effects of this agent, the much greater variations in DPPI content appear to play an even more important role in determining sensitivity to this agent.

DPPI is expressed at higher levels in spleen or PBL of all mammalian species than in most tissues that are not enriched in cells of bone marrow origin (McDonald and Barrett, 1986). However, human kidney has also been reported to contain significant levels of DPPI activity (McDonald and Barrett, 1986). A number of renal carcinoma cell lines have been screened for DPPI content and sensitivity to the toxic effects of Leu-Leu-OMe. Whereas DPPI content varies widely among these lines, renal tumor cell lines with DPPI levels comparable to that present in Leu-Leu-OMe-sensitive myeloid cell lines have been identified. Nevertheless, such renal carcinoma cells are not killed by exposure to concentrations of Leu-Leu-OMe that are 5 to 10 fold higher than those at which efficient killing of monocytes or tumor lines of myeloid cell origin is observed. Differences in rates of Leu-Leu-OMe uptake do not appear to account for differences in sensitivity to this agent. The mechanisms underlying resistance of human renal cells to Leu-Leu-OMe remain to be fully delineated. However, it has recently been appreciated that unique properties of cells with cytolytic potential may play a role in Leu-Leu-OMe toxicity. This may explain the resistance of renal cell lines to Leu-Leu-OMe despite the comparable content of DPPI.

When Leu-Leu-OMe-sensitive NK cells or myeloid tumor cell lines are examined by electron microscopy at various time points after exposure to Leu-Leu-OMe, the first morphologic abnormality noted is swelling and rupture of electron-dense lysosomal granules. However, in addition to these anticipated lysosomal changes, distortion of nuclear architecture and other changes characteristic of apoptosis are also observed very early in the process of Leu-Leu-OMe-mediated killing of these cells. When cytotoxic lymphocytes or myeloid cells are preincubated with the specific DPPI inhibitor Gly-Phe-CHN$_2$, both lysosomal and nuclear morphology remain normal after exposure to Leu-Leu-OMe. As shown in Figure 4, when human U937 myeloid tumor cell lines are treated with Leu-Leu-OMe, early release of both cytosolic ^{51}Cr and soluble ^{3}H-thymidine-labeled DNA fragments is observed, whereas antibody and complement treatment of these cells causes only ^{51}Cr release. When these cells are preincubated with Zn^{2+}, a known inhibitor of endonuclease activity, Leu-Leu-OMe-induced release of both ^{51}Cr and ^{3}H-thymidine is prevented, whereas no effect on antibody + C-induced ^{51}Cr release is noted (Thiele and Lipsky, 1991). Other studies have demonstrated that Zn^{2+} preincubation also protects a variety of myeloid and CTL cell lines from the toxic effects of Leu-Leu-OMe and prevents Leu-Leu-OMe-mediated depletion of human $CD16^+$ NK cells (Thiele and Lipsky, 1991). However, Zn^{2+} has been found to have no inhibitory effect on Leu-Leu-OMe uptake or intracellular conversion to (Leu-Leu)$_n$-OMe metabolites by these cells (Thiele and Lipsky, 1991). Moreover, Zn^{2+} does not inhibit lysis of RBC or nucleated targets mediated by extracellular production of DPPI-generated metabolites of Leu-Leu-OMe (Thiele and Lipsky, 1991). Thus, killing of cytotoxic lymphocytes and myeloid cells by Leu-Leu-OMe appears to be dependent upon generation of metabolites with membranolytic properties, but cell death induced by this process does not involve simple lysis of the plasma membrane. Rather, intracellular production of DPPI-generated (Leu-Leu)$_n$-OMe metabolites appears to trigger apoptosis of cells with cytolytic potential. It is likely that the activation of the cytolytic machinery by (Leu-Leu)$_n$-OMe-mediated disruption of cytolytic granules plays a role in the induction of the death of cytotoxic cells. The presence of lytic granules, therefore, may contribute to the selective nature of toxicity mediated by this agent.

Selective Immunosuppressive Effects of Leu-Leu-OMe

While all of the factors responsible for selective loss of myeloid cells and lymphocytes with cytolytic potential following exposure to Leu-Leu-OMe remain to be elucidated, the presence of DPPI-enriched lysosomal granules within cells of these lineages clearly plays a central role in this process. Unlike other granule proteins, such as perforin or the granzymes, that appear only to be expressed by activated cytotoxic lymphocytes, a variety of findings suggest that enhanced expression of DPPI is a reliable marker of the precursors of these granular effector cells. The mean DPPI content of purified human $CD8^+$ peripheral blood T cells is > threefold higher than that of $CD4^+$ T cells (Thiele and Lipsky, 1990a). Leu-Leu-OMe treatment of human or murine T-lymphocyte populations eliminates T cells with the highest DPPI content, with approximately two-thirds of $CD8^+$ and 5–10% of $CD4^+$ T cells being killed (Thiele and Lipsky, 1986; Thiele et al., 1987a; Thiele and Lipsky, 1990a, 1990b). As shown in Figure 5, Leu-Leu-OMe-mediated depletion of DPPI-enriched T cells

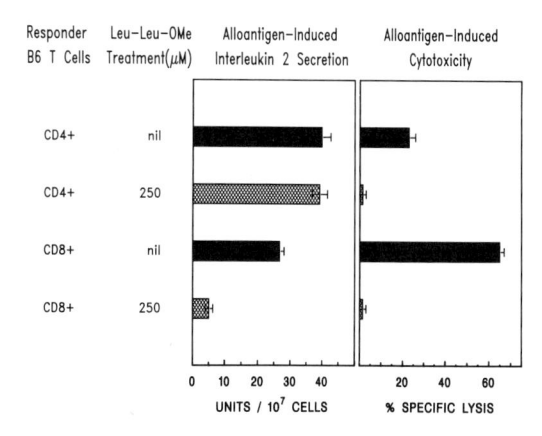

FIGURE 5. Leu-Leu-OMe treatment of murine spleen cells ablates $CD4^+$ and $CD8^+$ CTL, but $CD4^+$ T-cell alloantigen-induced IL-2 responses are preserved. $CD4^+$ T-cell-enriched or $CD8^+$ T-cell-enriched B6 (H-2^b) spleen cells were incubated for 15 min at 22°C in the presence or absence of 250 μM Leu-Leu-OMe and then cultured with T-cell-depleted B6D2F1 (H-2$^{b/d}$) spleen cells for 24 hr or 5 days before assessment of IL-2 production or allospecific lysis of B6D2F1 B cell blasts as previously detailed (Thiele et al. 1987a).

removes both $CD8^+$ and $CD4^+$ precursors of allospecific CTL. In contrast, interleukin-2 (IL-2) production by alloantigen-stimulated $CD4^+$ T cells is largely preserved. Of note, Leu-Leu-OMe treatment substantially reduces alloantigen-induced IL-2 production by $CD8^+$ T cells, thus suggesting that both cytokine-producing and cytolytic $CD8^+$ T cells are enriched in DPPI, Leu-Leu-OMe-mediated depletion of DPPI-enriched human PBL also removes cells responsible for suppressing mitogen- or antigen-induced generation of immunoglobulin (Ig) secreting cells while B-cells and T-helper cell activity necessary for these responses is preserved (Thiele and Lipsky, 1986). Thus, Leu-OMe or Leu-Leu-OMe treatment of peripheral blood mononuclear cells has proven to be an efficient technique for providing purified populations of B cells and T-helper cells devoid of cells capable of suppressing primary *in vitro*-activated Ig responses (Thompson et al., 1984; Borrebaeck, 1988).

The utility of Leu-Leu-OMe-mediated depletion of DPPI-enriched CTL has also been demonstrated in models of *in vivo* alloimmune responses. As demonstrated by the experiment detailed in Figure 6, depletion of Leu-Leu-OMe-sensitive CTL is effective in modulating the courses of graft versus host disease (GVHD). In this experiment, C57BL/6 × CBA F1 (B6 × CBAF1) recipient mice were

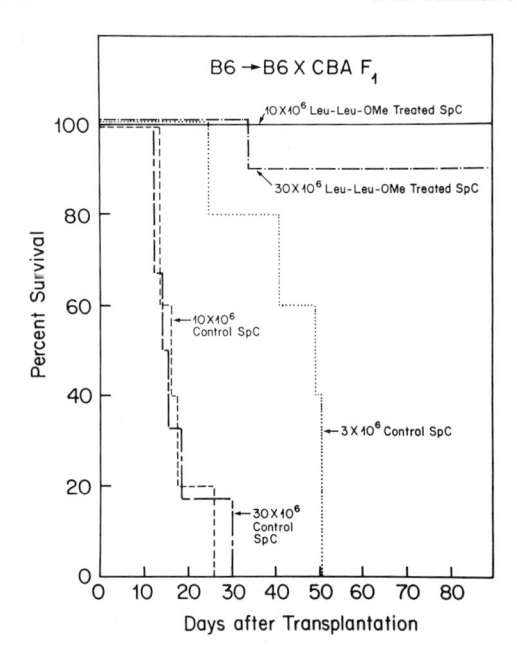

FIGURE 6. Leu-Leu-OMe treatment of B6 spleen cells prevents lethal GVHD in irradiated B6 × CBAF1 recipients. After irradiation (900 cGy), all B6 × CBAF1 mice received 5 × 10^6 anti-Thy-1 + C and Leu-Leu-OMe treated bone marrow cells in addition to the indicated numbers of spleen cells that had been preincubated for 15 min at 22°C in the presence or absence of 250 μM Leu-Leu-OMe. (Reprinted with permission of The Journal of Immunology from Thiele et al., 1988.)

lethally irradiated and then transplanted with 5 × 10^6 T- and NK-depleted C57BL/6 (B6) bone marrow cells and graded numbers of control or Leu-Leu-OMe-treated B6 spleen cells. No significant GVHD developed in B6 × CBAF1 mice reconstituted only with T- and MK-depleted B6 bone marrow (data not shown), whereas transfer of 3-30 × 10^6 control B6 spleen cells resulted in uniformly lethal GVHD. In contrast, recipients of 10-30 × 10^6 Leu-Leu-OMe-treated B6 spleen cells exhibited excellent long-term survival. As other studies have shown that in such semiallogeneic class I + II MHC and multiple non-MHC histocompatibility antigen-disparate strain combinations, lethal GVHD is prevented by depletion of all donor T cells, but not by depletion of CD4$^+$ or CD8$^+$ T-cell subsets alone (Korngold and Sprent, 1985), these results suggest that the marked survival advantage conferred by Leu-Leu-OMe treatment of donor cells relates to the capacity for removing both CD4$^+$ and CD8$^+$ CTL precursors. Additional studies performed in class I + II MHC disparate murine strain combinations have demonstrated that Leu-Leu-OMe treatment of donor cells has no deleterious effects on rates of engraftment and permits establishment of long-lived immunocompetent chimeras (Thiele et al., 1987b; Blazar et al., 1990).

Additional studies have demonstrated that transfer of Leu-Leu-OMe-resistant donor T cells is associated with extensive intestinal GVHD lesions in some class I + II MHC-disparate strain combinations (Thiele et al., 1989 and with lethal GVHD in the class II MHC only-disparate B6→B6 × bm12F1 strain combination (Thiele et al., 1988). However, in multiple strain combinations with class I + II MHC disparities, isolated class I MHC disparities or non-HMC histocompatibility antigen only disparities, Leu-Leu-OMe treatment of donor cells prevents lethal GVHD (Thiele et al., 1987a, 1988; Thiele and Lipsky, 1989). Thus, whereas Leu-Leu-OMe-sensitive, DPPI-enriched donor T cells are not responsible for generation of all aspects of GVHD, removal of these precursors of CTL is associated with dramatic survival benefits in the vast majority of allogeneic murine strain combinations.

To assess whether DPPI-enriched CTL played a similar role in the evolution of organ allgraft rejection, the experiments detailed in Figure 7 were performed. Adult B6 mice were thymectomized, lethally irradiated, transplanted with T cell-depleted bone marrow, and treated with anti-CD4 and anti-CD8 monoclonal antibodies. As shown by the results detailed in Figure 7, such adult-thymectomized, T-cell-depleted mice were generally unable to reject class I + II MHC and multiple non-MHC histocompatibility antigen-disparate B6 × DBA/2F1 (B6D2F1) skin grafts. However, following transfer of 7 × 10^7 syngeneic B6 spleen cells, rapid rejection of B6D2F1 skin grafts was observed. When B6 donor cells were treated with Leu-Leu-OMe prior to transfer, B6D2F1 skin graft rejection was significantly delayed. Although all recipients of Leu-Leu-OMe-treated B6 spleen cells eventually rejected B6D2F1 skin grafts, subsequent analysis of the adult-thymectomized, T-cell-depleted mice utilized in these experiments has demonstrated that they contain a population of host-derived, thymus-independent CTL precursors that are expanded after transfer of Leu-Leu-OMe-resistant T-helper cells. Thus, the results detailed in Figure 7 do not distinguish between a requisite role for CTL in skin allograft rejection and the possibility that in the presence of class I + II MHC differences, T-cell cytokine responses alone can mediate skin allograft rejection. Nevertheless, these results do suggest that

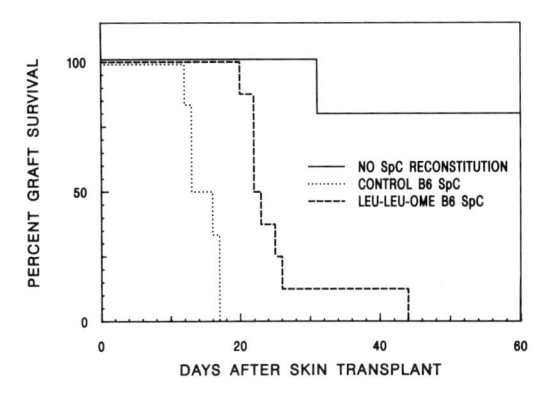

FIGURE 7. Survival of B6D2F1 (H-2$^{b/d}$) skin grafts applied to adult-thymectomized, T-cell-depleted B6 (H-2^2) mice in the presence or absence of reconstitution with control or Leu-Leu-OMe treated B6 spleen cells. Four weeks after treatment of adult-thymectomized, lethally irradiated, T-cell-depleted, bone-marrow-reconstituted mice with anti-L3T4 and anti-Lyt2, mice were infused with 70 × 10^6 control B6 spleen cells (····), 70 × 10^6 Leu-Leu-OMe B6 spleen cells (···), or no spleen cells (———). B6D2F1 skin grafts were applied within 24 hr. Survival of skin grafts in recipients of Leu-Leu-OMe treated spleen cells was significantly longer than that in mice reconstituted with control B6 spleen cells ($p < 0.001$ by Mann-Whitney nonparametric analysis).

therapeutic modalities designed to selectively ablate CTL responses may by efficacious in modulating the course of organ allograft rejection.

Summary

Leu-Leu-OMe, an agent that target cells enriched in the granule enzyme DPPI, has been demonstrated to serve as an immunosuppressive agent capable of selectively ablating CTL and ameliorating the course of GVHD and organ allograft rejection. These results suggest that other agents that target the unique granular effector mechanism utilized by CTL also may function as useful therapeutic agents.

References

Blazar BR, et al. (1990): Pretreatment of murine donor grafts with L-leucyl-L-leucine methyl ester: Elimination of graft-versus-host disease without detrimental effects on engraftment. *Blood* 75: 798-805

Bleackley RC, et al. (1988): The isolation and characterization of a family of serine protease genes expressed in activated cytotoxic T lymphocytes. *Immunol Rev* 103: 5-19

Borrebaeck CAK (1988): Human mAbs produced by primary in vitro immunization. *Immunol Today* 9: 355

Golding H, Singer A (1985): Specificity, phenotype, and precursor frequency of primary cytolytic T lymphocytes specific for class II major histocompatibility antigens. *J Immunol* 135: 1610

Goldman R, Kaplan A (1973): Rupture of rat liver lysosomes mediated by L-amino acid esters. *Biochem Biophys Acta* 318: 205

Jacobson S, et al. (1984): Measles virus-specific T4$^+$ human cytotoxic T cell clones are restricted by class II HLA antigens. *J Immunol* 133: 754

Korngold R, Sprent J (1985): Surface markers of T cells causing lethal graft-vs-host disease to class I vs class II H-2 differences. *J Immunol* 135: 3004

Lanier LL, et al. (1986): Natural killer cells: Definition of a cell type rather than a function. *J Immunol* 137: 2735-2739

McDonald JK, Barrett AJ (1986): *Mammalian Proteases: A Glossary and Bibliography*. London: Academic Press

Odake S, et al. (1991): Human and murine cytotoxic T lymphocyte serine proteases: Subsite mapping with peptide thioester substrates and inhibition of enzyme activity and cytolysis by isocoumarins. *Biochemistry* 30: 2217-2227

Peters PJ, et al. (1991): Cytotoxic T lymphocyte granules are secretory lysosomes, containing both perforin and granzymes. *J Exp Med* 173: 1099-1109

Poe M, et al. (1991): The enzymatic activity of human cytotoxic T-lymphocyte granzyme A and cytolysis mediated by cytotoxic T-lymphocytes are potently inhibited by a synthetic antiprotease, FUT-175. *Arch Biochem Biophys* 284: 215-218

Redelman D, Hudig D (1980): The mechanism of cell-mediated cytotoxicity. I. Killing by murine cytotoxic T lymphocytes requires cell surface thiols and activated proteases. *J Immunol* 124: 870-878

Reeves JP, et al. (1981): Intracellular disruption of rat heart lysosomes by leucine methyl ester: Effects on protein degradation. *Proc Natl Acad Sci USA* 78: 4426

Thiele DL, Lipsky PE (1985a): Modulation of human natural killer cell function by L-leucine methyl ester: Monocyte-dependent depletion from human peripheral blood mononuclear cells. *J Immunol* 134: 786-793

Thiele DL, Lipsky PE (1985b): Regulation of cellular function by products of lysosomal enzyme activity: Elimination of human natural killer cells by a dipeptide methyl ester generated from L-leucine methyl ester by monocytes or polymorphonuclear leukocytes. *Proc Natl Acad Sci USA* 82: 2468-2472

Thiele DL, Lipsky PE (1986): The immunosuppressive activity of L-leucyl-L-leucine methyl ester: Selective ablation of cytotoxic lymphocytes and monocytes. *J Immunol* 136: 1038-1048

Thiele DL, Lipsky PE (1989): The impact of major histocompatibility complex (MHC) and non-MHC antigenic differences on graft-versus-host disease mediated by helper T cells. *Trans Assoc Am Physicians* CII: 59–67

Thiele DL, Lipsky PE (1990a): Mechanism of L-leucyl-L-leucine methyl ester-mediated killing of cytotoxic lymphocytes: Dependence on a lysosomal thiol protease, dipeptidyl peptidase I, that is enriched in these cells. *Proc Natl Acad Sci USA* 87: 83–87

Thiele DL, Lipsky PE (1990b): The action of leucyl-leucine methyl ester on cytotoxic lymphocytes requires uptake by a novel dipeptide-specific facilitated transport system and dipeptidyl peptidase I-mediated conversion to membranolytic products. *J Exp Med* 172: 183–194

Thiele DL, Lipsky PE (1991): Apoptosis is an essential step in leucyl-leucine methyl ester (Leu-Leu-OMe) mediated killing of cytotoxic lymphocytes and myeloid cells. *FASEB J* 5: A937

Thiele DL, et al. (1983): Phenotype of the accessory cell necessary for mitogen-stimulated T and B cell responses in human peripheral blood: Delineation by its sensitivity to the lysosomotropic agent, L-leucine methyl ester. *J Immunol* 131: 2282

Thiele DL, et al. (1987a): Lethal graft-vs-host disease across major histocompatibility barriers: Requirements for leucyl-leucine methyl ester sensitive cytotoxic T cells. *J Immunol* 138: 51–57

Thiele DL, et al. (1987b): Leucyl-leucine methyl ester treatment of donor cells permits establishment of immunocompetent parent →F1 chimeras that are selectively tolerant of host alloantigens. *J Immunol* 139: 2137–2142

Thiele DL, et al. (1988): Lethal graft-vs-host disease induced by a class II MHC antigen only disparity is not mediated by cytotoxic T cells. *J Immunol* 141: 3377–3382

Thiele DL, et al. (1989): Intestinal graft-vs-host disease is initiated by donor T cells distinct from classic cytotoxic T lymphocytes. *J Clin Invest* 84: 1947–1956

Thompson PA, et al. (1984): Regulation of human B cell proliferation by prostaglandin E_2. *J Immunol* 133: 2446–2453

Tschopp J, Nabholz M (1990): Perforin-mediated target cell lysis by cytolytic T lymphocytes. *Annu Rev Immunol* 8: 279–302

VIII
Functions of Cytotoxic Cells *In Vivo*

33
Role of CD8⁺ αβ T Cells in Respiratory Infections Caused by Sendai Virus and Influenza Virus

Peter C. Doherty, William Allan, Maryna Eichelberger, Sam Hou, Jacqueline M. Katz, Rudolf Jaenisch, and Maarten Zijlstra

Introduction

Current understanding of the part played by CD8⁺ αβ T cells in respiratory virus infections is based on findings from a spectrum of approaches. These include direct analysis of the patterns of lymphocyte involvement in the virus-infected lung (McDermott et al., 1987; Pena-Cruz et al., 1989; Allan et al., 1990; Openshaw, 1991; Eichelberger et al., 1991b), adoptive transfer experiments utilizing either bulk immune T-cell populations or cell lines (Leung and Ada, 1992; Lukacher et al., 1984, 1986; Ada and Jones, 1986; Taylor and Askonas, 1986; Kast et al., 1986; Cannon et al., 1987; Askonas et al., 1988; Mackenzie et al., 1989), *in vivo* depletion with monoclonal antibody (mAb) (Lightman et al., 1987; Waldmann, 1989; Allan et al., 1990; Eichelberger et al., 1991b), and the use of H-2 mutant or genetically manipulated mice (de Waal et al., 1983; Eichelberger et al., 1991b).

This chapter summarizes the situation for the influenza and Sendai virus pneumonia models that are being analyzed in our laboratories. It complements a more extensive recent review of the roles of αβ and γδ T-cell subsets in viral immunity (Doherty et al., 1992), and adds new information that has been generated over the past few months. Some of the points depend on results that are currently being prepared for publication and that are neither provided for referenced. The aim is to use the available information to make as contemporary a synthesis as possible of the nature of CD8⁺ αβ T-cell involvement in viral pneumonias.

The Influenza and Sendai Models and the Development of the CD8⁺ αβ T-cell Response

Influenza in the mouse is something of an artifact. Mice do not become infected in nature with influenza A viruses, and even with the adapted strains used in the laboratory, the virus does not usually spread between infected and normal animals housed in the same cage. However, murine influenza pneumonia can be lethal, depending on the virulence and dose of the virus used (Kawaoka, 1991). Also, this experimental system has been analyzed in great detail, from the aspect of both T-cell specificity and *in vivo* effector function (Bennink et al., 1982; Ada and Jones, 1986; Townsend et al., 1986; Askonas et al., 1988; Braciale and Braciale, 1991; Yewdell, 1991). Given that *in vivo* experiments with genuine mouse pathogens (no matter how well contained) generate considerable unease in the minds of colleagues, influenza tends to be a convenient system for testing hypotheses in viral immunology. The experimental system has provided a great deal of insight into the nature of cell-mediated immunity.

The murine parainfluenza type 1 virus, Sendai virus, spreads very rapidly within mouse colonies and is the main cause of virus-induced respiratory disease in this species. Thus, in analyzing the immune response to these two viruses, we are comparing a situation where the virus and host have coevolved (Sendai) with a laboratory animal model system (influenza). Both are negative-strand RNA viruses, which use somewhat different infection and replication strategies but are ultimately lytic for respiratory epithelial cells. The pathogenesis of the resultant disease processes presumably reflects this divergence. However, at least on superficial analysis, the host response is similar in many respects, and the nature of CD8[+] T-cell involvement looks to be broadly comparable (Eichelberger et al., 1991b; Hou et al., 1992).

The C57BL/6J (B6) mice used in these experiments (Allan et al., 1990) are anesthetized and given a sublethal dose of the virus intranasally (i.n.). This results in extensive replication in the respiratory epithelium, with minimal evidence of infection of the regional, mediastinal lymph node (MLN) in influenza (Allan et al., 1990; Eichelberger et al., 1991c). Similarly, virus titers in the MLN of Sendai virus-infected mice are at least 1000-fold lower than those in the lung (Hou et al., 1992). Both viral RNA and mRNA can, however, be detected in the MLN of mice with influenza pneumonia using the polymerase chain reaction (PCR) (Eichelberger et al., 1991c). The MLN enlarge enormously within 2 to 3 days of infection due to massive, nonspecific recruitment of T cells and B cells from blood. The CD4:CD8 ratios (Figure 1a) remain at about 1:1 (influenza) or 2:1 (Sendai) throughout the course of infection. We think that this is where the primary CD8[+] cytotoxic T-lymphocyte (CTL) response is being generated in these virus infections, though we have no evidence to exclude the involvement of the bronchus-associated lymphoid tissue (BALT).

Effector CD8[+] CTL are generally not very apparent in the regional MLN, though CD8[+] CTL precursors (CTLp) are readily demonstrated following *in vitro* restimulation under bulk culture or limiting dilution analysis (LDA) conditions (Allan et al., 1990). Potent effector CTL are, however, present in the bronchoalveolar lavage (BAL) populations recovered from anesthetized, exsanguinated mice (Figure 2). The thinking is that the CTL response develops in the MLN, but because there are relatively few infected (or antigen-presenting) cells in the lymph node, most CTLp exit before becoming fully functional effectors. The CTLp that localize to the site of virus growth in the lung then acquire a capacity for lytic activity, becoming prominent at the time (days 7–10 postinfection) that the virus is being eliminated (Figure 2).

The T-cell infiltrate sampled by BAL is dominated by large CD8[+] lymphoblasts. The CD4:CD8 ratio in both infections is 1:2 to 1:3 (Figure 1b), with few B lymphocytes being present. The CD8[+] T cells found at the peak of Sendai pneumonia are LFA-1[1], CD44[+], Mel14[−], and CD45RB[−], with the intensity of LFA-1 staining decreasing rapidly after the virus is cleared. Many of the CD8[+] T cells that are recruited to the lung may not be virus-specific

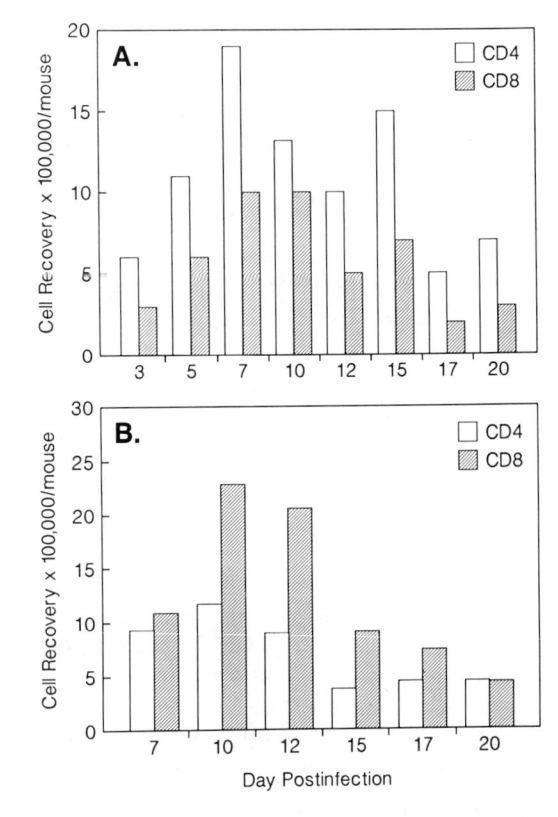

FIGURE 1. The kinetics of T-cell recruitment to the lymph node (A) and lung (B) are shown for B6 mice infected i.n. with 120 mouse ID$_{50}$ of Sendai virus (Hou et al., 1992). The numbers (per mouse) of CD4[+] and CD8[+] T cells recovered (A) from the mediastinal lymph node (MLN) and (B) by bronchoalveolar lavage (BAL) were calculated from total cell counts and the percent of cells detected by fluorescence-activated cell sorter (FACS) analysis.

CTL. The frequency of virus-specific CTLp in the BAL from mice with influenza pneumonia is < 1:100 (Allan et al., 1990). However, this may underestimate the prevalence of virus-specific effectors, as it is not clear whether the CD8[+] CTL are terminally differentiated or can be expanded further under LDA conditions. Current experiments are concentrating on defining the spectrum of T-cell receptor (TCR) gene usage in these responses, part of the idea being to separate actively and passively involved lymphocyte populations.

The overall pattern is thus that the CD8[+] T cell is the more prominent $\alpha\beta$ TCR[+] lymphocyte in the lung following infection with both influenza A viruses and Sendai virus (Figures 1b and 2). This is a comfortable finding, as most virus growth is in epithelial cells which would normally be class I

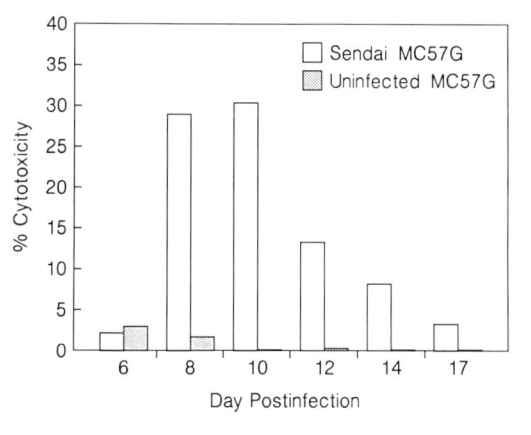

FIGURE 2. The level of class I MHC-restricted CTL activity was measured by specific ^{51}Cr release (E/T ratio 15:1) using class I MHC$^+$ class II MHC$^-$ virus-infected and normal MC57G target cells. The lymphocytes were obtained from the lungs of Sendai virus-infected B6 mice by BAL, then depleted of macrophages by adherence to plastic (Allan et al., 1990; Hou et al., 1992).

major histocompatibility complex (MHC)$^+$ class II MHC$^-$. Dominance of the inflammatory process by CD8$^+$ T cells (many of which may not be virus-specific) is also in accord with earlier, detailed studies of T-cell recruitment and localization in the murine lymphocyte choriomeningitis virus (LCM) model (Doherty et al., 1990).

Adoptive Transfer Experiments and the Roles of CD4$^+$ and CD8$^+$ T Cells

The concept has become established in the influenza literature that immune CD4$^+$ T cells promote severe inflammation and immunopathology in the virus-infected mouse lung, while the CD8$^+$ population induces less marked pneumonitis and clears the virus (Ada and Jones, 1986). This hypothesis is based on adoptive transfer experiments using bulk, memory T-cell populations that were stimulated *in vitro* and then injected into normal mice that had been infected i.n. with virus 24 hr previously.

Comparable adoptive transfer studies with CD4$^+$ and CD8$^+$ influenza-immune T-cell lines did not highlight this difference in capacity to induce immunopathology, and showed that both T-cell subsets could function to clear virus (Lukacher et al., 1984, 1986; Askonas et al., 1988; Scherle et al., 1992). Furthermore, when two influenza A viruses were given i.n. simultaneously, the CD4$^+$ and CD8$^+$ cell lines eliminated only the virus for which they were specific (Lukacher et al., 1984, 1986). Thus the CD4$^+$ T cells did not seem to be functioning via some nonspecific effector mechanism, such as macrophage activation. More recent experiments indicate that the CD4$^+$ T cells are of the TH1 type (Graham et al., 1991) and that their protective effect may be mediated via help for virus type-specific B lymphocytes (Scherle et al., 1992). The broad alternative is obviously that lymphokines/cytokines produced locally in the inflamed lung are inducing the virus-infected epithlial cells to express class II MHC glycoproteins, which then present viral peptides to the CD4$^+$ effectors (Pober and Cotran, 1991).

Adoptive transfer studies using nu/nu mice as recipients in the Sendai model indicated a requirement for the presence of both CD4$^+$ and CD8$^+$ T cells to promote recovery (Iwai et al., 1988; Kast et al., 1986). These experiments, which were done with established cell lines, were considered to support the idea of a need for CD4$^+$ T-cell help in the generation and maintenance of the CD8$^+$ response.

Depletion of the CD4$^+$ and CD8$^+$ T Cells with Monoclonal Antibodies

Elimination of the CD4$^+$ population by *in vivo* treatment with the GK1.5 mAb abrogates the virus-specific immunoglobulin G (IgG) response in both the Sendai and the influenza models (Lightman et al., 1987; Allan et al., 1990; Eichelberger et al., 1991c; Hou et al., 1992). The virus-specific CD8$^+$ T-cell response is, however, not greatly reduced in magnitude, and virus clearance from lung and lymph node is minimally delayed (Allan et al., 1990). The total numbers of influenza-specific CTLp in lung and regional lymph node may be reduced by, at most, about 20%. This lack of a mandatory need for class II MHC-restricted T-cell help for the virus-specific CD8$^+$ CTL response *in vivo* has now been found for a number of experimental systems using mAb depletion (reviewed in Doherty et al., 1992), negative selection protocols (Bennink and Doherty, 1979), and mice with a CD4 gene disruption (Rahemtulla et al., 1991). Conversely, the presence of CD4$^+$ helpers (or added lymphokines) is required for the development of influenza-specific CD8$^+$ CTL from primed CTLp under *in vitro* culture conditions (Palladino et al., 1991), perhaps because the generally smaller amount of interleukin-2 (IL-2) produced by the $CD8^+$ T cells is less diffused (and thus utilized more economically) in the lymph node than in a tissue culture flask.

The more surprising finding is that depletion of the $CD8^+$ population by *in vivo* treatment with the 2.43.1 mAb also caused a relatively slight delay in virus clearance in mice with influenza pneumonia: those that were treated with the mAb recovered normally (Eichelberger et al., 1991b). There was no evidence of the severe immunopathology that had been associated with the adoptive transfer of primed $CD4^+$ T cells into naive mice (Ada and Jones, 1986).

The effect of removing the $CD8^+$ population is greater in Sendai virus infection, with elimination of the pathogen from the lung being delayed by several days and death occurring in about 20% of the animals (Hou et al., 1992). Even so, the majority of the antibody-treated mice survived and cleared the virus, though another group treated concurrently with mAb to CD4 and CD8 failed to deal with the infection and succumbed. Both $CD4^-8^-$ $\alpha\beta$ and $\gamma\delta$ T cells are found in the BAL of the doubly depleted mice, but are unable to reduce virus titers.

The conclusion from the mAb-depletion studies is thus that either $CD4^+$ or $CD8^+$ T cells can terminate the pneumonia caused by these two RNA viruses, though the $CD8^+$ population promotes much more rapid clearance of Sendai virus. The problem with this approach is, however, that it is difficult to ensure that all the T cells of a particular subset have indeed been removed (Rosenberg et al., 1991). Culture of lymphocytes from the CD8-depleted mice sometimes reveals the presence of residual, virus-specific effector function for class I MHC^+ class II MHC^- target cells. It is also possible that some $\alpha\beta$ T cells may modulate surface CD4 or CD8 expression and thus escape elimination. Virus immune T cells do not necessarily need to engage CD8 to mediate lysis, as the "high-affinity" effectors generated *in vivo* may not be greatly inhibited by including mAb to CD8 at the time of assay.

Experiments with $\beta2$-m $(-/-)$ Transgenic Mice

The $\beta2$-m $(-/-)$ transgenics (Zijlstra et al., 1990), which lack mature $CD8^+$ $\alpha\beta$ T cells and functional class I MHC glycoproteins, show about the same capacity to deal with both viruses as the mice that are depleted with the mAb to CD8 (Eichelberger et al., 1991b; Hou et al., 1992). There is comparatively little effect on clearance of influenza virus, while the elimination of Sendai virus is substantially delayed, and there is an increase in mortality. We used a low dose of Sendai virus with these (129 × B6) F3 transgenics, as 129 mice are very susceptible to infection with this virus (Parker et al., 1978). It is likely that challenge with a higher titer of Sendai virus would have killed all the $\beta2$-m $(-/-)$ mice, but not the $\beta2$-m $(+/-)$ heterozygote controls. This would be in accord with earlier findings that the $H-2K^{bm1}$ mutant, which does not make a $CD8^+$ CTL response to Sendai virus, is much more susceptible than congenic B6 mice to infection with this virus (de Waal et al., 1983).

There was also a hint that the $\beta2$-m $(-/-)$ mice may be able to generate a minimal set of class I MHC-restricted CTLp (Hou et al., 1992). Such effector CTL were never detected on direct assay of the BAL population, but restimulation of MLN cells taken late in the course of the infectious process led to some lysis of virus-infected class I MHC^+ class II MHC^- targets that was not seen for the uninfected controls. The lymphocytes were stimulated *in vitro* with virus-infected B6 cells, which are class I MHC^+, and $\beta2$-m could also be provided from the fetal calf serum in the cultures. We have not yet tested whether this effect is reproducible, or if it is mediated by $CD8^+$ T cells. It is also possible that, especially in the absence of negative selection to tolerizing class I MHC glycoproteins in the $\beta2$-m $(-/-)$ thymus, a subset of $CD4^+$ T cells might show a class I MHC-restricted reactivity pattern. Even if a few such class I MHC-restricted CTLp are generated, it is doubtful that they would be able to promote the clearance of virus because the $\beta2$-m $(-/-)$ mice lack the class I MHC-restriction element *in vivo* (Zijlstra et al., 1990). However, in order to be fully satisfied that $CD8^+$ effectors were not involved in terminating the infection in the $\beta2$-m $(-/-)$ transgenics, we repeated the challenge with $\beta2$-m $(-/-)$ mice that were also treated with the 2.43.1 mAb to CD8. They all survived and cleared virus from the lung. Class I MHC-restricted CTLp could not be detected in either the BAL or MLN populations. We have also shown that class II MHC-restricted $CD4^+$ CTL are generated in these Sendai virus-infected $\beta2$-m $(-/-)$ mice (Hou et al., 1992).

Conclusions

There seems little doubt that mice can clear influenza A viruses and Sendai virus in the absence of

CD8$^+$ T cells, though this lymphocyte subset is probably the main mediator of virus elimination from infected tissues in normal animals (Doherty et al., 1992). The infectious process is certainly prolonged in the Sendai model, and the incidence of lethal disease is increased (Hou et al., 1992). The outcome of infection with any lytic virus can be considered to reflect a race between the development of the effective host response and the growth of the pathogen.

The experiments of F. Lehmann-Grube and J. Löhler (personal communication) indicate that the same strain of β2-m ($-/-$) mice that we have used in the experiments described here (Zijlstra et al., 1990) cannot clear murine LCMV. A somewhat contrary result has been obtained by J. Frelinger and colleagues (Chapter 13), who used β2-m ($-/-$) mice from a different source and showed variable, but generally delayed, elimination of LCMV (Muller et al., 1992). Adult β2-m ($-/-$) mice also failed to clear a low dose of Moloney murine leukemia virus, in contrast to the ($+/-$) controls (R. Jaenisch and M. Zijlstra, unpublished). Another study with mutant mice that do not express the CD8 molecule established that both the acute clearance of LCMV and the immunopathology in the central nervous system that is characteristic of this disease are greatly inhibited (Fung-Leung et al., 1991). However, the nonlytic LCMV which establishes a persistent infection in the absence of an immune T-cell response (Buchmeier et al., 1980), is eventually eliminated from the CD8$^-$ transgenic mice (Fung-Leung et al., 1991).

The overall impression is thus that the absence of a CD8$^+$, class I MHC-restricted $\alpha\beta$ T-cell response is likely to delay virus clearance, with this effect varying depending on the nature of the particular infectious process. The protective immune response in such CD8-depleted mice depends on the involvement of virus-specific CD4$^+$ T cells. The CD4$^+$ lymphocytes may function as CTL effectors to eliminate class II MHC$^+$ virus-infected target cells (Muller et al., 1992), or as helpers for virus-specific B cells and (perhaps) for other components of the host response such as CD4$^-$ 8$^-$ $\alpha\beta$ or $\gamma\delta$ T cells (Scherle and Gerhard, 1986; Carding et al., 1990; Eichelberger et al., 1991a, 1991b; Scherle et al., 1992).

We have known for some time that, even in normal mice, CD8$^+$ T cells are not the only mediators of virus clearance from solid tissues. This is obviously the case for neurons, which cannot be induced to express class I MHC glycoproteins under any circumstances (Joly et al., 1991), and which clear alphavirus infection by an antibody-dependent mechanism (Levine et al., 1991; Griffin et al., 1992). The various components of the immune system may both play different parts in the normal host response and provide alternative effector mechanisms when a particular lymphocyte subset is missing. One function that does not seem to be readily compensated is CD4$^+$ T-cell help for B cells. (Eichelberger et al., 1991c; Rahemtulla et al., 1991; Hou et al., 1992).

Acknowledgments

Maarten Zijlstra died suddenly early this year. We greatly regret this tragic loss of an intellectually incisive and generous young colleague.

References

Ada GL, Jones PD (1986): The immune response to influenza infection *Curr Top Microboil Immunol* 128: 1–54

Allan W, Tabi Z, Cleary A, Doherty PC (1990): Cellular events in the lymph node and lung of mice with influenza. Consequences of depleting CD4+ T cells. *J Immunol* 144: 3980–3986

Askonas BA, Taylor PM, Esquivel F (1988): Cytotoxic T cells in influenza infection. *Ann NY Acad Sci* 532: 230–237

Bennink JR, Doherty PC (1979): Different rules govern help for cytotoxic T cells and for B cells. *Nature* 276: 829–831

Bennink JR, Yewdell JW, Gerhard W (1982): A viral polymerase involved in recognition of influenza virus-infected cells by a cytotoxic T-cell clone. *Nature* 296: 75–76

Braciale TJ, Braciale VL (1991): Antigen presentation: Structural themes and functional variations. *Immunol Today* 12: 124–129

Buchmeier MJ, Welsh RM, Dutko FJ, Oldstone MB (1980): The virology and immunobiology of lymphocytic choriomeningitis virus infection *Adv Immunol* 30: 275–331

Cannon MJ, Stott EJ, Taylor G, Askonas BA (1987): Clearance of persistent respiratory syncytial virus infections in immunodeficient mice following transfer of primed T cells. *Immunology* 62: 133–138

Carding SR, Allan W, Kyes S, Hayday A, Bottomly K, Doherty PC (1990): Late dominance of the inflammatory process in murine influenza by $\gamma\delta$+ T cells. *J Exp Med* 172: 1225–1231

de Waal LP, Kast WM, Melvoid RW, Melief CJ (1983): Regulation of the cytotoxic T lymphocyte response against Sendai virus analyzed with H-2 mutants. *J Immunol* 130: 1090–1096

Doherty PC, Allan JE, Lynch F, Ceredig R (1990): Dissection of an inflammatory process induced by CD8+ T cells. *Immunol Today* 11: 55–59

Doherty PC, Allan W, Eichelberger M, Carding SR

(1992): Roles of $\alpha\beta$ and $\gamma\delta$ T cell subsets in viral immunity. *Annu Rev Immunol* 10: 123-151

Eichelberger M, Allan W, Carding SR, Bottomly K, Doherty PC (1991a): Activation status of the CD4−8− $\gamma\delta$-T cells recovered from mice with influenza pneumonia. *J Immunol* 147: 2069-2074

Eichelberger M, Allan W, Zijlstra M, Jaenisch R, Doherty PC (1991b): Clearance of influenza virus respiratory infection in mice lacking class I major histocompatibility complex-restricted CD8+ T cells. *J Exp Med* 174: 875-880

Eichelberger MC, Wang M, Allan W, Webster RG, Doherty PC (1991c): Influenza virus RNA in the lung and lymphoid tissue of immunologically intact and CD4-depleted mice. *J Gen Virol* 72: 1695-1698

Fung-Leung W-P, Kundig TM, Zinkernagel RM, Mak TW (1991): Immune response against lymphocytic choriomeningitis virus infection in mice without CD8 expression. *J Exp Med* 174: 1425-1429

Graham MB, Braciale VL, Braciale TJ (1991): Analysis of differential in vitro and in vivo effector functions of influenza specific TH1 and TH2 clones. *FASEB J* 5: 1338 (abstr)

Griffin DE, Levine B, Tyor WR, Irani DN (1992): The immune response in viral encephalitis. *Sem Immunol* 10:

Hou S, Doherty PC, Zijlstra M, Jaenisch R, Katz JM (1992): Delayed clearance of Sendai virus in mice lacking class I MHC-restricted CD8+ $\alpha\beta$ T cells. *J Immunol* 149: 1319-1325

Iwai H, Machii K, Otsuka Y, Ueda K (1988): T cells subsets responsible for clearance of Sendai virus from infected mouse lungs. *Microbiol Immunol* 32: 305-315

Joly E, Mucke L, Oldstone MB (1991): Viral persistence in neurons explained by lack of major histocompatibility class I expression. *Science* 253: 1283-1285

Kast WM, Bronkhorst AM, de Waal LP, Melief CJ (1986): Cooperation between cytotoxic and helper T lymphocytes in protection against lethal Sendai virus infection. Protection by T cells in MHC-restricted and MHC-regulated; a model for MHC-disease associations. *J Exp Med* 164: 723-738

Kawaoka Y (1991): Equine H7N7 influenza A viruses are highly pathogenic in mice without adaptation: Potential use as an animal model. *J Virol* 65: 3891-3894

Leung KN, Ada GL (1982): Different functions of subsets of effector T cells in murine influenza virus infection. *Cell Immunol* 67: 312-324

Levine B, Hardwick BJ, Trapp BD, Crawford TO, Bollinger RC, Griffin DE (1991): Antibody-mediated clearance of alphavirus infection from neurons. *Science* 254: 860

Lightman S, Cobbold S, Waldmann H, Askonas BA (1987): Do L3T4+ T cells act as effector cells in protection against influenza virus infection. *Immunology* 62: 139-144

Lukacher AE, Braciale VL, Braciale TJ (1984): In vivo effector function of influenza virus-specific cytotoxic T lymphocyte clones is highly specific. *J Exp Med* 160: 814-826

Lukacher AE, Morrison LA, Braciale VL, Braciale TJ (1986): T lymphocyte function in recovery from experimental viral infection: The influenza model. In: *Mechanisms of Host Resistance to Infectious Agents, Tumors and Allografts* pp. 233-254

Mackenzie CD, Taylor PM, Askonas BA (1989): Rapid recovery of lung histology correlates with clearance of influenza virus by specific CD8+ cytotoxic T cells. *Immunology* 67: 375-381

McDermott MR, Lukacher AE, Braciale VL, Braciale TJ, Bienenstock J (1987): Characterization and in vivo distribution of influenza-virus-specific T-lymphocytes in the murine respiratory tract. *Am Rev Respir Dis* 135: 245-249

Muller D, Koller BH, Whitton JL, LaPan K, Brigman KK, Frelinger JA (1992): MHC class I deficient mice kill LCM virus-infected cells using CD4+ MHC class II restricted T cells (submitted)

Openshaw PJ (1991): Pulmonary epithelial T cells induced by viral infection express T cell receptors alpha/beta. *Eur J Immunol* 21: 803-806

Palladino G, Scherle PA, Gerhard W (1991): Activity of CD4+ T-cell clones of type 1 and 2 in generation of influenza virus-specific cytotoxic responses in vitro. *J Virol* 65: 6071-6076

Parker JC, Whiteman MD, Richter CB (1978): Susceptibility of inbred and outbred mouse strains to Sendai virus and prevalence of infection in laboratory rodents. *Infect Immun* 19: 123-130

Pena-Cruz V, Reiss CS, McIntosh K (1989): Sendai virus infection of mice with protein malnutrition. *J Virol* 63: 3541-3544

Pober JS, Cotran RS (1991): Immunologic interactions of T lymphocytes with vascular endothelium. *Adv Immunol* 50: 261-301

Rahentulla A, Fung-Leung WP, Schilham MW, Kundig TM, Sambhara SR, Narendran A, Arabian A, Wakeham A, Paige CJ, Zinkenagel RM, Miller RG, Mak TW (1991): Normal development and function of CD8+ cells but markedly decreased helper cell activity in mice lacking CD4. *Nature* 353: 180-184

Rosenberg AS, Munitz TI, Maniero TG, Singer A (1991): Cellular basis of skin allograft rejection across a class I major histocompatibility barrier in mice depleted of CD8+ T cells in vitro. *J Exp Med* 173: 1463-1471

Scherle PA, Gerhard W (1986): Functional analysis of influenza-specific helper T cell clones in vivo. T cells specific for internal viral proteins provide cognate help for B cell responses to hemagglutinin. *J Exp Med* 164: 1114-1128

Scherle PA, Palladino G, Gerhard W (1992): Mice can recover from pulmonary influenza virus infection in the absence of class I-restricted cytotoxic T cells. *J Immunol* 148: 212-217

Taylor PM, Askonas BA (1986): Influenza nucleoprotein-specific cytotoxic T-cell clones are protective in vivo. *Immunology* 58: 417-420

Townsend AR, Rothbard J, Gotch FM, Bahadur G, Wraith D, McMichael AJ (1986): The epitopes of influenza nucleoprotein recognized by cytotoxic T lymphocytes can be defined with short synthetic peptides. *Cell* 44: 959–968

Waldmann H (1989): Manipulation of T-cell responses with monoclonal antibodies. *Annu Rev Immunol* 7: 407–444

Yewdell J (1991): Molecular and cellular biology of antigen processing and presentation. *Cancer Cells* 3: 278–282

Zijlstra M, Bix M, Simister NE, Loring JM, Raulet DH, Jaenisch R (1990): Beta 2-microglobulin deficient mice lack CD4-8+ cytolytic T cells. *Nature* 344: 742–746

34
CD4$^+$ and CD8$^+$ Cytolytic T Lymphocyte Recognition of Viral Antigens

Vivian Lam Braciale

Background

An enormous amount of work by many laboratories has brought us to our present understanding of the T cell-mediated immune response to viruses. To be reviewed here are some observations from our own work on the cytotoxic T-lymphocyte (CTL) responses to influenza virus which provides a glimpse of some of the issues addressed by laboratories in this area as well as the evolution of the direction of the field of viral immunology. Since the mid-1970s, the study of the cell-mediated immune responses to viruses has provided insight into the nature of T-lymphocyte recognition, the pathways of foreign antigen processing and presentation, and the biology of the cytotoxic T effector subset. It was first appreciated in studies with lymphocytic choriomeningitis virus (LCMV) by Zinkernagel and Doherty (1974) that T lymphocytes recognize viral antigens in a major histocompatibility complex (MHC)-restricted fashion. This led the way to studies dealing with the elucidation of the T-cell receptor, one that would seemingly need to be as diverse as the B-cell receptor, as well as studies of the MHC class I products in an effort to divine how the T cell can recognize both foreign antigen and the self-MHC product.

Groups working with ectromelia and influenza soon demonstrated that virus-specific T lymphocytes could be protective in the murine model (Blanden, 1974; Yap et al., 1978). Antiviral CTL could be derived from lungs of infected mice as well as from spleen (Yap and Ada, 1978). For the most part, the antiviral CTL were so considered because they demonstrated cytotoxic effector function *in vitro* using the ^{51}Cr release assay (Brunner et al., 1968), but the actual *in vivo* mechanism of protection by the antiviral CTL is still unkown. At that time various areas of investigation included determining parameters of target cell recognition, immune response (Ir) gene effects of viral antigen recognition, and isolation of antiviral CTL. It was shown that antiviral CTL could recognize early viral proteins and thereby lyse target cells before the release of progeny virion particles (Koszinowski and Ertl, 1976; Zinkernagel and Althage, 1977). Thus, for certain viruses, the early recognition of infected cells can prevent release and dissemination of new virus particles. Martz and Howell (1989) recently put forth a hypothesis that CTL may halt virus production via degradation of viral nucleic acids prior to CTL-induced lysis. It was also soon appreciated that antiviral T cells displayed cross-reactive recognition for viruses that are related but serologically non-cross-reactive (Braciale, 1977; Doherty et al., 1977; Effros et al., 1977; Zweerink et al., 1977; Lin and Askonas, 1980). This implies that vaccines might be generated to cross-reactive T-cell epitopes for viruses that do not share cross-reactive B-cell epitopes.

In Vitro Properties of Class I MHC-Restricted, Influenza-Specific CTL Clones

Our initial interests in the study of the CTL response to influenza virus resided in determining the fine specificity of recognition of influenza, as compared with the B-cell response, as well as the further characterization of the virus-specific CTL as an effector *in vitro* and *in vivo*. At this time, many groups saw that CTL clones would be useful for identifying the T-cell receptor; for understanding alloreactivity and MHC-restricted recognition of tumor cells, minor histocompatibility antigens, hapten-modified cells, and soluble protein antigens; and for characterizing T-cell subsets, e.g., cytotoxic effectors, delayed-type hypersensitivity (DTH) effectors, helpers, and suppressors. We were fortunate to isolate, characterize, and propagate stable clonal populations of T lymphocytes from influenza virus-primed mice. These clones were obtained according to the procedures for the derivation of alloreactive T-cell clones (Fathman and Hengartner, 1978; Glasebrook and Fitch, 1979). The cytotoxic T-cell clones we isolated were primarily CD8+, Thy-1+, MHC-restricted, and virus-specific both in cytotoxic assays and in proliferation assays (Braciale et al., 1981a). This latter assay convinced us that these clonal CTL were probably not terminally differentiated end-stage effectors, as are antibody-producing B cells, and set the stage for studies of requirements for T-cell activation and proliferation using homogeneous clonal populations.

A subset of clones demonstrated alloreactive recognition of uninfected target cells in CTL assays or of irradiated allogeneic splenocytes in proliferation assays (Braciale et al., 1981b). Also, several of these CD8+ CTL clones were minor lymphocyte stimulatory (Mls)-reactive in proliferation assays (Braciale and Braciale, 1981). Both observations imply that MHC-restricted foreign antigen-specific CTL overlap with the alloreactive CTL and Mls loci-reactive T lymphocytes.

The antiviral CTL are stable with regard to their cytolytic activity on a per cell basis as a function of time after the antigenic stimulus (Andrew et al., 1985), which is different from that described for heterogeneous alloreactive CTL populations (MacDonald et al., 1974) and for some alloreactive CTL clones (Howe and Russell, 1983). The clones are absolutely dependent on the presence of an exogenous source of interlenkin-2 (IL-2) in the culture medium, which was provided in the form of supernatant from concanavalin A (Con A)-activated rat splenocyte cultures. Detailed studies of the requirement for IL-2 and expression of high-affinity IL-2 receptors on these T cells revealed that the clones are receptive to the proliferative stimulus of IL-2 only for the 3 to 5 day period following antigenic stimulation, and that despite the expression of moderate to high levels of high-affinity IL-2 receptors, the T cells late after antigenic stimulation are refractory to the IL-2 stimulus (Churilla and Braciale, 1987). Whether interferon-γ (IFN-γ) is required as a cytolytic differentiation factor for these cells remains unknown at present. These CTL clones produce abundant IFN-γ as a result of antigenic stimulation (Henkel et al., 1987).

CD4+, Class II MHC-Restricted Influenza-Specific CTL Clones

The isolation and characterization of human CTL directed to influenza virus led to the unexpected observation that CTL with *in vitro* cytotoxic activity were CD4+ and produced IL-2 in an autocrine fashion (Kaplan et al., 1984). The Epstein-Barr virus (EBV)-immortalized B lymphoblastoid cell lines which were used as targets express class II as well as class I MHC molecules. Indeed, these cloned human CTL were class II MHC-restricted and virus-specific. Others reported human CD4+ CTL clones specific for herpes simplex virus and measles (Jacobson et al., 1984; Yasukawa and Zarling, 1984) and alloreactive human CTL clones (e.g., Meurer et al., 1982). Although there were reports of class II MHC-restricted alloreactive T-cell clones in the mouse (e.g., Pierres et al., 1982), the evidence for class II MHC-restricted murine CTL specific for viral antigens was lacking. Upon reexamination of class II MHC-restricted T-cell clones for cytotoxic activity on B lymphoma cells which express class II MHC molecules, we readily observed MHC-restricted, virus-specific lysis of infected target cells (Lukacher et al., 1985). In the ensuing years, there have been many reports of CD4+ CTL in the human, yet few have been reported in the mouse (Browning et al., 1990; Thomas et al., 1990). The importance of murine antiviral CD4+ CTL may increase following recent observations that they may be protective in the absence of CD8+ CTL in β_2 microglobulin-lacking mice (Eichelberger et al., 1991).

Using a modification of the iododeoxyuridine-125 (^{125}IUdR) release assay developed by Russell (Russell et al., 1980), we were able to detect cytolysis of influenza-infected target cells by both class I and class II MHC-restricted CTL clones in as little as 30 min after CTL–target cell contact (Maimone et al., 1986). No bystander killing was demonstrable in the standard 6-hr assay. Others have reported "bystander" killing by class II MHC-restricted T cells and inferred that therefore class II MHC-restricted CTL carry out the cytotoxic function via release of a soluble cytotoxic factor (Tite and Janeway, 1984). Lymphotoxin has been suggested as the cytotoxic factor (Ruddle et al., 1987). In many of the reports where "bystander" killing has been observed, the length of the assay generally exceeds the standard 4 to 6-hr ^{51}Cr-release assay. It has also been suggested that the longer assay of 16 to 18-hr duration may be more physiologically relevant than the short assays. With regard to this point and as mentioned previously, there would be an advantage for CTL to recognize and lyse virally infected cells in a rapid manner in order to prevent release of progeny virions. In addition, virally infected target cells usually die by 12–18 hr in the absence of CTL.

Stimulation of CD4$^+$ and CD8$^+$ Clones with Monoclonal Anti-TCR Antibodies, Superantigens, and Plant Lectins

Along another line of investigation, we wished to determine how T cells are activated to cytolytic effector function and stimulated to proliferate, and whether there are differences in the requirements. An obvious difference can be deduced from the difference in time frame of the readout: class I and class II MHC-restricted CTL do not require IL-2 for cytolysis but do require it for proliferation. In the presence of monoclonal antibody to the T-cell receptor (TCR) (Henkel et al., 1987) or to CD3 (Braciale, unpublished observations), the clones demonstrate nonspecific killing of irrelevant target cells and are not blocked from killing of specific target cells. This killing of nonspecific target cells requires adhesion of CTL to the target cell, which can be enhanced by gentle cocentrifugation of CTL and target cells in the assay wells. Certain T-cell mitogens, Con A and phytohemagglutinin (PHA), can stimulate the clones to proliferate as well as mediate killing of nonspecific target cells. "Superantigens" such as SEB (staphylococcal enterotoxin B) as well as Mls locus antigens have been so named due to the ability to stimulate a large subset of T cells (White et al., 1989). In general, these Mls-reactive T cells and the SEB-reactive T cells seem to utilize a limited set of Vβ in their TCR. Another way to look at "superantigens" is to consider them T-cell mitogens that are selective for certain T-cell subsets based on recognition of sites on the TCR. In this way they may stimulate T cells via their TCR in a manner similar to the lectins Con A and PHA, or the monoclonal antibodies directed to the TCR $\alpha\beta$ chains or to the CD3 chains. Our panel of CD4$^+$ and CD8$^+$ antiviral CTL can mediate nonspecific cytolysis in the presence of Con A, SEB, and anti-TCR or anti-CD3 antibodies on a variety of target cells independent of MHC restriction. In addition, class II MHC$^-$ cells can be lysed in the presence of SEB by SEB-reactive CD8$^+$ or CD4$^+$ CTL clones (Braciale et al., 1991).

In Vivo Function of Long-term In Vitro CD8$^+$ and CD4$^+$ Influenza-Specific CTL Clones

These long-term virus-specific CTL were used to determine whether clonal CTL populations, as was shown for heterogeneous virus-specific CTL, would be adequate to provide protection in vivo in an infectious influenza model in mice. It was possible that the CD8$^+$ CTL clone might not protect in vivo because (1) one might need to co-adoptively transfer primed virus-specific CD4$^+$ T cells to provide IL-2, (2) they might not home to the appropriate site, or (3) a mixture of CTL clones directed to several viral antigens might be required. In vivo studies of the homing of these long-term cultured T cells indicated that the CTL can home to the infected lung (McDermott et al., 1987). Studies of in vivo protection showed that for 10 LD$_{50}$ infectious virus doses, 10^6 CD8$^+$ clone T cells directed to the influenza hemagglutinin or 10^6 CD8$^+$ clone T cells directed to an internal conserved viral protein, nucleoprotein, were able to protect in the absence of additional antiviral CD4$^+$ T cells (Lukacher et al., 1984). In addition, irradiated clones were also protective, although higher CTL input numbers were required. The in vivo effector activity of the cloned CTL remained antigen-specific and MHC-restricted. In the case where mice were dually infected with two different strains of influenza virus, a clone with cross-reactive recognition could clear both strains of influenza to promote recovery. A strain-specific CTL clone, however, could clear only the specific virus strain, and since there was no "bystander" clearance of the irrelevant strain, the mice went on to die. This would suggest that in the acute time frame studied, protection may have been mediated by direct CTL contact with infected cells, and that release of IFN-γ and other lymphokines by the CTL had very localized effects. Any nonspecific host defence mechanisms that might limit the spread or production of the unrecognized virus strain were mobilized too late to deal with the overwhelming viral infection in the lungs.

To determine the in vivo effector function of the CD4$^+$ CTL subset, class II MHC-restricted clones were adoptively transferred into mice given

10 LD$_{50}$ infectious virus by the intranasal route. The results paralleled those seen for class I MHC-restricted CTL clones with regard to maintenance of MHC-restriction, virus specificity, and lack of "bystander" protection (Lukacher et al., 1986). These CTL also produce abundant IFN-γ, as was seen for the CD8$^+$ CTL. Like their human counterpart, these CD4$^+$ CTL make IL-2 in an autocrine fashion and therefore do not require an exogenous source of IL-2 at the time of antigenic stimulation. Therefore, as was seen for the class I MHC-restricted T cells, any secreted IFN-γ or cytotoxic factors produced by these CTL were not effective in this experimental model for clearance of the unrecognized virus.

Frequency of CD4$^+$ and CD8$^+$ Antiviral CTL

Influenza-specific human CTL, derived from peripheral blood, isolated and propagated in the presence of an exogenous source of IL-2, are generally CD4$^+$ and class II MHC-restricted. In the murine model, however, influenza—specific CTL, derived from primed splenocytes, cloned and propagated in the presence of exogenous IL-2, are predominantly CD8$^+$ and class I MHC-restricted. It was suggested that the few CD4$^+$ CTL clones were a highly selected subset of class II MHC-restricted T cells that survived the cell culture conditions or possibly were artifacts of long-term *in vitro* culture. To determine the relative frequency of CD4$^+$ CTL present in spleens of mice primed with infectious influenza virus, limiting dilution analyses of cytotoxic effector function as well as MHC-restriction and cell surface phenotyping were carried out (Figure 1). Such studies revealed that CD4$^+$ and CD8$^+$ CTL were present in influenza-primed spleens at approximately the same frequency ($\sim$1/10,000), and that restimulation with infectious virus *in vitro* yielded mostly the CD8$^+$ class I MHC-restricted CTL, while restimulation with UV-inactivated virus yielded predominantly CD4$^+$ class II MHC-restricted CTL (Morrison et al., 1988). It was thus clear that CD4$^+$ CTL were not an extremely rare T-effector subset with regard to recognition of viral antigens. This indicated that the form of the antigen has a major impact on the T-cell subsets elicited, which is important with regard to vaccine design and led to the examination of viral antigen processing and presentation to the CD4$^+$ and CD8$^+$ CTL clones.

Viral Antigen Processing and Presentation for Recognition by Class I and Class II MHC-Restricted CTL Clones

It was possible to compare the presentation of infectious and UV-inactivated virus to both class I and class II MHC-restricted CTL subsets by using the same assay. Although the proliferation assay could have been employed, the time frame would have involved days between the recognition event and the readout of ^{3}H-thymidine (^{3}H-TdR) incorporation as a measure of the rate of DNA synthesis in the activated T cells. In addition, one population is dependent on exogenous IL-2 while the other is not. The cytotoxicity assay provided an excellent opportunity to compare the presentation of viral antigen to the CTL clones using a short assay as well as a homogeneous target cell population which expressed both the class I and class II MHC restriction molecules.

Table 1 summarizes the similarities and differences in the processing and presentation of antigen in the context of class I and class II MHC molecules (Morrison et al., 1986). The fact that synthetic peptides, usually 8 to 20 amino acid residues in length, can mimic T-cell epitopes for antiviral class I MHC-restricted CTL was first reported by Townsend (Townsend et al., 1986a) for influenza nucleoprotein in the mouse and subsequently reported for many other viral antigens, e.g., influenza hemagglutinin (Braciale et al., 1987a), influenza matrix protein in the human (Gotch et al., 1987), human immunodeficiency virus-1 (HIV-1) gag protein (Nixon et al., 1988), GP2 of LCMV (Oldstone et al., 1988), murine cytomegalovirus (CMV) pp89 (Del et al., 1988), and SV40 T antigen (Tevethia, 1990). The finding that *de novo* viral protein synthesis is required for class I MHC presentation holds for most viruses, but there are circumstances in which viral protein synthesis is not required. If viral proteins can gain access to the cell cytoplasm, then they apparently can be presented in the context of class I MHC moelcules (Hosaka et al., 1988). This may also account for early observations that viruses with fusion activity, e.g., Sendai virus, were shown to sensitize target cells for CTL recognition in the absence of viral protein synthesis (Gething et al., 1978).

The expression of individual viral gene products in target cells can be accomplished by establishing permanent cell lines expressing transfected viral genes (Braciale et al., 1984; Townsend et al., 1984)

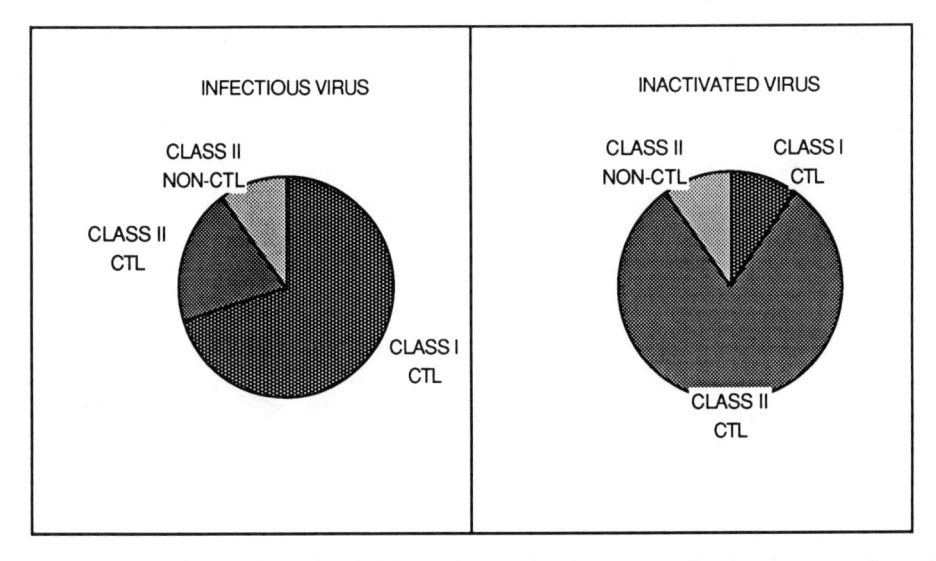

FIGURE 1. Splenocytes from mice primed 2–3 weeks previously were restimulated *in vitro* in a limiting dilution analysis with either irradiated stimulator splenocytes infected with influenza virus or irradiated splenocytes treated with UV-inactivated A/JAP/57 virions. Individual cultures were tested for cytolysis. The pie slices indicate the proportion and phenotype of culture wells responding to the infectious or inactivated virus stimulus. The apparent precursor frequency of class I CTL to infectious virus was approximately 1/10,000. The apparent precursor frequency of class II CTL in response to inactivated virions was approximately 1/10,000 as well. (Adapted from data in Morrison et al., 1988)

or by using viral expression vectors to transiently express viral genes in target cells. Studies employing the latter method, primarily using recombinant vaccinia viruses (Yewdell et al., 1988), have shown that expressed truncated or mutated viral gene products (Townsend et al., 1986b; Sweetser et al., 1988) are recognized by CD8$^+$ CTL. Indeed, the intracellular expression of the viral sequence

TABLE 1. Antigen presentation to Class I and Class II CTL

Treatment of ^{51}Cr-labeled A20 B lymphoma target cells	Recognition of A20 B lymphoma cells in ^{51}Cr-release assays	
	Class I MHC-restricted CTL clones	Class II MHC-restricted CTL clones
Uninfected	−	−
Infected with influenza virus[a]	+	+
Pulsed with UV-inactivated influenza viruse[a]	−	+
Infected with influenza virus + emetine[b]		
Infected with influenza virus + chloroquine[c]	+	−
Pulsed with UV-inactivated A/JAPAN/57 virus + chloroquine[c]	−	−
Pulsed with soluble bromelain-cleaved A/JAPAN/57 HA	−	+
Infected with recombinant vaccinia virus encoding influenza HA	+	−
Pulsed with synthetic peptide mimic of HA epitopes	+	+

[a] A/JAPAN/57 strain.
[b] Inhibitor of protein synthesis.
[c] Lysosomotropic agent.
Adapted from Braciale et al., 1987b.

encoding the T-cell epitope, a "minigene," in the target cell allows for CTL recognition (Gould et al., 1989; Sweetser et al., 1989; Whitton and Oldstone, 1989).

From these types of studies it was appreciated that more information on the cellular compartments where antigen can be processed, e.g., proteolytically cleaved, as well as localization of where the antigenic fragments may be bound to the class I or class II MHC molecules would be required in order to dissect where the differences in class I and class II MHC processing and presentation pathways may lie. The characterization of mutant cell lines defective in class I (Alexander et al., 1989; Townsend et al., 1989) or class II MHC antigen presentation (Stockinger et al., 1989, Mellins et al., 1990) will help clarify some of these issues. To further complicate matters, certain viruses encode proteins that can inhibit antigen presentation, e.g., adenovirus (Burgert et al., 1987; Cox et al., 1990). Since the direct comparison of antigen presentation to class I and class II MHC-restricted CTL was performed with a limited population of target cell lines, it is important to investigate these pathways for several cell types and several viruses in both the mouse and the human. Such studies should extend our picture of how certain cells process and present antigen, thereby affecting the balance of immune effector subsets elicited. At present there are some results emerging that point to apparent exceptions to the defined class I and class II MHC presentation pathways. Most notable are results for human class II MHC-restricted CTL where endogenous viral proteins appear to be recognized (Braciale et al., 1989; Jaraquemada et al., 1990; Nuchtern at al., 1990; Polydefkis et al., 1990), as well as for some CD4+ CTL in the mouse (Thomas et al., 1990). Certain distinctions in the two antigen presentation pathways remain. Antigen presentation in the context of class II MHC moelcules requires access to a low-pH endosomal compartment. Viral proteins expressed in the cytoplasm are not presented by class II MHC. These results hold true for acute viral infections. Recognition of viral proteins expressed in latently infected cells, self proteins, and tumor antigens, which may be self or viral, may not be subject to these distinctions.

A current summary of the antigen processing and presentation pathways is that for class I MHC presentation, there is an apparent symbiotic relationship between peptide binding and conformational changes in class I MHC molecules which permits their transit from the endoplasmic reticulum (ER) to expression on the cell surface (Townsend et al., 1989). For class II MHC presentation, there is a requirement for peptide to be bound by class II MHC molecules in a low-pH compartment prior to expression (for newly synthesized class II MHC molecules) or reexpression (for recycled class II MHC molecules) on the cell surface. Recent reports have described putative proteolytic machinery in the cytoplasm (Martinez and Monaco, 1991) as well as candidate peptide transporters (Monaco et al., 1990; Spies and DeMars, 1991) to account for antigenic peptide delivery to the ER. The elution of viral peptides from class I MHC molecules (Van Bleek and Nathenson, 1990) and the expression of minigenes encoding T-cell epitopes of differing length are being employed to determine the size requirements of endogenously processed antigen.

Acknowledgments

The work described in this chapter was made possible by the collaboration and support of Dr. Thomas Braciale, the hard work of our technicians and a talented group of graduate students and postdoctoral fellow who trained in our laboratories, and the generosity of colleagues who provided various critical reagents.

References

Alexander J, et al., (1989): Differential transport requirements of HLA and H-2 class I glycoproteins. *Immunogenetics* 29: 380–388

Andrew ME, et al., (1985): Stable expression of specific cytotoxicity by cytotoxic T lymphocyte clones. *J Immunol* 135: 3520–3523

Blanden RV (1974): T cell response to viral and bacterial infection. *Transplant Rev* 19: 56

Braciale TJ (1977): Immunologic recognition of influenza virus-infected cells. I. Generation of a virus strain-specific and a crossreactive subpopulation of cytotoxic T cells in the response to type A influenza viruses of different subtypes. *Cell Immunol* 33: 423

Braciale TJ, et al., (1981a): Heterogeneity and specificity of cloned lines of influenza virus-specific cytotoxic T lymphocytes. *J Exp Med* 153: 910–923

Braciale TJ, et al. (1981b): Simultaneous expression of H-2 restricted and alloreactive recognition by a cloned line of influenza virus-specific cytotoxic T lymphocytes. *J Exp Med* 153: 1371–1376

Braciale TJ, et al. (1984): CTL recognition of the influenza A/JAP/305/57 hemagglutinin gene product expressed by DNA mediated gene transfer. *J Exp Med* 159: 341–354

Braciale TJ, et al. (1987a): On the role of the transmembrane anchor sequence of influenza hemagglutinin in target cell recognition by class I MHC-restricted, hemagglutinin-specific cytolytic T lymphocytes. *J Exp Med* 166: 678–692

Braciale TJ, et al., (1987b): Antigen presentation pathways to class I and class II MHC-restrcited T lymphocytes. *Immunol Rev* 98: 95–114

Braciale TJ, et al. (1989): T lymphocyte recognition of a membrane glycoprotein. In: *Immunological Recognition*, Watson JD, Inglis Jr, eds. (Cold Spring Harbor Symposia on Quantitative Biology, Vol. 59) Cold Spring Harbor, NY: Cold Spring Harbor Press

Braciale VL, et al. (1991): Staphylococcal enterotoxin B triggers the cytolytic effector function of both class I and class II MHC-restricted T cell clones. *FASEB J* 5: A1336

Braciale VL, Braciale TJ (1981) M1s locus recognition by a cloned line of H-2 restricted influenza virus-specific cytotoxic T lymphocytes. *J Immunol* 127: 859–862

Browning MJ, et al. (1990): Cytolytic T lymphocytes from the BALB/c-H-2dm2 mouse recognize the vesicular stomatitis virus glycoprotein and are restricted by class II MHC antigens. *J Immunol* 145: 985–994

Brunner KT, et al., (1968): Quantitative assay of the lytic action of immune lymphoid cells on ^{51}Cr-labeled allogeneic target cells in vitro; inhibition by isoantibody and by drugs. *Immunology* 14: 181

Burgert HG, et al. (1987): "E3/19K" protein of adenovirus type 2 inhibits lysis of cytolytic T lymphocytes by blocking cell-surface expression of histocompatibility class I antigens. *Proc Natl Acad Sci USA* 84: 1356–1360

Churilla AM, Braciale VL (1987): Lack of IL 2-dependent proliferation despite significant expression of high-affinity IL 2 receptor on murine cytolytic T lymphocyte clones late after antigenic stimulation. *J Immunol* 138: 1388–1345

Cox JH, et al. (1990): Antigen presentation requires transport of MHC class I molecules from the endoplasmic reticulum. *Science* 247: 715–718

Del VM, et al. (1988): Molecular basis for cytolytic T-lymphocyte recognition of the murine cytomegalovirus immediate-early protein pp89. *J Virol* 62: 3965–3972

Doherty PC, et al. (1977): Heterogeneity of the cytotoxic response of thymus derived lymphocytes after immunization with influenza viruses. *Proc Natl Acad Sci USA* 74: 1209–1213

Effros RB, et al. (1977): Generation of both crossreactive and virus-specific T-cell populations after immunization with serologically distinct influenza A viruses. *J Exp Med* 145: 557

Eichelberger M, et al. (1991): Clearance of influenza virus respiratory infection in mice lacking class I major histocompatibility complex—restricted CD8$^+$ T cells. *J Exp Med* 174: 875–880

Fathman CG, Hengartner H (1978): Clones of alloreactive T-cells. *Nature* 272: 617

Gething MJ, et al. (1978): Fusion of Sendai virus with the target cell membrane is required for T cell cytotoxicity. *Nature* 274: 689

Glasebrook Al, Fitch FW (1979): T-cell lines which cooperate in generation of specific cytolytic activity. *Nature* 278: 171

Gotch F, et al. (1987): Cytotoxic T lymphocytes recognize a fragment of influenza virus matrix protein in association with HLA-A2. *Nature* 326: 881–882

Gould K, et al. (1989): A 15 amino acid fragment of influenza nucleoprotein synthezied in the cytoplasm is presented to class I restricted cytotoxic T lymphocytes. *J Exp Med* 170: 1051–1056

Henkel TJ, et al. (1987): Anti-T cell receptor antibodies fail to inhibit specific lysis by CTL clones but activate lytic activity for irrelevant targets. *J Immunol* 138: 1221–1228

Hosaka Y, et al. (1988): Recognition of noninfectious influenza virus by class I-restricted murine cytotoxic T lymphocytes. *J Immunol* 140: 606–610

Howe RC, Russell JH (1983): Isolation of alloreactive CTL clones with cyclical changes in lytic activity. *J Immunol* 131: 2141

Jacobson S, et al. (1984): Measles virus-specific T4+ human cytotoxic T cell clones are restricted by class II HLA antigens. *J Immunol* 133: 754

Jaraquemada D, et al. (1990): An endogenous processing pathway in vaccinia virus-infected cells for presentation of cytoplasmic antigens to class II-restricted T cells. *J Exp Med* 172: 947–954

Kaplan DR, et al. (1984): Influenza virus-specific human cytotoxic T cell clones: Heterogeneity in antigenic stimulation. *Cell Immunol* 88: 193–206

Koszinowski U, Ertl H (1976): Role of early viral surface antigens in cellular immune response to vaccinia virus. *Eur J Immunol* 6: 679

Lin L-Y, Askonas BA (1980): Cross reactivity for different type A influenza viruses of a cloned T-killer cell line. *Nature* 288: 164

Lukacher AE, et al. (1984): *In vivo* function of influenza virus-specific cytotoxic T lymphocyte clones is highly specific. *J Exp Med* 160: 814–826

Lukacher AE, et al. (1985): Expression of specific cytolytic activity by H-21 region-restricted influenza virus-specific T lymphocyte clones. *J Exp Med* 162: 171–187

Lukacher AE, et al. (1986): T lymphocyte function in recovery from experimental viral infection: The influenza model. In: *Mechanisms of Host Resistance to Infectious Agents, Tumors, and Allografts: A Conference in Recognition of the Trudeau Institute Centennial*, Steinman RM, North RJ, eds. New York: Rockefeller University Press

MacDonald HR, et al. (1974): Generation of cytotoxic T lymphocytes in vitro. II. Effect of repeated exposure to alloantigens on the cytotoxic activity of longterm mixed leukocyte cultures. *J Exp Med* 140: 718

Maimone MM, et al. (1986): Features of target cell lysis by class I and class II MHC-restricted cytolytic T lymphocytes. *J Immunol* 137: 3639–3643

Martinez CK, Monaco JJ (1991): Homology of proteasome subunits to a major histocompatibility complex-linked LMP gene. *Nature* 353: 664–667

Martz E, Howell DM (1989): CTL: Virus control cells first and cytolytic cells second? DNA fragmentation, apoptosis and the Prelytic halt hypothesis. *Immunol Today* 10: 79–86

McDermott MR, et al. (1987): Characterization and *in vivo* distribution of influenza virus-specific T-lymphocytes in the murine respiratory tract. *Am Rev Respir Dis* 135: 245–249

Mellins E, et al. (1990): Defective processing and presentation of exogenous antigens in mutants with normal HLA class II genes. *Nature* 343: 71–74

Meurer SC, et al. (1982): Clonal analysis of human cytotoxic T lymphocytes: T4+ and T8+ effector T cells recognize products of different major histocompatibility complex regions. *Proc Natl Acad Sci USA* 79: 4395

Monaco JJ, Cho S, Attaya M (1990): Transport protein genes in the murine MHC: Possible implications for antigen processing. *Science* 250: 1723–1726

Morrison LA, et al. (1986): Differences in antigen presentation to MHC class I- and class II-restricted influenza virus-specific cytolytic lymphocyte clones. *J Exp Med* 163: 903–921

Morrison LA, et al. (1988): Antigen form influences induction and frequency of influenza specific class I and class II MHC-restricted cytolytic T lymphocytes. *J Immunol* 141: 363–368

Nixon DF, et al. (1988): HIV-1 gag-specific cytotoxic T lymphocytes defined with recombinant vaccinia virus and synthetic peptides. *Nature* 336: 484–487

Nuchtern JG, et al. (1990): Class II MHC molecules can use the endogenous pathway of antigen presentation. *Nature* 343: 74–76

Oldstone MB, et al. (1988): Fine dissection of a nine amino acid glycoprotein epitope, a major determinant recognized by lymphocyte choriomeningitis virus-specific class I-restricted H-2Db cytotoxic T lymphocytes. *J Exp Med* 168: 559–570

Pierres A, et al. (1982): Characterization of an Lyt-1+ cytolytic T-cell clone specific for a polymorphic domain of the I-Ak molecule. *Scand J Immunol* 15: 619

Polydefkis M, et al. (1990): Anchor sequence-dependent endogenous processing of human immunodeficiency virus 1 envelope glycoprotein gp160 for CD4+ T cell recognition. *J Exp Med* 171: 875–887

Ruddle NH, et al. (1987): Purified lymphotoxin (LT) from class I restricted CTLs and class II restricted cytolytic helpers induce target cell DNA fragmentation. In: *Membrane-Mediated Cytotoxicity*, Bonavida B, Collier RJ, eds. New York: Alan R Liss

Russell JH, et al. (1980): Mechanisms of immune lysis. I. Physiological distinction between target cell death mediated by cytotoxic T lymphocytes and antibody plus complement. *J Immunol* 124: 1100

Spies T, DeMars R (1991): Restored expression of major histocompatibility class I molecules by gene transfer of a putative peptide transporter. *Nature* 351: 323–324

Stockinger B, et al. (1989): A role of Ia-associated invariant chains in antigen processing and presentation. *Cell* 56: 683–689

Sweetser MT, et al. (1988): Class I MHC-restricted recognition of cells expressing a gene encoding a 41 amino acid product of the influenza hemagglutinin. *J Immunol* 141: 3324–3328

Sweetser MT, et al. (1989): Class I but not class II MHC-restricted T cells recognize preprocessed endogenous antigen. *Nature* 342: 180–182

Tevethia SS (1990): Recognition of simian virus 40 T antigen by cytotoxic T lymphocytes. *Mol Biol Med* 7: 83–96

Thomas DB, et al. (1990): The role of the endoplasmic reticulum in antigen processing. N-glycosylation of influenza hemagglutinin abrogates CD4+ cytotoxic T cell recognition of endogenously processed antigen. *J Immunol* 144: 2789–2794

Tite JP, Janeway CR, Jr (1984): Cloned helper T cells can kill B lymphoma cells in the presence of specific antigens: Ia restriction and cognate vs noncognate interactions in cytolysis. *Eur J Immunol* 14: 878

Townsend ARM, et al., (1984): Cytotoxic T cell recognition of the influenza nucleoprotein and hemagglutinin expressed in transfected L cells. *Cell* 39: 13

Townsend ARM, et al. (1986a): The epitopes of influenza nucleoprotein recognized by cytotoxic T lymphocytes can be defined with short synthetic peptides. *Cell* 44: 959

Townsend ARM, et al. (1986b): Cytotoxic T lymphocytes recognize influenza hemagglutinin that lacks a signal sequence. *Nature* 324: 575

Townsend ARM, et al. (1989): Association of class I major histocompatibility heavy and light chains induced by viral peptides. *Nature* 340: 443–448

Van Bleek GM, Nathenson SG (1990): Isolation of an endogenously processed immunodominant viral peptide from the class I H-2Kb molecule. *Nature* 348: 213–216

White J, et al. (1989): The V beta-specific superantigen staphylococcal enterotoxin B: Stimulation of mature T cells and clonal deletion in neonatal mice. *Cell* 56: 27–35

Whitton JL, Oldstone MB (1989): Class I MHC can present an endogenous peptide to cytotoxic T lymphocytes. *J Exp Med* 170: 1033–1038

Yap KL, Ada GL (1978): Cytotoxic T cells in the lungs of mice infected with an influenza A virus. *Scand J Immunol* 8: 413

Yap KL, et al. (1978): Transfer of specific cytotoxic T lymphocytes protects mice inoculated with influenza virus. *Nature* 273: 238

Yasukawa M, Zarling JM (1984): Human cytotoxic T cell clones directed against herpes simplex virus-infected cells. I. Lysis restricted by HLA class II MB and DR antigens. *J Immunol* 133: 422

Yewdell J, et al. (1988): Use of recombinant vaccinia viruses to examine cytotoxic T lymphocyte recognition of individual viral proteins. *Adv Exp Med Biol* 239: 151–161

Zinkernagel RM, Althage A (1977): Antiviral protection by virus immunocytotoxic T cells: Infected target cells are lysed before infectious virus progeny is assembled. *J Exp Med* 145: 644

Zinkernagel RM, Doherty PC (1974): Activity of sensitized thymus-derived lymphocytes in lymphocytic choriomeningitis reflects immunological surveillance against altered self components. *Nature* 251: 547

Zweerink HL, et al. (1977): Cytotoxic T cells kill influenza virus infected cells but do not distinguish between serologically distinct type A viruses. *Nature* 267: 354

35
Can CTL Control Virus Infections Without Cytolysis? The Prelytic Halt Hypothesis

Eric Martz

There is excellent evidence that T cells are required for control of most virus infections and that cloned cytotoxic T lymphocytes (CTL) are sufficient for such control (reviewed in Martz and Howell, 1989; see also Moskophidis et al., 1989). In three cases that have been tested, control of one virus fails to control a second virus in a mixed infection (Lukacher et al., 1984; MacIntyre et al., 1985; Scherle et al., 1992). This evidence for control by a bystander-sparing effector action is consistent with control by the contact-dependent lytic action of CTL. It speaks against a nonspecific CTL effector mechanism such as release of interferons or other antireplicating cytokines, or activation of phagocytes by cytokines released by CTL and other cells, at least to the extent that the actions of such mediators would be "long-range," spilling into the compartment infected with bystander virus.

It is widely assumed that CTL halt virus infections by lysing infected cells. However, this is not the only possibility. We have recently delineated three hypotheses: **lytic halt, lytic inactivation, and prelytic halt** (Martz and Howell, 1989, where references will be found supporting many of the undocumented assertions to follow). That CTL can halt virus replication by lysing the host cell (**lytic halt**) is clearly true, but is this the only, or even the major, pathway of virus control *in vivo*?

Few studies have been reported in which CTL have been tested for **agents capable of directly inactivating virions** (Kaufmann, 1988; Joag and Young, 1989). This probably deserves further exploration. Since it is clear that CTL are heterogeneous in their effector capabilities, such studies should be conducted with CTL generated *in vivo* against virus infections. In particular, CTL should be examined for virus-inactivating agents such as the defensins

employed by neutrophils (Gabay et al., 1989; Ganz et al., 1990).

If CTL lysed target cells by a complement-like or purely perforin-mediated necrotic mechanism (see my overview of the CTL mechanism in Chapter 3, this volume, which includes explanation and references for this and other undocumented statements made here), there would be little reason to suggest alternatives to lytic halt. However, CTL induce **apoptosis** in targets, preceded by massive fragmentation of target cell DNA. The **DNA fragmentation** raises the possibility of halting virus replication by a prelytic or nonlytic mechanism (Clouston and Kerr, 1985; Lehmann-Grube et al., 1988; Sellins and Cohen, 1989; Martz and Howell, 1989). Cells undergoing apoptosis break into initially sealed vesicles that are targeted for phagocytosis (Savill et al., 1990; Kerr and Harmon, 1991), which may prevent release of virions and facilitate their inactivation by phagocytes (Clouston and Kerr, 1985).

CTL may induce DNA fragmentation by a **mechanism distinct** from that involved in other apoptotic death pathways, notably programmed cell death. In contrast with the latter, CTL may be able to induce DNA fragmentation in target cells in which transcription and translation have been blocked, and in some target cells they induce DNA fragmentation much faster (a few minutes). However, as I have concluded elsewhere in this volume, the requirement for transcription and translation needs more careful scrutiny.

I have summarized a list of more than 15 physical and chemical means of inducing cell lysis that do not induce DNA fragmentation (Martz and Howell, 1989; see also Martin et al., 1990). However, since that review was completed, it has become clear that there are **several chemical cytolysins that can induce rapid, prelytic DNA fragmenta-**

tion similar to that induced by CTl, at least in certain cell lines. The original report of such activity by valinomycin (Allbritton et al., 1988) has been confirmed (Zanovello et al., 1990). Of great interest, the rapid prelytic DNA fragmentation induced by valinomycin **does not require transcription or translation** (Cohen and Smith, 1990; Ojcius et al., 1991). Conceivably, valinomycin activates the same pathway used by CTL. Tumor necrosis factor (TNF)-induced DNA fragmentation also does not require transcription or translation (Gromkowski et al., 1989; Ojcius et al., 1991). Exposure of mouse thymocytes to 43°C for 1 hr induces prelytic DNA fragmentation independently of translation (Sellins and Cohen, 1991a). Diphtheria toxin, an inhibitor of translation, induces prelytic DNA fragmentation in U937 but not K562 cells; other inhibitors of translation fail to have this effect (Chang et al., 1989). Certain human leukemic cell lines (HL-60, CEM-C7) seem especially sensitive to induction of DNA fragmentation independently of translation by a variety of stimuli including blocking translation itself (Baxter et al., 1989; Bansal et al., 1990; Martin et al., 1990; Lennon et al., 1991; Martin and Cotter, 1991). Some other inducers of DNA fragmentation require translation or have not been tested (Kolber et al., 1990; Nagle et al., 1990; Waring et al., 1990; Zanovello et al., 1990). Further studies of the mechanism of induction of translation-independent, rapid prelytic DNA fragmentation may provide clues about the mechanism used by CTL.

It is becoming clear that cells and cell lines of **hematopoietic origin are especially sensitive** to induction of DNA fragmentation and apoptosis. This may be related to the need to prevent the proliferation of autoreactive or neoplastic clones. Nonhematopoietic cell lines show much less dramatic DNA fragmentation when attacked by CTL (Howell and Martz, 1987a; Sellins and Cohen, 1991b). Thus, to the extent that DNA fragmentation may be utilized by CTL to control virus replication, it will be crucial in future work to compare results between target cells of hematopoietic vs. nonhematopoietic origin.

Some viruses accumulate assembled virions in the host cell prior to virus-induced host cell lysis. CTL-mediated lysis of the host cell stops further production of virions, but what happens to those virions already assembled? We wished to test whether the prelytically assembled virions were released by the CTL unharmed and capable of spreading infection, or whether the CTL was somehow capable of inactivating the virions (**lytic inactivation**). We had enormous difficulty in finding a virus that met the technical criteria required to make a definitive test of this hypothesis. In the end, the only virus for which we were able to obtain an unequivocal answer was reovirus (Howell and Martz, 1987b). Reovirus was not inactivated when its host cell was lysed by CTL; it was released into the supernatant as infectious virions. However, note that possible inactivation of apoptotic-body-contained virions by phagocytes was not tested in this *in vitro* system. This result should not be generalized from reovirus, a doubly-coated, cytoplasmically replicating RNA virus, to other viruses, especially nuclearly-replicating DNA viruses. Hence, further testing of the lytic inactivation hypothesis, including the possible roles of neutrophils and macrophages, is needed.

CTL-induced prelytic fragmentation of nucleic acids (RNA has not been tested) would seem capable of halting virus replication prelytically. Indeed, the possibility should be considered that **cytolysis, rather than being the major antivirus effector mechanism of CTL, may be only a minor mechanism among several**. The heavy reliance on ^{51}Cr release to assay for CTL function began in 1967, many years before CTL were recognized as being virus control cells. The reliance on ^{51}Cr, coupled with the refinement of techniques for generating CTL of high *lytic* activity, and the predominant use of target cells exquisitely sensitive to *lysis*, has had the advantage of allowing investigators to get home for dinner. Conceivably, however, it has inadvertently led us away from effector mechanisms of greater importance in controlling virus replication. *In vivo*, the most common target cells for virus infection may be nonhematopoietic cells relatively resistant to CTL-mediated *lysis*, but susceptible to other CTL-mediated virus control mechanisms.

Recently, we reported what may be the **first quantitation of CTL-induced prelytic halt** of virus replication (Martz and Gamble, 1992). We selected a DNA virus that replicates in the nucleus, Herpes simplex virus (HSV) type I, believed to be controlled by CTL (Martin et al., 1988; Minagawa et al., 1988; Bonneau and Jennings, 1989). We attempted to imitate conditions more likely to occur *in vivo*. We avoided the usual high effector-to-target cell ratios (E/T), short incubations, and lysis-sensitive target cells. We used an unestablished, uncloned mouse fibroblast-like cell line; such cells are about 20-fold less sensitive to CTL-mediated lysis than P815 cells. CTL were generated *in vivo* to avoid possible *in vitro* artifacts. CTL were able to induce some DNA fragmentation in these cells; HSV alone neither induced DNA fragmentation nor prevented CTL-induced DNA fragmentation. After infecting these cells with HSV, we exposed them to allospecific CTL at low E/T for overnight incubation, and monitored the prelytic effect of the CTL on virus yield.

A modest amount of prelytic halt of virus replication was detected. A twofold reduction in yield occurred when the CTL has lysed less than 10% of the cells; a 75-90% reduction accompanied 15-30% lysis. The mechanism of this reduction could have involved well-known cytokines. Also, the results obtained did not require an action of the CTL distinct from their lytic activity, damage from which obviously could impair the rate of virus synthesis prelytically.

In retrospect, this study was less than optimal in several respects. In view of the existence of multiple CTL effector mechanisms and CTL heterogeneity, CTL generated *in vivo* in an antiviral response would be preferable to allospecific CTL. Also, the strain of HSV we employed was highly lytic itself, unlike some other strains (Hemady et al., 1989). The lysis induced by HSV infection alone began at about 10 hr and seriously limited the time window in which we could monitor prelytic effects of CTL. A virus with less cytopathic effect would allow the CTL more time to produce an impressive loss of virus yield.

One study of the control of HIV replication by CD8+ effector cells was consistent with a nonlytic control mechanism (Wiviott et al., 1990; Walker et al., 1991). However, the data presented were equally consistent with a purely lytic control mechanism, since the results obtained can be quantitatively predicted by a lytic control model (see discussion in Martz and Gamble, 1992).

Understanding the full range of virus control mechanisms employed by CTL and other effector cell populations will require an open-minded re-evaluation of their antivirus activities independent from their cytolytic activities.

Acknowledgment

I am grateful to the Cellular Biology Program of the National Science Foundation for grant support.

References

Allbritton NL, Verret CR, Wolley RC, Eisen HN (1988): Calcium ion concentrations and DNA fragmentation in target cell destruction by murine cloned cytotoxic T lymphocytes. *J Exp Med* 167: 514-527

Bansal N, Houle AG, Melnykovych G (1990): Dexamethasone-induced killing of neoplastic cells of lymphoid derivation: Lack of early calcium involvement. *J Cell Physiol* 143: 105-109

Baxter GD, Smith PJ, Lavin MF (1989): Molecular changes associated with induction of cell death in a human T-cell leukaemia line: Putative nucleases identified as histones. *Biochem Biophys Res Commun* 162: 30-37

Bonneau RH, Jennings SR (1980): Modulation of acute and latent herpes simplex virus infection in C57BL/6 mice by adoptive transfer of immune lymphocytes with cytolytic activity. *J Virol* 63: 1480-1484

Chang MP, Bramhall J, Graves S, Bonavida B, Wisnieski BJ (1989): Internucleosomal DNA cleavage precedes diphtheria toxin-induced cytolysis. Evidence that cell lysis is not a simple consequence of translation inhibition. *J Biol Chem* 264: 15261-15267

Clouston WM, Kerr JFR (1985): Apoptosis, lymphocytotoxicity and the containment of viral infections. *Med Hypoth* 18: 399-404

Cohen JJ, Smith PA (1990): Apoptosis induced by the potassium ionophore valinomycin: Potassium flux is not involved. *FASEB J* 4: A1707

Gabay JE, Scott RW, Campanelli D, Griffith J, Wilde C, Marra MN, Seeger M, Nathan CF (1989): Antibiotic proteins of human polymorphonuclear leukocytes [published erratum appears in Proc Natl Acad Sci USA 1989 Dec; 86(24): 101033]. *Proc Natl Acad Sci USA* 86: 5610-5614

Ganz T, Selsted ME, Lehrer RI (1990): Defensins. *Eur J Haematol* 44: 1-8

Gromkowski SH, Yagi J, Janeway CA, Jr (1989): Rapid, prelytic DNA fragmentation can be triggered by TNF. *Eur J Immunol* 19: 1709

Hemady R, Opremcak EM, Zaltas M, Berger A, Foster CS (1989): Herpes simplex virus type-1 strain influence on chorioretinal disease patterns followng intracameral inoculation in lgh-1 disparate mice. *Invest Ophthalmol Vis Sci* 30: 1750-1757

Howell DM, Martz E (1987a): The degree of CTL-induced DNA solubilization is not determined by the human vs. mouse origin of the target cell. *J Immunol* 138: 3695-3698

Joag S, Young JD (1989): The absence of direct antimicrobial activity in extracts of cytotoxic lymphocytes. *Immunol Lett* 22: 195-198

Kaufmann SHE (1988): CD8+ T lymphocytes in intracellular microbial infections. *Immunol Today* 9: 168-174

Kerr JFR, Harmon BV (1991): Definition and incidence of apoptosis: An historical perspective. In: *Apoptosis: The Molecular Basis of Cell Death*. Tomei D, Cope F, eds. New York: Cold Spring Harbor Laboratory Press

Kolber MA, Broschat KO, Landa-Gonzalez B (1990): Cytochalasin B induces cellular DNA fragmentation. *FASEB J* 4: 3021-3027

Lehmann-Grube F, Moskophidis D, Loehler J (1988): Recovery from acute virus infection. Role of cytotoxic T lymphocytes in the elimination of lymphocytic choriomeningitis virus from spleens of mice. *Ann NY Acad Sci* 532: 238-256

Lennon SV, Martin SJ, Cotter TG (1991): Dose-dependent induction of apoptosis in human tumour cell lines by widely diverging stimuli. *Cell Prolif* 24: 203–214

Lukacher AE, Braciale VL, Braciale TJ (1984): In vivo effector function of influenza virus-specific cytotoxic T lymphocyte clones is highly specific. *J Exp Med* 160: 814–826

MacIntyre KW, Bukowski JF, Welsh RM (1985): Exquisite specificity of adoptive immunization in aernavirus-infected mice. *Antiviral Res* 5: 299–305

Martin SJ, Cotter TG (1991): Ultraviolet B irradiation of human leukaemia HL-60 cells in vitro induces apoptosis. *Int J Radiat Biol* 59: 1001–1016

Martin S, Cantin E, Rouse BT (1988): Cytotoxic T lymphocytes. Their relevance in herpesvirus infections. *Ann NY Acad Sci* 532: 257–272

Martin SJ, Lennon SV, Bonham AM, Cotter TG (1990): Induction of apoptosis (programmed cell death) in human leukemic HL-60 cells by inhibition of RNA or protein synthesis. *J Immunol* 145: 1859–1867

Martz E, Gamble SR (1992): How do CTL control virus infections? Evidence for prelytic halt of *Herpes simplex Virol Immunol* 5: 81–91

Martz E, Howell DM (1989): CTL: Virus control cells first and cytolytic cells second? DNA fragmentation, apoptosis, and the prelytic halt hypothesis. *Immunol Today* 10: 79–86

Minagawa H, Sakuma S, Mohri S, Mori R, Watanabe T (1988): Herpes simplex virus type 1 infection in mice with sever combined immunodeficiency (SCID). *Arch Virol* 103: 73–82

Moskophidis D, Fang L, Gossmann J, Lehmann-Grube F (1989): mechanism of recovery from acute virus infection. IX. Clearance of lymphocytic choriomeningitis (LCM) virus from the feet of mice undergoing LCM virus-specific delayed-type hypersensitivity reaction. *J Gen Virol* 70: 3305–3316

Nagle WA, Soloff BL, Moss AJ, Jr, Henle KJ (1990): Cultured Chinese hamster cells undergo apoptosis after exposure to cold but nonfreezing temperatures. *Cryobiology* 27: 439–451

Ojcius DM, Zychlinsky A, Zheng LM, Young JD (1991): Ionophore-induced apoptosis: Role of DNA fragmentation and calcium fluxes. *Exp Cell Res* 197: 43–49

Savill J, Dransfield I, Hogg N, Haslett C (1990): Vitronectin receptor-mediated phagocytosis of cells undergoing apoptosis. *Nature* 343: 170–173

Scherle PA, Palladino G, Gerhard W (1992): Mice can recover from pulmonary influenza virus infection in the absence of class I-restricted cytotoxic T cells. *J Immunol* 148: 212–217

Sellins KS, Cohen JJ (1989): Polyomavirus DNA is damaged in target cells during cytotoxic T-lymphocyte-mediated killing. *J Virol* 63: 572–578

Sellins KS, Cohen JJ (1991a): Hyperthermia induces apoptosis in thymocytes. *Radiat Res* 126: 88–95

Sellins KS, Cohen JJ (1991b): Cytotoxic T lymphocytes induce different types of DNA damage in target cells of different origins. *J Immunol* 147: 795–803

Walker CM, Erikson AL, Hsueh FC, Levy JA (1991): Inhibition of human immunodeficiency virus replication is acutely infected CD4+ cells by CD8+ cells involves a noncytotoxic mechanism. *J Virol* 65: 5921–5927

Waring P, Egan M, Braithwaite A, Mullbacher A, Sjaarda A (1990): Apoptosis induced in macrophages and T blasts by the mycotoxin sporidesmin and protection by Zn^{2+} salts. *Int J Immunopharmacol* 12: 445–457

Wiviott LD, Walker CM, Levy JA (1990): CD8+ lymphocytes suppress HIV production by autologous CD4+ cells without eliminating the infected cells from culture. *Cell Immunol* 128: 628–634

Zanovello P, Bronte V, Rosato A, Pizzo P, di Virgilio F (1990): Responses of mouse lymphocytes to extracellular adenosine 5'-triphosphate (ATP). II. Extracellular ATP causes cell type-dependent lysis and DNA fragmentation. *J Immunol* 145: 1545–1550

36
Immunologic Control of *Toxoplasma gondii* Infection by CD8+ Lymphocytes: A Model for Class I MHC-Restricted Recognition of Intracellular Parasites

Ricardo T. Gazzinelli, Eric Denkers, Frances Hakim, and Alan Sher

Interactions of CD8+ T Lymphocytes with Parasitic Organisms

Immunity against parasites (protozoa and helminths) has traditionally been thought to depend exclusively on CD4+ helper functions controlling B-cell (i.e., antibody) and lymphokine-mediated effector mechanisms. It is only within the last decade that CD8+ lymphocytes have been demonstrated to display protective activity against parasitic infections. As would be predicted, most CD8+-dependent effector mechanisms are directed against intracellular parasites infecting host cells expressing class I major histocompatibility complex (MHC) products. Nevertheless, there are several examples of CD8+ cells interacting with extracellular parasites such as tachyzoites of *Toxoplasma gondii* (Khan et al., 1988) or schistosomula of *Schistosoma mansoni* (Butterworth et al., 1979). In the latter system, recognition is dependent on the presence of adsorbed MHC molecules on the parasite surface (Sher et al., 1978). While most examples of CD8+-dependent killing are directed against protozoa parasitizing host cells, the intracellular environments of the affected organisms differ considerably. Because of this, the study of parasite–CD8+ interactions provides some fascinating as well as highly relevant models for studying processing and presentation of foreign antigens by class I molecules.

Reproducible parasite-specific CD8+ cytotoxicity was first observed in cattle vaccinated against or undergoing remission of infection with *Theileria parva* (Emery et al., 1981). Infection with the protozoan is initiated when sporozoites deposited by the tick vector enter host T lymphocytes, escape from endocytic vacuoles, and develop within the cytoplasm as schizonts. It thus might be expected to employ an intracellular processing pathway similar to viruses and thus induce potentially protective CD8+ responses. The observed CD8+-mediated lysis is restricted by class I MHC gene products (Morrison et al., 1987) and correlates kinetically *in vivo* with the clearance of schizonts after challenge of animals immunized by infection and cure (Morrison and Godderis, 1990). While the induction of CD8+ effector function is a major approach in the development of an immunization method against this economically important parasite of Africa, the existence of MHC restriction *in vivo* and a high degree of strain specificity are serious obstacles limiting the potential effectiveness of a CD8+-dependent vaccine.

Trypanosoma cruzi infects a wide variety of host tissues, and in common with *Theileria parva* is able to escape from the parasitophorous vacuole into the cytoplasm of its target cells using a specific hemolysin (Andrews et al., 1990). Indeed, CD8+ depletion has been shown to markedly increase susceptibility to acute infection and abolish the protective immunity induced by an attenuated vaccine (Tarleton, 1990). Moreover, recent observations (Tarleton et al., 1992) indicate that β_2-microglobulin-deficient mice (produced by homologous recombination in a mouse strain normally highly resistant to *T. cruzi*) are highly susceptible to infection and die during the acute stage of the disease. Nevertheless, no *in vitro* correlates of the CD8+-dependent effector function have yet been described.

Induction of CD8+ effector activity is a major approach for vaccination against malaria. Immunization of laboratory animals or humans with irradiated sporozoites (the infective stage of plasmodium) results in solid protection against challenge with unattenuated parasites. *In vivo* depletion studies revealed that a major component of the protective response is CD8$^+$-dependent (Schofield et al., 1987; Kumar et al., 1988; Weiss et al., 1988). Moreover, CD8$^+$ cells from immune donors can lyse infected hepatocytes *in vitro*, and adoptive transfer of a cytolytic CD8$^+$ T-cell clone directed against the circumsporozoite (CS) protein (the major surface antigen on the infective parasite stage) confers protective immunity against sporozoite challenge (Hoffman et al., 1989; Romero et al., 1989). Nevertheless, there is considerable unexplained variability in this system, since other CD8$^+$ clones directed against the same epitope fail to transfer protection (Romero et al., 1990), and in some instances CD4$^+$ clones can also confer resistance against sporozoites (Tsuji et al., 1990). More importantly, vaccination with CS protein or an appropriate peptide containing the CTL epitope has failed to induce solid immunity against sporozoites, despite the induction of CTL activity. A possible explanation for the discrepancy in the results of vaccination with attenuated parasites and CS epitopes is that protection requires both recognition of multiple antigens and highly specific methods of antigen presentation (Khusmith et al., 1991). It should be noted that in contrast to *Theileria* and *T. cruzi*, the intrahepatic forms of plasmodia live within parasitophorous vacuoles, and therefore parasite epitopes must be transported across the vacuole membrane before associating with class I molecules for presentation to CD8$^+$ cells.

Role of CD8$^+$ Lymphocytes in Resistance to *T. gondii* Infection

The work of our laboratory on parasite–CD8$^+$ interactions has focused on the functions of this T-lymphocyte subset in immunity against *Toxoplasm gondii*. This protozoan is responsible for one of the major opportunistic infections encountered in AIDS patients. *T. gondii* is normally controlled by the host immune system, resulting in an asymptomatic, chronic infection maintained by dormant cysts. However, in immunosuppressed individuals the cysts rupture, resulting in reactivated infection. Cell-mediated rather than humoral immunity is thought to be the major host factor preventing reactivation of chronic infection as well as determining acquired resistance to reinfection (Frenkel and Taylor, 1982; Frenkel, 1988).

Mice vaccinated with ts-4, a temperature-sensitive strain of *T. gondii* (Pfefferkorn and Pfefferkorn, 1976), have been shown to be highly resistant to challenge infection with virulent parasites. Suzuki and Remington (1988) found that transfer of either immune CD4$^+$ or CD8$^+$ lymphocytes from vaccinated to naive mice conferred protection against challenge with a virulent strain of *T. gondii*. In addition, they showed that protection could be adoptively transferred more effectively by the combination of both CD4$^+$ and CD8$^+$ T-cell subsets. We have confirmed and extended these results by showing that *in vivo* depletion of CD8$^+$, but not CD4$^+$, lymphocytes at the time of challenge abrogates resistance in vaccinated mice. Nevertheless, a role for CD4$^+$ lymphocytes in resistance is evident from the fact that depletion of this T-cell subset before and during vaccination abrogates resistance induced by vaccination (Gazzinelli et al., 1991b). Therefore, we hypothesize that CD4$^+$ lymphocytes are involved in specific CD8$^+$ lymphocyte induction through interleukin-2 (IL-2) production and that CD8$^+$ T cells are the major effector cells either through interferon-γ (IFN-γ) production (Susuki and Remington, 1990) or by cytotoxic T-lymphocyte (CTL) activity (Hakim et al., 1991; Subauste et al., 1991) against tachyzoite-infected targets.

Both CD4$^+$ and CD8$^+$ T-cell subsets are important in natural resistance to new *T. gondii* infections. Thus, mice which are normally resistant to *T. gondii* infection became more susceptible, as judged by numbers of cysts in the brain, when depleted of the CD8$^+$ T-cell subset. Moreover, the class I H-2L locus was found to regulate the number of brain cysts which form following peroral infection with *T gondii*, with L^d and L^b conferring resistance and susceptibility, respectively (Brown and Mcleod, 1990). In addition, cyst numbers are amplified in the brains of B6-C-H-2^{bm12} mice (Brown and Mcleod, 1990), which have a mutation in the IA^b gene encoding the b-1 chain of their class II glycoproteins on antigen-presenting cells. Cyst numbers are also increased in mice depleted of CD4$^+$ lymphocytes (Araujo, 1991). These observations are consistent with the importance of helper T cells in the induction of parasite-specific CD8$^+$ effector lymphocytes.

In chronic toxoplasmosis, CD8$^+$ lymphocytes play a major role in preventing reactivation, but the function of CD4$^+$ cells is controversial. While long-term depletion of CD4$^+$ cells results in reactivation of an established infection (Vollmer et al.,

1987), short-term depletions have little or no effect on encephalitis and cyst numbers in the brain (Israelski et al., 1989). The result of long-term depletion of CD4+ cells in resistance can be explained by a secondary effect on the generation of effector cells such as CD8+ lymphocytes. Thus, resistance to chronic *T. gondii* infection is only partially affected by a major defect in CD4+ lymphocytes induced by a murine retrovirus (LP-BM5 MuLV). In these mice, further depletion of CD8+ lymphocytes is required to fully reactivate chronic toxoplasmosis. In contrast, infection with this retrovirus dramatically affects resistance to newly acquired infection, reinforcing the important role of CD4+ lymphocytes in the initial generation of a protective response against this protozoan (Gazzinelli et al., 1992b). Similar results were observed in mice chronically infected with *T. gondii* and then depleted of CD4+ cells and/or CD8+ cells with specific monoclonal antibody (mAb). Thus, depletion of both CD4+ and CD8+ lymphocytes rather than CD4+ cells alone is required to induce reactivation of infection (Gazzinelli et al., 1992a).

The mechanism by which CD8+ lymphocytes mediate resistance is unknown. However, antibody to IFN-γ eliminates the protective effect of adoptively transferring immune CD8+ cells (Suzuki and Remington, 1990). In addition, IFN-γ-reactive antibodies both induce fatal acute disease in mice challenged with a normally avirulent strain of *T. gondii* (Suzuki et al., 1988) and reactivate chronic infection (Suzuki et al., 1989). The explanation for the dual requirement of CD8+ T cells and IFN-γ is unclear, although several mechanisms have been proposed. These include (1) activation and generation of CD8+ effector cells by IFN-γ; (2) up-regulation of MHC class I antigens on host cells, therefore promoting CD8+ target cell recognition; (3) IFN-γ activation of macrophages, which kill parasites released from infected cells lysed by CD8+ CTL; and (4) a direct inhibitory effect of IFN-γ on intracellular parasite growth (Pfefferkorn, 1984).

In Vitro Effector Functions of *T. gondii*-specific CD8+ Lymphocytes

That CD8+ cells from immune hosts can limit the survival of *T. gondii in vitro* is now supported by a large body of evidence. The first description of CD8+ effector function against *T. gondii* demonstrated direct killing of free tachyzoites in a non-MHC-restricted manner. Similar activity was reported for both murine and human CD8+ effector cells (Khan et al., 1988; Khan et al., 1990). Yano et al. (1989) described the existence of CD8+ clones derived from chronically infected patients that were found to kill *T. gondii*-infected lymphoma targets in a MHC-restricted manner. These authors also demonstrated that peripheral blood lymphocytes (PBL) from an acutely infected individual displayed a low but significant degree of CD8-dependent cytotoxicity against human leukocyte antigen (HLA)-matched infected targets. We have confirmed the presence of CTL activity in seropositive donors using *T. gondii* soluble antigen-pulsed autologous Epstein-Barr virus (EBV)-transformed B-cell lines as targets (M. Clerici and R. T. Gazzinelli, unpublished observations).

In the murine system, we have recently described an MHC-restricted CTL activity directed against bone marrow-derived macrophages exposed to toxoplasma (described below) (Hakim et al., 1991). The existence of CTL directed against infected cells as opposed to free tachyzoites is also supported by recent observations by Subauste et al. (1991) employing infected tumor cells as targets. CD8+ lymphocytes from immune mice also release IFN-γ when stimulated with syngeneic infected host cells and the requirements for cytokine secretion closely follow those associated with CTL activity (Gazzinelli et al., 1991a).

Some key aspects of the MHC-restricted interaction between CD8+ lymphocytes and host cells exposed to toxoplasma which have been deduced from our studies are described below.

T-Cell Subset and MHC Restriction Element(s) Involved in the Killing of Infected Host Cells by Splenocytes from ts-4-Vaccinated Mice

After *in vitro* stimulation with irradiated tachyzoites, splenocytes from mice vaccinated with ts-4 display strong CTL activity against target cells infected with *T. gondii* or pulsed with the soluble parasite antigen (Figure 1). The *in vivo* induction of CTL requires live parasites, implying that active infection is a requisite for CTL generation. The cytotoxicity of effector cells is measured using as targets syngeneic bone marrow macrophages cultured with macrophage colony-stimulating factor (CSF)-containing culture supernatant from L cells. Such macrophages represent a homogeneous population of antigen-presenting cells expressing high levels of class I and no class II surface

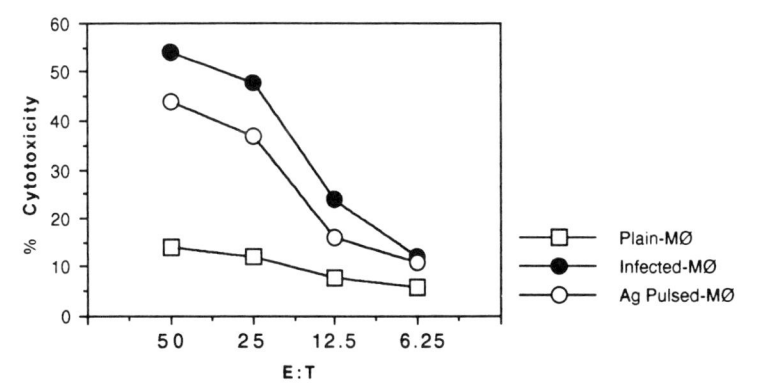

FIGURE 1. Specific killing of *T. gondii*-infected or antigen-pulsed bone marrow macrophages by spleen cells from ts-4 vaccinated mice. Spleen cells from ts-4 vaccinated mice were stimulated *in vitro* with irradiated tachyzoites and then incubated with plain, infected, or antigen-pulsed bone marrow macrophages.

molecules (De Libero and Kaufmann, 1986) (see Chapter 47, this volume). *T. gondii*-infected class I-bearing fibroblasts can be lysed by the same effector cells.

Before *in vitro* stimulation, depletion of either CD4$^+$ or CD8$^+$ lymphocytes blocks killing activity against *T. gondii*-infected host cells (Figure 2A). In contrast, after *in vitro* stimulation, depletion of CD4$^+$ lymphocytes has no effect on cytotoxicity, whereas depletion of CD8$^+$ lymphocytes totally eradicates cytolytic activity (Figure 2B). This suggests that CD4$^+$ lymphocytes are required to generate CD8$^+$ CTL effectors. The experiments shown in Figure 2A support this hypothesis, since addition of recombinant murine IL-2 restores CTL activity eliminated by CD4$^+$ depletion before *in vitro* stimulation. The CD8$^+$ activity is restricted by genes mapping in the D/L end of the MHC locus (Hakim et al., 1991) in BALB/c mice.

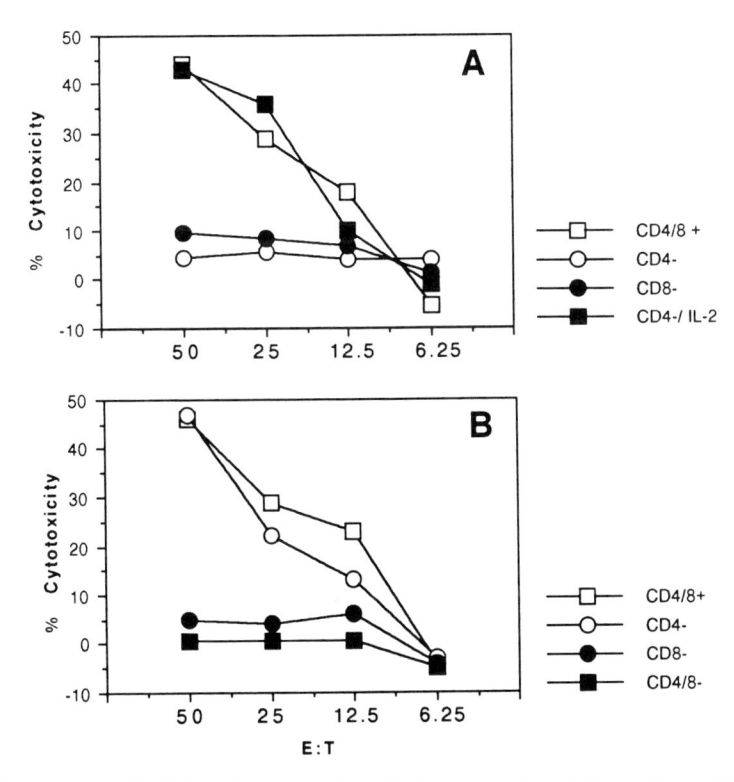

FIGURE 2. Requirement for CD4$^+$ lymphocytes or IL-2 in the generation of *T. gondii*-specific CD8$^+$-mediated cytolysis. Spleen cells from vaccinated mice were depleted of CD4$^+$ and/or CD8$^+$ subsets, either before (A) or after (B) *in vitro* stimulation with irradiated tachyzoites. Panel A shows that exogenous IL-2 restores the CTL activity of CD4$^+$-depleted spleen cells from vaccinated mice.

Pathways Involved in Class I-restricted *T. gondii* Antigen Presentation

In the classical model of antigen presentation for intracellular pathogens such as viruses or certain bacteria, antigens are proteolyzed in the cytosol, and the resulting peptides are transported to the lumen of the endoplasmic reticulum (ER) where they associated with newly synthesized MHC class I molecules. This complex is then transported to the cell surface via the Golgi. The intracellular protozoa *T. parva* and *T. cruzi*, which live in the host cell cytoplasm, probably utilize a similar or identical pathway. In contrast, plasmodium liver stages and *T. gondii* tachyzoites are sequestered within a parasitophorous vacuole, yet are clearly able to elicit highly efficient CD8$^+$ lymphocyte responses. Since these parasites grow in an environment isolated from host cell cytoplasm, antigen presentation may require the transport of parasite proteins or peptides from the parasitophorous vacuole to the host cell cytoplasm where they can access the classical pathway.

In addition to intracellular *T. gondii* infection, exogenous soluble antigen can also access the class I presentation pathway of bone marrow macrophages and thereby sensitize these targets for lysis (Figure 3). This is not simply due to low-M_r antigenic peptides associating with surface MHC class I molecules, since lytic activity co-resolves with proteins of $M_r \geq 12,000$ by gel filtration (Denkers et al., submitted). Thus, active processing of the antigenic proteins must be required. We have found that CD8$^+$ killing of *T. gondii*-infected cells but not the antigen-pulsed cells is inhibited when the targets are treated with brefeldin A, a drug which blocks protein transport from the ER to the Golgi. This observation suggests that extracellular antigen binds to class I molecules within close proximity of the cell surface, rather than to newly synthesized class I molecules in the ER. Such antigen processing could occur either via extracellular proteolysis (shown in Figure 3) or by endocytosis and intracellular degradation of the parasite antigens in class I-bearing vacuoles.

Although the presentation pathways in infected and antigen-pulsed cells are distinct (Figure 3), the peptides presented are likely to be similar or identical. Thus, lymphocytes restimulated *in vitro* with ts-4 lyse targets preincubated with soluble antigen, and conversely, lymphocytes restimulated with soluble antigen lyse infected targets. Moreover, the killing activity of antigen-pulsed macrophages, measured by ^{51}Cr release, is blocked by addition of nonradiolabeled *T. gondii*-infected targets (Figure 4). These results suggest that the same effector population reacts with both infected and antigen-loaded targets, implying that similar antigenic peptides bind MHC class I molecules in infected and antigen-loaded macrophages.

Target Antigens of CD8$^+$ Cytolysis

The capability of using exogenous *T. gondii* antigen to sensitize cells for lysis offers a powerful means to directly identify target antigens of CD8$^+$ T cells. To this end, we tested fractionated *T. gondii* proteins for their ability to induce cytolytic activity in our CTL assay, and found a preferential association of lytic activity with the antigens remaining in solution after centrifugation at $100,000 \times g$ (Denkers et al., in press). This suggests that the major target antigens are soluble rather than membrane-bound proteins. Moreover, when these soluble proteins were further resolved by Mono Q and Mono S fast protein liquid chromatography (FPLC) ion exchange chromatography, most of the cytolytic activity was associated with a single fraction (Denkers et al., in press). Thus, the spectrum of antigens recognized by *T gondii*-specific CD8$^+$ cells appears to be relatively restricted. In addition, our studies indicate that P30, a tachyzoite-specific antigen which is a target in extracellular killing of *T. gondii* by CD8$^+$ lymphocytes (Khan et al., 1991) is not present in the parasite fractions which sensitize host cell for lysis.

Summary and Conclusions

Because of their broad recognition of class I-bearing target cells and their dual cytolytic and IFN-γ-producing microbicidal activity, CD8$^+$ T lymphocytes appear to be ideal effector cells against intracellular parasites. As summarized in this chapter, CD8$^+$ cells have now been recognized to play an important role in immunity against intracellular protozoa, and in certain cases, the induction of CD8$^+$ activity represents a major approach toward vaccination against these pathogens.

As stressed above, the study of *T. gondii*–CD8$^+$ interactions is highly relevant to human disease and additionally offers easily manipulated *in vivo* and *in vitro* models. Indeed, the broad host and tissue range of this parasite and its now well-characterized intracellular cycle (Joiner et al., 1990) make *T. gondii*-infected cells a convenient and powerful model for studies of class I-associated antigen pro-

ALTERNATE PATHWAYS OF ANTIGEN PROCESSING FOR PRESENTATION OF TOXOPLASMA EPITOPES TO CD8+ T CELLS

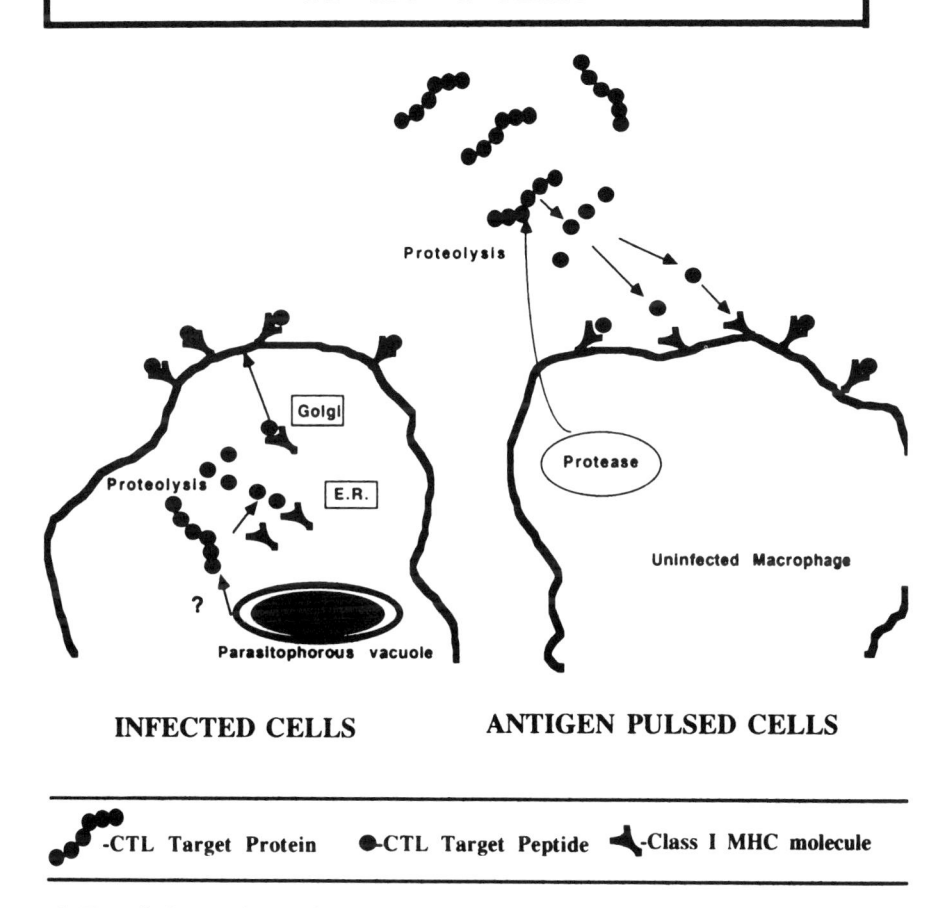

FIGURE 3. Two distinct pathways for *T. gondii* antigen presentation in the context of MHC class I antigens. This figure illustrates two different pathways involved in *T. gondii* antigen presentation by parasite-infected (left side) or antigen-pulsed (right side) bone marrow macrophages. An alternative pathway involved in the antigen presentation by pulsed cells (not shown in this figure) may involve antigen endocytosis, intracellular processing, and binding to MHC class I molecules.

cessing. As suggested by the data summarized in this chapter, toxoplasma antigens can be presented by macrophage class I molecules as a result of either exogenous or endogenous processing, the latter mechanism probably involving transport of tachyzoite antigens across the parasitophorous vacuole. Our current work is focused on the elucidation of these two processing pathways.

Because of the highly effective host cellular response against the parasite, most infections with *T. gondii* are asymptomatic. However, in AIDS as well as in other immunocompromised states, reactivation of chronic toxoplasmosis results in excessive cellular destruction, often leading to severe morbidity and mortality (Frenkel, 1988). Our studies suggest that a deficiency in CD8$^+$ lymphocyte function may be critical to the reactivation of chronic toxoplasmosis observed in immunocompromised individuals. Indeed, it is interesting to note that toxoplasma-induced encephalitis is a late phenomenon in AIDS (Luft et al., 1984) and therefore may be linked to the loss of CD8 function, rather than the earlier deficiencies in CD4-dependent responses (Shearer et al., 1986). Thus, we feel that further studies of toxoplasma–CD8$^+$ lymphocyte interactions should be pertinent in elucidating the pathogenesis of this important opportunistic infection.

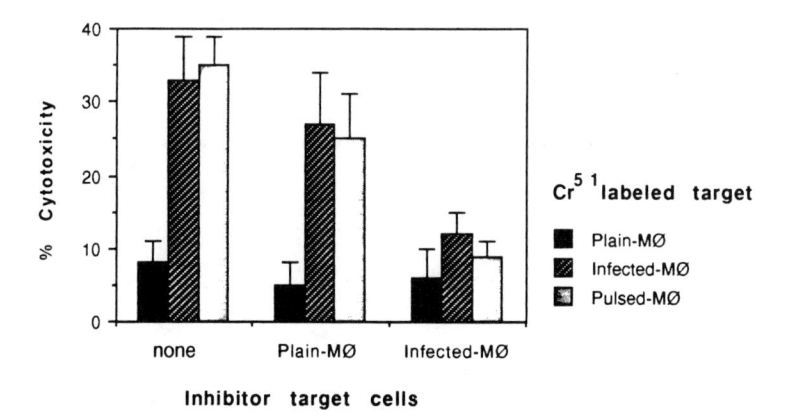

FIGURE 4. *T. gondii*-infected cells specifically inhibit killing of both infected and Ag-pulsed bone marrow macrophages. Plain or infected bone marrow macrophages were used at a ratio of 25:1 (cold/Cr^{51}-labeled targets) to inhibit Cr^{51} release induced by spleen cells from vaccinated mice stimulated *in vitro* with irradiated tachyzoites.

Acknowledgments

We thank Gene Shearer, Ronald Germain, and David Margulies for helpful discussions.

References

Andrews NW, Abrams CK, Stalin SL, Griffiths G (1990): A *T. cruzi*-secreted protein immunologically related to the complement component C9: Evidence for membrane pore-forming activity at low pH. *Cell* 61: 1277–1287

Araujo FG (1991): Depletion of L3T4$^+$ (CD4$^+$) T lymphocytes prevents development of resistance to *Toxoplasma gondii* in mice. *Infect Immun* 59: 1614–1619

Brown CR, Mcleod R (1990): Class I MHC and CD8$^+$ T cells determine cyst number in *T. gondii* infection. *J Immunol* 145: 3438–3441

Butterworth AE, Vadas MA, Martz E, Sher A (1979): Cytolytic T lymphocytes recognize alloantigens on schistosomula of *Schistosoma mansoni*, but fail to induce damage. *J Immunol* 122: 1314–1321

De Libero G, Kaufmann SHE (1986): Antigen-specific Lyt-2$^+$ cytolytic T lymphocytes from mice infected with the intracellular bacterium *Listeria monocytogenes*. *J Immunol* 137: 2688–2694

Denkers EY, Gazzinelli RT, Hieny S, Caspar P, Sher A (In press): Bone marrow macrophages process exogenous *Toxoplasma gondii* polypeptides for recognition by parasite-specific cytolytic T lymphocytes. *J Immunol*

Emery DL, Eugui EM, Nelson RT, Tenywa (1981): Cell mediated immune responses of *Theileria parva* (East coast fever) during immunization and lethal infections in cattle. *Immunology* 43: 233–335

Frenkel JK (1988): Pathophysiology of toxoplasmosis. *Parasitol Today* 4: 273–278

Frenkel JK, Taylor DW (1982): Toxoplasmosis in immunoglobulin M-suppressed mice. *Infect Immun* 38: 360–367

Gazzinelli RT, Hakim FT, Denkers E, Hieny S, Shearer G, Sher A (1991a): Vaccinated mice display CD8$^+$ dependent IFN-γ production and cytolytic activity against host cells infected with *T. gondii*. *FASEB* 5: A1338

Gazzinelli RT, Hakim FT, Hieny S, Shearer GM, Sher A (1991b): Synergisitic role of CD4$^+$ and CD8$^+$ T lymphocytes in IFN-γ production and protective immunity induced by an attenuated *Toxoplasma gondii* vaccine. *J Immunol* 146: 286–292

Gazzinelli R, Xu Y, Hieny S, Cheever A, Sher A (1992a): Simultaneous depletion of CD4$^+$ and CD8$^+$ T lymphocytes is required to reactivate chronic infection with *Toxoplasma gondii*. *J Immunol* 149: 175–180

Gazzinelli RT, Hartley JW, Fredrickson TN, Chattopadhyay SK, Sher A, Morse III HC (1992b): Opportunistic infections and retrovirus-induced immunodeficiency: Studies of acute and chronic infections with *Toxoplasma gondii* in mice infected with LP-BM5 murine leukemia viruses. *Infection and Immunity* 60: 4394–4401

Hakim FT, Gazzinelli RT, Denkers E, Hieny S, Shearer GM, Sher A (1991): CD8$^+$ T cells from mice vaccinated against *Toxoplasma gondii* are cytotoxic for parasite-infected or antigen-pulsed host cells. *J Immunol* 147: 2310–2316

Hoffmann SL, Isenbarger D, Long GW, Sedagah M, Szarfman A, Waters L, Hollingdale MR, van der Meide P, Finbloom D, Ballou WR (1989): Cloned cytotoxic T cells recognize an epitope in the circumsporozoite protein and protect against malaria. *Nature* 341: 323–326

Israelski DM, Araujo FG, Conley FK, Suzuki Y, Sharma S, Remington JS (1989): Treatment with anti-L3T4 (CD4) monoclonal antibody reduces the inflammatory response in toxoplasmic encephalitis. *J Immunol* 142: 954–958

Joiner KA, Fuhrman SA, Miettinen HM, Kasper LH, Mellman I (1990): *Toxoplasma gondii*: Fusion competence of parasitophorous vacuoles in Fc receptor-transfected fibroblasts. *Science* 249: 641–646

Khan IA, Smith KA, Kasper LH (1988): Induction of antigen-specific parasiticidal cytotoxic T cells splenocytes by a major membrane tachyzoite antigen (P30) of *Toxoplasma gondii*. *J Immunol* 141: 3600–3605

Khan IA, Smith KA, Kasper LH (1990): Induction of antigen-specific human cytotoxic T cells by *Toxoplasma gondii*. *J Clin Invest* 85: 1879–1886

Khan IA, Ely KH, Kasper LH (1991): Purified parasite antigen (p30) mediated CD8+ T cell immunity against fatal *Toxoplasma gondii* infection in mice. *J Immunol* 147: 3501–3506

Khusmith S, Charoenvit Y, Kumar S, Sedegah M, Beaudoin RL, Hoffman SL (1991): Protection against malaria by vaccination with sporozoite surface protein 2 plus CS protein. *Science* 252: 715–718

Kumar SL, Miller LH, Quakyi IA, Keister DB, Houghten RH, Maloy WL, Moss B, Berzofsky JA, Good MF (1988): Cytotoxic T cells specific for the circumsporozoite protein of *Plasmodium falciparum*. *Nature* 334: 258–260

Luft BJ, Brooks RG, Conley FK, McCabe RE, Remington JS (1984): Toxoplasmic encephalitis in patients with acquired immune response deficiency syndrome. *JAMA* 252: 913–917

Morrison WI, Godderis BM (1990): Cytotoxic T cells in immunity to *Theileria parva* in cattle. *Curr Top Microbiol Immunol* 155: 79–93

Morrison WI, Godderis BM, Teale AJ, Groocock CM, Kemp SJ, Stagg DA (1987): Cytotoxic T cells elicited in cattle challenged with *Theileria parva* (Muguga): Evidence for restriction by class I MHC determinants and strain specificity. *Parasite Immunol* 9: 563–578

Pfefferkorn ER (1984): Interferon-γ blocks the growth of *Toxoplasma gondii* in human fibroblasts by inducing the host cells to degrade tryptohan. *Proc Natl Acad Sci USA* 81: 908–912

Pfefferkorn ER, Pfefferkorn LC (1976): *Toxoplasma gondii*: Isolation and preliminary characterization of temperature-sensitive mutants. *Exp Parasitol* 39: 365–376

Romero P, Maryanski JL, Cordey AS, Corradin G, Nussenzweig RS, Zavala F (1990): Isolation and characterization of protective cytolytic T cells in a rodent malaria system. *Immunol Lett* 25: 27–32

Romero P, Maryanski JL, Corradin G, Nussenzweig RS, Nussenzweig V, Zavala F (1989): Cloned cytotoxic T cells recognize an epitope in the circumsporozoite protein and protect against malaria. *Nature* 341: 323–326

Schofield L, Villaquiran J, Ferreira A, Schellekens H, Nussenzweig RS, Nussenzweig V (1987): Gamma-Interferon, CD8+ T cells and antibodies required for immunity malaria sporozoites. *Nature* 330: 664–666

Shearer GM, Bernstein DC, Tung KST, Via CS, Redfield R, Salahuddin SZ, Gallo RC (1986): A model for the selective loss of major histocompatibility complex self-restricted T cell immune responses during the development of acquired immune deficiency syndrome (AIDS). *J Immunol* 137: 2514–2521

Sher A, Hall B, Vadas MA (1978): Acquisition of murine histocompatibility complex gene products by schistosomula of *Schistosoma mansoni*. *J Exp Med* 148: 46–57

Subauste CS, Koniaris AH, Remington JS (1991): Murine CD8+ cytotoxic T lymphocytes lyse *Toxoplasma gondii* infected cells. *J Immunol* 147: 3955–3959

Suzuki Y, Remington JS (1988): Dual regulation of resistance against *Toxoplasma gondii* infection by Lyt-2+ and L3T4+ T cells in mice. *J Immunol* 140: 3943–3950

Suzuki Y, Remington JS (1990): The effect of an anti-IFN-γ antibody on the protective effect of Lyt-2 immune T cells against toxoplasmosis in mice. *J Immunol* 144: 1954–1956

Suzuki Y, Conley FK, Remington JS (1989): Importance of endogenous IFN-γ for prevention of toxoplasmic encephalitis in mice. *J Immunol* 143: 2045–2050

Suzuki Y, Orellana MA, Schreiber RD, Remington JS (1988): Interferon-γ: The major mediator of resistance against *T. gondii*. *Science* 240: 516–519

Tarleton RL (1990): Depletion of CD8+ T cells increases susceptibility and reverses vaccine-induced immunity in mice infected with *Trypanosoma cruzi*. *J Immunol* 144: 717–724

Tarleton RL, Koller BH, Latour A, Postan M (1992): Susceptibility of β_2-microglobulin deficient mice to *Trypanosoma cruzi* infection. *Nature* 356: 338–340

Tsuji M, Romero P, Nussenzweig RS, Zavala F (1990): CD4+ cytolytic T cell clone confers protection against murine malaria. *J Exp Med* 172: 1353–1357

Vollmer LT, Waldor MK, Steinman L, Conley FK (1987): Depletion of CD4+ lymphocytes with monoclonal antibody reactivates toxoplasmosis in the central nervous system: A model of superinfection in AIDS. *J Immunol* 138: 3737–3741

Weiss W, Sedegah M, Beaudoin RL, Miller LH, Good MF (1988): CD8+ T cells (cytotoxic/suppressors) are required for protection in mice immunized with malaria sporozoites. *Proc Natl Acad Sci USA* 85: 573–576

Yano A, Aosai F, Ohta M, Hasekura H, Sugane K, Hayashi S (1989): Antigen presentation by *Toxoplasma gondii*-infected cells to CD4+ proliferative T cells and CD8+ cytotoxic cells. *J Parasitol* 75: 411–416

37
Antigen-Specific Suppression of Antibody Responses by Class II MHC-Restricted CTL

Nobukata Shinohara

Introduction

Mature T cells are known to be capable of suppressing immune responses in an antigen-specific manner. However the mechanism of suppression is not known. Despite many reports indicating the existence of "suppressor T cells" capable of secreting suppressor factors with affinity for native antigens, no evidence on a molecular level is available. Recent studies on class II-restricted cytotoxic T lymphocytes (CTL) raised the possibility that such CTL might explain antigen-specific suppression of antibody responses by T cells. Thus, CTL might play a role in the regulation of immune responses as well as the elimination of pathological cells.

Class II-Restricted CTL

Class II-Restricted CTL Are Not Rare

It is generally thought that CTL belong to the $CD8^+$ population of mature T cells and recognize cellular or viral protein antigens presented on the class I major histocompatibility complex (MHC) molecules. However, analyses of virus-specific as well as allospecific CTL revealed that a considerable number of class II-specific CTL consisting of both $CD8^+$ and $CD4^+$ cells exist (Vidovic et al., 1981; Shinohara and Kojima, 1984; Shinohara et al., 1987, 1988; Golding and Singer, 1985; Morrison et al., 1985). The magnitude of the class II-specific CTL response is comparable to that of the class I-specific response. Furthermore, pretreatment of the cells with antibodies to CD8 or CD4 together with complement revealed that class II-specific CTL consisted of both $CD8^+$ and $CD4^+$ cells. Generally, $CD8^+$ cells constitute a major portion of allo-class II-specific CTL responses. It was demonstrated that the precursor frequency of allogeneic class II-specific CTL is comparable with that of class I-specific allogeneic CTL (Golding and Singer, 1985). As described later, class II-restricted CTL were also found to be induced by immunization with soluble protein antigens.

$CD8^+$ Class II-Specific CTL

In allogeneic responses, $CD8^+$ cells constitute a large portion of class II-specific CTL (Vidovic et al., 1981; Shinohara, 1987; Shinohara et al., 1988). This is a conspicuous exception to the general correlation between the surface phenotype of T cells and the class of MHC molecule they recognize (Swain, 1983). Four lines of evidence indicated that the target molecules of such CTL are genuine class II molecules (Shinohara et al., 1988a): (1) The recognition of the target antigen is not influenced by allelic differences at class I loci. (2) Antibodies to the target class II molecules can block the recognition. (3) The CTL failed to recognize the target class II molecule with a mutation on either polypeptide chain. (4) A negative target can be converted to a positive target by transfection with the two structural genes of the target class II molecule. Interestingly, the recognition of the target cells by such $CD8^+$ class II-specific CTL is totally insensitive to the blocking effect of anti-CD8 antibodies, suggesting the CD8-independent nature of the recognition (Shinohara and Kojima, 1984; Shinohara et al., 1988a). $CD8^+$ class II-restricted CTL are also known to be induced upon immunization with exogenous protein antigens such as keyhole limpet hemocyanin (KLH) (Shinohara et al., 1988b) and hemagglutinin of influenza virus (Hioe and Hinshaw, 1989), but are difficult to detect in responses to some other antigens.

The origins of such CTL with a discordant combination of phenotype and specificity is totally unknown. The contemporary general consensus explains that immature $CD8^+4^+$ thymocytes undergo differentiation into $CD8^+$ or $CD4^+$ cells when they have been interacted with self class I or class II MHC molecules, respectively. Thus, all $CD8^+$ cells are expected to have class I restriction. We envision three possibilities to account for the existence of $CD8^+$ class II-restricted CTL. The first possibility is that the observed class II reactivity is an accidental cross-reaction of $CD8^+$ T cells that have been selected by self class I molecules. Although this possibility predicts the existence of a symmetrical population of cells, i.e., $CD4^+$ class I-restricted T cells, with a comparable frequency, such cells are extremely rare, if they exist at all. The second possibility assumes asymmetrical positive selection of $CD4^+8^+$ cells. In this model the CD8 molecule is not involved in signal transduction upon positive selection, while the CD4 molecule is. Thus, when the cells receive signals through the T-cell receptor (TCR) complex and the CD4 molecule, they differentiate into $CD4^+$ cells, whereas they undergo differentiation into $CD8^+$ cells if the signal is delivered only through the TCR complex. In such a situation, a portion of cells bearing TCR with high affinity for the class II molecule might differentiate into $CD8^+$ cells. The last possibility is the existence of an extrathymic differentiation pathway where immature T cells differentiate into $CD8^+$ T cells without experiencing thymic positive selection.

$CD4^+$ Class II-Restricted CTL

Certain $CD4^+$ T cells have been known to be capable of killing class II-positive B cells in the presence of the relevant antigen. Nevertheless, it was claimed that such $CD4^+$ T cells lyse neighboring cells by nondirectionally releasing cytotoxic factors such as lymphotoxin and tumor necrosis factor (TNF) upon recognition of specific antigen, and that such killing is totally different from specific and directional target cell lysis of $CD8^+$ CTL (Tite and Janeway, 1984; Ozaki et al., 1987). However, recent studies clearly demonstrated that $CD4^+$ cells can lyse specific targets without involving neighboring nonspecific bystander cells (Maimone et al., 1986; Takayama et al., 1991). It was also shown that such target cell lysis was also accompanied by fragmentation of DNA of the target cells. Thus there appears to be no reason to discriminate $CD4^+$ cells from CTL. Interestingly, many such $CD4^+$ CTL have been found not to express perforin mRNA, indicating the existence of a perforin-independent killing mechanism.

CTL Response to Soluble Protein Antigens

Class I-Restricted Soluble Protein-Specific CTL

Since exogeneous soluble proteins are generally internalized and presented on class II molecules, immunization with soluble protein results in induction of class II-restricted T cells, including CTL. However, there are a small number of examples showing induction of class I-restricted CTL by immunization with soluble proteins. Staerz et al. (1987) reported induction of ovalbumin (OVA)-specific class I-restricted CTL by immunization with a soluble form of the antigen. These CTL could lyse target cells with the appropriate H-2 haplotype only when peptide fragments of the antigen were present. A native form of the protein could not sensitize the target cells for specific lysis, showing that the target cells could not process the extracellularly administered antigen for class I presentation.

Class II-Restricted Soluble Protein-Specific CTL

There is much evidence for the induction of $CD4^+$ class II-restricted T cells with cytolytic activity in response to antigenic stimulation by soluble proteins (Tite et al., 1985; Shinohara et al., 1988; Bourgault et al., 1989). In addition, many cloned $CD4^+$ helper T cells were shown to be capable of lysing class II-positive cells in the presence of the specific antigen (Tite and Janeway, 1984; Ozaki et al., 1987; Watanabe et al., 1987) as mentioned in the preceding section. Furthermore, a few studies reported induction of $CD8^+$ class II-restricted CTL in response to antigenic stimulation by soluble proteins (Shinohara et al., 1988b; Hioe and Hinshaw, 1989).

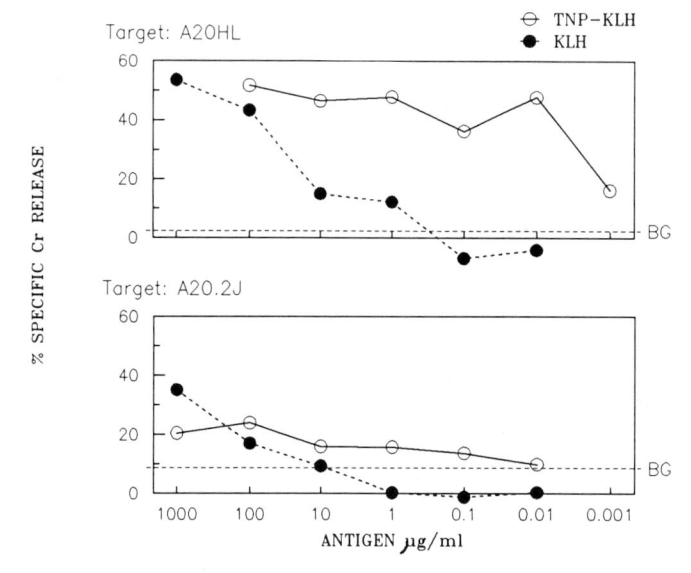

FIGURE 1. Preferential killing of antigen-reactive B cells by soluble antigen-specific class II-restricted CTL. An I-E^d-restricted, KLH-specific $CD4^+$ CTL clone of BALB/c origin, BK1, was tested on class II positive B cell tumor A20.2J or on A20HL, a transfectant expressing anti-TNP IgM on the surface. E/T ratio was 16:1. (From Shinohara et al., 198b.)

Specific Lysis of B Cells by Soluble Protein-Specific, Class II-Restricted CTL

B Cells as Specific Antigen Presenters

Generally, B lymphocytes are poor antigen-presenting cells of soluble antigens. However, when they encounter antigens reactive with their surface immunoglobulin (Ig) receptors, they become extremely efficient antigen presenters (Watanabe et al., 1986; Liano and Abbas, 1987), being far more efficient than versatile macrophages. Such a unique feature of B cells is believed to result from a specialized endocytotic machinary of these cells: they are designed to internalize molecules bound on the surface Ig receptors. Consequently, B cells reactive with a given antigen are preferential targets for cognate interaction with T cells reactive with the complex of class II and peptide fragments derived from the same antigen molecule. Such a characteristic of B cells explains the classical observations that B cells can receive help from T-helper cells when the B-cell determinant (haptenic determinant) and the T-cell determinant (carrier determinant) are on the same molecule. Similarly, class II-restricted, soluble protein-specific CTL preferentially lyse B-cell clones with Ig receptors reactive with the relevant antigen. Figure 1 depicts preferential killing of antigen-reactive B cells by class II-restricted soluble antigen-specific CTL. Note the tremendous difference between KLH and TNP-KLH in the minimal effective concentration required to sensitize the trinitrophenyl (TNP)-reactive B cells for specific lysis by I-E^d-restricted KLH-specific CTL.

Antigen Processing Is Necessary for Class II-Restricted CTL

As is the case with helper T cells, the target molecule on B cells recognized by class II-restricted, soluble antigen-specific CTL is the complex of class II and processed antigen rather than native antigen trapped on the Ig receptor. this conclusion was derived from three lines of evidence: (1) Monoclonal antibodies to class II molecules blocked specific target cell lysis. (2) Antibodies to native antigen and to Ig failed to inhibit recognition. (3) A variety of processing inhibitors inhibited sensitization of target cells by the antigen (Shinohara, unpublished observations). In the case of class I-restricted CTL, external administration of native antigens was totally ineffective, and only peptide fragments could sensitize the target cells (Staerz et al., 1987). There has been, however, one report (Barnaba et al., 1990) showing that antigen-reactive B cells presented exogenous antigen on class I molecules. At present it is not clear whether such a type of class I presentation is common or not.

Carrier-Specific Suppression of AntiHapten Antibody Responses by CTL

Theoretical considerations

The concentration of specific antigens required to

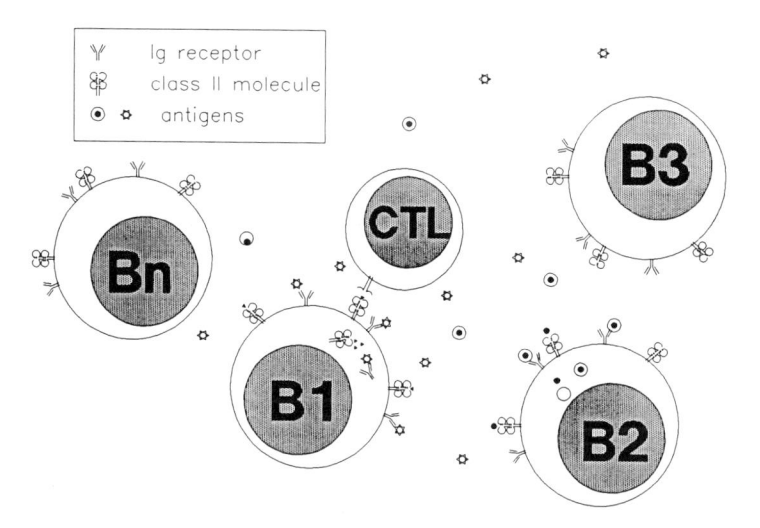

FIGURE 2. Schematic diagram of proposed mechanism of antigen-specific suppression of antibody responses by soluble antigen-specific class II-restricted CTL.

sensitize nonspecific B cells for killing by class II-restricted soluble antigen-specific CTL is considerably high (100 μg–1 mg/ml). Since it is rather difficult to imagine that such high concentrations can be attained by exogenous antigens *in vivo*, the primary targets of class II-restricted CTL *in vivo* are most likely B-cell clones reactive with the relevant antigen molecule (schematic diagram shown in Figure 2). Therefore, it was predicted that such CTL might be capable of antigen-specific suppression of antibody production. In this model it is essential that the haptenic determinant and a peptide stretch to be presented on the class II molecule reside on a single molecular complex in order for such suppression to be operative, because of the requirement for the ability of B cells to specifically trap the antigen. This type of suppression reminds us of the classical observations on carrier-specific suppression of anti-hapten antibody responses, which used to be accounted for by suppressor T cells.

In Vitro Suppression of Antibody Responses by Class II-Restricted Soluble Antigen-Specific CTL

When cloned class II-restricted soluble antigen-specific CTL were incorporated into *in vitro* secondary antibody responses, carrier-specific suppression of anti-hapten antibody production could be reproduced (Shinohara et al., 1991). Figure 3 demonstrates the result of one such experiment. A KLH-specific I-E^d-restricted CD4$^+$ CTL clone exerted a profound suppressive effect on anti-TNP antibody production in the *in vitro* secondary response to TNP-KLH, whereas it did not suppress anti-TNP-OVA response. An OVA-specific I-A^d-restricted CTL clone caused suppression of reciprocal specificity. The suppressive effects by these CTL clones were genetically restricted, i.e., they failed to suppress antibody response of lymphocytes with an incompatible H-2 haplotype. The observed carrier-specificity of suppression could be bypassed by adding appropriate TNP-conjugated protein in culture. Thus, OVA-specific CTL clones suppressed anti-TNP antibody response to TNP-KLH when TNP-OVA was added. Even under this condition, antibody production to KLH was not suppressed, reflecting the failure of KLH-specific B cells to efficiently present OVA-derived peptides. These results are consistent with the speculation that soluble antigen-specific CTL exert antigen-specific suppression of antibody responses by specifically killing B cells capable of trapping and presenting the relevant antigen molecules.

Similar carrier-specific suppression of anti-hapten antibody responses by soluble antigen-specific, class II-restricted T cells has been reported by other groups (Bottomly et al., 1983; Asano and Hodes, 1985). Although antigen-specific lytic activity of the T cells was not shown in these reports, it appears likely that those cells were actually capable of lysing class II-positive antigen-specific B cells in the presence of the antigen.

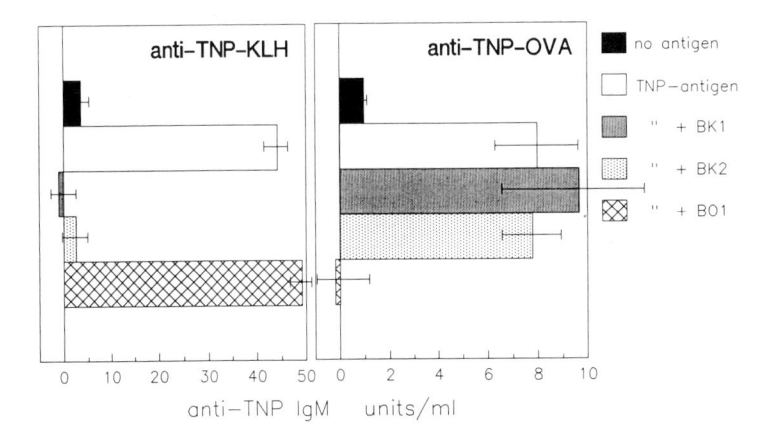

FIGURE 3. CTL-mediated specific suppression of anti-hapten antibody responses. Spleen cells of BALB/c mice immunized with TNP-KLH or TNP-OVA were restimulated *in vitro*. CTL added to the cultures were BK1, BK2 (I-E^d-restricted KLH specific clones), and BO1 (I-A^d-restricted OVA-specific clone). Anti-TNP antibody in day 9 culture supernatant was determined by ELISA assay. (From Shinohara et al., 1991.)

Suppression of Antibody Responses by Class I-Restricted Soluble Antigen-Specific CTL

Yefenof et al. (1990) demonstrated suppression of antihapten antibody response by class I-restricted soluble antigen-specific CTL. They showed that KLH-primed CD8$^+$ cells were capable of lysing TNP-binding B cells and caused suppression of anti-TNP antibody production of TNP-KLH primed cells. The killing was restricted by class I rather than class II MHC molecules. The mechanism of antigen presentation on class I molecules by these B cells was not elucidated. Thus, the specificity of suppression in terms of the antigen reactivity of B cells was not clear either. If class I presentation of the antigens was due to passive binding of peptide fragments resulting from an extracellular proteolytic process in culture, as was the case for OVA-specific class I-restricted CTL (Staerz et al. 1987), one would not expect specific elimination of B cells reactive with the antigen. A recent report by Barnaba et al. (1990) demonstrated specific presentation of hepatitis B virus envelope antigen on class I molecules by the antigen-specific B cells. It appears possible that certain proteins internalized by binding to Ig undergo the class I presentation pathway. If that is the case, class I-restricted CTL would also be capable of carrier-specific suppression of production of antibody directed to the same molecule.

Concluding Remarks

Evidence that class II-restricted, soluble antigen-specific CTL can mediate antigen-specific, genetically restricted suppression of antibody responses has been presented. These observations raised a possibility that CTL might play a role as regulators of immune responses besides their role as eliminators of pathological cells. Although at present the true biological implication of these observations is not clear, it is quite likely that a major portion of the observations on T-cell-mediated antigen-specific suppression of antibody responses can be accounted for by the natural occurrence of this type of CTL upon immunization. Nevertheless, it is not clear in what kind of situation such specific suppression is necessary. One attractive possibility is that they might be involved in the surveillance of autoantibody-producing B cells. If we assume the existence of a special subset of CTL which recognize self protein-derived peptides on class II molecules, B-cell clones reactive with self proteins would be eliminated as soon as they presented the antigens. This type of serveillance would allow autoantibody-producing B cells to survive unless they encountered the antigen. The normal lymphocyte population is known to contain B cells capable of producing autoantibodies, despite the fact that they do not mount responses even when immunized with cross-reactive foreign antigens. Although such a possibility is attractive, we have not, so far, obtained evidence for the existence of self peptide-specific class II-restricted CTL. Further and more extensive studies are required to elucidate the physiological significance of these CTL.

References

Asano Y, Hodes R (1985): T cell regulation of B cell activation Lyt-1$^+$,2$^-$ T cells modify the MHC-restricted function of heterogenous and cloned T suppressor cells. *J Exp Med* 162: 413–426

Barnaba V, Franco A, Alberti A, Benvenuto R, Balsano F (1990): Selective killing of hepatitis B envelope antigen-specific B cells by class I-restricted, exogenous antigen-specific T lymphocytes. *Nature* 345: 258–260

Bottomly K, Kaye J, Jones B, Jones F III, Janeway CA, Jr (1983): A cloned, antigen-specific, Ia-restricted Lyt-1$^+$2$^-$ T cell with suppressive activity. *J Col Cell Immunol* 1: 42–49

Bourgault I, Gomez A, Comrad E, Picard F, Levy JP (1989): A virus-specific CD4$^+$ cell-mediated cytolytic activity revealed by CD8$^+$ cell elimination regularly develops in uncloned human antiviral cell lines. *J Immunol* 142: 252–256

Golding H, Singer A (1985): Specificity, phenotype, and precursor frequency of primary cytolytic T lymphocytes specific for class II major histocompatibility antigens. *J Immunol* 135: 1610–1615

Hioe CE, Hinshaw VS (1989): Induction and activity of class II-restricted, Lyt-2$^+$ cytolytic T lymphocytes specific for the influenza H5 hemagglutinin. *J Immunol* 142: 2482–2488

Liano D, Abbas AK (1987): Antigen presentation by hapten-specific B lymphocytes. V. Requirements for activation of antigen-presenting B cells. *J Immunol* 139: 2562–2566

Maimone MM, Morrison LA, Braciale VL, Braciale TJ (1986): Features of target cell lysis by class I and class II MHC-restricted cytolytic T lymphocytes. *J Immunol* 137: 3639–3643

Morrison LA, Braciale VL, Braciale TJ (1985): Expression of H-2I region-restricted cytolytic activity by an Lyt2$^+$ influenza virus-specific T lymphocyte clone. *J Immunol* 135: 3691–3696

Ozaki S, York-Jolley J, Kawamura H, Berzofsky JA (1987): Cloned protein antigen-specific, Ia-restricted T cells with both helper and cytolytic activities: Mechanisms of activation and killing. *Cell Immunol* 105: 301–316

Shinohara N (1987): Class II antigen-specific murine cytolytic T lymphocytes (CTL). I. Analysis of bulk populations and establishment of Lyt-2$^+$L3T4$^-$ and Lyt-2$^-$L3T4$^+$ bulk CTL lines. *Cell Immunol* 107: 395–407

Shinohara N, Kojima M (1984): Mouse alloantibodies capable of blocking cytotoxic T cell function. V. The majority of *I* region-specific CTL are Lyt-2$^+$ but are relatively resistant to anti-Lyt-2 blocking. *j Immunol* 132: 578–583

Shinohara N, Hozumi N, Watanabe M, Bluestone JA, Johnson-Leva R, Sachs DH (1988a): Class II antigen-specific murine cytolytic T lymphocytes (CTL). II. Genuine class II specificity of Lyty-2$^+$ CTL clones. *J Immunol* 140: 30–36

Shinohara N, Huang Y, Muroyama A (1991): Specific suppression of antibody responses by soluble protein-specific, class II-restricted cytolytic T lymphocyte clones. *Eur J Immunol* 21: 23–27

Shinohara N, Watanabe M, Sachs DH, Hozumi N (1988b): Killing of antigen-reactive B cells by class II-restricted, soluble antigen-specific CD8$^+$ cytotoxic lymphocytes. *Nature* 334: 481–484

Staerz UD, Karasuyama H, Garner AM (1987): Cytotoxic T lymphocytes against a soluble protein. *Nature* 329: 449–451

Swain SL (1983): T cell subsets and the recognition of MHC class. *Immunol Rev* 74: 129–141

Takayama H, Shinohara N, Kawasaki A, Someya Y, Hanaoka S, Kojima H, Yagita H, Okumara K (1991): Antigen-specific directional target cell lysis by perforin-negative T lymphocyte clones. *Int Immunol* (in press)

Tite JP, Janeway CA, Jr (1984): Cloned helper T cells can kill B-lymphoma cells in the presence of specific antigen.: Ia-restriction and cognate vs. non-cognate interactions in cytolysis. *Eur J Immunol* 14: 878–886

Tite HP, Powell MB, Ruddle NH (1985): Protein-antigen specific Ia-restricted cytolytic T cells: Analysis of frequency, target cell susceptibility, and mechanism of cytolysis. *J Immunol* 135: 25–33

Vidovic D, Juretic A, Nagy ZA, Klein J (1981): Lyt phenotypes of primary cytotoxic T lymphocytes generated across the A and E region of the H-2 complex. *Eur J Immunol* 11: 499–504

Watanabe M, Wegman DR, Ochi A, Hozumi N (1986): Antigen presentation by a b-cell line transfected with cloned immunoglobulin heavy- and light-chain genes specific for a defined hapten. *Proc Natl Acad Sci USA* 83: 5247–5251

Watanabe M, Yoshizawa M, Hozumi N (1987): Cytotoxic function of a cloned helper T cell line. *Immun Lett* 15: 133–139

Yefenof E, Zehavi-Feferman R, Guy R (1990): Control of primary and secondary antibody responses by cytotoxic T lymphocytes specific for a soluble antigen. *Eur J Immunol* 20: 1849–1853

38
The Immunosenescence of Cytolytic T Lymphocytes (CTL): Reduction of Pore-Forming Protein and Granzyme Levels

Eda T. Bloom and Judith A. Horvath

Introduction

The decline in immunologic vigor that accompanies advancing age has been recognized for many years, and immunologic senescence has been the focus of two recent and extensive reviews (Thoman and Weigle, 1989; Miller, 1990). Although all immunologically active cells, including T cells, B cells, macrophages, and natural killer (NK) cells, appear to exhibit age-related alterations in one or more functions, T cells are widely considered to be the most vulnerable to the potentially deleterious effects of aging. Thymic involution, morphological evidence for which can be observed soon after the immune system functionally matures, is thought to be intimately connected with the subsequent decline in T-cell function (Hirokawa et al., 1990). Although evidence suggests that extrathymic T-cell development is also altered with advanced age (Gorczynski and Chang, 1984), the mechanisms are not yet understood.

The age-related functional decline in cytolytic T lymphocyte (CTL) effector cell activity was first observed nearly two decades ago (Menon et al., 1974; Weksler and Hutteroth, 1974; Shigemoto et al., 1975). Because CTL have been shown to be involved in protection against viral and other intracellular infections and against certain tumors, as well as in allograft responses, the senescent decline in CTL activity has important consequences, as evidenced by the increased morbidity and mortality observed among aged individuals due to infectious and malignant diseases. This chapter will first focus on current knowledge regarding the age-related decline in CTL activity, specifically with regard to "normal" aging, omitting, for the sake of brevity, information from studies in autoimmune and immunodeficient models. Most of the evidence has been obtained in murine model systems, unless otherwise stated. The chapter will then briefly review selected related immunologic activities vulnerable to aging, which may suggest additional potential mechanisms for the decline. Finally, it will present our recent data regarding the effect of age on the production of pore-forming protein (PFP, perforin, or cytolysin) and granule-associated serine esterases, and will discuss some implications of these findings.

The Age-Related Decline in CTL Activity

During the mid-1970s a number of reports appeared demonstrating that CTL function is reduced with advancing age. In 1973 Kishimoto et al. (1973) reported that cell-mediated immunity, as assessed by the ability of aged bone marrow cells to induce graft-versus-host disease in semiallogeneic recipients, declined in aged mice. Shortly thereafter, the same group reported that the cell-mediated cytotoxic response to allogeneic cells *in vitro* decreases with age (Shigemoto et al., 1975). In 1974 Menon et al. (1974) reported the results of examining alloimmunity across major histocompatibility complex (MHC) barriers, *in vivo* utilizing skin graft rejection, and *in vitro* by testing for cytolytic activity of splenocytes from alloimmunized mice by assessing ^{51}Cr release by specific targets. They found an 85% decrease in cytolytic activity between 2 and 18 months of age in long-lived mice, correlating with a delayed rejection of primary skin grafts. Furthermore, by measuring the rate of attachment of effector cells to specific target cells, these investigators concluded that the immune cells did not make multiple "serial attachments" to several target cells and suggested that the decline in lytic activity with age is due to fewer "sensitized cells". This latter conclusion may be subject to question, because current evidence suggests that, in general, lytic effector cells do, indeed, recycle and can lyse several target cells. At about the same time, the age-related decline in CTL activity was recognized in humans (Weksler and Hutteroth, 1974), and soon thereafter was confirmed in other strains of mice (Goodman and Makinodan, 1975; Hirano and Nordin, 1976). Since the initial descriptions of the age-related decrease in CTL activity, and with the increasing recognition of its broad scope and far-reaching implications, considerable effort has been expended to understand the mechanisms responsible for the decline.

Effector Cell Characteristics

There are several differences between CTL from aged and young individuals. For example, the affinity for specific target cells of CTL from aged mice has been reported to be reduced when generated at low stimulator cell concentrations, but not when equivalent numbers of stimulator and responder cells are co-cultured (Zharhary and Gershon, 1981).

That CTL from aged mice exhibit increased cross-reactivity, or decreased MHC restriction, was originally demonstrated by Kruisbeek and Steinmeier in 1980 (Kruisbeek and Steinmeier, 1980). In the light of more current knowledge, this might suggest alterations in T-cell repertoire or increases in MHC-unrestricted mechanisms, such as lymphokine-activated killer (LAK) cell activity. Both of these possibilities have been investigated and demonstrated to occur. Experimental results of examining the effect of normal aging on the T-cell repertoire have demonstrated decreased diversity (Chang and Gorczynski, 1984), increased degeneracy (Helfrich et al., 1989), and decreased T-cell receptor (TCR) density (Asano et al., 1990) on alloimmune murine CTL. The induction of LAK activity has been shown to increase or show little change with age in mice both *in vitro* (Kawakami and Bloom, 1987; Kawakami and Bloom, 1988; Bloom and Kubota, 1989) and *in vivo* (Ho et al., 1990), although one report of decreased LAK activity with age, depending upon the incubation period with interleukin-2, has appeared (Bubenik, et al., 1987).

The frequencies of T-cell subsets, including $CD8^+$ T cells, effector cells in MHC class I-restricted CTL, are generally believed to remain relatively stable with aging in virtually all species examined (Utsuyama and Hirokawa, 1987; Thoman and Weigle, 1989, and Miller, 1990, for reviews). Although the density of the CD8 accessory molecule may be altered by aging (Utsuyama and Hirokawa, 1987), it has been suggested that the cellular distribution and density of the CD8 and LFA-1 accessory molecules may play little role in the age-related decline in murine CTL activity (Jenski, 1990). Interestingly, Jenski (1990) further reported that increased alloantigen density was required to stimulate CTL from aged mice, consistent with both the requirement for high stimulator/effector cell ratios (Zharhary and Gershon, 1981) and the decreased density of TCR expression (Asano et al., 1990) with advanced age, as discussed above.

Functional declines in $CD8^+$ T-effector cells (Komuro et al., 1990; Bloom, 1991) in class I-restricted alloreactive CTL and $CD4^+$ effector cells in class II-restricted alloreactive CTL (Rosenberg et al., submitted) have been observed in mice. We have demonstrated that mRNA levels of granule-associated serine esterases (granzymes) and pore-forming protein (PFP, perforin, or cytolysin) are reduced with advancing age in class I MHC alloantigen-stimulated, $CD8^+$ T cells (Bloom et al., 1990; Bloom, 1991). Available evidence suggests that aging alters CTL effector cell activity, and affects multiple aspects of that function.

Antigen Presentation

CTL directed against viral antigens are decreased with advanced age in humans (Bloom et al., 1989) and mice (Effros and Walford, 1983a; Effros and Walford, 1983b; Effros and Walford, 1984), and the decrease is related to the severity of infection (Bender et al., 1991). Effros and Walford (1984) have reported that decreased efficiency of antigen presentation accounts for at least part of the age-related decline in influenza-specific secondary CTL, although they did not rigorously identify the antigen-presenting cell(s). More recently, others have reported that aged splenic adherent cells but not B cells are defective in stimulating autoreactive T cells (Seth et al., 1990). In contrast, others have observed that antigen presentation may not contribute to diminished alloimmune reactivity with age (Gottesman et al., 1985). In general, most evidence to date suggests that macrophage and B-cell antigen-presenting activities probably change little with age (reviewed by Miller, 1990).

Immunoregulatory Circuits

Age-related alterations in a number of suppressor and helper mechanisms have been implicated in the decline in CTL activity with aging. In one of the earlier reports, Bach (1979) observed that both decreased helper cell and increased suppressor cell activities contributed to the age-related decline in the generation of alloantigen-specific cytolytic activity. Subsequently, Gottesman and colleagues (1984, 1988) demonstrated an age-related *de*crease in allospecific T suppressor cells, while observing an increase in natural suppressor cells effective for suppressing proliferation but not CTL generation. The T-cell subpopulation of the allospecific suppressor cell was not identified. Recently, others have observed an age-related decrease in the capability to produce tumor-specific CD4$^+$ suppressor cells (Dunn and North, 1991). We could find no evidence to support an increase with age in allospecific CD4$^+$ down-regulatory activity in the results of experiments in which titrated numbers of CD4$^+$ T cells from aged mice were mixed with constant numbers of CD4$^+$ cells from young mice and assessed for their ability to support CTL generation in CD8$^+$ cells from young mice (Bloom, 1991; and unpublished observations). Therefore, little evidence exists to support the notion that increased suppressor activity with age is responsible for declining CTL activity.

In contrast, a great deal of evidence exists for the contribution of decreased T-helper cell activity to the age-related decrease in the generation of CTL. In particular, the production of interleukin-2 (IL-2), a major cytokine produced by a subset of CD4$^+$ T cells, is markedly and universally decreased with age, and the expression of IL-2 receptors has been reported to undergo marked age-related alterations in many, but not all, systems (see reviews by Thoman and Weigle, 1989; Miller, 1990). IL-2 has a central role in CTL generation, and it has been reported that the age-related decline can be completely (Miller and Stutman, 1981) or significantly (Thoman and Weigle, 1982; Effros and Walford, 1983b; Thoman and Weigle, 1985) corrected *in vitro* by exogenous IL-2. Zharhary and co-workers (1984), using limiting dilution analyses and bulk cultures, demonstrated that the reconstitution of the CTL response by exogenous IL-2 was highly variable among individual aged animals. We have shown, by removal of cells with the asialo GM$_1^+$ marker prior to allostimulation, that the predominant cell affected by attempts to reconstitute CTL activity is probably an LAK precursor cell (Bloom and Kubota, 1989). We have also recently observed that the MHC class II-restricted CTL response by CD4$^+$ T cells cannot be reconstituted either by mitogen stimulated lymphokine-containing supernatant or by exogenous recombinant IL-2 (Rosenberg et al., submitted). Therefore, although the profound age-related decrease in IL-2 production undoubtedly contributes to the decline in CTL activity, it does not account for the entire phenomenon, and it is clear that other age-related alterations in CD4$^+$ T-helper cells contribute prominently to the age-related decline in alloantigen-specific CTL activity (Bloom, 1991). Interestingly, although other cytokines have been reported to exhibit age-related alterations in production (e.g., IL-1, IL-3, interferons, etc.), conflicting reports have appeared regarding most of them (see reviews by Thoman and Weigle, 1989; Miller, 1990).

CTL Precursors (pCTL)

A number of studies have examined the pCTL frequencies in aged compared to young individuals using limiting dilution analyses (LDA). Goodman and Makinodan (1975) first reported, using LDA *in vivo* by transferring limiting dilutions of spleen cells into young irradiated (600 R) recipients and subsequently assessing splenic CTL, that the frequency of CTL "precursor units" is diminished with age in long-lived mice, but that this decrease does not account for the entire age-related effect. Subsequent LDA were performed *in vitro*. Zharhary and Gershon (1981) later observed decreased frequency of pCTL "units" among spleen cells of aged mice stimulated *in vitro* with semiallogeneic cells, and the frequency could be restored to young levels with IL-2 in some, but not all, individual mice. Most subsequent investigations included cytokine-rich conditioned medium in the LDA, and therefore measurements were interpreted to signify individual pCTL rather than "precursor units," which would also contain helper and accessory cells. LDA of alloreactive pCTL revealed a 10-fold decrease in aged compared to young mice (Nordin and Collins, 1983), and consistent data were subsequently obtained for pCTL among polyclonal activated T cells (Miller, 1984). However, "moderately" aged mice of the short-lived AKR strain exhibited no evidence of decreased frequencies of virus-specific pCTL (Wegmann et al., 1991). Moreover, whether pCTL in the bone marrow of very old long-lived mice exhibit a decrease with advanced age depends upon the time interval used for polyclonal activation (Sharp et al., 1990). Therefore, most evidence suggests that the frequency of pCTL is reduced in aged mice, although results may vary, depending upon the model system and conditions used.

Miller (1984) also measured the frequency of precursors of IL-2-producing cells when stimulated with antigen, and as with pCTL, found an age-related reduction. He further reported that levels of activity generated per responsive cell, whether cytolytic or helper, remain unaffected by aging, speculating that aging reduces the proportion of responsive T cells but not the activity per cell, a theory which has since become known as the "mosaic" theory of aging (Thoman and Weigle, 1989). However, a more recent report that cloned naive cells from aged mice express less cytolytic activity than those from young upon stimulation with alloantigen (Jenski, 1990) suggests that immunologically active cells, and CTL in particular, from aged mice have reduced levels of activity on a per cell basis. In addition, using a single-cell cytotoxicity assay, we found that the frequency of antigen-specific target-binding cells is not altered with age, while the frequency of lytic cells among the target-binding cells is diminished (Bloom et al., 1988). Although conflicting observations have been reported (Gottesman and Edington, 1990), these results suggest that alloantigen-stimulated spleen cells from aged mice contain cells that can undergo a portion of the events along the vector leading to mature CTL activity, at a similar frequency to those from young. Such nonlytic but target-binding cells would have escaped detection by LDA because LDA utilizes target cell destruction (^{51}Cr release) as an endpoint. The existence of such "incompletely" capable cells, as demonstrated by single-cell assay and by clonal analyses, suggests a potential need to redefine the "mosaic" theory of aging in terms of the window used to assess activity.

Related Immunologic Activities Altered by Aging

Recent findings in at least two areas have potential implications for the age-related decline in CTL activity. Ernst et al. (1990) reported that the distribution of certain CD4$^+$ T-cell subsets, defined by the CD45RB and Pgp-1 (CD44) markers, are altered in aged mice, with the CD4$^+$CD45RB^{lo}Pgp-1^{hi} (i.e., memory subset) more strongly represented with increased age. In agreement with this, they found that CD4$^+$ T cells from aged mice express different cytokine secretion patterns from young mice. In particular, an age-related increase in IL-4 was noted. The age-related alteration in CD4$^+$ T-cell subsets is further strengthened by the consistent observations made in humans (Pirruccello et al., 1989). However, the mechanisms for the shift in subset distribution remain unclear, since the proliferative responses of the CD45RB^{lo} *and* CD45RB^{hi} subsets of cells are diminished with age (Nagelkerken et al., 1991).

A second area that has been explored recently is that of age-related changes in signal transduction mechanisms in T cells. The subject has been reviewed recently (Miller, 1990), and therefore few details will be given here. Suffice it to say that altered Ca^{2+} flux (Miller et al., 1987; Philosophe and Miller, 1989; Miller et al., 1989; Philosophe and Miller, 1990; Grossmann et al., 1991), defects in phosphatidyl inositol metabolism (Proust et al., 1987), and deficient protein kinase C (PKC) translocation (Fong and Makinodan, 1990) have been reported in T cells from aged mice. Deficient Ca^{2+} flux has also been observed in T cells from aged humans (Chopra et al., 1987). The implications of altered signal transduction mechanisms with age in regard to CTL activity have not been explored.

Reduction of Pore-Forming Protein and Granzyme Levels with Age

PFP and granzymes are released upon granule exocytosis, one process thought to be involved in cell-mediated cytolysis (Henkart, 1985; Young and Cohn, 1988; Young et al., 1988; Peters et al., 1991; Tschopp and Nabholz, 1991). Thorough attention is given to the potential roles of PFP and granzymes in lysis by CTL elsewhere in this volume (Chapter 14). Having previously demonstrated an age-related alteration in the lytic mechanism or maturation thereof (Bloom et al., 1988), we utilized a murine model of alloimmunity, in which CTL are stimulated *in vitro* by mixed lymphocyte culture (MLC), to investigate the effect of aging on PFP and granzyme production. We have reported an age-related reduction in the levels of mRNA transcribed by genes encoding for PFP and granzymes B and C in CD8$^+$ T cells stimulated in MLC (Bloom et al., 1990; Bloom, 1991). Data presented here demonstrate that stimulator cell-induced release of granzyme A and intracellular PFP levels are also reduced by aging. We further report that at least a portion of this reduction appears to be a consequence of the existence of smaller or fewer granules per cell that contain these molecules.

Enriched splenic T cells were obtained from individual young adult (3–5 months) and aged (24–

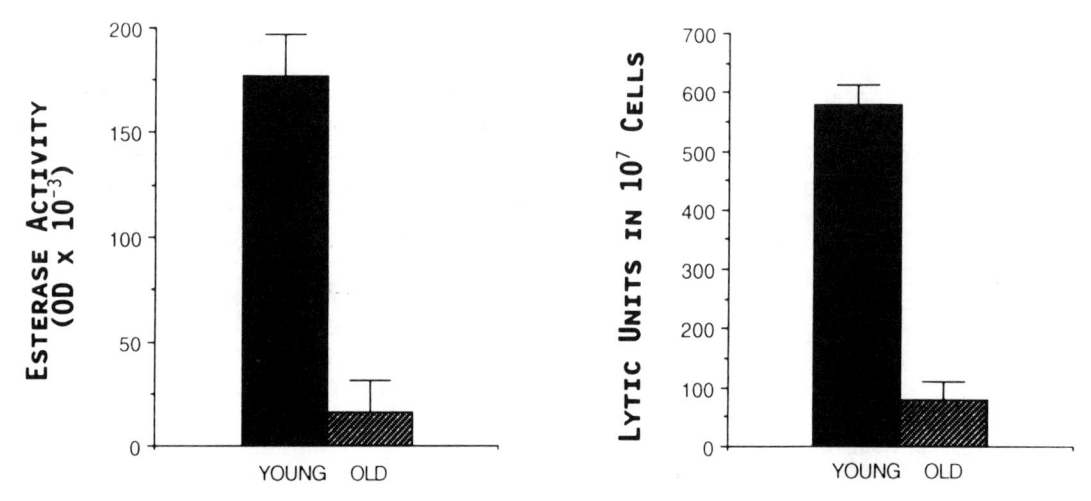

FIGURE 1. Released BLT esterase activity is reduced in CTL generated from old compared to young mice. The data represent the means of BLT esterase activity in supernatants from cultures of four young and four old mice each. Lytic activity is shown for the same mice. Error bars are SEM. BLT, esterase and lytic activity is greater in young than in old mice both with $p < 0.0025$.

28 months) female BALB/cNNia mice (H-2^d) and were co-cultured for 5 days with equivalent numbers of irradiated spleen cells from female C57BL/6NNia mice (H-2^b). We have previously reported the MHC-restricted nature of this system (Bloom et al., 1988), and routine tests for cytolysis of LAK-sensitive target cells verified this. Cytolytic activity was measured using ^{51}Cr-labeled EL-4 (H-2^b) target cells, and expressed as lytic units (LU) in 10^7 effector cells (Bloom and Korn, 1983). After culture, viable cells were collected by centrifugation over ficoll-hypaque. Supernatants were collected from MLC and tested for N-α-benzyloxycarbonyl-L-lysine thiobenzyl (BLT)-esterase activity, generally believed to be mediated by granzyme A. Cells were assessed for lytic activity. The results, displayed in Figure 1, clearly showed that release of BLT esterase was reduced in MLC-stimulated cells from aged compared to young mice. This reduction corresponded well with the parallel reduction in lytic activity measured on cells recovered from the same cultures.

Direct assessment of PFP activity by isolation and measurement of its lytic activity is not feasible for stimulated T cells from individual mice because of the large numbers of cells required. We therefore utilized monoclonal antibodies against murine PFP (P1-8 and P1-9, kindly supplied by Prof. Ko Okumura, Juntendo University, Tokyo, Japan), and measured PFP by immunocytochemical and enzyme-linked immunosorbent assay (ELISA) techniques using immunoperoxidase as described (Kawasaki et al., 1990). Immunocytochemistry slides were examined at 100× magnification, and a typical staining pattern of P1-8 on T cells from young and aged mice stimulated in primary MLC is shown in Figure 2. Staining with control, isotype-matched immunoglobulin G_{2a} (IgG_{2a}) was consistently and uniformly negative, with absolutely no stain visible (Figure 2A and B), whereas T cells stimulated in primary MLC exhibited a granular pattern of stain (Figure 2C–F). Because of the absolute absence of granular staining obtained using the negative control immunoglobulin, cells exhibiting even one granule were considered to be PFP$^+$. Staining of stimulated cells from young mice (Figure 2E and F) was more intense and generally displayed more PFP$^+$ cells compared to those from aged mice (Figure 2C and D). Stimulated T cells obtained from 28–29 individual aged and young mice were assessed for PFP$^+$ cells and cytolytic activity, and the results, displayed in Figure 3A, show that fewer MLC-stimulated cells from aged than young mice expressed PFP. Although this difference was highly statistically significant, the magnitude of the difference was remarkably less than that of the lytic activity (approximately two-fold compared to more than sevenfold, respectively).

Because the difference in PFP$^+$ cells between MLC-stimulated T cells from aged and young mice was only marginal, and because individual T cells from aged mice exhibited many fewer and more lightly stained or smaller granules per cell compared to CTL from young mice, the total PFP levels in cell lysates from individual young mice

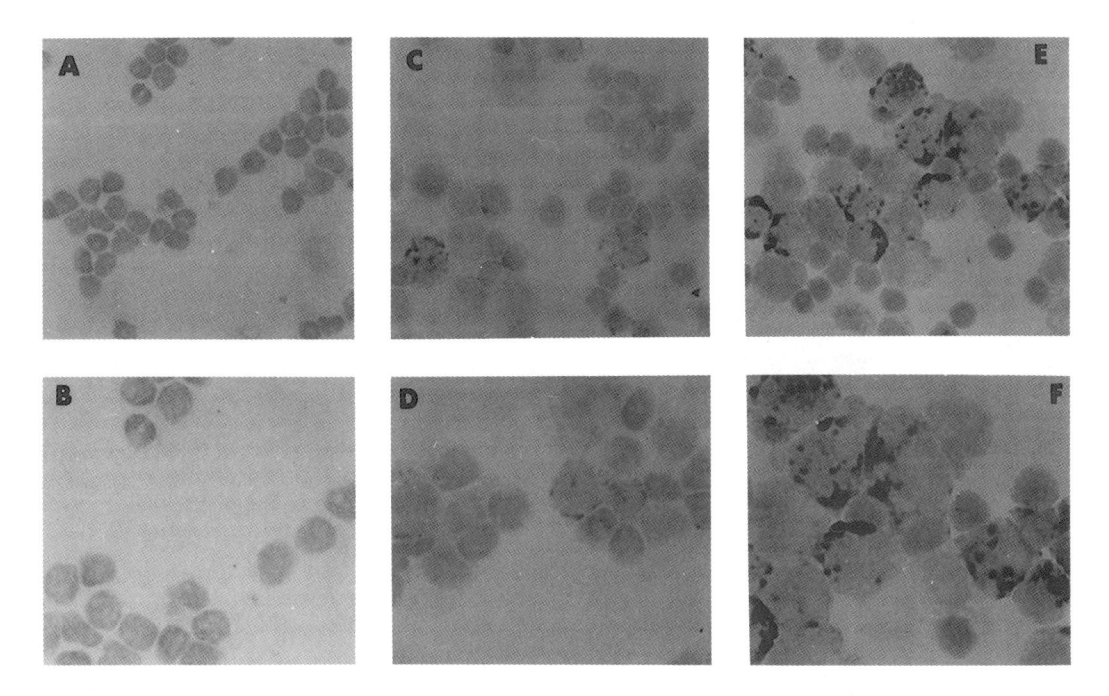

FIGURE 2. Immunocytochemical staining with monoclonal antibody to PFP (C–F) or isotype matched control IgG$_{2a}$ (A and B). A, B, E, and F are typical of MLC-stimulated cells from young mice; C and D are representative of those from old mice. Magnification was 40× for panels, A, C, and E, and 63× for panels B, D, and F.

were quantitated by ELISA. Data were normalized to a titration curve generated using lysates of CTLL.R8 (a PFP- and granzyme-producing CTL line, and a gift of Dr J.D.-E. Young, Rockefeller University, New York), and results are expressed as CTLL.R8 equivalents. The results, displayed in Figure 3B, show that the total PFP content of stimulated T cells from young mice was more than 12-fold that in stimulated cells from aged mice, substantially greater than the approximately twofold difference in the proportion of PFP$^+$ cells. These data demonstrate that the fewer and lighter-staining PFP$^+$ granules observed among T cells or purified CD8$^+$ stimulated cells from spleens of old compared to young mice (refer to Figure 2) are associated with greatly reduced levels of total cellular PFP as well as with diminished CTL activity.

The immunocytochemical staining patterns suggested that the stimulated T cells from aged mice may contain fewer and smaller cytolytic granules than those from young mice. In order to estimate the granularity of MLC-stimulated T cells from young and aged mice, we measured uptake of quinacrine, preloading of cells with which has been used to measure the degranulation of intracellular acidophilic granules (Kolber and Henkart, 1988), such as those that contain PFP and serine esterases (Ojcius et al., 1991; Peters et al., 1991). Replicate samples were double-stained with quinacrine plus phycoerythrin (PE)-labeled antibodies to either the CD8 or CD4 markers. The results, shown in Table 1, are consistent with the occurrence of an age-related reduction of the acidic intracellular compartment in CD8$^+$ CTL generated from the spleens of old compared to young mice. Such a reduction was not seen in CD4$^+$ T cells. We have already reported that the lytic effector cells comprise only CD8$^+$ T cells in this system (Bloom, 1991). It should be cautioned that quinacrine stains all acidophilic compartments, and from these data alone we cannot conclude definitively that there is a reduction in size or number of PFP-containing granules with age. Nevertheless, the different staining patterns achieved when PFP was assessed by immunocytochemistry in stimulated T cells from young and aged mice also support this notion.

Implications for understanding the lytic mechanism can be inferred from the observations that a large difference between MLC-stimulated cells

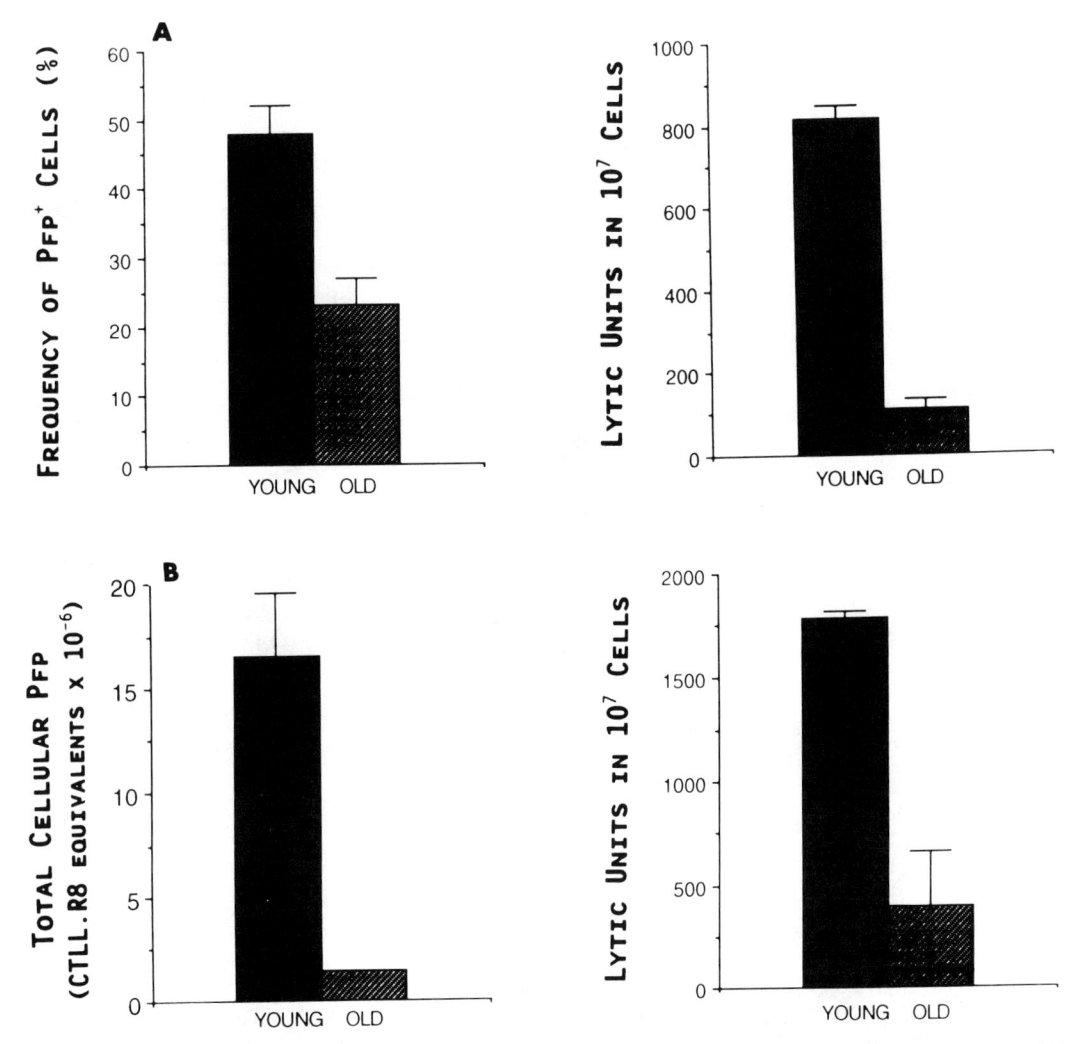

FIGURE 3. PFP levels are reduced in MLC-stimulated cells from aged mice compared to young. (A) Frequencies of PFP$^+$ cells in 29 young and 28 aged mice, and the mean lytic activities mediated by these same cell populations. Young > old for both determinations ($p < 0.0001$). (B) Mean total PFP content in lysates of 10^6 MLC-stimulated T cells from three each of young and old mice, and the mean lytic activities mediated by these same cell populations. PFP content is expressed as equivalents of CTLL.R8 cells, and determined relative to a titration of cell R8 cell lysates. Young > aged ($p < 0.05$). In this experiment, the frequencies of PFP$^+$ cells were 60% ± 15%, and 27% ± 4%, for young and aged mice, respectively ($p > 0.05$). All error bars are SEM.

from young and aged mice was observed in lytic activity (more than sevenfold at the population level, and about fourfold at the single-cell level (Bloom et al., 1988) and total PFP content (about 12-fold), while a much smaller difference was observed in the frequency in PFP$^+$ cells. Therefore, CTL generated from T cells from aged mice contain less PFP on a per cell basis and mediate greatly reduced cytotoxicity compared to those from young mice, i.e., the average PFP content per cell is reduced among effector cell populations exhibiting reduced lytic activity. This suggests the notion that the distribution of PFP and/or PFP-containing granules among the potential lytic effector cells may be important for conferring lytic activity. It is not possible to gather from our data whether this might mean that a threshold level of PFP would be required to inflict lethal damage, or whether decreased levels of PFP might reduce the ability of CTL to recycle and lyse additional target cells.

TABLE 1. Estimate of acidophilic granules in T-cell subsets of MLC-stimulated cells by quinacrine uptake suggests reduced granularity of CTL from aged compared to young mice[a]

Age	Quinacrine[+] Proportion[b] of CD8[+] T cells[c]	Quinacrine[+] Proportion of CD4[+] T cells[c]
Young	55 ± 1[d]	34 ± 1[e]
Old	46 ± 3	35 ± 3

[a] Data are means of three individual old and three individual young mice.
[b] Percentage of quinacrine[+] cells within the CD8[+] or CD4[+] population.
[c] Percentage of CD8 or CD4 cells among all viable cells recovered after MLC stimulation of splenic T cells did not differ between young and aged mice, $p > 0.05$, by t test.
[d] Young significantly different from old, $p < 0.05$.
[e] Young and old not significantly different, $p > 0.05$.

However, as reviewed above, although the results of early studies suggested that the ability of CTL from aged mice to recycle may not be comprised, these studies also demonstrated no recycling in CTL from young mice (Menon et al., 1974), and therefore may be open to question in the light of more recent knowledge. Also, the capacity for recycling may be related to the capability for *de novo* production of lytic mediators and formation of new granules.

It is noteworthy that the demonstration that PFP expression is reduced with advanced age on a per cell basis also supports our earlier demonstration that the proportion of killer cells is reduced with aging while the frequency of cells able to bind target cells remains unaffected (Bloom et al., 1988). These findings support the existence of stimulated T cells that have undergone at least partial differentiation toward becoming active CTL, thereby suggesting that not all effector cells from aged individuals can maintain the level of activity of those from young. Therefore, in view of information that can be obtained by examining single effector cells, these data confirm the need to reinterpret or modify the "mosaic" theory if aging, in terms of levels of differentiation.

References

Asano Y, Komuro T, Kuko M, Sano K, Tada T (1990): Age-related degeneracy of T cell repertoire: Influence of the aged environment on T cell allorecognition. *Gerontology* 36 (Suppl 1): 3–9

Bach M-A (1979): Influence of aging on T-cell subpopulations involved in the in vitro generation of allogeneic cytotoxicity. *Clin Immunol Immunopathol* 13: 220–230

Bender BS, Johnson MP, Small PA (1991): Influenza in senescent mice: Impared cytotoxic T-lymphocyte activity is correlated with prolonged infection. *Immunology* 72: 514–519

Bloom ET (1991): Functional importance of CD4[+] and CD8[+] cells in CTL activity and associated gene expression. Impact on the age-related decline in lytic activity. *Eur J Immunol* 21: 1013–1017

Bloom ET, Korn EL (1983): Quantification of natural cytotoxicity by human lymphocyte subpopulations isolated by density: Heterogeneity of the effector cells. *J Immunol Methods* 58: 323–335

Bloom ET, Kubota LF (1989): Effect of IL-2 in vitro on CTL generation in spleen cells of young and old mice: Asialo GM1[+] cells are required for the apparent restoration of the CTL response. *Cell Immunol* 119: 73–84

Bloom ET, Crim JA, Siegel JP (1989): Evaluation of influenza A virus-specific cell mediated immune responses in the elderly. *Clin Res* 37: 406A

Bloom ET, Kubota LF, Kawakami K (1988): Age-related decline in the lethal hit but not the binding stage of cytotoxic T-cell activity in mice. *Cell Immunol* 114: 440–446

Bloom ET, Umehara, H, Bleackley RC, Okumura K, Mostowski H, Babbit JT (1990): Age-related decrement in CTL activity is associated with decreased levels of mRNA encoded by two CTL-associated serine esterase genes and the perforin gene in mice. *Eur J Immunol* 20: 2309–2316

Bubeník J, Cinader B, Indrová M, Koh SW, Chou C-T (1987): Lymphokine-activated killer (LAK) cells: I. Age-dependent decline of LAK cell-mediated cytotoxicity. *Immunol Lett* 16: 113–119

Chang M-P, Gorczynski RM (1984): Peripheral (somatic) expansion of the murine cytotoxic T lymphocyte repertoire. I. Analysis of diversity in recognition repertoire of alloreactive T cells derived from the thymus and spleen of adult or aged DBA/2J mice. *J Immunol* 133: 2375–2380

Chopra RK, Nagel JE, Chrest FJ, Adler WH (1987): Impaired phorbol ester and calcium ionophore induced proliferation of T cells from old humans. *Clin Exp Immunol* 70: 456–462

Dunn PL, North RJ (1991): Effect of advanced aging on ability of mice to cause regression of an immunogenic lymphoma in response to immunotherapy based on depletion of suppressor T cells. *Cancer Immunol Immunother* 33: 421–423

Effros RB, Walford RL (1983a): Diminished T-cell response to influenza virus in aged mice. *Immunology* 49: 387–392

Effros RB, Walford RL (1983b): The immune response of aged mice to influenza: Diminished T-cell proliferation, interleukin 2 production and cytotoxicity. *Cell Immunol* 81: 298–305

Effros RB, Walford RL (1984): The effect of age on the antigen-presenting mechanisms in limiting dilution precursor cell frequency analysis. *Cell Immunol* 88: 531–539

Ernst DN, Hobbs MV, Torbett BE, Glasebrook A, Rehse

MA, Bottomly K, Hayakawa K, Hardy RR, Weigle WO (1990): Differences in the expression profiles of CD45RB, Pgp-1, and 3G11 membrane antigens and in the patterns of lymphokine secretion by splenic CD4+ T cells from young and aged mice. *J Immunol* 145: 1295-1302

Fong TC, Makinodan T (1990): Preferential enhancement by 2-mercaptoethanol of IL-2 responsiveness of T blast cells from old over young mice is associated with potentiated protein kinase C translocation. *Immunol Lett* 20; 149-154

Goodman SA, Makinodan T (1975): Effect of age on cell-mediated immunity in long-lived mice. *Clin Exp Immunol* 19: 533-542

Gorczynski RM, Chang M-P (1984): Peripheral (somatic) expansion of the murine cytotoxic T lymphocyte repertoire. II. Comparison of diversity in recognition repertoire of alloreactive T cells in spleen and thymus of young or aged DBA/2J mice transplanted with bone marrow from young or aged donors. *J Immunol* 133: 2381-2389

Gottesman SRS, Edington J (1990) Proliferative and cytotoxic immune functions in aging mice: V. Deficiency in generation of cytotoxic cells with normal lytic function per cell as demonstrated by the single cell conjugation assay. *Aging: Immunol Infec Dis* 2: 19-29

Gottesman SRS, Edington JM, Thorbecke GJ (1988): Proliferation and cytotoxic immune functions in aging mice. IV. Effects of suppressor cell populations from aged and young mice. *J Immunol* 140: 1783-1790

Gottesman SRS, Walford RL, Thorbecke GJ (1984): Proliferative and cytotoxic immune functions in aging. II. Decreased generation of specific suppressor cells in alloreactive cultures. *J Immunol* 133: 1782-1787

Gottesman SRS, Walford RL, Thorbecke GJ (1985): Proliferative and cytotoxic immune functions in aging mice. III. Exogenous interleukin-2 rich supernatnat only partially restores alloreactivity in vitro. *Mech Ageing Dev* 31: 103-113

Grossmann A, Maggio-Price L, Jinneman JC, Rabinovitch PS (1991): Influence of aging on intracellular free calcium and proliferation of mouse T-cell subsets from various lymphoid organs. *Cell Immunol* 135: 118-131

Helfrich BA, Segre M, Segre D (1989): Age-related changes in the degeneracy of the mouse T-cell repertoire. *Cell Immunol* 118: 1-9

Henkart PA (1985): Mechanisms of lymphocyte-mediated cytotoxicity. *Annu Rev Immunol* 3: 31-58

Hirano T, Nordin AA (1976): Age-associated decline in the in vitro development of cytotoxic lymphocytes in NZB mice. *J Immunol* 117: 1093-1098

Hirokawa K, Utsuyama M, Kasai M (1990): Role of the thymus in aging of the immune system. In: *Biomedical Advances in Aging*, Goldstein AL, ed. New York, Plenum, pp. 375-384

Ho S-P, Kramer KE, Ershler WB (1990): Effect of host age upon interleukin-2-mediated anti-tumor responses in a murine fibrosarcoma model. *Cancer Immunol Immunother* 31: 146-150

Jenski LJ (1990): Accessory molecules and antigen requirements for young and aging cytotoxic lymphocytes. *Mech Aging Dev* 55: 107-122

Kawakami K, Bloom ET (1987): Lymphokine-activated killer cells and aging in mice: Significance for defining the precursor cell. *Mech Ageing Dev* 41: 229-240

Kawakami K, Bloom ET (1988): Lymphokine-activated killer cells derived from murine bone marrow: Age-associated difference in precursor cell populations demonstrated by response to interferon. *Cell Immunol* 116: 163-171

Kawasaki A, Shinkai Y, Kuwana Y, Furuya A, Iigo Y, Itoh S, Yagita H, Okumura K (1990): Perforin, a pore-forming protein detectable by monoclonal antibodies, is a functional marker for killer cells. *Int Immunol* 2: 677-684

Kishimoto S, Shigemoto S, Yamamura Y (1973): Immune response in aged mice. Change of cell-mediated immunity with aging. *Transplantation* 15: 455-459

Kolber MA, Henkart PA (1988): Quantitation of secretion by rat basophilic leukemia cells by measurements of quinacrine uptake. *BBA* 939: 459-466

Komuro T, Sano K, Asano Y, Tada T (1990): Analysis of age-related degeneracy of T-cell repertoire—localized functional failure in CD8+ T-cells. *Scand J Immunol* 32: 545-553

Kruisbeek AM, Steinmeier FA (1980): Alloreactive cytotoxic T lymphocytes from aged mice express increased lysis of autologous and third-party target cells. *J Immunol* 125: 858-864

Menon M, Jaroslow BN, Koesterer R (1974): The decline of cell-mediated immunity in aging mice. *J Gerontol* 29: 499-505

Miller RA (1984): Age associated decline inprecursor frequency for different T cell-mediated reactions, with preservation of helper or cytotoxic effect per precursor cell. *J Immunol* 132: 63-68

Miller RA (1990): Aging and the immune response. In: *Handbook of the Biology of Aging*, Schneider EL, Rowe JW, eds. San Diego: Academic Press, pp. 157-180

Miller RA, Jacobson B, Weil G, Simons ER (1987): Diminished calcium influx in lectin-stimulated T cells from old mice. *J Cell Physiol* 132: 337-342

Miller RA, Philosophe B, Ginis I, Weil G, Jacobson B (1989): Defective control of cytoplasmic calcium concentration in T lymphocytes from old mice. *J Cell Physiol* 138: 175-182

Miller RA, Stutman O (1981): Decline in aging mice, of the anti-2,4,6-trinitrophenyl (TNP) cytotoxic T cell response attributable to loss of Lyt 2−, interleukin 2-producing helper cell function. *Eur J Immunol* 11: 751-756

Nagelkerken L, Hertogh-Huijbregts A, Dobber R, Dräger A (1991): Age-related changes in lymphokine production related to a decreased number of CD45RB^{hi} CD4+ T cells. *Eur J Immunol* 21: 273-281

Nordin AA, Collins GD (1983): Limiting dilution analysis of alloreactive cytotoxic precursor cells in aging mice. *J Immunol* 131: 2215-2218

Ojcius DM, Zheng LM, Sphicas EC, Zychlinsky A,

Young JD-E (1991): Subcellular localization of perforin and serine esterase in lymphokine-activated killer cells and cytotoxic T cells by immunogold labeling. *J Immunol* 146: 4427–4432

Peters PJ, Borst J, Oorschot V, Fukuda M, Krähenbühl O, Tschopp J, Slot JW, Geuze JH (1991): Cytotoxic T lymphocyte granules are secretory lysosomes, containing both perforin and granzymes. *J Exp Med* 173: 1099–1109

Philosophe B, Miller RA (1989): T lymphocyte heterogeneity in old and young mice: Functional defects in T cells selected for poor calcium signal generation. *Eur J Immunol* 19: 695–699

Philosophe B, Miller RA (1990): Diminished calcium signal generation in subsets of T lymphocytes that predominate in old mice. *J Gerontol Biol Sci* 45: B87–B93

Pirruccelo SJ, Collins M, Wilson JE, McManus BM (1989): Age-related changes in naive and memory CD4+ T cells in healthy human children. *Clin Immunol Immunopathol* 52: 341–345

Proust JJ, Filburn CR, Harrison SA, Buchholz MA, Nordin AA (1987): Age-related defect in signal transduction during lectin activation of murine T lymphocytes. *J Immunol* 139: 1472–1478

Rosenberg AS, Sechler JMG, Horvath JA, Maniero TG, Bloom ET: Assessment of alloreactive T cell subpopulations of aged mice *in vivo*. CD4+ but not CD8+ T cell mediated responses decline with advanced age. Submitted.

Seth A, Nagarkatti M, Nagarkatti PS, Subbarao B, Udhayakumar V (1990): Macrophages but not B cells from aged mice are defective in stimulating autoreactive T cells *in vitro*. *Mech Ageing Dev* 52: 107–124

Sharp A, Kukulansky T, Malkinson Y, Globerson A (1990): The bone marrow as an effector T cell organ in aging. *Mech Ageing Dev* 52: 219–233

Shigemoto S, Kishimoto S, Yamamura Y (1975): Change of cell mediated cytotoxicity with aging. *J Immunol* 115: 307–309

Thoman ML, Weigle WO (1982): Cell-mediated immunity in aged mice: An underlying lesion in IL 2 synthesis. *J Immunol* 128: 2358–2361

Thoman ML, Weigle WO (1985): Reconstitution of *in vivo* cell-mediated lympholysis responses in aged mice with interleukin 2. *J Immunol* 134: 949–952

Thoman ML, Weigle WO (1989): The cellular and subcellular bases of immunosenescence. *Adv Immunol* 46: 221–261

Tschopp J, Nabholz M (1991): Perforin-mediated target cell lysis by cytolytic T lymphocytes. *Annu Rev Immunol* 8: 279–302

Utsuyama M, Hirokawa K (1987): Age-related changes of splenic T cells in mice—A flow cytometric analysis. *Mech Ageing Dev* 40: 89–102

Wegmann KW, McMaster JS, Green WR (1991): Mechanism of nonresponsiveness to AKR/Gross leukemia virus in AKR.H-2^b:Fv-1^b mice. An analysis of precursor cytotoxic T lymphocyte frequencies in young versus moderately aged mice. *J Immunol* 146: 2469–2477

Weksler ME, Hutteroth TH (1974): Impaired lymphocyte function in aged humans. *J Clin Invest* 53: 99

Young JD-E, Cohn ZA (1988): How killer cells kill. *Sci Am* (Jan): 38–44

Young JD-E, Liu C-C, Persechini PM, Cohn ZA (1988): Perforin-dependent and -independent pathways of cytotoxicity mediated by lymphocytes. *Immunol Rev* 103: 161–202

Zharhary D, Gershon H (1981): Allogeneic T cytotoxic reactivity of senescent mice: Affinity for target cells and determination of cell number. *Cell Immunol* 60: 470–479

Zharhary D, Segev Y, Gershon HE (1984): T-cell cytotoxicity and aging: Differing causes of reduced response in individual mice. *Mech Ageing Dev* 25: 129–140

39
Bone Marrow Graft Rejection as a Function of T_{NK} Cells

Gunther Dennert

The Phenomenon of Acute Bone Marrow Graft Rejection

Almost three decades ago, G. Cudkowicz (Cudkowicz and Cosgrove, 1961) made the curious observation that irradiated (A × B)F1 hybrid mice transplanted with parental bone marrow from either parent A or parent B may acutely reject one but not the other graft. The observation was unexpected, as transplantation antigens are known to be inherited codominantly (Snell, 1953) and therefore parental marrow grafts should not be subject to immunological rejection mechanisms. He subsequently postulated that the antigenic determinants recognized during the rejection are expressed on parental and not on F1 hybrid cells and hence must be inherited noncodominantly (Cudkowicz and Stimpfling, 1964). The antigens were named hemopoietic histocompatibility (Hh) antigens (Cudkowicz, 1971; Cudkowicz and Lotzova, 1973) and genetic studies revealed that they often encoded close to the major histocompatibility complex (MHC) region. For example, the resistance of (C57BL/6 × C3H)F1 mice to C57BL/6 marrow was mapped to the D region of the MHC (Cudkowicz and Stimpfling, 1964b). Perhaps not surprisingly, the phenomenon of acute parental marrow graft rejection in F1 hybrid mice, also called hybrid resistance (HR), found its counterpart in the ability of irradiated mice to acutely reject allogeneic marrow grafts (Cudkowicz and Bennett, 1971a). Again, the specificity of acute allogeneic marrow graft rejection pointed to antigenic determinants that were encoded close to the MHC region. As examples of this, the rejection of $H-2^d$ marrow by C57BL/6 mice maps to the D region of the MHC (Cudkowicz and Stimpfling, 1964b; Cudkowicz, 1975a), and the rejection of $H-2^k$ marrow by strain 129 mice maps to the K region (Cudkowicz and Warner, 1979). The two mouse strains 129 and C57BL/6 which are of $H-2^b$ haplotype, provide an interesting experimental model because they show opposing responsiveness to $H-2^d$ and $H-2^k$ marrow grafts (Cudkowicz, 1975a; Cudkowicz and Warner, 1979). C57BL/6 fails to acutely reject $H-2^k$ marrow, and 129 is a nonresponder for $H-2^d$ marrow.

The NK Hypothesis of Acute Marrow Graft Rejection

Inasmuch as the noncodominant expression of antigens responsible for marrow graft rejection was surprising, the nature of the rejection mechanism itself appeared to be quite unconventional. The ability to reject marrow grafts matures at the age of about 3 weeks (Cudkowicz and Bennett, 1971a,b) and was found to be resistant to a single high dose of ionizing irradiation (Cudkowicz and Bennett, 1971a,b), suggesting that for the rejection to occur, no sensitization phase is necessary. The rejection is very rapid, i.e., 12–24 hr after transplantation, which would allow little time for sensitization. The logical conclusion from these observations was that common T-cell mediated mechanisms, responsible for the rejection of solid tissues, are not involved. In direct support of this, it was found that athymic nude mice (Cudkowicz, 1975b) or thymectomized mice (Cudkowicz and Bennett, 1971a) are able to acutely reject marrow grafts.

The question as to which type of effector cell may be causing acute marrow graft rejection appeared to be solved by the suggestion of Kiessling et al. (1977) that natural killer (NK) cells are involved.

At about that time an interesting new mouse mutant was described, the beige mouse. This mouse lacks NK activity (Roder and Duwe, 1979) and therefore presented itself as an animal model in which to study marrow graft rejection or the lack thereof. The results indeed showed that beige mice are not able to acutely reject marrow grafts (Kaminsky and Cudkowicz, 1980) and that normal mice depleted of NK cells by antibodies specific for the antigens ASGM1 and NK1 also lose their ability to reject marrow grafts acutely (Okumura et al., 1982; Lotzova et al., 1983). Further support for the notion that NK cells are the responsible effectors came from experiments in which cell clones with NK phenotype (Dennert, 1980) were shown to reconstitute NK-deficient mice to reject marrow grafts acutely (Warner and Dennert, 1982).

The NK hypothesis of acute marrow graft rejection was not without its problems, however. Doubts in our laboratory about the involvement of NK cells arose from experiments with the above-mentioned cloned cells with NK activity and phenotype (Dennert, 1980). These cells, when injected into NK-deficient mice, reconstitute specific responsiveness (Warner and Dennert, 1982), contrary to the fact that their *in vitro* cytolytic activity is not specific for Hh antigens. Another interesting observation contradicting the involvement of NK cells was that bone marrow graft rejection can be competitively inhibited by tumor cells syngeneic to the marrow graft, but not by cells that are good targets for NK cells (Daley and Nakamura, 1984). This strongly supported the contention that a specific recognition process must take place, a conclusion already drawn from the genetic specificity of the rejection. It therefore became evident that any hypothesis of acute marrow graft rejection involving NK cells as effectors had to explain how these cells are able to specifically recognize a marrow graft.

There are, of course, several ways that NK cells could specifically recognize bone marrow. For one, they may possess specific receptors not yet identified or they could cooperate with other cells expressing specific receptors. Alternatively, they could make use of receptors acquired from other cells. In pursuit of these possibilities, we focused our attention on the well-documented fact that the NK cells possess Fc receptors and are able to perform antibody-dependent cell-mediated cytotoxicity (ADCC) *in vitro* (Pollack and Kraft, 1977). We hypothesized that NK cells utilize antibody in the specific rejection of marrow grafts (Figure 1), in which case responder mice should possess the predicted antibody, whereas nonresponders should not. The prediction was tested in various strain combinations, and it was shown that transfer of serum from C57BL/6 responders into 129 nonresponder mice enables the latter to reject $H-2^d$ marrow (Warner and Dennert, 1985). Moreover, it was demonstrated that marrow graft rejection induced by serum transfer has the predicted specificity (D region), is caused by antibody, and involves a cell-mediated mechanism (Warner and Dennert, 1985).

An important prediction from the antibody hypothesis was that specific antibody should be detectable not only in mice able to reject allogeneic marrow but also in F1 hybrid mice able to reject parental grafts. Repeated attempts in our laboratory to demonstrate in F1 hybrid mice antibody able to induce parental marrow graft rejection were unsuccessful (Warner and Dennert, 1985 and unpublished results), although others have reported the tentative identification of such antibody (Cibotti et al., 1989). What also emerged from our experiments was that antibody-mediated rejection of marrow grafts is a relatively weak mechanism that cannot fully account for the strength of either allogeneic or parental marrow graft rejection (Dennert et al., 1986). It therefore became apparent that there are still major gaps in our understanding of the mechanisms involved in acute marrow graft rejection, and it seemed prudent then to approach the question as to what type of cell is responsible in a more fundamental fashion.

T_{NK} Cells as Major Effectors in Bone Marrow Graft Rejection

It is well documented that (C57BL/6 × C3H)F1 mice reject parental C57BL/6 and allogeneic BALB/c marrow grafts and that mice treated with a high dose of cyclophosphamide (Cudkowicz and Bennett, 1971a) lose the ability to reject either graft. This provided an assay system to identify cells able to cause rejection in an adoptive cell transfer system. Using this *in vivo* cell transfer system, we demonstrated that nylon wool nonadherent spleen cells from responder mice induce specific marrow graft rejection in nonresponder recipients (Yankelevich et al., 1989). Treatment of cells prior to injection with specific antibody and complement revealed the tentative phenotype of these cells as $NK1^+$ $ASGM1^+$ $CD4^-$ $CD8^-$ $Thy1^-$ $CD3^+$ TCR^+ $J11d2^-$ $Pgp1^+$. In support of this, purification of $NK1^+$ $CD3^+$ $CD4^-$ $CD8^-$ cells by fluorescence-activated cell sorting enabled transfer of responsiveness into nonresponder recipients

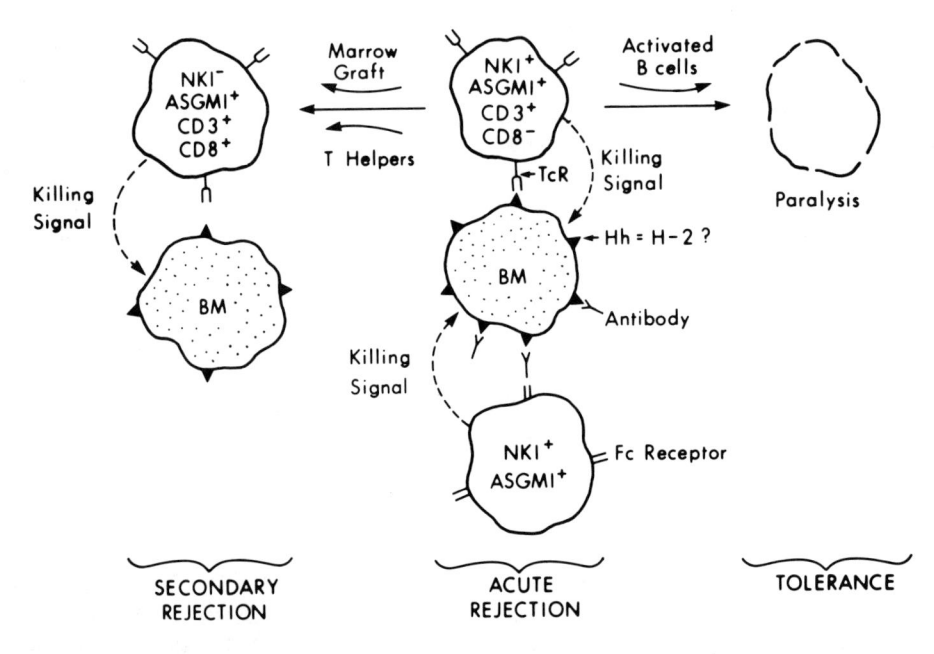

FIGURE 1. Effector mechanisms responsible for marrow graft rejection. NK1$^+$ TCR$^-$ effector cells possessing Fc receptors cause acute marrow graft rejection in some model systems. What the relationship of these cells if any to NK1$^+$ CD3$^+$ cells might be is not clear. NK1$^+$ CD3$^+$ cells can be paralyzed by a high dose of activated B cells. However, in response to a marrow graft they differentiate into NK1$^-$ CD8$^+$ cytotoxic T cells a reaction that apparently requires the presence of T helper cells. The NK1$^-$ CD8$^+$ cells are responsible for the delayed and secondary rejection of marrow grafts.

(Yankelevich et al., 1989). It therefore became apparent that acute marrow graft rejection is, to a major extent, due to cells with NK phenotype that express TCR (Figure 1).

This result raised the next question: What are the antigens that they recognize, and what is the fate of these cells once they are stimulated? Because the cells responsible for the rejection appear to be relatively radiation-resistant, it seemed reasonable to assay irradiated mice that had been transplanted with allogeneic marrow for the presence of these cells. We observed that irradiated C57BL/6 mice transplanted with H-2^d marrow generate cytotoxic effector cells specific for the graft. The phenotype of these cells was determined to be ASGM1$^+$ NK1$^-$ CD8$^+$ Thy1$^+$, i.e., they are cytotoxic T cells (CTL) (Figure 1) expressing the NK cell marker ASGM1 (Dennert et al., 1990). Assay of the cytolytic specificity of these cells revealed that they are H-2D^d-specific, i.e., they express a specificity similar to that of the acute rejection mechanism. This, as well as the fact that they are ASGM1$^+$, gave rise to the hypothesis that the ASGM1$^+$ NK1$^+$ CD8$^-$ Thy1$^-$ CD3$^+$ cells responsible for acute marrow graft rejection differentiate into ASGM1$^+$ NK1$^-$ CD8$^+$ Thy1$^+$ CD3$^+$ CTL. Support for this possibility came from the demonstration that NK1$^+$ CD3$^+$ cells purified by fluorescence-activated cell sorting and adoptively transferred into recipient mice differentiate into NK1$^-$ CD8$^+$ CTL in response to an allogeneic marrow graft (Dennert et al., 1990).

It is noteworthy to mention at this point that the presence of proliferating T cells of recipient origin had been demonstrated in irradiated mice transplanted with allogeneic bone marrow (Melchner and Bartlett, 1983). Moreover, the participation of these cells in the later stages of the rejection process had been suggested (Dennert et al., 1985). It was also known that *in vitro* stimulation of F1 hybrid spleen cells with parental stimulator cells induces CTL with antiparental specificity (Nakano et al., 1981). Therefore the participation of CTL in marrow graft rejection was well documented. However, it was thought at that time that NK- and CTL-mediated effects are independent events, whereas now it appears that the cells responsible for acute marrow graft rejection differentiate into CTL during the process of marrow graft rejection (Figure 1).

Recognition Specificity of T_{NK} Cells

The identification of CTL as descendants of the NK1$^+$ CD3$^+$ cells that are responsible for the acute rejection of marrow grafts raises the important question as to what the antigenic determinants might be that are recognized by these cells. It is well documented that the T-cell receptor (TCR) on conventional T cells is restricted to interaction with MHC antigens. Therefore one would predict that Hh antigens should be related to MHC antigens, and recent observations seem to indicate that this may be the case.

One of these observations arose during experiments in which we attempted to induce tolerance to parental marrow grafts in F1 hybrid mice. In these experiments it was observed that injection into F1 hybrid recipients of a very high dose of irradiated, lipopolysaccharide (LPS)-activated, parental B cells leads to induction of specific tolerance to a subsequent parental marrow graft (Figure 1). Unresponsiveness was found to be due to functional elimination of the cells responsible for HR (Nowicki et al., 1990). A surprising and unexpected observation was that tolerance to the parental marrow graft was not only inducible by parental cells but also by syngeneic (!) F1 hybrid cells (Nowicki et al., 1990). This result suggested that the antigenic determinants recognized by the HR mechanism are essentially self antigens and that the HR mechanism is in fact autoreactive (Nowicki et al., 1990). In further support of this interpretation, it was demonstrated that the HR mechanism is able to eliminate syngeneic stem cells *in vivo* (Nowicki et al., 1990).

Demonstration of the autoreactivity of the HR mechanism is of importance for several reasons. For one, it seriously questions the concept of non-codominant inheritance of Hh antigens and thereby opens the possibility that Hh antigens are in fact related to MHC antigens. In support of this are results in which it was shown that the CTL generated in irradiated C57BL/6 mice transplanted with H-2^d marrow can be H-2L^d-specific (Nowicki et al., 1990). Moreover, H-2L^d has been suggested to be one of the important antigens recognized in acute marrow graft rejection (Morgan and McKenzie, 1981). But perhaps the most convincing evidence for identity of Hh and MHC antigens was recently provided by Öhlen et al. (1989), who show that bone marrow from C57BL/6 mice carrying the H-2D^d transgene is rejected by C57BL/6 recipients. This suggests that MHC class I antigens can serve as determinants in acute marrow graft rejection.

Why Are T_{NK} Cells Autoreactive and What Is Their Physiological Function?

An important conclusion from our suggestion that the T_{NK} cells responsible for acute marrow graft rejection are MHC-restricted is that hybrid and perhaps also alloresistance are due to an autoreactive immune mechanism. This raises the question as to why these cells are autoreactive and what their physiological function might be. The observation that the HR mechanism has a down-regulatory effect on hemopoiesis (Nowicki et al., 1990) may suggest that it has a control function in hemopoiesis and perhaps immune responses. The previous observation that mice with genetic predisposition to develop autoimmunity tend to generate an abnormally high number of colony-forming unit–spleen (CFU–S) after sublethal irradiation and that there is a strong association between increase in the number of endogenous CFU–S and development of generalized autoimmunity (Scribner and Steinberg, 1988) would be consistent with this possibility.

It is interesting to note here that nude mice are able to specifically reject allogeneic marrow grafts in an acute response (Kaminsky and Cudkowicz, 1980) and hence should possess T_{NK} cells. It therefore appears that the development of T_{NK} cells does not require thymic selection of TCR specificities.

Conclusions

There is overwhelming evidence that cells expressing the NK1$^+$ ASGM1$^+$ phenotype are the main effectors responsible for the acute rejection of allogeneic and parental marrow grafts in mice. In at least some models of acute allogeneic marrow graft rejection, antibody plays a role, presumably by providing a specific recognition structure for Fc receptor-bearing NK effector cells (Figure 1). While it had originally been hypothesized that antibody-mediated graft rejection is the main mechanism by which marrow transplants are rejected, it is now clear that NK1$^+$ ASGM1$^+$ CD3$^+$ cells are the important effectors responsible for the rejection of both allogeneic and parental marrow grafts. The fact that this cell possesses TCR would predict that it recognizes MHC antigens on the grafted marrow cells, for which some evidence has been presented. Expression of these antigens not only on parental but also on F1 hybrid cells and the observation that H-2 transgenes code for antigens that cause acute marrow graft rejection seriously questions the existence of noncodominantly inherited antigens, i.e., Hh antigens, as responsible for the rejection.

Antigenic stimulation of the $NK1^+$ $ASGM1^+$ $CD3^+$ cell, we have shown, leads to two effects: its paralysis, i.e., induction of tolerance, or its differentiation into a CTL. Paralysis appears to require injection of large numbers of activated parental B cells. In contrast, stimulation of the cell by a bone marrow graft induces its differentiation into CTL. It seems likely that this latter pathway requires the participation of T-cell help, as nude mice are able to acutely reject marrow grafts yet fail to generate CTL. It is intriguing that during differentiation of $NK1^+$ effector cells into CTL, the NK1 antigen is lost, the ASGM1 antigen is preserved, and CD8 appears on these cells (Figure 1). This, of course, shows that both NK1 and ASGM1 are CTL differentiation antigens and are not unique to NK cells lacking TCR.

The finding that cells with TCR are responsible for marrow graft rejection raises two interesting questions for future experimentation. One is how the TCR repertoire of T_{NK} cells is generated, as its selection apparently does not require a functional thymus. The other is what the physiological function of these cells might be. A plausible hypothesis can be based on the observation that the HR mechanism is autoreactive. One could therefore postulate that these cells are endowed with important down-regulatory functions in hemopoiesis and immunity.

Acknowledgments

This work was supported by National Institutes of Health grants CA 39623 and CA 37706.

References

Cibotti R, Churaqui E, Bolonaki-Tsilivakos D, Scott-Algara O, Halle P, Kosmatopoulos K (1989): Specific inhibition of hybrid resistance in F1 hybrid mice pretreated with parent-strain spleen cells. *J Immunol* 143: 3484–3491

Cudkowicz G (1971): Genetic control of bone marrow graft rejection. I. Determinant-specific difference of reactivity in two pairs of inbred mouse strains. *J Exp Med* 134: 281–293

Cudkowicz G (1975a): Genetic control of resistance to allogeneic and xenogenic bone marrow grafts in mice. *Transplant Proc* 7: 155–159

Cudkowicz G (1975b): Rejection of bone marrow allografts by irradiated athymic nude mice. *Proc Am Assoc Cancer Res* 16: 170–177

Cudkowicz G, Bennett M (1971a): Peculiar immunobiology of bone marrow allografts. I. Graft rejection by irradiated responder mice. *J Exp Med* 134: 83–102

Cudkowicz G, Bennett M (1971b): Peculiar immunobiology of bone marrow allografts. II. Rejection of parental grafts by resistant F1 hybrid mice. *J Exp Med* 134: 1513–1528

Cudkowicz G, Cosgrove GE (1961): Modified homologous disease following transplantation of parental bone marrow and recipient liver into irradiated F1 mice. *Transplant Bull* 27: 90–94

Cudkowicz G, Lotzova E (1973): Hemopoietic cell-defined components of the major histocompatibility complex of mice: Identification of responsive and unresponsive recipients for bone marrow transplants. *Transplant Proc* 5: 1399–1405

Cudkowicz G, Stimpfling JH (1964a): Deficient growth of C57BL marrow cells transplanted in F1 hybrid mice. Association with the histocompatibility-2 locus. *Immunology* 7: 291–306

Cudkowicz G, Stimpfling JH (1964b): Lack of expression of parental isoantigen(s) in F1 hybrid mice. *Immunology* 7: 291–306

Cudkowicz G, Warner JF (1979): Natural resistance of irradiated 129-strain mice to bone marrow allografts: Genetic control by the H-2 region. *Immunogenetics* 8: 13–26

Daley JP, Nakamura I (1984): Natural resistance of lethally irradiated F1 hybrid mice to parental marrow grafts is a function of H2/Hh-restricted effectors. *J Exp Med* 159: 1132–1148

Dennert G (1980): Cloned lines of natural killer cells. *Nature (Lond)* 287: 47–49

Dennert G, Anderson CG, Warner JF (1985): T killer cells play a role in allogeneic bone marrow graft rejection but not in hybrid resistance. *J Immunol* 135: 3729–3734

Dennert G, Anderson CG, Warner J (1986): Induction of bone marrow allograft rejection and hybrid resistance in nonresponder recipients by antibody: Is there evidence for a dual receptor interaction in acute marrow graft rejection. *J Immunol* 136: 3981–3986

Dennert G, Knobloch C, Sugawara S, Yankelevich B (1990): Evidence for differentiation of $NK1^+$ cells into cytotoxic T cells during acute rejection of allogeneic marrow grafts. *Immunogenetics* 31: 161–168

Kaminsky S, Cudkowicz G (1980): Immune response to dextran isomaltosyl oligosaccharide-protein conjugates in CBA/N mice. *Fed Proc* 39: 466

Kiessling R, Hochman PS, Haller O, Shearer GM, Wigzell H, Cudkowicz G (1977): Evidence for a similar or common mechanism for natural killer cell activity and resistance to hemopoietic grafts. *Eur J Immunol* 7: 655–663

Lotsova E, Savary CA, Pollack SB (1983): Prevention of

rejection of allogeneic bone marrow transplants by NK1.1 antiserum. *Transplant* 35: 490–494

Melchner HV, Bartlett PV (1983): Mechanisms of early allogeneic marrow graft rejection. *Immunol Rev* 71: 31–56

Morgan GM, McKenzie KPL (1981): Implication of the H-2L locus in hybrid histocompatibility (Hh-1). *Transplantation* 31: 417–422

Nakano K, Nakamura I, Cudkowicz G (1981): Generation of F1 hybrid cytotoxic T lymphocytes specific for self H-2. *Nature (Lond)* 298: 559–563

Nowicki M, Yankelevich B, Kikly K, Dennert G (1990): Induction of tolerance to parental marrow grafts in F1 hybrid mice. Evidence for recognition of self-antigens. *J Immunol* 144: 47–52

Öhlen C, Kling G, Höglund P, Hansson M, Scangos G, Biebrich C, Jay G, Kärre K (1989): Prevention of allogeneic bone marrow graft rejection by H-2 transgene in donor mice. *Science* 246: 666–670

Okumura K, Habu S, Shimamura K (1982): The role of asialo GM1+ cells in the resistance to transplants of bone marrow or other tissues. In: *NK Cells and Other Natural Effector Cells*, Herberman RB, ed. New York: Academic Press

Pollack SB, Kraft DS (1977): Effector cells which mediate antibody-dependent cell-mediated cytotoxicity. *Cell Immunol* 34: 1–9

Roder JC, Duwe AK (1979): The beige mutation in the mouse selectively impairs natural killer cell function. *Nature (London)* 278: 451–453

Scribner CL, Steinberg AD (1988): The role of splenic colony-forming units in autoimmune disease. *Clin Immunol Immunopathol* 49: 133–142

Snell GD (1953): The genetics of transplantation. *J Natl Cancer Inst* 14: 691–700

Warner JF, Dennert G (1982): Effects of a cloned cell line with NK activity on bone marrow transplant, tumor development and metastasis *in vivo*. *Nature (Lond)* 300: 31–34

Warner JF, Dennert G (1985): Bone marrow graft rejection as a function of antibody-directed natural killer cells. *J Exp Med* 161: 563–576

Yankelevich B, Knobloch C, Nowicki M, Dennert G (1989): A novel cell type responsible for marrow graft rejection in mice. T cells with NK phenotype cause acute rejection of marrow grafts. *J Immunol* 142: 3423–3430

40
Class I MHC Antigens and the Control of Virus Infections by NK Cells

Raymond M. Welsh, Paul R. Rogers, and Randy R. Brutkiewicz

Due to their ability to lyse virus-infected cells before the release of mature virions, cytotoxic lymphocytes are among the most important host defense mechanisms against viral infections. The two major forms of cytotoxic lymphocytes are the natural killer (NK) cells and the cytotoxic T lymphocytes (CTL), which peak in response at early and late stages of infection, respectively. Mouse NK cells, defined here as $CD3^-$, TCR^-, $NK1.1^+$, asialo $GM1^+$ lymphocytes, are normally slowly dividing cells that lyse a limited range of "NK-sensitive" targets, but the interferon (IFN)-α/β induced during virus infection augments their cytolytic potential, enabling them to lyse most types of targets, and, most likely in concordance with other growth factors, stimulates their proliferation (Welsh, 1986). NK cells accumulate at sites of virus infection and may represent up to 20% of the infiltrating leukocyte population by 3 days postinfection (McIntyre and Welsh, 1986; Natuk and Welsh, 1987). Concomitantly, clones of CTL precursors recognizing viral peptides displayed on class I major histocompatibility complex (MHC) antigens proliferate and differentiate in response to IL-2 and IFN-γ and become detectable at later (day 6–10) stages of infection (Welsh, 1986; Townsend and Bodmer, 1989). At this later time the NK cell response diminishes, due both to a decrease in the levels of virus, which stimulate the IFN-α/β response, and due to suppressive effects of transforming growth factor-β produced predominantly by activated macrophages (Su et al., 1991).

The evidence demonstrating the antiviral properties of cytotoxic lymphocytes during infection is now substantial. The regulatory roles of NK cells in viral infection have most convincingly been established for murine cytomegalovirus (MCMV), whose replication is enhanced in mice having low NK cell activity as a result of young age, genetic background (e.g., beige), or depletion by immunochemical (e.g., anti NK1.1, anti asialo GM_1) or pharmacological (hydrocortisone, cyclophosphamide) agents (reviewed in Welsh, 1986; Welsh and Vargas-Cortes, 1992). Genetic resistance of mice to MCMV maps closely to the NK1.1 locus on chromosome 6 (Scalzo et al., 1990). Adoptive transfer of spleen NK cells or NK cells derived from lymphokine-activated killer (LAK) cell cultures protects suckling mice from MCMV (Bukowski et al., 1985, 1988). Antiviral roles for NK cells against MCMV have also been demonstrated in mice with severe combined immunodeficiency (SCID), indicating the effectiveness of NK cells in the complete absence of functional B and T cells (Welsh et al., 1991). NK cells may also be of great significance in human cytomegalovirus (HCMV) infection. A patient at the University of Massachusetts Medical Center with a complete and selective NK cell immunodeficiency has presented with exceptionally severe cases of HCMV, varicella zoster virus, and herpes simplex virus-1 (HSV-1) (Biron et al., 1989). Other studies have also correlated low NK cell responses with severe cases of HCMV (Quinnan et al., 1982) and HSV (Lopez et al., 1983).

NK Cell Recognition

The mechanism of NK cell recognition of, and triggering by, target cells is poorly understood, but recent developments have indicated that the presence on the target cells of class I MHC molecules identical to those on the NK cells confers some degree of resistance to NK cells (Ljunggren and Karre, 1990). Whether these MHC molecules obscure an NK cell target antigen or deliver a negative signal to the NK cell, causing it to leave its target before triggering, is not currrently known. Recent evidence indicates that NK cells are educated by bone marrow cells to be negatively regulated by self MHC, and therefore to recognize "absence of self" (Liao et al., 1991; Hoglund et al., 1991). A question not yet comprehensively addressed is whether the "absence of self" hypothesis is relevant to viral infections. The remainder of this review will focus on this topic.

Virus-Induced NK Cells Display "Absence of Self" Reactivity

The target specificities of NK cells taken from lymphocytic choriomeningitis virus (LCMV)-infected mice have been extensively analyzed (Welsh, 1978; Welsh et al., 1979; Kiessling and Welsh, 1980). It first should be noted that these NK cells are highly activated, with target cell ranges very comparable to those of NK1.1$^+$ "LAK" cells, and that *both* syngeneic and allogeneic targets can be lysed by them. However, our early work done in the 1970s showed a clear allogeneic preference in the lysis of freshly isolated peritoneal macrophages (Figure 1). This allogeneic preference was not observed against thymocyte targets (Kiessling and Welsh, 1980), but subsequent work showed that, in contrast to macrophages, thymocytes expressed very low levels of class I MHC antigens (Figure 2). A tenet of the "absence of self" hypothesis is that target cells must express a critical level of class I MHC antigens to be protected from NK cells (Ljunggren and Karre, 1990), and the high levels of lysis of class I-antigen-negative target cells by mouse NK cells was first shown against teratocarcinoma cells with LCMV-induced NK cells (Welsh, 1978). Thus, the high levels of lysis of class I antigen-negative teratocarcinoma cells, the allogeneic preference for high class I antigen-expressing macrophages, and the lack of allogeneic preferences for low class I antigen-expressing thymocytes are all consistent with the "absence of self" hypothesis.

IFN-Mediated Protection of Target Cells and Class I Antigen Expression

A paradox of the NK cell system is that IFN, in addition to activating the cytolytic and proliferative properties of NK cells, *protects* target cells from NK cell-mediated lysis (Trichieri and Santoli, 1978). This protection has elegantly been shown by Karre and co-workers to be related to the ability of IFN to enhance the expression of class I antigens on cells. Among the various biological properties of IFN is its ability to stimulate the transcription of messages for both the class I α chain and for β_2-microglobulin. Studies with a variety of class I Ag-negative cell lines indicated that IFN could not protect them from NK cells. However, IFN-induced protection could be restored to a β_2-microglobulin-negative cell line after transfection with a gene for β_2-microglobulin (Ljunggren et al., 1990; Ljunggren and Karre, 1990). Thus, by up-regulating the transcription of genes for class I antigens, IFN can simultaneously protect cells from NK cell-mediated lysis but render them more sensitive to CTL, which recognize these antigens (Bukowski and Welsh, 1985a).

IFN Protection and Class I Antigen Expression *In Vivo*

The normal host is exposed to only low levels of endogenous IFN production, and with the exception of the mature cells of the immune system, most cells in the body express very low levels of class I MHC antigens. The weak cytotoxic potential of endogenous NK cells may help prevent these low class I-expressing targets from being lysed. However, viral infections stimulate the production of very high levels of IFN, which, in addition to activating the NK cell response (Welsh, 1978), cause a major up-regulation of class I MHC antigens on tissues throughout the body (Bukowski and Welsh, 1986). Figure 2 depicts the frequency of high class I MHC antigen-expressing thymocytes at different days after LCMV infection. Of significance is that cells such as thymocytes taken from mice before virus infections are sensitive to lysis by NK cells but not to CTL, whereas cells removed from the body several days after infection are resistant to NK cells but are much more sensitive to allospecific CTL (Hanson et al., 1980; Bukowski and Welsh, 1986). Thus, *the IFN induced during a virus infection conditions the cells in the body to become resistant to NK cells as they become increasingly sensitive to CTL*. This may be of profound importance, because (1) without protection of the target cells, the highly activated NK cells would attack uninfected normal tissue; and (2) without up-regulation of class I MHC antigens on tissues normally expressing low levels, the CTL

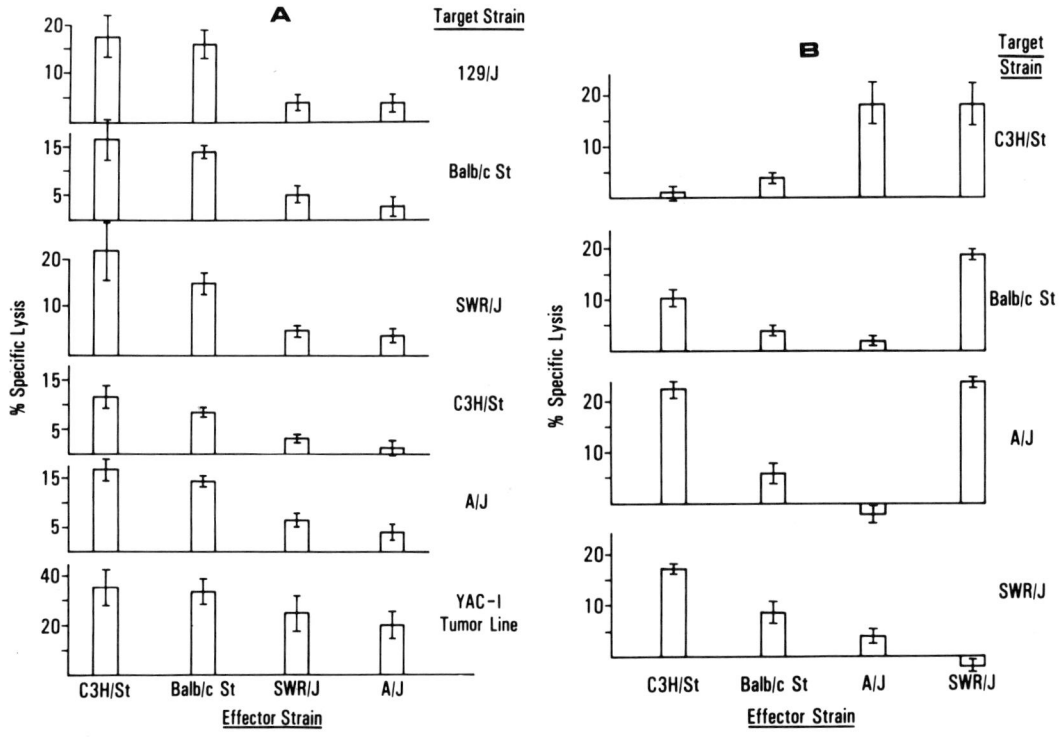

FIGURE 1. Influence of allogenicity on the sensitivity of thymocytes and peritoneal macrophages to LCMV-induced activated NK cells. Activated NK cells in spleen leukocyte populations taken from mice 3 days after LCMV infection were tested for cytotoxicity against thymocytes (A) at an effector-to-target ratio of 15:1 or against adherent peritoneal cells (macrophages) (B) at an effector-to-target ratio of 50:1. Thymocytes and macrophages were from uninfected mice. (Reprinted with permission of the *International Journal of Cancer* from Kiessling and Welsh, 1980.)

would not be able to clear the infection. Additional proof that IFN-mediated protection is a relevant *in vivo* finding comes from our studies showing that cells treated *in vitro* with IFN are rejected by NK cells from implanted virus-infected mice more slowly than are untreated cells (Welsh et al., 1981).

Means by Which Viral Alterations of Class I MHC Regulation Could Influence Susceptibility to NK Cells

It first should be stated that many factors are involved in the triggering of NK cells after binding to a target, and viral infections could influence many different stages of effector cell–target cell interaction. For instance, it has been demonstrated that certain viral glycoproteins can directly trigger NK cells (Harfast et al., 1980; Casali et al., 1981; Arora et al., 1984), and many cytopathic viruses can so alter the membrane integrity of target cells that they will not bind to NK cells (Welsh and Hallenbeck, 1980). Consequently, viruses can make cells more susceptible or more resistant to NK cells or not affect susceptibility at all. Here we will review only the work relating the virus-induced effects on class I MHC expression to surveillance by NK cells.

Global Inhibition in Protein Synthesis

The inhibition of class I antigen expression has been reported in infections with many viruses, including members of the adeno- (Andersson et al., 1985; Friedman and Ricciardi, 1988), retro- (Scheppler et al., 1989), rhabdo- (Hecht and Summers, 1972), and herpes- (Jennings et al., 1985; Mellencamp et al., 1991) virus groups. The most highly cytopathic viruses inhibit the expression of MHC antigens and many other host proteins. Virus infec-

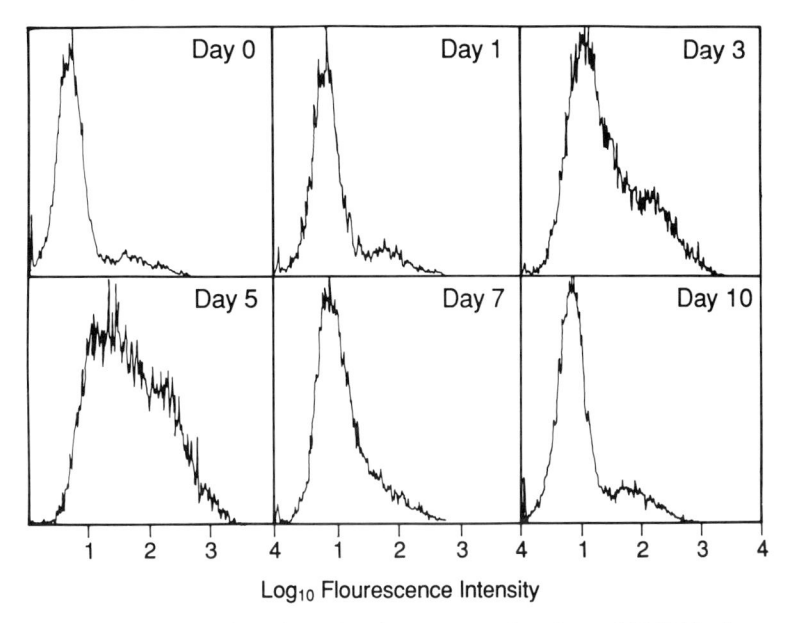

FIGURE 2. Expression of class I MHC antigens on thymocytes taken from C57BL/6 mice at different days after LCMV infection. Under these conditions of infection, peak IFN production is on days 2 and 3, and peak virus production is on days 4 and 5, after which it is rapidly cleared.

tions can by diverse mechanisms inhibit host protein synthesis at the levels of transcription or translation. Sometimes this reduction in class I antigen expression is associated with enhanced susceptibility to NK cell-mediated lysis, but this is often not the case, as the inhibition in synthesis of other proteins may impair NK cell recognition or triggering.

Prevention of IFN-Mediated Protection

Viral infections, by virtue of their abilities to interfere with cellular RNA and protein synthesis, can interfere with the function of IFN, which requires cellular RNA and protein synthesis to enhance MHC expression and to protect targets from NK cells. Thus, a virus infection could prevent a target cell from being protected by IFN, which would protect the uninfected cells in the body (Trinchieri and Santoli, 1978). Support for this hypothesis comes from studies with MCMV, which is an NK-sensitive virus, and with LCMV, whose infection is completely uninfluenced by NK cells. MCMV is a cytopathic virus which inhibits the ability of IFN to protect targets, whereas the relatively noncytopathic LCMV fails to prevent IFN-mediated protection (Bukowski and Welsh, 1985b).

Virus-induced inhibition of IFN-mediated up-regulation of class I antigens is not always due to global inhibitions in cellular protein synthesis (see next section) and can sometimes be due to more subtle mechanisms. It is proposed that certain murine retrovirus-infected cells contain cis-acting, negative regulatory factors that inhibit IFN-γ-induced up-regulation of H-2D^k but not H-2K^k transcription (Rich et al., 1990).

Selective Inhibition of Class I Antigen Expression

Some viruses have evolved mechanisms to inhibit the expression of class I antigens in a rather selective manner. HIV-1 and pseudorabies virus by unknown mechanisms each inhibit class I antigen expression under conditions where other cellular proteins are unaffected (Scheppler et al., 1989; Mellencamp et al., 1991). The adenovirus 12 E1A gene product is reported to inhibit the transcription of class I mRNA (Friedman and Ricciardi, 1988). Further, a 19-kDa adenovirus E3 gene product complexes with the class I α chain in the cytoplasm and prevents its expression on the cell surface (Andersson et al., 1985). Adenoviruses that down-regulate class I MHC expression have been reported to render cells more susceptible to lysis by NK cells (Cook et al., 1987; Dawson et al., 1989). HCMV encodes a class I α-chain-like molecule which can complex with β_2-microglobulin, and cell-encoded class I antigens in HCMV-infected cells are trapped in the cytoplasm, perhaps because of competition by the viral class I antigen for binding to β_2-microglobulin, which is required for efficient transport of class I antigen to the cell surface (Browne et al., 1990). The question of whether this down-regulation of cellular class I expression is relevant to the high sensitivity of HCMV-infected cells to NK cells (Borysiewicz et al., 1985) should be addressed.

Retroviruses may either up- or down-regulate MHC class I antigen expression through a variety of mostly poorly understood mechanisms (Flyer et al., 1985; Rich et al., 1990; Scheppler et al., 1989). The presence of an oncogene may influence the outcome, as Moloney sarcoma virus has been shown to inhibit an up-regulation of class I antigen expression induced by Moloney leukemia virus (Flyer et al., 1985). Recent work has indicated that the expression of certain protooncogenes, such as c-myc, reduces class I antigen expression and renders cells more sensitive to NK cells (Versteeg et al., 1988, 1989). This is consistent with a wide body of data indicating that rapidly proliferating cells, which express high levels of protooncogenes, are highly susceptible to NK cells. Delivery into the cells of oncogenes by viruses or the virus-induced activation of cellular protooncogenes may therefore contribute to class I down-regulation and enhanced sensitivity to NK cells.

Alterations in Class I Antigens Expressed

Recent evidence suggests that NK cells may be negatively regulated not only by target cell class I antigens *per se*, but by the class I antigens presenting endogenous host peptides. Target cells transfected with some, but not all, class I MHC antigens can be rendered more resistant to lysis, and exon-shuffling and site-directed mutagenesis experiments have defined residues on the α_1 domain, which couches the antigen-binding groove, to be important for resistance to killing (Storkus et al., 1991). Incubation of high class I-expressing, NK-resistant cell lines with viral peptides capable of being presented by the class I molecules renders the targets more sensitive to NK cells (Chadwick et al., 1991). Hence, replacement of endogenous peptides with virus-encoded peptides can render a cell sensitive to NK cells.

We have examined the vaccinia virus (VV) infection of L-929 cells and have found that during a

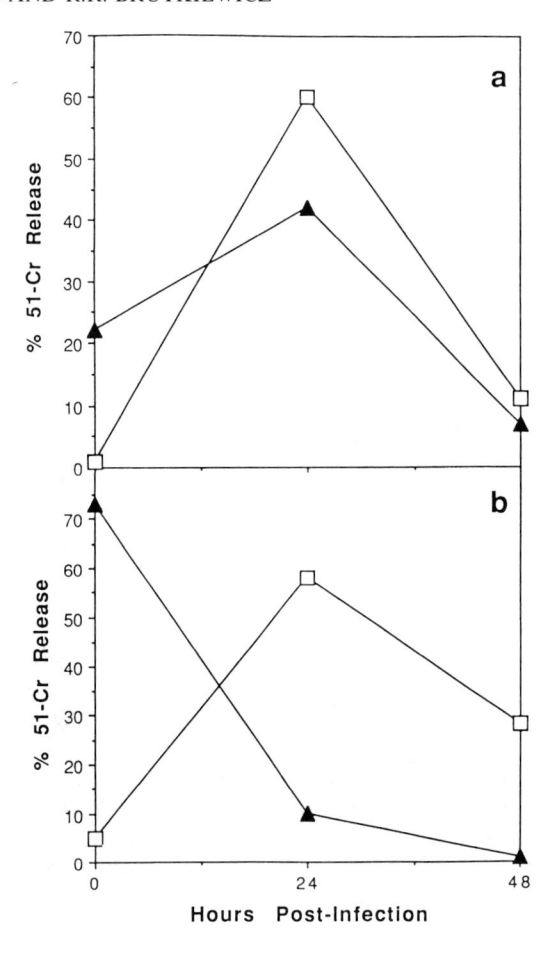

FIGURE 3. Correlation of the NK cell sensitivity of VV-infected target cells with resistance to allospecific CTL. (a) Lysis of VV-infected L-929 cells by VV-specific CTL from the spleens of C3H mice 7 days after infection with VV (□) and by activated NK cells from the spleens of C3H mice 3 days after infection with LCMV (▲). (b) Lysis of VV-infected L-929 cells by VV-specific CTL (□) and by anti-H-2^k allospecific CTL taken from the spleens of irradiated (750 rad) C3H mice 5 days after inoculation with BALB/c mouse splenocytes (▲).

discrete time period after infection there is an enhanced sensitivity or "window of vulnerability" to NK-cell-mediated lysis (Figure 3). During this time period there is substantial expression of class I antigens on the cell surface, yet the cells have completely lost their sensitivity to allospecific CTL. As allospecific CTL recognize allogeneic class I antigens in the context of presenting endogenous peptides, we intrepret this to indicate that the viral infection has changed the nature of peptide presentation, possibly by the insertion of viral peptides. After this "window of vulnerability," the cells become resistant to NK cells, allospecific CTL, and virus-specific CTL, but this is likely due to a global loss of cell surface molecules late in infection with this cytopathic virus.

NK-Sensitive and NK-Resistant Viruses

Our examination of a wide variety of viruses has indicated that many viruses are sensitive to NK cells, whereas others are resistant (Welsh et al., 1990). It is easy to understand why a virus should be resistant to NK cells, as this should give the virus a selective advantage *in vivo*. In fact, when an NK-sensitive variant of Pichinde virus was inoculated into SCID mice, a persistent infection followed, and virus recovered from these mice was NK-resistant (Welsh et al., 1991). This indicates that, in the absence of adaptive host response mechanisms, selective pressure favors the NK-resistant virus. This leads one to question why a virus should ever be NK-sensitive. This could only be explained if NK sensitivity conferred a selective advantage at some point during infection. If "NK sensitivity" is associated with reductions in class I antigen expression, then the selective advantage becomes obvious, as there should be impaired recognition and clearance by CTL. This hypothesis will be worth testing with NK-sensitive and -resistant variants of a given virus.

Acknowledgments

This work was upported by USPHS research grants AI17672, CA34461, and AR35506.

References

Andersson M, Paabo S, Nilsson T, Peterson PA (1985): Impaired intracellular transport of class I MHC antigens as a possible means for adenovirus to evade immune surveillance. *Cell* 43: 215–222

Arora DJS, Houde M, Justewicz DM, Mandeville R (1984): In vitro enhancement of human natural killer cell mediated cytotoxicity by purified influenza virus glycoproteins. *J Virol* 52: 839–845

Biron CA, Byron KS, Sullivan JS (1989): Severe herpes virus infections in an adolescent without natural killer cells. *New Engl J Med* 320: 1731–1735

Borysiewicz LK, Rodgers B, Morris S, Graham S, Sissons JGP (1985): Lysis of human cytomegalovirus infected fibroblasts by natural killer cells: Demonstration of an interferon-dependent component requiring expression of early viral proteins and characterization of effector cells. *J Immunol* 134: 2695–2701

Browne H, Smith G, Beck S, Minson T (1990): A complex between the MHC class I homologue encoded by human cytomegalovirus and β_2 microglobulin. *Nature* 347: 770–772

Bukowski JF, Welsh RM (1985a): Interferon enhances the susceptibility of virus-infected fibroblasts to cytotoxic T cells. *J Exp Med* 161: 257–262

Bukowski JF, Welsh RM (1985b): Susceptibility of virus-infected targets to natural killer cell-mediated lysis in vitro correlates with natural killer cell-mediated antiviral effects in vivo. *J Immunol* 135: 3537–3541

Bukowski JF, Welsh RM (1986): Enhanced susceptibility to cytotoxic T lymphocytes of target cells isolated from virus-infected or interferon-treated mice. *J Virol* 59: 735–739

Bukowski JF, Warner JR, Dennert G, Welsh RM (1985): Adoptive transfer studies demonstrating the antiviral effect of NK cells in vivo. *J Exp Med* 161: 40–52

Bukowski JF, Yang H, Welsh RM (1988): The antiviral effect of lymphokine activated killer (LAK) cells. 1. Characterization of the effector cells mediating prophylaxis. *J Virol* 62: 3642–3648

Casali P, Sissons JGP, Buchmeier MJ, Oldstone MBA (1981): In vitro generation of human cytotoxic lymphocytes by virus. Viral glycoproteins induce nonspecific cell-mediated cytotoxicity without release of interferon. *J Exp Med* 154: 840–855

Chadwick BS, Sambhara SR, Roder JC, Miller RG (1991): Effect of MHC-class-I-binding peptides on recognition by NK cells. *Nat Immun Cell Growth Regul* 10: 127 (abstract)

Cook JL, May DL, Lewis AM, Walker TA (1987): Adenovirus E1A gene induction of susceptibility to lysis by natural killer cells and infected macrophages in infected rodent cells. *J Virol* 61: 3510–3520

Dawson JR, Storkus WJ, Patterson EB, Cresswell P (1989): Adenovirus inversely modulates target cell class I MHC antigen expression and sensitivity to natural killing. In: *Natural Killer Cells and Host Defense*, Ades EW, Lopez C, eds. Basel: S Karger, pp. 156–159

Flyer DC, Burakoff SJ, Faller DV (1985): Retrovirus-induced changes in major histocompatibility complex antigen expression influence susceptibility to lysis by cytotoxic T lymphocytes. *J Immunol* 135: 2287–2292

Friedman DJ, Ricciardi RP (1988): Adenovirus type 12 E1A gene represses accumulation of MHC class I mRNAs at the level of transcription. *Virology* 165: 303–305

Hansson M, Kiessling R, Andersson B, Welsh RM (1980): Effect of interferon and interferon inducers on the NK sensitivity of normal mouse thymocytes. *J Immunol* 125: 2225–2231

Harfast B, Orvell C, Alsheikhly A, Andersson T, Perlmann P, Norrby E (1980): The role of viral glycoproteins in mumps virus-dependent lymphocyte-mediated cytotoxicity in vitro. *Scand J Immunol* 11: 391–400

Hecht TT, Summers DF (1972): Effect of vesicular stomatitis virus infection on histocompatibility antigens of L cells. *J Virol* 10: 578–585

Hoglund P, Glas R, Ohlen C, Ljunggren H-G, Karre K (1991): Alteration of the natural killer cell repertoire in H-2 transgenic mice: Specificity of rapid lymphoma cell clearance determined by the H-2 phenotype of the cell. *J Exp Med* 174: 327–334

Jennings SR, Rice PL, Kloszewski ED, Anderson RW, Thompson DL, Tevethia SS (1985): Effect of herpes simplex virus types 1 and 2 on surface expression of class I major histocompatibility complex antigens on infected cells. *J Virol* 56: 757–766

Kiessling R, Welsh RM (1980): Killing of normal cells by natural killer cells: Evidence for two patterns of genetic regulation of lysis. *Int J Cancer* 25: 611–615

Liao N-S, Bix M, Zilstra M, Jaenisch R, Raulet D (1991): MHC class I deficiency: Susceptibility to natural killer (NK) cells and impaired NK activity. *Science* 253: 199–202

Ljunggren H-G, Karre K (1990): In search of the 'missing self': MHC molecules and NK cell recognition. *Immunol Today* 11: 237–244

Ljunggren H-G, Sturmhofel K, Wolpert E, Hammerling G, Karre K (1990): Transfection of β_2-microglobulin restores IFN-mediated protection from natural killer cell lysis in YAC-1 lymphoma variants. *J Immunol* 145: 380–386

Lopez C, Kirkpatrick D, Reid SE, Fitzgerald PA, Pitt J, Pahwa S, Ching CY, Smithwick EM (1983): Correlation between low natural killing of fibroblasts infected with herpes simplex virus type 1 and susceptibility to herpesvirus infections. *J Infect Dis* 147: 1030–1035

McIntyre KW, Welsh RM (1986): Accumulation of natural killer and cytotoxic T large granular lymphocytes during virus infection. *J Exp Med* 164: 1677–1681

Mellencamp MW, O'Brien PCM, Stevenson JR (1991): Pseudorabies virus-induced suppression of major histocompatibility complex class I antigen expression. *J Virol* 65: 3365–3368

Natuk RJ, Welsh RM (1987): Accumulation and chemotaxis of natural killer/large granular lymphocytes to sites of virus replication. *J Immunol* 138: 877–883

Quinnan GV, Kirmani N, Rook AH, Manischewitz JF, Jackson L, Moreschi G, Santos GW, Saral R, Burns WH (1982): Cytotoxic T cells in cytomegalovirus infection: HLA-restricted T-lymphocyte and non-T-lymphocyte cytotoxic responses correlate with recovery from cytomegalovirus infection in bone marrow transplant recipients. *New Engl J Med* 307: 7–13

Rich RF, Gaffney KJ, White HD, Green WR (1990): Differential up-regulation of H-2D versus H-2K class I major histocompatibility expression by interferon-gamma: Evidence against a trans-acting factor. *J Interferon Res* 10: 505–514

Scalzo AA, Fitzgerald NA, Simmons A, LaVista AB, Shellam GR (1990): CMV-1, a genetic locus that controls murine cytomegalovirus replication in the spleen. *J Exp Med* 171: 1469–1483

Scheppler JA, Nicholson JKA, Swan DC, Ahmed-Ansari A, McDougal JS (1989): Down-modulation of MHC-I in a CD4[+] T cell line, CEM-E5, after HIV-1 infection. *J Immunol* 143: 2858–2866

Storkus WJ, Salter RD, Alexander J, Ward FE, Ruiz RE, Cresswell P, Dawson JR (1991): Class I-induced resistance to natural killing: Identification of nonpermissive residues in HLA-A2. *Proc Natl Acad Sci USA* 88: 5989–5992

Su HC, Ishikawa R, Leite-Morris KA, Braun L, Biron CA (1991): Role of TGF-beta in regulating NK cell proliferation. *Nat Immun Cell Growth Regul* 10: 147–148 (abstract)

Townsend A, Bodmer H (1989): Antigen recognition by class I-restricted T lymphocytes. *Annu Rev Immunol* 7: 601–624

Trinchieri G, Santoli D (1978): Antiviral activity induced by culturing lymphocytes with tumor derived or virus-transformed cells. Enhancement of natural killer activity by interferon and antagonistic inhibition of susceptibility of target cells to lysis. *J Exp Med* 147: 1314–1333

Versteeg R, Noodermeer IA, Kruse-Walters M, Reiter DJ, Schrier P (1988): c-myc down-regulates class I HLA expression in human melanomas. *EMBO J* 7: 1023–1029

Versteeg R, Peltenburg LTC, Plomp AC, Schrier PI (1989): High expression of the c-myc oncogene renders melanoma cells prone to lysis by natural killer cells. *J Immunol* 143: 4331–4337

Welsh RM (1978): Cytotoxic cells induced during lymphocytic choriomenigitis virus infection of mice: 1. Characterization of natural killer cell induction. *J Exp Med* 148: 163–181

Welsh RM (1986): Regulation of virus infections by natural killer cells. *Nat Immun Cell Growth Regul* 5: 169–199

Welsh RM, Hallenbeck LA (1980): Effect of virus infections on target cell susceptibility to natural killer cell-mediated lysis. *J Immunol* 124: 2491–2497

Welsh RM, Vargas-Cortes M (1992): Natural killer cells in virus infection. In: *The Natural Immune System: The Natural Killer Cell*, Lewis CE, Gee JOM, eds. Oxford University Press, pp. 108–150

Welsh RM, Brubaker JO, Vargas-Cortes M, O'Donnell CL (1991): Natural killer (NK) cell response to virus infections in mice with severe combined immunodeficiency. The stimulation of NK cells and the NK cell-dependent control of virus infections occur independently of T and B cell function. *J Exp Med* 173: 1053–1063

Welsh RM, Dundon PL, Eynon EE, Brubaker JO, Koo GC, O'Donnell CL (1990): Demonstration of the antiviral role of natural killer cells in vivo with a natural killer cell-specific monoclonal antibody (NK 1.1). *Nat Immun Cell Growth Regul* 9: 112–120

Welsh RM, Karre K, Hansson M, Kunkel LA, Kiessling RW (1981): Interferon-mediated protection of normal and tumor target cells against lysis by mouse natural killer cells. *J Immunol* 126: 219–225

Welsh RM, Zinkernagel RM, Hallenbeck LA (1979): Cytotoxic cells induced during lymphocytic choriomeningitis virus infection of mice. II. Specificities of the natural killer cells. *J Immunol* 122: 475–481

41
Clinical Trials of Immunotherapy of Cancer Utilizing Cytotoxic Cells

Stephen E. Ettinghausen and Steven A. Rosenberg

Standard therapies for patients with cancer have traditionally included surgery, chemotherapy, and radiation. Since the 1960s, however, there has been increasing interest in the use of a biologic or immunologic therapy to treat cancer in man (Rosenberg et al., 1989a; Rosenberg, 1991). Within the realm of biologic therapy, the greatest therapeutic promise lies with manipulations of the cellular arm of the immune system, since cellular host mechanisms are responsible for elimination of transplanted tissues bearing foreign proteins (Rosenberg and Terry, 1977).

In cellular adoptive immunotherapy, cells possessing specific or nonspecific antitumor reactivity are passively administered to the tumor-bearing host. Subsequently, these transferred cells mediate their anticancer effects directly via cytotoxic mechanisms or indirectly through host components (Rosenberg et al., 1989a; Rosenberg, 1991). Initially, attempts to utilize such therapies in humans were limited by the inability to identify tumor-reactive cells and to expand such populations to large enough numbers to achieve clinically meaningfully results (Rosenberg and Terry, 1977). However, advances in two areas of cellular immunologic research facilitated the development of immunotherapy and allowed its eventual progression into clinical trials. First, the discovery of T-cell growth factor (TCGF) in 1976 provided a means to expand specific lymphocyte populations that exhibited tumor reactivity (Morgan et al., 1976). However, the minute amounts of TCGF available limited the possible levels of cell expansion and the use of the reactive cells in potential human trials. In 1984, large quantities of the lymphokine, renamed interleukin-2 (IL-2), become available after the gene for IL-2 was cloned and expressed to high levels in *E. coli* (Fujita et al., 1983;

Taniguchi et al., 1983; Rosenberg et al., 1984). Subsequently, recombinant IL-2 (rIL-2) was purified to homogeneity and the biologic and immunologic activity of the recombinant material was found to be similar to the naturally occurring form (Rosenberg et al., 1984; Wang et al., 1984). Considerable strides were also made in the general understanding of T-cell biology, including T-cell activation, proliferation, differentiation, and distribution patterns *in vivo* (Rosenberg, 1991). The elucidation of the mechanisms of antigen recognition, processing, and presentation by accessory cells and details of effector cell function also guided the development of immunotherapeutic protocols.

Nonspecific Adoptive Immunotherapy with Lymphokine-Activated Killer (LAK) Cells: *In Vitro* Studies

During attempts to expand specifically reactive lymphocytes from murine tumors, Yron et al. (1980) found that culture of normal splenocytes in high-dose IL-2 induced the development of nonspecifically cytotoxic cells capable of lysing fresh, syngeneic, weakly immunogenic, and nonimmunogenic sarcomas in short-term chromium release assays. These lytic cells, which were termed lymphokine-activated killer (LAK) cells, would kill fresh tumor cells but not fresh normal cells (Yron et al., 1980; Rosenstein et al., 1984; Lefor and Rosenberg, 1991).

Subsequent investigations of the LAK cell phenomenon in the human system demonstrated that similar, nonspecifically cytotoxic cells could be generated by incubation of peripheral blood lym-

TABLE 1. *In vitro* cytolytic spectrum of human LAK cells

	Effector cells[a] [% lysis ($\pm$ SEM)][b]			
	LAK cells		Fresh PBL	
Target cells	40:1	10:1	60:1	15:1
Fresh tumor				
Sarcoma (autologous)	75 ± 6	73 ± 1	−4 ± 10	−9 ± 1
Sarcoma	88 ± 3	78 ± 1	7 ± 2	1 ± 1
Sarcoma	57 ± 3	48 ± 3	−8 ± 1	−12 ± 2
Sarcoma	85 ± 1	70 ± 5	9 ± 8	2 ± 5
Sarcoma	98 ± 1	87 ± 1	1 ± 2	1 ± 2
Sarcoma	64 ± 2	54 ± 7	−4 ± 3	−11 ± 8
Sarcoma	67 ± 3	57 ± 2	−6 ± 4	−3 ± 1
Colon cancer	62 ± 1	41 ± 3	−16 ± 1	−14 ± 1
Adrenal cancer	68 ± 2	41 ± 3	−16 ± 1	−20 ± 2
Esophageal cancer	78 ± 5	62 ± 2	0 ± 4	−1 ± 1
Pancreatic cancer	28 ± 5	17 ± 2	5 ± 1	−2 ± 1
Fresh PBL				
Sarcoma patient 1	4 ± 1	9 ± 6	−9 ± 3	−8 ± 2
Sarcoma patient 2	9 ± 2	5 ± 3	−5 ± 1	−3 ± 3
Cultured tumor				
K562 erythroleukemia	105 ± 6	89 ± 2	46 ± 3	15 ± 1
Daudi lymphoma	85 ± 5	101 ± 18	2 ± 2	3 ± 3

[a]Effector cells were obtained from a patient with sarcoma. LAK cells were generated in a 5-day incubation with rIL-2. NK activity is generated by fresh peripheral blood lymphocytes (PBL).
[b]Results of a 4-h ^{51}Cr-release assay.
From Rayner et al., 1985.

phocytes (PBL) in media containing TCGF or IL-2 (Lotze et al., 1981; Grimm et al., 1982, 1983b; Grimm and Rosenberg, 1984; Rayner et al., 1985). Human LAK cells could effectively lyse fresh autologous and allogeneic sarcomas and carcinomas, cultured normal or tumor cell lines, but not fresh PBL or lung fibroblasts (Lotze et al., 1981; Grimm et al., 1982, 1983b; Grimm and Rosenberg, 1984; Rayner et al., 1985). While able to produce low levels of lysis of cultured, NK-sensitive tumor lines, fresh PBL were unable to lyse fresh, noncultured tumor targets, as shown in Table 1 (Rayner et al., 1985). Additional *in vitro* studies of the LAK cell phenomenon were carried out in murine and human systems and are summarized in Table 2.

Nonspecific Adoptive Immunotherapy with LAK Cells: Preclinical Murine Studies

Extensive investigations of murine LAK cells *in vivo* have utilized models of pulmonary and hepatic metastases from a variety of early passage, chemically induced, weakly immunogenic, or nonimmunogenic carcinomas and sarcomas (Mazumder and Rosenberg, 1984; Mulé et al., 1984, 1985, 1986b; Papa et al., 1986; Eisenthal et al., 1988b). To induce metastases in mice, single-cell suspensions of tumor were injected intravenously via the lateral tail vein or intrasplenically during a laparotomy; tumor cells lodged and subsequently grew as metastases in the lung and liver, respectively. By 3 days following tumor injection when established micrometastases were present, therapy was begun. The adoptive transfer of LAK cells on days 3 and 6 in combination with repetitive injections of rIL-2 given between days 3 and 8 significantly reduced the number of lung or liver metastases when enumerated 13 to 17 days after tumor injection (Mazumder and Rosenberg, 1984; Mulé et al., 1984, 1985, 1986a, 1986b; Papa et al., 1986; Lafreniere and Rosenberg, 1985a, 1985b, 1986). Figure 1 presents the results of therapy of 3-day-old lung metastases using the combination of LAK cells and rIL-2. Therapy with LAK cells and rIL-2 also prolonged survival as compared to control treated mice Lafreniere and Rosenberg, 1985a; Mulé et al., 1986b). A direct

TABLE 2. Summary of *in vitro* characteristics of murine and human lymphokine-activated killer (LAK) cells

1.	Lymphokine-activated killer (LAK) cells are derived from lymphocytes incubated for 3–5 days in high concentrations of rIL-2. As functionally defined, LAK cells are capable of lysis of fresh syngeneic, autologous, allogeneic or xenogeneic sarcomas, carcinomas or melanomas in short-term ^{51}Cr- release assays (Yron et al., 1980; Lotze et al., 1981; Grimm et al., 1982, 1983a, 1983b; Grimm and Rosenberg, 1984; Rosenstein et al., 1984; Rayner et al., 1985; Lefor and Rosenberg, 1991).
2.	rIL-2 alone is sufficient for development of LAK activity; no exposure to the target antigen is required (Grimm et al., 1983b; Grimm and Rosenberg, 1984).
3.	LAK cells may be generated from peripheral blood lymphocytes (PBL), spleen, thymus, bone marrow, or thoracic duct lymphocytes (Grimm et al., 1982, 1983a).
4.	Irradiation of lymphocytes before culture in rIL-2 prevents development of LAK activity (Grimm et al., 1982).
5.	When incubated in rIL-2, PBL from normal volunteers or cancer patients develop similar levels of LAK activity (Grimm et al., 1982; Rayner et al., 1985).
6.	Murine LAK cell precursors reside within the null cell population of peripheral blood mononuclear cells (PBMC). Precursors lack Thy-1.2 and Ia antigens and are surface immunoglobulin-negative. Most LAK cell activity resides within a precursor splenocyte population bearing surface asialo GM1 characteristic of NK cells (Yang et al., 1986).
7.	Following incubation in rIL-2, murine LAK effector cells become Thy-1.2 antigen- and Fc receptor-positive while remaining Ia antigen-negative (Yang et al., 1986).
8.	LAK cell activity can be dissociated from NK-induced lysis, since certain immunodeficient mice can generate LAK- but not NK-mediated cytotoxicity (Andriole et al., 1985).
9.	Human LAK cell progenitors are derived from the null cell fraction of PBMC; purification of null cells by depletion of T and B cells and monocytes from PBMC results in significant augmentation of LAK activity. Although lesser degrees of LAK activity are found in other lymphocyte populations, the most potent LAK lysis is found in precursors that express CD11b (C3bi complement receptor), CD16 (Fc gamma receptor for IgG found on neutrophils, NK cells, and monocytes), and CD56 (NKH-1 found on NK cells and CTL subsets). Human LAK effector cells are primarily CD3-negative (Grimm et al., 1983a; Itoh et al., 1985; Ortaldo et al., 1986; Phillips and Lanier, 1986; Roberts et al., 1987; Skibber et al., 1987; Fox and Rosenberg, 1989).
10.	LAK cells can mediate antibody-dependent cellular cytotoxicity (ADCC); the lytic capacity of LAK cells may be potentiated up to 100-fold by use of antibody directed against specific antigens expressed on the tumor target. ADCC activity of LAK cells is mediated via LAK cell Fc receptors (Shiloni et al., 1987; Eisenthal et al., 1988b).

dose-response relationship could be demonstrated between the number of LAK cell infusions and the number of LAK cells per infusion, as well as with the dose of rIL-2 (Mulé et al., 1984, 1985; Lafreniere and Rosenberg, 1985a, 1985b, 1986; Mulé et al., 1986a, 1986b; Papa et al., 1986). Substitution of LAK cells with fresh or cultured (without IL-2) splenocytes failed to affect the number of metastases (Mulé et al., 1984, 1985; Lafreniere and Rosenberg, 1985b).

Table 3 details additional results from the preclinical investigations of immunotherapy with LAK cells and rIL-2.

Clinical Immunotherapeutic Trials of LAK Cells and IL-2

Based upon therapeutic successes with LAK cells and rIL-2 in murine models, this immunotherapeutic approach was applied to patients with metastatic cancer in the Surgery Branch of the National Cancer Institute (NCI). The initial clinical studies of LAK cells and rIL-2 were conducted in separate trials to establish their individual toxicities and safety (Lotze et al., 1980; Mazumder et al., 1984; Rosenberg, 1984; Lotze et al., 1985a, 1985b). Prior to the development of recombinant DNA techniques and the consequent availability of rIL-2, a purified Jurkat-derived IL-2 was administered to 12 patients with advanced cancer in a phase I trial (Lotze et al., 1985a). No anticancer effect was observed in any of the patients who received up to 2 mg of Jurkat IL-2 per week by bolus or continuous infusion (Lotze et al., 1985a). Subsequently, IL-2 became available and this cytokine entered into phase I evaluation in 20 patients with various malignancies in the Surgery Branch (Lotze et al., 1985b). Recombinant IL-2 was found to be tolerated as a single bolus up to 7.2 million IU/kg, while 21,600 IU/kg/hr was the highest dose that could be administered by continuous intravenous infusion before major life-threatening side effects developed (Lotze et al., 1985b). The rIL-2-associat-

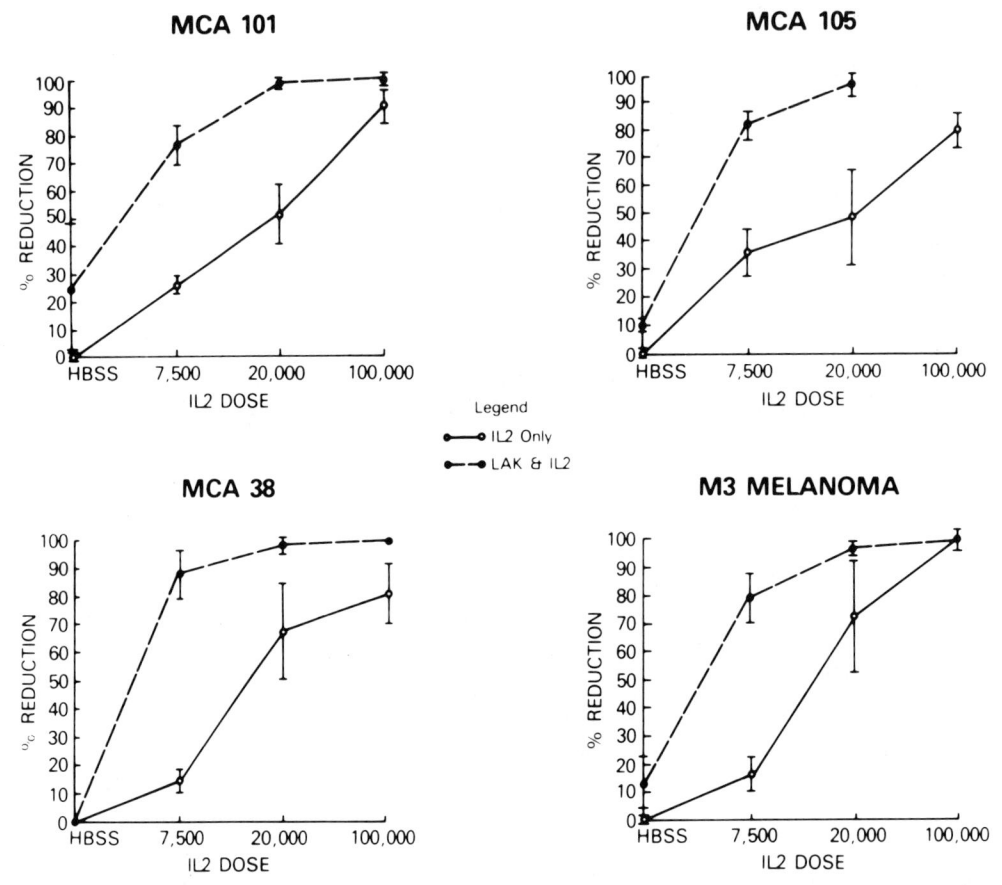

FIGURE 1. Effect of LAK cells and escalating doses of rIL-2 for treatment of 3-day lung metastases in mice bearing two sarcomas (MCA 101 and 105), a colon adenocarcinoma (MCA-38), and a melanoma (M-3). The figure depicts the pooled results of multiple experiments (MCA-101, $n = 5$; MCA-105, $n = 4$; MCA-38, $n = 4$; M-3, $n = 4$). Each point represents the mean percentage reduction of metastases for all experiments. The combination of LAK cells (2 injections of 10^8 cells on days 3 and 6) and rIL-2 (days 3–8) mediated a significantly greater increase in the percent reduction of lung metastases when compared with rIL-2 alone or LAK cells alone. (From Papa et al., 1986.)

ed toxicities have been well described and are similar to those observed with LAK cells and rIL-2, which are described below.

As early as the 1960s, various investigators had undertaken several clinical trials utilizing the adoptive transfer of lymphocytes to cancer patients; the details of these protocols and the results have been reviewed elsewhere (Rosenberg and Terry, 1977). All of these studies were limited by the inability to generate large numbers of autologous cells with antitumour activity for use in cancer therapy. In the Surgery Branch, trials of cellular therapy in patients with a variety of disseminated malignancies began in 1980 shortly after the LAK cell phenomenon was first noted (Lotze et al., 1980; Mazumder et al., 1984; Rosenberg, 1984).

In four separate trials, different cellular therapies were studied, including long-term cultured PBL, phytohemagglutinin (PHA)-activated killer (PAK) cells, PAK cells with preinfusion cyclophosphamide (to eliminate suppressor cells), and PAK cells with PHA-activated monocytes (Lotze et al., 1980; Mazumder et al., 1984; Rosenberg, 1984). Table 4 summarizes the experience in the Surgery Branch with these early studies as well as later adoptive immunotherapeutic trials.

In 1984, with the availability of rIL-2 for clinical use, six patients received LAK cells generated using the recombinant material by *in vitro* culture of thoracic duct lymphocytes or PBL obtained by repeated leukaphereses (Rosenberg, 1984, 1991). Although no patient in these early trials exhibited

TABLE 3. Summary of preclinical trials of immunotherapy using LAK cells and rIL-2 in mice

1. The administration of LAK cells in combination with multiple injections of rIL-2 to tumor-bearing mice can significantly reduce the number of lung or liver metastases from a variety of early-passage, weakly immunogenic sarcomas, melanomas, and carcinomas. The combination therapy imparts a survival benefit when compared with control treated mice (Mazumder and Rosenberg, 1984; Mulé et al., 1984; Lafreniere and Rosenberg 1985a, 1985b; Mulé et al., 1985; Lafreniere and Rosenberg, 1986; Mulé et al. 1986a, 1986b; Papa et al., 1986).
2. Immunotherapy with LAK cells and rIL-2 can reduce tumor load and prolong survival of mice with peritoneal carcinomatosis (Ottow et al., 1987a, 1987b).
3. LAK cells generated from splenocytes harvested from tumor-bearing mice are as effective as LAK cells derived from "naive" animals (Mazumder and Rosenberg, 1984).
4. Irradiation of LAK cells prior to adoptive transfer inhibits their *in vivo* efficacy; LAK cells proliferate *in vivo* under the effect of exogenously administered rIL-2 and, if extracted from lungs and liver, maintain their lytic capability as tested in *in vitro* assays (Ettinghausen et al., 1985; Mulé et al., 1985).
5. When administered with rIL-2, LAK cells derived from allogeneic splenocytes can significantly reduce established pulmonary and hepatic metastases (Mulé et al., 1985; Shiloni et al., 1986).
6. When given in combination with rIL-2, LAK cells are effective in the tumor-bearing host immunosuppressed by total body irradiation or total depletion of T cells by thymectomy, lethal irradiation, and reconstitution with T-cell-depleted bone marrow ("B" mice) (Mulé et al., 1984).
7. Residual lung metastases which remain after treatment of tumor-bearing mice with LAK cells and rIL-2 may be excised and are still susceptible to LAK lysis *in vitro* and *in vivo* if reinjected into "naive" mice. Failure of immunotherapy with LAK cells does not appear to result from the emergence of LAK-resistant tumor clones; instead treatment failures may be related to problems with *in vivo* trafficking of LAK cells or local tumor-induced immunosuppression (Mulé et al., 1986a).
8. The adoptive transfer of tumor-specific monoclonal antibodies and LAK cells mediates antibody-dependent cellular cytotoxicity *in vivo*; the adoptive transfer of tumor-specific monoclonal antibodies plus LAK cells and rIL-2 can significantly impact on lung micrometastases (Eisenthal et al., 1988a).

any anticancer response, the overall safety of *ex vivo* activated mononuclear cell administration was established.

Since the animal data predicted that combination therapy would have the greatest likelihood of therapeutic efficacy, patient trials with LAK cells and rIL-2 were initiated following Food and Drug Administration (FDA) approval and began in November 1984 (Rosenberg et al., 1985a, 1987, 1989b). Patients with metastatic cancer who had failed all standard therapies, had received no other treatment within the previous 30 days, and were in otherwise good health were accepted into therapeutic trials. As outlined in Figure 2, the schema for therapy utilized a 3- to 5-day period of bolus rIL-2 administration (720,000 IU/kg) given intravenously every 8 hr followed by a 2- to 3-day rest period. The early trials of rIL-2 demonstrated that shortly after initiation of rIL-2 infusion a marked decline in all mononuclear cell populations (including CD3, CD4, CD8, Leu 7, Leu M3, Leu 10, and Leu DR positive cells as well as LAK cell precursors) was observed in peripheral blood (Lotze et al., 1985b). By 24 to 72 hr after completion of rIL-2, a dramatic rebound above baseline levels of all phenotypes could be demonstrated (Lotze et al., 1985b). Taking advantage of this profound rise in circulating LAK cell precursors, daily leukaphereses were then performed over 4 to 5 days to harvest peripheral blood mononuclear cells (PBMC). Up to 2×10^{11} cells were obtained by this method and large-scale methods were required to generate LAK cells (Muul et al., 1986, 1987a; Aebersold et al., 1988). After a 3- to 4-day incubation of lymphocytes in 2.3-l roller bottles or 1 to 3-l polyolefin bags containing media with 6000–9000 IU of rIL-2/ml, LAK cells were returned to the patients along with bolus rIL-2 at a similar dose and schedule as prior to the leukaphereses. Recombinant IL-2 was continued for up to 15 doses but was usually discontinued earlier due to associated toxicity. A summary of 286 courses of immunotherapy administered to 180 patients receiving LAK cells and rIL-2 is presented in Table 5. As shown in Table 6, the majority of patients had metastatic renal cell carcinoma, melanoma, and colorectal cancer and demonstrated an objective (complete or partial) response rate of 35, 21, and 13%, respectively. A complete response was achieved in approximately 10% of patients with melanoma and renal cell cancer. Over half of the patients with non-Hodgkin's lymphoma have shown significant responses, although only seven patients received this therapy in this initial study. In several patients, the durability of these responses has translated into prolonged survival. As shown in Table 7, of 14 patients who achieved a complete response, nine have remained free of disease for 10 to 75+ months. Examples of patients

TABLE 4. Clinical studies of cellular adoptive immunotherapy in the Surgery Branch, National Cancer Institute

Year	Cells administered	No. of patients	Reference	Findings
1980	Long-term cultured peripheral blood lymphocytes (PBL)	3	Lotze et al., 1980b	Small numbers (up to 5×10^8) of long-term cultured PBL could be safely infused in humans
1981	Phytohemagglutinin-activated killer (PAK) cells	10	Mazumder et al., 1984	Large numbers (up to 1.7×10^{11}) of activated killer cells, obtained from up to 15 successive leukaphereses, could be infused safely in humans
1982	PAK cells plus cyclophosphamide	6	Rosenberg, 1984	Activated killer cells could be safely infused in conjunction with high-dose cyclophosphamide (50 mg/kg)
1983	PAK cells plus activated macrophages	5	Rosenberg, 1984	Activated killer cells plus PHA-activated macrophages up to 6.6×10^{10} total cells could be safely infused
1984	Lymphokine-activated killer (LAK) cells	6	Rosenberg, 1984	LAK cells (activated with recombinant IL-2) could be safely infused in humans
1985	LAK cells plus recombinant IL-2	25	Rosenberg et al., 1985a	Regression of metastatic cancer of a variety of types in some patients
1988	Tumor-infiltrating lymphocytes (TIL) plus IL-2	20	Rosenberg et al., 1988a	Complete and partial regression of melanoma
1989	LAK cells plus IL-2 or IL-2	307	Rosenberg et al., 1987, 1989b	Complete and partial regression of cancer of several histological types
1990	Gene-modified TIL (neo R gene)	10	Rosenberg et al., 1990;	TIL modified by insertion of gene coding for neomycin resistance
1991	Gene-modified TIL (TNF and neo R genes)	4	Rosenberg, 1992a	TIL modified by insertion of genes coding for tumor necrosis factor and neomycin resistance

Adapted from Rosenberg, 1991.

with melanoma and renal cell carcinoma responding to therapy are shown in Figures 3-5.

In order to further evaluate the efficacy of LAK cells and IL-2, six Medical Centers designated by the NCI as the IL-2/LAK Working Group (ILWG) initiated similar protocols with the combined adoptive immunotherapy. After treatment of the first 83 patients with renal cell carcinoma, melanoma, and colorectal cancer, the ILWG reported objective response rates of 16, 19, and 16%, respectively (Hawkins, 1989; Rosenberg, 1992a). The lower response rate for patients with renal cell cancers was presumed to be due to the greater tumor burden in patients selected for therapy by the ILWG.

Other groups have evaluated combined immunotherapy with LAK cells and rIL-2 using variations in the original Surgery Branch regimen (West et al., 1987; Fisher et al., 1988; Schoof et al., 1988; Dutcher et al., 1989; Hawkins, 1989; Negrien et al., 1989; Paciucci et al., 1989; Stahel et al., 1989; Thompson et al., 1989; Bar et al., 1990; Gaynor et al., 1990; Dutcher et al., 1991; Weiss et al., 1992). As presented in Table 8, these trials have yielded results similar to those reported from the Surgery Branch for patients with renal cell carcinoma and melanoma.

Therapy-associated toxicity has been substantial in all trials and has limited the duration of therapy in the great majority of patients (Rosenberg et al., 1985a, 1987; West et al., 1987; Fisher et al., 1988; Schoof et al., 1988; Dutcher et al., 1989; Hawkins, 1989; Negrier et al., 1989; Paciucci et al., 1989; Rosenberg et al., 1989b; Stahel et al., 1989; Thompson et al., 1989; Bar et al., 1990; Gaynor et al., 1990; Dutcher et al., 1991; Weiss et al., 1992). The major side effects are due to the concomitant administration of rIL-2 and are the result of a vascular leak syndrome associated with fluid and colloid loss into tissues with consequent hypotension, tachycardia, oliguria, and weight gain (Lotze et al., 1986; Rosenstein et al., 1986; Ettinghausen et

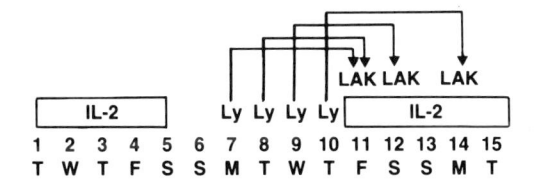

FIGURE 2. Schema of clinical protocol for treatment of patients with metastatic cancer using LAK cells and rIL-2 in the Surgery Branch, National Cancer Institute. IL-2 was administered from day 1 or 2 through the middle of day 5 at 720,000 IU/kg intravenously every 8 hrs followed by a 2-day rest period. Lymphocytophereses were carried out on days 7–10 with reinfusions of LAK cells on days 11, 12, and 14 (as shown by arrows). IL-2 was given concurrently with LAK cell infusions at 720,000 IU/kg every 8 hrs up to 15 doses or as limited by toxicity. (Adapted from Rosenberg, 1991.)

al., 1988; Table 9). Other toxicities have included fever, chills, skin rash, pruritus, mucositis, nausea, vomiting, diarrhea, hepatopathy, azotemia, disorientation, hallucinations, hypothyroidism, anemia, and thrombocytopenia (Belldegrun et al., 1987; Denicoff et al., 1987; Ettinghausen et al., 1987; Gaspari et al., 1987; Atkins et al., 1988; Cotran et al., 1988; Lee et al., 1988; Ogribene et al., 1988; Saris et al., 1988; Schwartzentruber et al., 1988; Webb et al., 1988; Belldegrun et al., 1989; Denicoff et al., 1989; Lee et al., 1989; Bock et al., 1990; Kragel et al., 1990; Schwartzentruber et al., 1991b; Table 9). All side effects have been rapidly reversible upon discontinuation of therapy. Treatment-related mortalities have been approximately 1–3%.

The mechanism by which LAK cells and rIL-2 exert their effect appears to be similar to that in murine models of metastases (Rosenberg et al., 1985b; Rosenberg, 1992a). Serial histologic sections of tumor-bearing mice undergoing immunotherapy showed progressively increasing numbers of activated lymphocytes infiltrating tumor deposits. In microscopic sections taken at later time points, necrosis of tumor cells became evident with subsequent regression of metastatic foci. Lymphocytes recovered from lungs and livers of these mice could be shown to be derived from the transferred LAK cells and could mediate characteristic LAK activity *in vitro* (Ettinghausen et al., 1985). Sequential biopsies of subcutaneous metastases in patients with melanoma demonstrated a dramatic lymphoid infiltration of tumors with subsequent necrosis and dropout of melanoma cells. Figure 6 shows the histologic appearance of subcutaneous metastases from melanoma before and after therapy with LAK cells and rIL-2.

TABLE 5. Summary of adoptive immunotherapy treatments with LAK cells and rIL-2

Treatment	No. of courses[a]
Total	286
Dose IL-2[b] (U/kg × 10^{-3})	
200	1
100	233
30	21
20	9
10	22
Number of IL-2 doses	
1–10	24
11–20	181
21–30	52
31–40	12
41+	17
Cumulative IL-2 dose[b] (U/kg × 10^{-3})	
1–500	17
501–1000	50
1001–2000	183
2001–3000	29
3001–4000	6
4001–5000	1
Number of cell doses	
1–5	249
6–10	34
11–14	3
Cumulative cells (× 10^{-10})	
0.1–1.0	1
1.1–5.0	35
5.1–10.0	116
10.1–15.0	78
15.1+	56

[a] Number of patients receiving first course = 180.
[b] One unit of rIL-2 is the equivalent of 7.2 International Units (IU).
Adapted from Rosenberg, 1991.

Regional Administration of LAK Cells

Studies of the *in vivo* distribution of LAK cells in murine models and sarcoma patients with lung metastases have shown that the transferred cells

TABLE 6. Summary of adoptive immunotherapy of patients with advanced cancer using LAK cells and rIL-2 (as of March, 1991)

Diagnosis	Assessable[a]	CR	PR	CP + PR (%)
Renal cell cancer	72	8	17	35
Melanoma	48	4	6	21
Colorectal cancer	30	1	3	13
Non-Hodgkin's lymphoma	7	1	3	57
Breast cancer	1	0	0	0
Sarcoma	6	0	0	0
Lung cancer	5	0	0	0
Other[b]	9	0	0	0
Total	178	14	29	24

[a]Includes all treated patients except one lost to follow-up and one death due to therapy.
[b]One patient each with cancer of brain, esophagus, ovary, testes, thyroid, gastrinoma, unknown primary, and two patients with Hodgkin's lymphoma.
CR, Complete response; PR, partial response.
Adapted from Rosenberg, 1992a.

rapidly localize to the lungs, but within 24 hr redistribute to the liver and spleen without significant tracking to tumor foci (Lotze et al., 1980). In order to optimize the distribution of LAK cells into sites of metastatic deposits, several investigators have utilized the regional or local administration of LAK cells (Jacob et al., 1986; Ingram et al., 1987; Shimizu et al., 1987; Steis et al., 1987; Yasumoto et al., 1987; Merchant et al., 1988; Yoshida et al., 1988; Barba et al., 1988; Urba et al., 1989; Nitta et al., 1990). Direct infusion of LAK cells and rIL-2 into the peritoneal cavity has been attempted for treatment of peritoneal carcinomatosis from ovarian and colorectal cancers (Steis et al., 1987; Urba et al., 1989). Response rates of up to 70% have been reported (Steis et al., 1987). Other groups have utilized regional immunotherapeutic approaches for treatment of malignant pleural effusions and meningeal carcinomatosis with symptomatic improvement and objective responses (Shimuzu et al., 1987; Yasumoto et al., 1987). Therapy of recurrent gliomas has been approached by a number of different local techniques, including direct injection of LAK cells and rIL-2 into the bed of tumor resection, application of LAK cells contained within a plasma clot onto the resection site, infusion via Ommaya reservoir postoperatively, and application of LAK cells in combination with bispecific antibodies (directed against CD3 and glioma) into brain tissue adjacent to the resection site (Jacob et al., 1986; Ingram et al., 1987; Barba et al., 1988; Merchant et al., 1988; Yoshida et al., 1988; Nitta et al., 1990). Table 10 summarizes the results of regional therapy of malignant brain tumors with LAK cells and IL-2.

These studies were largely preliminary, and few conclusions about the efficacy of these regional treatment approaches can be drawn.

Specific Adoptive Immunotherapy with Tumor-Infiltrating Lymphocytes *In Vitro* Studies

In an effort to improve the results of immunotherapy, cell populations with more specific tumor reactivity were sought. Studies shifted to the investigation of mononuclear cell infiltrates residing within murine and human neoplasms. The initial attempts to expand and isolate these tumor-infiltrating lymphocytes (TIL) were performed by culturing enzymatic digests of tumors with TCGF (Yron et al., 1980). The culture conditions promoted the eventual outgrowth of a pure population of T cells, while after an initial growth period, tumor cells regressed and died under the negative selection pressure. After rIL-2 became available, additional experiments demonstrated that the growth of TIL was solely dependent on IL-2 and that TIL could be reproducibly generated from a variety of weakly immunogenic murine sarcomas, melanomas, and carcinomas (Rosenberg et al., 1986; Spiess et al., 1987).

In contrast to LAK cells, murine TIL could recognize unique antigens associated with their tumor of origin (Barth et al., 1990, 1991a, 1991b). As shown in Table 11, chromium-release assays demonstrated specific lysis by TIL of the tumor from which they were grown but not of other his-

TABLE 7. Duration of responses (months) in patients receiving LAK cells and rIL-2 (as of March, 1991)

Diagnosis	Complete responses	Partial responses
Renal cell cancer	51+, 47+, 27, 15, 13, 11, 9, 6	45+, 38+, 19+, 13, 11, 11, 9, 7, 7, 6, 6, 6, 6, 3, 2
Melanoma	75+, 56+, 42+, 13	41, 6, 6, 3, 2, 2
Colorectal cancer	21	11, 6, 6, 2
Non-Hodgkin's lymphoma	10	55+, 31+, 14

Adapted from Rosenberg, 1992a.

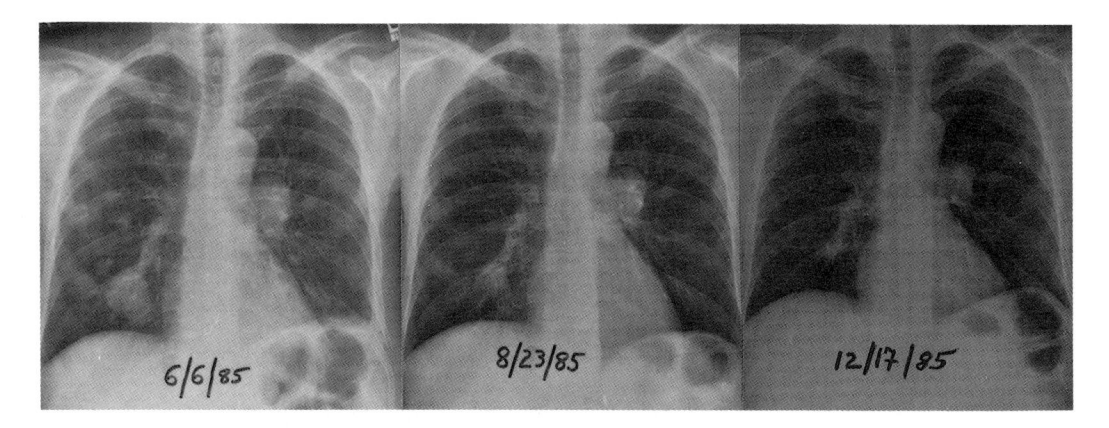

FIGURE 3. Serial chest x-rays showing partial regression of multiple pulmonary metastases after one cycle of LAK cells and rIL-2 in a patient with renal cell carcinoma. (From Rosenberg et al., 1985a.)

tologically similar or dissimilar tumor targets (Barth et al., 1990, 1991a, 1991b). Moreover, coculture of TIL with a panel of tumors showed that murine TIL specifically released interferon (IFN)-γ and/or tumor necrosis factor (TNF)-α only in the presence of the tumor of origin (Barth et al., 1990, 1991a, 1991b). A summary of the *in vito* characteristics of murine TIL is presented in Table 12.

In humans, TIL have been grown from approximately 80% of cancers, including melanomas, renal cell carcinoma, colorectal cancer, breast cancer, squamous head and neck cancer, sarcoma, neuroblastoma, lymphomas, and other neoplasms

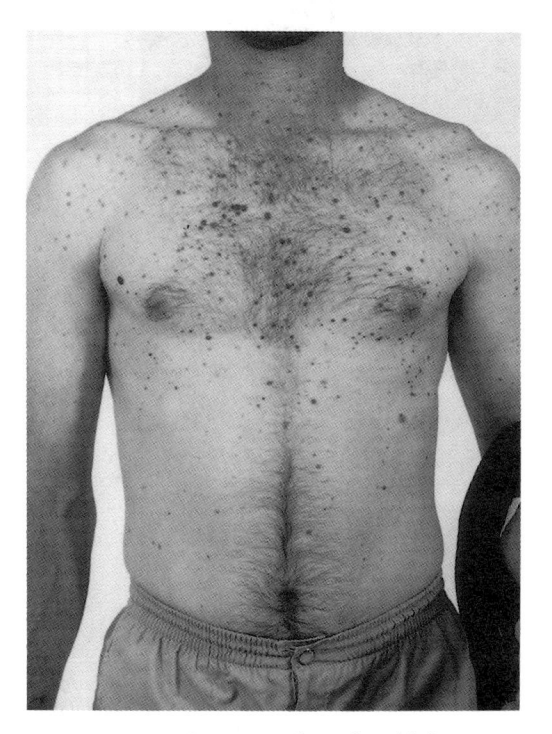

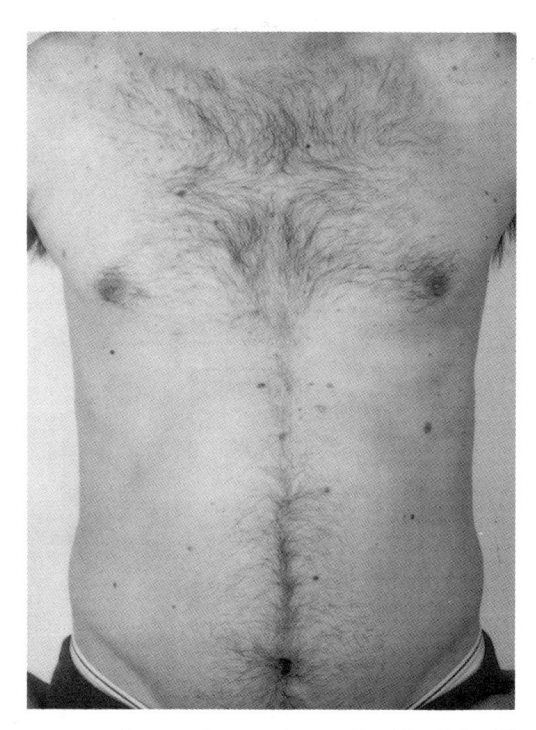

FIGURE 4. Complete regression of multiple cutaneous metastases from melanoma in a patient treated with LAK cells and rIL-2. (A) pretreatment; (B) posttreatment.

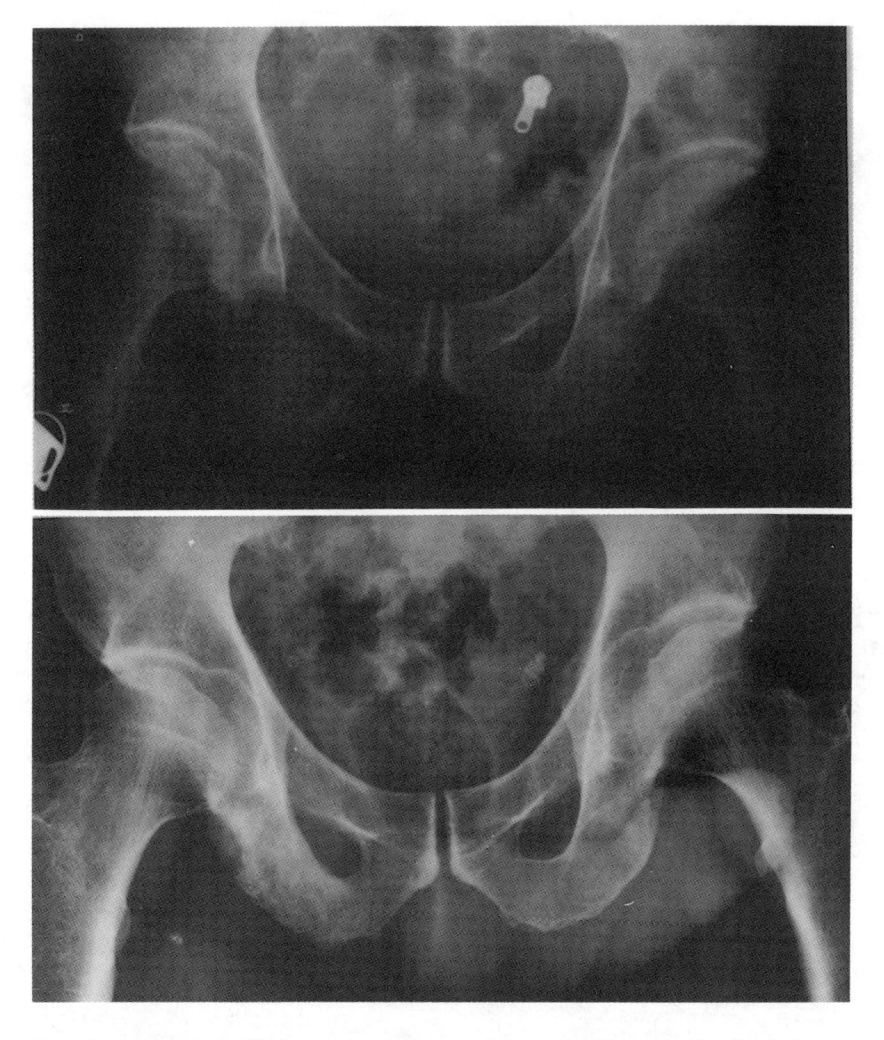

FIGURE 5. Complete regression of bony metastases (to pubic ramus) from renal cell carcinoma in a patient treated with LAK cells and rIL-2. (From Rosenberg, 1992a.)

(Galili et al., 1979; Itoh et al., 1986; Kurnick et al., 1986; Paine et al., 1986; Whiteside et al., 1986a, 1986b; Heo et al., 1987; Muul et al., 1987b; Rabinowich et al., 1987; Belldegrun et al., 1988; Heo et al., 1988; Itoh et al., 1988; Kuppner et al., 1988; Saito et al., 1988; Radrizzani et al., 1989; Tagaki et al., 1989; Topalian et al., 1989; Balch et al., 1990; Haas et al., 1990; Rivoltini et al., 1992). The growth conditions and kinetics for generation of human TIL are similar to those in murine systems, although higher concentrations of IL-2 (6000 IU/ml) are usually required to expand human TIL populations (Topalian et al., 1989). Although TIL grown from many human tumors are frequently nonspecific in their lytic spectrum, TIL derived from human melanomas develop specific cytolytic activity for their autologous tumor but not for allogeneic melanoma in approximately 1/3 to 2/3 of TIL populations (Muul et al., 1987b; Topalian et al., 1989). As shown in Figure 7, TIL derived from three separate melanoma patients mediated lysis of only the melanoma of origin but not of allogeneic melanomas (Topalian et al., 1989). Although TIL isolated from colorectal and breast cancers have not demonstrated lytic specificity, recognition by TIL of unique tumor antigens has been suggested by the production of cytokines by TIL on exposure to their tumor of origin (Schwartzentruber et al., 1991a). Figure 8 demonstrates specific secretion of granulocyte-macrophage colony-stimulating factor (GM-CSF), TNF-α, and IFN-γ by TIL derived from breast cancer during coincubation with autologous tumor but not by other tumors or autologous normal cells (Schwartzentruber et al., 1991a).

TABLE 8. Treatment of patients with advanced cancer using LAK cells plus IL-2

Authors	Year	IL-2 dose and schedule	LAK cells × 10⁻¹⁰ (mean No.)	Diagnosis	Total (No. patients)	CR (No. patients)	PR (No. patients)	CR + PR(%)
West et al.	1987	1 to 7 × 10⁶ U/m²/day CI	NA					
				Melanoma	10	0	5	50
				Renal	6	0	3	50
				Colon	13	0	0	0
Fisher et al.	1988	100,000 U/kg q8h	7.0	Renal	32	2	3	16
Schoof et al.	1988	30,000 U/kg q8h	4.3	Renal	10	1	4	50
				Melanoma	9	1	4	56
				Colorectal	4	0	0	0
Dutcher et al.	1989	100,000 U/kg q8h	8.9	Melanoma	32	1	5	19
Thompson et al.	1989	1 × 10⁶ U/m² q8h	3.4	Melanoma	4	0	0	0
				Renal	3	0	0	0
				Colorectal	4	0	0	0
		3 × 10⁶ U/m²/day CI	4.3	Melanoma	4	0	0	0
				Renal	8	1	0	13
Paciucci et al.	1989	1 to 5 × 10⁶ U/m²/day CI	5.6	Melanoma	5	0	1	20
				Renal	9	0	1	20
				Colorectal	1	0	1	100
Negrier et al.	1989	3 × 10⁶ U/m²/day	1.2	Renal	51	5	9	27
Stahel et al.	1989	3 × 10⁴ U/kg q8h	5.1	Renal	14	0	3	21
				Melanoma	7	0	1	14
				Colon	2	0	0	0
Hawkins	1989	10⁵ U/kg q8h		Melanoma	32	1	5	19
				Colorectal	19	1	2	16
Bar et al.	1990	18 × 10⁶ IU/m²/day CI	8.3	Melanoma	50	1	6	14
Gaynor et al.	1990	18–27 × 10⁶ IU/m²/day CI	14.3–16.0	Melanoma	30	0	1	3
				Renal	25	2	2	16
Dutcher et al.	1991	22.5 × 10⁶ IU/m²/day CI	16	Melanoma	33	0	1	3
Weiss et al.	1992	22.5 × 10⁶ IU/m²/day CI	16.3	Renal	48	2	5	15
		5.94 × 10⁵ IU/kg q8h	13.7	Renal	46	3	6	20

NA, Not available.
CI, continuous infusion; CR, complete response; PR, partial response.
Adapted from Rosenberg, 1991.

MHC-restricted TIL derived from melanomas exhibit lysis of some allogeneic melanomas suggesting recognition by TIL of a target shared by autologous and allogeneic tumors (Darrow et al., 1991; Hoy et al., 1991) Human leukocyte antigen (HLA) typing of TIL and melanomas demonstrates that in cases of allogeneic tumor lysis, the TIL and tumor target frequently share at least one HLA restriction element (Darrow et al., 1989; Hom et al., 1991). Systematic testing of panels of HLA-typed TIL and melanomas show that HLA-A2 and HLA-A24 are common restriction determinants (Hom et al., 1991). As demonstrated in Table 13, HLA-A2⁺ TIL 553 mediated lysis of its autologous melanoma and 14 of 15 HLA-A2⁺ matched allogeneic melanomas, while no significant lysis was generated against non-HLA-matched melanomas (Hom et al., 1991). Although recognition of shared MHC molecules is required for lysis, TIL recognition of melanoma antigen is also necessary, since autologous Epstein-Barr virus (EBV)-transformed B cell lines bearing common MHC determinants are not lysed by TIL (Hom et al., 1991).

A summary of the *in vitro* studies of human TIL is presented in Table 14.

Specific Adoptive Immunotherapy with TIL: Preclinical Trials

In murine models of lung and liver micrometastases from sarcomas, melanomas, and carcinomas, TIL can significantly reduce the numbers of metastases

TABLE 9. Toxicity of treatment with IL-2

No. of patients	652
No. of courses	1039
Chills	399
Pruritus	180
Necrosis	5
Anaphylaxis	1
Mucositis	30
Alimentation not possible	4
Nausea and vomiting	666
Diarrhea	596
Hyperbilirubinemia, maximum (mg/dl)	
2.1–6.0	547
6.1–10.0	179
10.1+	83
Oliguria	
< 80 ml/8 hr	347
< 240 ml/24 hr	42
Weight gain (% body weight)	
0.0–5.0	377
5.1–10.0	436
10.1–15.0	175
15.1–20.0	38
20.1+	13
Elevated creatinine, maximum (mg/dl)	
2.1–6.0	637
6.1–10.0	85
10.1+	10
Hematuria (gross)	2
Edema (symptomatic nerve or vessel compression)	17
Tissue ischemia	2
Respiratory distress	
Not intubated	67
Intubated	41
Bronchospasm	9
Pleural effusion (requiring thoracentesis)	17
Somnolence	114
Coma	33
Disorientation	215
Hypotension (requiring pressors)	508
Angina	22
Myocardial infarction	6
Arrhythmias	78
Anemia requiring transfusion (no. of units transfused)	
1–5	377
6–10	95
11–15	24
16+	14
Thrombocytopenia, minimum (/μl)	
< 20,000	131
20,001–60,000	361
60,001–100,000	285
Central line sepsis	63
Death	10

From Rosenberg et al., 1989b.

as compared with controls (Rosenberg et al., 1986; Spiess et al., 1987). Although optimal results are seen when TIL are given with repetitive injections of rIL-2, an immunotherapeutic effect can also be mediated by TIL alone (Spiess et al., 1987). TIL titration studies demonstrated a direct dose-response relationship between the *in vivo* therapeutic efficacy of TIL and the number of TIL transferred, as well as the amount of rIL-2 injected (Spiess et al., 1987; Figure 9). In comparative studies as presented in Figure 9, TIL were 50 to 100 times more potent on a per cell basis when compared with LAK cells (Spiess et al., 1987). In 8- to 14-day-old lung and liver metastasis models, therapy with TIL and rIL-2 in conjunction with cyclophosphamide could mediate significant regression of tumor and could prolong overall survival (Rosenberg et al., 1986). The results of other preclinical studies of TIL in murine models are presented in Table 15.

Clinical Trials with Specific Adoptive Therapy Utilizing TIL

Based upon promising results of murine studies utilizing TIL in the treatment of advanced cancers, a pilot trial of escalating doses of cyclophosphamide and TIL with rIL-2 was initiated in the Surgery Branch of the NCI in 12 patients with disseminated cancer (Topalian et al., 1988). TIL were expanded from various sites of tumor including subcutaneous, nodal, breast, and visceral areas (spleen, kidney). Cyclophosphamide (up to 50 mg/kg) was administered prior to TIL infusion, since animal models predicted that immunosuppression prior to adoptive therapy would yield the optimal results with advanced disease (Rosenberg et al., 1986). Varying numbers of TIL (0.8 to 23×10^{10} cells) and doses of rIL-2 (72,000 to 7.2 million IU/kg I.V. thrice daily) were administered to the patients. Phenotypic analysis of the transferred TIL demonstrated that > 80% of the cells were OKT3/Leu 4 positive in 10 of 12 patients. Leu 3-expressing TIL exceeded Leu 2 positive cells in 11 of 15 cultures. A partial response to therapy was observed in two patients—one with melanoma and the other with renal cell carcinoma. Subsequently, a phase II protocol utilizing TIL and high dose rIL-2 was initiated for patients with metastatic melanoma (Rosenberg et al., 1988a). This disease was selected for study because lytic specificity was frequently demonstrated in TIL derived from

(A)

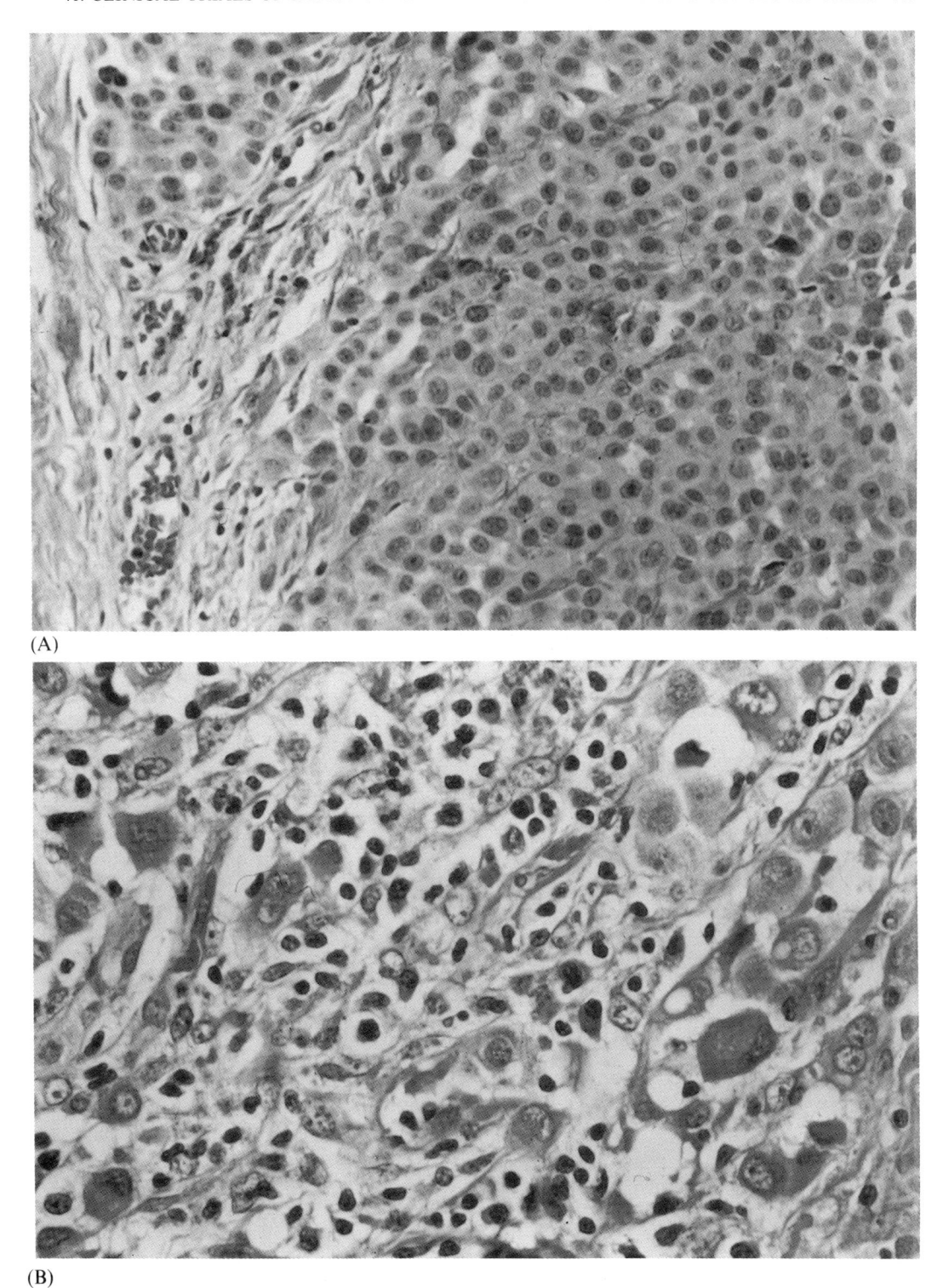

(B)

FIGURE 6. Light microscopic appearance of subcutaneous metastases from a melanoma patient treated with LAK cells and rIL-2. (A) pretreatment; (B) posttreatment. Histologic sections demonstrate extensive infiltration of tumor by activated lymphocytes associated with widespread tumor cell dropout and necrosis. (From Rosenberg, 1992a.)

TABLE 10. Regional therapy of malignant brain tumors with LAK cells and IL-2

Author	Year	Treatment	No. patients	Comments
Jacobs et al.	1986	Direct injection of LAK cells into tissue surrounding tumor cavity	15	No measurable tumor at time of treatment
Ingram et al.	1987	Plasma clot containing LAK cells placed into resected tumor bed (PHA + IL-2 used in cell incubation)	55	No measurable tumor at time of treatment; survival data pending
Merchant et al.	1988	Direct injection of LAK cells into tissue surrounding the tumor cavity during craniotomy	20	No measurable tumor at time of treatment; survival data pending
Yoshida et al.	1988	Direct injection of LAK cells into recurrent cavity with 50 to 400 units IL-2; multiple treatments	23	Tumor regression in six tumor patients
Barba et al.	1988	Direct injection of LAK cells plus IL-2 into tumor or resected tumor cavity using an Ommaya reservoir	9	One partial response. Neurological side effects secondary to cerebral edema
Nitta et al.	1990	Direct injection of LAK cells into tumor site after "debulking" in conjunction with anti-CD3-antiglioma bispecific antibody	10	"Regression" of tumor in four patients. No recurrence in any patient at 8 to 18 months follow-up

From Rosenberg, 1991.

melanomas but not in TIL derived from other histologies. In the treatment regimen, patients received cyclophosphamide (25 mg/kg; the optimally tolerated dose demonstrated in the phase I pilot trial) followed 36 hr later by TIL (1.3 to 75 × 10^{10} cells) accompanied by rIL-2 (720,000 IU/kg) I.V. every 8 hr up to 15 doses or as limited by toxicity. The results of the phase II trial are presented in Table 16. Four of 11 (36%) patients previously treated with rIL-2 showed an objective response to the combination of TIL and rIL-2, as compared with

TABLE 11. Cytolytic specificity of tumor-infiltrating lymphocytes (TIL) for tumor of origin

	MCA tumor target ($LU_{30}/10^7$ cells)[a]			
Effector	102	203	205	207
LAK cells	250	160	280	80
TIL 203	<1	1850	<1	<1
TIL 207	<1	<1	<1	590

[a] One lytic unit (LU) defined as the number of effector cells required to mediate 30% lysis of 10^4 fresh MCA sarcoma cells.
Adapted from Barth et al., 1990.

15 of 39 (38%) patients who had not received previous rIL-2-based immunotherapy (Rosenberg et al., 1988a; Rosenberg, 1992a). Although not proven in a randomized trial, the results of this study suggested that the response rate of patients receiving TIL and rIL-2 was higher than those treated in previous trials with LAK cells and rIL-2 (20 to 25% response rate) (Rosenberg, 1992a).

Considerable work has been undertaken to identify factors associated with TIL or tumor that might predict response to IL-2-based immunotherapy. Aebersold et al. found that *in vitro* cytolytic activity of TIL for autologous tumor was significantly higher in responding than in nonresponding patients (Aebersold et al., 1991). Other studies have evaluated MHC class I and II expression on tumors and its impact on response to immunotherapy (Rubin et al., 1989). Although pretreatment expression of MHC class II molecules on tumors was not a predictor of later clinical response, posttreatment biopsies of tumor demonstrated HLA-DR antigen on >50% of cells in 7 of 7 regressing tumors but in only 3 of 10 nonresponding metastases. Responding lesions were infiltrated by CD4 and CD8 T-cell subpopulations and macrophages.

TABLE 12. *In vitro* characteristics of murine TIL

1. TIL may be generated by the incubation in rIL-2-containing media of enzymatic digests of a variety of weakly immunogenic sarcomas, carcinomas, and melanoma (Rosenberg et al., 1986; Spiess et al., 1987; Barth et al., 1990, 1991a, 1991b).
2. TIL effector populations are CD3-, CD8-, and Thy-1-positive. TIL do not express the asialo GM1 antigen (Spiess et al., 1987; Barth et al., 1990, 1991a, 1991b).
3. TIL can often specifically lyse the tumor of origin but not other methylcholanthrene-induced sarcoma cells. *In vitro* stimulation of TIL with its tumor of derivation frequently results in the specific release of TNF-α and/or IFN-γ (Barth et al., 1990, 1991a, 1991b).
4. TIL with enhanced cytolytic potency may be grown from lymphocytes separated from bulk tumor digests using immunomagnetic beads coated with Thy-1.2 antibodies. Immunobeaded TIL can then be expanded to large numbers with maintenance of antitumor reactivity if cultured in media containing low doses of rIL-2 (5–20 U/ml) and stimulated repetitively with antigen using irradiated tumor of origin (Yang et al., 1990).

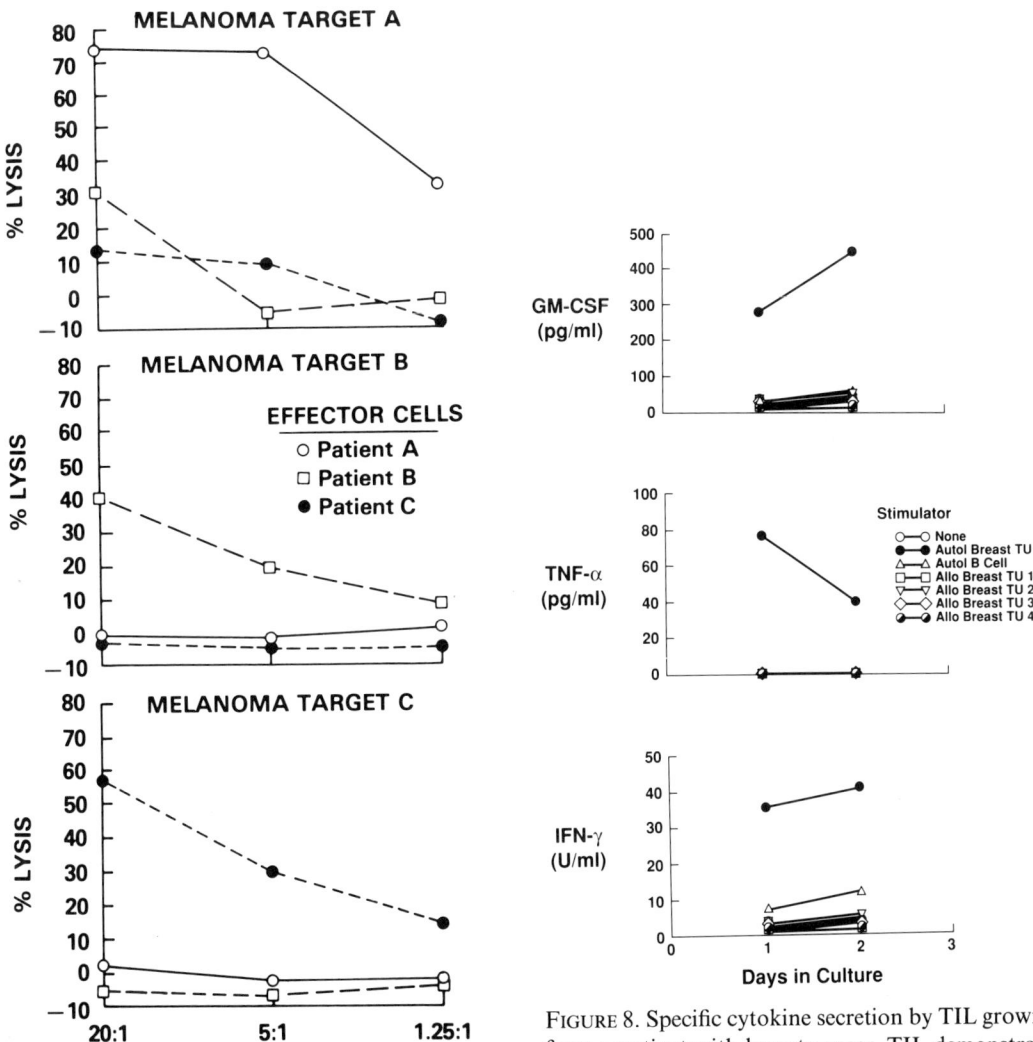

FIGURE 7. Specific lysis of autologous but not allogeneic melanomas by TIL grown from freshly excised metastatic melanoma in 3 patients (patient A, ○; patient B, □; patient c, ●). (From Muul et al., 1987b.)

FIGURE 8. Specific cytokine secretion by TIL grown from a patient with breast cancer. TIL demonstrated recognition of unique tumor antigens by secretion of GM-CSF, TNF-α, and IFN-γ after a 1- or 2-day incubation with irradiated autologous tumor, but not with four irradiated allogeneic breast tumors or an irradiated autologous B-cell line. (From Schwartzentruber et al., 1991a.)

TABLE 13. Cytolytic activity of 553 TIL against HLA-matched and unmatched targets

% lysis[a]	Autologous	Unmatched	HLA-A2[b]	HLA-A23	HLA-B40	HLA-B44	HLA-Cw3	HLA-Cw4
> 10	553 19 ± 4(4)[c]		864 15 ± 4			889 15 ± 4	(697)[d]	(882)
	553-mel[e] 29 ± 7(3)		836 24 ± 3(3)			878 13 ± 5(3)	(697mel)	(677)
			836-mel 12			(809)	(526-mel)	(677-mel)
			822 16 ± 2(3)			(549)		(501-mel)
			809 26 ± 4(3)					
			697 29 ± 7(3)					
			697-mel 60 ± 3(3)					
			677 40					
			677-mel 57					
			624-mel 47 ± 8					
			551 25 ± 3					
			549 20 ± 7					
			526-mel 58					
≤ 10		888 5	890 10 ± 1(3)	893 2		586-mel 1	(851)[f]	397-mel 6
		865 2	851[f] 6 ± 3			537-mel 6	(697-EBV)	(501-EBV)
		796 0	697-EBV[g] 3			(893)		
		790 1	551-EBV					
		Daudi 3 ± 1(18)	501-EBV 4					

[a]Lysis in 4-hr ^{51}Cr-release at E/T = 40:1. Values ≤ 10% are not considered significant.
[b]Target donor HLA Ag matching TIL effector HLA Ag.
[c]Target cells, mean % lysis ± SEM (n) = 1 where no SEM is indicated, or (n) = 2 unless otherwise specified. Target cells are fresh cryopreserved melanomas unless otherwise indicated.
[d]Targets matching more than one TIL HLA locus are indicated in parentheses.
[e]"Mel", culture melanoma line.
[f]Fresh sarcoma target.
[g]"EBV," EBV-transformed B-cell line.
From Hom et al., 1991.

Attempts to correlate TIL phenotype and *in vivo* TIL efficacy have not been successful (Topalian et al., 1988).

Kradin et al. have utilized TIL-based immunotherapy for 28 patients with metastatic melanoma, renal cell carcinoma, and non-small-cell carcinoma of the lung (Kradin et al., 1989). TIL were expanded from tumor fragments incubated in rIL-2 and later in media containing PHA with irradiated PBMC. While not utilizing pretreatment cyclophosphamide, the therapeutic regimen included up to 7 infusions of TIL at lower numbers (approximately 10^{10} cells) and IL-2 administered by continuous infusion (at 1–3 × 10^6 U/m^2/24 hr). Objective partial responses lasting 3–14 months were observed in 2 patients with renal cell cancer (29%) and three with melanoma (23%). None of the 8 patients with lung cancer responded.

During trials of TIL for treatment of cancer at the NCI, studies were also conducted to evaluate the ability of TIL to target to sites of tumor (Fisher et al., 1989; Griffith et al., 1989). In this protocol, 18 patients with metastatic melanoma received TIL labeled with ^{111}In just prior to adoptive transfer followed by rIL-2 injections. Five patients received radiolabeled autologous PBL. Radionuclide scintigraphy demonstrated TIL localization to the lungs, liver, and spleen during the initial 2 hr. Over the next 1 to 9 days, TIL were found to concentrate in sites of tumor in 13 of 18 patients, while PBL localized to tumor deposits in only 1 of 4 patients (Figure 10). Sequential biopsies of tumor and adjacent normal tissues demonstrated preferential localization (up to 3- to 40-fold) of labeled TIL to tumors over time (Figure 11). The demonstration of tumor-specific homing patterns of TIL laid the groundwork for the idea that TIL might be used as vehicles with which to deliver anticancer substances to the tumor site. With advances in molecular biological techniques, manipulation of TIL by the insertion of genes coding for therapeutic molecules became a logical strategy for improvement of immunotherapeutic approaches to cancer treatment.

TABLE 14. *In vitro* characteristics of human TIL

1. TIL may be grown from approximately 80% of human cancers, including melanomas, renal cell carcinomas, colorectal cancers, breast cancers, and other neoplasms (Galili et al., 1979; Itoh et al., 1986; Kurnick et al., 1986; Paine et al., 1986; Whiteside et al., 1986a, 1986b; Heo et al., 1987; Muul et al., 1987b; Rabinowich et al., 1987; Belldegrun et al., 1988; Heo et al., 1988; Itoh et al., 1988; Kuppner et al., 1988; Saito et al., 1988; Radrizzani et al., 1989; Tagaki et al., 1989; Topalian et al., 1989; Balch et al., 1990; Haas et al., 1990; Rivoltini et al., 1992).
2. Approximately 33 to 66% of TIL derived from melanomas display the ability to lyse the melanoma of origin, but not allogeneic melanomas (Muul et al., 1987; Topalian et al., 1989).
3. TIL derived from colorectal and breast cancers and sarcomas do not generally develop specific lysis of their autologous tumor; renal cell carcinoma TIL may occasionally demonstrate lytic specificity (Galili et al., 1979; Itoh et al., 1986; Kurnick et al., 1986; Paine et al., 1986; Whiteside et al., 1986a, 1986b; Heo et al., 1987; Muul et al., 1987; Rabinowich et al., 1987; Belldegrun et al., 1988; Heo et al., 1988; Itoh et al., 1988; Kuppner et al., 1988; Saito et al., 1988; Radrizzani et al., 1989; Tagaki et al., 1989; Topalian et al., 1989; Balch et al., 1990; Haas et al., 1990; Rivoltini et al., 1992).
4. Nonlytic TIL may demonstrate specific recognition of unique tumor antigens by secretion of cytokines during coincubation with autologous tumor but not autologous normal cells or allogeneic tumors (Schwartzentruber et al., 1991a).
5. TIL are predominantly $CD3^+$; the effector populations usually express CD4 or CD8 phenotypes or are mixtures of both cell types. Over time, the development of lytic specificity by melanoma TIL can be correlated with expression of CD3, CD8, as well as HLA-DR determinants (Muul et al., 1987; Topalian et al., 1989).
6. Recognition of tumor antigen by melanoma TIL is mediated through the T-cell receptor and is MHC class I-restricted; antibodies to class I and CD3 effectively block specific lysis by TIL (Topalian et al., 1989).
7. Preincubation of tumor cells with IFN-γ or TNF-α alone or in combination augments lysis by TIL (Stotter et al., 1992).
8. The expansion of melanoma TIL in IL-2 and IL-4 can lead to an increase in lytic specificity (Kawakami et al., 1988).
9. MHC-restricted TIL derived from melanoma may demonstrate lysis of allogeneic melanomas through the recognition of shared tumor antigens if the TIL and melanoma targets share a crucial common HLA determinant. HLA-A2 and -A24 are common restriction elements for tumor target recognition by melanoma TIL (Darrow et al., 1989; Hom et al., 1991).
10. HLA-A2 restricted-melanoma TIL recognize shared tumor antigens and lyse HLA-A2-negative melanoma targets after transfection of the gene for HLA-A2 into the targets (Kawakami et al., 1992).

Gene Therapy Utilizing TIL in Specific Adoptive Immunotherapy

Gene therapy involves the transfer of exogenous genetic information to a patient's cells so as to alter cellular functions to correct an inborn genetic error or impart a new function to the cells (Rosenberg, 1992a).

The initial manipulations of TIL in humans involved the insertion of the gene coding for neomycin phosphotransferase, thereby conferring upon these cells the ability to grow in otherwise toxic, neomycin-containing media (Kasid et al., 1990). More importantly, the neomycin-resistance (*neoR*) gene transferred into TIL served as a cellular marker which would distribute to progeny TIL during cell expansion (Kasid et al., 1990; Rosenberg, 1992a). This genetic marker could then facilitate the investigation of tissue distribution, tumor localization, and *in vivo* survival of TIL. Prior studies of *in vivo* distribution patterns of TIL labeled with ^{111}In had been limited by the short half-life of ^{111}In (2.8 days) and the spontaneous release of the radioisotope from cells (Fisher et al., 1989; Griffith et al., 1989). Moreover, potential injury to the labeled TIL existed from continuous autoirradiation by ^{111}In (Rosenberg, 1992a). Lastly, preliminary efforts to insert the *neoR* gene into human TIL could establish the safety and efficacy of the molecular biological techniques and of the resulting genetically altered cells in preparation for future protocols involving therapeutic manipulations of human cells (Rosenberg, 1992a).

The gene insertion protocols centered on the use of retrovirally mediated gene transfer, since physical methods yielded an unacceptably low and inefficient transfection rate, while retroviral vectors could mediate gene transfer with efficiencies up to 50% (Giboa et al., 1986; Eglitis and Anderson, 1988; Kasid et al., 1990; Kniegler, 1990; Rosenberg, 1992a, 1992b). Genetic transduction of human TIL was accomplished by the use of a modified Moloney murine leukemia retrovirus containing the gene encoding for neomycin phosphotrans-

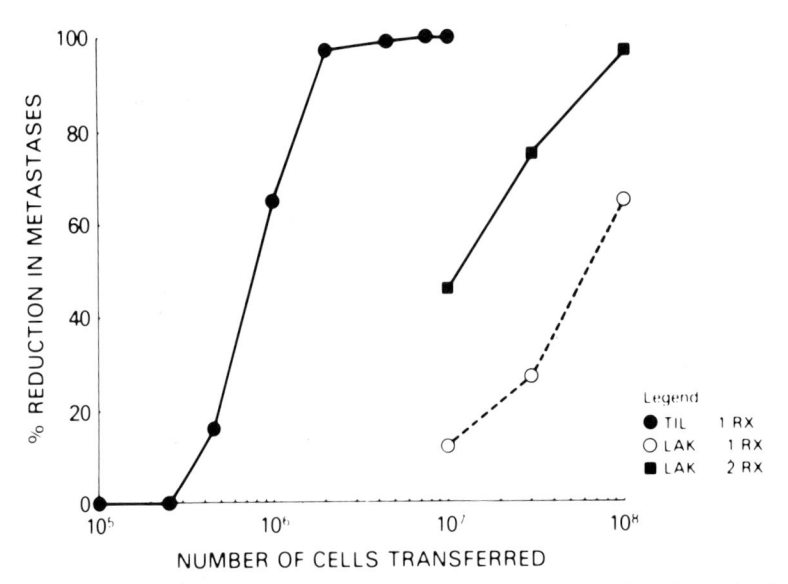

FIGURE 9. Comparison of TIL and LAK cells administered with rIL-2. Mice bearing 3-day lung metastases from MCA 105 received 2×10^6 TIL (1 treatment) or 10^8 LAK cells (1 or 2 treatments). TIL appear to be 50 to 100 times more potent in reducing pulmonary metastases when compared with LAK cells. (From Spiess et al., 1987.)

ferase (Miller and Buttimore, 1986; Miller and Rosman, 1989; Kasid et al., 1990). The PA317 packaging cell line was used to generate the retroviral vector, which lacked the viral *gag*, *pol*, and *env* genes (Giboa et al., 1986; Miller and Buttimore, 1986; Eglitis and Anderson, 1988; Miller and Rosman, 1989; Kasid et al., 1990; Kriegler, 1990; Rosenberg, 1992b). Consequently, infection of TIL by the retrovirus could result in transfer of the neomycin resistance gene into TIL without the risk of generation of new viral particles within the infected cells (Giboa et al., 1986; Miller and Buttimore, 1986; Eglitis and Anderson, 1988; Miller and Rosman, 1989; Kasid et al., 1990; Kriegler, 1990;

TABLE 15. Preclinical studies of TIL in the murine system

---

1. Adoptive transfer of TIL to tumor-bearing mice can mediate regression of lung and liver micrometastases from sarcomas, carcinomas, and melanomas (Rosenberg et al., 1986; Spiess et al., 1987).
2. A direct dose-response relationship may be demonstrated between the number of TIL and the amount of rIL-2 administered in the immunotherapeutic effect (Rosenberg et al., 1986; Spiess et al., 1987).
3. TIL are effective *in vivo* even when transferred without subsequent injections of rIL-2 (Spiess et al., 1987).
4. Addition of cyclophosphamide to immunotherapy with TIL and rIL-2 can successfully treat advanced (8- to 14-day old) lung and liver (Rosenberg et al., 1986).
5. On a per cell basis, TIL are 50 to 100 times more potent than LAK cells (Spiess et al., 1987).
6. TIL can be isolated from mice cured of micrometastases up to 6 weeks to 3 months after cell transfer (Alexander and Rosenberg, 1990, 1991).
7. TIL and rIL-2 can mediate significant tumor regressions when administered to mice immunosuppressed by pretherapy total body irradiation or cyclophosphamide. These results suggest that a fully immunocompetent host immune system is not required for *in vivo* TIL function (Rosenberg et al., 1986).
8. Application of local radiation to a tumor site provides a synergistic effect when combined with TIL and rIL-2 immunotherapy (Cameron et al., 1990).
9. The best predictor of *in vivo* antitumor efficacy of TIL is specific cytokine secretion by TIL *in vitro*. IFN-γ secretion appears to be a better predictor of therapeutic activity than TNF-α production by TIL. Nonlytic TIL which secrete IFN-γ when cocultured with the tumor of origin are effective in a micrometastatic model (Barth et al., 1991).
10. TIL can be raised from tumors lacking MHC class I surface molecules following the transfection and expression of the gene for class I in the tumor (Weber and Rosenberg, 1990).
11. Concurrent treatment of mice bearing lung metastases with IFN-α augments the immunotherapeutic effect of TIL and rIL-2 by up-regulation of MHC surface determinants (Rosenberg et al., 1988b).

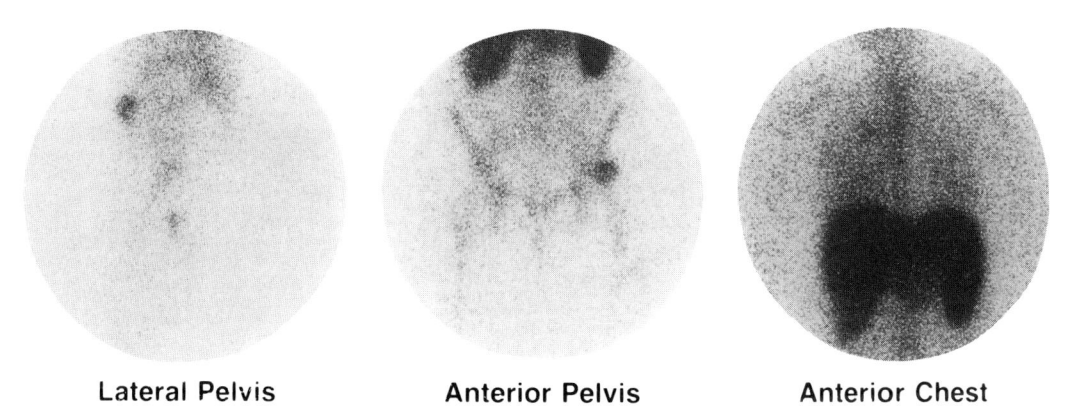

Lateral Pelvis   **Anterior Pelvis**   **Anterior Chest**

FIGURE 10. Radionuclear scans of the pelvis and anterior chest of a patient with left groin and right anterior chest wall metastases from melanoma. Images taken 5 days after TIL transfer show localization of ^{111}In-labeled TIL to the sites of tumor. (From Fisher et al., 1989.)

Rosenberg, 1992b). Extensive studies showed that the *neoR* gene could be successfully transduced into human TIL and be expressed as demonstrated by both Southern transfer analysis and continued growth of transduced TIL in media containing the neomycin analog, G418 (Kasid et al., 1990). Furthermore, phenotypic characteristics and *in vitro* function of TIL were not significantly altered following transduction (Kasid et al., 1990). Lastly, exhaustive safety testing with *in vitro* assays and in primates demonstrated that there was little or no risk to patients and no risk to health care personnel or the public from use of the gene-transduced TIL (Kasid et al., 1990; Culver et al., 1991b).

A pilot trial of *neoR*-transduced TIL was begun in May 1989 for patients with widely metastatic melanoma (Rosenberg et al., 1990). A total of 10 patients have been treated to date. None have

TABLE 16. TIL treatment of patients with melanoma[a]

	NR	Objective response (PR + CR)	PR + CR (all)	%
No previous IL-2				
IL-2 + CY	17	11	11/28	39
IL-2 (no CY)	7	4	4/11	36
Previous IL-2				
IL-2 + CY	3	3	3/6	50
IL-2 (no CY)	4	1	1/5	20

[a]Excludes patients with brain metastases at start of therapy and two patients who received low-dose rIL-2 (all NR).
NR, No response; PR, partial response; CR, complete response; CY, cyclophosphamide.
From Rosenberg et al., 1988a; Rosenberg, 1992a.

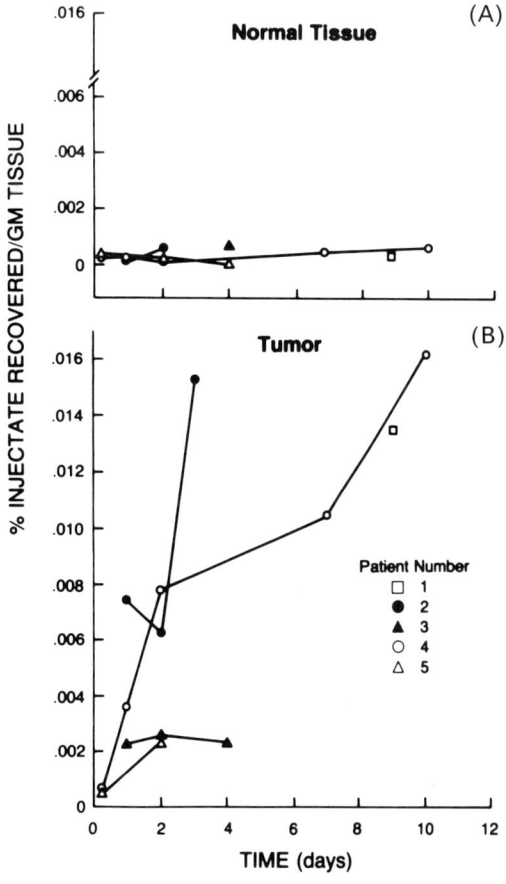

FIGURE 11. Preferential localization of TIL labeled with ^{111}In to subcutaneous melanoma metastases (B) over normal skin (A) in 5 patients (○, ●, △, ▲, □). The percent of injectate per gram of tumor and normal tissue is shown. In 3 of 5 patients, localization of TIL increased over time. (From Fisher et al., 1989.)

TABLE 17. Characteristics of patients with metastatic melanoma receiving neomycin-resistant TIL

Patient No.	Age (years)/sex	Primary site	Prior treatment	Harvested tumor Site	Size (cm)	Total No. cells obtained ($\times 10^{-7}$)	No. cells to start culture ($\times 10^{-7}$)[a]	% lymphocytes	% tumor cells	Sites of assessable disease
1	52/M	Neck	Wide local excision	Lymph node	$4 \times 4 \times 2$	33	12	15	85	Lung, liver, spleen
2	46/F	Finger	Amputation finger Lymph node dissection	Lymph node	$5 \times 5 \times 3$	157	50	70	30	Lymph nodes, intramuscular
3	42/M	Back	Wide local excision Lymph node dissection Melanoma vaccine IL-2/IFN-α	Lymph node Subcutaneous Subcutaneous	$6 \times 5 \times 4$ $2 \times 2 \times 2$ $2 \times 2 \times 2$	205	120	31	69	Lung, subcutaneous
4	41/M	Chest	Wide local excision Lymph node dissection IL-2/IFN-α	Subcutaneous Subcutaneous	$2 \times 1 \times 1$ $5 \times 4 \times 4$	41	4	39	61	Lung, liver, lymph nodes, subcutaneous, brain
5	26/F	Arm	Wide local excision	Subcutaneous	$2 \times 2 \times 2$ to $5 \times 4 \times 2$	71	15	16	84	Lung, lymph nodes, subcutaneous

[a]Denotes the number of cells used to start the cultures that were ultimately administered to the patient.
From Rosenberg et al., 1990.

TABLE 18. Characteristics of infused neomycin-resistant TIL

Patient No.	Cells transduced day of transduction	No. cells transduced ($\times 10^{-8}$)	Multiplicity of infection[a]	Total days of growth	Fold expansion[b]	Doubling time[c] (days)	Cells infused ($\times 10^{-10}$)	Estimate of cells transduced (%)	Cycles administered	IL-2 doses to patient
1	No	—	—	36, 37	63,400	2.5	22.8	—	1	7
	Yes/13	1.8	2.3	60	16,100	7.5	7.1	1	2	
2	No	—	—	65	209	3.5	0.2	—	1	15
	Yes/19	2.5	1.3	65	5,030	3.5	13.2	11	1	
3[d]										
A	No	—	—	35	874	2.8	10.8	—	1	13
	Yes/12	1.8	1.6	48	35,100	2.5	14.5	1	2	
B	No	—	—	35	324	2.3	3.5	—	1	
	Yes/12	1.4	1.6	48	21,400	3.5	5.5	1	2	
4	No	—	—	36	8,470	2.0	26.0	—	1	10
	Yes/16	1.8	1.7	36	9,480	2.0	3.0	4	1	
5	No	—	—	30	18,900	2.0	15.0	—	1	7
	Yes/8	0.6	1.6	30	5,250	2.3	6.2	10	1	

[a]Ratio of number of virions to number of TIL during the transduction procedure.
[b]Calculated fold expansion of cells administered. Not all cultured cells were given; some cells were diverted for experimental studies or lost to contamination. Nontransduced cells shown here are those in conjunction with the transduced cells administered to the patient.
[c]Varied during culture growth; doubling time at time of final cell infusion is presented here.
[d]Two separate cultures transduced and expanded separately. Culture A was started in AIM-V medium; culture B was started in RPMI 1640 plus 105 human serum.
From Rosenberg et al., 1990.

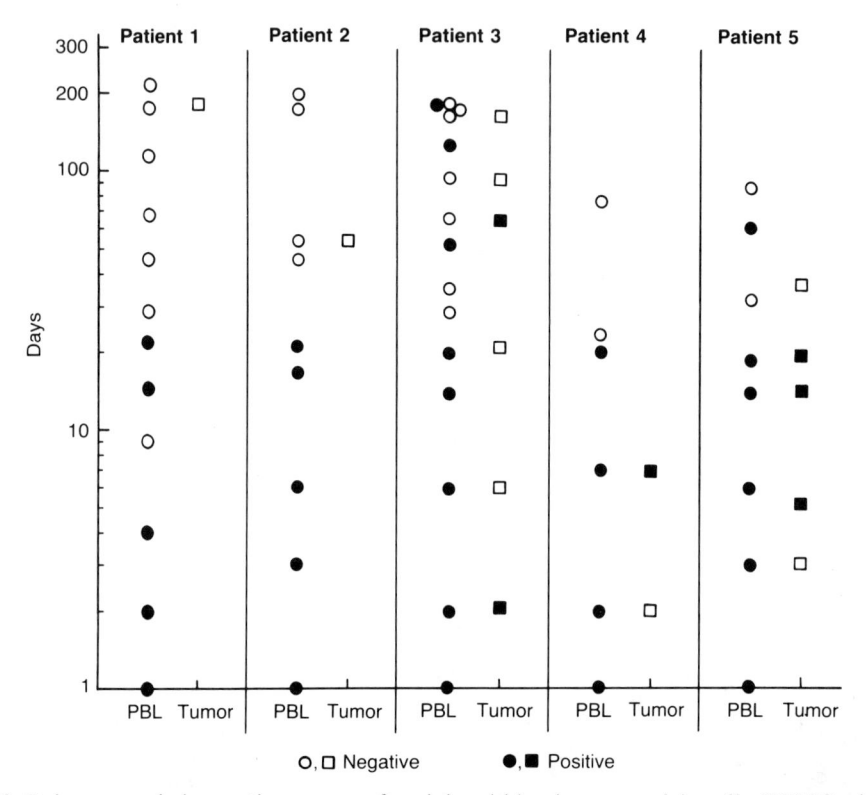

FIGURE 12. Polymerase chain reaction assays of peripheral blood mononuclear cells (PBMC, circles) and tumor biopsies (squares) obtained from 5 patients following infusion of TIL transduced with the gene for neomycin resistance. Open symbols: negative results; closed symbols: positive results. (From Rosenberg et al., 1990.)

demonstrated any toxicity referable to the use of the genetically altered TIL. No patient was exposed to transduced TIL contaminated by helper virus, as shown by $S^+L^-/3T3$ amplification assays or to viral genes encoding reverse transcriptase or capsid or envelope proteins. The characteristics of the initial five patients are shown in Table 17. Standard methods were employed to expand TIL from tumor digests *in vitro*. When TIL were actively proliferating, usually by days 8 to 16, 6 to 25 × 10^8 TIL were removed for transduction (Table 18). In the first five patients, 3.3 to 14.5 × 10^{10} transduced TIL were infused with rIL-2 approximately 1 to 2 months after the initial tumor harvest. As shown in Figure 12, gene-modified TIL could be isolated from peripheral blood and tumor up to 189 and 64 days after infusion, respectively, as determined by results of polymerase chain reaction assays. Two of the initial five patients treated with *neoR*-TIL exhibited significant tumor regression, including one patient with a complete response now maintained for almost 3 years.

With the demonstration that transduced human cells could be successfully generated and safely infused into patients, efforts turned toward gene modification for potential therapeutic benefit. In the first of two areas of study, investigators initiated a protocol to treat children with severe combined immunodeficiency, a disease caused by adenosine deaminase (ADA) deficiency (Calver et al., 1991a; Anderson, 1992). To correct this metabolic defect, the gene coding for the missing enzyme was retrovirally transduced into autologous lymphocytes obtained by leukaphereses.

At the same time, other investigations were begun to modify TIL by insertion of genes encoding molecules that would enhance the antitumor effect of TIL. The initial efforts were directed towards the use of the gene coding for the molecule TNF-α, which had been cloned successfully in 1985 (Wang et al., 1985). Early work in mice demonstrated that systemically administered TNF-α at doses of 400 µg/kg could mediate the regression of established hepatic and subcutaneous tumors (Asher et al., 1987). A phase I trial of escalating TNF-α in 39 patients with advanced cancer revealed that the maximally tolerated dose of TNF in humans was 8 to 10 μg/kg (Rosenberg et al., 1989b). Even at these toxic doses, no antitumor effect was observed (Rosenberg et al., 1989b). However, the mouse model predicted that a 40- to 50-fold increase in the TNF concentration at the local tumor site might result in antitumor responses in humans (Rosenberg, 1992a). With this rationale and the previous demonstration that TIL selectively localized to tumor, attempts were made to insert the gene for TNF into TIL. A retroviral vector was selected in which the TNF gene was promoted by the viral long terminal repeat and the neomycin phosphotransferase gene driven by the SV40 early promoter (Rosenberg, 1992a). Following demonstration of successful transduction and expression of the TNF and neomycin-resistance genes into TIL and the safety of the retrovirally altered TIL, a pilot trial of TNF-transduced TIL was initiated in January 1991 for patients with metastatic melanoma (Rosenberg, 1992a). The protocol schema called for twice weekly administration of TIL, beginning at 10^8 cells and escalating by threefold steps to 3×10^{11} cells or to the maximally tolerated number of TIL. Once that level had been reached, the number of TIL infused would be decreased 10-fold and the TIL infusions combined with rIL-2 injections at 180,000 IU/kg I.V. every 8 hr. Although seven patients have been treated to the present time, it is too early to make any assessment of the efficacy of these TNF-transduced TIL.

In the future, additional genetic modifications of TIL may further improve the antitumor efficacy of TIL (Rosenberg, 1992a). Insertion of other cytokines into TIL, such as IFN, may produce up-regulation of MHC molecules and augment the reactivity of TIL. Transduction of genes encoding IL-6, IL-1-α, or RANTES may modulate TIL response to tumors (Rosenberg, 1992a). Other genetic modifications of TIL include possible insertion of genes coding for the Fc receptor to increase antibody dependent cellular cytotoxicity, the IL-2 receptor to increase TIL sensitivity to rIL-2, or a chimeric T-cell receptor to increase specificity of TIL by the addition to the constant region of the T-cell receptor the variable region of monoclonal antibodies directed at tumor-specific antigens (Gross et al., 1989; Rosenberg, 1992a).

Another avenue for possible advancement in gene therapy is the genetic modification of tumors as a way to increase their immunogenicity. Several investigators have demonstrated that the insertion of genes encoding TNF-α, IL-2, IFN-α, IL-4 and GM-CSF into cancer cells can result in host-mediated elimination of genetically altered tumors in mice (Tepper et al., 1989; Fearon et al., 1990; Gansbacher et al., 1990a, 1990b; Leg et al., 1990; Asher et al., 1991; Blankenstein et al., 1991; Colombo et al., 1991; Golumbek et al., 1991; Teng et al., 1991). These findings form the basis of the clinical protocol in the Surgery Branch, NCI, for patients with advanced cancer in which the gene for TNF-α or IL-2 is inserted into a patient's tumor *in vitro*. Following selection and expansion of the gene-transduced tumor cells, the cells are injected subcutaneously and intradermally into the thigh of the patient. After 3 weeks of growth, the vaccination sites are operatively removed together with several draining lymph nodes from the inguinal region. The lymphocytes from the draining nodes are then expanded *in vitro* for later infusion into the patient together with rIL-2 (Rosenberg, 1992a). This clinical protocol was initiated in October, 1991.

References

Aebersold P, Carter CS, Hyatt C, et al. (1988): A simplified automated procedure for generation of human lymphokine-activated killer cells for use in clinical trials. *J Immunol Methods* 112: 1–7

Aebersold P, Hyatt C, Johnson S, et al. (1991): Lysis of autologous melanoma cells by tumor, infiltrating lymphocytes: Association with clinical response. *J Natl Cancer Inst* 83: 932–937

Alexander RB, Rosenberg SA (1990): Long term survival of adoptively-transferred tumor infiltrating lymphocytes in mice. *J Immunol* 145: 1615–1620

Alexander RB, Rosenberg SA (1991): Adoptively transferred tumor-infiltrating lymphocytes can cure established metastatic tumor in mice and persist long-term in vivo as functional memory T lymphocytes. *J Immunother* 10: 389–397

Anderson WF (1992): Human gene therapy. *Science* 256: 808–813

Andriole GL, Mulé JJ, Hansen CT, et al. (1985): Evidence that lymphokine-activated killer cells and natural killer cells are distinct based on an analysis of congenitally immunodeficient mice. *J Immunol* 135: 2911–2913

Asher AL, Mulé JJ, Kasid A, et al. (1991): Murine tumor cells transduced with the gene for tumor necrosis factor-γ: Evidence for paracrine immune effects of tumor necrosis effect against tumor. *J Immunol* 146: 3227–3234

Asher AL, Mulé JJ, Reichert CM, et al. (1987): Studies of the anti-tumor efficacy of systematically administered recombinant tumor necrosis factor against several murine tumors in vivo. *J Immunol* 138: 963–974

Atkins MB, Mier JW, Parkinson DR, et al. (1988): Hypothyroidism after treatment with interleukin-2 and lymphokine-activated killer cells. *N Engl J Med* 318: 1557–1563

Balch CM, Riley LB, Bae TJ, et al. (1990): Patterns of human tumor infiltrating lymphocytes in 120 human cancers. *Arch Surg* 125: 200–205

Bar M, Sznol M, Atkins MB, et al. (1990): Metastatic melanoma treated with combined bolus and continuous infusion interleukin-2 and lymphokine activated killer cells. *J Clin Oncol* 8: 1138–1147

Barba D, Saris SC, Holder C, et al. (1988): Immunotherapy of human glial tumors: Report of multiple dose intratumoral infusions of lymphokine-activated killer cells and interleukin-2. *J Neurosurg* 70: 175–182

Barth RJ, Jr, Bock SN, Mulé JJ, et al. (1990): Unique murine tumor associated antigens identified by tumor infiltrating lymphocytes. *J Immunol* 144: 1531–1537

Barth RJ, Mulé JJ, Asher AL, et al. (1991a): Identification of unique murine tumor associated antigens by tumor infiltrating lymphocytes using tumor specific secretion of interferon-gamma and tumor necrosis factor. *J Immunol Methods* 140: 269–279

Barth RJ, Jr, Mulé JJ, Spiess PJ, et al. (1991b): Interferon gamma and tumor necrosis factor have a role in tumor regressions mediated by murine CD8$^+$ tumor infiltrating lymphocytes. *J Exp Med* 173: 647–658

Belldegrun A, Muul LM, Rosenberg SA (1988): Interleukin-2 expanded tumor-infiltrating lymphocytes in human renal cell cancer: Isolation, characterization, and anti-tumor activity. *Cancer Res* 48: 206–214

Belldegrun A, Webb DE, Austin HA, et al. (1987): Effects of interleukin-2 on renal function in patients receiving immunotherapy for advanced cancer. *Ann Intern Med* 106: 817–822

Belldegrun A, Webb DE, Austin HA, et al. (1989): Renal toxicity of interleukin-2 administration in patients with metastatic renal cell cancers: Effect of pretherapy nephrectomy. *J Urol* 141: 499–503

Blankenstein T, Quin Z, Uberla K, et al. (1991): Tumor suppression after tumor cell-targeted tumor necrosis factor alpha gene transfer. *J Exp Med* 173: 1047–1052

Bock SN, Lee RE, Fisher B, et al. (1990): A prospective randomized trial evaluating prophylactic antibiotics to prevent triple-lumen catheter-related sepsis in patients treated with immunotherapy. *J Clin Oncol* 8: 161–169

Cameron RB, Spiess PJ, Rosenberg SA (1990): Synergistic antitumor activity of tumor infiltrating lymphocytes, interleukin-2 and local tumor irradiation: Studies on the mechanism of action. *J Exp Med* 171: 249–263

Colombo MP, Gerrari G, Stoppacciaro A, et al. (1991): Granulo-cyte colony-stimulating factor gene transfer suppresses tumorigenicity of a murine adenocarcinoma in vivo. *J Exp Med* 173: 889–897

Cotran RS, Pober JS, Gimbrone MS, Jr, et al. (1988): Endothelial activation during interleukin-2 (IL-2) immunotherapy: A possible mechanism for the vascular leak syndrome. *J Immunol* 140: 1883–1888

Culver KW, Anderson WF, Blaese RM (1991a): Lymphocyte gene therapy. *Hum Gene Ther* 2: 107–109

Culver K, Cornetta K, Morgan R, et al. (1991b): Lymphocytes as cellular vehicles for gene therapy in mouse and man. *Proc Natl Acad Sci USA* 88: 3155–3159

Darrow TL, Slingluff CL, Seigler HF (1989): The role of class I antigens in recognition of melanoma cells by tumor-specific cytotoxic T lymphocytes: Evidence for shared tumor antigens. *J Immunol* 142: 3329–3335

Denicoff KD, Durkin TM, Lotze MT, et al. (1989): The neuroendocrine effects of interleukin-2 treatment. *J Clin Endocrinol Metab* 69: 402–410

Denicoff KD, Rubinow DR, Papa MZ, et al. (1987): The neuropsychiatric effects of the treatment with interleukin-2 and lymphokine activated killer cells. *Ann Intern Med* 107: 293–300

Dutcher JP, Creekmore S, Weiss GR, et al. (1989): A phase II study of interleukin-2 and lymphokine activated killer cells in patients with metastatic malignant melanoma. *J Clin Oncol* 7: 477–485

Dutcher JP, Gaynor ER, Boldt DH, et al. (1991): A phase II study of high-dose continuous infusion interleukin-2 with lymphokine-activated killer cells in patients with metastatic melanoma. *J Clin Oncol* 9: 641–648

Eglitis MA, Anderson WF (1988): Retroviral vectors for introduction of genes into mammalian cells. *Bio Techniques* 6: 608–614

Eisenthal A, Cameron RB, Uppenkamp I, et al. (1988a): Effect of combined therapy with lymphokine-activated killer cells, interleukin-2 and specific monoclonal antibody on established B16 melanoma lung metastases. *Cancer Res* 48: 7140–7145

Eisenthal A, Shiloni E, Rosenberg SA (1988b): Characterization of IL-2 induced murine cells which exhibit ADCC activity. *Cell Immunol* 115: 257–272

Ettinghausen SE, Lipford EH, Mulé JJ, et al. (1985): Recombinant interleukin-2 stimulates in vivo proliferation of adoptively transferred lymphokine activated killer (LAK) cells. *J Immunol* 135: 3623–3635

Ettinghausen SE, Moore JG, White DE, et al. (1987): Hematologic effects of immunotherapy with lymphokine activated killer cells and recombinant interleukin-2 in cancer patients. *Blood* 69: 1654–1660

Ettinghausen SE, Puri RK, Rosenberg SA (1988): Increased vascular permeability in organs mediated by the systemic administration of lymphokine-activated killer cells and recombinant interleukin-2 in mice. *J Natl Cancer Inst* 80: 177–188

Fearon ER, Pardoll DM, Itaya T, et al. (1990): Interleukin-2 production by tumor cells bypasses T helper function in the generation of an antitumor response. *Cell* 60: 397–403

Fisher B, Packard BS, Read EJ, et al. (1989): Tumor localization of adoptively transferred indium-111 labeled tumor infiltrating lymphocytes in patients with metastatic melanoma. *J Clin Oncol* 7: 250–61

Fisher RI, Coltman CA, Doroshow JH, et al. (1988): Metastatic renal cancer treated with interleukin-2 and lymphokine activated killer cells. A phase II clinical trial. *Ann Int Med* 108: 518–523

Fox B, Rosenberg SA (1989): Heterogeneous lymphokine-activated killer cell precursor populations. *Cancer Immunol Immunother* 29: 115

Fujita T, Takaoka C, Matsui H, et al. (1983): Structure of the human interleukin-2 gene. *Proc Natl Acad Sci USA* 80: 7437–7441

Galili U, Vanky F, Rodriquez L, et al. (1979): Activated T lymphocytes within human solid tumors. *Cancer Immunol Immunother* 6: 129–133

Gansbacher B, Bannerji R, Daniels B, et al. (1990a): Retroviral vector-mediated γ-interferon gene transfer into tumor cells generates potent and long lasting antitumor immunity. *Cancer Res* 50: 7820–7825

Gansbacher B, Zier K, Daniels B, et al. (1990b): Interleukin-2 gene transfer into tumor cells abrogates tumorigenicity and induces protective immunity. *J Exp Med* 172: 1217–1224

Gaspari AA, Lotze MT, Rosenberg SA, et al. (1987): Dermatologic changes associated with interleukin-2 administration. *JAMA* 258: 1624–1629

Gaynor ER, Weiss GR, Margolin KA, et al. (1990): Phase I study of high-dose continuous-infusion recombinant interleukin-2 and autologous lymphokine activated killer cells in patients with metastatic or unresectable malignant melanoma and renal cell carcinoma. *J Natl Cancer Inst* 82: 1397–1402

Giboa E, Eglitis MA, Kantoff PW, et al. (1986): Transfer and expression of cloned genes using retroviral vectors. *Bio Techniques* 4: 504–512

Golumbek PT, Lazenby AJ, Levitsky HI, et al. (1991): Treatment of established renal cancer by tumor cells engineered to secrete interleukin-4. *Science* 254: 713–716

Griffith KD, Read EJ, Carrasquillo CS, et al. (1989): In vivo distribution of adoptively transferred indium-111 labeled tumor infiltrating lymphocytes and peripheral blood lymphocytes in patients with metastatic melanoma. *J Natl Cancer Inst* 81: 1709–1717

Grimm EA, Rosenberg SA (1984): The human lymphokine-activated killer cell phenomenon. In: *Lymphokines*, Pick E, Candy M, eds. New York: Academic Press, Vol 9, p. 279

Grimm EA, Mazumder A, Zhang HZ, et al. (1982): Lymphokine-activated killer cell phenomenon: Lysis of natural killer-resistant fresh solid tumor cells by interleukin-2 activated autologous human peripheral blood lymphocytes. *J Exp Med* 155: 1823–1841

Grimm EA, Ramsey KM, Mazumder A, et al. (1983a): Lymphokine-activated killer cell phenomenon. II. Precursor phenotype is serologically distinct from peripheral T lymphocytes, memory cytotoxic thymus-derived lymphocytes, and natural killer cells. *J Exp Med* 157: 884–897

Grimm EA, Robb RJ, Roth JA, et al. (1983b): Lymphokine-activated killer cell phenomenon. III. Evidence that IL-2 is sufficient for direct activation of peripheral blood lymphocytes into lymphokine-activated killer cells. *J Exp Med* 158: 1356–1361

Gross G, Waks T, Eshhar Z (1989): Expression of immunoglobulin-T cell receptor chimeric molecules as functional receptors with antibody-type specificity. *Proc Natl Acad Sci USA* 86: 10023–10028

Haas GP, Solomon D, Rosenberg SA (1990): Tumor infiltrating lymphocytes from non renal urological malignancies. *Cancer Immunol Immunother* 30: 342–350

Hawkins MT (1989): PPO updates IL-2/LAK. *Princ Pract Oncol* 3: 1–14

Heo DL, Whiteside TL, Johnson JT, et al. (1987): Long-term interleukin-2-dependent growth and cytotoxic activity of tumor-infiltrating lymphocytes from human squamous cell carcinomas of the head and neck. *Cancer Res* 47: 6353–6362

Heo DS, Whiteside TL, Kanbour A, et al. (1988): Lymphocytes infiltrating human ovarian tumors. I. Role of Leu-19 (NKH1)-positive recombinant IL-2 activated cultures of lymphocytes infiltrating human ovarian tumors. *J Immunol* 140: 4042–4049

Hom SS, Topalian SL, Simoni ST, et al. (1991): Common expression of melanoma tumor-associated antigens recognized by human tumor-infiltrating lymphocytes: Analysis by HLA restriction. *J Immunother* 10: 153–164

Ingram M, Jacques S, Freshwater DB, et al. (1987): Salvage immunotherapy of malignant glioma. *Arch Surg* 122: 1483–1486

Itoh K, Platsoucas CD, Balch CM (1988): Autologous tumor-specific cytotoxic T lymphocytes in the infiltrate of human metastatic melanomas: Activation by interleukin-2 and autologous tumor cells and involvement of the T cell receptor. *J Exp Med* 168: 1419–1441

Itoh K, Tilden AB, Balch CM (1986): Interleukin-2 activation of cytotoxic T-lymphocyte infiltrating into human metastatic melanomas. *Cancer Res* 46: 3011–3017

Itoh K, Tilden AB, Kumagai K, et al. (1985): Leu 11 positive lymphocytes with natural killer activity are precursors of recombinant interleukin-2 induced activated killer cells. *J Immunol* 134: 802

Jacob SK, Wilson DJ, Kornblith PL, et al. (1986): Interleukin-2 or autologous lymphokine-activated killer cell treatment of malignant glioma: Phase I trial. *Cancer Res* 46: 2101–2104

Kasid A, Morecki S, Aebersold P, et al. (1990): Human gene transfer: Characterization of human tumor-infiltrating lymphocytes as vehicles for retroviral-mediated gene transfer in man. *Proc Natl Acad Sci USA* 87: 473–477

Kawakami Y, Rosenberg SA, Lotze MT (1988): Interleukin-4 promotes the growth of tumor-infiltrating lymphocytes cytotoxic for human autologous melanoma. *J Exp Med* 168: 2183–2191

Kawakami Y, Zakut R, Topalian SL, et al. (1992): Shared human melanoma antigens: Recognition by tumor infiltrating lymphocytes in HLA-A2. 1-transfected melanomas. *J Immunol* 148: 638–643

Kradin RL, Kurnick JT, Lazarus DS, et al. (1989): Tumour-infiltrating lymphocytes and interleukin-2 in treatment of advanced cancer. *Lancet* 1: 577–580

Kragel AH, Travis WD, Steis RG, et al. (1990): Myocarditis or acute myocardial infarction associated with interleukin-2 therapy for cancer. *Cancer* 66: 1513–1516

Kriegler M (1990): *Gene Transfer and Expression. A Laboratory Manual.* New York: Stockton Press, pp. 1–242

Kuppner MC, Hamou MF, de Tribolet N (1988): Immunohistological and functional analyses of lymphoid infiltrates in human glioblastomas. *Cancer Res* 48: 6926–6932

Kurnick JT, Kradin RL, Blumberg R, et al. (1986): Functional characterization of T lymphocytes propagated from human lung carcinoma. *Clin Immunol Immunopathol* 38: 367–380

Lafreniere R, Rosenberg SA (1985a): Adoptive immunotherapy of murine hepatic metastases with lymphokine activated killer (LAK) cells and recombinant interleukin-2 (RIL-2) can mediate the regression of both immunogenic and non-immunogenic sarcomas and an adenocarcinoma. *J Immunol* 135: 4273–4280

Lafreniere R, Rosenberg SA (1985b): Successful immunotherapy of murine experimental hepatic metastases with lymphokine activated killer cells and recombinant interleukin-2. *Cancer Res* 45: 3735–3741

Lafreniere R, Rosenberg SA (1986): A novel approach to the generation and identification of experimental hepatic metastases in a murine model. *J Natl Cancer Inst* 76: 309–322

Lee RE, Gaspari AA, Lotze MT, et al. (1988): Interleukin-2 and psoriasis. *Arch Dermatol* 124: 1811–1815

Lee RE, Lotze MT, Skibber JM, et al. (1989): Cardiorespiratory effects of immunotherapy with interleukin-2. *J Clin Oncol* 7: 7–20

Lefor AT, Rosenberg SA (1991): The specificity of lymphokine activated killer (LAK) cells in vitro: Fresh normal murine tissues are resistant to LAK-mediated lysis. *J Surg Res* 50: 15–23

Ley V, Roth C, Langlade-Demoyen P, et al. (1990): A novel approach to the induction of specific cytolytic T cells in vivo. *Res Immunol* 141: 855–863

Lotze MT, Frana LW, Sharrow SO, et al. (1985a): In vivo administration of purified human interleukin-2. I. Half-life and immunologic effects of the Jurkat cell line-derived interleukin-2. *J Immunol* 134: 157–166

Lotze MT, Grimm EA, Mazumder A, et al. (1981): Lysis of fresh and cultured autologous tumor by human lymphocytes cultured in T-cell growth factor. *Cancer Res* 41: 4420–4425

Lotze MT, Line BR, Mathisen DJ, et al. (1980): The in vivo distribution of autologous human and murine lymphoid cells grown in T cell growth factor (TCGF): Implications for the adoptive immunotherapy of tumors. *J Immunol* 125: 1487–1493

Lotze MT, Matory YL, Ettinghausen SE, et al. (1985b): In vivo administration of purified human interleukin-2. II. Half-life, immunologic effects, and expansion of peripheral lymphoid cells in vivo with recombinant IL-2. *J Immunol* 135: 2865–2875

Lotze MT, Matory YL, Rayner AA, et al. (1986): Clinical effects and toxicity of interleukin-2 in patients with cancer. *Cancer* 58: 2764–2772

Mazumder A, Rosenberg SA (1984): Successful immunotherapy of NK-resistant established pulmonary melanoma metastases by the intravenous adoptive transfer of syngeneic lymphocytes activated in vitro by interleukin-2. *J Exp Med* 159: 495–507

Mazumder A, Eberlein TJ, Grimm EA, et al. (1984): Phase I study of the adoptive immunotherapy of human cancer with lectin activated autologous mononuclear cells. *Cancer* 53: 896–905

Merchant RE, Merchant LH, Cook SHS, et al. (1988): Intralesional infusion of lymphokine-activated killer (LAK) cells and recombinant interleukin-2 (rIL-2) for the treatment of patients with malignant brain tumor. *Neurosurgery* 23: 725–732

Miller A, Buttimore C (1986): Redesign of retrovirus packaging cell lines to avoid recombinant leading to helper virus production. *Mol Cell Biol* 6: 2895–2902

Miller AD, Rosman GK (1989): Improved retroviral vectors for gene transfer and expression. *Bio Techniques* 7: 980–986

Morgan DA, Ruscetti FW, Gallo RG (1976): Selective in vitro growth of T lymphocytes from normal bone marrow. *Science* 193: 1007–1008

Mulé JJ, Ettinghausen SE, Spiess PJ, et al. (1986a): The anti-tumor efficacy of lymphokine-activated killer cells and recombinant interleukin-2 in vivo: An analysis of survival benefit and mechanisms of tumor escape in mice undergoing immunotherapy. *Cancer Res* 46: 676–683

Mulé JJ, Shu S, Rosenberg SA (1985): The anti-tumor efficacy of lymphokine activated killer cells and recombinant interleukin-2 in vivo. *J Immunol* 135: 646–652

Mulé JJ, Shu S, Schwarz SL, et al. (1984): Adoptive immunotherapy of established pulmonary metastases with LAK cells and recombinant interleukin-2. *Science* 225: 1487–1489

Mulé JJ, Yang J, Shu S, et al. (1986b): The anti-tumor efficacy of lymphokine-activated killer cells and recombinant interleukin-2 in vivo. Direct correlation between reduction of established metastases and cytolytic activity of lymphokine activated killer cells. *J Immunol* 136: 3899–3909

Muul LM, Director EP, Hyatt CL, et al. (1986): Large scale production of human lymphokine activated killer cells for use in adoptive immunotherapy. *J Immunol Methods* 88: 265–275

Muul LM, Narson-Burchenal K, Carter CS, et al. (1987a): Development of an automated closed system for generation of human lymphokine-activated killer (LAK) cells for use in adoptive immunotherapy. *J Immunol Methods* 101: 171–181

Muul LM, Spiess PJ, Director EP, et al. (1987b): Identification of specific cytolytic immune responses against autologous tumor in humans bearing malignant melanoma. *J Immunol* 138: 989–995

Negrier S, Phillip T, Stoter G, et al. (1989): Interleukin-2 with or without LAK cells in metastatic renal cell carcinoma: A report of a European multicentre study. *Eur J Cancer Clin Oncol* 25: S21–S28

Nitta T, Sato K, Yagita H, et al. (1990): Preliminary trial of specific targeting therapy against malignant glioma. *Lancet* 335: 368–371

Ognibene FP, Rosenberg SA, Lotze M, et al. (1988): Interleukin-2 administration causes reversible hemodynamic changes and left ventricular dysfunction similar to those seen in septic shock. *Chest* 94: 750–754

Ortaldo JR, Mason A, Overton R (1986): Lymphokine-activated killer cells: Analysis of progenitors and effectors. *J Exp Med* 164: 1193

Ottow RT, Eggermont AM, Steller EP, et al. (1987a): The requirements for successful immunotherapy of intraperitoneal cancer using interleukin-2 and lymphokine-activated killer cells. *Cancer* 60: 1465–1473

Ottow RT, Stellar EP, Sugarbaker PH, et al. (1987b): Immunotherapy of intraperitoneal cancer with interleukin-2 and lymphokine-activated killer cells reduces tumor load and prolongs survival in murine models. *Cell Immunol* 104: 366–376

Paciucci PA, Holland JF, Glidewell O, et al. (1989): Recombinant interleukin-2 continuous infusion and adoptive transfer of recombinant interleukin-2 activated cells in patients with advanced cancer. *J Clin Oncol* 7: 869–878

Paine JT, Handa H, Yamasaki T, et al. (1986): Immunohistochemical analysis of infiltrating lymphocytes in central nervous system tumors. *Neurosurgery* 18: 766–772

Papa MZ, Mulé JJ, Rosenberg SA (1986): Antitumor efficacy of lymphokine-activated killer cells and recombinant interleukin-2 in vivo: Successful immunotherapy of established pulmonary metastases from weakly immunogenic and nonimmunogenic murine tumors of three distinct histological types. *Cancer Res* 46: 4973–4978

Phillips JH, Lanier LL (1986): Dissection of the lymphokine-activated killer phenomenon: Relative contribution of peripheral blood natural killer cells and T lymphocytes to cytolysis. *J Exp Med* 164: 814

Rabinowich H, Cohen R, Bruderman I, et al. (1987): Functional analysis of mononuclear cells infiltrating into tumors: Lysis of autologous human tumor cells by cultured infiltrating lymphocytes. *Cancer Res* 47: 173–177

Radrizzani M, Gambacorti-Passerini C, Parmiani G, et al. (1989): Lysis by interleukin-2 stimulated tumor-infiltrating lymphocytes of autologous and allogeneic tumor target cells. *Cancer Immunol Immunother* 28: 67–73

Rayner AA, Grimm EA, Lotze MT, et al. (1985): Lymphokine-activated killer (LAK) cells: Analysis of factors relevant to the immunotherapy of human cancer. *Cancer* 55: 1327

Rivoltini L, Avienti F, Orazi A, et al. (1992): Phenotypic and functional analysis of lymphocytes infiltrating pediatric tumors, with a characterization of the tumor phenotype. *Cancer Immunol Immunother* 34: 241–251

Roberts K, Lotze MT, Rosenberg SA (1987): Separation and functional studies of the human lymphokine-activated killer cell. *Cancer Res* 47: 4366

Rosenberg SA (1984): Immunotherapy of cancer by the systemic administration of lymphoid cells plus interleukin-2. *J Biol Resp Mod* 3: 501–511

Rosenberg SA (1991): Adoptive cellular therapy: Clinical applications. In: *Biologic Therapy of Cancer*, DeVita VT, Hellman S, Rosenberg SA, eds. Philadelphia: J.B. Lippincott, Chapter 12, pp. 214–236

Rosenberg SA (1992a): The immunotherapy and gene therapy of cancer. *J Clin Oncol* 10: 180–199

Rosenberg SA (1992b): Gene therapy of cancer. In. *Important Advances in Oncology*, Devita VT, Hellman S, Rosenberg SA, eds. Philadelphia: J.B. Lippincott pp. 17–38

Rosenberg SA, Terry W (1977): Passive immunotherapy of cancer in animals and man. *Adv Cancer Res* 25: 323–388

Rosenberg SA, Aebersold P, Cornetta K. et al. (1990): Gene transfer into humans—Immunotherapy of patients with advanced melanoma, using tumor-infiltrating lymphocytes modified by retroviral gene transduction. *N Engl J Med* 323: 570–578

Rosenberg SA, Grimm EA, McGrofan M, et al. (1984): Biological activity of recombinant human interleukin-2 produced in *Escherichia coli*. *Science* 223: 1412–1415

Rosenberg SA, Longo DL, Lotze MT (1989a): Principles and applications of biologic theory. In: *Cancer, Principles and Practices of Oncology*, DeVita VT, Hellman, S, Rosenberg, SA, eds. Philadelphia: J.B. Lippincott, pp. 1342–1398

Rosenberg SA, Lotze MT, Muul LM, et al. (1985a): Observations on the systemic administration of autologous lymphokine-activated killer cells and recombinant interleukin-2 to patients with metastatic cancer. *N Engl J Med* 313: 1485–1492

Rosenberg SA, Lotze MT, Muul LM, et al. (1987): A progress report on the treatment of 157 patients with advanced cancer using lymphokine activated killer cells and interleukin-2 or high dose interleukin-2 alone. *N Engl J Med* 316: 889–905

Rosenberg SA, Lotze MT, Yang JC, et al. (1989b): Experience with the use of high-dose interleukin-2 in the treatment of 652 cancer patients. *Ann Surg* 210: 474–485

Rosenberg SA, Mulé JJ, Spiess PJ, et al. (1985b): Regression of established pulmonary metastases and subcutaneous tumor mediated by the systemic administration of high-dose recombinant interleukin-2. *J Exp Med* 161: 1169–1188

Rosenberg SA, Packard BS, Aebersold PM, et al. (1988a): Use of tumor-infiltrating lymphocytes and interleukin-2 in the immunotherapy of patients with metastatic melanoma. *N Engl J Med* 319: 1676–1680

Rosenberg SA, Schwarz S, Spiess P (1988b): Combination immunotherapy of cancer: Synergistic anti-tumor interactions of interleukin-2, alpha-interferon and tumor infiltrating lymphocytes. *J Natl Cancer Inst* 80: 1393–1397

Rosenberg SA, Spiess P, Lafreniere R (1986): A new approach to the adoptive immunotherapy of cancer with tumor infiltrating lymphocytes. *Science* 233: 1318–1321

Rosenstein M, Ettinghausen SE, Rosenberg SA (1986): Extravasation of intravascular fluid mediated by the systemic administration of recombinant interleukin-2. *J Immunol* 137: 1735-1742

Rosenstein M, Yron I, Kaufman Y, et al. (1984): Lymphokine-activated killer cells: Lysis of fresh syngeneic natural killer-resistant murine tumor cells by lymphocytes cultured in interleukin-2. *Cancer Res* 44: 1946-1953

Rubin JT, Elwood LJ, Rosenberg SA, et al. (1989): Immunohistochemical correlates of response to recombinant interleukin-2 based immunotherapy in humans. *Cancer Res* 49: 7086-7092

Saito T, Tanaka R, Yoshida S, et al. (1988): Immunohistochemical analysis of tumor-infiltrating lymphocytes and major histocompatibility antigens in human gliomas and metastatic brain tumors. *Surg Neurol* 29: 435-442

Saris SC, Rosenberg SA, Friedman RB, et al. (1988): Penetration of recombinant interleukin-2 across the blood cerebrospinal fluid barrier. *J Neurosurg* 69: 29-34

Schoof DD, Gramolini BA, Davidson DL, et al. (1988): Adoptive immunotherapy of human cancer using low-dose recombinant interleukin-2 and lymphokine-activated killer cells. *Cancer Res* 48: 5007-5010

Schwartzentruber D, Lotze MT, Rosenberg SA (1988): Colonic perforation: An unusual complication of therapy with high-dose interleukin-2. *Cancer* 62: 2350-2353

Schwartzentruber DJ, Topalian SL, Mancini MJ, et al. (1991a): Specific release of granulocyte-macrophage colony-stimulating factor, tumor necrosis factor-a, and IFN-γ by human tumor-infiltrating lymphocytes after autologous tumor stimulation. *J Immunol* 146: 153-164

Schwartzentruber DJ, White DE, Zweig MH, et al. (1991b): Thyroid dysfunction associated with immunotherapy for patients with cancer. *Cancer* 68: 2384-2390

Shiloni E, Eisenthal A, Sachs D, et al. (1987): Antibody dependent cellular cytotoxicity mediated by murine lymphocytes activated in recombinant interleukin-2. *J Immunol* 138: 1992-1998

Shiloni E, Lefreniere R, Mulé JJ, et al. (1986): Effect of immunotherapy with allogeneic lymphokine-activated killer (LAK) cells and recombinant interleukin-2 (RIL-2) on established pulmonary and hepatic metastases in mice. *Cancer* 46: 5633-5640

Shimizu K, Okamoto Y, Miyao Y, et al. (1987): Adoptive immunotherapy of human meningeal gliomatosis and carcinomatosis with LAK cells and recombinant interleukin-2. *J Neurosurg* 66: 519-521

Skibber JM, Lotze MT, Uppenkamp I, et al. (1987): Identification and expansion of human lymphokine-activated killer cells: Implications for the immunotherapy of cancer. *J Surg Res* 42: 613

Spiess PJ, Yang JC, Rosenberg SA (1987): In vitro antitumor activity of tumor-infiltrating lymphocytes expanded in recombinant interleukin-2. *J Natl Cancer Inst* 79: 1067-1075

Stahel RA, Sculier J, Jost LM, et al. (1989): Tolerance and effectiveness of recombinant interleukin-2 (r-met Hu IL-2 (ala 125) and lymphokine-activated killer cells in patients with metastatic solid tumors. *Eur J Cancer Clin Oncol* 25: 965-972

Steis R, Bookman M, Clark J, et al. (1987): Intraperitoneal lymphokine activated killer (LAK) cell and interleukin-2 (IL-2) therapy for peritoneal carcinomatosis toxicity, efficacy, and laboratory results (abstract). *Proc Annu Meet Am Soc Clin Oncol* 6: 250

Stotter H, Weibke EA, Tomita S, et al. (1992): Cytokines after target cell susceptibility to lysis: II. Evaluation of tumor infiltrating lymphocytes. *J Immunol* 148: 638-643

Tagaki S, Chen K, Schwarz R, et al. (1989): Functional and phenotypic analysis of tumor-infiltrating lymphocytes isolated from human primary and metastatic liver tumors and cultured in recombinant interleukin-2. *Cancer* 63: 102-111

Taniguchi T, Matsui H, Fujita T, et al. (1983): Structure and expression of a cloned cDNA for human interleukin-2. *Nature* 303: 305-307

Teng MN, Park BH, Koeppen HKW, et al. (1991): Long-term inhibition of tumor growth by tumor necrosis factor in the absence of cachexia or T-cell immunity. *Proc Natl Acad Sci USA* 88: 3535-3539

Tepper RL, Pattengale P, Leder P (1989): Murine interleukin-4 displays potent anti-tumor active in vivo. *Cell* 57: 503-512

Thompson JA, Lee DJ, Lindgren CG, et al. (1989): Influence of schedule of interleukin-2 administration on therapy with interleukin-2 and lymphokine activated killer cells. *Cancer Res* 49: 235-240

Topalian SL, Solomon D, Avis FP, et al. (1988): Immunotherapy of patients with advanced cancer using tumor-infiltrating lymphocytes and recombinant interleukin-2: A pilot study. *J Clin Oncol* 6: 839-853

Topalian SL, Solomon D, Rosenberg SA (1989). Tumor-specific cytolysis by lymphocytes infiltrating human melanomas. *J Immunol* 142: 3714-3725

Urba WJ, Clark JW, Steis RG, et al. (1989): Intraperitoneal lymphokine-activated killer cell/interleukin-2 therapy in patients with intraabdominal cancer: Immunologic considerations. *J Natl Cancer Inst* 81: 602-611

Wang AM, Creasy AA, Ladner MB, et al. (1985): Molecular cloning of the complementary DNA for human tumor necrosis factor. *Science* 228: 149-154

Wang A, Lu S-D, Mark DF (1984): Site-specific mutagenesis of the human interleukin-2 gene: Structure-function analysis of the cysteine residues. *Science* 224: 1431-1433

Webb DE, Austin HA, Belldegrun A, et al. (1988): Metabolic and renal effects of interleukin-2 immunotherapy for metastatic cancer. *Clin Nephrol* 30: 141-145

Weber JS, Rosenberg SA (1990): Effects of murine tumor class I major histocompatibility complex expression on antitumor activity of tumor infiltrating lymphocytes. *J Natl Cancer Inst* 82: 755-761

Weiss GR, Margolin KA, Aronson FR, et al. (1992): A randomized phase II trial of continuous infusion interleukin-2 or bolus injection interleukin-2 plus lymphokine activated killer cells for advanced renal cell carcinoma. *J Clin Oncol* 10: 275–281

West WH, Tauer KW, Yannelli JR, et al. (1987): Constant-infusion recombinant interleukin-2 in adoptive immunotherapy of advanced cancer. *N Engl Med* 316: 898–905

Whiteside TL, Miescher S, Hurlimman J, et al. (1986a): Clonal analysis and in situ characterization of lymphocytes infiltrating human breast carcinomas. *Cancer Immunol Immunother* 23: 169–178

Whiteside TL, Miescher S, Hurlimann J, et al. (1986b): Separation, phenotyping, and limiting dilution analysis of T-lymphocytes infiltrating human solid tumors. *Int J Cancer* 38: 803–811

Yang JC, Mulé J, Rosenberg SA (1986): Murine lymphokine-activated killer (LAK) cells: Phenotypic characterization of the precursor and effector cells. *J Immunol* 137: 715–722

Yang JC, Perry-Lalley D, Rosenberg SA (1990): An improved method for growing murine tumor infiltrating lymphocytes with in vivo antitumor activity. *J Biol Response Md* 9: 149–159

Yasumoto K, Miyazaki K, Nagashima A, et al. (1987): Induction of lymphokine activated killer cells by intrapleural instillations of recombinant interleukin-2 in patients with malignant pleurisy due to lung cancer. *Cancer Res* 47: 2184–2187

Yoshida S, Tanaka R, Takai N, et al. (1988): Local administration of autologous lymphokine-activated killer cells and recombinant interleukin-2 to patients with malignant brain tumors. *Cancer Res* 48: 5011–5016

Yron I, Wood TA, Spiess PJ, et al. (1980): In vitro growth of murine T cells. V. The isolation and growth of lymphoid cells infiltrating syngeneic solid tumors. *J Immunol* 125: 238–245

IX
Macrophage-Mediated Cytotoxicity

42
Macrophage-Mediated Cytotoxicity

Penelope J. Duerksen-Hughes and Linda R. Gooding

Introduction

Macrophages are among the most versatile and multifunctional of cells. Their phagocytic ability is an essential part of the nonspecific immune defense against microorganisms, and macrophages are also required for the specific immune responses, mediated by T and B cells, to occur. In addition, macrophages manufacture and release a wide variety of substances with many different functions (for a review, see Nathan, 1987). Perhaps the most remarkable thing about macrophages is their ability to differentiate between normal and transformed cells, and between normal and infected cells, and to selectively leave the normal cells unharmed while killing the transformed or infected cells (Fidler and Schroit, 1988). Thus, macrophages not only can discriminate between self and nonself, but also can often distinguish between normal and abnormal self. This chapter will examine the cytotoxic activity of macrophages against transformed and infected cells both *in vivo* and *in vitro*.

Macrophage Activation

Macrophage functions vary depending on what state the macrophage is in, and these states can be modulated by various stimuli, generally referred to as Macrophage-Activating Factors (MAF). Generally speaking, these activating agents fall into one of two classes: microorganisms or their products, and lymphokines. Upon receiving one or more signals from MAF, macrophages will respond by becoming activated (reviewed by Adams and Hamilton, 1984). Historically, activated macrophages have been defined as those possessing antimicrobial or antitumor activity. Included in this definition is the acquired ability to perform such functions as binding to tumor cells and releasing reactive oxygen intermediates (Adams and Hamilton, 1984).

Activation is generally considered to take place in two steps, with a priming step (Marino and Adams, 1982; Pace et al., 1983) followed by a triggering step (Pace and Russell, 1981; Meltzer, 1981). Often, the priming agent used is interferon-γ (IFN-γ) and the triggering agent lipopolysaccharide (LPS). Alternatively, LPS can fulfill both roles when present at a higher (e.g., μg/ml as opposed to ng/ml) concentration (Gifford and Lohmann-Mathes, 1986; also see Hamilton et al., 1986 on IFN-γ/LPS synergy). IFN-γ is thought to activate macrophages via a pathway involving alterations of protein kinase c (PK_c) potential and the concentration of calcium. LPS-induced changes include the breakdown of phosphatidylinositol-bisphosphate (PIP_2), with the subsequent release of inositol trisphosphate (IP_3) and diacylglyceride followed by an influx of calcium and protein phosphorylation. LPS also induces the expression of a number of cellular genes (reviewed by Hamilton and Adams, 1987; Adams and Hamilton, 1987). Other events thought to be involved in activation involve diacylglycerol (DAG) accumulation independent of potentiated phosphatidylinositol metabolism (Sebaldt et al., 1990), Na^+/H^+ exchange and alterations in intracellular pH (reviewed by Uhing and Adams, 1989), and modulation of the stability of various mRNAs (Yu et al., 1990).

Macrophages can also be activated by monoclonal antibodies to Mac-1 (complement receptor type 3) (Ding et al., 1987), maleylated bovine serum albumin (BSA) (Somers et al., 1987; Haberland et al., 1989), peroxidases (Wei et al., 1986; Lefkowitz et al., 1989), muramyl dipeptide (Chedid et al.,

1979), and bacillus Calmette-Guérin (BCG) (Adams, 1980). Tumor necrosis factor (TNF) has also been shown to participate in activation (Drapier et al., 1988) and may synergize with IFN-γ and LPS in induction of nitric oxide synthesis (Amber et al., 1991). An alternative explanation for the effect of TNF in some systems is that the cytokine does not render blood monocytes tumoricidal, but binds to its own receptor on the effector cells and can hence produce lysis of TNF-sensitive cells by a "carryover" mechanism (Nii and Fidler, 1990).

Additionally, when liposomes containing immunomodulators are phagocytosed by rodent macrophages, efficient activation can occur, with these macrophages able to prolong the life of mice with metastatic tumors (Poste et al., 1979; Fidler et al., 1981; Fidler, 1985, Fidler et al., 1989; Utsugi et al., 1991a; Dinney et al., 1991). Such activated macrophages are also active against herpes simplex virus-2 (HSV-2)-infected cells (Koff et al., 1983; 1984; 1985). This mode of activation could have therapeutic significance, since such liposomes could be given to patients, thus activating their own macrophages. In the human system, liposome-activated blood monocytes form clusters around melanoma cells, and this clustering is followed by the development of specific points of contact between the monocytes and the tumor cells (Bucana et al., 1983). Recently, studies have demonstrated the ability of liposomes containing either lipophilic muramyl tripeptide or a new synthetic analog of lipoprotein, N-hexadecanoyl-S-[2(R)3-didodecanoyloxypropyl]-L-cysteinyl-L-alanyl-D-isoglutaminyl-glycyl-taurine sodium salt (CGP 31362) to activate human blood monocytes (Utsugi et al., 1991b). When liposomes containing muramyl tripeptide phosphatidylethanolamine were administered to human patients, their blood monocytes became cytotoxic for tumor cells (Kleinerman et al., 1989).

Activation is probably more complex than we currently envision (see Adams and Hamilton, 1984), but at present there is a marked shortage of reagents or markers that can be used to analyze the activation process. Commonly, activation is monitored by measuring specific macrophage functions, including the generation of reactive oxygen intermediates, such as H_2O_2 (Nathan et al., 1979a, 1979b), and reactive nitrogen intermediates, such as nitric oxide (Ding et al., 1988), or by the killing of tumor cells (Pace and Russell, 1981; Meltzer, 1981). Other markers noted include morphology and cell biology, secretory activity, cell surface antigen expression, and extoenzyme expression (Lambert and Paulnock, 1989, and references therein).

Another approach has been to determine the ability of the macrophage to bind the lectin I-B$_4$ from *Griffonia simplicifolia* (GSI-B$_4$) (Maddox et al., 1982). In this system, stimulated (thioglycollate-elicited) peritoneal macrophages but not resident macrophages were shown to react with the lectin. The binding could be inhibited by the presence of methyl α-D-galactopyranoside (Me α-D-Gal).

More recently, monoclonal antibodies reacting with the surface of macrophages at different stages of activation have been described, which can be used to define the steps of activation more specifically (Paulnock and Lambert, 1990). For example, TM-1 is expressed on RAW 264.7 (a macrophage-like cell line) cells that have been primed by IFN-γ or γ radiation. TM-2 is expressed on RAW 264.7 cells primed by IFN-γ but not by γ radiation alone, and TM-3 is detectable on RAW 264.7 cells primed by either IFN-γ or γ radiation, after subsequent triggering of the primed cells with LPS. It also appears that activation for the various functions may not necessarily be identical. For example, activation for kill of virus-infected cells shows different characteristics, in terms of the amounts of activators needed and the length of time macrophages are exposed to those activators, than does activation for kill of tumor cells (LeBlanc, 1989).

Macrophages can also be negatively regulated. The transforming growth factor-β family (TGF-β_1, TGF-β_2, and TGF-β_3) (Tsunawaki et al., 1988; Ding et al., 1990) as well as another factor, macrophage-deactivating factor (MDF), derived from tumor cells (Ding et al., 1990), have been shown to inhibit induction of macrophage nitric oxide synthesis by IFN-γ. In addition, several cytokines, including MDF (Szuro-Sudol and Nathan, 1982; Tsunawaki and Nathan, 1986; Srimal and Nathan, 1990), TGF-β_1 and TGF-β_2 (Tsunawaki et al., 1989), IL-4 (Lehn et al., 1989; Abramson and Gallin, 1990), and calcitonin and calcitonin gene-related peptide (Nong et al., 1989), can block the IFN-γ-mediated enhancement of the capacity of macrophages to release reactive oxygen intermediates. TGF-β_1 was also shown to inhibit the activation of RAW 264.7 cells for cell killing (Haak-Frendscho et al., 1990).

Prostaglandins of the E series have also been associated with suppression of tumoricidal properties (Schultz et al., 1978; Taffet and Russell, 1981; Figueiredo et al., 1990). However, a recent report does not support the hypothesis that prostaglandin E$_2$ (PGE$_2$) functions in a negative feedback mechanism to inhibit the tumoricidal properties of macrophages (Utsugi and Fidler, 1991). Activated macrophages cultured *in vitro* will lose their ability to undergo a respiratory burst over time. This may be due to altered signal transduction, as the concentration of various monovalent cations affects the kinetics of this process (Kitagawa and Johnston, 1986). Since it is the activated macrophages which have been shown to have tumoricidal and antiviral activity, this chapter will focus on the ability of activated macrophages to recognize and destroy tumor and virus-infected cells.

Antitumor Activity

Several types of evidence indicate that macrophages can act against transformed cells *in vivo*. The vast majority of both spontaneous and transplanted tumors contain infiltrating macrophages which can account for up to 65% of the cells derived from disaggregation of solid tumors (reviewed by Russell et al., 1980). This may suggest an attempt by the host to eliminate the tumor using these macrophages. Also, when Urban and Schreiber selected for progressor variants of UV-induced tumors *in vivo*, cells derived from these tumors turned out to have selective resistance to macrophage cytolysis *in vitro* as well (Urban and Schreiber, 1983). Similarly, our laboratory found that SV40-induced tumors selected for the ability to progress *in vivo* display decreased susceptibility to macrophage killing *in vitro* (Chapes and Gooding, 1985). Conversely, *in vitro* selection of cells that resist macrophage killing also selects for cells that grow progressively *in vivo*, while macrophage-sensitive cells from the same population fail to progress *in vivo* (Chapes and Gooding, 1985).

Additional evidence for the tumoricidal activity of macrophages comes from observations made by Cook and his co-workers (1980). While SV40-transformed mouse and rat cells are normally sensitive to macrophage cytolysis, SV40-transformed hamster cells are relatively resistant. Since the SV40 virus is oncogenic in hamsters while nononcogenic in rats and mice, Cook et al. proposed that susceptibility or resistance of transformed cells to macrophage cytolysis may be the basis for the species-specific oncogenicity of SV40. When comparing the characteristics of 13 DNA virus-transformed cell lines, these workers also reported an association between the level of the transformed cells' resistance to lysis by nonspecific host effector cells and the oncogenicity of the transforming virus (Cook et al., 1982). It should be noted that these *in vivo* effects may be due either to cytostatic or to cytotoxic activity.

Work done over the last several years by Fidler and co-workers has focused on the ability of liposomes containing immunomodulators to activate macrophages (see Macrophage Activation, above). Such activated macrophages are able to prolong the life of mice with metastatic tumors, and this therapy is currently being used in human clinical trials (reviewed by Whitworth et al., 1989/1990). This type of work may ultimately provide the best *in vivo* evidence for macrophage tumoricidal activity.

In addition to this *in vivo* evidence, many laboratories have also studied selective cytotoxicity for tumor cell lines *in vitro* (for example, see Fidler and Schroit, 1988; Evans and Alexander, 1972; Hibbs, 1974b; Adams and Nathan, 1983; Fidler and Kleinerman, 1984; Cooke et al., 1987; Cook et al., 1989). These studies show clearly that activated macrophages can recognize and destroy transformed cells. It should be noted, however, that not all transformed lines are sensitive to activated macrophages (Chapes and Gooding, 1985).

Antivirus Activity

Macrophages play an important role in defending the host against invading viruses. They can inhibit virus replication and infectivity (Mogenson, 1985; Morahan et al., 1985) as well as lyse infected cells (Stott et al., 1975; Morahan et al., 1985). Macrophages are widely distributed in the body and are located in areas that interface with the external environment (such as the lungs) and in organs that contact the circulating blood (such as the liver). Hence, macrophages are often among the first cells to contact an invading virus (Morahan et al., 1985; Mogenson, 1985). Furthermore, when viruses infect cells, a number of chemotactic stimuli are released which can recruit macrophages to the site of inflammation (Allison, 1974).

The macrophages of newborn mice are not fully functional at birth, and these newborns are susceptible to herpes simplex virus (HSV) while adults are not. When macrophages from syngeneic adults were injected into newborn mice, the newborns were protected from intraperitoneal challenge by HSV, and proteose-peptone-stimulated macrophages were more efficient in conferring protection than were unstimulated macrophages (Hirsch et al., 1970; Allison, 1974). Macrophages are capable of lysing reovirus type 1-infected cells (Letvin et al., 1982), as well as HSV-2-infected cells (Koff et al., 1983, 1984, 1985). Cells infected with vesicular stomatitis virus and Venezuelan equine encephalitis virus were also shown to be susceptible to killing by activated macrophages (LeBlanc, 1989), as were cells infected with human adenoviruses (Zachariades, Wold, and Gooding, unpublished).

When Ginsburg and his co-workers (Ginsberg et al., 1989, and references therein) studied adenovirus infections of epithelial cells lining the bronchi and bronchioles of cotton rats, they found that the onset of viral multiplication was followed by a progressively increasing infiltration of lymphocytes as well as macrophages/monocytes. Two phases of an immune response were observed, the first of which consisted of a lymphocyte and monocyte/macrophage intraalveolar and interstitial infiltration, as well as a scattering of polymorphonuclear leukocytes. Their results in a mouse model were similar (Ginsberg et al., 1991).

The antiviral effects of macrophages are modulated at least in part by secreted products, such as TNF. The effects of such soluble mediators are discussed more fully below (see Mechanism).

Selectivity

Macrophages are able to distinguish transformed or infected cells from normal cells; however, the basis for this selectivity remains unknown. Clearly, since some macrophages can kill and others cannot (i.e., activated versus nonactivated), and since some target cells are killed and others are not (tumorigenic versus normal), the observed selectivity is dependent on characteristics of both the target and the effector cells. Experiments done by Fidler and co-workers (Fidler and Kleinerman, 1984) showed that when tumor cells, normal cells, and activated monocytes were co-cultured, only the tumor cells were lysed.

A number of studies have shown that macrophage selectivity for cytolysis is independent of any known tumor or histocompatibility transplantation antigens (Fidler, 1978) and does not correlate with either the cells' doubling time or expression of endogenous C-type viruses (Fidler, 1978). Additionally, the susceptibility of cells to macrophage killing was determined by a mechanism distinct from those regulating the recognition of target cells by natural killer (NK) cells (Wiltrout et al., 1982; Nestle et al., 1984), in that sensitivity to macrophages did not correlate with sensitivity to NK cells. Furthermore, when cells transformed with SV40 or polyoma virus which exhibit a temperature-dependent transformed phenotype were tested for sensitivity, the cells were equally sensitive at both the permissive and the nonpermissive temperatures (Fidler, 1978; Fidler et al., 1978; Fidler and Schroit, 1988). This suggests that the mechanism by which macrophages recognize transformed cells does not involve temperature-dependent phenotypic characteristics such as the expression of cell-surface high-molecular-weight glycoproteins, Forssman antigen, surface changes for lectin agglutination, expression of SV40 T antigen, low saturation density, or density-dependent inhibition of DNA synthesis (Fidler, 1978; Fidler et al., 1978; Fidler and Schroit, 1988). It should be noted, however, that in all the studies mentioned above, an excellent correlation between sensitivity to macrophages and tumorigenic potential was observed.

Selectivity for macrophage killing could be exerted at at least two sites, binding or sensitivity to the lytic machinery itself. It has been shown that activated macrophages do bind selectively to tumor cells. Raz et al., (1977) showed that malignant lymphoma cells bound to macrophages at a level three- to five-fold higher than that of normal lymphocytes. Also, Adams and his co-workers (Somers et al., 1983) showed by scanning electron microscopy that two to five tumor targets clustered around fully activated (tumoricidal) macrophages as compared to very few tumor cells found at the periphery of inflammatory (noncytolytic) macrophages. This binding required divalent cations and trypsin-sensitive surface molecules on the macrophages. Intact energy metabolism, microfilaments, and microtubules were required to produce this binding, and radioactively labeled tumor targets could be competitively removed by the addition of excess unlabeled tumor cells (Somers et al., 1983). Further work by this laboratory (Somers et al., 1986) has demonstrated that the binding of tumor cells by activated macrophages proceeds through two distinct stages, characterized by their strength of binding. Studies by Marino and Adams (1980a, 1980b) also point to a specific binding. BCG-activated macrophages selectively bound tumor targets but not normal cells, and this interaction was dependent on the presence of divalent cations and trypsin-sensitive structures on the macrophages. Binding was temperature-dependent and required living, metabolically active macrophages. Binding in this sytem correlated with cytolysis.

Some evidence exists that phosphatidylserine-containing structures may participate in this binding (Fidler and Schroit, 1988), as liposomes containing phosphatidylserine were phagocytosed 5 to 10 times faster than liposomes of the same size and configuration consisting only of phosphatidylcholine. Furthermore, recognition of phosphatidylserine in the membranes of aged red blood cells may play a role in their removal (Tanaka and Schroit, 1983; Schroit et al., 1984; Schroit et al.,

1985; Fidler and Schroit, 1988). Our laboratory has found that after killing of adenovirus-infected targets by secreted TNF, the cellular remnants and nuclei are phagocytosed and destroyed by macrophages in an integrin-dependent fashion (Zachariades, Wold, and Gooding, unpublished). Therefore, integrin-mediated recognition may also play a role in binding and selectivity. As a precedent for this, macrophage recognition of aged neutrophils and lymphocytes has been shown to be integrin-mediated (Savill et al., 1990).

Another way that selectivity can be exerted is at the level of secreted mediators. In many cases, tumor or infected cells display a different sensitivity to macrophage-secreted mediators. Tumor necrosis factor (TNF), for example, was named for its ability to cause necrosis of experimental tumors, and also has selective antiviral properties (reviewed by Old, 1985; Beutler and Cerami, 1989; Fiers, 1991). Nitric oxide also is toxic for some cells but not for others (Klostergaard and Leroux, 1989; Klostergaard et al., 1991), while cytolytic proteases secreted by activated macrophages are reported to selectively harm neoplastic cells (Adams, 1980; Adams et al., 1980).

The macrophage–tumor cell interaction possesses an additional level of complexity not seen in other cytolytic situations, because tumor cells release substances that can affect the macrophage and influence the outcome of the interaction. For example, tumor cells were shown to induce the release of TNF by pyran copolymer-elicited murine peritoneal macrophages (Hasday et al., 1990). Also, work in our laboratory has shown that factors secreted by SV40-transformed targets can affect macrophage migration and cytolysis. At least one factor, produced by a macrophage-resistant target, inhibits both migration and cytolysis, while a second factor or factors, produced by a sensitive target, enhances macrophage kill of resistant targets (Laster et al., 1988; Laster and Gooding, 1990). MDF produced by tumor cells suppresses macrophage functions (Ding et al., 1990), as does p15e, a retrovirus product (Cianciolo et al., 1981). Other systems have been described where products secreted from target cells modulate macrophage functions in both a positive (Kleinerman et al., 1984) and a negative manner (Hammond et al., 1974; Bigazzi et al., 1975; Snyderman and Pike, 1976). Other mechanisms may affect the sensitivity of a given target cell. For example, IFN-γ, which activates macrophages, may also play a role in modulating the sensitivity of tumor cells to macrophage killing (Leu et al., 1991).

To summarize, the selectivity observed appears to be a complex function dependent on several aspects. These include the ability of the target cell to bind to macrophages, the intrinsic sensitivity of the target to various macrophage-derived mediators, and the ability of the target cell to modulate macrophage function.

Mechanism

Over the years, many laboratories have attempted to understand the way in which macrophages kill their targets. No clear, single mechanism has been defined, and it seems certain now that there is no single mechanism. Rather, macrophages possess a number of cytolytic mechanisms that can interact with each other and with the target cell, such that each macrophage–target cell interaction is unique. In addition to this intrinsic variability, difference in experimental design among different laboratories have created a literature that is difficult to synopsize.

Variations in the results reported for macrophage cytolysis have come from several sources: in the types of effector cells used, in the activation protocol followed, in the types of target cells, and in the assays used to measure function. Much of the work in this field has been done with either human monocytes or with macrophages elicited from the mouse peritoneal cavity. Human blood is a relatively convenient source of material, while in a murine system, the quantity of material available from the peritoneal cavity is much greater than that available from the circulatory system. Macrophage-like cell lines such as RAW 264.7 have also been used extensively, and some work has been done with macrophage hybridoma clones (Higuchi et al., 1990).

Activation protocols differ, as explained above (Macrophage Activation), and a variety of target cells have been used, from cultured cell lines such as P815 to normal, primary cells. As described above (Selectivity), the target chosen for a particular study will significantly affect the results observed.

A number of different assays have been utilized to measure function. Fortunately, when five different protocols were compared, including three that depended on the release of a radioisotope, one that depended on cell counting, and one that depended on flow cytometric quantification of remaining viable tumor cells after exposure to activated macrophages, good agreement was found among them (Russell et al., 1986). It is important to keep these sources of variation in mind when comparing the results from different laboratories.

Macrophages exert their cytolytic effects through two major general mechanisms: direct cell–cell contact, and through the action of the more than 100 different molecules secreted by the activated macrophage (Nathan, 1987). These two mechanisms are not necessarily mutually exclusive, and lysis may depend on cooperation between one or more soluble factor(s) and membrane-triggered events. In the past, not all studies have differentiated between the two forms of killing. For example, in experiments where macrophages are simply incubated with potential targets, killing of the targets could be due either to direct cell-to-cell contact and/or to effects mediated by secreted molecules. As an example of the need for careful analysis to differentiate between these two general mechanisms, we find that although macrophage killing of virus-infected murine fibroblasts, as measured by DNA degradation, appears to require cell-to-cell contact, the process actually occurs in two steps: the first involves killing of the targets by secreted molecules, and the second involves phagocytosis of the remains by the macrophages and the degradation of the cellular DNA (Zachariades, Wold, and Gooding, unpublished).

Not all susceptible targets are killed by the same mechanism, and more than one mechanism can be operative for a given target cell. For example, Higuchi and co-workers (1990) showed that for two macrophage-sensitive cell lines, L-P3 cells were killed primarily by TNF, and P815 cells were killed primarily by nitric oxide. These workers were also able to show synergy between the two mechanisms in lysis of Meth A, another macrophage-sensitive line. We find that contact-dependent killing of an SV40-transformed target involves both TNF and nitric oxide, in that interference with either mechanism results in suppression of cell-mediated cytolysis. However, attempts to lyse these targets by incubating them with both TNF and nitric oxide-producing agents were unsuccessful, suggesting that an additional factor may be necessary for macrophage-mediated killing (Duerksen-Hughes et al., submitted).

In some cases, target cells may be sensitive to several macrophage-secreted products, and the multiple lytic pathways may be redundant. In this situation, blocking one pathway, even of a mechanism cytolytic to a particular target cell, may not appear to be effective, as other mechanisms will continue to lyse the cells. Ichinose et al., (1988) showed that A375 cells, derived from a human melanoma, are sensitive to interleukin (IL)-1_α, IL-1_β, and TNF, and their lysis by each of these cytokines can be blocked by the appropriate antiserum. When fixed plasma membranes from activated monocytes were used as the effectors, each of these antisera provided only partial protection, while combining anti-IL-1 and anti-TNF sera provided complete protection. However, when these antisera, either alone or in combination, were added to cocultures of activated monocytes and targets, little if any protection was observed, indicating that some other function of the activated monocyte was lysing the target. In support of this conclusion, a variant of A375 that was resistant to both TNF and Il-1 was lysed by activated monocytes.

Thus, the activated macrophage possesses multiple, sometimes redundant, cytolytic mechanisms and each macrophage–target cell interaction probably contains elements unique to the individual target. However, in spite of this complexity, there are a number of common elements involved in macrophage-mediated cytolysis which have been studied individually. These cytolytic factors and their mechanisms of action are described in the next section.

Factor-Mediated Cytotoxicity

Hydrogen Peroxide and Other Reactive Oxygen Intermediates

Activated macrophages undergo an enhanced respiratory burst which results in the production of H_2O_2 as well as other reactive oxygen intermediates, such as the superoxide anion O_2^- (Nathan et al., 1979a, 1979b; Johnston and Kitagawa, 1985; Sasada et al., 1983). The enzyme responsible for the production of these cytotoxic mediators is NADPH oxidase. Upon activation by LPS and stimulation by phorbol myristate acetate (PMA) or phagocytosis, the NADPH oxidase present in macrophages undergoes alterations in its K_m and V_{max}. Also, the concentration of NADPH is increased. Together, these changes can account for a 2.2- to 3.5-fold increase in the calculated velocity of the oxidase (Sasada et al., 1983). H_2O_2 may react with myeloperoxidase to produce hypochlorous acid, which is cytotoxic (Clark and Klebanoff, 1979; Slivka et al., 1980; Clark and Szot, 1981). It should be noted, however, that competence for secretion of H_2O_2 does not always correlate with macrophage cytolytic ability (Cohen et al., 1982).

Nitric Oxide

Nitric oxide is produced by activated macrophages

from the guanidino nitrogen of arginine, with citrulline as the other product (Hibbs et al., 1987a; Hibbs et al., 1987b; Hibbs et al., 1988; Marletta et al., 1988; Stuehr and Nathan, 1989; Stuehr et al., 1989a). The enzyme responsible, NO synthase (sometimes referred to as arginine deiminase) is induced in macrophages by IFN-γ and LPS. The enzyme is heat-sensitive and exhibits substrate stereospecificity (Stuehr et al., 1989b). NADPH, FAD, FMN, tetrahydrobiopterin, and reduced thiol are all required as cofactors (Kwon et al., 1989; Stuehr et al., 1990; Stuehr et al., 1989b; Stuehr et al., 1991a; Tayeh and Marletta, 1989). The purified enzyme is active as a dimer of two 130-kDa subunits and exhibits a K_m of 2.8 μM for arginine and 0.3 μM for NADPH (Stuehr et al., 1991a). NO synthase is inhibited by arginine analogs (Hibbs et al., 1987b) as well as by inhibitors of flavoproteins (Stuehr et al., 1991b). The mechanisms of nitric oxide generation includes N^ω-hydroxy-L-arginine as an intermediate (Stuehr et al., 1991c), and it is known that the ureido oxygen of the L-citrulline products derives from dioxygen and not from water (Kwon et al., 1990).

Nitric oxide chelates iron and is thought to act by abstracting iron from iron-sulfur centers in a variety of enzymes (Lancaster and Hibbs, 1990; Hibbs et al., 1984; 1988), including those involved in mitochondiral respiration, such as NADH: ubiquinone oxidoreductase, succinate:ubiquinone oxidoreductase, and the citric acid enzyme aconitase (Drapier and Hibbs, 1986; 1988, and references therein). Due to the inactivation of these enzymes, the target cell suffers from a lack of ATP, which will ultimately lead to cell death. Nitric oxide can also affect the macrophage itself, inhibiting metabolism (Drapier and Hibbs, 1988). Macrophages compensate for this by increasing the activity in their glycolytic pathway. Nitric oxide has also been identified as endothelium-derived relaxing factor (EDRF) (Stuehr et al., 1989a), which functions to relax smooth muscle (Moncada et al., 1988) by modulating the activity of cGMP (reviewed by Marletta, 1989).

Nitric oxide has been shown to be instrumental in causing cytostasis [*Mycobacterium leprae* (Adams et al., 1991), *Cryptococcus neoformans* (Granger et al., 1988; 1990), *Toxoplasma gondii* (Adams et al., 1990)], or cytolysis [*Leishmania major* (Liew et al., 1990a, 1990b; Green et al., 1990), *Schistosoma mansoni* (James and Glavin, 1989)] in several microorganisms. In addition, it is operative in macrophage-mediated lysis of tumor cells (Hibbs et al., 1987b, 1988). Nitric oxide may also synergize with other macrophage-secreted products. Higuchi and co-workers (1990) showed that one particular cell line, Meth A, was killed primarily by nitric oxide, and that that killing could be enhanced by TNF. However, this system does not appear to be responsible for lysis of all targets (Klostergaard and Leroux, 1989).

Cytolytic Proteases

Activated macrophages secrete proteases which are lytic to several types of tumor cells, including leukemias, lymphomas, sarcomas, and carcinomas (Adams, 1980; Adams et al., 1980; Adams and Johnson, 1982). These proteases are maximally active at neutral pH, and their activity is inhibited by compounds such as bovine pancreatic trypsin inhibitor and diisopropylfluorophosphate, suggesting that they are serine protease(s) (Adams, 1980; Adams et al., 1980). Their mode of action is unknown, but it is suggested that H_2O_2 acts syngergistically with the protease in producing target injury (Adams et al., 1981; Adams and Johnson, 1982). Reidarson and co-workers (1982) have observed specific target cell binding by a macrophage cytotoxin, which was itself, or was associated with, protease activity. Additionally, the binding of maleylated proteins to their receptors has been associated with the release of three neutral proteases: neutral casenases, plasminogen activator, and cytolytic proteinase (Johnson et al., 1982).

Tumor Necrosis Factor

TNF is a multifunctional molecule, secreted by activated macrophages, that has been implicated in roles as diverse as antitumor activity, inflammatory and immune responses, and antiviral activities (reviewed by Old, 1985; Beutler and Cerami, 1986, 1989; Fiers, 1991). The secreted form of TNF is active as a trimer of three 17-kDa subunits (Aggarwal et al., 1985; Smith and Baglioni, 1987; Jones et al., 1989). TNF is itself cytolytic for some tumor cell lines (Urban et al., 1986; Zeigler-Heitbrock et al., 1986; Nissen-Meyer et al., 1987); hence, its synthesis and release can satisfactorily account for some macrophage-mediated cytolysis. For TNF-sensitive lines, TNF is, in fact, the prime mediator of lysis (Philip and Epstein, 1986; Feinman et al., 1987; Wilson et al., 1989).

The exact molecular mechanism by which TNF kills cells is unknown. However, certain aspects have been defined. TNF binds to specific receptors on the cell surface and is internalized and degraded

(Mosselmans et al., 1988; Tartaglia et al., 1991). This occurs in both sensitive and resistant cells. Of the two TNF receptors identified, only TNF-R1 (55 kDa) mediates cytotoxicity (Tartaglia et al., 1991). When normally resistant cells are treated with cyclohexamide or other inhibitors of protein synthesis, these cells become sensitive to TNF, implying that cellular protein synthesis is essential for the resistant phenotype [Ruff and Gifford, 1981; also see Shepard and Lewis (1988) to review mechanisms of resistance]. Pretreatment of cells with TNF can protect them from subsequent killing by TNF plus cyclohexamide, suggesting that TNF can induce the synthesis of cellular protective proteins (Hahn et al., 1985). One of these induced proteins is manganous superoxide dismutase (Wong et al., 1989). The exact lytic events are unknown, but it has been suggested that they may be due to aberrant mitochondrial energy metabolism, leading to the accumulation of reactive intermediates. It may be that in some transformed cell lines, the protective mechanisms are defective, rendering cells sensitive to TNF alone.

Secreted TNF has demonstrated antiviral properties (Mestan et al., 1986; Wong and Goeddel, 1986) and is instrumental in the kill of virus-infected and -transformed cells. We and others (Cook et al., 1987; Gooding et al., 1988) have shown that infection with adenovirus mutants induces sensitivity to TNF and therefore to activated macrophages in normally resistant cells. Transformation with the adenovirus E1A *onc* gene also induces TNF sensitivity (Chen et al., 1987; Duerksen-Hughes et al., 1989). TNF is directly lytic for other virus-infected cells as well (Paya et al., 1988; Wong and Goeddel, 1986). Presumably, in these cases, virus-directed functions modulate the cellular pathways leading to either lysis by or protection from TNF, and hence their susceptibility to TNF and activated macrophages.

Interleukin-1

IL-1 is another cytokine produced by activated macrophages. Lachman and co-workers (1986) have shown that recombinant IL-1$_\beta$ is cytotoxic for human melanoma cells, which are not sensitive to TNF. Soluble IL-1 is cytocidal for several additional cell lines, such as L929 (Onazaki et al., 1985), which are also killed by α-lymphotoxin. These workers were also able to show that the cytocidal activity of IL-1 only partially overlapped the target cell selectivity of α-lymphotoxin.

Cell–Cell-Mediated Cytotoxicity

In addition to mediating cell killing by releasing soluble mediators, macrophages kill by contact-dependent mechanisms. When cell contact-dependent killing occurs, the process can be thought of as occurring in three steps: migration of the macrophage toward the target, binding between the two cells, and finally, lysis of the target. In our laboratory, video analysis of macrophage migration showed that factors released by sensitive or resistant targets could affect the migration of macrophages toward their targets (Laster et al., 1988). Once the macrophages have arrived at the target cells, binding must occur for lysis to take place, and macrophages are, in fact, capable of selectively binding to transformed cells (Adams and Johnson, 1982; Somers et al., 1983; 1986) (see also Selectivity, above).

As in the case of the soluble mediators, there appears to be more than one mechanism by which the actual lysis can occur. In some cases, cytotoxicity may be the result of the transfer of lysosomal enzymes (Hibbs, 1974a). In support of this concept, Bucana and co-workers (1976) showed that destruction of tumor cells by BCG-stimulated macrophages was associated with exocytosis of lyosomes from the macrophage and endocytosis of the lysosomes by the targets.

Molecules associated with the surfaces of activated macrophages, such as TNF and IL-1, also play essential roles in cell contact-dependent killing. TNF is found both in the cytoplasm and associated with the membranes of murine macrophages (Chensue et al., 1988). Decker and co-workers have shown that TNF-dependent cytolysis could occur in the absence of any measurable secreted TNF and by paraformaldehyde-fixed, activated macrophages (Decker et al., 1987), indicating that surface-associated TNF was itself cytolytic. Bakouche et al. (1988) found that TNF exists as a 17-kDa molecule bound to its own receptor, and also showed that purified plasma membranes from LPS-activated human monocytes could lyse L929 cells. Luettig and co-workers (1989) demonstrated the existence and cytotoxicity of two forms of membrane-associated TNF: as an integral membrane protein, and as the secreted protein bound to its own receptor.

TNF is synthesized initially as a 26-kDa, membrane-associated molecule. Pulse-chase experiments have shown that this is the precursor to the secreted, 17-kDa molecule (Kriegler et al., 1988;

Jue et al., 1990). By transfecting genes that code either for the cleavable, wild-type protein or for a mutant, noncleavable (and therefore nonsecretable) form of the protein into NIH3T3 cells, Kriegler and his co-workers were able to show that the membrane-associated form of TNF is cytotoxic for TNF-sensitive targets (Perez et al., 1990). Therefore, the membrane-associated form of TNF can kill targets in a cell–cell contact-dependent mechanism. Work done in our laboratory using the transfected cell lines described above has shown that of several cell lines tested, cells sensitive to the 26-kDa molecule are also sensitive to the 17-kDa molecule (Day and Gooding, unpublished).

Like TNF, IL-1 is found as both a membrane-associated and a secreted molecule, and both human monocytes and murine macrophages stimulated with LPS can express the membrane-associated forms of the cytokine (Kurt-Jones et al., 1985; 1986; Matsushima et al., 1986; Bakouche et al., 1987; Conlon et al., 1987). Fixed membranes from human monocytes that had been activated with LPS or desmethyl muramyl dipeptide were cytotoxic to A375 melanoma cells and HT-29 colon cancer cells but not to actinomycin D-treated L929 cells (a TNF-sensitive cell line), and the monocyte cytotoxicity was inhibited by antibodies to IL-1 but not by antibodies to TNF (Okubo, 1989). Ichinose and co-workers (1988) have also provided evidence that membrane-bound Il-1 can lyse tumor targets. However, it has been suggested that "membrane-associated IL-1" could be caused by leakage from inadequately fixed cells, and that it is the soluble factor which actually mediates cytotoxicity (Suttles et al., 1990).

These different mechanisms of cell–cell contact-dependent killing do not necessarily work independently of each other or of the soluble molecules released by the activated macrophage. In our laboratory, we have shown that an SV40-transformed target (E8) is sensitive to activated macrophages. Lysis can be inhibited by anti-TNF antibodies or by agents that interfere with nitric oxide generation and function [Fe(II) and N^GMMA], showing that both TNF and nitric oxide participate in its lysis. However, the killing is contact-dependent, suggesting that yet a third signal may be involved (Duerksen-Hughes et al., submitted). The observed requirement for actual cell-to-cell contact could be due to at least two possibilities, which are not mutually exclusive. In one case, actual binding of a surface-associated ligand to a surface-associated receptor could trigger events in the target cell which would culminate in cytolysis. In the other, products secreted by macrophages could be the effector molecules, but due to concentration requirements or to a short life of these molecules (i.e., nitric oxide), very close proximity of the target and effector cells is necessary for cytolysis. In another study, work done by Klostergaard et al., (1990) showed that while some targets (i.e., L929) could be killed both by activated macrophages and soluble TNF, other targets (i.e., EMT-6) were killed only by activated macrophages and not by soluble TNF. In both these cases, anti-TNF blocked lysis. Thus, EMT-6 cells may respond to macrophages in a manner similar to that of E8 cells.

It seems clear that killing can be due to a complex interaction between surface events and soluble mediators. This is in contrast to CTL killing, where antibodies to soluble mediators such as perforin do not inhibit lysis (Reynolds et al., 1987; Shiver and Henkart, 1991).

Summary and Conclusions

Macrophages play a central role in eliminating virus-infected and transformed cells. Upon exposure to various agents, macrophages become activated and capable of carrying out many secretory and contact-mediated functions related to cytotoxicity. Often, the macrophages will selectively bind their targets and lyse them in a contact-dependent manner. However, any ligands involved in this binding have yet to be definitively identified. Another mode of action is by the secretion of various effectors, such as proteases, reactive oxygen intermediates, reactive nitrogen intermediates, and TNF. These various modes of action are not mutually exclusive and may synergize with each other. It appears clear that macrophages have been equipped with an amazing variety of potential methods by which unwanted cells, as well as infectious organisms, can be recognized and eliminated.

Since the stage of macrophage activation (and hence lytic capacity) can be modulated by microorganisms or their products, lymphokines produced by immune cells, and products secreted by tumor cells, there is an intriguing possibility, as work from Fidler and his colleagues shows, that these immunomodulators may some day be used in tumor therapy by activating macrophages *in situ*.

References

Abramson SL, Gallin JI (1990): IL-4 inhibits superoxide production by human mononuclear phagocytes. *J Immunol* 144: 625–630

Adams DO (1980): Effector mechanisms of cytolytically activated macrophages. I. Secretion of neutral proteases and effect of protease inhibitors. *J Immunol* 124: 286-292

Adams DO, Hamilton TA (1984): The cell biology of macrophage activation. *Annu Rev Immunol* 2: 283-318

Adams DO, Hamilton TA (1987): Molecular transductional mechanisms by which IFN-γ and other signals regulate macrophage development. *Immunol Rev* 97: 5-27

Adams DO, Johnson WJ (1982): Activation of murine mononuclear phagocytes for destroying tumor cells: Analysis of effector mechanisms and development. *Adv Exp Med Biol* 155: 707-720

Adams DO, Nathan CF (1983): Molecular mechanisms in tumor cell-killing by activated macrophages. *Immunol Today* 4: 166-170

Adams LB, Franzblau SG, Vavrin Z, Hibbs JB, Jr, Krahenbuhl JL (1991): L-Arginine-dependent macrophage effector functions inhibit metabolic activity of *Mycobacterium leprae*. *J Immunol* 147: 1642-1646

Adams LB, Hibbs JB, Taintor RR, Krahanbuhl JL (1990): Microbiostatic effect of murine-activated macrophages for *Toxoplasma gondii*: Role for synthesis of inorganic nitrogen oxides from L-arginine. *J Immunol* 144: 2725-2729

Adams DO, Johnson WJ, Fiorito E, Nathan CF (1981): Hydrogen peroxide and cytolytic factor can interact synergistically in effecting cytolysis of neoplastic targets. *J Immunol* 127: 1973-1977

Adams DO, Kao K-J, Farb R, Pizzo SV (1980): Effector mechanisms of cytolytically activated macrophages. II. Secretion of a cytolytic factor by activated macrophages and its relationship to secreted neutral proteases. *J Immunol* 124: 293-300

Aggarwal BB, Kohr WJ, Hass PE, Moffat B, Spencer SA, Henzel WJ, Bringman TS, Nedwin GE, Goeddel DV, Harkins RN (1985): Human tumor necrosis factor: Production, purification and characterization. *J Biol Chem* 260: 2345-2354

Allison AC, (1974): On the role of mononuclear phagocytes in immunity against viruses. *Prog Med Virol* 18: 15-31

Amber IJ, Hibbs JB, Jr, Parker CJ, Johnson BB, Taintor RR, Vavrin Z (1991): Activated macrophage conditioned medium: Identification of the soluble factors inducing cytotoxicity and the L-arginine dependent effector mechanism. *J Leukocyte Biol* 49: 610-620

Bakouche O, Brown D, Lachman LB (1987): Subcellular localization of human monocyte interleukin 1: Evidence for an inactive precursor molecule and a possible mechanism for IL-1 release. *J Immunol* 138: 4249-4255

Bakouche O, Ichinose Y, Heicappell R, Fidler IJ, Lachman LB (1988): Plasma membrane-associated tumor necrosis factor: A non-integral membrane protein possibly bound to its own receptor. *J Immunol* 140: 1142-1147

Beutler B, Cerami A (1986): Cachectin and tumour necrosis factor as two sides of the same biological coin. *Nature* 320: 584-588

Beutler B, Cerami A (1989): The biology of cachectin/ TNF—a primary mediator of the host response. *Annu Rev Immunol* 7: 625-655

Bigazzi PE, Yoshida T, Ward PA, Cohen S (1975): Production of lymphokine-like factors (cytokines) by Simian virus 40-infected and Simian virus 40-transformed cells. *Am J Pathol* 80: 69-77

Bucana C, Hoyer LC, Hobbs B, Breesman S, McDaniel M, Hanna MG, Jr (1976): Morphological evidence for the translocation of lysosomal organelles from cytotoxic macrophages into the cytoplasm of tumor target cells. *Cancer Res* 36: 4444-4458

Bucana CD, Hoyer LC, Schroit AJ, Kleinerman E, Fidler IJ (1983): Ultrastructural studies of the interaction between liposome-activated human blood monocytes and allogeneic tumor cells *in vitro*. *Am J Pathol* 112: 101-111

Chapes SK, Gooding LR (1985): Evidence for the involvement of cytolytic macrophages in rejection of SV40-induced tumors. *J Immunol* 135: 2192-2198

Chedid L, Carelli L, Audibert F (1979): Recent developments concerning muramyl dipeptide, a synthetic immunoregulating molecule. *J Reticuloendothel Soc.* 26: 631

Chen MJ, Holskin B, Strickler J, Gorniak J, Clark MA, Johnson PJ, Mitcho M, Shalloway D (1987): Induction by E1A oncogene expression of cellular susceptibility to lysis by TNF. *Nature* 330: 581-583

Chensue SW, Remick DG, Shmyr-Forsch C, Beals TF, Kunkel SL (1988): Immunohistochemical demonstration of cytoplasmic and membrane-associated tumor necrosis factor in murine macrophages. *Am J Pathol* 133: 564-572

Cianciolo G, Hunter J, Silva J, Haskill JS, Snyderman R (1981): Inhibitors of monocyte responses to chemotaxins are present in human cancerous effusions and react with monoclonal antibodies to the $p_{15}(E)$ structural protein of retroviruses. *J Clin Invest* 68: 831-844

Clark RA, Klebanoff SJ (1979): Role of the myeloperoxidase-H_2O_2-halide system in concanavalin A-induced tumor cell killing by human neutrophils. *J Immunol* 122: 2605-2610

Clark RA, Szot S (1981): The myeloperoxidase-hydrogen peroxide-halide system as effector of neutrophil-mediated tumor cell cytotoxicity. *J Immunol* 126: 1295-1301

Cohen MS, Taffet SM, Adams DO (1982): The relationship between competence for secretion of H_2O_2 and completion of tumor cytotoxicity by BCG-elicited murine macrophages. *J Immunol* 128: 1781-1785

Conlon PJ, Grabstein KH, Alpert A, Prickett KS, Hopp TP, Gillis S (1987): Localization of human mononuclear cell interleukin 1. *J Immunol* 139: 98-102

Cook JL, Hibbs JB, Lewis AM (1980): Resistance of simian virus 40-transformed hamster cells to the cytolytic effect of activated macrophages: A possible factor in species-specific viral oncogenicity. *Proc Natl Acad Sci USA* 77: 6773-6777

Cook JL, Hibbs JB, Lewis AM, Jr (1982): DNA virus-transformed hamster cell-host effector cell interactions:

Level of resistance to cytolysis correlated with tumorigenicity. *Int J Cancer* 30: 795–803

Cook JL, May DL, Lewis AM, Jr, Walker TA (1987): Adenovirus E1A gene induction of susceptibility to lysis by natural killer cells and activated macrophages in infected rodent cells. *J Virol* 61: 3510–3520

Cook JL, May DL, Wilson BA, Holskin B, Chen M-J, Shalloway D, Walker TA (1989): Role of tumor necrosis factor-α in E1A oncogene-induced susceptibility of neoplastic cells to lysis by natural killer cells and activated macrophages. *J Immunol* 142: 4527–4534

Decker T, Lohmann-Matthes M-L, Gifford GE (1987): Cell-associated tumor necrosis factor (TNF) as a killing mechanism of activated cytotoxic macrophages. *J Immunol* 138: 957–962

Ding A, Nathan CF, Graycar J, Derynck R, Stuehr DJ, Srimal S (1990): Macrophage deactivating factor and transforming growth factors-β1, β2, and -β3 inhibit induction of macrophage nitrogen oxide synthesis by IFN-γ. *J Immunol* 145: 940–944

Ding AH, Nathan CF, Stuehr DJ (1988): Release of reactive nitrogen intermediates and reactive oxygen intermediates from mouse peritoneal macrophages: Comparison of activating cytokines and evidence for independent production. *J Immunol* 141: 2407–2412

Ding A, Wright SD, Nathan C (1987): Activation of mouse peritoneal macrophages by monoclonal antibodies to Mac-1 (complement receptor type 3). *J Exp Med* 165: 733–749

Dinney CPN, Bucana CD, Utsugi T, Fidler IJ, von Eschenbach AC, Killion JJ (1991): Therapy of spontaneous lung metastasis of murine renal adenocarcinoma by systemic administration of liposomes containing the macrophage activator CGP 31362. *Cancer Res* 51: 3741–3747

Drapier J-C, Hibbs JB, Jr. (1986): Murine cytotoxic activated macrophages inhibit aconitase in tumor cells: Inhibition involves the iron-sulfur prosthetic group and is reversible. *J Clin Invest* 78: 790–797

Drapier J-C, Hibbs JB, Jr (1988): Differentiation of murine macrophages to express nonspecific cytotoxicity for tumor cells results in L-arginine-dependent inhibition of mitochondrial iron-sulfur enzymes in the macrophage effector cells. *J Immunol* 140: 2829–2838

Drapier J-C, Wietzerbin J, Hibbs JB, Jr (1988): Interferon-γ and tumor necrosis factor induce the L-arginine-dependent cytotoxic effector mechanism in murine macrophages. *Eur J Immunol* 18: 1587–1592

Duerksen-Hughes P, Wold WSM, Gooding LR (1989): Adenovirus E1A renders infected cells sensitive to cytolysis by tumor necrosis factor. *J Immunol* 143: 4193–4200

Evans R, Alexander P (1972): Mechanism of immunologically specific killing of tumour cells by macrophages. *Nature* 236: 168–170

Feinman R, Henriksen-DeStefano D, Tsujimoto M, Vicek J (1987): Tumor necrosis factor is an important mediator of tumor cell killing by human monocytes. *J Immunol* 138: 635–640

Fidler IJ (1978): Recognition and destruction of target cells by tumoricidal macrophages. *Isr J Med Sci* 14: 177–191

Fidler IJ (1985): Macrophages and metastasis—a biological approach to cancer therapy: Presidential Address. *Cancer Res* 5: 4714–4726

Fidler IJ, Kleinerman ES (1984): Lymphokine-activated human blood monocytes destroy tumor cells but not normal cells under cocultivation conditions. *J Clin Oncol* 2: 937–943

Fidler IJ, Schroit AJ (1988): Recognition and destruction of neoplastic cells by activated macrophages: Discrimination of altered self. *Biochim Biophys Acta* 948: 151–173

Fidler IJ, Fan D, Ichinose Y (1989): Potent in situ activation of murine lung macrophages and therapy of melanoma metastases by systemic administration of liposomes containing muramyltripeptide phosphatidylethanolamine and interferon gamma. *Invasion Metastasis* 9: 75–88

Fidler IJ, Roblin RO, Poste G (1978): *In vitro* tumoricidal activity of macrophages against virus-transformed lines with temperature-dependent transformed phenotypic characteristics. *Cell Immunol* 38: 131–146

Fidler IJ, Sone S, Fogler WE, Barnes ZL (1981): Eradication of spontaneous metastases and activation of alveolar macrophages by intravenous injection of liposomes containing muramyl dipeptide. *Proc Natl Acad Sci USA* 78: 1680–1684

Fiers W (1991): Tumor necrosis factor: Characterization at the molecular, cellular and *in vivo* level. *FEBS Lett* 285: 199–212

Figueiredo F, Uhing RJ, Okonogi K, Gettys TW, Johnson SP, Adams DO, Prpic V (1990): Activation of the cAMP cascade inhibits an early event involved in murine macrophage Ia expression. *J Biol Chem* 265: 12317–12323

Gifford GE, Lohmann-Matthes M-L (1986): Requirement for the continual presence of lipopolysaccharide for production of tumor necrosis factor by thioglycollate-induced peritoneal murine macrophages. *Int J Cancer* 38: 135–137

Ginsberg HS, Lundholm-Beauchamp U, Horswood RL, Pernis B, Wold WSM, Chanock RM, Prince GA (1989): Role of early region 3 (E3) in the pathogenesis of adenovirus disease. *Proc Natl Acad Sci USA* 86: 3823–3827

Ginsberg HS, Moldawer LL, Sehgal PB, Redington M, Kilian PL, Chanock RM, Prince GA (1991): A mouse model for investigating the molecular pathogenesis of adenovirus pneumonia. *Proc Natl Acad Sci USA* 88: 1651–1655

Gooding LR, Elmore LW, Tollefson AE, Brady HA, Wold WSM (1988): A 14,700 MW protein from the E3 region of adenovirus inhibits cytolysis by tumor necrosis factor. *Cell* 53: 341–346

Granger DL, Hibbs JB, Jr, Perfect JR, Durack DT (1988): Specific amino acid (L-arginine) requirement for the microbiostatic activity of murine macrophages. *J Clin Invest* 81: 1129–1236

Granger DL, Hibbs JB, Jr, Perfect JR, Durack DT (1990):

Metabolic fate of L-arginine in relation to microbiostatic capability of murine macrophages. *J Clin Invest* 85: 264–273

Green SJ, Meltzer MS, Hibbs JB, Jr, Nacy CA (1990): Activated macrophages destroy intracellular *Leishmania major* amastigotes by an L-arginine-dependent killing mechanism. *J Immunol* 144: 278–283

Haak-Frendscho M, Wynn TA, Czuprynski CJ, Paulnock D (1990): Transforming growth factor-β1 inhibits activation of macrophage cell line RAW 264.7 for cell killing. *Clin Exp Immunol* 82: 404–410

Haberland ME, Tannenbaum CS, Williams RE, Adams DO, Hamilton TA (1989): Role of the maleyl-albumin receptor in activation of murine peritoneal macrophages in vitro. *J Immunol* 142: 855–862

Hahn T, Toker L, Budilovsky S, Aderka D, Eshhar Z, Wallach D (1985): Use of monoclonal antibodies to a human cytotoxin for its isolation and for examining the self-induction of resistance to this protein. *Proc Natl Acad Sci USA* 82: 3814–3818

Hamilton TA, Adams DO (1987): Molecular mechanisms of signal transduction in macrophages. *Immunol Today* 8: 151–158

Hamilton TA, Jansen MM, Somers SD, Adams DO (1986): Effects of bacterial lipopolysaccharide on protein synthesis in murine peritoneal macrophages: Relationship to activation for macrophage tumoricidal function. *J Cell Physiol* 128: 9–17

Hammond ME, Roblin RO, Dvorak AM, Selvaggio SS, Black PH, Dvorak HF (1974): MIF-like activity in simian virus 40-transformed 3T3 fibroblast cultures. *Science* 185: 955–957

Hasday JD, Shah EM, Lieberman AP (1990): Macrophage tumor necrosis factor—a release is induced by contact with some tumors. *J Immunol* 145: 371–379

Hibbs JB, Jr (1974a): Heterocytolysis by macrophages activated by *Bacillus Calmette-Guerin*: Lysosome exocytosis into tumor cells. *Science* 184: 468–471

Hibbs JB, Jr (1974b): Discrimination between neoplastic and non-neoplastic cells *in vitro* by activated macrophages. *J Natl Cancer Inst* 53: 1487–1492

Hibbs JB, Jr, Taintor RR, Vavrin Z (1984): Iron depletion: Possible cause of tumor cell cytotoxicity induced by activated macrophages. *Biochem Biophys Res Commun* 123: 716–723

Hibbs JB, Jr, Taintor RR, Vavrin Z (1987a): Macrophage cytotoxicity: Role for L-arginine deiminase and imino nitrogen oxidation to nitrite. *Science* 235: 473–476

Hibbs JB, Jr, Taintor RR, Vavrin Z, Rachlin EM (1988): Nitric oxide: A cytotoxic activated macrophage effector molecule. *Biochem Biophys Res Commun* 157: 87–94

Hibbs JB, Jr, Vavrin Z, Taintor RR (1987b): L-Arginine is required for expression of the activated macrophage effector mechanism causing selective metabolic inhibition in target cells. *J Immunol* 138: 550–565

Higuchi M, Higashi N, Taki H, Osawa T (1990): Cytolytic mechanisms of activated macrophages: Tumor necrosis factor and L-arginine-dependent mechanisms act synergistically as the major cytolytic mechanisms of activated macrophages. *J Immunol* 144: 1425–1431

Hirsh MS, Zisman B, Allison AC (1970): Macrophages and age-dependent resistance to herpes simplex virus in mice. *J Immunol* 104: 1160–1165

Ichinose Y, Bakouche O, Tsao JY, Fidler IJ (1988): Tumor necrosis factor and IL-1 associated with plasma membranes of activated human monocytes lyse monokine-sensitive but not monokine-resistant tumor cells whereas viable activated macrophages lyse both. *J Immunol* 141: 512–518

James SL, Glavin J (1989): Macrophage cytotoxicity against schistosomula of *Schistosoma mansoni* involves arginine-dependent production of reactive nitrogen intermediates. *J Immunol* 143: 4208–4212

Johnson WJ, Pizzo SV, Imber MJ, Adams DO (1982): Receptors for maleylated proteins regulate secretion of neutral proteases by murine macrophages. *science* 218: 574–576

Johnston, RB, Jr, Kitagawa S (1985): Molecular basis for the enhanced respiratory burst of activated macrophages. *Fed Proc* 44, 2927–2932

Jones EY, Stuart DI, Walker NPC (1989): Structure of tumour necrosis factor. *Nature* 338: 225–228

Jue D-M, Sherry B, Luedke C, Manogue KR, Cerami A (1990): Processing of newly synthesized cachectin/tumor necrosis factor in endotoxin-stimulated macrophages. *Biochemistry* 29: 8371–8377

Kitagaua S, Johnston RB (1986): Deactivation of the respiratory burst in activated macrophages: Evidence for alteration of signal transduction. *J Immunol* 136: 2605–2612

Kleinerman ES, Murray JL, Snyder JS, Cunningham JE, Fidler IJ (1989): Activation of tumoricidal properties in monocytes from cancer patients following intravenous administration of liposomes containing muramyl tripeptide phosphatidylethanolamine. *Cancer Res* 49: 4665–4670

Kleinerman ES, Zicht R, Sarin PS, Gallo RC, Fidler IJ (1984): Constitutive production and release of a lymphokine with macrophage-activating factor activity distinct from γ-interferon by a human T-cell leukemia virus-positive cell line. *Cancer Res* 44: 4470–4475

Klostergaard J, Leroux MF (1989): L-Arginine independent macrophage tumor cytotoxicity. *Biochem Biophys Res Commun* 165: 1262–1266

Klostergaard J, Leroux ME, Hung M-C (1991): Cellular models of macrophage tumoricidal effector mechanisms in vitro: Characterization of cytolytic responses to tumor necrosis factor and nitric oxide pathways in vitro. *J Immunol* 147: 2802–2808

Klostergaard J, Stoltje PA, Kull FC, Jr (1990): Tumoricidal effector mechanisms of murine BCG-activated macrophages: Role of TNF in conjugation-dependent and conjugation-independent pathways. *J Leukocyte Biol* 48: 220–228

Koff WC, Fidler IJ, Showalter SD, Chakrabarty MK, Hampar B, Ceccorulli LM, Kleinerman ES (1984): Human monocytes activated by immunomodulators in liposomes lyse herpesvirus-infected but not normal cells. *Science* 224: 1007–1009

Koff WC, Showalter SD, Chakrabarty MK, Hampar B,

Ceccorulli LM, Kleinerman ES (1985): Human monocyte-mediated cytotoxicity against herpes simplex virus-infected cells: Activation of cytotoxic monocytes by free and liposome-encapsulated lymphokines. *J Leukocyte Biol* 37: 461–472

Koff WC, Showalter SD, Seniff DA, Hampar B (1983): Lysis of Herpesvirus-infected cells by macrophages activated with free or liposome-encapsulated lymphokine produced by a murine T cell hybridoma. *Infect. Immun.* 42: 1067–1072

Kriegler M, Perez C, DeFay K, Albert I, Lu SD (1988): A novel form of TNF/cachectin is a cell surface cytotoxic transmembrane protein: Ramifications for the complex physiology of TNF. *Cell* 53: 45–53

Kurt-Jones EA, Beller DI, Mizel SB, Unanue ER (1985): Identification of a membrane-associated interleukin 1 in macrophages. *Proc Natl Acad Sci USA* 82: 1204–1208

Kurt-Jones EA, Virgin HW IV, Unanue ER (1986): *In vivo* and *in vitro* expression of macrophage membrane interleukin 1 in response to soluble and particulate stimuli. *J Immunol* 137: 10–14

Kwon NS, Nathan CF, Gilker C, Griffith OW, Matthews DE, Stuehr DJ (1990): L-Citrulline production from L-arginine by macrophage nitric oxide synthase: The ureido oxygen derives from dioxygen. *J Biol Chem* 265: 13442–13445

Kwon NS, Nathan CF, Stuehr DJ (1989): Reduced biopterin as a cofactor in the generation of nitrogen oxides by murine macrophages. *J Biol Chem* 264: 20496–20501

Lachman LB, Dinarello CA, Llansa ND, Fidler IJ (1986): Natural and recombinant human interleukin 1-β is cytotoxic for human melanoma cells. *J Immunol* 136: 3098–3102

Lambert LE, Paulnock DM (1989): Differential induction of activation markers in macrophage cell lines by interferon-γ. *Cell Immunol* 120: 401–418

Lancaster JR, Jr, Hibbs JB, Jr (1990): EPR demonstration of iron-nitrosyl complex formation by cytotoxic activated macrophages. *Proc Natl Acad Sci USA* 87: 1223–1227

Laster SM, Gooding LR (1990): Evidence that a target-derived soluble factor is necessary for the selective lysis of SV40-transformed fibroblasts by activated mouse macrophages. *J Immunol* 144: 1438–1443

Laster SM, Wood JG, Gooding LR (1988): Target-induced changes in macrophage migration may explain differences in lytic sensitivity among simian virus 40-transformed fibroblasts. *J Immunol* 141: 221–227

LeBlanc PA (1989): Macrophage activation for cytolysis of virally infected target cells. *J Leukocyte Biol* 45: 345–352

Lefkowitz DL, Mone J, Mills K, Hsieh T-C, Lefkowitz SS (1989): Peroxidases enhance macrophage-mediated cytotoxicity via induction of tumor necrosis factor. *Proc Soc Exp Biol Med* 190: 144–149

Lehn M, Weiser WY, Engelhorn S, Gillis S, Remold HG (1989): IL-4 inhibits H_2O_2 production and antileishmanial capacity of human cultured monocytes mediated by IFN-γ. *J Immunol* 143: 3020–3024

Letvin NL, Kauffman RS, Finberg R (1982): An adherent cell lyses virus-infected targets: Characterization, activation, and fine specificity of the cytotoxic cell. *J Immunol* 129: 2396–2401

Leu RW, Leu NR, Shannon BJ, Fast DJ (1991): IFN-γ differentially modulates the susceptibility of L1210 and P815 tumor targets for macrophage-mediated cytotoxicity: Role of macrophage-target interaction coupled to nitric oxide generation, but independent of tumor necrosis factor production. *J Immunol* 147: 1816–1822

Liew FY, Li Y, Millott S (1990a): Tumor necrosis factor-α synergizes with IFN-γ in mediating killing of *Leishmania major* through the induction of nitric oxide. *J Immunol* 145: 4306–4310

Liew FY, Millott S, Parkinson C, Palmer RMJ, Moncada S (1990b): Macrophage killing of *Leishmania* parasite in vivo is mediated by nitric oxide from L-arginine. *J Immunol* 144: 4794–4797

Luettig B, Decker T, Lohmann-Matthes M-L (1989): Evidence for the existence of two forms of membrane tumor necrosis factor: An integral protein and a molecule attached to its receptor. *J Immunol* 143: 4034–4038

Maddox DE, Shibata S, Goldstein IJ (1982): Stimulated macrophages express a new glycoprotein receptor reactive with *Griffonia simplicifolia* I-B_4 isolectin. *Proc Natl Acad Sci USA* 79: 166–170

Marino PA, Adams DO (1980a): Interaction of Bacillus Calmette-Guerin-activated macrophages and neoplastic cells *in vitro*: I. Conditions of binding and its selectivity. *Cell Immunol* 54: 11–25

Marino PA, Adams DO (1980b): Interaction of Bacillus Calmette-Guerin-activated macrophages and neoplastic cells *in vitro*: II. The relationship of selective binding to cytolysis. *Cell Immunol* 54: 26–35

Marino PA, Adams DO (1982): The capacity of activated murine macrophages for augmented binding of neoplastic cells: Analysis of induction by lymphokine containing MAF and kinetics of the reaction. *J Immunol* 128: 2816–2823

Marletta MA (1989): Nitric oxide: Biosynthesis and biological significance. *Trends Biol Sci* 14: 488–492

Marletta MA, Yoon PS, Iyengar R, Leaf CD, Whishnok JS (1988): Macrophage oxidation of L-arginine to nitrite and nitrate: Nitric oxide is an intermediate. *Biochemistry* 27: 8706–8711

Matsushima K, Taguchi M, Kovacs EJ, Young HA, Oppenheim JJ (1986): Intracellular localization of human monocyte associated interleukin 1 (IL-1) activity and release of biologically active IL-1 from monocytes by trypsin and plasmin. *J Immunol* 136: 2883–2891

Meltzer MS (1981): Macrophage activation for tumor cytotoxicity: Characterization of priming and triggering signals during lymphokine activation. *J Immunol* 127: 179–183

Mestan J, Digel W, Mittnacht S, Hillen H, Blohm D, Moller A, Jacobsen H, Kirchner H (1986): Antiviral effects of recombinant tumour necrosis factor *in vitro*. *Nature* 323: 816–819

Mogensen SC (1985): Genetic aspects of macrophage involvement in natural resistance to virus infections. *Immunol Letts* 11: 219–224

Moncada S, Palmer RMJ, Higgs EA (1988): The discovery of nitric oxide as the endogenous nitrovasodilator. *Hypertension* 12: 365–372

Morahan PS, Connor JR, Leary KR (1985): Viruses and the versatile macrophage. *Br Med Bull* 41: 15–21

Mosselmans R, Hepburn A, Dumont JE, Fiers W, Galand P (1988): Endocytic pathway of recombinant murine tumor necrosis factor in L-929 cells. *J Immunol* 141: 3096–3100

Nathan CF (1987): Secretory products of macrophages. *J Clin Invest* 79: 319–326

Nathan CF, Brukner LH, Silverstein SC, Cohn ZA (1979a): Extracellular cytolysis by activated macrophages and granulocytes. I. Pharmacologic triggering of effector cells and the release of hydrogen peroxide. *J Exp Med* 149: 84–99

Nathan CF, Silverstein SC, Brukner LH, Cohn ZA (1979b): Extracellular cytolysis by activated macrophages and granulocytes. II. Hydrogen peroxide as a mediator of cytotoxicity. *J Exp Med* 149: 100–113

Nestel FP, Carson PR, Wiltrout RH, Kerbel RS (1984): Alterations in sensitivity to nonspecific cell-mediated lysis associated with tumor progression: Characterization of activated macrophage- and natural killer cell-resistant tumor variants. *J Natl Cancer Inst* 73: 483–490

Nii A, Fidler IJ (1990): The incubation of human blood monocytes with tumor necrosis factor-α leads to lysis of tumor necrosis factor-sensitive but not resistant tumor cells. *Lymphokine Res* 9: 113–124

Nissen-Meyer J, Austgulen R, Espevik T (1987): Comparison of recombinant tumor necrosis factor and the monocyte-derived cytotoxic factor involved in monocyte-mediated cytotoxicity. *Cancer Res* 47: 2251–2258

Nong Y-H, Titus RG, Ribeiro JMC, Remold HG (1989): Peptides encoded by the calcitonin gene inhibit macrophage function. *J Immunol* 143: 45–49

Okubo A, Sone S, Tanaka M, Ogura T (1989): Membrane-associated interleukin 1α as a mediator of tumor cell killing by human blood monocytes fixed with paraformaldehyde. *Cancer Res* 49: 265–270

Old LJ (1985): Tumor necrosis factor (TNF). *Science* 230: 630–632

Onozaki K, Matsushima K, Aggarwal BB, Oppenheim JJ (1985): Human interleukin 1 is a cytocidal factor for several tumor cell lines. *J Immunol* 135: 3962–3968

Pace JL, Russell SW (1981): Activation of mouse macrophages for tumor cell killing. I. Quantitative analysis of interactions between lymphokine and lipopolysaccharide. *J Immunol* 126: 1863–1867

Pace JL, Russell SW, Torres BA, Johnson HM, Gray PW (1983): Recombinant mouse γ interferon induces the priming step in macrophage activation for tumor cell killing. *J Immunol* 130: 2011–2013

Paulnock DM, Lambert LE (1990): Identification and characterization of monoclonal antibodies specific for macrophages at intermediate stages in the tumoricidal activation pathway. *J Immunol* 144: 765–773

Paya CV, Kenmotsu N, Schoon RA, Liebson PJ (1988): Tumor necrosis factor and lymphotoxin secretion by human natural killer cells leads to antiviral cytotoxicity. *J Immunol* 141: 1989–1995

Perez C, Albert I, DeFay K, Zachariades N, Gooding L, Kriegler M (1990): A nonsecretable cell surface mutant of tumor necrosis factor (TNF) kills by cell-to-cell contact. *Cell* 63: 251–258

Philip R, Epstein LB (1986): Tumor necrosis factor as immunomodulator and mediator of monocyte cytotoxicity induced by itself, γ-interferon and interleukin-1. *Nature* 323: 86–89

Poste G, Kirsh R, Fogler WE, Fidler IJ (1979): Activation of tumoricidal properties in mouse macrophages by lymphokines encapsulated in liposomes. *Cancer Res* 39: 881–892

Raz A, Inbar M, Goldman R (1977): A differential interaction *in vitro* of mouse macrophages with normal lymphocytes and malignant lymphoma cells. *Eur J Cancer* 13: 605–615

Reidarson TH, Granger GA, Klostergaard J (1982): Inducible macrophage cytotoxins. II. Tumor lysis mechanism involving target cell-binding proteases. *J Natl Cancer Inst* 69: 889–894

Reynolds CW, Reichardt D, Henkart M, Millard P, Henkart P (1987): Inhibition of NK and ADCC activity by antibodies against purified cytoplasmic granules from rat LGL tumors. *J Leukocyte Biol* 42: 642–652

Ruff MR, Gifford GE (1981): Tumor necrosis factor. *Lymphokines* 2: 235–272

Russell SW, Gillespie GY, Pace JL (1980): Evidence for mononuclear phagocytes in solid neoplasms and appraisal of their nonspecific cytotoxic capabilities. *Contemp Top Immunobiol* 10: 143–166

Russell SW, Pace JL, Varesio L, Akporiaye E, Blasi E, Celado A, Schreiber RD, Schultz RM, Stevenson AP, Stewart CC, Stewart SJ (1986): Comparison of five short-term assays that measure nonspecific cytotoxicity mediated to tumor cells, by activated macrophages. *J Leukocyte Biol* 40: 801–813

Sasada M, Pabst MJ, Johnston RB, Jr (1983): Activation of mouse peritoneal macrophages by lipopolysaccharide alters the kinetic parameters of the superoxide-producing NADPH oxidase. *J Biol Chem* 258: 9631–9635

Savill J, Dransfield I, Hogg N, Haslett C (1990): Vitronectin receptor-mediated phagocytosis of cells undergoing apoptosis. *Nature* 343: 170–173

Schroit AJ, Madsen JW, Tanaka Y (1985): *In vivo* recognition and clearance of red blood cells containing phosphatidylserine in their plasma membranes. *J Biol Chem* 260: 5131–5138

Schroit AJ, Tanaka Y, Madsen J, Fidler IJ (1984): The recognition of red blood cells by macrophages: Role of phosphatidylserine and possible implications of membrane phospholipid asymmetry. *Biol Cell* 51: 227–238

Schultz RM, Pavlidis NA, Stylos WA, Chirigas MA

(1978): Regulation of macrophage tumoricidal function: A role for prostaglandins of the E series. *Science* 202: 320–321

Sebaldt RJ, Prpic V, Hollenbach PW, Adams DO, Uhing RJ (1990): IFN-γ potentiates the accumulation of diacylglycerol in murine macrophages. *J Immunol* 145: 684–689

Shepard HM, Lewis GD (1988): Resistance of tumor cells to tumor necrosis factor. *J Clin Immunol* 8: 333–341

Shiver JW, Henkart PA (1991): A noncytotoxic mast cell tumor line exhibits potent IgE-dependent cytotoxicity after transfection with the cytolysin/perforin gene. *Cell* 64: 1175–1181

Slivka A, Buglio AFL, Weiss SJ (1980): A potential role for hypochlorous acid in granulocyte-mediated tumor cell cytotoxicity. *Blood* 55: 347–350

Smith RA, Baglioni C (1987): The active form of tumor necrosis factor is a trimer. *J Biol Chem* 262: 6951–6954

Snyderman R, Pike MC (1976): An inhibitor of macrophage chemotaxis produced by neoplasms. *Science* 192: 370–372

Somers SD, Hamilton TA, Adams DO (1987): Maleylated bovine serum albumin triggers cytolytic function in selected populations of primed murine macrophages. *J Immunol* 139: 1361–1368

Somers SD, Mastin JP, Adams DO (1983): The binding of tumor cells by murine mononuclear phagocytes can be divided into two qualitatively distinct types. *J Immunol* 131: 2086–2093

Somers SD, Whisnant CC, Adams DO (1986): Quantification of the strength of cell–cell adhesion: The capture of tumor cells by activated murine macrophages proceeds through two distinct stages. *J Immunol* 136: 1490–1496

Srimal S, Nathan C (1990): Purification of macrophage deactivating factor. *J Exp Med* 171: 1347–1361

Stott EJ, Probert M, Thomas LH (1975): Cytotoxicity of alveolar macrophages for virus-infected cells. *Nature* 255: 710–712

Stuehr DJ, Nathan CF (1989): Nitric oxide: A macrophage product responsible for cytostasis and respiratory inhibition in tumor target cells. *J Exp Med* 169: 1543–1555

Stuehr DJ, Cho HJ, Kwon NS, Weise MF, Nathan CF (1991a): Purification and characterization of the cytokine-induced macrophage nitric oxide synthase: An FAD- and FMN-containing flavoprotein. *Proc Natl Acad Sci USA* 88: 7773–7777

Stuehr DJ, Fasehun OA, Kwon NS, Gross SS, Gonzalez JA, Levi R, Nathan CF (1991b): Inhibition of macrophage and endothelial cell nitric oxide synthase by diphenyleneiodonium and its analogs. *FASEB J* 5: 98–103

Stuehr DJ, Gross SS, Sakuma I, Levi R, Nathan CF (1989a): Activated murine macrophages secrete a metabolite of arginine with the bioactivity of endothelium-derived relaxing factor and the chemical reactivity of nitric oxide. *J Exp Med* 169: 1011–1020

Stuehr DJ, Kwon NS, Gross SS, Thiel BA, Levi R, Nathan CF (1989b): Synthesis of nitrogen oxides from L-arginine by macrophage cytosol: Requirement for inducible and constitutive components. *Biochem Biophys Res Commun* 161: 420–426

Stuehr DJ, Kwon NS, Nathan CF (1990): FAD and GSH participate in macrophage synthesis of nitric oxide. *Biophys Biochem Res Commun* 168: 558–565

Stuehr DJ, Kwon NS, Nathan CF, Griffith OW, Feldman PL, Wiseman J (1991c): N$^\omega$-hydroxy-L-arginine is an intermediate in the biosynthesis of nitric oxide from L-arginine. *J Biol Chem* 266: 6259–6263

Suttles J, Carruth LM, Mizel SB (1990): Detection of IL-1α and IL-1β in the supernatants of paraformaldehyde-treated human monocytes: Evidence against a membrane form of IL-1. *J Immunol* 144: 170–174

Szuro-Sudol A, Nathan CF (1982): Suppression of macrophage oxidative metabolism by products of malignant and nonmalignant cells. *J Exp Med* 156: 945–961

Taffet SM, Russell SW (1981): Macrophage-mediated tumor cell killing: Regulation of expression of cytolytic activity by prostaglandin E. *J Immunol* 126: 424–427

Tanaka Y, Schroit AJ (1983): Insertion of fluorescent phosphatidylserine into the plasma membrane of red blood cells: Recognition by autologous macrophages. *J Biol Chem* 258: 11335–11343

Tartaglia LA, Weber RF, Figari IS, Reynolds C, Palladino MA, Jr, Goeddel DV (1991): The two different receptors for tumor necrosis factor mediate distinct cellular responses. *Proc Natl Acad Sci USA* 88: 9292–9296

Tayeh MA, Marletta MA (1989): Macrophage oxidation of L-arginine to nitric oxide, nitrite, and nitrate: Tetrahydrobiopterin is required as a cofactor. *J Biol Chem* 264: 19654–19658

Tsunawaki S, Nathan CF (1986): Macrophage deactivation: Altered kinetic properties of superoxide-producing enzyme after exposure to tumor cell-conditioned medium. *J Exp Med* 164: 1319–1331

Tsunawaki S, Sporn M, Ding A, Nathan C (1988): Deactivation of macrophages by transforming growth factor-β. *Nature* 334: 260–262

Tsunawaki S, Sporn M, Nathan C (1989): Comparison of transforming growth factor-β and a macrophage deactivating polypeptide from tumor cells: Differences in antigenicity and mechanism of action. *J Immunol* 142: 3462–3468

Uhing RJ, Adams DO (1989): Molecular events in the activation of murine macrophages. *Agents Actions* 26: 9–14

Urban JL, Schreiber H (1983): Selection of macrophage-resistant progressor tumor variants by the normal host: Requirement for concomitant T-cell mediated immunity. *J Exp Med* 157: 642–656

Urban JL, Shepard HM, Rothstein JL, Sugarman BJ, Schreiber H (1986): Tumor necrosis factor: A potent effector molecule for tumor cell killing by activated macrophages. *Proc Natl Acad Sci USA* 83: 5233–5237

Utsugi T, Fidler IJ (1991): Prostaglandin E$_2$ does not inhibit tumoricidal activity of mouse macrophages against adherent tumor cells. *J Immunol* 146: 2066–2071

Utsugi T, Dinney CPN, Killion JJ, Fidler IJ (1991a): In situ activation of mouse macrophages and therapy of spontaneous renal cell cancer metastasis by liposomes containing the lipopeptide CPT 31362. *Cancer Immunol Immunother* 33: 375–381

Utsugi T, Nii A, Fan D, Pak CC, Denkins Y, van Hoogevest P, Fidler IJ (1991b): Comparative efficacy of liposomes containing synthetic bacterial cell wall analogues for tumoricidal activation of monocytes and macrophages. *Cancer Immunol Immunother* 33: 285–292

Wei RQ, Lefkowitz SS, Lefkowitz DL, Everse J (1986): Activation of macrophages by peroxidases. *Proc Soc Exp Biol Med* 182: 515–521

Whitworth PW, Pak CC, Esgro J, Kleinerman ES, Fidler IJ (1989/1990): Macrophages and cancer. *Cancer Metastasis Rev* 8: 319–351

Wilson KM, Siegal G, Lord EM (1989): Tumor necrosis factor-mediated cytotoxicity by tumor-associated macrophages. *Cell Immunol* 123: 158–165

Wiltrout RH, Brunda MJ, Holden HT (1982): Variation in selectivity of tumor cell cytolysis by murine macrophages, macrophage-like cell lines and NK cells. *Int J Cancer* 30: 335–342

Wong GHW, Goeddel DV (1986): Tumour necrosis factors α and β inhibit virus replication and synergize with interferons. *Nature* 323: 819–822

Wong GHW, Elwell JH, Oberley LW, Goeddel DV (1989): Manganous superoxide dismutase is essential for cellular resistance to cytotoxicity of tumor necrosis factor. *Cell* 58: 923–931

Yu S-F, Koerner TJ, Adams DO (1990): Gene regulation in macrophage activation: Differential regulation of genes encoding for tumor necrosis factor, interleukin-1, JE, and KC by interferon-γ and lipopolysaccharide. *J Leukocyte Biol* 48: 412–419

Ziegler-Heitbrock HWL, Moller A, Linke RP, Haas JG, Rieber EP, Riethmuller G (1986): Tumor necrosis factor as effector molecule in monocyte mediated cytotoxicity. *Cancer Res* 46: 5947–5952

X
Methods

43
The ^{51}Cr-Release Assay for CTL-Mediated Target Cell Lysis

Eric Martz

Introduction

Strengths and Weaknesses

Two of the best ways to quantitate cytolysis are by dye exclusion and ^{51}Cr release. Each has strengths and weaknesses.

Dye exclusion involves exposing cells to a protein stain such as trypan blue or eosin. The dye is excluded by living cells and stains lysed cells. If the cells have been lysed for too long, they may disintegrate and not be recognizable as "dead cell bodies." This problem does not arise with ^{51}Cr. Quantitating by dye exclusion involves tedious microscopic counting. This has the advantage of high accuracy: you can distinguish 90% dead from 99% dead, and you can usually be sure whether a given cell has lysed or not. It has the disadvantage of low precision (high statistical uncertainty), since one rarely has the patience to count more than a few hundred events (cells) per sample. The method is tiring, induces headaches, and limits a single investigator to a maximum of a couple of dozen tests per day. For cell-mediated killing assays, it requires that you have a microscopic way of distinguishing effectors from targets. In many cases, cell size differences suffice (e.g. Brunner et al., 1966; Martz, 1975). Like all microscopic methods, you "see" a lot about what is happening to the cells.

^{51}Cr release has complementary strengths. It is easily capable of high precision since thousands of events (radioactive counts) can be quantitated per sample. However, it is inherently inaccurate, since the percentage of the ^{51}Cr released at 100% lysis is difficult to determine reliably and hence uncertain. It is quite incapable of distinguishing between 90 and 99% lysis, although the agreement between replicate samples can be excellent. It enables the automated quantitation of hundreds of samples per day. On the other hand, if you don't look in the microscope, you may overlook situations that would otherwise be obvious.

No other radioisotopic agent has proved as convenient, versatile, and reliable as ^{51}Cr for general quantitation of cytolysis. ^{3}H-Thymidine is a close second (see method elsewhere in this volume); it has lower spontaneous release, but requires a growing target cell (e.g., it cannot label normal thymocytes) and involves the extra cost, toxicity, and waste disposal problems of liquid scintillation. Other agents may be useful in special circumstances (Sanderson, 1976a) but typically have much higher spontaneous release rates: ^{14}C-Nicotinamide (50% spontaneous release in about 7 hr, Martz et al., 1974), ^{14}C-amino isobutyric acid (Burakoff et al., 1975; Thelestam and Möllby, 1976), ^{111}In (low spontaneous release but half-life 2.8 days: Wiltrout et al., 1978; Ravdin et al., 1980; Shortman and Wilson, 1981), ^{99m}Tc (6-hr half life: Gillespie et al., 1973), and ^{86}Rb (Henney, 1973; Burakoff et al., 1975; Martz, 1976a; Russell and Dobos, 1983). Much work has utilized release of endogenous lactate dehydrogenase (LDH) as a nonradioactive measure of lysis; this is more attractive now that the results can be read in a 96-well tray in an enzyme-linked immunosorbent assay (ELISA) reader (Korzeniewski and Callewaert, 1983), but may be problematic in cytotoxic T lymphocyte (CTL) assays since the CTL may also release some LDH.

History

I have reviewed ^{51}Cr history elsewhere, where references to the following studies can be found (Martz, 1976b). Briefly, as a label for living cells, ^{51}CrO$_4^{2-}$ was first applied to erythrocytes in 1950

CYTOTOXIC CELLS
M. Sitkovsky and P. Henkart, Editors
© 1993 Birkhäuser Boston

and to ascites tumor cells in 1958. Arnold Sanderson first used it to assay complement-mediated lysis in 1964, Göran Holm and Peter Perlmann for cell-mediated lysis in 1967, and Ted Brunner in 1968 and Gideon Berke in 1969 for CTL-mediated lysis. ^{51}Cr release correlates well with trypan blue staining (Berke, 1977), loss of cloning ability, and release of macromolecules. Not surprisingly, in sufficient amounts, uptake of chromate has adverse long-term effects on cells (Wright and Bonavida, 1987); however, minimal levels of ^{51}Cr sufficient to monitor lysis of populations of cells can permit cloning of the labeled cells in soft agar (Martz, 1977, and unpublished data). ^{51}Cr has been used to monitor lysis of schistosomula worms (Butterworth et al., 1977) and bacteria (Fierer et al., 1974).

Principles of the ^{51}Cr-Release Assay

Chemistry

For reasons which are not clear, most kinds of cells take up Na_2CrO_4 when it is supplied in trace amounts. Once taken up, the Cr^{6+} in chromate is believed to be reduced by cell constituents to Cr^{3+}. This species binds tightly to multivalent anions present in the cells, such as citrate and amino acids, and the resulting complexes are released slowly if at all as long as the cells remain viable. Breakage of the plasma membrane and lysis release the majority of the isotope complexes as soluble material with an average apparent molecular weight of less than 4000 daltons (Martz, 1976b; Sanderson, 1976b; Berke, 1977). Thus, the ratio of soluble to cell-associated ^{51}Cr is roughly proportional to the percentage of cells that have lysed (see below for details).

Nonreutilization

In order for soluble ^{51}Cr to reflect cell lysis, it is important that the soluble ^{51}Cr released from lysing cells not be taken up again (reutilized) by any remaining viable cells. That this is the case can be easily demonstrated by incubating viable cells with ^{51}Cr released from labeled cells (e.g., by freezing and thawing). This can be contrasted with the behaviour of many other soluble cytoplasmic substances, and it is one of the key properties of ^{51}Cr which makes it so useful. For example, ^{3}H-thymidine, ^{14}C-nicotinamide (Martz et al., 1974), or ^{14}C-amino isobutyric acid (Burakoff et al., 1975), when released by dying cells, are readily reutilized unless excess "cold" material is provided. Inorganic tracers such as ^{86}Rb (a potassium analog) have the same problem.

Spontaneous Release

A ^{51}Cr-labeled cell population, incubated in culture medium, releases between 1 and 10% of the total isotope per hour, depending on the cell type and conditions. Initially, some of this "spontaneous" release is due to the lysis of a subpopulation of cells, probably as a result of damage during preparation. (The percentage of spontaneous ^{51}Cr release which is due to "spontaneous" cell death can be determined with cells doubly labeled with ^{51}Cr and ^{3}H- or ^{14}C-thymidine plus DNAse). With incubations longer than a couple of hours, much of the spontaneous release may represent slow leakage from viable cells. Spontaneous release usually limits the usefulness of the ^{51}Cr-release assay to a maximum of about 18 hr. (If longer incubations are needed, see the section in this volume on use of ^{3}H-thymidine to quantitate cytolysis.)

"Resting" lymphocytes (e.g., splenocytes) have a higher spontaneous release rate for ^{51}Cr than do tumor cells or mitogen-induced lymphoblasts. However, the spontaneous release rate can be greatly diminished by reducing the incubation temperature a few degrees, e.g. from 37 to 32°C, a reduction which often has a lesser effect on the cytolytic process being studied and thus improves the signal/noise ratio (Balk and Mescher, 1981).

Physics and Counting

When a ^{51}Cr atom decays, the nucleus, which contains 24 protons, captures an electron, thereby converting a proton to a neutron, and becoming the stable, naturally abundant isotope of vanadium, ^{51}V. The disappearance of an inner orbital electron initiates a chain reaction of electrons falling into lower orbitals, which in turn so disturbs the atom that an average of about 5 low-energy electrons, termed Auger electrons, are thrown off per decay. One hundred percent of the decays of ^{51}Cr emit Auger electrons. These electrons range in energy from a few electron volts to 10 keV or higher, with an average near that of the beta particles emitted by ^{3}H, around 5 keV; in liquid scintillation, ^{51}Cr cannot be resolved from ^{3}H.

The half-life of ^{51}Cr is 27.8 days. This is long enough that corrections for decay are not necessary for the several-hour interval needed to count a batch of experimental samples. It does, however, limit the usefulness of a batch of ^{51}Cr to about 3 months from the date of purchase.

Nine percent of the ^{51}Cr decay events emit a gamma ray with an average energy of 320 keV. This allows ^{51}Cr to be counted in a gamma counter, a simpler and less expensive procedure than liquid scintillation. The gamma counting efficiency is typically between 2 and 5%, depending on the size of the crystal detector. ^{51}Cr can be resolved nicely by a gamma spectrometer from lower- (e.g., ^{125}I or ^{131}I) or higher- (e.g., ^{86}Rb) energy gamma emitters should double or triple labeling be useful.

^{51}Cr can be counted with about fivefold higher efficiency by liquid scintillation, based on the Auger electrons and low-energy x rays emitted, using the standard ^{3}H channel. It is possible to resolve mixtures of ^{51}Cr and ^{3}H by counting each sample in both a gamma spectrometer and a liquid scintillation spectrometer (see Chapter 44, this volume).

Safety

Every person who will use ^{51}Cr should first be trained in the safe use and disposal of radioactive materials. ^{51}Cr is less hazardous to work with than other common isotopes, such as ^{3}H, ^{14}C, or ^{125}I, because relatively small amounts are needed, it has a short physical half-life (28 days), it is not concentrated in any organ, and label released from cells tends to be rapidly eliminated from the body. Nevertheless, common sense and good laboratory procedures dictate that unnecessary exposure be avoided (minimize time and maximize distance) and that ingestion (or aerosol inhalation) of the isotope into the body should be strictly prevented.

If the released (reduced) form of ^{51}Cr is inadvertently taken into the body, the vast majority will be promptly eliminated in the urine. As chromate (before being used to label cells), it will combine irreversibly with proteins and hence have a longer residence time in the body. Many people are concerned about the gamma emissions, which are dramatically penetrating, a substantial percentage easily going through a 2-inch stone bench top or a human body. In fact, were isotope to enter the body, some of the gamma rays would pass out without being captured in the body, doing no damage at all. On the other hand, the energy of the Auger electrons would be entirely captured within a few cell diameters of the emission point. Since the Auger electrons are roughly 50-fold more numerous than the gamma rays, they represent a substantial source of biological damage from internal isotope.

A geiger counter is often used to check for ^{51}Cr contamination, e.g., of the benchtop or hands. Geiger counters are about 1000-fold less sensitive than gamma spectrometers at detecting ^{51}Cr. Many geiger counters can barely detect 0.2 μCi of ^{51}Cr; however, this amount emits 60,000 gamma rays per minute and roughly 3,000,000 Auger electrons per minute! Therefore, you should wash your hands very well, even when the geiger counter shows nothing above background.

A typical experiment with a single type of target cell uses about 150 μCi to label cells. Typically, less than one-third of this is taken up, so after the first rinse, one is working with ≤ 50 μCi. If this is on the bench an average of 2 feet from your body, the body receives 0.043 mR/hr from the gamma radiation. For a typical experiment, the total necessary exposure time at this distance would not exceed 4 hr, or a total dose of 0.17 mR. Thus, even if one were to perform five such experiments per week, every week of the year, the total cumulative exposure, 44 mR would be less than one-tenth of the legally allowable maximum annual whole body exposure for a pregnant woman, 500 mR/year, which is one-tenth of that allowed for other adults.

Labeling Target Cells

Media

For production of cells in culture, a bicarbonate-buffered complete growth medium (**MB** for medium bicarbonate) is preferred, since bicarbonate is the physiological buffer. This of course requires that the proper [CO_2] be maintained in the gas phase, typically 5% CO_2 in air.

When bicarbonate-buffered media are used for bench-top (0% CO_2) experimental manipulations of cells in small volumes of media, the HCO_3^- dissociates to $OH^- + CO_2$, and the CO_2 is lost irreversibly to the air. This raises the pH of the medium to > 8.5 in a matter of minutes, which is highly undesirable for nearly all purposes. The maximum pH can be reduced somewhat by the inclusion of 10 mM N-2-hydroxyethylpiperazine-N'-2 ethanesulfonic acid (HEPES) (expensive!), but despite this, if the medium contains bicarbonate buffer, the pH will be unstable and undefined. The potencies of some drugs (e.g., tertiary amine anesthetics) can be altered significantly by a pH change as small as 0.1 unit.

For these reasons, it is recommended that media used for benchtop, short-term experiments (especially rapid events such as CTL–target adhesion, programming for lysis, etc.) be bicarbonate-free. For many years, we have employed Liebowitz-15 medium for this purpose, a complete, defined bicarbonate-free medium, which is commercially readily available. We supplement it with 10 mM HEPES, since it contains only 1.8 mM phosphate, and 5 mM (1 g/l) glucose, since it has only galactose and pyruvate as energy sources and CTL are known to require glucose for purposes other than an energy source (MacLennan and Golstein, 1978; MacDonald and Cerottini, 1979). We will refer to this type of bicarbonate-free medium as **MNB** (medium no bicarbonate).

Serum

We routinely use 5 to 10% fetal calf or calf bovine serum in both MB and MNB. The serum does not prevent effective labeling, and it is very beneficial to cell health. When cells contact serum-free solid surfaces (such as test tubes and pipets), they adhere rapidly and irreversibly, even in the cold (Grinnell, 1976), probably as a result of physical denaturation of the cell surface proteins against the solid. Handling and centrifugation of cells in serum-free medium often results in dramatically reduced viability and/or loss of cells. If it is necessary to use serum-free medium for a special purpose, precoating the test tubes and pipets with serum (which forms a molecular monolayer within minutes) and rinsing them with serum-free buffer will greatly improve cell yield. Pure serum is 4 to 7% protein; about 60% of serum protein is albumin. Hence, 10% serum is about 0.6% protein. Purified albumin or gelatin, 0.1 to 1%, can be used as a protective protein in place of serum.

Labeling Procedure

We have had best results labeling cells overnight under optimal culture conditions. This is convenient, since it minimizes the time required to prepare labeled cells on the following day when the experiment is performed. However, cells can also be satisfactorily labeled with a short incubation (less than an hour). Both methods are given below.

Procedures are given below for cells that grow in suspension (most lymphoid cell lines, hybridomas, myelomas) and also for cells that grow as attached monolayers. The procedure given for monolayered cells is for the case in which a monodisperse *suspension* of the labeled cells is needed for experimental use. Most experimental purposes are better served with a suspension of target cells, since the volume of experimental medium can be smaller, effector cells can be well mixed with targets, and effector–target contact can be maximized by centrifugation.

When handling cell suspensions, remember that cells sediment rapidly at unit gravity. Always vortex cell suspensions (and examine them carefully to be sure the cells are not aggregated) before removing a sample aliquot.

For lowest spontaneous release rates, use healthy exponentially growing cells. Cells from a crowded culture which is entering the stationary phase will have higher spontaneous release. Most cell lines which grow unattached are healthiest between 1 and 5×10^5 cells/ml. Monolayers are healthiest from 2×10^4 to 10^5 cells/sq cm depending on the cell line.

A. Overnight method (recommended)

This method has generally given us healthier cells, with lower spontaneous release rates, than the short-term method.

1A. *For cells which do not attach*: On the late afternoon of the day before labeled cells are needed, seed a 75 sq cm cell culture flast with 6×10^6 cells (harvested from an exponentially growing culture) and bring the total volume to approx. 30 ml with fresh MB culture medium (giving 2×10^5 cells/ml).

Most cell lines can be harvested in healthy condition at densities of $5-10 \times 10^5$ cells/ml. Seeding at this density provides room for the 0.5–1.5 cell divisions which will occur during overnight culture, depending on the characteristics of the cell line employed.

For attached monolayer cells: Seed one or more 75 sq cm flasks in advance so that the monolayers are well established, but still growing, by the late afternoon of the day before the labeled cells are needed. Add the customary amount of MB (e.g., 15 ml/flask).

2A. Add to the flask sterile sodium chromate (^{51}Cr) in isotonic saline, and mix. Use 5 µCi/ml, thus 150 µCi per flask of unattached cells or 75 µCi for a monolayer flask. Place a radioactive warning label on the flask.

We have good results with New England Nuclear Na$_2$^{51}CrO$_4$ NEZ-030S (about 500 mCi/mg), which comes at 1 mCi/ml in sterile isotonic saline. It comes in an injection vial, and the isotope is transferred to the culture flask with a sterile 1-ml syringe fitted with a 1.5-inch number 22 needle. Wear gloves and watch carefully for spilled droplets (especially when pulling the needle out of the injection vial—remember to inject a volume of air equal to the volume of fluid you withdraw).

3A. Incubate the flask(s) horizontally in your CO$_2$ incubator overnight.

Gamma radiation slightly above background will be emitted through the incubator wall. No one should work continuously immediately adjacent to the incubator. Such exposure would be within permissible limits but is unnecessary.

4A. *Unattached cells*: The next morning you should have $3-5 \times 10^5$ cells/ml (depending on the growth rate of the cell line), giving a total of about 10^7 cells. Resuspend the cells and pour the contents of the flask into a 50-ml centrifuge tube.

Attached cells: Suspend the cells with your usual procedure [e.g., trypsin ethylenediaminetetraacetate (EDTA)]. Add serum to neutralize the trypsin, and transfer to a centrifuge tube.

B. Short incubation method (an alternative to A):

This method is not recommended for sticky (monolayering) cells because the cells will aggregate during the incubation.

1B. Place 10^7 cells in a 12×75 mm test tube. Centrifuge (1500 rpm 3 min) and discard the supernatant. Vortex the pellet to disperse it in the residual volume, add 0.5 ml fresh MNB, and vortex again.

2B. Add 150 µCi Na$_2$^{51}CrO$_4$ (typically 0.15 ml of 1 mCi/ml stock), vortex.

See note for step 2A.

3B. Incubate the cell suspension for 15–60 min in a 37°C water bath. Vortex every 10 min during the incubation.

Uptake will be approximately linearly related to time for this time range. Incubations longer than 1 hr tend to reduce cell health and increase spontaneous release. See note for step 3A.

If the cells are allowed to sediment into a pellet during this incubation, the cells on the bottom will be starved for oxygen and nutrients, and will likely give higher spontaneous release.

It has been reported that uptake of ^{51}CrO$_4^{2-}$ is energy-independent, and occurs well at room temperature (Sanderson, 1976b). Incubations at room temperature should allow high densities of cells without loss of viability, but we have not tested this recommendation.

4B. Remove from water bath, add 2 ml MNB, vortex.

Both methods:

5. Centrifuge at 1500 rpm for 3 min. Discard the supernatant as liquid radioactive waste. (If you wish, save 1 ml of this supernatant to count so you can calculate the total cpm added, and the percentage taken up. Uptake should be 5–30% of the added isotope.) Vortex the pellet to resuspend the cells in the residual fluid, then add a few milliliters of MNB and (if starting with a larger volume) transfer to a 12×75 ml transparent test tube.

The transparent tube allows you to see, after you add the few milliliters of MNB, whether any lumps of undispersed cells remain. Vortexing in the small residual volume produces a higher average shear stress and breaks the pellet into monodisperse cells more easily than vortexing after you add medium. Difficulty in resuspending the cells can result from either (i) DNA released by a small percentage of dead cells forming a sticky gel, or (ii) spontaneous homotypic adhesion formation between healthy cells.

(i) DNA gel is easily dissolved by adding a few tiny crystals of pure crystalline DNAse I (e.g. Sigma D-5025), and incubating a few minutes at 37°C. Extracellular DNAse does not harm healthy cells.

(ii) Spontaneous homotypic adhesion can usually be prevented by handling the cells on ice ($<10°$ during centrifugations), or by including 10 mM EDTA in the MNB.

6. Fill the tube half-full of MNB, vortex, and centrifuge, discarding the supernatant as radioactive waste. Vortex in the residual volume to resuspend. Repeat this centrifugation for a total of three centrifugations since the labeling incubation was completed.

The cells typically take up roughly 1/5 of the added ^{51}Cr. The goal of the rinsing is to reduce the extracellular ^{51}Cr to less than 1% of the cell-associated ^{51}Cr. Each centrifugation cycle dilutes the extracellular ^{51}Cr by at least 10-fold. Two centrifugations may be sufficient; three is conservative. If you are concerned about the effectiveness of your rinsing, save the last supernatant in a tube and count it and the tube containing the rinsed cells in the gamma counter. The supernatant should have $\leq 10\%$ of the cpm in the cells.

For cells that do not stand up well to repeated centrifugation, an alternative method is a single centrifugation through a 1-step density gradient. The cell suspension in MNB is layered over a few milliliters of 100% serum. The supernatant must be removed carefully to avoid contaminating the pellet with unincorporated ^{51}Cr.

7. Count the cells (e.g., in a hemocytometer with a microscope) and adjust to the desired cell density.

The cells should be monodisperse. If you have lots of clumps, see note above under step 5.

If the labeled cells are held for more than 30 min before the experiment is started, you may wish to give them a final rinse immediately before use to remove any ^{51}Cr which leaked out during the holding period.

Labeled cells can be held on ice or at room temperature for hours. At these temperatures, metabolism is reduced, so it is not necessary to keep the cells suspended with frequent vortexing.

Increasing the cpm/Cell

The above procedure gives roughly 0.1 gamma cpm/cell, which is good for most purposes. If you need a microassay with very few cells/assay, remember that you can increase the cpm/cell fivefold merely by counting with liquid scintillation. Uptake levels near 1 gamma cpm/cell (5 liquid scintillation cpm/cell) can be obtained. To increase uptake, use the short-term labeling method. Reduce the volume of MNB during labeling so that the μCi/ml is increased, and incubate for a full 60 min (or even 90 min if the cells can withstand it). Alternatively, after 45 min of labeling, you can rinse the cells and give them another 45 min labeling in fresh medium with fresh ^{51}Cr.

Design of the CTL Assay

Overview

CTL and ^{51}Cr-labeled target cells are mixed and centrifuged together to maximize intercellular contact. The pellet is then incubated at 37°C to allow the CTL to lyse the target cells. At the end of the incubation, it may be desirable to centrifuge the assay to ensure that all cell-associated ^{51}Cr is in the pellet; this can usually be omitted for 96-well plate assays if the plate has not been jarred. If the assay is done in a small volume (e.g., 150 μl) in a test tube, 2 ml phosphate-buffered saline (PBS) needs to be added at this time to expand the volume to enable the released ^{51}Cr to be decanted; this addition mixes the contents and hence the tubes must be centrifuged. The released soluble ^{51}Cr is separated from the cell pellet by decanting (for test tubes) or removing a portion (typically half) of the supernatant with a pipet (wells).

The Assay Container

CTL assays are optimal in round-bottomed test tubes (12 × 75 mm) or round-bottomed 96-well plates when 10^4 to 10^5 target cells are used per assay. During the centrifugation just before the main incubation, the cells slide down the round bottom into a pellet and intercellular contact is maximized; this may not happen reliably in flat-bottomed vessels. The round-bottomed vessel makes a thin, disc-shaped pellet (in plastic tubes with truly round bottoms) or a thin, crescent-shaped pellet (in glass tubes with a lumpy bottom from the tube sealing during manufacture). The thinness provides adequate nutrient/waste exchange.

V-bottom wells make a more compact pellet. This may smother the cells on the bottom, reducing CTL efficiency, when larger numbers of cells/well are used (near 10^5). However, for microassays with as few as 10^3 cells/well, V-shaped wells are optimal, and round-bottomed wells will give less intercellular contact and less efficient killing. V-wells also have a convenient angle in the wall against which the pipet tip can be stopped when removing the released ^{51}Cr; this helps avoid disturbing the pellet with the pipet tip.

Cells per Assay Tube/Well

The 95% confidence interval around a mean observed value is approximately twice the standard error of the mean, 2(SEM), and the 99% confidence interval, 3(SEM). When counting events occurring randomly in time (radioactive decays) or space (cells in a hemocytometer), a special case applies: the SEM is equal to the standard deviation is equal to the square root of the number of events counted.

Cumulative pipeting errors in typical CTL assays result in SEMs of at least a few percent. A reasonable goal for the counting of the radioisotope is to have the uncertainty resulting from the sampling error in the counting process alone not exceed 1%, so that it does not significantly increase the uncertainty from other unavoidable sources such as pipeting. This uncertainty goal can be relative to the maximum scale value (100% ^{51}Cr release); that is, the uncertainty for the lowest count in the experiment can be 1% of the maximum release, not 1% of the count itself. Hence, we want sufficient gross counts in our maximum release controls to make the SEM < 1%. This dictates that the maximum release controls should be counted long enough to give 10,000 gross counts, so that the SEM = $\sqrt{N} = \sqrt{10,000} = 100 = 1\%$. In practice, satisfactory data can usually be obtained with half this number of gross counts.

The above labeling procedures give roughly 0.1 cpm/cell. Typically, you may wish to count each sample for not more than 3 min in the gamma counter. Thus, to get 6000 counts in the maximum release control, you need 2000 cpm, or 2×10^4 cells/tube. In fact, this is only a fewfold above the minimum suitable for CTL assays in round-bottomed tubes (see The Assay Container, above).

Taking these considerations and other design requirements of your experiment into account, you can determine a suitable number of cells/tube.

Volume per Assay Tube/Well

For CTL assays, total volumes per well/tube of 100 to 300 μl work well, provided incubations over 15 min or so are done in a humid atmosphere to minimize evaporation. Smaller assay volumes are workable if careful attention is paid to aliquot delivery and mixing.

Consider an effector to target cell ratio of 50. With 2×10^4 target cells/assay, the assay contains 10^6 cells. If the assay volume is 100 μl, the nominal cell density is 10^7 cells/ml. Most cell types are unable to grow above 10^6 cells/ml in long-term culture due to nutrient depletion and waste accumulation. However, at 10^7 cells/ml, CTL can function adequately as long as the incubation does not exceed a few hours.

Controls for Spontaneous Release

Spontaneous release is ideally the release from the targets in an assay treated identically to the experimental in every way except that mock effector cells are substituted for effector cells. The best mock effector cells are typically CTL which fail to recognize the target cells. A simpler method of obtaining mock effector cells is to replace the effector cells with an equal number of unlabeled target cells or another convenient similar type of nonkiller cell. The simplest and most often used control for spontaneous release is to omit the effector cells altogether (keeping the total volume of medium the same as when effector cells are present). This latter control should always be included so that in case the mock effector cells have a cytotoxic effect, it will be obvious. The addition of mock effectors usually *reduces* slightly the amount of spontaneous release.

Controls for Release at 100% Lysis

Maximum release means the amount of ^{51}Cr released when 100% of the cells are lysed *by the mechanism being studied*. The following truth cannot be overemphasized:

Uncertainty in the release at 100% lysis is by far the largest uncertainty in a properly conducted ^{51}Cr-release assay.

This is because the amount of ^{51}Cr released following the lysis of a cell depends upon: (1) the type of cell; (2) the history of the cell, including the manner in which it was labeled and the time interval since labeling; (3) the chemical environment prior to and at the time of lysis; (4) the events that induce lysis; and (5) the time interval between lysis and the separation of soluble from insoluble ^{51}Cr.

When a cell is lysed by a CTL, by mechanical means, or by complement, a variable minority of the ^{51}Cr remains insoluble. This ^{51}Cr is likely bound to the cytoskeletal shell and nuclear remnant, which are initially insoluble after breakdown of the plasma membrane. Thus, if the total ^{51}Cr taken up (determined by counting unlysed cells) is used as a 100% basis for calculating corrected release, CTL will be unable to produce 100% corrected release (unless the supernatant removal is delayed until far beyond the actual time of lysis, which permits autolysis of the initially insoluble remnants).

It is common practice to lyse a control sample of cells in sodium dodecyl sulfate (which truly solubilizes the entire cell) and use this as the 100% basis. This is acceptable as long as the resulting corrected percentage release values are regarded as roughly proportional (by an unknown factor) to the percentage of the target cells lysed, rather than equating them to the latter. If the investigator wishes to make a somewhat more accurate estimate of the actual percentage of the target cells lysed, then the control used as the 100% basis should produce 100% lysis (verified by trypan blue) by a mechanism similar to the mechanism being studied.

The most convenient, quick, and reliable method of lysing 100% of a cell sample by mechanical damage is to freeze and thaw it rapidly (the ice crystals break the plasma membrane). In order to do this quickly and guarantee that the freeze penetrates the entire sample, a low temperature can be used, such as a $-80°C$ freezer or a dry ice bath. A single cycle of freezing and thawing lyses 100% of the cells, and the subsequent release, at least from P815 cells, occurs instantly at 0°C (unpublished data) and hence does not require any enzymatic or other temperature-dependent process.

Different kinds of lysis release different percentages of the cell-associated ^{51}Cr. CTL release more than does freezing and thawing (Martz and Benacerraf, 1973; Sanderson, 1976b), and complement or nonionic detergent may release yet different amounts depending on the cell type. Freezing and thawing renders the nuclear DNA more accessible to DNAse than does Triton X-100 detergent, presumably because the ice crystals damage the nuclear envelope (Howell and Martz, 1988). Since it is impossible to lyse 100% of the targets in exactly the way lysis occurs in the experimental assays, the choice of a basis for 100% is arbitrary. We have usually used freeze-thaw because of its reliability at lysing 100% of the cells, speed, and convenience.

Perhaps the most accurate, though still somewhat uncertain, estimates of the percentage of targets lysed can be made by determining the *total releasable* ^{51}*Cr* for each assay tube individually. In this procedure (Martz and Benacerraf, 1973), after the removal of the *experimentally released* isotope, the cell pellet is frozen and thawed, PBS is added, the sample is centrifuged, and a second supernatant is removed to determine the *remaining releasable* isotope. The total releasable is calculated as the sum of the experimentally released and the remaining releasable.

The total releasable ^{51}Cr, defined above, varies according to conditions 1–5 listed above. It can be as low as 50%, but it is typically 75–90%. Cells that have a high nucleus-to-cytoplasm ratio (e.g., resting lymphocytes) release less than lymphoblasts or tumor cells, as would be expected.

The total ^{51}Cr releasable from short-term labeled P815 mouse mastocytoma cells by freezing and thawing has been determined for a variety of conditions (Martz and Benacerraf, 1973). In the absence of CTL, it starts out about 70%, decreases to around 60% during the first 2 hr of culture, and then stabilizes. The total releasable is increased by CTL-mediated killing, and eventually exceeds the initial value by as much as 20%. Addition of anti-CTL antibody plus complement reduces total releasable target ^{51}Cr, sometimes by more than 20%. EDTA increases total releasable by up to 20%. If incubation continues for many hours after the initial linear phase of CTL-induced lysis is completed, the CTL-released ^{51}Cr gradually approaches 100% of the total releasable. Postlytic autolysis is presumed gradually to solubilize the ^{51}Cr, which is insoluble immediately after lysis. These effects illustrate the considerable uncertainty about the level of ^{51}Cr release which represents 100% lysis in ordinary assays.

A Basic CTL Assay Procedure

The procedure below is a basic one suitable for simple titrations of lytic activity by varying the effector-to-target cell ratio. Controls should include spontaneous release (e.g., CTL omitted) and a freeze-thaw of the targets to estimate maximum releasable. It would be informative also to count the pellets of the freeze-thaw control tubes for summing with the supernatants to give the total ^{51}Cr/tube. Each assay should be done in triplicate. Label each replicate set with a single number, distinguishing individual tubes with primes, e.g., 1, 1′, 1″. Order the relevant controls before each experimental set. The tubes need not be capped at any point during the assay.

1. To each 12 × 75 mm test tube, add 100 μl MNB containing 3 × 10^4 ^{51}Cr-labeled target cells.
2. Add 100 μl MNB containing the desired number of CTL. For spontaneous-release controls, this aliquot should contain mock CTL or simply MNB.
3. Mix the CTL with the targets by vortexing.
4. Centrifuge for 2 min at 1500 rpm. Handle the tubes gently from this time until the end of step 5 to avoid disturbing the pellets.

    In none of the centrifuges we have used does centrifugation of uncapped tubes produce a radioactive aerosol. The first time you use a new centrifuge, test for this by wiping the inside of the centrifuge afterwards with a tissue and counting the tissue in the gamma counter. It should have no radioactivity.

5. Incubate at 37°C in a water bath for the desired times. The water need cover only the filled portion of each tube. If the times exceed 15 min, cover the water bath to keep the air over the tubes humidified to minimize evaporation. When removing the cover, be careful not to allow condensation to drip into the tubes.
6. Remove the tubes from the water bath and add 2 ml PBS to each tube. (This can be squirted in vigorously to mix the contents.)
7. Centrifuge for 5 min at 1500 rpm.
8. Decant the supernatant into a separate tube having the same label in a different color.

    The tubes which receive the supernatants need not be chemically clean, only nonradioactive. We rinse them in tap water and recycle them. Preloading the supernatant tubes with

0.1 ml 1% sodium azide (caution: toxic!) prevents growth of microorganisms during gamma counting. This not only makes the tubes smell better, but makes effective rinsing easy.

9. Stopper and count the supernatant tubes in a gamma counter. Ideally, set the counting time per sample sufficient to get a gross count of 5000 counts in the freeze-thaw supernatants. (Save the pellets in a freezer. If peculiarities are seen in the supernatant counts, you may wish to count the pellets to determine whether the total amount of radioactivity added per tube was uniform as intended.)

Interpretation of the Results

Calculating the Percentage Corrected Release

Terminology: t is total releasable ^{51}Cr, the basis you select to define as 100% lysis (e.g. ^{51}Cr released by freeze-thaw); c is the control value for spontaneous release; e is the release in an experimental tube, e.g., one containing CTL.

In order to determine the ^{51}Cr released by the CTL, the spontaneous release must be subtracted. The result has often been called *specific release*, but this may incorrectly be taken to imply immunological specificity; hence, it is best called *corrected release*. In this simplest method

1. percentage corrected release = $100(e - c)/t$.

Investigators have usually been dissatisfied with the above correction because the corrected release values can never reach 100%. It underestimates the release that would be observed in the ideal case in which spontaneous release was zero. On the assumption that, were spontaneous release zero, the CTL would have lysed the same fraction of targets among those which spontaneously released ^{51}Cr as among those which did not, one may calculate

2. percentage corrected release = $100(e - c)/(t - c)$.

This assumption is likely valid when most of the spontaneous release represents leakage from living cells (clearly the case with labels that have higher spontaneous release than ^{51}Cr, such as ^{86}Rb, ^{14}C-nicotinamide, or ^{14}C-amino isobutyric acid: Burakoff et al., 1975; Martz et al., 1974; discussed in an appendix in Martz, 1976a). It is less likely valid when spontaneous release represents spontaneous lysis of a subpopulation of the targets (probably often the case with ^{51}Cr).

Which formula is used is not much of an issue for ^{51}Cr as long as spontaneous release is less than 20%. However, when spontaneous release is in the 20 to 35 % range, formula (1) is highly recommended as less likely to be misleading. In view of the fact that total releasable ^{51}Cr can be as low as 50% of total uptake, **when spontaneous release exceeds 35% of total uptake, estimates of lysis based on ^{51}Cr release are likely to be meaningless**. That is, the changes in release may not represent lysis at all, but could easily represent nothing more than changes in the chemical associations of the ^{51}Cr within the cell debris.

In a situation where spontaneous ^{51}Cr release cannot be kept below 35% of total uptake, the best solution would be to use ^{3}H-thymidine to estimate cytolysis (see Chaper 44, this volume). If the target cell cannot be labeled with ^{3}H-thymidine, ^{51}Cr-release data are least likely to be misleading if corrected release is calculated as a percentage of total freeze-thaw releasable ^{51}Cr determined for each individual tube as explained above.

Lytic Activity and Lytic Units

It seems impossible to derive a lytic activity value from ^{51}Cr-release assays which is independent of the assay conditions. Hence, the original definition of a lytic unit (Cerottini and Brunner, 1974) still seems the best way to quantitate lytic activity, although it is inherently entirely dependent on the assay conditions. Thus, it can only be used to compare values obtained under strictly identical assay conditions.

A *lytic unit* is defined as the *number of effector cells* required to lyse 30% of the target cells under specified assay conditions. *Lytic activity* is then typically expressed as lytic units per million cells.

Reliable Interpretation of the Results

In view of the uncertainties detailed above which are inherent in estimating the percentage of target cells lysed from ^{51}Cr release data, results must be considered carefully and interpreted with caution (or tossed out altogether) where appropriate.

The high reproducibility of ^{51}Cr-release assays can give a false sense of security. Despite the fact that a 5% difference in corrected release may be statistically significant, it cannot be assumed to represent a difference in cytolysis. It could well represent an effect on the solubility of the ^{51}Cr attached to cell debris immediately after lysis, including spontaneous lysis. In the same vein, the difference between 75, 90 and 99% *lysis* simply cannot be determined by ^{51}Cr release, due to the uncertainty in the amount released at 100% lysis.

In view of these limitations, prudence compels the following guidelines for ordinary ^{51}Cr assays:

1. **Regardless of statistical significance, only "large" differences ($\geq 20\%$ of total uptake) can be relied upon to signify differences in cytolysis.**
2. **Release values above 70% of total uptake should not be relied upon to signify additional cytolysis.**
3. **When presenting corrected release data, always specify the calculation used and the range of values determined for spontaneous release.**

Acknowledgments

Thanks to Gideon Berke, who diagnosed the reasons for the failure of my first ^{51}Cr-release assay in Boston in 1970, to Kandula Sastry for explaining Auger electrons, to James Tocci for help with safety, and to the National Science Foundation for grant support.

References

Balk SP, Mescher MF (1981): Cytolytic T lymphocyte mediated chromium-51 release versus spontaneous release from blast and spleen cell targets at low temperatures. *Cell Immunol* 65: 201–205

Berke G (1977): Comparative analysis of single cell and population events in T lymphocyte-mediated cytolysis. In: *International Symposium on Tumor-Associated Antigens and Their Specific Immune Response*, Spreafico F, ed. New York: Academic Press

Brunner KT, Mauel J, Schindler R (1966): In vitro studies of cell-bound immunity; cloning assay of the cytotoxic action of sensitized lymphoid cells on allogeneic target cells. *Immunology* 11: 499–506

Burakoff SJ, Martz E, Benacerraf B (1975): Is the primary complement lesion insufficient for lysis? Failure of cells damaged under osmotic protection to lyse in EDTA or at low temperature after removal of osmotic protection. *Clin Immunol Immunopathol* 4: 108–126

Butterworth AE, Remold HG, Houba V, David JR, Franks D, David PH, Sturrock RF (1977): Antibody-dependent eosinophil-mediated damage to ^{51}Cr-labeled schistosomula of Schistosoma mansoni: Mediation by IgG, and inhibition by antigen-antibody complexes. *J Immunol* 118: 2230

Cerottini J-C, Brunner KT (1974): Cell-mediated cytotoxicity, allograft rejection, and tumor immunity. *Adv Immunol* 18: 67–132

Fierer J, Finley F, Braude AI (1974): Release of ^{51}Cr-endotoxin from bacteria as an assay of serum bactericidal activity. *J Immunol* 112: 2184

Gillespie GY, Barth RF, Gobuty A (1973): A new radiosotopic microassay of cell-mediated immunity utilizing technetium-99m labeled target cells. *Proc Soc Exp Biol Med* 142: 378–382

Grinnell F (1976): Biochemical analysis of cell adhesion to a substratum and its possible relevance to cell metastasis. In: *Membranes and Neoplasia: New Approaches and Strategies*, Marchesi VT, ed. New York: Alan R. Liss (*Prog Clin Biol Res* 9: 227–236)

Henney C (1973): Studies on the mechanism of lymphocyte-mediated cytolysis II. The use of various target cell markers to study cytolytic events. *J Immunol* 110: 73–84

Howell DM, Martz E (1988): Evidence against damage to the nuclear envelope of the target cell in the induction of nuclear disintegration by cytolytic T lymphocytes. *J Immunol* 140: 689–692

Korzeniewski C, Callewaert DM (1983): An enzyme-release assay for natural cytotoxicity. *J Immunol Methods* 64: 313–320

MacDonald HR, Cerottini J-C (1979): Inhibition of T cell-mediated cytolysis by 2-deoxy-D-glucose: Dissociation of the inhibitory effect from glycoprotein synthesis. *Eur J Immunol* 9: 466–470

MacLennan ICM, Golstein P (1978): Requirement for hexose, unrelated to energy provision, in T cell-mediated cytolysis at the lethal hit stage. *J Exp Med* 147: 1551–1567

Martz E (1975): Early steps in specific tumor cell lysis by sensitized mouse T-lymphocytes I. Resolution and characterization. *J Immunol* 115: 261–267

Martz E (1976a): Early steps in specific tumor cell lysis by sensitized mouse T lymphocytes II. Electrolyte permeability increase in the target cell membrane concomitant with programming for lysis. *J Immunol* 117: 1023–1027

Martz E (1976b): Sizes of isotopically-labelled molecules released during lysis of tumor cells labelled with ^{51}Cr and [^{14}C]nicotinamide. *Cell Immunol* 26: 313–321

Martz E (1977): Mechanism of specific tumor cell lysis by alloimmune T-lymphocytes: Resolution and characterization of discrete steps in the cellular interaction. *Contemp Top Immunobiol* 7: 301–361

Martz E, Benacerraf B (1973): An effector-cell independent step in target cell lysis by sensitized mouse lymphocytes. *J Immunol* 111: 1538–1545

Martz E, Burakoff SJ, Benacerraf B (1974): Interruption of the sequential release of small and large molecules from tumor cells by low temperature during cytolysis mediated by immune T-cells or complement. *Proc Natl Acad Sci USA* 71: 177–181

Ravdin JI, Croft BY, Guerrant RL (1980): Cytopathogenic mechanisms of Entamoeba histolytica. *J Exp Med* 152: 377–390

Russell JH, Dobos CB (1983): Accelerated ^{86}Rb$^+$ (K$^+$) release from the cytotoxic T lymphocyte is a physiologic event associated with delivery of the lethal hit. *J Immunol* 131: 1138–1148

Sanderson CJ (1976a): The mechanism of T cell mediated cytotoxicity I. The release of different cell components. *Proc Roy Soc B* 192: 221–239

Sanderson CJ (1976b): The uptake and retention of chromium by cells. *Transplant* 21: 526–529

Shortman K, Wilson A (1981): A new assay for cytotoxic

lymphocytes, based on a radioautographic readout of ^{111}In release, suitable for rapid, semi-automated assessment of limit-dilution cultures. *J Immunol Meth* 43: 135

Thelestam M, Möllby R (1976): Cytotoxic effects on the plasma membrane of human diploid fibroblasts—a comparative study of leakage tests. *Med Biol* 54: 39–49

Wiltrout RH, Frost P, Cummings GD (1978): Isotope-release cytotoxicity assay with the use of indium-111: Advantage over chromium-51 in long-term assays. *J Natl Cancer Inst* 61: 183–188

Wright SC, Bonavida B (1987): Studies on the mechanism of natural killer cell-mediated cytotoxicity. VII. Functional comparison of human natural killer cytotoxic factors with recombinant lymphotoxin and tumor necrosis factor. *J Immunol* 138: 1791–1798

44
DNA Fragmentation and Cytolysis Assayed by ^{3}H-Thymidine

Eric Martz

The present note has two purposes: first, to recommend the use of ^{3}H-thymidine over iododeoxyuridine-125 (^{125}IUdR) for assays of DNA fragmentation; and second, to point out the advantages of using the same ^{3}H-thymidine label to assay cytolysis (instead of using double label with ^{51}Cr). We have recently published data in which both DNA fragmentation and lysis were quantitated with ^{3}H-thymidine (Martz and Gamble, 1992).

Duke and Cohen (1992) have recently provided an excellent and comprehensive guide to methods for assaying DNA fragmentation and apoptosis, including staining with acridine orange for determination of the apoptotic index, quantitation of DNA fragmentation by diphenylamine for unlabeled cells or by radioisotopic release from labeled cells, and demonstration of internucleosomal cleavage of DNA by agarose gel electrophoresis. The reader interested in details of these basic techniques is directed there.

The verification that DNA fragmentation involves the type of internucleosomal cleavage characteristic of apoptosis requires agarose gel electrophoresis to visualize the typical "ladder" pattern of nucleosomal monomers and oligomers, multiples of approximately 200 bp.

Quantitation of DNA fragmentation, on the other hand, cannot be accomplished with the usual agarose gel electrophoresis methods. Indeed, fragmentation of an insignificantly small percentage of the total DNA could make an impressive ladder pattern on gels. For quantitation, the fraction of DNA rendered soluble must be determined.

The basic principle of this assay is that DNA which has suffered extensive double-strand breakage is rendered soluble, while intact chromatin is insoluble by ordinary low-speed centrifugation. In order to ensure that prelytically solubilized DNA (often the most important fraction) is not trapped inside intact plasma membranes, or inside the nuclear envelope, the assay employs a special solubilization buffer. This buffer must include Triton X-100 to destroy phospholipid bilayers, and should also be hypotonic and include ethylenediametetraacetic acid (EDTA), to explode nuclear remnants and release any fragmented DNA (Russell, 1983; Duke and Cohen, 1988; see "TTE" buffer in Duke and Cohen, 1992; see also legend to Table 1 in Martz and Howell, 1989). Unless the cells involved are nonreplicating, this is most easily accomplished by prelabeling the cell DNA.

^{3}H-Thymidine Avoids Radiotoxicity of ^{125}IUdR and Subpopulation Artifacts

^{125}IUdR has been used to quantitate DNA fragmentation in the majority of published studies; ^{3}H-thymidine has been used in the minority. However, ^{3}H-thymidine has several compelling advantages, which argue for its use in future studies.

The typical use of ^{125}IUdR has involved a short labeling pulse (about 1 hr). Since even rapidly cycling mammalian cells have a DNA synthesis cycle not less than 7 hr, this pulse labels only a small fraction of the DNA. Moreover, it labels only the cells which happen to be in the S phase of the cell cycle during the pulse, typically a minority. Hence, this type of labeling is undesirable because the results could conceivably differ between the minority of the cells/DNA labeled, and the majority which is therefore invisible by this technique.

This problem can be minimized by labeling with ^{3}H-thymidine, provided the exposure to label extends over approximately one generation time for the cells and culture conditions employed. Typically, label can be introduced into a cell culture about 1 day before the cells are harvested for use in the assay.

The second major problem with ^{125}I is radiotoxicity. A sufficiently high level of any radioisotope will fragment DNA by itself. However, this is much easier to achieve inadvertently with ^{125}I than with ^{3}H. At equal cpm/sample, ^{125}I is about 100-fold more radiotoxic than ^{3}H! ^{125}I decays by electron capture, and only 7% of the disintegrations emit a gamma ray. Consequently, a gamma spectrometer typically counts ^{125}I with only a few percent efficiency. Every ^{3}H disintegration releases a beta particle, and in liquid scintillation, ^{3}H can be counted with 30% efficiency. Thus, one needs about 10 times as many ^{125}I disintegrations as ^{3}H to get the same cpm/sample. However, although only 7% of the ^{125}I disintegrations release a gamma ray, 100% of the disintegrations release a shower of > 10 low-energy electrons (Auger electrons). These electrons are extremely damaging to DNA, since, unlike the much higher-energy gamma rays, the Auger electron energy is absorbed in the immediate vicinity of their emission. In conjunction with the difference in counting efficiency, then, the overall radiotoxicity of ^{125}I is roughly 100-fold that of ^{3}H at the same cpm/sample.

In view of the unavoidably higher DNA damage associated with ^{125}I labeling, it would seem advisable to use ^{3}H when the goal is to detect DNA fragmentation induced by CTL. Most previous studies in which ^{125}I was employed showed controls indicating that the ^{125}I alone (e.g., in the absence of CTL) did not fragment the DNA. Nevertheless, the ^{125}I-induced damage may "prime" the DNA for fragmentation by other means such as CTL.

In addition to the above considerations, ^{125}IUdR is considerably more expensive that ^{3}H-thymidine, and has a much shorter half-life (60 days compared to 12 years).

Quantitation of Cytolysis with ^{3}H-Thymidine

For many purposes, ^{51}Cr is a highly satisfactory way to quantitate cytolysis. However, ^{51}Cr release above 70% is incapable of distinguishing 70% lysis from 90% lysis from 99% lysis (see Chapter 43, this volume), and spontaneous release limits its use to assays of a fraction of a day in duration. Moreover, resolution of ^{51}Cr from ^{3}H in double-label experiments poses significant, though not insoluble, problems (see below).

A ^{3}H-thymidine-based assay for cytolysis suffers from none of the above disadvantages of ^{51}Cr. In a series of experiments using mouse fibroblast-like cells, spontaneous lysis quantitated with ^{3}H-thymidine was 4% at 8 hr, and 5-8% at 12-14 hr (Martz and Gamble, 1992). The major reason for preferring ^{51}Cr or ^{125}I in many cases is that the liquid scintillation fluid required for counting ^{3}H is expensive to purchase and dispose of, flammable, toxic, and unpleasant to work with. However, if incubations of 10 hr or more are needed, or if it is necessary to determine high levels of lysis accurately, ^{3}H-thymidine is preferable. If ^{51}Cr is used for lysis, it is easier to resolve ^{125}I than ^{3}H from ^{51}Cr in double-label experiments (see below). However, if you have decided to use ^{3}H-thymidine to determine DNA fragmentation, you can avoid the complications of double labeling with ^{51}Cr by using the ^{3}H-thymidine to quantitate lysis in addition to DNA fragmentation. Unlike double labeling, this method requires that replicate samples be processed separately for lysis vs. DNA fragmentation (see below).

In order for soluble ^{3}H-thymidine accurately to reflect the percentage of cells that have lysed, the following two criteria must be met. First, DNA fragmented prelytically within unlysed cells must not be soluble. This means that Triton X-100 must not be used in the lysis assay, so that the DNA of all unlysed cells will be contained within their plasma membranes (regardless of its degree of fragmentation) and will be insoluble by low-speed centrifugation. Second, 100% of the DNA of every lysed cell must be soluble. Cytolysis itself may solubilize none of the DNA, depending on what induced it (Martz and Howell, 1989). Therefore, DNAse is added prior to harvest to ensure solubilization of 100% of the DNA of all nuclei exposed to the culture medium by lysis of the plasma membrane.

In practice, 1 hr before harvesting a sample to determine cytolysis, a small volume of medium is added containing DNAse I sufficient to solubilize all exposed DNA. Incubation is continued at 37°C for the final hour in the presence of the DNAse I (which requires the Mg^{2+} normally present in culture media). Extracellular DNAse does no harm to viable cells, to my knowledge; it certainly does not solubilize their DNA. At the end of this last hour of incubation, the sample is centrifuged (e.g., $500 \times g$ for 3 min), and soluble ^{3}H is determined in the supernatant by liquid scintillation counting.

The proper time of incubation and concentration of DNAse should be determined for each type of target cell being used. For mouse fibroblast-like cells lysed with 0.1% Triton X-100 (which does not inhibit DNAse I: Howell and Martz, 1988), we found that exposure to 100 μg/ml DNAse I (Type IV, Sigma Chemical Co., St. Louis, MO) solubilized 87% of the DNA, a near-plateau value. (Use of freeze-thaw to lyse the cells before titrating the DNAse is not recommended, as this method renders the DNA more sensitive to DNAse, probably by damaging the nuclear envelope: Howell and Martz, 1988).

If the cells are doubly labeled with ^{51}Cr and a DNA label (either ^{125}IUdR or ^{3}H-thymidine), it is possible to process a single sample into three fractions from which can be calculated the percentage of cells lysed and the percentage of DNA fragmented (see Duke and Cohen, 1992 for details). Alternatively, replicate samples can be prepared, with one being processed for lysis and one for DNA fragmentation. When ^{3}H is used for both assays, replicate tubes *must* be used, since lysis and DNA fragmentation must be determined without and with DNAse, respectively.

To determine CTL-induced DNA fragmentation, replicate samples *to which no DNAse was added* are harvested using the hypotonic EDTA Triton "TTE" buffer mentioned above to determine soluble (fragmented) DNA. It is important that no exogenous DNAse be added to these samples, since DNAse I is not inhibited by Triton X-100 detergent (Howell and Martz, 1988) and will therefore tend to solubilize DNA not solubilized by the CTL. In the absence of CTL and DNAse, lysis by freeze-thaw followed by TTE buffer, or by TTE buffer alone, should result in < 5% soluble ^{3}H (prior to incubation with CTL). Mouse fibroblast-like cells incubated without CTL for 8 hr gave 3% spontaneous DNA fragmentation and 4% spontaneous lysis by the ^{3}H-thymidine method (Figure 1 in Martz and Gamble, 1992). Some cell lines have higher levels of spontaneous DNA fragmentation; we observed up to 12% in 2 hr in some hematopoietic cell lines (Howell and Martz, 1987).

Additional Technical Considerations

Any assay employing thymidine-labeled cells should include (after labeling and rinsing of the cells is completed) excess unlabeled thymidine to prevent reincorporation of released isotope. We use cold thymidine at 40 μM (10 μg/ml) during the assay incubation.

Beware that some cell lines, notably EL4, are extremely sensitive to thymidine blockade of DNA synthesis (Stadecker et al., 1977). The amount of thymidine present during labeling is many orders of magnitude below the amount needed for blockage of even the most sensitive cell lines. However, unlabeled thymidine used at 10 μg/ml would be sufficient for blockage of EL4; since it is about 10^6-fold more concentrated than the thymidine used during labeling, a lower concentration could be used with EL4 or other sensitive cell lines. For most cells, 40 μM is safe. Spleen cell proliferation to mitogens requires 3 mM thymidine for blockade; P815 and L1210 cell lines are blocked at 200 μM (Stadecker et al., 1977).

Because the digestion of any unfragmented DNA from lysed cells requires about 1 hr of incubation, the ^{3}H-thymidine assay is not suitable for total incubation times of less than several hours. When lysis must be determined for incubations of less than a few hours, ^{51}Cr is preferable.

Note that merely adding liquid scintillation fluid to ^{3}H-labeled cells may not count the ^{3}H efficiently, evidently due to inadequate solubilization of the DNA. In order to determine the total ^{3}H present per sample, the DNA must first be dissolved by adding 0.1% sodium dodecyl sulfate (SDS), then mixed with liquid scintillation fluid. This gives a substantially higher count than when SDS is omitted.

When labeling cells with ^{3}H-thymidine, it is best to verify that the conditions used do not inhibit cell proliferation significantly. It is easy to add enough ^{3}H-thymidine to stop cell proliferation due to radiotoxicity. However, lower levels which do not stop proliferation are adequate for most purposes, giving for example 5000 cpm in a sample containing 2×10^4 cells. We label $1-2 \times 10^6$ mouse fibroblasts overnight with 4 μCi ^{3}H-thymidine in 15 ml medium. For lymphoid cell lines, we label at about 2×10^5/ml, using 0.1–0.2 μCi/ml overnight, giving about 0.1 cpm/cell.

You may wish to use medium without phenol red for ^{3}H protocols; the phenol red quenches liquid scintillation, reducing counting efficiency by roughly twofold.

In the event that you prefer to use ^{51}Cr for lysis with ^{3}H-thymidine for DNA fragmentation, careful planning is necessary to permit satisfactory resolution of the two isotopes. The problem is that in liquid scintillation spectrometry the Auger electrons released by ^{51}Cr are indistinguishable from the beta particles released by ^{3}H. The solution is to prepare the samples for liquid scintillation, then count each sample once in a gamma counter for ^{51}Cr, and once in the liquid scintillation counter for the sum of ^{3}H + ^{51}Cr. The "spillover" of ^{51}Cr into the liquid scintillation count (determined by counting a ^{3}H-free sample of ^{51}Cr in both counters) is then subtracted to yield the ^{3}H count. (^{3}H produces no counts whatsoever in the gamma counter.) Planning is required, since ^{51}Cr typically gives several fold more counts in liquid scintillation than in the gamma counter. Thus, if the ratio of ^{51}Cr to ^{3}H is too high, the ^{3}H signal cannot be accurately determined by subtracting the much larger ^{51}Cr "spillover." As an example, assume that you will have 500 gamma cpm ^{51}Cr per sample (counted for 5 min per sample). This amount of ^{51}Cr will give about 2000 cpm in liquid scintillation. The worst case will be a sample which has 100% lysis and 5% of DNA soluble. In order to see the ^{3}H signal accurately above the 2000 cpm from ^{51}Cr, the ^{3}H signal should be at least 500 cpm. Since this is only 5% of the total ^{3}H, the goal should be to have at least 10,000 cpm of ^{3}H per sample. Thus, it is necessary to label with ^{3}H just below the level that blocks cell division due to radiotoxicity, and to label very lightly with ^{51}Cr.

Acknowledgments

I am grateful to Rick Duke and John Cohen for a pre-publication copy of their methods article, Kandula Sastry for explaining Auger electrons, and to the Cellular Physiology Program of the National Science Foundation for grant support.

References

Duke RC, Cohen JJ (1988): The role of nuclear damage in lysis of target cells by cytotoxic T lymphocytes. In: *Cytolytic Lymphocytes and Complement: Effectors of the Immune System*, Podack ER, ed. Boca Raton, FL: CRC Press, 2: 35–37

Duke RC, Cohen JJ (1992): Cell death and apoptosis. In: *Current Protocols in Immunology*, Coligan JE, Kruisbeek AM, Margulies DH, Shevach EM, Strober W, eds. Brooklyn, NY: Greene Publishing Associates (in press)

Howell DM, Martz E (1987): The degree of CTL-induced DNA solubilization is not determined by the human vs. mouse origin of the target cell. *J Immunol* 138: 3695–3698

Howell DM, Martz E (1988): Evidence against damage to the nuclear envelope of the target cell in the induction of nuclear disintegration by cytolytic T lymphocytes. *J Immunol* 140: 689–692

Martz E, Gamble SR (1992): How do CTL control virus infections? Evidence for prelytic halt of *Herpes simplex*. *Virol Immunol* (in press)

Martz E, Howell DM (1989): CTL: Virus control cells first and cytolytic cells second? DNA fragmentation, apoptosis, and the prelytic halt hypothesis. *Immunol Today* 10: 79–86

Russell, JH (1983): Internal disintegration model of cytotoxic lymphocyte-induced target damage. *Immunol Rev* 72: 97–118

Stadecker MJ, Calderon J, Karnovsky ML, Unanue ER (1977): Synthesis and release of thymidine by macrophages. *J Immunol* 119: 1738–1743

45
The JAM Test: An Assay of Cell Death

Polly Matzinger

Cytotoxic T lymphocytes (CTL) appear to have at least two different ways of killing their targets: they drill holes in the target cell membrane, and they can also trigger the targets to commit suicide. For 30 years, ^{51}Cr-release has been a reasonably efficient and useful method to measure membrane disintegration, but until very recently, the assessment of target cell suicide has been a slow, laborious, and inefficient process. Here I describe the JAM Test (Matzinger, 1991) a method that measures the DNA fragmentation that accompanies cell suicide. As a general measure of apoptosis, it is fast, quantitative, and easy to do. As a method for measuring the activity of CTL, it surpasses ^{51}Cr release in several different ways. It is faster, more sensitive, more convenient, safer, more flexible, and cheaper. Its main disadvantage is that the targets must be in growth phase.

Materials and Methods

For a complete comparison of the JAM Test and ^{51}Cr-release assays, see Matzinger (1991). Here I will describe only how to do the JAM Test itself and give a few examples of killing assays done this way in our laboratory.

General Overview

In every respect except the labeling and harvesting of targets, testing killers by the Jam Test is done in the same way as by ^{51}Cr release. Killer cells are plated into microtiter plates and titrated in two- or threefold dilutions. The targets are added and the plates are then incubated at 37°C for a period of time before harvesting.

The differences are that the targets are labeled with ^{3}H-Thymidine (^{3}H-TdR), instead of ^{51}Cr, and that the *cells* in the wells are harvested rather than the supernatant. The harvesting step is done essentially as in a proliferation assay, in that the contents of the wells are harvested onto fiberglass filters. Since DNA is physically trapped in these filters, rather than chemically bound, small pieces of fragmented DNA from dead targets are washed through, leaving only the intact DNA from living cells. The associated radioactivity thus corresponds to the number of living targets remaining in each well.

To calculate percent specific killing, I substituted into the standard formula used for ^{51}Cr-release assays, which is

$$\frac{E - Sp}{T - Sp} \times 100$$

Where
E = label released in the presence of killers (in cpm)
S = spontaneous amount of label released in the absence of killers
T = total releasable label.

For the JAM Test, the amount of label retained (and captured by the filter) is a measure of living cells, and the amount of label lost (as above) reflects the amount of target death. Therefore if
R = label retained in the presence of killers (in cpm)
S = label retained in the absence of killers (spontaneous)
T = total retainable label
one can substitute into the standard formula thus:

$$\% \text{ specific DNA loss} = \frac{(T - R) - (T - S)}{T - (T - S)} \times 100$$

which becomes:

$$\% \text{ specific killing} = \frac{S - R}{S} \times 100$$

T (the total incorporated counts) drops out of the formula. It should not be neglected, however, because its relationship to S, the amount of DNA spontaneously retained by the targets over the course of the assay, is a measure of how healthy the targets are. S should never be less than 80% of T over a 4-hr assay. I have found that it is generally much better than that.

Media

I use the following (but any of the media routinely used for cell culture can be used):

For lymphocyte culture (killer generation and target growth): Iscove's modified Dulbecco's modified Eagle's medium (IMDM, Biofluids, Bethesda, MD) plus 10% fetal calf serum (FCS) plus glutamine, 5×10^{-5} M 2-mercaptoethanol (2-ME), and antibiotics.

For handling cells: I use IMDM without sodium bicarbonate (the tonicity is made up with additional NaCl, Biofluids) plus 1% FCS. This is a complete medium, buffered for use in air rather than 5% CO_2, in which cells can happily sit at room temperature for hours.

For cell lysis assays: IMDM plus glutamine and 5% FCS. Because of its superior buffering capacity, IMDM maintains a fairly constant pH during large assays and lengthy manipulations.

Cells

Growth and labeling of target cells: I have used two sorts of targets, tumor lines and concanavalin A (Con A) blasts. Their growth and labeling are slightly different.

Growth:

P815: are normally grown in plastic tissue culture flasks (Costar, Falcon, etc.) at 37°C, 5% CO_2 under standard conditions. The day before the assay, I subculture growing cells into about 50% new medium (leaving the old, conditioned medium so as not to disturb them too much) in 24-well tissue culture plates at about 2.5–5×10^5 cells/ml.

Con A blasts: I culture 2×10^6 fresh spleen cells [or 3×10^6 thawed spleen cells which had been frozen in 90% FCS, 10% dimethylsulfoxide (DMSO)] in 2 ml/well with 2 µg/ml Con A. I label 30–48 hr later (this gives good labeling and low spontaneous death). The blasts are at their very best when labeled at about 30 hr for 3–6 hr before use, but both the timing and length of labeling can be varied extensively.

Labeling

Without disturbing the target cells, I add 100 µl ^{3}H-thymidine/well to a final concentration of 2.5–5 µCi/ml (9.25×10^4–18.5×10^4 Bq/ml) and allow them to continue incubating at 37°C until use. The targets can be labeled for varying lengths of time from 3 hr to overnight. Three to four hours is best for targets that grow rapidly, such as tumor lines, because otherwise they incorporate too much label and begin to die from the radioactivity. I have found that each new target cell type should be tested for its sensitivity to the label. I do this by plating duplicate plates of targets and comparing the number of counts incorporated (by harvesting a sample at time 0) and the number of counts retained after 4 hr of culture. The cells should not lose more than 20% of their label over 4 hr. Most types of targets can take a wide range of conditions, but a few are finicky.

Just before the assay, *gently* pellet the cells, wash once, and resuspend to about 1×10^5 cells/ml in assay medium. Since temperature changes can cause cells to apoptose, the targets are best kept at room temperature for the handling procedures.

Killer Cells

Cytotoxic cells are raised under standard conditions. There are, however, almost as many different "standard" ways of generating killer cells as there are laboratories doing it. I find more efficient killing, per input cell, from killers raised in 2-ml cultures in 24-well plates (Falcon, Costar, Nunc, etc.) at responder-to-stimulator ratios of 2:1 or 4:1 than from killers raised in flasks or in microtiter plates.

For bulk cultures of fresh cells: Spleen cells are cultured at 4×10^6 cells/2-ml well in 24-well plates with irradiated (3000 rads) spleen stimulator cells (2×10^6 cells/well) for 5 days. The cells are then harvested, pooled, pelleted, resuspended in assay medium, and plated (150 µl/well) into 200-µl round-bottomed microwells (96-well plates, flat or V-bottomed wells can also be used. I find that the round-bottomed wells are the easiest to use and harvest). Fifty microliters from each of these wells is then serially diluted into 100 µl medium to set up a titration of threefold dilutions. In general, I set up two culture wells (8×10^6 responders) for each target that the killers will be tested against. After 5 to 7 days, the two wells are pooled and resuspended in 300 µl for the assay. For example, for the anti-H-Y killers depicted in Figure 2, four wells were set up against male stimulators and, after the 5-day culture period, the killers were spun and resuspended to 0.6 ml of assay medium for plating in duplicate against male and female targets. In this way, the highest responder-to-target ratio is about 133:1, calculated from the number of spleen cells originally cultured.

For limiting dilution assays: Lymph node or spleen cells are cultured in 0.2-ml round-bottomed microwells at various numbers of cells/well against 1×10^6 irradiated spleen cell stimulators in medium containing 5% Con A Sup (supernatant from 2-day cultures of 5×10^6 Lewis rat spleen cells/ml plus 5 µg/ml Con A). To neutralize the Con A, 100 mM α-methyl mannoside was added to the stock Con A Sup after harvesting. After 7 days the plates were flicked to remove the medium, and resuspended in 100 µl assay medium (which leaves space to add 100 µl of targets). For the assay in Figure 3, the spleen cell responders were enriched for T cells by panning on plates coated with sheep anti-mouse Ig.

The Assay Itself

When the plates containing killers are ready, I wash the targets (once), resuspend them to 10^5/ml, and plate 100 µl/well into three sorts of wells:

1. Those containing killers;
2. Those containing 100 µl of medium only, to measure spontaneous retention (usually on the same plates as those above);
3. Those containing 100 µl of medium only, to measure the value of the total incorporated/releasable label (on a different plate, to be harvested immediately or frozen immediately for later harvesting).

The plates are then incubated for 1 to 4 hr at 37°C in a humidified atmosphere at 5% CO_2. To harvest, one can aspirate the cells and their medium onto fiberglass filters using any of the cell harvesters currently used for harvesting cells for proliferation assays. The filters are then dried, wetted with liquid scintillation fluid, and counted in a liquid scintillation counter, exactly as for proliferation assays. They can also be frozen and harvested later. In fact, we often freeze the plates as a way of stopping the reaction when we have very large assays to harvest.

We find that, with an LKB Betaplate Beta counter and a Brandell (Gaithersberg, MD) 96-well harvester, harvesting and counting the assay is fast (it takes about 1.5 min to harvest each plate, and the Betaplate, counting six samples at a time, can count a plate in as little as 12.5 min when set for 30 sec/sample). The Betaplate also has the advantage that it does not require scintillation vials or caps, and uses very little scintillation fluid. Thus, as well as being easy to use, it generates far less radioactive waste than conventional counting methods.

Examples

Kinetics of Killing

Figure 1 shows the rate at which killing can be measured by the JAM Test. The assay was a standard alloreactive T-cell response measuring the lysis of BALB/c Con A blast targets by alloreactive B6 anti-BALB/c cytotoxic T lymphocytes (CTL) from a 5-day culture. The assay was set up in nine replicate plates, one of which was harvested every half hour. Panel A, depicting specific lysis, is a typical representation of CTL titration curves at different times, and panel B, percent total incorporated label, shows the amount of retained counts at different times as a proportion of the original counts incorporated, without subtracting the spontaneous death. In panel A we see that greater than 65% killing appears as early as 1 hr at a responder-to-target ratio (R/T) of 100:1, (a level not usually seen by ^{51}Cr-release until much later), and that even higher levels of kill can be achieved with 10-fold fewer responders by 2.5 hr. Panel A also shows that JAM plates, unlike ^{51}Cr release plates, can be frozen. Plates frozen after 3.5 hr incubation and harvested the next day give the same results as the plates harvested immediately at the 3.5-hr time point: a bonus for anyone doing many plates or lengthy assays. We now routinely freeze large assays as a means of stopping the reaction. Panel B shows that the spontaneous death rate is fairly low, reaching about 14% at 3–3.5 hours. Thus, at an R/T of 100:1, the level of killing is easily distinguished from the spontaneous death as early as 1 hr into the assay, and even at ratios of 5:1, there is a difference between test and control wells at 1 hr that becomes clearly significant by 2.5 hr.

There are two reasons for this increased sensitivity compared to ^{51}Cr-release assays. The first is that DNA disintegration occurs very quickly and that cytoplasmic leakage, on which the ^{51}Cr-release assay relies, takes time. In fact, according to Duke, apoptosis and complete DNA degradation occur

45. THE JAM TEST: AN ASSAY OF CELL DEATH

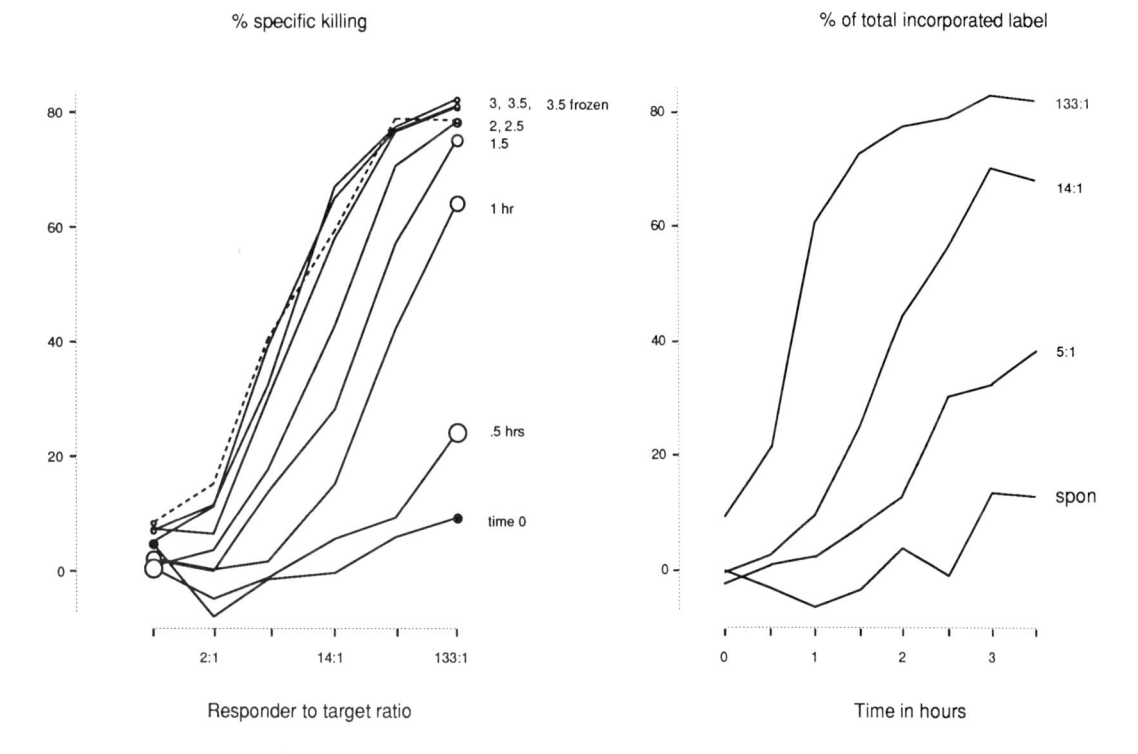

FIGURE 1.

within half an hour of the lethal hit. The second reason for the sensitivity is the harvesting procedure. In ^{51}Cr-release assays, targets that die near the end of the incubation period will not be counted as dead, since they will not have released much label, and even those that have died early may not have released all of their cytoplasmic contents. In the JAM Test, the cells are smashed up in the process of harvesting and their entire contents are loaded onto the filters. Thus it is not necessary to wait for anything to leak out, and almost all dead cells read out as dead.

Example of a "Weaker" Response

Figure 2 shows the anti-H-Y (left panel) and anti-MHC (right panel) responses of four B6 female mice that were primed once (thin lines) or twice (heavier lines) to H-Y. The JAM Test picks up the small differences between the two sets of mice showing that boosted mice respond somewhat better to H-Y than unboosted mice, and that the boost is specific, since the level of MHC responses is the same for all four.

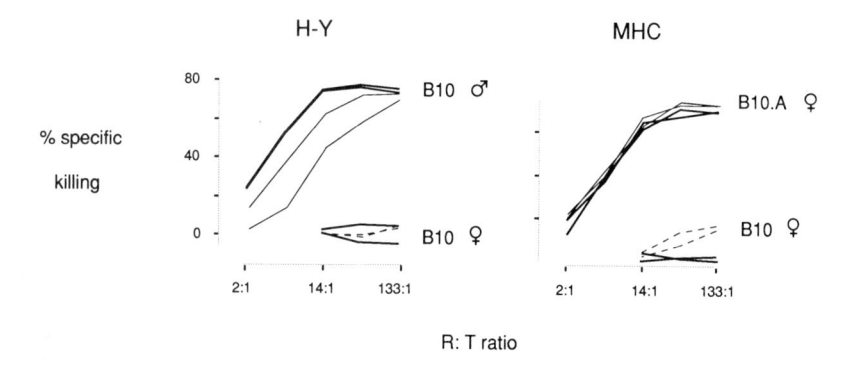

FIGURE 2.

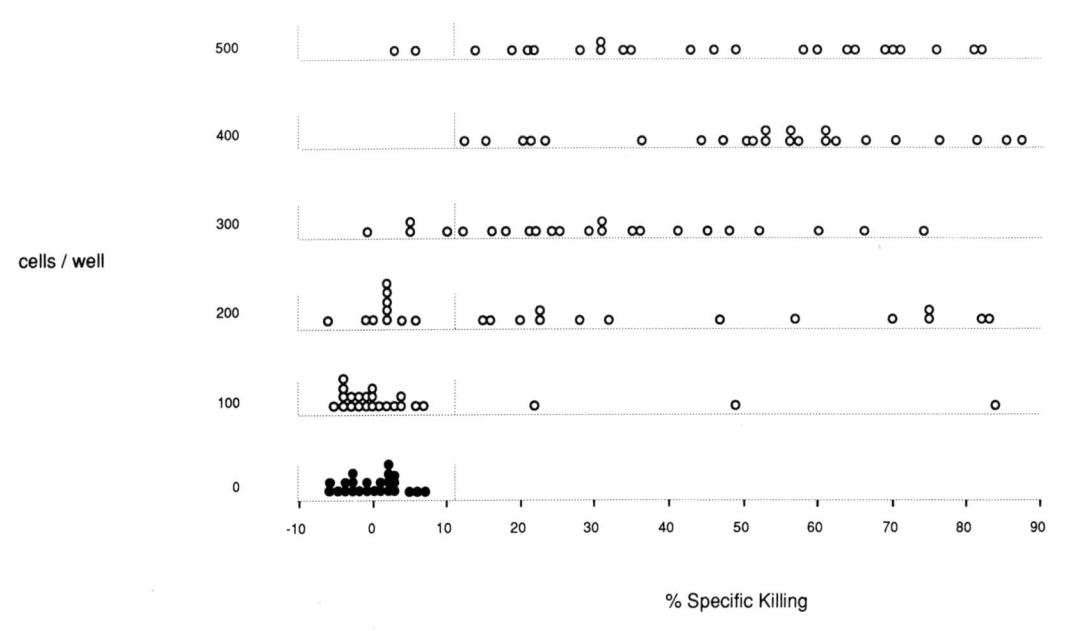

FIGURE 3.

Example of a Limiting Dilution Assay

Figure 3 shows the results of a limiting dilution experiment testing B6 anti-BALB/c killers on P815 targets. The assay was run for 3.5 hr, and the plates were frozen before harvesting. The frequency estimate for this test was 1/260, counting as positive each well that gave killing greater than 3 standard deviations from the mean of the spontaneous death (vertical lines at 12%). Clearly the JAM Test allows good discrimination between strongly positive, weakly positive, and negative wells. In this respect I have found it to be consistently better than ^{51}Cr release, where weakly positive wells are difficult to distinguish from negatives.

Summary

The JAM Test is clearly a simple assay with several benefits.

1. It is fast and easy to set up; targets that have been ^{3}H-TdR labeled overnight require very few manipulations (count viable cells, wash once, resuspend to desired concentration) and less than ½ hr to prepare.
2. It is sensitive; specific lysis can be seen with incubation periods of 1 to 1.5 hr and/or with small numbers of killer cells. This makes the JAM Test ideal for limiting dilution assays, where each well contains a very small number of effector cells, and for any other test where the number of specific effectors might be limited (for example, an estimate of the effectors found locally in biopsies of a tumor or graft site).
3. It is not expensive; since ^{3}H-TdR has a very long half-life, we use a shipment until it is used up. Thus we no longer routinely throw away unused, half-decayed radioactive labels, and in time, the ^{3}H-TdR may be replaced by DNA labels that are tagged with completely non-radioactive substances.
4. It is also convenient in other respects. Many targets can be labeled for vastly different lengths of time (4 to 18 hr) without apparent harm. Two to three consecutive tests can be run in one day, very large assays can routinely be run, and the plates can be frozen for future harvest and analysis. The same equipment and materials used for proliferation assays can be used for the JAM Test, and it lends itself well to automated harvesting.

A Couple of Things to Watch Out for

Since its inception, I have received several comments on the JAM Test and the problems people have had with it. The most common problem has been overlabeling of the targets. Cells that incorporate too much ^{3}H die during the assay without the addition of killers, so it is best to work out the conditions for each cell line or cell preparation so that the targets are growing happily and are labeled to no more than about 0.5 to 4 counts per cell. A second is harvesting. When there are a lot of killers plus targets per well, a lot of intact DNA will be collected on the filters. This will tend to trap the small pieces of degraded DNA. Harvesting rinse cycles should therefore be extended so that the small pieces of degraded DNA will be washed through. We use a manually operated Brandell harvester, which gives very good results. Automatic harvestors should be programmed for long rinses.

Finally, one must remember that the JAM Test measures only one type of cell death, namely, death by apoptosis. Other forms of death, and death of nondividing cells, will need other assays.

The assay is dedicated to the memory of Tangy Matzinger.

References

Matzinger P (1991): The JAM Test: A simple assay for DNA fragmentation and cell death. *J Immunol Methods* 145: 185–192

46
Target Cell Detachment Assay

Scott I. Abrams and John H. Russell

Conventional assays used for the measurement of cell-mediated cytotoxicity involve interactions between effector lymphocytes and antigen-bearing targets maintained and/or manipulated into a suspension culture. However, *in vivo*, immunologic reactions also involve T lymphocytes responding to targets that are bound to each other or to an extracellular matrix. To that end, we have developed an *in vitro* murine model, which more closely mimics such an *in vivo* situation, to explore the functional consequences that result from T cell-tissue or tumor interaction (Russell et al., 1988).

We have modified the conventional ^{51}Cr-release cytotoxic assay using adherent target cells (fibroblasts, macrophages, tumor cells) which are maintained, radiolabeled, and functionally tested in their native conformation as monolayer cultures. As discussed in Chapter 19, we have identified and functionally characterized in detail a unique lymphoid activity expressed by both $CD8^+$ and $CD4^+$ T-cell lines and clones which results in the detachment of viable target cells from their substratum early during the effector-target interaction, which is distinct and potentially independent of the lytic response (Abrams et al., 1989, 1991, 1992; Russell et al., 1988).

The following protocol describes the T-cell-target cell detachment assay, which also simultaneously measures lytic activity (Figure 1). Adherent fibroblasts and tumor cells are seeded at 2×10^4 cells/well in 200 μl of culture medium in sterile 96-well, flat-bottomed microtiter plates and incubated at 37°C for 18 to 24 hr before the assay to ensure stable and confluent monolayer cultures. Due to the comparatively smaller size of macrophages and their inability to replicate under these culture conditions, they are plated at 5×10^4 cells/well. Macrophages are derived from peritoneal exudate cells of normal mice previously injected with thioglycollate medium (enhances cell yields) or concanavalin A (Con A) [augments adherence properties and major histocompatibility complex (MHC) Class I and II antigen expression via interferon-γ (IFN-γ) induction].

The next day, the monolayers are washed 2 to 3 times by aspiration and warm culture medium to remove unbound cells using an ordinary 6" Pasteur pipette, previously curved at the tip, situated against the wall of the well. After washing, monolayers are radiolabeled in 50 μl of warm culture medium containing 200 to 300 μCi/ml of ^{51}Cr and incubated for 1.5 hr at 37°C. Monolayers are then washed 3 times with 100 μl of warm culture medium to remove unincorporated radioactivity and incubated for an additional 0.5 to 1 hr at 37°C before washing 3 more times, which helps to reduce the background or spontaneous release of ^{51}Cr. Radiolabeled targets are mixed with effector T cells (1, 2, or 4×10^5 cells/well, for example) in 200 μl of culture medium, the plate centrifuged ($450 \times g$, 2 min) to initiate cell contact, and incubated at 37°C for various time intervals up to 18 hr. Parallel cultures prepared without effectors, but containing culture medium alone or 0.2% Triton X-100 (or 0.1 N NaOH), provide essential information on spontaneous isotope leakage and maximum isotope uptake, respectively. Due to the nature of the assay design, precise effector/target ratios are determined by enumerating target cells (fibroblasts, tumors) released and recovered from control wells using ethylenediaminetetraacetic acid (EDTA) or by counting nuclei of detergent-lysed macrophage cultures.

The method by which samples are collected and processed here at the end of the reaction further distinguishes it from that of the ordinary lytic

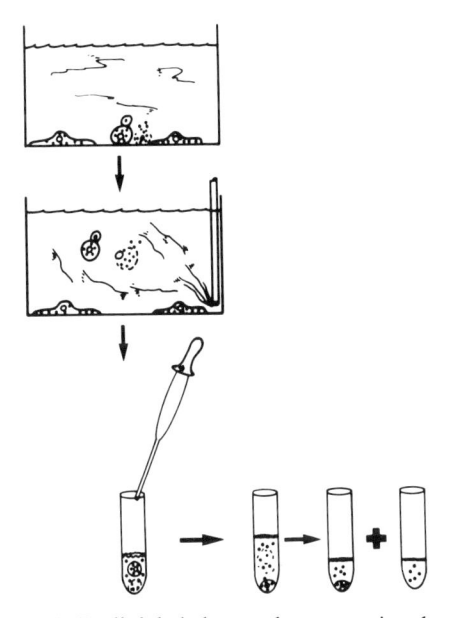

FIGURE 1. Radiolabeled monolayers are incubated with effector T cells at 37°C in a 96-well, flat-bottomed microtiter plate. Target cells, which originally are elongated and attached firmly to their substratum, round-up (depicted as the larger circle) and detach after a specific binding interaction with effector T cells (depicted as the smaller circle). Target cytolysis may also occur as a consequence of the interaction and, as depicted here, resulting in a fragmented, floating cell. At the end of the culture period, a pipette is placed against the side of the well with the tip near, but not touching, the bottom. The cell suspension is vigorously pipetted six times, and the medium and nonadherent cellular fractions (live and dead cells) are collected in a tube containing 1 ml PBS with 1 mM EDTA. The cell mixture is vortexed and centrifuged, and one-half of the supernatant fluid is removed and transferred to a fresh tube. Finally, radioactivity in both tubes is quantitated, and the magnitude of target cell detachment and lysis is determined by the formulas described in the text.

assay, which simply measures radioactivity released into the supernatant fluid by dead and damaged targets. In the combined detachment-lytic assay, both supernatant fluid and nonadherent cellular fractions are collected and evaluated (Figure 1). A Pasteur pipette is situated against the side of the well, with the tip near, but not contacting the bottom. The cell suspension is vigorously agitated 6 times to dislodge loosely adherent cells, collected and transferred to a 12 × 75 mm tube containing 1 ml of cold phosphate-buffered saline (PBS) with 1 mM EDTA. The tube is vortexed, centrifuged, and one-half of the supernatant fluid removed and transferred to a second 12 × 75 mm tube. Finally, both tubes are counted in a traditional γ-counter.

As with the conventional ^{51}Cr-release cytotoxic assay, supernatant or soluble-associated radioactivity provides an estimate of target cell death. Here, it is determined as twice the radioactivity in the tube containing one-half of the supernatant fluid divided by the total radioactivity incorporated into the monolayer. The magnitude of target cell detachment, which reflects intact as well as damaged cells, is determined as the sum of radioactivity in both tubes divided by the total radioactivity incorporated into the monolayer. Spontaneous release of ^{51}Cr in the absence of effectors is approximately the same for both detached and soluble fractions at any given time point and usually < 15% in a short-term (4 to 6 hr) assay and < 40% in a longer-term (18 hr) assay. The data are standardized to percent specific release for both detached (i.e., supernatant fluid + pellet) and soluble (i.e., supernatant fluid only) radioactivity and are determined by the equation:

% specific release

$$= \frac{\% \text{ experimental release} - \% \text{ spontaneous release}}{100 - \% \text{ spontaneous release}}$$

$$\times 100.$$

Results are expressed as the mean ± SEM of triplicate cultures. The mathematical difference between the percent specific detachment and the percent specific lysis represents the percent specific detachment of viable target cells. To evaluate, compare, and elucidate the nature of different forms of immune damage, we have adapted this assay for radiolabeling with iododeoxyuridine-125 (^{125}IUdR) (which measures nuclear, rather than cytoplasmic, changes within the adherent target cell) and, further, for examining interactions between adherent target cells and Ab + C, tumor necrosis factor (TNF)-α/β, or granules which contain perforin (cytolysin) activity (Abrams et al., 1989, 1991, 1992; Russell et al., 1988).

References

Abrams SI, McCulley DE, Meleedy-Rey P, Russell JH (1989): Cytotoxic T lymphocyte (CTL)-induced loss of target cell adhesion and lysis involve common and separate CTL signaling pathways. *J Immunol* 142: 1789–1796

Abrams SI, Russell JH (1991): CD4$^+$ T lymphocyte-induced target cell detachment. A model for T cell-mediated lytic and nonlytic inflammatory processes. *J Immunol* 146: 405–413

Abrams SI, Wang R, Munger WL, Russell JH (1993): Detachment and lysis of adherent target cells by CD4$^+$ T cell clones involve multiple effector mechanisms *Cell Immunol.* (in press)

Russell JH, Musil L, McCulley DE (1988): Loss of adhesion. A novel and distinct effect of the cytotoxic T lymphocyte-target interaction. *J Immunol* 140: 427–432

47
Protocol for Assaying CTL Activity Against *Toxoplasma gondii*

Ricardo T. Gazzinelli, Eric Denkers, Frances Hakim, and Alan Sher

Parasites and Antigen Preparation

Tachyzoites of the ts-4 strain of *T. gondii* are maintained by *in vitro* passage in human foreskin fibroblasts at 33°C. *T. gondii* soluble antigens (STAg) are prepared from parasites obtained by peritoneal lavage, 3 days after infection of Swiss Webster mice with $5-10 \times 10^6$ RH strain tachyzoites. Tachyzoites are concentrated by centrifugation ($1500 \times g$, 10 min) and repeatedly sonicated until no intact parasites remain. After centrifugation at $10,000 \times g$ for 30 min, the supernatant is collected, dialyzed against PBS, and stored at $-80°C$.

Target Cells

Bone marrow macrophages are obtained from femur and tibia bones of 8- to 12-week-old female mice. Cells are washed, resuspended in marrow culture media, and plated at 10×10^6 cells/10 ml on 100-mm non-tissue-culture treated Petri dishes (Falcon). Marrow culture media consist of Dulbecco's modified Eagle's medium (DMEM) (GIBCO) supplemented with 10% fetal calf serum (FCS), nonessential amino acids, sodium pyruvate, and penicillin/streptomycin, plus 30% supernatant from confluent cultures of L929 fibroblasts, as a source of macrophage-colony-stimulating factor (CSF). After 5 to 7 days of culture at 37°C, 10% CO_2, plates are gently washed to remove nonadherent cells. Adherent cells are then either pulsed with STAg (37°C) or infected with ts-4 (37°C) overnight, washed to remove excess antigen or extracellular parasites, and used as targets in the cytotoxic assay.

In Vitro Generation and Assay of CTL

Splenic responder cells (5×10^6) from control or mice vaccinated with ts-4 are cultured for 5–7 days in the presence of $0.5-1 \times 10^6$ irradiated (15,000 rad) tachyzoites, in DMEM supplemented with 10% FCS, plus nonessential amino acids, glutamine, sodium pyruvate, 5×10^{-5} M 2-mercaptoethanol (2-ME) and pencillin/streptomycin. Assays are performed in parallel, with untreated, *T. gondii*-infected and antigen-pulsed bone marrow macrophages as target cells. After exposure to parasite antigens, the target cells are incubated with ^{51}Cr, washed, and then used to measure the cytotoxicity of *T. gondii*-specific lymphocytes. Effector cells are assayed in triplicate at four different effector/target (E/T) ratios in a 4-hr ^{51}Cr-release assay. The percentage of specific lysis is calculated as $100 \times$ (experimental release − spontaneous release)/(maximum release − spontaneous release).

48
Granule Exocytosis Assay of CTL Activation

Michail V. Sitkovsky

Granule exocytosis assay for cytotoxic T lymphocyte (CTL) activation has become an important additional method of evaluating activating stimuli and of studying cell surface receptors and transmembrane signaling. It also provides the opportunity to study the effects of enhancing and inhibiting agents on the exocytic response by predictably manipulating the intensity of granule exocytosis which can be triggered by immobilized monoclonal antibody (mAb) or by the antigen (Ag)-bearing target cell. This assay is based on the ability of CTL to release cytoplasmic granules into the culture medium that contain the easily detectable enzyme serine esterase (Pasternack et al., 1986; Takayama and Sitkovsky, 1987). We describe here the granule exocytosis assay in response to immobilized anti-T-cell receptor (TCR) mAb and to Ag-bearing target cells. For more details the reader is directed to (Taffs and Sitkovsky (1991).

Anti-TCR mAb-Induced Exocytosis

This assay starts with preparation of microtiter plates with immobilized antibodies. After incubation of CTL in Ab-coated wells the supernatants are harvested for detection of secreted serine esterase. The substrate (BLT) with coloring agent (DTNB) is added to supernatants, and the enzyme activity is measured spectrophotometrically.

Materials

Phospate-buffered saline (PBS) pH 7.2 to 7.4;
monoclonal Ab to TCR (5 to 10 μg/ml, most convenient; most universal are mAb to CD3 ε chain (Leo et al., 1987) or clonotypic anti-Vβ8-TCR mAb F23.1 (Staerz et al., 1985).
culture medium (e.g. RPMI-10);
Ficoll-purified, washed CTL which are suspended in complete culture medium;
1% Triton X-100;
20 mM BLT (N-α-benzyloxycarboxyl-L-lysine thiobenzyl ester) stock (Calbiochem);
22 mM DTNB [5,5'-dithio-bis(2-nitrobenzoic acid) stock (Pierce)];
0.1 M phenylmethanesulfonyl flouride (PMSF) in dimethylsulfoxide (DMSO; Sigma);
96-well round-bottomed microtiter plates, made from flexible polyvinyl chloride (Dynatech #001-010-2401);

The preparation of the 96-well round-bottomed microtiter plates should be planned keeping in mind the need for triplicate samples to reserve cells for background, total, and Ab-induced exocytosis.

1. Add 50 μl of 5 μg/ml (or other dilutions of Ab to each well designed for Ab immobilization and wait at least 30 min at room temperature to allow immobilization of Ab.
2. Aspirate solutions of Ab from wells and wash them with 100 μl of RPMI-10 medium. Immediately after the last wash and the removal of all liquid from the well, add 100,000 CTL in 50 μl of complete RPMI-10 medium to all wells, expect those that are dedicated to the determination of total granule release.
3. Add 40 μl of complete RPMI-10 medium and 10 μl of 1% Triton X-100 in the wells for the determination of total granule release. All wells have a final volume of 100 μl.
4. After 4 hr incubation in a CO$_2$ incubator at 37°C (times can be varied depending on the responsiveness of a particular clone or the goal of the

experiment), the plates should be centrifuged 5 min using a plate-carrier rotor (Sorvall H-1000B) at 1100 rpm (200 × g) at 4°C and 50 μl of culture supernatant transferred to 12 × 75 mm tubes.

5. To measure the amount of secreted enzyme, prepare 50 ml of serine esterase substrate (BLT) solution by adding to a 50-ml conical tube:
500 μl 20 mM BLT stock (0.02 mM final);
500 μl 22 mM DTNB stock (0.22 mM final);
500 μl 1% Triton X-100 (0.01% final),
48.5 ml PBS

6. Add 950 μl of BLT substrate to each 50-μl aliquot of culture supernatant and incubate 20 min at 37°C in a water bath.

7. After removal of the tubes to an ice bath, immediately add 1 μl of 0.1 M PMSF to each tube. Add 1.0 ml PBS for a final concentration of 0.5 mM PMSF in a final volume of 2 ml.

8. Calculate percentage values for antibody-induced secretion of esterase according to the following formula: % secretion = 100 × $(E-B)/(T-B)$, where E is the mean absorbance value for antibody-stimulated cell supernatants, B is the background or mean absorbance value from CTL in untreated wells, and T is the total enzyme content.

Induction of Granule Exocytosis by Target Cells

In many experimental situations, it is desirable to compare the abilities of different target cells to activate CTL through TCR triggering by the antigen. Granule exocytosis triggered by TC is also a powerful evidence of CTL-TC conjugate formation.

1. To 50 μl of 2 × 10^6 cells/ml CTL suspended in complete RPMI-10 medium in all wells of 96-well round-bottom microtiter plate, add 50 μl of 2 × 10^5 target cell suspension in the same medium but only in the wells reserved for testing TC-induced exocytosis. Then add 40 μl complete medium and 10 μl of 1% Triton X-100 to wells designated for determination of total release and 50 μl complete medium to wells designated for determination of background release. The assay's application could be limited if BLT-esterase content of TC is high. It is recommended to vary target cell number from 5 × 10^4 to 4 × 10^5 cells/well to determine the optimum stimulation level.

2. Incubate plates 4 hr at 37°C. Proceed with enzyme assay as described for anti-TCR mAb induced exocytosis in steps 6 to 10 of the protocol for Ab-induced exocytosis.

References

Leo O, Foo M, Sachs DH, Samelson LE, Bluestone JA (1987): Identification of a monoclonal antibody specific for a murine T3 polypeptide. *Proc Natl Acad Sci USA* 84: 1374–1378

Pasternack M, Verret CR, Liu MA, Eisen HN (1986): Serine esterase in cytolytic T lymphocytes. *Nature* 322: 740–743

Staerz UD, Rammensee H-G, Benedetto JD, Bevan MJ (1985): Characterization of a murine monoclonal antibody specific for an allotypic determinant on T cell antigen receptor. *J Immunol* 134: 3994–4000

Taffs RE, Sitkovsky MV (1991): Granule enzyme exocytosis assay for CTL activation. In: *Current Protocols in Immunology*. Coligan J, Kruisbeek AM, Margulies DH, Shevach EM, Strober W, eds. Unit 3.16

Takayama H, Sitkovsky MV (1987): Antigen receptor-regulated exocytosis in cytotoxic T lymphocytes. *J Exp Med* 166: 725–743

49
Measurement of Cytolysin Hemolytic Activity

Pierre Henkart

Purpose

Determine the lytic functional activity of cytolysin (perforin) in granule and cell extracts. Red cell lysis is measured by optical absorbance of supernatant hemoglobin.

Reagents

1. Dilution buffer [0.2% polyethylene glycol (PEG) 6000 in phosphate-buffered saline (PBS)]. Use a PBS containing no calcium (not Dulbecco's) and dissolve PEG 6000 (Sigma) to 0.2% (w/v). The PEG helps prevent adsorption of highly diluted protein to the plastic surface.
2. Red blood cell (RBC) suspension. Make up a 0.2% (v/v) suspension of washed human or sheep red cells in balanced salt solution containing 0.01 M N-2-hydroxyethylpiperazine-N'-ethanesulfonic acid (HEPES), pH 7.4, and 2 mg/ml bovine serum albumin (BSA).

Procedure

1. In a U-bottom microtiter plate, add 100 μl of dilution buffer to a row of wells for each sample. To the first well, add sample and buffer to a volume of 200 μl. Mix and carry out twofold serial dilutions of the samples so that the volumes become 100 μl. Include control wells with dilution buffer alone and with 0.1% Triton X-100 as the total.
2. Add 100 ml of RBC suspension to each well. Incubate plate at 37°C for 10 min. Spin plate at 1500 rpm, 5 min. Remove 150 μl of supernatant and transfer to a flat-bottomed enzyme-linked immunosorbent assay (ELISA)-type microtiter plate, avoiding bubbles. Read absorbance of wells at 420 nm in an ELISA reader.
3. Calculate the dilution of cytolysin giving 50% lysis graphically or by means of a program. We have defined 1 unit/ml as that final concentration in the assay well giving 50% lysis; thus the sample activity in units/ml is the reciprocal of this dilution. We have found a data logger on the ELISA reader and the use of spreadsheets for calculations very useful.

Cautions

1. Many substances inhibit cytolysin activity, including those found in cell extracts (e.g., membranes of any kind) and in serum (lipoproteins). Thus, if you fail to find activity, cytolysin could be present but its activity masked by inhibitors.
2. Make sure that the cytolysin sample is not exposed to calcium before the assay. All extraction and granule purification steps should be carried out in calcium-free buffers containing 1 mM EGTA. Cytolysin is a potent mediator which is dependent on protein conformation and should be kept cold until assayed. Preparations can be frozen for periods of weeks.

References

Henkart PA, Millard PJ, Reynolds CW, Henkart MP (1984): Cytolytic activity of purified cytoplasmic granules from cytotoxic rat LGL tumors. *J Exp Med* 160: 75

Millard PJ, Henkart MP, Reynolds CW, Henkart PA (1984): Purification and properties of cytoplasmic granules from cytotoxic rat LGL tumors. *J Immunol* 132: 3197

Yue CC, Reynolds CW, Henkart PA (1987): Inhibition of cytolysin activity in granules of large granular lymphocytes by lipids: Evidence for a membrane insertion mechanism of lysis. *Mol Immunol* 24: 647

50
SPDP Crosslinking of Antibodies to Form Heteroconjugates Mediating Redirected Cytotoxicity

David M. Segal

Brief Description

Succinimidyl-3-(2-pyridyldithiol)-propionate (SPDP) is used to heterocrosslink two different antibodies or their Fab fragments. Crosslinking is through a disulfide bond, and is therefore reducible. Variations on this procedure crosslink antibodies via nonreducible thioether linkages.

Reagents and Sources

SPDP (Pierce or Pharmacia) $M_r = 312$
Absolute ethanol
PD10 (disposable Sephadex G25) column (Pharmacia)
Cytochrome C (Sigma)
Dithiothreitol (DTT) (Sigma)
Iodoacetamide (Sigma)
1 M NaCl, 1 M sodium acetate, pH 4.5 (reducing buffer)
Borate-buffered saline (0.15 M NaCl, 0.01 M sodium borate, pH 8.5) (BBS)
A 1.6 × 90 cm Ultrogel AcA 34 column (IBF Biotechnics, Savage, MD) or an FPLC unit equipped with a Superose 12 HR 10/30 column (Pharmacia)
CX10 immersible membranes (Millipore) or Centricon 10 Microconcentrators (Amicon) for concentrating samples

Method

Start with pure unaggregated antibodies at 10 mg/ml or higher, in BBS. Use equal amounts of each antibody; 5 mg of each is convenient. Dissolve a weighed amount of solid SPDP in absolute ethanol at about 2 mg/ml. Make sure it is totally dissolved, and add a 4-fold molar excess of SPDP to each antibody (42 μg of SPDP for 5 mg of intact antibody). For Fab fragments, use a 3-fold molar excess. Allow the SPDP to react with the antibodies for about half an hour at room temperature. During this time, equilibrate a PD10 column with BBS.

Take one of the antibodies and add 0.1 volume of reducing buffer. Then add 1–2 mg DTT followed by enough cytochrome C to give a strong red color. The amounts of DTT and cytochrome C are not crucial. When the cytochrome C and DTT are dissolved, pass the sample over the PD10 column, eluting in BBS, and collect the red peak in a minimum volume. Immediately add the reduced antibody (red peak) to the non-reduced, SPDP-modified second antibody. Incubate for about 4 hr at 37°C. During this time the two proteins are crosslinking by disulfide exchange. At higher protein concentrations, the crosslinking reaction will be more rapid and will give higher-molecular-weight aggregates and higher yields of crosslinked material. Therefore try to keep the protein concentration as high as possible.

After incubation, add 1–2 mg of iodoacetamide to the crosslinked protein. Apply the sample to the AcA 34 column. Separate and pool polymerized and monomeric material. Concentrate each fraction with a CX10 immersible membrane. Determine volumes and concentrations of each sample. A reasonable recovery after concentration is 50% of protein applied to the column, with one-half to two-thirds being crosslinked material. Store samples in BBS to prevent bacterial growth.

Alternative Procedures

1. To produce nonreducible thioether bonds, react one antibody with SPDP, as described above, and the second antibody with an active ester maleimide compound (at about a 4-fold molar excess) such as *m*-maleimidobenzoyl-*N*-hydroxysuccinimide ester (MBS, Pierce). Reduce the SPDP-antibody and mix with the MBS-antibody, as above. The rest of the procedure is the same.

2. After incubation of the two antibodies (crosslinking step) and addition of iodoacetamide, concentrate material to about 0.2–0.4 ml in a Centricon 10 microconcentrator. Then fractionate on a Superose 12 FPLC column in BBS. Collect 1-ml fractions and save the crosslinked material without further concentration.

Cautions

Make sure the starting protein is pure and in BBS. Amine contaminants will interfere with the SPDP coupling reaction. Try to keep the protein concentrations high, especially at the crosslinking step. Store the proteins in BBS. If the crosslinked antibodies are stored at relatively high concentrations, they can be diluted directly into medium and used without dialysis.

Reference

Titus JA, Garrido MA, Hecht TT, Winkler DF, Wunderlich JR, Segal DM (1987): Human T cells targeted with anit-T3 crosslinked to anti-tumor antibody prevent tumor growth in nude mice. *J Immunol* 138: 4018–4022

51
Derivatization of Cells with Antibody

Anna Ratner and William R. Clark

This technique is very useful for redirected lysis of target cells which are otherwise not recognized by a particular CTL. Target cells are derivatized with antibody specific either for the α- or β-chain of the T-cell receptor (TCR), or for a member of the CD-3 complex of TCR-associated proteins. Useful control antibodies are class I or CD-8, neither of which should induce target cell lysis.

1. Make up the heterobifunctional reagent *N*-succinimidyl 3-(2-pyridyldithio)propionate (SPDP) to 10 mM in absolute ethanol. It is stable in that form for at least 2 weeks, stored at 4°C.
2. React the antibody (1–2 ml) with about a 20-fold molar excess of SPDP for 1 hr at room temperature. If a precipitate forms, the amount of SPDP added needs to be lowered. Dialyze against phosphate-buffered saline (PBS) overnight. The product, SPDP-Ab, can be stored at −70°C indefinitely.
3. Pretreat target cells with 10–20 mM dithiothreitol (DTT) in PBS for 1 hr at room temperature to generate free sulfhydryl groups at the cell surface. Wash the targets three times with PBS.
4. Add 100–200 μl of the modified antibody (SPDP-Ab) to a pellet of the treated target cells, resuspend, and allow to react for 1 hr at room temperature with occasional shaking.
5. Wash the derivatized targets three times with PBS and use immediately.

If sheep red blood cells (SRBC) are to be derivatized, additional sulfhydryl groups must be introduced onto the SRBC surface prior to reduction with DTT. To do so, react 1 ml 50% SRBC suspension [washed in BBS, pH 7.0 (see below)] with 182 mg dithidiglycolate, dissolved first in 1 ml of 2 M NaOH and then diluted to 25 ml with BBS, pH 6.0 (see below). After thorough mixing, 250 mg 1-ethyl-3--(3-dimethylaminopropyl)carbodiimide (EDCI) in 2.5 ml water is added, and the reaction is allowed to proceed at room temperature for 30 min. The SRBC are then washed 4× with PBS, and resuspended in 25 ml. Total available disulfide groups are reduced by addition of 1 ml 1 M DTT (dissolved in PBS) and incubation for 1 hr at room temperature on a rotating mixer. The SRBC are then washed 4× again with PBS, and resuspended in about 1 ml to make a 50% suspension to which is added the modified antibody (SPDP-Ab), usually 200 μl SPDP-Ab to 100 μl of 50% modified SRBC in small Eppendorf tubes. This is allowed to react overnight at room temperature on a rotating mixer. Cells are washed 4× with PBS before use.

BBS: 0.167 M H_3BO_3
0.137 M NaCl
0.022 M NaOH
dilute 1:12 in 0.147 M NaCl
pH to 6.0 or 7.0 with 1 M HCl

This technique is adopted from Jou and Bankert (1981): *Proc Natl Acad Sci USA* 78: 2493–2496

52
Mixed Lymphocyte Culture for the Generation of Allospecific CTL

Eda T. Bloom

Introduction

The mixed lymphocyte culture (MLC) is an effective means by which to generate allospecific cytolytic T lymphocytes (CTL). Detailed methods and discussions of methodological considerations have been published for setting up MLC for use in assessing the proliferative repsonse to alloantigenic stimulation (Bradley, 1980; Kruisbeek and Shevach, 1991) and for inducing CTL (Grabstein, 1980; Wunderlich and Shearer, 1991). The preparation of cultures for the generation of CTL is virtually identical to that used to assess proliferative responses. The exception is that cells recovered after culture are assessed for lytic activity, and therefore MLC used to generate CTL often contain larger numbers of cells than those used to assess proliferative responses.

Virtually every parameter of the preparation of MLC is amenable to a variety of modifications that still enable recovery of highly specific and active CTL. Therefore, the prescribing of a precise protocol would be somewhat presumptuous and misleading. With this in mind, I will describe the procedure that we currently use and include brief discussions of a few of the possible permutations that can be used successfully.

A Method for Stimulation of Murine Alloimmune CTL by MLC In Vitro

A.1. Materials for Culture

a. Responder cells: Lymphocytes from spleen, lymph node, or thymus of nonimmunized mice, prepared by standard methods. Purified T cells (e.g., by negative selection with antibody or nylon wool) or, in some cases, T-cell subpopulations can also be used. After preparation, and whenever possible, cells should be maintained on ice until use in culture.

b. Stimulator cells: Allogeneic spleen cells containing all leukocyte populations, which have been irradiated or treated with mitomycin C (MMC), or depleted of T cells. We generally use 2000 rad for irradiation, although other investigators have used up to 10,000 rad. MMC treatment can be accomplished as described (Kruisbeek and Shevach, 1991; Wunderlich and Shearer, 1991), by treating cells at 5×10^7/ml with a final concentration of 50 μg/ml MMC in the dark for 20 min at 37°C. Cells should be washed thoroughly and maintained on ice until use in culture.

c. Multiple-well tissue-culture plates or tissue-culture flasks: The vessel size is selected on the basis of the volume of cells to be cultured, which is selected based on the number of CTL needed. We have used 6 well (5–10 ml cultures) to 24 well (~1 ml cultures) plates, and 25 cm² flasks (8–15 ml cultures) to 175 cm² flasks (50+ ml cultures) with equivalent results. (Some investigators have used 96 well plates to generate CTL, and have assessed lytic activity by adding target cells directly to the same plates, without adjusting responder cell concentration prior to assay.)

d. Cell-culture medium: RPMI-1640 containing 10% fetal calf serum (heat-inactivated), 2 mM L-glutamine, 50 μg/ml gentamycin sulfate, and 50 μM 2-mercaptoethanol (2-ME). Modifications of RPMI [e.g., including nonessential amino acids, pyruvate, different antibiotics, N-2-hydroxyethylpiperazine-N'-ethanesulfonic acid (HEPES) buffer, etc.] or other media may also be used.

A.2. Materials for Cell Harvest

a. Centrifuge tubes: 15 ml or 50-ml sizes, depending upon the size of the MLC, and various-sized pipets.

b. Ficoll-hypaque (F/H) solution (optional): Standard F/H at a density of 1.077 gm/ml can be used. Although this is generally used for the isolation of human mononuclear cells, it is suitable for removal of dead mouse cells as well. If F/H is used, phosphate-buffered saline (PBS) or Hanks' balanced salt solution (HBSS) will also be needed.

c. Medium: RPMI-1640 or other medium with additives as discussed above, except without 2-ME.

B.1. Procedure: Setup of MLC

a. Adjust responder and stimulator cells to 1×10^7/ml in medium containing 2-ME (A.1.c above), and mix in equal proportions in a tissue culture vessel of suitable size. The ratio of responder to stimulator cells, and the concentrations of each cell type to be used, exhibit variation and often flexibility, both of which may depend upon the antigenic system under investigation. Therefore, ideally the optimal conditions should be determined for each system using a checkerboard titration design, using a general range of initial concentrations of $1-10 \times 10^6$ cells/ml. It should be noted that for certain investigations, e.g., to maximize differences, use of suboptimal conditions may be desirable.

b. Incubate culture for 5 days at 37°C in a humidified atmosphere containing 5% CO_2, under sterile conditions but allowing air exchange. All of these conditions may be altered to achieve appropriate results. For example, depending upon the antigen difference, shorter or longer intervals may be more desirable. Some investigators find 39°C a preferred temperature, and the CO_2 concentration may need to be modified depending upon the buffer system in the culture medium.

B.2. Procedure: Cell Harvest

a. Harvest nonadherent cells from MLC using appropriate-size pipet to transfer cells into an appropriate-size centrifuge tube. Pellet cells at $200 \times g$ for 10 min.

b. If high viability (85–95%) of recovered cells is desirable, centrifuge cells, prior to pelleting, over a F/H cushion (12 ml for a 50-ml tube, 4 ml for a 15-ml tube) at $400 \times g$ for 15 min. The viable cells are recovered from the interface and washed twice more in PBS or HBSS. It should be noted that high viability is not required for the demonstration of lytic activity in a ^{51}Cr-release assay.

c. The final pellet is suspended in a volume of medium (section A.2.c) depending upon the original culture volume. After counting and assessing the cell viability, cells are suspended at concentrations of $\sim 1-5 \times 10^6$ viable cells/ml and reserved on ice for assay by ^{51}Cr release or other methods. (If 96-well plates are used for MLC, cells are generally not collected, and cell concentrations are not readjusted prior to assay of lytic activity. In this case, presentation and interpretation of ^{51}Cr-release data should make it clear that effector cell concentrations and resultant effector/target ratios were estimated from the original cell concentrations used to establish the MLC.)

Conclusion

The method described above provides a framework for the generation of murine CTL in MLC. Some modifications may need to be made to suit the system at hand, or to use human cells. Additional discussions of problems, as well as of related issues such as antigenic modification of stimulator cells, polyclonal activation, and redirected lysis can be found elsewhere (Wunderlich and Shearer, 1991; Grabstein, 1980).

References

Bradley LM (1980): Cell proliferation. In: *Selected Methods in Cellular Immunology*, Mishell BB, Shiigi SM, eds. New York: W.H. Freeman, pp. 153–172

Grabstein K (1980): Cell-mediated cytolytic responses. In: *Selected Methods in Cellular Immunology*, Mishell BB, Shiigi SM, eds. New York: W.H. Freeman pp. 124–137

Kruisbeek AM, Shevach E (1991): Proliferative assays for T cell function. In: *Current Protocols in Immunology*, Coligan JE, Kruisbeek AM, Margulies DH, Shevach EM, Strober W, eds. New York: Greene Publishing Associates and Wiley-Interscience. pp. 3.12.1–3.12.14

Wunderlich J, Shearer G (1991): Induction and measurement of cytotoxic T lymphocyte activity. In: *Current Protocols in Immunology*, Coligan JE, Kruisbeek AM, Margulies DH, Shevach EM, Strober W, eds. New York: Greene Publishing Associates and Wiley-Interscience, pp. 3.11.1–3.11.15

53
Generation of CD4+ and CD8+ Antiinfluenza CTL and Assay of In Vitro Cytotoxicity

Vivian Lam Braciale

Generation of Antiinfluenza-Specific Cytolytic T Lymphocytes (CTL)

Mice that are primed *in vivo* at 6–12 weeks of age by immunizing intravenously with 100–300 hemagglutinating units (HAU) of virus in allantoic fluid are used as spleen donors three or more weeks afterwards. Immune spleen cells (100×10^6) are stimulated *in vitro* by co-culture with 20×10^6 virus-infected, irradiated (2000 rad) naive splenocytes in 75-cm^2 flasks containing 40 ml of minimal essential medium (MEM) (GIBCO) plus 10% fetal bovine serum (FBS) and 5×10^{-5}M 2-mercaptoethanol (2-ME) (Braciale et al., 1982).

Isolation of Antiinfluenza CTL Clones

These *in vitro*-stimulated cells (day 7) or immune splenocytes obtained directly from immune mice are centrifuged over Isopaque-Ficoll (Davidson and Parish, 1975) to remove red cells and dead cells. They are plated at 10^2–10^4 cells/well together with 10^6 infected, irradiated sygeneic stimulators in 96-well flat-bottomed plates. The medium is Iscove's modified Dulbecco's MEM (GIBCO 430–2200) plus 5×10^{-5}M 2-ME, 10% FBS, and 10% crude factor containing interleukin-2 (IL-2) [rat concanavalin A (Con A) supernatant prepared according to Lafferty and Woolnough (1977)]. Positive cultures, as visualized by inverted scope, are propagated and expanded up to 5-ml cultures in 6-well plates with weekly restimulation. Both CD8+ and CD4+ CTL can be obtained in this manner, but the CD8+ clones are most frequent. In order to increase the frequency of CD4+ CTL, the *in vitro* stimulus should be irradiated, syngeneic splenocytes plus 100 HAU ultraviolet (UV)-inactivated virus (Morrison et al., 1988).

In Vitro Assay of Cytolysis by Antiviral CTL

CD8+ CTL (days 4–7 after antigenic stimulation) are assayed on influenza-infected target cells, e.g., P815 (H-2^d), EL-4 (H-2^b), L929 (H-2^k), class I major histocompatibility complex (MHC)-transfected L cells, A20 B lymphoma (H-2^d). Target cells are infected with influenza virus and ^{51}Cr-labeled for 1–2 hr at 37°C or overnight at room temperature. The targets are washed and incubated with cloned CTL at various effector:target ratios for 6 hr at 37°C. Usually CTL are plated with 10^4 target cells per well in 96-well plates, and 100 µl of supernatant is harvested and counted on a gamma counter. Alternatively, target cells can be labeled with iododeoxyuridine-125 ^{125}IUdR-infected, and incubated with CTL for 1 to 4 hr at 37°C. 1% Triton (final concentration) is added per well at the end of the assay prior to centrifugation of the plate and harvesting of 100 µl of supernant.

CD4+ CTL are assayed on class II MHC-expressing cells, e.g., A20 B lymphoma (H-2^d), class II MHC-transfected L cells, Epstein-Barr virus (EBV)-transformed autologous B lymphoblastoid cell lines (for human CTL). ^{51}Cr-labeled target cells are infected with virus or treated with UV-inactivated virus. Other conditions of assay are identical to that for the CD8+ CTL.

Assays using peptide-treated target cells are as described in Braciale et al. (1987). Peptide at various doses is added to assay wells for the entire assay. For studies involving peptide competion as a functional measure of peptide MHC binding, the competitor target cell is first preincubated with the competitor peptide for 15 min. The cells are washed twice, then the sensitizing peptide (peptide which is known to be recognized by the CTL clone at a known concentration) is added at the start of the assay.

References

Braciale TJ, Braciale VL, Andrew ME (1982): Cloned continuous lines of H-2 restricted influenza virus-specific CTL: Probes of T lymphocyte specificity and heterogeneity. In: *Isolation, Characterization and Utilization of T Lymphocytes*, Fathman CG, Fitch FW, eds. New York: Academic Press

Braciale TJ, Braciale VL, Winkler M, Stroynowski I, Hood L, Sambrook J, Gething M-J (1987): On the role of the transmembrane anchor sequence of influenza hemagglutinin in target cell recognition by class I MHC-restricted, hemagglutinin-specific cytolytic T lymphocytes. *J Exp Med* 166: 678–692

Davidson WF, Parish CR (1975): A procedure for removing red cells and dead cells from lymphoid cell suspensions. *J Immunol Methods* 7: 291

Lafferty KJ, Woolnough J (1977): The origin and mechanism of the allograft reaction. *Immunol Rev* 35: 231

Morrison LA, Braciale VL, Braciale TJ (1988): Antigen form influences induction and frequency of influenza specific class I and class II MHC-restricted cytolytic T lymphocytes. *J Immunol* 141: 363–368

54
Generation of Antigen-Specific Murine CTL Under Weakly Immunogenic Conditions

Sergey G. Apasov

The two-step generation of allospecific cytotoxic T cell lymphocytes (CTL) is an effective method to obtain highly active antigen-specific CTL. The first step involves the *in vivo* induction of CTL precursors (pCTL) followed by *in vitro* incubation of these precursors as the second step for the development of pCTL into CTL. The method was introduced (Wagner et al., 1976; Starzinski-Powitz et al., 1976) to produce CTL recognizing the hapten-conjugated syngeneic structures of the major histocompatibility complex (MHC). Later this approach was successfully used to develop CTL recognizing minor alloantigens (Czitrom and Gascoigne, 1983) or to generate CTL in the system where the donor and recipient differ by point mutations in MHC class I molecules (Brondz et al., 1987).

This procedure facilitates the study of the immune response against weakly immunogenic antigens by sensitization of T cells *in vivo* to obtain the antigen-specific MHC-restricted CTL when sensitization *in vitro* by mixed lymphocyte culture (MLC) or *in vivo* immunizations (intraperitoneal or subcutaneous) has failed. In addition, this approach permits investigation of the mechanism of antigen-induced CTL differentiation and the role of different lymphokines, since the development of noncytotoxic precursors into CTL takes place *in vitro* under easily controlled conditions. The establishment of long-term cytotoxic cell lines and clones from CTL generated by this procedure and subsequent stimulation with antigen and interleukin-2 (IL-2) is another convenient application of the method.

Induction of pCTL *In Vivo*

Induction of pCTL starts by injecting the footpads of mice with a suspension of cells bearing the antigen of interest. The suspension of donor splenic cells for injection (e.g., allogeneic with minor antigens, syngeneic hapten-modified, etc.) has to be prepared in phosphate-buffered saline (PBS) or in a serum-free medium. After removing red blood cells from the suspension with red blood cell-lysing buffer, $1-2 \times 10^7$ irradiated donor cells are injected into both hind footpads of the responder animals in a total volume of $40\,\mu$l under anesthesia. Groups of 4–6 mice are used per experiment. The use of nonirradiated donor cells is recommended to increase the immune response when the chosen antigen is of very low immunogenicity. Secondary immunization with irradiated cells 10–14 days after primary immunization is another way to enhance the immune response.

Four days after the last immunization, the animals are sacrificed and the popliteal and inguinal lymph nodes are removed using sterile technique. A lymph node suspension is prepared by standard procedures and washed twice by centrifugation. Usually $1.5-2 \times 10^7$ lymph node cells can be recovered from one mouse by this way.

Development of CTL *In Vitro*

Pooled lymph node cells from the immunized animals are resuspended in complete growth medium (RPMI 1640 medium supplemented with 5% FCS and other standard additives). At this step CD8$^+$ cells have no detectable cytotoxic activity. The development of active CTL requires 2 to 3 days of cell culture at 37°C and 5–7% CO_2. The culture can be maintained in 24-well plastic plates ($2.5-3 \times 10^6$ cells/ml and 1.5–2 ml per well), in 96-well plates ($2-4 \times 10^6$ cells/ml; 0.2 ml per well), or in upright 25-cm² culture flasks (1–2 × 10⁶ cells/ml in 13–18 ml per flask).

After 2 or 3 days the cultured cells are collected and cytotoxic activity can be detected in a standard ^{51}Cr-release assay. Usually about 60% of the initial amount of the cells can be recovered from 3 days' culture.

In Vitro Culture of Purified CD8⁺ pCTL

If necessary, CD8⁺ pCTL can be isolated from the total lymph node cell population immediately after the sensitization set by the removal of B cells and CD4⁺ cells using standard cell separation techniques such as panning or complement lysis. Development of the purified pCTL into highly cytotoxic CTL takes place using the same culture conditions as described above with the addition of IL-2 and/or other lymphokines. We recommend adding 1–5% culture supernatant of concanavalin A (Con A)-stimulated rat splenocytes as a source of lymphokines. Recombinant IL-2 is also sufficient to develop CTL; however, its use may result in the development of background nonspecific cytotoxic activity.

Materials

Ethanol 70%
Sterile (PBS), pH 7.2–7.4
Sterile tuberculin syringe with fine hypodermic needles
RPMI 1640, heat inactivated FCS (Gibco, Biofluids etc.), supplement for full growth medium
Plastic: Petri tissue culture dishes, 24- or 96-well plates, flat or round bottom, 25-cm² vented culture flask
Mice: 4–8 weeks old

References

Brondz BD, Khan ES, Chervonsky AV, Isacova VR, Apasov SG, Blandova ZK (1987): Differential genetic requirement for *in vivo* and *in vitro* induction of T-killer and T-suppressor cells in the mutant H-2K^b system and the cross-reactivity of the T-killer clones. *Exp Clin Immunogenet* 4: 211–221

Czitrom AA, Gascoigne NRJ (1983): Primary T-cell responses to minor alloantigens I. Characterization of cytotoxic effector cells generated from regional lymph nodes after immunization in the footpad. *Immunology* 50: 121–129

Gascoigne NRJ, Crispe IN (1984): Differential expression of the repertoire in *in vitro* and *in vivo* responses to minor alloantigens. *Immunogenetics* 19: 511–517

Starzinski-Powitz A, Pfizenmaier K, Rollingoff M, Wagner H (1976): In vivo sensitization of T cells to hapten-conjugated syngeneic structures of major histocompatibility complex. I. Effect of *in vitro* culture upon generation of cytotoxic T lymphocytes. *Eur J Immunol* 6: 799–805

Wagner H, Starzinski-Powitz A, Pfizenmaier K, Rollingoff M (1976): Regulation of T cell-mediated cytotoxic allograft responses. II. Evidence for antigen-specific suppressor T cells. *Eur J Immunol* 6: 873–878

55
Commercial Liposomes and Electroporation Can Deliver Soluble Antigen for Class I Presentation in CTL Generation

Weisan Chen and James McCluskey

Major histocompatibility complex (MHC) class I antigens present processed antigens derived almost exclusively from the cellular cytoplasm (see Chapter 5, this volume). Accordingly, extracellular soluble antigens are unusual targets for class I-restricted cytotoxic T lymphocytes (CTL) and little is known regarding mechanisms of self-tolerance to soluble antigens in the CD8$^+$ T-cell compartment. In addition, vaccines designed to elicit CD8$^+$ class I-restricted CTL usually require immunization with live attenuated virus, rather than with purified antigens. Such vaccines are contraindicated in patients who are immune-suppressed or who have a history of life-threatening hypersensitivity. Moreover, vaccination with attenuated viruses can cause postvaccine illness, which is undesirable in large vaccination programs. Therefore, strategies which might generate class I-restricted CTL by immunization with soluble (recombinant) protein would be very valuable in relation to practical vaccine development and in probing the scope of self-tolerance by CD8$^+$ T lymphocytes.

Soluble protein can be introduced into the cytoplasm of living cells using osmotic loading methods in which cells are incubated in high concentrations of protein and then transferred to a relatively hypotonic environment (Moore et al., 1988). Following hypotonic challenge, small pinosomes containing the soluble protein are thought to burst or release their contents into the cytoplasm, whereupon they become accessible to the class I presentation pathway. However, this method requires quite high concentrations of soluble antigen, and in our hands tends to give variable levels of class I-antigen loading.

For this reason, we have explored two new methods of introducing soluble antigens into the class I pathway: electroporation and fusion with antigen-loaded artificial lipid vesicles (liposomes). Electroporation has been widely used as a method for delivering DNA into cultured cells for transfection studies; however, this method is also suitable for introducing proteins into live cells (Mir et al., 1988; Lambert et al., 1990). The electrical disturbance is believed to transiently reorient components of the plasma membrane, creating nanometer-size pores (electropores) through which exogenous molecules can enter the cytoplasm (Zimmerman et al., 1982).

Recently, several groups have described the use of cationic or pH-sensitive liposomes as a vehicle for delivering soluble antigen into the class I pathway (Monte and Szoka, 1989; Harding et al., 1991, Reddy et al., 1991; Nair et al., 1992). Preparing these liposomes from mixtures of different lipids can be troublesome, however, and so we describe here a method of sensitizing live cells for class I-restricted presentation of soluble antigen delivered by either commercially available liposomes or by electroporation.

Methods

For development of these techniques we used the murine OVA/K^b antigen combination in which the hen ovalbumin (OVA) peptide 257–264 forms a minimal determinant presented by H-2K^b molecules to CD8$^+$ T cells in H-2^b mice. T hybridomas recognizing the OVA/K^b complex were used to read out antigen presentation *in vitro* on a number of combinations of antigen-presenting cells (APC) and antigen-delivery conditions.

Antigen-Presenting Cells

The thymoma cell line EL-4 (H-2^b) (class II

negative), mouse L cells ($H\text{-}2^k$) transfected with the gene encoding $H\text{-}2K^b$, and C57BL/6 spleen cells ($H\text{-}2^b$) were tested as APC. Cell culture was carried out in DME containing 10% fetal calf serum (FCS), antibiotics, glutamine, and 5×10^{-5} M 2-mercaptoethanol. Spleen cells were prepared freshly and irradiated with 3000 rad prior to antigen loading.

Antigen

OVA (grade VI, Sigma, St. Louis, MO) was freshly dissolved in either phosphate-buffered saline (PBS) containing 10 mM $MgCl_2$ for electroporation (electroporation buffer) or DME without FCS for liposome loading.

Antigen Loading

Electroporation

Fresh spleen cells were prepared as a single-cell suspension which was washed once in DME lacking any FCS and then resuspended in 0.4 ml of electroporation buffer containing various concentrations of soluble antigen (OVA). The range of OVA concentration that gave linear T-cell responses following electroporation was 0.1–5 mg/ml. Cells were incubated on ice in a volume of 0.4–0.8 ml for 10 min in a BioRad electroporation curvette and then electroporated using a BioRad Gene Pulser followed by a further 10-min incubation on ice. Cells were washed once prior to testing for in vitro antigen presentation. In general, optimal conditions for electroporation varied between cell types. Prior to selecting the precise electroporation conditions, cell types were tested by trypan blue exclusion in a "killing curve" experiment to establish conditions that resulted in 50% cell death within 4–6 hr of electroporation. Optimum electroporation conditions were then determined empirically around the LD_{50} settings. The electroporation optimum for the cells described here was 0.45 kV/ 250–500 μF.

Liposome Antigen Loading

Commercial liposomes capable of delivering soluble antigen to class I molecules include DOTAP {N-[1-(2,3-dioleoyloxy)propyl)]N,N,N-trimethylammoniummethylsulfate} (Boehringer-Mannheim, Biochemica, Cat. No. 1202375); Lipofectin [1:1 mixture of N-[1-(2,3-dioleoyloxy)propy]-N,N,N-trimethylammonium choloride (DOTMA) and dioleoyl phosphatidylethanolamine] (BRL, Cat. No. 8292SA); and Transfectam (cationic lipid with lipospermine head groups) (Promega, Cat. No. E1231). For antigen loading, liposomes (100–200 μg/ml) are incubated with antigen (100 μg/ml–5 mg/ml) at room temperature for 10 min in medium lacking any FCS. This mixture is then added directly to the APC pellet (5×10^6 spleen cells; or $1\text{–}2 \times 10^6$ L cell transfectants or EL-4 cells) for 10 min at room temperature. We have observed that washing of liposome/Ag-pulsed cells prior to setup of antigen presentation assays results in a greater response than that observed when liposomes are left in the coculture. DOTAP gave better results than Lipofectin. It also seems likely that for some antigens the liposome–antigen mixture may need to be incubated for longer than 10 min, and as long as overnight.

CTL Priming

For priming CTL responses *in vivo*, we have used the above electroporation and liposome protocols followed by reinjection of 5×10^7 antigen-pulsed spleen cells into the tail vein of a syngeneic mouse. In the $H\text{-}2^b$/OVA system a single *in vivo* priming followed by a conventional *in vitro* restimulation is sufficient to raise CTL in C57BL/6 mice. Additional *in vivo* priming may be necessary in other antigen systems. *In vivo* priming directly subcutaneously with cell-free liposome–antigen mixtures may be sufficient for priming CTL responses to certain antigens.

Acknowledgements

We are grateful to Dr. F Carbone (Monash University, Melbourne, Australia) for the generous gift of T hybridomas and antigens. This work was supported by the NHMRC Australia, Flinders Medical Research Foundation, Arthritis Foundation of Australia and Australian Red Cross Society, S.A. Division.

References

Harding CV, Collins DS, Kanagawa O, Unanue ER (1991): Liposome-encapsulated antigens engender lysosomal processing for class II presentation and cytosolic processing for class I presentation. *J Immunol* 147: 2860–2863

Lambert H, Pankov R, Gauthier J, Hancock R (1990): Electroporation-mediated uptake of proteins into mammalian cells. *Biochem Cell Biol* 68: 729–734

Mir LM, Banoun H, Paoletti C (1988): Introduction of definite amounts of nonpermeant molecules into living cells after electropermeabilization: Direct access to the cytosol. *Exp Cell Res* 175: 15–25

Monte PD, Szoka JR (1989): Effect of liposome encapsulation on antigen presentation *in vitro. J Immunol* 142: 1437–1443

Moore MW, Carbone FR, Bevan MJ (1988): Introduction of soluble protein into the class I pathway of antigen processing and presentation. *Cell* 54: 777–785

Nair S, Zhou F, Huang L, Reddy R, Huang L, Rouse BT (1992): Soluble proteins delivered to dendritic cells via pH-sensitive liposomes induce primary cytotoxic T lymphocyte responses *in vitro. J Exp Med* 175: 609–612

Reddy R, Zhou F, Huang L, Carbone F, Bevan M, Rouse BT (1991): pH sensitive liposomes provide an efficient means of sensitizing target cells to class I restricted CTL recognition of a soluble protein. *J Immunol Methods* 141: 157–163

Zimmerman U (1982): Electric field mediated fusion and related phenomena. *Biochim Biophys Acta* 694: 227–277

56
Stimulation of CTL on Antibody-Coated Plates

Anna Ratner and William R. Clark

This technique is useful for a number of purposes. Antibodies to either the T-cell receptor (TCR) or to the associated CD-3 molecules are effective, and antibodies such as class I or CD-8 are good controls. We have used this method quite successfully to induce CTL degranulation. It also causes cells to release tumor necrosis factor-α (TNF-α), if they produce it. CTLs stimulated by TCR or CD-3 antibody on plates also show greatly enhanced lysis of specific and nonspecific target cells.

For best results we have found it useful to first attach a primary antibody to the surface of the plate, and then add the specific antibody. For example, if the CTL are to be stimulated by an antibody to the CD-3 molecule, such as 2C11, which is a hamster–mouse hybridoma, we would coat the plate first with a goat anti-hamster antibody ("primary"), and then add the 2C11 antibody ("secondary").

1. First dilute the primary antibody (e.g., goat anti-hamster), to 1 μg/ml in phosphate-buffered (PBS). Coat a flat-bottom 96-well (100 μl/well) ELISA plate with the primary antibody by incubation overnight at 4°C.

2. Wash the plate with PBS/2% calf serum (CS) three times.

3. Dilute the secondary antibody, 2C11, to 10 μg/ml in PBS/2%CS. Add the secondary antibody to the plate (100 μl/well), and incubate 4 hr at 37°C.

4. Wash the plate with PBS/2% CS three times.

5. Add the effector cells at 2×10^6/ml.

6. Incubate for 3–18 hr, as needed to harvest supernates, or add labeled target cells directly to the plate for an "enhanced" ^{51}Cr-release assay.

57
CTL Recognition of Purified MHC Antigens and Other Cell Surface Ligands

Matthew F. Mescher, Paul Champoux, and Kevin P. Kane

T-cell receptor binding to peptide antigen–major histocompatibility complex (MHC) complexes or MHC alloantigen clearly plays the central role in activating T-cell responses. However, as discussed elsewhere in this volume, a number of additional receptor–ligand interactions, including LFA-1/ICAM, CD8/class 1, and binding of VLA to fibronectin or laminins, can either be required to initiate response or contribute to increased response levels. Both the types and amounts (surface densities) of ligands present on a target cell can determine the levels of responses, and may also determine whether a particular response occurs (e.g., proliferation vs. production of cytokines).

Defining the roles of individual receptor–ligand pairs is difficult when antigen-bearing cells are used; the types and amounts of ligands cannot be varied in a controlled way, and there is the potential for contributions from additional, unknown receptors and ligands. In contrast, use of purified membrane protein ligands allows precise control of the interactions which can occur, and this approach has been effective for defining qualitative and quantitative contributions of various receptors to adhesion, transmembrane signal generation, and functional responses. In addition, this approach provides a means of studying the parameters that influence peptide binding to class I to form the antigenic complex and has recently been extended to manipulation of cytotoxic T lymphocyte (CTL) response levels in vivo.

Effective CTL interaction with an antigen-bearing surface requires that contact occur over a relatively large surface area. Although MHC proteins can readily be incorporated into unilamellar liposomes by detergent dialysis procedures, antigen on these small ($< 0.2 \mu$m) vesicles is very inefficient in activating responses. In contrast, the same antigen displayed on larger surfaces triggers T-cell response levels comparable to those elicited by intact target cells (Goldstein and Mescher, 1986). Incorporation of MHC proteins onto large surfaces for effective presentation to T cells has been accomplished by several different methods (Nakanishi et al., 1983; Brian and McConnell, 1984; Coeshott et al., 1986; Goldstein and Mescher, 1986; Quill and Schwartz, 1987; Kane et al., 1989a; Kozlowski et al., 1991). Below are summarized the methods we have used in our laboratory that have proven to be simple, reliable, and reproducible for studying normal and cloned CTL.

Immobilization in Microtiter Plate Wells

MHC proteins can be readily purified from detergent lysates of cells or isolated membranes by affinity chromatography on monoclonal antibody (mAb) columns, and procedures for this have been described in detail (Mescher et al., 1983; Parham, 1983). Detergent is included in the elution buffer to keep the protein in solution; we have routinely used deoxycholate (DOC), but other nondenaturing detergents can be used. Purifications typically yield a preparation having a protein concentration of 20 to 250 μg/ml in 0.5% DOC, 10 mM Tris, pH 8.2 with 20 to 150 mM NaCl, depending on the elution conditions required for the particular antigen. Steps for immobilization in microtiter wells for CTL functional assays or enzyme-linked immunosorbent assay (ELISA) determinations are:

Step 1. Affinity-purified MHC antigen is diluted in a tube to a final concentration of 2 μg/ml in Dulbecco's phosphate-buffered saline (D-PBS).

Step 2. Immediately after dilution and mixing,

0.1 ml of the antigen solution is placed into wells of Falcon microtest III flexible assay plates (Becton Dickinson, Oxnard, CA). When antigen titrations are done, appropriate dilutions are prepared and the desired amount of antigen in 0.1 ml added to wells. Alternatively, dilutions can be done in the wells. The desired volume of D-PBS is first put into the well and a volume of antigen solution then added so that the total volume is 0.1 ml.

Step 3. Plates are incubated at room temperature for 1.5 hr to allow binding of the antigen.

Step 4. Antigen solutions are removed and the wells washed twice with D-PBS (0.15 ml/wash). Then 0.15 ml of 1% bovine serum albumin in D-PBS (BSA/D-PBS) is added and the plates are incubated for 0.5 hr at room temperature to block unreacted sites on the plastic.

Step 5. The BSA/D-PBS solution is removed and the plates are washed four times with D-PBS (0.15 ml/wash).

The antigen-bearing wells are now ready for addition of CTL to assay for binding or response (Kane et al., 1989a, 1989b; Mescher et al., 1991). Parallel wells can be prepared and the immobilized antigen quantitated in a standard two-step ELISA using an appropriate first antibody followed by enzyme-linked second antibody. (Note: the same procedure described above can be used to monitor purification of MHC antigens by diluting individual fractions eluted from the affinity column, immobilizing, and doing ELISA or CTL response assays.)

MHC antigens and other membrane proteins can be immobilized on plastic in the presence of detergents at concentrations that completely inhibit binding of nonmembrane proteins (Kane et al., 1989a). Binding is comparable using Falcon microtest III (Becton Dickinson) or Immulon I or II (Dynatech Laboratories, Alexandria, VA) plates, and lower but still substantial using Corning "Cell Well" plates (Corning, NY). DOC does not affect the extent of binding at final concentrations in the well of up to 0.02% (but inhibits binding by about 50% at 0.09% final concentration). Binding is affected very little by octyl-β-glucoside concentrations up to 0.2%. In contrast, Triton X-100 is very inhibitory, with 50% inhibition of binding occurring at about 0.002% final concentration.

The amount of MHC protein immobilized in the well is a linear function of the amount added, up to about 0.2 μg/well, and binding saturates at higher input levels. Using the procedure described above, about one-third of the added protein binds, as estimated using radiolabeled class I.

Immobilization of class I at 0.04 to 0.05 μg/well yields a surface density comparable to that on normal spleen cells and stimulates optimal responses by allogeneic CTL. When D-PBS containing 2% fetal calf serum (FCS) is added to the wells and the plates stored at 4°C, immobilized class I is stable for at least several days as measured by ELISA or CTL response assays. The procedure described above has been used successfully to immobilize functionally active (by ELISA and CTL binding and/or response assays) murine class I and II MHC proteins, TLA, HLA-A2, and murine ICAM, indicating that this method will be generally useful for cell surface membrane proteins.

Peptide–MHC Antigenic Complex Formation

Recognition of purified class I by allogeneic CTL, at least in some cases, probably involves recognition of peptide antigens which remain bound during the purification (Heath et al., 1991). Exogenous peptide antigens can also be bound to purified, immobilized class I proteins to form antigenic complexes, as evidenced by CTL stimulation (Kane et al., 1989c; Kane et al., 1991). Complex formation is dependent on exogenous β_2-microglobulin (β_2m) (Kane et al., 1992). In the protocol below, this is provided by the FCS present during incubation with peptide. The appropriate class I protein is first immobilized as described above, and pulsed with peptide by the following steps:

Step 1. Peptide at an appropriate concentration in D-PBS containing 2% FCS is added to microtiter wells (0.1 ml/well) in which the class I restriction protein has been immobilized.

Step 2. The plate is covered to prevent evaporation and incubated overnight (12–16 hr) at 37°C.

Step 3. Unbound peptide is removed and the wells washed three times with BSA/D-PBS (0.15 ml/wash).

CTL can then be added to the wells and their binding and response measured. Conditions described above have been found to yield optimal antigenic complex formation using several class I/peptide combinations including K^d/NP(147–158R$^-$), D^b/NP(365–380), and K^b/OVA(253–276), with peptide concentrations ranging from 1 to 60 μM. Experiments have shown that antigenic complex formation requires the presence of exogenous β_2m, supplied by the FCS used in the above procedure (Kane et al., 1991). Either mouse or human sera can be substituted for the FCS.

In addition to providing free β_2m, the presence of serum increases the stability of the immobilized class I during prolonged incubation at 37°C. It appears likely that components in the serum block sites on the plastic that are not effectively blocked by BSA, and that lead to denaturation of the class I. Comparable stability is obtained if serum is replaced by BSA/D-PBS containing 50 μM dimyristoylphosphatidylcholine (DMPC). In this case, the requirement for exogenous β_2m can then be provided by addition of 2 μg/ml purified β_2m (Sigma Chem. Co.).

Coimmobilization of Antigen with Additional Ligands

The effects of additional surface interactions can be studied using microtiter wells in which both the MHC protein and an additional ligand(s) have been immobilized (Kane and Mescher, 1990). This can be done using a two-step procedure in which the first ligand is incubated in the well as described above, unbound protein is removed, and the wells are washed two times with D-PBS (0.15 ml/wash). The second ligand solution is then added and incubated for an additional 1.5 hr at room temperature, and the wells are washed and blocked with BSA as described above.

When the ligand to be coimmobilized with the MHC protein is not a membrane protein, it should be bound to the plastic in the first incubation step, followed by immobilization of the MHC protein. Reversing the order results in decreased recovery of active MHC antigen and greater variability in the level of binding of the other ligand. Provided that the proteins are immobilized at less than saturating levels, the extent of immobilization of each is a linear function of the amount added to the wells. Again, ELISA determinations using the appropriately specific antibodies with parallel sets of wells can be used to determine the levels of each of the coimmobilized ligands. Coimmobilization of class I antigens with nonmembrane proteins (immunoglobulins and fibronectin) and other membrane proteins [class I (nonantigen) and ICAM] has been successfully used to study the effects of these additional ligands on stimulation of CTL by the antigen (Kane and Mescher, 1990, and unpublished results).

Cell-Sized Microspheres

Additional experimental flexibility is provided by incorporation of antigen and other ligands onto cell-size microspheres. Originally, this was accomplished by a 2-day dialysis procedure to incorporate class I and lipid onto 5 μm diameter silica beads having covalently attached C18 alkyl chains (Goldstein and Mescher, 1986; Goldstein and Mescher, 1987). More recently, we have used 5-μm latex microspheres. Antigen can be incorporated onto these in a rapid, reproducible manner and they are quantitatively as effective for CTL stimulation as the antigen-bearing silica microspheres (Mescher, 1992). The procedure for preparation includes the following steps:

Step 1. 5-μm diameter sulfate polystyrene microspheres (Interfacial Dynamics Corp., Portland, OR), supplied in distilled water, are diluted with D-PBS to give a suspension having 1.5×10^8 particles/ml.

Step 2. Affinity-purified antigen is diluted with D-PBS to a concentration of 5 μg/ml.

Step 3. To prepare 1.5×10^6 antigen-bearing microspheres, 0.01 ml of bead suspension is mixed with 0.3 ml of antigen solution in a tube. The sample is then incubated for 1.5 hr at room temperature on a rotating platform adjusted to a speed that maintains the beads in suspension.

Step 4. 0.5 ml of 1% BSA in D-PBS is then added and incubation is continued for 0.5 hr at room temperature.

Step 5. The beads are pelleted by low-speed centrifugation (1000 rpm for 5 min) and washed three times with BSA/D-PBS.

Following preparation, the antigen-coated microspheres can be handled and used in essentially the same way as antigen-bearing stimulator cells. Particle number can be determined using a hemocytometer and antigen-quantitated by ELISA or by flow cytometry. Dose dependences (particle numbers) for assays or CTL function are quite comparable for microspheres and stimulator cells.

Using the radio indicated above (0.75 μg class I per 10^6 particles) yields microspheres having the same amount of antigen as an equivalent number of normal spleen cells, based on ELISA determinations. Varying the protein-to-particle ratio used for preparation results in a corresponding variation in the density of antigen on the resulting beads. DOC concentrations up to 0.2% do not interfere with binding of class I to the beads, allowing use of antigen preparations that are at low concentrations. Although we have not yet examined incorporation of multiple ligands onto these microspheres, it is likely that a two-step procedure as described above for microtiter wells will be effective. It should be mentioned that fluorescent sulfate latex microspheres are also available (Interfacial

Dynamics Corp.), providing additional flexibility in the design of experiments.

References

Brian AA, McConnell HM (1984): Allogeneic stimulation of cytotoxic T cells by supported planar membranes. *Proc Natl Acad Sci USA* 81: 6159–6163

Coeshott CM, Chesnut RW, Kubo RT, Grammer SF, Jenis DM, Grey HM (1986): Ia-specific mixed leukocyte reactive T cell hybridomas: Analysis of their specificity by using purified Class II MHC molecules in a synthetic membrane system. *J Immunol* 136: 2832–2838

Goldstein SAN, Mescher MF (1986): Cell-size, supported artificial membranes (pseudocytes): Response of precursor cytotoxic T lymphocytes to Class 1 proteins. *J Immunol* 137: 3383–3392

Goldstein SAN, Mescher MF (1987): Cytotoxic T cell activation by Class I protein on cell-size artificial membranes: Antigen density and Lyt-2/3 function. *J Immunol* 138: 2034–2043

Heath WR, Kane KP, Mescher MF, Sherman LA (1991): Alloreactive T cells discriminate among a diverse set of endogenous peptides. *Proc Natl Acad Sci USA* 88: 5101–5105

Kane KP, Mescher MF (1990): Antigen recognition by T cells: Quantitative effects of augmentation by antibodies providing accessory adhesion. *J Immunol* 144: 824–829

Kane KP, Champoux P, Mescher MF (1989a): Solid-phase binding of class I and class II MHC proteins: Immunoassay and T cell recognition. *Mol Immunol* 26: 759–768

Kane KP, Sherman LA, Mescher MF (1989b): Molecular interactions required for triggering alloantigen-specific cytolytic T lymphocytes. *J Immunol* 142: 4153–4160

Kane KP, Sherman LA, Mescher MF (1991): Exogenous β2-microglobulin is required for antigenic peptide binding to isolated class I major histocompatibility complex molecules. *Eut J Immunol* 21: 2289–2292

Kane KP, Vitiello A, Sherman LA, Mescher MF (1989c): Cytolytic T lymphocyte response to isolated class 1 H-2 proteins and influenza peptides. *Nature* 340: 157–159

Kozlowski ST, Takeshita T, Boehncke W-H, Takahashi H, Boyd LF, Germain RN, Berzofsky JA, Margulies DH (1991): Excess β2-microglobulin promoting functional peptide association with purified soluble class I MHC molecules. *Nature* 349: 74–77

Mescher MF (1992): Surface contact requirements for activation of cytotoxic T lymphocytes. *J Immunol* 149: 2402–2405

Mescher MF, O'Rourke AM, Champoux P, Kane KP (1991): Equilibrium binding of cytotoxic T lymphocytes to class I antigen. *J Immunol* 147: 36–41

Mescher MF, Stallcup KC, Sullivan CP, Turkewitz AP, Herrmann SH (1983): Purification of murine MHC antigens by monoclonal antibody affinity chromatography. *Methods Enzymol* 92: 86–109

Nakanishi M, Brian AA, McConnell HM (1983): Binding of cytotoxic T-lymphocytes to supported lipid monolayers containing trypsinized H-2K^k. *Molec Immunol* 20: 1227–1231

Parham P (1983): Monoclonal antibodies against HLA products and their use in immunoaffinity purification. *Methods Enzymol* 92: 110–138

Quill H, Schwartz RH (1987): Stimulation of normal inducer T cell clones with antigen presented by purified Ia molecules in planar lipid membranes: Specific induction of a long-lived state of proliferative nonresponsiveness. *J Immunol* 138: 3704–3712

58
Use of Protease Inhibitors as Probes for Biological Functions: Conditions, Controls, and Caveats

Dorothy Hudig and James C. Powers

Introduction to Serine-Dependent Proteases

A new subfamily of serine proteases which are unique to lymphocytes has recently been discovered (Jenne et al., 1989). These proteases are induced when cytotoxic function is induced (Gershenfeld and Weissman, 1986; Lobe et al., 1986) and are stored in intracellular granules at low pH (Masson et al., 1990). When the granules are released during cytolysis, the proteases control cytolysis (Chapter 27, this volume) and may cleave perforin to activate pore formation. The proteases are also likely to produce localized mediators and systemic physiological signals (see review by Simon et al., 1989a). It has been suggested that lymphocyte proteases can cleave the HIV gp120/160 surface protein to promote viral entry into lymphocytes (Maraganore J, Biogen, personal communication), cleave secreted pro-interleukin-1_β (Sleath et al., 1990) to an active form, and arrest tumor cell growth (Sayers et al., 1992). To date, we have identified five different substrate specificities in rodents (Chapter 27, this volume), indicating the potential to cleave, activate, and inactivate numerous substrates. The human lymphocyte serine proteases are not as well characterized, though Asp-ase and tryptase activities have been purified (Poe et al., 1991; Fruth et al., 1987).

When a function is controlled by proteases, inhibition by protease inhibitors can help an investigator select which proteases to study and recover undegraded natural substrates. This chapter offers practical advice for use of protease inhibitors, starting with the design of the biological assay, proceeding to protease assays, and offering an overview of different types of serine protease inhibitors. It concludes with a discussion of how to proceed after initial success or failure and offers specific suggestions for the study of protease-dependent lymphocyte- and granule-mediated cytotoxicity.

Members of the serine-dependent family of proteases share biochemical properties which affect how they are inhibited. They are translated as pre-pro-enzymes and usually secreted as pro-enzymes (or zymogens). Upon cleavage to remove the propeptide, the new amino-terminal isoleucine forms a salt bridge with an internal aspartic acid residue, converting the inactive zymogen to an active protease. *For practical applictions, in zymogen form the protease is essentially nonsusceptible to protease inhibitors.*

After conversion from zymogen form, the active conformation is created which has the substrate binding region S_1 and reactive Ser and His residues that are targets for inhibition. These amino acid residues are in close proximity to the substrate-binding pocket, which binds the amino acid residue that will form the new carboxy terminus of one of the peptide cleavage products. By the chymotrypsinogen numbering system, the active site triad of amino acids is His_{57}, Asp_{102}, and Ser_{195} (Figure 1A). The Asp_{102} and His_{57} make the gamma-hydroxyl group of Ser_{195} an unusually reactive nucleophile which is essential for catalysis. This serine gives this protease family its name. The gamma-hydroxyl of Ser_{195} attacks the carbonyl of the scissile amide bond of the substrate. A tetrahedral adduct is formed which decomposes to form an acyl serine derivative of the enzyme with the peptide cleavage product. The acyl portion of the scissile substrate bond is covalently linked to the enzyme. The peptide cleavage bond from the amino side of the substrate's scissile bond is released as His_{57} pro-

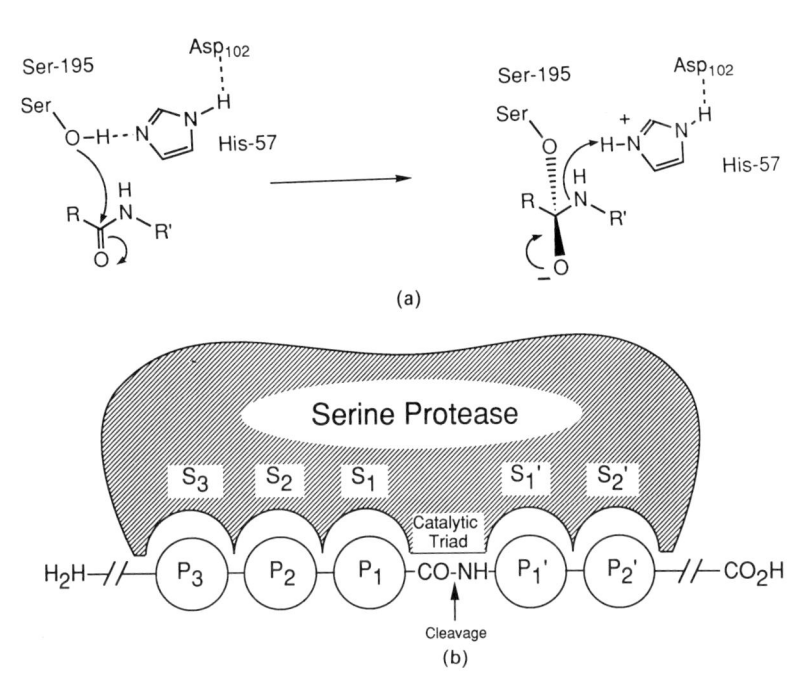

FIGURE 1. (A) Catalytic site of serine-dependent proteases. The active site serine, histidine, and aspartic acid amino acid residues are illustrated with the chymotrypsinogen numbering system. The R and R' groups represent the peptide chain of the substrate which interacts with portions of the enzyme's extended substrate binding site on both sides of the scissile bond. Primary specificity is usually determined by the amino acid residue which provides the carbonyl group of the scissile bond. (B) Numbering system for the substrate and the extended substrate-binding site of proteases (Schechter and Berger, 1967). The interaction of a serine protease with its natural substrate is illustrated. Individual amino acid residues (P) are shown interacting with corresponding subsites (S) of the enzyme. Primary specificity is usually determined by the S_1–P_1 interaction. Most synthetic serine protease substrates contain only sequences which interact with subsites on the carbonyl side (S subsites) of the scissile bond. The catalytic triad is composed of Ser_{195}, His_{57}, and Asp_{102}.

tonates the amino group of the scissile bond when the tetrahedral intermediate decomposes, thereby releasing one peptide product. The remaining acyl-enzyme intermediate contains the amino terminal fragment of the original substrate, with the P_1 amino acid residue inserted into the primary substrate binding pocket of the protease (Figure 1B). This acyl enzyme intermediate is subsequently hydrolyzed to regenerate an active enzyme and free the second peptide cleavage fragment. Because the active site aspartic residue is buried within the protease structure, the active site serine and histidine are the most frequent targets for inhibitors. *Irreversible inhibition at the serine residue generally occurs by acylation, sulfonylation, or phosphonylation, while irreversible inhibition at the histidine usually occurs as a result of alkylation.*

The pH and the presence of natural substrates also affect inhibition of serine proteases. Serine proteases have optimal activity between pH 7 and 8 and little activity below pH 6.5. At suboptimal pH values, these proteases are not affected by most serine protease inhibitors because the active site His is protonated and the active site is less reactive. Like most enzymes, serine proteases have high affinity for their natural substrates. If present during inhibition, natural substrates will compete with inhibitors and may be very protective, since they often have a much higher affinity than the inhibitor for the enzyme.

Design of Biological Assays to Detect Protease Functions

Use rate assays for the function. Often, not all protease activity is inhibited when using protease inhibitors. As a consequence, the physiological function of the protease may be slowed but not stopped. For example, blood clotting times can be greated extended when proteases are inactivated, but a measurable clot will form after several minutes at 37°C (Kam et al., 1988). If rates are not used, the inhibition can be completely overlooked. Similar difficulties are encountered in the arrest of lymphocyte granule-mediated lysis (Hudig et al., 1987). To slow a physiological function to a measurable rate, one can dilute the reactants (protease and natural substrate), and/or render the conditions suboptimal (such as lowering pH and temperature). For example, to measure rates of granule-mediated lysis, we dilute the granule extracts, perform the assays at room temperature with suboptimal calcium concentrations, and stop the assays by acidification at 0.5- or 1-min intervals (Hudig et al., 1988).

Inhibition of the protease(s) may not equal the inhibition of the function. Proteases are often present in great excess over the concentration needed to catalyze function. Thus a substantial fraction of protease can be inactivated without affecting function. Furthermore, if (1) new proteases can be activated from zymogens or (2) inhibitor–protease complexes dissociate spontaneously, active proteases can reappear after the initial inhibition reaction, particularly when the inhibitor is diluted or removed from the experimental system or is unstable in the reaction medium. Therefore, it is advisable to assay for function immediately after inhibition.

Optimize access of the inhibitors to the protease(s). Consider where a protease is likely to be when it is in its active form. To react with serine proteases inside lymphocyte granules, a reagent has to traverse both the plasma and the granule membranes. Small and uncharged inhibitors are most likely to cross membranes. Dimethyl sulfoxide may improve the passage for some inhibitors. The sequence of the biological processes studied may also affect access of the inhibitor. If a zymogen is only converted to an active protease immediately before it performs its physiological function, it may be essential to have the serine protease inhibitor present at the time the function is initiated.

Select inhibitors with aqueous stability. Many serine protease inhibitors have half-lives in water that are as short as a few minutes. These inhibitors can be destroyed by spontaneous decay during biological assays that require hours of incubation, such as lymphocyte-mediated lysis or inhibition of tumor cell growth. Half-lives depend on the type of inhibitor and on its substituents. Primary references should be consulted to ensure that the inhibitor is appropriate for the proposed use. In general, protein inhibitors such as aprotinin, soybean trypsin inhibitor, and α_1-antichymotrypsin are stable in aqueous buffers, while sulfonyl fluorides and isocoumarins are not.

Recognize the other conditions that can perturb the bioassay. For example, high concentrations of some protease inhibitors can alter electrostatic protein interactions, form hydrophobic micelles, or perturb pH. If the bioassay is affected by these parameters, data can be misinterpreted. For example, high concentrations of the chloride salts of amino acid esters will lower the pH of weak buffers. We have found that >10 mM concentrations of any of these reagents reduce lymphocyte granule-mediated lysis when the pH is not corrected.

Assays for the Serine Proteases

Good specific substrates are needed to measure individual proteases and to determine levels of inactivation. It is important to determine which inhibitors inactivate each protease, to rank the potency of the inhibitors, and to assay the extent of protease inactivation that occurs within the biological system. Only specific and highly reactive substrates are suitable for this purpose, although initially almost all serine proteases can be detected with the radioactive general inhibitor diisopropyl fluorophosphate (^{3}H-DFP), followed by sodium dodecyl sulfate–polyacrylamide gel electrophoresis (SDS–PAGE) separation and autoradiography (see Masson et al., 1986, for radiolabeled lymphocyte proteases). Nonspecific protein substrates, such as radiolabeled or dye-associated casein, are not useful for detection of individual proteases within a mixture and are cumbersome for routine assays of proteases. The most useful substrates are synthetic ones made for simple, fast, and convenient spectrophotometric assays.

To find the optimal synthetic substrates to measure protease activities, we recommend screening a "library" of potential substrates containing representative P_1 amino acids that fit in the primary substrate-binding pocket of the enzyme. During this screening, the presence of small, uncharged P_2 and P_3 amino acids such as Ala will usually accelerate the rate of substrate hydrolysis (Castillo et al., 1979). The introduction of charged or aromatic P_2 or P_3 amino acids could reduce the rate of hydrolysis if these amino acids are incompatible with the extended substrate-binding sites of the protease (Cho et al., 1984; Harper et al., 1984; Kam et al., 1987). Blocking the charge of the amino terminus of the substrate also reduces unfavourable secondary interactions, particularly with endopeptidases. For example, we screened the lymphocyte granule proteases with a set of Boc-Ala-Ala-[X]-SBzl peptides, where X was any one of 13 representative amino acids, and found five good substrates for different protease activities (Hudig et al., 1987; Odake et al., 1991; Hudig et al., 1991).

After proteases are purified, better substrates can be identified by selecting the optimal P_1 amino acid and varying the P_2 amino acid to find the best P_1-P_2 combination, while maintaining the P_3 residue (Ala) unchanged. After selection of P_1 and P_2, the P_3 amino acid can be varied to determine the best tripeptide sequence. Substrate reactivity for many proteases is not improved by further extension of a P_3-P_2-P_1 peptide. Many substrates suitable for initial protease screening are commercially available (see suppliers listed at the end of these publications: Powers and Harper, 1986, and Beynon and Bond, 1989).

To evaluate protease inhibitors, biochemical constants are more useful than ID_{50} determinations. ID_{50} estimates are often dependent upon the type of assay and time of incubation of the enzyme with the inhibitor and cannot be used as a basis to calculate the extent of inhibition in other experimental situations. Comparison of biochemical constants can help determine why some reagents block function where others do not, under conditions in which these reagents appear to equally and completely inactivate proteases. Detection of trace quantities of remaining enzymes can be extremely difficult. With lymphocyte proteases, greater than 95% inactivation is difficult to measure. However, residual active proteases can be calculated from (1) dissociation constants for reversible protease inhibitors and (2) second-order inhibition rate constants for irreversible protease inhibitors. These calculations can indicate biologically significant differences in remaining protease activity.

When feasible, determine the rate of cleavage of natural substrates. Cleavage of putative natural substrates can be detected by autoradiography of radiolabeled substrates or by Western immunoblots with antibodies to the substrates. When a relevant protease is inactivated, there will be a slower rate of natural substrate cleavage. To favor detection, the natural substrate should be at a concentration at least 10-fold greater than its K_m with the protease (if the K_m can be determined) and in 100-fold molar excess of the protease. If a reversible inhibitor is employed, a high concentration relative to the K_I should initially be used, because the natural substrate may have much greater affinity for the protease and thus successfully compete with the inhibitor.

Selection of Serine Protease Inhibitors

General vs. specific protease inhibitors. In theory, a general or broadly reactive serine protease inhibitor should block any function dependent upon serine protease activity. Representative general serine protease inhibitors are listed in Table 1. The general, mechanism-based isocoumarin serine protease inhibitor, 3,4-dichloroisocoumarin (Harper et al., 1985) is quite useful since it reacts much faster than either diisopropylfluorophosphate (DFP) or phenylmethylsulfonyl fluoride (PMSF). However, general inhibitors do not inactivate all serine proteases and react at different rates with different proteases. Specific serine protease inhibitors usually react much faster with their target proteases than general inhibitors. If general inhibitors do not affect biological function and/or protease activity, highly reactive inhibitors designed for elastase-like, chymotrypsin-like, and trypsin-like serine proteases should be also evaluated. If protease activity is still unaffected, it may be advisable to consider if the protease is not in the serine protease family but belongs to one of the other three classes of endoproteases (Barrett, 1986).

Types of serine protease inhibitors and their important properties. Serine protease inhibitors can be of low M_r, either synthetic or natural in origin, or can be macromolecular protein inhibitors. Within each of these two large divisions, there are several types of both reversible and irreversible reagents. Table 2 provides key features of commonly used types of inhibitors and examples for each type.

The properties summarized in this table are of practical importance. The reactivity of reversible inhibitors is indicated by the dissociation constant K_i, with smaller K_i values indicating better inhibition. The second-order inhibition rate constant ($k_{observed}/[I]$) applies to irreversible inhibitors and allows one to calculate the surviving fraction of enzyme after any time of pretreatment with any concentration of inhibitor. The constant is the rate of loss of enzyme activity per second ($k_{observed} = (\ln[E_t/E_0] s^{-1})$ divided by the molar concentration of inhibitor used ([I]). The larger the rate constant, the better the inhibitor. The dissociation rate is expressed as $t_{1/2}$, the time in which half the inhibitor-enzyme will dissociate, rather than the first-order rate of dissociation. Fast dissociation rates are of experimental importance because

TABLE 1. General or broadly reactive irreversible serine protease inhibitors

	Protease reactivity, $K_{obsd}/[I]$ in M-1 s-1				
Inhibitor	Trypsin	Chymotrypsin	Pancr. elastase	Aqueous 1/2-life	Comments
Low-M_r					
Diisopropylfluorophospate (DFP)	250[a]	16[a]	?[b]	1 hr[c]	Not all serine proteases react; volatile and potentially lethal
Phenylmethylsulfonylfluoride (PMSF)	4.5[d]	370[e]	2.7[e]	55 min, pH 7.5[f]	Not all serine proteases react; non volatile with little toxicity
3,4-Dichloroisocoumarin (DCL)	198[g]	570[g]	2500[g]	18 min, pH 7.5[g]	Most serine proteases react, except complement C2a and Bb; non-volatile; low toxicity
Macromolecular inhibitors	Trypsin	Chymotrypsin	Pancr. elastase	Aqueous 1/2-life	
Alpha-1-protease inhibitor (a1-PI)	7.3×10^{4}[h]	5.9×10^{6}[h]	6.5×7[h]	Months[h]	Not all serine proteases react; a1-PI inactivated by oxidation and low pH

[a] Cohen et al., 1967.
[b] Rate not provided, Naughton and Sanger, 1961.
[c] Beynon and Salvesen, 1989
[d] Farney and Gold, 1963.
[e] Lively and Powers, 1978.
[f] James, 1978.
[g] Harper et al., 1985.
[h] Beatty et al., 1980.

enzyme activity can be restored if the inhibitor is no longer present.

In the following section, we offer an overview of the advantages and disadvantages of each type of inhibitor. Storage conditions, maximum solubility and stability in aqueous solutions, and nonspecific reactivity are discussed. For a description of mechanisms, comparison of reactivities of numerous serine protease inhibitors, and primary references see Powers and Harper, 1986. For additional practical information, see Barrett and Salveseon (1986); Beynon and Bond (1989); Keesey (1987); and Methods in Enzymology 45 and 80, edited by Lorand in 1976 and 1981.

Low-M_r Natural and Synthetic Protease Inhibitors

Reversible Low-M_r Natural and Synthetic Protease Inhibitors Include Oligopeptides, Peptide Esters, Peptide Ketoesters, and Peptide Aldehydes

Model oligopeptides. Unblocked, short oligopeptides that contain the amino acid sequence of the natural substrate can effectively compete with the natural substrate. Unfortunately, they do not usually possess desirably low dissociation constants because (1) simple peptides do not bind well to endopeptidases, and (2) their charged amino and carboxy termini interact unfavorably with proteases. Although they are stable in aqueous solutions, they may be cleaved by the target protease. If they block function, it is very important to scramble the amino acid sequence of the peptide to control for nonspecific electrostatic and hydrophobic effects, which can also occur at the high (millimolar) concentrations at which model peptides react with serine proteases.

Peptide esters and amides are alternate substrates for proteases and act as competitive inhibitors with the natural substrate; their K_m value become K_i values in this capacity. The carboxy terminus of the esters is often blocked by an uncharged methyl or ethyl group which is likely to interact minimally with the $S_{1'}$ substrate binding site. The amide bond of nitroanilide and aminocoumarin substrates also blocks the often unfavorable charge of the carboxyl group. The peptide amino terminus may also be blocked to negate its charge and reduce unfavorable interactions with the protease. These peptide esters and amides frequently have lower, more favorable dissociation constants than free peptides and are better competitive, reversible inhibitors. They are likely to compete in about 0.1 millimolar concentrations when the right peptide sequence is presented. Many of these esters and amides are widely used as substrates and are commercially available.

Peptide alpha-ketoesters have much lower dissociation constants with serine proteases than peptide esters because they are transition-state analogs and form tetrahedral complexes which dissociate slowly from the enzymes. Their hydrolysis products, alpha-ketoacids, are also potent protease inhibitors. The dissociation constants are quite low (often μM or lower) for these very promising, stable reversible inhibitors. Unfortunately, these new synthetic reagents (Hori et al., 1985; and Powers, manuscript in preparation) are not yet commercially available.

Peptide aldehydes are natural products of fungi (e.g., chymostatin, leupeptin) and have also been synthesized using peptide sequences that are recognized by specific proteases [e.g., D-Phe-Pro-Arginal, which is an excellent thrombin inhibitor (Bajusz et al., 1978)]. These reagents form tetrahedral transition state-like complexes with serine proteases and are not hydrolyzed, though the Schiff's bases which they can form with amino groups are unstable to oxidation. The best inhibitor-serine protease combinations (e.g., trypsin and leupeptin) have K_i values of $\ll 1\ \mu$M. Peptide aldehydes can be slowly oxidized. They should be stored with desiccant, dissolved in anyhydrous dimethylsulfoxide (DMSO) under an inert gas (nitrogen, argon) and used within a week after making the DMSO solution.

Irreverisble Low-M_r Inhibitors Include Sulfonyl Fluorides, Peptide Phosphonates, Peptine Chloromethyl Ketones, and Isocoumarin Reagents

Sulfonyl fluoride serine protease inhibitors first gained popularity when phenylmethylsulfonyl fluoride (PMSF) was recognized as an effective general serine protease inhibitor that was safe for routine use. PMSF lacked the volatility of diisopropylfluorophosphate (DFP), which can be lethal to the experimenter due to its ability to inhibit acetylcholinesterase. Sulfonyl fluorides sulfonylate the active site serine and form stable inhibited derivatives. As a group, these inhibitors are only moderately reactive, with second order inhibition rate constants of $\sim 10\ \text{M}^{-1}\text{s}^{-1}$ for PMSF and no greater than $4000\ \text{M}^{-1}\text{s}^{-1}$ for similar compounds

TABLE 2. Characteristics of reversible and irreversible serine protease inhibitors

Type of inhibitor	Example	Representative enzyme inhibited	Reactivity with 2nd-order rate, $M^{-1} s^{-1}$
Synthetic and related inhibitors			
Reversible inhibitors			
Peptide esters	Ac-Tyr-OEt	Chymotrypsin	
Peptide α-ketoesters	Bz-DL-Phe-COOEt	Chymotrypsin	
		Cathepsin G	
Fungal aldehydes	Chymostatin	Chymotrypsin	
Irreversible inhibitors			
Sulfonyl fluoride	2-(N-(3-(Benzyloxycarbonylamino) propanoyl)amino)benzenesulfonyl fluoride	Chymotrypsin	3,300[d]
		RNK chymase	913[e]
Peptide chloromethylketones	Z-Gly-Leu-Phe-CH2Cl	Chymotrypsin	3[f]
		Cathepsin G	51[g]
		RNK chymase	86[h]
	D-Phe-Pro-Arg-CH2Cl	Trypsin	3.5×10^6[f]
Isocoumarin	3-Benzyloxy-4-cloro-7-nitroisocoumarin	Chymotrypsin	> 10,500[i]
		Cathepsin G	> 2,600[i]
		RNK chymase	2,010[j]
Peptide phosphonates	Z-Phe-Pro-Phe-P-(OPh)2	Chymotrypsin	17,000[k]
		Cathepsin G	5,100[k]
Protein inhibitors (natural)			
Reversible			
Plant inhibitors, Kunitz type	Soybean trypsin inhibitor, Kunitz (SBTI)	Trypsin	6×10^6[l]
Plant inhibitors, Bowman-Birk	Lima bean trypsin inhibitor	Trypsin	23,000[n]
		Chymotrypsin	~ 23,000
Aprotinin = basic pancreatic trypsin inhibitor, Kunitz	Aprotinin	Trypsin	
		Chymotrypsin	
Irreversible			
Serpins (plasma antiprotease)	α₂-Antiplasmin	Plasmin	3.8×10^7[p]
α₂-Macroglobulin (plasma)	α₂-Macroglobulin	Trypsin	25,000[q]

References
[a] Hein and Niemann, 1961.
[b] Hori et al., 1985.
[c] Umezawa and Aoyagi, 1977.
[d] Yoshimura et al., 1982.
[e] Ewoldt et al., submitted.
[f] Powers and Harper, 1986.
[g] Powers et al., 1977.
[h] Zunino et al., 1988.
[i] Harper and Powers, 1985.
[j] Hudig et al., 1989.
[k] Oleksyszyn and Powers, 1991.
[l] Luthy et al., 1973.
[m] Finkenstadt et al., 1974.
[n] Kran and Stevens, 1972.
[o] Fritz and Wunderer, 1983.
[p] Lijnen and Collen, 1986.
[q] Salvesen et al., 1981.

Representative enzyme K_i	Spontaneous dissociation (time for 50% restoration of represent, protease)	Aqueous stability of representative cpd half-lives	Range of reactivity of the inhibitor type with very reactive enzymes (2nd-order or K_i)
0.7 mM[a]	NA		~1 mM
0.28 μM[b]	NA		0.1 to 640 μM[b]
58 μM[b]			
0.3 μM[c]	NA	<1 week under inert gas	0.02–0.6 μM[c]
	NA	3.3 min	20–3,000 M^{-1}s^{-1} [d]
	NA	~1 day for related cpds[g]	3–86 M^{-1}s^{-1} [f]
	NA	<30 min at high pHs[f]	~1,000–3 × 10^6 M^{-1}s^{-1} [f]
	0.7 hr	17 min (Powers, unpublished)	~100–100,000 M^{-1}s^{-1} [f]
	0.05 hr		
	26 hr	>7 days	100–44,000 M^{-1}s^{-1} [k]
	3.2 days[m]	Very stable	
	0.5 to ⩾5 hr, depending on variant[m]	Very stable	
6 × 10^{-14} [o]	17 weeks	Stable even to boiling!	6 × 10^{-6} to 6 × 10^{-14} M
9 × 10^{-9} [o]	12 min		
	NA	Very stable	Usually ⩾100,000 M^{-1}s^{-1}
	Negligible; enzyme is "trapped"	Thioester is unstable	Reacts with all 4 classes of proteases

with substituents that enhance reactivity. These inhibitors are also hydrolyzed with half-lives of minutes. The half-lives become progressively shorter as pH values are increased above neutrality (James, 1978). Thus, the combination of only moderate reactivity and short half-lives in solution limits the utility of these reagents. In addition to PMSF, the somewhat more reactive, uncharged 4-(2-aminoethyl)benzenesulfonyl fluoride and positively charged (p-amidinophenyl)methanesulfonyl fluoride (Laura et al., 1980) are commercially available. Only the charged compound is soluble at greater than 1 mM in aqueous solution.

Peptide phosphonate inhibitors phosphonylate the active site serine (Oleksyszyn and Powers, 1991). They are very stable in water and, with a favorable match between the peptide sequence and the protease substrate-binding site, are excellent inhibitors with second-order inhibition rate constants $> 40{,}000\ M^{-1} s^{-1}$.

Peptide chloromethylketones are quite reactive inhibitors, with second-order inhibition rate constants that may be as large as $23{,}000{,}000\ M^{-1}s^{-1}$ (for D-Phe-Phe-Arg-CH$_2$Cl with kallikrein, Kettner and Shaw, 1981). They are relatively stable in aqueous solutions, with the exception of the compounds with Arg or Lys P$_1$ amino acids, which have short half-lives at pH values above neutrality (Kettner and Shaw, 1981). These reagents form a tetrahedral complex with the protease active site Ser and alkylate the active side His. Alkylation of the His gives a stable product that will not spontaneously dissociate, which means that the serine proteases will remain permanently inactivated. Because chloromethylketones are potent alkylating reagents, they can react nonspecifically with other protein nucleophiles, particularly sulfhydryl groups. Therefore, for biological experiments, chloromethyl ketones containing different peptide sequences are essential controls for this nonspecific reactivity.

Isocoumarin serine protease inhibitors are mechanism-based (so-called "suicide") inhibitors that are reactive only after being metabolized by active enzymes and thus have great specificity for active serine proteases. The iscoumarin ring is opened by attack of the active site Ser on the carbonyl of the isocoumarin ring, forming an acylated serine derivative which unmasks a previously concealed functional group that can alkylate the active site His or react with water to give a rearranged acyl enzyme (Harper et al., 1985; Powers et al., 1989). When His alkylation takes place, the proteases are stable toward reactivation. Proteases that are only acylated will recover activity spontaneously upon deacylation. The acyl-serine protease complexes dissociate with half-lives varying between hours and days. The enzyme activity of these complexes can be restored quickly by hydroxylamine treatment. Such restoration can recover both proteases and protease-dependent functions [e.g., granule proteases and lysis: (Hudig et al., 1991)]. Though isocoumarin reagents hydrolyze in water, they are reasonably stable when stored under anhydrous conditions and solubilized in anhydrous DMSO. The molarity of the active reagents should be verified at the time of use by measuring the absorbance of the isocoumarin ring. Despite the mechanism-based inhibition of those compounds, they may also exhibit some slow nonspecific reactivity, resulting in the acylation of Lys residues (Rusbridge and Beynon, 1990). In addition, isocoumarin reagents are less stable in the presence of sulfhydryl groups, including the free thiol group of albumin. Thus, in biological experiments, albumin should be omitted and treatment should be only as long as is necessary to block biological function.

Protein (Macromolecular) Serine Protease Inhibitors

Reversible protein (macromolecular) serine protease inhibitors include the Kunitz soybean trypsin inhibitor, Bowman-Birk plant inhibitors, and aprotinin (basic pancreatic trypsin inhibitor, Kunitz). These protein inhibitors are very stable in aqueous solutions. Because they can bind with high affinity to serine proteases, they can be bound to solid matrixes for affinity chromatography of the proteases.

Soybean trypsin inhibitor (Kunitz) is a 22,000 M$_r$ protein that has Arg in its reactive site (which binds to the S$_1$-substrate binding site of serine proteases). This inhibitor would be expected to interact primarily with trypsin-like proteases, but weaker inhibition of other serine proteases has been observed. The fact that commercial preparations (Sigma Chemical Co., St. Louis, MO) inactivate all five lymphocyte serine protease activities (Chapter 27, this volume) may indicate that the commercial preparations are contaminated with Bowman-Birk inhibitors.

Bowman-Birk plant serine protease inhibitors are also found in seeds. The constitute a large family of proteins, of which lima bean trypsin inhibitor (LBTI), M$_r$ 7000, is a representative member. Each Bowman-Birk molecule has two serine protease reactive sites. In the case of LBTI, one reactive site has Leu and the other Lys (Odani and Ikenaka,

1973) as primary binding sites for proteases. Thus the Bowman-Birk inhibitors have 2:1 stoichiometry of inactivation of proteases and can inactivate both chymotrypsin-like and trypsin-like proteases.

Aprotinin is the same protein as basic pancreatic trypsin inhibitor (Kunitz) and Trasylol (trade name). It is found in mast cells and is usually isolated from the lung or pancreas. This 6.5 kDa protein has Lys in its reactive site and binds with very high affinity to trypsin ($K_1 = 10^{-14}$). However, even though aprotinin has much higher affinity for trypsin, aprotinin columns will also bind neutrophil elastase and cathepsin G. These proteins can be eluted at higher pH values than trypsin (Baugh and Travis, 1976), so that this protein can be a very good matrix for protease purification.

Irreversible protein (macromolecular) serine protease inhibitors include the serpin (*ser*ine protease *in*hibitor) family of plasma proteins and α_2-macroglobulin.

The serpins are a family of related proteins that includes α_1-protease inhibitor (α_1PI), α_1-antichymotrypsin (α_1-X), antithrombin III (AT III), complement C1 esterase inactivator (C1 INA), and α_2-plasmin inactivator. They range in size from 52 (α_1PI) to 106 kDa (C1 INA) and are remarkably reactive with the serine proteases they inactivate, having second-order inhibition rate constants as great as 380,000 $M^{-1}s^{-1}$ (α_1-X with chymotrypsin: Schechter et al., 1989). The serpins bind and inactivate serine proteases with a 1:1 stoichiometry. In the case of α_1PI, a tetrahedral complex is formed between the protease active site serine and the serpin (Matheson et al., 1991). The serpin–protease complexes are stable in SDS–PAGE but can be dissociated with strong nucleophiles like hydroxylamine to regenerate active serine proteases. Specificity is determined by the P_1 amino acid in a reactive loop of the serpin: α_1-PI has Met in the P_1 position, α_1-X has Leu, and AT III has Arg in this position. The reagents are very reactive and essentially nontoxic. Caveats are, however, appropriate. The serpin reactivity should be checked before use and the purity checked after the first indication of efficacy. Serpin activity can be lost in preparation and the reactivity should be checked with common susceptible serine proteases. Serpins share low pI values and have similar M_rs, making it quite possible that serpin contaminants contribute to multiple bioactivity of some commercial preparations.

α_2-*Macroglobulin* (α_2-M) is not a serpin, but an extremely large ($\sim$ 720 kDa) plasma protein which is able to covalently trap serine proteases and the other three classes of proteases as well. The trapped proteases are still catalytically active, but substrate access is limited to small peptide substrates and proteins smaller than $\sim$ 12 kDa. Thus α_2-M will not inactivate protease hydrolysis of synthetic substrates, though it may block functions dependent upon cleavage of large natural substrates. The trapping of proteases is dependent upon covalent interaction with an essential thioester bond in α_2-M. Since this essential thioester bond is subject to decay, α_2-M reactivity should be verified at the time of use. Spectrophotometric measurement can be used to distinguish free proteases from α_2-M-trapped proteases. In these assays, an appropriate serpin is included to inactivate the free protease but not the trapped protease (e.g., lack of inactivation of trapped trypsin by α_1-PI). An excellent review of α_2-M is found in Barrett (1981).

Interpretation of Success or Failure in Blocking Function

If both proteases and biological function are blocked, a new project may be underway. Initially, if possible, the protease inhibitors should be inactivated and retested. Many synthetic inhibitors, such as isocoumarins, sulfonyl fluorides etc., can by hydrolyzed by extended incubation in buffers. Several serpins lose activity after acidification.

Inhibition should be demonstrated with several different types of inhibitors that have similar specificity, i.e., inactivate enzymes with common S_1 substrate specificities. In particular, within one type of compound, inhibitors with different primary specificities and reagents with different reactivities should affect function differently. Even if a protease is not yet identified to target in the biological system, there should be differences in the dose-titrations which will later correlate with efficacy of inhibition of the protease(s).

Subsequent identification of natural substrates can be arduous. Ideally, one should be able to return endogenous protease activity to the biological system and simultaneously restore blocked function. Later, one should demonstrate that (putative) natural substrates are hydrolyzed and that the biological function requires and correlates with this substrate hydrolysis. Ultimately, final proof will involve reconstitution of function with purified proteases and natural substrates.

If a biological function is not blocked by protease inhibitors, one should reassess the approach. First the investigator should consult the literature to determine how reactive the reagents are with the proteases for which they were initially designed and to ensure that very potent inhibitors were used for the initial screening. Then, the investigator should verify that the inhibitors remain active after incubation in the biological system. Often, this verification is simple. For example, when isocoumarin inhibitors are inactivated, the isocoumarin ring is opened and the characteristic isocoumarin absorption (330–375 nm) is lost. Other protease inhibitors can be assayed at the start and finish of the experiment for their ability to inactivate common proteases such as trypsin, chymotrypsin, and elastase. In addition, the reactivity of the protease inhibitors should be determined with the specific proteases found in the biological system. When the reactivity is defined by rate constants, the investigator can mathematically predict the fraction of surviving protease. If calculations predict survival of less than one molecule of active protease, failure to block function is compelling evidence against a role for *only the individual protease that was inactivated.* However, negative results can be inconclusive, since (1) protease–inhibitor complexes can dissociate, (2) anomalous behavior of low concentrations of active proteases could leave a few catalytic molecules, and (3) it is difficult to obtain complete inhibition with many synthetic inhibitors.

Special Applications

Lymphocyte-mediated cytolysis. Specific requirements and procedures apply to lymphocyte-mediated killing. In *in vitro* cytotoxicity assays, T and natural killer (NK) lymphocyte killing is dependent upon energy metabolism and lymphocyte receptor–target ligand interactions, but not upon *de novo* mRNA or protein synthesis. (Cellular induction of lytic function, however, requires all of these capabilities.) Gross toxicity of protease inhibitors for the effectors can be measured by the ability to exclude vital dyes such as trypan blue. Damaged targets usually leak radiolabel at an accelerated rate. To detect low levels of toxicity, we recommend assay of cell proliferation of mitogen-stimulated normal lymphocytes. Only reagents of minimal toxicity will permit entry from G_0 into G_1 and normal rates of cell division.

For killing, it is also essential that there be no effects of the reagents on lymphocyte receptor–target cell interactions. While extreme effects can be detected by microscopic examination of conjugates of effector lymphocytes and targets, the sample numbers are too small to detect modest effects. Flow cytometric analyses, in which effectors and targets are labeled with different fluors and two-color cytometric events are scored as conjugates of effectors and targets, are optimal for this evaluation. Chloromethylketone reagents at approximately millimolar concentrations disrupt conjugates (Ristow et al., 1983). It should be noted that conjugates are not always highly reproducible. Measurement of the sequelae of receptor interaction, such as increases in intracellular calcium concentrations, may be more practical and perhaps also more meaningful.

In addition, protease inhibitors may be slow but not stop lysis. Rates of lysis are needed and can be determined from data collected by terminating assays every few hours (e.g., Hudig et al., 1981).

The sequestration of lymphocyte serine proteases in acidic granules presents problems of access and efficacy for protease inhibitors. Reagents must cross cell membranes and pass through the reducing atmosphere of the cytoplasm. When they reach the granules, the acidic pH is unfavorable to serine protease activity, and therefore the proteases cannot be inhibited in this intracellular environment. A very useful procedure (Henkart et al., 1987) is to temporarily increase the pH of the granules so that the proteases can be inactivated. Ammonium chloride or monensin (which blocks the sodium pump) will raise the granule pH. DMSO can be used to increase the permeability of the cell membranes temporarily. Reagents such as PMSF (Henkart et al., 1987) or 3,4-dichloroisocoumarin (DCI) (Hudig et al., unpublished) can then cross the membranes and inactivate protease activity more effectively than if the pH were not raised. After the pretreatment of the lymphocytes, all reagents can be removed, returning the cells to near normal conditions. Furthermore, when the protease inhibitors are used without altering the granule pH, controls are established in which non-granule proteases are still affected. If a function is blocked only (1) when the lymphocytes are pretreated with inhibitors and additional reagents to raise pH, and not with inhibitors alone or pH-altering reagents alone, and (2) when granule serine protease activities are inactivated, protease control of function is strongly implicated. Some words of caution are in order. If any of the multiple proteases are in zymogen form inside the granules, they will not be susceptible to inhibition until activated.

Cytotoxicity mediated by lymphocyte granule extracts is easier to block, since there is ready access to the proteases. The nonpolar solvent concentration should be kept low (no more than 10% DMSO for example) to prevent loss of lytic activity. Low-M_r inhibitors can be removed from the granule extracts by desalting columns made of acrylic beads. Considerable lytic activity is lost upon dialysis, and almost all is lost upon contact with matrixes made of natural sugars such as Sephadex. To detect inhibition, rates of granule-mediated lysis are essential (Hudig et al., 1988). To stop granule-mediated lysis, virtually all protease molecules must be inactivated. Only substantial concentrations of the most highly reactive serine protease inhibitors will prevent lysis. Furthermore, natural substrates within the granules may protect the proteases from inhibition, so that treatment of dilute granule extracts is more effective than the treatment of concentrated extracts.

Perforin activity isolated from granules appears dependent upon serine proteases (Hudig et al., 1991; Chapter 27, this volume). Nonpolar solvents can destroy perforin activity. All attempts should be made to keep DMSO concentrations as low as possible (< 2% v/v) and pretreatment conditions as short as possible.

Overall Utility of Serine Protease Inhibitors

Serine protease inhibitors are useful reagents throughout the development of a research project. In the beginning, specific probes will help determine which proteases are most likely to be important for function, and therefore, which ones to purify. This discrimination would be the case for the proposed activation of proIL-1 and cleavage of HIV gp160 to promote viral entry. Often the same probes can be used to protect the natural substrates from degradation during isolation. Such protection would facilitate determination of the natural substrate when Asp-ase blocks tumor growth (Sayers et al., in press). Effective inhibitors can be used to distinguish the functional proteases after they are purified from a system containing multiple proteases. As in the case of the lymphocyte granule proteases, only a subset of the granule proteases may directly regulate cytotoxicity. Ultimately, initial probes or subsequently designed serine protease inhibitors will serve as therapeutic agents. Serine protease inhibitors that inactivate lysis are likely to reduce damage and ameliorate the symptoms produced by cytotoxic lymphoctyes in progressively degenerative autoimmune diseases.

Acknowledgments

We thank the predoctoral student, Gerald R. Ewoldt, and Dr. Terry Woodin of the University of Nevada, Reno, for their helpful suggestions and critique of the manuscript. The research was made possible by NIH grants CA38942 (D.H.), GM42212 (D.H. and J.C.P.), and HL29307 (J.C.P.).

Abbreviations Used

a_1P1, alpha-1-protease inhibitor; a_2M, alpha-1-macroglobulin; a_1X, alpha-1-antichymotrypsin; ADCC, antibody-dependent cell-mediated cytotoxicity; Asp-ase, protease activity cleaving Boc-Ala-Ala-Asp-SBzl; AT III, antithrombin III; BLT, Benzyloxycarbonylysine thiobenzyl ester; Boc, butyloxycarbonyl; DCI, 3,4-dichloroisocoumarin; DMSO, dimethylsulfoxide; IC, isocoumarin; NK, natural killer lymphocyte; OEt, ethyl ester; PMSF, phenylmethysulfonyl fluoride; SBzl, thiobenzyl ester; SBTI, soybean trypsin inhibitor, T cell, thymus-dependent lymphocyte; Z, benzyloxycarbonyl.

References

Bajusz S, Barabas E, Tolnay P, Szell E, Bagdy D (1978): Inhibition of thrombin and trypsin by tripeptide aldehydes. *J Pep Pro Res* 12: 217–221

Barrett AJ (1981): Alpha-2 Macroglobulin. *Methods Enzymol* 80: 737–754

Barrett AJ (1986): An introduction to the proteinases. In: *Proteinase Inhibitors*, Barrett AJ, Salvesen G, eds. Amsterdam: Elsevier

Barrett AJ, Salvesen G, eds. (1986): *Proteinase Inhibitors*. Amsterdam: Elsevier

Baugh RJ, Travis J (1976): Human leukocyte granule elastase: Rapid isolation and characterization. *Biochemistry* 15: 836–841

Beatty K, Bieth J, Travis J (1980): Kinetics of association of serine proteinases with native and oxidized α_1-proteinase inhibitor and α_1-antichymotrypsin. *J Biol Chem* 255: 3931–3934

Beynon RJ, Bond JS, eds. (1989): *Proteolytic Enzymes: A Practical Approach*. Oxford: IRL Press

Beynon RJ, Salvesen G (1989): Appendix III. Commercially available protease inhibitors. In: *Proteolytic Enzymes: A Practical Approach*, Beynon RJ, Bond JS, eds. Oxford: IRL Press

Castillo MJ, Nakajima K, Zimmerman M, Powers JC (1979): Sensitive substrates for human leukocyte and porcine pancreatic elastase: A study of the merits of various chromophoric and fluorgenic leaving groups in assays for serine proteases. *Anal Biochem* 99: 53–64

Cho K, Tanaka T, Cook RR, Kisiel W, Fujikawa K, Kurachi K, Powers JC (1984): Active-site mapping of bovine and human blood coagulation serine proteases using synthetic peptide 4-nitroanilide and thio ester substrates. *Biochemistry* 23: 644–650

Cohen JA, Oosterbaan RA, Berends F (1967): Organophosphorus compounds. *Methods Enzymol* 11: 686–702

Farney DE, Gold A (1963): Sulfonyl fluorides as inhibitors of esterases: 1. Rates of reaction with acetylcholinesterase, alpha-chymotrypsin and trypsin. *J Am Chem Soc* 85: 977–1000

Finkenstadt WR, Hamid MA, Mattis JA, Schrode J, Sealock RW, Wand D, Laskowski W, Jr (1974): In: *Bayer Symposium V "Proteinase Inhibitors,"* Fritz H, Tschesche H, Greene LJ, eds. New York: Springer-Verlag

Fruth U, Sinigaglia F, Schlesier M, Kilgus J, Kramer MD, Simon MM (1987): A novel serine proteinase (HuTSP) isolated from a cloned human CD8+ cytolytic T cell line is expressed and secreted by activated CD4+ and CD8+ lymphocytes. *Eur J Immunol* 17: 1625–1633

Gershenfeld HK, Weissman IL (1986): Cloning of a cDNA for a T cell-specific serine protease from a cytotoxic T lymphocyte. *Science* 232: 854–857

Harper JW, Powers JC (1985): Reaction of serine proteases with substituted 3-alkoxy-4-chloroisocoumarins and 3-alkoxy-7-amino-4-chloroisocoumarins: New reactive mechanism-based inhibitors. *Biochemistry* 24: 7200–7213

Harper JW, Cook RR, Roberts CJ, McLaughlin BJ, Powers JC (1984): Active site mapping of the serine proteases human leukocyte elastase, cathepsin G, porcine pancreatic elastase, rat mast cell proteases I and II, bovine chymotrypsin A alpha, and Staphylococcus aureus protease V-8 using tripeptide thiobenzyl ester substrates. *Biochemistry* 23: 2995–3002

Harper JW, Hemmi K, Powers JC (1985): Reaction of serine proteases with substituted isocoumarins: Discovery of 3,4-dichloroisocoumarin, a new general mechanism-based serine protease inhibitor. *Biochemistry* 24: 1831–1841

Hein GE, Niemannn C (1961): An interpretation of the kinetic behavior of model substrates of alpha-chymotrypsin. *Proc Natl Acad Sci USA* 47: 1341–1344

Henkart PA, Berrebi GA, Takahama H, Mungar WE, Sitkovsky M (1987): Biochemical and functional properties of serine esterases in acidic cytoplasmic granules of cytotoxic T lymphocytes. *J Immunol* 139: 2398–2405

Hori H, Yatsutake A, Minematsu Y, Powers JC (1985): Inhibition of human leukocyte elastase, porcine pancreatic elastase and cathepsin G by peptide ketones. In: *Peptides: Structure and Function*, Deber CM, Hruby VJ, Kopple KD, eds. Rockford, IL: Pierce Chemical Co

Hudig D, Allison NJ, Pickett TM, Kam C-M, Powers JC (1991): The function of lymphocyte proteases: Inhibition and restoration of granule-mediated lysis with isocoumarin serine protease inhibitors. *J Immunol* 47: 1360–1368

Hudig D, Callewaert DM, Redelman D, Gregg NJ, Krump M, Tardieu B (1988): Lysis by RNK-16 cytotoxic–lymphocyte granules: Rate assays and conditions to study control of cytolysis. *J Immunol Methods* 115: 169–177

Hudig D, Gregg NJ, Kam C-M, Powers JC (1987): Lymphocyte granule-mediated cytolysis requires serine protease activity. *Biochem Biophys Res Commun* 149: 882–888

Hudig D, Haverty T, Fulcher C, Redelman D, Mendelsohn J (1981): Inhibition of human natural cytotoxicity by macromolecular antiproteinases. *J Immunol* 126: 1569–1574

Hudig D, Powers JC, Allison NJ, Kam C-M (1989): Selective isocoumarin protease inhibitors block RNK-16 lymphocyte granule-mediated cytolysis. *Mol Immunol* 26: 793–798

James GT (1978): Inactivation of the proteinase inhibitor phenylmethylsulfonyl fluoride in buffers. *Anal Biochem* 86: 574–579

Jenne DE, Masson D, Zimmer M, Haefliger JA, Li WH, Tschopp J (1989): Isolation and complete structure of the lymphocyte serine protease granzyme G, a novel member of the granzyme multigene family in murine cytolytic T lymphocytes. Evolutionary origin of lymphocyte proteases. *Biochemistry* 28: 7953–7961

Kam CM, Fujikawa K, Powers JC (1988): Mechanism-based isocoumarin inhibitors for trypsin and blood coagulation serine proteases: New anticoagulants. *Biochemistry* 27: 2547–2557

Kam CM, McRae BJ, Harper JW, Niemann MA, Volanakis JE, Powers JC (1987): Human complement proteins D, C2 and B. Active site mapping with peptide thioester substrates. *J Biol Chem* 262: 3444–3451

Keesey J (1987): *Biochemical Information: A Revised Biochemical Reference Source*. Indianapolis: Boehringer Mannheim

Kettner C, Shaw E (1981): Inactivation of trypsin-like enzymes with peptides of arginine chloromethyl ketone. *Methods Enzymol* 80: 826–842

Krahn J, Stevens FC (1972): Lima bean protease inhibitor: Comparative study of the trypsin and chymotrypsin inhibitor activity of the four chromatographic variants. *FEBS Lett* 28: 313–316

Laura R, Robinson DJ, Bing DH (1980): (p-Amidinophenyl)methanesulfonyl fluoride, an irreversible inhibitor of serine proteases. *Biochemistry* 19: 4859–4864

Lijnen HR, Collen D (1986): Alpha-2-Antiplasmin. In: *Proteinase Inhibitors*, Barrett AJ, Salvesen G, eds. Amsterdam: Elsevier

Lively MO, Powers JC (1978): Specificity and reactivity of human granulocyte elastase and cathepsin G, porcine pancreatic elastase, bovine chymotrypsin and trypsin toward inhibition with sulfonyl fluorides. *Biochem Biophys Acta* 525: 171–179

Lobe CG, Finlay BB, Paranchych W, Paetkau VH, Bleackley RC (1986): Novel serine proteases encoded by two cytotoxic T lymphocyte-specific genes. *Science* 232: 858–861

Lorand L, ed. (1976): Proteolytic enzymes, Part B. *Methods Enzymol* 45

Lorand L, ed. (1981): Protoelytic enzymes, Part C. *Methods Enzymol* 80

Luthy JA, Praissman M, Finkenstadt WR, Laskowski M, Jr (1973): Detailed mechanism of interaction of bovine β-trypsin with soybean trypsin inhibitor (Kunitz) *J Biol Chem* 248: 1706–1766

Masson D, Nabholtz M, Estrade C, Tschopp J (1986): Granules of cytotoxic lymphocytes contain two serine esterases. *EMBO J* 5: 1595–1600

Masson D, Peters PJ, Geuze HJ, Borst J, Tschopp J (1990): Interaction of chondroitin sulfate with perforin and granzymes of cytolytic T-cells is dependent upon pH. *Biochemistry* 29: 11229–11235

Matheson NR, van Halbeek H, Travis J (1991): Evidence for a tetrahedral intermediate complex during serpin-proteinase interactions. *J Biol Chem* 266: 13489–13491

Naughton MA, Sanger F (1961): Purification and specificity of pancreatic elastase. *Biochem J* 78: 156–163

Odake S, Kam C-M, Narasimhan L, Poe M, Blake JT, Krahenbuhl O, Tschopp J, Powers JC (1991): Human and murine cytotoxic T lymphocyte serine proteases: Subsite mapping with peptide thioester substrates and inhibition of enzyme activity and cytolysis by isocoumarins. *Biochemistry* 30: 2217–2227

Odani S, Ikenaka T (1973): Studies on soybean trypsin inhibitors VIII. Disulfide bridges in soybean Bowman-Birk trypsin inhibitor. *J Biochem* 74: 857–860

Oleksyszyn J, Powers JC (1989): Irreversible inhibition of serine proteases by peptidyl derivatives of alpha-aminoalkylphosphonate diphenyl esters. *Biochem Biophys Res Commun* 161: 143–149

Oleksyszyn J, Powers JC (1991): Irreversible inhibition of serine proteases by peptide derivatives of (alpha-aminoalkyl)phosphonate diphenyl esters. *Biochemistry* 30: 485–493

Poe M, Blake JT, Boulton DA, Gammon M, Sigal NH, Wu JK, Zweerink HJ (1991): Human cytotoxic lymphocyte granzyme B: Its purification from granules and the characterization of substrate and inhibitor specificity: *J Biol Chem* 266: 98–103

Powers JC, Harper JW (1986): Inhibitors of serine proteinases. In: *Proteinase Inhibitors*, Barrett AJ, Salvesen G, eds. Amsterdam: Elsevier

Powers JC, Gupton BF, Harley AD, Nishino N, Whitley RJ (1977): Specificity of porcine pancreatic elastase, human leukocyte elastase and cathepsin G. Inhibition with peptide chloromethyl ketones. *Biochim Biophys Acta* 485: 156–166

Powers JC, Kam CM, Narasimhan L, Oleksyszyn J, Hernandez MA, Ueda T (1989): Mechanism-based isocoumarin inhibitors for serine proteases: Use of active site structure and substrate specificity in inhibitor design. *J Cell Biochem* 39: 33–46

Ristow SS, Starkey JR, Hass GM (1983): *In vitro* effects of protease inhibitors on murine natural killer cell activity. *Immunology* 48: 1–8

Rusbridge NM, Beynon RJ (1990): 3,4-Dichlorisocoumarin, a serine protease inhibitor, inactivates glycogen phosphorylase b. *FEBS Lett* 268: 133–136

Salvesen GS, Sayers CA, Barrett AK (1981): Further characterization of the covalent linking reaction of α_2 macroglobulin. *Biochem J* 195: 453–461

Sayers TJ, Wiltrout TA, Sowder R, Munger WL, Smyth MJ, Henderson LE (1992): Purification of a factor from the granules of a rat natural killer cell line (RNK) that affects tumor cell growth and morphology: Molecular identity with a granule serine protease (RNK-P1). *J Immunol* (in press)

Schechter I, Berger A (1967): On the size of the active site in proteases. I. Papain. *Biochem Biophys Res Commun* 27: 157–162

Schechter NM, Sprows JL, Schoenberger OL, Lazarus GS, Cooperman BS, Rubin H (1989): Reaction of human skin chymotrypsin-like proteinase chymase with plasma proteinase inhibitors. *J Biol Chem* 264: 21308–21315

Simon MM, Fruth U, Simon HG, Gay S, Kramer MD (1989): Evidence for multiple functions of T-lymphocytes associated serine proteinases. *Adv Exp Med Biol* 247A: 609–613

Umezawa H, Aoyagi T (1977): Activities of proteinase inhibitors of microbial origin. In: *Proteinases in Mammalian Cells and Tissues*, Barrett AJ ed. Amsterdam: Elsevier

Yoshimura T, Barker LN, Powers JC (1982): Specificity and reactivity of human leukocyte elastase, porcine pancreatic elastase, human granulocyte cathepsin G, and bovine pancreatic chymotrypsin with arylsulfonyl fluorides. Discovery of a new series of potent and specific irreversible elastase inhibitors. *J Biol Chem* 257: 5077–5084

Zunino SJ, Allison NJ, Kam C-M, Powers JC, Hudig D (1988): Localization, implications for function, and gene expression of chymotrypsin-like proteinases of cytotoxic RNK-16 lymphocytes. *Biochim Biophys Acta* 967: 331–340

59
The Murine T Cell-Specific Serine Proteinase-1: Cleavage Activity on Synthetic and Natural Substrates

M.D. Kramer, U. Vettel, K. Ebnet, and M.M. Simon

Introduction

The murine T-cell-associated serine proteinase-1 (MTSP-1; Simon et al., 1986), synonymously denominated BLT esterase (Pasternack and Eisen, 1985; Pasternack et al., 1986), granzyme A (Masson et al., 1986), and SE-1 (Young et al., 1986), is contained within cytoplasmic granules of activated T cells (Fruth et al., 1987; Kramer et al. 1989) and is released into the extracellular space by receptor-triggered secretory exocytosis (Takayama et al., 1987; Takahashi et al., 1991). MTSP-1 is a trypsin-like endopeptidase, as revealed on low-molecular-weight chromogenic peptide substrates (Kramer et al., 1986) and has an optimal activity at neutral to slightly alkaline pH (Simon et al., 1986). Its functional characteristics suggest MTSP-1 to be operative only in the extracellular milieu.

T-cell extravasation and migration into inflammatory foci are likely to involve hydrolytic enzymes able to solubilize extracellular matrices (ECM), in particular the subendothelial basement membrane (Savion et al., 1984). We have therefore tested the proteolytic activity of MTSP-1 containing cellular subfractions or of the purified enzyme on either purified matrix proteins (Simon et al., 1988; Simon et al., 1991) or whole basement membrane-like extracellular matrix generated by endothelial cells *in vitro* (Simon et al., 1987; Vettel et al., 1991). The methods for testing MTSP-1 activity on low-molecular-weight chromogenic peptide substrates, on isolated ECM proteins, and on *in vitro*-produced extracellular matrix are described.

Preparation of Cell Lysates or of Supernatants Containing the Antigen-Induced Secretory Products of T Cells and Purification of MTSP-1

Two methods for preparation of cell lysates were used: detergent lysis, and lysis by freezing and thawing. For preparation of detergent lysates, stimulated and nonstimulated *ex vivo*-derived T-cell populations (thymocytes, splenocytes, nylon-wool purified lymph node T cells) and monoclonal CD8+ and CD4+ T-cell lines were used (Kramer et al., 1986). Cells were washed twice in sterile phosphate-buffered saline (PBS) and lysed at a concentration of 5×10^6 cells/ml in 0.1 M Tris/HCl pH 8.5, 0.1% Triton X-100 for 30 min on ice. For preparation of freeze/thaw lysates, cells are washed twice in sterile PBS, adjusted to a density of 5×10^6/ml in 0.1 M Tris/HCl pH 7.0, and lysed by freezing at $-70°C$ and thawing three times. Particulate material in both preparations is precipitated by centrifugation at $10,000 \times g$ for 5 min, and the clear supernatants are taken as enzyme preparations and stored at $-20°C$.

For preparation of supernatants of monoclonal CD8+ T-cell lines, the cells are washed twice in PBS and are incubated in round-bottomed, 96-well microtiter plates at a density of 4×10^5 cells/well together with 4×10^5 appropriate MTSP-1 negative target cells in a total volume of 0.2 ml. Medium is RPMI 640 without phenol red (Gibco) supplemented with 0.01 M and N-2-hydroxyethyl-piperazine-N'-2-ethanesulfonic acid (HEPES) and

1 mg/ml bovine serum albumin (Boehringer Mannheim). The plates are incubated at 37°C for 3 hr, and afterwards the supernatant is removed, centrifuged at 10,000 × g, and immediately frozen at −20°C.

MTSP-1 is purified from cell lysates in two steps by a modification of our previously published protocol (Simon et al., 1986). $5 \times 10^8 - 1 \times 10^9$ cells are lysed in 10 mM Tris/HCl, pH 7.5, 0.1% Triton X-100 at a density of 5×10^7 cells/ml. The lysates are subjected to ammonium sulfate precipitation at 60% saturation. Supernatant thereof, which contains more than 90% of total MTSP-1 activity, is extensively dialyzed against 10 mM BIS/Tris/HCl pH 6.0 [BIS-TRIS = bis[2-hydroxyethyl]-iminotris[hydroxymethyl]methane; 2-bis[2-hydroxyethyl]amino-2[hydroxymethyl]-1,3-propanediol (Sigma)] (2 × 2000 ml, 1 hr and 2 hr) and 50 mM BIS/Tris/HCl pH 6.0 (2 × 2000 ml overnight and 2 hr) at 4°C, filtrated (0.2 μm pore size), and applied to a Mono S cation exchange column (Mono S HR5/5, Pharmacia). The column is eluted by a combined step/linear gradient ranging from 0 to 2.0 M NaCl in 50 mM BIS/Tris/HCl pH 6.0. Elution of MTSP-1 occurs at 800–840 mM NaCl. Fractions containing MTSP-1 activity are pooled and dialyzed against PBS pH 7.0. Purification is ascertained by sodium dodecyl sulfate–polyacrylamide gel electrophoresis (SDS–PAGE) under reducing and nonreducing conditions followed by silver staining (Ansorge, 1985) or by immunoblotting using an MTSP-1-specific monoclonal antibody (Kramer et al., 1989). Purified MTSP-1 is stored at −70°C.

Chromogenic Assay for Amidolytic Activity of MTSP-1

Enzyme-containing preparations are tested for amidolytic activity as follows: 100 μl of test samples are titrated in serial dilutions in 0.1 M Tris/HCl, pH 8.5, and are mixed with 100 μl (3×10^{-4} M) of the chromogenic peptide substrate HD-prolyl-phenylalanyl-arginyl-nitroanilide (Bachem) in flat-bottomed 38microtiter plates. The absorbance at 405 nm is directly determined from the plate after 1 hr at 37°C by using an enzyme-linked immunosorbent assay (ELISA) reader. An increase in the absorbance of 0.01/hr at 37°C is defined as 1 unit of enzyme activity. The pH dependence of amidolytic activity of purified MTSP-1 is shown in Figure 1.

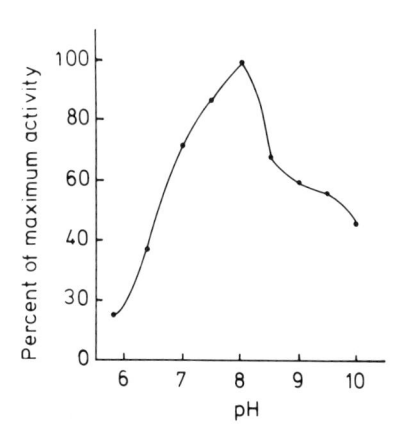

FIGURE 1. pH dependence of MTSP-1 activity on the substrate HD-prolyl-phenylalanyl-arginyl-*para*-nitroanilide. The buffers used at a final concentration of 50 mM were phosphate buffer (pH 5.8–7.4) and Tris/HCl (pH 7.5–10.0). Reprinted with permission of Oxford University Press from Simon MM, et al. (1986): Purification and characterization of a T cell specific serine proteinase (TSP-1) from cloned cytolytic T lymphocytes. *EMBO J* 5: 3267–3274.

Degradation of High-Molecular-Weight Matrix Proteins by Purified MTSP-1

The capability of MTSP-1 to degrade high-molecular-weight matrix proteins is studied by admixing purified MTSP-1 to preparations of purified matrix proteins, such as human fibronectin (Simon et al., 1988) or human collagen types I–VI (Simon et al., 1991). The cleavage products are analyzed by SDS–PAGE and silver staining. As control MTSP-1 is inactivated by incubation with the selective MTSP-1 inhibitor prolyl-phenylalanyl-arginyl-chloromethyl-ketone (PFR-CK; Simon et al., 1989) at 0.1 mM for 30 min at 37°C.

Degradation of fibronectin

Purified human fibronectin (10 μg) is incubated with a fixed concentration of purified MTSP-1 (1 μg) in 100 μl of 100 mM Tris/HCl, pH 8.5, at 37°C. The enzyme reaction is terminated after different intervals of time (0–120 min) by adding reducing SDS–PAGE sample buffer. Fibronectin digests are then subjected to SDS–PAGE under reducing conditions. The time-dependent appearance of distinct fibronectin fragments is shown in Figure 2.

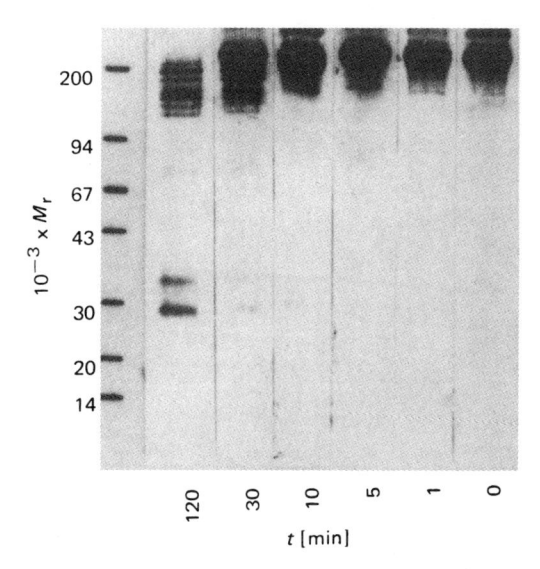

FIGURE 2. Time dependence of fibronectin degradation by purified MTSP-1. Purified human plasma fibronectin (10 μg) was mixed with purified MTSP-1 (1 μg) and incubated in 100 μl 100 mM Tris/HCl pH 8.5 for different intervals of time at 37°C. Afterwards the reaction mixture was analyzed by SDS–polyacrylamide gel electrophoresis (10–15% gradient separating gel) and silver staining. Distinct fibronectin split products with progressively smaller molecular weight are generated in a time-dependent manner. Reprinted with permission from: Simon et al., 1988.

Degradation of collagen type IV

Native collagens type I–VI were derived via extraction of appropriate human tissues by limited pepsin digestion and further purification and characterization by established procedures (Furuto and Miller, 1987). Two micrograms of soluble collagens are incubated with 1 μg MTSP-1 in 20 μl 100 mM Tris/HCl, pH 8.5, containing 10 μg/ml heparin for 8 hr at 27°C. The degradation products are analyzed by SDS–PAGE (7.5% separating gel) under reducing conditions followed by silver staining. Under these conditions, only collagen type IV, but none of the collagens type I, II, III, V, and VI, are sensitive to proteolytic cleavage by MTSP-1. Degradation of collagen type IV is shown in Figure 3.

Preparation of [^{35}S]O$_4$-labeled Extracellular Matrix Produced by Endothelial Cells *In Vitro*

Cultures from bovine corneal endothelial cells are established from bovine eyes as described (Gospodarowicz et al., 1977). Cells are cultured in Dulbecco's modified Eagle's medium (DMEM H-16; Gibco) supplemented with 10% bovine calf serum, 5% fetal calf serum, penicillin (50 U/ml), and streptomycin (50 μg/ml) at 37°C in a humidified atmosphere of 10% CO$_2$. Basic fibroblast growth factor (bFGF) is added to a final concentration of 100 ng/ml every other day during the phase of active cell growth.

Corneal endothelial cells are plated at an initial density of 4×10^4 cells per 35-mm dish (Falcon) and maintained as described above, except that

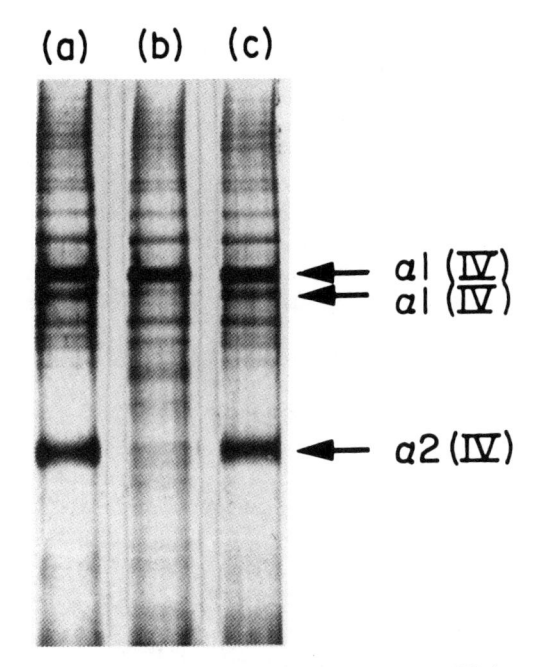

FIGURE 3. Degradation of collagen type IV by MTSP-1. Two micrograms collagen type IV was incubated with 1 μg MTSP-1 in 20 μl 100 mM pH 8.5, containing 10 μg/ml heparin for 8hr at 27°C. The products were analyzed on a 7.5% SDS–polyacrylamide gel under reducing conditions. Lane a: Collagen type IV incubated with buffer alone. Lane b: Collagen type IV incubated with MTSP-1. Lane c: Collagen type IV incubated with PFR-CK-treated MTSP-1. The two α1(IV) chains of 140 and 120 kDa, respectively, as well as the α2(IV) chain of 80 kDa are marked by arrows. Reprinted with permission of Blackwell Scientific Publications from Simon MM et al. (1991): CD8$^+$ activated T cells secrete MTSP-1, a collagen-degrading enzyme: Potential pathway for the migration of lymphocytes through vascular basement membranes. *Immunology* 73: 117–119.

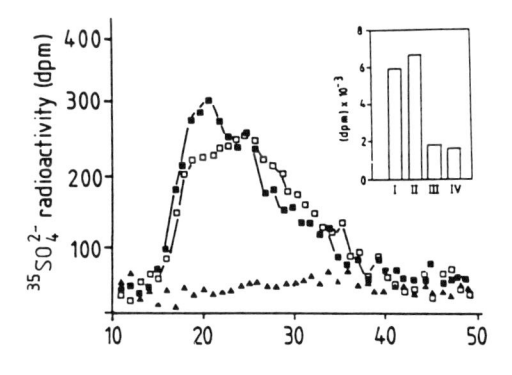

FIGURE 4. Degradation of [^{35}S]O$_4$-labeled corneal endothelial ECM by purified MTSP-1. Purified MTSP-1 (150 U/ml in 1.5 ml 0.1 M Tris/HCl, pH 7.0/35 mm dish) was incubated with sulfate-labeled ECM (48 hr, 37°C, pH 7.0, humidified atmosphere) either in the absence (open rectangles) or presence of heparin (20 μg/ml, closed rectangles) or PFR-CK (0.5 mM, closed triangles). The resulting degradation products were separated by gel filtration (Sepharose 6B) and radioactivity in the fractions was determined by liquid scintillation counting. Inset: Total [^{35}S]O$_4^-$-radioactivity released by purified MTSP-1. I: MTSP-1 without additions. II: MTSP-1 in the presence of heparin (20 μg/ml). III: MTSP-1 in the presence of PFR-CK (0.5 mM). IV: Spontaneous release of radioactivity in the presence of buffer only. Degradation of 48 hr, at 37°C, pH 7.0. Reprinted with permission of VCH Verlagsgesellschaft from: Vettel et al., 1991.

Degradation of *In Vitro*-Produced Endothelial ECM and Analysis of Degradation Products

Labeled ECM dishes are washed twice with sterile PBS and then incubated with a total buffer volume of 1.5 ml 0.1 M Tris/HCl at pH 7.0 and 37°C in a humidified atmosphere of 7% CO$_2$. After 48 hr the supernatant is collected and centrifuged at 10,000 × g for 10 min. Five hundred microliters of the supernatant are subjected to gel filtration on Sepharose 6B columns (0.75 × 20 cm). See Figure 4. Samples are eluted with PBS/0.05% NaN$_3$, and 200-μl fractions are collected at a flow rate of 12 ml/hr. Radioactivity in the resulting fractions is determined by scintillation counting using Aquasol scintillation fluid (DuPont/NEN). The excluded volume (V_0) is marked by blue dextrane (Pharmacia) and the total included volume (V_t) by phenol red. The k_{av} is calculated according to the following formula:

$$k_{av} = \frac{V_e - V_0}{V_t - V_0}$$

where V_e is the elution volume, V_0 the void volume, and V_t the column volume.

References

Ansorge W (1985): Fast and sensitive detection of protein and DNA bands by treatment with potassium permanganate. *J Biochem Biophys Methods* 11: 13–20

Bar-Ner M, Kramer MD, Schirrmacher V, Ishai-Michaeli R, Fuks Z, Vlodavsky I (1985): Sequential degradation of heparan sulfate in the subendothelial extracellular matrix by highly metastatic lymphoma cells. *Int J Cancer* 35: 483–491

Fruth U, Prester M, Golecki J, Hengartner H, Simon HG, Kramer MD, Simon MM (1987): The T cell-specific serine endopeptidase TSP-1 is associated with cytoplasmic granules of cytolytic T lymphocytes. *Eur J Immunol* 176: 613–622

Furuto DK, Miller EJ (1987); Isolation and characterization of collagens and procollagens. *Methods Enzymol* 144: 41–61

Gospodarowicz D, Delgado D, Vlodavsky I (1980): Permissive effect of the extracellular matrix on cell proliferation in vitro. *Proc Natl Acad Sci USA* 77: 4094–4098

Gospodarowicz D, Mescher AR, Birdwell C (1977): Stimulation of corneal endothelial cell proliferation in vitro by fibroblast and epidermal growth factors. *Exp Eye Res* 25: 75–89

Kramer MD, Binninger L, Schirrmacher V, Moll H,

---

sulfate-free medium is used and 5% dextran T-40 (Sigma) is included in the growth medium. Na$_2$[^{35}S]O$_4$ (540–590 mCi/mM) is added to the cultures 3 and 7 days after seeding (40 μCi/ml) without medium change. Five to seven days after reaching confluence (10–12 days after seeding), the cell layer is dissolved by exposure (10 min, 22°C) to 0.5% Triton X-100 in PBS (v/v), leaving the underlying extracellular matrix intact and firmly attached to the tissue culture dish (Gospodarowicz et al., 1980; Vlodavsky et al., 1980, 1983). The remaining nuclei and cytoskeletons are removed by a 2- to 3-min exposure to 0.025 N NH$_4$OH followed by 4 washes in PBS. The labeled ECM is then ready to be used in degradation experiments. Previous studies have shown that 70–75% of the total ECM-bound radioactivity is incorporated into heparan sulfate side chains of matrix proteoglycans (Bar-Ner et al., 1985).

Prester M, Nerz G, Simon MM (1986): Characterization and isolation of a trypsin-like serine protease from a long-term culture cytolytic T cell line and its expression by functionally distinct T cells. *J Immunol* 136: 4644–4651

Kramer MD, Fruth U, Simon H-G, Simon MM (1989): Expression of cytoplasmic granules with T cell-associated serine proteinase-1 activity in Ly-2+(CD8+) T lymphocytes responding to lymphocytic choriomeningitis virus in vivo. *Eur J Immunol* 19: 151–156

Masson D, Zamai M, Tschopp J (1986): Identification of granzyme A isolated from cytotoxic T-lymphocyte granules as one of the proteases encoded by CTL-specific genes. *FEBS Lett* 208: 84–88

Pasternack MS, Eisen HN (1985): A novel serine esterase expressed by cytotoxic T lymphocytes. *Nature* 314: 743–745

Pasternack MS, Verret CR, Liu MA, Eisen HN (1986): Serine esterase in cytolytic T lymphocytes. *Nature* 322: 740–743

Savion N, Vlodavsky I, Fuks Z (1984): Interaction of T lymphocytes and macrophages with cultured vascular endothelial cells: Attachment, invasion, and subsequent degradation of the subendothelial extracellular matrix. *J Cell Physiol* 118: 169–178

Simon MM, Hoschützky H, Fruth U, Simon H-G, Kramer MD (1986): Purification and characterization of a T cell specific serine proteinase (TSP-1) from cloned cytolytic T lymphocytes. *EMBO J* 5: 3267–3274

Simon MM, Kramer MD, Gay S (1991): CD8+ activated T cells secrete MTSP-1, a collagen-degrading enzyme: Potential pathway for the migration of lymphocytes through vascular basement membranes. *Immunology* 73; 117–119

Simon MM, Prester M, Kramer MD, Fruth U (1989): An inhibitor specific for the mouse T-cell associated serine proteinase-1 (TSP-1) inhibits the cytolytic potential of cytoplasmic granules but not of intact cytolytic T cells. *J Cell Biochem* 40: 1–13

Simon MM, Prester M, Nerz G, Kramer MD, Fruth U (1988): Release of biologically active fragments from human plasma fibronectin by murine T cell-specific serine proteinase-1 (MTSP-1). *Biol Chem Hoppe-Seyler* 369 (Suppl): 107–112

Simon MM, Simon H-G, Fruth U, Epplen J, Müller-Hermelink HK, Kramer MD (1987): Cloned cytolytic T-effector cells and their malignant variants produce an extracellular matrix degrading typsin like serine proteinase. *Immunology* 60: 219–230

Takahashi K, Nakamura T, Adachi H, Yagita H, Okumura K (1991): Antigen-independent T cell activation mediated by a very late activation antigen-like extracellular matrix receptor. *Eur J Immunol* 21: 1559–1562

Takayama H, Trenn C, Humphrey W, Bluestone JA, Henkart PA, Sitkovsky MV (1987): Antigen receptor-triggered secretion of a trypsin-type esterase from cytotoxic T lymphocytes. *J Immunol* 138: 566–569

Vettel U, Bar-Shavit R, Simon MM, Brunner G, Vlodavsky I, Kramer MD (1991): Coordinate secretion and functional synergism of T cell-associated serine proteinase-1 (MTSP-1) and endoglycosidase(s) of activated T cells. *Eur J Immunol* 21: 2247–2251

Vlodavsky I, Fuks Z, Bar-Ner M, Ariav Y, Schirrmacher V (1983): Lymphoma cell-mediated degradation of sulfated proteoglycans in the subendothelial extracellular matrix: Relationship to tumor cell metastasis. *Cancer Res* 43: 2704–2711

Vlodavsky I, Lui GM, Gospodarowicz D (1980): Morphological appearance, growth behavior and migratory activity of human tumor cells maintained on extracellular matrix versus plastic. *Cell* 19: 607–616

Young JDE, Leong LG, Liu CC, Damiano A, Wall DA, Cohn ZA (1986): Isolation and characterization of a serine esterase from cytolytic T cell granules. *Cell* 47: 183–194

60
Detection of Specific mRNAs by In Situ Hybridization

Gillian M. Griffiths, Susan Alpert, R. Jane Hershberger, Lishan Su, and Irving L. Weissman

RNA *in situ* hybridization

Materials

All glassware should be baked at Θ200°C overnight All solutions, except those containing Tris, should be treated with diethylpyrocarbonate (Sigma, D-5758) for at least 2 hr and then autoclaved. Gloves should be worn throughout.

Slides

Slides should be washed in Chromerge overnight, rinsed in deionized water, baked at 200°C overnight, and finally coated with poly-L-lysine (Sigma P-1399), 50 μg/ml, 10 mM Tris pH 8.

Radiolabeled RNA Probe

^{35}S-UTP (Amersham, SJ1303); RNA polymerases, e.g. SP6 and T7 (Promega); ribonucleotide buffer; 10 mM rATP, 10 mM rGTP, 10 mM rGTP, 10 mM rCTP HEPES, pH 7.5; SP6 buffer, 5×: 200 mM Tris-HCl, pH 7.9, 300 mM MgCl$_2$, 10 mM spermidine; RNasin (Promega); DNase I, 1 mg/ml (Worthington); yeast tRNA, type X-S (Sigma R-0128), 10 mg/ml; phenol/chloroform/iso-amyl alcohol (24:24:1); TE 10 mM Tris, ph 7.5, 1 mM ethylenediaminetetraacetic acid (EDTA); carbonate buffer; 400 mM NaHCO$_3$, 600 mM Na$_2$CO$_3$.

Fixation

Paraformaldehyde, EM grade (Polysciences 0380) 4% solution in phosphate-buffered saline (PBS). This should be used for only 1–4 days after preparation. In order to dissolve the paraformaldehyde, it is necessary to use calcium- and magnesium-free PBS and to heat to 60°C.

Hybridization

Proteinase K (Sigma P0390); dextran sulfate, MW ~500,000 (Pharmacia 17-0340-02); formamide (EM Science), deionized with resin 100 g/l (Bio-Rad, AG501-X8), filtered and twice crystallized by freezing at −20°C; 20% dextran sulfate/formamide: 6 g dextran sulfate in 30 ml (heat to 60°C). Store 4°C, dark; Denhardt's solution (50×): 1% ficol, 1% polyvinylpyrrolidone, 1% bovine serum albumin (BSA) (fraction V), filtered and stored at −20°C; dithiothreitol (DTT) (Sigma, D-0632): 15.4 mg in 50 μl. Freshly prepared.

Washings

Formamide (Mallinckrodt 3797, technical grade); 20× SSC: 175 g NaCl and 88.2 g sodium citrate in 800 ml water, adjusted to pH 7 with conc. HCl and brought to 1 l final volume; RNase A, type II-A (Sigma R-5000): 10 mg/ml in PBS; RNase T1, grade IV (Sigma R-8251).

Autoradiography

Emulsion, type NTB2 (Kodak 165-4433); developer, D19 (Kodak, 146-4593); fixer (Kodak 197-1746).

Preparation of Radiolabeled Probe Using T7 or SP6 RNA Polymerases

1. Dry 25 μl (250 μCi) of ^{35}S-UTP.
2. Add polymerase reaction mixture for 20 μl reaction:
   4 μl 5× enzyme buffer
   2 μl 100 mM DTT

2 µl ribonucleotide mix
2 µl BSA (5 mg/ml)
1 µl RNasin (20 u/ml)
0.5 µl polymerase (20 u/λ)
8.5 µl H$_2$O containing 4 µg probe, linearized to allow synthesis of insert only

Incubate for 45 min at 37°C for T7 and 40°C for SP6.

3. Make a second addition of enzyme (2–4 u) in enzyme buffer containing 1× ribonucleotides, and continue incubation for a further 30 min.
4. Treat with 1 µg DNase (with 0.5 µl Rnasin) to digest the template. 37°C, 15 min.
5. Bring volume to 200 µl, add 5 µl tRNA, and extract with 200 µl phenol/CHCl$_3$. Spin in microfuge 5 min, 4°C, remove upper phase, and count 1/100th on scintillation filter paper.
6. Precipitate labeled riboprobe by addition of 100 µl 7.5 M NH$_4$Ac and 800 µl ethanol. Mix and keep on dry ice for 15 min before spinning in a microfuge for 15 min at 4°C. Remove supernatant and resuspend pellet in 200 µl 10 mM Tris pH8, 1mM EDTA (TE). Count 1/100th on scintillation filter paper and calculate percent incorporation. This should be > 75%.
7. Size reduce the probe to a final length of 150–200 bp by alkaline lysis by addition of 30 µl carbonate buffer and incubation at 60°C for T min according to the equation:

$$T = \frac{L_0 - L_f}{0.11 \times L_0 \times L_f}$$

where L_0 and L_f are the initial and final lengths of the probe in kilobasepairs.

8. Terminate lysis by addition of 24 µl sodium acetate mix (292.5 µl 3M NaAc + 21.5 µl conc. acetic acid).
9. Add tRNA to a final concentration of 7.5 µg/1–2 × 10^6 cpm and reprecipitate with 950 µl ethanol.
10. Resuspend at 4 × 10^6 cpm/µl in TE. Heat at 90°C for 2–3 min and store at −70°C.

Fixation

1. Sections or cells on slides are fixed in 4% paraformaldehyde in PBS for 20 min at room temperature. Samples are then sequentially washed in 3× PBS and twice in 1× PBS for 5 min each, rinsed in water and dehydrated through graded alcohols (50%, 80%, 95%), and air dried. Samples can then be stored at 4°C or hybridized immediately.
2. Prior to hybridization, samples are treated with proteinase K at 1 µl/ml in 100 mM Tris-HCL, pH 8.0, 50 mM EDTA at 37°C for 30 min. Samples are then fixed again as in (1).
3. Slides are rinsed in 0.1 M triethanolamine and then incubated for 10 min in 0.1 M triethanolamine to which acetic anhydride is added to 0.25% while mixing. Slides are then rinsed in water and air dried.

Hybridization

1. Hybridization solution is prepared allowing approximately 15 µl per slide for a 20-mm^2 coverslip. 2 × 10^6 cpm of probe are prepared in 10% dextran sulfate, 50% formamide, 300 mM NaCl, 20 mM Tris-HCl, pH 7.5, 5 mM EDTA 1× Denhardt's solution, and 100 mM DTT.
2. Probe is added to the section and covered with a siliconized, baked coverslip which is sealed with rubber cement. Slides are then sealed in a plastic slide box containing filter paper saturated with 50% formamide/2 × saline sodium citrate (SSC). Hybridization is carried out for 17–20 hr at 45–50°C in a dry incubator.

Washing

1. Slides are immersed in washing solution (50% formamide, 2× SSC, 1mM EDTA, 10 mM DTT), and the coverslips can then be easily removed using (sterile) toothpicks and a pair of forceps. Washing is carried out at 54°C for 20 min in a glass slide dish. A magnetic stir bar is used to provide continual mixing.
2. The slides are rinsed twice in 2× SSC, 10 mM DTT prior to RNase digestion for 30 min at 37°C in 2× SSC, 1mM EDTA containing 20 µl/ml RNaseA1 and 1 unit/ml RNase T1.
3. Slides are then washed twice more in washing solution for 45 min each at 54°C. Slides are then rinsed in 2 × SSC prior to dehydration in graded ethanol (50%, 80%, and 95%) containing 300–400 mM ammonium acetate to help stabilize the RNA hybrids. Slides are then air dried.

Autoradiography

1. Slides are exposed, in the dark, in Kodak emulsion, diluted to 50% in 750 mM ammonium acetate. Emulsion is melted at 40–50°C for

about 1 hr, diluted, and used immediately. Slides are dipped twice and stood in a test-tube rack to dry completely (about 1–2 hr) before being placed in a light-tight box containing desiccant and stored in a desiccator during exposure. Exposure time varies from 3 days to 3 weeks depending on the abundance of the target mRNA.

2. Slides are developed, in the darkroom, in Kodak D-19 developer for 2.5 min at 15–16°C, stopped in 2% acetic acid for 30 sec, and fixed in Kodak fixer for 5 min at 15–16°C before extensive washing in running tap water (30 min). Slides can be counterstained with Giemsa or hematoxylin.

Keyword Index

This index was established according to the keywords supplied by the authors. Page numbers refer to the beginning of the chapter.

Adenosine triphosphate (ATP), 307, 314
Adenovirus, 400
Adhesion, 9, 65, 72
　assay, 498
　to substratum, 202
Adoptive immunotherapy, 407
Aging, 384
Alloantigen, 49, 488
　minor, 492
Antibody
　bispecific, 96, 485
　crosslinked, 485
　monoclonal, 49
　response, 378
Antigen
　major histocompatibility complex, 498
　presentation, 358, 370, 378, 494
　processing, 358, 378
　recognition, 72
　tumor, 49
　viral, 49
Antiviral CTL, 358, 490
Apoptosis, 9, 153, 196, 213, 263, 314, 472
Artificial membranes, 65, 498
Auto-immunity, 273
Avidity, 49

β2-microgobulin, 3, 58, 145, 351
Blood cell, red, 484
BLT-esterase, 3, 113, 166
Bystander lysis, 9

Calcineurin, 331
Calcium, 72, 196, 314, 321
Calmodulin, 331
cAMP-dependent protein kinase, 331
Cancer immunotherapy, 96
CD2, 84
CD3, 84
CD4, 145, 166
CD4$^+$
　CTL, 190, 351, 358
　T cell clones, isolation of, 490
CD8, 65, 166
CD8$^+$
　CTL, 190, 358, 370, 481
　T cell clones, isolation of, 490
CD16, 84
CD58 (LFA-3), 72

cDNA library, 237
Chymase, 295, 502
Collagen, 516
Conjugation, 9
Contact-induced lysis, 196
Cosignalling, 65
^{51}Cr (51-chromium), 457
　release assay, 5, 478
Crosslinked antibody, 485
CTL
　antiviral, 358, 490
　blasts, 113
　function, 366
　generation, 492
　hybridoma, 113
　in vivo, 113
　peritoneal exudate, 113
Cyclosporin A, 331
Cytokine, 128
Cytolysin (*see also* Perforin), 3, 153, 263, 484
　role in killing, 213
Cytolysis, 166, 190, 457, 468
Cytomegalovirus, 400
Cytoskeleton, 478
Cytotoxic T cells, 3, 96, 278, 494
Cytotoxicity
　assay, 490
　bystander, 190
　perforin-independent, 190
　targeted, 485
Degranulation, 321, 498
Detachment, 9, 478
Dipeptidyl peptidase I, 340
DNA
　fragmentation, 9, 153, 190, 213, 223, 251, 366, 468

Ecto-ATPase, 314
Electroporation, 494
ELISA, 498
Endonuclease activation, 213
Esterase
　BLT, 3, 113, 166
Exocytosis, 153, 166, 223, 263, 307, 331, 498
Extracellular matrix, 478, 516
Extravasation, 278

Fatty acids, 321
Fc receptor, 72, 84, 96

KEYWORD INDEX

Fibronectin, 516
Fragmentin, 263

Gene
  structure, 273
  therapy, 407
  "knock-out", 3
Granule, 153, 263, 340, 484
Granzymes, 153, 166, 223, 251, 384, 502
Granzyme A, 3, 263, 273, 516
Growth factors, 213

Haptens, 49
Hemolysis, 484
Herpes simplex, 366, 400
Histocompatibility, 145
Hybridoma
  CTL, 113
  hybrid, 96

Ig receptor, 378
Immune damage, 478
Immunopathology, 351
Immunosuppression, 340
Immunotherapy
  adoptive, 407
  cancer, 96
In vivo effector function, 358
Infection, viral, 278
Inflammation, 202, 351, 478
Influenza virus, 351
Interferon (IFN)
  IFN-α, 178
  IFN-γ, 331, 400, 407
Interleukin, 128
Interleukin-2, 138, 213, 407
Interleukin-12, 138
^{125}Iododeoxyuridine, 468
Ion channel, 314

Labels, 472
LAK, 128, 138, 407
Large granular lymphocyte, 223
Lethal hit, 321
Leucine methyl ester, 340
Leucyl-leucine methyl ester, 340
LFA-3 (CD58), 72
Limiting dilution, 128, 472
Lipofectin, 494
Liposomes, 494
Ly49, 84
Lymph node T-lymphocytes, 492
Lymphocyte
  activation, 65, 307, 331, 498
  cytolytic T, 113, 128, 178, 358
  large granular, 223
  lymph node, 492
  recruitment, 351
  subsets, 223
  tumor-infiltrating, 407
Lymphocytic choriomeningitis virus (LCMV), 400
Lymphokine-activated killer cells (LAK), 128, 138, 407
Lymphotoxin (TNF-β), 190
Lysis
  bystander, 9, 190
  contact-induced, 196
  osmotic, 314
  redirected, 166, 487, 497
Lysosomes, 340

Macrophage, 439
  activation, 439
  antitumor, 439
  antivirus, 439
  bone marrow, 481
  selectivity, 439
Marrow graft rejection, 394
Melanoma, 407
Membrane protein immobilization, 498
Membranes, artificial, 65, 498
Metastasis, 478
Methods, 457, 468, 472
MHC, 58, 145, 378, 400, 498
MHC-restricted, 378, 384
  class I, 58, 65, 400
  class II, 145
Microspheres, 498
Minor alloantigen, 492
Mixed leukocyte culture (MLC), 49, 128, 384, 488
Mutation, somatic, 49

Necrosis, 213
Nitric oxide, 439
NK cells, 3, 72, 84, 223, 278, 295, 340, 394, 400, 502
Nuclear
  degradation, 166
  disintegration, 202
  DNA release (NDR) activity, 263
Nucleolin, 263
Nucleus, 153

Oligonucleosomes, 263
Osmotic lysis, 314
Ovalbumen, 494

P56^{lck}, 65
P815 cells, 5
pCTL, 384
Peptide, 58
  binding, 58
  transport, 58
  -class I complex, 498
Perforin, 3, 113, 153, 166, 178, 196, 263, 295, 384, 484

KEYWORD INDEX

role in killing, 213
  expression, 190
    mRNA, 190
  -independent cytotoxicity, 190
Peripheral blood, 223
Peritoneal exudate CTL (PEL), 113, 196
Phosphatidylinositol, 72
Phospholipase, 321
Pichinde virus, 400
Pneumonia, 351
Prelytic halt hypothesis, 366
Priming *in vitro*, 494
Programmed cell death, 196, 202, 213
Protease, 58, 251, 278, 295, 482, 502, 516
  inhibitors, 295, 502
  serine, 278, 295, 502
Protein folding, 58
Protein kinase, 72, 331
Protein phosphatase, 331
Proteoglycan, 516
Purinergic receptors, 314

Radiotoxicity, 468
Receptor
  Fc, 72, 84, 96
  Ig, 378
  purinergic, 314
  T cell, 65, 96, 331
Redirected lysis, 166, 487, 497
Renal cell carcinoma, 407
Reovirus, 366
Retroviral transduction, 407
RNA
  alternation splicing, 72
  in situ hybridization, 273, 521

Secretion, 153
Sendai virus, 351
Serpin, 295, 502
Signaling, 321
  cosignaling, 65
  transmembrane, 65
Substrate, 251
Suicide, 9
Superantigen reactivity, 358
Supressor, 378

Toxoplasma, 370, 481
Transgenes, 49
Transgenic mice, 3
Transmembrane signalling, 65
Transplantation, 273
Tryptase, 295, 502
Tumor, 96
  antigens, 49
  necrosis factor, 407, 439
  necrosis factor (TNF-α), 138, 178, 190
  -infiltrating lymphocytes, 407

Viral
  antigens, 49
  infection, 96, 278
Virus
  adenovirus, 400
  clearance, 351
  control mechanisms, 366
  cytomegalovirus, 400
  herpes simplex, 366, 400
  influenza, 351
  lymphocytic choriomeningitis (LCMV), 400
  pichinde, 400
  reovirus, 366
  sendai, 351